100 Years of General Relativity – Vol. 1

NUMERICAL RELATIVITY

100 Years of General Relativity
ISSN: 2424-8223

Series Editor: Abhay Ashtekar *(Pennsylvania State University, USA)*

This series is to publish about two dozen excellent monographs written by top-notch authors from the international gravitational community covering various aspects of the field, ranging from mathematical general relativity through observational ramifications in cosmology, relativistic astrophysics and gravitational waves, to quantum aspects.

Published

Vol. 1 Numerical Relativity
 by Masaru Shibata (Kyoto University, Japan)

Forthcoming

Chern–Simons (Super)Gravity
by Mokhtar Hassaine (Universidad of Talca, Chile) &
Jorge Zanelli (Centro de Estudios Científicos, Chile)

100 Years of General Relativity – Vol. 1

NUMERICAL RELATIVITY

Masaru Shibata

Kyoto University, Japan

World Scientific

NEW JERSEY · LONDON · SINGAPORE · BEIJING · SHANGHAI · HONG KONG · TAIPEI · CHENNAI · TOKYO

Published by

World Scientific Publishing Co. Pte. Ltd.
5 Toh Tuck Link, Singapore 596224
USA office: 27 Warren Street, Suite 401-402, Hackensack, NJ 07601
UK office: 57 Shelton Street, Covent Garden, London WC2H 9HE

British Library Cataloguing-in-Publication Data
A catalogue record for this book is available from the British Library.

100 Years of General Relativity — Vol. 1
NUMERICAL RELATIVITY

Copyright © 2016 by World Scientific Publishing Co. Pte. Ltd.

All rights reserved. This book, or parts thereof, may not be reproduced in any form or by any means, electronic or mechanical, including photocopying, recording or any information storage and retrieval system now known or to be invented, without written permission from the publisher.

For photocopying of material in this volume, please pay a copying fee through the Copyright Clearance Center, Inc., 222 Rosewood Drive, Danvers, MA 01923, USA. In this case permission to photocopy is not required from the publisher.

ISBN 978-981-4699-71-6

In-house Editor: Song Yu

Preface: What is numerical relativity?

Status and prospect for the study of general relativity

Since Einstein formulated general relativity in 1915, hundred years passed. In the last hundred years, general relativity resolved several unsolved issues and also predicted new phenomena in nature [see Will (1993, 2014) for a comprehensive review]. Soon after the Einstein's success in formulating general relativity, the perihelion shift of Mercury, which was a serious mystery in nineteenth century, was precisely explained. Also, the bending angle of light passing near the sun was correctly predicted and subsequently this gravitational lensing effect was verified by the well-known observation by Eddington and his collaborators. General relativity also predicted the expansion of the universe, the presence of gravitational waves, and the Shapiro time delay effect. All these predictions were already verified through astronomical observations. In particular, the light bending and Shapiro time delay effects are confirmed within less than 0.1% error through several experiments in the solar system, indicating the validity and robustness of general relativity. Also, a long-term observation of the Hulse–Taylor binary pulsar (PSR B1913+16) proved the presence of gravitational waves and the observed luminosity of gravitational waves of this binary agrees with the prediction of general relativity within $\sim 0.2\%$ error [Weisberg et al. (2010)].

In addition, in the past 50 years, a large number of general relativistic objects and phenomena have been discovered in the universe. Neutron stars, binary neutron stars, black holes, and supermassive black holes are the representative objects in general relativity. The universe itself is also an object described by general relativity. Stellar core collapses of massive stars, supernova explosions, and gamma-ray bursts are representative high-energy phenomena in which general relativity plays an important role.

However, we have not yet fully understood general relativity, general relativistic objects, and general relativistic phenomena. To date, precise tests of general relativity have been performed primarily in the solar system, which is a weakly gravitating system from the view point of general relativity because the non-dimensional compactness parameter of the sun, $GM_\odot/c^2 R_\odot$, is very small, 4×10^{-6}, where G, c, $M_\odot$, and $R_\odot$ are the gravitational constant, the speed of light, the solar mass, and the solar radius, respectively. The properties of general relativity will be revealed in an object of the compactness $\sim 0.1-1$ such as black holes and neutron stars. General relativity has been also tested by observing the systems composed of two neutron stars [Stairs (2003, 2004); Lorimer (2008); Weisberg et al. (2010)]. However, the gravity in the observed systems is not strongly general relativistic, because the orbital radius is comparable to $R_\odot$, and thus, the compactness of the system is also much smaller than unity. No precise tests have been performed in the strongly general relativistic environments such as vicinities of black holes and neutron stars.

Moreover, although a number of neutron stars and black holes have been discovered in the past fifty years, we have not yet observed *their formation processes* in detail. Only one chance was closely encountered in the observation of supernova 1987A via the neutrino detection [Hirata et al. (1987); Bionta et al. (1987)]. However, the number of the detected neutrinos was too small to derive detailed information about the formation process of a neutron star or a black hole. In the formation of a neutron star or a black hole, a strongly general relativistic and time varying gravitational field would be realized. However, we have not yet directly observed such a violent phenomenon.

Einstein's equation, the basic equation of general relativity, is a highly nonlinear equation. Hence, the nonlinearity is a key property of general relativity. The nonlinear nature will be revealed in violently time varying and strongly gravitating phenomena, and therefore, to explore the detailed nature of general relativity, we have to observe dynamical phenomena in a strongly gravitating environment. In such events, it will also be feasible to test the validity of general relativity in a strongly general relativistic environment.

Observation of such phenomena will be realized through the gravitational-wave observation by laser-interferometric gravitational-wave detectors (see section 1.2.6). Gravitational waves detectable are emitted when a strongly gravitational field varies violently, e.g., in the stellar core collapse of massive stars and coalescences of binary neutron stars/black holes. One of the key properties of gravitational waves is that they are transparent to any matter, and hence, they can directly propagate from the source to us even if the source is densely surrounded by a matter field. This property contrasts well with that of electromagnetic waves which interact strongly with baryons and leptons. Hence, observed electromagnetic waves are usually emitted in a region far from the strongly gravitating object, and thus, do not carry direct information for the strongly general relativistic and dynamical objects. Neutrinos are also transparent to matter fields, but they are less transparent than gravitational waves. For example, the central region of supernova explosions is surrounded by a neutrino sphere where neutrinos are primarily emitted. Therefore, detecting gravitational waves is the best way for directly observing strongly gravitating and dynamical objects.

In addition to the gravitational-wave observation, observation tools in high-energy astrophysics, such as X-ray and gamma-ray satellites, have been significantly developed in the past four decades. One of the central subjects in this research field is to clarify the origin of gamma-ray bursts [Zhang and Meszáros (2004); Piran (2004); Nakar (2007)]. These are transient and high-energy phenomena that primarily emit gamma-rays, and they uniformly and frequently (every ~ 10 hours) occur in the universe. The duration is $0.01-1000$ s, the typical variation timescale is milliseconds, and the total energy emitted is estimated to be typically $\sim 10^{51}$ ergs, which is comparable to the total energy emitted by the sun in its lifetime $\sim 10^{10}$ yrs. Furthermore, they are not repeaters.

Because the short variation timescale implies that they are compact, the origin of gamma-ray bursts is inferred to be a compact object. In addition, the fact that they emit huge total energy in a short timescale suggests that the energy source would be associated with gravitational binding energy. Finally, the fact that they are not repeaters indicates that the central engine should be formed in a dynamical phenomenon. All these evidences strongly suggest that the central engine would be composed of a stellar-mass black hole surrounded by a dense and hot torus which are formed in a dynamical phenomenon such as stellar core collapse or merger of compact neutron star binaries. This indicates that we are approaching the observation of a strongly general relativistic and dynamical phenomena through the observation of gamma-ray bursts.

Numerical relativity required

To theoretically clarify the properties of strongly general relativistic and dynamical phenomena in nature as well as to explain observational results for high-energy astrophysical phenomena, it is often necessary to solve Einstein's equation and associated matter field equations. However, these equations are nonlinear partial-differential equations and hence it is impossible to solve them analytically for general problems. This implies that a numerical computation is necessary in this research field. Numerical relativity is a field of theoretical physics in which Einstein's equation is solved using computers to clarify the nature of general relativity, the properties of astrophysical compact objects such as black holes and neutron stars, and the high-energy phenomena in the universe such as stellar core collapses and supernova explosions.

Numerical relativity has been an active field in general relativity and astrophysics since 1990s. One of the reasons for this is that projects of the gravitational-wave detection using laser-interferometric detectors became in earnest in those days (see section 1.2.6). Advanced gravitational-wave detectors such as LIGO, VIRGO, and KAGRA (formerly LCGT) will be in operation sequentially from 2015. These detectors will detect gravitational waves for the first time, and subsequently, they will contribute a lot to observing general relativistic and dynamical phenomena. For the detection of gravitational-wave signals and subsequent extraction of physical information from the signals, however, a precise prediction of gravitational waveforms (i.e., theoretical templates) is required. Gravitational-wave detections will be confirmed only when the cross correlation of the signal with a theoretical template is sufficiently high. Thus, the theoretical templates have to be highly accurate, because a systematic inaccuracy in them can reduce the value of the cross correlation and lead to the failure of the detection. Theoretical templates of gravitational waves from highly nonlinear (strongly general relativistic) phenomena can be derived only by numerical-relativity simulations. This fact has been stimulating the activity of this field significantly.

High-energy astrophysics has also been developed significantly, and consequently, new unsolved issues have been raised frequently. As mentioned above, the central engine of gamma-ray bursts is likely to be composed of a stellar-mass black hole surrounded by a dense and hot accretion torus. However, its progenitor and the formation process are still unsolved. It is quite difficult to directly observe the formation process by detecting electromagnetic signals because they cannot directly carry the information of the central object. Future possible detections of gravitational waves and neutrinos may constrain the formation mechanism, but probably, the information will be a limited one. To understand the formation process taking into account the limited observational data, theoretical studies are obviously necessary. For this purpose, numerical-relativity simulations are also required.

Motivated by these two facts, numerical relativity has been significantly developed in the last two decades. In particular, in the last decade, a variety of scientific simulations for the coalescence of binary neutron stars, binary black holes, and black hole-neutron star binaries have been performed. A detailed physical simulation for the supernova core collapse leading to the formation of a neutron star and a black hole has been also reported in the last five years. These simulations become feasible due to the significant progress in formulations and numerical techniques in numerical relativity. The purpose of this volume is to describe these established formulations and techniques and to report the knowledge obtained from the numerical simulations performed so far.

Outline of this volume

This volume is organized in the following order. Chapter 1 describes the backgrounds of numerical relativity; basics of general relativity, gravitational waves, compact objects, and general relativistic astrophysics, as well as unsolved issues in these fields are described. Basic equations of gravitational waves and gravitational-wave sources are described in detail, because computing gravitational waves from a variety of gravitational-wave sources is the primary subject in numerical relativity. If a reader is familiar with these topics, he/she does not have to read chapter 1. On the other hand, if a reader is a beginner of general relativity and astrophysics, and if he/she would like to study the topics described in chapter 1 from the basics in depth, we recommend them to refer to well-known textbooks, e.g., Misner *et al.* (1973); Wald (1984); Schutz (1985); Poisson (2004) for general relativity, to Thorne (1987) for gravitational waves, and to Zel'dovich and Novikov (1971); Shapiro and Teukolsky (1983) for relativistic astrophysics.

Chapters 2–7 constitute the part I of this volume in which we describe the formulations and numerical techniques often used in numerical relativity. In particular, chapters 2–5 are devoted to describing the core parts of numerical relativity. In chapter 2, formulations in numerical relativity are summarized, paying particular attention to 3+1 (and $N+1$ with N the spatial dimension) formulations and their modified versions in general relativity. The 3+1 formulations for numerical relativity are also summarized in the latest textbooks by Alcubierre (2008); Baumgarte and Shapiro (2010); Gourgoulhon (2012). The introductory part in our volume is written in a similar manner to those in these previous textbooks. However, the method of descriptions would not be always the same. We will be grateful if the readers could feel a new flavor from our volume.

In chapter 3, we introduce numerical methods, focusing on finite-differencing methods, for a numerical solution of the evolution equations (hyperbolic equations) in the 3+1 ($N+1$) formulations. For the help of the students who would like to make a numerical-relativity code by themselves, finite-differencing schemes are described in detail. In addition, testbed problems, which should be employed for test simulations to confirm the reliability of a new numerical code, are described with typical results of several test simulations.

In chapter 4, we describe matter field equations in general relativity: scalar-field equations, equations for collisionless particles, hydrodynamics and magneto-hydrodynamics equations, and radiation transfer equations. First, basic equations together with their properties are summarized. Then, we pay attention to numerical methods for a solution of hydrodynamics/magnetohydrodynamics equations together with the testbed problems.

In chapter 5, we describe the methods for a solution of the initial-value problems in general relativity. In the 3+1 ($N+1$) formulations of general relativity, the constraint equations are present, and they have to be always solved for preparing an initial condition. The typical procedure for a solution of these equations together with several known analytic solutions are summarized in this chapter. In addition, we describe how to define energy, linear momentum, and angular momentum of asymptotically flat systems, which characterize properties of the initial condition.

Chapters 6 and 7 are devoted to describing the often-used necessary tools for analyzing numerical-relativity results. As already mentioned in this preface, one of the most important roles in numerical relativity is to derive accurate gravitational waves emitted in general relativistic astrophysical phenomena. Thus, in chapter 6, we show the representative methods for extracting gravitational waves from geometric variables. In chapter 7, we summarize the tools for analyzing black holes in numerical relativity. In particular, we describe in detail about apparent horizon, which is usually determined for confirming the presence of a black hole and for analyzing the properties of the black hole.

Chapters 8–11 constitute the part II of this volume, in which we review the representative numerical-relativity simulations and the knowledge obtained from these simulations. In chapter 8, one of the highlights of this volume, we introduce a variety of numerical-relativity simulations for the coalescence of binary compact objects, binary neutron stars, black hole-neutron star binaries, and binary black holes. In the last decade, a great amount of new knowledge for the merger process and the fates after the merger of these binary systems have been obtained. These provide valuable predictions for the forthcoming gravitational-wave observation as well as for the high-energy astrophysical observation. We will report the current status in this research field. Since the author has been deeply working in the neutron-star binaries, the focus is in particular on neutron star-neutron star and black hole-neutron star binaries.

In chapter 9, we describe the current status for the general relativistic simulations of black-hole formation, which is also one of the main subjects in numerical relativity and computational astrophysics. Although this field has a long history since 1960s, we focus only on describing the latest progress of multi-dimensional, fully general relativistic simulations. Specifically, we will describe the mechanism for the stellar core collapse to a black hole, collapse of neutron stars to a black hole, collapse of a supermassive star to a black hole, and primordial black hole formation.

Beside the merger of binary compact objects and stellar core collapse to a black hole, there are several important subjects in numerical relativity and astrophysics; exploring non-radial instability of neutron stars and accretion disks/torus surrounding a black hole. If these instability turn on, the neutron star and accretion disks or torus are deformed, being a strong emitter of gravitational waves. The dynamical evolution for these objects could be a high-energy transient phenomenon that may be observable by astronomical telescopes. Thus, exploring the instability of general relativistic objects in detail is also an important issue in numerical relativity. In chapter 10, we describe some of the progress for these subjects.

There are also several new frontiers in numerical relativity. One of the most exciting fields is higher-dimensional numerical relativity. Since a possibility of production of mini black holes in huge particle accelerators such as CERN Large Hardon Collider (LHC) was pointed out, studies of the black-hole formation have been important: If our space is a 3-brane in large [Arkani-Hamed *et al.* (1998)] or warped [Randall and Sundrum (1999b,a)] extra dimensions, the Planck energy could be of $O(\text{TeV})$ that may be accessible by LHC. Thus, in the presence of the extra dimensions, mini black holes may be produced and its evidence may be detected. Numerical relativity is the unique approach for clarifying their formation process quantitatively, and in the past decade, the higher-dimensional numerical relativity has been an active field in *high-energy* physics [Cardoso *et al.* (2012)]. Chapter 11 introduces several numerical results in this research field.

Acknowledgments

This volume could not have been written without the encouragement, valuable discussions, and insights provided by my colleagues and collaborators. In particular, I am deeply indebted to Satoru Ikeuchi, Takashi Namakura, Misao Sasaki, and Stuart Shapiro for their mentoring. I am also grateful to Kenta Hotokezaka, Kyo-hei Kawaguchi, Kenta Kiuchi, Koutarou Kyutoku, Hiroki Nagakura, Hiroyuki Nakano, Yuichiro Sekiguchi, Yudai Suwa, and Hirotada Yoshino for reading a part of this volume, for giving me valuable comments, and in addition, for correcting a part of this volume, and to Sho Fujibayashi, Taku Watanabe, and Yasuho Yamashita for checking a part of this volume. I also thank Hideki Asada, Luca Baiotti, Thomas Baumgarte, Alessandra Buonanno, Matt Duez, Yoshiharu Eriguchi, Toni Font, John Friedman, Naoteru Gouda, Eric Gourgoulhon, Izumi Hachisu, Kunihito Ioka, Yuk-Tang Liu, Kei-ichi Maeda, Pedro Montero, Ken-ichi Nakao, Takashi Okamura, Hirotada Okawa, Ken-ichi Oohara, Luciano Rezzolla, Tetsuya Shiromizu, Takeru Suzuki, Hideyuki Tagoshi, Fumio Takahara, Mariko Takahara, Masaomi Tanaka, Takahiro Tanaka, Keisuke Taniguchi, Koji Uryū, Shoichi Yamada, Tetsuro Yamamoto, Shijun Yoshida, and Takashi Yoshida for the fruitful discussions. Finally, I am grateful to my family, Keiko and Yūki, for their encouragement.

Although we checked the content of this volume for many times, we have no doubt that there would be still many typos and mistakes. Whenever we find a typo or mistake, we will list them in the following website: http://www2.yukawa.kyoto-u.ac.jp/~masaru.shibata/NRbook.html.

Masaru Shibata
December 15, 2014

Contents

Preface   v

1. Preliminaries for numerical relativity   1
   1.1 Brief introduction of general relativity 1
      1.1.1 Einstein's equation 1
      1.1.2 Nature of Einstein's equation 3
   1.2 Gravitational waves 4
      1.2.1 Linearized Einstein's equation 5
      1.2.2 Propagation of gravitational waves 6
      1.2.3 Generation of gravitational waves 7
      1.2.4 Gravitational-wave luminosity 10
      1.2.5 Gravitational waves from a binary 16
      1.2.6 Gravitational-wave detectors 18
   1.3 Black holes 22
      1.3.1 Four dimensional black holes 25
      1.3.2 Properties of four-dimensional black holes 30
      1.3.3 Higher-dimensional black holes 38
   1.4 Neutron stars 40
      1.4.1 Formation of neutron stars 41
      1.4.2 Basic properties of neutron stars 42
      1.4.3 Hydrostatic equations for cold neutron stars 43
      1.4.4 Cold neutron-star equations of state 45
      1.4.5 Structure and sequences of neutron stars 50
      1.4.6 Supramassive and hypermassive neutron stars 51
      1.4.7 Finite-temperature equations of state for high-density matter 55
      1.4.8 Binary neutron stars 57
   1.5 Sources of gravitational waves 62
      1.5.1 Inspiral of binary compact objects 63
      1.5.2 Merger of binary neutron stars 69
      1.5.3 Merger of binary black holes 74

		1.5.4	Merger of black hole-neutron star binaries	77
		1.5.5	Gravitational collapse and core collapse supernova	82
	1.6	Matched filtering techniques for gravitational-wave data analysis		103

Methodology   105

2. Formulation for initial-value problems of general relativity   107
　　2.1 Formulations based on spacetime foliation 109
　　　　2.1.1 Foliation of spacetime 109
　　　　2.1.2 Derivative operator 111
　　　　2.1.3 Extrinsic curvature 112
　　　　2.1.4 The Gauss and Codazzi equations 114
　　　　2.1.5 Constraint and evolution equations 115
　　　　2.1.6 Evolution of constraint equations 119
　　　　2.1.7 Conformal decomposition 120
　　2.2 Gauge conditions 123
　　　　2.2.1 Required conditions for the gauge choice 123
　　　　2.2.2 Maximal slicing 129
　　　　2.2.3 Harmonic gauge 137
　　　　2.2.4 Issues for black hole evolution and excision 139
　　　　2.2.5 Minimal distortion gauge 140
　　　　2.2.6 Dynamical gauge 142
　　2.3 Formulations in numerical relativity 148
　　　　2.3.1 Linearized standard 3+1 formulation 149
　　　　2.3.2 Linearized equations for purely conformal decomposition formulation 153
　　　　2.3.3 BSSN formulation 155
　　　　2.3.4 Generalized harmonic formulation 160
　　　　2.3.5 Other formulations 162
　　2.4 Formulations in axisymmetric spacetime 166
　　　　2.4.1 Introduction 167
　　　　2.4.2 Regularity conditions 168
　　　　2.4.3 Formulation with quasi-isotropic gauge 172
　　　　2.4.4 Formulation with quasi-radial gauge 176
　　　　2.4.5 NCSA formulation 179
　　　　2.4.6 (2+1)+1 formulation 180
　　2.5 Cartoon method 186
　　　　2.5.1 Cartoon method for axisymmetric spacetime 186
　　　　2.5.2 Cartoon method for higher-dimensional spacetime: Modified cartoon method 189

	2.6	Formulations for asymptotically de Sitter and Friedmann spacetime 192
		2.6.1 Asymptotically de Sitter spacetime 192
		2.6.2 Asymptotically Friedmann spacetime 194

3. Numerical methods for a solution of Einstein's evolution equation 197

    3.1 Solving hyperbolic equations 197
        3.1.1 Finite-differencing for spatial derivatives 198
        3.1.2 Time integration 200
        3.1.3 Boundary condition at outer boundaries 206
        3.1.4 Courant–Friedrichs–Lewy condition 209
        3.1.5 Numerical experiments 211
    3.2 Handling advection terms 216
    3.3 Adaptive mesh refinement 219
        3.3.1 Why necessary? 219
        3.3.2 Spatial interpolation in the buffer zone 222
        3.3.3 Time integration scheme in the buffer zone 225
        3.3.4 Restriction 229
        3.3.5 Kreiss–Oliger dissipation 229
    3.4 Testing numerical-relativity code: Vacuum spacetime 230
        3.4.1 Examining convergence property 230
        3.4.2 Propagation of linear gravitational waves 231
        3.4.3 Evolution of black holes 234

4. Matter equations in general relativity 243

    4.1 Scalar fields 243
    4.2 Collisionless particles 244
    4.3 3+1 form of stress-energy conservation: $\nabla_a T^a{}_b = 0$ 248
    4.4 Hydrodynamics 249
        4.4.1 Basic equations 249
        4.4.2 Properties of hydrodynamics equations 252
    4.5 Hydrodynamics with microphysics 258
    4.6 Electromagnetohydrodynamics 260
        4.6.1 Definitions 261
        4.6.2 Ideal magnetohydrodynamics 266
        4.6.3 Properties of the ideal magnetohydrodynamics equations . 269
        4.6.4 Force-free electromagnetodynamics 272
    4.7 Radiation transfer and radiation hydrodynamics 275
        4.7.1 Boltzmann's equation 277
        4.7.2 Moment formalism 285
        4.7.3 Leakage scheme 293

4.8 Numerical methods for hydrodynamics and
    magnetohydrodynamics: Handling transport term 295
    4.8.1 Monotonicity preserving 297
    4.8.2 Godunov's theorem 297
    4.8.3 Circumventing Godunov's theorem 300
    4.8.4 Total variation diminishing 303
    4.8.5 Reconstruction of numerical flux at cell interfaces 304
    4.8.6 Approximate Riemann solvers 308
4.9 Other ingredients in numerical hydrodynamics and
    magnetohydrodynamics 313
    4.9.1 Solving normalization relation for four velocity 313
    4.9.2 Constrained transport 314
    4.9.3 Implicit-explicit scheme 317
4.10 Testing hydrodynamics and magnetohydrodynamics codes 319
    4.10.1 Riemann shock-tube problems in hydrodynamics 319
    4.10.2 Riemann shock-tube problems in magnetohydrodynamics . 321
    4.10.3 Bondi flow 324
    4.10.4 Stationary accretion torus around a black hole 328
4.11 Testing a numerical-relativity code with matter 328
    4.11.1 Stability of neutron stars 328
    4.11.2 Oscillation of neutron stars 332
    4.11.3 Collapse of unstable neutron stars 335

5. Formulations for initial data, equilibrium, and quasi-equilibrium   337
    5.1 Properties of initial-data equations 337
    5.2 York–Lichnerowicz formulation 339
    5.3 Mass, linear momentum, and angular momentum 340
        5.3.1 ADM mass, linear momentum, and angular momentum in
              the Hamiltonian formulation 341
        5.3.2 Komar mass and angular momentum 345
        5.3.3 Virial relation 348
        5.3.4 Irreducible mass 351
    5.4 Initial data for pure gravitational waves 352
    5.5 Initial data for black holes 353
        5.5.1 Time symmetric case 353
        5.5.2 Time asymmetric case 358
    5.6 Isenberg–Wilson–Mathews/Conformal thin-sandwich formulation   365
        5.6.1 Isenberg–Wilson–Mathews formulation 366
        5.6.2 Conformal thin-sandwich formulation 367
        5.6.3 Handling black-hole horizon 368
        5.6.4 IWM plus conformal transverse-tracefree formulation . . . 372

		5.6.5	Using excision initial data of black holes in the time evolution 373
	5.7	Formulation of equilibrium and quasi-equilibrium 374
		5.7.1	Axisymmetric equilibria 375
		5.7.2	Quasi-equilibria for non-axisymmetric spacetime 387
	5.8	Numerical methods for a solution of elliptic equations 396
		5.8.1	Finite-differencing schemes 396
		5.8.2	Spectral and pseudo-spectral methods 401

6. Extracting gravitational waves　　　　　　　　　　　　　　　　409

   6.1 Gauge-invariant wave extraction 410
   6.2 Extraction using a complex Weyl scalar 415
       6.2.1 Overview of the Newman–Penrose formalism 416
       6.2.2 The gauge freedom of the Newman–Penrose quantities and classification of spacetime 418
       6.2.3 Weyl tensor in the 3+1 formulation 420
       6.2.4 Perturbation on type D spacetime 421
       6.2.5 Ψ_4 and outgoing gravitational waves 422
       6.2.6 Waveforms and luminosity 424
   6.3 Quadrupole formula 427

7. Finding black holes　　　　　　　　　　　　　　　　　　　　　429

   7.1 Apparent horizon 429
       7.1.1 Spherically symmetric case 431
       7.1.2 Axially symmetric case 432
       7.1.3 Non-axially symmetric case 435
       7.1.4 Finding apparent horizons for simple initial data 437
   7.2 Event horizon 441
   7.3 Analyzing horizons 443

Applications　　　　　　　　　　　　　　　　　　　　　　　　445

8. Coalescence of binary compact objects　　　　　　　　　　　　　447

   8.1 Binary neutron stars: Brief introduction 448
   8.2 Binary neutron stars: Quasi-equilibrium states and sequences . . . 449
       8.2.1 Velocity field 450
       8.2.2 Gravitational-field equations 453
       8.2.3 Numerical results 454
   8.3 Binary neutron stars: Numerical simulations 461
       8.3.1 Equations of state for numerical simulations 461
       8.3.2 Eccentricity reduction for initial condition 464

		8.3.3	Merger process and remnants	464

        8.3.3   Merger process and remnants 464
        8.3.4   Gravitational waves 486
        8.3.5   Dynamical mass ejection and electromagnetic counterparts 502
  8.4   Black hole-neutron star binaries: Quasi-equilibrium states and
        sequences 514
        8.4.1   Formulations 514
        8.4.2   Two types of quasi-equilibrium sequences 516
        8.4.3   Endpoint of sequences 518
  8.5   Black hole-neutron star binaries: Numerical simulations 521
        8.5.1   Methodology of simulations and equations of state 521
        8.5.2   Merger process 523
        8.5.3   Properties of the remnants 537
        8.5.4   Gravitational waves 545
        8.5.5   Fourier spectrum of gravitational waves 549
        8.5.6   Dynamical mass ejection and electromagnetic counterparts 558
  8.6   Binary black holes: Quasi-equilibrium states 563
        8.6.1   Effective-potential method 564
        8.6.2   Quasi-equilibrium with physical horizon boundary
                conditions 567
  8.7   Binary black holes: Numerical simulations 571
        8.7.1   Two approaches for evolving binary black holes 572
        8.7.2   Inspiral and merger processes 573
        8.7.3   Modeling gravitational waveforms and their spectrum . . 585

9. Gravitational collapse to a black hole   591
  9.1   Collapse of a supramassive neutron star to a black hole 591
        9.1.1   The criterion for the quasi-radial collapse 592
        9.1.2   The dynamical stability and final fate of gravitational
                collapse 595
  9.2   Collapse of a rotating supermassive star to a supermassive
        black hole 599
        9.2.1   Equation of state for massive stars 599
        9.2.2   Equilibrium and stability of rotating supermassive stars . . 602
        9.2.3   The final state of rapidly rotating supermassive stars . . . 606
  9.3   Stellar core collapse of massive stars to a black hole 609
        9.3.1   Path to the gravitational collapse of massive stars 611
        9.3.2   Numerical-relativity simulation of rotating massive stars to
                a black hole 614
  9.4   Formation of a primordial black hole 635

10. Non-radial instability and magnetohydrodynamics instability   641
  10.1 Non-axisymmetric instability of rapidly rotating neutron stars . . . 641

- 10.2 Non-axisymmetric instability of torus surrounding black hole ... 646
- 10.3 Magnetohydrodynamical instability of neutron stars 649
  - 10.3.1 Magnetohydrodynamics instability 649
  - 10.3.2 Numerical simulations 656

11. Higher-dimensional simulations 671
    - 11.1 Gregory–Laflamme instability of black string 671
    - 11.2 Black hole collisions and scattering 674
      - 11.2.1 Black hole head-on collision in $D=5$ 674
      - 11.2.2 Off-axis collision of two black holes 675
    - 11.3 Bar-mode instability of Myers–Perry black holes 678
      - 11.3.1 Studies for the instability of Myers–Perry black holes ... 678
      - 11.3.2 Numerical simulations 680

12. Conclusion 687

Appendix A Killing vector and Frobenius' theorem 689
- A.1 Killing vector fields 689
- A.2 Frobenius' theorem 691

Appendix B Numerical relativity in spherical symmetry 693
- B.1 Spatially conformally flat gauge 694
- B.2 Zero shift 695
  - B.2.1 Evolving γ_{rr} and $\gamma_{\theta\theta}$ 695
  - B.2.2 Maximal slicing 696
- B.3 Radial gauge 697
- B.4 Misner–Sharp–May–White and Hernandez–Misner formulations 698

Appendix C Decomposition by spherical harmonics 701
- C.1 Spherical harmonics 701
  - C.1.1 Scalar spherical harmonics 701
  - C.1.2 Vector spherical harmonics 702
  - C.1.3 Tensor spherical harmonics 703
  - C.1.4 Decomposition of spatial tensor by spherical harmonics .. 705
  - C.1.5 Spin-weighted spherical harmonics 705
- C.2 Tensor harmonics decomposition of spatial tensors and linear gravitational waves 706
- C.3 Vector harmonics decomposition of spatial vectors 709

Appendix D Lagrangian and Hamiltonian formulations of general relativity 711
- D.1 Action principle for gravitational fields 711
- D.2 Action principle for general relativistic fluids 714

	D.3	Gravitational-field action in $N+1$ variables	715
	D.4	ADM formulation from the variation principle	717
	D.5	Hamiltonian formulation	718

Appendix E Solutions of Riemann problems in special relativistic hydrodynamics — 721

	E.1	Riemann problems for pure hydrodynamics in special relativity	721
	E.2	Solutions of shock-tube problem in hydrodynamics	723
		E.2.1 Riemann invariants and solution for rare-faction waves	723
		E.2.2 Jump conditions at shock waves	725
	E.3	Wall shock problem in hydrodynamics	727
	E.4	Solutions of Riemann problems in ideal magnetohydrodynamics	727
		E.4.1 Solutions for shocks	728
		E.4.2 Solutions for Riemann problems	729
		E.4.3 Solutions for nonlinear Alfvén waves	731

Appendix F Landau–Lifshitz pseudo tensor — 733

Appendix G Laws of black hole and apparent horizon — 737

	G.1	Raychaudhuri's equation	738
	G.2	Expansion, shear, and rotation on event horizons	739
	G.3	Zeroth law	741
	G.4	First law	742
		G.4.1 Stationary and axisymmetric case	743
		G.4.2 Helically symmetric case	748
	G.5	Apparent horizon and dynamical horizon	751
	G.6	Isolated horizon	754

Appendix H Post-Newtonian results for coalescing compact binaries — 757

	H.1	Solutions for circular orbits	758
	H.2	Luminosity, emission rate of linear and angular momenta, amplitude of gravitational waves	760
	H.3	Adiabatic approximations for the orbital evolution, and Taylor approximants	763
	H.4	Effective one-body approach	766

Bibliography — 769

Index — 809

Notation

Unless otherwise stated, we always use the following notation throughout this volume:

g_{ab} : spacetime metric

γ_{ab} : spatial metric

$\tilde{\gamma}_{ab}$: conformal spatial metric

∇_a : spacetime covariant derivative associated with g_{ab}

D_a : spatial covariant derivative associated with γ_{ab}

$\tilde{D}_a$: spatial covariant derivative associated with $\tilde{\gamma}_{ab}$

η_{ab} : flat spacetime metric

f_{ab} : flat spatial metric

$^{(D)}R_{abcd}$: D-dimensional spacetime Riemann tensor

$^{(4)}R_{abcd}$: four-dimensional spacetime Riemann tensor

$^{(D)}R_{ab}$: D-dimensional spacetime Ricci tensor

$^{(4)}R_{ab}$: four-dimensional spacetime Ricci tensor

G_{ab} : spacetime Einstein tensor

R_{abcd} : spatial Riemann tensor

R_{ab} : spatial Ricci tensor

K_{ab} : extrinsic curvature on spatial hypersurfaces

n_a : future-directed unit normal timelike vector field satisfying $\gamma_{ab}n^b = 0$

T_{ab} : spacetime stress-energy tensor

$\rho_{\rm h} = T_{ab}n^a n^b$

$J_c = -T_{ab}\gamma^a{}_c n^b$

$S_{cd} = T_{ab}\gamma^a{}_c \gamma^b{}_d$

ρ : rest-mass density of baryon

u^a : four velocity of fluid

ε : specific internal energy of fluid

h : specific enthalpy of fluid

P : pressure of fluid

Ψ : velocity potential

$\mathscr{L}_k$: Lie derivative with respect to a vector k^a

$\epsilon_{abcd\cdots}$: completely antisymmetric tensor with $\epsilon_{0123\cdots} = \sqrt{-g}$

and ∂_μ denotes the partial derivative. D and N denote the spacetime and spatial dimensions, respectively.

Throughout this volume, we adopt the $(-, +, +, +, \cdots, +)$ metric signature. We basically follow an abstract index notation of the textbook of Wald (1984); Latin indices, $a, b, c, \cdots$, and h on a tensor do not represent components but are parts of the notation for the tensor itself. Any equation involving tensors (on D-dimensional spacetime or N-dimensional space) that employs the Latin indices represents a relation between tensors. On the other hand, Greek indices $\alpha, \beta, \gamma, \cdots$ on a spacetime tensor represent spacetime components, as in the usual convention. In addition, we employ a notation that Latin indices $i, j, k, \cdots$ on a spatial tensor represent spatial components. This notation is often employed in the field of numerical relativity. Unless otherwise stated, we employ the geometrical units in which $c = 1 = G$ where c and G are the speed of light and gravitational constant, respectively. Thus, in four dimensions, the length dimension (and time dimension) is equivalent to the mass dimension, and in higher dimensions, the length dimension is equivalent to the (mass)$^{1/(D-3)}$ dimension. k_B and $\hbar$ are Boltzmann's constant and reduced Planck's constant, respectively.

$[ab \cdots cd]$ for the subscripts denotes that the antisymmetric relation is taken, e.g., as

$$T_{[ab]} = \frac{1}{2!}\Big(T_{ab} - T_{ba}\Big), \quad T_{[abc]} = \frac{1}{3!}\Big(T_{a[bc]} + T_{b[ca]} + T_{c[ab]}\Big),$$

while $(ab \cdots cd)$ for the subscripts denotes that the symmetric relation is taken, e.g., as

$$T_{(ab)} = \frac{1}{2!}\Big(T_{ab} + T_{ba}\Big), \quad T_{(abc)} = \frac{1}{3!}\Big(T_{a(bc)} + T_{b(ca)} + T_{c(ab)}\Big).$$

Chapter 1

Preliminaries for numerical relativity

Before we describe the details for the formulations and numerical methods of numerical relativity, in this chapter, we shall touch on general relativity, gravitational waves, general relativistic objects such as black holes, neutron stars, and binary neutron stars, and sources of gravitational waves, to which special attention is paid in numerical relativity and to which we often refer in the subsequent chapters.

1.1 Brief introduction of general relativity

1.1.1 *Einstein's equation*

The purpose of numerical relativity is to clarify the nature of spacetime dynamics and dynamical evolution of general relativistic objects, by numerically solving Einstein's equation

$$G_{ab} = 8\pi T_{ab}, \tag{1.1}$$

where G_{ab} and T_{ab} denote the D-dimensional Einstein tensor and the stress-energy tensor for a matter field, respectively. G_{ab} is defined from the Ricci tensor by

$$G_{ab} = \overset{(D)}{R}_{ab} - \frac{1}{2}g_{ab}\overset{(D)}{R}, \tag{1.2}$$

where $\overset{(D)}{R}_{ab}$ is the D-dimensional Ricci tensor and $\overset{(D)}{R}$ is the Ricci scalar defined by $\overset{(D)}{R}_{ab}g^{ab}$. g_{ab} is the spacetime metric, which determines the invariant spacetime interval (distance) between two nearby points according to

$$ds^2 = g_{\mu\nu}dx^\mu dx^\nu. \tag{1.3}$$

Here dx^μ denotes the infinitesimal difference of the spacetime coordinates x^μ.

The covariant derivative ∇_a is defined as an operator that maps a tensor field of type (k, l) to a tensor field of type $(k, l+1)$ as $\nabla_a T^{b_1 b_2 \cdots b_k}{}_{c_1 c_2 \cdots c_l}$, and satisfies $\nabla_a g_{bc} = 0$. The covariant derivative of the tensor is related to the ordinary derivative with the introduction of connection coefficients (Christoffel symbols), $\overset{(D)}{\Gamma}{}^a{}_{bc}$, which are written in a coordinate basis

$$\overset{(D)}{\Gamma^{\alpha}}_{\beta\gamma} = \frac{1}{2}g^{\alpha\mu}\Big(\partial_{\beta}g_{\mu\gamma} + \partial_{\gamma}g_{\mu\beta} - \partial_{\mu}g_{\beta\gamma}\Big), \tag{1.4}$$

where the symmetry relation, $\overset{(D)}{\Gamma^{\alpha}}_{\beta\gamma} = \overset{(D)}{\Gamma^{\alpha}}_{\gamma\beta}$, is satisfied for this. The connection coefficients are not tensor. This is found, e.g., from the fact that they vanish in a local Lorentz frame at which $g_{ab} = \eta_{ab}$ where η_{ab} denotes the flat, Minkowski metric.

Geodesic equations for a test particle with a four vector u^a, which moves in some spacetime of metric g_{ab}, are written as

$$u^b \nabla_b u^a = 0 \quad \text{or} \quad u^\nu \partial_\nu u^\mu + \Gamma^\mu{}_{\nu\sigma} u^\nu u^\sigma = 0. \tag{1.5}$$

In local Lorentz frames, this reduces to $u^\nu \partial_\nu u^\mu = 0$. This implies that a local observer never measures the strength of the gravitational field.

The Ricci tensor is defined from the Riemann tensor $\overset{(D)}{R}_{abcd}$ as

$$\overset{(D)}{R}_{ac} = \overset{(D)}{R}_{abcd} g^{bd}. \tag{1.6}$$

The Riemann tensor is defined by

$$\overset{(D)}{R}_{abc}{}^d \omega_d = (\nabla_a \nabla_b - \nabla_b \nabla_a)\omega_c, \tag{1.7}$$

where ω^a denotes an arbitrary vector field. Using the connection coefficients, this is written in a coordinate basis as

$$\overset{(D)}{R}_{\mu\nu\sigma}{}^\lambda = \partial_\nu \Gamma^\lambda{}_{\mu\sigma} - \partial_\mu \Gamma^\lambda{}_{\nu\sigma} + \Gamma^\alpha{}_{\mu\sigma}\Gamma^\lambda{}_{\nu\alpha} - \Gamma^\alpha{}_{\nu\sigma}\Gamma^\lambda{}_{\mu\alpha}. \tag{1.8}$$

Here, the Riemann tensor satisfies symmetry relations, $\overset{(D)}{R}_{abcd} = -\overset{(D)}{R}_{bacd} = -\overset{(D)}{R}_{abdc} = \overset{(D)}{R}_{cdab}$ and $\overset{(D)}{R}_{[abc]d} = 0$. In general relativity (with $D \geq 4$), the Riemann tensor, which cannot be erased by any coordinate transformation in contrast to connection coefficients, plays a role for measuring the strength of the gravitational field through the geodesic deviation equation

$$a^a_{\text{mid}} = u^c \nabla_c (u^b \nabla_b X^a) = \overset{(D)}{R}_{bcd}{}^a X^b u^c u^d, \tag{1.9}$$

where u^a is a timelike vector field which obeys equation (1.5) and X^a is a spatial vector field orthogonal to u^a. Here, the acceleration, a^a_{mid}, measures the relative acceleration, i.e., tidal force, between two neighboring observers for which the four velocity is u^a. Thus, if $\overset{(D)}{R}_{abcd} \neq 0$, $a^a_{\text{tid}} \neq 0$, and hence, the observers feel a tidal force. Because the Riemann tensor plays a central role for measuring the gravitational-field strength, we often use a curvature invariant like $\overset{(D)}{R}_{abcd} \overset{(D)}{R}{}^{abcd}$ for determining the absolute field strength.

The Ricci tensor is the trace part of the Riemann tensor as shown in equation (1.6). On the other hand, the tracefree part of the Riemann tensor is referred to as the Weyl tensor, which is defined by

$$\overset{(D)}{C}_{abcd} = \overset{(D)}{R}_{abcd} - \frac{1}{D-2}\left(g_{ac}\overset{(D)}{R}_{bd} - g_{ad}\overset{(D)}{R}_{bc} - g_{bc}\overset{(D)}{R}_{ad} + g_{bd}\overset{(D)}{R}_{ac}\right)$$
$$+ \frac{1}{(D-1)(D-2)}(g_{ac}g_{bd} - g_{ad}g_{bc})\overset{(D)}{R}. \qquad (1.10)$$

The Weyl tensor is often referred to as the conformal tensor because it is invariant under the conformal transformation $g_{ab} \to \Psi^2 \bar{g}_{ab}$ as

$$\overset{(D)}{C}_{abc}{}^d = \bar{C}_{abc}{}^d, \qquad (1.11)$$

where $\bar{C}_{abc}{}^d$ is the D-dimensional Weyl tensor with respect to $\bar{g}_{ab}$ and Ψ is a conformal factor.

Einstein's equation shows that the Ricci tensor vanishes in vacuum $T_{ab} = 0$. However, the Weyl tensor may not vanish even in the vacuum. In this sense, the Weyl tensor may be regarded as a curvature of purely geometrical origin. It is also worthy to note that the Weyl tensor vanishes for $D \le 3$; for $D = 3$, the Riemann tensor is simply written by the Ricci tensor. This fact simplifies some of manipulation in computing complex Weyl scalars for $D = 4$ (see section 6.2).

The Einstein tensor satisfies a relation resulting from the Bianchi identity as

$$\nabla^a G_{ab} = 0. \qquad (1.12)$$

This equation is derived by contracting the original form of the Bianchi identity, written as $\nabla_{[e}\overset{(D)}{R}_{ab]cd} = 0$. Equation (1.12) implies that a conservation equation for T_{ab} is satisfied as

$$\nabla^a T_{ab} = 0, \qquad (1.13)$$

which is the basic equation for matter fields. Equations for several matter systems will be described in chapter 4. A unique property of general relativity is that we have to solve Einstein's equation and matter equations simultaneously; for evolving Einstein's equation, T_{ab} has to be known while for evolving the matter equation, g_{ab} has to be known.

1.1.2 Nature of Einstein's equation

First, we touch on the so-called gauge freedom in general relativity. General relativity is a covariant theory, and hence, any coordinate system may be employed when one solves this equation. This fact implies that there are D degrees of freedom for choosing the D-dimensional spacetime coordinates. Since g_{ab} is a tensor, it obeys the following relation for the coordinate transformation $x^\mu \to x'^\mu$:

$$g_{\alpha'\beta'} = \frac{\partial x^\alpha}{\partial x'^{\alpha'}} \frac{\partial x^\beta}{\partial x'^{\beta'}} g_{\alpha\beta}. \qquad (1.14)$$

This shows that by an appropriate choice of the coordinates, we set at least D components of $g_{\alpha'\beta'}$ to be a desired form, e.g., $g_{tt} = -1$ and $g_{tk} = 0$. This is called the gauge freedom and for solving Einstein's equation, we have to specify a gauge condition.

Second, it should be mentioned that Einstein's equation is a hyperbolic-type equation. The Riemann tensor is composed of partial derivatives of the metric $g_{\mu\nu}$ up to second order together with nonlinear terms in $g_{\mu\nu}$ and $\partial_\sigma g_{\mu\nu}$. Thus, Einstein's equation is equivalent to a coupled system of nonlinear second-order partial differential equations for the metric components $g_{\mu\nu}$. Because the metric has the Lorentz signature, these equations have a hyperbolic nature like a free scalar wave equation

$$\Box \phi = (-\partial_t^2 + \Delta_f)\phi = 0, \tag{1.15}$$

where Δ_f is the flat-space Laplacian.

The hyperbolic nature of Einstein's equation is clearly found in a special gauge condition in the following manner (see also appendix F). First we define a quantity

$$\mathcal{G}^{\mu\nu} := \sqrt{-g}\, g^{\mu\nu}. \tag{1.16}$$

As shown in Landau and Lifshitz (1962), using $\mathcal{G}^{\mu\nu}$, Einstein's equation is written in the form

$$\partial_\alpha \partial_\beta \left(\mathcal{G}^{\mu\nu} \mathcal{G}^{\alpha\beta} - \mathcal{G}^{\mu\alpha} \mathcal{G}^{\nu\beta} \right) = -16\pi(-g)\left(T^{\mu\nu} + t_{LL}^{\mu\nu}\right), \tag{1.17}$$

where $t_{LL}^{\mu\nu}$ is the so-called Landau–Lifshitz pseudo tensor, which is composed only of more than second-order terms of $\mathcal{G}^{\mu\nu}$ and its first derivative, $\partial_\sigma \mathcal{G}^{\mu\nu}$.

As we mentioned above, there is the gauge freedom in general relativity and we have to give a gauge condition. To show the hyperbolic nature of Einstein's equation, an often-used gauge is the harmonic gauge defined by

$$\Box_g x^\mu = 0, \quad \text{and hence,} \quad \partial_\nu \mathcal{G}^{\mu\nu} = 0, \tag{1.18}$$

where

$$\Box_g = \frac{1}{\sqrt{-g}} \partial_\alpha \left(\sqrt{-g}\, g^{\alpha\beta} \partial_\beta \right). \tag{1.19}$$

Then, equation (1.17) is written to

$$\sqrt{-g}\, \Box_g \mathcal{G}^{\mu\nu} = -16\pi(-g)\left(T^{\mu\nu} + t_{LL}^{\mu\nu}\right) + \left(\partial_\alpha \mathcal{G}^{\nu\beta}\right) \partial_\beta \mathcal{G}^{\mu\alpha}. \tag{1.20}$$

Thus, equation (1.20) shows that Einstein's equation may be viewed as a coupled system of $D(D+1)/2$ nonlinear second-order partial differential hyperbolic equations for $\mathcal{G}^{\mu\nu}$.

1.2 Gravitational waves

Equation (1.20) shows that Einstein's equation may be viewed as second-order partial differential hyperbolic equations for $\mathcal{G}^{\mu\nu}$. This indicates that gravitational waves, i.e., a slightly curved ripple in spacetime which propagates with the speed of light, exist as a solution of Einstein's equation. This facts are clearly found in a linear perturbation analysis for Einstein's equation. In the following, we will describe the linear perturbation theory paying attention only to the four-dimensional case ($D = 4$) for simplicity: In four dimensions, there are only two modes of gravitational waves, but in higher dimensions, there are more modes and the analysis becomes much more complicated.

1.2.1 Linearized Einstein's equation

We consider weakly gravitating spacetime which only slightly deviates from the flat Minkowski spacetime. In such a case, the metric components are written as

$$g_{\mu\nu} = \eta_{\mu\nu} + \epsilon h_{\mu\nu}, \tag{1.21}$$

where ϵ is a small dimensionless parameter and $h_{\mu\nu}$ denotes a perturbed metric. The basic equation for linearized Einstein's theory is derived by substituting equation (1.21) into Einstein's equation and then by taking the terms into account up to the first order in ϵ. The resulting Einstein tensor is written as

$$G_{\mu\nu} = \frac{\epsilon}{2}\left(-\Box\psi_{\mu\nu} + \partial_\alpha\partial_\mu\psi^\alpha_{\ \nu} + \partial_\alpha\partial_\nu\psi^\alpha_{\ \mu} - \eta_{\mu\nu}\partial_\alpha\partial_\beta\psi^{\alpha\beta}\right), \tag{1.22}$$

where

$$\psi_{\mu\nu} := h_{\mu\nu} - \frac{1}{2}\eta_{\mu\nu}\eta^{\alpha\beta}h_{\alpha\beta}, \tag{1.23}$$

and the superscripts of $\psi_{\alpha\beta}$ are raised by $\eta^{\mu\nu}$ as $\psi^\alpha_{\ \nu} = \eta^{\alpha\mu}\psi_{\mu\nu}$. Equating $G_{\mu\nu}$ in equation (1.22) with $8\pi T_{\mu\nu}$ in the assumption that $T_{\mu\nu}$ is of order ϵ, we obtain linearized Einstein's equation.

The next task is to rewrite equation (1.22) by using a gauge freedom. We denote an infinitesimal coordinate variation by $\epsilon\xi^\mu$. Then, for an infinitesimal coordinate transformation $x^\mu \to x^\mu - \epsilon\xi^\mu$, $h_{\mu\nu}$ and $\psi_{\mu\nu}$ are transformed as

$$h_{\mu\nu} \to \bar{h}_{\mu\nu} = h_{\mu\nu} + \partial_\mu\xi_\nu + \partial_\nu\xi_\mu, \tag{1.24}$$

$$\psi_{\mu\nu} \to \bar{\psi}_{\mu\nu} = \psi_{\mu\nu} + \partial_\mu\xi_\nu + \partial_\nu\xi_\mu - \eta_{\mu\nu}\partial_\alpha\xi^\alpha, \tag{1.25}$$

where $\xi_\mu = \eta_{\mu\nu}\xi^\nu$. We then impose the following gauge condition:

$$\eta^{\alpha\beta}\partial_\alpha\bar{\psi}_{\beta\mu} = 0. \tag{1.26}$$

This is often called the Lorentz gauge condition which agrees with the harmonic gauge (1.18) in the framework of linearized Einstein's theory. For four degrees of freedom of ξ^μ, we have four conditions, and hence, it is possible to satisfy this condition.

In the Lorentz gauge condition, the linearized Einstein tensor is written as

$$G_{\mu\nu} = -\frac{\epsilon}{2}\Box\psi_{\mu\nu}. \tag{1.27}$$

Hence, linearized Einstein's equation reduces to linear second-order partial-differential hyperbolic equations

$$\Box\psi_{\mu\nu} = -16\pi\epsilon^{-1}T_{\mu\nu}. \tag{1.28}$$

Again we supposed that $T_{\mu\nu}$ is of order ϵ in the linearized theory.

1.2.2 Propagation of gravitational waves

First, we consider propagation of gravitational waves in vacuum, i.e., $T_{\mu\nu} = 0$, by solving

$$\Box \psi_{\mu\nu} = 0. \tag{1.29}$$

Equation (1.29) appears to have 10 components. However, we have already imposed the Lorentz gauge condition, and hence, there are actually only six degrees of freedom for $\psi_{\mu\nu}$. Furthermore, the Lorentz gauge condition is preserved in a coordinate transformation by ξ^μ that satisfies

$$\Box \xi^\mu = 0. \tag{1.30}$$

This property is essentially the same as that of the Lorentz gauge condition in Maxwell's theory of electromagnetism. Because of the presence of the additional four degrees of freedom, the number of the true degrees of freedom can be reduced to $6 - 4 = 2$. These two modes correspond to gravitational-wave modes, so-called + and × modes.

Specifically, for plane waves, we can write the components of $\psi_{\mu\nu}$ as

$$\psi_{\mu\nu} = \begin{pmatrix} 0 & 0 & 0 & 0 \\ 0 & h_+ & h_\times & 0 \\ 0 & h_\times & -h_+ & 0 \\ 0 & 0 & 0 & 0 \end{pmatrix}, \tag{1.31}$$

where the matrix components are displayed in the order t, x, y, z, and we assumed that gravitational waves propagate toward $+z$-direction. Here, h_+ and $h_\times$ are functions of $t - z$. Finally, the linearized metric components are obtained by $h_{\mu\nu} = \psi_{\mu\nu}$ because $\psi_{\mu\nu}$ is tracefree, $\psi_{\mu\nu} \eta^{\mu\nu} = 0$.

Equation (1.31) tells us that gravitational waves have the tracefree and transverse nature. Here, "transverse" implies that they do not have the components for the propagation direction, z. The tracefree and transverse nature of gravitational waves are not clear in general coordinate conditions. In general coordinates, we have to extract the gravitational-wave components in an appropriate manner. In linearized Einstein's theory, a well-known prescription is to define a projection operator $P_i{}^k = \delta_i{}^k - \bar{n}_i \bar{n}^k$, where $\bar{n}^i$ denotes a unit vector pointing the propagation direction, and to operate it for the spatial components of h_{ij} as

$$h_{ij}^{\rm GW} = \left[P_i{}^k P_j{}^l - \frac{1}{2} P_{ij} P^{kl} \right] h_{kl}. \tag{1.32}$$

With this projection, the components in the propagation direction as well as the time components become automatically zero, and the transverse nature is achieved. It is also easy to check that the tracefree condition is automatically guaranteed for $h_{ij}^{\rm GW}$.

Then, how could we detect gravitational waves? The straight way is to observe the relative acceleration of two nearby test particles, i.e., to measure the gravitational tidal force. For two nearby freely falling test particles, this acceleration is

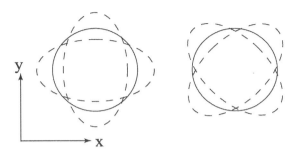

Fig. 1.1 Schematic figure for the deviation of space found from the motion of test particles, when gravitational waves propagate in the perpendicular direction of this page surface. The solid circles denote the location of the point particles before gravitational waves reach, and the dashed elliptical curves show the deviation of their positions by gravitational waves. The left and right figures show the deviations by h_+ and $h_\times$, respectively.

governed by the geodesic deviation equation (1.9). If the two test particles are initially at rest in a global inertial coordinate system of $\eta_{\mu\nu}$, we have the following equations of motion for the metric (1.31)

$$\ddot{x} = \frac{1}{2}\left(\ddot{h}_+ x + \ddot{h}_\times y\right), \qquad \ddot{y} = \frac{1}{2}\left(\ddot{h}_\times x - \ddot{h}_+ y\right). \tag{1.33}$$

Figure 1.1 displays how the test particles are forced to be moved by gravitational waves. The left and right panels of figure 1.1 show the particle motions by h_+ and $h_\times$, respectively. In this figure, we suppose that gravitational waves propagate in the direction perpendicular to this page surface. The unique feature is that the particles, which initially constitute a circle, are forced to constitute an ellipse: Gravitational waves cause a quadrupole deviation of space. Another important fact is that the wave phases of $+$ and $\times$ modes differ by $45°$.

1.2.3 *Generation of gravitational waves*

Next, we derive a solution of equation (1.28) in the presence of nonzero stress-energy tensor. For simplicity, we hereafter employ Cartesian coordinates for the spatial ones. Then, equation (1.28) is viewed as a system of 10 independent hyperbolic equations and each of 10 components can be solved by a Green function method as

$$\psi_{\mu\nu}(t, x^i) = 4\int \frac{T_{\mu\nu}(t - |x^i - y^i|, y^j)}{|x^i - y^i|} d^3y. \tag{1.34}$$

Gravitational waves are defined only in a wave zone, which is a region with a distance from the gravitational-wave source larger than the wavelength of gravitational waves, λ. Thus, we consider the solution (1.34) only for $r = |x^i| > \lambda$. In addition, we suppose that the stress-energy tensor is not zero only for a central region with the extent R. Because the wavelength is usually much longer than R, the wave zone is located for $r \gg R$. Hence, supposing that $|x^i| \gg |y^i|$, equation (1.34) is written

as
$$\psi_{\mu\nu}(t, x^i) = \frac{4}{r} \int T_{\mu\nu}(t - |x^i - y^i|, y^j) d^3y. \tag{1.35}$$

In addition, we have $|x^i - y^i| \approx r - \sum_i x^i y^i / r$ and $|y^i| \lesssim R$. Thus, we can perform a Taylor expansion around $t - r$ as

$$T_{\mu\nu}(t - |x^i - y^i|, y^j) = T_{\mu\nu}(t_{\text{ret}}, y^j) + \frac{\sum_i x^i y^i}{r} \partial_t T_{\mu\nu}(t_{\text{ret}}, y^j) + \cdots, \tag{1.36}$$

where $t_{\text{ret}} := t - r$ denotes a retarded time in the Minkowski spacetime. We note that if $T_{\mu\nu}$ is not absent for a distant zone, the prescription described above cannot be used.

Now, let τ denote a characteristic time scale for the variation of $T_{\mu\nu}$. Then, the ratio of the second to the first terms of equation (1.36) is of order $R/c\tau$, where R/τ denotes the order of the characteristic velocity of the gravitational-wave source and we recover c. Therefore, the Taylor expansion performed in equation (1.36) is equivalent to the expansion in terms of v/c. For neglecting the higher-order terms of v/c, we here assume that the characteristic velocity of the source is much smaller than the speed of light. Then, we have

$$\psi_{\mu\nu}(t, x^i) = \frac{4}{r} \int T_{\mu\nu}(t_{\text{ret}}, y^j) d^3y. \tag{1.37}$$

Here, we employ the stress-energy tensor of an ideal fluid for $T_{\mu\nu}$, neglecting the terms of higher-order in v/c. In its lowest order, we have $T_{tt} = T^{tt} = \rho$, $T_{ti} = T_t{}^i = -\rho v^i$ where ρ and $v^i(= v_i)$ are the rest-mass density and the three velocity of the fluid. First, we derive solutions of equation (1.36) for $(\mu, \nu) = (t, t)$ and (t, x^k) components with x^k Cartesian coordinates as

$$\psi_{tt}(t, x^i) = \frac{4}{r} \int \rho \, d^3y = \frac{4M}{r}, \tag{1.38}$$

$$\psi_{tk}(t, x^i) = -\frac{4}{r} \int \rho v^k \, d^3y = -\frac{4P_k}{r}. \tag{1.39}$$

Here, ρ and v^k in the integrals are functions of t_{ret}. However, the integrations in equations (1.38) and (1.39) lead to the mass M and the linear momentum P_k of the system, which are conserved quantities. Therefore, ψ_{tt} and ψ_{tk} are time-independent and do not have the components of gravitational waves.

The remaining components are derived from the following identities, resulting from $\partial_\mu T^{\mu\nu} = 0$ that holds in the lowest-order in ϵ:

$$T^{ij} = \frac{1}{2} \left[\partial_k \left(T^{ki} x^j + T^{kj} x^i \right) + \partial_t \left(T^{ti} x^j + T^{tj} x^i \right) \right], \tag{1.40}$$

$$T^{ti} = \partial_k \left(T^{kt} x^i \right) + \partial_t T^{tt} x^i. \tag{1.41}$$

Then,

$$\psi^{ij} = \frac{2}{r} \int d^3x \left[\partial_k \left(T^{ki} x^j + T^{kj} x^i \right) + \partial_k \partial_t \left(T^{kt} x^i x^j \right) + \partial_t^2 T^{tt} x^i x^j \right]. \tag{1.42}$$

Again, in the lowest order in v/c, $T^{tt} = \rho$. By performing integration by parts and by discarding the surface integrals that result from the integral of total divergence terms, we finally have

$$\psi^{ij} = \frac{2}{r} \frac{d^2}{dt^2} \int d^3 x \rho x^i x^j = \frac{2}{r} \ddot{I}_{ij}(t_{\text{ret}}), \qquad (1.43)$$

where I_{ij} denotes the quadrupole moment of the system.

From ψ^{ij}, ingredients of gravitational waves have to be extracted using the projection tensor [see equation (1.32)]:

$$h_{ij}^{\text{GW}} = \frac{2}{r} \frac{d^2 \mathcal{I}_{ij}^{\text{TT}}}{dt^2} = \left[P_i{}^k P_j{}^l - \frac{1}{2} P_{ij} P^{kl} \right] \frac{2}{r} \frac{d^2 \mathcal{I}_{kl}}{dt^2}. \qquad (1.44)$$

This is the so-called quadrupole formula of gravitational waves. Here, $\mathcal{I}_{ij}$ is the tracefree part of I_{ij} defined by

$$\mathcal{I}_{ij} := I_{ij} - \frac{1}{3} \delta_{ij} \sum_k I_{kk}. \qquad (1.45)$$

Equation (1.44) shows that gravitational waves are not emitted from spherically symmetric systems for which $\mathcal{I}_{ij} = 0$. It also shows that gravitational waves are emitted only for a dynamical system in which $\ddot{I}_{ij} \neq 0$.

A word of caution is appropriate here. The quadrupole formula (1.44) in this section was derived in the framework of the linearized Einstein's theory. However, the linearized theory cannot be applied to a system composed of self-gravitating objects. This is found from the equation of motion, $\partial_\mu T^{\mu\nu} = 0$, used for this theory, in which no information of curved spacetime (i.e., gravitational field) is involved in the linear order. Therefore, the quadrupole formula derived here cannot be in principle used for a system composed of self-gravitating objects such as binary black holes and neutron stars. However, what is really nice is that the quadrupole formula derived for a self-gravitating system using the post-Newtonian approximation agrees with equation (1.44) as far as $v/c \ll 1$, as shown rigidly by Chandrasekhar and Esposito (1970)[1]. Thus, the derivation of the quadrupole formula here is incomplete for the general self-gravitating systems, but the result can be, fortunately, applied to a variety of systems as far as the velocity of the source can be assumed to be slow.

Equation (1.44) indicates that the observed amplitude of gravitational waves depends on the direction of the observer because of the presence of the projection operator. Thus, there always exist the most optimistic direction for the observation of gravitational waves. We often refer to the maximum amplitude $h_{\max}$ to clarify how strong the object is as a source of gravitational waves. Thus, it is useful to show the expression for $h_{\max}$. For the case that the source is an axisymmetric system, $l = 2$ and $m = 0$ spherical harmonics mode is the dominant one for gravitational

---

[1] The post-Newtonian approximation is an approximate formulation of general relativity in which all the equations are expanded by a parameter v/c assuming that it is much smaller than unity. See also appendix H.

waves. The amplitude of gravitational waves for this mode becomes maximum when the observer is located on the equatorial plane which is perpendicular to the symmetry axis. Let z and ϖ be the coordinates along the symmetry axis and the cylindrical direction, respectively. Then,

$$h_{\max} = \frac{1}{2D_{\rm obs}}\left(2\ddot{I}_{zz} - \ddot{I}_{\varpi\varpi}\right) = \frac{1}{2D_{\rm obs}}\left(2\ddot{I}_{zz} - \ddot{I}_{xx} - \ddot{I}_{yy}\right), \tag{1.46}$$

where $D_{\rm obs}$ is a hypothetical distance from the source. For the case that the system is non-axisymmetric and the rotational axis agrees with the z-axis, $l = |m| = 2$ spherical harmonics modes are the dominant ones for gravitational waves. The amplitude for these modes becomes maximum when the observer is located along the z-axis. Then, if we employ Cartesian coordinates, we have

$$h_{\max} = \frac{1}{D_{\rm obs}}\left(\ddot{I}_{xx} - \ddot{I}_{yy}\right) \quad \text{or} \quad h_{\max} = \frac{2}{D_{\rm obs}}\ddot{I}_{xy}. \tag{1.47}$$

1.2.4 *Gravitational-wave luminosity*

One of the subtle problems in general relativity is to define energy and momentum of gravitational waves. In general relativity, we cannot locally define the energy and the momentum of gravitational fields, and hence, this is also the case for gravitational waves. By contrast, it is possible to globally define the energy and the momentum of the system in asymptotically flat spacetime; the frequently used ones in numerical relativity are the Arnowitt–Deser–Misner (ADM) mass and momentum (see section 5.3). This indicates that the energy and the momentum as well as the energy and momentum fluxes of gravitational waves can be defined in a non-local manner. Isaacson (1968a,b) established in his shortwave (high-frequency) formalism how to define the energy and the momentum of gravitational waves. Here, we outline his method of defining an effective stress-energy tensor of gravitational waves [see also section 35 of Misner *et al.* (1973)].

We focus on gravitational waves propagating in vacuum background spacetime and suppose that the metric components can be split into a background part (averaged part), $\overset{(b)}{g}_{\mu\nu}$, and a perturbation part, $h_{\mu\nu}$, as

$$g_{\mu\nu} = \overset{(b)}{g}_{\mu\nu} + h_{\mu\nu}. \tag{1.48}$$

Here, $\overset{(b)}{g}_{\mu\nu}$ does not have to be the flat Minkowski metric. We further suppose that the following properties hold: (i) the amplitude of the metric perturbation $h_{\mu\nu}$, which we write as $\mathcal{A}$, is much smaller than that of the background metric, $\overset{(b)}{g}_{\mu\nu}$, for which the order of magnitude is assumed to be unity; (ii) the curvature scale length of $\overset{(b)}{g}_{\mu\nu}$ is denoted by $\mathcal{R}$, and $|\partial_\alpha \overset{(b)}{g}_{\mu\nu}| \sim |\overset{(b)}{g}_{\mu\nu}|/\mathcal{R}$; (iii) the scale, on which $h_{\mu\nu}$ varies, $\bar{\lambda} := \lambda/2\pi$, is much shorter than $\mathcal{R}$, and $|\partial_\alpha h_{\mu\nu}| \sim |h_{\mu\nu}|/\bar{\lambda}$. Here, λ denotes the typical wavelength of gravitational waves. Then, the following order-of-magnitude relations hold,

$$\partial_\alpha \partial_\beta \overset{(b)}{g_{\mu\nu}} = O\left(1/\mathcal{R}^2\right),$$
$$\partial_\alpha \partial_\beta h_{\mu\nu} = O\left(\mathcal{A}/\lambda^2\right),$$
$$(\partial_\alpha h_{\mu\nu})\partial_\beta h_{\sigma\lambda} = O\left(\mathcal{A}^2/\lambda^2\right), \tag{1.49}$$

and it is found that the Ricci tensor is expanded as

$$R_{\mu\nu} = \overset{(b)}{R_{\mu\nu}} + \overset{(1)}{R_{\mu\nu}} + \overset{(2)}{R_{\mu\nu}} + O\left(\mathcal{A}^3/\lambda^2\right), \tag{1.50}$$

where $\overset{(1)}{R_{\mu\nu}}$ and $\overset{(2)}{R_{\mu\nu}}$ are first- and second-order in $h_{\mu\nu}$, and their magnitudes are of $O\left(\mathcal{A}/\lambda^2\right)$ and $O\left(\mathcal{A}^2/\lambda^2\right)$, respectively. $\overset{(b)}{R_{\mu\nu}}$ is the Ricci tensor associated with $\overset{(b)}{g_{\mu\nu}}$, and its magnitude is of $O\left(1/\mathcal{R}^2\right)$.

In the shortwave formalism, Einstein's equation is solved for each order separately. Because $\overset{(1)}{R_{\mu\nu}}$ is supposed to be only one term of $O\left(\mathcal{A}/\lambda^2\right)$ in Einstein's equation, we need to solve

$$\overset{(1)}{R_{\mu\nu}} = 0. \tag{1.51}$$

One then splits the remaining components of the Ricci tensor into a part that is free of fluctuations of scale λ and the other part that is composed of the fluctuations. This split is achieved by averaging over a scale much larger than λ, yielding

$$\overset{(b)}{R_{\mu\nu}} + \langle \overset{(2)}{R_{\mu\nu}} \rangle = O\left(\mathcal{A}^3/\lambda^2\right). \tag{1.52}$$

Here, $\langle \overset{(1)}{R_{\mu\nu}} \rangle = 0$ even in the absence of equation (1.51) [see equation (1.62)]. Note that equation (1.52) implies that $\mathcal{R}$ should be of order $\lambda/\mathcal{A}$, and we find that $\overset{(b)}{R_{\mu\nu}}$ should be of order of $\mathcal{A}^2/\lambda^2$. Hence, an effective Einstein's equation may be written as

$$\overset{(b)}{R_{\mu\nu}} - \frac{1}{2}\overset{(b)}{g_{\mu\nu}}\overset{(b)}{R} = 8\pi T_{\mu\nu}^{\mathrm{GW}}, \tag{1.53}$$

where $T_{\mu\nu}^{\mathrm{GW}}$ is the effective stress-energy tensor of gravitational waves defined by

$$T_{\mu\nu}^{\mathrm{GW}} := -\frac{1}{8\pi}\left(\langle \overset{(2)}{R_{\mu\nu}} \rangle - \frac{1}{2}\overset{(b)}{g_{\mu\nu}}\langle \overset{(2)}{R} \rangle\right), \tag{1.54}$$

and $\overset{(2)}{R} = \overset{(2)}{R_{\mu\nu}}\overset{(b)}{g^{\mu\nu}}$.

The explicit forms of $\overset{(1)}{R_{\mu\nu}}$ and $\overset{(2)}{R_{\mu\nu}}$ are calculated in the following manner. Let $\overset{(b)}{\nabla}_a$ be the covariant derivative operator associated with $\overset{(b)}{g_{ab}}$, and we define a connection tensor, $C^c{}_{ab}$, by

$$\nabla_a \omega_b = \overset{(b)}{\nabla}_a \omega_b - C^c{}_{ab}\omega_c, \tag{1.55}$$

where ω^c denotes an arbitrary vector field. The same manipulation as for deriving the connection coefficient yields,

$$C^c{}_{ab} = \frac{1}{2} g^{cd} \left(\overset{(b)}{\nabla}_a g_{bd} + \overset{(b)}{\nabla}_b g_{ad} - \overset{(b)}{\nabla}_d g_{ab} \right)$$

$$= \frac{1}{2} \left(\overset{(b)}{g^{cd}} - h^{cd} \right) \left(\overset{(b)}{\nabla}_a h_{bd} + \overset{(b)}{\nabla}_b h_{ad} - \overset{(b)}{\nabla}_d h_{ab} \right), \quad (1.56)$$

where we used the relation

$$g^{cd} = \overset{(b)}{g^{cd}} - h^{cd} + O(h^2), \quad h^{cd} = \overset{(b)}{g^{ac}} \overset{(b)}{g^{bd}} h_{ab}. \quad (1.57)$$

For the following manipulation, we define

$$\overset{(1)}{C}{}^c{}_{ab} := \frac{1}{2} \overset{(b)}{g^{cd}} \left(\overset{(b)}{\nabla}_a h_{bd} + \overset{(b)}{\nabla}_b h_{ad} - \overset{(b)}{\nabla}_d h_{ab} \right), \quad (1.58)$$

$$\overset{(1)}{C}_{cab} := \frac{1}{2} \left(\overset{(b)}{\nabla}_a h_{bc} + \overset{(b)}{\nabla}_b h_{ac} - \overset{(b)}{\nabla}_c h_{ab} \right). \quad (1.59)$$

The definition of the Riemann tensor yields

$$R_{abc}{}^d = \overset{(b)}{R}_{abc}{}^d + \overset{(b)}{\nabla}_b C^d{}_{ac} - \overset{(b)}{\nabla}_a C^d{}_{bc} + C^e{}_{ac} C^d{}_{be} - C^e{}_{bc} C^d{}_{ae}, \quad (1.60)$$

and thus, the Ricci tensor is written as

$$R_{ac} = \overset{(b)}{R}_{ac} + \overset{(b)}{\nabla}_b C^b{}_{ac} - \overset{(b)}{\nabla}_a C^b{}_{bc} + C^e{}_{ac} C^b{}_{be} - C^e{}_{bc} C^b{}_{ae}. \quad (1.61)$$

From this equation, we have

$$\overset{(1)}{R}_{ac} = \frac{1}{2} \left(h_{cb|a}{}^{|b} + h_{ab|c}{}^{|b} - h_{ac|b}{}^{|b} - h_{|ac} \right), \quad (1.62)$$

$$\overset{(2)}{R}_{ac} = - \left(h^{bd} \overset{(1)}{C}_{dac} \right)_{|b} + \left(h^{bd} \overset{(1)}{C}_{dbc} \right)_{|a} + \overset{(1)}{C}{}^d{}_{ac} \overset{(1)}{C}{}^b{}_{bd} - \overset{(1)}{C}{}^d{}_{bc} \overset{(1)}{C}{}^b{}_{ad}, \quad (1.63)$$

where $|a$ denotes $\overset{(b)}{\nabla}_a$ and $h = h_{ac} \overset{(b)}{g^{ac}}$.

The final task is to obtain $T_{\mu\nu}^{\rm GW}$ by taking the average of $\overset{(2)}{R}_{\mu\nu}$. Because we focus only on the terms of $O\left(\mathcal{A}^2/\lambda^2\right)$, the averaging should be performed keeping the terms of this order while discarding the higher-order terms. For example, we may replace $h_{ab|cd}$ to $h_{ab|dc}$ in this approximation because $\overset{(b)}{R}_{abcd}$, which appears in this replacement, is a higher-order quantity by the order $(\lambda/\mathcal{R})^2$. We may also discard the terms of the form $\langle (\cdots)_{|a} \rangle$, because the fractional error by this discarding is of order $\lambda/\mathcal{R}$. This also implies that one can freely integrate by parts discarding the total divergence term, i.e., $\langle AB_{|ab}\rangle = -\langle A_{|b}B_{|a}\rangle$ where A and B are some tensors. A straightforward calculation using these rules and $\overset{(1)}{R}_{\mu\nu} = 0$, we find

$$\langle \overset{(2)}{R} \rangle = 0, \qquad (1.64)$$

and

$$T^{\text{GW}}_{\mu\nu} = \frac{1}{32\pi}\Big\langle \psi_{\alpha\beta|\mu}\psi^{\alpha\beta}{}_{|\nu} - \frac{1}{2}\psi^\alpha{}_{\alpha|\mu}\psi^\beta{}_{\beta|\nu} - \psi^{\alpha\beta}{}_{|\beta}(\psi_{\alpha\mu|\nu} + \psi_{\alpha\nu|\mu})\Big\rangle, \qquad (1.65)$$

where $\psi_{\mu\nu}$ here is defined by

$$\psi_{\mu\nu} := h_{\mu\nu} - \frac{1}{2}\overset{(b)}{g}_{\mu\nu}h. \qquad (1.66)$$

It is straightforward to show that equation (1.65) is gauge-invariant within the accuracy of $O\left(\mathcal{A}^2/\lambda^2\right)$: This can be shown by substituting the following into it

$$h_{\mu\nu} = \hat{h}_{\mu\nu} - \overset{(b)}{\nabla}_\mu \xi_\nu - \overset{(b)}{\nabla}_\nu \xi_\mu, \qquad (1.67)$$

where ξ^μ denotes the gauge variable, then by appropriately performing integration by parts, and by finally using $\overset{(1)}{R}_{\mu\nu} = 0$ to erase the terms coupled with ξ^μ. It is also trivial to show that

$$\overset{(b)}{\nabla}{}^\mu T^{\text{GW}}_{\mu\nu} = 0, \qquad (1.68)$$

holds within the accuracy of $O\left(\mathcal{A}^2/\lambda^2\right)$.

For a transverse gauge (similar to the Lorentz gauge), $\psi^{\alpha\beta}{}_{|\beta} = 0$, the third and fourth terms in equation (1.65) vanish. In addition, $\psi_{\mu\nu}$ can be set to be tracefree by an appropriate choice of a gauge condition, and thus, the second term can be erased. Then, we have an often-seen relation

$$T^{\text{GW}}_{\mu\nu} = \frac{1}{32\pi}\Big\langle (\psi_{\alpha\beta|\mu})\psi^{\alpha\beta}{}_{|\nu}\Big\rangle. \qquad (1.69)$$

Alternatively, if we take into account the leading order term [the term of $O(r^{-2})$] for $r \to \infty$, equation (1.65) reduces to

$$T^{\text{GW}}_{\mu\nu} = \frac{1}{32\pi}\Big\langle (\partial_\mu h_{ij})\partial_\nu h^{ij}\Big\rangle. \qquad (1.70)$$

Here and in the following, we set that the background geometry is flat for $r \to \infty$ because the deviation from the flat geometry is usually a tiny correction.

In the rest of this section, we describe equations using Cartesian coordinates, x^k. The energy flux of gravitational waves per unit time and unit area along a direction of x^k is written as

$$F^{\text{GW}}_k = -T^{\text{GW}}_{tk} = -\frac{1}{32\pi}\sum_{i,j}\Big\langle \dot{h}^{\text{GW}}_{ij}\partial_k h^{\text{GW}}_{ij}\Big\rangle. \qquad (1.71)$$

Here, h^{GW}_{ij} is the gravitational-wave component of h_{ij}, and $\langle\cdots\rangle$ denotes taking an average for a time duration, which is at least several periods of gravitational-wave cycles. This averaging is necessary, because in general relativity, we cannot locally define the energy and the energy flux: A non-local definition is required for obtaining these quantities.

Using F_k^{GW}, the energy flux per solid angle, $d\Omega$, is defined by

$$\frac{dE}{dt d\Omega} = F_k^{\mathrm{GW}} r^2 \bar{n}^k = \frac{r^2}{32\pi} \sum_{i,j} \langle \dot{h}_{ij}^{\mathrm{GW}} \dot{h}_{ij}^{\mathrm{GW}} \rangle, \tag{1.72}$$

where $\bar{n}^k$ is a unit normal to a sphere of radius r, and we used the fact that h_{ij}^{GW} is a function of the retarded time $t - r$ (as well as x^i). In the lowest order in v/c (i.e., in the quadrupole order),

$$\frac{dE}{dt d\Omega} = \frac{1}{8\pi} \sum_{i,j} \left\langle \frac{d^3 \mathcal{I}_{ij}^{\mathrm{TT}}}{dt^3} \frac{d^3 \mathcal{I}_{ij}^{\mathrm{TT}}}{dt^3} \right\rangle, \tag{1.73}$$

where we substituted equation (1.44). By integrating it for the entire sphere, we finally have the luminosity in the quadrupole formula as

$$\frac{dE}{dt} = \frac{1}{5} \sum_{i,j} \left\langle \frac{d^3 \mathcal{I}_{ij}}{dt^3} \frac{d^3 \mathcal{I}_{ij}}{dt^3} \right\rangle. \tag{1.74}$$

By the same procedure, the linear- and angular-momentum emission rates are defined by

$$\frac{dP_k}{dt} = \frac{r^2}{32\pi} \oint d\Omega \bar{n}_k \sum_{i,j} \left\langle \dot{h}_{ij}^{\mathrm{GW}} \dot{h}_{ij}^{\mathrm{GW}} \right\rangle, \tag{1.75}$$

$$\frac{dJ_k}{dt} = \frac{r^2}{32\pi} \oint d\Omega \sum_{i,j,l,m} \epsilon_{klm} x^l \left\langle \dot{h}_{ij}^{\mathrm{GW}} \partial_m h_{ij}^{\mathrm{GW}} \right\rangle, \tag{1.76}$$

where ϵ_{ijk} is the completely antisymmetric spatial tensor with $\epsilon_{xyz} = 1$ here. The emission rate of the angular momentum in the lowest order in v/c is also written only by the quadrupole moment as

$$\frac{dJ_k}{dt} = -\frac{2}{5} \sum_{i,j,l} \epsilon_{kij} \left\langle \frac{d^2 \mathcal{I}_{il}}{dt^2} \frac{d^3 \mathcal{I}_{jl}}{dt^3} \right\rangle. \tag{1.77}$$

The emission rate of the linear momentum in the lowest order in v/c is not written only by the quadrupole moment because the $d\Omega$ integration in equation (1.75) vanishes if only the quadrupole part in h_{ij}^{GW} is taken into account. Actually, the leading-order contribution to the linear-momentum flux comes from coupling terms between the quadrupole moment and next-order higher-multipole moments. Fitchett (1983) first derived such a formula, and subsequently, a concise formula was derived by Wiseman (1992); Blanchet et al. (2005) as

$$\frac{dP_i}{dt} = \left\langle \frac{2}{63} \sum_{j,k} \frac{d^3 \mathcal{I}_{jk}}{dt^3} \frac{d^4 \mathcal{I}_{ijk}}{dt^4} + \frac{16}{45} \sum_{j,k,l} \epsilon_{ijk} \frac{d^3 \mathcal{I}_{jl}}{dt^3} \frac{d^3 J_{kl}}{dt^3} \right\rangle, \tag{1.78}$$

where

$$\mathcal{I}_{ijk} = \int d^3 x \rho \left[x^i x^j x^k - \frac{1}{5} r^2 \left(\delta^{ij} x^k + \delta^{ik} x^j + \delta^{jk} x^i \right) \right], \qquad (1.79)$$

$$J_{ij} = \frac{1}{2} \int d^3 x \rho \sum_{k,l} \left(\epsilon_{jkl} x^i x^k v^l + \epsilon_{ikl} x^j x^k v^l \right). \qquad (1.80)$$

Before closing this section, we show that the amplitude of gravitational waves is in general very small. Let f, T, and D_{obs} be the frequency of gravitational waves, the time duration for their emission, and the distance from the source, respectively. Using equations (1.44) and (1.74), the order of magnitude of the gravitational-wave luminosity is estimated as

$$\Delta E T^{-1} \sim c^3 G^{-1} (h D_{\text{obs}})^2 f^2. \qquad (1.81)$$

Here, we recover c and G to clarify the dimension. We write $T = N/f$ where N denotes the wave cycles and thus larger than unity, and $\Delta E = \epsilon_E M c^2 (\epsilon_E \ll 1)$. Then, we obtain

$$h \sim \frac{1}{D_{\text{obs}}} \left[\epsilon_E \frac{GM}{cfN} \right]^{1/2}$$

$$\sim 10^{-17} \left(\frac{10\,\text{kpc}}{D_{\text{obs}}} \right) \left(\frac{\epsilon_E}{10^{-2}} \right)^{1/2} \left(\frac{M}{10 M_\odot} \right)^{1/2} \left(\frac{f}{1\,\text{kHz}} \right)^{-1/2} N^{-1/2}. \qquad (1.82)$$

As an example, we here consider the case that a star of mass $10 M_\odot$ collapses to a black hole in our Galaxy ($D_{\text{obs}} \sim 10\,\text{kpc}$). The frequency is $\sim 1\,\text{kHz}$ [cf. equation (1.234)], $N = O(1)$, and ϵ_E is at best 1%. Then, $h \sim 10^{-17}$, and hence, the amplitude is quite small. During the gravitational-wave propagation, the distance between two test particles varies as shown in equation (1.33). The possible longest straight line taken on the surface of the earth would be $\sim 100\,\text{km}$, and thus, we here consider this straight line. When gravitational waves of amplitude 10^{-17} pass through the earth, the distance of $100\,\text{km}$ line varies by $\sim 10^{-10}\,\text{cm}$, which is about 1% of the atomic radius. This shows that the detection of gravitational waves requires a precise measurement of a tiny distance. However, the detection of gravitational waves will be achieved in the very near future, as we introduce in section 1.2.6.

Equation (1.44) can also be used for estimating the order of magnitude of the gravitational-wave amplitude. Let R, τ, and v be the characteristic length scale, the characteristic variation time scale, and the characteristic velocity for the source, respectively. Setting $|\ddot{I}_{ij}| \sim MR^2/\tau^2 \sim Mv^2$, we have

$$h \propto \delta_{\text{as}} \left(\frac{GM}{c^2 D_{\text{obs}}} \right) \left(\frac{v}{c} \right)^2 \quad \text{or} \quad h \propto \delta_{\text{as}} \left(\frac{R}{D_{\text{obs}}} \right) \left(\frac{GM}{c^2 R} \right) \left(\frac{v}{c} \right)^2, \qquad (1.83)$$

where δ_{as} denotes the degree of anisotropy of the source which is smaller than unity. $\mathcal{C} := GM/c^2 R$ is a dimensionless parameter (the so-called compactness parameter) which is smaller than unity. This parameter is about 1 for black holes and

0.13–0.25 for neutron stars (see also section 1.4.2). Therefore, equation (1.83) shows that strong gravitational waves are emitted when the source is compact, its motion is very fast, close to the speed of light, and the source is significantly deformed (highly non-axisymmetric), $\delta_\text{as} \sim 1$.

1.2.5 *Gravitational waves from a binary*

Close binary systems composed of neutron stars and/or black holes are among the most promising sources of gravitational waves because they could realize the state of $GM/Rc^2 \gtrsim 0.1$, $v/c \gtrsim 0.1$, and $\delta_\text{as} \sim 1$ (see also section 1.5 for other reasons). In this section, we derive the gravitational-wave luminosity and waveforms from them using the quadrupole formula together with the point-particle approximation and simplifying that the binary motion is determined by Newtonian gravity. Although a variety of general-relativistic effects have to be taken into account for the quantitative estimate of gravitational waves [Blanchet (2014)], the simple analysis presented here is sufficient for an approximately quantitative estimate.

Let m_1 and m_2 be the masses of two stars that are approximated by the point particles. The total mass is written by $m = m_1 + m_2$, and we suppose $m_1 \geq m_2$. These two stars are supposed to be in a circular orbit because such orbits are likely to be realized for strong gravitational-wave sources (see section 1.4.8). Choosing the center of mass as the origin and assuming that the orbital plane is the equatorial plane ($z = 0$), the orbital positions for these two stars can be written as

$$(x_1, y_1) = [\ a_1 \cos(\Omega t),\ a_1 \sin(\Omega t)],$$
$$(x_2, y_2) = [-a_2 \cos(\Omega t), -a_2 \sin(\Omega t)], \tag{1.84}$$

where we assumed that at $t = 0$, two stars were located on the x-axis and Ω denotes the Keplerian angular velocity, $\sqrt{m/a^3}$, with a the orbital separation. a_1 and a_2 are written as $a_1 = am_2/m$ and $a_2 = am_1/m$, respectively.

The nonzero components of the quadrupole moment of the system are

$$I_{xx} = \mu a^2 \cos^2(\Omega t),\quad I_{yy} = \mu a^2 \sin^2(\Omega t),\quad I_{xy} = \mu a^2 \cos(\Omega t)\sin(\Omega t), \tag{1.85}$$

where μ is the reduced mass defined by $m_1 m_2/m$. Using equation (1.44), the amplitude of gravitational waves measured by an observer at a distance from the source, D_obs, along the z-axis (rotation axis) is

$$h_\text{max} = \frac{4\mu a^2 \Omega^2}{D_\text{obs}} = \frac{4m\mu}{D_\text{obs} a} = \frac{4\mu(\pi f m)^{2/3}}{D_\text{obs}}, \tag{1.86}$$

where f is the gravitational-wave frequency written by

$$f = \frac{\Omega}{\pi} = \frac{1}{\pi}\sqrt{\frac{m}{a^3}}. \tag{1.87}$$

Note that f is determined by the time variation rate of the quadrupole moment of the system, and thus, it is $\Omega/(2\pi) \times 2$. If the observer is not located on the z-axis, the amplitude observed is smaller than that in equation (1.86).

Equation (1.74) yields the luminosity of gravitational waves as

$$\left(\frac{dE}{dt}\right)_{\text{GW}} = \frac{32}{5}\left(\frac{\mu}{m}\right)^2\left(\frac{m}{a}\right)^5 = \frac{32}{5}\left(\frac{\mu}{m}\right)^2(\pi fm)^{10/3}. \tag{1.88}$$

From equation (1.77), we also find $(dJ_z/dt)_{\text{GW}} = (dE/dt)_{\text{GW}}/\Omega$. Here, the total energy of the binary system in Newtonian gravity is $E = -m\mu/(2a)$. If the binary is assumed to dissipate its energy by the gravitational-wave emission adiabatically, a is varied (fixing m and μ), and thus,

$$\frac{dE}{dt} = \frac{m\mu}{2a^2}\frac{da}{dt}. \tag{1.89}$$

Equating $-(dE/dt)_{\text{GW}}$ and dE/dt yields an evolution equation for a as

$$\frac{da}{dt} = -\frac{64}{5}\frac{m^2\mu}{a^3}. \tag{1.90}$$

This is easily integrated to give

$$a^4 = a_0^4 - \frac{256m^2\mu}{5}t, \tag{1.91}$$

where a_0 is the orbital separation at $t = 0$. Thus, the binary merges at (the time for $a = 0$ is)

$$t_{\text{gw}} = \frac{5a_0^4}{256m^2\mu} \approx 1.2 \times 10^8 \left(\frac{a_0}{10^{11}\,\text{cm}}\right)^4 \left(\frac{m}{2.8M_\odot}\right)^{-3} \left(\frac{m}{4\mu}\right) \text{ yrs}. \tag{1.92}$$

This gives an approximate lifetime of the compact binary system at an orbital separation a_0. We often call this time scale the gravitational-radiation reaction time scale of the binary. This time scale is much shorter than the age of the universe ($\approx 1.38 \times 10^{10}$ yrs) if the system is compact enough as $a_0 \lesssim 3 \times 10^{11}$ cm for $m \sim 3M_\odot$.

The ratio of t_{gw} to the binary orbital period at $a = a_0$, $P_0 = 2\pi\sqrt{a_0^3/m}$, is

$$\frac{t_{\text{gw}}}{P_0} = \frac{5}{512\pi}\left(\frac{a_0}{m}\right)^{5/2}\left(\frac{m}{\mu}\right) \approx 1.1\left(\frac{a_0}{6m}\right)^{5/2}\left(\frac{m}{4\mu}\right). \tag{1.93}$$

Because $\mu \leq m/4$, t_{gw} is found to be always longer than P_0 for $a_0 \geq 6m$. Here,

$$6m = 25\left(\frac{m}{2.8M_\odot}\right) \text{ km}. \tag{1.94}$$

Thus, for binary neutron stars with the total mass $\sim 2.8M_\odot$ and each stellar radius ~ 10 km, the merger occurs when the orbital period becomes comparable to the gravitational-radiation reaction time scale. For binary black holes, the stable circular orbits are likely to present only for $a \geq a_{\text{ISCO}}$ where $a_{\text{ISCO}} = 1\text{–}9m$ (see section 1.3.2.1). Thus, for most of stable orbits (except for very close orbits), the binary also evolves in the adiabatic manner.

To summarize, the binary compact objects evolve by the emission of gravitational waves. In a distant orbit $a \gg 6m$, the orbital separation decreases adiabatically. With the decrease of a, the evolution time scale becomes shorter and shorter, and near the orbit of $a \sim 6m$, the gravitational-radiation reaction time scale becomes as short as the orbital period, implying that the radial approaching velocity steeply increases. Finally, the merger occurs. This is the universal picture for the coalescence of binary compact objects.

We note that the luminosity of gravitational waves and the solution of the equations of motion have been already derived up to a high post-Newtonian order [Blanchet (2014)] (see also appendix H). Thus, the orbital evolution of inspiraling compact binaries can be now more accurately predicted. However, the picture introduced in this section still approximately holds and provides us a guideline.

Before closing this section, we note that in unequal-mass binaries, linear momentum is carried away by gravitational waves. Fitchett (1983) derived the lowest-order emission rate in v/c for circular and elliptic orbits [see also Wiseman (1992); Kidder (1995); Blanchet et al. (2005) for higher post-Newtonian formulas. See also appendix H]. For circular orbits in the x-y plane, the emission rate is written as

$$\frac{dP_i}{dt} = \frac{464}{105} \left(\frac{m}{a}\right)^{11/2} f_{\text{fitchett}}(q) \left[-\sin(\Omega t), \cos(\Omega t), 0\right], \tag{1.95}$$

where

$$f_{\text{fitchett}}(q) = \frac{q^2(1-q)}{(1+q)^5}. \tag{1.96}$$

$f_{\text{fitchett}}(q)$ is the maximum for $q := m_2/m_1 \approx 0.38$, and for $m_1 = m_2$, it vanishes, i.e., no emission of the linear momentum for equal-mass binaries (but see Kidder (1995) in the presence of spins; see also appendix H).

1.2.6 *Gravitational-wave detectors*

For detecting gravitational waves, we have to measure the gravitational tidal force caused by gravitational waves. One of the most promising methods to achieve the detection is to employ a laser interferometer for which the conceptual figure is displayed in figure 1.2. Broadly speaking, a laser interferometer is composed of a laser, a beam splitter, two long-arm vacuum cavities, suspended mirrors, and a photodetector [for details of laser-interferometric gravitational-wave detectors, see, e.g., Creighton and Anderson (2011)]. The typical arm length of the vacuum duct is $3-4$ km for ground-based gravitational-wave detectors. For planned space-craft detectors, the vacuum duct is not necessary and the arm length is much longer as $\sim 10^6$ km.

Laser photons are injected toward the beam splitter at which photons are separately injected toward two long arms of the interferometer. At two ends of the long arms, suspended mirrors, which are implemented to behave as test masses, are placed. Photons injected from the laser are reflected at these mirrors and come eventually back to the beam splitter, at which laser photons coming from two different directions are adjusted to interfere with each other to suppress the fraction of photons entering the photo detector as small as possible. Suppose that the laser

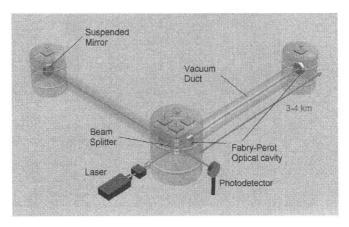

Fig. 1.2 Schematic figure for a laser interferometric gravitational-wave detector (by the courtesy of M. Fujimoto and H. Ishizaki at National Astronomical Observatory of Japan).

interferometer is placed in the $x-y$ plane and gravitational waves propagate along the z-axis direction. Here, we implicitly assume that the wave length of gravitational waves, which we aim to detect, is $\gtrsim 100$ km and much longer than the arm length. As illustrated in figure 1.1, during the propagation of gravitational waves in the detector, its arm length along the x and y axes varies with time. Here, the variation is anisotropic: For example, if the arm length along the x axis increases, that along the y axis decreases. This reduces the degree of the interference at the beam splitter. Therefore, in the presence of gravitational wave, an additional fraction of photons come into the photo detector. If the amplitude of gravitational waves is sufficiently high and the number of the photons are sufficiently large, we can achieve a gravitational-wave detection.

In the near future, there will be five advanced ground-based detectors for which the arm length is 3–4 km. Three of them are LIGO in USA and India; two of three are composed of 4 km arm-length interferometers placed at Hanford in Washington State (see figure 1.3) and at Livingstone in Louisiana state in USA. The 2 km-arm detector, which used to be placed at Hanford, is planned to be moved to India to construct the LIGO-India. One of the five detectors is Italian–French VIRGO, which is composed of a 3 km arm-length detector placed at Cascina, Italy. The last one is Japanese KAGRA (formerly LCGT), which is also a 3 km arm-length detector placed at Kamioka (under Mt. Ikenoyama), at which a well-known neutrino detector, super Kamiokande, is also placed (see figure 1.4).

Figure 1.5 depicts some of designed sensitivities (noise levels) of advanced LIGO and KAGRA. The sensitive curve for initial LIGO, which is by a factor of about 15 poorer than that for advanced LIGO, was already achieved in 2006 [Abbott et al. (2009)]. The planned sensitivity level of advanced VIRGO is approximately

Fig. 1.3 A 4 km-arm laser-interferometric gravitational-wave detector LIGO at Hanford, USA. The picture is taken from http://www.ligo.caltech.edu.

Fig. 1.4 A schematic figure for a 3 km-arm laser-interferometric gravitational-wave detector KAGRA (formerly LCGT) at Kamioka, Japan. The figure is taken from http://www.iccr.u-tokyo.ac.jp/gr/plans.html.

identical with those of advanced LIGO and KAGRA [Accadia *et al.* (2011); Kuroda (2010)].

The best sensitivity of these advanced detectors is $\sim 3 \times 10^{-23}$ at a frequency band $f \sim 100$ Hz. They typically have a high sensitivity in the range between ~ 10 and several thousand Hz. One of the promising sources of gravitational waves for these advanced detectors is coalescing compact binaries composed of black holes and/or neutron stars (see section 1.5.1). By the advanced detectors, these gravitational waves will be detected if the coalescences happen within the distance of

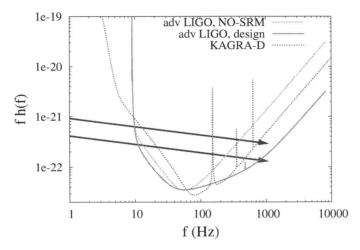

Fig. 1.5 Designed sensitivities of advanced LIGO and KAGRA. "adv LIGO, NO-SRM" and "design" denote a possible early configuration and the designed sensitivity of advanced LIGO, respectively (data for them are taken from https://dcc.ligo.org/cgi-bin/DocDB/ShowDocument?docid=2974). "KAGRA-D" denotes the official designed sensitivity of KAGRA (taken from http://gwcenter.icrr.u-tokyo.ac.jp/researcher/parameters). Although the sensitivity curves are usually defined in units of $Hz^{-1/2}$, we multiply $f^{1/2}$ to obtain the curves of dimensionless units which can be directly compared with the gravitational-wave amplitude. Two slowly declining lines with an arrow denote effective gravitational-wave amplitudes from inspiraling equal-mass binary neutron stars (or binary black holes) at a hypothetical distance of 100 Mpc (740 Mpc for binary black holes); see equation (1.202). The upper and lower lines, respectively, denote the amplitudes for the most optimistic source direction and orbital inclination, and for the average over the random orientation, random inclination angle of the orbital plane, and random polarization. Masses of neutron stars and black holes are assumed to be $1.35 M_\odot$ and $10 M_\odot$, respectively, in this plot.

several hundred Mpc from the Earth. Note that the limit distance for the detection depends on the sky direction and the inclination of the binary orbital plane to the line of sight [Schutz and Tinto (1987)].

It should also be mentioned that in the future, 10 km scale detectors, called Einstein Telescope, may be in operation [Hild *et al.* (2008, 2010)]. This project is now proposed as a future plan. The sensitivity of this detector will be by a factor of ~ 10 better than the advanced detectors with a wider detection band between 1 Hz and 10 kHz.

When we try to detect gravitational waves by the ground-based interferometric detectors, one of the serious noise sources is a seismic noise for a low-frequency band of $f \lesssim 10$ Hz: see the noise curve of the low-frequency band in figure 1.5. This shows it impossible to detect low-frequency gravitational waves with $f \ll 10$ Hz by ground-based detectors, and suggests us to construct a detector in space when one wants to detect low-frequency gravitational waves.

The most well-known space mission for this purpose was the LISA project by ESA and NASA. After NASA withdrew from this project, ESA replanned the so-

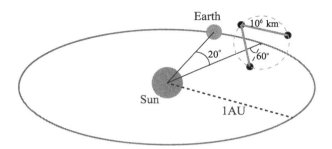

Fig. 1.6 Schematic figure for eLISA's orbital configuration.

called eLISA project, and in 2013, this project was approved for the launch in 2034. The eLISA project plans to launch three compact drag-free spacecrafts, which will be placed into the same heliocentric orbit as the Earth occupies but will follow 20° behind the Earth (see figure 1.6). The spacecrafts will fly at the vertex of an equilateral triangle that will be inclined at an angle of 60° to the Earth's orbital plane. The planned triangle's arm length is 1×10^6 km.

The planned noise curve is depicted in figure 1.7. eLISA will have a good sensitivity in the frequency range between ~ 1 and 100 mHz. The best sensitivity is approximately 3×10^{-21} at $f \sim 10$ mHz. The most promising sources for the space interferometers are coalescing supermassive black holes of mass $\sim 10^6 - 10^7 M_\odot$ and a number of nearby galactic binary systems. With the eLISA's sensitivity, gravitational waves from the binary supermassive black holes will be detected with high signal-to-noise ratio > 100 as shown in figure 1.7 (see also section 1.5) and will provide us rich information on supermassive black holes and galactic centers.

1.3 Black holes

Clarifying the formation and evolution processes of black holes and black-hole binaries is the central subject in numerical relativity. Stellar-mass black holes are believed to be formed as an end product of the evolution of very massive stars with initial mass larger than $\sim 20 M_\odot$. They should be also formed after the merger of two neutron stars and after a substantial mass accretion onto a neutron star in binary systems with an ordinary star or a white dwarf. Stellar-mass black holes are observed in our Galaxy and in the nearby galaxies such as Large Magellanic Cloud. To date, about 50 black holes including candidates have been observed in binary systems with ordinary stars [McClintock and Remillard (2006); Özel et al. (2010)]. For some black holes, the mass is determined to be larger than $5 M_\odot$ with a small error [Özel et al. (2010)], and hence, they are confirmed to be really black holes, because neutron stars of such high mass cannot exist [Shapiro and Teukolsky (1983)]. The formation and subsequent evolution of a non-spherical black hole in

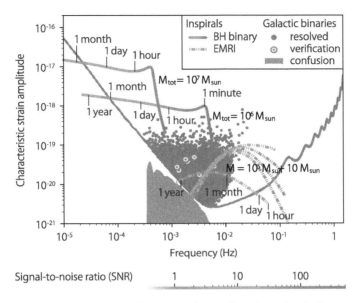

Fig. 1.7 eLISA's projected noise level $h_{\rm rms}$ and the predicted amplitude for several promising sources such as coalescing binary supermassive black holes and resolved galactic binary systems. "Confusion" shows the predicted superposition of gravitational waves from unresolved galactic binaries. The figure is taken from the report of eLISA Consorthium (2013) (see figure 13 of this reference for details).

the stellar core collapse have been poorly understood yet and their study is a central subject of high-energy astrophysics, gravitational-wave astrophysics, and numerical relativity (see section 9.3 for a numerical-relativity study).

Candidates of supermassive black holes with mass larger than $\sim 10^6 M_\odot$ are often observed in the central region of galaxies [e.g., Rees (1996); Magorrian et al. (1998); Ferrarese and Ford (2005)]. Now, we believe that each galaxy usually possesses a supermassive black hole in their center, although it is not still easy to strictly confirm their presence due to their high compactness and the resulting difficulty of observation for their vicinity. However, there are at least two observationally robust evidences which show that supermassive black holes really exist. One is the supermassive black hole located in our Galactic (Milky Way) center. A long-term observation of ordinary stars orbiting the vicinity of the center of our Galaxy clarifies that hidden mass in the central region of our Galaxy is about $4 \times 10^6 M_\odot$ [Schödel et al. (2002)]. Such huge mass hidden in a very small volume less than $(0.01\,{\rm pc})^3$ could be explained only by the presence of a supermassive black hole. Miyoshi et al. (1995) found the other robust evidence for a galaxy called NGC 4258 which is about 7 Mpc far from our Galaxy. This is a spiral galaxy like our Galaxy. Its quite unique property is that the galactic plane is parallel to our line of sight. This is ideal for measuring the orbital velocity of the objects in the galactic plane. Miyoshi *et al.* (1995) observed molecular clouds around the central region in which there are many maser sources. They measured the Doppler shift of a characteristic wavelength of the maser spectrum lines and determined the orbital velocity of the molecular clouds at an orbital radius $\sim 0.13\,\mathrm{pc}$ around the central region. Again, there are not sufficiently massive bright sources inside this radius. This results in the conclusion that there must be a supermassive black hole of mass about $3.6 \times 10^7 M_\odot$.

Supermassive black holes are believed to be formed in an early epoch of the universe in galactic centers [e.g., Rees (1996); Magorrian *et al.* (1998); Ferrarese and Ford (2005)]. However, the detailed formation process is still poorly known. During its formation, anyhow, general relativistic gravity and relativistic motion are likely to be associated. Thus, clarifying its formation process is also one of the most important subjects in numerical relativity (see section 9.2 for a scenario).

There may exist another type of black holes for which the formation process is different from those mentioned above. One is a primordial black hole which might be formed in a very early epoch of the universe (typically radiation-dominant epoch) with age $t < 1\,\mathrm{s}$ [Carr and Hawking (1974)]. Such a black hole may be formed from a super-horizon-scale curvature fluctuation of a high amplitude (see section 9.4). Specifically, a primordial black hole could be formed when the length scale of a high-amplitude fluctuation becomes smaller than the cosmological horizon scale that is proportional to the age of the universe t. Thus, the mass of the primordial black hole is proportional to the formation epoch, and for example, if it is formed at $t \sim 10^{-6}$ s, the mass is of order $M_\odot$. The study for the formation of the primordial black holes is also an issue in numerical relativity.

Recently, a possible formation mechanism of a mini black hole in particle accelerators was proposed. The hypothesis in this scenario is that our world could not be four-dimensional spacetime but higher-dimensional spacetime [Arkani-Hamed *et al.* (1998); Antoniadis *et al.* (1998); Randall and Sundrum (1999b,a)]. The higher-dimensional spacetime hypothesis is often required in string theories. If this is indeed the case, the Planck energy is not necessarily $\sim 10^{19}\,\mathrm{GeV}$ but could be of $O(\mathrm{TeV})$ that may be accessible with particle accelerators such as CERN Large Hadron Collider (LHC) in operation, because the gravitational constant in higher dimensions is not known. In the presence of the extra dimensions, mini black holes of very small mass energy $\gtrsim \mathrm{TeV}$ may be produced during the particle collision in the accelerators [Argyres *et al.* (1998); Banks and Fischler (1999); Dimopoulos and Landsberg (2001)] because the "true" Planck energy may be confined inside a black-hole horizon. Clarifying the formation process of mini black holes in high-velocity particle collisions in higher-dimensional spacetime becomes a new subject in numerical relativity (see chapter 11).

As briefly described here, there are many subjects associated with the formation and the evolution of black holes in numerical relativity. In the subsequent chapters of this volume, I will describe these topics in more detail. Thus in the rest of this section, fundamental properties of black holes are summarized for the preparation of the description of the following chapters. More detailed properties of four-dimensional black holes are also summarized in appendix G.

1.3.1 Four dimensional black holes

First, we review four-dimensional black holes. The line element of non-spinning black holes in the Schwarzschild coordinates is written as [e.g., Hawking and Ellis (1973); Wald (1984)]

$$ds^2 = -\left(1 - \frac{2M}{r}\right)dt^2 + \left(1 - \frac{2M}{r}\right)^{-1}dr^2 + r^2 d\Omega, \tag{1.97}$$

where M is the mass, r is the areal radius that defines the area of the $r = $ const sphere as $4\pi r^2$, and $d\Omega$ is the line element of the 2-dimensional unit sphere. This black-hole spacetime is static in the sense that it possesses a timelike Killing vector field

$$\xi^a := \left(\frac{\partial}{\partial t}\right)^a, \tag{1.98}$$

and in addition, ξ^a is orthogonal to the spatial ($t = $ const) hypersurface, satisfying Frobenius' theorem (see appendix A.2)

$$\epsilon_{abcd}\xi^b \nabla^c \xi^d = 0. \tag{1.99}$$

Here, ϵ_{abcd} is the completely antisymmetric tensor with $\epsilon_{tr\theta\varphi} = \sqrt{-g} = r^2 \sin\theta$ for the Schwarzschild coordinates.

The event horizon of the Schwarzschild black hole is placed at $r = 2M$ with its area $A_\mathrm{H} = 16\pi M^2$. The curvature invariant of the Schwarzschild black hole is [e.g., Landau and Lifshitz (1962)]

$$^{(4)}R_{abcd}\,^{(4)}R^{abcd} = \frac{48M^2}{r^6}, \tag{1.100}$$

which blows up at the origin, implying that $r = 0$ is the physical singularity (whereas the singularity of $g_{rr} = \infty$ at $r = 2M$ is a coordinate singularity, not physical one). This physical singularity is enclosed by the event horizon, and if we focus only on our world (the region outside the horizon), it does not matter to us.

The line element of spinning black holes (Kerr black holes) in the Boyer–Lindquist coordinates (the most well-known coordinates for spinning black holes) is written as [Boyer and Lindquist (1967)]

$$ds^2 = -\left(1 - \frac{2Mr}{\Sigma}\right)dt^2 - \frac{4Mar\sin^2\theta}{\Sigma}dt d\varphi + \frac{\Sigma}{\Delta}dr^2 + \Sigma d\theta^2 + \frac{\Xi}{\Sigma}\sin^2\theta d\varphi^2, \tag{1.101}$$

where a is the spin parameter[2] with which the angular momentum of the black hole is written as $J = Ma$. Σ, Δ, and Ξ are defined by

$$\Sigma := r^2 + a^2 \cos^2\theta, \quad \Delta := r^2 - 2Mr + a^2,$$
$$\Xi := (r^2 + a^2)\Sigma + 2Ma^2 r \sin^2\theta. \tag{1.102}$$

---

[2]Only in this section, a denotes the black-hole spin (not the orbital separation of binaries). A dimensionless spin parameter defined by a/M is denoted by $\hat{a}$ throughout this volume.

The event horizon of the Kerr black hole is placed at an outer surface that satisfies $\Delta = 0$, i.e., $r = r_+ = M + \sqrt{M^2 - a^2}$ and its area is

$$A_{\rm H} = \oint_{r=r_+} \sqrt{g_{\theta\theta}g_{\varphi\varphi}}d\theta d\varphi = 8\pi M\left(M + \sqrt{M^2 - a^2}\right). \tag{1.103}$$

Thus, for the event horizon to be present, the absolute magnitude of the spin parameter has to be smaller than M ($|a| \leq M$). Equation (1.103) shows that for a given value of mass M, the area of the spinning black hole is always smaller than that of $a = 0$.

The Kerr black-hole spacetime is stationary, axisymmetric in the sense that it possesses timelike and spacelike Killing vector fields [e.g., Wald (1984)]

$$\xi^a := \left(\frac{\partial}{\partial t}\right)^a, \quad \varphi^a := \left(\frac{\partial}{\partial \varphi}\right)^a, \tag{1.104}$$

and $\xi^a \varphi_a \neq 0$ for $a \neq 0$.

For the Kerr metric, $g_{tt} \neq 0$ on the horizon surface, but the lapse function (which is derived from $1/\sqrt{-g^{tt}}$; see chapter 2 for its definition) vanishes. The lapse function of the Kerr metric in the Boyer–Lindquist coordinates is written as

$$\alpha = \alpha_{\rm K} = \sqrt{\frac{\Sigma\Delta}{\Xi}}. \tag{1.105}$$

The curvature invariant of the Kerr black hole is [e.g., Poisson (2004)]

$${}^{(4)}R_{abcd}\,{}^{(4)}R^{abcd} = \frac{48M^2(r^2 - a^2\cos^2\theta)(r^4 - 14a^2\cos^2\theta + a^4\cos^4\theta)}{\Sigma^6}. \tag{1.106}$$

This blows up at $r = 0$ and $\theta = \pi/2$, which is the physical singularity. This physical singularity is again enclosed by the event horizon.

The line elements of Schwarzschild and Kerr black holes are often written in other coordinates, like isotropic and quasi-isotropic coordinates. In the isotropic coordinates, the line element of non-spinning black holes is written as

$$ds^2 = -\left(\bar{r} - \frac{M}{2}\right)^2 \left(\bar{r} + \frac{M}{2}\right)^{-2} dt^2 + \left(1 + \frac{M}{2\bar{r}}\right)^4 \left(d\bar{r}^2 + \bar{r}^2 d\Omega^2\right), \tag{1.107}$$

where $\bar{r}$ is related to the radial coordinate of the Schwarzschild black hole by

$$r = \left(1 + \frac{M}{2\bar{r}}\right)^2 \bar{r}, \quad \bar{r} = \frac{1}{2}\left(r - M + \sqrt{r(r - 2M)}\right). \tag{1.108}$$

The event horizon is placed at $\bar{r} = M/2$ in this coordinate.

The line element of spinning black holes in the quasi-isotropic coordinates is written as [Krivan and Price (1998)]

$$ds^2 = -\frac{\Sigma\Delta}{\Xi}dt^2 + \frac{\Xi}{\bar{r}^2\Sigma}\left[\frac{\Sigma^2}{\Xi}\left(d\bar{r}^2 + \bar{r}^2 d\theta^2\right) + \bar{r}^2 \sin^2\theta\left(-\frac{2Mar}{\Xi}dt + d\varphi\right)^2\right], \tag{1.109}$$

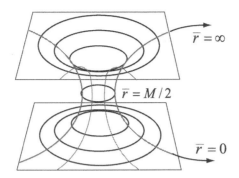

Fig. 1.8 Schematic embedding diagram of the Schwarzschild geometry in isotropic coordinates. The spatial hypersurface with $t = $ const on the equatorial plane ($\theta = \pi/2$) is embedded as a 2-dimensional surface in a 3-dimensional flat space. The minimum surface is the surface of the event horizon, $\bar{r} = M/2$.

where $\bar{r}$ is the radial coordinate in the quasi-isotropic coordinates and r is that in the Boyer–Lindquist ones. These are related by

$$r = \bar{r}\left(1 + \frac{M+a}{2\bar{r}}\right)\left(1 + \frac{M-a}{2\bar{r}}\right), \quad \bar{r} = \frac{r - M + \sqrt{\Delta}}{2}. \tag{1.110}$$

Here, the definitions of Σ, Δ, and Ξ are the same as those in equation (1.102). The event horizon is placed at $\bar{r} = \sqrt{M^2 - a^2}/2$ in this case.

Figure 1.8 displays a schematic embedding diagram for the topology of a spatial hypersurface of a non-spinning black hole in isotropic coordinates. The minimum surface labeled by $\bar{r} = M/2$ is the surface of the event horizon, and the regions with $\bar{r} > M/2$ and $\bar{r} < M/2$ denote our world and other world that is causally disconnected by us, respectively (see figure 1.9). The throat connecting two world sheets is sometimes called Einstein-Rosen-bridge or Schwarzschild wormhole throat [Misner et al. (1973)]. The two world sheets are symmetric with respect to a coordinate transformation $\bar{r} \to (M/2)^2/\bar{r}$. For Kerr spacetime, the region with $\bar{r} < \sqrt{M^2 - a^2}/2$ in quasi-isotropic coordinates denotes the other world which is symmetric with our world of $\bar{r} > \sqrt{M^2 - a^2}/2$, and $\bar{r} = 0$ denotes the spatial infinity of the other world. This fact is also found by a coordinate transformation of $\bar{r} \to \sqrt{M^2 - a^2}/2\bar{r}$: With this transformation, the line element is unchanged. These metric functions are free of physical singularities, although the coordinate singularity is present at $\bar{r} = 0$. The singularity-free nature of this metric is useful in numerical-relativity simulations of black-hole spacetime, and this type of the metric is often used as initial data (see section 5.5.2.2).

One of the important properties of the isotropic and quasi-isotropic coordinates is that t and $\bar{r}$ are always the timelike and spacelike coordinates, respectively. However, the spacetime region inside the event horizon is not covered by these coordinates (only regions I and III, i.e., outside the horizons, are covered in figure 1.9).

Another coordinates often used in numerical relativity are the Kerr–Schild coordinates (the original form of the Kerr metric), in which the line element of spinning black holes is written as [Kerr (1963)]

$$ds^2 = -d\bar{t}^2 + dx^2 + dy^2 + dz^2$$
$$+ \frac{2Mr^3}{r^4 + a^2z^2}\left(\frac{r(xdx + ydy) - a(xdy - ydx)}{r^2 + a^2} + \frac{zdz}{r} + d\bar{t}\right)^2, \quad (1.111)$$

where the relation between these coordinates and Boyer–Lindquist coordinates is as follows [e.g., Hawking and Ellis (1973)]:

$$x + iy = (r + ia)\sin\theta \exp\left[i\int\left(d\varphi + \frac{a}{\Delta}dr\right)\right],$$
$$z = r\cos\theta, \quad \bar{t} = t - r + \int \frac{r^2 + a^2}{\Delta}dr. \quad (1.112)$$

The relation between r and (x, y, z) are written as

$$r^4 - \left(x^2 + y^2 + z^2 - a^2\right)r^2 - a^2z^2 = 0. \quad (1.113)$$

In this coordinate system, the physical singularity ($r = 0$) is located on a ring shape denoted by $x^2 + y^2 = a^2$ and $z = 0$, and the event horizon becomes a spheroid denoted by

$$\frac{x^2 + y^2}{r_+^2 + a^2} + \frac{z^2}{r_+^2} = 1. \quad (1.114)$$

One of the most remarkable properties of the Kerr–Schild coordinates is that $\bar{t} = \text{const}$ hypersurfaces are always spacelike, and in addition, they penetrate the horizon. Furthermore, there is no singular behavior of the metric on the event horizon (although each spacelike hypersurface hits the physical singularity at $r = 0$).

For analyzing the global structure of non-spinning black holes, the Kruskal–Szekeres coordinates [Kruskal (1960)] are quite useful. This coordinate system covers the entire manifold of the black-hole spacetime in a well-defined manner in contrast to other coordinates mentioned above. Also it is free of the coordinate singularity on the event horizon. The line element in the Kruskal–Szekeres coordinates takes the form

$$ds^2 = \frac{32M^3 e^{-r/2M}}{r}\left(-dT^2 + dX^2\right) + r^2 d\Omega, \quad (1.115)$$

where the relation between (t, r) and (T, X) is given by

$$\left(\frac{r}{2M} - 1\right)e^{r/2M} = X^2 - T^2, \quad \frac{t}{4M} = \tanh^{-1}\left(\frac{T}{X}\right). \quad (1.116)$$

A spacetime diagram of the Kruskal–Szekeres extension is shown in figure 1.9. In this diagram, the timelike coordinate T is plotted vertically and a spacelike coordinate is plotted horizontally. Region I corresponds to the region of our world with $r > 2M$, i.e., the region outside the event horizon where we live. Region II is inside the event horizon $r < 2M$. Regions III ($r > 2M$) and IV ($r < 2M$) denote

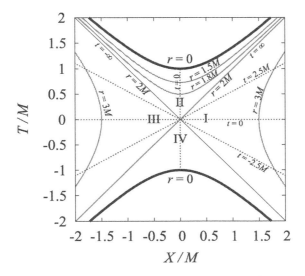

Fig. 1.9 A Kruskal–Szekeres diagram for the spacetime structure of a non-spinning black hole. The spatial surfaces of $r = 0$ are the physical singularities and the null surfaces of $r = 2M$ are the event horizons. Region I corresponds to our world with $r > 2M$, i.e., the region outside one of the event horizons. Region II is inside the event horizon $r < 2M$. Regions III ($r > 2M$) and IV ($r < 2M$) denote the other world and inside of the white hole, respectively. For the entire region, T is the timelike coordinate and X is a spacelike coordinate. Spatial hypersurfaces of $t = $ const in the Schwarzschild coordinates are described by tilted dashed lines in the regions I and III. The vertical line of $t = 0$ in the regions II and IV denotes the minimal surface of the wormhole throat if the symmetric spatial hypersurface with respect to this plane is chosen (see, e.g., section 2.2.2).

the other world and inside of the white hole, respectively. For the entire region, T is the timelike coordinate and X is a spacelike coordinate.

The causal structure of the extended Schwarzschild spacetime is easily seen in the Kruskal–Szekeres coordinates: The radially-directing null geodesics are denoted by the $X = \pm T$, i.e., 45° diagonal lines in the Kruskal–Szekeres coordinates. The physical singularity at $r = 0$ is found to be an extended region at $X = \pm\sqrt{T^2 - 1}$. It is found that the physical singularity at $r = 0$ has a spacelike structure which exists in the future of region II and in the past of region IV. The Kruskal–Szekeres diagram will be used for describing several slicing conditions (the condition for choosing spatial hypersurfaces) for non-spinning black-hole spacetime in section 2.2.

The Kruskal–Szekeres coordinates clarify the causal structure of non-spinning black-hole spacetime. However, we have another useful set of coordinates, double-null coordinates system, defined by

$$U = \tan^{-1}(T - X) \quad \text{and} \quad V = \tan^{-1}(T + X). \tag{1.117}$$

By this transformation, null rays are mapped onto $U = $ const or $V = $ const. Hence, 45° diagonal lines in the T-X plane are transformed to 45° diagonal lines in the $\tilde{T}$-$\tilde{X}$ plane where $\tilde{T} = (U + V)/2$ and $\tilde{X} = (V - U)/2$. In addition, the entire region

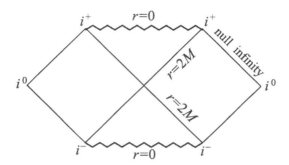

Fig. 1.10 The Penrose–Carter diagram for the spacetime structure of non-spinning black holes. $i^{\pm}$ and i^0 denote the future and past timelike infinities and the spatial infinity, respectively.

of the spacetime with $-\infty < T < \infty$ and $-\infty < X < \infty$ is mapped onto a finite domain composed of $-\pi/2 \leq U \leq \pi/2$ and $-\pi/2 \leq V \leq \pi/2$. For example, the future and past physical singularities are denoted by $U + V = \pi/2$ and $U + V = -\pi/2$, respectively. Such a transformation enables us to draw a compact diagram like figure 1.10: This type of the diagram is called the Penrose–Carter diagram or conformal diagram.

It is a good exercise to consider spacelike hypersurfaces of the Eddington–Finkelstein coordinates [Finkelstein (1958)] in the Kruskal–Szekeres diagram. The line element of the Eddington–Finkelstein coordinates is obtained by setting $a = 0$ for equation (1.111) as

$$ds^2 = -\left(1 - \frac{2M}{r}\right)d\bar{t}^2 + \frac{4M}{r}d\bar{t}dr + \left(1 + \frac{2M}{r}\right)dr^2 + r^2 d\Omega, \quad (1.118)$$

where $\bar{t} = t + 2M \log|r/2M - 1|$. For this case, one finds that $\bar{t} = $ const hypersurfaces are always spacelike and penetrate the horizon hitting the physical singularity at $r = 0$. Thus, $\bar{t}$ can be used as a time coordinate for covering spatial hypersurfaces for $r > 0$.

1.3.2 Properties of four-dimensional black holes

In this section, we summarize several important properties of four-dimensional black holes, which play a key role in the analysis of numerically generated spacetime.

1.3.2.1 Circular orbits around black holes

First, the properties of a test particle orbiting a black hole with a circular orbit are summarized [see, e.g., Shapiro and Teukolsky (1983) for more detailed description]. The orbit is determined by the geodesic equation for a given black-hole geometry. For the Schwarzschild black hole, we have a timelike Killing vector field and spacelike ones associated with rotational symmetries around the center. Because the rotational symmetries are present and thus in the Schwarzschild background, the orbital plane never changes, we may focus only on orbits on the equatorial plane of $\theta = \pi/2$ without the loss of generality. For this case, the relevant Killing vector fields that characterize the test particle motion are

$$\xi^a = \left(\frac{\partial}{\partial t}\right)^a \quad \text{and} \quad \varphi^a = \left(\frac{\partial}{\partial \varphi}\right)^a, \quad (1.119)$$

and the relevant conserved quantities for the test particle, i.e., energy and angular momentum, are derived, respectively, as

$$E = -m\xi^a u_a \quad \text{and} \quad \ell = m\varphi^a u_a, \quad (1.120)$$

where u^a is the four velocity of the test particle and m is its mass. These two conserved quantities together with an initial condition completely determine the orbit. For a circular orbit with its orbital radius r, we have

$$E = m\frac{r - 2M}{\sqrt{r(r - 3M)}} \quad \text{and} \quad \ell = m\frac{\sqrt{Mr}}{\sqrt{r - 3M}}. \quad (1.121)$$

The angular velocity defined by $\Omega_K := u^\varphi / u^t$ is

$$\Omega_K = \sqrt{\frac{M}{r^3}}. \quad (1.122)$$

The circular orbit is allowed only outside the innermost stable circular orbit (ISCO), for which the radius is $r = 6M$. At the ISCO, $E/m = \sqrt{8/9}$ and $\ell/m = 2\sqrt{3}M$, and hence, $E/m > \sqrt{8/9}$ and $\ell/m > 2\sqrt{3}$ for $r > 6M$. The binding energy at the ISCO is $(1 - \sqrt{8/9})m \approx 0.057m$. The presence of the ISCO means that if the specific angular momentum of a test particle, ℓ/m, is smaller than $2\sqrt{3}M$, it has to fall into the black hole.

For Kerr black holes, the geodesic equation is much more complicated than that for the Schwarzschild case. However, it is also known that the geodesic equation is integrable due to a geometrically special property of the Kerr geometry: For Kerr spacetime, in addition to the Killing vector fields shown in equation (1.119), the following Killing tensor field is present [Walker and Penrose (1970)]

$$K_{\mu\nu} = 2\Sigma \hat{l}_{(\mu} \hat{n}_{\nu)} + r^2 g_{\mu\nu}, \quad (1.123)$$

where $\hat{l}^a$ and $\hat{n}^a$ are null vector fields defined by

$$\hat{l}^a := \left(\frac{r^2 + a^2}{\Delta}\right)\left(\frac{\partial}{\partial t}\right)^a + \left(\frac{a}{\Delta}\right)\left(\frac{\partial}{\partial \varphi}\right)^a + \left(\frac{\partial}{\partial r}\right)^a, \quad (1.124)$$

$$\hat{n}^a := \left(\frac{r^2 + a^2}{2\Sigma}\right)\left(\frac{\partial}{\partial t}\right)^a + \left(\frac{a}{2\Sigma}\right)\left(\frac{\partial}{\partial \varphi}\right)^a - \frac{\Delta}{2\Sigma}\left(\frac{\partial}{\partial r}\right)^a. \quad (1.125)$$

Hence, in addition to E and ℓ, we have the third conserved quantity (called the Carter constant) [Carter (1968)],

$$Q = mK_{ab}u^a u^b. \quad (1.126)$$

These three constants of motion determine the orbit of the test particle.

The Carter constant primarily determines the orbital motion in the θ direction. If $C := Q + (\ell - aE)^2 \neq 0$, the test particle precesses around the equatorial plane, $\theta = \pi/2$. For $C = 0$, the test particle moves only on the equatorial plane. Because the equatorial orbits reflect black-hole spin effects in the most remarkable manner, we here pay attention only to them. In this case, we do not have to consider the Carter constant independently, because it is written by E and ℓ. For circular orbits of the orbital radius r, these two constants are written as [Bardeen et al. (1972)]

$$E = m \frac{r^2 - 2Mr \pm a\sqrt{Mr}}{r\sqrt{r^2 - 3Mr \pm 2a\sqrt{Mr}}}, \tag{1.127}$$

$$\ell = \pm m \frac{\sqrt{Mr}\left(r^2 \mp 2a\sqrt{Mr} + a^2\right)}{r\sqrt{r^2 - 3Mr \pm 2a\sqrt{Mr}}}, \tag{1.128}$$

where the plus and minus signs refer, respectively, to corotating and counterrotating orbits with respect to the black-hole spin. The angular velocity is written by

$$\Omega_K = \frac{\sqrt{M}}{r^{3/2} \pm a\sqrt{M}}. \tag{1.129}$$

Thus, the angular velocity is written in a form different from the usual Keplerian form because of the presence of the black-hole spin. Equation (1.129) shows that in the presence of a spin corotating (counterrotating) with the orbital motion, the angular velocity is smaller (larger) than that with no spin for a given orbital radius.[3] This fact may be described in the terminology of post-Newtonian approximations as follows: In the presence of a spin, there exists a coupling effect between the spin and the orbital angular momentum (spin-orbit coupling effect). The correction force associated with this is proportional to $\mathbf{S} \cdot \mathbf{L}$ where $\mathbf{S}$ and $\mathbf{L}$ denote the vectors of the black-hole spin and the orbital angular momentum, respectively [e.g., Kidder et al. (1993); Kidder (1995)]. Thus, for the corotating (counterrotating) case, the additional repulsive (attractive) force is yielded. In the presence of the repulsive (attractive) force, the centrifugal force has to be weaker (stronger) for maintaining a circular orbit for a given gravitational force determined by the monopole component, and therefore, the orbital angular velocity has to be smaller (larger). This spin-orbit coupling effect often plays an important role in the evolution of binary black holes and black hole-neutron star binaries.

The radius of the ISCO depends also strongly on the black-hole spin [Bardeen et al. (1972)]:

$$r_{\text{ISCO}} = M\left[3 + Z_2 \mp \sqrt{(3 - Z_1)(3 + Z_1 + 2Z_2)}\right] =: M\hat{r}_{\text{ISCO}}. \tag{1.130}$$

Here

$$Z_1 = 1 + (1 - \hat{a}^2)^{1/3}\left[(1 + \hat{a})^{1/3} + (1 - \hat{a})^{1/3}\right], \quad Z_2 = \sqrt{3\hat{a}^2 + Z_1^2},$$

---

[3] In the Boyer–Lindquist coordinates, the circumferential length at r on the equatorial plane is $2\pi r$ irrespective of a. Thus, r here has a physical meaning as the circumferential radius.

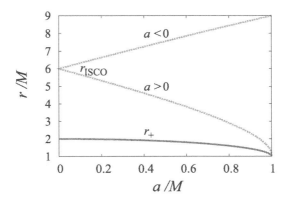

Fig. 1.11 The radii of the event horizon and the ISCO in the Boyer–Lindquist radial coordinate as functions of a/M. $a > 0$ and $a < 0$ imply that the test particle corotates and counterrotates with the black-hole spin, respectively, on the equatorial plane.

and $\hat{a} := a/M$: a dimensionless spin parameter. Figure 1.11 shows the location of the ISCO and the event horizon together: For $a = 0$, $r_{\rm ISCO} = 6M$. However, for $a \to \pm M$, $r_{\rm ISCO}$ approaches M for the corotating case and $9M$ for the counterrotating case.

Figure 1.12 displays E/m and $\ell/(Mm)$ at the ISCO as functions of a/M. For the corotating orbits, E/m and ℓ/m at the ISCO are decreasing functions of a/M while for the counterrotating case, they are increasing functions of $|a|/M$. It is remarkable that toward $a \to M$ for the corotating orbit, E/m and ℓ/m at the ISCO steeply decrease. The smallest values are achieved at $a = M$ as $E/m = \sqrt{1/3}$ and $\ell/m = M$, which are much smaller than those for $a = 0$. Thus, the binding energy there is quite large as $\left(1 - \sqrt{1/3}\right)m \approx 0.423m$. The small value of ℓ shows that a test particle is less subject to the capture by the black hole. Thus, the higher corotating black-hole spin, in particular the spin with a/M close to the unity, enhances the formation of a disk around the black hole and also the formation of a close binary with a star. In addition, this spin effect enhances the possibility of tidal disruption of a star orbiting the black hole, because a closer orbit, at which the tidal force of the black hole is stronger, is allowed for a high black-hole spin. The effect of the black-hole spin for the tidal problem will be highlighted when we describe the coalescence of black hole-neutron star binaries (see section 8.5).

1.3.2.2 Static observer and angular velocity of black holes

Locally non-rotating observers in stationary and axisymmetric spacetime were originally defined in Bardeen (1970); Bardeen et al. (1972). This family of the observers is at rest with respect to $t = $ constant hypersurfaces. Thus, their four-velocity, u^a, is proportional to $\nabla^a t$, and thus, $u^t \propto g^{tt}$, $u^\varphi \propto g^{t\varphi}$, and $u^r = u^\theta = 0$ for the metric (1.101) and/or (1.109). Then, the angular velocity of such observers is

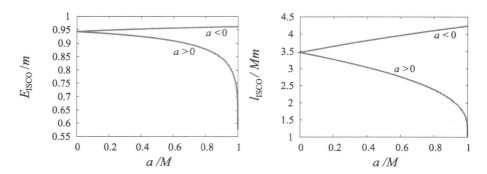

Fig. 1.12 E/m and ℓ/Mm at the ISCO (of equatorial orbits) as functions of a/M.

defined by
$$\Omega_{\text{static}} = \frac{g^{t\varphi}}{g^{tt}} = -\frac{g_{t\varphi}}{g_{\varphi\varphi}}. \tag{1.131}$$

In the limit that the orbital radius approaches the event horizon, $r \to r_+$, this angular velocity becomes a constant written by
$$\Omega_{\text{H}} := \frac{a}{r_+^2 + a^2} = \frac{a}{2Mr_+}. \tag{1.132}$$

This Ω_{H} is often referred to as the angular velocity of spinning (Kerr) black holes.

1.3.2.3 Quantities associated with the horizon

Here, we summarize the quantities of the event horizon. As mentioned already, its area is written as
$$A_{\text{H}} = 8\pi M \left(M + \sqrt{M^2 - a^2} \right). \tag{1.133}$$

Hawking (1971) proved the second law (the so-called area theorem) for black-hole spacetime. This states that the area of the black-hole horizons never decreases in the predictable spacetime (see also appendix G). Associated with this fact, "irreducible" mass is often defined by
$$M_{\text{irr}}^2 = \frac{A_{\text{H}}}{16\pi} = \frac{M\left(M + \sqrt{M^2 - a^2}\right)}{2}. \tag{1.134}$$

Using this, M is written by
$$M^2 = M_{\text{irr}}^2 + \frac{J^2}{4M_{\text{irr}}^2}. \tag{1.135}$$

This is called the Christodoulou's formula [Christodoulou (1970)].

The circumferential lengths around the equatorial surface and a meridian curve are, respectively, written by [e.g., Smarr (1973b)]
$$C_{\text{eq}} = 4\pi M \quad \text{and} \quad C_{\text{p}} = 4 \int_0^{\pi/2} d\theta \sqrt{r_+^2 + a^2 - a^2 \sin^2 \theta}. \tag{1.136}$$

Thus, the mass for the (four-dimensional) Kerr spacetime may be determined from C_{eq}. It is worthy to note that $C_{\text{p}}/C_{\text{eq}}$ is a monotonically decreasing function of a/M; for $a = 0$, it is unity and for $a = M$, it is ≈ 0.608. This implies that the black-hole spin may be determined from $C_{\text{p}}/C_{\text{eq}}$.

The Ricci scalar on the horizon surface is written as [Smarr (1973b)]

$$\overset{(2)}{R} = 2 \frac{(r_+^2 + a^2)(r_+^2 - 3a^2 \cos^2 \theta)}{(r_+^2 + a^2 \cos^2 \theta)^3}. \tag{1.137}$$

For $a = 0$, it is uniform on the horizon, while for the Kerr case, it is not. The ratio of the minimum value at the pole to the maximum value at the equator is a monotonically decreasing function of a/M. Thus, the black-hole spin may be also determined from this ratio [Lovelace et al. (2008)]. This quantity is also useful for determining the direction of the black-hole spin axis (the minima of the Ricci scalar are located along the spin axis).

It is well-known that using Ω_{H}, a Killing vector field, which is normal to the event horizon, can be constituted as

$$\chi^a = \xi^a + \Omega_{\text{H}} \varphi^a. \tag{1.138}$$

Since the event horizon is a null surface and χ^a is normal to the horizon, we have $\chi^a \chi_a = 0$ on the horizon; $\chi^a \chi_a$ is constant on the event horizon. This means that $\nabla_a(\chi^c \chi_c)$ does not have components along the event horizon, and hence, it is normal to the event horizon. Thus, on the event horizon, there exists a function, κ, such that

$$\nabla^a(\chi^c \chi_c) = -2\kappa \chi^a. \tag{1.139}$$

Here κ is called the surface gravity of the black hole and is a constant on the horizon written as [Bardeen et al. (1973)]

$$\kappa = \frac{\sqrt{M^2 - a^2}}{2M \left(M + \sqrt{M^2 - a^2} \right)}. \tag{1.140}$$

Using κ, A_{H}, Ω_{H}, and J, the black-hole mass is written as (see also appendix G)

$$M = \frac{\kappa}{4\pi} A_{\text{H}} + 2\Omega_{\text{H}} J. \tag{1.141}$$

This is called the Smarr relation [Smarr (1973a)]. It plays a role in deriving a mechanical relation of black holes (see section 1.3.2.4).

1.3.2.4 Theorems for black holes

A number of remarkable theorems that clarify global properties of black-hole spacetime have been proved since the discovery of the Kerr black hole. We here list several important theorems and laws which play a crucial role in numerical relativity, without any derivation and proof [see Bardeen et al. (1973) for the description of the four laws of black hole mechanics; see also appendix G].

The second law of black-hole mechanics (the area theorem of black holes) proved by Hawking (1971) has played a quite important role for expecting the results of dynamical evolution of black-hole spacetime. For example, consider a head-on collision of two non-spinning black holes of each mass M_1 and M_2. In this case, the final outcome is also a non-spinning black hole for which we write the final mass as $M_{\rm f}$. The area theorem states that the total area has to increase, namely,

$$16\pi M_{\rm f}^2 \geq 16\pi \left(M_1^2 + M_2^2\right). \tag{1.142}$$

Possible energy emitted from the system by gravitational waves is estimated to give

$$\Delta E = M_1 + M_2 - M_{\rm f} \leq M_1 + M_2 - \sqrt{M_1^2 + M_2^2}. \tag{1.143}$$

For the equal mass case $M_1 = M_2$, the efficiency, $\Delta E/(M_1 + M_2)$, could be maximum, and in this case,

$$\Delta E \leq 2M_1 \left(1 - 2^{-1/2}\right) \approx 2M_1 \times 0.29. \tag{1.144}$$

Thus we find that more than 29% of the total energy cannot be emitted from the system.

The following first law of black-hole mechanics (which is often called the first law of black-hole thermodynamics) also plays an important role (see also appendix G):

$$dM = \frac{\kappa}{8\pi}dA_{\rm H} + \Omega_{\rm H}dJ. \tag{1.145}$$

Here, dM, $dA_{\rm H}$, and dJ denote small variations of M, $A_{\rm H}$, and J for two neighboring black-hole states. The area theorem constrains the evolution of black holes as $dA_{\rm H} \geq 0$, and the third law of black holes states $\kappa > 0$. Thus,

$$dM \geq \Omega_{\rm H}dJ. \tag{1.146}$$

Consider the case in which gravitational waves of a monotonic angular frequency, $\omega(> 0)$, are emitted from the black-hole spacetime. In this case, $dM = \omega dJ(< 0)$ according to equations (1.72) and (1.76), and thus, equation (1.146) is written as

$$|dM| \leq \Omega_{\rm H}|dJ|. \tag{1.147}$$

This implies that the emission is possible only for the case $\omega \leq \Omega_{\rm H}$. Thus, gravitational waves may carry energy and angular momentum from Kerr black holes, if the angular frequency of gravitational waves is lower than $\Omega_{\rm H}$. The emission of such gravitational waves is called super-radiance and the condition (1.146) or (1.147) is referred to as the super-radiance condition [Bekenstein (1973); Teukolsky and Press (1974)].

In four dimensions, it is known that the resonant angular frequency [the so-called quasi-normal mode frequency, e.g., Chandrasekhar and Detweiler (1975); Leaver (1985); Schutz and Will (1985)] of any black hole is larger than $\Omega_{\rm H}$. Hence, the spontaneous emission of gravitational waves does not occur and black holes are stable. However, this is not the case for higher-dimensional black holes (see section 11.3).

1.3.2.5 Tetrad

There are three sets of tetrad basis vectors, $e^a_{(\alpha)}$, that are often used for Kerr spacetime in the Boyer–Lindquist coordinates. The first one is a canonical orthonormal tetrad introduced by Carter (1968)

$$e^a_{(0)} = \frac{1}{\sqrt{\Delta\Sigma}} \left[(r^2 + a^2) \left(\frac{\partial}{\partial t}\right)^a + a \left(\frac{\partial}{\partial \varphi}\right)^a \right],$$

$$e^a_{(1)} = \sqrt{\frac{\Delta}{\Sigma}} \left(\frac{\partial}{\partial r}\right)^a, \quad e^a_{(2)} = \frac{1}{\sqrt{\Sigma}} \left(\frac{\partial}{\partial \theta}\right)^a,$$

$$e^a_{(3)} = \frac{1}{\sqrt{\Sigma}\sin\theta} \left[a\sin^2\theta \left(\frac{\partial}{\partial t}\right)^a + \left(\frac{\partial}{\partial \varphi}\right)^a \right]. \tag{1.148}$$

The second one is the so-called locally non-rotating frame [Bardeen et al. (1972)]

$$e^a_{(0)} = \sqrt{\frac{\Xi}{\Delta\Sigma}} \left[\left(\frac{\partial}{\partial t}\right)^a + \frac{2Mar}{\Xi} \left(\frac{\partial}{\partial \varphi}\right)^a \right],$$

$$e^a_{(1)} = \sqrt{\frac{\Delta}{\Sigma}} \left(\frac{\partial}{\partial r}\right)^a, \quad e^a_{(2)} = \frac{1}{\sqrt{\Sigma}} \left(\frac{\partial}{\partial \theta}\right)^a,$$

$$e^a_{(3)} = \frac{\sqrt{\Sigma}}{\sqrt{\Xi}\sin\theta} \left(\frac{\partial}{\partial \varphi}\right)^a, \tag{1.149}$$

where it is worthy to note that $e_{(0)a} = -\alpha \nabla_a t$ for this case with α, the lapse function. These two sets of the tetrad satisfy

$$g^{ab} = -e^a_{(0)} e^b_{(0)} + e^a_{(1)} e^b_{(1)} + e^a_{(2)} e^b_{(2)} + e^a_{(3)} e^b_{(3)}, \tag{1.150}$$

and $e^a_{(\alpha)} e_{(\beta)a} = \eta_{\alpha\beta}$. The third one is the Kinnersley's null tetrad [Kinnersley (1969)]

$$e^a_{(1)} = \frac{1}{\Delta} \left[(r^2 + a^2) \left(\frac{\partial}{\partial t}\right)^a + \Delta \left(\frac{\partial}{\partial r}\right)^a + a \left(\frac{\partial}{\partial \varphi}\right)^a \right],$$

$$e^a_{(2)} = \frac{1}{2\Sigma} \left[(r^2 + a^2) \left(\frac{\partial}{\partial t}\right)^a - \Delta \left(\frac{\partial}{\partial r}\right)^a + a \left(\frac{\partial}{\partial \varphi}\right)^a \right],$$

$$e^a_{(3)} = \frac{1}{\sqrt{2}(r + ia\sin\theta)} \left[ia\sin\theta \left(\frac{\partial}{\partial t}\right)^a + \left(\frac{\partial}{\partial \theta}\right)^a + \frac{i}{\sin\theta} \left(\frac{\partial}{\partial \varphi}\right)^a \right], \tag{1.151}$$

and $e^a_{(4)}$ is the complex conjugate of $(e_3)^a$. Here, all these basis vectors are null. These satisfy

$$g^{ab} = -e^a_{(1)} e^b_{(2)} - e^a_{(2)} e^b_{(1)} + e^a_{(3)} e^b_{(4)} + e^a_{(4)} e^b_{(3)}, \tag{1.152}$$

and

$$-e^a_{(1)} e_{(2)a} = e^a_{(3)} e_{(4)a} = 1,$$

$$e^a_{(1)} e_{(3)a} = e^a_{(1)} e_{(4)a} = e^a_{(2)} e_{(3)a} = e^a_{(2)} e_{(4)a} = 0. \tag{1.153}$$

1.3.3 Higher-dimensional black holes

In higher-dimensional spacetime, there are many black objects [e.g., see a review by Emperan and Reall (2008)]. Here, black objects imply that their horizon topology is not always spherical, and it may be, e.g., a torus. This situation is completely different from that in four-dimension spacetime in which the uniqueness theorem proves that the Kerr solution is the unique solution for stationary axisymmetric black holes of no electric charge [e.g., Wald (1984) for a review]. Reviewing a variety of the black objects is far beyond the scope of this volume, and thus, we will focus only on the higher-dimensional version of the Schwarzschild and Kerr black holes in the following.

The line element of non-spinning black holes (the so-called Schwarzschild–Tangherlini solution) in any spatial dimension is written as [Tangherlini (1963)]

$$ds^2 = -\left(1 - \frac{\mu}{r^{D-3}}\right)dt^2 + \left(1 - \frac{\mu}{r^{D-3}}\right)^{-1}dr^2 + r^2 d\Omega_{D-2}, \quad (1.154)$$

where $d\Omega_{D-2}$ is the line element of a $(D-2)$-dimensional unit sphere, μ is a mass parameter related to the black-hole mass (see below), and r is an areal radius, which defines the hyper-area of the $(D-2)$-dimensional hypersphere, $r^{D-2}\Omega_{D-2}$, with

$$\Omega_{D-2} = \frac{2\pi^{(D-1)/2}}{\Gamma[(D-1)/2]}. \quad (1.155)$$

Here, $\Gamma[z]$ denotes the Gamma function. We note that the mass dimension is equivalent to the dimension of (length)$^{D-3}$ in higher dimensions in the geometrical units $c = G_D = 1$ where G_D is the D-dimensional gravitational constant. The event horizon of the black hole is placed at $r = \mu^{1/(D-3)} =: r_g$ with its hyper-area $A_H = \mu^{(D-2)/(D-3)}\Omega_{D-2}$.

As in the four-dimensional case, the line element in isotropic coordinates is derived to yield

$$ds^2 = -\left(\frac{\bar{r}^{D-3} - \mu/4}{\bar{r}^{D-3} + \mu/4}\right)^2 dt^2 + \left(1 + \frac{\mu}{4\bar{r}^{D-3}}\right)^{4/(D-3)}[d\bar{r}^2 + \bar{r}^2 d\Omega_{D-2}], \quad (1.156)$$

where $\bar{r}$ is a radial coordinate related to r by

$$\frac{\bar{r}^{D-3}}{\mu/4} = \frac{r^{D-3} - \mu/2}{\mu/2} + \sqrt{\left(\frac{r^{D-3} - \mu/2}{\mu/2}\right)^2 - 1}. \quad (1.157)$$

The event horizon in the $\bar{r}$ coordinate is located at $\bar{r}^{D-3} = \mu/4$.

The line element of a spinning black hole with one spin (a class of the so-called Myers–Perry black hole [Myers and Perry (1986)]) in any spatial dimension is written as

$$ds^2 = -dt^2 + \frac{\mu}{r^{D-5}\Sigma}(dt - a\sin^2\theta d\varphi)^2 + \frac{\Sigma}{\Delta}dr^2 + \Sigma d\theta^2$$
$$+ (r^2 + a^2)\sin^2\theta d\varphi^2 + r^2\cos^2\theta d\Omega^2_{D-4}, \quad (1.158)$$

where μ and a are mass and spin parameters, respectively, and

$$\Sigma = r^2 + a^2 \cos^2\theta \quad \text{and} \quad \Delta = r^2 + a^2 - \mu r^{5-D}. \tag{1.159}$$

For this black hole, only one spin parameter exists, although there may be more than two spin parameters in general for higher-dimensional spacetime as shown by Myers and Perry (1986).

The location of the event horizon, r_+, is determined from the equation $\Delta = 0$. For $D = 5$, the location of the event horizon is $r_+ = \sqrt{\mu - a^2}$, and thus, a has to be smaller than $\sqrt{\mu}$ for the presence of the event horizon; otherwise, a naked singularity appears. Note that for $a = \sqrt{\mu}$, not the event horizon but a naked singularity appears (the situation is different from the four-dimensional case). For $D \geq 6$, the event horizon can be present for any value of $a < \infty$, in contrast to the four- and five-dimensional cases: "Ultra-spinning" black-hole solutions are possible.

The $(D-2)$-dimensional area of the event horizon for spinning black holes is

$$A_\mathrm{H} = \mu r_+ \Omega_{D-2}. \tag{1.160}$$

In the limit $a \to \sqrt{\mu}$ for $D = 5$ and $a \to \infty$ for $D \geq 6$, r_+ approaches zero. This means that A_H also approaches zero. This fact suggests that the ultra-spinning black hole would be unstable to a transition to an object with a higher value of A_H, because the area theorem states that the area has to increase, and hence, the state with a larger area is likely to be more stable [Emperan and Myers (2003)].

The mass and the angular momentum of spinning black holes are

$$M = \frac{(D-2)\Omega_{D-2}\mu}{16\pi}, \tag{1.161}$$

$$J = \frac{2}{D-2} M a. \tag{1.162}$$

As in the four-dimensional case, circumferential lengths around black holes have important information. The circumferential length for $\theta = \pi/2$ is

$$C_\mathrm{eq} = 2\pi \frac{\mu}{r_+^{D-4}}. \tag{1.163}$$

Since r_+ depends on a, C_eq for $D \geq 5$ is a function not only of the mass parameter but also of the spin by contrast to the four-dimensional case. The meridian length is written by

$$C_\mathrm{p} = 4 \int_0^{\pi/2} d\theta \sqrt{r_+^2 + a^2 \cos^2\theta}. \tag{1.164}$$

The ratio $C_\mathrm{p}/C_\mathrm{eq}$ is a monotonically decreasing function of $a/\mu^{1/(D-3)}$ as in the four-dimensional case. However, the limiting value for it with $a \to \sqrt{\mu}$ for $D = 5$ and $a \to \infty$ for $D \geq 6$ is zero in this case. This is due to the fact that r_+ approaches zero for this limit. Thus, for the rapidly spinning case, the black hole approaches a flattened, pancake-like object. For $D \geq 6$, such a black hole is often referred to as a black membrane [Emperan and Myers (2003)].

As in the four-dimensional case, the line elements are written in isotropic or quasi-isotropic coordinates, for which the radial coordinate is defined by

$$\bar{r} = r_+ \exp\left[\pm \int_{r_+}^{r} \frac{dR}{\sqrt{R^2 + a^2 - \mu R^{5-D}}}\right], \quad (1.165)$$

where the plus and minus signs are adopted for the regions $\bar{r} \geq r_+$ and $\bar{r} \leq r_+$, respectively. This coordinate is useful in numerical-relativity simulations (see section 11.3).

Kerr black holes in four dimensions are known to be stable irrespective of the magnitude of the spin [Press and Teukolsky (1973); Whiting (1989)]. By contrast, higher-dimensional black holes are known to be often unstable if their spin is sufficiently high [Emperan and Myers (2003); Dias et al. (2009); Shibata and Yoshino (2010a)]. This is likely to be also the case for highly deformed black objects [Choptuik et al. (2003c); Lehner and Pretorius (2010)]. Clarifying the stability property and the final fate of the unstable black objects is one of the interesting subjects in higher-dimensional numerical relativity (see chapter 11).

1.4 Neutron stars

Clarifying the formation and evolution processes of neutron stars and neutron-star binaries is the very central subject in numerical relativity. To date, more than 2000 neutron stars have been observed [e.g., Lyne and Graham-Smith (2005)] since the first discovery by Bell et al. in 1967 [Hewish et al. (1968)].[4] The neutron stars have shown a wide variety of high-energy phenomena (see section 1.4.2). In addition, six binary systems composed of two neutron stars have already been confirmed [Lorimer (2008)] since the first discovery by Hulse and Taylor (1975). The orbits of the six binary neutron stars are sufficiently compact and hence they will merge within the age of the universe (Hubble time) ≈ 13.8 Gyrs ($= 1.38 \times 10^{10}$ yrs) (see table 1.2). Binary neutron stars have been an invaluable experimental field for the tests of general relativity [Stairs (2003, 2004)]. Furthermore, the coalescence of binary neutron stars in close orbits is the promising source for gravitational-wave detectors (see, e.g., section 1.2.6) and for the central engine of gamma-ray bursts [Eichler et al. (1989); Narayan et al. (1992); Zhang and Meszáros (2004); Piran (2004); Nakar (2007)]. In this section, we outline a typical formation process of neutron stars, properties of neutron stars, and binary neutron stars.

---

[4]According to a lecture by R. N. Manchester at Kyoto in June of 2013, the total number of the observed pulsars is more than 2200: See http://www2.yukawa.kyoto-u.ac.jp/∼ykis2013.ws/conference/conference.php?mode=AbstractInvited.

1.4.1 Formation of neutron stars

Neutron stars are typically formed through the gravitational collapse of an iron core or an oxygen-neon-magnesium core of evolved massive stars (see also section 1.5.5) and subsequent supernova explosion. Some of neutron stars are also likely to be formed as a result of the accretion-induced collapse of massive white dwarfs. In the following, we will outline the typical formation scenario of neutron stars.

Massive stars of their initial mass larger than $\sim 8 M_\odot$ are believed to evolve through a series of nuclear burning, and eventually ignite carbon [e.g., Kippenhahn and Weigert (1994); Woosley *et al.* (2002)]. If the initial mass is between $\sim 8 M_\odot$ and $10 M_\odot$, a core composed of oxygen, neon, and magnesium is finally formed. Then, if the initial mass is larger than $\sim 9 M_\odot$, radial instability against the gravitational collapse, due to electron captures of these heavy nuclei and resulting pressure depletion, is believed to set in. If the initial mass is larger than $10 M_\odot$, the nuclear burning continues until iron are produced in the core. Then, radial instability against the gravitational collapse, associated with electron captures and partial photo-dissociation of the iron core, [e.g., Shapiro and Teukolsky (1983)], is believed to set in.

After the gravitational collapse, broadly speaking, there are three possibilities [e.g., Woosley *et al.* (2002); Smartt (2009); see also section 9.3]. Here, we assume that the massive stars have solar metallicity. If the initial mass is smaller than a critical value $M_{\rm crit} \sim 15\text{–}20 M_\odot$, it is believed that a neutron star is formed after a supernova explosion. If the mass is larger than the critical value, on the other hand, a black hole would be formed.[5] If the initial mass is not very large, the collapse does not result directly in a black hole. Rather, a shock, which is formed after the gravitational collapse and proto-neutron star formation, initially propagates outward (see, e.g., section 1.5.5); i.e., for a while, a proto-neutron star is alive. However, the shock eventually stalls because its energy is consumed for the photo-dissociation of iron surrounding the proto-neutron star and it is eventually defeated by the ram pressure of the matter infalling from the outer part of the progenitor star. In such a case, a *fall-back collapse* to a black hole occurs. In this scenario, after a substantial amount of mass falls onto the proto-neutron star, it collapses to a black hole. The third possibility is that a black hole is *directly* formed during the gravitational collapse. This could occur if the iron core mass is very large (see also section 9.3).

In the supernova explosion, a hot proto-neutron star with the maximum temperature $\sim 10\,{\rm MeV}$ is formed [Bethe (1990)]. Although the self-gravity of the proto-neutron star is sustained primarily by the pressure associated with the nuclear force (see below for more details), the thermal pressure also contributes to a fraction of the total pressure, and hence, the radii of the proto-neutron stars are larger than

---

[5]The final fate could depend not only on the initial mass but also on the metallicity and presence or absence of its companion [Woosley *et al.* (2002); Smartt (2009)]. In the text we describe a typical scenario ignoring the complexity. See also section 9.3.

those of cold neutron stars subsequently formed after cooling. The order of the gravitational binding energy of the proto-neutron stars is

$$\sim \frac{M_{\rm PNS}^2}{R_{\rm PNS}} \sim 1.7 \times 10^{53} \left(\frac{M_{\rm PNS}}{1.4 M_\odot}\right)^2 \left(\frac{R_{\rm PNS}}{30\,{\rm km}}\right)^{-1} {\rm ergs}, \quad (1.166)$$

where $M_{\rm PNS}$ and $R_{\rm PNS}$ are the mass and the radius of the proto-neutron stars. The order of the gravitational binding energy is comparable to the total thermal energy.

The hot proto-neutron stars subsequently emit copious neutrinos with the total luminosity $\sim 10^{52}$ ergs/s, and cool down in ~ 10 s. After the substantial emission of neutrinos, the effect of the thermal pressure becomes negligible (i.e., the thermal energy of nucleons and electrons becomes much smaller than the Fermi energy of them), and a cold neutron star sustained by the pressure associated with the repulsive force in the nuclear matter (basically among nucleons; neutrons and protons) is formed, if a supernova explosion successfully occurs. Here, "cold" means that the Fermi energy of nucleons is much higher than their thermal energy. The typical value of the Fermi energy of neutrons is estimated as [Shapiro and Teukolsky (1983)]

$$E_{\rm F} \sim 100\,{\rm MeV} \left(\frac{\rho}{2\rho_{\rm nuc}}\right)^{2/3}, \quad (1.167)$$

where $\rho_{\rm nuc}$ is the nuclear density $\approx 2.8 \times 10^{14}\,{\rm g/cm}^3$. Thus, as long as the temperature is much lower than 10^{12} K, the thermal pressure plays a tiny role in cold neutron stars.

1.4.2 Basic properties of neutron stars

We believe that central density, mass, and radius of neutron stars are typically $10^{15}\,{\rm g/cm}^3$, $1.3\text{--}1.4 M_\odot$, and $10\text{--}15$ km, respectively, although the typical values of these fundamental quantities are still uncertain. For example, the mass of neutron stars in binary neutron stars is in a narrow range $1.23\text{--}1.44 M_\odot$ (see table 1.2), while neutron stars may have high mass as $M_{\rm NS} \approx 2 M_\odot$ like PSR J1614-2230 and PSR J0348+432 [Demorest et al. (2010); Antoniadis et al. (2013)]. One of the most serious problems is that the typical radius has not been known yet, because the accurate determination of the neutron-star radius by astronomical observations is not easy [but see, e.g., Lattimer (2012); Steiner et al. (2013) for a review on one of the promising efforts], and also, the equation of state for high-density nuclear matter has not been accurately determined by nuclear experiments and nuclear theory yet (see section 1.4.4).

Observations of pulsars (more specifically, the measurement of their spin period and its change rate) have indicated that most of neutron stars are strongly magnetized with the typical magnetic-field strength $\sim 10^{11}\text{--}10^{13}$ G and with the typical spin period $0.1\text{--}1$ s [see, e.g., figure 8.8 of Lyne and Graham-Smith (2005) or figure 3 of Lorimer (2008)]. Some special class, the so-called recycled pulsar, has a much weaker magnetic field $\sim 10^8\text{--}10^9$ G. The neutron stars in this class are usually rapidly spinning with the spin period ~ 1–$10\,\mathrm{ms}$ [Lyne and Graham-Smith (2005)], and hence, they are often called recycled millisecond pulsars. Such rapid spin is believed to be achieved by mass accretion from a normal binary companion. There is another class for neutron stars, the so-called magnetars (soft gamma-ray repeaters or anomalous X-ray pulsars), for which the typical magnetic-field strength is $\sim 10^{14}$–$10^{15}\,\mathrm{G}$. The neutron stars in this class are always slowly spinning with the spin period ~ 1–$10\,\mathrm{s}$ [Woods and Thompson (2006)]. Resolving the origin of such neutron stars of a wide range of the strong magnetic field is an important subject of relativistic astrophysics and numerical relativity.

Neutron stars are composed primarily of neutrons, and the fraction of protons is much smaller than neutrons. The fundamental reason for this is that the density is so high that the chemical potential of degenerate electrons is much larger than the mass difference between neutron and proton $(m_n - m_p)c^2 = 1.293\,\mathrm{MeV}$: The energy of the Fermi surface of electrons is of order $100\,\mathrm{MeV}$ [comparable to E_F of equation (1.167); Shapiro and Teukolsky (1983)], which is much larger than $1.293\,\mathrm{MeV}$. Therefore, neutrons can be produced during the gravitational collapse of a massive stellar core via the inverse β-decay process of protons,

$$p + e^- \to n + \nu_e, \tag{1.168}$$

until the Fermi energy of neutrons becomes as high as that of electrons. Due to essentially the same reason, exotic particles like mesons and hyperons may appear in the core of high-mass neutron stars when the density and resulting Fermi energy of neutrons are high enough to overcome mass differences (see section 1.4.4).

1.4.3 *Hydrostatic equations for cold neutron stars*

All the neutron stars except for the proto-neutron stars and the merger remnants of binary neutron stars can be accurately modeled as a cold state. Here, the term of the "cold state" is used for referring to the state with zero temperature and zero entropy. The main pressure source of such cold neutron stars is the two-body repulsive force among nucleons. The pressure P and the specific internal energy ε for the cold neutron stars are written as functions of the rest-mass density ρ.

The structure of the cold neutron stars is determined by a set of hydrostatic equations called Tolman–Oppenheimer–Volkoff equations [Tolman (1939); Oppenheimer and Volkoff (1939)] (often referred to as TOV equations), if they are not rotating and their magnetic-field strength is not extremely high ($\ll 10^{18}\,\mathrm{G}$). For deriving the hydrostatic equations, the simplest choice of the line element is

$$ds^2 = -e^{2\Phi}dt^2 + \left(1 - \frac{2m}{r}\right)^{-1} dr^2 + r^2 d\Omega, \tag{1.169}$$

where Φ and m are functions of r, which is the areal radius. From Einstein's equation, we obtain

$$\frac{dm}{dr} = 4\pi r^2 \rho(1+\varepsilon), \tag{1.170}$$

$$\frac{d\Phi}{dr} = \frac{m}{r^2}\left(1 + \frac{4\pi P r^3}{m}\right)\left(1 - \frac{2m}{r}\right)^{-1}, \tag{1.171}$$

where we supposed that the stress-energy tensor is written as

$$T_{\mu\nu} = (\rho + \rho\varepsilon + P)u_\mu u_\nu + Pg_{\mu\nu}, \tag{1.172}$$

with vanishing spatial components of the four velocity, $u_i = 0$, and thus, $e^\Phi u^t = 1$. From the conservation equation of the stress-energy tensor (1.13), a hydrostatic equation is derived as

$$\frac{dP}{dr} = -\rho h \frac{d\Phi}{dr}, \tag{1.173}$$

where $h := 1 + \varepsilon + P/\rho$ is the relativistic specific enthalpy. A neutron-star model is obtained by integrating these ordinary differential equations (1.170), (1.171), and (1.173) from the origin for a given value of the central density, ρ_c, to the stellar surface.

For isentropic fluids with $s =$ constant where s is the specific entropy, the first law of thermodynamics is written as

$$d\varepsilon = -P d\left(\frac{1}{\rho}\right) \quad \text{or} \quad dh = \frac{dP}{\rho}. \tag{1.174}$$

Thus, equation (1.173) is integrated to yield

$$\Phi = \ln h + \text{const.} \tag{1.175}$$

This may be used instead of equation (1.173) for isentropic fluids.

Because of the presence of Birkoff's theorem [e.g., Wald (1984)], the geometry outside any spherical star should be identical to the Schwarzschild geometry (1.97). Thus, in the chosen coordinate system, m for $r \to \infty$ agrees with the gravitational mass of the system, M. Strictly speaking, the mass determined from m should be referred to as Arnowitt–Deser–Misner (ADM) mass (see section 5.3.1.1). Due to this fact, the gravitational mass of the system is calculated by integrating equation (1.170) as

$$M = 4\pi \int_0^R \rho(1+\varepsilon) r^2 dr, \tag{1.176}$$

where R is a stellar radius defined in the coordinate r. R in this definition is often called the circumferential radius, because the circumferential physical length around the stellar surface is $2\pi R$.

The baryon rest mass is different from M in general relativity [see equation (4.45) for the definition of the rest mass], and it is written as

$$M_0 := \int_0^R \rho u^t \sqrt{-g}\, dr d\theta d\varphi = 4\pi \int_0^R \rho \left(1 - \frac{2m}{r}\right)^{-1/2} r^2 dr, \tag{1.177}$$

where g is the determinant of $g_{\mu\nu}$. The binding energy of the star is thus given by

$$E_{\rm b} = M - M_0. \tag{1.178}$$

Here, the binding energy is defined by the sum of the internal energy and the gravitational potential energy.

Because Birkoff's theorem holds, the asymptotic behavior of Φ should be

$$\Phi \to -\frac{M}{r}, \quad \text{and thus,} \quad \frac{d\Phi}{dr} \to \frac{M}{r^2}. \tag{1.179}$$

Hence, the gravitational mass is also calculated from Φ. Strictly speaking, the mass determined from Φ should be referred to as the Komar mass (see section 5.3.2). The equality of the ADM mass and the Komar mass gives a constraint relation, the so-called virial relation. After some manipulation, the virial relation for spherical stars is written in the integral form as (see also section 5.3.3)

$$4\pi \int (\rho + \rho\varepsilon + 3P) e^{\Phi} \left(1 - \frac{2m}{r}\right)^{-1/2} r^2 dr = 4\pi \int \rho(1+\varepsilon) r^2 dr. \tag{1.180}$$

This relation has to be satisfied for any spherical star in equilibrium. Confirming this relation for a star numerically obtained is useful for checking the reliability of the numerical calculation.

In numerical relativity, it is often convenient to derive a configuration of spherical stars in isotropic coordinates for which the line element is written by

$$ds^2 = -e^{2\Phi} dt^2 + \psi^4 \left(d\bar{r}^2 + \bar{r}^2 d\Omega\right), \tag{1.181}$$

where $\bar{r}$ is the isotropic radial coordinate [cf. equation (1.107)] and ψ is a function of $\bar{r}$. Here, the point is that the spatial line element is written in the conformally flat form, in which the procedure for the solution of initial data is simplified (see section 5.2).

From Einstein's equation, it is also easy to obtain the basic equations for ψ and $\chi := e^{\Phi} \psi$ as

$$\frac{1}{\bar{r}^2} \frac{d}{d\bar{r}} \left(\bar{r}^2 \frac{d\psi}{d\bar{r}}\right) = -2\pi \rho (1+\varepsilon) \psi^5, \tag{1.182}$$

$$\frac{1}{\bar{r}^2} \frac{d}{d\bar{r}} \left(\bar{r}^2 \frac{d\chi}{d\bar{r}}\right) = 2\pi \left[\rho(1+\varepsilon) + 6P\right] \chi \psi^4. \tag{1.183}$$

These two equations together with the hydrostatic equation (1.173) constitute the basic equations of spherical stars in the isotropic coordinates.

1.4.4 Cold neutron-star equations of state

For a given cold equation of state, $P = P(\rho)$ and $\varepsilon = \varepsilon(\rho)$, a sequence of neutron-star models (more specifically, the gravitational mass and the circumferential radius as functions of the central density) is obtained by solving the TOV equations described

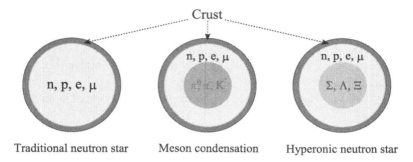

Fig. 1.13 Schematic figures for three representative possibilities of neutron stars. The left panel describes a model that the inner region of neutron stars are composed simply of neutrons, protons, electrons, and muons. The middle and right panels illustrate models that the core of neutron stars is composed of mesons (pions and kaons) and hyperons (Σ, Λ, and Ξ), respectively.

in the previous section.[6] Neutron-star models have been computed for a wide variety of hypothetical equations of state derived in various nuclear-physics theories. The reason why we have to consider various possibilities for the equation of state is that the definite equation of state for the high-density nuclear matter has not been known yet [Lattimer and Prakash (2004, 2007); Lattimer (2012)]. The specific reasons for this are that (i) the two-body force between nucleons, which is determined by the strong interaction, is still poorly known if the separation between two nucleons is smaller than $\sim 10^{-13}$ cm; (ii) for the high-density range with $\rho \gtrsim 3 \times 10^{14}$ g/cm^3, not only the two-body force but also many-body forces such as three-body force play an important role, and for such a state, it is not theoretically easy to determine the physical state of many-body systems; (iii) for extremely high density, exotic particles such as mesons (e.g., pions and kaons) and hyperons may appear. Because the properties of the interaction among the ordinary nucleons and the exotic particles, and the criterion for the appearance of mesons and hyperons are poorly known for the high-density range $\rho \gtrsim 10^{15}$ g/cm^3, many possible states are still considered (see figure 1.13 for three representative possibilities).

Figure 1.14 shows the pressure as a function of the number density of nucleons for various hypothetical equations of state ("cold" equations of state with $T = 0$ and $s = 0$). Each equation of state is computed assuming a hypothetical model of the nuclear forces. This figure illustrates the fact that the pressure above the nuclear density, $\rho \approx 2.8 \times 10^{14}$ g/cm^3 (or the number density ≈ 0.17 fm^{-3}), is still uncertain with the factor of ~ 5. This implies that for theoretical studies of neutron stars and neutron-star binaries, we have to keep in mind that there are a wide range of possible (unconstrained) equations of state.

If a wide variety of hypothetical equations of state could be described in a simple and systematic manner, that would be quite helpful for theoretical studies

---

[6]In general relativity, not only an equation of state $P = P(\rho)$ but also a relation $\varepsilon = \varepsilon(\rho)$ [or $h(\rho)$] has to be provided for computing an equilibrium state, in contrast to the Newtonian case.

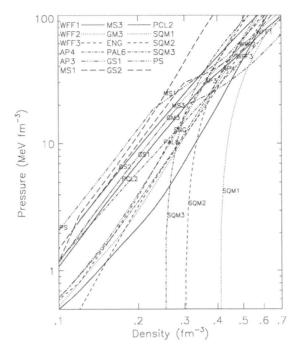

Fig. 1.14 Pressure as a function of number density of nucleons for various hypothetical equations of state. "fm" denotes 10^{-13} cm and $0.1\,\text{fm}^{-3}$ means that the rest-mass density is approximately $1.66 \times 10^{14}\,\text{g/cm}^3$. The figure is taken from Lattimer and Prakash (2001).

of phenomena associated with neutron stars. Read *et al.* (2009a) showed that a number of hypothetical cold equations of state can be accurately and systematically modeled by a piecewise polytropic equation of state [see also Özel and Psaltis (2009) for a similar idea]. This is a phenomenologically parametrized equation of state, which reproduces many hypothetical cold equations of state at high density only with a small number of polytropic constants κ_i and indices n_i ($i = 0, 1, 2, \cdots, n$) as

$$P = \kappa_i \rho^{1+1/n_i} \quad \text{for} \quad \rho_i \leq \rho < \rho_{i+1} \quad (0 \leq i \leq n), \tag{1.184}$$

where $n + 1$ is the number of the pieces used for parametrizing an equation of state. κ_i is the polytropic constant and n_i is the polytropic index for each piece. ρ_i denote values of the boundary density at which the values of κ_i and n_i change. Here, $\rho_0 = 0$ and $\rho_{n+1} \to \infty$. It is also possible to match to an equation of state for subnuclear density in terms of piecewise polytrope. However, the equation of state for the low-density range is not very essential for determining the structure of neutron stars, and hence, it may be described simply by a single piece composed only of κ_0 and $n_0 \approx 3$.

At each boundary density, $\rho = \rho_i$ ($i = 1, \cdots, n$), the pressure is required to be continuous, i.e., $\kappa_i \rho_{i+1}^{1/n_i} = \kappa_{i+1} \rho_{i+1}^{1/n_{i+1}}$. Thus, if we give κ_0, n_i, and ρ_i ($i = 1, \cdots, n$), the equation of state is totally determined. In this equation of state, the first law of thermodynamics (1.174) for $T = 0$ holds, and thus, ε and h are also determined except for the choice of the integration constants, which are fixed by the continuity condition of ε (hence equivalently h) at each value of ρ_i.

Read et al. (2009a) illustrated that 4-piece equations of state ($n = 3$) can approximately reproduce a number of hypothetical equations of state within $\sim 5\%$ error for the pressure at a given value of density for the wide density range, if the parameters are chosen appropriately. They proposed that n_0 and κ_0 are fixed to be

$$1/n_0 = 0.35692395, \qquad (1.185)$$

$$\kappa_0 = 3.99873692 \times 10^{-8} \text{ (the cgs unit)}. \qquad (1.186)$$

Then, they chose n_1, n_2, n_3, and the pressure at $\rho = \rho_2 = 5 \times 10^{14}\,\mathrm{g/cm^3}$ (denoted by p_1) as 4 free parameters. Here, n_1, n_2, and n_3 are polytropic indices for the density $\rho_1 \leq \rho \leq \rho_2 = 5 \times 10^{14}\,\mathrm{g/cm^3}$, $\rho_2 \leq \rho \leq \rho_3 = 1 \times 10^{15}\,\mathrm{g/cm^3}$, and $\rho \geq \rho_3$, respectively. It is worthy to note that the choice of p_1 as one of the free parameters stems from the fact that it has a strong correlation with the neutron-star radius of canonical mass $\sim 1.4 M_\odot$, as pointed out by Lattimer and Prakash (2001): For larger values of p_1, radii of spherical neutron stars are larger.

Parameters of the piecewise polytrope for several representative hypothetical equations of state are listed in table 1.1. Here, only stiff equations of state, with which the maximum allowed mass of spherical neutron stars is larger than $\sim 2.0 M_\odot$, can be realistic because of the presence of high-mass neutron stars with mass $\sim 2 M_\odot$ [Demorest et al. (2010); Antoniadis et al. (2013)]. APR4 was derived by a variational method with modern nuclear potentials [Akmal et al. (1998)] for the hypothetical components composed of neutrons, protons, electrons, and muons; SLy was derived by using an effective potential approach of the Skyrme type [Douchin and Haensel (2001)] for the hypothetical components composed of neutrons, protons, electrons, and muons; MS1 was derived by a relativistic mean-field theory for the hypothetical components composed of neutrons, protons, electrons, and muons again [Müller and Serot (1996)]; PS was derived using a potential approach incorporating effects of a pion condensation [Pandhripande and Smith (1975)]; H3 and H4 were derived by a relativistic mean-field theory including effects of hyperons [Glendenning and Moszkowski (1991); Lackey et al. (2006)], and they are, respectively, soft and stiff equations of state. Here, the stiffness can be controlled by a choice of parameters in the mean-field theory; ALF2 is a hybrid equation of state which assumes an ordinary nuclear matter for a low-density range and a quark matter for a high-density range with the transition density $\sim 3\rho_{\mathrm{nuc}}$.

Broadly speaking, there are two types of hypothetical equations of state. One is based on the variational-method calculation (i.e., APR4; SLy is similar to this). For this type, the pressure at ρ_2 (i.e., p_1) is relatively low, while for $\rho \gtrsim \rho_3$, the pressure becomes high and enables to sustain a high-mass neutron star with mass exceeding $2M_\odot$. Thus, for this equation of state, the radius of canonical-mass ($\sim 1.3 M_\odot - 1.4 M_\odot$ according to the observation of Galactic binary neutron stars; see table 1.2)

Table 1.1 Parameters of a piecewise polytrope for several equations of state [Read et al. (2009a)]. p_1 is shown in units of dyn/cm^2. $M_{\mathrm{max},0}$ and $R_{1.4}$ are the maximum mass of spherical neutron stars in units of $M_\odot$ and the circumferential radius of $1.4 M_\odot$ neutron stars in units of km for a given equation of state, respectively.

EOS	$\log_{10}(p_1)$	Γ_1	Γ_2	Γ_3	$M_{\mathrm{max},0}$ ($M_\odot$)	$R_{1.4}$ (km)
APR4	34.269	2.830	3.445	3.348	2.20	11.1
SLy	34.384	3.005	2.988	2.851	2.06	11.4
ALF2	34.616	4.070	2.411	1.890	1.99	12.4
H3	34.646	2.787	1.951	1.901	1.79	13.5
H4	34.669	2.909	2.246	2.144	2.03	13.6
MS1	34.858	3.224	3.033	1.325	2.77	14.4
PS	34.671	2.216	1.640	2.365	1.74	15.5

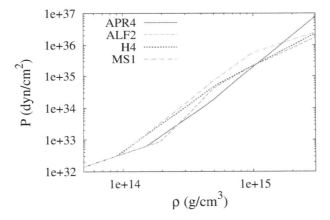

Fig. 1.15 Pressure as a function of the rest-mass density for four representative hypothetical equations of state in a piecewise polytrope approximation for them.

neutron stars is relatively small (see table 1.1). Here, we note that for a small value of p_1, Γ_2 and/or Γ_3 have to be large (~ 3) to enhance the pressure for a high-density range because the maximum mass of spherical neutron stars, $M_{\mathrm{max},0}$, for the hypothetical equations of state has to be larger than $\sim 2 M_\odot$. The other type is based on relativistic mean-field theories (i.e., MS1 and H4). For this type, p_1 is high, and hence, the radius of canonical-mass neutron stars becomes large, $> 13\,\mathrm{km}$. On the other hand, the adiabatic index for high-density ranges $\rho \gtrsim \rho_2$ or $\rho \gtrsim \rho_3$ is small, and the pressure does not increase with the increase of the density as steeply as that for APR4 and SLy. ALF2 is a mixture of these two types: For $\rho \lesssim \rho_2$, it is similar to APR4, while for the higher-density range, it is similar to MS1 and H4.

Figure 1.15 plots the pressure as a function of the rest-mass density for four piecewise polytropic equations of state. As already mentioned, APR4 has relatively small pressure for $\rho_1 \leq \rho \lesssim \rho_3$, while it has high pressure for $\rho \gtrsim \rho_3$. By contrast, H4 and MS1 have pressure higher than APR4 for $\rho \lesssim \rho_3$, while the equations of state become softer for a high-density range $\rho \gtrsim \rho_3$. ALF2 has small pressure for

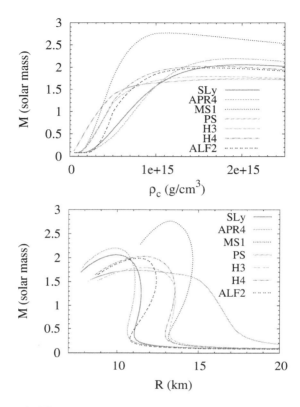

Fig. 1.16 Upper panel: The gravitational mass M as a function of the central density ρ_c for cold spherical neutron stars with selected equations of state listed in table 1.1. Lower panel: The gravitational mass M as a function of the circumferential radius R.

$\rho \leq \rho_2$ as in the case of APR4, but for $\rho_2 \lesssim \rho \leq \rho_3$, the pressure is higher than that for APR4. For $\rho \geq \rho_2$ the pressure of ALF2 is as high as that for H4. These properties are well reflected in the process and remnants of binary-neutron-star mergers (see section 8.3.3 on this topic).

1.4.5 *Structure and sequences of neutron stars*

Using piecewise polytropic equations of state such as those listed in table 1.1, it is quite easy to systematically study the structure of neutron stars by numerically solving the TOV equations. Often referred relations among the global quantities of the neutron-star models are those between M and R and between M and ρ_c, which characterize the properties of neutron stars and their dependence on the equations of state clearly. Here, M, R, and ρ_c are the gravitational mass, the circumferential radius, and the central density, respectively.

Figure 1.16 plots M as a function of ρ_c (upper panel) and R (lower panel) for seven hypothetical equations of state listed in table 1.1. This figure shows the important properties of neutron stars summarized as follows:

- The maximum density is by a factor of ~ 2–4 larger than the nuclear density $\rho_{\rm nuc}$ for the canonical neutron-star mass $M_{\rm NS} = 1.3$–$1.4 M_\odot$.
- For any equation of state, there is the maximum mass ($M_{\rm max,0} \sim 1.5 M_\odot$–$2.5 M_\odot$) but its value depends strongly on the equations of state.
- For any given equation of state, the circumferential radius is approximately constant within $\sim 1\,{\rm km}$ fluctuation for a wide mass range, $\sim M_\odot$–$2 M_\odot$ except for special equations of state (e.g., PS EOS). The radius depends also strongly on the equations of state.
- Compactness parameter of neutron stars, defined by $\mathcal{C} := M/R$, is typically 0.13–0.25.

The first fact shows that the typical neutron stars have a high value of the central density, and due to this reason, the properties of the neutron stars are still poorly known.

The second fact implies that neutron stars with their mass larger than the maximum mass cannot be in equilibrium. Hence, a compact star with mass larger than $M_{\rm max,0}$ has to collapse to a black hole. It should be noted that there are two branches for a given value of the mass lower than the maximum mass. The neutron stars in the higher-density branch (or the smaller radius) are known to be unstable against gravitational collapse to a black hole [Zel'dovich and Novikov (1971); Shapiro and Teukolsky (1983)] while those for the other branch are stable. Neutron stars in nature have to be on the stable branch. At the maximum mass, the central density becomes ~ 1–$2 \times 10^{15}\,{\rm g/cm}^3$. Thus, if a high-density state with $\rho_c \gtrsim 2 \times 10^{15}\,{\rm g/cm}^3$ is realized, the neutron star will collapse to a black hole.

The third fact implies that for a given equation of state, neutron stars have a typical radius irrespective of their mass. Thus, determining the equation of state for neutron stars is approximately equivalent to determining their radius. The typical radius depends on the equations of state and is in the range between ~ 10 and $\sim 16\,{\rm km}$. Thus, constraining the neutron-star radius within $1\,{\rm km}$ error will significantly constrain the possible equations of state.

1.4.6 Supramassive and hypermassive neutron stars

Neutron stars in nature are in general rotating. If the rotation velocity is large (say, larger than $\sim 30\%$ of the Keplerian velocity at the equator, $\sqrt{M/R}$ where R is the equatorial circumferential radius), the effect of the rotation plays an appreciable role for determining the structure of the neutron stars. For a solution of a rotating equilibrium state and its sequence, we have to solve partial differential equations because the system is not spherically symmetric (see section 5.7.1).

Rotating neutron stars in equilibrium were actively studied in 1980s and 1990s

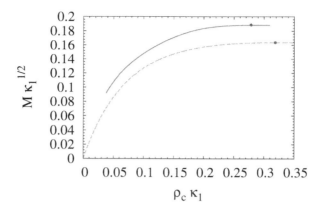

Fig. 1.17 The gravitational mass M as a function of the central density ρ_c for neutron stars with a simple polytropic equation of state, $P = \kappa_1 \rho^2$ (the polytropic index $n = 1$). The solid and dashed curves show the relations for maximally rigidly rotating and non-rotating neutron stars, respectively. The filled circles mark the maximum mass along each sequence. The density and the mass are shown in the polytropic units; dimensionless quantities are plotted for each axis in the geometrical units $c = 1 = G$ with $\kappa_1 = 1$. For having a quantity in physical units, we only need to give a particular value for κ_1.

[see, e.g., a review article by Stergioulas (2003) and a textbook by Friedman and Stergioulas (2013)], and now, it is not difficult to compute any equilibrium even if the rotation velocity is as high as the Keplerian velocity at the equator. In this section, we briefly summarize the properties for sequences of rapidly rotating neutron stars.

First, we focus on rigidly rotating neutron stars that have a uniform angular velocity. Figure 1.17 displays a typical relation of the gravitational mass (i.e., ADM mass and Komar Mass) as a function of the central density for a given equation of state: In this example, we simply choose a polytropic equation of state $P = \kappa_1 \rho^2$ with the polytropic index $n = 1$. Here, κ_1 is the polytropic constant. The solid and dashed curves denote the relations for the maximally rigidly rotating and non-rotating cases, respectively. Here, for the maximally rigidly rotating configuration, the rotation velocity at the equator is equal to the Keplerian velocity $\sqrt{M/R}$. This figure shows that for rigidly rotating neutron stars, the maximum allowed mass (denoted by $M_{\mathrm{max,spin}}$ in the following) can be larger than that for spherical neutron stars (denoted by $M_{\mathrm{max,0}}$), and $M_{\mathrm{max,spin}}$ is by $\approx 15\%$ larger than $M_{\mathrm{max,0}}$ in this case. It is known that this relation ($M_{\mathrm{max,spin}}/M_{\mathrm{max,0}} = 1.15$–$1.20$) holds for many possible equations of state for neutron stars [Cook et al. (1994a)]. An important fact here is that rotating neutron stars with mass larger than $M_{\mathrm{max,0}}$ can exist. Such massive neutron stars are called *supramassive neutron stars* [Cook et al. (1992, 1994b)]. The reason for the fact that $M_{\mathrm{max,spin}}$ is larger than $M_{\mathrm{max,0}}$ is that the centrifugal force contributes to sustaining the additional self-gravity of the supramassive neutron star. A supramassive neutron star can be formed e.g., when a substantial amount of matter is accreted onto a neutron star from a normal binary companion, as in the formation of recycled millisecond pulsars.

Supramassive neutron stars with the central density smaller than a critical value are dynamically stable against gravitational collapse [Friedman *et al.* (1988)]. Here, the critical density depends on the mass and the angular momentum of the neutrons star. Supramassive neutron stars, however, could become eventually unstable against gravitational collapse in the presence of a dissipation process by which their angular momentum is dissipated and centrifugal force is weakened. In reality, rotating neutron stars in nature are usually pulsars, which lose their angular momentum through the electromagnetic-wave emission. Thus, supramassive neutron stars will collapse to a black hole in a dissipation time scale, although they can live for a finite lifetime that is much longer than their dynamical time scale and rotation period. Here, a characteristic lifetime of a pulsar due to the magnetic-dipole radiation is calculated by [Shapiro and Teukolsky (1983)]

$$\tau_{\text{chr}} = \frac{3IP_{\text{spin}}^2 c^3}{2\pi^2 B_{\text{p}}^2 R^6}$$

$$\approx 4 \times 10^{15} \sec \left(\frac{I}{10^{45}\,\text{g}\cdot\text{cm}^2}\right)\left(\frac{B_{\text{p}}}{10^{12}\,\text{G}}\right)^{-2}\left(\frac{R}{10^6\,\text{cm}}\right)^{-6}\left(\frac{P_{\text{spin}}}{1\,\text{s}}\right)^2, \quad (1.187)$$

where I is the moment of inertia, B_{p} is the poloidal magnetic-field strength at the magnetic pole, R is the neutron-star radius, and P_{spin} is the rotation period, respectively. Thus, even for an extremely large magnetic-field strength $B_{\text{p}} \sim 10^{15}$ G and a rapid rotation $P_{\text{spin}} \sim 1$ ms, the lifetime, $\sim 4 \times 10^3$ s, is much longer than the rotation period.

In the presence of differential rotation, even higher neutron-star mass can be sustained by an enhanced rapid rotation [Baumgarte *et al.* (2000)]: If the angular velocity in an inner region of a neutron star is sufficiently larger than that near the equatorial surface, the centrifugal force in the central region is enhanced, and the mass can be larger than $M_{\text{max,spin}}$. For some special rotational profile, the mass can be twice as large as $M_{\text{max,spin}}$, as first shown by Baumgarte *et al.* (2000) (see figure 1.18). Such differentially rotating neutron stars with mass larger than $M_{\text{max,spin}}$ are called *hypermassive neutron stars* [Baumgarte *et al.* (2000)].[7] Numerical simulations [e.g., see a pioneer work by Baumgarte *et al.* (2000); Shibata *et al.* (2000b)] showed that hypermassive neutron stars can be dynamically stable against gravitational collapse. By contrast, hypermassive neutron stars are often unstable against non-axisymmetric deformation, because their ratio of rotational kinetic energy to gravitational potential energy (denoted by β in figure 1.18) can be larger

---

[7]Precisely speaking, the definition of the supramassive and hypermassive neutron stars should be done in terms of the rest mass: Supramassive neutron stars have to have the rest mass exceeding the maximum-allowed rest mass of spherical neutron stars for a given equation of state. In the same way, hypermassive neutron stars have to have the rest mass exceeding the maximum-allowed rest mass of rigidly rotating neutron stars. However, even when we define these massive stars in terms of the gravitational mass, the difference is minor, and hence, we are not very careful in distinguishing the two definitions in this volume.

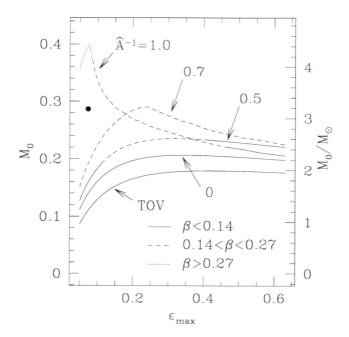

Fig. 1.18 The baryon rest mass (M_0) as a function of the maximum energy density ($\epsilon_{\max}$) for differentially rotating neutron stars with a simple polytropic equation of state, $P = \kappa_1 \rho^2$ ($n = 1$). $\hat{A}$ denotes a degree of differential rotation [which is the ratio of ϖ_0 in equation (5.212) to the equatorial stellar radius], and for the larger value of $\hat{A}^{-1}$, the angular velocity has steeper profiles. $\hat{A}^{-1} = 0$ implies that the angular velocity is uniform. For a special choice of $\hat{A} \sim 1$, a high-mass equilibrium can be constructed. $\beta := T_{\rm rot}/|W|$ denotes the ratio of the rotational kinetic energy $T_{\rm rot}$ to the absolute value of the gravitational potential energy W, and for $\beta \gtrsim 0.27$ and $0.14 \lesssim \beta \lesssim 0.27$, the neutron star would be dynamically and secularly unstable against non-axisymmetric deformation (see section 1.5.5.4). In this figure, the mass (in the left axis) and the energy density are shown in units of $\kappa_1 = 1$, and the mass in the right axis are shown for a particular choice of κ_1 with which the maximum rest mass of non-rotating neutron stars is $\approx 2 M_\odot$. We note that for the gravitational mass, the profiles are qualitatively the same as that for the rest mass except for the fact that the gravitational mass is smaller than the rest mass by $\sim 10\%$. The figure is taken from Baumgarte et al. (2000).

than critical values for the onset of the so-called bar-mode instability [e.g., chapter 7 of Shapiro and Teukolsky (1983); see also section 10.1]. Thus, hypermassive neutron stars of very high mass $\sim 3 M_\odot$ can exist in nature, but they would have a non-axisymmetric configuration.

In the presence of a dissipation process or an angular-momentum transport process, hypermassive neutron stars could be eventually unstable against gravitational collapse. We here note that since the mass of hypermassive neutron stars is larger than $M_{\rm max,spin}$, they will be unstable against gravitational collapse, when the degree of the differential rotation approaches zero, even with no angular-momentum dissipation. For example, in the presence of a viscous effect, the angular momentum will be transported from the inner region to the outer region inside a hypermassive neutron star, if the angular velocity in the inner region is larger than that in the outer region. When the degree of the differential rotation is decreased below a critical degree, the gravitational collapse will set in [Duez et al. (2004)]. A similar process works also by magnetohydrodynamics effects [Duez et al. (2006a); see also section 10.3.2.1].

Hypermassive neutron stars are likely to be formed after the merger of binary neutron stars (see sections 1.5.2 and 8.3), for which the typical total mass would be $2.6-2.8M_\odot$ (see table 1.2). As we mentioned in the above paragraph, the formed hypermassive neutron stars cannot live forever. Rather, they will eventually collapse to a black hole by some dissipation or angular-momentum transport mechanism. Exploring possible scenarios for the dynamical evolution of hypermassive neutron stars are one of the central issues in numerical relativity (see section 8.3).

1.4.7 Finite-temperature equations of state for high-density matter

Proto-neutron stars formed after the stellar core collapse of a massive star and massive neutron stars formed after the merger of binary neutron stars could be very hot. The temperature rises to ~ 10 MeV for the stellar core collapse [Bethe (1990)] and $\sim 50-100$ MeV for the merger of binary neutron stars [Sekiguchi et al. (2011b,a)]. For such high temperature, thermal pressure becomes one of the important pressure sources. In addition, neutrinos interact strongly with nucleons and electrons at the high temperature because the interaction cross section is proportional to the square of the neutrino energy. Thus, it takes a long time scale for neutrinos to escape from the hot neutron stars; neutrinos are trapped, that does not occur for cold neutron stars. Because they are captured inside the neutron stars, neutrinos are in degenerate states. This prevents the inverse-beta decay to neutrons and thus increases the fraction of protons and electrons, enhancing the contribution of the electron degenerate pressure. As these examples show, the finite-temperature effects play an important role for determining the structure, components, and evolution of hot neutron stars. For studying the properties of such hot neutron stars, equations of state incorporating the finite-temperature effects are necessary.

Equations of state for the high-density and high-temperature matter are usually constructed in a tabulated form [Lattimer and Swesty (1991); Shen et al. (1998a,b, 2011c,b,a); Hempel et al. (2012)]. In them, thermodynamical quantities such as pressure, internal energy, and entropy as well as fractions of protons, neutrons, and helium are written as functions of rest-mass density, ρ, temperature, T, and electron fraction per baryon, Y_e, e.g., $P = P(\rho, T, Y_e)$.

The importance of the finite-temperature effects for the structure of hot neutron stars can be explored by constructing sequences of spherical neutron stars. Figure 1.19 plots the gravitational mass as a function of the central density for a Shen equation of state [Shen et al. (1998a)] for setting the specific entropy, s, and

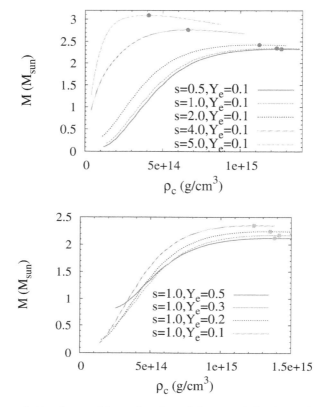

Fig. 1.19 The gravitational mass M as a function of the central density ρ_c for a Shen equation of state [Shen *et al.* (1998a)] for $Y_e = 0.1$ with $s/k_B = 0.5, 1, 2, 4$, and 5 (upper), and for $s = k_B$ with $Y_e = 0.1, 0.2, 0.3$, and 0.5 (lower). The filled circles denote the maximum mass along each sequence. Figures were kindly supplied by Y. Sekiguchi.

electron fraction, Y_e, to be constant for a wide range. We note for reference that the specific entropy of iron cores of massive stars is of order $s \sim k_B$ [Bethe (1990)]. We also note that the temperature of the model neutron stars with the maximum mass and $s = k_B$ is $\sim 30-40$ MeV for $Y_e = 0.5-0.1$.

Figure 1.19 shows that the equilibrium sequence is modified by the values of $s(\gtrsim k_B)$ and Y_e. In particular, for $s/k_B > 1$, the mass of the neutron star for a given value of the central density is significantly increased. The maximum temperature for the neutron star with the maximum mass along the sequence of $s/k_B = 5$ and $Y_e = 0.1$ is about 100 MeV. This suggests that if the temperature of a neutron star rises to ~ 100 MeV with $Y_e \sim 0.1$, a neutron star with mass $3M_\odot$ does not collapse to a black hole. Such a high temperature state may be achieved by dynamical phenomena in general relativity such as mergers of binary neutron stars. For these dynamical phenomena, the finite-temperature effect could play a crucial role for the evolution of the system (see section 8.3.3.4).

Table 1.2 Binary neutron stars observed to date, that will merge within the age of the universe $\approx 13.8\,\mathrm{Gyrs}$. P_{orb}: Orbital period in units of day, e: Orbital eccentricity, m: Total mass, m_1: Pulsar's mass, m_2: Companion's mass, and t_{gw}: Emission time scale of gravitational waves in units of 10^8 yrs [see, e.g., equation (1.93) for $e = 0$]. Note that the semi-major axis length, a, is estimated by $a \approx 1.8 \times 10^{11} \left(\dfrac{P_{\mathrm{orb}}}{0.3\,\mathrm{day}}\right)^{2/3} \left(\dfrac{m}{2.7 M_\odot}\right)^{1/3}$ cm, and thus, the periastron radius, $a(1 - e)$, is typically comparable to the solar radius $\approx 7.0 \times 10^{10}$ cm. References 1: Weisberg et al. (2010), 2: Stairs (2004), 3: Jacoby et al. (2006), 4: Lyne et al. (2004), 5: Faulkner et al. (2005); Ferdman et al. (2014), and 6: Lorimer et al. (2006); Kasian (2012). See also Lattimer (2012) and Ferdman et al. (2013) for a summary of the observed compact binary neutron stars. See also http://stellarcollapse.org/nsmasses for the latest mass measurement.

PSR	P_{orb}	e	m	m_1	m_2	t_{gw}	Ref.
B1913+16	0.323	0.617	2.828	1.441	1.387	3.0	1
B1534+12	0.421	0.274	2.678	1.333 ± 0.001	1.345 ± 0.001	27	2
B2127+11C	0.335	0.681	2.71	1.36 ± 0.01	1.35 ± 0.01	2.2	3
J0737-3039	0.102	0.088	2.587	1.338 ± 0.001	1.249 ± 0.001	0.86	4
J1756-2251	0.320	0.181	2.570	1.341 ± 0.007	1.230 ± 0.007	17	5
J1906+746	0.166	0.085	2.61	1.29 ± 0.01	1.32 ± 0.01	3.1	6

1.4.8 Binary neutron stars

Compact binaries composed of neutron stars and/or black holes are formed as an end product of a binary system composed of two massive stars. Compact binary systems may be also formed through a three-body interaction of stars in a dense stellar system such as globular clusters.[8] Among them, only binary neutron stars have been observed in our Galaxy by radio observations [Stairs (2003); Lorimer (2008)]. Here, we briefly describe observational facts and their typical evolution scenario.

To date, there exist six confirmed binary neutron stars observed in our Galaxy (see table 1.2). All these binaries have relatively compact orbits with their orbital period shorter than half a day. For them, it is feasible to measure the individual mass of binary components by the precise determination of relativistic effects in the pulsar timing observation [Lorimer (2008)]: It should be emphasized that the precise measurement of the relativistic effects of binary neutron stars [e.g., chapter 16 of Shapiro and Teukolsky (1983)] is the key to identify that they are really composed of two neutron stars. There are several other candidates of binary neutron stars of a wider orbit with the orbital period longer than a day. For them, it is difficult to determine the individual mass because the relativistic effects cannot be measured precisely.

All the binary neutron stars shown in table 1.2 will merge within the age of the universe $\approx 13.8\,\mathrm{Gyrs}$ (Gyrs= 10^9 yrs) because the orbital radius decreases due to the

---

[8] A binary neutron star of PSR B2127+11C (see table 1.2) is an example that is likely to be formed by a three-body encounter. However, the formation rate for the three-body encounter is inferred to be smaller than that through the evolution of massive binaries; see, e.g., Phinney (1991).

gravitational-wave emission (see section 1.2.5). Their lifetime may be approximately found in a theoretical calculation using a formula derived by Peters (1964). He gave the equations governing the evolution of the orbital semi-major axis a and eccentricity e by the gravitational-wave emission for binaries in Keplerian orbits as

$$\frac{da}{dt} = -\frac{64m_1m_2m}{5a^3(1-e^2)^{7/2}}\left(1+\frac{73}{24}e^2+\frac{37}{96}e^4\right), \qquad (1.188)$$

$$\frac{de}{dt} = -\frac{304m_1m_2me}{15a^4(1-e^2)^{5/2}}\left(1+\frac{121}{304}e^2\right), \qquad (1.189)$$

where m_1 and m_2 are masses of the binary components with $m = m_1 + m_2$. For deriving these equations, Keplerian equations of motion together with the gravitational-radiation reaction based on the quadrupole formula are used. Integrating equations (1.188) and (1.189) from the present values of (a,e) to the time when $a = 0$, thus, yields the lifetime of the binary, $t_{\rm gw}$ (see table 1.2). An important finding of Peters (1964) was that a is analytically written as a function of e as

$$a(e) = \frac{c_0 e^{12/19}}{1-e^2}\left[1+\frac{121}{304}e^2\right]^{870/2299}, \qquad (1.190)$$

where c_0 is determined by the initial conditions $a = a_0$ and $e = e_0$. Equation (1.190) implies that e decreases more steeply than a, and the orbit quickly approaches a circular one when a decreases due to the gravitational-wave emission. Note that substituting equation (1.190) into equation (1.189), an ordinary differential equation for e is obtained, and thus, the time evolution for e and a can be obtained by a straightforward integration.

The number of the observed binary neutron stars is much smaller than the number of observed neutron stars. The fraction of the observed compact binary neutron stars with $t_{\rm gw} \ll 13.8\,{\rm Gyrs}$ is less than 1% of the total number of the observed neutron stars. This fact implies that the merger of binary neutron stars is a rare event in our Galaxy. The reason for this is likely to result from the facts that (i) the formation of binaries composed of two massive stars, which are most promising progenitors of binary neutron stars (but see footnote 7), is an event rarer than that for single massive stars or binaries composed of a massive star and a less-massive star and/or (ii) binary massive stars do not always result in binary compact objects. In the following, we will touch on the second point (see figure 1.20).

Massive-star binaries will experience supernova events twice during an interactive evolutionary path. Initially, the binary is composed of two massive stars. The more massive star evolves in a shorter time scale and eventually forms a neutron star after the first supernova explosion, unless its initial mass is very large. In the supernova event, roughly half of the total mass is ejected from the system, and thus, the orbital motion would be significantly perturbed. However, the companion star is much more massive than the formed neutron star and thus can control the dynamical motion of the system. Hence, the binary is not likely to be disrupted

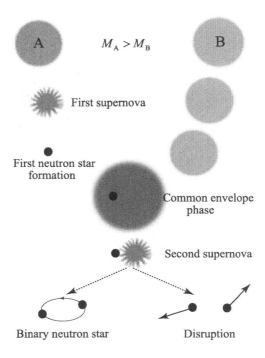

Fig. 1.20 A schematic formation scenario of binary neutron stars.

in the first supernova. In the subsequent evolution, the binary will experience the second supernova after a common envelope phase through which the binary orbital separation is appreciably reduced (while avoiding the merger) and a compact binary would be formed [e.g., Postnov and Yungelson (2006); Lorimer (2008) for a review]. In the second supernova explosion, again, a large amount of mass is ejected from the system. In contrast to the first supernova, the fraction of the ejected mass is much larger than half of the total mass of the system. This suggests that the binary is subject to disruption. Of course, some fraction of the system may remain to be a binary and eventually form a system of a binary neutron star. This would be in particular the case if the binary orbital separation before the second supernova explosion is sufficiently small and the binding between two stars is sufficiently strong. However, with a not small probability, in particular for less compact binaries, the binary may be disrupted. This is the possible reason that binary neutron stars may not be frequently formed.

However, we still have several binary neutron stars for which $t_{\rm gw} \ll 13.8\,{\rm Gyrs}$. This fact has a quite important meaning for the observation of gravitational waves, because this observational result indicates that a non-negligible number of the merger of binary neutron stars would occur in our Galaxy: Suppose that a binary neutron star of lifetime 0.1 Gyrs, such as PSR J0737-3039, is observed. If we assume

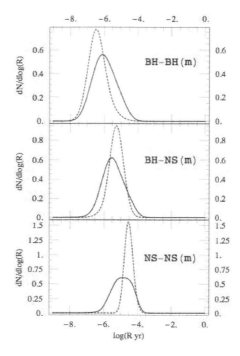

Fig. 1.21 A plot of the probability distributions for the black hole-black hole (upper), black hole-neutron star (center), and neutron star-neutron star (lower) merger rates per our-Galaxy-equivalent galaxy. The dashed and solid curves denote the results by two different models. This figure is taken from Kalogera *et al.* (2007); see this reference for details.

that stars and binary systems have been formed with a uniform rate in the history of our Galaxy, then, it is natural to conclude that the merger rate for this type of binary neutron stars would be at least 1 per every 0.1 Gyrs. In reality, the expected merger rate in our Galaxy should be much larger than this value, because there are many binary systems that cannot be observed due to the following several reasons; the intrinsic luminosity of neutron stars as a pulsar could be too weak for a signal to reach the Earth; the distance from the Earth could be so large that the signal could be too weak to be observed; the pulsars do not isotropically emit the radio pulse but emit toward a particular direction, and thus, the signal could not reach the Earth; there are some absorbers between the Earth and the pulsars, and hence, the signals might be absorbed before reaching the Earth. All these uncertainties are not known precisely, and therefore, the researchers of the merger rate have to rely on statistical studies, based on the limited observational results.

Statistical studies for the merger rate of binary compact objects have been performed by many authors [e.g., Phinney (1991); Narayan *et al.* (1991); Voss and Tauris (2003); Kalogera *et al.* (2004, 2007); O'Shaughnessy *et al.* (2008, 2010); Dominik *et al.* (2012)]. Figures 1.21 and 1.22 show a result of the statistical stud-

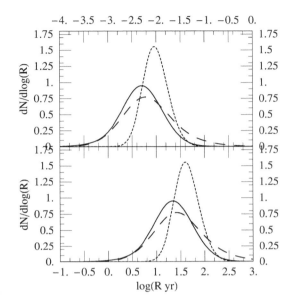

Fig. 1.22 Ranges of expected detection rates for initial (upper) and advanced (lower) LIGO for black hole-black hole (dashed), black hole-neutron star (solid), and neutron star-neutron star (dotted) binaries, based on models that satisfy observational constraints and under the assumptions outlined in Kalogera *et al.* (2007). This figure is taken from Kalogera *et al.* (2007).

ies by Kalogera *et al.* (2007). Figure 1.21 shows a probability distribution of the merger rates per our-Galaxy-equivalent galaxy. For deriving the curves in this figure, a number of assumptions were made and the results of population synthesis studies were used. These assumptions and input models for the population synthesis studies contain large uncertainties, and thus, have been updated year by year. Hence, the event rates have been modified frequently, although the order of the magnitude does not change. Describing the details for this statistical calculation is far beyond this volume, and thus, we recommend the readers who are interested in the details to refer to, e.g., Postnov and Yungelson (2006); Kalogera *et al.* (2007); Dominik *et al.* (2012).

Figure 1.21 shows that the most probable Galactic merger rate of binary neutron stars is one every $\sim 10^5$ yrs, although the error-bar is large; it is still possible that the merger rate is less than one every $\sim 10^6$ yrs for the most conservative case. The important fact is, however, that this conservative number is still an appreciable value [Phinney (1991); see also below].

For compact black-hole binaries, we do not have any observational results, and thus, a purely theoretical modeling is necessary to infer the merger rates. For the theoretical modeling, a population synthesis study is usually employed. This study follows the evolution of massive-star binaries, making a number of assumptions for complicated physical processes and treating them in a simple manner: Here, the physical processes include mass and angular momentum transports in the binary system, common envelope formation and its evolution, and evolution of accretion disks around the first-born neutron star. Although the uncertainty for their results seems to be quite high, the results usually predict that the most probable merger rate of black-hole binaries in our Galaxy is one every 10^6–10^7 yrs.

The result shown in figure 1.21 is the probability distribution for our-Galaxy-type galaxies. There are a huge number of galaxies in the universe, and thus, the total merger rate in the universe should be quite high. From the view point of the gravitational-wave observation, the most interesting prediction is associated with the detection rate of gravitational waves by gravitational-wave detectors. For example, advanced LIGO will be able to detect gravitational waves from the coalescence of binary neutron stars with total mass $2.8 M_\odot$ up to the distance of ~ 450 Mpc (this is the case when the source direction and the inclination angle of the binary orbital plane with respect to our lines of sight are most optimistic for the detection). Because the average number density of the galaxies in the universe is $\sim 0.01/\text{Mpc}^3$, there are $\sim 10^6$ galaxies inside the sphere of radius 300 Mpc around our Galaxy. This suggests that if the average merger rate in one galaxy is one every 10^5 yrs, ~ 10 detections per year may be expected for advanced LIGO. Figure 1.22 shows a predicted probability distribution for the detection number by initial and advanced LIGO. This predicts that advanced LIGO will detect ~ 40 gravitational-wave events per year for binary neutron stars in the most probable detection rate. For black-hole binaries, the predicted number is slightly smaller. However, ~ 20 events per year may be expected in the most probable detection rate both for binary black holes and black hole-neutron star binaries.

We note that the results shown in figures 1.21 and 1.22 were based on one of many theoretical models. As mentioned above, there are many uncertainties in the theoretical modeling and in the statistical studies, and thus, it is safe to consider that the real event rates could be changed by a factor of 10. However, even for a conservative estimate in which the event rate would be by a factor of 10 lower than the best estimate that is shown in figure 1.22, gravitational waves from binary neutron stars will be detected by advanced LIGO every year. This is the encouraging prediction for the gravitational-wave detection. Anyway, gravitational-wave detectors will clarify the real event rates in the near future. This fact is also very important in the field of stellar population synthesis.

1.5 Sources of gravitational waves

As shown in section 1.2, a non-stationary and non-spherical object emits gravitational waves. For a large amount of the emission, the system has to be moving rapidly and sufficiently compact. These conditions imply that gravitational-wave sources are composed of general relativistic objects in dynamical spacetime.

Speaking from the view point of detecting them, the gravitational-wave events have to occur frequently. In addition, it is desirable that the signal of gravitational waves can be accurately predicted and/or the event can be confirmed by other observational means, e.g., by observing an electromagnetic counterpart event. Unless these two conditions are satisfied, it will be quite difficult to confirm the detection of gravitational waves for a possible observed event. This section summarizes the properties of the gravitational-wave sources that could satisfy the conditions listed here.

1.5.1 *Inspiral of binary compact objects*

Compact binary systems composed of neutron stars and black holes in close quasi-circular orbits (inspiral orbits) are among the most promising sources of ground-based gravitational-wave detectors. In particular, coalescing binary neutron stars are considered to be the best sources. There are three reasons for this: (i) statistical studies (see figure 1.22) indicate that gravitational waves from them will be detected by advanced gravitational-wave detectors every year even for a conservative estimate; (ii) gravitational waveforms in the inspiral stage can be predicted with a high accuracy by post-Newtonian calculations [Blanchet (2014)], and thus, the extraction of the signal from a noisy data stream can be efficiently performed using the theoretical templates (see section 1.6); (iii) electromagnetic counterparts that will be generated during the merger and will be observed by large-scale optical and radio telescopes are likely to be accompanied with the gravitational-wave signal [e.g., Li and Paczyński (1998); Metzger *et al.* (2010); Nakar and Piran (2011)].

Before describing the features of gravitational waveforms from coalescing binary neutron stars, we outline their evolution process. Binary neutron stars will merge within the age of the universe and could be a source of ground-based gravitational-wave detectors, if their orbital period is short enough, i.e., less than about one day; see table 1.2. For a specific example, we here consider the evolution of the system of PSR J0737-3039 (see figure 1.23). For this binary, the orbital period is 2.4 hour(≈ 8640 s) with a small eccentricity ($e \approx 0.088$). Thus, nearly monotone gravitational waves with ~ 0.23 Hz are continuously emitted now. Because of this gravitational-wave emission, the orbital separation gradually decreases (each neutron star in this binary inspirals toward each other). Because the decrease time scale of the eccentricity is shorter than that of the semi-major axis, the eccentricity will approach zero during the inspiral and in the late inspiral stage, the orbit is approximately circular. After the gravitational-wave emission for $\sim 10^8$ yrs, two neutron stars will eventually merge.

The frequency of gravitational waves from a binary in a circular orbit is given, in Newtonian gravity, by

$$f = \frac{1}{\pi}\sqrt{\frac{m}{a^3}} = 10.5 \left(\frac{m}{2.8 M_\odot}\right)^{1/2} \left(\frac{a}{700\,\text{km}}\right)^{-3/2}\text{Hz}, \qquad (1.191)$$

where m and a are the total mass and the orbital separation, respectively [see

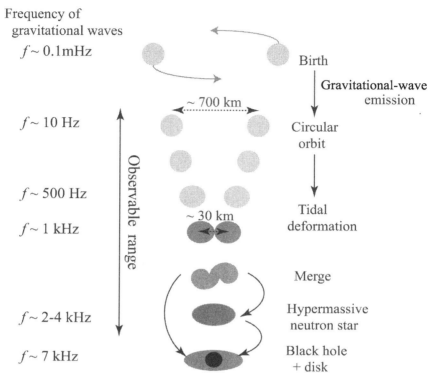

Fig. 1.23 A typical evolution scenario of binary neutron stars in close orbits and characteristic gravitational-wave frequencies. See section 8.3 for more details.

equation (1.87)]. In the following, we assume $m = 2.8 M_\odot$ because it is a typical value of the total mass for binary neutron stars. As shown in figure 1.5, ground-based advanced detectors will have a good sensitivity for $f \gtrsim 10$ Hz. This implies that gravitational waves from binary neutron stars can be detected when the orbital separation becomes less than ~ 700 km. Equation (1.92) gives the lifetime of binary neutron stars of $m = 2.8 M_\odot$ as ~ 880 s at $f = 10$ Hz. Thus, advanced gravitational-wave detectors will observe last ~ 15 minutes of binary neutron stars.

As shown by equation (1.93), the emission time scale of gravitational waves is longer than the orbital period up to a very close orbit just before the merger. Hence, the binaries evolve in an adiabatic manner for most of their life. On the other hand, various effects break the adiabatic nature of the evolution when the orbital separation decreases to ~ 30–40 km (the corresponding frequency of gravitational waves is $f \approx 1.2$–0.8 kHz). First, the gravitational-wave emission time scale becomes as short as the orbital period. Second, the general relativistic attractive force is so strong that the stable circular orbit could disappear for a close orbit (see section 1.3.2). Finally, each neutron star is tidally elongated. This modifies the attractive force between two neutron stars, because by the tidal deformation, a quadrupole moment is induced for each star [Rasio and Shapiro (1992); Lai et al. (1993a,b, 1994b,a,c)]. For example, in Newtonian gravity, the gravitational potential by a star of mass m_1 and of tracefree quadrupole moment I_{ij} becomes

$$V(r) = -\frac{m_1}{r} - \frac{3 I_{ij} \hat{x}^i \hat{x}^j}{2r^3}, \qquad (1.192)$$

where $\hat{x}^i := x^i/r$ with r, the radial distance from the center of the star. Thus, due to the correction by the tidal deformation, the additional attractive force is induced. Here, I_{ij} is induced by the tidal force from the companion star, and hence, it is proportional to r^{-3}. This implies that the attractive force associated with the tidal deformation is by $O(r^{-5})$ weaker than the leading part, and thus, for a distant orbit, it is negligible. Specifically, the tidal part is of $O[(R/r)^5]$ of the leading part, where R is the stellar radius in isolation. Thus, for a close orbit $a \sim 3R$, the attractive force due to the tidal part becomes appreciable enough that the stable circular orbit could disappear [Rasio and Shapiro (1992); Lai et al. (1994b)].

All these three effects mentioned in the above paragraph can enhance the radial approaching velocity of binary neutron stars, and hence, for $a \lesssim 30$–40 km, two neutron stars will start merging. We note that the typical orbital period at the onset of the merger is 2 ms. This is usually much shorter than the spin period of each neutron star (unless the neutron star is the so-called recycled millisecond pulsar[9]). Thus, the spin of each neutron star is not likely to modify this scenario significantly. By contrast, the equation of state of neutron stars, which is still poorly known (see section 1.4.4), can affect the quantitative details of the late inspiral motion, because it determines the neutron-star radius, R, which is the key quantity for determining the attractive force due to the tidal deformation. For a stiff equation of state, i.e., for a large value of R, the tidal force plays an important role for a distant orbit, say at $a \sim 40$ km. Clarifying the dependence of the orbital motion for the close orbits on the equation of state is one of the most important subjects in numerical relativity (e.g., see section 8.3.4.1).

After the onset of the merger, a rotating black hole or a massive and rapidly rotating neutron star (namely a supramassive or hypermassive neutron star [Baumgarte et al. (2000)]; see section 1.4.6) is formed as a remnant. The details about the merger process, properties of the merger remnant, and subsequent evolution of the remnant are described in section 1.5.2 briefly and in section 8.3 in detail.

The evolution process of inspiraling binary black holes and black hole-neutron star binaries is qualitatively the same as that of binary neutron stars except for the final merger stage. However, because the typical mass of black holes is likely to be much larger than $3M_\odot$, the values of the characteristic frequency are quantitatively different from those of binary neutron stars. For example, for hypothetical total

---

[9]In a typical scenario for the formation of binary neutron stars, one of two neutron stars is a recycled pulsar that was spun up by the mass accretion from its companion [Belczynski et al. (2002); Kalogera et al. (2007)]. However, the observation data for the binary neutron stars show that the resulting spin period is $\gtrsim 20$ ms, and thus, indicate that the spin-up was not significant enough to result in a millisecond pulsar; see, e.g., the listed date of Lorimer (2008).

mass of the system $m = 10M_\odot$, the orbital separation at $f = 10\,\mathrm{Hz}$ is $a \approx 1100\,\mathrm{km}$, and at an approximate ISCO radius (assuming a small black-hole spin), $a = 6m \approx 90\,\mathrm{km}$, the frequency is $f \approx 430\,\mathrm{Hz}$. The lifetime from $f = 10\,\mathrm{Hz}$ to the onset of the merger for an equal-mass binary of $m = 10M_\odot$ is about two minutes. The merger processes are also different from those of binary neutron stars. These topics are described in section 1.5.3 for binary black holes and in section 1.5.4 for black hole-neutron star binaries.

In the rest of this section, we will describe an approximate waveform and a spectrum of gravitational waves from inspiraling compact binaries in quasi-circular orbits, assuming the Keplerian motion for two bodies and using the quadrupole formula (1.86). For a compact binary just before the merger, the orbital velocity, v, becomes 20–40% of the speed of light, and hence, the Newtonian approximation together with the quadrupole formula cannot provide quantitatively accurate results for the orbital motion and gravitational waves. Post-Newtonian calculations incorporating the terms up to a quite high order in v/c have to be performed for the quantitative study [e.g., see Blanchet (2014) for a review and also appendix H]. However, it is still possible to derive a qualitative result for the evolution of the inspiral stage in the Newtonian approximation with the quadrupole formula.

In the quadrupole formula, the amplitude of gravitational waves as a function of time at a detector is written as

$$h(t) = 4Q_h \frac{\mathcal{M}}{D_{\mathrm{obs}}} [\pi F(t) \mathcal{M}]^{2/3} \cos \Phi(t), \quad \Phi(t) = \int 2\pi F(t) dt, \quad (1.193)$$

where $Q_h (> 0)$ is a function of the orientation angles of the gravitational-wave source with respect to the detector plane, the inclination angle of the binary orbital plane to the line of sight, and the polarization of gravitational waves. The value of Q_h is smaller than unity with the average $1/\sqrt{5} \approx 0.447$ [Schutz and Tinto (1987); Thorne (1987)] (by averaging over the random orientation, the random inclination angle of the orbital plane, and the random polarization). $F(t)$ and $\Phi(t)$ are the gravitational-wave frequency and phase as functions of time, respectively. $\mathcal{M}$ is the so-called *chirp mass* defined by

$$\mathcal{M} := \mu^{3/5} m^{2/5}, \quad (1.194)$$

where μ is the reduced mass. The chirp mass is $1.2188 M_\odot$ for $m_1 = m_2 = 1.4 M_\odot$ and $8.7055 M_\odot$ for $m_1 = m_2 = 10 M_\odot$. It is worthy to note that $\mathcal{M}$ is only the parameter of mass dimension that appears in the quadrupole formula. Thus, by a gravitational-wave detection, $\mathcal{M}$ is one of the primary parameters to be determined.

Due to the gravitational-wave emission, the orbital separation, a, decreases and the frequency, $F(t)$, increases. Using equation (1.87), the evolution equation for $F(t)$ is written as

$$\frac{dF}{dt} = \frac{96}{5\pi \mathcal{M}^2} [\pi F(t) \mathcal{M}]^{11/3}, \quad (1.195)$$

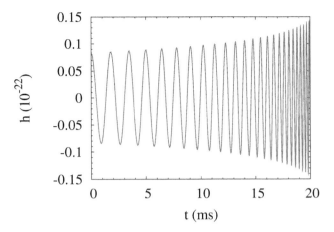

Fig. 1.24 Gravitational waveforms ("chirp" signal) from an equal-mass binary neutron star with the total mass $m = 2.8 M_\odot$ for a hypothetical distance to the source $D_{\rm obs} = 140$ Mpc. Gravitational waves are assumed to be observed from the axis perpendicular to the orbital plane (i.e., most optimistic direction). The vertical and horizontal axes denote the amplitude in units of 10^{-22} and the time in units of milliseconds, respectively.

where we used $d\ln F/dt = -(3/2) d\ln a/dt$. By integrating equation (1.195) and using equation (1.193), we obtain

$$\Phi(t) = \Phi_0 - \frac{1}{16} \left[\pi F(t) \mathcal{M}\right]^{-5/3}, \qquad (1.196)$$

where Φ_0 is a constant.

Then, $h(t)$ is written as a function of time. Figure 1.24 plots a gravitational waveform of $h(t)$ as a function of time for an inspiraling equal-mass binary neutron star with $m = 2.8 M_\odot$ for the duration of ≈ 20 ms before the onset of the merger. This illustrates that the frequency and the amplitude increase with the decrease of the orbital separation. This type of gravitational waveforms is called the *chirp signal*.

Next, we describe the property of the gravitational-wave spectrum defined by

$$\tilde{h}(f) := \int h(t) e^{2\pi i f t} dt. \qquad (1.197)$$

Assuming that the emission time scale of gravitational waves is much longer than the orbital period,[10] the stationary-phase approximation may be used for evaluating the integral of equation (1.197). Then, we obtain [e.g., see Creighton and Anderson (2011)]

$$\tilde{h}(f) = \left(\frac{5}{24\pi^3}\right)^{1/2} \frac{Q_h}{D_{\rm obs}} f^{-7/6} (\pi \mathcal{M})^{5/6} e^{i\psi(f)}, \qquad (1.198)$$

$$\psi(f) = 2\pi t_0 f + \psi_0 + \frac{3}{128} (\pi \mathcal{M} f)^{-5/3}, \qquad (1.199)$$

---

[10] Note that this assumption is violated for the late inspiral orbits just before the merger.

where ψ_0 is a constant and t_0 denotes the coalescence time. Note that $\tilde{h}(f)$ is a quantity of time dimension. It is often useful to define an effective amplitude of gravitational waves, $h_{\rm eff} := |f\tilde{h}(f)|$, which is dimensionless and may be considered as the amplitude of gravitational waves for a given value of f.

Here, it is interesting to compare $h_{\rm eff}$ with the maximum amplitude of $h(t)$, $h_{\rm max}$, for a given value of frequency $f = F(t)$:

$$\frac{h_{\rm eff}}{h_{\rm max}} = \sqrt{\frac{5}{384\pi}} (\pi \mathcal{M} f)^{-5/6} \approx 82 \left(\frac{f}{10\,{\rm Hz}}\right)^{-5/6} \left(\frac{\mathcal{M}}{1.2188 M_\odot}\right)^{-5/6}$$

$$\approx 12 \left(\frac{f}{100\,{\rm Hz}}\right)^{-5/6} \left(\frac{\mathcal{M}}{1.2188 M_\odot}\right)^{-5/6}. \quad (1.200)$$

This equation shows that $h_{\rm eff}$ is larger than $h_{\rm max}$ if the binary is in an adiabatic inspiral orbit, i.e., $f \lesssim 1\,{\rm kHz}$. In particular for non-close orbits, $f \ll 1\,{\rm kHz}$, the amplification factor, $h_{\rm eff}/h_{\rm max}$, is much larger than unity. This is reasonable because gravitational waves of a given value of f are accumulated in the integration for the Fourier spectrum if the gravitational-radiation reaction time scale is longer than the orbital period. Specifically, the amplification factor is approximately written as $\sqrt{N}$ where N denotes a number of cycle of gravitational waves for a given value of f, $N \sim ft_{\rm gw}$, where $t_{\rm gw}$ is defined in equation (1.92) and is proportional to $F(t)(dF/dt)^{-1}$.

This accumulation mechanism of the wave cycle has a significant impact on the detection of gravitational waves: The value of $h_{\rm max}$ for a binary neutron star at a hypothetical distance of $100\,{\rm Mpc}$ is

$$h_{\rm max} = 3.6 \times 10^{-23} Q_h \left(\frac{\mathcal{M}}{1.2188 M_\odot}\right)^{5/3} \left(\frac{100\,{\rm Mpc}}{D_{\rm obs}}\right) \left(\frac{f}{100\,{\rm Hz}}\right)^{2/3}. \quad (1.201)$$

This value is too small for gravitational waves to be detected by advanced ground-based detectors such as advanced LIGO; see its sensitivity curve of figure 1.5. However, the effective amplitude is by a factor of $\sqrt{N}$ larger than $h_{\rm max}$ as

$$h_{\rm eff} = 4.3 \times 10^{-22} Q_h \left(\frac{\mathcal{M}}{1.2188 M_\odot}\right)^{5/6} \left(\frac{100\,{\rm Mpc}}{D_{\rm obs}}\right) \left(\frac{f}{100\,{\rm Hz}}\right)^{-1/6}. \quad (1.202)$$

Hence, the value of $h_{\rm eff}$ is larger than the noise level of advanced detectors: Compare slowly declining lines (for which $Q_h = 1$ and $1/\sqrt{5}$ are assumed) with the noise curve in figure 1.5; this figure illustrates that for the source distance of several $100\,{\rm Mpc}$, gravitational waves will be detected by advanced gravitational-wave detectors. We here emphasize that for a *long-term observation* of gravitational waves for which the observation time is longer than $t_{\rm gw}$, the output in the detectors for the amplitude is $h_{\rm eff}$ not $h_{\rm max}$. This fact makes compact binaries promising sources of gravitational waves for the advanced detectors.

As mentioned earlier, although the analysis of the gravitational waveforms and spectrum in this section is qualitatively valid, it is not quantitatively accurate. In particular, for the late inspiral stage just before the merger, there are many elements to be taken into account; general relativistic effects play an important role; a dynamical approach is necessary because the adiabatic approximation breaks down; and finite-size effects become appreciable and point-particle approximation becomes invalid. In the relatively early stage of the inspiral, the last two elements are not important, but general relativistic corrections have to be taken into account even for $f \sim 10\,\mathrm{Hz}$. For such a stage, the orbital evolution and resulting gravitational waveform depend not only on the chirp mass but also on the reduced mass and the spin of each object. The study for this stage has been extensively performed in the framework of post-Newtonian approximations [Blanchet (2014)] (see also appendix H). Because the gravitational waveforms depend on the reduced mass and the spin, we have the possibility to determine the mass and the spin of two objects by analyzing gravitational waves in the relatively early inspiral stage.

The last two points become crucial in particular for the orbit with $a \lesssim 3R$. For such a stage, usual post-Newtonian approximations together with the point-particle approximation cannot be employed. We have to fully take into account the dynamical and finite-size effects of compact objects together with general relativistic gravity. The unique approach for the study of this stage is numerical relativity.

1.5.2 Merger of binary neutron stars

As schematically described in figure 1.23, the outcome formed after the merger of binary neutron stars is either a rotating black hole or a massive (typically supra-massive or hypermassive) neutron star. This depends strongly on the total mass of the system and the equation of state for the high-density nuclear matter, which is still poorly known (see section 1.4.4).

The readers may well wonder why the massive neutron star can be formed for the typical total mass of the system $\sim 2.6\text{--}2.8 M_\odot$ that would be much larger than the maximum allowed mass of spherical neutron stars in isolation for typical equations of state (see section 1.4.5). The reason for this is that the remnant formed soon after the merger is hot and rapidly rotating, and thus, it has significant thermal pressure and centrifugal force. Here, the high thermal energy comes from shock heating at the collision of two neutron stars. As a consequence of these effects, the maximum allowed mass can be increased by several ten percents [e.g., Shen et al. (1998a); Baumgarte et al. (2000)], resulting in the possible formation of a remnant massive neutron star.

The typical gravitational waveform during the prompt formation of a black hole, obtained in our numerical-relativity simulation, is displayed in figure 1.25. We display the waveform only during the formation of the black hole omitting the inspiral waveform. This gravitational waveform is characterized by a ringdown

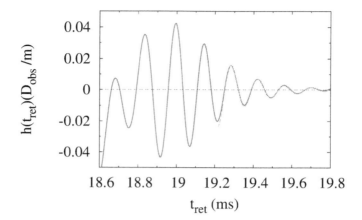

Fig. 1.25 Gravitational waves associated with a black-hole formation process in the merger of a binary neutron star observed along the axis perpendicular to the orbital plane. $D_{\rm obs}$ and $m(=2.7M_\odot)$ are a hypothetical distance from the source and total mass of the binary, respectively. The solid curve is the result of a numerical-relativity simulation and the dashed curve is the fitting one obtained by using the frequency of the dominant quasi-normal mode [see equation (1.204)]. Note that the prompt black-hole formation for $m = 2.7M_\odot$ occurs only when the equation of state of neutron stars is soft and would not be very realistic (see section 8.3.3).

oscillation described by

$$h(t) \propto \text{Re}\left[e^{i\omega_{\rm QNM}t}\right], \tag{1.203}$$

where $\omega_{\rm QNM} := \omega_r + i\omega_i$ is the angular frequency of a quasi-normal mode of the formed black hole [see, e.g., Chandrasekhar and Detweiler (1975); Leaver (1985); Schutz and Will (1985) for the quasi-normal modes of black holes]. The real part of $\omega_{\rm QNM}$ is in particular important because it gives the highest angular frequency in the Fourier spectrum; i.e., the amplitude of the Fourier spectrum exponentially damps for $\omega > \omega_r$.

ω_r and ω_i depend only on the mass and the spin of the formed black hole. Berti et al. (2009) derived an accurate fitting formula for them except for the extremely rapidly spinning case, and ω_r and ω_i for the fundamental quadrupole modes (i.e., for $l = m = 2$) are written as

$$\omega_r M_{\rm BH} \approx \left[1.5251 - 1.1568(1-\hat{a})^{0.1292}\right],$$
$$\omega_i \approx \frac{1}{2}\omega_r \left[0.700 + 1.4187(1-\hat{a})^{-0.4990}\right]^{-1}, \tag{1.204}$$

where $M_{\rm BH}$ and $\hat{a}$ denote the mass and dimensionless spin of the formed black hole. In the merger of binary neutron stars, the typical value of $\hat{a}$ is 0.6–0.8 (the details will be described in section 8.3). This implies that the characteristic frequency defined by $\omega_r/(2\pi)$ is ≈ 6.0–$7.0(2.7M_\odot/M_{\rm BH})$ kHz. Indeed, the ringdown oscillation shown in figure 1.25 for $t_{\rm ret} \gtrsim 19.3$ ms is fitted well with the parameters determined from the horizon as $M_{\rm BH} \approx 2.636 M_\odot$ and $\hat{a} \approx 0.806$.

A detection of this ringdown oscillation by gravitational-wave detectors will imply our direct observation for the black-hole formation. However, the characteristic frequency $\sim 6\text{--}7\,\text{kHz}$ is too high, and moreover, the amplitude at a hypothetical distance of $D_\text{obs} = 100\,\text{Mpc}$ is unfortunately much lower than the sensitivity level of the advanced detectors (compare with figure 1.5):

$$h = 5.2 \times 10^{-23} \left(\frac{hD_\text{obs}/m}{0.04}\right) \left(\frac{D_\text{obs}}{100\,\text{Mpc}}\right)^{-1} \left(\frac{m}{2.7 M_\odot}\right). \quad (1.205)$$

(See figure 1.25 that shows that the maximum value of hD_obs/M is ~ 0.04.) Thus, only for the case that the merger event happens in a nearby galaxy ($D_\text{obs} \lesssim 100\,\text{kpc}$) quite fortunately, the detection of ringdown gravitational waves will be possible.

A massive neutron star, if it is formed as a remnant, is hot and rapidly rotating. It is also highly deformed and has a non-axisymmetric figure; see section 8.3.3 for the details. Because of the rapid rotation and the non-axisymmetric configuration, quasi-periodic gravitational waves with the typical frequency $2\text{--}4\,\text{kHz}$ are emitted from the remnant massive neutron stars; see figure 1.26. The frequency depends strongly on the equation of state, which is still poorly known, as well as on the remnant mass. Because of the angular-momentum dissipation by the gravitational-wave emission and outward angular-momentum transport to its envelope, the remnant massive neutron star will lose its angular momentum gradually. Due to this, the degree of non-axisymmetry is decreased, and eventually, the gravitational-wave emission ceases (see section 1.5.5.4). However, the emission time scale, typically $\sim 10\text{--}100\,\text{ms}$, is much longer than the rotation period $\sim 1\,\text{ms}$. This implies that gravitational waves with approximately the same frequency are emitted for $\sim 10\text{--}100$ cycles, and in the detectors, the signal is accumulated, as in the chirp signal (see section 1.5.1). Hence, a peak with a high amplitude is seen in the Fourier spectrum (see section 8.3.4 for details).

Gravitational waves in the final inspiral and merger stages have rich information for the properties of neutron stars. First, gravitational waves from binary neutron stars in close orbits carry the information of the tidal deformation as already mentioned. The degree of the tidal deformation at a given orbital radius (or orbital frequency) depends strongly on the equation of state or the radius of the neutron stars. Specifically, the evolution of the phase $\Phi(t)$ in the final inspiral stage depends strongly on the equation of state. Thus, by analyzing gravitational waves in this stage for which $f \sim 500\text{--}1000\,\text{Hz}$, it may be possible to constrain their equation of state. Second, quasi-periodic gravitational waves from remnant massive neutron stars will carry the information of the equation of state, because, as already mentioned, the characteristic frequency of these gravitational waves depends strongly on the equation of state. If they are detected by advanced detectors and their frequency is determined, the equation of state for the high-density nuclear matter will be strongly constrained. These facts have been discovered and clarified in numerical-relativity simulations. The details will be described in section 8.3.

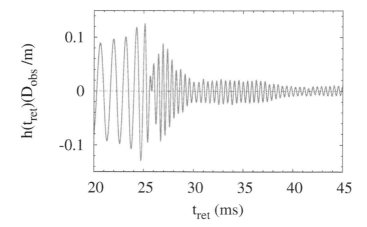

Fig. 1.26 Gravitational waves from a remnant massive neutron star formed after the merger of a binary neutron star observed along the axis perpendicular to the orbital plane. $D_{\rm obs}$ and $m(=2.7M_\odot)$ are a hypothetical distance from the source and the total mass, respectively.

The mergers of binary neutron stars (and also black hole-neutron star binaries) are also promising sources for high-energy astrophysical phenomena such as gamma-ray bursts [Eichler et al. (1989); Narayan et al. (1992); Zhang and Meszáros (2004); Piran (2004); Nakar (2007)], because a class of the merger events is likely to result in the formation of a black hole surrounded by a massive, dense, and high-temperature torus (see descriptions in sections 8.3 and 8.5). An analytic modeling [e.g., Di Matteo et al. (2002); Kohri and Mineshige (2002); Chen and Beloborodov (2007)] and numerical simulations [e.g., Lee et al. (2005); Setiawan et al. (2006); Shibata et al. (2007)] showed that from a dense ($\rho \gtrsim 10^{11}$ g/cm^3) and hot ($T \gtrsim 10$ MeV) disk or torus surrounding a stellar-mass black hole with mass $\sim 3M_\odot$, copious neutrinos are generated in the accretion disk/torus due to its high density and high temperature, and thus, a neutrino-dominated accretion flow [named by Popham et al. (1999); Narayan et al. (2001)] is formed. Here, a hot state is realized due to the fact that the viscous accretion (viscous heating) time scale is quite short for such a compact system. Although copious neutrinos are generated in the disk/torus, all of them are not emitted freely but a fraction of neutrinos could be trapped because the density and the temperature of the disk/torus are so high that free escape may not be allowed (the mean-free path of neutrinos becomes shorter than the size of the disk/torus for its inner part) [Di Matteo et al. (2002)]. Hence, the neutrino luminosity is not as high as that originally reported by Popham et al. (1999), but still it could be $\gtrsim 10^{53}$ ergs/s for a sufficiently high mass accretion rate $dM/dt \gtrsim M_\odot$/s. For such a high value of the luminosity, the neutrino pair annihilation could be enhanced significantly. If this occurs efficiently in a low-density region near the black hole (e.g., near the inner surface of the accretion disk/torus [Setiawan et al. (2006); Birkl et al. (2007); Zalamea and Beloborodov (2011)]), a fireball composed of pho-

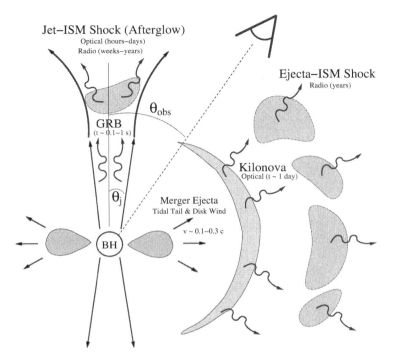

Fig. 1.27 A schematic picture for possible electromagnetic counterparts of mergers of binary neutron stars and black hole-neutron star binaries. Following the merger, a massive and hot disk/torus surrounding the central black hole is formed. Rapid accretion lasting $\lesssim 1\,\mathrm{s}$ could power a collimated relativistic jet, which produces a short-duration gamma-ray burst. Due to relativistic beaming, the gamma-ray emission is restricted to observers within $\theta_j \ll 1$, where θ_j is the half-opening angle of the jet. Non-thermal afterglow emission results from the interaction of the jet with the surrounding circumburst medium. Optical afterglow emission is observable on time scales up to $\sim$ days–weeks by observers with viewing angles of $\sim 2\theta_j$. Radio afterglow emission is observable from all viewing angles (i.e., quasi-isotropic) once the jet decelerates to mildly relativistic speeds on a time scale of weeks–months (written as "Jet-ISM Shock"), and can also be produced on time scales of years from sub-relativistic ejecta interacting with interstellar matter (written as "Ejecta-ISM Shock"). Short-lived quasi-isotropic optical emission lasting $\sim$ several days (written as Kilonova) can also accompany the merger, powered by the radioactive decay of heavy elements synthesized in the ejecta. The figure is taken from Metzger and Berger (2012).

tons and electron-positron pairs is likely to be produced. Such a fireball is believed to launch gamma-ray bursts. Although the scenario is plausible, this hypothesis, i.e., the formation of a black hole surrounded by a massive hot disk/torus, subsequent copious neutrino emission, and high neutrino pair annihilation rate, has to be proved. For pursuing this problem theoretically, numerical relativity is probably the most promising approach (see sections 8.3 and 8.5.)

The mergers of binary neutron stars (and black hole-neutron star binaries) could also emit transient electromagnetic signals observable, if a substantial amount of the material is ejected from the system during the merger process [see a schematic figure 1.27 and for the pioneer proposal of scenarios, see, e.g., Li and Paczyński (1998); Kurkarni (2005); Metzger *et al.* (2010); Nakar and Piran (2011)]. The sources of one possible signal are radioactive *r*-process nuclei [Burbidge *et al.* (1957)], which could be produced from the neutron-rich material in the merger ejecta [Li and Paczyński (1998); Kurkarni (2005); Metzger *et al.* (2010); Goriely *et al.* (2011); Metzger and Berger (2012); Barnes and Kasen (2013); Kasen *et al.* (2013); Tanaka and Hotokezaka (2013)]. A fraction of the unstable nuclei produced subsequently decays in a short time scale and could heat up the ejecta, which emits ultra-violet, visible, and infrared light that may be observable by optical telescopes like LSST (Large Synoptic Survey Telescope). This radioactively powered transient phenomena is called "kilonova" or "macronova".

Another possible signal could be generated during the free expansion and the subsequent Sedov phase of the ejecta as a result of sweeping-up interstellar medium and forming blast waves [Nakar and Piran (2011)]. In this process running, the shocked material at the blast waves could amplify magnetic fields, which subsequently accelerate particles that emit synchrotron radiation in the radio-wave band, for a hypothetical amplification of the electromagnetic field and a hypothetical electron injection.

In these two scenarios, the electromagnetic signals are likely to be emitted in a quasi-isotropic manner. Therefore, they should be observed from any viewing angles. This property is in clear contrast to gamma-ray bursts, which can be observed only from a particular direction (see figure 1.27).

A coincident detection of gravitational waves and these electromagnetic counterparts is particularly important for confirming the detection of gravitational waves from the merger of binary compact objects and for obtaining the information of the gravitational-wave source that cannot be found only from the gravitational-wave signal [e.g., Kochanek and Piran (1993)]. Here, a solid confirmation will be achieved if the observed electromagnetic signals are what we expect to be observed. This implies that theoretical studies for the electromagnetic signals are as important as those for the gravitational-wave signals. For this purpose, numerical relativity also plays a crucial role (see sections 8.3 and 8.5).

1.5.3 *Merger of binary black holes*

The merger of binary black holes is a highly nonlinear but relatively simple phenomenon, because the merger simply results in a new black hole. The new black hole soon after its formation is not in a stationary state. It subsequently settles to a stationary state after the emission of gravitational waves associated with the ringdown oscillation of quasi-normal modes [see equations (1.203) and (1.204)]. The merger process of coalescing binary black holes was first studied by Pretorius (2005a, 2006) in his numerical-relativity simulation. Figure 1.28 plots the gravitational waveform in the late inspiral and merger stages of an equal-mass binary non-spinning black

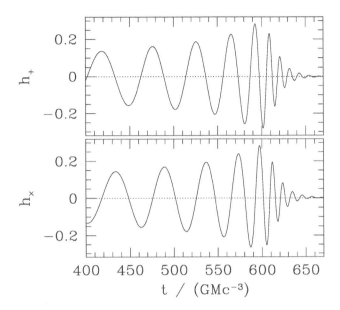

Fig. 1.28 Gravitational waves from the late inspiral and merger of an equal-mass binary non-spinning black hole. The computation was done by F. Pretorius. M is the total mass of the system. The horizontal axis shows the time in units of GM/c^3. h_+ and $h_\times$ denote the plus and cross modes of gravitational waves observed from the axis perpendicular to the orbital plane and these amplitudes are shown in units of $D_{\rm obs}/M$. Gravitational waves for $t/(GM/c^3) \gtrsim 600$ denote the ringdown waveforms associated with a fundamental quasi-normal mode. The data for this figure was kindly provided by F. Pretorius in 2006.

hole computed by him. This shows that gravitational waves are composed of the chirp, short-term merger, and ringdown signals.

The characteristic frequency of the ringdown waveforms is found from equation (1.204). For the merger of equal-mass binary black holes of no spins, it is well established that the final spin is $\hat{a} \approx 0.686$ [Buonanno et al. (2007a); Scheel et al. (2009)]. Assuming that each mass of the binary is $10 M_\odot$, the final mass of the formed black hole is approximately $19 M_\odot$ ($\sim 5\%$ of the total mass is radiated by gravitational waves). For such parameters, the oscillation frequency of the quasi-normal mode (the real part of the quasi-normal mode) is $\approx 900\,{\rm Hz}$, which is much lower than that for binary-neutron-star mergers in which it is $6-7\,{\rm kHz}$. Figure 1.28 shows that the amplitude of ringdown gravitational waves is approximately $h(D_{\rm obs}/m) \sim 0.1$. For a hypothetical distance of $D_{\rm obs} = 100\,{\rm Mpc}$, thus, the gravitational-wave amplitude (for the most optimistic case) is

$$h \approx 9 \times 10^{-22} \left(\frac{h D_{\rm obs}/m}{0.1} \right) \left(\frac{D_{\rm obs}}{100\,{\rm Mpc}} \right)^{-1} \left(\frac{m}{19 M_\odot} \right). \qquad (1.206)$$

This is much larger than that for binary-neutron-star mergers [compare with equation (1.205)]. We note that the amplitude increases and the characteristic frequency decreases as the total mass increases. Therefore, if $m \gtrsim 20 M_\odot$, the signal of ringdown gravitational waves will be a possible target detectable by advanced gravitational-wave detectors (see figure 1.5), and thus, the merger of high-mass black hole binaries is one of the most promising sources, probably the best source, for observing a formation process of a new black hole and a black hole in dynamical spacetime.

The wave shape of figure 1.28 is valid for any black-hole mass due to the scale-free nature of the vacuum system. For example, we may consider an equal-mass supermassive binary black hole for which each mass is $10^6 M_\odot$. In this case, the frequency of the quasi-normal mode becomes $\approx 9.0\,\text{mHz}$ and for a hypothetical distance of $D_{\text{obs}} = 10\,\text{Gpc}$,

$$h \approx 9 \times 10^{-19} \left(\frac{hD_{\text{obs}}/m}{0.1} \right) \left(\frac{D_{\text{obs}}}{10\,\text{Gpc}} \right)^{-1} \left(\frac{m}{1.9 \times 10^6 M_\odot} \right). \quad (1.207)$$

The frequency is so low that gravitational waves cannot be detected by ground-based detectors. However, these gravitational waves are promising sources for space interferometers such as eLISA. Taking a look at figure 1.7, it is found that such a signal can be detected by eLISA-type space interferometers with the signal-to-noise ratio $\gtrsim 10^2$.

There are increasing evidences that supermassive black holes exist in the central region of many of galaxies [e.g., Rees (1996); Magorrian et al. (1998); Ferrarese and Ford (2005)]. It is also believed that the mergers of galaxies have often occurred in the history of the universe. During a merger process of two galaxies, the supermassive black holes at their centers will eventually form a supermassive binary black hole. There are still many debates about whether such binaries can merge within the Hubble time. If we assume that the merger of supermassive binary black holes of mass $m \sim 10^6 M_\odot – 10^7 M_\odot$ may frequently occur in the center of merged galaxies and if a space interferometer is in operation, gravitational waves will be detected with an extremely high signal-to-noise ratio (see figure 1.7). They will be the invaluable sources for observing the formation process of a new black hole, the nonlinear nature of the black-hole merger, and behavior of a black hole in dynamical spacetime.

Gravitational waves from coalescing binary black holes could also carry rich information about nonlinear general relativistic phenomena, if black holes are rapidly spinning. In binaries composed of spinning black holes, the orbital plane in general precesses (see figure 1.29). This is in contrast with the orbit of binary neutron stars in which each neutron star is unlikely to be rapidly spinning. The precession of the orbital plane is induced primarily by the spin-orbit and spin-spin coupling effects [e.g., Kidder et al. (1993); Apostolatos et al. (1994); Kidder (1995); Faye et al. (2006)]. This occurs when the directions of the black-hole spin and orbital angular momentum vectors do not coincide. The time scale of the precession is in general proportional to a^3 which is longer than the orbital period ($\propto a^{3/2}$) but shorter than the gravitational-wave emission time scale ($\propto a^4$). Here, a is the

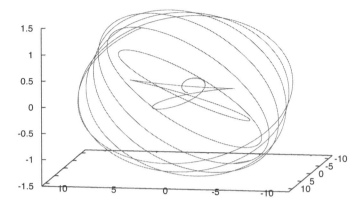

Fig. 1.29 Time evolution for the relative trajectory of spinning black holes calculated by Campanelli *et al.* (2009) in a post-Newtonian approximation.

orbital separation. Thus, for close orbits, the ratio of the precession time scale to the orbital period approaches unity, and the precession becomes a major feature of the orbital motion. Associated with this effect, the amplitude of gravitational waves emitted from binary spinning black holes could be significantly modulated. Detecting these gravitational waves enables us to observe a nonlinear general relativistic effect. For extracting the information of a large modulation from the signal of gravitational waves, it is necessary to prepare accurate theoretical templates. Numerical relativity plays a crucial role for this task (see section 8.7 for details).

A unique property in the merger of binary spinning black holes, which was recently discovered, is that the remnant black holes could have a large kick (recoil) velocity [e.g., Baker *et al.* (2006c); González *et al.* (2007b); Campanelli *et al.* (2007b); González *et al.* (2007a); Schnittman *et al.* (2008); Buchman *et al.* (2012)]. This kick velocity is induced by the back reaction (recoil) of the linear-momentum emission of gravitational waves [Fitchett (1983)]. The kick velocity for binaries of non-spinning black holes is at most 175 km/s [González *et al.* (2007b); Buchman *et al.* (2012)], and this maximum is achieved for unequal-mass binaries of a special mass ratio $\sim 1/3$. However, for binaries of spinning black holes, it may be faster than 1000 km/s, which is larger than the escape velocity of typical galaxies (several hundred km/s). Thus, after the merger of binary black holes, their remnant may escape from the galaxy. This suggests in particular an interesting possibility for the merger of supermassive black holes, because in some galactic centers, no supermassive black hole may exist, if the corresponding galaxies experienced the merger in their history.

1.5.4 *Merger of black hole-neutron star binaries*

After the merger of black hole-neutron star binaries, a variety of remnants is likely to be possible by contrast to those for binary black holes. The final fates of black hole-neutron star binaries are broadly separated into two categories; (i) a neutron star is tidally disrupted before it is swallowed by its companion black hole; (ii) a neutron star is simply swallowed by its companion black hole. There may be the third possibility, in which the so-called stable mass transfer occurs after the onset of mass shedding of the neutron star. Here, the mass shedding is a phenomenon that the neutron-star matter is stripped from its inner edge by the tidal field of its companion black hole, and in the stable mass transfer, this stripping occurs continuously for a time scale longer than the orbital period, avoiding the decrease of the orbital separation due to the gravitational-radiation reaction. Although this might be possible, numerical-relativity simulations performed so far have not found this phenomenon.

The final fate of a neutron star in a binary with a black hole is determined primarily by the mass and the spin of the companion black hole, the neutron-star mass, and the compactness of the neutron star, $\mathcal{C} := M_{NS}/R_{NS}$, where M_{NS} is the neutron-star mass and R_{NS} is the circumferential radius of the neutron star in isolation. According to nuclear-physics theories for the high-density matter, the compactness, $\mathcal{C}$, is in the range between ~ 0.13 and ~ 0.25 [Lattimer and Prakash (2001, 2004)] for the typical neutron stars of mass $1.2-1.5\,M_\odot$. For the case that the black-hole mass is small enough or the neutron-star radius is large enough, the neutron star will be tidally disrupted before it is swallowed by the black hole, irrespective of the black-hole spin. A necessary (but not sufficient) condition for this will be semi-quantitatively derived as follows.

Tidal disruption of a neutron star occurs after the onset of mass shedding by black-hole tidal force. Thus, the onset of the mass shedding is a necessary condition for tidal disruption. The mass shedding of a neutron star sets in when the tidal force of the black hole at a neutron-star inner surface is stronger than neutron-star's self-gravity. This condition is approximately (assuming Newtonian gravity) written as

$$\frac{M_{BH}(c_R R_{NS})}{a^3} \gtrsim \frac{M_{NS}}{(c_R R_{NS})^2}, \qquad (1.208)$$

where M_{BH} is the black-hole mass and a is the orbital separation again. We introduce a function, c_R, which denotes the degree of tidal deformation of the neutron star; we suppose that the semi-major axis is elongated as $c_R R_{NS}$. It is a function of a and is larger than unity; its dependence on a depends on the equation of state of the neutron star. The left-hand side of equation (1.208) approximately denotes the tidal force by the black hole at the neutron-star inner surface and the right-hand side is the self-gravitational force of the neutron star there.

We emphasize here that equation (1.208) is the condition for the onset of *mass shedding*, strictly speaking. The tidal disruption occurs after a substantial amount of mass is stripped from the neutron-star surface, during the decrease of the orbital separation due to the gravitational-radiation reaction. Thus, the tidal disruption should set in for a smaller orbital separation (larger orbital angular velocity) than

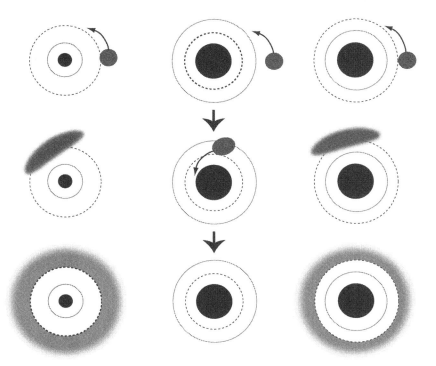

Fig. 1.30 Schematic pictures for three types of merger process of black hole-neutron star binaries. Left column: a neutron star is tidally disrupted and the extent of the tidally disrupted material is as large as or larger than the black-hole circumferential length. This is the case when the black-hole mass (or ratio of the black-hole mass to the neutron-star mass, Q) is not very large. Middle column: a neutron star is not tidally disrupted and simply swallowed by a black hole. This is the case when the black-hole mass (or Q) is large enough. Right column: a neutron star is tidally disrupted and the extent of the tidally disrupted material in the vicinity of the black-hole horizon is smaller than the black-hole circumferential length. This is the case when the black-hole mass is not small but its spin is large and corotating with the orbital motion; in this case, the surface area of the horizon is smaller than that with no spin. The filled black sphere denotes the black hole, the blue distorted ellipsoid denotes the neutron star, the solid (red) circle is the location of ISCO, and the dashed circle is the location of the tidal disruption limit, respectively. Note that tidal disruption should occur outside the ISCO.

that derived from equation (1.208). We also note that the neutron-star radius is assumed to depend weakly on the neutron-star mass (see figure 1.16). If the radius quickly *increases* with mass loss by the mass shedding, the tidal disruption may occur soon after the onset of the mass shedding.

Assuming that the binary is in a circular orbit with the Keplerian angular velocity, equation (1.208) may be written in terms of the angular velocity $\Omega(=\sqrt{m/a^3}$ with $m = M_{\rm BH} + M_{\rm NS})$ as

$$\Omega^2 \geq c_{\rm R}^{-3} \frac{M_{\rm NS}}{R_{\rm NS}^3}\left(1 + Q^{-1}\right), \qquad (1.209)$$

where Q denotes the mass ratio defined by $M_{\rm BH}/M_{\rm NS}$. A general relativistic numerical study for quasi-equilibrium states of black hole-neutron star binaries derives a more quantitative result [Taniguchi *et al.* (2007, 2008)] (see also section 8.4) as

$$\Omega^2 \geq C_\Omega^2 \left(\frac{M_{\rm NS}}{R_{\rm NS}^3}\right)(1+Q^{-1}), \qquad (1.210)$$

where the value of the dimensionless constant, C_Ω, is $\lesssim 0.3$ for a system composed of a neutron star with the irrotational velocity field and with $n=1$ polytropic equation of state, and of a non-spinning black hole. The small value of C_Ω (<1) indicates that mass shedding is enhanced by a significant tidal deformation (i.e., $c_R > 1$) and/or possibly by a general relativistic effect. Equation (1.210) indicates that the frequency of gravitational waves at mass shedding is

$$f = \frac{\Omega}{\pi} \geq 1.0\,{\rm kHz} \left(\frac{C_\Omega}{0.3}\right) \left(\frac{M_{\rm NS}}{1.4\,M_\odot}\right)^{1/2} \left(\frac{R_{\rm NS}}{12\,{\rm km}}\right)^{-3/2} \sqrt{1+Q^{-1}}. \qquad (1.211)$$

Thus, tidal disruption is likely to occur at higher frequency $f \gtrsim 1\,{\rm kHz}$ for a hypothetical neutron star of $M_{\rm NS} = 1.4\,M_\odot$ and $R_{\rm NS} = 12\,{\rm km}$ irrespective of the black-hole mass.

To the above paragraph, we have implicitly assumed that orbits with arbitrary orbital separations might be possible for the binary system. However, black hole-neutron star binaries should have an ISCO determined by the general relativistic effect, and hence, we should impose the condition that the mass shedding (and the tidal disruption) has to occur before the binary orbit reaches the ISCO in this analysis. According to a post-Newtonian analysis [e.g., Blanchet (2014) for a review], the dimensionless orbital compactness parameter, $m\Omega$, at the ISCO is ~ 0.10 for systems composed of a non-spinning black hole with $1 \leq Q \leq 5$. Note that the tidal-deformation effect could reduce this value by $\sim 10\text{--}20\%$ [Taniguchi *et al.* (2007, 2008)]. In the presence of the ISCO, the condition for the onset of mass shedding is written by

$$\mathcal{C}_{\rm ISCO} := m\Omega_{\rm ISCO} \geq m\Omega \geq C_\Omega \left(\frac{M_{\rm NS}}{R_{\rm NS}}\right)^{3/2} (1+Q)^{3/2} Q^{-1/2}, \qquad (1.212)$$

where $\Omega_{\rm ISCO}$ is the angular velocity at the ISCO. $\mathcal{C}_{\rm ISCO} \sim 0.1$ for a system composed of a non-spinning black hole, and could be up to ~ 0.5 for a spinning black hole, because the orbital radius at the ISCO around spinning black holes is much smaller than $6M_{\rm BH}$ (see section 1.3.2). Thus, the system composed of a large black-hole spin could be more subject to tidal disruption.

Equation (1.212) is rewritten to

$$\left(\frac{\mathcal{C}_{\rm ISCO}}{C_\Omega}\right)^{2/3} \frac{Q^{1/3}}{1+Q} \geq \frac{M_{\rm NS}}{R_{\rm NS}} = \mathcal{C}, \qquad (1.213)$$

where $Q^{1/3}/(1+Q)$ is a monotonically decreasing function of Q for $Q > 1/2$. This estimate suggests that tidal disruption of a neutron star by a non-spinning black hole (i.e., $\mathcal{C}_\text{ISCO} \sim 0.1$) could occur *for a binary of a relatively low mass ratio of* $Q \lesssim 5$ for $\mathcal{C} \sim 0.14$ and $Q \lesssim 2$ for $\mathcal{C} \sim 0.20$ assuming $C_\Omega \sim 0.3$. For a rapidly spinning black hole (in particular, for the case that the directions of the black-hole spin and orbital angular momentum vectors are parallel), the mass ratio allowed for tidal disruption could be much larger; e.g., for $\hat{a} \gtrsim 0.9$, Q may be larger than 10 even for $\mathcal{C} \sim 0.20$. Equation (1.213) also shows that the conditions for the onset of mass shedding and for tidal disruption depend strongly on the compactness of the neutron star. Figure 1.30 schematically summarizes the conditions for whether tidal disruption sets in or not. To sum up, the tidal-disruption condition is determined by the relation among the circumferential length of the black-hole horizon, the ISCO radius, and the neutron-star radius.

In the above simple estimate, the effect of tidal deformation of neutron stars to the orbital motion is not taken into account. As a result of the tidal deformation, the gravitational force between two objects is modified, so are the orbital evolution and the criterion for tidal disruption, as we mentioned in section 1.5.1. Rasio and Shapiro (1992); Lai *et al.* (1993a, 1994b,a); Lai and Wiseman (1996) found that the two-body attractive force is strengthen by the effect of the tidal deformation and, by this, the orbital separation of the ISCO is increased and also gravitational waveforms in the late inspiral stage are modified.

Gravitational waveforms in the merger stage depend strongly on the final fate. Figure 1.31 plots two typical waveforms for the late inspiral and merger stages. The upper panel plots the typical waveform for the case that the neutron star is tidally disrupted by its companion black hole. In this case, the amplitude of gravitational waves quickly damps during the late inspiral stage, because the degree of the non-axial symmetry of the system significantly decreases after the tidal disruption. The bottom panel plots the typical waveform for the case that the neutron star is not tidally disrupted by its companion black hole. In this case, the merger waveform is characterized by a ringdown oscillation associated with the fundamental quasi-normal mode of the formed black hole, and is qualitatively the same as that in the merger of binary black holes (compare with figure 1.28).

For the case that tidal disruption occurs, the highest gravitational-wave frequency is recorded at the moment of the tidal disruption. As shown in equation (1.211), this frequency depends strongly on the radius (or compactness) of the neutron star (for given values of mass and spin of the black hole and mass of the neutron star). This suggests that the neutron-star radius may be constrained by observing gravitational waves emitted at the tidal disruption. In reality, the frequency depends also on other parameters, especially strongly on the black-hole spin and inclination angle of this spin to the orbital plane. However, if these quantities are determined, the neutron-star radius may be strongly constrained. Exploring this possibility is one of the major subjects in numerical relativity. More details based on the results of numerical-relativity simulations will be described in section 8.5.

Tidal disruption could also accompany a short gamma-ray burst and a mass ejec-

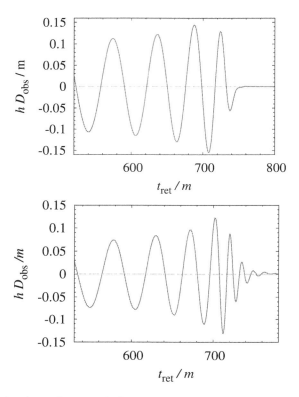

Fig. 1.31 Gravitational waveforms in the late inspiral and merger stages of binaries composed of a non-spinning black hole and a neutron star. For the upper panel, $Q = 2$ and $\mathcal{C} = 0.145$. For such small values of Q and $\mathcal{C}$, the neutron star is tidally disrupted during the late inspiral stage, and the gravitational-wave amplitude quickly damps at the disruption (see for $t_{\rm ret}/m \gtrsim 740$). For the bottom panel, $Q = 5$ and $\mathcal{C} = 0.145$. For such a large value of Q, the neutron star is not tidally disrupted and simply swallowed by the black hole. In this case, the merger waveform is characterized by a ringdown oscillation associated with the fundamental quasi-normal mode of the formed black hole (see figure 1.28). Note that m denotes the total mass, and for $Q = 2$ and $Q = 5$, plausible values are $m \approx 4 M_\odot$ and $\approx 8 M_\odot$, respectively, and thus, the units of the time are ≈ 0.02 ms and 0.04 ms in the upper and lower panels.

tion that emits an observable electromagnetic signal. As described in section 1.5.2, exploring such electromagnetic counterparts is as important as exploring gravitational waveforms in numerical relativity (see section 8.5).

1.5.5 *Gravitational collapse and core collapse supernova*

The gravitational collapse and subsequent supernova explosion of massive stellar cores are among the promising sources of gravitational waves, and a variety of studies have been performed on this subject. Several types of burst gravitational waves are expected to be emitted in the collapse and explosion processes. In the following, we will first outline the standard scenario of the gravitational collapse

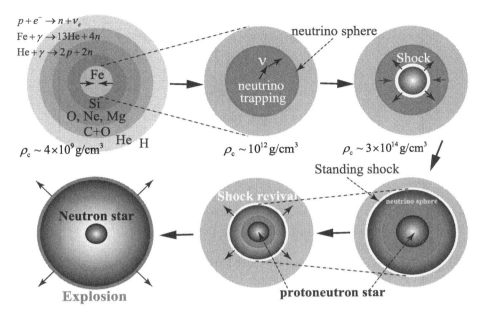

Fig. 1.32 Standard scenario for the processes of gravitational collapse and subsequent supernova explosion. In the final stage of a massive star with its initial mass $\gtrsim 10 M_\odot$, the iron core collapses to a proto-neutron star. A shock wave is formed at the surface of the proto-neutron star, and then, propagates outward. However, the shock stalls in the middle of the expansion. Some mechanism such as neutrino heating should activate the shock to cause the supernova explosion, although the exact mechanism has not yet been resolved.

and the supernova explosion; see figure 1.32 [see Bethe (1990) if the readers want to study these processes from the basics in depth]. Then, we will briefly summarize the possible gravitational-wave signals that could be emitted during the collapse and the explosion.

The gravitational collapse of a massive star with its initial mass $\gtrsim 10 M_\odot$ is triggered by the partial photo-dissociation of iron in its core as

$$^{56}\text{Fe} + \gamma \to 13 ^4\text{He} + 4n. \tag{1.214}$$

For the case that the initial mass of a massive star is slightly smaller than $10 M_\odot$ (in the range $\sim 9\text{--}10 M_\odot$), the following electron capture process of heavy nuclei triggers the collapse of the core composed of oxygen, neon, and magnesium,

$$(Z, A) + e^- \to (Z-1, A) + \nu_e. \tag{1.215}$$

Here, A and Z denote the mass and charge numbers of the heavy nuclei. These two processes reduce thermal energy and electron degenerate pressure from the core, and as a result, the core is destabilized, leading to the gravitational collapse.

At the onset of the collapse, the central density, temperature, and specific entropy are approximately $\sim 4 \times 10^9 \text{ g/cm}^3$, $\sim 10^{10}$ K, and $\sim k_\text{B}$ [e.g., Bethe (1990)].

Then, the dominant pressure source is degenerate electrons, because the gas, radiation, and electron-degenerate pressures are written, respectively, by

$$P_{\text{gas}} = 5.9 \times 10^{25} \text{dyn/cm}^2 \left(\frac{\rho}{4 \times 10^9 \text{ g/cm}^3}\right) \left(\frac{T}{10^{10} \text{ K}}\right) \left(\frac{A}{56}\right)^{-1}, \quad (1.216)$$

$$P_{\text{rad}} = 2.6 \times 10^{25} \text{dyn/cm}^2 \left(\frac{T}{10^{10} \text{ K}}\right)^4, \quad (1.217)$$

$$P_e = 2.8 \times 10^{27} \text{dyn/cm}^2 \left(\frac{\rho}{4 \times 10^9 \text{ g/cm}^3}\right)^{4/3} \left(\frac{Y_e}{26/56}\right)^{4/3}, \quad (1.218)$$

where we assumed that the core is composed of iron, electrons, and photons. This indicates that the electron capture process, (1.215), plays an important role. However, other pressure sources are also not negligible, and thus, the depletion of the thermal energy by the photo-dissociation of heavy nuclei also plays an important role for destabilizing the core.

In an early stage of the gravitational collapse for which the central density ρ_c is smaller than $\sim 10^{11}$ g/cm^3, neutrinos produced by the electron capture freely escape from the star and subtract thermal energy from the collapsing core. Furthermore, the electron capture reduces the number of degenerate electrons (and thus, the neutronization proceeds). Due to these effects, the collapse is accelerated. However, for $\rho_c \gtrsim 3 \times 10^{11}$ g/cm^3, neutrinos cannot freely escape from the core because the cross section of neutrinos for a coherent scattering with heavy nuclei is quite large: At such high density, the time scale for a neutrino to escape from the core by the diffusion process becomes longer than the dynamical time scale of the system, $\sim \rho_c^{-1/2}$. Then, neutrinos are effectively trapped [Sato (1975a,b)]. As a result of this neutrino trapping, neutrinos are in degenerate states. This blocks the further electron capture, and hence, the electron fraction per baryon, Y_e (and thus, the number ratio of protons to neutrons) is approximately frozen in the collapsing core after the onset of the neutrino trapping (see figure 9.11). In addition, the Fermi energy of degenerate neutrinos gradually increases during the collapse. Since the cross section between baryons and neutrinos is approximately proportional to the square of the neutrino energy, the opacity of neutrinos is enhanced. This effect is also important for the continuous neutrino trapping during the collapse. In the presence of the neutrino trapping, any efficient cooling process is absent. Thus, for $\rho_c \gtrsim 3 \times 10^{11}$ g/cm^3, the collapse proceeds in an (approximately) adiabatic manner.

The stellar core collapse is halted when the central density exceeds the nuclear density ($\rho_{\text{nuc}} \approx 2.8 \times 10^{14}$ g/cm^3): For such high density, the repulsive force associated with the strong interaction among nucleons is strong enough to significantly enhance the nuclear-matter pressure. Then, a core bounce occurs and a proto-neutron star is formed. At the bounce, strong sound waves propagate outwards, and when they arrive at a region for which the inward flow is supersonic, a shock, which is the key for the supernova explosion, is formed and starts propagating outwards. The subsequent evolution of the shock wave and the formed proto-neutron star as well as possible emission mechanisms of gravitational waves will be described as follows.

First, we briefly summarize the current understanding for the core-collapse supernova explosion mechanism. To date, the precise explosion mechanism of this type of supernovae is still poorly known, although a variety of numerical simulations have been performed in the past ~ 50 years since the first work by Colgate and White (1966). A consensus reached so far is that the explosion cannot occur for the iron core collapse,[11] if the supernova process proceeds in the spherically symmetric manner. This fact is based on a number of spherical symmetric simulations performed independently by several groups incorporating detailed microphysics and neutrino radiation transfer, that have shown the failure of the supernova explosion [see, e.g., Rampp and Janka (2000); Mezzacappa et al. (2001); Bruenn et al. (2001); Liebendörfer et al. (2001b); Sumiyoshi et al. (2005) for the latest works; but see also Wilson (1985); Bruenn (1985) for the pioneer works]. In all the spherically symmetric simulations, the shock formed at the core bounce stalls during its propagation within ~ 100 ms and becomes a standing accretion shock. This is due to the effects that (i) during the propagation of the shock, a substantial amount of its thermal energy is consumed to photo-dissociate iron that is the main component of the infalling matter in the early time after the core bounce and (ii) neutrinos carry the thermal energy of the shock-heated matter located behind the shock (i.e., the thermal pressure inside the shock radius is reduced). Subsequently, the heating by neutrinos, which are emitted from the proto-neutron star, sustains the standing accretion shock for a while. Up to this stage, the process is believed to be universal irrespective of the geometrical symmetry imposed. However, the neutrino heating alone is not efficient enough to drive the explosion in spherical symmetry, and thus, as the accretion of the matter from the outer part of the progenitor star proceeds, the standing accretion shock eventually goes back to the proto-neutron star; the explosion fails. This strongly suggests that some other ingredients such as non-spherical processes are necessary for the successful explosion. Because the asymmetry would be the key for the explosion, gravitational waves are likely to be emitted during the gravitational collapse and the supernova explosion processes.

There are several candidates for the key mechanisms for inducing the non-spherical deformation. One is the rotation that the progenitor has. By this, the collapsing core is deformed by the centrifugal force, in particular at the core bounce at which the maximum central density is achieved. Thus, the rotation plays an important role for enhancing the emissivity of gravitational waves at the core bounce (see section 1.5.5.1). However, it should be noted that this mechanism could play an important role only when the progenitor core is rapidly rotating. For ordinary supernovae, the progenitors may not be rapidly rotating because the spin period

---

[11] For the collapse of the oxygen-neon-magnesium core, this is not the case [Kitaura et al. (2006)] because the progenitor star has a less compact structure only with a small amount of the envelope mass that can be exploded relatively easily.

of the observed new-born neutron stars is $\gtrsim 10$ ms, which is much longer than the possible minimum one $\lesssim 1$ ms [Lyne and Graham-Smith (2005)] [see also Woosley *et al.* (2005)]. Alternatively, for the collapse of massive cores that eventually produce a central engine of a gamma-ray burst composed of a rotating black hole surrounded by a massive torus, a massive and rapidly rotating proto-neutron star may be formed. For such a case, high-amplitude gravitational waves associated with the rapid rotation may be emitted.

The most important and universal mechanism for inducing non-spherical motion is neutrino-driven convection, which is believed to be always active in the post bounce stage because the proto-neutron star and the surrounding region (inside the standing accretion shock) often possess a negative entropy gradient and a negative lepton-fraction gradient due to neutrino emission processes [e.g., Burrows and Fryxell (1992); Janka (2001) and section 1.5.5.2]. In the presence of these negative gradients, criteria for the onset of the convective instability, i.e., the Schwarzschild criterion for the pure entropy gradient or the Ledoux criterion for the entropy and composition gradients, could be satisfied [see, e.g., chapter 6 of Kippenhahn and Weigert (1994) for several kinds of the convective instability in the context of stellar physics]. By the convection, hot material is carried from the inner region to the outer one in the proto-neutron star. This could increase the temperature of the neutrino sphere and could enhance the neutrino luminosity (see figure 1.34). Also, the convection that occurs just behind the standing accretion shock could supply a significant power to this stalled shock (see section 1.5.5.2), and in addition, could activate the non-spherical motion of the fluid outside the proto-neutron star, contributing to a substantial amount of the gravitational-wave emission and the anisotropic emission of neutrinos. This anisotropic neutrino emission also plays a role for the emission of gravitational waves (see section 1.5.5.3).

The effects of the neutrino-driven convection have been investigated by non-spherically symmetric (primarily axisymmetric) simulations in Newtonian gravity since the pioneer works by Burrows and Fryxell (1992); Herant *et al.* (1994). These simulations showed that the convection indeed plays a key role in activating the standing accretion shock. However, they also showed that the heating time scale by neutrinos enhanced by the convection is not as short as the advection time scale of the matter infalling onto the proto-neutron star [Buras *et al.* (2006)]. Thus, only by the effects of the convection and the neutrino heating, the standing accretion shock does not revive.

Another novel mechanism, which was discovered in 2000s and has been actively studied since then, is the so-called standing accretion shock instability (SASI) [Blondin *et al.* (2003); Marek and Janka (2009)]. This is a different mechanism from convection, and it could help supplying energy to the standing accretion shock (see section 1.5.5.2 for its mechanism). This also could induce highly non-spherical motion inside the standing accretion shock and could trigger the emission of gravitational waves. The numerical simulations have shown [e.g., Burrows *et al.* (2007); Marek and Janka (2009); Suwa *et al.* (2010) for the pioneer works] that the SASI could significantly help the standing accretion shock to revive and could lead to a success in the supernova explosion, although subsequent three-dimensional numerical simulations suggest that the asymmetry would not be very pronounced in the absence of the axisymmetry [see, e.g., Takiwaki *et al.* (2012); Hanke *et al.* (2013) for the pioneer self-consistent simulations including the neutrino radiation transfer]. Overall, the numerical simulations performed so far suggest that the SASI could be one of the important mechanisms for the supernova explosion. Nevertheless, this may not be the final answer because the numerical simulations have shown that (i) the explosion can occur only for a restricted class of progenitor models [see, e.g., Ugliano *et al.* (2012) for the possible reason], and for many models (even for low-mass models), the standing accretion shock does not revive; (ii) even for the case that the explosion occurs, its energy is often by one-order of magnitude smaller than the typical explosion energy observed $\sim 10^{51}$ ergs. Moreover, fully general relativistic simulations with a sophisticated treatment of the neutrino radiation transfer have not been done yet (but see Kuroda *et al.* (2012); Ott *et al.* (2013) for the first numerical-relativity simulations with approximate cooling and heating by neutrinos).

To summarize, the exact explosion mechanism has not yet been known, although it is very likely that certain non-spherical mechanisms are the keys for the explosion. We note that all the numerical simulations so far have been performed with an approximate treatment of some of physical processes. For example, in most of the multi-dimensional numerical simulations, neutrino radiation transfer is solved only in an approximate manner and general relativity is not strictly taken into account. In addition, due to the restriction of the computational resources, most of the simulations have been done in the assumption of axial symmetry. Hence, we should keep in mind that this field is still in progress. The final answer has to be awaited until a more detailed simulation taking into account all the important physical processes is performed.

1.5.5.1 *Gravitational waves in the core bounce*

After the core collapse of a massive star, a proto-neutron star is formed. At its formation, the proto-neutron star quasi-radially oscillates in a quasi-periodic manner because the strong restoring force due to the repulsive nuclear force induces such an oscillation. If the collapse occurs in a non-spherical manner, gravitational waves are emitted by this oscillation (see figure 1.33 for typical waveforms). The order of the gravitational-wave amplitude is estimated approximately by the quadrupole formula (1.44) with

$$\ddot{I}_{ij}\Big|_{\max} \sim \kappa_{\mathrm{I}} \delta_{\mathrm{as}} M R^2 (2\pi f)^2, \qquad (1.219)$$

where M and R are the mass and the radius of the proto-neutron star, and f is the typical frequency of quasi-periodic gravitational waves. κ_{I} is a dimensionless quan-

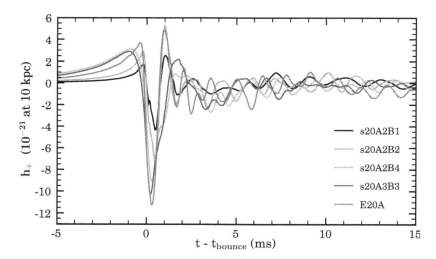

Fig. 1.33 Typical gravitational waveforms emitted in the core-bounce stage. The figure is taken from Ott et al. (2007a).

tity $\approx \sum_{i=x,y,z} I_{ii}/MR^2$, which is determined by the density profile of the proto-neutron star and is of $O(0.1)$. δ_{as} denotes a degree of non-spherical deformation of the proto-neutron star. Assuming that the collapse proceeds in an axisymmetric manner, the gravitational-wave amplitude observed in the most optimistic direction is written as

$$h_{\max} \sim 10^{-21} \left(\frac{10\,\mathrm{kpc}}{D_{\mathrm{obs}}}\right)\left(\frac{\kappa_{\mathrm{I}}}{0.1}\right)\left(\frac{\delta_{\mathrm{as}}}{0.01}\right)\left(\frac{M}{1.4M_\odot}\right)\left(\frac{R}{20\,\mathrm{km}}\right)^2\left(\frac{f}{1\,\mathrm{kHz}}\right)^2, \quad (1.220)$$

where D_{obs} is a hypothetical distance to the source and δ_{as} is defined here by $|1 - 2I_{zz}/(I_{xx} + I_{yy})| (\leq 1)$. This deformation parameter is zero in spherical symmetry and can become $O(0.1)$ when the collapsing core is rapidly rotating and the spheroidal deformation is enhanced. Thus, gravitational waves associated with the core bounce could be an important source in particular for a rapidly rotating progenitor. This has been confirmed by a variety of numerical simulations, since the first work by Müller (1982) [see also the general relativistic works, e.g., by Dimmelmeier et al. (2002); Shibata and Sekiguchi (2004, 2005b); Ott et al. (2007a); Dimmelmeier et al. (2007, 2008)].

The typical value of f is estimated from the dynamical time scale of the proto-neutron star as

$$f \sim \frac{1}{2\pi}\sqrt{\frac{M}{R^3}} \approx 770\,\mathrm{Hz}\left(\frac{M}{1.4M_\odot}\right)^{1/2}\left(\frac{R}{20\,\mathrm{km}}\right)^{-3/2}. \quad (1.221)$$

Equations (1.220) and (1.221) show that gravitational waves may be in a sensitive band of advanced detectors if an event occurs in our Galaxy (see figure 1.5).

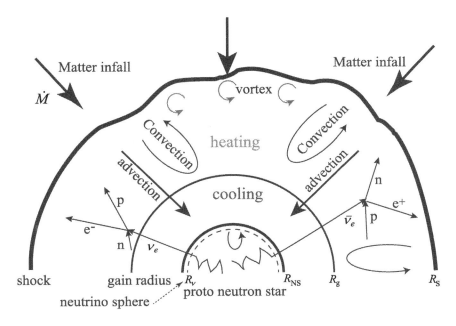

Fig. 1.34 A schematic figure, summarizing major processes that occur during the evolution of the standing accretion shock formed after the core bounce. See the text for details.

1.5.5.2 *Gravitational waves from convective motion and the resulting g-mode oscillation of proto-neutron stars*

In the first part of this subsection, we will describe a popular and promising mechanism of the supernova explosion, the so-called neutrino-heating mechanism, in more detail, and subsequently, summarize possible emission mechanisms of gravitational waves in this scenario.

The structure composed of a proto-neutron star and matter surrounding it formed after the formation of the standing accretion shock was clearly depicted by Janka (2001) (see figure 1.34 for a slightly modified version). During this stage, the stellar matter falls from the outer region onto the stalled standing accretion shock at radius $R_\mathrm{S}(\approx 100\text{--}200\,\mathrm{km})$ with a mass accretion rate $\dot{M}$ and with an approximately free-fall velocity ($\sim 0.1c$). The standing accretion shock is usually deformed due to hydrodynamics instability (see below). After the deceleration at the shock, the infalling matter is advected towards the proto-neutron star with the velocity much smaller than the free-fall one. The radius of the proto-neutron star, $R_\mathrm{NS}(\approx 20\text{--}40\,\mathrm{km})$, is defined by a radius at which the density steeply decreases. R_NS is approximately as large as the radius of the neutrino sphere denoted by R_ν. For $r < R_\nu$, the neutron-star matter is optically thick to neutrinos, and hence, a thermal and chemical equilibrium approximately holds. Here r denotes the distance from the center.

Heating by neutrinos emitted from the vicinity of the neutrino sphere balances cooling by the neutrino emission at the gain radius $R_g (\approx 50\text{--}100\,\text{km})$ located outside the neutrino sphere: The cooling is dominant for $R_\nu < r < R_g$, while for $r > R_g$, the heating is dominant. The reason that the gain radius appears is qualitatively explained as follows: For $r > R_\nu$, the heating rate is proportional to the flux of neutrinos emitted from the proto-neutron star, and thus, it is proportional to r^{-2}. On the other hand, the cooling rate is approximately proportional to T^6 with T the temperature of hot matter surrounding the proto-neutron star. Outside the neutrino sphere, T is known to be broadly proportional to r^{-1}, and hence, the cooling rate is approximately proportional to r^{-6}. These facts imply that in an inner region, the cooling would be dominant while the heating could be dominant for an outer region. In particular just outside the gain radius, the neutrino heating most efficiently works, because the neutrino flux is larger for the inner region.

In the situation described above, the convective motion is activated in two regions. One is located between the gain radius and the standing accretion shock. This is triggered by the buoyancy of the matter in the heating-dominant region just outside the gain radius. This convection is usually called neutrino-driven convection. By this convection, the layer between the gain radius and the standing accretion shock is mixed. This could help the revival of the shock propagation. The other region is located in the vicinity of the neutrino sphere. This convection is caused by a gradient of the entropy and lepton fractions. The negative entropy gradient is generated by the copious neutrino emission in a very early stage just after the shock wave propagates outward following the core bounce. This plays a role only for the early stage of the proto-neutron star evolution. On the other hand, the lepton fraction gradient is generated by the neutrino emission near the neutrino sphere that removes the lepton number. The convection induced by the former and latter mechanisms are called *prompt convection* and *proto-neutron-star convection*, respectively. These convection activities carry hot matter from the inside to the outside of the proto-neutron star, and enhances the neutrino luminosity. However, the effect by them is relatively weaker than the neutrino-driven convection which occurs behind the stalled standing accretion shock, because the time duration for the prompt and proto-neutron star convection is rather short.

As already outlined, the convective activity alone is unlikely to launch a supernova explosion. The SASI [Blondin *et al.* (2003); Marek and Janka (2009)] is one of the likely key mechanisms for reactivating the stalled shock. This instability is characterized by globally non-spherical motion and deformation of the standing accretion shock. Axisymmetric numerical simulations [Burrows *et al.* (2007); Marek and Janka (2009); Suwa *et al.* (2010)] showed that the primary deformation modes are dipole modes. This suggested that the SASI could enhance the gravitational-wave emission. Murphy *et al.* (2009); Yakunin *et al.* (2010); Müller *et al.* (2012) indeed showed that this could be the case. However, the subsequent three-dimensional simulations showed that the degree of the asymmetry by the

SASI is not as high as that in the axisymmetric simulations [Takiwaki *et al.* (2012); Hanke *et al.* (2013)], although we have to await more detailed three-dimensional simulations in the future.

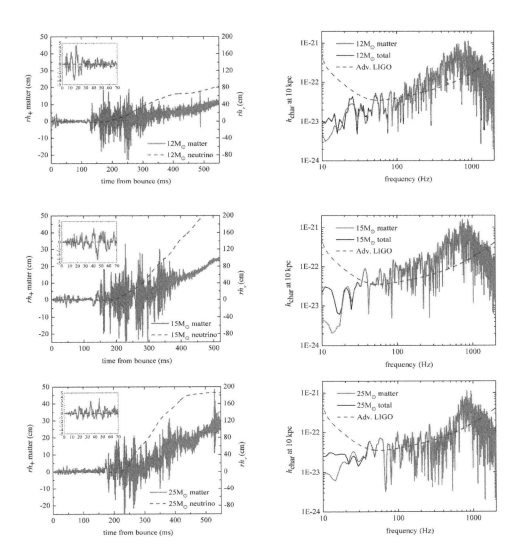

Fig. 1.35 Left panels: Gravitational waveforms emitted in the core collapse and subsequent evolution for non-rotating progenitors of initial mass 12 (upper), 15 (middle), and $25M_\odot$ (bottom panels). The dashed and solid curves denote gravitational waves emitted by the neutrino emission and matter motion, respectively. The inset shows the gravitational waveforms in an early phase. The vertical axis shows $hD_{\rm obs}$ in units of cm; i.e., the amplitude of gravitational waves observed along the most optimistic direction is $h \approx 10^{-21} \, (hD_{\rm obs}/31\,{\rm cm}) \, (10\,{\rm kpc}/D_{\rm obs})$. Right panel: Gravitational-wave spectrum for a hypothetical distance of $D_{\rm obs} = 10\,{\rm kpc}$. This figure is taken from Yakunin *et al.* (2010).

The gravitational-wave emission is enhanced by the anisotropic motion associated with the convection and the SASI mentioned above [e.g., Burrows and Hayes (1996); Müller and Janka (1997); Müller et al. (2004); Murphy et al. (2009); Scheidegger et al. (2010); Yakunin et al. (2010); Müller et al. (2012, 2013)]. Figure 1.35 displays typical gravitational waveforms and their spectra in the core collapse of non-rotating progenitors and in the subsequent evolution of proto-neutron stars, computed by Yakunin et al. (2010). As they described clearly in their numerical simulations, there are four stages for the gravitational-wave emission associated with the matter motion. (1) A weak prompt signal is emitted in the first ~ 100 ms after the core bounce. This is generated by the prompt convection activated inside proto-neutron stars and by the infalling matter for which the trajectory is anisotropic because of the anisotropy induced by the convection. (2) A quiescent stage is followed after the prompt signal. The reason for this is that the prompt convection ceases and no other activity is strong enough during this stage. (3) A subsequent long-term signal is emitted until the explosion is achieved (or collapse to a black hole). This main part of the signal is induced primarily by a g-mode oscillation of the proto-neutron star [Müller et al. (2013)] which is excited by the infalling matter that hits the proto-neutron star in an anisotropic manner due to the SASI and convective motion. (4) When the explosion is succeeded, the matter explodes in an anisotropic manner with a certainly large velocity. Then, the components of I_{ij} have nonzero values for which the order is typically

$$\frac{G}{c^4}\delta M_{\text{exp}} v_{\text{exp}}^2 \sim 100 \, \text{cm} \left(\frac{\delta M_{\text{exp}}}{0.1 M_\odot}\right) \left(\frac{v_{\text{exp}}}{0.1 c}\right)^2, \quad (1.222)$$

where δM_{exp} denotes the mass that contributes to the anisotropic explosion. This results in a memory-type signal associated with the matter motion.

In addition to gravitational waves induced by the matter motion, memory-type gravitational waves associated with the anisotropic neutrino emission may be emitted. The dashed curves in the left panels of figure 1.35 show such gravitational waveforms. This topic will be touched in section 1.5.5.3.

The strongest signal is excited in the stage (3), and hence, we describe more details on this. As clarified by Müller et al. (2013), this signal is induced by the g-mode oscillation excited by the SASI and convective motion. Indeed, the dominant frequency of gravitational waves $\sim 500-1000$ Hz (see the Fourier spectrum of figure 1.35) agrees approximately with the Brunt–Väisälä frequency of the proto-neutron star that is the frequency associated with the buoyancy (i.e., g-mode) in the convectively stable region. Thus, gravitational waves in this stage are emitted by a coherent oscillation of the proto-neutron star. Then, the order of the gravitational-wave amplitude is estimated by equations (1.44) and (1.219) as

$$h_{\text{max}} \sim 10^{-21} \left(\frac{10 \, \text{kpc}}{D_{\text{obs}}}\right) \left(\frac{\kappa_{\text{I}}}{0.1}\right) \left(\frac{\delta_{\text{as}}}{0.1}\right) \left(\frac{\delta M}{0.1 M_\odot}\right) \left(\frac{R}{20 \, \text{km}}\right)^2 \left(\frac{f}{1 \, \text{kHz}}\right)^2, \quad (1.223)$$

where δM denotes a part of the mass of the proto-neutron star that contributes to the g-mode oscillation. Several numerical simulations in the framework of Newtonian gravity [Burrows and Hayes (1996); Müller and Janka (1997); Müller et al. (2004); Murphy et al. (2009); Scheidegger et al. (2010); Yakunin et al. (2010); Müller et al. (2012)] and in an approximate treatment of general relativity [Müller et al. (2013)] indeed have shown that the typical amplitude is very approximately given by equation (1.223).

The right panels of figure 1.35 show the gravitational-wave spectrum for the waveforms shown in the left panels. The spectrum universally has a peak at 700–800 Hz. This peak frequency is determined by the g-mode oscillation of the proto-neutron stars. Since the spectrum peak is determined by the g-mode of the proto-neutron stars, the gravitational-wave spectrum depends sensitively on the stellar structure. Müller et al. (2013) showed that the g-mode frequency derived in their Newtonian simulations is significantly different from that in their (approximate) general relativistic simulations. This implies that for deriving an accurate gravitational waveform and spectrum, a numerical-relativity simulation is absolutely necessary in the future.

1.5.5.3 Gravitational waves by anisotropic neutrino emission

Epstein (1978) first pointed out that gravitational waves are emitted not only by the bulk matter motion but also by the anisotropic emission of a huge amount of neutrinos. A computation of gravitational waves emitted by neutrinos in a multi-dimensional hydrodynamics simulation was first performed by Burrows and Hayes (1996); Müller and Janka (1997).

For a hypothetical observation distance $D_{\rm obs}$, the gravitational-wave amplitude by the neutrino emission is written as [Müller and Janka (1997)]

$$h_{ij}^{\rm TT}(x^i, t) = \frac{4}{D_{\rm obs}} \int_{-\infty}^{t-D_{\rm obs}} dt' \int d\Omega' \frac{(\bar{n}_i \bar{n}_j)^{\rm TT}}{1 - \cos\theta} \frac{dL_\nu(\Omega', t')}{d\Omega'}, \qquad (1.224)$$

where θ is the angle between the line of sight toward the observer and the direction Ω' of the neutrino emission, and $dL_\nu(\Omega', t')/d\Omega'$ denotes the direction-dependent neutrino luminosity (the energy radiated at time t' per unit time and per unit solid angle). The angular coordinate Ω' is measured by the source frame. $\bar{n}_i$ is a unit normal spatial vector in the observer frame. The meaning of TT is the same as in equation (1.44). Equation (1.224) implies that a neutrino pulse, which was emitted at $t - D_{\rm obs}$ and has passed through a detector at t, induces a change in the gravitational-wave amplitude. Thus, the gravitational waveform associated with this effect does not have a oscillatory one. Rather, the amplitude increases or decreases in a monotonic manner, if neutrinos are emitted to a particular direction in an anisotropic manner. Therefore, gravitational waves associated with the neutrino emission primarily have the so-called memory feature. This implies that there is no characteristic frequency except for that associated with the duration of the neutrino emission.

An order estimate for the gravitational-wave amplitude is given from equation

(1.224) by

$$h_{\max} \sim 2 \times 10^{-20} \left(\frac{10\,\text{kpc}}{D_{\text{obs}}}\right) \left(\frac{\Delta E_\nu}{10^{52}\,\text{ergs}}\right) \left(\frac{\epsilon_\nu}{0.1}\right), \tag{1.225}$$

where ΔE_ν and ϵ_ν denote the total amount of the energy emitted by neutrinos and a degree of anisotropy of the neutrino emission. This suggests that the amplitude of gravitational waves is as high as that emitted by the core bounce and convective motion. The dashed curves in the left panels of figure 1.35 depict an example of the gravitational waveforms induced by the anisotropic neutrino emission. This clearly shows that gravitational waves emitted after the core bounce are composed of two major ingredients; those by the matter motion and by the neutrino emission.

As shown by the order estimates in sections 1.5.5.1, 1.5.5.2, and 1.5.5.3, three ingredients induce the emission of gravitational waves of approximately the same order of amplitude. As shown in the right panels of figure 1.35, a characteristic feature of the spectrum is that it has a peak at $f \lesssim 1\,\text{kHz}$ together with a rather broad shape for $f \lesssim 500\,\text{Hz}$, if the gravitational-wave signal by the neutrino emission is strong. This broad feature is associated with the memory-type gravitational-wave emission and universally seen in any numerical simulations [e.g., Müller et al. (2004); Murphy et al. (2009); Yakunin et al. (2010); Scheidegger et al. (2010)].

1.5.5.4 Gravitational waves from rapidly rotating proto-neutron stars

Rapidly rotating stars are subject to non-axisymmetric rotational instability. An exact treatment for this exists only for incompressible equilibrium fluids in Newtonian gravity [see, e.g., Chandrasekhar (1969); Shapiro and Teukolsky (1983)]. For these configurations, global rotational instability arises from non-radial toroidal modes $e^{im\varphi}$ ($m = \pm 1, \pm 2, \ldots$) when $\beta := T_{\text{rot}}/W$ exceeds a certain critical value. Here φ is the azimuthal coordinate and T_{rot} and W are the rotational kinetic energy and gravitational potential energy, respectively. For the incompressible and rigidly rotating model, the fastest growing mode is the so-called bar mode with $|m| = 2$. For the case that the degree of differential rotation is quite high, the $|m| = 1$ mode could become the fastest growing mode [Centrella et al. (2001)].

There exist two different mechanisms and corresponding time scales for the bar-mode instability. Rigidly rotating, incompressible stars in Newtonian theory are *secularly* unstable to bar-mode formation when $\beta \geq \beta_s \approx 0.14$ [Chandrasekhar (1969)]. However, this instability can grow only in the presence of some dissipative mechanism, like gravitational radiation, and the growth rate of the bar-mode perturbation is determined by the dissipative time scale, which is usually much longer than the dynamical time scale of the system [e.g., Chandrasekhar (1970); Friedman and Schutz (1978)]. By contrast, *dynamical* instability to bar-mode deformation sets in when $\beta \geq \beta_d \approx 0.27$ [Chandrasekhar (1969)]. This instability is independent of any dissipative mechanisms, and the growth rate is determined purely by the hydrodynamical time scale of the system.

The secular instability in compressible stars, both rigidly and differentially rotating, has been analyzed numerically within linear perturbation theory, by means of a variational principle with trial functions, by solving the eigenvalue problem, or by other approximate means. For relativistic stars, the critical value of β_s depends on the compactness M/R of the star (where M is the gravitational mass and R the circumferential radius at the equator), on the rotation law, and on the dissipative mechanism. The gravitational-radiation driven instability occurs for smaller rotation rates, i.e. for values $\beta_s < 0.14$, in general relativity: For extremely compact stars [Stergioulas and Friedman (1998)] or strongly differentially rotating stars [Karino and Eriguchi (2003)], the critical value can be as small as $\beta_s < 0.1$. By contrast, viscosity drives this instability for higher rotation rates $\beta_s > 0.14$ as the configurations become more compact [Bonazzola et al. (1996)].

Determining the onset of the dynamical bar-mode instability, as well as the subsequent evolution of an unstable star, requires a numerical simulation for fully nonlinear hydrodynamics. A large number of simulations performed in Newtonian theory [e.g., Tohline et al. (1985); Houser et al. (1994); New et al. (2000)] and in general relativity [e.g., Shibata et al. (2000a); Baiotti et al. (2007)] have shown that β_d depends only very weakly on the stiffness of the equation of state. Once a bar has developed, the formation of spiral arms plays an important role for redistributing the angular momentum and forming a core-halo structure. This topic will be revisited in section 10.1.

In some numerical simulations, it has been shown that, similar to the onset of the secular instability, β_d can be much smaller than 0.27 for stars with a high degree of differential rotation [Tohilne and Hachisu (1990); Centrella et al. (2001); Shibata et al. (2003)]. For this case, the degree of global deformation for the unstable stars is not very high, and a moderately elliptic figure or a slightly asymmetric configuration with a $|m| = 1$ mode is an outcome. In addition, the growth time scale of the dynamical instability is relatively long; up to the saturation of the growth, $\gtrsim 10$ dynamical time scale is necessary. Thus, the resulting slightly deformed star could not be a strong burst emitter of gravitational waves but a long-term emitter of them.

In a naive consideration, we may expect that a core collapse of massive stars may lead to the formation of a rapidly rotating proto-neutron star with $\beta \sim 0.3$. The reason for this is that the parameter β increases approximately in proportional to R^{-1} during the stellar core collapse because $T_{\rm rot}$ and W are approximately proportional to $MR^2\Omega^2 \propto R^{-2}$ and M^2/R, respectively. Here, we supposed that Ω would be roughly proportional to R^{-2} in the assumption of the conservation of specific angular momentum for each fluid element. During the collapse, the core radius decreases from $\sim 2000\,\rm km$ to $\sim 20\,\rm km$, and hence β may increase by about two orders of magnitude. Thus, a moderately rotating progenitor-star core may yield a rapidly rotating proto-neutron star which may reach the criterion for the onset of the dynamical instability. Similar arguments hold for the accretion induced collapse of white dwarfs to neutron stars and for the merger of binary white dwarfs to neutron stars. Motivated by this expectation, three-dimensional simulations for rapidly rotating cores were performed with a simplified equation of state [Rampp et al. (1998); Shibata and Sekiguchi (2005b)]. These works suggested a possibility that the dynamical instability could indeed set in for an extremely rapidly rotating progenitor.

For a rapidly rotating proto-neutron star that is unstable against non-axisymmetric (bar-mode) instability, the angular velocity is approximately equal to the Keplerian velocity. Thus, equation (1.219) may be replaced by

$$\ddot{I}_{ij}\bigg|_{\max} \sim 4\kappa_I M R^2 \Omega^2 \sim 4\kappa_I \frac{M^2}{R}, \qquad (1.226)$$

Then, the amplitude of gravitational waves is approximately estimated, giving

$$h_{\max} \sim 10^{-22}\left(\frac{10\,\text{Mpc}}{D_{\text{obs}}}\right)\left(\frac{\kappa_I}{0.1}\right)^2\left(\frac{M}{1.4M_\odot}\right)^2\left(\frac{R}{30\,\text{km}}\right)^{-1}, \qquad (1.227)$$

with the frequency

$$f = \frac{\Omega}{\pi} \approx \frac{1}{\pi}\sqrt{\frac{M}{R^3}} = 840\,\text{Hz}\left(\frac{M}{1.4M_\odot}\right)^{1/2}\left(\frac{R}{30\,\text{km}}\right)^{-3/2}. \qquad (1.228)$$

This suggests that these gravitational waves may be detected if the event happens in a nearby galaxy with $D_{\text{obs}} \lesssim 10\,\text{Mpc}$ (cf. figure 1.5).

However, the latest numerical simulations with a realistic equation of state [Dimmelmeier et al. (2007); Ott et al. (2007a); Dimmelmeier et al. (2008)] have shown that this possibility is not very likely for the ordinary supernova collapse. The reason for this is that for very rapidly rotating progenitors, the collapse of the core is hung up before its central density exceeds the nuclear density, implying that the spin-up is not sufficiently achieved. For the collapse with a special rotation configuration (e.g., with a high degree of differential rotation) of a progenitor, a large value of $\beta \sim 0.25$ may be achieved. However, it is not clear whether such a special configuration is natural or not. Rather, Ott et al. (2007a) discovered a different type of dynamical instability, which sets in for a proto-neutron star with a high degree of differential rotation but with a low value of $\beta_d \sim 0.1$ as suggested by Tohilne and Hachisu (1990); Centrella et al. (2001); Shibata et al. (2003) (see below and section 10.1).

Alternatively, dynamical instability may occur for the collapse of more massive progenitors leading to a massive proto-neutron star that eventually collapses to a black hole after a long-term accretion process. A reason for this is that the very massive progenitors are often inferred to be rapid rotators in nature; e.g., the progenitor of a system composed of a spinning black hole surrounded by a torus that is believed to be the central engine of gamma-ray bursts should be massive and rotating with sufficiently high angular momentum. This issue has not be explored yet in numerical relativity, and thus, this is one of the issues in the future.

As already mentioned, dynamical instability could set in for the ordinary core collapse if the proto-neutron star core has a high degree of differential rotation. For

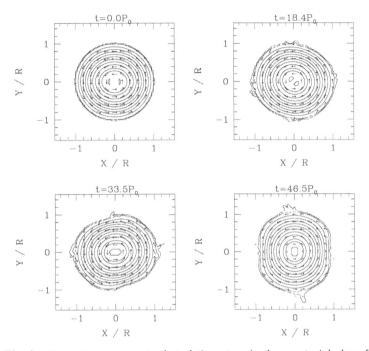

Fig. 1.36 The density contour curves at selected time steps in the equatorial plane for a rotating neutron star with a high degree of differential rotation. Here, P_0 is $2\pi/\Omega_0$ with Ω_0 being the angular velocity at the rotation axis. R denotes the equatorial stellar radius referred to as $R_{\rm eq}$ in the text. The contour curves are drawn for $\rho/\rho_{\max} = 0.95, 0.9, 0.8, 0.7, 0.6, 0.5, 0.4, 0.3, 0.2, 0.1$ 0.01 and 0.001, where $\rho_{\max}$ denotes the maximum density at each time slice. The dashed curves are plotted for $\rho/\rho_{\max} = 0.01$ and 0.001. The figure is taken from Shibata et al. (2003).

such a case, the resulting value of β for the proto-neutron star may not have to be very large [Shibata et al. (2003)]. Ott et al. (2007a) indeed found that this is the case [see also Ott et al. (2005); Scheidegger et al. (2010) for simulations in Newtonian gravity]. For this type of the dynamical instability, the resulting proto-neutron star is not highly deformed (see figure 1.36). Thus, the amplitude of gravitational waves for each period is by $\sim O(0.1)$ smaller than that in equation (1.227). However, the small amplitude implies that the emission time scale (dissipation time scale by gravitational waves) is longer. Thus, a large number of wave cycles could be accumulated (see figure 1.37). Then, the effective amplitude of gravitational waves would be significantly enhanced as in the case of inspiraling compact binaries (see section 1.5.1). In the following, we will approximately estimate the effective amplitude following Lai and Shapiro (1995).

As illustrated in figure 1.38, we suppose that an unstable proto-neutron star deforms to be an ellipsoidal figure, and then, it emits gravitational waves for a time scale much longer than the rotation period. The degree of the deformation gradually decreases in the gravitational-wave emission time scale. Here, the rotational motion

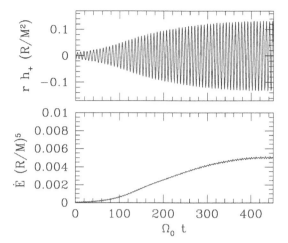

Fig. 1.37 Gravitational waves emitted by the differentially rotating neutron star shown in figure 1.36 as a function of $\Omega_0 t$. Gravitational-wave amplitude is shown in units of $M^2/rR_{\rm eq}$ and the luminosity of gravitational waves $\dot{E}$ is shown in units of $(M/R_{\rm eq})^5$. Here, $R_{\rm eq}$ (R in the figure) is the equatorial radius, and r is a hypothetical distance to the source. This figure is taken from Shibata *et al.* (2003).

for the ellipsoidal star is composed of two ingredients; one is the internal motion and the other is the figure rotation. Gravitational waves are emitted by this figure rotation and their characteristic frequency is determined by $\Omega_{\rm fig}/\pi$ with $\Omega_{\rm fig}$ being the angular velocity of the ellipsoidal figure which is as large as or slightly less than the Keplerian velocity $\sqrt{M/R_{\rm eq}^3}$ with $R_{\rm eq}$, the equatorial radius. A large number of numerical simulations have shown that at the onset of the dynamical instability, the figure-rotation mode is certainly excited and a large fraction of the rotational kinetic energy is composed of this rotation mode. The subsequent evolution of the rotating star is determined by the hydrodynamical angular-momentum transport process and the gravitational-radiation reaction. We will describe how the slightly-deformed rotating neutron star evolves, ignoring the hydrodynamical process, as follows.

Using equation (1.74), the luminosity of gravitational waves is approximately estimated as

$$\dot{E}_{\rm GW} \sim \epsilon_{\rm gw}^2 (\kappa_{\rm I} M R_{\rm eq}^2 \Omega_{\rm fig}^3)^2, \qquad (1.229)$$

where $\epsilon_{\rm gw}$ is a parameter of order 0.1, which approximately denotes the ellipticity. Then, the time scale of the gravitational-wave emission is approximately estimated as

$$t_{\rm gw} \sim \frac{T_{\rm rot}}{\dot{E}_{\rm GW}} \sim \epsilon_{\rm gw}^{-2}(MR_{\rm eq}^2\Omega_{\rm fig}^4)^{-1}$$
$$\sim 73\,{\rm s}\,\left(\frac{\epsilon_{\rm gw}}{0.1}\right)^{-2}\left(\frac{\kappa_{\rm I}}{0.1}\right)^{-1}\left(\frac{M}{1.4M_\odot}\right)^{-1}\left(\frac{R_{\rm eq}}{30\,{\rm km}}\right)^{-2}\left(\frac{f}{1\,{\rm kHz}}\right)^{-4}, \qquad (1.230)$$

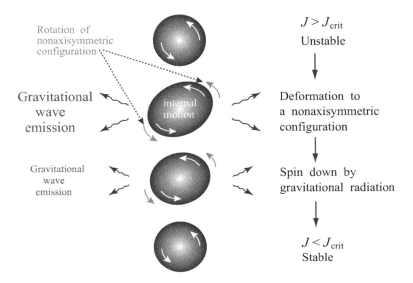

Fig. 1.38 A schematic figure for the evolution of non-axisymmetric (bar-mode) deformation of a rapidly rotating proto-neutron star.

where we simply set $T_{\rm rot} \sim \kappa_{\rm I} M R_{\rm eq}^2 \Omega_{\rm fig}^2/2$. Since the rotation period is much shorter than 1 s, we find that the time scale, $t_{\rm gw}$, is much longer than the rotation period.

The characteristic frequency of gravitational waves is assumed to be $\sim 1\,{\rm kHz}$ according to equation (1.228). Assuming that the kinetic energy would be primarily dissipated by the gravitational-wave emission, the accumulated cycles of gravitational waves N for a given value of the frequency are estimated by

$$N := f t_{\rm gw}$$
$$\approx 7.3 \times 10^4 \left(\frac{\kappa_{\rm I}}{0.1}\right)^{-1} \left(\frac{\epsilon_{\rm gw}}{0.1}\right)^{-2} \left(\frac{M}{1.4 M_\odot}\right)^{-1} \left(\frac{R_{\rm eq}}{30\,{\rm km}}\right)^{-2} \left(\frac{f}{1\,{\rm kHz}}\right)^{-3}. \quad (1.231)$$

As described in section 1.5.1, the effective amplitude of gravitational waves may be defined by $h_{\rm eff} := N^{1/2} h$ where h denotes the characteristic amplitude of gravitational waves for each cycle. Using this relation, we find

$$h_{\rm eff} \sim \frac{4(\kappa_{\rm I} M R_{\rm eq}^2 f)^{1/2}}{D_{\rm obs}}$$
$$\sim 10^{-21} \left(\frac{\kappa_{\rm I}}{0.1}\right)^{1/2} \left(\frac{M}{1.4 M_\odot}\right)^{1/2} \left(\frac{R_{\rm eq}}{30\,{\rm km}}\right) \left(\frac{f}{1\,{\rm kHz}}\right)^{1/2} \left(\frac{100\,{\rm Mpc}}{D_{\rm obs}}\right), \quad (1.232)$$

where we note that $\epsilon_{\rm gw}$ is canceled out; this effective amplitude does not depend on the degree of the deformation. Equation (1.232) suggests that gravitational waves from proto-neutron stars of a high degree of differential rotation of mass $\sim 1.4 M_\odot$

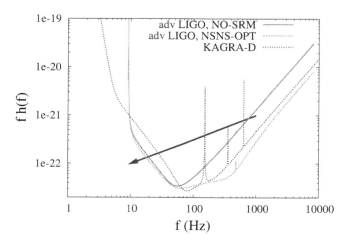

Fig. 1.39 A predicted spectrum of gravitational waves emitted from a rapidly rotating and non-axisymmetric proto-neutron star for a hypothetical distance of 100 Mpc (the thick line with arrow) together with designed sensitivities of advanced LIGO and KAGRA (see the caption of figure 1.5).

and of radius $\gtrsim 30$ km at a hypothetical distance of $\lesssim 100$ Mpc could be a source for advanced gravitational-wave detectors, if other dissipation processes are negligible (see figure 1.39). We note that the similar long-term gravitational-wave emission could also occur for secularly unstable proto-neutron stars [Lai and Shapiro (1995)].

For deriving a spectrum of gravitational waves, we have to also follow the evolution for all the relevant modes of the gravitational-wave emission. To do this, numerical simulations have to be done, but it has not been performed yet [but see Ou et al. (2004); Shibata and Karino (2004) for an effort]. Lai and Shapiro (1995) performed a semi-analytical numerical calculation employing an ellipsoidal approximation of the deformed proto-neutron star. In their model, the configuration and the velocity field of the star are modeled by a finite number of parameters, which are evolved by solving ordinary differential equations. This method cannot provide quantitatively accurate results but qualitative features can be explored by it. Figure 1.39 schematically displays a predicted spectrum of gravitational waves emitted by a rapidly rotating star with non-axisymmetric (bar-mode) deformation. In this prediction, the gravitational-wave frequency gradually decreases with time, and the effective amplitude decreases with lowering f in proportional to $f^{1/2}$, as described in equation (1.232).

1.5.5.5 Gravitational waves in the formation of black holes

For the case that the progenitor core is massive enough, supernova explosion of the massive star will fail and a black hole will be formed eventually, after the gravitational collapse and substantial matter accretion from the outer region. There are also several possibilities that produce a black hole in the gravitational collapse.

One is the collapse of a very massive, metal-poor star for which the mass is much larger than $100 M_\odot$. For such a star, the evolution process is different from ordinary massive stars because its core has a high entropy per baryon number [Bond et al. (1984)]. The core will be unstable against the gravitational collapse in the oxygen-core phase due to the instability associated with the pressure depletion by the pair creation of electron-positron pairs, which results from high temperature ($\sim 10^{10}$ K) and relatively low density (less than 10^7 g/cm^3). After the onset of the pair instability, there are two possibilities [Fryer et al. (2001)]: If the mass of the oxygen core is not high enough, an explosive oxygen burning occurs after the gravitational collapse, leading to a pair-instability explosion. For this case, there will be no remnant at the center after the explosion. On the other hand, if the mass is high enough, the oxygen burning does not supply the power sufficient to halt the collapse, and the gravitational collapse continues up to the formation of a black hole. Clarifying the process of this type of gravitational collapse is one of the interesting issues in numerical relativity [Nakazato et al. (2006); Sekiguchi and Shibata (2011)] (see also section 9.3).

The other possibility for the direct formation of a black hole is that the collapse of a supermassive star of mass larger than $\sim 10^5 M_\odot$ to a supermassive black hole [Rees (1984)]. It is not clear whether a supermassive star was really formed in an early epoch of the universe. However, if it was formed, such a high-mass gas cloud subsequently shrink by radiative cooling in a short time scale (much shorter than the Hubble time), and eventually, general relativistic instability against the radial collapse sets in leading to the formation of a supermassive black hole [see, e.g., Baumgarte and Shapiro (1999a)]. Alternatively, an extremely high-mass main sequence star with mass $\gtrsim 10^5 M_\odot$ may be formed in an early epoch of the universe [Hosokawa et al. (2013)], if an extremely high accretion rate could be preserved in a massive star formation epoch. Such an extremely high-mass star may collapse to a supermassive black hole after the end of the hydrogen burning due to the pair instability of the helium or oxygen core. The resulting black hole could be a seed of supermassive black holes. Thus, the study for the collapse process of supermassive stars and extremely high-mass stars is also a subject of numerical relativity [Shibata and Shapiro (2002)] (see also section 9.2).

As in the merger of binary black holes, gravitational waves emitted at the formation of a black hole would be characterized by a ringdown oscillation associated with a quasi-normal mode of the formed black hole. In this case, the dominant mode would be $l = 2$ and $m = 0$ mode because the collapse would occur approximately in an axisymmetric manner. To date, however, in the context of the gravitational collapse to a black hole, detailed features of the gravitational waveform have not been known yet. This is mainly due to the reason that gravitational waves from the formation of a black hole can be computed only by a numerical-relativity simulation but it is not technically easy to extract gravitational waves of small amplitude accurately.

To date, only one work has been performed by Ott *et al.* (2011) for the collapse of ordinary massive stars in the context of the so-called failed supernova. This work shows that the waveform in the black-hole formation phase is indeed characterized by a ringdown oscillation. The amplitude is $\sim 10^{-20}$ for an event of a hypothetical distance of $D_{\text{obs}} = 10\,\text{kpc}$ with the typical frequency $\sim 3-4\,\text{kHz}$.

Gravitational waveforms for the ringdown phase of a formed black hole can be modeled by equation (1.203). Here, ω_r and ω_i for the fundamental quadrupole mode with $l = 2$ and $m = 0$ are written approximately as [Berti *et al.* (2009)]

$$\omega_r M_{\text{BH}} \approx [0.4437 - 0.0739(1 - \hat{a})^{0.3350}],$$
$$\omega_i \approx \frac{1}{2}\omega_r[4.000 - 1.955(1 - \hat{a})^{0.1420}]^{-1}. \qquad (1.233)$$

Thus, the frequency depends weakly on the black-hole spin and is determined primarily by the black-hole mass;

$$f_r = \frac{\omega_r}{2\pi} \approx \begin{cases} 3.0\,\text{kHz}\left(\dfrac{M_{\text{BH}}}{4M_\odot}\right)^{-1} & \text{for } \hat{a} = 0, \\[1em] 3.6\,\text{kHz}\left(\dfrac{M_{\text{BH}}}{4M_\odot}\right)^{-1} & \text{for } \hat{a} = 1. \end{cases} \qquad (1.234)$$

Assuming that the amplitude is approximately proportional to $M_{\text{BH}}/D_{\text{obs}}$, the result by Ott *et al.* (2011) may be rescaled for different black-hole mass. In reality, the amplitude is likely to depend strongly on the black-hole spin, because it determines the degree of the non-spherical deformation, but we here suppose that the formed black hole has a spin as large as that in Ott *et al.* (2011). In these assumptions, we may guess that for the formation of black holes of mass $10^3 M_\odot$ and $10^6 M_\odot$, the amplitude of gravitational waves would be

$$h \sim \begin{cases} 10^{-21}\left(\dfrac{10\,\text{Mpc}}{D_{\text{obs}}}\right)\left(\dfrac{M_{\text{BH}}}{10^3 M_\odot}\right) & \text{for } M_{\text{BH}} = 10^3 M_\odot, \\[1em] 10^{-21}\left(\dfrac{10\,\text{Gpc}}{D_{\text{obs}}}\right)\left(\dfrac{M_{\text{BH}}}{10^6 M_\odot}\right) & \text{for } M_{\text{BH}} = 10^6 M_\odot, \end{cases} \qquad (1.235)$$

with the corresponding frequency $f \sim 12-15\,\text{Hz}$ and $12-15\,\text{mHz}$, respectively. When plotting these values in figures 1.5 and 1.7, we find that a signal from the black-hole formation of mass $\sim 10^3 M_\odot$ may be detected by the advanced detectors for an event of a hypothetical distance $\sim 10\,\text{Mpc}$, and those from the supermassive black-hole formation of $\lesssim 10^6 M_\odot$ may be detected by space interferometers for a cosmological event. This suggests that formation processes of a variety of black holes may be observed by the gravitational-wave detectors in the future.

1.6 Matched filtering techniques for gravitational-wave data analysis

For the detection of gravitational waves, the data analysis plays a crucial role. A reason for this is that gravitational-wave signals, which could be detected, do not come frequently, and moreover, the typical signal-to-noise ratio of the signals is expected to be ~ 10 for the best-sensitivity frequency band. This implies that we are required to extract a signal, which suddenly comes, from a very noisy data stream. Finding the signal in the noisy data stream may be compared to finding a small-size treasure lost in the sea in the windy day. Thus, for detecting gravitational waves, in addition to improving the sensitivity of the detectors, we have to make an effort of efficiently extracting the signal; an efficient data analysis technique has to be developed.

An efficient extraction of gravitational-wave signals is feasible when we know a priori the waveform of the corresponding signals. In the presence of a prediction for gravitational waveforms (we call it *template*), the signal could be efficiently extracted by taking a cross correlation between the data stream and the template, because the noise is primarily Gaussian and thus taking the cross correlation with the template of the signal enables us to remove the major part of the noise. The well-known data analysis technique, in which theoretical templates of gravitational waves are prepared and used, is called a matched filtering technique [Thorne (1987); Schutz (1991); Creighton and Anderson (2011)].

The process of the matched filtering is summarized as follows. First, let $h(t)$ and $n(t)$ be data streams of a gravitational-wave signal and noise of a detector, respectively. Thus, the total data stream in the detector is $s(t) = h(t) + n(t)$. In each process of the matched filtering, we take a cross correlation between $s(t)$ and a theoretical template $q(t)$. Here, the cross correlation is defined by

$$c(t) := (s,q)(t) = \int_{-\infty}^{\infty} s(t')q(t-t')dt'$$
$$= \int_{-\infty}^{\infty} \hat{s}(f)\hat{q}^*(f)\exp(2\pi i f t)df, \quad (1.236)$$

where $\hat{s}$ and $\hat{q}$ are the Fourier transform of s and q, and $\hat{q}^*$ is the complex conjugate of $\hat{q}$. Assuming that the noise is completely random, Gaussian, and stationary in average, the expectation value of c, written as $\langle c(t) \rangle$, is $(h,q)(t)$. The variance of $c(t)$ due to the noise becomes

$$\sigma^2 := \left\langle [c(t) - \langle c(t) \rangle]^2 \right\rangle = \frac{1}{2}\int_{-\infty}^{\infty} S_n(f)|\hat{q}(f)|^2 df, \quad (1.237)$$

where $S_n(f)$ is called the one-sided spectrum density, and defined by

$$\langle \hat{n}(f)\hat{n}(f') \rangle = \frac{1}{2}S_n(f)\delta(f-f'). \quad (1.238)$$

Note that the sensitivity curves in figure 1.5 correspond to $\sqrt{S_n(f)f}$.

Now, the ratio of $c(t)$ to σ denotes the signal-to-noise ratio at each time:

$$\frac{S}{N}(t) := \frac{c(t)}{\sigma}. \tag{1.239}$$

In addition, setting $\hat{q}(f) = 2\hat{g}(f)/S_n(f)$ yields

$$\frac{S}{N}(t) = \frac{(h|ge^{-2\pi ift})}{(g|g)^{1/2}}, \tag{1.240}$$

where

$$(h|g) := 4\int_0^\infty \frac{\hat{h}(f)\hat{g}(f)^*}{S_n(f)}df. \tag{1.241}$$

S/N defined in equation (1.240) becomes maximum when $\hat{g}$ is equal to $\hat{h}e^{2\pi ift}$. Usually, this maximum value, $(g|g)^{1/2}$, is called the signal-to-noise ratio. It should be emphasized that this maximum value is achieved only when we know the correct waveform $q(t)$ [or $g(t)$] prior to the detection. If the template is not correct, the maximum signal-to-noise ratio is not achieved in the data analysis. In particular, if the phase part of the template is not calculated very accurately, we will lose the signal-to-noise ratio significantly. The reason for this is that the phase cancellation in the integral of equation (1.241) is quite serious. This implies that the templates have to be accurate, in particular in phase.

The loss of the signal-to-noise ratio is the serious problem in the gravitational-wave *detection*. As already mentioned in the first paragraph of this section, the expected signal-to-noise ratio for the predicted gravitational-wave sources is at most 20–30, and typically ~ 10. The detection of gravitational waves will be confirmed only for the case that the signal-to-noise ratio is sufficiently high as $\gtrsim 8$. Therefore, if the loss is serious, we will fail to detect many gravitational-wave signals, which will not frequently come.

Preparing accurate templates are also crucial for extracting physical information of gravitational-wave sources from the detected signal, i.e., in the *parameter estimation*. In the matched filtering technique, one takes cross correlations between the detected signal and many theoretical templates. Then, the physical parameters are determined by finding a template that results in the maximum value of the signal-to-noise ratio. Here, if the templates were not accurate, we would fail to correctly determine the properties of the gravitational-wave source. More specifically, the error resulting from the inaccuracy of the templates at least has to be smaller than the statistical error associated with the detector noise.

All these facts tell us that in the gravitational-wave detection project, highly accurate theoretical templates have to be prepared. What do we need to do for this issue? The answer is to perform the theoretical calculation or computation of gravitational waveforms by accurately solving Einstein's equation. For most of the gravitational-wave sources, the theoretical calculation cannot be analytically done. Therefore, we have to perform numerical-relativity simulations!

PART 1
Methodology

Chapter 2

Formulation for initial-value problems of general relativity

The primary purpose of numerical relativity is to clarify the nature of nonlinear dynamics for a variety of spacetime by numerically solving Einstein's equation. For this, we need to solve Einstein's equation as an initial-value problem (we need to evolve spatial hypersurfaces by solving Einstein's equation forward in time). For this procedure, we first have to reformulate Einstein's equation to a form with which the evolution of geometric quantities forward in time is feasible. As already seen in section 1.1.2, Einstein's equation is basically a hyperbolic equation. However, to explicitly clarify its hyperbolic nature, a well-defined formulation is necessary.

The most popular way for deriving an initial-value formulation of general relativity is based on the so-called Arnowitt–Deser–Misner (ADM) formulation [Arnowitt *et al.* (1962)] or $N+1$ formulation [York (1979)] (here $N = D-1$ denotes the spatial dimensionality and "1" denotes the time). In this approach, the time and the space are explicitly decomposed. Thus, it provides a clear geometrical interpretation of the foliation of spacetime: Spacelike hypersurfaces are successively chosen to fill the spacetime, and geometrical quantities on each spacelike hypersurface can be evolved forward in time with no ambiguity. In addition, in this approach, the freedom of the gauge choice, which is associated with the general covariance in general relativity, is totally preserved. However, for the ADM-type formulation to be employed in numerical computation, we need reformulation, because the original ADM and $N+1$ formulations are not suitable for stable numerical computations (see section 2.3.3).

The other popular way to explicitly write Einstein's equation in an initial-value formulation is to employ a special gauge condition as introduced in section 1.1.2. A generalized formulation of this line in numerical relativity is called generalized harmonic (GH) formulation, which is derived in slightly modified harmonic coordinates. As described in section 2.3.4, in the GH approach, the gauge freedom is largely consumed for deriving 10 components of hyperbolic equations. Only a part of the degrees of freedom, for which the geometrical meaning is not very clear, remains. Nevertheless, this is a quite robust formulation for stable numerical computation, as a number of numerical experiments have demonstrated [e.g., Pretorius (2005a)].

Although the GH approach is one option, the notion of the spacetime foliation is still quite useful in numerical relativity, because this guarantees the evolution of spatial hypersurfaces forward in time. This usefulness can be reminded by a counterexample that an unsuitable choice of coordinates in general relativity, which is not based on the spacetime foliation, often leads to pathological behavior in a strongly gravitating region. Then, a well-known example is the Schwarzschild coordinates for non-spinning black holes described in equation (1.97). For this, t behaves as the time outside the event horizon $r > 2M$. However, inside the event horizon, it becomes a spacelike coordinate. This is simply due to the bad choice of the coordinates. To avoid this failure, it is better to foliate the spacetime to a series of spacelike hypersurfaces before fixing the coordinates.

The final merit in the formulations based on the spacetime foliation is that dynamical variables are naturally introduced. Consider the evolution of particles in Newtonian gravity. Their motion is determined by the equations of motion with an initial condition for the position x_p^i and the velocity v_p^i where p denotes the label of the particles. Specifically, the equations of motion for a particle p are

$$\frac{dx_p^i}{dt} = v_p^i \quad \text{and} \quad \frac{dv_p^i}{dt} = f_p^i\left(x_q^j, v_q^j\right), \tag{2.1}$$

where f_p^i denotes a force vector. In the spacetime-foliation approach such as ADM formulation, the dynamical variables such as x_p^i and its time derivative are naturally introduced. This fact is helpful for straightforward evolution of spatial hypersurfaces forward in time. Due to this reason, we describe mainly the formulation based on the spacetime foliation in this chapter.

Before going ahead, it will be helpful for the readers to have an analogy between the formulations for Einstein's equation and for solving massless scalar fields, ϕ, which obey the equation

$$\Box \phi = 0. \tag{2.2}$$

In the GH formulation, Einstein's equation is written into 10 components of scalar-wave equations like equation (2.2). In the $N+1$ formulation, we rewrite equation (2.2) into the form

$$\dot{\phi} = \eta, \quad \dot{\eta} = \Delta_\text{f} \phi, \tag{2.3}$$

where Δ_f denotes the flat spatial Laplacian. It is found that equation (2.3) is similar to equation (2.1): (ϕ, η) correspond to (x^i, v^i).

Finally, we may rewrite equation (2.2) into the form

$$\dot{\phi} = \eta, \quad \dot{\eta} = \sum_i \partial_i \Pi_i, \quad \dot{\Pi}_i = \partial_i \eta. \tag{2.4}$$

In this case, all the equations have only the first-order derivatives. This type of formulations is called hyperbolic formulations. The merit in this formulation is that it is straightforward to derive characteristic speeds of the field propagation, which can be used for integrating the equations in a physical manner. Use of the characteristic speeds for the accurate numerical computation is popular in numerical fluid hydrodynamics (see chapter 4). Although the strategy is well motivated and resulting formulations are often written in a mathematically elegant manner, this line of the formulations has not provided a success in numerical relativity (i.e., for a solution of geometric equations) because an unphysical growth of numerical errors is not controlled in the known hyperbolic formulations. Thus, in this volume, we do not touch on the hyperbolic formulations. For the readers who are interested in the hyperbolic formulations, we refer to the references; e.g., Reula (1998); Bona and Massó (1992); Bona et al. (1995); Anderson and York (1999); Kidder et al. (2001); Alcubierre (2008).

2.1 Formulations based on spacetime foliation

In this section, we describe formulations of Einstein's equation based on the spacetime foliation, i.e., ADM and $N+1$ formulations. In the following, we will implicitly assume to consider globally hyperbolic spacetime because well-defined evolution of spatial hypersurfaces in the initial-value formulations is feasible only in this case.

2.1.1 *Foliation of spacetime*

We can foliate globally hyperbolic spacetime by spacelike hypersurfaces (Cauchy surfaces), Σ_t, parametrized by a global time function t. Here, each hypersurface is assumed not to intersect each other. Let n^a be a future-directed timelike unit vector field normal to the hypersurfaces Σ_t: At each point on Σ_t, n^a is normal to it (i.e., $n_a \propto \nabla_a t$) and satisfies $n^a n_a = -1$. Then, the spacetime metric g_{ab} induces a spatial metric γ_{ab} on each hypersurface by

$$\gamma_{ab} = g_{ab} + n_a n_b, \qquad (2.5)$$

where $n^a \gamma_{ab} = 0$. $\gamma^a_{\ b}$ can be used for mapping a spacetime vector (and tensor), V^a, to a spatial vector on a spatial hypersurface by $\gamma^a_{\ b} V^b$.

Let t^a be a timelike vector field on the spacetime which is the tangent to the time axis, $t^a = (\partial/\partial t)^a$, and $t^a \nabla_a t = t^t = 1$. Note that t^a is not always normal to Σ_t, and thus, it has components both on Σ_t and along n^a (see figure 2.1). We decompose t^a into two components as

$$\alpha := -t^a n_a, \quad \beta^b := t^a \gamma_a^{\ b}. \qquad (2.6)$$

Here, α and β^a are called the lapse function and the shift vector, respectively. Obviously, $\beta^a n_a = 0$. Using these quantities, t^a is written as

$$t^a = \alpha n^a + \beta^a. \qquad (2.7)$$

Note that n_a can be written as $n_a = -\alpha \nabla_a t$.

Each time axis is a spacetime curve of a constant value of spatial coordinates, which can be arbitrarily chosen (e.g., the origin on $\Sigma_{t+\Delta t}$ is arbitrary; see figure 2.1). Here, we have N degrees of freedom for choosing the spatial coordinates on each

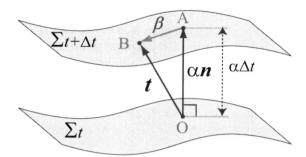

Fig. 2.1 Schematic figure that shows the meaning of the lapse function and the shift vector. The lapse function determines the physically proper time ($\alpha\Delta t$) between two points on two neighboring spatial hypersurfaces, Σ_t and $\Sigma_{t+\Delta t}$, along a timelike vector αn^a which is normal to Σ_t: Points O and A are connected by αn^a in this example. The shift vector specifies the difference between αn^a and t^a which determines the direction of the time axis for each spatial point. In a coordinate basis, points O and B have the same spatial coordinates while the spatial coordinates for O and A are different by $-\beta^i$.

hypersurface Σ_t. Thus, choosing the time axis is equivalent to choosing the spatial coordinates on $\Sigma_{t+\Delta t}$ at each time step. If the time axis is parallel to n^a, $\beta^a = 0$ and otherwise $\beta^a \neq 0$; see equation (2.7). This illustrates that the choice of spatial coordinates is equivalent to the choice of β^a. Therefore, β^a denotes the freedom of the spatial coordinate choice on each $\Sigma_{t+\Delta t}$, or equivalently, the freedom for the choice of the direction of the time axis.

The meaning of the lapse function is found in the following manner. For two neighboring hypersurfaces Σ_t and $\Sigma_{t+\Delta t}$ (see figure 2.1), a timelike vector connecting two points on Σ_t and $\Sigma_{t+\Delta t}$ along n^a is written as $\alpha n^a \Delta t$. The norm of this vector is $-(\alpha\Delta t)^2$, implying that the spacetime proper distance (proper time) between two points is $\alpha\Delta t$: see figure 2.1. Therefore, at each point of Σ_t, α determines the proper distance between two neighboring hypersurfaces along a timelike normal vector. Because we have a degree of freedom for choosing the way of foliating the spacetime, α is freely chosen.

Since α and β^a are related to the degrees of freedom for the coordinate choice, they are not dynamical variables such as ϕ in equation (2.2). Thus, the induced metric γ_{ab}, the last components in g_{ab}, is the dynamical variable that should obey an evolution equation derived from Einstein's equation.

For the help of manipulations, we here describe several useful relations. The time and space components of n^a, n_a, and γ_{ab} are written as

$$n^\mu = \left(\frac{1}{\alpha}, \frac{-\beta^i}{\alpha}\right) \text{ and } n_\mu = (-\alpha, \mathbf{0}), \qquad (2.8)$$

and

$$\gamma_{\mu\nu} = \begin{pmatrix} \beta^k \beta_k & \beta_j \\ \beta_i & \gamma_{ij} \end{pmatrix}, \qquad (2.9)$$

where $\beta_i = \gamma_{ij}\beta^j$ and $\beta^\mu = (0, \beta^i)$. On the other hand, $\gamma^{\mu\nu}$ does not have time components, i.e., $\gamma^{tt} = \gamma^{ti} = 0$.

Using α, β^k, and γ_{ij}, spacetime metric components are written as

$$g_{\mu\nu} = \begin{pmatrix} -\alpha^2 + \beta^k\beta_k & \beta_j \\ \beta_i & \gamma_{ij} \end{pmatrix}, \qquad (2.10)$$

and

$$g^{\mu\nu} = \begin{pmatrix} -\dfrac{1}{\alpha^2} & \dfrac{\beta^j}{\alpha^2} \\ \dfrac{\beta^i}{\alpha^2} & \gamma^{ij} - \dfrac{\beta^i\beta^j}{\alpha^2} \end{pmatrix}. \qquad (2.11)$$

From equation (2.10), it is easily found that the determinant of $g_{\mu\nu}$, g, is written as $-\alpha^2\gamma$ where γ is the determinant of γ_{ij}.

2.1.2 Derivative operator

In the spacetime-foliation formulations, we consider the evolution of spatial hypersurfaces forward in time. As found in section 2.1.1, the dynamical variable in this approach is the spatial metric, γ_{ab}. Thus, basic equations should be written in terms of γ_{ab}. As a first step toward developing a covariant formulation with respect to γ_{ab}, we have to define the covariant derivative associated with γ_{ab}, which maps spatial tensors into spatial tensors. We can construct this derivative by projecting all indices in a D-dimensional covariant derivative ∇_a into Σ_t as

$$D_a T_{a_1 a_2 \ldots}{}^{b_1 b_2 \ldots} = \gamma_a{}^c \gamma_{a_1}{}^{c_1} \gamma_{a_2}{}^{c_2} \cdots \gamma^{b_1}{}_{d_1} \gamma^{b_2}{}_{d_2} \cdots \nabla_c T_{c_1 c_2 \ldots}{}^{d_1 d_2 \ldots}, \qquad (2.12)$$

where D_a is the covariant derivative associated with γ_{ab}, and $T_{a_1 a_2 a_3 \ldots}{}^{b_1 b_2 b_3 \ldots}$ is an arbitrary tensor on spatial hypersurfaces Σ_t. Using equation (2.12), the covariant derivative of γ_{ab} is shown to vanish as

$$\begin{aligned} D_a \gamma_{bc} &= \gamma_a{}^d \gamma_b{}^e \gamma_c{}^f \nabla_d(g_{ef} + n_e n_f) \\ &= \gamma_a{}^d \gamma_b{}^e \gamma_c{}^f (n_e \nabla_d n_f + n_f \nabla_d n_e) = 0, \end{aligned} \qquad (2.13)$$

where we used $\nabla_d g_{ef} = 0$ and $n_e \gamma_b{}^e = 0$. Thus, D_a is the unique derivative operator associated with γ_{ab}.

From equation (2.13), the spatial connection coefficients are derived as

$$\Gamma^i{}_{jk} = \frac{1}{2}\gamma^{il}\left(\partial_j \gamma_{kl} + \partial_k \gamma_{jl} - \partial_l \gamma_{jk}\right). \qquad (2.14)$$

The spatial Riemann tensor associated with γ_{ab} is defined by

$$R_{abc}{}^d \omega_d = (D_a D_b - D_b D_a)\omega_c, \qquad (2.15)$$

where ω^a is a spatial vector field on Σ_t. In a coordinate component, the spatial Riemann tensor is written as in the spacetime case as

$$R_{ijk}{}^l = \partial_j \Gamma^l{}_{ik} - \partial_i \Gamma^l{}_{jk} + \Gamma^m{}_{ik} \Gamma^l{}_{jm} - \Gamma^m{}_{jk} \Gamma^l{}_{im}, \qquad (2.16)$$

and the spatial Ricci tensor is $R_{ac} = R_{abc}{}^b = R_{abcd}\gamma^{bd}$.

The Riemann tensor may be written in the form

$$R_{abc}{}^d = \tilde{R}_{abc}{}^d + \tilde{D}_b C^d{}_{ac} - \tilde{D}_a C^d{}_{bc} + C^e{}_{ac} C^d{}_{be} - C^e{}_{bc} C^d{}_{ae}, \tag{2.17}$$

where $\tilde{R}_{abc}{}^d$ and $\tilde{D}_a$ are, respectively, the Riemann tensor and the covariant derivative associated with $\tilde{\gamma}_{ab}$ which is a spatial metric different from γ_{ab}. $C^c{}_{ab}$ is a connection tensor defined by

$$D_a \omega_b = \tilde{D}_a \omega_b - C^c{}_{ab} \omega_c, \tag{2.18}$$

namely,

$$C^c{}_{ab} = \frac{1}{2} \gamma^{cd} \left(\tilde{D}_a \gamma_{bd} + \tilde{D}_b \gamma_{ad} - \tilde{D}_d \gamma_{ab} \right), \tag{2.19}$$

and

$$C^k{}_{ij} = \Gamma^k{}_{ij} - \tilde{\Gamma}^k{}_{ij}, \tag{2.20}$$

where $\tilde{\Gamma}^k{}_{ij}$ is a connection coefficient associated with $\tilde{\gamma}_{ab}$.

2.1.3 Extrinsic curvature

As found in section 2.1.1, γ_{ab} is the dynamical variable that corresponds to x_p^i in the example of equation (2.1). In the similar manner, we would like to define a variable that corresponds to the velocity v_p^i, i.e., a quantity related to the time derivative of the spatial metric on spatial hypersurfaces Σ_t. A well-defined covariant "time-derivative" of the spatial metric is the extrinsic curvature, which is defined by

$$K_{ab} := -\gamma_a{}^c \nabla_c n_b = -\frac{1}{2} \mathscr{L}_n \gamma_{ab}. \tag{2.21}$$

Here, K_{ab} is a symmetric spatial tensor, i.e., $K_{ab} = K_{ba}$ and $K_{ab} n^b = 0$. Note that the symmetric relation of K_{ab} is guaranteed by Frobenius' theorem (see appendix A). $\mathscr{L}_n$ denotes the Lie derivative with respect to n^a, and satisfies the following relation:

$$\begin{aligned}\mathscr{L}_n \gamma_{ab} &:= n^c \nabla_c \gamma_{ab} + \gamma_{ac} \nabla_b n^c + \gamma_{bc} \nabla_a n^c \\ &= n^c \nabla_c (g_{ab} + n_a n_b) + g_{ac} \nabla_b n^c + g_{bc} \nabla_a n^c \\ &= \gamma_a{}^c \nabla_c n_b + \gamma_b{}^c \nabla_c n_a = 2 \gamma_a{}^c \nabla_c n_b. \end{aligned} \tag{2.22}$$

We note that it is straightforward to show $\gamma_b{}^c \nabla_c n_a = \gamma_a{}^c \nabla_c n_b$, because

$$\gamma_\beta{}^\sigma \nabla_\sigma n_\alpha = \gamma_\alpha{}^\mu \gamma_\beta{}^\sigma \nabla_\sigma n_\mu = -\gamma_\alpha{}^\mu \gamma_\beta{}^\sigma \overset{(D)}{\Gamma^\nu{}_{\sigma\mu}} n_\nu. \tag{2.23}$$

The Lie derivative associated with n^a may be written as

$$\alpha \mathscr{L}_n \gamma_{ab} = \mathscr{L}_t \gamma_{ab} - \mathscr{L}_\beta \gamma_{ab}. \tag{2.24}$$

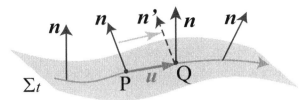

Fig. 2.2 Concept of the extrinsic curvature. $\boldsymbol{n}$ denotes the timelike unit normal vector field n^a of a spatial hypersurface Σ_t, and $\boldsymbol{n}'$ denotes a timelike vector that is obtained by the parallel transportation of $\boldsymbol{n}$ from P to Q along a spatial geodesic for which the tangent vector is $\boldsymbol{u}$. The extrinsic curvature denotes the degree of the difference $\boldsymbol{n}' - \boldsymbol{n}$ which does not vanish if the spatial hypersurface is embedded in a bended manner for given spacetime.

Here, to derive equation (2.24), we use equation (2.7) and $\gamma_{ab}n^b = 0 = \gamma_{ab}(t^b - \beta^b)$. We note that $\mathscr{L}_t \gamma_{ab}$ is the time derivative of γ_{ab}, $\partial_t \gamma_{ab}$, in the coordinate basis. $\mathscr{L}_\beta \gamma_{ab}$ is written as

$$\mathscr{L}_\beta \gamma_{ab} = \beta^c \nabla_c \gamma_{ab} + \gamma_{ac} \nabla_b \beta^c + \gamma_{bc} \nabla_a \beta^c. \tag{2.25}$$

Thus,

$$\begin{aligned}\mathscr{L}_\beta \gamma_{ij} &= \beta^k \partial_k \gamma_{ij} + \gamma_{ik} \partial_j \beta^k + \gamma_{jk} \partial_i \beta^k \\ &= \beta^k D_k \gamma_{ij} + \gamma_{ik} D_j \beta^k + \gamma_{jk} D_i \beta^k \\ &= \gamma_{ik} D_j \beta^k + \gamma_{jk} D_i \beta^k = D_j \beta_i + D_i \beta_j,\end{aligned} \tag{2.26}$$

and hence, the extrinsic curvature is related to the time derivative of the spatial metric by

$$\mathscr{L}_t \gamma_{ij} = -2\alpha K_{ij} + D_i \beta_j + D_j \beta_i. \tag{2.27}$$

Therefore, K_{ab} is employed as the conjugate variable of γ_{ab}, and equation (2.27) is usually employed as the evolution equation for γ_{ab}. For the following manipulations, we note that equation (2.27) may be rewritten in the form,

$$\mathscr{L}_t \gamma^{ij} = 2\alpha K^{ij} - (D^i \beta^j + D^j \beta^i). \tag{2.28}$$

In addition to its usefulness in numerical simulations, the extrinsic curvature has a clear geometric meaning as its name shows. Let u^a be a spatial vector field tangent to a geodesic on a spatial hypersurface Σ_t (see figure 2.2). Then, from equation (2.21), we have

$$u^a K_{ab} = -u^c \nabla_c n_b. \tag{2.29}$$

This implies that if n^a is not parallel-transported along a spatial geodesic, for which u^a is the tangent vector field, K_{ab} does not vanish (see figure 2.2). Nonzero K_{ab} is realized when the spatial hypersurface is embedded in a bended manner for given spacetime. Thus, K_{ab} is a measure of the bending of Σ_t.

Before closing this section, we describe a useful relation. From equation (2.21), we find that K_{ab} is written as
$$K_{ab} = -n_a a_b - \nabla_a n_b, \qquad (2.30)$$
where a_b denotes an acceleration, $a_b := n^c \nabla_c n_b$, which is a spatial vector written by
$$a_b = D_b \ln \alpha. \qquad (2.31)$$
Thus,
$$\nabla_a n_b = -K_{ab} - n_a D_b \ln \alpha. \qquad (2.32)$$
This relation will be used for many times in the manipulation of section 2.1.4.

2.1.4 The Gauss and Codazzi equations

Up to the previous sections, we have defined the dynamical variables (γ_{ab}, K_{ab}). The next step is to derive equations for them from Einstein's equation, which are composed originally of the Ricci tensor associated with g_{ab}. Thus, specifically, the next step is to write the spacetime Ricci tensor in terms of (γ_{ab}, K_{ab}). Such relations are called the Gauss and Codazzi equations (often called the Gauss-Codazzi equations).

The Gauss equation is derived from equation (2.15). Let ω^a be a spatial vector field satisfying $\omega_a n^a = 0$. Then, the double spatial covariant derivative of ω_a is written as
$$\begin{aligned} D_a D_b \omega_c &= \gamma_a{}^f \gamma_b{}^g \gamma_c{}^h \nabla_f \left(\gamma_g{}^d \gamma_h{}^e \nabla_d \omega_e \right) \\ &= \gamma_a{}^f \gamma_b{}^d \gamma_c{}^e \nabla_f \nabla_d \omega_e + \gamma_a{}^f \gamma_b{}^d \gamma_c{}^h \left(\nabla_f \gamma_h{}^e \right) \left(\nabla_d \omega_e \right) \\ &\quad + \gamma_a{}^f \gamma_b{}^g \gamma_c{}^e \left(\nabla_f \gamma_g{}^d \right) \left(\nabla_d \omega_e \right) \\ &= \gamma_a{}^f \gamma_b{}^d \gamma_c{}^e \nabla_f \nabla_d \omega_e - K_{ac} K_b{}^d \omega_d - n^d (\nabla_d \omega_e) \gamma_c{}^e K_{ab}, \end{aligned} \qquad (2.33)$$
where we used
$$\begin{aligned} \nabla_c \gamma_a{}^b &= \nabla_c \left(g_a{}^b + n_a n^b \right) = n_a \nabla_c n^b + n^b \nabla_c n_a \\ &= -n_a \left(K_c{}^b + n_c a^b \right) - n^b (K_{ac} + n_c a_a), \end{aligned} \qquad (2.34)$$
$$\gamma_c{}^d \gamma_a{}^e \nabla_d \gamma_e{}^b = \gamma_c{}^d \gamma_a{}^e n^b \nabla_d n_e = -n^b K_{ac}, \qquad (2.35)$$
$$\gamma_b{}^d n^e \nabla_d \omega_e = -\gamma_b{}^d \omega_e \nabla_d n^e = \omega_e K_b{}^e, \qquad (2.36)$$
and equation (2.32). Then,
$$\begin{aligned} (D_a D_b - D_b D_a) \omega_c &= \gamma_a{}^f \gamma_b{}^g \gamma_c{}^e \left(\nabla_f \nabla_g - \nabla_g \nabla_f \right) \omega_e + \left(K_{bc} K_a{}^d - K_{ac} K_b{}^d \right) \omega_d \\ &= \left(\gamma_a{}^f \gamma_b{}^g \gamma_c{}^e \overset{(D)}{R}{}_{fge}{}^d + K_{bc} K_a{}^d - K_{ac} K_b{}^d \right) \omega_d. \end{aligned} \qquad (2.37)$$
This relation has to hold for any spatial vector field ω^d. Therefore,
$$R_{abcd} = \gamma_a{}^e \gamma_b{}^f \gamma_c{}^g \gamma_d{}^h \overset{(D)}{R}{}_{efgh} + K_{ad} K_{bc} - K_{ac} K_{bd}. \qquad (2.38)$$

This is called the Gauss equation.

The Codazzi equation is derived from the following identity:

$$\gamma_a{}^d \gamma_b{}^e \gamma_f{}^c (\nabla_d \nabla_e - \nabla_e \nabla_d) n^f = -\gamma_a{}^d \gamma_b{}^e \gamma_f{}^c \overset{(D)}{R}_{deg}{}^f n^g. \tag{2.39}$$

Using equation (2.30), we find

$$\gamma_a{}^d \gamma_b{}^e \gamma_f{}^c \nabla_d \nabla_e n^f = \gamma_a{}^d \gamma_b{}^e \gamma_f{}^c \nabla_d \left(-K_e{}^f - n_e a^f \right)$$
$$= -D_a K_b{}^c - \gamma_a{}^d \gamma_b{}^e a^c \nabla_d n_e = -D_a K_b{}^c + a^c K_{ab}. \tag{2.40}$$

Hence, equation (2.39) is rewritten as

$$D_a K_{bc} - D_b K_{ac} = -\gamma_a{}^d \gamma_b{}^e \gamma_c{}^f \overset{(D)}{R}_{defg} n^g. \tag{2.41}$$

This is called the Codazzi equation.

2.1.5 *Constraint and evolution equations*

Basic equations for (γ_{ab}, K_{ab}) are derived from the Gauss and Codazzi equations. There are two types of equations in them. The first type is the constraint equation in which the second time derivatives of γ_{ab} (or the first time derivatives of K_{ab}) do not appear. Remembering that Einstein's equation is a nonlinear wave equation for g_{ab} (see section 1.1.2), the dynamical equation for γ_{ab} should include the second time derivative of γ_{ab}. Hence, the constraint equations are not the equations that describe the evolution of the spatial geometries. The other type is the evolution equation that includes the second time derivative of γ_{ab}. This equation describes the dynamical evolution of the spatial geometries.

Before going ahead, we give a spacetime-decomposition relation of the stress-energy tensor as

$$T_{ab} = \rho_{\mathrm{h}} n_a n_b + J_a n_b + J_b n_a + S_{ab}, \tag{2.42}$$

where

$$\rho_{\mathrm{h}} := T_{ab} n^a n^b, \tag{2.43}$$

$$J_a := -T_{bc} n^b \gamma^c{}_a, \tag{2.44}$$

$$S_{ab} := T_{cd} \gamma^c{}_a \gamma^d{}_b. \tag{2.45}$$

First, we derive two constraint equations. The Hamiltonian constraint is derived from the Gauss equation (2.38) by contracting it (by multiplying the spatial metric) twice to give a scalar equation. The first contraction by γ^{bd} yields

$$R_{ac} = \gamma_a{}^e \gamma^{bd} \gamma_c{}^h \overset{(D)}{R}_{ebhd} + K_{ab} K_c{}^b - K_{ac} K$$
$$= \gamma_a{}^e \gamma_c{}^h \left(\overset{(D)}{R}_{eh} + \overset{(D)}{R}_{ebhd} n^b n^d \right) + K_{ab} K_c{}^b - K_{ac} K, \tag{2.46}$$

where $K = K_a{}^a$. By the second contraction, we have

$$R = \overset{(D)}{R} + 2\,\overset{(D)}{R}_{bd} n^b n^d + K_{ab}K^{ab} - K^2 = 2\,G_{ab}n^a n^b + K_{ab}K^{ab} - K^2. \quad (2.47)$$

Using Einstein's equation, $G_{ab}n^a n^b$ is written as $8\pi T_{ab}n^a n^b = 8\pi\rho_{\mathrm{h}}$. Then, we obtain the Hamiltonian constraint equation for (γ_{ab}, K_{ab}) as

$$R - K_{ab}K^{ab} + K^2 = 16\pi\rho_{\mathrm{h}}. \quad (2.48)$$

The momentum constraint is derived from the Codazzi equation. Multiplying γ^{ac} to equation (2.41) yields

$$\begin{aligned}
D_a K_b{}^a - D_b K &= -\gamma_b{}^e \left(g^{df} + n^d n^f\right) \overset{(D)}{R}_{defg} n^g \\
&= -\gamma_b{}^e \overset{(D)}{R}_{eg} n^g \\
&= -8\pi \gamma_b{}^e T_{eg} n^g,
\end{aligned} \quad (2.49)$$

and thus,

$$D_a K_b{}^a - D_b K = 8\pi J_b. \quad (2.50)$$

Equation (2.50) is called the momentum constraint equation.

The Hamiltonian and momentum constraints include only the spatial metric, extrinsic curvature, and their spatial derivatives, as mentioned already. They are the conditions that have to be satisfied on each spatial hypersurface Σ_t. In other words, to embed a spatial hypersurface Σ_t in spacetime, these constraints have to be satisfied. These are similar to the energy and momentum conservation laws for the system of particles that also do not include the time derivative of the velocity.

The evolution equation, which is used for evolving K_{ab} forward in time, is derived from equation (2.46). This is found from the fact that the term

$$\gamma_a{}^e \gamma_c{}^h \overset{(D)}{R}_{ebhd} n^b n^d,$$

includes a term of the time derivative of K_{ab}. To find this fact, we start from the following relation:

$$\overset{(D)}{R}_{abc}{}^d n_d = (\nabla_a \nabla_b - \nabla_b \nabla_a) n_c. \quad (2.51)$$

Using equation (2.32) with $a_a = D_a \ln \alpha$,

$$\begin{aligned}
\nabla_a \nabla_b n_c &= -\nabla_a(K_{bc} + n_b D_c \ln \alpha) \\
&= -\nabla_a K_{bc} + (K_{ab} + n_a D_b \ln \alpha)D_c \ln \alpha - n_b \nabla_a D_c \ln \alpha.
\end{aligned} \quad (2.52)$$

Thus,

$$\gamma_a{}^e \gamma_c{}^h \overset{(D)}{R}_{ebhd} n^b n^d = \gamma_a{}^e \gamma_c{}^h n^d \left(-\nabla_e K_{dh} + \nabla_d K_{eh}\right) + \frac{1}{\alpha} D_a D_c \alpha. \quad (2.53)$$

Here,
$$-\gamma_a{}^e\gamma_c{}^h n^d \nabla_e K_{dh} = \gamma_a{}^e K_{cd}\nabla_e n^d = -K_{cd}K_a{}^d, \tag{2.54}$$

and
$$\gamma_a{}^e\gamma_c{}^h n^d \nabla_d K_{eh} = \gamma_a{}^e\gamma_c{}^h \left(\mathscr{L}_n K_{eh} - K_{eb}\nabla_h n^b - K_{bh}\nabla_e n^b\right)$$
$$= \mathscr{L}_n K_{ac} + 2K_{ab}K_c{}^b. \tag{2.55}$$

As a result, we obtain
$$\gamma_a{}^e\gamma_c{}^h \overset{(D)}{R}_{ebhd} n^b n^d = \mathscr{L}_n K_{ac} + K_{ab}K_c{}^b + \frac{1}{\alpha}D_a D_c \alpha, \tag{2.56}$$

and equation (2.46) is rewritten to
$$\mathscr{L}_n K_{ac} = R_{ac} - \gamma_a{}^e\gamma_c{}^h \overset{(D)}{R}_{eh} - 2K_{ab}K_c{}^b + KK_{ac} - \frac{1}{\alpha}D_a D_c \alpha. \tag{2.57}$$

Using Einstein's equation,
$$\gamma_a{}^e\gamma_c{}^h \overset{(D)}{R}_{eh} = 8\pi\gamma_a{}^e\gamma_c{}^h \left(T_{eh} - \frac{1}{D-2}g_{eh}T_b{}^b\right) = 8\pi\left(S_{ac} - \frac{1}{D-2}\gamma_{ac}T_b{}^b\right), \tag{2.58}$$

and
$$\alpha\mathscr{L}_n K_{ac} = \mathscr{L}_t K_{ac} - \mathscr{L}_\beta K_{ac}, \tag{2.59}$$
$$\mathscr{L}_\beta K_{ac} = \beta^b D_b K_{ac} + K_{ab}D_c\beta^b + K_{bc}D_a\beta^b, \tag{2.60}$$

we finally obtain the evolution equation of K_{ab} as
$$\mathscr{L}_t K_{ac} = \alpha R_{ac} - 8\pi\alpha\left[S_{ac} - \frac{1}{D-2}\gamma_{ac}(S_b{}^b - \rho_h)\right] + \alpha\left(-2K_{ab}K_c{}^b + KK_{ac}\right)$$
$$- D_a D_c \alpha + \beta^b D_b K_{ac} + K_{ab}D_c\beta^b + K_{bc}D_a\beta^b, \tag{2.61}$$

where we used $T_a{}^a = S_a{}^a - \rho_h$ and we note that abstract indices in equations (2.60) and (2.61) are used only for the spatial tensor. For the later convenience, we also note the equations
$$\mathscr{L}_t K_a{}^c = \alpha R_a{}^c - 8\pi\alpha\left[S_a{}^c - \frac{1}{D-2}\gamma_a{}^c\left(S_b{}^b - \rho_h\right)\right] + \alpha K K_a{}^c$$
$$- D_a D^c \alpha + \beta^b D_b K_a{}^c + K_b{}^c D_a \beta^b - K_a{}^b D_b \beta^c, \tag{2.62}$$

and
$$\mathscr{L}_t \hat{K}_{ac} = \alpha\sqrt{\gamma}R_{ac} - 8\pi\alpha\sqrt{\gamma}\left[S_{ac} - \frac{1}{D-2}\gamma_{ac}\left(S_b{}^b - \rho_h\right)\right] - \frac{2\alpha}{\sqrt{\gamma}}\hat{K}_{ab}\hat{K}_c{}^b$$
$$- \sqrt{\gamma}D_a D_c \alpha + D_b\left(\beta^b \hat{K}_{ac}\right) + \hat{K}_{ab}D_c\beta^b + \hat{K}_{bc}D_a\beta^b, \tag{2.63}$$

where $\hat{K}_{ab} := \sqrt{\gamma}K_{ab}$ is a weighted extrinsic curvature. Equations (2.62) and (2.63) are often employed in spherically and axially symmetric numerical-relativity simulations (see section 2.4 and appendix A.2).

Table 2.1 Number of degrees of freedom in geometric variables.

	Total	Constraints	Gauge	Dynamical one for γ_{ij}
General case	$(D-1)D$	D	D	$(D-3)D/2$
$D=4$ case	12	4	4	2

To summarize, there are three equations, the Hamiltonian constraint, momentum constraint, and evolution equations. These are derived by operating $n^a n^b$, $n^a \gamma^b{}_c$, and $\gamma^a{}_c \gamma^b{}_d$, respectively, to Einstein's equation. Each of them has 1, $D-1$, and $D(D-1)/2$ components, and the total number is $D(D+1)/2$, which agrees with the total number of components of Einstein's equation in D-dimension. Equation (2.61) together with equation (2.27) determines the time evolution of all the components of (γ_{ab}, K_{ab}). This implies that the constraint equations are redundant for their evolution. Thus, if (γ_{ab}, K_{ab}) are precisely evolved by their evolution equations, the constraint equations have to be satisfied automatically. This point will be discussed in section 2.1.6.

In numerical simulations, the evolution equations are solved in a coordinate basis. In numerical relativity, equations (2.27) and (2.61) are often written in terms of a coordinate basis as follows:

$$(\partial_t - \beta^k \partial_k)\gamma_{ij} = -2\alpha K_{ij} + \gamma_{ik}\partial_j \beta^k + \gamma_{jk}\partial_i \beta^k, \tag{2.64}$$

$$(\partial_t - \beta^k \partial_k)K_{ij} = \alpha R_{ij} - 8\pi\alpha \left[S_{ij} - \frac{1}{D-2}\gamma_{ij}\left(S_k{}^k - \rho_{\rm h}\right)\right]$$
$$+ \alpha\left(-2K_{ik}K_j{}^k + KK_{ij}\right)$$
$$- D_i D_j \alpha + K_{ik}\partial_j \beta^k + K_{jk}\partial_i \beta^k. \tag{2.65}$$

Equations (2.64) and (2.65) for $D=4$ are often referred to as the standard 3+1 formulation [York (1979)].

Because they are often used in the later sections, we also write the equations for the contraction of equations (2.64) and (2.65):

$$(\partial_t - \beta^k \partial_k)\sqrt{\gamma} = \sqrt{\gamma}\left(-\alpha K + \partial_k \beta^k\right), \tag{2.66}$$

$$(\partial_t - \beta^k \partial_k)K = \alpha R - 8\pi\alpha \left[S_k{}^k - \frac{D-1}{D-2}\left(S_k{}^k - \rho_{\rm h}\right)\right] + \alpha K^2 - D_i D^i \alpha$$

$$= 8\pi\alpha \frac{S_k{}^k + (D-3)\rho_{\rm h}}{D-2} + \alpha K_{ij} K^{ij} - D_i D^i \alpha. \tag{2.67}$$

Here the Hamiltonian constraint was used for deriving the second line of equation (2.67).

Finally, we count the number of real dynamical degrees of freedom in the geometric variables (γ_{ab}, K_{ab}). There are in total $D(D-1)$ components for them. However, the constraint equations impose D relations among (γ_{ab}, K_{ab}), thus effectively reducing the number of freely specifiable functions on Σ_t to $D(D-2)$. In addition, we have D degrees of gauge freedom for choosing α and β^a. Thus, the number of the dynamical degrees of freedom is even less as $D(D-2) - D = D(D-3)$.

This is the total number for the freely specifiable functions for (γ_{ab}, K_{ab}). Dividing by two, we conclude that γ_{ab} has $D(D-3)/2$ dynamical degrees of freedom at each spatial point. This is the same as the number of the gravitational-wave mode; i.e., for $D = 4$, it is two which corresponds to plus and cross modes of gravitational waves. We summarize each number of degrees of freedom in Table 2.1.

2.1.6 *Evolution of constraint equations*

In initial-value formulations of Einstein's equation, one usually evolves spatial hypersurfaces in the following manner. The first step is to prepare an initial condition on an initial spatial hypersurface $\Sigma_{t=0}$ that satisfies the constraint equations. Then, the system is evolved by solving the evolution equations for (γ_{ab}, K_{ab}) appropriately giving α and β^a. Here, (γ_{ab}, K_{ab}) have to satisfy the constraint equations on each spatial hypersurface of $t > 0$. Now, a question is whether the constraint equations are automatically satisfied, when the evolution equations are solved. The answer is yes: This is guaranteed in general relativity.

To show that the constraints are preserved, we define

$$A_{ab} := G_{ab} - 8\pi T_{ab}, \tag{2.68}$$

and write it in the form

$$A_{ab} = H_0 n_a n_b + H_a n_b + H_b n_a + H_{ab}, \tag{2.69}$$

where

$$H_0 := A_{ab} n^a n^b, \quad H_a := -A_{bc} n^b \gamma^c{}_a, \quad H_{ab} := A_{cd} \gamma^c{}_a \gamma^d{}_b. \tag{2.70}$$

Einstein's equation gives $A_{ab} = 0$, and the Hamiltonian and momentum constraints correspond to $H_0 = 0$ and $H_a = 0$, respectively. We may assume that at $t = 0$, $H_0 = 0 = H_a$ are satisfied because we can always prepare the initial condition that satisfies the constraint equations. On the other hand, the evolution equation is described by $H_{ab} = 0$. Thus, we here should show that if the condition, $H_{ab} = 0$, is satisfied, the constraints, $H_0 = 0$ and $H_a = 0$, are guaranteed to be satisfied for $t > 0$.

The preservation of the constraint equations are found by deriving the evolution equations for H_0 and H_a. This is achieved using $\nabla_a A^a{}_b = 0$. Here, this relation holds due to the Bianchi identity, $\nabla_a G^a{}_b = 0$, and to the assumption that the matter equation, $\nabla_a T^a{}_b = 0$, is solved precisely. A straightforward manipulation yields

$$\mathscr{L}_t H_0 = \beta^a D_a H_0 + \alpha K H_0 - 2 H^a D_a \alpha - \alpha D_a H^a + \alpha H_{ab} K^{ab}, \tag{2.71}$$

$$\mathscr{L}_t H_a = -H_0 D_a \alpha + \alpha K H_a + \beta^b D_b H_a + H_b D_a \beta^b - D_b \left(\alpha H^b{}_a \right). \tag{2.72}$$

Equations (2.71) and (2.72) show that if $H_0 = 0$ and $H_a = 0$ on a spatial hypersurface Σ_t (this is the case at $t = 0$), and if the evolution equation is solved precisely, $H_{ab} = 0$, the right-hand sides of these equations vanish. Therefore, H_0 and H_a remain zero on the subsequent spatial hypersurfaces.

2.1.7 Conformal decomposition

The conformal decomposition is an often-used prescription to factor out a scalar component (say, the primary gravitational-potential part) from the spatial metric in numerical relativity. A conformal decomposition formulation was originally developed by Lichnerowicz (1944) and by York (1971, 1972, 1973) for initial-data problems in general relativity (see also chapter 5). Subsequently, the conformal decomposition has been also employed for reformulating evolution equations in the 3+1 and N+1 formulations. In the following sections, we will often describe formulations in numerical relativity based on the conformal decomposition. As a step for them, we here summarize a conformal decomposition formulation of the N+1 formulation.

In the conformal decomposition formulation, we define the following variables as the fundamental quantities for an appropriately-chosen conformal factor, ψ:

$$\tilde{\gamma}_{ab} := \psi^{-4/(N-2)} \gamma_{ab}, \tag{2.73}$$

$$\tilde{A}_{ab} := \psi^{-4/(N-2)} \left(K_{ab} - \frac{1}{N} \gamma_{ab} K \right), \tag{2.74}$$

$$K := \gamma^{ab} K_{ab}. \tag{2.75}$$

In the often-used formulation, ψ is defined by

$$\psi := (\gamma/f)^{(N-2)/4N}, \tag{2.76}$$

where $\gamma = \det(\gamma_{ab})$ as before and $f = \det(f_{ab})$ with f_{ab} being the flat spatial metric in a chosen coordinate system; in Cartesian coordinates, $\det(\tilde{\gamma}_{ab})$ is assumed to be unity. We note that subscripts of $\tilde{A}_{ij}$ and $\tilde{A}^{ij}$ (and any other quantities with tilde) are raised and lowered by the conformal spatial metric $\tilde{\gamma}^{ij}$ and $\tilde{\gamma}_{ij}$, respectively.

In the conformal decomposition formulation, there are $N(N+1)+2$ components for the variables $\tilde{\gamma}_{ab}$, $\tilde{A}_{ab}$, ψ, and K. Since we increased two additional components, ψ and K, two new constraints have to be introduced. One is the tracefree condition for $\tilde{A}_{ab}$ as

$$\tilde{A}_{ab} \tilde{\gamma}^{ab} = 0. \tag{2.77}$$

For the definition of (2.76), the following relation is the other constraint:

$$\det(\tilde{\gamma}_{ab}) = \det(f_{ab}). \tag{2.78}$$

We note that in general, equation (2.78) may be replaced, e.g., by $\det(\tilde{\gamma}_{ab}) = F(x^i)$ where $F(x^i)$ is a function of spatial coordinates. For such a case, the definition of ψ is different from equation (2.76). In the following, we will always suppose equation (2.76) for ψ (except in section 2.4).

Associated with the conformal metric, $\tilde{\gamma}_{ab}$, we define the covariant derivative $\tilde{D}_a$ that satisfies

$$\tilde{D}_a \tilde{\gamma}_{bc} = 0. \tag{2.79}$$

The relation between D_a and $\tilde{D}_a$ is written in equation (2.18), and in the present context, the connection tensor $C^c{}_{ab}$ is

$$C^c{}_{ab} = \frac{2}{(N-2)\psi}\left(\delta^c{}_b \tilde{D}_a\psi + \delta^c{}_a \tilde{D}_b\psi - \tilde{\gamma}_{ab}\tilde{D}^c\psi\right). \tag{2.80}$$

Equation (2.17) gives a relation for the Ricci tensor in the form

$$R_{ac} = \tilde{R}_{ac} + R^\psi_{ac}, \tag{2.81}$$

where $\tilde{R}_{ac}$ is the Ricci tensor with respect to $\tilde{\gamma}_{ab}$, and

$$\begin{aligned}R^\psi_{ac} &= \tilde{D}_b C^b{}_{ac} - \tilde{D}_a C^b{}_{bc} + C^d{}_{ac}C^b{}_{bd} - C^e{}_{bc}C^b{}_{ae} \\ &= -\frac{2}{\psi}\tilde{D}_a\tilde{D}_c\psi + \frac{2N}{(N-2)\psi^2}(\tilde{D}_a\psi)\tilde{D}_c\psi \\ &\quad - \frac{2}{(N-2)\psi^2}\tilde{\gamma}_{ac}\left[\psi\tilde{\Delta}\psi + (\tilde{D}_b\psi)\tilde{D}^b\psi\right],\end{aligned} \tag{2.82}$$

with $\tilde{\Delta} = \tilde{D}_a\tilde{D}^a$.

The basic equations in the conformal decomposition formulation are written as

$$\mathscr{L}_t\tilde{\gamma}_{ab} = -2\alpha\tilde{A}_{ab} + \tilde{D}_a\tilde{\beta}_b + \tilde{D}_b\tilde{\beta}_a - \frac{2}{N}\tilde{\gamma}_{ab}\tilde{D}_c\beta^c, \tag{2.83}$$

$$\begin{aligned}\mathscr{L}_t\tilde{A}_{ab} &= \psi^{-4/(N-2)}\left[\alpha\left(R_{ab} - \frac{1}{N}\gamma_{ab}R_c{}^c\right) - \left(D_aD_b\alpha - \frac{1}{N}\gamma_{ab}D_cD^c\alpha\right)\right] \\ &\quad + \alpha\left(K\tilde{A}_{ab} - 2\tilde{A}_{ac}\tilde{A}_b{}^c\right) \\ &\quad + (\tilde{D}_a\beta^c)\tilde{A}_{bc} + (\tilde{D}_b\beta^c)\tilde{A}_{ac} - \frac{2}{N}(\tilde{D}_c\beta^c)\tilde{A}_{ab} + \beta^c\tilde{D}_c\tilde{A}_{ab} \\ &\quad - 8\pi\alpha\psi^{-4/(N-2)}\left(S_{ab} - \frac{1}{N}\gamma_{ab}S_c{}^c\right),\end{aligned} \tag{2.84}$$

$$\mathscr{L}_t\psi = -\frac{(N-2)\psi}{2N}\left(\alpha K - \tilde{D}_a\beta^a\right) + \mathscr{L}_\beta\psi, \tag{2.85}$$

$$\mathscr{L}_t K = \alpha\left[\tilde{A}_{ab}\tilde{A}^{ab} + \frac{1}{N}K^2\right] - D_aD^a\alpha + \frac{8\pi\alpha}{N-1}\left[(N-2)\rho_{\rm h} + S_c{}^c\right] + \mathscr{L}_\beta K, \tag{2.86}$$

where $\tilde{\beta}_a = \tilde{\gamma}_{ab}\beta^b$ and we used the Hamiltonian constraint to write equation (2.86). We also note that for deriving equation (2.83), we used a relation

$$\psi^{-4/(N-2)}\left(D_a\beta_b + D_b\beta_a - \frac{2}{N}\gamma_{ab}D_c\beta^c\right) = \tilde{D}_a\tilde{\beta}_b + \tilde{D}_b\tilde{\beta}_a - \frac{2}{N}\tilde{\gamma}_{ab}\tilde{D}_c\tilde{\beta}^c, \tag{2.87}$$

where $\tilde{\beta}^a = \beta^a$. The typical forms in the Cartesian coordinate basis are written as

$$(\partial_t - \beta^l\partial_l)\tilde{\gamma}_{ij} = -2\alpha\tilde{A}_{ij} + \tilde{\gamma}_{ik}\partial_j\beta^k + \tilde{\gamma}_{jk}\partial_i\beta^k - \frac{2}{N}\tilde{\gamma}_{ij}\partial_k\beta^k, \tag{2.88}$$

$$\begin{aligned}(\partial_t - \beta^l\partial_l)\tilde{A}_{ij} &= \psi^{-4/(N-2)}\left[\alpha\left(R_{ij} - \frac{1}{N}\gamma_{ij}R_k{}^k\right) - \left(D_iD_j\alpha - \frac{1}{N}\gamma_{ij}D_kD^k\alpha\right)\right] \\ &\quad + \alpha\left(K\tilde{A}_{ij} - 2\tilde{A}_{ik}\tilde{A}_j{}^k\right) + (\partial_i\beta^k)\tilde{A}_{kj} + (\partial_j\beta^k)\tilde{A}_{ki} - \frac{2}{N}(\partial_k\beta^k)\tilde{A}_{ij} \\ &\quad - 8\pi\alpha\psi^{-4/(N-2)}\left(S_{ij} - \frac{1}{N}\gamma_{ij}S_k{}^k\right),\end{aligned} \tag{2.89}$$

$$(\partial_t - \beta^l \partial_l)\psi = -\frac{(N-2)\psi}{2N}\left(\alpha K - \partial_k \beta^k\right), \tag{2.90}$$

$$(\partial_t - \beta^l \partial_l)K = \alpha\left[\tilde{A}_{ij}\tilde{A}^{ij} + \frac{1}{N}K^2\right] - D_i D^i \alpha + \frac{8\pi\alpha}{(N-1)}\left[(N-2)\rho_{\rm h} + S_k^{\ k}\right]. \tag{2.91}$$

As shown in equation (2.81), R_{ab} is decomposed into two parts: Ricci tensor associated with $\tilde{\gamma}_{ab}$ and R_{ab}^{ψ} composed of ψ. We here describe some manipulation for rewriting $\tilde{R}_{ab}$. Using the flat spatial metric f_{ab} and the covariant derivative associated with f_{ab}, $\overset{(0)}{D}_c$, $\tilde{R}_{ab}$ is written in the form

$$\tilde{R}_{ab} = \overset{(0)}{D}_c \overset{(0)}{\Gamma}{}^c{}_{ab} - \overset{(0)}{\Gamma}{}^d{}_{ac} \overset{(0)}{\Gamma}{}^c{}_{bd}, \tag{2.92}$$

where $\overset{(0)}{\Gamma}{}^c{}_{ab}$ is a connection tensor defined by

$$\overset{(0)}{\Gamma}{}^c{}_{ab} = \frac{1}{2}\tilde{\gamma}^{cd}\left(\overset{(0)}{D}_a\tilde{\gamma}_{bd} + \overset{(0)}{D}_b\tilde{\gamma}_{ad} - \overset{(0)}{D}_d\tilde{\gamma}_{ab}\right), \tag{2.93}$$

and $\overset{(0)}{\Gamma}{}^c{}_{ac} = 0$ for $\tilde{\gamma} = f$, which is used for writing equation (2.92). Then, $\tilde{R}_{ab}$ is written in the form

$$\tilde{R}_{ab} = -\frac{1}{2}\tilde{\gamma}^{cd}\left(\overset{(0)}{D}_c\overset{(0)}{D}_d\tilde{\gamma}_{ab} - \overset{(0)}{D}_a\overset{(0)}{D}_d\tilde{\gamma}_{bc} - \overset{(0)}{D}_b\overset{(0)}{D}_d\tilde{\gamma}_{ac}\right)$$
$$+ \left(\overset{(0)}{D}_c\tilde{\gamma}^{cd}\right)\overset{(0)}{\Gamma}_{d,ab} - \Gamma^d{}_{ac}\Gamma^c{}_{bd}, \tag{2.94}$$

where $\overset{(0)}{\Gamma}_{c,ab} = \tilde{\gamma}_{cd}\overset{(0)}{\Gamma}{}^d{}_{ab}$. In Cartesian coordinates, $\overset{(0)}{D}_i = \partial_i$ and $\overset{(0)}{\Gamma}{}^k{}_{ij} = \tilde{\Gamma}^k{}_{ij}$, which is the connection coefficient associated with $\tilde{\gamma}_{ij}$.

The Hamiltonian and momentum constraint equations (2.48) and (2.50), are written in the forms

$$\tilde{\Delta}\psi = \frac{(N-2)\psi}{4(N-1)}\tilde{R} - \frac{4(N-2)\pi}{N-1}\rho_{\rm h}\psi^{(N+2)/(N-2)}$$
$$- \frac{(N-2)\psi^{(N+2)/(N-2)}}{4(N-1)}\left(\tilde{A}_{ab}\tilde{A}^{ab} - \frac{N-1}{N}K^2\right), \tag{2.95}$$

$$\psi^{-2N/(N-2)}\tilde{D}_a\left[\psi^{2N/(N-2)}\tilde{A}^a{}_b\right] - \frac{N-1}{N}\tilde{D}_b K = 8\pi J_b, \tag{2.96}$$

where we used equations (2.81) and (2.82) to derive

$$R = \psi^{-4/(N-2)}\left[\tilde{R} - \frac{4(N-1)}{(N-2)\psi}\tilde{\Delta}\psi\right]. \tag{2.97}$$

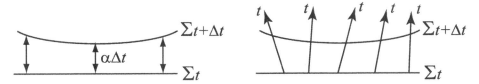

Fig. 2.3 Left panel schematically shows that the proper time between two spatial hypersurfaces Σ_t and $\Sigma_{t+\Delta t}$ is different at each spatial point and controlled by $\alpha(x^i)$. Right panel schematically shows that the time axis determined by the shift vector is chosen arbitrarily at each spatial point.

2.2 Gauge conditions

In the formulation based on the spacetime foliation, the gauge freedom is totally preserved; the lapse function α and the shift vector β^a can be freely chosen. We have to appropriately fix these degrees of freedom by imposing a gauge condition that enables successful evolution of the spacetime we are interested in. However, an appropriate gauge condition is not trivial at all, and also, it depends on the problem concerned in general. The choice of the appropriate gauge condition has been one of the most important issues in numerical relativity. In this section, we summarize several representative gauge conditions proposed and used so far.

2.2.1 *Required conditions for the gauge choice*

As described in section 2.1.1, the lapse function determines the proper time between two spacetime points on two neighboring spatial hypersurfaces. Thus, it controls how the proper time is elapsing forward in time *at each spatial point* on a spatial hypersurface (see the left panel of figure 2.3). In other words, it determines the configuration of a series of spatial hypersurfaces embedded in a spacetime manifold. For this reason, we often refer to the condition for the lapse function as the *time slicing* condition.

The shift vector determines the direction of the time axis *at each spatial point* on a spatial hypersurface (see the right panel of figure 2.3). This implies that the relative position of two neighboring time axes on spatial hypersurfaces changes with time. Thus, the shift vector controls how we choose the spatial coordinates on each spatial hypersurface. For this reason, we often refer to the condition for the shift vector as the *spatial gauge* condition. In the following, we describe the fundamental ideas for fixing these degrees of freedom.

2.2.1.1 *Avoiding the appearance of coordinate singularities*

There are several points to be kept in mind for the choice of the gauge condition in numerical relativity. The first point, which is required both for the lapse function and the shift vector, is that we must not choose a bad gauge condition in which *coordinate singularities* appear. A well-known bad example, irrespective of the dimensionality D, is the geodesic gauge in asymptotically flat spacetime, in which

$$\alpha = 1 \quad \text{and} \quad \beta^i = 0. \tag{2.98}$$

For $\alpha = 1$, the hypersurface normal vector field n^a satisfies $n^b \nabla_b n^a = D^a \ln \alpha = 0$. Thus, n^a satisfies a geodesic equation. With $\beta^a = 0$, in addition, n^a is tangent to time axes. This means that t^a also satisfies the same geodesic equation. For this case, time axes, which initially intersect a spatial hypersurface at different spatial points, could focus each other, and they can often collide at a spacetime point, resulting in a coordinate singularity. In the following, we will describe this property in a specific example.

In vacuum spacetime with $T_{ab} = 0$, equation (2.67) in the geodesic gauge is written as

$$\partial_t K = K_{ij} K^{ij}. \tag{2.99}$$

To show the bad property of this condition even in weakly gravitating spacetime, we consider the case in which gravitational waves of a tiny amplitude initially located around the central region propagate away toward null infinity. Gravitational waves are transverse-tracefree parts of the spatially geometric quantities (see section 1.2). Thus, we decompose K_{ij} as

$$K_{ij} = A_{ij} + \frac{1}{N} \gamma_{ij} K, \tag{2.100}$$

where A_{ij} is the tracefree part of K_{ij}, satisfying $A_{ij} \gamma^{ij} = 0$. Then, equation (2.99) is written as

$$\partial_t K = A_{ij} A^{ij} + \frac{1}{N} K^2. \tag{2.101}$$

This equation implies that $\partial_t K > 0$ in the presence of nonzero A_{ij}. For simplicity we consider the case that $K = 0$ at $t = 0$. This implies that $K > 0$ for $t > 0$. After gravitational waves go away, the equation reduces approximately to

$$\partial_t K = \frac{1}{N} K^2 \quad \text{or} \quad \partial K^{-1} = -\frac{1}{N}. \tag{2.102}$$

Integrating this equation yields the following solution for all the spatial points:

$$K^{-1} = K_0^{-1} - \frac{t - t_0}{N}, \tag{2.103}$$

where $K_0(x^i)$ (> 0) is $K(x^i)$ at $t = t_0$ with t_0 being larger than the time scale for the propagation of gravitational waves. This shows that K diverges at $t = t_0 + N/K_0$; the first divergence occurs at a point at which K_0 is maximum. Equation (2.66) in the geodesic gauge is written as

$$\partial_t \ln \sqrt{\gamma} = -K. \tag{2.104}$$

Thus, for $t \to t_0 + N/K_0$ and $K \to \infty$, γ goes to zero. This obviously shows that a coordinate singularity appears in the finite time.

The divergence of K implies that the focusing singularity appears [Smarr and York (1978a)] because of the bad choice of the gauge condition. Remember that K is written as $-\nabla_a n^a$. Namely, $-K$ is the expansion of n^a. In the geodesic gauge, n^a agrees with t^a. Hence, the divergence of K at a spacetime point implies that several time axes focus at the spacetime point, and the coordinate system breaks down.

Coordinate singularities also appear in a bad choice of the shift vector. This fact is understood by studying asymptotically flat spacetime that has an angular momentum. We here consider a Kerr black hole for which the metric is given by equation (1.101). One of the characteristic features in the Kerr metric is that it has a nonzero shift vector for the φ component. With this choice of the shift vector, the stationary form of the Kerr metric is preserved for given spatial hypersurfaces parametrized by t. This fact is found by evolving the Kerr metric by equation (2.64) choosing the three-metric of the Kerr black hole as the initial condition. Here, α, β^i, γ_{ij}, and K_{ij} for a Kerr black hole are written as

$$\alpha = \sqrt{\frac{\Delta \Sigma}{\Xi}}, \quad \beta^\varphi = -\frac{2Mar}{\Xi},$$

$$\gamma_{rr} = \frac{\Sigma}{\Delta}, \quad \gamma_{\theta\theta} = \Sigma, \quad \gamma_{\varphi\varphi} = \frac{\Xi}{\Sigma}\sin^2\theta,$$

$$K_{r\varphi} = \frac{Ma\sin^2\theta}{(\Xi\Delta)^{1/2}\Sigma^{3/2}}\left[3r^4 + r^2a^2(1+\cos^2\theta) - a^4\cos^2\theta\right],$$

$$K_{\theta\varphi} = -\frac{2Ma^3 r\Delta^{1/2}}{\Xi^{1/2}\Sigma^{3/2}}\cos\theta\sin^3\theta, \qquad (2.105)$$

where we list up only the nonzero components (see section 1.3.1). The extrinsic curvature does not vanish in the presence of an angular momentum of the system. This fact may be inferred from the fact that for a rotating star, solutions of the momentum constraint are not trivial because $J_\varphi \neq 0$. For the Kerr metric, the relation $2\alpha K_{ij} = D_i\beta_j + D_j\beta_i$ is satisfied, and hence, $\partial_t \gamma_{ij} = 0$. However, for the zero shift vector, $\partial_t \gamma_{ij} \neq 0$, and hence, $\gamma_{r\varphi}$ and $\gamma_{\theta\varphi}$ deviate from zero. These induced nonzero components correspond to unphysical gauge modes.

The zero-shift gauge condition causes a more serious problem for systems with angular momenta as indicated from the following analysis. Consider two lines which initially constitute the x-axis ($\varphi = 0$ and π) and y-axis ($\varphi = \pi/2$ and $3\pi/2$) in the Boyer–Lindquist coordinates. Then, we consider a group of static observers located on these lines; each observer is at rest and stationary in these coordinates, i.e., the spatial components of the observer's four velocity is zero, $u^i = 0$. In the 3+1 formulation, the three velocity of test particles, $v^i = dx^i/dt = u^i/u^t$, is written by (see section 4.4.1)

$$v^i = -\beta^i + \frac{\gamma^{ij}u_j}{u^t} = -\beta^i + \frac{\alpha\gamma^{ij}u_j}{\sqrt{1+\gamma^{ij}u_i u_j}}. \qquad (2.106)$$

Thus, for the static observers in the Kerr spacetime with the shift vector given by equation (2.105), we have $u_r = 0 = u_\theta$, and u_φ (the specific angular momentum of

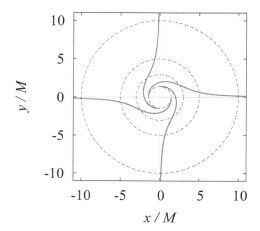

Fig. 2.4 The configuration of the coordinate curves at $t = 10M$ (solid curves), which are initially $x = 0$ and $y = 0$ lines, for the case that the zero shift gauge condition is employed for Kerr spacetime with the Boyer–Lindquist coordinates. In the Kerr metric, the lines are unchanged whereas in the zero shift gauge, it is tightly wound around the origin. The dotted circles are plotted for $r = r_+$, $3M$, $5M$, and $10M$. The spin parameter in this example is $a = 0.9M$.

the observers) is determined by

$$\beta^\varphi = \frac{\alpha \gamma^{\varphi\varphi} u_\varphi}{\sqrt{1 + \gamma^{\varphi\varphi}(u_\varphi)^2}}. \tag{2.107}$$

Then, we consider the zero-shift gauge condition fixing u_φ and α. For the short-term evolution, $\gamma^{\varphi\varphi}$ may be considered to be approximately unchanged. Because β^φ in the Boyer–Lindquist coordinates is negative, v^φ in the zero-shift gauge becomes negative. This implies that the static observers defined above move to the negative direction of φ in the new coordinates. In other words, the coordinate lines rotate in the positive direction of φ with the rotational angular velocity $d\varphi/dt = -\beta^\varphi$, if we consider the static observers to be fiducial observers.

Figure 2.4 displays the variation of the profiles of the two coordinate curves (which were originally located on $x = 0$ and $y = 0$) after the time interval of $\Delta t = 10M$. For drawing this figure, we assume that α, γ_{ij}, and K_{ij} are unchanged for simplicity. This shows that two lines are tightly wound around the origin, i.e., the coordinate curves are highly distorted. The degree of this distortion monotonically increases with time. Associated with this, some components of γ_{ij} monotonically increase (or decrease) and eventually diverge (or be zero). Namely, a coordinate singularity appears in the future. Moreover, γ_{ij} is dominated by an unphysical gauge mode associated with the rotational distortion. For such a situation, the physical content of the spacetime geometry is obscured by the unphysical component of large amplitude.

Figure 2.5 displays bird's-eye views of a numerical-relativity result for $\tilde{\gamma}_{xx} := \gamma_{xx}\gamma^{-1/3}$ on the equatorial plane for Kerr spacetime with $a = 0.6M$ in the zero-

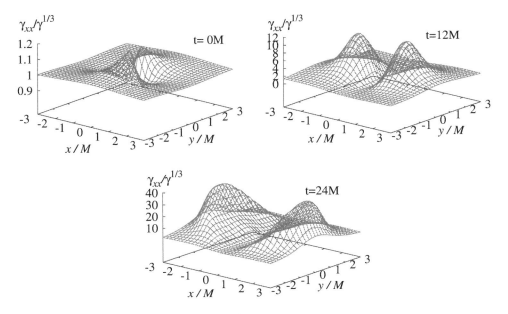

Fig. 2.5 Bird's-eye views of $\tilde{\gamma}_{xx} := \gamma_{xx}\gamma^{-1/3}$ on the equatorial plane for the evolution of Kerr spacetime with $a = 0.6M$ in the zero-shift gauge condition at $t/M = 0$, 12, and 24. Note that the maximum value steeply increases from ~ 1.2 to 40, showing that the metric is dominated by an unphysical content associated with the coordinate distortion.

shift gauge condition at $t/M = 0$, 12, and 24. For this simulation, we initially prepared the Kerr metric in quasi-isotropic coordinates described in section 1.3.1 and employed the dynamical slicing condition (2.158) with (2.160) (see section 2.2.6). In the Kerr metric with an appropriate shift vector, $\tilde{\gamma}_{xx}$ relaxes to a stationary state in a dynamical times scale $\sim 10M$ (see section 3.4.3.2). However, with the zero-shift vector, its maximum value monotonically and rapidly increases from ~ 1.2 to 40 at $t = 24M$, showing that the metric is dominated by an unphysical content associated with the coordinate distortion. The numerical simulation eventually crashed at $t \approx 30M$ for this case, because it was not possible to accurately compute the evolution of the metric with such a large unphysical value and steep gradient. We note that by current numerical-relativity implementations, it is not a problem to follow the Kerr metric for a time scale longer than $10^3 M$ stably and accurately (see section 3.4.3.2). However, for a bad choice of the gauge condition, the lifetime of simulations is much shorter than $100M$, as this example shows. This result clearly illustrates that an appropriate choice of the gauge condition is the key for the successful simulation in numerical relativity.

From the above two examples, we have learned the following lessons: (i) We should not choose $\alpha = 1$ and $\beta^i = 0$ (and also other gauges in which α and β^i are constant) for any dynamical and asymptotically flat spacetime, even if the system is only weakly gravitating. A simple choice does not work. It is necessary to choose gauge conditions physically and geometrically motivated in which α and β^i are adjusted to be suitable for the corresponding spacetime. (ii) For spacetime with angular momentum, the zero-shift gauge condition $\beta^i = 0$ does not work, because the rotation of an object induces coordinate distortion around it. This suggests that the most important role of the spatial gauge condition will be to suppress the coordinate distortion. (iii) We should employ a gauge condition such that the metric in (physically) stationary spacetime remains to be time-independent with it: The gauge condition has to guarantee the stationarity of stationary spacetime in the dynamical evolution. Otherwise, the geometric quantities may be contaminated by unphysical gauge contents. If the unphysical contents dominate the physical one, they may prevent us from extracting the physical information from the geometric quantities.

These three lessens appropriately constrain the choice of the gauge condition. However, they are the minimum requirements for the gauge choice. There is an additional requirement in the presence of black holes, as described in the next section.

2.2.1.2 Avoiding black hole singularities

The lapse function often plays a crucial role for the successful simulation in numerical relativity when black holes are formed and/or present. The reason for this is that black holes possess physical singularities inside their event horizon. If we allow spatial hypersurfaces to proceed forward in time with no braking, a spatial hypersurface Σ_t eventually hits the physical singularity. At this singularity, the spacetime curvature diverges, and such divergence is reflected in some of geometric quantities as divergence. In any numerical computation, it is not possible to handle an extremely large number, and hence, when a spatial hypersurface is too close to the physical singularity, the computation crashes. Such a situation has to be prohibited if we want to perform a long-term simulation of black holes.

A typical bad example for simulating black-hole spacetime is again the geodesic gauge condition. We here consider to foliate Schwarzschild spacetime into a series of spatial hypersurfaces with the initial condition of the Schwarzschild slice at $t = 0$ where t is the Schwarzschild coordinate time. In the geodesic gauge condition, the timelike unit normal vector field, n^a, satisfies geodesic equations and the proper time τ is equivalent to the coordinate time. Thus, by integrating the geodesic equations with respect to τ for a series of the initial radius $r = R_0$ ($\geq 2M$) with $dr/d\tau = 0$ at $\tau = 0$, we find the evolution of spatial hypersurfaces. Denoting τ by a cycloid parameter η as

$$\tau = \left(\frac{R_0^3}{8M}\right)^{1/2} (\eta + \sin \eta), \tag{2.108}$$

for a given value of R_0, the relation between t and r in this gauge condition is

$$r = \frac{R_0}{2}(1 + \cos \eta), \qquad (2.109)$$

$$t = 2M \left[\ln \left| \frac{(R_0/2M - 1)^{1/2} + \tan(\eta/2)}{(R_0/2M - 1)^{1/2} - \tan(\eta/2)} \right| \right.$$
$$\left. + \left(\frac{R_0}{2M} - 1 \right)^{1/2} \left\{ \eta + \frac{R_0}{4M}(\eta + \sin \eta) \right\} \right]. \qquad (2.110)$$

This shows that at $\tau = \pi M$ (i.e., for a particle of $r = R_0 = 2M$ at $t = 0$), the spatial hypersurface first hits the physical singularity at $r = 0$.

To avoid the situation that the hypersurface hits or approaches closely physical singularities, it is necessary to appropriately control the lapse function. Specifically, the elapse of the proper time should be forced to be zero in the vicinity of the physical singularities by adjusting the magnitude of the lapse function. If a slicing condition has such a property, we refer to it as a condition that possesses a *singularity-avoidance* property. In the following sections, we will describe several slicing conditions that have the singularity-avoidance property.

2.2.2 Maximal slicing

The maximal slicing condition is defined by

$$K = 0 = \partial_t K. \qquad (2.111)$$

This slicing condition is expected to have several required properties for slicing of asymptotically flat spacetime.[1] First, several stationary spacetime including stationary black holes and stationary neutron-star spacetime (see section 1.4.3) satisfy the condition (2.111) for a standard gauge condition. Thus, with this slicing condition, a number of stationary spacetime can be observed as stationary one; one of the three conditions described in section 2.2.1.1 is satisfied.

The second good sign is found from equation (2.66), which yields the following equation for $K = 0$,

$$\partial_t \sqrt{\gamma} = \partial_i \left(\sqrt{\gamma} \beta^i \right). \qquad (2.112)$$

This shows that the proper volume element $\sqrt{\gamma}$ satisfies a continuity equation in maximal slicing. This suggests that γ would not behave in a bad manner as far as a regular shift vector is chosen; e.g., for a special case $\beta^i = 0$, $\gamma = \text{const}$, and hence, as far as the initial condition for γ_{ab} is regular, γ is regular forever. Near physical singularities, γ often goes to zero as found from several metric forms of black holes with singularity non-avoidance slicing (see section 1.3.1). Equation (2.112) suggests that with the maximal slicing condition, such bad behavior could be avoided, and hence, it is likely to have a singularity-avoidance property.

---
[1] For other types of spacetime, see, e.g., section 2.6.

A good sign is also found from equation (2.67), which yields an elliptic equation for the lapse function for $K = 0$ as

$$D_i D^i \alpha = 8\pi\alpha \frac{S_k{}^k + (N-2)\rho_\mathrm{h}}{N-1} + \alpha A_{ij} A^{ij}, \qquad (2.113)$$

where A_{ij} is the tracefree part of K_{ij} ($A_{ij}\gamma^{ij} = 0$). Combining the Hamiltonian constraint in the form of equation (2.95), equation (2.113) may be written to

$$\tilde{\Delta}(\alpha\psi) = \frac{N-2}{4(N-1)}\alpha\psi\tilde{R} + \frac{4\pi}{N-1}\left[(N-2)\rho_\mathrm{h} + 2S_k{}^k\right]\alpha\psi^{(N+2)/(N-2)}$$
$$+ \frac{3N-2}{4(N-1)}\alpha\psi^{(N+2)/(N-2)} \tilde{A}_{ij}\tilde{A}^{ij}. \qquad (2.114)$$

This indicates that α and $\alpha\psi$ should be smooth functions as far as the spatial metric is regular, because they are solutions of these elliptic equations. Thus, the appearance of coordinate singularities associated with this slicing condition is expected to be avoided as far as the regular spacetime is concerned. This property is also expected from the fact that $K = 0$ is equivalent to $\nabla_a n^a = 0$, which implies that the focusing of the timelike unit normal vector field (and hence, the appearance of an irregular region on spatial hypersurfaces) is prohibited.

In 1970s and 80s, several important works were done for studying the properties of maximal slicing. These works indeed show that spatial hypersurfaces with maximal slicing have required properties in numerical relativity. The most important fact of maximal slicing is that it has a singularity-avoidance property for spherically symmetric black-hole spacetime not only in 4 dimensions but also in higher dimensions. This implies that a series of spatial hypersurfaces asymptotically approaches the so-called limit hypersurface before a hypersurface hits the physical singularity of the black holes. At the limit hypersurface, the lapse function becomes zero inside event horizons, i.e., the elapse of the proper time is stopped even if the coordinate time proceeds. This fact was shown by Estabrook *et al.* (1973) for the 4-dimensional case and by Nakao *et al.* (2009) for the higher-dimensional cases. A series of spatial hypersurfaces in maximal slicing is derived in the same manner for any dimensionality without loss of generality. In the following, we describe how to derive such hypersurfaces.

There are infinite numbers of a set of maximal-sliced spatial hypersurfaces, depending on the boundary condition of the lapse function that obeys an elliptic equation. For example, a series of spatial hypersurfaces labeled by the Schwarzschild time coordinate is the most well-known example. For this case, the boundary condition is $\alpha = 0$ on the event horizon; e.g., tilted lines in regions I and III in figure 1.9 denote such spatial hypersurfaces. Namely, the region inside the event horizon (regions II and IV according to the definition of figure 1.9) is not covered in this case. In this maximal slicing, not only K but also $K_\mathrm{T} := K_\theta{}^\theta + K_\varphi{}^\varphi$ is zero (hence, $K_r{}^r = 0$ in this case). Therefore, it is often called by a different name as *polar slicing* (see section 2.4.4). In the following, we will consider black-hole spacetime with the wormhole topology as depicted in figure 1.8, and for determining each spatial hypersurface, we impose the symmetric boundary condition at the minimal surface of the wormhole throat (i.e., the vertical line of $t = 0$ in regions II and IV in figure 1.9). Namely, we consider the slicing that covers a wide domain which extends to regions I–III (or I, IV, and III) in figure 1.9.

A method of analysis for maximal slicing of spherically symmetric black holes was developed originally by Estabrook *et al.* (1973), and can be used for any spherically symmetric black-hole spacetime irrespective of the dimensionality D [Nakao *et al.* (2009)]. Thus, we consider spherically symmetric black holes of arbitrary dimensionality. We start from the Schwarzschild–Tangherlini metric as [Tangherlini (1963)] (see section 1.3.3)

$$ds^2 = -\left[1 - \left(\frac{r_g}{r}\right)^{D-3}\right] dt_{\rm s}^2 + \left[1 - \left(\frac{r_g}{r}\right)^{D-3}\right]^{-1} dr^2 + r^2 d\Omega_{D-2}, \quad (2.115)$$

where $t_{\rm s}$ denotes the "time" coordinate in the Schwarzschild–Tangherlini spacetime ($t_{\rm s}$ is not always timelike). $d\Omega_{D-2}$ is the line element of $(D-2)$-dimensional unit sphere, as before. Following Estabrook *et al.* (1973), the line element of the black hole in the maximal slicing condition is written as

$$ds^2 = -\alpha^2 dt^2 + f(dr + f^{-1}\beta dt)^2 + r^2 d\Omega_{D-2}, \quad (2.116)$$

where α, β, and f are functions of r and t with t being the new time coordinate, which we require to be always timelike. α, β, and f are determined by Einstein's equation. Specifically, we only need to consider the maximal slicing condition $K = 0 = \partial_t K$, the Hamiltonian and momentum constraints, and the evolution equation of K_{rr} because of the spherical symmetry. The conditions $K = 0$ and $\partial_t K = 0$, respectively, yield

$$-\partial_t \ln f + \beta f^{-1} \partial_r \ln \left[\beta^2 f^{-1} r^{2(D-2)}\right] = 0, \quad (2.117)$$

$$\left[\partial_r^2 + \frac{D-2}{r}\partial_r - \frac{1}{2}(\partial_r \ln f)\partial_r\right]\alpha = \frac{D-2}{r^2}\left[(D-3)(f-1) + r\partial_r \ln f\right]\alpha. \quad (2.118)$$

The Hamiltonian and momentum constraints, respectively, yield

$$(D-1)\alpha^{-2}\beta^2 f^{-1} = (D-3)(f-1) + r\partial_r \ln f, \quad (2.119)$$

$$\partial_r \ln\left(\alpha^{-1}\beta f^{-1} r^{D-2}\right) = 0, \quad (2.120)$$

where we used $\alpha K_{\theta\theta} = D_\theta \beta_\theta = -\gamma_{rr}\Gamma^r{}_{\theta\theta}\beta = r\beta$, and $K_r{}^r = -(D-2)K_\theta{}^\theta$. Finally, the evolution equation of K_{rr} results in

$$\partial_t \ln(\alpha^{-1}\beta) = \left[(2D-5)\beta f^{-1} + (D-3)\alpha^2 \beta^{-1}(f-1)\right] r^{-1}$$

$$+ 3f^{-1}\partial_r \beta + \frac{\alpha^2 \beta^{-1} - 4\beta f^{-1}}{2}\partial_r \ln f - (\beta f^{-1} + \alpha^2 \beta^{-1})\partial_r \ln \alpha. \quad (2.121)$$

Equation (2.120) is immediately solved to yield

$$\beta = T(t)\alpha f r^{-(D-2)}, \qquad (2.122)$$

where $T(t)$ is a function only of t. Substituting equation (2.122) into equation (2.119), we obtain a linear first-order equation for f^{-1} as

$$\partial_r \left(r^{D-3} f^{-1} \right) = (D-3) r^{D-4} - (D-1) T(t)^2 r^{-D}, \qquad (2.123)$$

for which the solution is

$$f^{-1} = 1 - \left(\frac{r_c(t)}{r} \right)^{D-3} + \frac{T(t)^2}{r^{2(D-2)}}, \qquad (2.124)$$

where $r_c(t)$ is a function of t at this stage.[2]

Eliminating β from equation (2.117) and using equation (2.124), we obtain

$$\partial_r \left(\alpha f^{1/2} \right) = f^{3/2} \left[\frac{r}{2T(t)} \partial_t \left(r_c^{D-3} \right) - r^{-(D-2)} \partial_t T(t) \right]. \qquad (2.125)$$

On the other hand, eliminating β from equation (2.121) and rewriting the result with help of equation (2.124), we find that $\partial_t r_c = 0$. Because r_c should be r_g for $T(t) = 0$, we find $r_c = r_g$. Then, equation (2.125) is integrated to yield

$$\alpha = \left[1 - \left(\frac{r_g}{r} \right)^{D-3} + \frac{T(t)^2}{r^{2(D-2)}} \right]^{1/2}$$
$$\times \left[1 + \frac{\partial_t T(t)}{r_g^{D-3}} \int_0^{r_g/r} dx\, x^{D-4} \left(1 - x^{D-3} + T(t)^2 r_g^{-2(D-2)} x^{2(D-2)} \right)^{-3/2} \right]. \qquad (2.126)$$

Equations (2.122), (2.124), and (2.126) show that if an arbitrary function $T(t)$ is given, solutions for α, β, and f are determined. Here, $T(t)$ is determined by specifying boundary conditions of spatial hypersurfaces. Following Estabrook *et al.* (1973), we impose a condition in which each spatial hypersurface has an inversion symmetry with respect to the minimal surface of the wormhole throat.

The location of the minimal surface of the wormhole throat is denoted by $t_s = 0$ in the coordinates of the metric (2.115) (see figure 1.9). Here, we note that t_s is a spatial coordinate in the black-hole interior. On the other hand, r is timelike in the black-hole interior and the coordinate value at the minimal surface of the throat decreases with the elapse of the proper time from $r = r_g$ to $r = 0$ (see figure 1.9). In contrast to t_s, t in the metric (2.116) is the timelike coordinate in any region. In particular, the inner product of $(\partial/\partial t)^a$ and $(\partial/\partial t_s)^a$ is zero at $t_s = 0$ when we

---

[2] Since f diverges at a radius where $f^{-1} = 0$, the metric (2.116) has a coordinate singularity. However, as long as the roots are not degenerate, this singularity does not cause any serious problem because it can be erased by simply employing a new coordinate $\tilde{r} = \int dr f^{1/2}$. However, in the presence of the double root, the coordinate singularity has a fundamental problem as discussed in section 2.2.4.

impose the inversion symmetric condition. Thus, the condition for $T(t)$ is derived from

$$\left(\frac{\partial}{\partial t_s}\right)^a \nabla_a t = \frac{\partial t}{\partial t_s} = 0, \quad \text{at the minimal surface.} \tag{2.127}$$

The relation between t and t_s is derived from the relations of the coordinate transformation as follows:

$$\frac{\partial t_s}{\partial t} = \alpha f^{1/2}, \tag{2.128}$$

$$\frac{\partial t_s}{\partial r} = -f^{1/2} T r^{-(D-2)} \left[1 - \left(\frac{r_g}{r}\right)^{D-3}\right]^{-1}. \tag{2.129}$$

Equation (2.129) is solved to give

$$t_s = T r_g^{-(D-3)} \int_{r_g/r}^{X(T)} \frac{x^{D-4}}{(x^{D-3} - 1)\sqrt{1 - x^{D-3} + T(t)^2 r_g^{-2(D-2)} x^{2(D-2)}}} dx, \tag{2.130}$$

which also satisfies equation (2.128) provided that the function $X(T)$ satisfies

$$\frac{dX}{dT} = T(t)^{-1} X^{-(D-4)} (X^{D-3} - 1)\sqrt{1 - X^{D-3} + T(t)^2 r_g^{-2(D-2)} X^{2(D-2)}}$$

$$\times \left[\frac{r_g^{D-3}}{\partial_t T(t)} + \int_0^X \frac{x^{D-4}}{\left[1 - x^{D-3} + T(t)^2 r_g^{-2(D-2)} x^{2(D-2)}\right]^{3/2}} dx\right]. \tag{2.131}$$

In the inversion symmetric boundary condition at the the minimal surface of the wormhole throat, $X(T)$ has to be determined by the smallest value of r (the value of r at the minimal surface of the wormhole throat) to which we refer as $r = r_{\min}$. Thus, $X(T) = r_g/r_{\min}$. From the requirement (2.127), the value of $X(T)$ has to be a solution of

$$F(x) = 1 - x^{D-3} + T(t)^2 r_g^{-2(D-2)} x^{2(D-2)} = 0. \tag{2.132}$$

Then, there are at most two real roots for equation (2.132). $X(T)$ is the smaller root, because for the larger root, the integrand of equation (2.131) becomes imaginary and thus unphysical.

As we mentioned above, $r_{\min}$ decreases with the elapse of the proper time. Thus, $X(T)$ increases with the elapse of the proper time. To know how $X(T)$ proceeds with time, we have to determine the relation between $T(t)$ and t. This is derived by requiring that t agrees with t_s for $r \to \infty$. From equation (2.130), this condition is written as

$$t = T r_g^{-(D-3)} \int_0^{X(T)} \frac{x^{D-4}}{(x^{D-3} - 1)\sqrt{1 - x^{D-3} + T(t)^2 r_g^{-2(D-2)} x^{2(D-2)}}} dx. \tag{2.133}$$

If equation (2.132) does not have a degenerate double root, t is finite, whereas if it has, the integral of (2.133) diverges, i.e., $t = \infty$ is reached, although T and X are

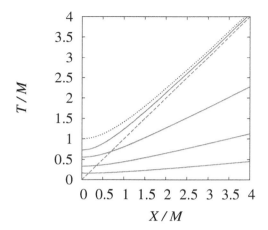

Fig. 2.6 A series of spatial hypersurfaces in maximal slicing is plotted on a Kruskal–Szekeres diagram for the four-dimensional non-spinning black hole. The hypersurfaces are plotted by the thick solid curves at $T/T_\infty = 0.3, 0.6, 0.9$, and 1. The top one is the limit hypersurface ($r = 1.5M$). The dotted curve and dashed line denote the physical singularity ($r = 0$) and the event horizon ($r = 2M$), respectively (cf. figure 1.9).

still finite. This means that the elapse of the proper time along a series of spatial hypersurfaces stops before the physical singularity ($X = \infty$) is reached, although the elapse of the coordinate time proceeds up to infinity. Such the spatial hypersurface, which a series of spatial hypersurfaces in maximal slicing asymptotically approach, is called the *limit hypersurface*.

Because the double root of equation (2.132) is also the root of $dF/dx = 0$, we can easily derive the maximum value of $T(t)$ as

$$T_\infty^2 = T(t = \infty)^2 = \frac{D-3}{2(D-2)} \left[\frac{D-1}{2(D-2)} \right]^{(D-1)/(D-3)} r_g^{2(D-2)}. \quad (2.134)$$

Then, the minimum value of $r_{\min}$, $r_{\lim}$, is given by

$$r_{\lim} = \left[\frac{D-1}{2(D-2)} \right]^{1/(D-3)} r_g. \quad (2.135)$$

Thus, maximally sliced hypersurfaces approach a spatial hypersurface distant from the physical singularity located at $r = 0$. This shows that maximal slicing has the singularity-avoidance property. Figure 2.6 plots selected spatial hypersurfaces in maximal slicing on a Kruskal–Szekeres diagram. The thick solid curves denote the spatial hypersurfaces, and the top one denotes the limit hypersurface (the dotted curve is the physical singularity, $r = 0$). This clearly shows that the limit hypersurface is away from the physical singularity.

One of the most important properties of maximal slicing is that the lapse function inside the event horizon quickly approaches zero. Figure 2.7 shows the profiles

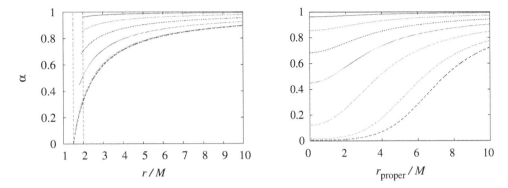

Fig. 2.7 Evolution of the profile of α as a function of r (left panel) and as a function of r_{proper} (right panel) for $T/T_\infty = 0.2$, 0.4, 0.6, 0.8, 0.98, 0.9998, and 0.99998. Note that the values of α decrease with time. A four-dimensional black hole is considered here. The dashed vertical lines in the left panel are $r = r_{\text{lim}}(= 1.5M < 2M)$ and $r = 2M$, respectively.

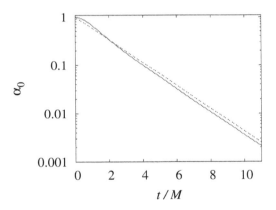

Fig. 2.8 Evolution of the value of α at the minimal surface of the throat ($r = r_{\text{min}}$) as a function of t. The dashed tilted line is $e^{-t/\tau}$ where $\tau = (3/2)^{3/2}M$, which is derived by Petrich et al. (1985).

of α at selected time slices as a function of r (left panel) and as a function of the proper distance (right panel) defined by

$$r_{\text{proper}} = \int_{r_{\text{min}}}^{r} dr' f(r')^{1/2}. \tag{2.136}$$

Here, a four-dimensional black hole is considered but the qualitative feature is independent of the dimensionality. The left panel shows that α approaches zero only at the minimal surface of the wormhole throat, $r = r_{\text{min}}$. However, the right panel shows that the region with $\alpha \approx 0$ expands when the spatial hypersurfaces approach the limit hypersurface. This illustrates the fact that for a wide region

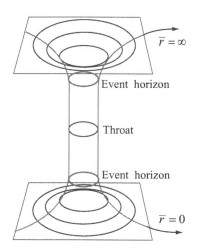

Fig. 2.9 Schematic embedding diagram of the Schwarzschild geometry near the limit hypersurface for which a long cylindrical topology appears. The spatial hypersurface with $t = $ const on the equatorial plane $\theta = \pi/2$ is embedded as a two-dimensional surface in a three-dimensional flat space. The minimum surface of the wormhole throat is located at $r = r_{\min}$ and it is not the surface of the event horizon in this case.

inside the event horizon, α approach zero (see the next paragraph). Figure 2.8 plots the value of the lapse function at $r = r_{\min}$ as a function of t. This shows that α approaches zero in an exponential manner with respect to the coordinate time [Smarr and York (1978a); Petrich et al. (1985)]. This implies that even for $t \to \infty$, the proper time calculated by $\int \alpha dt$ is still finite.

In the limit hypersurface, $r = r_{\lim}$ is satisfied everywhere inside the event horizon. From equation (2.126), we find $\alpha = 0$ inside the event horizon in the limit hypersurface. This reconfirms that the elapse of the proper time of spatial hypersurfaces stops inside the event horizon in maximal slicing. It is also worthy to note that the limit hypersurface is the static slice. Thus, the metric form in this hypersurface is one of several forms for the exact solution of non-rotating black holes.

In the above example, maximal slicing is applied to vacuum spacetime. It is not clear only from this example whether maximal slicing also has a singularity-avoidance property for non-vacuum spacetime. Eardley and Smarr (1978) explored spherically symmetric gravitational collapse of pressure-free dust in maximal slicing and found that as far as the degree of inhomogeneity of the collapsing dust is not very high, maximal slicing has a strong singularity-avoidance property. A semi-analytic solution for the uniform dust collapse [Oppenheimer and Snyder (1939)] in maximal slicing was also given by Petrich et al. (1985) and shows that maximal slicing has a singularity-avoidance property.

One remarkable point of maximal-sliced hypersurfaces near the limit hypersurface of non-spinning black holes is that a proper length of the wormhole throat (e.g., proper length between inner and outer event horizons) increases with the elapse of the proper time, and at the limit hypersurface, the proper length diverges. Figure 2.9 schematically depicts this fact. This stretching effect is understood by the fact that the proper distance between the minimal surface of the wormhole throat and a point along constant angular coordinates is written by

$$\int_{r_{\min}}^{r} f(y)^{1/2} dy = \int_{r_{\min}}^{r} \left[1 - (r_g/y)^{D-3} + T^2/y^{2(D-2)}\right]^{-1/2} dy, \quad (2.137)$$

where $r > r_{\min}$. This integral diverges in the logarithmic manner at the limit hypersurface. This means that with maximal slicing, the region inside the event horizon is not resolved in numerical simulations. This problem is discussed in section 2.2.4 in more detail.

2.2.3 Harmonic gauge

The harmonic gauge condition is defined by

$$\Box_g x^\alpha = 0, \quad (2.138)$$

and thus,

$$\partial_\beta \left(g^{\alpha\beta} \sqrt{-g}\right) = 0, \quad (2.139)$$

where we supposed to use Cartesian coordinates for x^i. In the $N+1$ formulation, the harmonic gauge condition settles to the following evolution equations for α and β^i:

$$(\partial_t - \beta^k \partial_k)\alpha = -\alpha^2 K, \quad (2.140)$$

$$(\partial_t - \beta^k \partial_k)\beta^i = -\frac{\alpha}{\sqrt{\gamma}} \partial_j \left(\alpha \sqrt{\gamma} \gamma^{ij}\right). \quad (2.141)$$

Thus, the equations are completely different from those of maximal slicing and minimal distortion gauge condition (see section 2.2.5) in which equations are elliptic-type. In the harmonic gauge condition, all the metric functions, α, β^i, and γ_{ij}, obey hyperbolic equations. From the viewpoint of numerical computations, it is a desired property because the computational costs for solving hyperbolic equations in initial-value problems are in general much less expensive than for solving elliptic equations in a boundary-value problem.

As shown in section 1.1.2, the hyperbolic nature of Einstein's equation is explicitly found in this gauge condition. Thus, it is natural to expect that this condition is suitable for following the propagation of gravitational waves, and for numerically extracting gravitational waves in the wave zone. Furthermore, axisymmetric stationary black holes are written to be in a stationary form in harmonic coordinates [Cook and Scheel (1997)]. This indicates that with an appropriate choice of the initial condition, coordinate distortion is always suppressed, and also, stationary spacetime are described in a stationary manner in the harmonic gauge. Then, a remaining issue is whether it has a singularity-avoidance property in the presence of black holes (see also section 2.3.4 for other possible issues). In the following, we will show a *weak* singularity-avoidance property for harmonic slicing.

Although the harmonic gauge condition determines the equations both for the lapse function and the shift vector, we do not have to employ both conditions together. For example, we may combine the harmonic slicing condition (2.140) with other spatial gauge condition. Motivated by this idea, Bona and Massó (1988, 1992) explored the property of the harmonic slicing condition and showed that it has a weak singularity-avoidance property [Bona and Massó (1988); Alcubierre (2003)]. Following them, we here describe this property choosing the zero-shift vector for simplicity. For the case $\beta^k = 0$, equation (2.140) may be written as

$$\partial_t \left(\alpha \gamma^{-1/2} \right) = 0, \tag{2.142}$$

where we used $\alpha K \sqrt{\gamma} = -\partial_t \sqrt{\gamma}$. Thus, for given spatial coordinates x^i, we have the relation

$$\alpha = A(x^i)\sqrt{\gamma}, \tag{2.143}$$

where $A(x^i)$ is a function of x^i and is determined by the initial condition of $\alpha/\sqrt{\gamma}$. We here suppose that $A(x^i)$ is a regular function.

Bona and Massó (1988) supposed that at a physical singularity that hypothetically appears at a proper time τ_S and at spatial coordinates x^i, the spatial volume element $\sqrt{\gamma}$ vanishes and the proper-time derivative of $\sqrt{\gamma}$ is bounded. Thus,

$$\sqrt{\gamma(\tau_S)} = 0, \quad |\partial_\tau \sqrt{\gamma}|_{\tau=\tau_S} < C, \tag{2.144}$$

where C is a finite constant. The above assumptions for γ imply that $\sqrt{\gamma}$ behaves as $(\tau_S - \tau)^n$ with $n \geq 1$ in the Laurent expansion for $\tau \to \tau_S$. This is equivalent to assuming that K does not diverge faster than $1/\sqrt{\gamma}$ near the physical singularity.

Variation of the proper time along constant spatial coordinates is written as $d\tau = \alpha dt$, and hence,

$$\Delta t = \int dt = \int_0^{\tau_S} \alpha^{-1} d\tau \to \infty. \tag{2.145}$$

This means that the physical singularities defined by equation (2.144) are not reached in a finite coordinate time, t; the harmonic slicing condition has a singularity-avoidance property.

However, this singularity-avoidance property is weaker than that of maximal slicing, because spatial hypersurfaces eventually hit the physical singularity for $t \to \infty$ in harmonic slicing: The hypersurfaces may be close to the physical singularity for a sufficiently large value of t. In other words, in this slicing condition, there is no limit hypersurface that exists for maximal slicing. Near the physical singularities, γ is close to zero and absolute values of some of geometrical quantities are likely to be very large. In numerical computations, such a small and/or large number cannot be evolved accurately; e.g., by numerical errors, γ could be smaller than zero. This suggests that the harmonic slicing condition might not have a desired singularity-avoidance property in the context of numerical relativity.

2.2.4 Issues for black hole evolution and excision

Maximal slicing has a singularity-avoidance property, and hence, it could be well-suited for evolving black-hole spacetime. However, there is a problem raised by this choice as described in the following: With maximal slicing, the proper distance from event horizon to the center of the wormhole throat on a spatial hypersurface steeply increases when the spatial hypersurfaces approach the limit hypersurface, as shown in equation (2.137). Eventually the distance becomes infinite (see figure 2.9). This implies that the proper distance between two neighboring grid points located inside the event horizon monotonically and quickly increases, and hence, the grid resolution becomes monotonically poor. Specifically, this effect is reflected in geometric quantities in a pathological manner: The magnitude of some components of the spatial metric increases quickly near the limit hypersurface leading to divergence, i.e., a coordinate singularity appears (see for example the behavior of $\tilde{r} = \int f^{1/2} dr$ in section 2.2.2 which diverges at the limit hypersurface). Because of this pathology, numerical computations often crashed in the past simulations for the case that spatial hypersurfaces approach the limit hypersurface. This phenomenon is called *horizon sucking* or *horizon stretching*. Thus, when we employ the maximal slicing condition or a slicing condition that has a singularity-avoidance property, an additional prescription is necessary for avoiding this problem.

The most popular idea for avoiding the coordinate singularities (as well as physical singularities) is to excise a stretched region inside black-hole horizons from computational regions and to solve basic equations imposing boundary conditions on the excised surfaces. This idea was originally proposed by Unruh [as cited in Thornburg (1987)] and first implemented by Seidel and Suen (1992) in their spherically symmetric simulation. Numerical implementations based on this idea are called *black-hole excision* algorithm. If the excised regions are located inside apparent horizons (see section 7.1 for the apparent horizon), it would be guaranteed that any information cannot escape from them, because the apparent horizons are located inside black-hole event horizons for globally hyperbolic spacetime. Thus, we should impose an ingoing boundary condition at the excised boundaries. However, it is not easy to impose the exact ingoing boundary condition for rather arbitrarily excised regions. As a consequence of imposing approximate conditions, numerical computations could crash or numerical accuracy be poor near the black holes by an accumulation of the numerical error. For a successful excision algorithm, we need an appropriate formulation and gauge conditions: The formulation and the gauge conditions have to guarantee that all the information inside the black hole never escape from the black hole, and also, any numerical instability does not set in. We will describe one of the most popular conditions that have the desired properties in section 2.2.6.

Even if a slicing condition has only a weak singularity-avoidance property, it enables us to perform stable evolution of black-hole spacetime by using a black-hole excision algorithm, i.e., by excising a region inside the black-hole horizon that will eventually hit the black-hole physical singularity. Numerical experiments have shown that this is possible with the generalized harmonic formulation in which a slicing condition similar to the harmonic slicing condition (that has only a weak singularity-avoidance property) is employed. We will outline this in section 2.3.4.

2.2.5 Minimal distortion gauge

The idea of minimal distortion gauge was first proposed by Smarr and York as a radiation gauge in general relativity [Smarr and York (1978b,a)], which is suitable for suppressing the accumulation of the coordinate distortion and for extracting gravitational waves from geometric quantities in a wave zone. Let h_{ij} be a linear perturbation component of the conformal spatial metric, $\tilde{\gamma}_{ij} - f_{ij}$. Here, we suppose $\tilde{\gamma} = f$ and h_{ij} is tracefree, $h_{ij}f^{ij} = 0$, in the linear perturbation level. (For simplicity, we hereafter suppose to use Cartesian coordinates in this section.) In general, the magnitude of h_{ij} depends on the chosen gauge condition, and to extract gravitational-wave components from h_{ij} in a straightforward manner, the magnitude of gauge components in h_{ij} should be as small as possible. For this purpose, a transverse-tracefree gauge condition imposing $\sum_i \partial_i h_{ij} = 0$ is likely to be useful. However, this condition cannot be written in a covariant way because $\tilde{D}_i \tilde{\gamma}^{ij}$ are trivially zero and a quantity such as $D_i \tilde{\gamma}^{ij}$ does not yield the transverse gauge.

Instead of directly imposing a transverse-tracefree gauge condition, Smarr and York (1978b) proposed the following condition,

$$D_i \left[\gamma^{-1/N} \partial_t \left(\gamma^{1/N} \gamma^{ij} \right) \right] = 0, \qquad (2.146)$$

or in the conformal decomposition formulation,

$$\tilde{D}_i \left(\sqrt{\tilde{\gamma}} \partial_t \tilde{\gamma}^{ij} \right) = 0. \qquad (2.147)$$

This is usually referred to as the minimal distortion gauge condition. In the wave zone, this condition may be written as

$$\partial_t \left(\sum_i \partial_i h_{ij} \right) = 0. \qquad (2.148)$$

Thus, if an initial condition satisfies $\sum_i \partial_i h_{ij} = 0$, the transverse-tracefree condition is preserved in this gauge condition.

Using the evolution equation for $\tilde{\gamma}^{ij}$, equation (2.147) is written to a vector elliptic equation for $\tilde{\beta}^k$ as

$$\tilde{D}^i \tilde{D}_i \tilde{\beta}_j + \frac{N-2}{N} \tilde{D}_j \tilde{D}_i \tilde{\beta}^i + \tilde{R}_{jk} \tilde{\beta}^k + \frac{1}{\sqrt{\tilde{\gamma}}} \tilde{D}^i \sqrt{\tilde{\gamma}} \left(\tilde{D}_i \tilde{\beta}_j + \tilde{D}_j \tilde{\beta}_i - \frac{2}{N} \tilde{\gamma}_{ij} \tilde{D}_k \tilde{\beta}^k \right)$$
$$- 2\tilde{A}_{ij} \tilde{D}^i \alpha - \frac{2(N-1)}{N} \alpha \tilde{D}_j K = 16\pi \alpha J_j, \qquad (2.149)$$

where $\tilde{\beta}_k = \tilde{\gamma}_{kl} \beta^l$ and $\tilde{\beta}^k = \beta^k$. Accordingly, the minimal distortion gauge provides the condition for the shift vector.

Smarr and York (1978b) showed that this gauge condition is also derived by minimizing the following action I with respect to β^k on each spatial hypersurface

$$I = \int d^N x (\partial_t \tilde{\gamma}_{ij})(\partial_t \tilde{\gamma}_{kl}) \tilde{\gamma}^{ik} \tilde{\gamma}^{jl} \sqrt{\gamma}. \tag{2.150}$$

Thus, by choosing the minimal distortion gauge condition, the global change rate of $\tilde{\gamma}_{ij}$ *based on the action I* is minimized in every spatial hypersurface. Thus, the change rate of the coordinate distortion is globally minimized. This is the reason why it is called the minimal distortion gauge.

Here, the following point is worthy to note. In the minimal distortion gauge condition described above, the *change rate* of coordinate distortion is minimized. Thus, if large coordinate distortion is present in an initial condition, it is badly preserved. Therefore, a good initial choice for the spatial gauge condition is necessary when the minimal distortion gauge condition is employed.

Since I is not the unique action that defines the global distortion, we may consider an alternative minimal distortion gauge condition by slightly changing the definition of the action. Shibata (1999c) proposed another action I' as

$$I' = \int d^N x (\partial_t \tilde{\gamma}_{ij})(\partial_t \tilde{\gamma}_{kl}) \tilde{\gamma}^{ik} \tilde{\gamma}^{jl}. \tag{2.151}$$

This corresponds to defining a certain action in the conformal space in which the determinant of the metric $\det(\tilde{\gamma}_{ij})$ is unity (assuming that Cartesian coordinates are employed). In this case, by taking the variation of I' with respect to $\tilde{\beta}^k$, we obtain a slightly different gauge condition,

$$\tilde{D}_i(\partial_t \tilde{\gamma}^{ij}) = 0, \tag{2.152}$$

or for $\tilde{\beta}^k$,

$$\tilde{D}^i \tilde{D}_i \tilde{\beta}_j + \frac{N-2}{N} \tilde{D}_j \tilde{D}_i \tilde{\beta}^i + \tilde{R}_{jk} \tilde{\beta}^k$$
$$- 2\tilde{A}_{ij}\left(\tilde{D}^i \alpha - \frac{\alpha}{\sqrt{\gamma}} \tilde{D}^i \sqrt{\gamma}\right) - \frac{2(N-1)}{N} \alpha \tilde{D}_j K = 16\pi \alpha J_j. \tag{2.153}$$

A merit of this condition is that a coupling term between $\tilde{\beta}^k$ and $\tilde{D}^i \sqrt{\gamma}$ is absent, and hence, the equation for $\tilde{\beta}^k$ is slightly simpler than equation (2.149), although it is still a complicated vector elliptic equation.

Although it is desirable to adopt a shift vector in which I or I' is *exactly* minimized, this may not be always necessary. We may employ a slightly different shift vector in which I or I' is *approximately* minimized. Using this idea, Shibata (1999c) proposed a gauge condition by simplifying equation (2.153) as

$$\delta_{ij}\Delta_{\mathrm{f}}\beta^i + \frac{N-2}{N}\beta^k{}_{,kj}$$
$$- 2\tilde{A}_{ij}\left(\tilde{D}^i \alpha - \frac{\alpha}{\sqrt{\gamma}} \tilde{D}^i \sqrt{\gamma}\right) - \frac{2(N-1)}{N} \alpha \tilde{D}_j K = 16\pi \alpha J_j. \tag{2.154}$$

Here, $\Delta_{\rm f}$ is the flat-space Laplacian. In this case, equation (2.154) can be rewritten into simple elliptic equations for a vector P_i and a scalar η using the following decomposition,

$$\beta^j = \delta^{ji} \left[\frac{3N-2}{4(N-1)} P_i - \frac{N-2}{4(N-1)} \left(\eta_{,i} + P_{k,i} x^k \right) \right], \qquad (2.155)$$

where P_i and η satisfy

$$\Delta_{\rm f} P_i = S_i, \qquad \Delta_{\rm f} \eta = -S_i x^i, \qquad (2.156)$$

and

$$S_i = 16\pi \alpha J_i + 2\tilde{A}_{ij} \left(\tilde{D}^j \alpha - \frac{2N\alpha}{(N-2)\psi} \tilde{D}^j \psi \right) + \frac{2(N-1)}{N} \alpha \tilde{D}_i K. \qquad (2.157)$$

Here, we assumed to use Cartesian coordinates for performing the decomposition (2.155). Shibata and Uryū (2000, 2002) showed that this approximately-minimal distortion gauge condition works well in their early simulations of binary neutron star mergers.

2.2.6 Dynamical gauge

From the purely physical point of view, the maximal slicing and minimal distortion gauge conditions are excellent as the slicing and spatial gauge conditions, respectively. However, to impose these conditions, we have to numerically solve elliptic equations. In multi-dimensional numerical simulations, solving elliptic equations in a boundary-value problem is computationally expensive. In addition, we need an appropriate boundary condition for solving elliptic equations. Such boundary conditions are easily found for spatial infinity (and hence finding the boundary condition does not matter in the absence of black holes). However, in the presence of black holes, it is often needed to excise a region inside black-hole horizons as described in section 2.2.4. For the excised boundaries, we have to impose a boundary condition, but it is not trivial what type of boundary conditions should be imposed. By contrast, for hyperbolic equations, boundary conditions are in a sense trivial because we know that any information cannot escape from the inside of the excised regions that are located inside horizons. Thus, an ingoing boundary condition should be imposed (although imposing a precise ingoing boundary condition is not a trivial issue). Motivated by these considerations, dynamical gauge conditions for the lapse function and shift vector were extensively explored in 1990s and 2000s [e.g., Bona et al. (1995); Alcubierre et al. (2001b, 2003)]. In particular, the so-called *moving puncture gauge condition* is now established as one of the standard gauge conditions in numerical relativity. Thus, in this section, we describe this gauge condition.

2.2.6.1 *Slicing*

For deriving an equation of dynamical slicing conditions, the equation of harmonic slicing (2.140) is referred to as a starting point. As we described in section 2.2.3, the harmonic slicing condition has only a weak singularity-avoidance property [Bona and Massó (1988); Alcubierre (2003)]. Bona *et al.* (1995) then modified the equation of harmonic slicing to the following form that covers a wide class of slicing conditions:

$$(\partial_t - \beta^k \partial_k)\alpha = -\alpha^2 f(\alpha) K. \tag{2.158}$$

Here $f(\alpha)$ is an arbitrary function of α and $f = 1$ for harmonic slicing. The geodesic slicing is then included as a sub-class of $f = 0$. For finding a good choice of f, numerical experiment is the unique way, and has been extensively performed. Eventually, Anninos *et al.* (1995c) (after an extensive exploration for dynamical evolution of spherical black holes by Bernstein) discovered that the following particular choice of f is suitable for evolving the black-hole spacetime

$$f = \frac{1}{\alpha}. \tag{2.159}$$

They found that this slicing is similar to maximal slicing near its limit hypersurface (i.e., for the case that the lapse function approaches zero). This slicing condition is often referred to as $1+\log$ slicing because for $\beta^k = 0$ with $\alpha K = -\partial_t \ln \sqrt{\gamma}$, it leads to the condition $\alpha = C(x^i) + \ln \sqrt{\gamma}$ or $\alpha = \alpha_0 + \ln \sqrt{\gamma/\gamma_0}$ where $C(x^i)$, α_0, and γ_0 are functions of spatial coordinates that are determined by the initial condition.

More detailed subsequent numerical exploration [Alcubierre *et al.* (2003)] discovered that the best choice of f in four-dimensional spacetime ($D = 4$) was not the form of equation (2.159), but

$$f = \frac{2}{\alpha}. \tag{2.160}$$

In this case (with $\beta^k = 0$), α is written as $C(x^i) + \ln \gamma$ or $\alpha_0 + \ln(\gamma/\gamma_0)$. We note that $\alpha_0 \leq 1$ and $\gamma_0 \geq 1$ are usually satisfied for the initial condition of astrophysical objects. The dynamical slicing condition with this form of f is now standard slicing in numerical relativity with $D = 4$. (For the higher-dimensional case, the factor may be smaller than 2 although it should be larger than 1 [Shibata and Yoshino (2010b,a)].)

The dynamical slicing conditions with equations (2.159) and (2.160) are expected to have a strong singularity-avoidance property because $\alpha = 0$ is reached for a finite value of $\gamma > 0$ and $\gamma < \gamma_0$; e.g., for $\beta^k = 0$, $\alpha = 0$ is reached when $\gamma = \gamma_0 \exp(-2\alpha_0)$ for equation (2.159) and $\gamma = \gamma_0 \exp(-\alpha_0)$ for equation (2.160). Hence, a limit hypersurface, which is distant from physical singularities, is likely to be reached in the finite proper time.

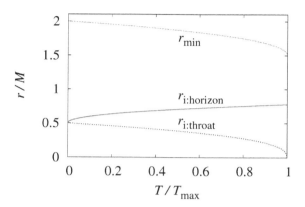

Fig. 2.10 Evolution of $r_{\min}$ and locations of the minimal surface of the wormhole throat and the event horizon in the isotropic radial coordinate (denoted by $r_{\mathrm{i:throat}}$ and $r_{\mathrm{i:horizon}}$) for $D=4$. $T_{\max} = T_\infty = 3\sqrt{3}M^2/4$ at which the limit hypersurface is reached, and the isotropic radial coordinate at the horizon at $T=T_\infty$ is $\approx 0.779M$.

In the presence of such a limit hypersurface, we may expect that it agrees with the limit hypersurface in the maximal slicing condition, irrespective of the dimensionality. This expectation comes from the fact that $\partial_\tau \alpha = 0$ on the limit hypersurface with the choice of $\beta^k = 0$ where τ is the proper time along each time axis. This implies that $\partial_t \alpha / \alpha = 0$ along each time axis and hence $K = 0$ according to equation (2.158) with equation (2.159) or (2.160). Indeed, this is the case as numerically confirmed in Brügmann et al. (2008) for $D=4$ and in Yoshino and Shibata (2009) for $D=5$. Thus, in the dynamical slicing conditions, the maximal slicing condition (or the condition similar to it for $\beta^k \neq 0$) is asymptotically achieved, and therefore, they have excellent properties that maximal slicing has.

However we already learned in section 2.2.2 that in maximal slicing, horizon stretching occurs inside the event horizon and the grid resolution there becomes poor. Nevertheless, the dynamical slicing works well for evolving black-hole spacetime in numerical relativity. What is the reason? The answer is as follows [also see Hannam et al. (2007)].

Since numerical simulations, in which dynamical slicing is employed, are usually performed in the coordinate system similar to isotropic coordinates, we first rewrite the line element of a non-spinning black hole with maximal slicing in the form [compare it with equation (2.116)]

$$ds^2 = -\alpha^2 dt^2 + \psi^4(d\bar{r} + \beta dt)^2 + \psi^4 \bar{r}^2 d\Omega, \qquad (2.161)$$

where $\bar{r}$ is the isotropic radial coordinate. We focus only on a four-dimensional black hole but the following result does not depend on the dimensionality qualitatively. The relation between r in equation (2.116) and $\bar{r}$, and the functional form of ψ are found from

$$r = \psi^2 \bar{r}, \qquad f^{1/2} dr = \psi^2 d\bar{r}. \qquad (2.162)$$

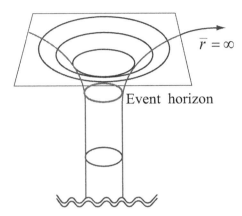

Fig. 2.11 Schematic embedding diagram of Schwarzschild geometry near the limit hypersurface. All the domains in the other world sheet (the causally disconnected region from our world) are denoted by a point $\bar{r} = 0$ in isotropic coordinates and are effectively excised. Compare with figure 2.9.

Here, we consider maximal slicing in the inversion symmetric boundary condition at the minimal surface of the wormhole throat as in section 2.2.2, and thus, $f^{-1} = 1 - 2M/r + T^2/r^4$ with $0 \leq T^2 \leq 27M^4/16$. We suppose that at $t = 0$ (and $T = 0$), the event horizon is located at $r = 2M$ and $\bar{r} = M/2$. Note that at $t = 0$, $\bar{r} = 0$ denotes the spatial infinity in the other world sheet, i.e., region III of figure 1.9. As shown in section 2.2.2, with the elapse of the proper time, spatial hypersurfaces approach the limit hypersurface which is reached at $T = \sqrt{27}M^2/4$. The radial coordinate of the minimal surface of the wormhole throat, $r_{\min}$, decreases from $r = 2M$ toward $3M/2 (= 3r_g/4)$; see figure 2.10. With this change, the locations of the minimal surface of the throat and the event horizon in the isotropic radial coordinate $\bar{r}$ also change as shown in figure 2.10. Here, particularly remarkable is that the location of the minimal surface of the wormhole throat approaches $\bar{r} = 0$. Remember that $\bar{r} = 0$ denotes the location of the spatial infinity in the other world sheet (region III of figure 1.9) for $T < T_\infty$. However, in the limit hypersurface, all the points in region III (as well as a half of region II) are degenerate into $\bar{r} = 0$: This implies that region III is effectively excised (see figure 2.11 as a schematic picture). In addition, the proper distance from $\bar{r} = 0$ to a point with $\bar{r} \neq 0$ diverges even if the point is located inside the event horizon. Thus, the domain with $\bar{r} = 0$ is causally disconnected from other places with $\bar{r} > 0$. Therefore, surprisingly, without any special numerical treatment, an excision is automatically achieved! [see Hannam et al. (2007); Brügmann et al. (2008); Baumgarte and Naculich (2007)].

At $\bar{r} = 0$, $\psi = \sqrt{r/\bar{r}}$ diverges because $r = 3M/2 (\neq 0)$ (finite) there. Thus, it is not possible to follow the evolution of ψ at $\bar{r} = 0$ in numerical computation. However, even if we do not evolve ψ at $\bar{r} = 0$ in the vicinity of the limit hypersurface, it does not matter because this point is inside the event horizon and not causally connected with the regions concerned. The only task is to avoid a numerical problem due to the divergence at $\bar{r} = 0$. For this, we only need to simply excise a small domain in the vicinity of $\bar{r} = 0$.

The desired behavior near the limit hypersurface described above is realized both in maximal and dynamical slicing. However, in real computation, dynamical slicing obviously has advantage. The reason for this is that for imposing maximal slicing, we have to solve an elliptic equation for α or for $\alpha\psi$. For example, let us consider equation (2.114) for $\alpha\psi$. Because ψ at $\bar{r} = 0$ diverges at the limit hypersurface, it is not clear whether the right-hand side of equation (2.114) is finite or not at $\bar{r} = 0$ in the presence of K_{ij} or matter source terms. Thus, we need to employ an excision algorithm in this case. However, the boundary condition on the excised surface is not clear as already mentioned. By contrast, if we solve the evolution equation for α with an excision algorithm, the divergence problem at $\bar{r} = 0$ is not serious because the point is causally disconnected: We may impose any arbitrary boundary condition as long as it is physically acceptable (e.g., we may set that ψ has a moderately large value as 10 at $\bar{r} = 0$). Indeed, numerical experiments show that this simple treatment is acceptable.

Actually, we can fix the problem of the divergence at $\bar{r} = 0$ by simply choosing an appropriate variable instead of ψ; e.g., for $D = 4$, the popular choice is ψ^{-4} or ψ^{-2}. With a suitable formulation and choice of the variables, it is possible to guarantee that no divergence appears for the chosen variables. In this case, dynamical slicing also has advantage to guarantee the absence of divergence. This point will be revisited in section 2.3.3.1.

2.2.6.2 Spatial gauge: dynamical shift

As we mentioned in section 2.2.5, the minimal distortion gauge or its variant will have many excellent properties. However, we do not want to solve any elliptic equation because the procedure for its solution in numerical relativity requires an expensive cost in computation, and moreover, we have to impose an appropriate boundary condition, which is not clear in the presence of black holes as already mentioned in the previous sections. Due to these reasons, it has been desired to discover a good spatial gauge condition for which the equation can be written in a dynamical form. Specifically, a number of effort has been paid for modifying the elliptic equation of the minimal distortion gauge to a dynamical form.

Broadly speaking, there are two possibilities: to rewrite the vector elliptic equation for the shift vector into a parabolic-type or a hyperbolic-type equation as

$$\partial_t \beta^k = C \tilde{D}_j \left(\sqrt{\gamma} \partial_t \tilde{\gamma}^{jk} \right), \tag{2.163}$$

or

$$(\partial_t^2 + C_1 \partial_t) \beta^k = C_2 \tilde{D}_j \left(\sqrt{\gamma} \partial_t \tilde{\gamma}^{jk} \right), \tag{2.164}$$

where C, C_1, and C_2 are functions to be appropriately chosen. Note that we also have an option for replacing the partial derivative ∂_t to $\partial_t - \beta^l \partial_l$. If the quasi-stationarity for the shift vector, such as $\partial_t \beta^k \approx 0$ and $\partial_t^2 \beta^k \approx 0$, could be reached in a short time scale, equation (2.163) or (2.164) can be an alternative of the minimal distortion gauge condition in which $\tilde{D}_j \left(\sqrt{\gamma} \partial_t \tilde{\gamma}^{jk} \right) = 0$.

However, the parabolic-type equation (2.163) was soon found to be unsuitable in numerical simulations because the quasi-stationarity is not likely to be reached in a short time scale. This is understood in the following analysis: For a parabolic equation, the time scale for achieving a stationary solution is approximately estimated as $t \sim R_c^2/C$ where R_c is a characteristic length scale of the system. Since C is an arbitrary function, one might think that C should be chosen to be large enough to shorten this time scale. However, this is not possible because of the presence of the so-called Courant-Friedrichs-Lewy condition (see section 3.1.4), which constrains the possible time-step interval in numerical simulations, approximately as $\Delta t \leq (\Delta x)^2/C$ where Δt and Δx are the time-step interval and the grid spacing, respectively. Hence, the required time-step number is constrained by $t/\Delta t \geq (R/\Delta x)^2$ irrespective of C. This shows that for a well-resolved simulation (for small values of Δx), the required value of $t/\Delta t$ is significantly large (i.e., computational costs become huge).

The remaining possibility is to choose a hyperbolic-type equation such as equation (2.164). The first practical proposal was done by Alcubierre et al. (2001b, 2003), and since the success of binary-black-hole-merger simulations [Campanelli et al. (2006a); Baker et al. (2006a)], such hyperbolic shift conditions have been employed together with the Baumgarte–Shapiro–Shibata–Nakamura (BSSN) formulation (see section 2.3.3) not only for binary black holes but also for neutron star binaries and stellar core collapse. Currently, the most popular condition for the "$\tilde{\Gamma}^i$ version" of the BSSN formulation is written in the following form [van Meter et al. (2006); Brügmann et al. (2008)],

$$(\partial_t - \beta^k \partial_k) \beta^i = \frac{3}{4} B^i,$$
$$(\partial_t - \beta^k \partial_k) B^i = (\partial_t - \beta^k \partial_k) \tilde{\Gamma}^i - \eta_B B^i, \tag{2.165}$$

where B^i is an auxiliary variable. Here, $\tilde{\Gamma}^i = -\overset{(0)}{D}_k \tilde{\gamma}^{ki}$ and its evolution equation includes the terms schematically written as $\overset{(0)}{D}{}^k \overset{(0)}{D}_k \beta^i + (1 - 2/N) \overset{(0)}{D}{}^i \overset{(0)}{D}_k \beta^k$ [e.g., equation (2.223)] which is similar to the left-hand side of the equation for the minimal distortion gauge, (2.149). Thus, β^i obeys a vector hyperbolic equation. The factor 3/4 comes from the fact that the characteristic speed in four-dimensional spacetime becomes approximately the speed of light. This is changed for $D \geq 5$ [Yoshino and Shibata (2009); Shibata and Yoshino (2010b,a)] to $N/(2N-2)$ if the characteristic speed should be set to the speed of light. η_B is a constant, which is introduced to suppress an unphysical oscillation of the shift vector. This dynamical shift condition has been examined for a variety of spacetime including binary black holes, binary neutron stars, black hole-neutron star binaries, and rotating stellar core collapse. All these simulations have been successfully performed, indicating that the coordinate distortion is preserved to be small in this gauge condition. Since this spatial gauge condition together with dynamical slicing works well for spacetime with moving black holes in the puncture formulation (see section 2.3.3.1), it is often called *moving puncture gauge*.

For the "F_i version" of the BSSN formulation, Shibata (2003b) proposed a dynamical shift condition in the form

$$\partial_t \beta^i = \tilde{\gamma}^{ij} \left(F_j + \Delta t \partial_t F_j \right). \tag{2.166}$$

Here, $F_i = \sum_k \partial_k \tilde{\gamma}_{ki}$, and its evolution equation also includes the terms of a vector Laplacian of β^k [e.g., equation (2.220)]. This dynamical shift condition also has been examined for a variety of spacetime including binary black holes, binary neutron stars, black hole-neutron star binaries, and rotating stellar core collapse by the Japanese group [e.g., Shibata and Uryū (2000); Yamamoto *et al.* (2008)], and successful numerical results have been reported.

It is not easy to precisely state the reason that these dynamical shift conditions work well. The reason may be inferred from the similarity with the dynamical slicing condition. For dynamical slicing with $\partial_t \alpha = -2\alpha K$, $\partial_t K = 0$ is achieved in a short time scale, leading to $\partial_t \alpha = 0 = \alpha$ near the limit hypersurface for non-spinning black-hole spacetime (here we assume $\beta^k = 0$). For the dynamical shift condition, it seems that the similar evolution occurs: For the $\tilde{\Gamma}^i$ version of the dynamical shift, $\partial_t B^i = 0$ and $\partial_t \beta^i = 0$ are achieved in a short time scale, and for the F_i version, $\partial_t F_i = 0$ is achieved. Then, the shift vector satisfies a vector elliptic equation that is similar to the equation of the minimal distortion gauge, (2.149).

What we should emphasize here is that the dynamical gauge conditions have been developed primarily through a large number of numerical experiments. Only by a mathematical consideration, it would not be possible to find this excellent gauge condition. This great success indicates that numerical experiments are essential in numerical relativity.

2.3 Formulations in numerical relativity

The standard 3+1 (and N+1) formulation is a well-defined initial-value formulation in general relativity. However, it is not suitable for numerical-relativity simulations. The reason for this is that with this formulation, a numerical error, even if its magnitude is sufficiently small initially, grows steeply with time and eventually the computation crashes. In numerical relativity, numerical errors associated with discretization are inevitably generated. However, if their magnitude does not grow with time (i.e., it is preserved to be in a small level), and in addition, it decreases toward zero with improving the grid resolution, we can accept the approximate numerical solution. Here, the second point is often referred to as the *convergence property*, which guarantees us to obtain an exact solution in principle by an extrapolation, taking a limit of infinitesimal grid spacing. This provides a foundation of numerical simulations. However, if the magnitude of the numerical errors grows with no bound, the convergence property cannot be satisfied (i.e., the foundation of the numerical computation is lost), and moreover, the computation usually crashes. Hence, the standard 3+1 (and N+1) formulation cannot be employed in numerical relativity. We need other formulations in which numerical errors are maintained to be small and the convergence property is satisfied.

The main subject of this section is to describe two formulations that have been proven to be robust by a number of numerical experiments. Before describing these formulations, we show a specific reason that the standard 3+1 formulation is not suitable in numerical relativity using a linear perturbation analysis. This indicates that the standard N+1 formulation is also unsuitable in numerical relativity.

2.3.1 Linearized standard 3+1 formulation

In this section, we will show that the standard 3+1 formulation is not suitable even for numerically evolving small-amplitude gravitational waves in vacuum for a long time duration. This implies that even for spacetime of a tiny gravity, it is not suitable.

The linearized equations in the standard 3+1 formulation are derived by expanding α and γ_{ab} as

$$\alpha = 1 + \epsilon A, \quad \text{and} \quad \gamma_{ab} = f_{ab} + \epsilon h_{ab}, \qquad (2.167)$$

where f_{ab} is the flat spatial metric, ϵ is a small parameter, and A and h_{ab} are perturbed metric components. In addition, we suppose that β^a and K_{ab} are quantities of order ϵ; thus, we write

$$\beta^a = \epsilon B^a, \quad \text{and} \quad K_{ab} = \epsilon k_{ab}, \qquad (2.168)$$

where B^a and k_{ab} are perturbed quantities. The subscripts of the perturbed quantities are raised and lowered by the flat spatial metric, f^{ab} and f_{ab}, respectively.

Collecting the terms up to the first order in ϵ, we obtain the following equations from the standard 3+1 evolution equations, (2.64) and (2.65):

$$\partial_t h_{ab} = -2k_{ab} + \overset{(0)}{D_a} B_b + \overset{(0)}{D_b} B_a, \qquad (2.169)$$

$$\partial_t k_{ab} = -\frac{1}{2}\left[\Delta_{\mathrm{f}} h_{ab} + \overset{(0)}{D_a}\overset{(0)}{D_b} h - \left(\overset{(0)}{D_a}\overset{(0)}{D^c} h_{bc} + \overset{(0)}{D_b}\overset{(0)}{D^c} h_{ac}\right)\right] - \overset{(0)}{D_a}\overset{(0)}{D_b} A. \qquad (2.170)$$

Here, $h = f^{ab} h_{ab}$, $\overset{(0)}{D_a}$ is the covariant derivative associated with f_{ab}, and $\Delta_{\mathrm{f}} = \overset{(0)}{D^a}\overset{(0)}{D_a}$. The abstract index $a, b, \cdots$ is here used only for spatial components. Eliminating k_{ab} from equations (2.169) and (2.170) yields a second-order hyperbolic equation for h_{ab} as

$$\partial_t^2 h_{ab} = \Delta_{\rm f} h_{ab} + \overset{(0)}{D_a} \overset{(0)}{D_b} h - \left(\overset{(0)}{D_a} \overset{(0)}{D^c} h_{bc} + \overset{(0)}{D_b} \overset{(0)}{D^c} h_{ac} \right)$$
$$+ 2 \overset{(0)}{D_a} \overset{(0)}{D_b} A + \overset{(0)}{D_a} \dot{B}_b + \overset{(0)}{D_b} \dot{B}_a, \qquad (2.171)$$

where $\dot{B}_a = \partial_t B_a$. Equation (2.171) shows that h_{ab} obeys a wave-like equation composed of a d'Alembertian operator $\Box h_{ab} = (-\partial_t^2 + \Delta_{\rm f}) h_{ab}$, although there are several additional terms (second, third, and fourth terms in its right-hand side). These terms cause a problem breaking a strong hyperbolicity [see, e.g., Alcubierre (2008) for a description of the strong hyperbolicity].

In addition to the evolution equations, we have the Hamiltonian and momentum constraint equations in the linear approximation, which are written, respectively, as

$$\Delta_{\rm f} h - \overset{(0)}{D^a} \overset{(0)}{D^c} h_{ac} = 0, \qquad (2.172)$$

$$\overset{(0)}{D^a} k_{ab} - \overset{(0)}{D_b} k_a{}^a = 0. \qquad (2.173)$$

These will be also used later.

For identifying the origin of the problem, it is useful to perform an analysis using a harmonics-decomposition method, in which we expand the variables by

$$h_{ab} = \Psi f_{ab} + \overset{(0)}{D_a} \overset{(0)}{D_b} \Phi + 2 \overset{(0)}{D_{(a}} V_{b)} + h_{ab}^{\rm tt}, \qquad (2.174)$$

$$k_{ab} = \Psi_{\rm k} f_{ab} + \overset{(0)}{D_a} \overset{(0)}{D_b} \Phi_{\rm k} + 2 \overset{(0)}{D_{(a}} U_{b)} + k_{ab}^{\rm tt}, \qquad (2.175)$$

$$B_a = \overset{(0)}{D_a} B + \omega_a, \qquad (2.176)$$

where Ψ, Φ, $\Psi_{\rm k}$, $\Phi_{\rm k}$, and B as well as A are scalars, V_a, U_a, and ω_a are vectors, and $h_{ab}^{\rm tt}$ and $k_{ab}^{\rm tt}$ are tensors. Here, the vector and tensor are defined as the quantities that, respectively, satisfy the conditions

$$\overset{(0)}{D_a} V^a = 0, \qquad (2.177)$$

$$\overset{(0)}{D^a} h_{ab}^{\rm tt} = 0 \text{ and } f^{ab} h_{ab}^{\rm tt} = 0. \qquad (2.178)$$

Thus, a tensor field in three spatial dimensions is composed of two scalar components, two vector components, and two tensor components. Using this decomposition, we can decompose two basic equations (2.169) and (2.170) as

$$\dot{\Psi} = -2\Psi_{\rm k},$$
$$\dot{\Phi} = -2\Phi_{\rm k} + 2B,$$
$$\dot{V}_a = -2U_a + \omega_a,$$
$$\dot{h}_{ab}^{\rm tt} = -2k_{ab}^{\rm tt}, \qquad (2.179)$$

and

$$\dot{\Psi}_k = -\frac{1}{2}\Delta_f \Psi,$$
$$\dot{\Phi}_k = -\frac{1}{2}\Psi - A,$$
$$\dot{U}_a = 0,$$
$$\dot{k}_{ab}^{tt} = -\frac{1}{2}\Delta_f h_{ab}^{tt}. \tag{2.180}$$

Equations (2.179) and (2.180) show that the tensor part composed of (h_{ab}^{tt}, k_{ab}^{tt}) and one of the scalar parts (Ψ, Ψ_k) constitute wave equations, $\Box h_{ab}^{tt} = 0$ and $\Box \Psi = 0$. On the other hand, for other components, the type of the equations depends on the gauge condition.

The constraint equations (2.172) and (2.173) are written in the forms

$$\Delta_f \Psi = 0, \tag{2.181}$$
$$\Delta_f U_a = 0 \quad \text{and} \quad \overset{(0)}{D}_a \Psi_k = 0. \tag{2.182}$$

This suggests that $\Psi = 0$, $U_a = 0$, and $\Psi_k = 0$ should be satisfied in asymptotically flat spacetime. However, if we solve only the evolution equations, as that is usually the case in numerical relativity, the constraint equations are not satisfied precisely. This is a possible origin of the problem concerned here.

If all the basic equations obey the wave equation like $\Box h_{ab}^{tt} = 0$, the strong hyperbolicity is guaranteed for the system, and we may suppose that there is no problem for deriving a solution in numerical simulations, even if a small error associated with a constraint violation is present. The reason for this is that the small error propagates toward the infinity (disperses away) if the basic equations obey the wave equation. However, in the presence of a non-wave equation, e.g., if one of equations does not contain any spatial derivatives and the time evolution for the variable concerned is determined only by the local information, it could not be suitable for numerical simulations. A simple but informative example is obtained for the equation of Φ with the gauge conditions $A = 0$ and $B^a = 0$. In this case, the relevant equations are

$$\dot{\Phi} = -2\Phi_k \quad \text{and} \quad \dot{\Phi}_k = -\frac{1}{2}\Psi. \tag{2.183}$$

As we mentioned above, Ψ is zero if the Hamiltonian constraint is satisfied. However, a small error is inevitably generated in numerical simulations. Because Ψ obeys a wave equation, the error for it eventually disperses away. However, in the presence of its error, Φ_k deviates from zero in general. If it is a slowly varying function or a constant (this is the case after Ψ relaxes to zero), $|\Phi|$ grows (approximately) linearly with time. The problem is that there is no term that prevents the growth or enforces the propagation of this unphysical mode, since the evolution of Φ is determined only by local information.

The system of the evolution equations, which results in this type of undesirable solutions, is called a weakly hyperbolic system [e.g., Alcubierre (2008)]. We refer to a problem, which is defined in such a weakly hyperbolic system, as a not-well-posed problem. By contrast, if a system is strongly hyperbolic, the problem becomes well posed as long as the linear perturbation equations are concerned; undesirable growth of a numerical error does not appear at least for the linear perturbation level. Thus, employing a strongly hyperbolic system is a minimum necessary requirement for stably and accurately performing a numerical simulation.

Here, several words of caution are appropriate to avoid possible misunderstanding in this problem. First, the strong hyperbolicity is a necessary condition for setting a well-posed *numerical* problem: To guarantee that numerical errors do not grow steeply, the strong hyperbolicity is required. Thus, it is not the matter of physics but the matter of numerical computation. Second, the strong hyperbolicity is not the sufficient condition for stable numerical simulations, when we solve fully nonlinear Einstein's equation. The reason for this is that many nonlinear terms present in the nonlinear equation could change the stability property of the system. It would be perhaps impossible to find a well-posed formulation in numerical relativity, if one is trapped only on the notion of the hyperbolicity of the linear perturbation system. The notion of the hyperbolicity should be used only for excluding a weakly hyperbolic system in numerical relativity. For finding a really good formulation, numerical experiments are always required because tests for the highly nonlinear formulations by analytic calculations are not feasible.

The problem of the hyperbolicity found in the standard 3+1 formulation can be resolved if we employ a special class of gauge conditions. As one of the useful gauge conditions, the harmonic gauge condition introduced in section 2.2.3 is well known. In the linear approximation, the gauge condition is written as

$$\dot{A} = -3\Psi_k - \Delta_f \Phi_k, \tag{2.184}$$

$$\dot{B} = -A - \frac{1}{2}\Psi + \frac{1}{2}\Delta_f \Phi, \tag{2.185}$$

$$\dot{\omega}_a = \Delta_f V_a. \tag{2.186}$$

Substituting these gauge conditions into equations (2.179) and (2.180), we find, of course, that all the components of h_{ab} obey the wave equations as

$$\Box \Psi = 0, \quad \Box \Phi = 0, \quad \Box V_a = 0, \quad \Box h_{ab}^{tt} = 0. \tag{2.187}$$

The success of the GH formulation is based partly on this fact. We note that even in other gauge conditions, e.g., in which the right-hand side of equation (2.185) is composed only of $\Delta_f \Phi / 2$, the problem associated with Φ mentioned above could be resolved.

However, if we have to rely on the special choice of the gauge condition, the merit of the $3+1$ formulation (and $N+1$ or any spacetime-foliation formulation) is lost, because we lose the freedom for choosing the gauge condition: We note again that the special gauge condition chosen is not guaranteed to be a good condition for all the problems. To preserve the merit of choosing arbitrary gauge conditions, therefore, we have to modify the standard 3+1 (and N+1) formulations.

2.3.2 Linearized equations for purely conformal decomposition formulation

Before describing formulations well-suited in numerical relativity, we point out that the (purely) conformal decomposition formulation derived in section 2.1.7 is *not* suitable in numerical relativity. Rather, it introduces an even worse property as seen in the following.

For deriving linearized equations for the conformal decomposition formulation for $D=4$ ($N=3$), we write α and β^i as in section 2.3.1 [see equations (2.167) and (2.168)]. In addition, $\tilde{\gamma}_{ab}$, $\tilde{A}_{ab}$, ψ, and K are written as

$$\tilde{\gamma}_{ab} = f_{ab} + \epsilon \tilde{h}_{ab}, \quad \tilde{A}_{ab} = \epsilon \bar{A}_{ab}, \quad \psi = 1 + \epsilon \phi, \quad K = \epsilon k, \tag{2.188}$$

where ϵ is again a small parameter, and $\tilde{h}_{ab}$, $\bar{A}_{ab}$, ϕ, and k are perturbed quantities.

The linearized equations for the conformal decomposition formulation, (2.83)–(2.86), are

$$\partial_t \tilde{h}_{ab} = -2\bar{A}_{ab} + \overset{(0)}{D}_a B_b + \overset{(0)}{D}_b B_a - \frac{2}{3} f_{ab} \overset{(0)}{D}_c B^c, \tag{2.189}$$

$$\partial_t \bar{A}_{ab} = -\frac{1}{2}\left[\Delta_f \tilde{h}_{ab} - \left(\overset{(0)}{D}_a \overset{(0)}{D}{}^c \tilde{h}_{bc} + \overset{(0)}{D}_b \overset{(0)}{D}{}^c \tilde{h}_{ac}\right) - \frac{1}{3} f_{ab}\left(\Delta_f \tilde{h} - 2\overset{(0)}{D}{}^c \overset{(0)}{D}{}^e \tilde{h}_{ce}\right)\right]$$
$$- 2\left(\overset{(0)}{D}_a \overset{(0)}{D}_b \phi - \frac{1}{3} f_{ab} \Delta_f \phi\right) - \left(\overset{(0)}{D}_a \overset{(0)}{D}_b A - \frac{1}{3} f_{ab} \Delta_f A\right), \tag{2.190}$$

$$\partial_t \phi = -\frac{1}{6}\left(k - \overset{(0)}{D}_a B^a\right), \tag{2.191}$$

$$\partial_t k = -\Delta_f A. \tag{2.192}$$

Here, we left $\tilde{h}$ because the constraint $\tilde{h} = 0$ is not guaranteed to be satisfied. As done in section 2.3.1, we decompose geometric quantities into scalar, vector, and tensor components as

$$\tilde{h}_{ab} = Q f_{ab} + \overset{(0)}{D}_a \overset{(0)}{D}_b \Phi + 2 \overset{(0)}{D}_{(a} V_{b)} + h_{ab}^{\text{tt}}, \tag{2.193}$$

$$\bar{A}_{ab} = Q_{\text{k}} f_{ab} + \overset{(0)}{D}_a \overset{(0)}{D}_b \Phi_{\text{k}} + 2 \overset{(0)}{D}_{(a} U_{b)} + A_{ab}^{\text{tt}}. \tag{2.194}$$

ϕ and k as well as the perturbed lapse function A are scalar components, and the shift vector is expanded as equation (2.176). Then, from equations (2.189) and (2.191), we have

$$\dot{Q} = -2Q_k - \frac{2}{3}\Delta_f B, \tag{2.195}$$

$$\dot{\Phi} = -2\Phi_k + 2B, \tag{2.196}$$

$$\dot{V}_a = -2U_a + \omega_a, \tag{2.197}$$

$$\dot{h}_{ab}^{tt} = -2A_{ab}^{tt}, \tag{2.198}$$

$$\dot{\phi} = \frac{1}{6}\left(-k + \Delta_f B\right), \tag{2.199}$$

and from equations (2.190) and (2.192),

$$\dot{Q}_k = -\frac{1}{3}\Delta_f\left(\frac{1}{2}\Delta_f \Phi + Q - 2\phi - A\right), \tag{2.200}$$

$$\dot{\Phi}_k = \frac{1}{2}(\Delta_f \Phi + 2Q - 4\phi - 2A), \tag{2.201}$$

$$\dot{U}_a = 0, \tag{2.202}$$

$$\dot{A}_{ab}^{tt} = -\frac{1}{2}\Delta_f h_{ab}^{tt}, \tag{2.203}$$

$$\dot{k} = -\Delta_f A. \tag{2.204}$$

The Hamiltonian and momentum constraint equations in the linear approximation are

$$\Delta_f \phi = -\frac{1}{4}\Delta_f Q, \tag{2.205}$$

$$Q_k + \Delta_f \Phi_k - \frac{2}{3}k = 0 \quad \text{and} \quad \Delta_f U_a = 0, \tag{2.206}$$

and new constraints (2.78) and (2.77) are

$$3Q + \Delta_f \Phi = 0, \tag{2.207}$$

$$3Q_k + \Delta_f \Phi_k = 0, \tag{2.208}$$

which are derived by taking the trace of equations (2.193) and (2.194). Note that to derive the Hamiltonian constraint (2.205), equation (2.207) was used.

Equations (2.196) and (2.201) show that the problem could again occur for the equations of (Φ, Φ_k). This is found by erasing Φ_k from these equations, that yields an *elliptic* equation for spacetime coordinates as

$$\ddot{\Phi} = -\Delta_f \Phi + 2\dot{B} - (2Q - 4\phi - 2A). \tag{2.209}$$

This equation is not suitable in the initial-value formulation at all.[3] Therefore, the purely conformal decomposition formulation makes the situation even worse in numerical relativity. People often refer to the BSSN formulation (see section 2.3.3) as a "conformal formulation". However, this is a really bad naming, which leads people unfamiliar with this field to misunderstanding.

---

[3] Again, a special choice of the gauge condition such as harmonic gauge condition could resolve this pathology.

2.3.3 BSSN formulation

As we see in the previous two sections 2.3.1 and 2.3.2, the standard 3+1 (and N+1) formulation and the (purely) conformal decomposition formulation are not suitable in numerical relativity. Reformulation is necessary if we want to employ a formulation based on the spacetime foliation.

The origin of the undesirable property both in the standard 3+1 formulation and the conformal decomposition formulation is associated with the presence of an unphysical scalar mode composed of (Φ, Φ_k), which arises from two terms in the Ricci tensor,

$$\overset{(0)}{D}_a \overset{(0)}{D^c} h_{bc} + \overset{(0)}{D}_b \overset{(0)}{D^c} h_{ac}$$

in the standard 3+1 (and N+1) formulation, and

$$\overset{(0)}{D}_a \overset{(0)}{D^c} \tilde{h}_{bc} + \overset{(0)}{D}_b \overset{(0)}{D^c} \tilde{h}_{ac}$$

in the conformal decomposition formulation. In the following, we will show how to fix this problem by reformulating the standard 3+1 formulation.

Following Shibata and Nakamura (1995), we start from the conformal decomposition formulation. We emphasize, however, that this problem can be also resolved even if we start from the standard 3+1 formulation as shown by Nakamura [in Nakamura et al. (1987)]. Using the mode decomposition (2.193), we have

$$F_a := \overset{(0)}{D^c} \tilde{h}_{ac} = \overset{(0)}{D}_a(Q + \Delta_f \Phi) + \Delta_f V_a. \qquad (2.210)$$

Thus, the scalar mode Φ, which is the origin of the problem, is included in $\overset{(0)}{D^c} \tilde{h}_{ac} = F_a$. However, this scalar mode is excluded from $\dot{F}_a$, if we use the momentum constraint equation (2.206) together with the evolution equations (2.195), (2.196), and (2.197), as

$$\begin{aligned}\dot{F}_a &= \overset{(0)}{D}_a(\dot{Q} + \Delta_f \dot{\Phi}) + \Delta_f \dot{V}_a \\ &= \overset{(0)}{D}_a\left(-2Q_k - 2\Delta_f \Phi_k + \frac{4}{3}\Delta_f B\right) + \Delta_f(-2U_a + \omega_a) \\ &= \frac{4}{3}\overset{(0)}{D}_a(-k + \Delta_f B) + \Delta_f(-2U_a + \omega_a).\end{aligned} \qquad (2.211)$$

This suggests that if we introduce a new auxiliary variable F_a and evolve it independently of the equations for $\tilde{\gamma}_{ij}$, $\tilde{A}_{ij}$, ψ, and K, the problem found in the previous two sections 2.3.1 and 2.3.2 could be resolved. This is an answer for the reformulation of the standard 3+1 formulation.

The robustness of introducing such an auxiliary variable was first discovered by Nakamura [in Nakamura et al. (1987)]. Originally, he considered to reformulate the standard 3+1 formulation without the conformal decomposition. Then, Shibata and Nakamura (1995) derived a new formulation with $F_a := f^{bc}\overset{(0)}{D}_c \tilde{\gamma}_{ab}$ starting from the conformal decomposition formulation. Subsequently, Baumgarte and Shapiro

(1999b) proposed a new auxiliary variable, $\tilde{\Gamma}^a := -\overset{(0)}{D}_b \tilde{\gamma}^{ab}$ with which the basic equation for the auxiliary variable has a slightly simpler form than that derived by Shibata and Nakamura (1995), although the basic idea for the reformulation was not changed.

In the formulation with the new auxiliary variable, total number of the dynamical variables is $17 (= 12 + 5)$, which are composed of

$$\tilde{\gamma}_{ab}, \quad \tilde{A}_{ab}, \quad \psi, \quad K, \quad F_a \text{ or } \tilde{\Gamma}^a, \tag{2.212}$$

and in addition to the Hamiltonian and momentum constraint equations, we have five new components of the constraint equations as

$$\det(\tilde{\gamma}_{ab}) = \det(f_{ab}), \tag{2.213}$$

$$\tilde{A}_{ab} \tilde{\gamma}^{ab} = 0, \tag{2.214}$$

$$F_a - f^{bc} \overset{(0)}{D}_c \tilde{\gamma}_{ab} = 0 \quad \text{or} \quad \tilde{\Gamma}^a + \overset{(0)}{D}_c \tilde{\gamma}^{ac} = 0. \tag{2.215}$$

Now this type of the formulation is called Baumgarte–Shapiro–Shibata–Nakamura (BSSN) formulation. In the following, we describe the basic equations in the BSSN formulation derived originally in Shibata and Nakamura (1995) in more detail.

In the original BSSN formulation, equations (2.83), (2.85), and (2.86) are used with no change. However, using F_a, $\tilde{R}_{ab}$ in equation (2.84) is rewritten as

$$\tilde{R}_{ab} = -\frac{1}{2} \left[\tilde{\gamma}^{cd} \overset{(0)}{D}_c \overset{(0)}{D}_d \tilde{\gamma}_{ab} - \overset{(0)}{D}_a \left(f^{cd} \overset{(0)}{D}_d \tilde{\gamma}_{bc} \right) - \overset{(0)}{D}_b \left(f^{cd} \overset{(0)}{D}_d \tilde{\gamma}_{ac} \right) \right.$$
$$\left. - \delta \tilde{\gamma}^{cd} \left(\overset{(0)}{D}_a \overset{(0)}{D}_d \tilde{\gamma}_{bc} + \overset{(0)}{D}_b \overset{(0)}{D}_d \tilde{\gamma}_{ac} \right) \right] + \left(\overset{(0)}{D}_c \tilde{\gamma}^{cd} \right) \overset{(0)}{\Gamma}_{d,ab} - \overset{(0)}{\Gamma}{}^d{}_{ac} \overset{(0)}{\Gamma}{}^c{}_{bd}$$
$$= -\frac{1}{2} \left[\Delta_f \tilde{\gamma}_{ab} - \overset{(0)}{D}_a F_b - \overset{(0)}{D}_b F_a + \delta \tilde{\gamma}^{cd} \left(\overset{(0)}{D}_c \overset{(0)}{D}_d \tilde{\gamma}_{ab} - \overset{(0)}{D}_a \overset{(0)}{D}_d \tilde{\gamma}_{bc} - \overset{(0)}{D}_b \overset{(0)}{D}_d \tilde{\gamma}_{ac} \right) \right]$$
$$+ \left(\overset{(0)}{D}_c \tilde{\gamma}^{cd} \right) \overset{(0)}{\Gamma}_{d,ab} - \overset{(0)}{\Gamma}{}^d{}_{ac} \overset{(0)}{\Gamma}{}^c{}_{bd}, \tag{2.216}$$

where $\delta \tilde{\gamma}^{ab} := \tilde{\gamma}^{ab} - f^{ab}$. Then, the evolution equation of F_a is derived from equation (2.83) as

$$\partial_t F_a = \partial_t \left(f^{bc} \overset{(0)}{D}_c \tilde{\gamma}_{ab} \right)$$
$$= f^{bc} \overset{(0)}{D}_c \left(-2\alpha \tilde{A}_{ab} + \tilde{D}_a \tilde{\beta}_b + \tilde{D}_b \tilde{\beta}_a - \frac{2}{3} \tilde{\gamma}_{ab} \tilde{D}_d \beta^d \right). \tag{2.217}$$

As we show in equation (2.211), the crucial step is to use the momentum constraint equation to rewrite $f^{bc} \overset{(0)}{D}_c \tilde{A}_{ab}$ to the form

$$f^{bc} \overset{(0)}{D}_c \tilde{A}_{ab} = -\overset{(0)}{D}_c \left(\delta \tilde{\gamma}^{bc} \tilde{A}_{ac} \right) + \overset{(0)}{\Gamma}{}^c{}_{ab} \tilde{A}^b{}_c - 6 \left(\overset{(0)}{D}_b \ln \psi \right) \tilde{A}^b{}_a + \frac{2}{3} \overset{(0)}{D}_a K + 8\pi J_a, \tag{2.218}$$

where $\overset{(0)}{\Gamma}{}^c{}_{ab}$ was defined in equation (2.93). Then, we have

$$\partial_t F_a == 2\alpha \left[\overset{(0)}{D}_b \left(\delta \tilde{\gamma}^{bd} \tilde{A}_{ad} \right) - \overset{(0)}{\Gamma}{}^c{}_{ab} \tilde{A}^b{}_c + 6 \left(\overset{(0)}{D}_b \ln \psi \right) \tilde{A}^b{}_a - \frac{2}{3} \overset{(0)}{D}_a K - 8\pi J_a \right]$$
$$- 2\tilde{A}_{ab} f^{bc} \overset{(0)}{D}_c \alpha + f^{bc} \overset{(0)}{D}_c \left(\tilde{D}_a \tilde{\beta}_b + \tilde{D}_b \tilde{\beta}_a - \frac{2}{3} \tilde{\gamma}_{ab} \tilde{D}_d \beta^d \right). \quad (2.219)$$

Thus, we may say that the momentum constraint is written to the evolution equation for F_a. In a coordinate basis, equation (2.219) is often written as

$$\left(\partial_t - \beta^k \overset{(0)}{D}_k \right) F_i = 2\alpha \left[\delta \tilde{\gamma}^{kj} \overset{(0)}{D}_j \tilde{A}_{ik} + \left(\overset{(0)}{D}_j \tilde{\gamma}^{kj} \right) \tilde{A}_{ik} - \frac{1}{2} \tilde{A}^{jl} \overset{(0)}{D}_i \tilde{\gamma}_{jl} \right.$$
$$\left. + 6 \left(\overset{(0)}{D}_k \ln \psi \right) \tilde{A}^k{}_i - \frac{2}{3} \overset{(0)}{D}_i K \right]$$
$$- 2 f^{jk} \left(\overset{(0)}{D}_k \alpha \right) \tilde{A}_{ij} + f^{jl} \left(\overset{(0)}{D}_l \beta^k \right) \overset{(0)}{D}_k \tilde{\gamma}_{ij}$$
$$+ f^{jk} \overset{(0)}{D}_k \left(\tilde{\gamma}_{il} \overset{(0)}{D}_j \beta^l + \tilde{\gamma}_{jl} \overset{(0)}{D}_i \beta^l - \frac{2}{3} \tilde{\gamma}_{ij} \overset{(0)}{D}_l \beta^l \right) - 16\pi\alpha J_i. \quad (2.220)$$

Baumgarte and Shapiro (1999b) modified the original version of the BSSN formulation. This is done by employing $\tilde{\Gamma}^a$ instead of F_a. In this case, $\tilde{R}_{ab}$ is written as

$$\tilde{R}_{ab} = -\frac{1}{2} \left[\tilde{\gamma}^{cd} \overset{(0)}{D}_c \overset{(0)}{D}_d \tilde{\gamma}_{ab} - \tilde{\gamma}_{ac} \overset{(0)}{D}_b \tilde{\Gamma}^c - \tilde{\gamma}_{bc} \overset{(0)}{D}_a \tilde{\Gamma}^c \right.$$
$$\left. + \left(\overset{(0)}{D}_a \tilde{\gamma}^{cd} \right) \overset{(0)}{D}_c \tilde{\gamma}_{bd} + \left(\overset{(0)}{D}_b \tilde{\gamma}^{cd} \right) \overset{(0)}{D}_c \tilde{\gamma}_{ad} - \tilde{\Gamma}^c \overset{(0)}{D}_c \tilde{\gamma}_{ab} \right] - \tilde{\Gamma}^d{}_{ac} \tilde{\Gamma}^c{}_{bd}. \quad (2.221)$$

Then, the evolution equation for $\tilde{\Gamma}^a$, which is derived from $\partial_t \tilde{\Gamma}^a = -\overset{(0)}{D}_b \partial_t \tilde{\gamma}^{ab}$, is simplified compared with that for F_a, because of the presence of a relation

$$\overset{(0)}{D}_b \left(\tilde{D}^a \beta^b + \tilde{D}^b \beta^a - \frac{2}{3} \tilde{\gamma}^{ab} \tilde{D}_d \beta^d \right)$$
$$= \tilde{\gamma}^{bc} \overset{(0)}{D}_b \overset{(0)}{D}_c \beta^a + \frac{1}{3} \tilde{\gamma}^{ac} \overset{(0)}{D}_c \overset{(0)}{D}_b \beta^b - \tilde{\Gamma}^c \overset{(0)}{D}_c \beta^a + \beta^c \overset{(0)}{D}_c \tilde{\Gamma}^a + \frac{2}{3} \tilde{\Gamma}^a \overset{(0)}{D}_c \beta^c. \quad (2.222)$$

The resulting evolution equation for $\tilde{\Gamma}^a$ is written in a slightly simple form as

$$\left(\partial_t - \beta^b \overset{(0)}{D}_b \right) \tilde{\Gamma}^a = -2\tilde{A}^{ab} \overset{(0)}{D}_b \alpha$$
$$+ 2\alpha \left[\tilde{\Gamma}^a{}_{bc} \tilde{A}^{bc} - \frac{2}{3} \tilde{\gamma}^{ab} \overset{(0)}{D}_b K - 8\pi \tilde{\gamma}^{ab} J_b + 6 \left(\overset{(0)}{D}_b \ln \psi \right) \tilde{A}^{ab} \right]$$
$$+ \tilde{\gamma}^{bc} \overset{(0)}{D}_b \overset{(0)}{D}_c \beta^a + \frac{1}{3} \tilde{\gamma}^{ac} \overset{(0)}{D}_c \overset{(0)}{D}_b \beta^b - \tilde{\Gamma}^b \overset{(0)}{D}_b \beta^a + \frac{2}{3} \tilde{\Gamma}^a \overset{(0)}{D}_b \beta^b. \quad (2.223)$$

Although the computational cost is slightly reduced, no substantial difference has been found between two versions by Shibata and Nakamura and by Baumgarte and Shapiro. Indeed, both formulations have provided numerical results for which the accuracy is essentially the same [e.g., Yamamoto et al. (2008)].

2.3.3.1 BSSN-puncture formulation

Before 2005, the BSSN formulation had not achieved a success for the long-term evolution of black-hole spacetime. However, it was substantially improved by Campanelli et al. (2006a) [see also Baker et al. (2006a)]. By their modification, stable and accurate computation for black-hole spacetime is well founded in the BSSN formulation with an appropriate choice of gauge conditions.

Before describing how they modified the BSSN formulation, we remind the reader of the line element of a Schwarzschild black hole of mass M in isotropic coordinates

$$ds^2 = -\alpha^2 dt^2 + \psi^4(dx^2 + dy^2 + dz^2), \qquad (2.224)$$

where $\alpha = (1-M/2r)/(1+M/2r)$ and $\psi = 1+M/2r$. Note that in these coordinates, $\tilde{\gamma}_{ij} = \delta_{ij}$, $\tilde{A}_{ij} = 0$, $K = 0$, and $\beta^i = 0$. One might think that the weak point of this coordinate system for the use in numerical relativity is that the conformal factor ψ diverges at $r = 0$. However, this is nothing but a coordinate singularity because the curvature invariant is zero at $r = 0$ (see also section 1.3). This suggests that by a locally appropriate coordinate transformation or other prescription, this divergence may be removed. Nevertheless, in the original BSSN formulation, ψ diverges at $r = 0$, and in the absence of a prescription, the simulation is prohibited. What Campanelli et al. (2006a) pointed out that by defining $\chi := \psi^{-4}$ or $W := \psi^{-2}$ and rewriting the basic equations in terms of this new variable,[4] the problem mentioned above could be fixed. Indeed, with the new variable (e.g., for W), the basic equations are rewritten as

$$(\partial_t - \beta^l \partial_l)\tilde{\gamma}_{ij} = -2\alpha \tilde{A}_{ij} + \tilde{\gamma}_{ik}\partial_j \beta^k + \tilde{\gamma}_{jk}\partial_i \beta^k - \frac{2}{3}\tilde{\gamma}_{ij}\partial_k \beta^k, \qquad (2.225)$$

$$(\partial_t - \beta^l \partial_l)\tilde{A}_{ij} = \left[\alpha W^2 \left(R_{ij} - \frac{1}{3}\gamma_{ij}R_k^{\ k}\right) - W^2\left(D_i D_j \alpha - \frac{1}{3}\gamma_{ij}\Delta\alpha\right)\right]$$
$$+ \alpha\left(K\tilde{A}_{ij} - 2\tilde{A}_{ik}\tilde{A}_j^{\ k}\right) + \left(\partial_i \beta^k\right)\tilde{A}_{kj} + \left(\partial_j \beta^k\right)\tilde{A}_{ki} - \frac{2}{3}\left(\partial_k \beta^k\right)\tilde{A}_{ij}$$
$$- 8\pi\alpha W^2 \left(S_{ij} - \frac{1}{3}\gamma_{ij}S_k^{\ k}\right), \qquad (2.226)$$

$$(\partial_t - \beta^l \partial_l)W = \frac{W}{3}\left(\alpha K - \partial_k \beta^k\right), \qquad (2.227)$$

---

[4] In the original paper by Campanelli et al. (2006a), they employed $\chi = \psi^{-4}$. W was proposed by Marronetti et al. (2008).

$$(\partial_t - \beta^l \partial_l) K = \alpha \left[\tilde{A}_{ij} \tilde{A}^{ij} + \frac{1}{3} K^2 \right] - W^2 \left(\tilde{\Delta} \alpha - \frac{\partial_i W}{W} \tilde{\gamma}^{ij} \partial_j \alpha \right)$$
$$+ 4\pi \alpha \left(\rho_{\rm h} + S_k{}^k \right), \tag{2.228}$$

$$(\partial_t - \beta^l \partial_l) \tilde{\Gamma}^i = -2\tilde{A}^{ij} \partial_j \alpha + 2\alpha \left[\tilde{\Gamma}^i_{jk} \tilde{A}^{jk} - \frac{2}{3} \tilde{\gamma}^{ij} \partial_j K - 8\pi \tilde{\gamma}^{ik} J_k - 3 \frac{\partial_j W}{W} \tilde{A}^{ij} \right]$$
$$- \tilde{\Gamma}^j \partial_j \beta^i + \frac{2}{3} \tilde{\Gamma}^i \partial_j \beta^j + \frac{1}{3} \tilde{\gamma}^{ik} \partial_k \partial_j \beta^j + \tilde{\gamma}^{jk} \partial_j \partial_k \beta^i, \tag{2.229}$$

where for the decomposition $R_{ij} = \tilde{R}_{ij} + R^W_{ij}$, we have

$$W^2 R^W_{ij} = W \tilde{D}_i \tilde{D}_j W + \tilde{\gamma}_{ij} \left(W \tilde{D}_k \tilde{D}^k W - 2 \tilde{D}_k W \tilde{D}^k W \right). \tag{2.230}$$

Here, $0 \leq W \leq 1$ and $W = 0$ at $r = 0$ for the metric (2.224). Equations (2.225)–(2.229) together with equation (2.230) show that there is no divergent term for the Schwarzschild metric in the isotropic coordinates. Even for the case that $\tilde{\gamma}_{ij}$, $\tilde{A}_{ij}$, and K are not zero, all the terms are finite if the lapse function, α, is zero at points where $W = 0$. This suggests that for an appropriate choice of the slicing condition in which $\alpha = 0$ at the points of $W = 0$, the black-hole spacetime can be evolved in the BSSN formulation. This can be indeed done for the case that we choose the moving puncture gauge condition described in section 2.2.6. Since 2005, many simulations along this line have been performed for a variety of black-hole spacetime. Now, the choice of W (or χ) is the standard in the BSSN formulation.

BSSN-puncture formulation was generalized for arbitrary spatial dimensions by Yoshino and Shibata (2009). The basic equations are qualitatively the same as those for three spatial dimensions but some of coefficients are changed. Specifically, they are (in Cartesian coordinates)

$$(\partial_t - \beta^l \partial_l) \tilde{\gamma}_{ij} = -2\alpha \tilde{A}_{ij} + \tilde{\gamma}_{ik} \partial_j \beta^k + \tilde{\gamma}_{jk} \partial_i \beta^k - \frac{2}{N} \tilde{\gamma}_{ij} \partial_k \beta^k, \tag{2.231}$$

$$(\partial_t - \beta^l \partial_l) \tilde{A}_{ij} = \chi \left[-\left(D_i D_j \alpha - \frac{1}{N} \gamma_{ij} D_k D^k \alpha \right) \right.$$
$$\left. + \alpha \left(R_{ij} - \frac{1}{N} \gamma_{ij} R \right) - 8\pi \alpha \left(S_{ij} - \frac{1}{N} \gamma_{ij} S_k{}^k \right) \right]$$
$$+ \alpha \left(K \tilde{A}_{ij} - 2 \tilde{A}_{ik} \tilde{A}^k{}_j \right)$$
$$+ \tilde{A}_{ik} \partial_j \beta^k + \tilde{A}_{jk} \partial_i \beta^k - \frac{2}{N} \tilde{A}_{ij} \partial_k \beta^k, \tag{2.232}$$

$$(\partial_t - \beta^l \partial_l) \chi = \frac{2}{N} \chi \left(\alpha K - \partial_i \beta^i \right), \tag{2.233}$$

$$(\partial_t - \beta^l \partial_l) K = -D_i D^i \alpha + \alpha \left(\tilde{A}^{ij} \tilde{A}_{ij} + \frac{1}{N} K^2 \right)$$
$$+ \frac{8\pi \alpha}{N-1} \left[(N-2) \rho_{\rm h} + S_k{}^k \right], \tag{2.234}$$

$$(\partial_t - \beta^l \partial_l)\tilde{\Gamma}^i = -2\tilde{A}^{ij}\partial_j\alpha$$
$$+2\alpha\left[\tilde{\Gamma}^i{}_{jk}\tilde{A}^{jk} - \frac{N-1}{N}\tilde{\gamma}^{ij}\partial_j K - 8\pi\tilde{\gamma}^{ij}J_j - \frac{N}{2}\frac{\partial_j\chi}{\chi}\tilde{A}^{ij}\right]$$
$$-\tilde{\Gamma}^j\partial_j\beta^i + \frac{2}{N}\tilde{\Gamma}^i\partial_j\beta^j + \frac{N-2}{N}\tilde{\gamma}^{ik}\partial_k\partial_j\beta^j + \tilde{\gamma}^{jk}\partial_j\partial_k\beta^i, \quad (2.235)$$

for the evolution equations, and

$$R + \frac{N-1}{N}K^2 - \tilde{A}_{ij}\tilde{A}^{ij} = 16\pi\rho_{\rm h}, \quad (2.236)$$

$$\chi^{N/2}\tilde{D}_i(\chi^{-N/2}\tilde{A}^i{}_j) - \frac{N-1}{N}\tilde{D}_j K = 8\pi J_j, \quad (2.237)$$

for the constraint equations. Here the following quantities are defined:

$$\tilde{\gamma}_{ij} = \chi\gamma_{ij} \quad [\det(\tilde{\gamma}_{ij}) = 1],$$
$$\tilde{A}_{ij} = \chi\left(K_{ij} - \frac{1}{N}\gamma_{ij}K\right). \quad (2.238)$$

2.3.4 Generalized harmonic formulation

The harmonic gauge has been the most popular gauge condition in mathematical relativity as well as in post-Newtonian and post-Minkowski approximations [Blanchet and Damour (1986); Blanchet (2014)], because they are useful to prove local existence of solutions of Einstein's equation [see chapter 10 of Wald (1984) for a review] and to directly integrate Einstein's equation in perturbation analyses. As described in section 1.1.2, the hyperbolicity of Einstein's equation is explicitly shown in this gauge (see section 2.2.3). The strongly hyperbolic nature of Einstein's equation in this gauge is also found in a linear perturbation analysis in section 2.3.1.

By contrast, the harmonic gauge had not been employed in numerical relativity until the middle of 2000s. This is due to their possible drawback associated with a fact that harmonic coordinates are solutions of the wave equation $\Box_g x^\mu = 0$, and the "time coordinate" may not remain to be globally timelike in general. This could cause a problem in numerical relativity. Also, if the gradients of all the spacetime coordinate directions fail to be linearly independent at a point, a coordinate singularity appears and the computation crashes [see, e.g., Alcubierre (1995); Alcubierre and Massó (1997) for such pathological examples]. On the other hand, harmonic gauges should have an excellent property in numerical relativity as mentioned in section 2.2.3. In particular, they are well suited for computing the propagation of gravitational waves in a wave zone, because all the equations obey the wave equation in this coordinate system.

For a practical use of the harmonic gauge in numerical relativity, Garfinkle (2002) proposed a modified version of the harmonic gauge for the first time. Subsequently, Pretorius (2005a,b) extended the idea of Garfinkle for simulating binary-black-hole spacetime. The modification by Pretorius (2005a,b) enabled him to successfully simulate the binary-black-hole merger for the first time. Now, the modified version of the harmonic gauge is called the generalized harmonic (GH) gauge, and the formulation based on this gauge condition is called generalized harmonic (GH) formulation. In the rest of this section, we outline this formulation following the original paper by Garfinkle (2002).

In the GH formulation, as in post-Newtonian approximations, Einstein's equation is written as

$$\overset{(D)}{R}_{ab} = 8\pi \left(T_{ab} - \frac{1}{D-2} g_{ab} T \right), \tag{2.239}$$

and rewrite this equation without foliating spacetime. The GH gauge condition satisfies

$$\Box_g x^\mu = \frac{1}{\sqrt{-g}} \partial_\alpha \left(\sqrt{-g} g^{\alpha\mu} \right) = H^\mu, \tag{2.240}$$

where H^μ is an arbitrary function that specifies the gauge condition. We may give an appropriate function to it or we may construct evolution equations for H^μ. For $H^\mu = 0$, the harmonic gauge is obtained. If H^μ is determined by an evolution equation, we have an additional constraint

$$\mathcal{C}^\mu = H^\mu - \Box_g x^\mu. \tag{2.241}$$

Using spacetime metric g_{ab} and H^μ, $\overset{(D)}{R}_{\alpha\beta}$ is written in the form

$$2\overset{(D)}{R}_{\alpha\beta} = -g^{\mu\nu} \partial_\mu \partial_\nu g_{\alpha\beta} - (\partial_\alpha H_\beta + \partial_\beta H_\alpha)$$
$$+ \mathcal{C}_\alpha{}^{\mu\nu} \mathcal{C}_{\mu\nu\beta} + \mathcal{C}_\beta{}^{\mu\nu} \mathcal{C}_{\mu\nu\alpha} - 2\overset{(D)}{\Gamma}{}^\mu{}_{\nu\alpha} \overset{(D)}{\Gamma}{}^\nu{}_{\mu\beta} + 2H_\nu \overset{(D)}{\Gamma}{}^\nu{}_{\alpha\beta}, \tag{2.242}$$

where $H_\mu = g_{\mu\nu} H^\nu$, $\mathcal{C}_{\alpha\mu\nu} := \partial_\alpha g_{\mu\nu}$, and $\mathcal{C}_\alpha{}^{\mu\nu} = g^{\mu\beta} g^{\nu\gamma} \mathcal{C}_{\alpha\beta\gamma}$. Equation (2.242) shows that equation (2.239) is a hyperbolic equation for $g_{\mu\nu}$, as shown in section 1.1.2.

The reason why the nonzero term of H^μ has to be considered is that this term allows us to change the behavior of the coordinates from the purely harmonic-gauge case, and thus, it could eliminate possible bad behavior like appearance of coordinate singularities in harmonic coordinates. In principle, it is possible to choose any gauge condition by an appropriate choice of H^μ. However, H^μ does not have a direct geometrical meaning in contrast to the lapse function, α, and the shift vector, β^i. It is therefore less clear how to choose the gauge condition of desired properties. Because any systematic way to find appropriate forms of H^μ has not been developed yet, it is currently necessary to perform numerical experiments of the try-and-error search for this term. For example, Pretorius (2005a) discovered a suitable choice for H^μ for evolving black-hole spacetime, following the idea proposed by Gundlach et al. (2005) that was developed in a different context. In his choice, H_t obeys a wave equation in which several arbitrary parameters are included, while the spatial components of H_i are chosen to be zero. The free parameters are determined in a try-and-error manner. The prescription of Pretorius (2005a) was further explored by Scheel *et al.* (2009) who performed a precise long-term numerical simulation for the inspiral and the merger of binary black holes.

The GH formulation is quite different from the BSSN formulation. In the original version of the GH formulation, the spacetime metric is only the component to be evolved. No conjugate momentum like extrinsic curvature in the 3+1 and N+1 formulations is present. Namely, the basic equations are written in the form of second-order hyperbolic equations. In subsequent works by Lindblom *et al.* (2006); Scheel *et al.* (2009); Anderson *et al.* (2008), first-order forms of the GH formulation were developed. In these formulations, in addition to the spacetime metric $g_{\alpha\beta}$, a time-derivative term $-n^\mu \partial_\mu g_{\alpha\beta}$ and the spatial derivative of $g_{\mu\nu}$, $\partial_i g_{\alpha\beta}$, are employed as quantities to be evolved. The new system is also strongly hyperbolic and allows a stable and long-term numerical simulation. Some of numerical codes based on the modified GH formulation have been applied to the coalescence of binary black holes and black hole-neutron star binaries (see sections 8.5 and 8.7).

Another fact to be stressed is that the singularity-avoidance property is not very strong in the time coordinate of the GH gauge (see section 2.2.3). Thus, in the formation or in the presence of a black hole, a black-hole excision prescription is necessary in this formulation (unless a special choice is made for H^t for significantly modifying the slicing condition).

2.3.5   Other formulations

Besides the BSSN and GH formulations, several formulations, which could be useful for the practical use in numerical relativity, have been proposed. One is a constrained formulation proposed by Bonazzola *et al.* (2004); Cordero-Carrión *et al.* (2009) and the others are the so-called Z4 formulations, which were originally proposed by Bona *et al.* (2003, 2004) and subsequently explored, e.g., in Gundlach *et al.* (2005); Bernuzzi and Hilditch (2010); Alic *et al.* (2012a); Hilditch *et al.* (2013). Among them, the CCZ4 [Alic *et al.* (2012a)] and Z4c formulations [Hilditch *et al.* (2013)] are getting popular because of its robustness for simulations of inspiraling neutron-star binaries, and hence, we here briefly refer to this type of the formulation. Because there are several variants for the basic equations in these formulations, we only describe their basic concept in our own manner.

The CCZ4 and Z4c formulations may be viewed as a modified version of the BSSN-puncture formulation. The motivation of the modification stems from the fact that in the BSSN-puncture formulation, the magnitude of the constraint violation grows slowly with time. This is the result from the presence of a mode with the zero propagation speed that stays at the original position. Although this is not the growing mode in the BSSN formulation, the accumulation of numerical errors due to the absence of the constraint propagation enhances the constraint violation. The idea of these Z4-type formulations is to suppress the growth of this constraint violation by adding an evolution equation for a new auxiliary variable, Z^a (in particular n^a part of Z^a), in the original BSSN formulation. Specifically, in the Z4 formulation, Einstein's equation is modified as

$$R_{ab} + \nabla_a Z_b + \nabla_b Z_a - \kappa_1 \left[n_a Z_b + n_b Z_a - (1 + \kappa_2) g_{ab} n_c Z^c \right]$$
$$= 8\pi \left(T_{ab} - \frac{1}{2} g_{ab} T \right), \tag{2.243}$$

where κ_1 and κ_2 are small coefficients associated with the constraint damping and they control the time scale for the damping of the constraint violation. This damping term is not essential in the Z4c formulations, and hence, we often drop these terms in the following. We also pay attention to $D = 4$ ($N = 3$) case for simplicity. The extension is quite straightforward for $D > 4$.

The 3+1 equations in the Z4 formulation are easily derived when we view the terms associated with Z^a as an additional stress-energy tensor as

$$8\pi T_{ab} \to 8\pi T_{ab} - (\nabla_a Z_b + \nabla_b Z_a - g_{ab} \nabla_c Z^c)$$
$$+ \kappa_1 \left(n_a Z_b + n_b Z_a + \kappa_2 g_{ab} n_c Z^c \right). \tag{2.244}$$

Then, ρ_h, J_a, and S_{ab} [see equations (2.43)–(2.45) for the definitions of them] are replaced as

$$8\pi \rho_h \to 8\pi \rho_h + n^a \nabla_a \Theta + K\Theta - D_a \hat{Z}^a + \hat{Z}^a D_a \ln \alpha$$
$$+ \kappa_1 (2 + \kappa_2) \Theta, \tag{2.245}$$
$$8\pi J_a \to 8\pi J_a + \Theta D_a \ln \alpha - D_a \Theta + \gamma^c{}_a n^b \nabla_b \hat{Z}_c + K_{ab} \hat{Z}^b - \kappa_1 \hat{Z}_a, \tag{2.246}$$
$$8\pi S_{ab} \to 8\pi S_{ab} + 2 K_{ab} \Theta - \left(D_a \hat{Z}_b + D_b \hat{Z}_a - \gamma_{ab} D_c \hat{Z}^c \right)$$
$$+ \gamma_{ab} \left(-K\Theta + n^c \nabla_c \Theta + \hat{Z}^c D_c \ln \alpha \right) - \kappa_1 \kappa_2 \Theta \gamma_{ab}, \tag{2.247}$$
$$4\pi (\rho_h + S_a{}^a) \to 4\pi (\rho_h + S_a{}^a) + 2 n^a \nabla_a \Theta + 2 \hat{Z}^a D_a \ln \alpha + \kappa_1 (1 - \kappa_2) \Theta, \tag{2.248}$$

where $\Theta := -n_a Z^a$ and $\hat{Z}^a := \gamma^a{}_b Z^b$. In the presence of these additional terms, the evolution equations for $\tilde{A}_{ij}$, K, and $\tilde{\Gamma}^i$ are modified, respectively, by the modification of S_{ab}, $\rho_h + S_a{}^a$, and J_a [see equations (2.225)–(2.229)]. As a result, in the right-hand side of the equations for K and $\tilde{\Gamma}^i$, the terms associated with the time derivative as $2\alpha n^a \nabla_a \Theta$ and $-2\alpha n^a \nabla_a \hat{Z}_b$ appear, respectively. Then, by defining new variables $\tilde{K} := K - 2\Theta$ and $\bar{\Gamma}^i := \tilde{\Gamma}^i + 2 \tilde{\gamma}^{ij} \hat{Z}_j$, we can obtain the evolution equations for $\tilde{K}$ and $\bar{\Gamma}^i$, for which no time derivative terms appear in the right-hand side.

The Hamiltonian and momentum constraint equations are also modified. In particular, the modification by the presence of Θ is crucial in this formulation. Thus, in the following, we set $\hat{Z}_a = 0$ for simplicity. Dropping also the constraint damping terms, the Hamiltonian and momentum constraint equations are written as

$$R - K_{ab} K^{ab} + K^2 = 16\pi \rho_h + 2 n^a \nabla_a \Theta + 2 K \Theta, \tag{2.249}$$
$$D_a K^a{}_b - D_b K = 8\pi J_b + \Theta D_b \ln \alpha - D_b \Theta, \tag{2.250}$$

and hence,

$$(\partial_t - \beta^l \partial_l)\Theta = \frac{1}{2}\alpha\left[R - \tilde{A}_{ij}\tilde{A}^{ij} + \frac{2}{3}K^2 - 16\pi\rho_{\rm h}\right] - \alpha K\Theta$$
$$= \alpha H_0 - \alpha K\Theta, \qquad (2.251)$$
$$\partial_t H_i = \partial_t(\Theta D_i \ln\alpha - D_i\Theta), \qquad (2.252)$$

where $H_0 = (R - K_{ab}K^{ab} + K^2 - 16\pi\rho_{\rm h})/2$ and $H_i = D_j K^j{}_i - D_i K - 8\pi J_i$: see equation (2.70) for the definition of H_0 and H_i. Now, remember the evolution equations for the constraints, (2.71) and (2.72), in the ADM formulation. The principle parts of these equations have the forms

$$\partial_t H_0 \sim -D_i H^i, \qquad \partial_t H_i \sim -D_j H^j{}_i = 0, \qquad (2.253)$$

where we used $H^i{}_j = 0$ supposing that the evolution equation for the spatial geometry is accurately solved. Equation (2.253) indicates that there is no propagation mode for the evolution equations of the constraints in the ADM formulation. However, in the Z4 formulation, the evolution equation for H_i is replaced by equation (2.252). For this, the principle part is

$$\partial_t H_i \sim -D_i(\partial_t \Theta) \sim -D_i H_0, \qquad (2.254)$$

where we used equation (2.251). As a result, H_0 and H_i obey wave equations, and therefore, the constraint violation can propagate away in this formulation.

As shown in the above paragraph, the crucial point of the Z4 formulation is the introduction of the new variable Θ and the equation for it. Thus, we here set $Z^a = -n^a\Theta$ with $\kappa_1 = \kappa_2 = 0$ to focus on the important terms. In this case, the basic equations are only slightly different from the original BSSN-puncture formulation as

$$(\partial_t - \beta^l \partial_l)\tilde{\gamma}_{ij} = -2\alpha\tilde{A}_{ij} + \tilde{\gamma}_{ik}\partial_j\beta^k + \tilde{\gamma}_{jk}\partial_i\beta^k - \frac{2}{3}\tilde{\gamma}_{ij}\partial_k\beta^k, \qquad (2.255)$$

$$(\partial_t - \beta^l \partial_l)\tilde{A}_{ij} = \left[\alpha W^2\left(R_{ij} - \frac{1}{3}\gamma_{ij}R_k{}^k\right) - W^2\left(D_i D_j\alpha - \frac{1}{3}\gamma_{ij}\Delta\alpha\right)\right]$$
$$+ \alpha\left[\tilde{K}\tilde{A}_{ij} - 2\tilde{A}_{ik}\tilde{A}_j{}^k\right] + (\partial_i\beta^k)\tilde{A}_{kj} + (\partial_j\beta^k)\tilde{A}_{ki} - \frac{2}{3}(\partial_k\beta^k)\tilde{A}_{ij}$$
$$- 8\pi\alpha W^2\left(S_{ij} - \frac{1}{3}\gamma_{ij}S_k{}^k\right), \qquad (2.256)$$

$$(\partial_t - \beta^l \partial_l)W = \frac{W}{3}\left(\alpha(\tilde{K} + 2\Theta) - \partial_k\beta^k\right), \qquad (2.257)$$

$$(\partial_t - \beta^l \partial_l)\tilde{K} = \alpha\left[\tilde{A}_{ij}\tilde{A}^{ij} + \frac{1}{3}(\tilde{K} + 2\Theta)^2\right] - W^2\left(\tilde{\Delta}\alpha - \frac{\partial_i W}{W}\tilde{\gamma}^{ij}\partial_j\alpha\right)$$
$$+ 4\pi\alpha(\rho_{\rm h} + S_k{}^k), \qquad (2.258)$$

$$(\partial_t - \beta^l \partial_l)\tilde{\Gamma}^i = -2(\tilde{A}^{ij} + \Theta \tilde{\gamma}^{ij})\partial_j \alpha$$
$$+ 2\alpha \left[\tilde{\Gamma}^i_{jk} \tilde{A}^{jk} - \frac{1}{3}\tilde{\gamma}^{ij}\partial_j(2\tilde{K} + \Theta) - 8\pi \tilde{\gamma}^{ik} J_k - 3\frac{\partial_j W}{W}\tilde{A}^{ij} \right]$$
$$- \tilde{\Gamma}^j_{\rm d}\partial_j \beta^i + \frac{2}{3}\tilde{\Gamma}^i_{\rm d}\partial_j\beta^j + \frac{1}{3}\tilde{\gamma}^{ik}\partial_k\partial_j\beta^j + \tilde{\gamma}^{jk}\partial_j\partial_k\beta^i, \qquad (2.259)$$
$$(\partial_t - \beta^l \partial_l)\Theta = \frac{1}{2}\alpha\left[R - \tilde{A}_{ij}\tilde{A}^{ij} + \frac{2}{3}(\tilde{K} + 2\Theta)^2 - 16\pi\rho_{\rm h}\right] - \alpha K \Theta, \qquad (2.260)$$

where $\tilde{\Gamma}^i_{\rm d} := -\overset{(0)}{D}_j \tilde{\gamma}^{ij}$ and the tracefree part of the extrinsic curvature is derived by $K = \tilde{K} + 2\Theta$. For $\Theta = 0$, the BSSN-puncture equations are recovered.

In the CCZ4 [Alic et al. (2012a)] and Z4c [Hilditch et al. (2013)] formulations, the basic equations are slightly different. For the CCZ4 case, the equations for Θ and $\hat{Z}^i$ are fully solved appropriately choosing κ_1 and κ_2. On the other hand, in the Z4c case, the following equations are chosen:

$$(\partial_t - \beta^l\partial_l)\tilde{\gamma}_{ij} = -2\alpha \tilde{A}_{ij} + \tilde{\gamma}_{ik}\partial_j\beta^k + \tilde{\gamma}_{jk}\partial_i\beta^k - \frac{2}{3}\tilde{\gamma}_{ij}\partial_k\beta^k, \qquad (2.261)$$
$$(\partial_t - \beta^l\partial_l)\tilde{A}_{ij} = \left[\alpha W^2 \left(R_{ij} - \frac{1}{3}\gamma_{ij}R_k^{\ k}\right) - W^2\left(D_i D_j \alpha - \frac{1}{3}\gamma_{ij}\Delta\alpha\right)\right]$$
$$+ \alpha\left[\left(\tilde{K}+2\Theta\right)\tilde{A}_{ij} - 2\tilde{A}_{ik}\tilde{A}_j^{\ k}\right]$$
$$+ \left(\partial_i\beta^k\right)\tilde{A}_{kj} + \left(\partial_j\beta^k\right)\tilde{A}_{ki} - \frac{2}{3}\left(\partial_k\beta^k\right)\tilde{A}_{ij}$$
$$- 8\pi\alpha W^2\left(S_{ij} - \frac{1}{3}\gamma_{ij}S_k^{\ k}\right), \qquad (2.262)$$
$$(\partial_t - \beta^l\partial_l)W = \frac{W}{3}\left(\alpha\left(\tilde{K}+2\Theta\right) - \partial_k\beta^k\right), \qquad (2.263)$$
$$(\partial_t - \beta^l\partial_l)\tilde{K} = \alpha\left[\tilde{A}_{ij}\tilde{A}^{ij} + \frac{1}{3}\left(\tilde{K}+2\Theta\right)^2\right] - W^2\left(\tilde{\Delta}\alpha - \frac{\partial_i W}{W}\tilde{\gamma}^{ij}\partial_j\alpha\right)$$
$$+ 4\pi\alpha(\rho_{\rm h} + S_k^{\ k}) + \alpha\kappa_1(1-\kappa_2)\Theta, \qquad (2.264)$$
$$(\partial_t - \beta^l\partial_l)\tilde{\Gamma}^i = -2\tilde{A}^{ij}\partial_j\alpha$$
$$+ 2\alpha\left[\tilde{\Gamma}^i_{jk}\tilde{A}^{jk} - \frac{1}{3}\tilde{\gamma}^{ij}\partial_j\left(2\tilde{K}+\Theta\right) - 8\pi\tilde{\gamma}^{ik}J_k - 3\frac{\partial_j W}{W}\tilde{A}^{ij}\right]$$
$$- \tilde{\Gamma}^j_{\rm d}\partial_j\beta^i + \frac{2}{3}\tilde{\Gamma}^i_{\rm d}\partial_j\beta^j + \frac{1}{3}\tilde{\gamma}^{ik}\partial_k\partial_j\beta^j + \tilde{\gamma}^{jk}\partial_j\partial_k\beta^i$$
$$- 2\alpha\kappa_1\left(\tilde{\Gamma}^i - \tilde{\Gamma}^i_{\rm d}\right), \qquad (2.265)$$
$$(\partial_t - \beta^l\partial_l)\Theta = \frac{1}{2}\alpha\left[R - \tilde{A}_{ij}\tilde{A}^{ij} + \frac{2}{3}\left(\tilde{K}+2\Theta\right)^2 - 16\pi\rho_{\rm h}\right] - \alpha\kappa_1(2+\kappa_2)\Theta. \qquad (2.266)$$

In this formulation, terms associated with Θ in the equation for $\tilde{A}_{ij}$, $\tilde{\Gamma}^i$, and Θ are artificially dropped from the original Z4 formulation to improve the accuracy (compare them with equations (2.255)–(2.260) excluding the constraint damping terms).

However, numerical experiments (for four spacetime dimensions) have shown that the slight difference among these formulations does not give serious damage in the numerical results, as far as the effect of the constraint propagation is appropriately taken into account.[5]

2.4 Formulations in axisymmetric spacetime

The formulations described in section 2.3 are popular in numerical relativity for spacetime with no symmetry. In a simulation for non-symmetric spacetime using such formulations, we usually employ Cartesian coordinates, because it is not technically easy to successfully perform a simulation in curvilinear coordinates due to the presence of coordinate singularities [but see Baumgarte *et al.* (2013) for a special technique]: In the curvilinear coordinates, it is not easy to guarantee the regularity conditions for tensor quantities at the coordinate singularities (see section 2.4.2 for the reason).[6]

The formulations and the numerical codes developed for general three-dimensional space written in Cartesian coordinates may be employed for any system of special symmetries including spherically and axially symmetric space without modification. However, it is a waste of computational costs to employ such codes for the spacetime of special symmetries. The cost can be significantly saved, if we explicitly impose the symmetry conditions using a curvilinear coordinate system. In this section, we introduce several numerical-relativity formulations for axially symmetric spacetime, which have been frequently used in the history of numerical relativity. The formulation depends on our choice of the gauge condition and on the choice of basic equations; i.e., whether the constraint equations are solved instead of the evolution equations for some of geometric components. Although there are a wide variety of possible formulations, we here focus only on describing several representative formulations.

Simulations for axisymmetric (and spherical symmetric) spacetime can be also performed in Cartesian coordinates, still imposing symmetries, by the so-called cartoon method, which was originally proposed by Alcubierre *et al.* (2001a). This method is quite different from those in this section and, in addition, it can be applied to a variety of spacetime, not only to axisymmetric spacetime but also to others. For this reason, we describe the cartoon method in a separate section (see section 2.5).

---

[5]In the Z4c formulation, the general covariance is violated. However, as far as the constraint violation is maintained to be small everywhere, the covariance violation would be only as serious as the small constraint violation. By contrast, if the constraint violation is quite large, it could be dangerous to use this formulation.

[6]For example, in spherical polar coordinates, $r = 0$, and $\theta = 0$ and π are the coordinate singularities. In this section, the coordinate singularities do not mean those caused by an inappropriate choice of gauge conditions.

2.4.1 Introduction

In this section, we describe numerical-relativity formulations for four-dimensional axisymmetric spacetime, supposing that curvilinear coordinates are used as a coordinate basis. For this case, first of all, the general form of the metric components depends on whether angular momentum of the system is present or not. In its absence, we can write the metric, e.g., in spherical polar coordinates (r, θ, φ), as

$$g_{\mu\nu} = \begin{pmatrix} -\alpha^2 + \beta^k\beta_k & \beta_r & \beta_\theta & 0 \\ * & \gamma_{rr} & \gamma_{r\theta} & 0 \\ * & * & \gamma_{\theta\theta} & 0 \\ * & * & * & \gamma_{\varphi\varphi} \end{pmatrix}, \qquad (2.267)$$

where $*$ denotes the relation of symmetry $g_{\mu\nu} = g_{\nu\mu}$. Namely, it is natural to set $\beta_\varphi = \beta^\varphi = 0$, $\gamma_{r\varphi} = \gamma_{\theta\varphi} = 0$, and $K_{r\varphi} = K_{\theta\varphi} = 0$ for this case. On the other hand, in the presence of nonzero angular momentum, full components of the metric have to be solved. In the following, we suppose that the system considered has angular momentum.

There are several possible formulations of Einstein's equation in axially symmetric numerical relativity:

(1) *Free-evolution formulation:* Employing a 3+1 formulation, all the components of γ_{ij} and K_{ij} (or new variables written by a combination of them) are evolved and some gauge conditions for α and β^i, which are freely chosen as in the usual 3+1 formulation, are imposed. This method was employed in the pioneer work by Smarr (1979) in 1970s and by Nakamura and his collaborators in 1980s using their (2+1)+1 formulation (see section 2.4.6) [Nakamura (1981); Nakamura and Sato (1981, 1982); Nakamura (1983); Nakamura et al. (1987)].

(2) *Partially-constrained evolution:* Employing a 3+1 formulation, some components of γ_{ij} and K_{ij} are evolved by evolution equations for them and the rest components are determined using the constraint equations, imposing some gauge conditions for α and β^i. This method has not been adopted yet.

(3) *Partially-constrained evolution with particular spatial gauge:* Employing a 3+1 formulation, but imposing particular relations among the components of γ_{ij} that lead to a spatial gauge condition for β^i, a part of the components of γ_{ij} are evolved. The lapse function is determined by a slicing condition. The constraint equations are not used or partially used. This type of formulation was most popular and was employed by Miyama (1981), Stark and Piran (1985); Piran and Stark (1986); Stark and Piran (1987), Abrahams and Evans (1992, 1993, 1994), Abrahams et al. (1992); Anninos et al. (1993); Brandt and Seidel (1995a,b); Anninos et al. (1995b,d), Shibata et al. (1994b), and Garfinkle and Duncan (2001).

(4) *Fully-constrained evolution with particular spatial gauge:* The same as (3) but the constraint equations are fully used. This method was employed by Evans (1985, 1986), Shapiro and Teukolsky (1991, 1992b,a); Abrahams et al. (1994a,b,

1995); Shapiro et al. (1995), Choptuik et al. (2003a,b, 2004), and Rinne (2008, 2010).

(5) *GH formulation:* Employing the GH formulation, 7 (for no angular momentum) or 10 (with angular momentum) evolution equations for the spacetime metric are solved. This method has been recently employed by Sorkin (2010) [see also Lehner and Pretorius (2010) for a similar context].

In the following, we will describe formulations of types (1), (3), and (4) by which great success was achieved in axisymmetric numerical relativity in 1970s, 1980s, and 1990s.

One of the most crucial points in numerical relativity performed with curvilinear coordinate bases is to guarantee regularity conditions for the geometric variables that have to be satisfied at coordinate singularities.[7] If the regularity conditions are violated, the computation usually crashes around the coordinate singularities; the accuracy becomes poor or the numerical stability is lost. The regularity conditions also prohibit employing a class of gauge conditions and numerical methods in which the regularity conditions are never satisfied as illustrated in detail by Bardeen and Piran (1983). By contrast, the regularity conditions are written in a simpler form in a good formulation, that helps imposing them in numerical computation. All these facts imply that guaranteeing the regularity conditions is the key for the success in axisymmetric numerical relativity. Therefore, before describing the specific formulations, we remind the readers of the regularity conditions in the next section.

2.4.2 *Regularity conditions*

As already mentioned, curvilinear coordinates have coordinate singularities such as $r = 0$, and $\theta = 0$ and π in spherical polar coordinates, and $\varpi = 0$ in cylindrical coordinates (ϖ, φ, z). At such coordinate singularities, vector and tensor components written in a curvilinear coordinate basis have to satisfy certain conditions in order for them to be regular. Otherwise, correct solutions for their basic equations cannot be obtained, and moreover, the computation often crashes. These conditions to be satisfied are referred to as regularity conditions.

Regularity conditions for scalars (represented by Q in the following) are trivial: They are derived from the requirement that Q is a regular function of $\varpi^2 = x^2 + y^2$ and z under the axisymmetric condition $\partial_\varphi Q = 0$.

Regularity conditions for tensors were described in detail by Bardeen and Piran (1983) for three-dimensional axisymmetric space. They are derived from the relation that the Lie derivative of a symmetric tensor represented by Q_{ab} (which is γ_{ab} or K_{ab} in the present context) vanishes as

$$\mathcal{L}_\varphi Q_{ab} = \varphi^c \nabla_c Q_{ab} + Q_{ac} \nabla_b \varphi^c + Q_{cb} \nabla_a \varphi^c = 0, \qquad (2.268)$$

---

[7]We here suppose that the same coordinate bases are used both for assigning the grid points and for describing the tensor components.

where $\varphi^c = (\partial/\partial\varphi)^c$ is the rotational Killing vector field associated with the axial symmetry, for which components in the Cartesian coordinate basis are $\varphi^x = -y$, $\varphi^y = x$, and $\varphi^z = 0$. In the Cartesian coordinate basis, six components of equation (2.268) are written as

$$\begin{aligned}
(-y\partial_x + x\partial_y)Q_{xx} + 2Q_{xy} &= 0, \\
(-y\partial_x + x\partial_y)Q_{yy} - 2Q_{xy} &= 0, \\
(-y\partial_x + x\partial_y)Q_{xy} + Q_{yy} - Q_{xx} &= 0, \\
(-y\partial_x + x\partial_y)Q_{zz} &= 0, \\
(-y\partial_x + x\partial_y)Q_{xz} + Q_{yz} &= 0, \\
(-y\partial_x + x\partial_y)Q_{yz} - Q_{xz} &= 0.
\end{aligned} \qquad (2.269)$$

The general regular solution for these equations will contain six independent functions, f_n ($n = 1 \cdots 6$), which depend only on ϖ^2 and z [i.e., $f_n(\varpi^2, z)$], and which can be expanded in non-negative integer powers of ϖ^2 and z near the origin $r = 0$ and the symmetric axis $\varpi = 0$. A specific form is written as

$$\begin{aligned}
Q_{xx} &= f_1 - 2f_2 xy + f_3 y^2, & Q_{yy} &= f_1 + 2f_2 xy + f_3 x^2, \\
Q_{xy} &= f_2(x^2 - y^2) - f_3 xy, & Q_{zz} &= f_4, \\
Q_{xz} &= f_5 x - f_6 y, & Q_{yz} &= f_5 y + f_6 x.
\end{aligned} \qquad (2.270)$$

The components in the spherical polar coordinate basis are

$$\begin{aligned}
Q_{rr} &= \sin^2\theta f_1 + \cos^2\theta f_4 + 2r\cos\theta \sin^2\theta f_5, \\
Q_{r\theta} &= r[\sin\theta \cos\theta (f_1 - f_4) + r\sin\theta(\cos^2\theta - \sin^2\theta)f_5], \\
Q_{\theta\theta} &= r^2[\cos^2\theta f_1 + \sin^2\theta f_4 - 2r\cos\theta \sin^2\theta f_5], \\
Q_{\varphi\varphi} &= r^2 \sin^2\theta [f_1 + r^2 \sin^2\theta f_3], \\
Q_{r\varphi} &= r\sin^2\theta [r\cos\theta f_6 + r^2 \sin^2\theta f_2], \\
Q_{\theta\varphi} &= r^2 \sin^2\theta [-r\sin\theta f_6 + r^2 \sin\theta \cos\theta f_2],
\end{aligned} \qquad (2.271)$$

and the components in the cylindrical coordinate basis are

$$\begin{aligned}
Q_{\varpi\varpi} &= f_1, & Q_{\varphi\varphi} &= \varpi^2[f_1 + \varpi^2 f_3], & Q_{\varpi\varphi} &= \varpi^3 f_2, \\
Q_{zz} &= f_4, & Q_{\varpi z} &= \varpi f_5, & Q_{\varphi z} &= \varpi^2 f_6.
\end{aligned} \qquad (2.272)$$

Note that in the presence of the reflection symmetry with respect to the equatorial plane ($z = 0$), f_1–f_4 have to be symmetric functions and f_5 and f_6 be asymmetric functions of z. Thus, for that case, f_1–f_4 are functions of z^2 and ϖ^2, and f_5 and f_6 are written as zf_5' and zf_6' where f_5' and f_6' are functions of z^2 and ϖ^2.

For vectors, the regularity conditions are derived in the same manner: In the spherical polar coordinate basis,

$$Q^r = r\sin^2\theta \hat{f}_1 + \cos\theta \hat{f}_3, \quad Q^\theta = \sin\theta(\cos\theta \hat{f}_1 - r^{-1}\hat{f}_3), \quad Q^\varphi = \hat{f}_2, \qquad (2.273)$$

and in the cylindrical coordinate basis,

$$Q^\varpi = \varpi \hat{f}_1, \quad Q^z = \hat{f}_3, \quad Q^\varphi = \hat{f}_2, \qquad (2.274)$$

where Q^a represents vectors. $\hat{f}_1-\hat{f}_3$ are functions of ϖ^2 and z, and can be expanded in non-negative integer powers of ϖ^2 and z near the origin $r=0$ and the symmetric axis $\varpi=0$.

We note that the regularity conditions in spherically symmetric space are soon found from equations (2.271) as

$$Q_{rr} = f_a(r^2) + r^2 f_b(r^2), \quad r^{-2}Q_{\theta\theta} = (r\sin\theta)^{-2}Q_{\varphi\varphi} = f_a(r^2),$$
$$Q_{r\theta} = Q_{r\varphi} = Q_{\theta\varphi} = 0, \qquad (2.275)$$

where we set $f_2 = f_6 = 0$ due to the absence of angular momentum of the system, and $f_1 = f_a + \varpi^2 f_b$, $f_3 = -f_b$, $f_4 = f_a + z^2 f_b$, and $f_5 = z f_b$ for imposing the spherical symmetry.

In numerical relativity, it is crucially important to satisfy the regularity conditions as precisely as possible to the spatial metric, because we have to evaluate their Ricci tensor which is composed of first and second spatial derivatives of the spatial metric and in any finite-differencing schemes the derivatives near the coordinate singularities depend entirely on the regularity conditions. Then, the analysis of the regularity conditions often indicates an important guideline for the gauge choice; we could find that some gauge conditions are not suitable in numerical relativity, because it is difficult to impose the regularity conditions with such gauges. This situation is found from equation (2.271) which obviously shows it quite complicated to impose the regularity conditions in spherical polar coordinates, in particular for Q_{rr}, $Q_{r\theta}$, and $Q_{\theta\theta}$, because they include three independent functions and moreover for each, θ-dependence is different. It may be in practice impossible to impose such multi θ-dependent regularity conditions at $r=0$. This fact strongly suggests that one should not choose a formulation in which all the components of the spatial metric are evolved with no assumption using spherical polar coordinates. Such a formulation could probably be chosen only when one employs cylindrical coordinates as the basis coordinates.

To circumvent the difficulty for imposing the regularity conditions in the spherical polar coordinate basis, a special spatial gauge condition such as $\gamma_{rr} = \gamma_{\theta\theta} r^{-2}$ has been often chosen to simplify the regularity conditions. In this example, equation (2.271) gives a relation near $r=0$ as

$$(f_1 - f_4)(\cos^2\theta - \sin^2\theta) = 4r\sin^2\theta\cos\theta f_5. \qquad (2.276)$$

This implies that $f_1 - f_4$ and f_5 have to be written as

$$f_1 - f_4 = 4 f_7 r^3 \sin^2\theta\cos\theta, \qquad (2.277)$$
$$f_5 = f_7 r^2 (\cos^2\theta - \sin^2\theta), \qquad (2.278)$$

where f_7 is a regular function. Then, the regularity conditions for γ_{rr} and $\gamma_{r\theta}$ become

$$\gamma_{rr} = \gamma_{\theta\theta} r^{-2} = f_1 - 2r^3\cos\theta\sin^2\theta f_7, \qquad (2.279)$$
$$\gamma_{r\theta} = r^4 \sin\theta f_7. \qquad (2.280)$$

Thus, the condition is significantly simplified. Although there are still θ-dependent terms in these conditions (for $f_7 \neq 0$), they are zero at $r = 0$, and the complexity for imposing the regular boundary condition is much less serious than that in the general form in which the conditions (2.271) have to be satisfied. To further simplify the regularity conditions, the additional condition $\gamma_{r\theta} = 0$ (or $\gamma^{r\theta} = 0$) is often chosen. Then, $f_7 = 0$, and thus, the regularity conditions for γ_{rr} and $\gamma_{\theta\theta} r^{-2}$ reduce to a simple one; e.g., $\partial_r \gamma_{rr} = 0$ at $r = 0$. Formulations with this type of the gauge conditions are introduced in sections 2.4.3, 2.4.4, and 2.4.5.

To explicitly guarantee the regularity conditions, it is also helpful to employ regularized variables in which information of the regularity conditions is taken into account. For example, Nakamura (1981) [see also Nakamura et al. (1980)] employed special variables defined from the spatial metric (in the cylindrical coordinate basis) and rewrote the basic equations using the new variables. Specifically, these variables are

$$\varpi^{-2}\left(\sqrt{\gamma_{\varpi\varpi}} - \varpi^{-1}\sqrt{\gamma_{\varphi\varphi}}\right), \quad \gamma_{\varpi z}(\varpi z)^{-1},$$
$$\left(K^{\varpi}_{\varpi} - K^{\varphi}_{\varphi}\right)\varpi^{-2}, \quad K^{z}_{\varpi}(\varpi z)^{-1}, \qquad (2.281)$$

where we note that the equatorial plane symmetry was assumed in their works. All these variables are smooth, satisfying $\partial_\varpi \cdots = 0$ near the rotation axis, $\varpi = 0$, and the boundary conditions can be imposed without taking into account the special behavior of the metric components near the coordinate singularities (as in the case that the Cartesian coordinate basis is employed). This prescription was crucial for their success for stable and reliable simulations of rotating stellar core collapses to a black hole.

Piran and Stark (1986); Stark and Piran (1987) employed the similar regularized variables in the spherical polar coordinate basis. For example, for the extrinsic curvature, they defined

$$K_1 := K_{(1)(1)}, \quad K_2 := \frac{K_{(1)(2)}}{\sin\theta}, \quad K_3 := \frac{K_{(1)(3)}}{\sin\theta},$$
$$K_+ := \frac{1}{2\sin^2\theta}\left(K_{(3)(3)} - K_{(2)(2)}\right), \quad K_\times := \frac{K_{(2)(3)}}{\sin^3\theta}, \qquad (2.282)$$

where $K_{(i)(j)} = e^a_{(i)} e^b_{(j)} K_{ab}$ and $e^a_{(i)}$ ($i = 1, 2, 3$) are orthonormal tetrad basis vectors defined appropriately. This choice is also helpful for guaranteeing the regularity conditions.

Before closing this section, we would like to point out that the regularity conditions at the coordinate singularities are much more complicated in spacetime of no symmetry. This is because we do not have any equation like equation (2.268). In this case, the regularity conditions depend not only on r and θ (or ϖ and z) but also on φ: For example, the regularity conditions become complicated functions composed of θ and φ at $r = 0$ if the spherical polar coordinate basis is employed [see, e.g., Stark (1989) for an example of explicit description of the regularity conditions]. Practically, it does not seem to be possible to impose such a condition precisely in numerical simulations. Indeed, in the history of numerical relativity, few simulation has been performed in curvilinear coordinate bases for non-axisymmetric spacetime [but see Baumgarte et al. (2013) for a special technique that may resolve this issue]. Almost all the works have been done using the Cartesian coordinate basis. The reason for this is that it is extremely difficult to impose the regularity conditions, if we employ the curvilinear coordinate basis.[8]

2.4.3 *Formulation with quasi-isotropic gauge*

The "quasi-isotropic-gauge" formulation was originally developed by Evans (1985), and extensively used by Shapiro and Teukolsky, and their collaborators in 1980s and 1990s [Shapiro and Teukolsky (1991, 1992b,a); Abrahams et al. (1994a,b, 1995); Shapiro et al. (1995)]. A similar formulation was also described and analyzed in detail by Bardeen and Piran (1983) to which we recommend to refer.

In this formulation, the line element is written in the form

$$ds^2 = -(\alpha^2 - \beta_i\beta^i)dt^2 + 2\beta_i dx^i dt$$
$$+\psi^4\left[e^{2\eta}(dr^2 + r^2 d\theta^2) + r^2(\sin\theta d\varphi + \xi d\theta)^2\right], \quad (2.283)$$

where ψ is a conformal factor that is one of the scalar components in the spatial metric. η and ξ are radiative variables that denote dynamical degrees of freedom, and carry information of gravitational waves (see below). This quasi-isotropic-gauge formulation is characterized by a choice of the metric form as $\gamma^{rr} = r^2\gamma^{\theta\theta}$ and $\gamma^{r\theta}(=\gamma_{r\theta}) = \gamma_{r\varphi} = 0$, and this formulation reduces to the isotropic formulation in spherically symmetric spacetime for which $\eta = \xi = 0$ (see appendix A.2). The determinant of the spatial metric is $\gamma = \psi^{12}e^{4\eta}r^4\sin^2\theta$, and below, we denote the determinant of the flat spatial metric $r^4\sin^2\theta$ by $f\,[=\det(f_{ab})]$. We note that in the choice of (2.283) for the metric form, the determinant of the conformal metric is not f but $e^{4\eta}f$.

The spatial gauge condition, i.e., equations for the shift vector, is determined by the three conditions $\gamma^{rr} = r^2\gamma^{\theta\theta}$, $\gamma^{r\theta} = \gamma_{r\varphi} = 0$. More precisely, the conditions for the shift vector are derived from the following evolution equations,

$$\partial_t\left(\gamma^{rr} - r^2\gamma^{\theta\theta}\right) = \partial_t\gamma^{r\theta} = \partial_t\gamma_{r\varphi} = 0, \quad (2.284)$$

which yield the following three equations, respectively:

$$2\alpha\left(K^{rr} - K^{\theta\theta}r^2\right) = 2\gamma^{rr}\left(\partial_r\beta^r - \partial_\theta\beta^\theta - \beta^r/r\right), \quad (2.285)$$
$$2\alpha K^{r\theta} = \gamma^{rr}\left(\partial_r\beta^\theta + r^{-2}\partial_\theta\beta^r\right), \quad (2.286)$$
$$2\alpha K_{r\varphi} = \gamma_{\varphi\varphi}\partial_r\beta^\varphi + \gamma_{\theta\varphi}\partial_r\beta^\theta. \quad (2.287)$$

---

[8]Precisely speaking, we here focus only on the case that the components of geometric variables are defined in curvilinear coordinate bases. Curvilinear coordinates may be used for assigning the grids on the computational domain, if the components of the geometric variables are defined in the Cartesian coordinate basis; see, e.g., Nakamura and Oohara (1989); Boyle et al. (2008); Scheel et al. (2009); Szilágyi et al. (2009) for such examples. For this mixed case, we may not have to worry about the coordinate singularities.

Equations (2.285) and (2.286) constitute simultaneous equations for β^r and β^θ. Evans (1985) proposed to rewrite these equations by setting

$$\beta^r = r^2 \partial_r \chi + r \partial_\theta \phi, \quad \beta^\theta = r \partial_r \phi - \partial_\theta \chi, \tag{2.288}$$

where χ and ϕ are auxiliary functions. Then, equations (2.285) and (2.286) yield elliptic equations for them in the forms

$$\left(\partial_r^2 + r^{-1}\partial_r + r^{-2}\partial_\theta^2\right)\chi = \frac{\alpha}{r^2 \gamma^{rr}}\left(K^{rr} - K^{\theta\theta}r^2\right), \tag{2.289}$$

$$\left(\partial_r^2 + r^{-1}\partial_r + r^{-2}\partial_\theta^2\right)\phi = \frac{2\alpha}{r\gamma^{rr}}K^{r\theta}. \tag{2.290}$$

Alternatively, Choptuik et al. (2003a,b) proposed to use the momentum constraint to derive the equations for β^r and β^θ by substituting equations (2.285) and (2.286) into it. In this formulation, the momentum constraint is automatically satisfied.

Equation (2.287) is an ordinary differential equation for determining β^φ, and the solution for it is nontrivial only for the system that possesses nonzero angular momentum. This equation should be radially integrated either from $r = 0$, where β^φ has a certain value, to $r = \infty$, or from $r = \infty$, where β^φ should be zero, to $r = 0$. In the former case, the value at $r = \infty$ obtained by the integration would be different from zero for an arbitrary trial constant at $r = 0$. Then, we subtract this constant from β^φ to set $\beta^\varphi = 0$ at $r = \infty$. The issue in this numerical computation is that the value of β^φ at $r = 0$ may depend on θ because the constant to be subtracted depends in general on θ (note that the radial integration has to be performed for a fixed value of θ). Alternatively, when the integration is performed from $r = \infty$ toward $r = 0$, the values of β^φ at $r = 0$ depend in general on θ. This pathology is due to a gauge ambiguity in this formulation: There is a degree of a further coordinate transformation as $\varphi \to \varphi + h(\theta, t)$ where $h(\theta, t)$ is an arbitrary function of θ and t and its magnitude is supposed to be very small. By this transformation, ξ and β^φ are transformed, respectively, as

$$\xi - \sin\theta \partial_\theta h, \quad \beta^\varphi - (\partial_t h - \beta^\theta \partial_\theta h). \tag{2.291}$$

This shows that we have a degree of freedom for further adding a correction associated with h to β^φ to achieve β^φ at $r = 0$ being independent of θ. However, after β^φ is corrected, we further need to perform a gauge transformation to ξ. This is a rather messy procedure because we have to integrate $\partial_t h$ to obtain h. A prescription to avoid this procedure is to evolve $\partial_r \xi$ (not ξ) as first pointed out by Bardeen and Piran (1983), because $\partial_r \xi$ is independent of h. In this prescription, ξ is derived by integrating $\partial_r \xi$ along r for each value of θ. This prescription was employed by Abrahams et al. (1994a) in the quasi-isotropic gauge condition, and also by Stark and Piran (1985); Piran and Stark (1986) in the quasi-radial gauge condition (see section 2.4.4).

The works in the quasi-isotropic-gauge formulation were always carried out employing the maximal slicing condition, $K = 0$. In this case, the lapse function obeys an elliptic equation as shown in section 2.2.2, and in this formulation, equation (2.114) is written as

$$\frac{1}{r^2}\partial_r\left[r^2\partial_r(\alpha\psi)\right] + \frac{1}{r^2}\partial_u\left[(1-u^2)\partial_u(\alpha\psi)\right]$$
$$= -\frac{\alpha\psi}{4}\left[\frac{1}{r}\partial_r(r\partial_r\eta) + \partial_\theta^2\eta + \frac{e^{-2\eta}}{4}(\partial_r\xi)^2\right]$$
$$+ 2\pi\alpha\psi^5 e^{2\eta}(\rho_\mathrm{h} + 2S_k{}^k) + \frac{7}{8}\alpha\psi^5 e^{2\eta} K_{ij}K^{ij}, \qquad (2.292)$$

where $u = \cos\theta$. We note that $\tilde{\Delta}(\alpha\psi) = e^{-2\eta}\Delta_\mathrm{f}(\alpha\psi)$ in the quasi-isotropic gauge, where Δ_f denotes the Laplacian for flat axisymmetric space. K in this formulation may be written in the form

$$K = K_r{}^r + K_\theta{}^\theta + K_\varphi{}^\varphi = K_r{}^r + K^{\theta\theta}(\gamma^{\theta\theta})^{-1} + K_{\varphi\varphi}(\gamma_{\varphi\varphi})^{-1}. \qquad (2.293)$$

Thus, the equation $K = 0$ is used for eliminating $K_\theta{}^\theta$ or $K^{\theta\theta}$ from several equations.

By this choice, the components of the extrinsic curvature to be solved are reduced to five. On the other hand, there are three components of the spatial metric to be solved. Thus, there are in total eight components to be solved. For evolving these eight components, one has two choices: evolve them by their evolution equations or evolve some of them (maximally four components) by the constraint equations. In the quasi-isotropic-gauge formulation, all the constraint equations have been frequently employed. The reason for this is that the constraint equations lead to non-evolution type equations that may be stably solved, precisely imposing the regularity conditions at the coordinate singularities. When four of eight components are solved by the constraint equations, there remain only four components that have to be evolved by the evolution equations. These four components could carry the information of gravitational waves. Thus, when the constraint equations are fully used, the evolution equations are employed only for evolving the degree of freedom of gravitational waves.

In the quasi-isotropic-gauge formulation, it is considered that η and ξ represent dynamical degrees of freedom: Broadly speaking, η and ξ may be considered as the even- and odd-parity radiative components, respectively. Equations for η and ξ are written in the forms

$$(\partial_t - \beta^r\partial_r - \beta^\theta\partial_\theta)\eta = -2\alpha K_e + \sin\theta\partial_\theta\left(\beta^\theta/\sin\theta\right), \qquad (2.294)$$
$$(\partial_t - \beta^r\partial_r - \beta^\theta\partial_\theta)\xi = -2\alpha K_o + \xi\sin\theta\partial_\theta\left(\beta^\theta/\sin\theta\right) + \sin\theta\partial_\theta\beta^\varphi, \qquad (2.295)$$

where

$$K_e := \frac{1}{2}\left[K_r{}^r + 2K^{\theta\theta}(\gamma^{\theta\theta})^{-1}\right] = -\frac{1}{2}\left(K_r{}^r + 2K_\varphi{}^\varphi + \frac{2\xi}{\sin\theta}K_\varphi{}^\theta\right), \qquad (2.296)$$

$$K_o := \xi\left[K_{\theta\varphi}(\gamma_{\theta\varphi})^{-1} - K_{\varphi\varphi}(\gamma_{\varphi\varphi})^{-1}\right] = \frac{e^{2\eta}K_\varphi{}^\theta}{\sin\theta}. \qquad (2.297)$$

Note that in the original formulation by Evans (1985), K_e is denoted by $-\lambda/2$, which agrees with the variable $-K_+$ of Bardeen and Piran (1983) for $K = 0$. K_e

and K_o are components that are conjugate with η and ξ, respectively, and they are zero in spherical symmetry with $K = 0$.

The Hamiltonian constraint is used for deriving the equation for ψ in the form

$$\frac{1}{r^2}\partial_r\left[r^2\partial_r\psi\right] + \frac{1}{r^2}\partial_u\left[(1-u^2)\partial_u\psi\right] = -\frac{\psi}{4}\left[r^{-1}\partial_r(r\partial_r\eta) + \partial_\theta^2\eta + \frac{e^{-2\eta}}{4}(\partial_r\xi)^2\right]$$
$$-2\pi\psi^5 e^{2\eta}\rho_\mathrm{h} - \frac{1}{8}\psi^5 e^{2\eta}K_{ij}K^{ij}. \quad (2.298)$$

This equation is not suitable for obtaining the unique solution of ψ because in its right-hand side the source terms proportional to ψ^5 (i.e., a positive power, 5, of ψ which is beyond unity) are present, as has been well known [e.g., Rinne (2008) for a review of the uniqueness of the solution for elliptic equations]. To reduce the power of ψ for the terms of ρ_h and $K_{ij}K^{ij}$, one often defined the rest-mass density $\rho_* := \rho\alpha u^t\sqrt{\gamma/f}\ (\propto \psi^6)$ and a weighted extrinsic curvature $\hat{K}_{ij} = K_{ij}\sqrt{\gamma/f}$, and solved the equations for ρ_* and $\hat{K}_{ij}$ [see equation (2.63)]; here, we note that the major part in ρ_h is ρ. With this prescription, the power of ψ that appears in the right-hand side of equation (2.298) becomes negative, and the solution of ψ is guaranteed to be unique.

Now, we turn our attention to the equations for the extrinsic curvature. In the case that the momentum constraint is fully used, three components of the extrinsic curvature are determined by it, and other two components are determined by the evolution equations. Then, we may well think it quite natural to solve the evolution equations for K_e and K_o while solving the momentum constraint for $K_r{}^r$, $K_\theta{}^r$, and $K_\varphi{}^r$, because K_e and K_o may be considered as the radiative variables that are conjugate with η and ξ. Indeed, K_o is usually determined by solving the evolution equation. Evans (1985) originally proposed to solve the evolution equation for K_e as well. In this case, we have to determine $K_1 := K_r{}^r$ (or $K_\mathrm{T} := K_\theta{}^\theta + K_\varphi{}^\varphi = -K_1$) and $K_2 := K_r{}^\theta r$ by using the simultaneous equations derived from two components of the momentum constraint. However, the use of this method is severely limited by the issue of imposing the regularity of the solution for K_1 and K_2, as described in Bardeen and Piran (1983). This fact is found by analyzing the r and θ components of the momentum constraint, which are, respectively, written as

$$\partial_r(K_r{}^r\sqrt{\gamma}) + \partial_\theta(K_r{}^\theta\sqrt{\gamma}) - \frac{1}{2}\sqrt{\gamma}K^{ij}\partial_r\gamma_{ij} = 8\pi J_r\sqrt{\gamma}, \quad (2.299)$$

$$\partial_r(K_\theta{}^r\sqrt{\gamma}) + \partial_\theta(K_\theta{}^\theta\sqrt{\gamma}) - \frac{1}{2}\sqrt{\gamma}K^{ij}\partial_\theta\gamma_{ij} = 8\pi J_\theta\sqrt{\gamma}. \quad (2.300)$$

Now, we consider these equations in the flat background with $J_i = 0$. Then, the equations may be written as

$$r^{-2}\partial_r(K_1 r^2) - r^{-1}\partial_u(K_2\sin\theta) + r^{-1}K_1 = 0, \quad (2.301)$$
$$r^{-2}\partial_r(K_2 r^3) - \partial_u(K_\theta{}^\theta\sin\theta) - K_\varphi{}^\varphi\cot\theta = 0. \quad (2.302)$$

Here, $K_\theta{}^\theta = (2K_e - K_1)/2$ and $K_\varphi{}^\varphi = -(2K_e + K_1)/2$. Thus, the equations reduce to a linear elliptic system for K_1 and K_2 for a given function of K_e. For finding the homogeneous solution with $K_e = 0$, we set

$$K_1 = a_l(r)P_l(u), \quad K_2 = -b_l(r)\sin\theta\partial_u P_l(u), \qquad (2.303)$$

where P_l is the l-th order Legendre function, and $a_l(r)$ and $b_l(r)$ are functions of r. Equations (2.301) and (2.302) then yield the homogeneous solution of a_l and b_l in the form $r^{\bar{\alpha}}$ where

$$\bar{\alpha} = -3 \pm \sqrt{l(l+1)/2}. \qquad (2.304)$$

Thus, for any value of l, a_l and b_l are not regular at $r = 0$ (for $l = 0$ and 1, $\bar{\alpha}$ is a negative integer and for $l = 2, 3, \cdots$, $\bar{\alpha}$ is not an integer). In the presence of gravity and $J_i \neq 0$, regular solutions K_1 and K_2 may be obtained. However, in the presence of numerical errors, the undesirable homogeneous solutions will contaminate the numerical solution, leading to the violation of the regularity conditions.

This problem can be avoided by employing K_2 as the dynamical evolution variable, as pointed by Bardeen and Piran (1983). In their series of works, Shapiro and Teukolsky (1991, 1992b,a) employed this method. In their simulations that paid attention to systems of no angular momentum, they solved K_1 and K_e (or $K_\varphi{}^\varphi$) using the momentum constraint while K_2 and K_o are evolved using the evolution equations.

Alternatively, Choptuik et al. (2003a,b) proposed to substitute equations (2.285) and (2.286) into the momentum constraint. This procedure results in elliptic equations for β^r and β^θ. The merit in this approach is that not only β^r and β^θ but also $K_r{}^r$ and $K_\theta{}^r$ are determined, because the momentum constraint are solved.[9]

Another alternative is that all the components of the evolution equation for the extrinsic curvature are solved, as done in Miyama (1981); Shibata et al. (1994b); Garfinkle and Duncan (2001) (for a vacuum system of no rotation). In this method, only the Hamiltonian constraint is solved. Although it was applied only to vacuum spacetime, this method is also shown to work well.

2.4.4 Formulation with quasi-radial gauge

The quasi-radial-gauge formulation was developed by Bardeen and Piran (1983) as an extension of the radial-gauge formulation for spherically symmetric spacetime (see appendix B.3), and it is often called Bardeen–Piran formulation. This formulation was used for one of the pioneer studies of stellar core collapse to a black hole by Stark and Piran (1985); Piran and Stark (1986). A remarkable advantage in this formulation is that it has relatively simple basic equations and, in addition, it is suitable for extracting gravitational waves in a local wave zone.

Bardeen and Piran (1983) wrote the line element in the form

$$ds^2 = -(\alpha^2 - \beta^i\beta_i)dt^2 + 2\beta_i dx^i dt + A^2 dr^2$$
$$+ B^{-2}r^2 d\theta^2 + r^2 B^2(\sin\theta d\varphi + \xi d\theta)^2, \qquad (2.305)$$

---

[9]Note that Choptuik et al. (2003a,b) employed cylindrical coordinates, but the resulting equations for the shift vector are qualitatively the same as in spherical polar coordinates.

where A, B, and ξ represent the spatial metric components. The area element for a $r = \text{const}$ surface is $\sqrt{\gamma_{\theta\theta}\gamma_{\varphi\varphi} - (\gamma_{\theta\varphi})^2} = r^2 \sin\theta$, and hence, minimal surfaces are not present as in the case of the radial gauge in spherical symmetry. This implies that hypersurfaces with a wormhole topology, such as shown in figure 1.8, are not described in this coordinate system.

As in the radial gauge in spherical symmetry, A is determined primarily by the mass of the system (e.g., remember the Schwarzschild metric). $\eta := B^2 - 1$ and ξ are not zero only in non-spherically symmetric spacetime, and denote the dynamical degrees of freedom (i.e., radiative variable) that carry information of gravitational waves as in the case of the quasi-isotropic-gauge formulation (see section 2.4.3).

The spatial gauge condition is determined by three conditions

$$\partial_t \left[\gamma_{\theta\theta}\gamma_{\varphi\varphi} - (\gamma_{\theta\varphi})^2\right] = 0, \quad \partial_t \gamma^{r\theta} = 0 = \partial_t \gamma_{r\varphi},$$

from which the equations for the shift vector is derived. The first equation [rewritten as $\partial_t(\gamma/\gamma_{rr}) = 0$] yields a simple relation

$$\alpha \left(K_\theta{}^\theta + K_\varphi{}^\varphi\right) = 2r^{-1}\beta^r + \partial_\theta \beta^\theta + \cot\theta \beta^\theta, \quad (2.306)$$

and the second and third equations are essentially the same as equations (2.286) and (2.287). This immediately implies that β^φ should be determined in the same manner as in the quasi-isotropic-gauge formulation. Eliminating β^r from equations (2.306) and $2\alpha K^{r\theta} = \gamma^{rr}\partial_r \beta^\theta + \gamma^{\theta\theta}\partial_\theta \beta^r$, a linear *parabolic* (not elliptic) equation for β^θ is derived as

$$\partial_u^2 \left[\hat{G}(1-u^2)\right] - \frac{2r}{A^2 B^2}\partial_r \hat{G} = -\partial_u(\alpha K_\text{T}) - \frac{4r\alpha}{A^2 B^2 \sin\theta} K_r{}^\theta, \quad (2.307)$$

where $\hat{G} := \beta^\theta/\sin\theta$ and $K_\text{T} = K_\theta{}^\theta + K_\varphi{}^\varphi$. If β^θ is determined from equation (2.307), β^r is immediately determined by equation (2.306). Thus, the structure of the equations for the shift vector is totally different from that in the quasi-isotropic-gauge formulation, in which the equations for β^r and β^θ obey two elliptic equations. Usually, solving elliptic equations is more time-consuming than solving parabolic equations, and thus, from the computational point of view, the quasi-radial-gauge formulation has an advantage in fixing the spatial gauge.

In this formulation, dynamical variables, η and ξ, are evolved using the evolution equations in the forms

$$\left(\partial_t - \beta^r \partial_r - \beta^\theta \partial_\theta\right)\eta = -2\alpha B^2 K_+ + B^2 \left(1 - u^2\right) \partial_u \hat{G}, \quad (2.308)$$

$$\left(\partial_t - \beta^r \partial_r - \beta^\theta \partial_\theta\right)\xi = -2\alpha\xi K_\times - \left(1 - u^2\right)\left(\xi\partial_u \hat{G} + \partial_u \beta^\varphi\right), \quad (2.309)$$

where

$$K_+ := \frac{1}{2}\left(K_\varphi{}^\varphi - K_\theta{}^\theta + 2\frac{\gamma_{\theta\varphi}}{\gamma_{\varphi\varphi}}K_\varphi{}^\theta\right), \quad (2.310)$$

$$K_\times := K_\varphi{}^\theta \frac{\gamma_{\theta\theta}\gamma_{\varphi\varphi} - (\gamma_{\theta\varphi})^2}{\gamma_{\theta\varphi}\gamma_{\varphi\varphi}}. \quad (2.311)$$

As in the quasi-isotropic-gauge formulation, the gauge ambiguity shown in equation (2.291) is present associated with the freedom of a coordinate transformation in φ. Thus, the solution for ξ again has to be carefully determined: As already mentioned in section 2.4.3, the most popular method is to evolve $\partial_r \xi$.

$\sqrt{\gamma_{rr}} = A$ can be determined by the Hamiltonian constraint which yields a parabolic equation as

$$\partial_r \left[r(1 - A^{-2}) \right] - A^{-1} \partial_u \left[B^2 (1 - u^2) \partial_u A \right]$$
$$= 8\pi \rho_h r^2 + \frac{r^2}{2} \left(K_{ij} K^{ij} - K^2 \right) + \frac{1}{2} \partial_u^2 \left[\eta(1 - u^2) \right]$$
$$+ \frac{r^2}{4A^2} \left[B^{-4} (\partial_r \eta)^2 + B^4 (\partial_r \xi)^2 \right]. \quad (2.312)$$

Remember the fact that an elliptic equation for ψ is derived in the quasi-isotropic-gauge formulation. Thus, from the computational point of view, the quasi-radial-gauge formulation has an advantage.

Bardeen and Piran (1983) proposed two slicing conditions: maximal slicing and *mixed slicing* (i.e., polar slicing mixed with maximal slicing). As shown in section 2.2.2, the maximal slicing condition, $K = 0$, yields an elliptic equation for α as

$$\frac{1}{Ar^2} \left[\partial_r (A^{-1} r^2 \partial_r \alpha) + \partial_u \{ (1 - u^2) B^2 A \partial_u \alpha \} \right]$$
$$= 4\pi \alpha \left(\rho_h + S_k^{\ k} \right) + \alpha K_{ij} K^{ij}. \quad (2.313)$$

On the other hand, the polar slicing condition is defined by $K_T = 0$. It is well-known that in spherical symmetry, this condition yields a first-order ordinary differential equation for α (see appendix B.3). In axial symmetry, the equation becomes a parabolic type equation for α, as in the case for A. Bardeen and Piran (1983) derived this equation from

$$\partial_t K_T = -\alpha R_r^{\ r} - (D_\theta D^\theta + D_\varphi D^\varphi) \alpha + 8\pi \alpha (S_r^{\ r} + \rho_h) + \alpha (K_{ij} K^{ij} - K K_r^{\ r})$$
$$+ \beta^l \partial_l K_T + K_i^{\ \theta} \partial_\theta \beta^i - K_\theta^{\ i} \partial_i \beta^\theta - K_\varphi^{\ i} \partial_i \beta^\varphi, \quad (2.314)$$

where the Hamiltonian constraint was used for deriving this. The resulting equation for $\partial_t K_T = 0$ is written as a parabolic equation for αA because of the following relation

$$\alpha R_r^{\ r} + (D_\theta D^\theta + D_\varphi D^\varphi) \alpha = \frac{2}{A^3 r} \partial_r (\alpha A) + \frac{A}{r^2} \partial_u \left[\frac{B^2}{A^2} (1 - u^2) \partial_u (\alpha A) \right]$$
$$- \frac{2\alpha}{r^2} \partial_u \left[(1 - u^2) B^2 \partial_u \ln A \right]$$
$$- \frac{\alpha}{2A^2} \left[B^{-4} (\partial_r \eta)^2 + B^4 (\partial_r \xi)^2 \right]. \quad (2.315)$$

An important property of the polar slicing condition, in particular for the use with the radial gauge condition, is that they have a singularity-avoidance property [see Petrich et al. (1986)]. This is expected from the fact that spatial hypersurfaces for non-spinning black-hole spacetime agree with those labeled by the Schwarzschild time coordinate in polar slicing with the radial gauge condition. This slicing condition enforces the spatial hypersurfaces to avoid penetrating event horizons (as expected from the Schwarzschild solution). Thus, the singularity-avoidance property is stronger than that of maximal slicing. Another important property is that the geometry asymptotically approaches the Schwarzschild metric in the far zone in this gauge condition. This situation is desirable for extracting gravitational waves from geometric quantities because it is easy to separate perturbed gravitational-wave components from the full metric [see Bardeen (1983) for a detailed analysis].

However, this slicing condition cannot be employed in the entire region of spatial hypersurfaces for axisymmetric problems. The reason for this is that with $K_T = 0$, solutions of the momentum constraint for some components of the extrinsic curvature become irregular [Bardeen and Piran (1983)]. Specifically, non-spherical components in the solution of the extrinsic curvature cannot be regular in the entire region of spatial hypersurfaces. In particular, the irregularity at the origin will cause a serious problem.

To overcome this issue, Bardeen and Piran (1983) proposed a mixed slicing condition which is composed of maximal slicing near the origin and of polar slicing for most of the region on spatial hypersurfaces except the vicinity of the origin. They also proposed to match two slicing conditions in an intermediate zone. This prescription is well-organized for preserving excellent properties of polar slicing and maximal slicing; extracting gravitational waves in the wave zone and singularity-avoidance property near black-hole horizons.

Stark and Piran (1985); Piran and Stark (1986) performed numerical simulations for rotating stellar core collapse using this quasi-radial-gauge formulation. They succeeded in computing formation and subsequent evolution of the collapsed object, and in addition, in extracting gravitational waves in the wave zone. They clarified that gravitational waves are characterized by a ringing oscillation associated with a quasi-normal mode of the formed black holes. This was the first success in numerical relativity for extracting gravitational waves emitted from collapsing objects. Also, they clarified that the emissivity of gravitational waves increases with the increase of the spin parameter of the formed black holes.

2.4.5 NCSA formulation

In 1990s, a numerical-relativity group in National Center for Supercomputing Applications (NCSA) performed a wide variety of simulations for black-hole spacetime including the collision of two black holes. They employed a formulation different from those described in two previous sections. The unique point of their formulation is in the choice of the line element in the form

$$ds^2 = -(\alpha^2 - \beta^k \beta_k)dt^2 + 2\beta_k dx^k dt + Adr^2 \\ + Br^2 d\theta^2 + Dr^2 \sin^2\theta d\varphi^2 + Fr^2 \sin\theta d\theta d\varphi, \qquad (2.316)$$

where A, B, D, and F represent metric components: They employed the gauge conditions $\gamma_{r\theta} = \gamma_{r\varphi} = 0$. We note that they employed spherical-polar-like coordinates different from the ordinary one; instead of using the ordinary radial coordinate r, they employed a coordinate as $r' \propto \ln r$. However, in the following, we describe their formulation simply using the ordinary radial coordinate.

Their spatial gauge condition yields the following relations:

$$2\alpha K_{r\theta} = \gamma_{rr}\partial_\theta \beta^r + \gamma_{\theta\theta}\partial_r \beta^\theta + \gamma_{\theta\varphi}\partial_r \beta^\varphi, \quad (2.317)$$

$$2\alpha K_{r\varphi} = \gamma_{\theta\varphi}\partial_r \beta^\theta + \gamma_{\varphi\varphi}\partial_r \beta^\varphi. \quad (2.318)$$

These two equations are the conditions for determining three components of the shift vector. For determining three components from two equations, they set [Bernstein et al. (1994); Brandt and Seidel (1995a)],

$$\beta^r = \partial_\theta \tilde{\Omega}, \qquad \beta^\theta = \partial_r \tilde{\Omega}, \quad (2.319)$$

where $\tilde{\Omega}$ is an auxiliary function. Eliminating $\partial_r \beta^\varphi$ from equations (2.317) and (2.318), an elliptic equation for $\tilde{\Omega}$ is derived as

$$\left[\gamma_{rr}\gamma_{\varphi\varphi}\partial_\theta^2 + (\gamma_{\theta\theta}\gamma_{\varphi\varphi} - \gamma_{\theta\varphi}^2)\partial_r^2\right]\tilde{\Omega} = 2\alpha(\gamma_{\varphi\varphi}K_{r\theta} - \gamma_{\theta\varphi}K_{r\varphi}). \quad (2.320)$$

Once $\tilde{\Omega}$ is determined, β^φ is obtained by integrating equation (2.318).

For other ingredients, their formulation is quite standard. For example, they employed the maximal slicing condition to derive the equation for α and they did not solve the constraint equations during numerical evolution; K_{ij} was evolved by solving the evolution equation. Details of their formulation and basic equations are found in Bernstein et al. (1994).

2.4.6 *(2+1)+1 formulation*

Geroch (1971) developed a dimensional reduction formulation of Einstein's equation in the presence of a Killing vector field ξ^a that satisfies Killing's equation (see appendix A.1)

$$\nabla_a \xi_b + \nabla_b \xi_a = 0. \quad (2.321)$$

This formulation has been used for generating new solutions of Einstein's equation [e.g., Wald (1984) for an outline of this].

Starting from this Geroch's formulation, Maeda et al. (1980) developed a new numerical-relativity formulation for axisymmetric spacetime setting $\xi^a = (\partial/\partial\varphi)^a$. This formulation is called a (2+1)+1 formulation. The merit in their (2+1)+1 formulation is that the terms composed of β^φ is naturally absorbed in new variables, which obey evolution equations in the Geroch's formulation. Since β^φ does not appear explicitly in this formulation, we do not have to consider the spatial gauge condition for β^φ: The quantities associated with β^φ are obtained in a gauge-invariant manner.

In his formulation, Geroch (1971) proposed to decompose symmetric spacetime in a 3+1 way where "1" is the direction of ξ^a and "3" here denotes three-dimensional *timelike* hypersurfaces Σ_φ which are labeled by φ. Like in the 3+1 formulation described in section 2.1, the induced metric, which satisfies $h_{ab}\xi^b = 0$, is defined by

$$h_{ab} := g_{ab} - \lambda^{-2}\xi_a\xi_b, \tag{2.322}$$

where

$$\lambda^2 := \xi^a\xi_a = g_{\varphi\varphi}. \tag{2.323}$$

λ^2 is positive because ξ^a is a spatial vector field. Here, we note that the determinant of the spacetime metric is written as $g = h\lambda^2$ where $h = \det(h_{ab}) < 0$. Associated with h_{ab}, the covariant derivative, denoted by $\overset{(h)}{D}_a$, is defined as in the 3+1 formulation.

In addition to h_{ab} and λ, the twist, which is a vector field orthogonal to ξ^a, is defined by

$$\omega_a := \epsilon_{abcd}\xi^b\nabla^c\xi^d, \tag{2.324}$$

where ϵ_{abcd} is the completely anti-symmetric tensor. Then, 10 degrees of freedom for g_{ab} are formally decomposed into the 6 + 1 + 3 degrees of freedom composed of h_{ab}, λ, and ω_a, respectively. In spherical polar coordinates, the components of ω^a are written as

$$\omega^t = -\lambda^3(-h)^{-1/2}(\partial_r V_\theta - \partial_\theta V_r), \tag{2.325}$$

$$\omega^r = \lambda^3(-h)^{-1/2}(\partial_t V_\theta - \partial_\theta V_t), \tag{2.326}$$

$$\omega^\theta = \lambda^3(-h)^{-1/2}(\partial_r V_t - \partial_t V_r), \tag{2.327}$$

where $V_\mu := g_{\mu\varphi}/g_{\varphi\varphi}$ and ω^φ is calculated from $\omega^a\xi_a = 0$. Thus, ω^t is associated with the spatial metric, and ω^r and ω^θ are associated with the time derivative of the spatial metric, i.e., extrinsic curvature, and β_φ.

The next task is to derive equations for λ, ω_a, and h_{ab}. The equation for λ is derived by operating $\nabla_a\nabla^a$ to equation (2.323). This yields

$$\lambda\nabla_a\nabla^a\lambda + (\nabla_a\lambda)\nabla^a\lambda = \xi_b\nabla_a\nabla^a\xi^b + (\nabla_a\xi_b)\nabla^a\xi^b$$
$$= -\overset{(4)}{R}_{ab}\xi^a\xi^b + (\nabla_a\xi_b)\nabla^a\xi^b, \tag{2.328}$$

where to rewrite the first term in the right-hand side, we used a formula for the Killing vector field, equation (A.8). The covariant derivative of the Killing vector field is written using equation (2.324) as

$$\nabla_a\xi_b = \frac{1}{2\lambda^2}\epsilon_{abcd}\xi^c\omega^d - \frac{1}{\lambda}\left(\xi_a\overset{(h)}{D}_b - \xi_b\overset{(h)}{D}_a\right)\lambda, \tag{2.329}$$

and thus,

$$(\nabla_a\xi_b)\nabla^a\xi^b = -\frac{\omega^a\omega_a}{2\lambda^2} + 2h^{ab}\left(\overset{(h)}{D}_a\lambda\right)\overset{(h)}{D}_b\lambda. \tag{2.330}$$

Using an identity

$$\nabla_a \nabla^a \lambda^2 = 2\lambda h^{ab} \overset{(h)}{D}_a \overset{(h)}{D}_b \lambda + 4h^{ab} \left(\overset{(h)}{D}_a \lambda \right) \overset{(h)}{D}_b \lambda, \tag{2.331}$$

we finally obtain the basic equation for λ,

$$h^{ab} \overset{(h)}{D}_a \overset{(h)}{D}_b \lambda = -\frac{1}{\lambda} \overset{(4)}{R}_{ab} \xi^a \xi^b - \frac{\omega^a \omega_a}{2\lambda^3}. \tag{2.332}$$

This is the evolution equation for λ.

Equations for ω_a are derived by taking the divergence and rotation of equation (2.324). The divergence yields

$$\nabla_a \omega^a = \epsilon_{abcd} (\nabla^a \xi^b) \nabla^c \xi^d - \epsilon_{abcd} \xi^b \overset{(4)}{R}{}^{cdae} \xi_e, \tag{2.333}$$

where we used equation (A.7). The second term in the right-hand side of equation (2.333) is zero because of an identity for the Riemann tensor

$$\overset{(4)}{R}_{[abc]d} = 0. \tag{2.334}$$

The first term in the right-hand side of equation (2.333) is evaluated using equation (2.329), and is written as

$$\frac{4}{\lambda} \omega^a \overset{(h)}{D}_a \lambda. \tag{2.335}$$

The left-hand side of equation (2.333) is written as

$$\frac{1}{\lambda} \overset{(h)}{D}_a (\lambda \omega^a), \tag{2.336}$$

because ω^a is a vector field on Σ_φ. Thus, we obtain

$$\overset{(h)}{D}_a \left(\lambda^{-3} \omega^a \right) = 0. \tag{2.337}$$

This gives the evolution equation for $\omega^a n_a$.

Using equations (2.321) and (A.7), the rotation of ω^a yields

$$\epsilon^{abcd} \nabla_a \omega_b = 2 \left(\xi^c \overset{(4)}{R}{}_e^d - \xi^d \overset{(4)}{R}{}_e^c \right) \xi^e. \tag{2.338}$$

This is rewritten to

$$\overset{(h)}{D}_a \omega_b - \overset{(h)}{D}_b \omega_a = 2\lambda \epsilon_{abc} \xi^e \overset{(4)}{R}{}_e^c, \tag{2.339}$$

where $\epsilon_{abc} := \lambda^{-1} \epsilon_{abcd} \xi^d$ is a three-dimensional completely anti-symmetric tensor. Equation (2.339) gives the evolution equation for the spatial components of ω_a and a constraint equation.

The equation for h_{ab} is derived essentially in the same manner as that used for deriving the Gauss equation (2.38). Let k^a be an arbitrary vector field on Σ_φ. Then,

$$\overset{(h)}{D_a}\overset{(h)}{D_b}k_c = h_a{}^d h_b{}^e h_c{}^f \nabla_d \nabla_e k_f - \left(h_a{}^d h_b{}^e \nabla_d \xi_e\right)\lambda^{-2} h_c{}^f \xi^g \nabla_g k_f$$
$$- (h_a{}^d h_c{}^f \nabla_d \xi_f)\lambda^{-2} h_b{}^g \xi^h \nabla_g k_h. \tag{2.340}$$

Using $\xi^a k_a = 0$ and $\mathscr{L}_\xi k_a = \xi^c \nabla_c k_a + k_c \nabla_a \xi^c = 0$, we have

$$(\overset{(h)}{D_a}\overset{(h)}{D_b} - \overset{(h)}{D_b}\overset{(h)}{D_a})k_c$$
$$= h_a{}^d h_b{}^e h_c{}^f (\nabla_d \nabla_e - \nabla_e \nabla_d) k_f + 2 h_a{}^d h_b{}^e (\nabla_d \xi_e)\lambda^{-2} h_c{}^f (\nabla_f \xi^g) k_g$$
$$+ \left[(h_a{}^d h_c{}^f \nabla_d \xi_f) h_b{}^g - (h_b{}^d h_c{}^f \nabla_d \xi_f) h_a{}^g\right]\lambda^{-2} (\nabla_g \xi^h) k_h, \tag{2.341}$$

where we used Killing's equation (2.321). Thus,

$$\overset{(h)}{R}_{abcd} = h_a{}^e h_b{}^f h_c{}^g h_d{}^h \left[\overset{(4)}{R}_{efgh} + \lambda^{-2}\left\{2\nabla_e \xi_f \nabla_g \xi_h + (\nabla_e \xi_g)\nabla_f \xi_h - (\nabla_f \xi_g)\nabla_e \xi_h\right\}\right], \tag{2.342}$$

where $\overset{(h)}{R}_{abcd}$ is the Riemann tensor with respect to h_{ab}. We note that the antisymmetric relations for $\overset{(h)}{R}_{abcd}$ are guaranteed to hold because of the presence of Killing's equation (2.321).

Contraction of equation (2.342) by h^{bd} yields

$$\overset{(h)}{R}_{ac} = h_a{}^e h_c{}^g \left[\overset{(4)}{R}_{eg} - \lambda^{-2}\overset{(4)}{R}_{efgh}\xi^f \xi^h + 3\lambda^{-2} h^{fh}\nabla_e \xi_f \nabla_g \xi_h\right], \tag{2.343}$$

where we used $h^{ab}\nabla_a \xi_b = 0$ and Killing's equation (2.321). Equation (A.7) yields

$$\overset{(4)}{R}_{efgh}\xi^f \xi^h = -\lambda \nabla_e \nabla_g \lambda - (\nabla_e \lambda)\nabla_g \lambda + (\nabla_e \xi_f)\nabla_g \xi^f. \tag{2.344}$$

Using equation (2.329), we also have

$$h_a{}^e h_c{}^g g^{fh}\nabla_e \xi_f \nabla_g \xi_h = \frac{1}{4\lambda^2}\left(\omega_a \omega_c - \omega_b \omega^b h_{ac}\right) + (\overset{(h)}{D_a}\lambda)\overset{(h)}{D_c}\lambda. \tag{2.345}$$

Substituting equations (2.344) and (2.345) into (2.343) finally yields

$$\overset{(h)}{R}_{ac} = h_a{}^e h_c{}^f \overset{(4)}{R}_{ef} + \frac{1}{\lambda}\overset{(h)}{D_a}\overset{(h)}{D_c}\lambda + \frac{1}{2\lambda^4}\left(\omega_a \omega_c - \omega_b \omega^b h_{ac}\right). \tag{2.346}$$

Substituting the matter terms into $\overset{(4)}{R}_{ef}$ using Einstein's equation, a three-dimensional Einstein-type equation is obtained with its source term composed of the matter, λ, and ω^a as

$$\overset{(h)}{R}_{ac} - \frac{1}{2}h_{ac}\overset{(h)}{R}_b{}^b = \bar{T}_{ac}, \tag{2.347}$$

where

$$\bar{T}_{ac} = 8\pi h_a{}^e h_c{}^g \left[T_{eg} + \frac{1}{\lambda^2} h_{eg} \left(T_{bd}\xi^b\xi^d - \frac{\lambda^2}{2}T \right) \right]$$
$$+ \frac{1}{\lambda}\overset{(h)}{D}_a \overset{(h)}{D}_c \lambda + \frac{1}{4\lambda^4}\left(2\omega_a\omega_c + \omega^b\omega_b h_{ac} \right), \quad (2.348)$$

and we used equation (2.332). This is the consequence of the dimensional reduction, which we often see.

Maeda et al. (1980) developed a (2+1)+1 formulation starting from this Geroch's formulation, by further performing 2+1 (two-dimensional space Σ_t + time t) decomposition to equation (2.347) as well as to equations (2.332), (2.337), and (2.339). First, they defined an induced metric on two-dimensional spatial hypersurface Σ_t by

$$H_{ab} = h_{ab} + n_a n_b, \quad (2.349)$$

where n^a is the unit timelike vector field normal to Σ_t defined on Σ_φ.

Associated with this 2+1 decomposition, two-dimensional covariant derivative and extrinsic curvature are defined. We denote them, respectively, by $\check{D}_A$ and χ_{AB}, where the capital Latin denotes two-dimensional spatial components. The lapse function, α, and the shift vector β^A are defined as in the 3+1 formulation, and then, $n^a = (\alpha^{-1}, -\beta^A\alpha^{-1})$ and $n_a = (-\alpha, 0)$. As in the 3+1 formulation, the evolution equation for H_{AB} is

$$\partial_t H_{AB} = -2\alpha\chi_{AB} + \check{D}_A\beta_B + \check{D}_B\beta_A. \quad (2.350)$$

In addition, they defined the following quantities:

$$\mathcal{K} := -n^a \overset{(h)}{D}_a\lambda, \qquad \Omega_A := H_A{}^a\omega_a, \qquad \Omega_\omega := n^a\omega_a,$$
$$\rho_{\rm h} := T_{ab}n^a n^b, \qquad J_\varphi := -T_{ab}n^a\xi^b, \qquad J_A := -T_{ab}n^a H^b{}_A,$$
$$S_A := T_{ab}\xi^a H^b{}_A, \qquad S_{AB} := T_{ab}H^a{}_A H^b{}_B, \qquad \mathcal{T} := \lambda^{-2}T_{ab}\xi^a\xi^b. \quad (2.351)$$

The definition of $\mathcal{K}$ is equivalent to the evolution equation of λ:

$$(\partial_t - \beta^A\partial_A)\lambda = -\alpha\mathcal{K}. \quad (2.352)$$

Equation for $\mathcal{K}$ is derived from equation (2.332) as

$$\left(\partial_t - \beta^A\partial_A\right)\mathcal{K} = \alpha\mathcal{K}\chi_A{}^A - (\check{D}_A\alpha)\check{D}^A\lambda - \alpha\check{D}_A\check{D}^A\lambda$$
$$- \frac{1}{2\lambda^3}\alpha\left(\Omega_A\Omega^A - \Omega_\omega^2\right) - 4\pi\alpha\lambda\left(\rho_{\rm h} - S_A{}^A + \mathcal{T}\right). \quad (2.353)$$

The Hamiltonian and momentum constraints, and the evolution equation for three-dimensional Einstein's equation (2.347) are written in the same way as in the 3+1 formulation as

$$\overset{(2)}{R} + (\chi_A{}^A)^2 - \chi_{AB}\chi^{AB} = 2\bar{T}_{ac}n^a n^c, \quad (2.354)$$
$$\check{D}_A\chi^A{}_B - \check{D}_B\chi_A{}^A = -\bar{T}_{ac}n^a H^c{}_B, \quad (2.355)$$
$$\partial_t\chi_{AB} = \alpha\left(\overset{(2)}{R}_{AB} + \chi_C{}^C\chi_{AB} - 2\chi_{AC}\chi_B{}^C\right) - \check{D}_A\check{D}_B\alpha + \mathcal{L}_\beta\chi_{AB}$$
$$- \alpha\overset{(h)}{R}_{ab}H^a{}_A H^b{}_B, \quad (2.356)$$

where $\overset{(2)}{R}_{AB}$ is the Ricci tensor associated with H_{AB} and $\overset{(2)}{R} = \overset{(2)}{R}{}_A^A$. Here, the source terms associated with $\bar{T}_{ab}$ in the right-hand side of equations (2.354)–(2.356) are written as

$$\bar{T}_{ab} n^a n^b = \frac{1}{\lambda}(\check{D}_A \check{D}^A \lambda - \mathcal{K}\chi_A{}^A) + \frac{1}{4\lambda^4}(\Omega_A \Omega^A + \Omega_\omega^2) + 8\pi\rho_{\rm h}, \qquad (2.357)$$

$$\bar{T}_{ab} n^b H^a{}_A = \frac{\Omega_\omega \Omega_A}{2\lambda^4} + \frac{1}{\lambda}\left(-\check{D}_A \mathcal{K} + \chi_A{}^B \check{D}_B \lambda\right) - 8\pi J_A, \qquad (2.358)$$

$$\overset{(h)}{R}{}_{ab} H^a{}_A H^b{}_B = \frac{1}{2\lambda^4}\left[\Omega_A \Omega_B - H_{AB}(\Omega_C \Omega^C - \Omega_\omega^2)\right] + \frac{1}{\lambda}\left(\check{D}_A \check{D}_B \lambda - \mathcal{K}\chi_{AB}\right)$$
$$+ 8\pi\left[S_{AB} - \frac{1}{2}H_{AB}\left(-\rho_{\rm h} + S_C{}^C + \mathcal{T}\right)\right]. \qquad (2.359)$$

The evolution equation for Ω_ω is derived from equation (2.337) as

$$\left(\partial_t - \beta^A \partial_A\right)\Omega_\omega = \lambda^3 \check{D}_A\left(\alpha\lambda^{-3}\Omega^A\right) + \alpha\chi_A{}^A\Omega_\omega - 3\alpha\mathcal{K}\Omega_\omega \lambda^{-1}. \qquad (2.360)$$

By operating $n^a H^b{}_B$ and $H^a{}_A H^b{}_B$, the evolution and constraint equations for Ω_A are derived from equation (2.339), respectively, as

$$\partial_t \Omega_A = \beta^B \check{D}_B \Omega_A + \Omega_B \check{D}_A \beta^B + \check{D}_A(\alpha\Omega_\omega) + 16\pi\alpha\lambda\epsilon_{AB}S^B, \qquad (2.361)$$

$$\check{D}_A \Omega_B - \check{D}_B \Omega_A = 16\pi\lambda\epsilon_{AB}J_\varphi, \qquad (2.362)$$

where $\epsilon_{AB} = \epsilon_{ABc}n^c$. To clarify its nature, the last equation may be written to an equation similar to Gauss' law for the electric field as

$$\lambda^{-1}\check{D}_A\left(\lambda E^A\right) = 16\pi J_\varphi, \qquad (2.363)$$

where $E^A = \epsilon^{AB}\Omega_B/\lambda$.

The basic equations for the (2+1)+1 formulation are summarized as follows: (1) Equations (2.352) and (2.353) determine the evolution of the $\varphi\varphi$ component. (2) Equations (2.350) and (2.356) determine the evolution of three spatial components, e.g., in the spherical polar coordinate basis, rr, $r\theta$, and $\theta\theta$ components. (3) Equations (2.360) and (2.361) determine the evolution of the mixed components such as $r\varphi$ and $\theta\varphi$ in the spherical polar coordinate basis. (4) Equations (2.354), (2.355), and (2.362) are the constraint equations. (5) In this formulation, any gauge conditions can be chosen for α and β^A, but β^φ does not appear explicitly.

The remarkable point associated with (3) is that there are in total only three components, $\Omega_\omega(=-\alpha\omega^t)$ and Ω_A (related to ω^r and ω^θ), to be evolved although in general four components, e.g., in the spherical polar coordinate basis, $\gamma_{r\varphi}$, $\gamma_{\theta\varphi}$, $K_{r\varphi}$, and $K_{\theta\varphi}$, have to be evolved. In addition, β^φ, which is incorporated in ω^a, does not explicitly appear. Namely, variables, which are invariant for the gauge transformation with respect to φ, are chosen in this formulation. In the quasi-isotropic-gauge and quasi-radial-gauge formulations, an ambiguity in the gauge fixing associated with the following transformation exists: $\varphi \to \varphi + h(\theta, t)$, where $h(\theta, t)$ is an arbitrary function of θ and t. By this, $\xi(= g_{\theta\varphi}\sin\theta/g_{\varphi\varphi})$ and β^φ are transformed as $\xi - \sin\theta\partial_\theta h$ and $\beta^\varphi - \partial_t h + \beta^\theta\partial_\theta h$. In the (2+1)+1 formulation, by contrast, all the variables are invariant in this coordinate transformation due to the choice of ω^a as variables [see equation (2.327)].

Nakamura (1981); Nakamura and Sato (1981); Nakamura (1983) employed this (2+1)+1 formulation for computing rotating stellar core collapse. They chose $\beta^A = 0$ conditions together with maximal slicing or a phenomenological slicing condition in which the functional form for α is a priori determined. Although the regularity conditions become complicated in the zero-shift gauge condition, they avoided possible problems by employing appropriate variables to be evolved for guaranteeing the regularity conditions, as shown in equation (2.281), as well as by employing the cylindrical coordinate basis. Using this formulation, Nakamura (1981) successfully performed rotating stellar core collapse to a black hole for the first time. His success was also the first one in numerical relativity that is applied to the system with nonzero angular momentum.

2.5 Cartoon method

2.5.1 *Cartoon method for axisymmetric spacetime*

Even when we employ Cartesian coordinates both for the coordinate basis and for assigning the grid points, it is still possible to perform a simulation for symmetric spacetime imposing the corresponding symmetry conditions in an appropriate manner. The cartoon method is a novel method proposed for the first time in this direction. This was proposed by Alcubierre *et al.* in 1999 (the corresponding paper was published in 2001 [Alcubierre *et al.* (2001a)]) for vacuum spacetime. Soon after the preprint of Alcubierre *et al.* (2001a) appeared, Shibata (2000) adopted the cartoon method for a system coupled with hydrodynamics and showed that this method is quite robust for simulating rotating stellar collapses stably and accurately.

As we describe in this section, the cartoon method enables us to develop a numerical-relativity code (a solver for Einstein's equation) for any symmetric spacetime quite efficiently, when one has a code for fully three-dimensional space constructed using Cartesian coordinates (x, y, z). The computational cost for a simulation performed in a cartoon code is approximately the same as that in a code employing cylindrical coordinates. The most remarkable merit in the cartoon method is that we do not have to worry about the regularity condition in any computational domain because we adopt Cartesian coordinates.

The first task to make an axisymmetric code from a three-dimensional Cartesian code is to restrict the computational domain to $0 \leq x \leq L$ and $-L_y \leq y \leq L_y$ as shown in figure 2.12. Here L denotes the location of the outer boundary along the x-axis, and L_y is a small value (see below). The z-axis is supposed to be the symmetry axis and the computational domain for the z-direction may be arbitrarily chosen, e.g., as $-L_z \leq z \leq L_z$ where $L_z \sim L$ typically. In the cartoon method, the evolution equations for geometric variables are solved only for the x-z plane of

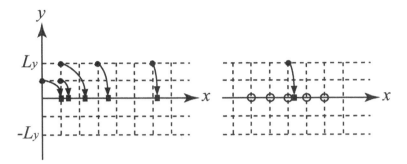

Fig. 2.12 Left: A x-y cross section of a three-dimensional hypersurface. The intersections of vertical and horizontal dashed lines denote the grid points. In axisymmetric space, each pair of a filled circle (at $y \neq 0$) and a square (at $y = 0$) connected by the arrow denotes the equivalent points. Thus, the value of any quantity on the circle points is found from that on the corresponding square points. Right: On the locations of the squares, there is in general no grid point. To assign the data there, an interpolation procedure is performed using the data at the nearby points (open circles). In this example, a fourth-order Lagrangian interpolation scheme is supposed to be employed.

$y = 0$; simulations are essentially performed on the two-dimensional subspace as in ordinary axisymmetric simulations.

Since a Cartesian grid is employed, we have to evaluate all the spatial derivatives in these coordinates. The derivatives associated with the x- and z-coordinates can be evaluated by finite differencing in a straightforward manner (see, e.g., section 3.1.1). On the other hand, we need an idea for evaluating the derivatives associated with y. Because evolution equations are solved only on the $y = 0$ plane, we do not explicitly evolve the data of $y \neq 0$ which are necessary to evaluate the derivatives associated with y. However, in the presence of axial symmetry, it is possible to determine the data on the points of $y \neq 0$ using the relations of this symmetry, as shown in the following.

Specifically, we need several grid points in the neighborhood of $y = 0$ for evaluating the derivatives associated with y at $y = 0$ by finite differencing. The number of the necessary points (i.e., the choice of L_y) depends on the choice of the order of accuracy in the finite differencing. If we employ the second-order finite differencing [see equation (3.7)], we need only three grid points, $y = 0$ and $\pm \Delta y$ (i.e., $L_y = \Delta y$), where Δy is the grid spacing, which is usually chosen to be equal to Δx. Then, the derivatives of a quantity Q associated y at $y = 0$ are evaluated as[10]

$$\partial_y Q = \frac{Q(\Delta y) - Q(-\Delta y)}{2\Delta y}, \tag{2.364}$$

$$\partial_{yy} Q = \frac{Q(\Delta y) - 2Q(0) + Q(-\Delta y)}{(\Delta y)^2}. \tag{2.365}$$

If one employs the fourth-order scheme [see equation (3.6)], we have to prepare the data on the additional two points $y = \pm 2\Delta_y$ (i.e., $L_y = 2\Delta y$).

---

[10]See section 3.1.1 for details of the spatial finite differencing.

The next task is to assign the data on the points of $y \neq 0$ using the axisymmetric relation. First, we define the cylindrical radius by $\varpi = \sqrt{x^2 + y^2}$ for a point (x,y) with $y \neq 0$. Two points (x,y) and $(\varpi, 0)$ are equivalent in axisymmetric spacetime (see figure 2.12). Then, for a scalar function (denoted by Q), the z-component of a vector function (denoted by Q^z), and the zz-component of a tensor function (denoted by Q^{zz}), the data at (x,y) are determined from the axisymmetric relation as

$$Q(x,y,z) = Q(\varpi, 0, z), \quad Q^z(x,y,z) = Q^z(\varpi, 0, z), \quad Q^{zz}(x,y,z) = Q^{zz}(\varpi, 0, z).$$

In general, the point $(\varpi, 0)$ is not located on any grid point along the x-axis. Thus, the values there have to be determined by an interpolation scheme using the nearby several grid points (see the right panel of figure 2.12). When the n-th-order finite differencing scheme is employed, we usually use the n-th-order Lagrangian interpolation; e.g., in the fourth-order interpolation, we use the points $i-2$, $i-1$, i, $i+1$, and $i+2$ where the point i is the nearest point of $x = \varpi$ with $x_i < \varpi$.

For other components, the data at (x,y) are also determined by the axisymmetric relation. However, in this case, we have to take into account the fact that the coordinate basis rotates for the transformation from (x,y) to $(\varpi, 0)$. Thus, we first determine the values of each component at $(\varpi, 0)$ using an appropriate interpolation scheme on the x-axis. We denote such an interpolated value, e.g., as $Q_{AB}^{(0)}$ for a tensor component where the capital Latin indices denote the components of x or y. For obtaining the data at (x,y), then, the following operation associated with the rotation of the coordinate basis has to be performed for the interpolated values:

$$Q_{AB} = \Lambda_A{}^C \Lambda_B{}^D Q_{CD}^{(0)}, \quad Q_{Az} = \Lambda_A{}^C Q_{Cz}^{(0)}, \quad Q_A = \Lambda_A{}^C Q_C^{(0)},$$

where

$$\Lambda_A{}^B = \begin{pmatrix} \cos\varphi & -\sin\varphi \\ \sin\varphi & \cos\varphi \end{pmatrix}, \tag{2.366}$$

and $\varphi = \varphi(x,y) = \tan^{-1}(y/x)$. Here, in the BSSN formulation, for example, Q_{ij}, Q_i, and Q denote $(\tilde{\gamma}_{ij}, \tilde{A}_{ij})$, $(F_i$ or $\tilde{\Gamma}^i, \beta^i)$, and (ψ, K, α), respectively,

Finally, we remark a necessary modification for the method of numerically handling Einstein's evolution equations, which are usually written in the form (see section 2.3.3)

$$\partial_t Q - \beta^k \partial_k Q = \text{right hand side}. \tag{2.367}$$

Here, Q denotes one of geometric variables to be evolved. In the case that the evolution equations are solved without imposing special symmetries, we usually handle the transport term, i.e., the second term in the left-hand side of equation (2.367), using an upwind scheme (see section 3.2). In the cartoon method, the same scheme should be employed for $\beta^x \partial_x Q$ and $\beta^z \partial_z Q$. However, for $\beta^y \partial_y Q$, employing the upwind scheme is physically inappropriate because the transportation should not occur in the direction of the rotational angle φ. This point is easily found when we write Einstein's equation in spherical polar or cylindrical coordinates, because such a transportation term is absent in axial symmetry. Rather, it is appropriate to rewrite $\beta^y \partial_y Q$ using the symmetry relation associated with the presence of the Killing vector field, $(\partial/\partial\varphi)^a$: For a scalar function of Q such as $Q = \phi$ and K or z-components of vector and tensor functions such as $\tilde{\gamma}_{zz}$ and $\tilde{A}_{zz}$, we find $\beta^y \partial_y Q = 0$. For other variables, we should use the following equation to derive the term $\beta^y \partial_y Q$; e.g., for a tensor, the relation

$$\mathscr{L}_\varphi Q_{ab} = 0$$

yields the following nontrivial relations along the x-axis:[11]

$$\partial_y Q_{xx} = -\frac{2Q_{xy}}{x}, \quad \partial_y Q_{xy} = \frac{Q_{xx} - Q_{yy}}{x}, \quad \partial_y Q_{yy} = \frac{2Q_{xy}}{x},$$
$$\partial_y Q_{xz} = -\frac{Q_{yz}}{x}, \quad \partial_y Q_{yz} = \frac{Q_{xz}}{x}. \quad (2.368)$$

These are often-seen terms that appear in curvilinear coordinates. Alternatively, we may simply use the cell-centered finite differencing to evaluate $\partial_y Q_{ab}$. The result of this finite differencing agrees with that derived by equation (2.368) within the numerical error associated with the finite differencing.

2.5.2 Cartoon method for higher-dimensional spacetime: Modified cartoon method

The cartoon method described in the previous subsection is an invaluable method for solving Einstein's equation in any dimension in the presence of some symmetries. In particular, it is a robust method for simulating higher-dimensional spacetime. In this section, we illustrate this fact focusing on the spacetime of $SO(n+1)$ symmetry where $n = D - 4$ denotes the dimensionality of the extra-dimensional space.[12] As in the previous section, we suppose that Einstein's equation for such spacetime is solved using Cartesian coordinates $(x, y, z, w_1, \cdots, w_n)$ with the BSSN formulation where w_q ($q = 1, 2, \cdots, n$) denote the extra-dimensional space coordinates. The $SO(n+1)$ symmetry is assumed to be imposed for the subspace composed of $(z, w_1, ..., w_n)$ coordinates. In the following, we suppose to solve Einstein's equation in the (x, y, z) hyperplane (i.e., $w_1 = w_2 = \cdots = w_n = 0$ hyperplane). This type of the symmetry is present when we consider, for example, a collision of black holes moving in the x-y plane (see section 11.2).

In a cartoon method for axisymmetric four-dimensional spacetime described in section 2.5.1, the evolution equations are solved in the (x, z) (i.e., $y = 0$) plane, and two or four additional grid points are prepared for $y \neq 0$ to evaluate the quantities associated with ∂_y and ∂_{yy} by finite differencing. In this method, the

---
[11] Equations in (2.368) cannot be used at $x = 0$ (on the symmetric axis). There, it is better to simply employ a finite-differencing scheme with respect to y for the y-derivative.
[12] For spacetime of other symmetries, we have to modify the method described in this section.

values of the geometric quantities on these extra grid points are determined using the symmetry relation associated with the rotational Killing vector field. In the higher-dimensional case, we also have to develop a method to evaluate the derivatives associated with the extra-dimensional coordinates, w_q, at $w_1 = w_2 = \cdots = w_n = 0$.

The original method applied to axisymmetric spacetime is straightforwardly applicable for the higher-dimensional case [Yoshino and Shibata (2009)]. However, it requires to take 3^n or 5^n times larger memory because we have to increase two or four grids for each w_q direction. In higher-dimensional simulations with a large value of n, the memory size could be too large for a simulation to be feasible.

To save the memory size used in numerical computation, a modified cartoon method developed by Shibata and Yoshino (2010a) is quite robust. In this method, the derivatives with respect to w_q are replaced to those with respect to z using the symmetry relations, without preparing additional grid points for the extra-dimensional directions. In the following, we describe this method denoting the scalar, vector, and tensor by Q, Q^i, and Q_{ij}, respectively, and decomposing the subscripts for the entire space $i, j, \cdots$ into a (small Latin that denotes x, y, and z), and w_q ($q = 1-n$). We also suppose that the BSSN formulation is employed in higher-dimensional simulations.

In the (x, y, z) hyperplane, the following relations hold because of the assumed isotropy:

$$Q_{ww} := Q_{w_1 w_1} = \cdots = Q_{w_n w_n}, \qquad (2.369)$$

and

$$Q^{w_q} = Q_{a w_q} = Q_{w_q w_r} = 0 \quad (q \neq r). \qquad (2.370)$$

Equation (2.370) implies that we also have the following relations in the (x, y, z) hyperplane:

$$Q^{w_q}{}_{,a} = Q_{b w_q, a} = Q_{w_q w_r, a} = 0 \quad (q \neq r), \qquad (2.371)$$

where $,a$ denotes the partial derivative with respect to x^a.

For other components, the derivatives associated with the extra-dimensional directions in the (x, y, z) hyperplane are evaluated in the following manner: For the scalar quantities,

$$Q_{,w_q} = Q_{,a w_q} = 0, \qquad Q_{,w_q w_r} = \frac{Q_{,z}}{z} \delta_{qr}, \qquad (2.372)$$

for the vector quantities,

$$Q^a{}_{,w_q} = Q^a{}_{,b w_q} = Q^{w_p}{}_{,w_q w_r} = 0,$$

$$Q^{w_q}{}_{,w_r} = \frac{Q^z}{z} \delta_{qr}, \quad Q^{w_q}{}_{,w_r a} = \left(\frac{Q^z}{z}\right)_{,a} \delta_{qr}, \quad Q^A{}_{,w_q w_r} = \frac{Q^A_{,z}}{z} \delta_{qr},$$

$$Q^z{}_{,w_q w_r} = \left(\frac{Q^z_{,z}}{z} - \frac{Q^z}{z^2}\right) \delta_{qr}, \qquad (2.373)$$

and for the tensor quantities,

$$Q_{ab,w_q} = Q_{w_p w_q, w_r} = 0, \qquad Q_{Aw_q, w_r} = \frac{Q_{Az}}{z} \delta_{qr},$$

$$Q_{zw_q, w_r} = \frac{Q_{zz} - Q_{ww}}{z} \delta_{qr},$$

$$Q_{AB, w_q w_q} = \frac{Q_{AB,z}}{z}, \qquad Q_{Az, w_q w_q} = \frac{Q_{Az,z}}{z} - \frac{Q_{Az}}{z^2},$$

$$Q_{zz, w_q w_q} = \frac{Q_{zz,z}}{z} - \frac{2}{z^2}(Q_{zz} - Q_{ww}),$$

$$Q_{w_q w_q, w_r w_r} = \frac{Q_{ww,z}}{z} + \frac{2}{z^2}(Q_{zz} - Q_{ww})\delta_{qr}. \qquad (2.374)$$

Here, the capital Latin A and B denote x or y (not z), and δ_{qr} is the Kronecker's delta. We did not evaluate Q_{ij,aw_q} and $Q_{ij,w_q w_r}$ ($q \neq r$) even if they do not vanish, because they do not appear in the BSSN formulation with $SO(n+1)$ symmetry: Note that the second derivatives appear only in the term $-(1/2)\tilde{\gamma}^{kl} \partial_k \partial_l \tilde{\gamma}_{ij}$ of equation (2.221) and $\tilde{\gamma}^{kl}$ satisfies the relation of equation (2.370). We also do not have to evaluate $Q_{aw_q,ij}$ and $Q_{w_q w_r, ij}$ ($q \neq r$), because they appear only in the aw_q and $w_q w_r$ components of equation (2.232) with equation (2.221) which do not have to be evolved.

Some of the above prescriptions can be used only for $z \neq 0$ in numerical computation because of the presence of the terms associated with $1/z$, although all the terms are actually regular. Thus, at $z = 0$, the following relations, which are found from $SO(n+1)$ symmetry, should be employed:

$$Q_{,w_q w_q} = Q_{,zz}, \quad Q^{w_q}_{,w_q} = Q^z_{,z}, \quad Q^A_{,w_q w_q} = Q^A_{,zz}, \quad Q^z_{,w_q w_q} = 0,$$

$$Q_{AB, w_q w_q} = Q_{AB, zz}, \qquad Q_{Az, w_q w_q} = 0, \qquad Q_{Aw_q, w_q} = Q_{Az, z},$$

$$Q_{zw_q, w_q} = 0, \qquad Q_{zz, w_q w_q} = Q_{ww, zz}, \qquad Q_{w_q w_q, w_q w_q} = Q_{zz, zz},$$

$$Q_{w_q w_q, w_r w_r} = Q_{w_q w_q, zz}. \qquad (2.375)$$

Here, the z-derivative is evaluated simply by finite differencing.

Finally, we remark the treatment of the advection terms such as $\beta^k \partial_k \tilde{\gamma}_{ij}$: Because $\beta^{w_q} = 0$ for the $w_q = 0$ hyperplane in the present case, the advection terms associated with β^{w_q} is always zero in the computational domain chosen in our method. Namely, we have only to evaluate the advection in the x-, y-, and z-directions.

With these prescriptions, all the terms composed of spatial derivatives associated with the extra-dimensional coordinates can be replaced to finite-differencing terms with respect to x^a or with no finite differencing. It is worthy to note, therefore, that total amount of the computational operation is only slightly larger than that for 3+1 simulations.

Because of the relations (2.369) and (2.370) that follow from $SO(n+1)$ symmetry, the implementation for the higher-dimensional contribution can be even simplified: We have only to evolve (i) the scalar equations, (ii) the "a" components of the vector equations, and (iii) the "ab" components and one of the $w_q w_q$ components of the tensor equations (i.e., the equations for $\tilde{\gamma}_{ij}$ and $\tilde{A}_{ij}$) in the BSSN formulation. Therefore, we need to increase only one component for each of $\tilde{\gamma}_{ij}$ and $\tilde{A}_{ij}$ irrespective of the dimensionality. For example, in the evolution equations, we often have terms such as $\beta^i{}_{,i}$ or $\tilde{\gamma}^{kl}\tilde{\gamma}_{ak,l}$, which are evaluated by

$$\beta^i{}_{,i} = \beta^a{}_{,a} + n\beta^{w_1}{}_{,w_1}, \qquad (2.376)$$

$$\tilde{\gamma}^{kl}\tilde{\gamma}_{ak,l} = \tilde{\gamma}^{cd}\tilde{\gamma}_{ac,d} + n\tilde{\gamma}^{w_1 w_1}\tilde{\gamma}_{aw_1,w_1}, \qquad (2.377)$$

where the prescriptions shown in equations (2.372)–(2.375) are used for evaluating the second terms in the right-hand side. These facts imply that once a five-dimensional code is implemented, it is quite straightforward to extend it to a code for $D \geq 6$, as in the case that we employ a curvilinear coordinate system. Furthermore, the code is free from the regularity problem.

2.6 Formulations for asymptotically de Sitter and Friedmann spacetime

We have described a number of formulations for asymptotically flat spacetime in the previous sections. The derived formulations may be used for other types of spacetime by slightly modifying the original one. In the following, we will describe formulations for asymptotically *non-flat* spacetime obtained by slightly modifying the BSSN formulation, restricting our attention only to the $D = 4$ case. We note that, for closed spacetime such as closed Friedmann universe, we need to employ a different formulation. For the readers who are interested in a formulation for closed universe, we refer, e.g., to Goldwirth and Piran (1989).

In the pure de Sitter and Friedmann spacetime, the spatial volume expands uniformly and isotropically. In the BSSN formulation, this fact can be extracted by writing the conformal factor in the form [Nakao *et al.* (1991b); Shibata *et al.* (1994b); Shibata and Sasaki (1999)]

$$\psi = \Psi a(t)^{1/2}, \qquad (2.378)$$

where $a(t)$ denotes the so-called scale factor that obeys the Friedmann equation for the cosmological expansion, and Ψ is a conformal factor in the comoving coordinate frame which is unity in the pure de Sitter and Friedmann spacetime. Only with this replacement, a formulation for the cosmological spacetime is derived and it is quite similar to that for asymptotically flat spacetime.

2.6.1 *Asymptotically de Sitter spacetime*

For deriving a 3+1 formulation of asymptotically de Sitter spacetime, it is convenient to replace the stress-energy tensor in the following form

$$8\pi T_{ab} \to 8\pi T_{ab} - \Lambda g_{ab}, \qquad (2.379)$$

where Λ is the positive cosmological constant. Thus, we only need to replace the matter source terms as

$$\rho_h \to \rho_h + \frac{\Lambda}{8\pi}, \qquad J_a \to J_a, \qquad S_{ab} \to S_{ab} - \frac{\Lambda}{8\pi}\gamma_{ab}. \tag{2.380}$$

Here, ρ_h, J_a, and S_{ab} in the right-hand side are defined only from the matter source.

With this preparation, the evolution equations in a BSSN formulation become

$$\left(\partial_t - \beta^l \partial_l\right)\tilde{\gamma}_{ij} = -2\alpha\tilde{A}_{ij} + \tilde{\gamma}_{ik}\partial_j\beta^k + \tilde{\gamma}_{jk}\partial_i\beta^k - \frac{2}{3}\tilde{\gamma}_{ij}\partial_k\beta^k, \tag{2.381}$$

$$\left(\partial_t - \beta^l \partial_l\right)\tilde{A}_{ij} = \Psi^{-4}a^{-2}\left[\alpha\left(R_{ij} - \frac{1}{3}\gamma_{ij}R_k^{\ k}\right) - \left(D_iD_j\alpha - \frac{1}{3}\gamma_{ij}\Delta\alpha\right)\right.$$
$$\left. - 8\pi\alpha\left(S_{ij} - \frac{1}{3}\gamma_{ij}S_k^{\ k}\right)\right] + \alpha\left(K\tilde{A}_{ij} - 2\tilde{A}_{ik}\tilde{A}_j^{\ k}\right)$$
$$+ \left(\partial_i\beta^k\right)\tilde{A}_{kj} + \left(\partial_j\beta^k\right)\tilde{A}_{ki} - \frac{2}{3}\left(\partial_k\beta^k\right)\tilde{A}_{ij}, \tag{2.382}$$

$$\left(\partial_t - \beta^l\partial_l + \frac{H}{2}\right)\Psi = \frac{\Psi}{6}\left(-\alpha K + \partial_k\beta^k\right), \tag{2.383}$$

$$\left(\partial_t - \beta^l\partial_l\right)K = \alpha\left[\tilde{A}_{ij}\tilde{A}^{ij} + \frac{1}{3}K^2\right] - \Delta\alpha - \alpha\Lambda + 4\pi\alpha\left(\rho_h + S_k^{\ k}\right), \tag{2.384}$$

$$\left(\partial_t - \beta^l\partial_l\right)\tilde{\Gamma}^i = -2\tilde{A}^{ij}\partial_j\alpha + 2\alpha\left[\tilde{\Gamma}^i_{\ jk}\tilde{A}^{jk} - \frac{2}{3}\tilde{\gamma}^{ij}\partial_jK + 6\frac{\partial_j\Psi}{\Psi}\tilde{A}^{ij} - 8\pi\tilde{\gamma}^{ij}J_j\right]$$
$$- \tilde{\Gamma}^j\partial_j\beta^i + \frac{2}{3}\tilde{\Gamma}^i\partial_j\beta^j + \frac{1}{3}\tilde{\gamma}^{ik}\partial_k\partial_j\beta^j + \tilde{\gamma}^{jk}\partial_j\partial_k\beta^i, \tag{2.385}$$

where $H := \dot{a}/a$ is the so-called Hubble parameter. We note that $W = \Psi^{-2}$ would be a better variable in the presence of a black hole. Here, rewriting the equation in terms of W is a straightforward task.

The Hamiltonian and momentum constraints are written in the forms

$$\tilde{\Delta}\Psi = \frac{\psi}{8}\tilde{R} - 2\pi\rho_h\Psi^5 a^2 - \frac{1}{4}\Lambda\Psi^5 a^2 - \frac{\Psi^5 a^2}{8}\left(\tilde{A}_{ij}\tilde{A}^{ij} - \frac{2}{3}K^2\right), \tag{2.386}$$

$$\tilde{D}_i(\Psi^6 \tilde{A}^i_{\ j}) - \frac{2}{3}\Psi^6\tilde{D}_jK = 8\pi J_j \Psi^6. \tag{2.387}$$

Equations (2.383) and (2.386) show that for asymptotically de Sitter spacetime with $\alpha \to 1$, $\beta^k \to 0$, $\Psi \to 1$, $\tilde{R} \to 0$, $\tilde{A}_{ij} \to 0$, $K = $ const, and $T_{\mu\nu} = 0$, K should be $-3H = -\sqrt{3\Lambda}$. We note that from the Friedmann equation, we obtain a well-known expansion law

$$a(t) = a_0 \exp\left(\sqrt{\Lambda/3}t\right), \tag{2.388}$$

where a_0 is a constant. Taking into account the asymptotic form of K, we set

$$K = -3H + \tilde{K}. \tag{2.389}$$

Note that slicing with $\tilde{K} = 0$ is called constant-mean-curvature slicing [Goldwirth and Piran (1989); Nakao et al. (1991a); Shibata et al. (1994b)]. Here, we consider a broader class for the slicing condition with $\tilde{K} \neq 0$.

Substituting equation (2.389) into the equation of the BSSN formulation, the evolution equations are rewritten as

$$\left(\partial_t - \beta^l \partial_l + 3\alpha H\right) \tilde{A}_{ij} = \Psi^{-4} a^{-2} \left[\alpha \left(R_{ij} - \frac{1}{3}\gamma_{ij} R_k^{\ k}\right) - \left(D_i D_j \alpha - \frac{1}{3}\gamma_{ij} \Delta\alpha\right)\right.$$
$$\left. -8\pi\alpha \left(S_{ij} - \frac{1}{3}\gamma_{ij} S_k^{\ k}\right)\right]$$
$$+\alpha \left(\tilde{K}\tilde{A}_{ij} - 2\tilde{A}_{ik}\tilde{A}_j^{\ k}\right)$$
$$+ \left(\partial_i \beta^k\right) \tilde{A}_{kj} + \left(\partial_j \beta^k\right) \tilde{A}_{ki} - \frac{2}{3}\left(\partial_k \beta^k\right) \tilde{A}_{ij}, \qquad (2.390)$$

$$\left(\partial_t - \beta^l \partial_l - \frac{H(\alpha-1)}{2}\right) \Psi = \frac{\Psi}{6}\left(-\alpha \tilde{K} + \partial_k \beta^k\right), \qquad (2.391)$$

$$\left(\partial_t - \beta^l \partial_l + 2\alpha H\right) \tilde{K} = \alpha \left[\tilde{A}_{ij}\tilde{A}^{ij} + \frac{1}{3}\tilde{K}^2\right] - \Delta\alpha + 4\pi\alpha \left(\rho_h + S_k^{\ k}\right), \qquad (2.392)$$

where the equations for $\tilde{\gamma}_{ij}$ and $\tilde{\Gamma}^i$ are the same as equations (2.381) and (2.385), respectively. Thus, the evolution equations for asymptotically de Sitter spacetime are written in a form quite similar to the equations in asymptotically flat spacetime. The terms which additionally appear as the modification are found only in the operator in the left-hand side and in the replacement from K to $\tilde{K}$.

The Hamiltonian and momentum constraints are rewritten in the forms

$$\tilde{\Delta}\Psi = \frac{\Psi}{8}\tilde{R} - 2\pi\rho_h \Psi^5 a^2 - \frac{\Psi^5 a^2}{8}\left(\tilde{A}_{ij}\tilde{A}^{ij} - \frac{2}{3}\tilde{K}^2 + 4H\tilde{K}\right), \qquad (2.393)$$

$$\tilde{D}_i \left(\Psi^6 \tilde{A}^i_{\ j}\right) - \frac{2}{3}\Psi^6 \tilde{D}_j \tilde{K} = 8\pi J_j \Psi^6. \qquad (2.394)$$

Again, the equations are similar to those for asymptotically flat spacetime.

The boundary conditions for the asymptotic region are

$$\alpha \to 1, \quad \beta^k \to 0, \quad \Psi \to 1, \quad \tilde{\gamma}_{ij} \to f_{ij}, \quad \tilde{A}_{ij} \to 0, \qquad (2.395)$$

for slicing in which $\tilde{K} \to 0$ in the asymptotic region. Thus, the boundary conditions are essentially the same as those for asymptotically flat spacetime.

2.6.2 Asymptotically Friedmann spacetime

Next, we describe a BSSN-like formulation in asymptotically Friedmann spacetime. For this spacetime, the cosmological expansion is driven by the presence of matter. Thus, we have to specify the stress-energy tensor for the first step. Because we consider the evolution of spacetime and matter in the cosmological context (for the early universe after inflation epoch), we write the stress-energy tensor in the following form

$$T_{ab} = (E+P)u^a u^b + P g_{ab}, \qquad (2.396)$$

where E, P, and u^a are the energy density, pressure, and four velocity of the matter, respectively. The equation of state is assumed to be in an adiabatic form

$P = (\Gamma - 1)E$ where Γ is the adiabatic constant for which we assume $1 \leq \Gamma \leq 2$: The condition, $\Gamma \leq 2$, comes from the causality condition that the sound velocity has to be smaller than the speed of light [see equation (4.68) for the sound velocity]. We typically consider the case of $\Gamma = 4/3$. In this equation of state, we have

$$\rho_\mathrm{h} = Ew^2 + P(w^2 - 1), \quad J_a = \Gamma Ew\bar{u}_a, \quad S_{ab} = \Gamma E\bar{u}_a\bar{u}_b + P\gamma_{ab}, \quad (2.397)$$

where $w := -u^a n_a = \alpha u^t$ and $\bar{u}_a = \gamma_a{}^b u_b$. We do not suppose the presence of the cosmological constant and the spatial curvature because their effects are not important in the early universe (after inflation epoch and before nucleosynthesis).

First, we consider the homogeneous and isotropic universe. Denoting the energy density and pressure for this case as $E_0(t)$ and $P_0(t)$, we have the well-known Friedmann equation [e.g., chapter 5 of Wald (1984)] as

$$\ddot{a} = -\frac{4\pi}{3}a(E_0 + 3P_0), \quad \dot{a}^2 = \frac{8\pi}{3}a^2 E_0, \quad (2.398)$$

where $a(t)$ is the scale factor as in the previous section. From these equations together with the adiabatic equation of state, the solutions of the scale factor and the energy density are obtained as

$$a(t) = a_0 t^{2/3\Gamma}, \quad E_0 = \frac{1}{6\pi\Gamma^2 t^2}, \quad (2.399)$$

where a_0 is a constant.

Next, we consider general asymptotically Friedmann spacetime. Following the prescription used in asymptotically de Sitter spacetime, we write $K = -3H + \tilde{K}$ where $H = \dot{a}/a = 2/(3\Gamma t)$ is the Hubble parameter which is time-dependent in this case. Then, the evolution equations in a BSSN formulation are written as

$$\left(\partial_t - \beta^l \partial_l\right) \tilde{\gamma}_{ij} = -2\alpha \tilde{A}_{ij} + \tilde{\gamma}_{ik}\partial_j\beta^k + \tilde{\gamma}_{jk}\partial_i\beta^k - \frac{2}{3}\tilde{\gamma}_{ij}\partial_k\beta^k, \quad (2.400)$$

$$\left(\partial_t - \beta^l \partial_l + 3\alpha H\right) \tilde{A}_{ij} = \Psi^{-4} a^{-2}\left[\alpha\left(R_{ij} - \frac{1}{3}\gamma_{ij}R_k{}^k\right) - \left(D_i D_j \alpha - \frac{1}{3}\gamma_{ij}\Delta\alpha\right)\right]$$

$$+ \alpha\left(\tilde{K}\tilde{A}_{ij} - 2\tilde{A}_{ik}\tilde{A}_j{}^k\right) + \left(\partial_i\beta^k\right)\tilde{A}_{kj} + \left(\partial_j\beta^k\right)\tilde{A}_{ki} - \frac{2}{3}\left(\partial_k\beta^k\right)\tilde{A}_{ij}$$

$$- 8\pi\alpha\Psi^{-4}a^{-2}\Gamma E\left(u_i u_j - \frac{1}{3}\tilde{\gamma}_{ij}\tilde{\gamma}^{kl}u_k u_l\right), \quad (2.401)$$

$$\left(\partial_t - \beta^l\partial_l - \frac{H}{2}(\alpha - 1)\right)\Psi = \frac{\Psi}{6}\left(-\alpha\tilde{K} + \partial_k\beta^k\right), \quad (2.402)$$

$$\left(\partial_t - \beta^l\partial_l + 2\alpha H\right)\tilde{K} = \alpha\left[\tilde{A}_{ij}\tilde{A}^{ij} + \frac{1}{3}\tilde{K}^2\right] - \Delta\alpha$$

$$+ 4\pi\left[2\alpha\Gamma E(w^2 - 1) + (3\Gamma - 2)\alpha(E - E_0) + 3\Gamma E_0(\alpha - 1)\right], \quad (2.403)$$

$$\left(\partial_t - \beta^l\partial_l\right)\tilde{\Gamma}^i = -2\tilde{A}^{ij}\partial_j\alpha + 2\alpha\left[\tilde{\Gamma}^i{}_{jk}\tilde{A}^{jk} - \frac{2}{3}\tilde{\gamma}^{ij}\partial_j\tilde{K} + 6\frac{\partial_j\Psi}{\Psi}\tilde{A}^{ij} - 8\pi\Gamma Ewu_j\tilde{\gamma}^{ij}\right]$$

$$- \tilde{\Gamma}^j\partial_j\beta^i + \frac{2}{3}\tilde{\Gamma}^i\partial_j\beta^j + \frac{1}{3}\tilde{\gamma}^{ik}\partial_k\partial_j\beta^j + \tilde{\gamma}^{jk}\partial_j\partial_k\beta^i. \quad (2.404)$$

The Hamiltonian and momentum constraints are written in the forms

$$\tilde{\Delta}\Psi = \frac{\Psi}{8}\tilde{R} - 2\pi(\rho_h - E_0)\Psi^5 a^2 - \frac{\Psi^5 a^2}{8}\left(\tilde{A}_{ij}\tilde{A}^{ij} - \frac{2}{3}\tilde{K}^2 + 4H\tilde{K}\right), \quad (2.405)$$

$$\tilde{D}_i\left(\Psi^6 \tilde{A}^i{}_j\right) - \frac{2}{3}\Psi^6 \tilde{D}_j \tilde{K} = 8\pi \Gamma E w u_j \Psi^6. \quad (2.406)$$

Therefore, it is found that for the homogeneous isotropic universe with $\alpha = 1$, $\beta^i = 0$, $\tilde{\gamma}_{ij} = \delta_{ij}$, $\tilde{A}_{ij} = 0$, $\tilde{K} = 0$, $\tilde{\Gamma}^i = 0$, $E = E_0$, $u_i = 0$, and $w = 1$, all the equations reduce to a trivial equation. This implies that the effect of the spacetime expansion is subtracted from the variables employed, and the equations are written in the form quite similar to those in asymptotically flat spacetime.

The boundary condition for the asymptotic region is the same as equation (2.395) for slicing in which $\tilde{K} \to 0$ in the asymptotic region. Thus, the boundary condition is essentially the same as that for asymptotically flat spacetime, although the behavior of the asymptotic region is slightly different from that in asymptotically flat spacetime [Shibata and Sasaki (1999)].

When performing a simulation in a realistic setting based on the standard cosmological scenario, we have to be careful in setting the initial condition. This issue is described in detail in section 9.4.

Before closing this section, we describe the hydrodynamics equations for the special choice of the equation of state $P = (\Gamma - 1)E$ (see sections 4.3 and 4.4 for the more general case). In this case, the basic equations are derived fully from $\nabla_a T^a{}_b = 0$, and the energy and Euler equations may be written in the forms (in general coordinate systems)

$$\partial_t\left(w\Psi^6 a^3 \sqrt{\tilde{\gamma}} E^{1/\Gamma}\right) + \partial_k\left(w\Psi^6 a^3 \sqrt{\tilde{\gamma}} E^{1/\Gamma} v^k\right) = 0, \quad (2.407)$$

$$\partial_t\left\{w\Psi^6 a^3 \sqrt{\tilde{\gamma}}(E+P)u_j\right\} + \partial_k\left\{w\Psi^6 a^3 \sqrt{\tilde{\gamma}}(E+P)v^k u_j\right\}$$
$$= -\alpha\Psi^6 a^3 \sqrt{\tilde{\gamma}}\partial_j P + w\Psi^6 a^3 \sqrt{\tilde{\gamma}}(E+P)\left[-w\partial_j\alpha + u_k\partial_j\beta^k - \frac{u_k u_l}{2u^0}\partial_j\gamma^{kl}\right], \quad (2.408)$$

where

$$v^k := \frac{u^k}{u^0} = -\beta^k + \tilde{\gamma}^{kl}\frac{u_l}{\Psi^4 a^2 u^0}. \quad (2.409)$$

Here, the energy equation, (2.407), is written in a continuity-equation-like form. Of course, equation (2.407) can be rewritten to the form that is the same as the usual energy equation (see section 4.4).

Chapter 3

Numerical methods for a solution of Einstein's evolution equation

This chapter is devoted to describing several numerical methods for a solution of Einstein's evolution equations. In general, solving Einstein's equation is composed of two procedures to do. One is to solve Einstein's evolution equations, which are hyperbolic partial differential equations with an advection term associated with the shift vector. The other is to solve the constraint equations which are elliptic equations and are usually solved for providing an initial condition for numerical-relativity simulations. Therefore, numerical relativists need to solve both hyperbolic and elliptic partial differential equations. In this chapter, we focus only on describing typical methods for solving hyperbolic equations; in particular, we will describe several finite-differencing schemes. Methods for a solution of elliptic equations are described in chapter 5.8. Spectral and pseudo-spectral methods are getting popular in the field of numerical relativity since late 1990s [e.g., Bonazzola et al. (1999); Pfeiffer et al. (2003); Grandclément and Novak (2009); Boyle et al. (2007)], in particular for solving the constraint equations and vacuum Einstein's equation. We will also outline these methods in section 5.8.2.

In sections 3.1 and 3.2, we will summarize several numerical schemes for a solution of second-order and first-order wave equations. Then, an adaptive mesh refinement algorithm, which is one of the necessary ingredients in numerical relativity, will be described in section 3.3. In section 3.4, we will introduce several testbed problems in numerical relativity for vacuum spacetime, for which we often perform a numerical simulation to examine whether the solutions can be reproduced, for confirming the reliability of newly-developed numerical implementations.

3.1 Solving hyperbolic equations

As shown in section 2.3.1, Einstein's evolution equations can be written in simple multi-component wave equations in the linear approximation when a particular gauge condition such as harmonic gauge is employed or when basic equations are rewritten enforcing constraint equations (e.g., BSSN formulation; see section 2.3.3).

The resulting linear equation in vacuum is

$$\Box h_{ij} = 0, \tag{3.1}$$

where h_{ij} is the perturbed three metric. In Cartesian coordinates, this equation is equivalent to six-components scalar-wave equations schematically written as

$$\Box \phi_a = 0, \quad (a = 1, 2, \cdots, 6). \tag{3.2}$$

Although fully nonlinear Einstein's equation is much more complicated, $\Box h_{ij}$ is always the primarily part in this equation. This indicates that a robust method for a solution of equation (3.2) is one of the key ingredients for solving Einstein's evolution equations. Indeed, a large number of numerical experiments in the field of numerical relativity have shown that the finite-differencing schemes, which can be used for stably and accurately solving equation (3.2), can also be employed as a solver of Einstein's evolution equations. Thus, in this section, we focus on describing several finite-differencing methods for a numerical solution of equation (3.2).

3.1.1 *Finite-differencing for spatial derivatives*

First, we describe the standard method for evaluating the spatial partial derivatives in $\Delta_\mathrm{f} \phi_a$ where Δ_f denotes the flat spatial Laplacian as before. In the following, we will only consider the single scalar component, ϕ, for simplicity.

If ϕ is a continuous function around a point $x = x_0$ in the x-coordinate direction, it is possible to expand it by a Taylor series around $x = x_0$ as

$$\phi(x) = \phi_0 + \phi_0' \Delta x + \frac{1}{2!} \phi_0'' \Delta x^2 + \frac{1}{3!} \phi_0''' \Delta x^3 + \frac{1}{4!} \phi_0'''' \Delta x^4 + O(\Delta x)^5, \tag{3.3}$$

where $\Delta x = x - x_0$, and $\phi_0, \phi_0', \cdots, \phi_0''''$ denote the values of $\phi, \partial_x \phi, \cdots, \partial_x^4 \phi$ at $x = x_0$. In the absence of stress-energy tensor, the assumption that the metric function is continuous is justified, as far as we employ an appropriate gauge condition (in pathological gauge conditions, discontinuities could appear). In the presence of a nonzero stress-energy tensor, a discontinuity may be formed for more than second-order derivatives of the metric function (i.e., for the Ricci tensor), if the stress-energy tensor is not smooth. For example, in the presence of shocks for fluid quantities (e.g., rest-mass density), this is the case. In numerical simulations, however, it is impossible to capture the discontinuities for the stress-energy tensor completely sharply: Numerical solutions for such "discontinuities" are always rather smooth, and hence, in reality, sharp discontinuities are not present. Therefore, we usually suppose that the metric function for any numerical solution can be expanded by the Taylor series.

Finite-differencing equations for the metric functions are derived from the formula of the Taylor expansion (3.3). For the finite differencing, we first discretize a spatial domain into a large number of discrete cells. Here, we assign a value of physical quantities at a representative point of each cell, which we call it a *grid point*. As the grid points, vertexes or centers of the cells are usually chosen, and we call them vertex grid points and cell-centered grid points, respectively (see section 3.3.2 for more details). In the following, we will assume to use the vertex grid points. The value of partial derivatives has to be evaluated at the grid points to translate a partial differential equation into a finite-differencing form.

For simplicity, we assume that the grid points are uniformly distributed as $x_j = j\Delta x$, and consider the finite differencing at $x = x_0$. Then, equation (3.3) yields the following relations at $x_{\pm 2} = x_0 \pm 2\Delta x$ and $x_{\pm 1} = x_0 \pm \Delta x$,

$$\phi(x_{\pm 2}) = \phi_0 \pm 2\phi_0' \Delta x + 2\phi_0'' \Delta x^2 \pm \frac{4}{3}\phi_0''' \Delta x^3 + \frac{2}{3}\phi_0'''' \Delta x^4,$$

$$\phi(x_{\pm 1}) = \phi_0 \pm \phi_0' \Delta x + \frac{1}{2}\phi_0'' \Delta x^2 \pm \frac{1}{6}\phi_0''' \Delta x^3 + \frac{1}{24}\phi_0'''' \Delta x^4, \quad (3.4)$$

where we take into account the terms up to fourth order in Δx for deriving the fourth-order centered finite-differencing scheme, which is the most popular scheme in modern numerical relativity. For non-uniform grid spacing, the extension is straightforward, although Δx depends on the position, and thus, the resulting equations are more complicated.

Supposing that the values of ϕ at $x = x_{-2}$, x_{-1}, x_0, x_1, and x_2 are known, four unknown values ϕ_0', ϕ_0'', ϕ_0''', and ϕ_0'''' are determined by four equations in equation (3.4). Einstein's equation is composed of first- and second-order derivatives and we here only need ϕ_0' and ϕ_0'', which are given by

$$\phi_0' = \frac{-[\phi(x_2) - \phi(x_{-2})] + 8[\phi(x_1) - \phi(x_{-1})]}{12\Delta x}, \quad (3.5)$$

$$\phi_0'' = \frac{-[\phi(x_2) + \phi(x_{-2})] + 16[\phi(x_1) + \phi(x_{-1})] - 30\phi(x_0)}{12\Delta x^2}. \quad (3.6)$$

Thus, four and five points are used for the first- and second-order derivatives, respectively. We note that by definition, the error for these formulas is of $O(\Delta x^4)$. For the y- and z-coordinate directions, the finite-differencing equation can be written in the same manner; see, e.g., figure 3.1.

In the past, a second-order finite-differencing scheme was most popular in numerical relativity. If we do not evolve spacetime composed of a high-curvature region such as black-hole spacetime and if high accuracy is not required, the second-order scheme may be employed. In particular, for a beginner of numerical simulations, this scheme is recommendable because the derivatives of ϕ_0, ϕ_0' and ϕ_0'' are evaluated simply by

$$\phi_0' = \frac{\phi(x_1) - \phi(x_{-1})}{2\Delta x}, \quad \phi_0'' = \frac{\phi(x_1) + \phi(x_{-1}) - 2\phi(x_0)}{\Delta x^2}. \quad (3.7)$$

It is easy to check that the error for these finite-differencing schemes is of $O(\Delta x^2)$.

In numerical relativity, we have to evaluate first- and second-order partial derivatives of the spatial metric (γ_{ij} or $\tilde{\gamma}_{ij}$ and conformal factor in the BSSN formulation), first-order derivative of the extrinsic curvature, and first- and second-order derivatives of gauge variables α and β^i. The finite-differencing method for these variables is essentially the same as those described above, except for the advection term for

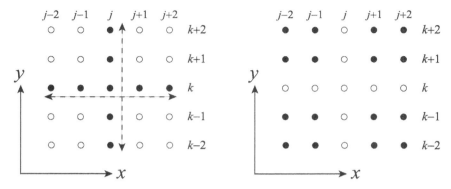

Fig. 3.1 Left: Method of the fourth-order centered finite differencing at a point (x, y) which is labeled by (j, k). For one direction, five points $j - 2 - j + 2$ or $k - 2 - k + 2$ (filled circles) are used. Right: For evaluation of $\partial_x \partial_y \gamma_{ij}$, sixteen points (filled circles) in the region between $j - 2$ and $j + 2$ and between $k - 2$ and $k + 2$ except for j and k are used; see equation (3.10).

which the methods will be described in section 3.2. For example, for γ_{ij}, the x derivatives at $x = x_0$ are evaluated in the fourth-order scheme by

$$\partial_x \gamma_{ij} = \frac{-[\gamma_{ij}(x_2) - \gamma_{ij}(x_{-2})] + 8[\gamma_{ij}(x_1) - \gamma_{ij}(x_{-1})]}{12 \Delta x}, \tag{3.8}$$

$$\partial_x^2 \gamma_{ij} = \frac{-[\gamma_{ij}(x_2) + \gamma_{ij}(x_{-2})] + 16[\gamma_{ij}(x_1) + \gamma_{ij}(x_{-1})] - 30 \gamma_{ij}(x_0)}{12 \Delta x^2}. \tag{3.9}$$

One new type, which is a characteristic term in general relativity, is $\partial_k \partial_l \gamma_{ij}$ for $k \neq l$. This term is evaluated, e.g., for $k = x$ and $l = y$, by

$$\partial_x \partial_y \gamma_{ij} = \frac{1}{12^2 \Delta x \Delta y}$$
$$\times \Big[-[-\{\gamma_{ij}(x_2, y_2) - \gamma_{ij}(x_{-2}, y_2)\} + 8\{\gamma_{ij}(x_1, y_2) - \gamma_{ij}(x_{-1}, y_2)\}]$$
$$+ [-\{\gamma_{ij}(x_2, y_{-2}) - \gamma_{ij}(x_{-2}, y_{-2})\} + 8\{\gamma_{ij}(x_1, y_{-2}) - \gamma_{ij}(x_{-1}, y_{-2})\}]$$
$$+ 8[-\{\gamma_{ij}(x_2, y_1) - \gamma_{ij}(x_{-2}, y_1)\} + 8\{\gamma_{ij}(x_1, y_1) - \gamma_{ij}(x_{-1}, y_1)\}]$$
$$- 8[-\{\gamma_{ij}(x_2, y_{-1}) - \gamma_{ij}(x_{-2}, y_{-1})\} + 8\{\gamma_{ij}(x_1, y_{-1}) - \gamma_{ij}(x_{-1}, y_{-1})\}] \Big]. \tag{3.10}$$

Thus, 4×4 points are used as a straightforward extension of the first-order partial derivative: see the right panel of figure 3.1.

3.1.2 Time integration

The next task for the numerical computation of equation (3.2) is to choose an appropriate scheme of the time integration. Here, as in the 3+1 formulation, we define a variable like the extrinsic curvature as

$$\eta := \partial_t \phi, \tag{3.11}$$

and then, we rewrite the wave equation in the form of a simultaneous equation as
$$\partial_t \phi = \eta, \quad \partial_t \eta = \Delta_f \phi. \tag{3.12}$$
If the right-hand sides of these equations are given at each time step, we may view this equation system as a simultaneous ordinary differential equation for the derivative by t. For a numerical solution of ordinary differential equations, there are several popular explicit schemes; Runge–Kutta, iterative Crank–Nicolson, and leapfrog schemes [e.g., Press *et al.* (1989)]. The Runge–Kutta scheme has second-, third-, fourth-, and higher-order schemes, while the Crank–Nicolson and leapfrog schemes are second-order accurate. For this reason, we will focus primarily on the description of the Runge–Kutta schemes in the following.

Before going ahead, we point out that implicit schemes are another options for the time integration. In these schemes, some or all quantities in the right-hand side of equation (3.12) are evaluated using quantities at an advanced [the $(n+1)$-th] time step for the time integration from n-th to $(n+1)$-th time step. In the typical implicit scheme, n-th quantities (denoted by ϕ^n and η^n) appear only for evaluating $\partial_t \phi$ and $\partial_t \eta$ by $(\phi^{n+1} - \phi^n)/\Delta t$ and $(\eta^{n+1} - \eta^n)/\Delta t$, where Δt is the time-step interval. In this case, the equations for $\phi^{n+1}_{j,k,l}$ and $\eta^{n+1}_{j,k,l}$, where the subscripts j, k, l denote the labels of spatial grid points, become a large matrix equation. For such equations, computational costs for a numerical solution are quite expensive.

The implicit scheme often becomes the best choice when one needs to solve a "stiff" equation such as radiation-transfer equations (see, e.g., section 4.7). Fortunately, Einstein's equation (as well as scalar-wave equations) is not a stiff equation, and hence, we do not have to employ an implicit scheme for this.

One of the points to be kept in mind for constructing a time integration scheme is that the method of the Taylor expansion is not directly used for achieving a higher-order accuracy. In principle, it is possible to evaluate time derivatives of ϕ and η by a finite differencing storing their values at several previous time steps and using the Taylor expansion, e.g., as
$$\partial_t \phi = \frac{1}{2\Delta t} \left(\phi^{n+1} - \phi^{n-1} \right), \quad \partial_t \phi = \frac{1}{6\Delta t} \left(2\phi^{n+1} + 3\phi^n - 6\phi^{n-1} + \phi^{n-2} \right). \tag{3.13}$$
However, this straightforward extension is not very popular except for the second-order scheme which is essentially the same as the leapfrog scheme (see below). The reasons for this are that (i) for higher-order schemes such as a fourth-order scheme, it is necessary to store a large amount of data for previous time steps and (ii) it is not easy to guarantee the *stability* of the time integration. The latter point will be revisited later in this section.

By contrast, in the Runge–Kutta schemes, we do not have to store a large amount of data for previous time steps, and furthermore, higher-order-accurate time integration schemes that enable us to stably perform a simulation can be constructed. To concisely describe the idea of the Runge–Kutta scheme, we write the simultaneous equation (3.12) in the form
$$\frac{d\boldsymbol{x}}{dt} = \boldsymbol{F}[\boldsymbol{x}, t], \tag{3.14}$$

where $\boldsymbol{x} = (\phi, \eta)$, and $\boldsymbol{F}$ denotes the right-hand sides of equation (3.12). We consider time integration of this equation from a time step $t = t_n$ to a time step $t = t_{n+1} = t_n + \Delta t$.

In the Runge–Kutta schemes, $\boldsymbol{x}$ at $t = t_{n+1}$ is formally written by

$$\boldsymbol{x}_{n+1} = \boldsymbol{x}_n + \sum_{i=1}^{p} w_i \boldsymbol{k}_i, \qquad (3.15)$$

where

$$\boldsymbol{k}_i = \Delta t \boldsymbol{F}\left[\boldsymbol{x}_n + \sum_{j=1}^{p} \alpha_{ij} \boldsymbol{k}_j, t + c_i \Delta t\right], \quad c_i = \sum_{j=1}^{p} \alpha_{ij} \quad (i = 1, 2, \cdots, p), \quad (3.16)$$

and w_i and p are weight factors and an integer, respectively, which are determined by the required order of accuracy (denoted by q). Then, α_{ij} and w_i are determined so as for $(\boldsymbol{x}_{n+1} - \boldsymbol{x}_n)/\Delta t$ to agree with the equation derived in the Taylor expansion up to the required order q as

$$\frac{1}{\Delta t}(\boldsymbol{x}_{n+1} - \boldsymbol{x}_n) = \frac{d\boldsymbol{x}}{dt} + O(\Delta t^q). \qquad (3.17)$$

Here, p can be equal to q for $q \leq 4$ (see below).

For the case that $\alpha_{ij} \neq 0$ for $i \leq j$, equation for $\boldsymbol{k}_i$ in (3.16) becomes implicit. Thus, we require $\alpha_{ij} = 0$ for $i \leq j$ to make the scheme explicit. In the explicit scheme,

$$\begin{aligned}
\boldsymbol{k}_1 &= \Delta t \boldsymbol{F}[\boldsymbol{x}_n, t], \\
\boldsymbol{k}_2 &= \Delta t \boldsymbol{F}\left[\boldsymbol{x}_n + \alpha_{21} \boldsymbol{k}_1, t + c_2 \Delta t\right], \\
\boldsymbol{k}_3 &= \Delta t \boldsymbol{F}\left[\boldsymbol{x}_n + \alpha_{31} \boldsymbol{k}_1 + \alpha_{32} \boldsymbol{k}_2, t + c_3 \Delta t\right], \\
&\vdots
\end{aligned} \qquad (3.18)$$

and

$$c_1 = 0, \quad c_2 = \alpha_{21}, \quad c_3 = \alpha_{31} + \alpha_{32}, \cdots. \qquad (3.19)$$

For this simple case, it is straightforward to derive the equations for α_{ij} and w_i up to the fourth-order scheme ($q = 4$).

In general, there are more than one solution for w_i and α_{ij} for a given value of q. Hence, there are infinite numbers for the option in the Runge–Kutta schemes. In the following, we will list the popular schemes.

(1) For the second-order Runge–Kutta scheme for which the accuracy of a numerical solution is of $O(\Delta t^2)$ (i.e., $q = 2$ and we choose $p = 2$), we obtain two conditions $w_1 + w_2 = 1$ and $w_2 \alpha_{21} = 1/2$ for three coefficients w_1, w_2, and α_{21}. There are two popular schemes. One is

$$\begin{aligned}
\boldsymbol{k}_1 &= \Delta t \boldsymbol{F}[\boldsymbol{x}, t], \\
\boldsymbol{k}_2 &= \Delta t \boldsymbol{F}\left[\boldsymbol{x} + \frac{1}{2}\boldsymbol{k}_1, t + \frac{1}{2}\Delta t\right], \\
\boldsymbol{x}(t + \Delta t) &= \boldsymbol{x}(t) + \boldsymbol{k}_2,
\end{aligned} \qquad (3.20)$$

where $w_1 = 0$, $w_2 = 1$, $c_2 = 1/2$, and $\alpha_{21} = 1/2$, and the other is

$$\boldsymbol{k}_1 = \Delta t \boldsymbol{F}[\boldsymbol{x}, t],$$
$$\boldsymbol{k}_2 = \Delta t \boldsymbol{F}[\boldsymbol{x} + \boldsymbol{k}_1, t + \Delta t],$$
$$\boldsymbol{x}(t + \Delta t) = \boldsymbol{x}(t) + \frac{1}{2}(\boldsymbol{k}_1 + \boldsymbol{k}_2), \qquad (3.21)$$

where $w_1 = 1/2$, $w_2 = 1/2$, $c_2 = 1$, and $\alpha_{21} = 1$.

(2) For the third-order Runge–Kutta scheme for which the accuracy of a numerical solution is of $O(\Delta t^3)$ (i.e., $q = 3$ and we choose $p = 3$), we obtain four conditions $w_1 + w_2 + w_3 = 1$, $c_2 w_2 + c_3 w_3 = 1/2$, $c_2^2 w_2 + c_3^2 w_3 = 1/3$, and $c_2 w_3 \alpha_{32} = 1/6$ for six coefficients w_1, w_2, w_3, c_2, c_3, and α_{32}. There are two popular schemes as

$$\boldsymbol{k}_1 = \Delta t \boldsymbol{F}[\boldsymbol{x}, t],$$
$$\boldsymbol{k}_2 = \Delta t \boldsymbol{F}\left[\boldsymbol{x} + \frac{1}{3}\boldsymbol{k}_1, t + \frac{1}{3}\Delta t\right],$$
$$\boldsymbol{k}_3 = \Delta t \boldsymbol{F}\left[\boldsymbol{x} + \frac{2}{3}\boldsymbol{k}_2, t + \frac{2}{3}\Delta t\right],$$
$$\boldsymbol{x}(t + \Delta t) = \boldsymbol{x}(t) + \frac{1}{4}(\boldsymbol{k}_1 + 3\boldsymbol{k}_3), \qquad (3.22)$$

and

$$\boldsymbol{k}_1 = \Delta t \boldsymbol{F}[\boldsymbol{x}, t],$$
$$\boldsymbol{k}_2 = \Delta t \boldsymbol{F}\left[\boldsymbol{x} + \frac{1}{2}\boldsymbol{k}_1, t + \frac{1}{2}\Delta t\right],$$
$$\boldsymbol{k}_3 = \Delta t \boldsymbol{F}[\boldsymbol{x} - \boldsymbol{k}_1 + 2\boldsymbol{k}_2, t + \Delta t],$$
$$\boldsymbol{x}(t + \Delta t) = \boldsymbol{x}(t) + \frac{1}{6}(\boldsymbol{k}_1 + 4\boldsymbol{k}_2 + \boldsymbol{k}_3). \qquad (3.23)$$

(3) For the fourth-order Runge–Kutta scheme for which the accuracy of a numerical solution is of $O(\Delta t^4)$ (i.e., $q = 4$ and we choose $p = 4$), we obtain eight conditions for ten coefficients w_1, w_2, w_3, w_4, c_2, c_3, c_4, α_{32}, α_{42}, and α_{43}. The most popular scheme is the classical Runge–Kutta scheme

$$\boldsymbol{k}_1 = \Delta t \boldsymbol{F}[\boldsymbol{x}, t],$$
$$\boldsymbol{k}_2 = \Delta t \boldsymbol{F}\left[\boldsymbol{x} + \frac{1}{2}\boldsymbol{k}_1, t + \frac{1}{2}\Delta t\right],$$
$$\boldsymbol{k}_3 = \Delta t \boldsymbol{F}\left[\boldsymbol{x} + \frac{1}{2}\boldsymbol{k}_2, t + \frac{1}{2}\Delta t\right],$$
$$\boldsymbol{k}_4 = \Delta t \boldsymbol{F}[\boldsymbol{x} + \boldsymbol{k}_3, t + \Delta t],$$
$$\boldsymbol{x}(t + \Delta t) = \boldsymbol{x}(t) + \frac{1}{6}(\boldsymbol{k}_1 + 2\boldsymbol{k}_2 + 2\boldsymbol{k}_3 + \boldsymbol{k}_4). \qquad (3.24)$$

There are infinite variants of the fourth-order Runge–Kutta scheme and a well-known variant is the Runge–Kutta 3/8 scheme as

$$k_1 = \Delta t F[x, t],$$

$$k_2 = \Delta t F\left[x + \frac{1}{3}k_1, t + \frac{1}{3}\Delta t\right],$$

$$k_3 = \Delta t F\left[x - \frac{1}{3}k_1 + k_2, t + \frac{2}{3}\Delta t\right],$$

$$k_4 = \Delta t F\left[x + k_1 - k_2 + k_3, t + \Delta t\right],$$

$$x(t + \Delta t) = x(t) + \frac{1}{8}(k_1 + 3k_2 + 3k_3 + k_4). \tag{3.25}$$

For examining that these schemes can yield a solution of the required order, it is convenient to consider the general solution of a simple ordinary differential equation

$$\frac{dx}{dt} = \zeta x, \tag{3.26}$$

where ζ is a complex constant. For this equation, the second-, third-, and fourth-order schemes, irrespective of the choice of w_i and α_{ij}, have to correctly yield the following equations,

$$x_{n+1} = x_n + a x_n + \frac{a^2}{2} x_n, \tag{3.27}$$

$$x_{n+1} = x_n + a x_n + \frac{a^2}{2} x_n + \frac{a^3}{6} x_n, \tag{3.28}$$

$$x_{n+1} = x_n + a x_n + \frac{a^2}{2} x_n + \frac{a^3}{6} x_n + \frac{a^4}{24} x_n, \tag{3.29}$$

where $a = \zeta \Delta t$. It is easily found that these relations are indeed satisfied for the Runge–Kutta schemes described above.

The time-step interval in the Runge–Kutta schemes cannot be taken to be arbitrarily large for guaranteeing the stability of numerical integration. To illustrate this fact, again, we consider the simplest equation (3.26) for evaluating the typical possible time-step interval. In particular, for the case that ζ is purely imaginary, the solution should be oscillatory one, that is similar to the solution for wave equations (3.2). Hence, we pay particular attention to this case. Equations (3.27)–(3.29) show that for the second-, third-, and fourth-order Runge–Kutta schemes, for each component of x_n, the following equation holds, respectively:

$$\frac{x_{n+1}}{x_n} = 1 + a + \frac{a^2}{2}, \tag{3.30}$$

$$\frac{x_{n+1}}{x_n} = 1 + a + \frac{a^2}{2} + \frac{a^3}{6}, \tag{3.31}$$

$$\frac{x_{n+1}}{x_n} = 1 + a + \frac{a^2}{2} + \frac{a^3}{6} + \frac{a^4}{24}. \tag{3.32}$$

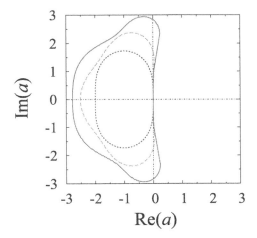

Fig. 3.2 Stable regions for the fourth-, third-, and second-order Runge–Kutta schemes (solid, dashed, and dotted curves). Each closed curve shows the marginally stable region and its inside is the stable region.

If $|x_{n+1}/x_n|$ is larger than unity, the amplitude of x increases with time in an exponential manner: i.e., the stability of the numerical time integration is lost. For obtaining a solution which is finite for $t \to \infty$, the condition, $|x_{n+1}/x_n| \leq 1$, has to be satisfied. This inequality provides us an allowed region for the complex number a, which is called a stable region for a.

A straightforward calculation by setting $a = re^{i\theta}$ yields the stable region for a. Figure 3.2 shows the stable regions for the second-, third-, and fourth-order schemes: The closed curves denote the marginally stable regions and their insides are the stable regions. The figure clearly shows that the fourth-order Runge–Kutta scheme has the widest stable region. In particular, along the imaginary axis, the stable region becomes quite wide for the fourth-order scheme, implying that we can take a large value of Δt. By contrast, for the second-order scheme, there is no stable region along the imaginary axis: It would not be possible to stably follow any oscillatory solution for this case (we will experimentally find that this is indeed the case in section 3.1.5). For the third- and fourth-order schemes, the stable regions along the imaginary axis are for $|a| < \sqrt{3}$ and for $|a| < 2\sqrt{2}$, respectively. This indicates that the value of Δt in the fourth-order scheme can be $\sqrt{8/3} \approx 1.6$ times as large as in the third-order scheme.

If we focus on the second-order schemes, there are other schemes besides the Runge–Kutta schemes. In the leapfrog scheme, we put η and ϕ at different time levels, $n + 1/2$ and n, respectively, while the spatial grids are identical for them. Then, the finite-differencing evolution equations for η and ϕ are written as

$$\eta^{n+1/2} = \eta^{n-1/2} + \Delta t \, \Delta_{\rm f} \phi^n, \tag{3.33}$$

$$\phi^{n+1} = \phi^n + \Delta t \, \eta^{n+1/2}. \tag{3.34}$$

This time integration scheme is second-order accurate and can yield a stable numerical solution by contrast to the second-order Runge–Kutta scheme.

In the Runge–Kutta schemes, η and ϕ are usually supposed to be evolved in the same integration scheme. For improving the stability property of the second-order Runge–Kutta scheme, we may change this policy. For example, the following scheme is also second-order accurate: In the first step, we evolve the equations from n-th level to $(n+1/2)$-th level as (note the right-hand side of the second equation)

$$\eta^{n+1/2} = \eta^n + \frac{1}{2}\Delta t\, \Delta_{\rm f}\phi^n, \tag{3.35}$$

$$\phi^{n+1/2} = \phi^n + \frac{1}{2}\Delta t\, \eta^{n+1/2}, \tag{3.36}$$

and in the next step, we employ

$$\eta^{n+1} = \eta^n + \Delta t\, \Delta_{\rm f}\phi^{n+1/2}, \tag{3.37}$$

$$\phi^{n+1} = \phi^n + \Delta t\, \eta^{n+1/2}, \tag{3.38}$$

or

$$\eta^{n+1} = \eta^n + \Delta t\, \Delta_{\rm f}\phi^{n+1/2}, \tag{3.39}$$

$$\phi^{n+1} = \phi^n + \frac{1}{4}\Delta t\left(\eta^n + 2\eta^{n+1/2} + \eta^{n+1}\right). \tag{3.40}$$

Here, for evolving η, a second-order Runge–Kutta scheme is employed, while for ϕ, a different scheme is chosen. In section 3.1.5, we will use these schemes referring to them as second-order modified Runge–Kutta schemes, version 1 and 2, respectively.

All the above examples show that there are many possible schemes even if we determine the order of accuracy: the difference appears only in the $(q+1)$-th-order truncation error for the q-th order accurate schemes. Here, one important caution is that we cannot always employ any scheme in numerical simulations: The chosen scheme has to enable us to *stably* perform a numerical simulation. It is important to emphasize that even if the order of accuracy is the same between two different schemes, the property of the stability could significantly change [see Teukolsky (2000) for a similar subject]. The reason for this is that the higher-order error of $O(\Delta t^{q+1})$ could determine the stability property. This point will be illustrated in section 3.1.5, showing that the purely second-order Runge–Kutta scheme does not enable us to perform stable time integration, while by the modified ones, it becomes feasible.

3.1.3 Boundary condition at outer boundaries

In numerical-relativity simulations, basic equations are usually solved in a finite computational domain unless a special coordinate system covering to spatial infinity is employed. Hence, outer boundaries are located at a region for which the distance from its center is finite. For simplicity, we here consider a computational domain

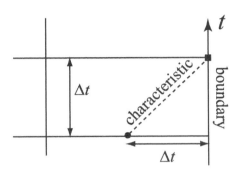 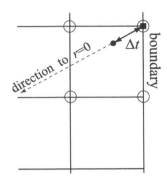

Fig. 3.3 Idea for imposing an outer boundary condition. The left panel shows a spacetime diagram composed of t (vertical axis) and r (horizontal axis), and the right panel shows a spatial domain near an outer boundary. For both figures, the filled square is the point for which the boundary condition is required and the filled circle is the point at which the value of $rQ = (r_o - \Delta t)Q(t - \Delta t, r_o - \Delta t)$ is equal to that at the square point, $r_o Q(t, r_o)$: see equation (3.44). The filled circle is not located on a grid point and the value there has to be determined by an interpolation procedure, using the values on the nearby grid points at a time level $t - \Delta t$ (see the open circles in the right panel).

composed of a Cartesian cube as $-L \leq x \leq L$, $-L \leq y \leq L$, and $-L \leq z \leq L$, where L is a finite value but it is much larger than the physical size of a system that we consider (e.g., the stellar radius or separation of two stars) and also it is larger than the wavelength of gravitational waves and electromagnetic waves.

In the presence of the outer boundaries at a finite distance from the center, we have to impose some boundary conditions. As described in section 2.3.1, geometric quantities are composed of scalar, vector, and tensor parts in general relativity, for which we have to impose different boundary conditions. Approximately speaking, the scalar part is composed of gravitational-potential and gauge parts, the vector part is composed primarily of gauge parts, and the tensor part is composed of gravitational waves. We note that for linear equations of the form (3.2), only the wave components are taken into account, but we here consider more general cases. With an appropriate choice of gauge conditions, gauge components can be excluded in the distant zone, and hence, we here pay attention to gravitational-potential and gravitational-wave parts. In the following, we will only consider four-dimensional asymptotically flat spacetime. Brief descriptions for other spacetime will be given in the last two paragraphs of this section.

The gravitational-potential part behaves at a large distance as $O(r^{-1})$ with a very slowly (or no) varying coefficient. The gravitational-wave part behaves at a large distance (in a wave zone) also as $O(r^{-1})$, coincidentally in four dimensions, but the coefficient is a function of a retarded time, approximately written as $r-t$. In geometrical quantities, these two modes are usually mixed, and we have to impose outer boundary conditions that are satisfied by these two modes. An often used prescription is to assume the following behavior for geometrical variables Q in the far zone,

$$Q = \frac{f(r-t)}{r}, \quad (3.41)$$

where $f(x)$ is a regular function of x and may be constant. Then, both conditions can be satisfied: For the gravitational-potential part, $f(r-t)$ should be approximately constant, and for the gravitational-wave part, it should be a non-constant function of $r-t$.

Specifically, the outer boundary condition is often written in the Sommerfeld form

$$\partial_t(rQ) + \partial_r(rQ) = 0. \quad (3.42)$$

Note that if Q does not vary with time, we automatically have the condition, $\partial_r(rQ) = 0$, from equation (3.42). Thus, equation (3.42) includes the condition that should be satisfied by the gravitational-potential part.

In numerical computations, it is not a trivial issue how to translate the boundary condition in the form of (3.42) into a finite-differencing form, because the finite-differencing condition has to enable us accurate and stable numerical simulations. One possibility is to rewrite it in a form, e.g., as

$$\frac{Q_l^n - Q_l^{n-1}}{\Delta t} + \frac{1}{r_l}\frac{r_l Q_l - r_{l-1} Q_{l-1}}{\Delta r_l} = 0, \quad (3.43)$$

where r_l denotes the l-th grid point of the radial coordinate, and here, r_l corresponds to the location of the outer boundary. In the case that spherical-polar coordinates are employed, this finite-differencing scheme is straightforwardly implemented. However, in numerical relativity of no assumption for symmetries, we usually employ Cartesian coordinates, and the grid points are located at (i, j, k) for (x, y, z). In this case, the finite differencing with respect to r cannot be taken in a straightforward manner.

A robust prescription is to determine the value of Q at the outer boundaries at time t by a physically-motivated interpolation [Shibata and Nakamura (1995)]. Equation (3.41) implies that near the outer boundaries, rQ is preserved along characteristic curves, $r - t =$const (see figure 3.3). Thus, the following equality should be satisfied

$$r_o Q(t, r_o) = (r_o - \Delta t) Q(t - \Delta t, r_o - \Delta t), \quad (3.44)$$

namely,

$$Q(t, r_o) = \left(1 - \frac{\Delta t}{r_o}\right) Q(t - \Delta t, r_o - \Delta t). \quad (3.45)$$

Here, r_o is a radial coordinate at the outer boundaries and $t - \Delta t$ denotes the time at the previous time step. In this method, we need only to substitute the value of $Q(t - \Delta t, r_o - \Delta t)$ for the boundary condition. A minor issue here is that at $r = r_o - \Delta t$, there is no grid point. Thus, the value there has to be determined by an interpolation procedure (see the right panel of figure 3.3). The interpolation is first-order accurate if nearby eight grid points are used, and it is second-order accurate if twenty-seven grid points are used.

The final remark about the outer boundary conditions for asymptotically flat spacetime is on the following point: As we describe in appendix C.2, the outer boundary condition written in the form of equation (3.41) is only approximate for gravitational waves, if r is finite: The approximation is justified only for $r \gg \omega^{-1}$ where ω is the angular velocity of gravitational waves. This means that the outer boundary conditions have to be imposed in the wave zone, which is usually far from the region that gravitational-wave sources are located. This situation motivates us to employ an adaptive mesh refinement algorithm (see section 3.3).

For higher-dimensional asymptotically flat spacetime, gravitational-potential parts fall off as r^{-N+2}, while gravitational-wave parts as $r^{-(N-1)/2}$, where N is the spatial dimensionality as before. This implies that in the wave zone (and far zone), the gravitational-wave parts are dominant. Assuming that the gravitational-potential parts are negligible, we only need to impose outer boundary conditions for the gravitational-wave parts. For this case, we can use the same boundary condition as that described above for $N = 3$, assuming the behavior of geometric quantities in the far zone as

$$Q = \frac{f(r-t)}{r^{-(N-1)/2}}. \tag{3.46}$$

For asymptotically flat Friedmann and de Sitter spacetime, essentially the same boundary condition may be used if we subtract the cosmological expansion part from geometric quantities appropriately (see section 2.6). For asymptotically anti de Sitter spacetime, by contrast, the situation is completely different. Much more complicated boundary conditions have to be imposed because its global structure is completely different from asymptotically flat and cosmological spacetime. For this topic, the numerical method for imposing the boundary conditions has not been established yet, although pioneer works such as Bizoń and Rostworowski (2011); Namtilan et al. (2012) have been reported.

3.1.4 Courant–Friedrichs–Lewy condition

For the stable numerical integration of wave equations, we have to appropriately choose the time-step interval, Δt. A stable time integration is feasible only for a small time-step interval limited by the grid spacing Δx as $\nu := \Delta t/\Delta x \lesssim 1$ for typical explicit time integration schemes. Such a constraint is called the Courant–Friedrichs–Lewy (CFL) condition, and ν is referred to as the Courant number or the CFL number.

The physical reason for the presence of the maximum permissible value of ν is often understood in the following manner. Consider discretized 1+1 flat spacetime described by coordinates t and x; see figure 3.4. n and j label the corresponding

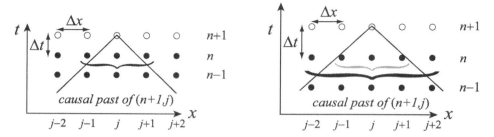

Fig. 3.4 Schematic figures for illustrating the notion of the CFL condition. In both panels, the circles denote discretized points of spacetime and solid oblique lines are characteristics determined by the speed of light. n and j denote the time and spatial discretized points. Left: $\Delta t/\Delta x$ is small enough to enable stable time integration. Right: $\Delta t/\Delta x$ is too large to enable stable time integration for second-order spatial finite differencing performed with three spatial points $j-1$, j, and $j+1$, but it is small enough to enable stable time integration for fourth-order spatial finite differencing performed with five points $j-2$, $j-1$, j, $j+1$, and $j+2$.

time and spatial discretized points. The quantities at $(n+1, j)$ are computed from the information at the n-th time level and at three points $j-1$, j, and $j+1$ in the second-order finite-differencing scheme, and at five points $j-2, \cdots, j+2$ in the fourth-order finite-differencing scheme. First, we consider the second-order spatial finite-differencing scheme. In this scheme, one assumes that the quantities at $(n+1, j)$ are determined from the information only in the domain $x_{j-1} \leq x \leq x_{j+1}$ at the n-th time level. In other words, we assume in this scheme that all the information in the domain $x_{j-1} \leq x \leq x_{j+1}$ are completely known (in the second-order accuracy) whereas the data for other domain may be arbitrary. Here, causality plays an important role. Because information propagates with the speed of light for the quantities that obey wave equations, all the data at $(n+1, j)$ have to be determined by the data in the causal past of the point $(n+1, j)$, i.e., in the region between two solid oblique lines in figure 3.4. If its causal past at the n-th time level is located between x_{j-1} and x_{j+1} (see the left panel of figure 3.4 for which $\Delta t < \Delta x$), all the necessary data are prepared in the second-order spatial finite differencing, and thus, we can consider that globally hyperbolic initial data are prepared for determining the data at $(n+1, j)$. By contrast, if a part of the causal past of the point $(n+1, j)$ at the n-th time level is located outside the domain $x_{j-1} \leq x \leq x_{j+1}$ (see the right panel of figure 3.4 for which $\Delta t > \Delta x$), we should consider that the global hyperbolicity is lost. This implies that the data at $(n+1, j)$ is not physically determined. For such a case, numerical instability takes place.

In the fourth-order spatial finite differencing, this restriction is relaxed. In this case, the quantities at $(n+1, j)$ are determined from the information for $x_{j-2} \leq x \leq x_{j+2}$ at the n-th time level. In other words, we assume in this scheme that all the information in the domain $x_{j-2} \leq x \leq x_{j+2}$ are prepared (in the fourth-order accuracy). Thus, if the causal past of the point $(n+1, j)$ is located in the domain $x_{j-2} \leq x \leq x_{j+2}$, the global hyperbolicity is guaranteed. This implies that Δt in the fourth-order scheme can be by a factor of 2 larger than that in the second-order scheme.

According to the CFL condition, the maximum permissible value of the CFL number is 1 for the second-order scheme and 2 for the fourth-order scheme. In reality, the possible maximum permissible values for it are smaller than these expected values in numerical-relativity simulations. The values experimentally found are, respectively, ~ 0.4 and ~ 0.7 for linear wave equations and smaller than these values for nonlinear problems. A partial reason for this is that in the presence of outer boundaries, the order of accuracy is reduced due to the lower-order-accurate boundary conditions, below the value predicted naively. Another reason is that for a large CFL number that is close to the maximum permissible value, the accuracy of a numerical solution is often worse than expected, in particular by a nonlinear coupling, which could give a significant effect in general relativity.

3.1.5 Numerical experiments

In the previous sections 3.1.1–3.1.3, we described how to solve linear wave equations of the form (3.2). As remarked in the last paragraph of section 3.1.2, all the numerical schemes presented in sections 3.1.1 and 3.1.2 do not always permit us a stable numerical integration. In this section, we will illustrate this fact by showing several results of a numerical experiment.

In this numerical experiment, we solve the wave equation for a single scalar field, $\Box \phi = 0$, in three-dimensional Cartesian coordinates (x, y, z). The computational domain is $-L \leq x \leq L$, $-L \leq y \leq L$, and $0 \leq z \leq L$, where the equatorial plane symmetry with respect to $z = 0$ is imposed. At the outer boundaries, $|x| = L$, $|y| = L$, and $z = L$, the boundary condition based on the information of characteristic curves is imposed, as described in section 3.1.3. Hereafter, we take the width of the wave packet λ as the units of length and time, and choose $L = 10\lambda$.

As an initial condition of this numerical experiment, we prepare a wave packet of the quadrupole configuration as

$$\phi = 0, \quad \eta = -2\left(x^2 - y^2\right) e^{-r^2/2}, \tag{3.47}$$

where $\lambda = 1$. The exact solution of the wave equation for this initial condition is easily obtained by the method described in appendix C.1.1, and thus, we can examine the accuracy of numerical solutions by comparing them with the exact solution. Specifically, we extract the scalar wave at $(x, y, z) = (L/2, 0, 0)$.

The numerical simulations were performed employing the following schemes:

(1) Fourth-order spatial finite differencing and fourth-order Runge–Kutta time-integration scheme are combined (referred to as "4th-4th").
(2) Second-order spatial finite differencing and second-order Runge–Kutta time-integration scheme are combined (referred to as "2nd-2nd").

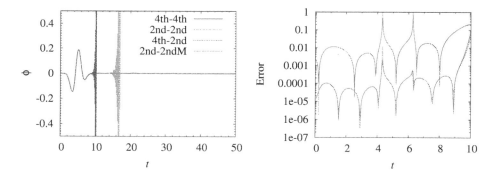

Fig. 3.5 Left: ϕ extracted at $(x, y, z) = (L/2, 0, 0)$. 4th-4th, 2nd-2nd, 4th-2nd, and 2nd-2ndM denote the results for fourth-order spatial finite differencing with the fourth-order Runge–Kutta scheme, for second-order spatial finite differencing with the second-order Runge–Kutta scheme, for fourth-order spatial finite differencing with the second-order Runge–Kutta scheme, and for second-order spatial finite differencing with the second-order modified Runge–Kutta scheme, respectively. Right: Error of ϕ as a function of time for $N = 100$ (solid curves), $N = 80$ (dashed curves), and $N = 64$ (dotted curves). The results are shown for the the schemes "4th-4th" (lower curves) and "2nd-2ndM" (upper curves). The errors for $N = 80$ and $N = 64$ are plotted multiplying $(N/100)^q$ where q is the order of accuracy, 2 or 4, and the curves of different values of N for each scheme approximately overlap, showing that the q-th order convergence is achieved.

(3) Fourth-order spatial finite differencing and second-order Runge–Kutta time-integration scheme are combined (referred to as "4th-2nd").
(4) Second-order spatial finite differencing and second-order modified Runge–Kutta time-integration scheme are combined (referred to as "2nd-2ndM").

In the numerical simulations, uniform spatial grids are prepared with the grid spacing $\Delta x = L/N$ where we take $N = 64$, 80, and 100 to examine the convergence property with respect to the grid resolution. The time-step interval for the time integration, Δt, is taken as $\nu = \Delta t / \Delta x = 1/2$ for the fourth-order time-integration scheme. For the second-order time-integration scheme, the value of ν is varied from 0.15 to 0.25, but the stability property discussed below does not depend on the value of ν in this case.

The left panel of figure 3.5 plots the time evolution of ϕ at $(x, y, z) = (L/2, 0, 0)$ for the numerical simulations by the four different numerical schemes with $N = 100$. As the second-order modified Runge–Kutta scheme, we here employed the version 1 [see equations (3.35)–(3.38) of section 3.1.2], but the version 2 yields essentially the same results. This figure shows that in the purely second-order Runge–Kutta scheme, the scalar field unphysically oscillates in an overstable manner, resulting in the divergence of the values of $|\phi|$ at $t = 10$–20. This is the case not only for the second-order spatial finite differencing but also for the fourth-order one. This illustrates that the second-order Runge–Kutta time-integration scheme cannot stably yield the solution irrespective of the scheme of the spatial finite differencing. By contrast, the fourth-order Runge–Kutta and second-order modified Runge–

Kutta time-integration schemes enable stable time evolution. Although we stopped the simulations at a finite time $t = 300$, they would stably continue for much longer time.

The right panel of figure 3.5 plots the time evolution of the error defined by

$$\left| 1 - \frac{\phi_{\text{numerical solution}}}{\phi_{\text{exact solution}}} \right|, \tag{3.48}$$

for three different grid resolutions. The errors for $N = 80$ and $N = 64$ are plotted multiplying $(N/100)^q$ where q is the order of accuracy, 2 or 4. The results are shown for the schemes "4th-4th" and "2nd-2ndM". Three curves for each scheme of this figure approximately overlap, showing that the error converges at the required orders (fourth and second orders) as long as the information of the outer boundary does not significantly affect the numerical results at $(L/2, 0, 0)$. We note that once the information from the outer boundary reaches the point of extracting ϕ (for $t \gtrsim 9$), the accuracy becomes poorer because a first-order scheme is employed for the outer boundary conditions. This figure also shows that the accuracy for the fourth-order scheme is approximately by $(\Delta x/\lambda)^2$ better than that for the second-order scheme.

Now, we describe the reason that purely second-order Runge–Kutta time-integration schemes yield unstable numerical results, while the other schemes enable us a stable numerical simulation. The best way for showing the stability property of a given scheme is the so-called von Neumann stability analysis [Press et al. (1989)]. Here, we consider this analysis for the $1+1$ wave equation

$$\partial_t^2 \phi = \partial_x^2 \phi. \tag{3.49}$$

For this equation, the general solution is written as $\phi = f_o(t - x) + f_i(t + x)$ where f_o and f_i are arbitrary functions. This implies that if we focus only on either an outgoing or ingoing wave solution, the wave packet simply propagates without changing its shape and the maximum amplitude should be constant. Taking into account this fact, we formally write a solution of ϕ on the discretized time and space points as

$$\phi(t_n, x_j) = \phi_0 \sum_k (A_k)^n e^{ikj\Delta x}, \tag{3.50}$$

where ϕ_0 is a constant (hereafter we set it to be unity), k is a wave number, n and j denote the discretized points for time and space, Δx is the grid spacing, and A_k is a complex which denotes the amplification factor of the amplitude. Note that the superscript for $(A_k)^n$ denotes the power of A_k (it does not mean the n-th time level). Because we consider an equation linear in ϕ, the equation for each value of k is identical, and thus, hereafter, we omit k.

As mentioned above, the amplitude of ϕ should be constant in the frame moving with a wave packet. Thus, the ratio of the amplitude of two neighboring time steps $|A^{n+1}/A^n| = |A|$ should be unity. Any numerical scheme has to guarantee this property as accurately as possible. If a numerical scheme yields a numerical solution with $|A| > 1$, the amplitude of the wave packet will be unphysically amplified, and therefore, such numerical schemes cannot be employed. Thus, a necessary condition for a numerical scheme to be employed is that it guarantees to always yield a numerical solution of $|A| \leq 1$.

Now, we analyze several second-order time-integration schemes. First, we consider the 2nd-2nd scheme for which the finite-differencing equations are written as

$$\eta_j^{n+1/2} = \eta_j^n + \frac{\Delta t}{2\Delta x^2}\left(\phi_{j+1}^n - 2\phi_j^n + \phi_{j-1}^n\right),$$
$$\phi_j^{n+1/2} = \phi_j^n + \frac{\Delta t}{2}\eta_j^n,$$
$$\eta_j^{n+1} = \eta_j^n + \frac{\Delta t}{\Delta x^2}\left(\phi_{j+1}^{n+1/2} - 2\phi_j^{n+1/2} + \phi_{j-1}^{n+1/2}\right),$$
$$\phi_j^{n+1} = \phi_j^n + \Delta t\,\eta_j^{n+1/2}. \qquad (3.51)$$

Eliminating $\eta^{n+1/2}$ and $\phi^{n+1/2}$ yields

$$\eta_j^{n+1} = \eta_j^n + \frac{\Delta t}{\Delta x^2}\left[\phi_{j+1}^n - 2\phi_j^n + \phi_{j-1}^n + \frac{\Delta t}{2}\left(\eta_{j+1}^n - 2\eta_j^n + \eta_{j-1}^n\right)\right],$$
$$\phi_j^{n+1} = \phi_j^n + \Delta t\left[\eta_j^n + \frac{\Delta t}{2\Delta x^2}\left(\phi_{j+1}^n - 2\phi_j^n + \phi_{j-1}^n\right)\right]. \qquad (3.52)$$

For the von Neumann analysis, we substitute the following relations into the above equations

$$\phi_j^n = (A_k)^n e^{ikj\Delta x}, \qquad \eta_j^n = \alpha_k(A_k)^n e^{ikj\Delta x}, \qquad (3.53)$$

where α_k is a complex number and we assume that the amplification factors of ϕ and η agree with each other. Then, equation (3.52) results in a simultaneous equation for A_k and α_k. Eliminating α_k from two equations, we obtain the equation for A_k as (hereafter we omit k)

$$\left(A - 1 + 2\nu^2\sin^2\theta\right)^2 + 4\nu^2\sin^2\theta = 0, \qquad (3.54)$$

where $\theta = k\Delta x/2$ and $\nu = \Delta t/\Delta x$. Here θ satisfies $0 < \theta \leq \pi/2$. Note that the above and following calculations are easily done if we remember the relations

$$\phi_{j+1}^n - 2\phi_j^n + \phi_{j-1}^n = -4A^n\sin^2\theta\,e^{ikj\Delta x},$$
$$\phi_{j+2}^n - 4\phi_{j+1}^n + 6\phi_j^n - 4\phi_{j-1}^n + \phi_{j-2} = 16A^n\sin^4\theta\,e^{ikj\Delta x}.$$

For two complex solutions of equation (3.54), the absolute values are identical as

$$|A|^2 = 1 + 4(\nu\sin\theta)^4 > 1. \qquad (3.55)$$

Since $|A| > 1$ in this scheme, the amplitude of the scalar field is amplified in an exponential manner, and thus, this scheme yields an unstable numerical solution. This agrees with the results shown in figure 3.5.

This unstable property is unchanged even if the order of spatial finite differencing is improved to the fourth order. In this 4th-2nd scheme, the resulting equation for A is slightly modified as

$$\left(A - 1 + 2\nu^2 C\right)^2 + 4\nu^2 C = 0, \qquad C = \sin^2\theta + \frac{\sin^4\theta}{3}, \qquad (3.56)$$

and thus, $|A|^2 = 1 + 4\nu^4 C^2 > 1$. Because $C \geq \sin^2 \theta$, the instability should develop more rapidly in this scheme than in the 2nd-2nd scheme. This agrees with the numerical result shown in the left panel of figure 3.5.

Next, we analyze the 2nd-2ndM scheme. In the first version, the second equation of (3.51) is rewritten as

$$\phi_j^{n+1/2} = \phi_j^n + \frac{\Delta t}{2} \eta_j^{n+1/2}. \tag{3.57}$$

In this scheme, after $\eta^{n+1/2}$ is determined, $\phi^{n+1/2}$ is updated. The von Neumann analysis, as done above, yields the equation for A as

$$\left(A - 1 + 2\nu^2 \sin^2 \theta\right)^2 + 4\nu^2 \sin^2 \theta \left(1 - \nu^2 \sin^2 \theta\right) = 0. \tag{3.58}$$

This gives two complex solutions of A which satisfies $|A| = 1$ (in the reasonable assumption $0 < \nu \sin \theta < 1$). Thus, this scheme yields a (marginally) stable numerical solution. The numerical experiments indeed show that this is the case (see figure 3.5).

In the second version of the 2nd-2ndM scheme, the second and fourth equations of (3.51) are rewritten, respectively, as

$$\phi_j^{n+1/2} = \phi_j^n + \frac{\Delta t}{2} \eta_j^{n+1/2},$$
$$\phi_j^{n+1} = \phi_j^n + \frac{\Delta t}{4} \left(\eta_j^{n+1} + 2\eta_j^{n+1/2} + \eta_j^n\right). \tag{3.59}$$

In this case, the equation for A is

$$A^2 - A \left(2 - 4\nu^2 \sin^2 \theta + \nu^4 \sin^4 \theta\right) + 1 - (\nu \sin \theta)^4 = 0, \tag{3.60}$$

and for two complex solutions of A (in the reasonable assumption of $0 < \nu \sin \theta \lesssim 0.955$),

$$|A|^2 = 1 - (\nu \sin \theta)^4 < 1. \tag{3.61}$$

Thus, a stable numerical solution is guaranteed to be obtained as long as ν satisfies a weak condition $0 < \nu \lesssim 0.955$. For the 4th-4th scheme, the similar result can be obtained, although the analysis is much more complicated.

All these properties for the stability also hold when integrating Einstein's evolution equations: Einstein's evolution equations can be stably integrated when we employ the 4th-4th scheme or 2nd-2ndM scheme. There are also other possibilities; for example, the third-order Runge–Kutta time-integration scheme together with the fourth-order spatial finite differencing is known to work well.

The final remark in this section is on computational costs. The computational cost for one-step time integration in the fourth-order Runge–Kutta scheme is approximately twice as expensive as that in the second-order one, because the number of the Runge–Kutta substeps is twice larger. However, the computational cost for the *physical-time integration* in the fourth-order Runge–Kutta scheme can be approximately as expensive as that in the second-order scheme, because the maximum permissible time-step interval, Δt, in the fourth-order scheme may be by a factor of ~ 2 larger than that in the second-order scheme and this property compensates the demerit of the larger number of Runge–Kutta substeps. Therefore, in numerical relativity, the fourth-order Runge–Kutta scheme is the recommendable time-integration scheme because the accuracy can be improved without significantly increasing the computational costs. We note that for the case that one implements an adaptive or fixed mesh refinement algorithm in a numerical-relativity code, a computational cost in the fourth-order scheme could be more expensive than that in the second-order scheme, because there are additional processes in the refinement algorithm. This point will be described in detail in section 3.3.

3.2 Handling advection terms

In the 3+1 (and N+1) formulation of Einstein's equation, the presence of the shift vector results in the advection term in the evolution equations, e.g., as (see sections 2.1.5 and 2.3.3)

$$\partial_t \gamma_{ij} - \beta^k \partial_k \gamma_{ij} = \text{other terms}. \tag{3.62}$$

We have to be careful in numerically handling this advection term because we cannot use the centered finite differencing for this term (see below).

We here consider numerical methods for solving a simple 1+1 first-order partial differential equation of the form [Toro (1999)]

$$\partial_t \phi + V \partial_x \phi = 0, \tag{3.63}$$

where V is a characteristic speed (i.e., in numerical relativity, the characteristic speed in the x^k direction is $-\beta^k$). In this section, we assume that ϕ is a continuous function because we consider the evolution of geometric quantities (in section 4.8, we will describe several methods for handling the advection term when ϕ could locally have a discontinuity). The general solution for equation (3.63) is easily written by $\phi = f(x - Vt)$ where f is an arbitrarily continuous function. Thus, the solution has a form that a profile of ϕ simply propagates with the propagation speed V preserving its profile and amplitude.

First of all, we show that usual centered finite-differencing schemes do not work for stably solving equation (3.63). For example, we consider the following second-order and fourth-order (spatially) finite-differencing schemes as in section 3.1.5:

$$\phi_j^{n+1/2} = \phi_j^n - \frac{\nu}{2} \left(\phi_{j+1}^n - \phi_{j-1}^n \right),$$
$$\phi_j^{n+1} = \phi_j^n - \nu \left(\phi_{j+1}^{n+1/2} - \phi_{j-1}^{n+1/2} \right), \tag{3.64}$$

for the second-order case, and

$$\phi_j^{n+1/2} = \phi_j^n - \frac{\nu}{24} \left[-\left(\phi_{j+2}^n - \phi_{j-2}^n \right) + 8 \left(\phi_{j+1}^n - \phi_{j-1}^n \right) \right],$$
$$\phi_j^{n+1} = \phi_j^n - \frac{\nu}{12} \left[-\left(\phi_{j+2}^{n+1/2} - \phi_{j-2}^{n+1/2} \right) + 8 \left(\phi_{j+1}^{n+1/2} - \phi_{j-1}^{n+1/2} \right) \right], \tag{3.65}$$

for the fourth-order case, where $\nu := V\Delta t/\Delta x$. Here, the second-order scheme is assumed for the time integration. The von Neumann analysis setting $\phi_j^n = A^n e^{ikj\Delta x}$ (see section 3.1.5) shows that for equation (3.64),

$$|A|^2 = 1 + \frac{(\nu \sin\theta)^4}{4} > 1, \tag{3.66}$$

and for equation (3.65),

$$|A|^2 = 1 + \frac{64(\nu \sin\theta)^4}{81}\left(1 - \frac{\cos\theta}{4}\right)^4 > 1, \tag{3.67}$$

where $\theta = k\Delta x$ in this case. Thus, the amplitude of ϕ increases exponentially with time (unstable numerical solutions are always yielded) irrespective of the order of the spatial finite-differencing scheme. We note that the results are qualitatively unchanged even when first-order time integration schemes are employed.

One of the prescriptions for overcoming this instability problem is to employ a modified time integration scheme. One simple prescription is to employ a leapfrog scheme [Press et al. (1989)] as

$$\phi_j^{n+1} = \phi_j^{n-1} - \nu\left(\phi_{j+1}^n - \phi_{j-1}^n\right). \tag{3.68}$$

In this case, the von Neumann analysis yields

$$A = -i\nu \sin\theta \pm \sqrt{1 - \nu^2 \sin^2\theta}, \quad \text{and} \quad |A|^2 = 1. \tag{3.69}$$

Thus, this scheme can yield physical numerical solutions as long as $\nu \sin\theta \leq 1$. This property is unchanged even if we employ the fourth-order centered spatial finite-differencing scheme.

The leapfrog scheme is, however, second-order accurate in time. If we would like to improve the accuracy for the time integration, this prescription has the limitation. Thus, the other robust prescription is necessary. One of the robust answers to this issue is to employ an upwind scheme. This scheme is popular not only in handling the β-advection terms but also in computational hydrodynamics (see section 4.8).

Again, we consider equation (3.63). In the upwind schemes, the rule of spatial finite differencing is changed depending on the sign of V; see figure 3.6 for its notion. For simplicity, we first consider the scheme which is first-order in time and space. In this case, the finite-differencing equation is written as

$$\phi_j^{n+1} = \phi_j^n - \nu \begin{cases} \phi_j^n - \phi_{j-1}^n & V \geq 0, \\ \phi_{j+1}^n - \phi_j^n & V \leq 0. \end{cases} \tag{3.70}$$

Thus, for the spatial finite differencing, a lopsided finite differencing is employed. Physically speaking, the information of characteristic curves are used in this finite-differencing scheme: Remember that the general solution of equation (3.63) is written as $f(x - Vt)$. Hence, the value of f at (x_0, t_0) is equal to that at $(x_0 - V\Delta t, t_0 - \Delta t)$. This implies that for $V > 0$, the finite differencing should be executed giving a weight in the side of $x < x_0$, while for $V < 0$, the weight should be given in the side of $x > x_0$.

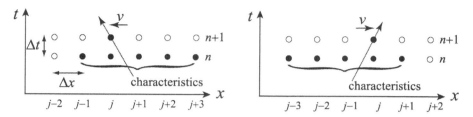

Fig. 3.6 Schematic figures for a fourth-order upwind scheme for $v < 0$ (left) and for $v > 0$ (right) where v denotes the characteristic speed. The filled circles at n-th time level are used for the upwind finite differencing for determining the value at $(n+1)$-th time level (filled circle). The dotted lines with arrows denote the characteristic curves.

The von Neumann analysis for equation (3.70) yields
$$A = 1 - |\nu|(1 - \cos\theta) - i\nu\sin\theta, \tag{3.71}$$
and thus,
$$|A|^2 = 1 - 2|\nu|(1 - |\nu|)(1 - \cos\theta). \tag{3.72}$$
Therefore, as long as $|\nu| \leq 1$, this scheme is guaranteed to yield stable numerical solutions. The point in this asymmetric finite differencing is that the term proportional to $-|\nu|$ appears in A [compare it with equations (3.66) and (3.67) in which there is no term linear in ν]. This guarantees that the unphysical amplification does not occur for small values of $|\nu|$.

The problem of this first-order scheme is that it is too dissipative: The amplitude of a wave packet damps in a short time scale unless we prepare a sufficiently high grid resolution (here a high grid resolution implies that the typical value of θ for a given problem is much smaller than unity and hence $1 - \cos\theta \ll 1$). This drawback can be improved when we employ higher-order schemes for the spatial finite differencing. In the second-order and fourth-order schemes, we may choose the following lopsided spatial finite differencing, respectively, as
$$\partial_x \phi = \frac{1}{2\Delta x} \begin{cases} 3\phi_j^n - 4\phi_{j-1}^n + \phi_{j-2}^n & V \geq 0, \\ -\phi_{j+2}^n + 4\phi_{j+1}^n - 3\phi_j^n & V \leq 0, \end{cases} \tag{3.73}$$
and
$$\partial_x \phi = \frac{1}{12\Delta x} \begin{cases} 3\phi_{j+1}^n + 10\phi_j^n - 18\phi_{j-1}^n + 6\phi_{j-2}^n - \phi_{j-3}^n & V \geq 0, \\ -3\phi_{j-1}^n - 10\phi_j^n + 18\phi_{j+1}^n - 6\phi_{j+2}^n + \phi_{j+3}^n & V \leq 0. \end{cases} \tag{3.74}$$
Here, we assume that ϕ is continuous up to the second- and fourth-order spatial derivatives, respectively. We depict a schematic figure of the fourth-order upwind finite-differencing scheme in figure 3.6.

The von Neumann analysis for these schemes shows that irrespective of the order of accuracy for the time integration,
$$|A|^2 = \begin{cases} 1 - 2|\nu|(1 - \cos\theta)^2 + O(\nu^2) & \text{second-order scheme,} \\ 1 - \dfrac{2|\nu|}{3}(1 - \cos\theta)^3 + O(\nu^2) & \text{fourth-order scheme.} \end{cases} \tag{3.75}$$

Thus, with increasing the order of accuracy, the power of the suppression factor $1 - \cos\theta$ increases and the degree of the unphysical damping can be reduced.

In numerical relativity, the most popular scheme is the fourth-order scheme shown in equation (3.74). In particular, it is experimentally found that employing the fourth-order scheme is crucial for evolving moving black holes [Campanelli *et al.* (2006a); Baker *et al.* (2006a)].

The upwind schemes described above can be used for a function continuous up to the required order. In compressible hydrodynamics for which the basic equations are similar to equation (3.63), shocks (discontinuities) are often formed, and hence, these unwind schemes are not applicable. For compressible hydrodynamics, schemes slightly different from them have to be employed in particular for accurately computing steeply varying quantities. Typical schemes for numerical hydrodynamics will be introduced in section 4.8.

3.3 Adaptive mesh refinement

3.3.1 *Why necessary?*

One of the most important roles in numerical relativity is to perform a large number of simulations for coalescing compact objects employing a wide range of binary parameters. For this purpose, an efficient scheme for the numerical simulation is absolutely required. For the two-body problems considered here, adaptive mesh-refinement (AMR) algorithm [Berger and Oliger (1984)] is well-suited, and the essential ingredient. First of all, we describe the reason for this.

In the two-body problems, there are three characteristic length scales; the radius of the compact objects, R, the orbital separation, ℓ, and a gravitational-wave length, λ. If we assume that the binary is in a circular orbit and has a Keplerian velocity, $\lambda \approx \pi(\ell^3/m_0)^{1/2}$ where m_0 is the total mass of the system. We have to accurately resolve these three scales which are in the relation $R < \ell < \lambda$, and typically, $R \ll \lambda$.[1] For accurately resolving each compact object, the grid spacing of size Δx in the vicinity of the compact objects has to be much smaller than R; according to the results of many numerical-relativity simulations, $R/\Delta x$ should be $\gtrsim 20$ for black holes and $\gtrsim 50$ for neutron stars. On the other hand, gravitational waves have to be extracted from geometric quantities in a wave zone, i.e., in the region that satisfies $r \gtrsim \lambda$. This implies that the size of the computational domain should be larger than λ (see chapter 6). By simply using a uniform Cartesian grid, the required grid number in one direction is $N_g = 2\lambda/\Delta x$ where the factor 2 comes from the fact that there are plus and minus directions in each axis direction of Cartesian

---

[1]We assume that the sizes of two compact objects are of the same order. This assumption is likely to be valid, as far as we consider binaries composed of stellar-mass objects. However, for binaries composed of, e.g., a very massive black hole of mass $\gg 10 M_\odot$ and a neutron star, this is not the case. If their scales are significantly different, we would have four different length scales to be resolved.

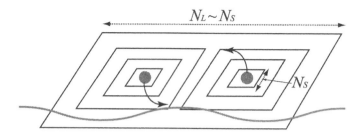

Fig. 3.7 Schematic figure of the structure of computational domains for simulating a binary system (denoted by the filled circles) in an AMR algorithm. The whole computational domain is composed typically of 10–15 subdomains each of which has a similar total grid number, denoted by N_S and $N_L \sim N_S$. The small high-resolution domains are prepared to resolve compact objects with a sufficiently small grid spacing while the large domains are prepared to follow the propagation of gravitational waves in the wave zone (see the thick wavy curve).

coordinates. Because of the facts $\ell \gtrsim 2R$ and $R > m_0$, the required value of N_g is larger than $\sim 10^3$. For following a binary inspiral motion from $\ell \sim 5R$, N_g has to be larger than 2×10^3. Even by supercomputers currently available for general users, it approximately takes at least a few months to perform a simulation when using such a huge grid number. This implies that it would not be feasible to perform a large number of simulations for a wide range of binary parameters in the uniform grid.

In the AMR algorithms (see figure 3.7 for a schematic figure of computational domains), one can change the grid spacing and the structure of the computational domains arbitrarily for different scales in order to guarantee the required grid resolutions for each scale and to significantly reduce the computational cost. For accurately resolving each compact object in a binary, we need to take a grid number of $N_S \sim 2R/\Delta x \gtrsim 100$ to cover the region in the vicinity of the compact objects. However for other regions, we do not have to take such small grid spacing, in particular, for a distant zone for which we need only to resolve gravitational-wave length. For following the propagation of gravitational waves in the distant zone, the required grid spacing is ~ 0.05–0.1λ which is by an order of magnitude larger than the grid spacing, Δx, required near the compact objects. Thus, by choosing an acceptably large grid spacing in the wave zone with the grid number $N_L \sim N_S \sim 100$, we can significantly save the grid number for covering a large computational region with $L \gtrsim \lambda$.

Furthermore, computational costs are significantly saved because we can employ a multi time-step-interval scheme in the AMR algorithm (see figure 3.8): The time-step interval for subdomains that cover distant zones with low grid resolutions can be taken to be much larger than the minimum time-step interval for the refined region, because their maximum permissible sizes are proportional to the grid spacing limited by the CFL condition (for each subdomain, not the time-step interval but

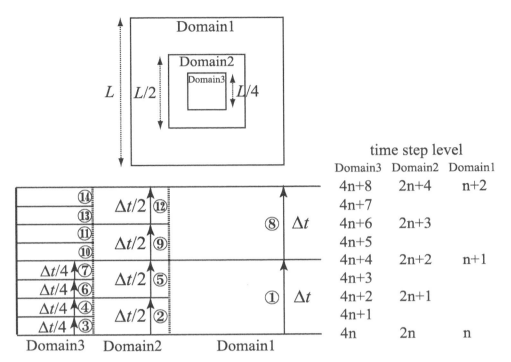

Fig. 3.8 Upper panel: Schematic figure of computational domains in an AMR scheme. In this example, the total computational domain is composed of three subdomains 1, 2, and 3 of size L, $L/2$, and $L/4$, respectively. Each subdomain is supposed to be covered by the same grid number, N, and thus, the grid spacing is $\Delta x = L/N$, $\Delta x/2$, and $\Delta x/4$, respectively. Lower panel: Time-step interval and order of time integration in an AMR algorithm. For the identical CFL number, the time-step interval is Δt, $\Delta t/2$, and $\Delta t/4$, respectively, where Δt denotes the largest interval. The numbers with the circle show the order of the time integration: First, the equations for subdomain 1 are evolved, second, the equations for subdomain 2 are evolved, third and fourth, the equations for subdomain 3 are evolved, etc. (see section 3.3.3 for details). The numbers in the right side, e.g., n, $n+1$, $n+2$, illustrate the time-step numbers of each subdomain with n an integer.

the CFL number can be identical; see section 3.1.4). A small time-step interval is necessary only in the vicinity of the compact objects. This implies that if we can save the total grid number for one axis direction by a factor of $F \sim N_g/N_S \sim 10$ compared with the uniform-grid case, the computational costs are saved by a factor of $F^4 \sim 10^4$, where the power, 4, comes not only from three spatial dimensions but also from time dimension. (For higher-dimensional spacetime, the factor of the saving is even larger.)

These reasons show that the AMR algorithms have to be employed. Indeed, all the major numerical-relativity groups have been employing them. In the following, we will outline typical AMR algorithms used in numerical-relativity communities [Schnetter et al. (2004); Brügmann et al. (2008); Yamamoto et al. (2008)].

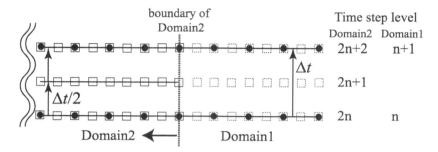

Fig. 3.9 Structure of the typical grid prolongation near an outer boundary of an AMR subdomain. Subdomains 1 and 2 are parent (larger domain size) and child (smaller domain size) subdomains, respectively. The filled circles and the open squares denote grid points of the parent and child subdomains, respectively. The dashed squares denote the prolonged grid points of the child subdomain, for which the data in general have to be determined by an interpolation procedure both in time and space. The numbers (n, $n+1$, $2n$, $2n+1$, and $2n+2$) in the right side illustrate the time-step numbers of each subdomain with n being an integer.

3.3.2 Spatial interpolation in the buffer zone

The most crucial process for implementing an AMR algorithm is to supply quantities in the vicinity of outer-boundary regions of each subdomain (in the following, we will call such regions the refinement boundaries or buffer zones). Here, the buffer zones are prolonged regions just outside the regular computation region, and usually they are not composed of single surface but a series of surfaces. The data in these zones have to be supplied from their "parent" subdomain by an interpolation procedure. Here, the parent subdomain completely encloses its "child" subdomain including the buffer zone: See, e.g., the upper panel of figure 3.8. For this figure, the parents of subdomains 2 and 3 are subdomains 1 and 2, respectively. On the other hand, the subdomains 2 and 3 are the children of the subdomains 1 and 2, respectively. The subdomains 2 and 3 are completely enclosed in the subdomains 1 and 2.

A method for supplying the data in the buffer zones can be understood by figure 3.9. In this figure, we display a spacetime structure of only two subdomains, a parent (subdomain 1) and a child (subdomain 2), for simplicity. In particular, we focus on a region near one of buffer zones of subdomain 2. The filled circles and the open squares denote grid points of the parent and child subdomains, respectively. The dashed squares in figure 3.9 denote the prolonged grid points of the child's buffer zone, which are necessary for evolving geometric quantities near the outer boundaries of the regular zone of subdomain 2 (i.e., in the left-hand side of the vertical dashed line). In the AMR algorithms, it is crucial to appropriately supply the data in the buffer zone by interpolating the data on the grid points of the parent subdomain *both in time and space* as illustrated in figure 3.9 (see section 3.3.3 for interpolation and integration schemes in time for the buffer zone).

Before going ahead, several words of explanation about the method for setting

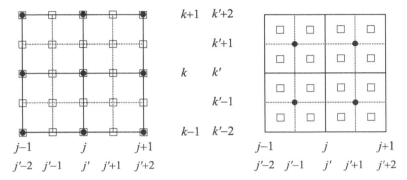

Fig. 3.10 Vertex grid (left) and cell-centered grid (right) in the case of two spatial dimensions. The squares and the circles denote the grid points for child and parent subdomains, and the dotted and solid lines denote coordinate lines for the child and parent subdomains, respectively.

grid points are necessary: There are two popular ways for preparing cells and grid points. One way is to set a vertex grid point (see the left panel of figure 3.10), in which the vertex points are assigned as the grid points; e.g., for the x-axis direction, the grid points are located at

$$x_j^p = j\Delta x_p, \quad \text{for the parent subdomain,}$$

$$x_{j'}^c = j'\frac{\Delta x_p}{2}, \quad \text{for the child subdomain,}$$

where j and j' denote a series of integer, and Δx_p is the grid spacing of the parent subdomain. The other popular option for the choice of grid points is the so-called cell-centered grid (see the right panel of figure 3.10), in which the grid points are assigned at

$$x_j^p = \left(j - \frac{1}{2}\right)\Delta x_p, \quad \text{for the parent subdomain,}$$

$$x_j^c = \left(j' - \frac{1}{2}\right)\frac{\Delta x_p}{2}, \quad \text{for the child subdomain.}$$

In the vertex grid, the grid points of the parent subdomain are overlapped with the grid points of the corresponding child subdomain. For this grid configuration, the interpolation and restriction processes (see section 3.3.4) are fairly easily executed. Thus, in the following, we will describe an AMR scheme assuming to employ the vertex grid.

In reality there are multi spatial dimensions. For the three spatial dimensions case, for example, there are four types for the grid location in the vertex grid setting. Denoting the grid points of parent and child subdomains by (x_j^p, y_k^p, z_l^p) and $(x_{j'}^c, y_{k'}^c, z_{l'}^c)$, the four types are classified as follows:

(1) All three coordinates agree (the grid points of the parent and child subdomains overlap each other): $x_j^p = x_{j'}^c$, $y_k^p = y_{k'}^c$, and $z_l^p = z_{l'}^c$. In this case, any

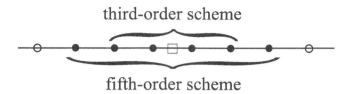

Fig. 3.11 Lagrangian interpolation scheme. For the interpolation at the position of the open square, six and four filled circles, respectively, are used in the fifth-order and third-order Lagrangian interpolation schemes for the choice of the uniform grid.

interpolation is not necessary for supplying the data in the buffer zone of the child subdomain.

(2) Two of three coordinates agree: $x_j^p = x_{j'}^c$, $y_k^p = y_{k'}^c$, and $z_l^p \neq z_{l'}^c$ or $x_j^p = x_{j'}^c$, $y_k^p \neq y_{k'}^c$, and $z_l^p = z_{l'}^c$ or $x_j^p \neq x_{j'}^c$, $y_k^p = y_{k'}^c$, and $z_l^p = z_{l'}^c$. In this case, for the disagreed direction, an interpolation procedure is necessary in the buffer zone of the child subdomain.

(3) Only one of three coordinates agrees: $x_j^p = x_{j'}^c$ or $y_k^p = y_{k'}^c$ or $z_l^p = z_{l'}^c$. In this case, for the disagreed two directions, an interpolation procedure is necessary.

(4) All three coordinates disagree: In this case, for all the three directions, an interpolation procedure is necessary.

We note that for the cell-centered grid, an interpolation procedure is necessary for all the grid points in the buffer zone of child subdomains.

The interpolation in the buffer zones is usually performed using a Lagrangian interpolation scheme. Thus, a fitting formula is constructed in terms of a polynomial function using the data on the grid points of the corresponding parent subdomain; e.g., the interpolation along the x direction for Q is done by

$$Q = \sum_{l=0}^{p} a_l x^l, \tag{3.76}$$

where a_l are coefficients, and p denotes the order of accuracy which depends on the scheme for solving Einstein's evolution equations. For example, when the finite-differencing scheme is fourth-order accurate, the interpolation scheme should be more than fourth-order accurate for preserving the fourth-order accuracy. In the vertex grid, an interpolation is necessary if the corresponding grid point of the child subdomain is in the middle of the grid points of the parent subdomain (see figure 3.11). If we employ a symmetric interpolation scheme, we have to employ even-number grid points of the parent subdomain for the Lagrangian interpolation. For preserving more than fourth-order accuracy, we have to employ at least fifth-order Lagrangian interpolation scheme with $p \geq 4$, and for performing the symmetric interpolation, we have to employ six grid points to determine six coefficients, a_l ($l = 0$–5: fifth-order scheme). If we employed only four grid points (i.e., $p = 3$) of the parent subdomain, the accuracy is third-order and such an interpolation scheme is available only for the case that a second-order scheme is employed for solving Einstein's evolution equations.

3.3.3 Time integration scheme in the buffer zone

In addition to the interpolation procedure for the spatial direction, we have to assign the data in the buffer zone for intermediate time steps. As figures 3.8 and 3.9 illustrate, the time-step interval of the child subdomain is typically half of the corresponding parent-subdomain's interval. This implies that the data for the buffer zone in the middle of two time-step levels of the parent subdomain have to be supplied by some way; e.g., the data of the dashed open squares at the time-step level of $\Delta t/2$ in figure 3.9 have to be assigned by an interpolation scheme in time. We furthermore need interpolation procedures in time for several *substep* levels because we usually employ a higher-order time integration scheme in which one time-step integration (for Δt) is composed of multi-substep procedures (see section 3.1.2: for example, there are four substep levels in the fourth-order Runge–Kutta scheme). There are several options for determining the data in these portions of the buffer zone.

Before describing popular schemes in numerical relativity, we outline the general prolongation procedures necessary for the buffer zone, assuming that we use a fourth-order Runge–Kutta scheme for the time integration, a fourth-order centered spatial finite-differencing scheme [see equations (3.8)–(3.10)] for evaluating spatial derivative terms except for the advection term, and a fourth-order lopsided upwind scheme for the advection term [see equation (3.74)] when solving Einstein's evolution equations. In the following, we will focus only on the buffer zone along the x-axis, because for other axis directions, the procedure is essentially the same. Below, the x-coordinate of the grid points is denoted by $x_j = x_0 + j\Delta x$ where j is a series of integer.

Because the fourth-order lopsided upwind scheme is supposed to be employed, the data at grid points $j - 3$, $j - 2$, $\cdots$, $j + 2$, and $j + 3$ could be necessary for evaluating the spatial derivatives with respect to x at a grid point j at each time-step level, although only one of two grid points of $j \pm 3$ is necessary as the shift vector points to a certain direction. In a fourth-order Runge–Kutta scheme, four procedures for four substeps are executed for one time-step interval Δt; see figures 3.12 and 3.13. If we do not want to do any interpolation in time in Δt, we have to supply the data at twelve grid points in the buffer zone before the first Runge–Kutta integration, because at each Runge–Kutta procedure, three grid points both in the left and right sides around the corresponding grid point j could be used; see the caption of figure 3.12 for details. On the other hand, if some interpolation in time is employed for substeps in the time-step interval Δt, we will be able to save the number of grid points in the buffer zone (see below).

Specifically, there are two popular schemes in numerical relativity. The scheme

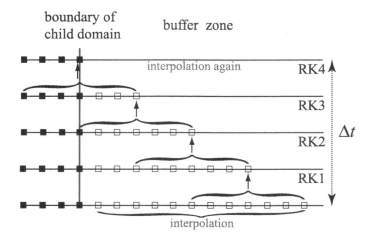

Fig. 3.12 Schematic figure of a time-integration scheme for the buffer zones in an AMR algorithm. The filled and open squares denote the regular grid points in a subdomain and the grid points in its buffer zone, respectively. In this scheme, twelve grid points are prepared in the buffer zone, and the data for these grid points are determined by an interpolation scheme using the data of its parent subdomain before the first substep of the Runge–Kutta procedure. At the first Runge–Kutta procedure, the data for the inner nine grid points in the buffer zone can be evolved without any special treatment in the fourth-order Einstein's equation solver. For other three outer grid points, however, it is not possible to integrate the basic equations because of the lack of the buffer-zone points, and hence, the data for such grid points are subsequently discarded. In the same way, the Runge–Kutta procedure is repeated for the next three substeps discarding three outer grid points at each substep. Then, the data for the regular zone (at the filled squares) can be evolved without any special treatment, although no data in the buffer zone can be evolved in the end. Note that the substep levels of the Runge–Kutta time integration, denoted by RK1–RK4, do not reflect the exact physical time levels.

shown in figure 3.12 was originally implemented in CARPET code, developed by Schnetter *et al.* (2004). In this scheme, before the first Runge–Kutta substep, the data at twelve grid points in the buffer zone are determined by a spatial interpolation using the data of its parent's subdomain. Then, the four-step Runge–Kutta time integration is fully performed during the time-step interval Δt. Specifically, at the first, second, and third Runge–Kutta procedures, the data at nine, six, and three grid points in the buffer zone are evolved as shown in figure 3.12. In this case, a spatial interpolation is needed only before the first Runge–Kutta substep. For this scheme, a time-interpolation procedure is necessary only when the corresponding time-step levels do not agree with its parent's time-step level; they are in the middle of its parent's time-step level (see the child's time-step level $2n+1$ in figure 3.14). In this time-step level, the interpolation in time is performed using the data of the parent subdomain at nearby several time-step levels. As described in figure 3.8, the time integration for the data of a parent subdomain is always performed *before* the time integration for its child subdomain. Thus, for determining the data at $2n+1$ time-step level for the child subdomain, the data at $n+1$, n,

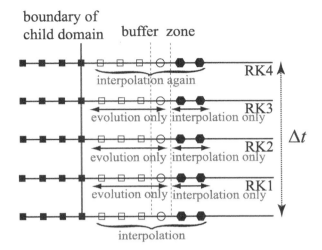

Fig. 3.13 Schematic figure for another time integration scheme for the buffer zones in an AMR algorithm which is different from that shown in figure 3.13. The filled squares denote the regular grid points in a subdomain, and the open squares, open circles, and filled hexagons are the grid points in its buffer zone, respectively. There are three types of the grid points in the buffer zone, for each of which different integration (and evolution) schemes are employed (see the text for details). Note again that the substep levels of the Runge–Kutta time integration, denoted by RK1–RK4, do not reflect the exact physical time levels.

··· time-step levels of the parent subdomain are used. For example, if one strictly requires the fourth-order accuracy in time, the data of the parent's subdomain at $n+1$, n, $n-1$, $n-2$, and $n-3$ time-step levels have to be used, storing the data at n, $n-1$, $n-2$, and $n-3$ time-step levels. For the case that the second-order accuracy is required, only the data at $n+1$, n, and $n-1$ time-step levels are necessary.

There is the other popular AMR scheme in numerical relativity developed by Brügmann et al. (2008) and Yamamoto et al. (2008) (BAM and SACRA codes, respectively). In this scheme, we need to prepare only six grid points along each axis direction in the buffer zone, as illustrated in figure 3.13. As in the scheme for CARPET code described above, a spatial interpolation procedure has to be performed before the first Runge–Kutta substep. During a series of the Runge–Kutta procedures, no interpolation is performed for four inner grid points in the buffer zone neighboring the boundary of the regular domain (see the open squares and the open circles of figure 3.13). Thus, the data for these grid points are determined by solving field equations as done for the data in the regular zone. Here, for evolving the data at the fourth grid point (denoted by the open circles), one grid point in the buffer zone may lack, if the shift vector points to a positive direction (for this location of the buffer zone). For such case, the lopsided upwind finite differencing is evaluated by the second-order upwind scheme [see equation (3.73)]. This prescription was proposed first by Brügmann et al. (2008). On the other hand, if the shift vector

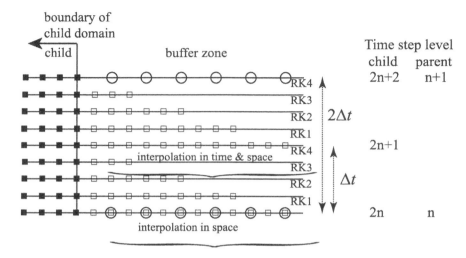

Fig. 3.14 More details about the AMR scheme shown in figure 3.12. The filled and open squares denote the regular grid points in a child subdomain and the grid points in its buffer zone, respectively, and the open circles denote the regular grid points in its parent subdomain. The interpolation in space is executed before the first Runge–Kutta substeps (or after the last Runge–Kutta substeps). The interpolation in time is executed at a time-step level (at $2n+1$ in this figure) at which its parent subdomain does not have the regular data. The numbers in the right side ($2n$, $2n+1$, $2n+2$, and n, $n+1$) illustrate the time-step levels for child and parent subdomains, respectively.

points to the opposite direction, the fourth-order lopsided scheme is used. One may think that the numerical accuracy would be deteriorated by this prescription. However, numerical experiments have shown that the deterioration due to this prescription is quite minor and the fourth-order accuracy is approximately preserved (see section 3.4 for demonstration).

The remaining two outer points (denoted by the hexagons in figure 3.13) are always determined purely by an interpolation procedure in time. This interpolation is performed using the data in its parent's subdomain. If we wanted to strictly preserve the fourth-order accuracy in time, we would have to store the data for four previous time-step levels, as we already mentioned. Brügmann et al. (2008); Yamamoto et al. (2008), however, employed a second-order accurate interpolation scheme for these grid points. In this case, the interpolation is performed using the data only for three time-step levels (for $2n+1$ time-step level of the child subdomain, $n+1$, n, and $n-1$ time-step levels of the parent subdomain are used; see figure 3.14). Numerical experiments also have shown that the deterioration caused by this is minor.

In section 3.4, we will present results of testbed simulations in numerical relativity for vacuum spacetime using the **SACRA** code in which the AMR scheme of Yamamoto et al. (2008) is implemented. The numerical results will indeed show that the six-points-buffer-zone method may be used for approximately preserving the fourth-order accuracy.

3.3.4 Restriction

In any AMR algorithm, a child subdomain have to be completely enclosed by its parent subdomain. Then, the data for the region overlapped by the child and parent subdomains should agree with each other. Here, for the overlapped region, the accuracy of the data in the child subdomain is better than that in the parent subdomain. Thus, the data in the child subdomain should be copied to those in the parent subdomain. This procedure is called *restriction*.

The restriction is always performed whenever the time-step level of the child subdomain agrees with that of its parent. For example, in figure 3.14, the time-step levels of them agree at $2n$ and $2n+2$ time-step levels of the child subdomain (at n and $n+1$ time-step levels of the parent subdomain). For such time-step levels, the restriction should be executed.

In the vertex grid, the grid points in the parent subdomain overlap with the corresponding grid points of the child subdomain for the overlapped region. Thus, the restriction procedure is done by a simple copying. In the cell-centered grid, on the other hand, an interpolation procedure (with a required accuracy) has to be done, because the grid points in the parent subdomain do not overlap with any of the child subdomain; see, e.g., figure 3.10.

3.3.5 Kreiss – Oliger dissipation

As we described in the previous two subsections, the data in the buffer zone are determined basically by an interpolation procedure in space and time. In addition, some of the data in the parent subdomains are determined by a restriction procedure. For these processes, a high-frequency numerical noise is often introduced in the numerical data. Usually, this does not give any serious damage in the numerical stability. However, the numerical data are often appreciably contaminated by the high-frequency noise. This could be a serious problem when one wants to accurately extract gravitational waves in the wave zone.

A popular prescription to suppress such high-frequency noise is to add an artificial dissipation term without reducing the order of the numerical accuracy. Here, as the most popular one, we introduce the Kreiss–Oliger-type dissipation term supposing that a fourth-order finite differencing scheme is employed in the numerical simulation.

Remember that in fourth-order accurate computations, one may add any term which is more than fifth-order (as far as the numerical stability is maintained). Then, the Kreiss–Oliger-type dissipation scheme recommends to modify geometric quantities as

$$Q_l \to Q_l - \sigma_{\mathrm{KO}}(\Delta x)^6 Q_l^{(6)}, \qquad (3.77)$$

where Q_l is a quantity in the l-th refinement level in an AMR algorithm, $Q_l^{(6)}$ is the sum of the sixth-order partial derivative along the x, y, and z axis directions, and σ_{KO} is a constant of $\sim 10^{-3}$. In the uniform grid, $Q_l^{(6)}$ at (x_0, y_0, z_0) is written as

$$Q_l^{(6)} = \frac{Q_l(x_3) + Q_l(x_{-3}) - 6[Q_l(x_2) + Q_l(x_{-2})] + 15[Q_l(x_1) + Q_l(x_{-1})] - 20 Q_l(x_0)}{\Delta x^6}$$
$$+ \text{ contribution from } y \text{ and } z \text{ directions}, \qquad (3.78)$$

where $x_k = x_0 + k\Delta x$ with $-3 \le k \le 3$.

3.4 Testing numerical-relativity code: Vacuum spacetime

When one develops a new numerical-relativity code, she/he first has to confirm the reliability to it. For this task, one of the best ways is to examine the convergence property of numerical solutions obtained by the new code, changing the grid resolution for a wide range. This method will be described in section 3.4.1.

The other way is to perform a simulation for a problem for which an analytic or semi-analytic solution is known and to confirm that the numerical solution agrees with the analytic or semi-analytic solution. There are several known analytic solutions for vacuum spacetime. One is for the propagation of (linear) gravitational waves of small amplitude in vacuum spacetime. The axisymmetric solution for this in four-dimensional spacetime is called Teukolsky waves [Teukolsky (1982)] [for the non-axisymmetric solution, see Nakamura (1984) and section C.2]. By a test simulation for reproducing this solution, we can examine the reliability of a numerical code about whether it can follow the propagation of gravitational waves and we can extract gravitational waves in the wave zone. Another solutions are for stationary black holes (Schwarzschild or Kerr black holes) and those for perturbed black holes. By the test simulations using these solutions, we can confirm that the code can evolve highly nonlinear spacetime. In sections 3.4.2 and 3.4.3, we will present some numerical results for these test problems.

3.4.1 *Examining convergence property*

The convergence test is a robust method not only for examining the reliability of a numerical code but also for understanding the accuracy of a numerical solution. The code in numerical relativity has usually constructed employing a second-order or fourth-order accurate scheme (besides deterioration by a minor source such as outer boundary conditions). This implies that the error of any quantity should be of order $(\Delta x)^p$ where Δx is the grid spacing, and $p = 2$ and 4 for the second-order and fourth-order accurate schemes, respectively. Specifically, for a quantity Q, the numerical result should behave as

$$Q_{\text{num}} = Q_0 + Q_p(\Delta x)^p + O\left[(\Delta x)^{p+1}\right], \tag{3.79}$$

where Q_{num} is a numerical solution and Q_0 is the (unknown) exact solution. It is an invaluable test to examine whether numerical solutions obtained in different grid-resolution runs indeed satisfy this predicted relation.

The convergence relation (3.79) can be also used for deriving the unknown exact solution, Q_0, from a series of numerical solutions. In the following, we will focus on a fourth-order accurate scheme and illustrate the procedure for this. In the fourth-order scheme, any quantity obtained in different grid-resolution runs should behave as equation (3.79) with $p = 4$. Here, we do not know Q_0 and Q_4. Moreover, we have to confirm that the numerical solutions indeed satisfy equation (3.79) because it is not trivially satisfied: We should not consider a priori that the power of Δx would be four. Hence, there are three unknown quantities (neglecting the higher-order error). Thus, we have to perform more than three simulations employing more than three grid resolutions. For such a series of the simulations, we have the simultaneous equations

$$Q_{\text{num}}(l) = Q_0 + Q_p(\Delta x_l)^p, \quad l = 1, 2, 3, \cdots, \tag{3.80}$$

where $Q_{\text{num}}(l)$ is the numerical result for $\Delta x = \Delta x_l$. From the three numerical results, we can determine Q_0, Q_p, and p. If we obtain more than four numerical results, we can further examine the degree of for the convergence property of equation (3.79), e.g., dispersion of the value of p can be estimated.

By contrast, it is not possible to derive Q_0, Q_p, and p only from two numerical solutions with two different grid resolutions, unless we assume the value of p as four. By one simulation, nothing is determined, but for Q_{num} which does not have clear physical meaning: It should be stressed that our purpose is not to derive Q_{num} but to derive Q_0. Simply obtaining the quantity Q_{num} is meaningless at all. Therefore, we have to perform at least two simulations with two different grid resolutions; but this is acceptable only when we have already confirmed the convergence property of a given code and a given problem. Strictly speaking, we have to always perform more than three simulations with more than three grid resolutions.

3.4.2 *Propagation of linear gravitational waves*

In this section, we describe a test simulation using a linear gravitational-wave solution written by equation (C.59) in appendix C.2, restricting our attention to the four-dimensional spacetime. In this solution, at $t = 0$, the perturbation for the conformal three-metric is zero, and hence, the three metric is written in a conformally flat form as $\gamma_{ab} = \psi^4 f_{ab}$ (see section 2.1.7). In addition, the trace and divergence of the extrinsic curvature are zero because of the transverse and tracefree nature of gravitational waves. This implies that the momentum constraint is automatically satisfied at $t = 0$ in the linear approximation level. Also, as far as the amplitude of gravitational waves is small, we may approximately set $\psi = 1$. Thus, the constraint equations are trivially satisfied at $t = 0$.

The linear gravitational-wave solution is derived in the assumption that $\alpha - 1$, β^k, and the trace parts of $\gamma_{ij} - f_{ij}$ and K_{ij} (i.e., ψ and K) are nonlinear quantities and their amplitude is negligible. In order for the gravitational-wave solution in appendix C.2 to be an approximate solution of Einstein's evolution equations, we have to employ a class of gauge conditions, in which the nonlinear nature of these quantities is guaranteed (i.e., their amplitude can be negligible). The simplest choice is the geodesic gauge $\alpha = 1$ and $\beta^k = 0$ (see section 2.2). However, as shown in section 2.2.1.1, this gauge does not enable us to perform a long-term stable numerical simulation. For simplicity, the zero shift gauge condition may be chosen, but then, the condition for α has to be a nontrivial one. As already described in section 2.2, several good slicing conditions such as maximal slicing (see section 2.2.2), harmonic slicing (see section 2.2.3), and dynamical slicing (see section 2.2.6) have been known. In all these conditions, the non-linearity of $\alpha - 1$ for the linear gravitational-wave solution is guaranteed. Indeed in the linear theory,

$$\Delta_f \alpha = 0, \quad \text{for maximal slicing,} \tag{3.81}$$

$$\partial_t \alpha = -\eta_\alpha K, \quad \text{for harmonic and dynamical slicing,} \tag{3.82}$$

with the evolution equation for K being

$$\partial_t K = -\Delta_f \alpha, \tag{3.83}$$

where $\eta_\alpha = 1$ (harmonic slicing) or 2 (dynamical slicing). Thus, in the test simulations, it is recommendable to choose one of these slicing conditions. Here, we employ the dynamical slicing condition for the test simulation.

As we described above, the test simulation is performed for the initial condition

$$\tilde{\gamma}_{ij} = f_{ij}, \quad \psi \approx 1, \quad K = 0,$$

and

$$\hat{A}_{ij} = \sum_{l,m} \begin{pmatrix} a_l Y_{lm} & b_l \partial_\theta Y_{lm} & b_l \partial_\varphi Y_{lm} \\ * & r^2(-a_l Y_{lm}/2 + f_l W_{lm}) & r^2 f_l X_{lm} \\ * & * & r^2(-a_l Y_{lm}/2 - f_l W_{lm})\sin^2\theta \end{pmatrix}, \tag{3.84}$$

where a_l, b_l, and f_l are functions of r ant t, which are written as

$$a_l = r^{l-2}\left(\frac{1}{r}\frac{\partial}{\partial r}\right)^l \frac{Q(t-r) + Q(t+r)}{r}, \quad b_l = \frac{1}{l(l+1)}\frac{1}{r}\partial_r(r^3 a_l),$$

$$f_l = \frac{1}{(l-2)(l+1)}\left[-\frac{1}{2}a_l + \partial_r b_l + \frac{2}{r}b_l\right], \tag{3.85}$$

and Q is a regular arbitrary function. One of the simplest choices is $Q(r) = Ae^{-r^2/(2r_0^2)}$ where r_0 is the length unit of the wave packet and A denotes the amplitude with $|A| \ll 1$ for linear gravitational waves. Setting $r_0 = 1$ (fixing the unit of the length as unity) for simplicity, the components of $\hat{A}_{ij}$ in Cartesian coordinates are written as

$$\hat{A}_{xx} = A_{12}e^{-r^2/2}\left[12 - 8y^2 - 16z^2 + z^2(x^2 + 3y^2 + 2z^2) + y^2(-x^2 + y^2)\right],$$
$$\hat{A}_{yy} = A_{12}e^{-r^2/2}\left[-12 + 8x^2 + 16z^2 - z^2(3x^2 + y^2 + 2z^2) + x^2(-x^2 + y^2)\right],$$
$$\hat{A}_{zz} = A_{12}e^{-r^2/2}\left(x^2 - y^2\right)\left(-8 + x^2 + y^2 + 2z^2\right),$$
$$\hat{A}_{xy} = A_{12}e^{-r^2/2}\,xy(x^2 - y^2),$$
$$\hat{A}_{xz} = A_{12}e^{-r^2/2}\,xz(12 - 2r^2 + x^2 - y^2),$$
$$\hat{A}_{yz} = A_{12}e^{-r^2/2}\,yz(-12 + 2r^2 + x^2 - y^2),$$

where $A_{12} = A/12$.

Figure 3.15 displays results of a testbed simulation for $A = \sqrt{2\pi/15} \times 10^{-6}$. In this simulation, SACRA code implementing a BSSN formulation and an AMR scheme was used [Yamamoto et al. (2008)] (see also section 3.3). Note that the SACRA code employs fourth-order accurate schemes in time and space for the major parts of the finite-differencing. The simulation was performed preparing four subdomains with different domain sizes but with the same number of the grid points labeled by N where N is an even integer. Each subdomain covers a region with $-L_i \leq x \leq L_i$, $-L_i \leq y \leq L_i$, and $0 \leq z \leq L_i$, where $i = 0, 1, 2,$ and 3 with $L_i = 19.2/2^i$ and thus the grid spacing in each subdomain is $\Delta x_i = 19.2/(2^i N)$. Note that the equatorial-plane symmetry is imposed. The CFL number is set to be 0.5 for the finer subdomains with $i \geq 2$ but for $i \leq 1$, the time-step interval Δt is set to be equal to that for $i = 2$ because the large CFL number often causes numerical instability around outer boundaries of the largest-size subdomain ($i = 0$) for which the fourth-order accuracy is not preserved due to the choice of an outer boundary condition (see section 3.1.3). The convergence of the numerical results is examined by varying N as 24, 30, and 36 in this test.

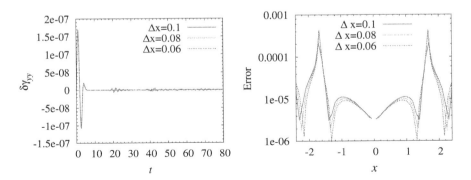

Fig. 3.15 Left: $\delta\gamma_{yy} = \gamma_{yy} - 1$ extracted at $x = 1.2$ for $N = 24$, 30, and 36 (the grid spacing for the smallest subdomain is $\Delta x = 0.1$, 0.08, and 0.06, respectively). Right: Errors of numerical results of $\gamma_{yy} - 1$ measured at $t = 2.4$ for $-2.4 \leq x \leq 2.4$ for $\Delta x = 0.1$, 0.08, and 0.06. The errors are plotted by multiplying $(0.06/\Delta x)^4$ to align the three curves.

The left panel of figure 3.15 shows $\gamma_{yy} - 1$ extracted at $x = 1.2$ for three different grid resolutions. During the propagation of the gravitational-wave packet, the metric oscillates but after it goes away, the amplitude settles approximately to zero. Because of the reflection of the wave at refinement boundaries and back to the inner subdomain, the gravitational-wave amplitude a bit deviates from zero in the late time ($t > 10$). However, this reflection is suppressed with improving the grid resolution. Because an appropriate formulation, appropriate gauge conditions, and an appropriate numerical scheme are employed, the simulation would be able to continue stably for a long time scale (although the simulations were stopped at $t \sim 100$ to save the computational time).

The right panel of figure 3.15 shows the error of $\gamma_{yy}-1$ along the x-axis measured at $t = 2.4$ for three different grid resolutions. The error is defined by

$$\left| 1 - \frac{(\gamma_{yy})_{\text{numerical solution}} - 1}{(\gamma_{yy})_{\text{analytic solution}} - 1} \right|. \tag{3.86}$$

Here, the analytic solution implies the exact solution for linearized Einstein's equation. The errors are plotted by multiplying $(0.06/\Delta x)^4$, and as a result, three curves approximately align. This implies that the fourth-order convergence is approximately achieved.

Before closing this section, we note that the simulation for non-linear gravitational waves can be also performed using the similar setting to that described in this section. For the case that the amplitude of gravitational waves is not small, we have to take into account the nonlinear effects when constructing the initial data. This can be easily incorporated to the momentum constraint for the conformally flat geometry with $K = 0$, by substituting the linear solution into $\hat{A}^a{}_b := \psi^6 \tilde{A}^a{}_b$ where $\tilde{A}_{ab}$ is the conformal tracefree extrinsic curvature [see equation (2.96)]. This is because for $K = 0$ and for the conformally flat geometry, the momentum constraint is written as $D_a \overset{(0)}{\hat{A}}{}^a{}_b = 0$. The remaining task is to satisfy the Hamiltonian constraint, which has to be solved at $t = 0$. For this, we should solve the equation for the conformal factor with a given function of $\hat{A}_{ab}$ as shown in equation (2.95) [see also Nakamura et al. (1987); Shibata and Nakamura (1995)].

3.4.3 *Evolution of black holes*

In this section, we illustrate the results of several testbed problems associated with black holes. In the first three subsections, we describe popular testbed problems for four-dimensional spacetime. In section 3.4.3.4, we introduce test problems for five-dimensional black-hole spacetime.

3.4.3.1 *Non-spinning black hole*

Evolving a non-spinning black hole is one of the simplest but robust testbed simulations. As described in section 2.2, for non-spinning black-hole spacetime, we know that spatial hypersurfaces eventually reach the limit hypersurface when we employ maximal slicing or dynamical slicing conditions [Estabrook et al. (1973); Hannam et al. (2007); Brügmann et al. (2008); Baumgarte and Naculich (2007)] (this property is qualitatively unchanged even for higher-dimensional black holes [Nakao et al. (2009)]). The metric in the limit hypersurface is written analytically if $K = 0$ is satisfied, and by comparing the numerical metric for a relaxed stationary state reached after a sufficiently long-term simulation with the analytic solution, the validity of the numerical code can be examined for highly nonlinear spacetime.

Baumgarte and Naculich (2007) derived an analytic solution of the limit hypersurface for a black hole of mass M in isotropic coordinates as

$$ds^2 = -\alpha^2 dt^2 + 2\beta_r dr dt + \psi^4 \left(dr^2 + r^2 d\Omega\right), \tag{3.87}$$

where

$$\alpha = \left(1 - \frac{2M}{R} + \frac{27M^4}{16R^4}\right)^{1/2}, \quad \beta^r = \frac{3\sqrt{3}M^2}{4\psi^6 r^2}, \quad \psi = \sqrt{\frac{R}{r}},$$

$$K_{rr} = -\frac{3\sqrt{3}M^2 \psi^4}{2\alpha R^3} \left(= -\frac{1}{2r^2}K_{\theta\theta} = -\frac{1}{(2r\sin\theta)^2}K_{\varphi\varphi}\right), \tag{3.88}$$

and R is the Schwarzschild radial coordinate related to r by

$$r = \left[\frac{2R + M + \sqrt{4R^2 + 4MR + 3M^2}}{4}\right]$$

$$\times \left[\frac{(4 + 3\sqrt{2})(2R - 3M)}{8R + 6M + 3\sqrt{8R^2 + 8MR + 6M^2}}\right]^{1/\sqrt{2}}. \tag{3.89}$$

In this testbed simulation, we prepare this exact solution and evolve it using an appropriate gauge condition. A simple gauge is a version of the dynamical gauge condition in the form (see section 2.2.6)

$$\partial_t \alpha = -2\alpha K, \tag{3.90}$$

$$\partial_t \beta^i = \frac{3}{4}B^i, \quad \partial_t B^i = (\partial_t - \beta^k \partial_k)\tilde{\Gamma}^i - \eta_B B^i, \tag{3.91}$$

with $B^i = 0$ at $t = 0$ and $\eta_B = 1/M$. Here, we suppose to use a BSSN formulation in this test. For this form, the metric of the exact solution remains unchanged approximately except for the truncation error associated with finite differencing. Thus, by comparing the numerical solution obtained after long-term evolution with the exact solution described above, we can find how accurate the numerical solution is.

Figure 3.16 displays the results of the comparison between numerical solutions of and exact solutions of α and $W(= \psi^{-2})$ along the x-axis. This figure plots the absolute values for the error of the numerical solutions defined by

$$\left|1 - \frac{\text{numerical solution}}{\text{exact solution}}\right|. \tag{3.92}$$

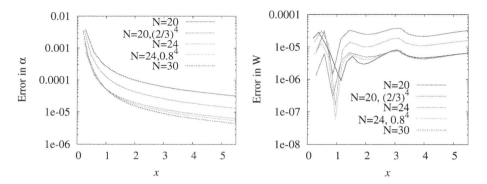

Fig. 3.16 Numerical errors for α (left) and W (right) measured at $t = 115.2M$ for a non-spinning black hole at the limit hypersurface. The results for three different grid resolutions are plotted. The error multiplied by $(N/30)^4$ as well as the absolute values of the error are plotted.

In this simulation, SACRA code implementing a BSSN formulation and an AMR scheme is used [Yamamoto et al. (2008)], preparing eight subdomains with different domain sizes but with the same number of the grid points labeled by N. Each subdomain covers a region with $-L_l \leq x \leq L_l$, $-L_l \leq y \leq L_l$, and $0 \leq z \leq L_l$, where $l = 0, 1, 2, \cdots$, and 7 with $L_l = 184.32M/2^l$ and $L_7 = 1.44M$. The grid spacing of each subdomain is $\Delta x_l = 184.32/(2^l N)$ and $\Delta x_7 = 0.06M \times (24/N)$. The CFL number is set to be 0.5 for the finer subdomains with $l \geq 3$ but for $l \leq 2$, the timestep interval Δt is set to be equal to that for $l = 3$. The reason for this treatment is the same as that for the propagation of linear gravitational waves. The convergence property of the numerical results is examined by varying N as 20, 24, and 30.

Figure 3.16 shows that the error is much smaller than 1% even with a small grid number $N = 20$, and furthermore, the fourth-order convergence is approximately achieved for the region far outside the horizon which is located at $x \approx 0.78M$. By contrast, the convergence is lost inside the horizon, although the error level is still small. This is reasonable, because at the center, the regularity is lost in the presence of a puncture. However, it is also confirmed that this irregular behavior does not give any serious damage for the region outside the horizon, because the information does not escape from its inside. Therefore, by this testbed simulation, we can confirm not only that the code can accurately follow black-hole spacetime for a long time scale but also that the coordinate singularity located inside the horizon does not matter for the evolution of its outside.

3.4.3.2 *Spinning black hole*

Evolving spinning black holes in general requires a more grid resolution than evolving non-spinning one for a given value of the black-hole mass and for a fixed gauge condition, because the area of the spinning black hole is smaller than that of the non-spinning one (see section 1.3.2). In particular, for the extremely spinning limit,

$\hat{a} \to 1$, the area steeply decreases, and hence, a high grid resolution is in general required. Here, $\hat{a}$ denotes the dimensionless spin parameter; see section 1.3.1. When we employ quasi-isotropic coordinates (see section 1.3.1) as is often the case in numerical relativity, this fact becomes even more severe, because the coordinate radius of the event horizon approaches zero for $\hat{a} \to 1$ as the horizon radius is $M\sqrt{1-\hat{a}^2}/2$ (unless a special coordinate system is employed [Liu et al. (2010)]). In numerical relativity, a rapidly spinning black hole is often formed after gravitational collapses and after the merger of binary neutron stars and black hole-neutron star binaries. Thus, it is important to know whether a new code can accurately evolve rapidly spinning black holes for a long time scale and to examine how fine grid resolution is necessary for evolving them. In the following, we will show that typical numerical-relativity codes can evolve rapidly spinning black holes stably, although the required grid resolution is very high when $\hat{a}$ approaches unity.

The testbed simulations were again performed by the **SACRA** code [Yamamoto et al. (2008)] preparing eight AMR subdomains. Each subdomain covers a region $-L_l \leq x \leq L_l$, $-L_l \leq y \leq L_l$, and $0 \leq z \leq L_l$, where $l = 0, 1, 2, \cdots$, and 7 with $L_l = 138.24M/2^l$ and $L_7 = 1.08M$, and the grid spacing of each subdomain is $\Delta x_l = 138.24/(2^l N)$ and $\Delta x_7 = 0.045M \times (24/N)$ with $N = 24$, 30, and 36. The CFL number is set to be 0.5 for the finer subdomains with $l \geq 3$ but for $l \leq 2$, the time-step interval Δt is set to be equal to that for $l = 3$. A dynamical gauge condition (the version preserving the advection term associated with the shift vector) was employed for the evolution of the lapse function and the shift vector (see section 2.2.6).

Figure 3.17 plots the numerical evolution of the area and axis ratio $C_\mathrm{p}/C_\mathrm{eq}$ of black-hole apparent horizons for the spin parameters $\hat{a} = 0$, 0.6, 0.8, and 0.9 with $N = 24$. Here, for the definitions of the area, C_p, and C_eq, see section 1.3.2.3 and for the definition of the apparent horizon, see section 7.1. The apparent horizon does not agree with the event horizon in general, but for the stationary black holes considered here, they should agree with each other.

Figure 3.17 shows that for $\hat{a} = 0$ and 0.6, the area and the axis ratio remain approximately constant up to $t = 1000M$. For $\hat{a} = 0.8$, by contrast, they increase but the error is smaller than 1% at $t = 1000M$. However, for $\hat{a} = 0.9$, the increase rate for the error is much larger than those for $\hat{a} = 0$, 0.6 and 0.8, and the error is much larger than 1% at $t = 1000M$. This implies that for $\hat{a} = 0$, 0.6, and 0.8, the grid spacing, $\Delta x = 0.045M$, is reasonably small (the grid resolution is good) for evolving the black holes for the time scale $1000M$. However, this grid resolution is not acceptable for evolving the black hole with $\hat{a} = 0.9$.

How fine grid resolution is necessary for evolving the black hole with $\hat{a} = 0.9$? For determining the required grid resolution, the testbed simulations were performed varying N from 24 to 36 (fixing the domain size). Figure 3.18 plots the evolution of the errors of the area and axis ratio. This shows that the error converges approximately at fourth order; by changing N from 24 to 36, the error decreases by a factor

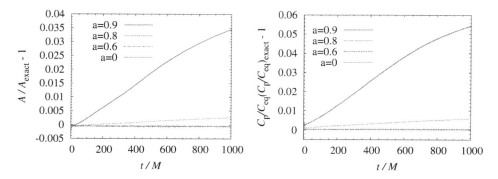

Fig. 3.17 Left: Evolution of the area of apparent horizons, A, for Kerr black holes with dimensionless spin parameters $\hat{a} = 0$, 0.6, 0.8, and 0.9 with the grid number $N = 24$ and the grid spacing $\Delta x = 0.045M$. $A/A_{\rm exact} - 1$ is shown where $A_{\rm exact}$ is the exact area of the Kerr black holes. Right: The same as the left panel but for the axis ratio $C_{\rm p}/C_{\rm eq}$ as a function of time (see section 1.3.2.3 for the definition of $C_{\rm p}$ and $C_{\rm eq}$). Note that the small error found even for $\hat{a} = 0$ indicates the typical size for the error associated with finding and analyzing apparent horizons. For this example, this error converges at second order because the apparent-horizon finder employed is second-order accurate in Δx.

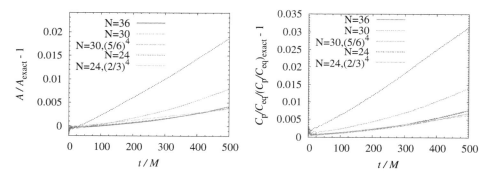

Fig. 3.18 Left: Evolution of the area of apparent horizons, A, for a Kerr black hole with spin parameter $\hat{a} = 0.9$ for three different grid resolutions, $N = 24$, 30, and 36. $A/A_{\rm exact} - 1$ is shown where $A_{\rm exact}$ is the exact area of the Kerr black hole. $(A/A_{\rm exact} - 1)(N/36)^4$ is approximately identical, implying that the fourth-order convergence is approximately achieved. Right: The same as the left panel but for the axis ratio $C_{\rm p}/C_{\rm eq}$ as a function of time. We note that for this example, an apparent-horizon finder, which is second-order accurate in Δx, was employed, and hence, quantities associated with the apparent horizon do not converge exactly at fourth order.

of $16/81 \sim 20\%$. Then, the error size at $t = 1000M$ is $\sim 1\%$. If we require that the error is within 1%, we have to evolve the black hole by $N = 36$, i.e., $\Delta x = 0.03M$. For evolving a black hole with $\hat{a} > 0.9$, the required grid resolution steeply increases; e.g., for $\hat{a} = 0.95$, it is $\Delta x \approx 0.02M$, which is approximately proportional to the initial horizon radius $M\sqrt{1 - \hat{a}^2}/2$.

As we remarked above, the required grid resolution could be decreased if we choose a smart coordinate [Liu et al. (2010)]. However, for dynamical spacetime, it is not easy to find a way getting a suitable coordinate automatically. Thus, in general, it is necessary to prepare a fine grid resolution (by a factor more than 2 finer than the grid resolution required for evolving non-spinning or slowly black holes) when a rapidly spinning black hole with $\hat{a} \sim 0.9$ is numerically evolved.

3.4.3.3 Collision of two non-spinning black holes

Simulations for a head-on collision of two non-spinning black holes was a primarily challenging issue in numerical relativity through 1970s to early 1990s [Smarr (1979); Anninos et al. (1993, 1995b,d)]. However, now, they are good testbed problems that can be performed by a personal computer when the initial separation of two black holes is not extremely large. The challenging problem in a past epoch always becomes a testbed problem eventually.

In this test, two black holes momentarily at rest are considered as the initial condition. This initial condition can be analytically given using the method by Lindquist (1963); Brill and Lindquist (1963) (see section 5.5.1.2) as

$$\gamma_{ij} = \psi^4 f_{ij}, \quad K_{ij} = 0, \tag{3.93}$$

where

$$\psi = 1 + \frac{m_1}{2r_1} + \frac{m_2}{2r_2}, \tag{3.94}$$

with r_i ($i = 1, 2$) being the coordinate distances from each black-hole center and m_i the mass parameters.

The collision process of two black holes is known to be quite simple as follows: Two black holes approach each other, merge to be an oscillating black hole, and the merged black hole eventually relaxes to a new static black hole. In the final ringdown phase, it is well-known that the new horizon shows a damping oscillation which is characterized mainly by primary quasi-normal modes [Anninos et al. (1993, 1995b,d)]. Specifically, the damping oscillation is appreciably found for a special non-spherical-mode, and for example, the axial ratio of the horizon C_p/C_{eq} [see equation (1.136)] shows such a damping oscillation (see the left panel of figure 3.19).

The oscillation angular frequency, ω_r, and the damping rate, ω_i, of quasi-normal modes were studied in detail in the framework of the black-hole perturbation theory [Chandrasekhar and Detweiler (1975); Leaver (1985); Schutz and Will (1985)]. The numerical studies of the quasi-normal modes for a non-spinning black hole by Chandrasekhar and Detweiler (1975); Leaver (1985) showed that they are

$$\omega_r = 0.3737 M_{\rm BH}^{-1}, \quad \omega_i = 0.0889 M_{\rm BH}^{-1}, \tag{3.95}$$

where $M_{\rm BH}$ is the mass of a black hole eventually formed. Thus, it is a good test to examine whether the damping oscillation curve agrees with a predicted curve of the form

$$\propto {\rm Re}\left(e^{i\omega_r t - \omega_i t}\right). \tag{3.96}$$

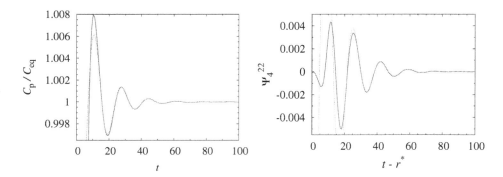

Fig. 3.19 Left: Time evolution of $C_{\rm p}/C_{\rm eq}$ after the merger of two black holes. Right: Waveform of $l = m = 2$ mode (the complex Weyl scalar, Ψ_4) as a function of the retarded time. For both panels, the solid and dashed curves are the numerical results and the predicted fitting curves, respectively. The unit of the time is M.

Figure 3.19 displays numerical results for a test simulation. In this test, two black holes are initially set at $x = \pm 0.5M$ and $m_1 = m_2 = M/2$ where M is approximately equal to the total mass of the system. The simulation was again performed by the SACRA code [Yamamoto et al. (2008)] preparing eight AMR subdomains with $N = 30$. Each subdomain covers a region $-L_l \leq x \leq L_l$, $-L_l \leq y \leq L_l$, and $0 \leq z \leq L_l$, where $l = 0, 1, 2, \cdots$, and 7 with $L_l = 153.6M/2^l$ and $L_7 = 1.2M$, and the grid spacing of each subdomain is $\Delta x_l = 153.6/(2^l N)$ and $\Delta x_7 = 0.04M$. The CFL number is set to be 0.5 for the finer subdomains with $l \geq 3$ but for $l \leq 2$, the time-step interval Δt is set to be equal to that for $l = 3$. A dynamical gauge condition was employed (see section 2.2.6).

The left panel of figure 3.19 plots the time evolution of $C_{\rm p}/C_{\rm eq}$ of the apparent horizon encompassing two black holes which is found for $t \gtrsim 1.9M$. The solid curve is the numerical result and the dashed curve is a fitting curve in the form (3.96) with equation (3.95). Here, the value of $M_{\rm BH}$ is determined by analyzing the area of the black hole finally formed (mass is equal to $\sqrt{A/16\pi}$; see section 1.3.1). This figure shows that in the final damping phase for $t \gtrsim 30M$, the numerical result is well fitted by the predicted curve. This agrees with the previously established numerical results by Anninos et al. (1993, 1995b,d), and hence, the code is validated.

The right panel of figure 3.19 plots the $l = m = 2$ mode of the outgoing component of complex Weyl scalars (the so-called Ψ_4; see section 6.2) as a function of the retarded time, which is closely related to gravitational waves emitted by the collision and the ringdown. Again, the solid curve is the numerical result and the dashed curve is a fitting curve. It is shown that the gravitational waveform in the final ringdown phase $(t - r^* \gtrsim 30M)$ is determined by the ringdown oscillation associated with the primary quasi-normal mode. This result shows that gravitational waves are accurately extracted in this code.

3.4.3.4 Test simulations in five dimensions

Test simulations, which are essentially the same as in four dimensions, may be performed for calibrating a higher-dimensional numerical-relativity code. Here, we describe test simulations using a Myers–Perry black hole [see equation (1.158)] and a collision of two non-spinning black holes, and illustrate successful results of the test simulations for a five-dimensional code (named SACRAND), in which the higher-dimensional version of the BSSN formulation and an AMR algorithm are implemented as in the four-dimensional case.

As in the four-dimensional spacetime, for numerical-relativity simulations, it is convenient to write the metric of the black holes in isotropic and quasi-isotropic coordinates [see equation (1.165)] in which there is no coordinate singularity. For the non-spinning black hole, the initial condition is momentarily static and is written in the conformally flat form as [see equation (5.83)]

$$\gamma_{ij} = \psi^2 f_{ij}, \quad K_{ij} = 0, \tag{3.97}$$

where

$$\psi = 1 + \frac{\mu}{4r^2}. \tag{3.98}$$

Here, μ is a parameter of mass dimension, which is related to the area and mass of the black hole by equations (1.160) and (1.161), respectively. The radius of the event horizon in this isotropic coordinate is denoted by $r_\mathrm{h} = \mu^{1/2}/2$.

As a straightforward extension of the single black-hole case, the initial condition for two equal-mass black holes in the momentarily static state ($K_{ij} = 0$) is written analytically in the conformally flat form. For this case, the conformal factor is written as

$$\psi = 1 + \frac{\mu'}{8r_1^2} + \frac{\mu'}{8r_2^2}. \tag{3.99}$$

Here, the meaning of r_1 and r_2 is the same as in equation (3.94). For the case of the two non-spinning black holes, the collision of two black holes will result in a non-spinning black hole of its mass determined approximately by $\mu \approx 2\mu'$.

Figure 3.20 shows the area of a black hole with the spin $a = 0.6\mu^{1/2}$ as a function of time. This test simulation was performed in the same manner as in section 3.4.3.1 with three grid resolutions, $N = 20, 24$, and 30, where the grid spacing in the finest AMR subdomain is $\Delta x/\mu^{1/2} = 0.96/N$. This figure illustrates that the code can evolve the black hole stably for a long time scale, and the accuracy is improved approximately at fourth order with the improvement of the grid resolution, as in the four-dimensional case.

The collision of two non-spinning black holes can be also performed as a test simulation as in section 3.4.3.3. In this test, two black holes are initially set at $x = \pm 0.5 r_\mathrm{h}$ and merge at $t \sim 1.6 r_\mathrm{h}$. Figure 3.21 plots the time evolution of $C_\mathrm{p}/C_\mathrm{eq}$ of the apparent horizon encompassing two black holes. The solid curve is the numerical result and the dashed curve is a fitting curve in the form (3.96). Cardoso *et al.*

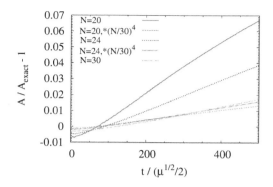

Fig. 3.20 Time evolution of the area of a spinning black hole with the spin $a = 0.6\mu^{1/2}$ in five dimensions. The area converges approximately at fourth order with the increase of N. We note that for this example, an apparent-horizon finder, which is second-order accurate in Δx, was employed, and hence, A does not converge exactly at fourth order.

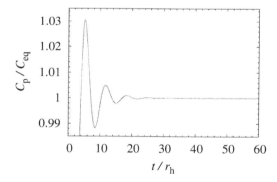

Fig. 3.21 Time evolution of $C_{\rm p}/C_{\rm eq}$ after the merger of two non-spinning black holes in five dimensions.

(2003) [see also Yoshino et al. (2006)] calculated the frequency of the quasi-normal mode of the fundamental quadrupole mode for five-dimensional non-spinning black holes as

$$\omega_r = 0.9477, \quad \omega_i = 0.2561 \quad \text{in units of } \mu = 1, \tag{3.100}$$

and hence, we adopt this value. This figure shows that in the final ringdown phase for $t \gtrsim 20 r_{\rm h}$, the numerical result is well fitted by the predicted curve, as in the four-dimensional case. By this procedure, the code is validated. This type of the test simulations can be easily done for higher-dimensional black holes with $D \geq 6$.

Chapter 4

Matter equations in general relativity

In the presence of a matter field, we have to solve equations of motion for it in addition to Einstein's equation. The basic equations for the matter field are derived primarily from equation (1.13), $\nabla^a T_{ab} = 0$. Here, the stress-energy tensor, T_{ab}, is written as a function of the matter field supposed in the chosen problem. In addition to this equation, several equations inherent in each matter field, which are often derived from conservation laws, have to be solved. In this chapter, we describe a variety of basic equations and several numerical methods for a solution of these equations. In sections 4.1 and 4.2, we will outline basic equations and typical numerical methods for solving equations of motion for scalar fields and swarms of collisionless particles. In the rest of this chapter, we will describe basic equations and numerical methods for a solution of compressible hydrodynamics, magnetohydrodynamics, and radiation hydrodynamics in detail. As in chapter 2, α, β^a, γ_{ab}, K_{ab}, and D_a denote the lapse function, shift vector, spatial metric, extrinsic curvature, and covariant derivative with respect to γ_{ab}, respectively, throughout this chapter.

4.1 Scalar fields

For simplicity, we here consider the basic equations for the case that only one single real scalar field is present. In the presence of multi scalar components, the equations are derived in essentially the same manner.

In the presence of a single scalar field, ϕ, the stress-energy tensor is written as

$$T_{ab} = (\nabla_a \phi)\nabla_b \phi - g_{ab}\left[\frac{1}{2}(\nabla_c\phi)\nabla^c\phi + V(\phi)\right], \qquad (4.1)$$

where $V(\phi)$ denotes a potential for ϕ.

From equation (1.13), the field equation for ϕ is derived as

$$\Box_g \phi = \frac{\partial V}{\partial \phi}, \qquad (4.2)$$

where $\Box_g$ is defined in equation (1.19). The left-hand side of equation (4.2) can be rewritten in several manners. Unless we define a new variable as in the generalized harmonic formulation for Einstein's equation (see section 2.3.4), equation (4.2) is written as

$$\Box_g \phi = \frac{1}{\sqrt{-g}} \partial_\mu \left(\sqrt{-g} g^{\mu\nu} \partial_\nu \phi \right) = g^{\mu\nu} \partial_\mu \partial_\nu \phi + H^\alpha \partial_\alpha \phi, \qquad (4.3)$$

where we recall that $H^\alpha = \partial_\mu(\sqrt{-g} g^{\alpha\mu})/\sqrt{-g}$ (see section 2.3.4). Thus, in this case, equation (4.2) is written in a hyperbolic equation.

Equation (4.2) may be written in a form similar to that in the $N+1$ formulation. For achieving that, we write

$$\begin{aligned}\Box_g \phi &= \frac{1}{\alpha\sqrt{\gamma}} \partial_\mu \left[\alpha\sqrt{\gamma}(\gamma^{\mu\nu} - n^\mu n^\nu) \partial_\nu \phi \right] \\ &= D_i D^i \phi + (D_i \ln \alpha) D^i \phi + (\nabla_\mu n^\mu) \Pi + n^\mu \partial_\mu \Pi, \end{aligned} \qquad (4.4)$$

where $\Pi := -n^\mu \partial_\mu \phi$. Thus, in a coordinate basis, equation (4.2) is written to a simultaneous equation as

$$(\partial_t - \beta^k \partial_k)\phi = -\alpha \Pi, \qquad (4.5)$$

$$(\partial_t - \beta^k \partial_k)\Pi = -\alpha D_i D^i \phi - (D_i \alpha) D^i \phi + \alpha K \Pi + \alpha \frac{\partial V}{\partial \phi}. \qquad (4.6)$$

Hence, the equations for (ϕ, Π) are written in the same form as those for (γ_{ij}, K_{ij}). This implies that the equations for the scalar field can be numerically solved in the same method as that for Einstein's equation (see section 2.1.1).

4.2 Collisionless particles

In numerical astrophysics, simulations for the formation of galaxies and star clusters are often performed approximating stars and dark-matter particles as point particles. In particular, these objects can be approximated by collisionless particles if the relaxation time scale by multiple two-body encounters is much longer than the crossing time (typical time scale of objects for traveling the size of their system) [Binney and Tremaine (1987)]. A simulation that follows a large number of particles in self-gravitating systems is often called N-body simulation. In general relativity, N-body simulations for the system composed of collisionless particles are also feasible in essentially the same method as in Newtonian gravity. After the pioneer works by Shapiro and Teukolsky (1985a,b), N-body simulations in general relativity have been performed by several groups [Shapiro and Teukolsky (1986, 1991, 1992b,a); Abrahams et al. (1994a,b); Shapiro et al. (1995); Shibata (1999a,c); Yamada and Shinkai (2011)], focusing in particular on exploring the nature of highly deformed and strongly gravitating objects. In this section, we outline a formulation and a numerical method for simulating N-body systems in general relativity.

The stress-energy tensor and the rest-mass density for a particle is written respectively as [Landau and Lifshitz (1962)]

$$T^{\mu\nu} = m \int d\tau \frac{\delta^{(4)}\left[x^\alpha - x^\alpha(\tau)\right]}{\sqrt{-g}} u^\mu u^\nu, \tag{4.7}$$

$$\rho = m \int d\tau \frac{\delta^{(4)}\left[x^\alpha - x^\alpha(\tau)\right]}{\sqrt{-g}}, \tag{4.8}$$

where m is the rest mass of the particle, $x^\alpha(\tau)$ denote its spacetime coordinates, u^μ is its four velocity, and $\delta^{(4)}$ is the delta function in four-dimensional spacetime. From equation (4.7), the geodesic equation

$$u^\mu \nabla_\mu u^\nu = 0, \tag{4.9}$$

is derived by operating ∇_μ to $T^{\mu\nu}$ and by performing integration by parts.

The stress-energy tensor for a swarm of collisionless particles is written as

$$\begin{aligned}T^{\mu\nu} &= \sum_{I=1}^N m_I \int d\tau \frac{\delta^{(4)}\left[x^\alpha - x_I^\alpha(\tau)\right]}{\sqrt{-g}} u_I^\mu u_I^\nu \\ &= \sum_{I=1}^N m_I \frac{\delta^{(3)}\left[x^j - x_I^j(t)\right]}{\alpha\sqrt{\gamma}} \left(\frac{u_I^\mu u_I^\nu}{u_I^t}\right),\end{aligned} \tag{4.10}$$

where m_I, x_I^j, and u_I^μ are their rest mass, position, and four-velocity, respectively. In this section, N is the total particle number. We used $d\tau = dt/u^t$ and performed the integration in terms of t to derive the second line of equation (4.10). The weighted rest-mass density $\rho_* := \rho u^t \alpha \sqrt{\gamma}$ is written in a simple form

$$\rho_* = \sum_{I=1}^N m_I \delta^{(3)}\left[x^j - x_I^j(t)\right], \tag{4.11}$$

from which the total rest mass of the system, M_0, is calculated as

$$M_0 = \int \rho_* d^3x = \sum_{I=1}^N m_I. \tag{4.12}$$

Equations of motion for each particle are derived from the geodesic equation (4.9). In a coordinate basis, the spatial component of the geodesic equation is written as

$$\frac{du_i}{dt} = -\alpha u^t \partial_i \alpha + u_j \partial_i \beta^j - \frac{u_j u_k}{2u^t} \partial_i \gamma^{jk}. \tag{4.13}$$

From u_i, u^t is determined from the normalization relation of the four-velocity, $u^a u_a = -1$, as

$$(\alpha u^t)^2 = 1 + \gamma^{ij} u_i u_j. \tag{4.14}$$

The relation between a coordinate-based three velocity $dx^i/dt := u^i/u^t$ and u_j is written in the form

$$\frac{dx^i}{dt} = -\beta^i + \frac{\gamma^{ij} u_j}{u^t}. \tag{4.15}$$

Equations (4.13) and (4.15) constitute the basic equations for the evolution of the particle position, x^i, and the specific momentum, u_i. These equations are ordinary differential equations, and hence, solutions for the equations of motion are much more easily obtained than that for hydrodynamic and magnetohydrodynamics equations. For integrating these ordinary differential equations, second- or fourth-order Runge–Kutta method is often used [e.g., Press *et al.* (1989); see also section 3.1.2].

In the 3+1 formulation, we also need to calculate $\rho_h = T_{ab}n^a n^b$, $J_i = -T_{ab}n^a \gamma^b{}_i$, and $S_{ij} = T_{ab}\gamma^a{}_i \gamma^b{}_j$ for the time integration of Einstein's evolution equations. These quantities are written as

$$\rho_h = \sum_{I=1}^{N} m_I (\alpha u_I^t) \gamma^{-1/2} \delta^{(3)} \left(x^k - x_I^k \right), \tag{4.16}$$

$$J_i = \sum_{I=1}^{N} m_I (u_i)_I \gamma^{-1/2} \delta^{(3)} \left(x^k - x_I^k \right), \tag{4.17}$$

$$S_{ij} = \sum_{I=1}^{N} m_I (u_i)_I (u_j)_I (\alpha u_I^t)^{-1} \gamma^{-1/2} \delta^{(3)} \left(x^k - x_I^k \right). \tag{4.18}$$

In the N-body problem, we have to provide gravitational-field quantities at each particle position to integrate the equations of motion. Also, to evaluate ρ_h, J_i, and S_{ij}, we have to approximately evaluate the delta function which cannot be directly handled in numerical simulations. The often-used scheme for the evaluation of the delta function is a particle-mesh method [e.g., Hockney and Eastwood (1988)]. In this method, gravitational-field quantities and their derivatives are represented approximately on discretized grid points, as is usually done in finite-differencing numerical schemes. Then, for integrating the equations of motion for particles, which are not located on the grid points in general, the field quantities and their derivatives at the particle positions have to be determined by some interpolation. For interpolating these quantities at the particle positions, the quantities at several grid points in the vicinity of the corresponding particles are chosen. On the other hand, to integrate gravitational-field equations, the quantities associated with the stress-energy of the particles such as ρ_h, J_i, and S_{ij} have to be provided for each grid point. This is achieved by assigning the contribution of each particle for these quantities to nearby grid points with an appropriate weight. For this assignment, we rewrite the delta function to some function as

$$\delta^{(3)} \left[x^k - x^k(t) \right] \to D \left[x^k - x^k(t) \right], \tag{4.19}$$

which satisfies

$$\int d^3 x D \left[x^k - x^k(t) \right] = 1. \tag{4.20}$$

Assuming that Cartesian coordinates, (x, y, z), are employed, the simplest choice for D is

$$D \left[x^k - x^k(t) \right] = \begin{cases} (\Delta x \Delta y \Delta z)^{-1} & \text{inside a cube in which particle is located,} \\ 0 & \text{otherwise,} \end{cases}$$

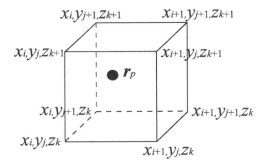

Fig. 4.1 Schematic figure for a particle in a Cartesian cube for which the edges are composed of the nearest eight grid points.

where $(\Delta x, \Delta y, \Delta z)$ denotes a set of grid spacing for each direction. In this example, a particle at $x^k(t) = (x_p, y_p, z_p)$ is assumed to be inside a cube of volume $\Delta x \Delta y \Delta z$ with $x \leq x_p \leq x + \Delta x$, $y \leq y_p \leq y + \Delta y$, and $z \leq z_p \leq z + \Delta z$.

One popular scheme often used for the interpolation is a cloud-in-cell (CIC) scheme [Hockney and Eastwood (1988)]. For three-dimensional space, the nearest eight grid points are employed for the interpolation in the typical CIC scheme. Let (x_i, y_j, z_k) be a discrete grid point in the Cartesian coordinates and the grid spacing is defined as $\Delta x_i = x_{i+1} - x_i$, $\Delta y_j = y_{j+1} - y_j$, and $\Delta z_k = z_{k+1} - z_k$. Suppose that a particle is located at $\bm{r}_p = (x_p, y_p, z_p)$ where $x_i < x_p < x_{i+1}$, $y_j < y_p < y_{j+1}$, and $z_k < z_p < z_{k+1}$ (see figure 4.1). Then, the value of a field variable Q at $\bm{r}_p$ in the CIC scheme is determined in the linear interpolation as

$$\left[\{f_x^+ Q_{i,j,k} + f_x^- Q_{i+1,j,k}\} f_y^+ + \{f_x^+ Q_{i,j+1,k} + f_x^- Q_{i+1,j+1,k}\} f_y^- \right] f_z^+$$
$$+ \left[\{f_x^+ Q_{i,j,k+1} + f_x^- Q_{i+1,j,k+1}\} f_y^+ + \{f_x^+ Q_{i,j+1,k+1} + f_x^- Q_{i+1,j+1,k+1}\} f_y^- \right] f_z^-, \quad (4.21)$$

where $f_x^+ = (x_{i+1} - x_p)/\Delta x_i$, $f_x^- = 1 - f_x^+$, $f_y^+ = (y_{j+1} - y_p)/\Delta y_j$, $f_y^- = 1 - f_y^+$, $f_z^+ = (z_{k+1} - z_p)/\Delta z_k$, and $f_z^- = 1 - f_z^+$. $Q_{i,j,k}$ denotes the value of Q at (x_i, y_j, z_k). In short, the weight factor is determined by a fraction of the volume in the cube.

In the same manner, we can employ a rule that a particle of rest mass m staying at $\bm{r}_p$ contributes to ρ_h, J_a, and S_{ab} of the nearest eight grid points. Specifically, it is convenient to consider $S_0 := \rho_h \sqrt{\gamma}$, $S_i := J_i \sqrt{\gamma}$, and $S_{ij}\sqrt{\gamma}$. In the typical CIC scheme, the contribution of a particle at $\bm{r}_p$ is determined by

$$(S_0)_{i+1/2 \mp 1/2, \ j+1/2 \mp 1/2, \ k+1/2 \mp 1/2} = \frac{f_x^{\pm} f_y^{\pm} f_z^{\pm}}{\Delta V_{ijk}} m \alpha u^t, \quad (4.22)$$

$$(S_l)_{i+1/2 \mp 1/2, \ j+1/2 \mp 1/2, \ k+1/2 \mp 1/2} = \frac{f_x^{\pm} f_y^{\pm} f_z^{\pm}}{\Delta V_{ijk}} m u_l, \quad (4.23)$$

$$(\sqrt{\gamma} S_{lm})_{i+1/2 \mp 1/2, \ j+1/2 \mp 1/2, \ k+1/2 \mp 1/2} = \frac{f_x^{\pm} f_y^{\pm} f_z^{\pm}}{\Delta V_{ijk}} m \frac{u_l u_m}{\alpha u^t}, \quad (4.24)$$

where $\Delta V_{ijk} = \Delta x_i \Delta y_j \Delta z_k$. The total amount of ρ_h, J_i, and S_{ij} at each grid point is computed by summing up the contribution from all the nearby particles.

We note that the CIC scheme is second-order accurate in Δx_i, Δy_j, and Δz_k. In higher-order schemes, the order of accuracy can be improved.

4.3 3+1 form of stress-energy conservation: $\nabla_a T^a{}_b = 0$

The conservation equation of T_{ab}, (1.13), yields equations of motion and an energy conservation equation in general relativity. For hydrodynamics and magnetohydrodynamics, it provides the Euler and energy equations. In modern numerical hydrodynamics and magnetohydrodynamics [since it was proposed by Martí et al. (1991) for the first time], these equations are written in a *conservative form*.[1] In this form, the momentum and energy of the system could be accurately conserved, and in addition, jump conditions such as Rankine–Hugoniot conditions across discontinuities [Landau and Lifshitz (1959)] are guaranteed to be captured with no special and artificial prescriptions. The 3+1 formulation of equation (1.13) is well suited for writing down the equations in the conservative form. Thus, in this section, we rewrite equation (1.13) in the 3+1 form without assuming specific forms of T_{ab}.

First of all, we remind the reader that T^{ab} can be decomposed into the form in the 3+1 formulation,

$$T^{ab} = \rho_h n^a n^b + J^a n^b + J^b n^a + S^{ab}, \qquad (4.25)$$

where $\rho_h := T_{ab} n^a n^b$, $J^c := -T_{ab} n^a \gamma^{bc}$, and $S^{cd} := T_{ab} \gamma^{ac} \gamma^{bd}$ as defined in **notation** page. For the following calculation, we also define

$$S_i := \sqrt{\gamma} J_i, \qquad S_0 := \sqrt{\gamma} \rho_h. \qquad (4.26)$$

We will describe the equations for these variables. We note that in this definition, we implicitly assume to use Cartesian coordinates. In curvilinear coordinates, an extra factor associated with the curved coordinates appears in γ. For this case, we should factor out the extra term from γ as $\gamma = \gamma_{\text{grav}} f$ where γ_{grav} is the determinant associated purely with the gravitational field and f is the determinant associated with the curvilinear coordinates; $f = \det(f_{ij})$ with f_{ij} the flat spatial metric.

For deriving the evolution equations for S_i and S_0, we consider the projections of $\nabla_b T^b{}_a = 0$ as

$$\gamma_i{}^a \nabla_b T^b{}_a = 0, \qquad (4.27)$$
$$n^a \nabla_b T^b{}_a = 0. \qquad (4.28)$$

Equation (4.27) results in the equations of motion such as Euler equation. For any

---

[1] For other forms of hydrodynamics and magnetohydrodynamics equations, see Font (2006) for a detailed review. In this volume, we only describe numerical approaches based on the conservative form.

stress-energy tensor, it is rewritten as

$$\begin{aligned} 0 = \gamma_k{}^a \nabla_b T^b{}_a &= \frac{1}{\sqrt{-g}} \left[\partial_t \left(T^t{}_k \sqrt{-g} \right) + \partial_j \left(T^j{}_k \sqrt{-g} \right) \right] - \frac{1}{2} T^{\mu\nu} \partial_k g_{\mu\nu} \\ &= \frac{1}{\alpha\sqrt{\gamma}} \left[\partial_t S_k + \partial_j \left(\alpha \sqrt{\gamma} S^j{}_k - \beta^j S_k \right) \right] \\ &\quad - \frac{1}{\alpha} \left(-\rho_{\rm h} \partial_k \alpha + J_i \partial_k \beta^i - \frac{1}{2} \alpha S_{ij} \partial_k \gamma^{ij} \right), \end{aligned} \quad (4.29)$$

where we used

$$\begin{aligned} T^{\mu\nu} \partial_k g_{\mu\nu} &= (\rho_{\rm h} n^\mu n^\nu + J^\mu n^\nu + J^\nu n^\mu + S^{\mu\nu}) \partial_k g_{\mu\nu} \\ &= \frac{1}{\alpha} \left(-2 \rho_{\rm h} \partial_k \alpha + 2 J_i \partial_k \beta^i - \alpha S_{ij} \partial_k \gamma^{ij} \right). \end{aligned} \quad (4.30)$$

Equation (4.29) is written to the equations of motion in the form

$$\partial_t S_k + \partial_j \left(\alpha \sqrt{\gamma} S^j{}_k - \beta^j S_k \right) = -S_0 \partial_k \alpha + S_i \partial_k \beta^i - \frac{1}{2} \alpha \sqrt{\gamma} S_{ij} \partial_k \gamma^{ij}. \quad (4.31)$$

Equation (4.28) is, for any stress-energy tensor, written as

$$\begin{aligned} 0 = n^a \nabla_b T^b{}_a &= \nabla_b (T^b{}_a n^a) - T^{ab} \nabla_b n_a \\ &= -\nabla_b \left(\rho_{\rm h} n^b + J^b \right) + T^{ab} (K_{ab} + n_a a_b) \\ &= -\frac{1}{\sqrt{-g}} \left[\partial_t S_0 + \partial_i \left(-S_0 \beta^i + \alpha S^i \right) \right] + S^{ij} K_{ij} - J^i a_i, \end{aligned} \quad (4.32)$$

where we used equation (2.30). Also, remember that $a_i = D_i \ln \alpha$. Thus, the energy equation is written as

$$\partial_t S_0 + \partial_i \left(-S_0 \beta^i + \alpha S^i \right) = \alpha \sqrt{\gamma} S^{ij} K_{ij} - S_i D^i \alpha. \quad (4.33)$$

In the following, equations (4.31) and (4.33) will be used for deriving the basic equations of hydrodynamics and magnetohydrodynamics.

4.4 Hydrodynamics

This section is devoted to describing hydrodynamics equations with a simple equation of state; we focus on the simplest case in which we need only to evolve rest-mass density, momentum density, and energy density (five components). In section 4.5, we will describe equations for more general cases in which microphysics components are taken into account.

4.4.1 Basic equations

In this section, we consider the perfect fluid with no dissipation (i.e., no viscosity, cooling, and heating), and write the stress-energy tensor in the form

$$T_{ab} = \rho h u_a u_b + P g_{ab}, \quad (4.34)$$

where the specific enthalpy, h, is written as $1 + \varepsilon + P/\rho$. In hydrodynamics, there are at least 5 components to be determined, ρ: the rest-mass density, u_i: spatial components of the four velocity, and P: the pressure (or ε: the specific internal energy). Here, P and ε are usually related by an equation of state to each other. In this section, we assume that the equation of state has the form

$$P = P(\rho, \varepsilon). \tag{4.35}$$

For the stress-energy tensor defined in equation (4.34), we have

$$S_i = \sqrt{\gamma}\rho w h u_i, \quad S_0 = \sqrt{\gamma}(\rho h w^2 - P), \quad S_{ij} = \rho h u_i u_j + P\gamma_{ij}, \tag{4.36}$$

where $w := -u^a n_a = \alpha u^t$.

The evolution equations for S_i and S_0 (i.e., the Euler and energy equations) are derived by substituting equations (4.36) into equations (4.31) and (4.33), respectively. The Euler equation is derived as

$$\partial_t S_i + \partial_k \left(S_i v^k + P\alpha\sqrt{\gamma}\delta^k_i \right) = -S_0 \partial_i \alpha + S_k \partial_i \beta^k - \frac{1}{2}\alpha\sqrt{\gamma}S_{jk}\partial_i\gamma^{jk}, \tag{4.37}$$

where $v^i := u^i/u^t$ denotes the three velocity in the laboratory frame, which is rewritten as

$$v^i = -\beta^i + \gamma^{ij}\frac{u_j}{u^t}. \tag{4.38}$$

For deriving equation (4.37), we used

$$\alpha\sqrt{\gamma}S^j{}_k - \beta^j S_k = S_k v^j + P\alpha\sqrt{\gamma}\delta^j{}_k, \tag{4.39}$$

which holds for the stress-energy tensor written in the form (4.34).

A word of caution is appropriate here: u_i is equal to $u_a \gamma^a{}_i$ whereas u^i (a spatial component of the four velocity) is not equal to $\gamma^{ij} u_j$ in general. This property is understood by the fact that for the unit timelike vector field normal to spatial hypersurfaces n^a, we know $n_i = 0$, but $n^i = -\beta^i/\alpha$ is not zero in general.

The energy equation is written in the form

$$\partial_t S_0 + \partial_i \left[S_0 v^i + P(v^i + \beta^i)\sqrt{\gamma} \right] = \alpha\sqrt{\gamma}S_{ij}K^{ij} - S_i D^i \alpha, \tag{4.40}$$

where we used

$$-S_0 \beta^i + \alpha S^i = P(v^i + \beta^i)\sqrt{\gamma} + S_0 v^i, \tag{4.41}$$

which holds for the stress-energy tensor of the form (4.34).

As mentioned above, there are five independent components in hydrodynamics. Thus, we have to solve one more equation in addition to the Euler and energy equations. That is the conservation equation for the rest-mass density ρ; the so-called continuity equation, which is written in the form

$$\nabla_a(\rho u^a) = 0. \tag{4.42}$$

In a coordinate basis, equation (4.42) is written as

$$\partial_t \left(\rho\sqrt{-g}u^t \right) + \partial_i \left(\rho\sqrt{-g}u^i \right) = 0. \tag{4.43}$$

For the weighted rest-mass density, $\rho_* := \rho\sqrt{-g}u^t = \rho w\sqrt{\gamma}$, we have
$$\partial_t \rho_* + \partial_i(\rho_* v^i) = 0. \tag{4.44}$$
Thus, the equation is written in the same form as that in flat spacetime, and we find that the continuity equation should be considered as the evolution equation for ρ_*. From equation (4.43) or (4.44), we can define the total rest mass as a conserved quantity as
$$M_0 := \int d^3x\, \rho_* = \int d^3x\, \rho\sqrt{-g}u^t = \int dV \rho\alpha u^t, \tag{4.45}$$
for which $\partial_t M_0 = 0$. Here, $dV = \sqrt{\gamma}d^3x$. We also note that S_i and S_0 are written in terms of ρ_* as
$$S_i = \rho_* h u_i,$$
$$S_0 = \rho_*\left(hw - \frac{P}{\rho w}\right).$$
Thus, hu_i and $e_0 := hw - P/(\rho w)$ are considered to be specific momentum and specific energy in general relativity, respectively. (Note that in the Newtonian limit, they are u_i and $1 + \varepsilon$, respectively.)

In addition to the five evolution equations, we have to determine the time component of the four velocity u^t or $w = \alpha u^t$. The equation for this is derived from the normalization condition of u^a, i.e., $u^a u_a = -1$, and written in the form
$$w^2 = 1 + \gamma^{ij} u_i u_j = 1 + \gamma^{ij}\frac{S_i S_j}{\rho_*^2 h^2}. \tag{4.46}$$
This equation, combined with an equation of state, is considered as an algebraic equation for w and h because at every time step, γ_{ij}, S_i, and ρ_* are updated by their evolution equations, and hence, they are given quantities for equation (4.46). Here, h is a function of ρ and ε through the equation of state of the form (4.35), and thus, it is not directly related to ρ_* and S_0 but a function of ρ_*, S_0, and w. Specifically, the following equation gives the relation among these quantities
$$e_0 = \frac{S_0}{\rho_*} = hw - \frac{P(h,w)\sqrt{\gamma}}{\rho_*}, \tag{4.47}$$
where e_0 is considered as a given quantity. We used the fact that P is a function of $\rho(= \rho_*/(w\sqrt{\gamma}))$ and h, and thus, $P = P(h,w)$ for given values of ρ_* and γ. For a given set of (ρ_*, S_i, S_0) and an equation of state, the algebraic equations (4.46) and (4.47) constitute a set of simultaneous equations for w and h or by eliminating w using equation (4.46), an algebraic equation for h is obtained.

As an illustration, we consider a Γ-law equation of state in the form
$$P = (\Gamma - 1)\rho\varepsilon, \tag{4.48}$$
where Γ is the adiabatic index and assumed to be a constant. Substituting this equation of state into the definition of h, we have $h = 1 + \Gamma\varepsilon$. Then, the equation of state may be written as
$$P = \frac{\Gamma - 1}{\Gamma}\rho(h - 1). \tag{4.49}$$

Substituting it into equation (4.47) yields

$$e_0 = hw - \frac{\Gamma - 1}{\Gamma} \frac{h(h-1)}{hw}. \tag{4.50}$$

Eliminating hw from equations (4.46) and (4.50), we obtain

$$e_0^2(h^2 + q^2) - \Gamma^{-2}[h^2 + (\Gamma - 1)h + \Gamma q^2]^2 = 0, \tag{4.51}$$

where $q^2 := \gamma^{ij} S_i S_j \rho_*^{-2}$. Thus, a fourth-order equation for h is derived, and it is not a difficult task to obtain a solution which satisfies $h \geq 1$.

For general equations of state, the resulting algebraic equation is not written in such a simple form. Thus, numerical computation is necessary for determining w and h. The typical numerical method for a solution is an iterative method such as Newton–Raphson method [Press et al. (1989)]; for this, see section 4.9.1.

When a solution of w and h is obtained, we can subsequently determine the so-called primitive variables, ρ, ε, and u_i (or v^i). Hence, solving the algebraic equation for w or h is equivalent to recovering these primitive variables. Thus this procedure is also called *primitive recovery*.

4.4.2 *Properties of hydrodynamics equations*

Equations (4.44), (4.37), and (4.40) constitute a set of the basic equations in general relativistic hydrodynamics. These are often written in a vector form

$$\partial_t \boldsymbol{Q}_a + \partial_k \boldsymbol{F}_a^k = \boldsymbol{S}_a, \tag{4.52}$$

where $\boldsymbol{Q}_a$, $\boldsymbol{F}_a^k$, and $\boldsymbol{S}_a$ are written in a five-components vector form defined by

$$\boldsymbol{Q}_a := \begin{pmatrix} \rho_* \\ S_i \\ S_0 \end{pmatrix}, \tag{4.53}$$

$$\boldsymbol{F}_a^k := \begin{pmatrix} \rho_* v^k \\ S_i v^k + P\alpha\sqrt{\gamma}\delta^k_i \\ S_0 v^k + P(v^k + \beta^k)\sqrt{\gamma} \end{pmatrix}, \tag{4.54}$$

$$\boldsymbol{S}_a := \begin{pmatrix} 0 \\ -S_0 \partial_i \alpha + S_k \partial_i \beta^k - \frac{1}{2}\alpha\sqrt{\gamma} S_{jk} \partial_i \gamma^{jk} \\ \alpha\sqrt{\gamma} S_{ij} K^{ij} - S^i D_i \alpha \end{pmatrix}. \tag{4.55}$$

The second term in the left-hand side of equation (4.52) determines the total amount of advection in hydrodynamics, and it is often referred to as advection term. The right-hand side of equation (4.52) is associated with gravitational forces. In numerical hydrodynamics, the accuracy and the stability in numerical computation are determined primarily by the numerical scheme chosen for handling the advection term.

In general relativistic hydrodynamics, the advection terms, for many cases, are evaluated using a scheme based on an *approximate Riemann solver*, which relies on the characteristic decomposition of the hydrodynamics equations [see Ibáñez and Martí (1999); Martí and Müller (2003); Font (2006) and references therein]. Here, the *Riemann solver* means that it can yield an accurate numerical solution for the Riemann problem (see section 4.8.6 and appendix E), and the exact Riemann solver is often called a Godunov-type scheme [Toro (1999)]. For constructing an approximate Riemann solver, we usually need to compute a *Jacobian matrix*, and then, to carry out a spectral decomposition for it. In the following, we will outline these procedures.

The advection term or the Jacobian matrix has the information of several characteristic speeds of hydrodynamics such as a bulk flow speed and propagation speeds of sound waves, which are associated with profiles of density and pressure as well as the fluid motion. The characteristic speeds then can be used for deriving a solution of the basic equations. This fact is easily understood by considering a simple scalar-wave equation like

$$\partial_t \phi + V \partial_x \phi = 0, \tag{4.56}$$

where V is a constant propagation velocity. For this scalar-field equation, the general solution is written as

$$\phi = F(x - Vt), \tag{4.57}$$

where F denotes an arbitrary function.

Equation (4.57) further shows that along each characteristic curve, $x - Vt = C$, $\phi = F(C)$ is constant. This suggests that even in general relativistic hydrodynamics in which the basic equations are much more complicated than equation (4.56), a solution or a semi-analytic solution could also be derived using the information of the characteristics.

In hydrodynamics, the Jacobian matrix for the k-th direction, $\boldsymbol{M}_{ab}^k$, is defined by

$$\boldsymbol{M}_{ab}^k := \frac{\partial \boldsymbol{F}_a^k}{\partial \boldsymbol{Q}_b} \qquad (k = x, y, z). \tag{4.58}$$

$\boldsymbol{M}_{ab}^k$ for each direction of k is a 5×5 matrix. Using equation (4.58), equation (4.52) is rewritten in the form

$$\partial_t \boldsymbol{Q}_a + \sum_{b=1}^{5} \boldsymbol{M}_{ab}^k \partial_k \boldsymbol{Q}_b = \boldsymbol{S}_a. \tag{4.59}$$

Comparing this with equation (4.56), it is found that $\boldsymbol{M}_{ab}^k$ indeed has the information of the characteristic speeds for the k-th direction.

General relativistic hydrodynamics equations are nonlinear equations, and hence, the method for a solution of the scalar-wave equation (4.56) may not be applied as it is. However, if we assume that $\boldsymbol{M}_{ab}^k$ is a constant matrix (which would be a good approximation at least locally) and consider equation (4.59) as a linear equation for $\boldsymbol{Q}_a$, we find that the method used for the scalar-wave equation can be extended for the hydrodynamics equations in a straightforward manner. Actually, the system of the hydrodynamics equations is quasilinear, and hence, a similar method can be used for the formal solution, as shown in the following.

The Jacobian matrix in general relativistic hydrodynamics is calculated and decomposed by Anile (1989); Font et al. (1994); Banyuls et al. (1997). We first follow the prescription of Font et al. (1994); Banyuls et al. (1997) for the calculation of $\boldsymbol{M}_{ab}^k$, in which it is written as

$$\boldsymbol{M}_{ab}^k = \sum_{c=1}^{5} \frac{\partial F_a^k}{\partial q_c} \frac{\partial q_c}{\partial Q_b} := \sum_{c=1}^{5} \boldsymbol{B}_{ac}^k \boldsymbol{C}_{bc}^{-1}, \quad (4.60)$$

where

$$\boldsymbol{q}_c := \begin{pmatrix} \rho \\ v^x \\ v^y \\ v^z \\ \varepsilon \end{pmatrix}, \quad (4.61)$$

is a vector composed of the so-called primitive variables, and

$$\boldsymbol{B}_{ac}^k := \frac{1}{\sqrt{\gamma}} \frac{\partial F_a^k}{\partial q_c}, \quad \boldsymbol{C}_{bc} := \frac{1}{\sqrt{\gamma}} \frac{\partial Q_b}{\partial q_c}. \quad (4.62)$$

Explicit forms for $\boldsymbol{C}_{ab}$ and $\boldsymbol{B}_{ab}^x$ are

$$\boldsymbol{C}_{ab} = \begin{bmatrix} w & \rho w^3 \dfrac{V_x}{\alpha^2} & \rho w^3 \dfrac{V_y}{\alpha^2} & \rho w^3 \dfrac{V_z}{\alpha^2} & 0 \\ \dfrac{h_1 w^2}{\alpha} V_x & \dfrac{\rho h w^2}{\alpha} \bar{f}_{xx} & \dfrac{\rho h w^2}{\alpha} \bar{f}_{xy} & \dfrac{\rho h w^2}{\alpha} \bar{f}_{xz} & \rho h_2 \dfrac{w^2 V_x}{\alpha} \\ \dfrac{h_1 w^2}{\alpha} V_y & \dfrac{\rho h w^2}{\alpha} \bar{f}_{xy} & \dfrac{\rho h w^2}{\alpha} \bar{f}_{yy} & \dfrac{\rho h w^2}{\alpha} \bar{f}_{yz} & \rho h_2 \dfrac{w^2 V_y}{\alpha} \\ \dfrac{h_1 w^2}{\alpha} V_z & \dfrac{\rho h w^2}{\alpha} \bar{f}_{xz} & \dfrac{\rho h w^2}{\alpha} \bar{f}_{yz} & \dfrac{\rho h w^2}{\alpha} \bar{f}_{zz} & \rho h_2 \dfrac{w^2 V_z}{\alpha} \\ h_1 w^2 - \chi & 2\dfrac{\rho h w^4}{\alpha^2} V_x & 2\dfrac{\rho h w^4}{\alpha^2} V_y & 2\dfrac{\rho h w^4}{\alpha^2} V_z & \rho h_2 w^2 - \rho \kappa \end{bmatrix}, \quad (4.63)$$

and

$$\boldsymbol{B}_{ab}^x = \begin{bmatrix} wv^x & \rho w(1 + w^2 V_x v^x \alpha^{-2}) & \rho w^3 V_y v^x \alpha^{-2} \\ \alpha^{-1} h_1 w^2 V_x v^x + \alpha\chi & \alpha^{-1}\rho h w^2 (F_{xx}^x + V_x) & \alpha^{-1}\rho h w^2 F_{xy}^x \\ \alpha^{-1} h_1 w^2 V_y v^x & \alpha^{-1}\rho h w^2 (F_{xy}^x + V_y) & \alpha^{-1}\rho h w^2 F_{yy}^x \\ \alpha^{-1} h_1 w^2 V_z v^x & \alpha^{-1}\rho h w^2 (F_{xz}^x + V_z) & \alpha^{-1}\rho h w^2 F_{yz}^x \\ h_1 w^2 v^x + \chi\beta^x & \rho h w^2 (1 + 2w^2 \alpha^{-2} V_x v^x) & 2\rho h w^4 \alpha^{-2} V_y v^x \end{bmatrix}$$

$$\begin{matrix} \rho w^3 V_z v^x \alpha^{-2} & 0 \\ \alpha^{-1}\rho h w^2 F_{xz}^x & \alpha^{-1}\rho h_2 w^2 V_x v^x + \rho\kappa\alpha \\ \alpha^{-1}\rho h w^2 F_{yz}^x & \alpha^{-1}\rho h_2 w^2 V_y v^x \\ \alpha^{-1}\rho h w^2 F_{zz}^x & \alpha^{-1}\rho h_2 w^2 V_z v^x \\ 2\alpha^{-2}\rho h w^4 V_z v^x & \rho h_2 w^2 v^x + \rho\kappa\beta^x \end{matrix} \Bigg], \quad (4.64)$$

where
$$V_i := \gamma_{ij}(v^j + \beta^j), \quad \bar{f}_{ij} := \gamma_{ij} + \frac{2w^2 V_i V_j}{\alpha^2}, \quad F^k{}_{ij} := v^k \bar{f}_{ij},$$

$$\chi := \left.\frac{\partial P}{\partial \rho}\right|_\varepsilon, \quad \kappa := \left.\frac{1}{\rho}\frac{\partial P}{\partial \varepsilon}\right|_\rho, \quad h_1 := 1 + \varepsilon + \chi, \quad h_2 := 1 + \kappa.$$

$\boldsymbol{B}^y_{ab}$ and $\boldsymbol{B}^z_{ab}$ are obtained by appropriate exchanges of subscripts and components among x, y, z of $\boldsymbol{B}^x_{ab}$.

The eigenvalues of the matrix $\boldsymbol{M}^k_{ab}$ (hereafter referred to as λ^k) correspond to the characteristic speeds of the fluid in the k-th direction, and are derived from the eigenvalue equation

$$\det\left(\boldsymbol{B}^k_{ab} - \lambda^k \boldsymbol{C}_{ab}\right) = 0. \tag{4.65}$$

By a lengthy but straightforward calculation, the solutions of equation (4.65) are derived as

$$\lambda^k = \lambda^k_\pm, \quad v^k \quad (v^k \text{ is the triple root}), \tag{4.66}$$

where

$$\lambda^k_\pm = \frac{1}{\alpha^2 - V^2 c_s^2}\left[v^k \alpha^2 (1 - c_s^2) - \beta^k c_s^2 (\alpha^2 - V^2)\right.$$
$$\left.\pm \alpha c_s \sqrt{(\alpha^2 - V^2)\{\gamma^{kk}(\alpha^2 - V^2 c_s^2) - (1 - c_s^2) V^k V^k\}}\right]$$
(no summation for k), \tag{4.67}

$V^k = \gamma^{kl} V_l = v^k + \beta^k$, $V^2 = V_i V^i$, and c_s is the sound velocity defined by

$$c_s^2 := \left.\frac{\partial P}{\partial \rho_{\text{ene}}}\right|_s = \frac{1}{h}\left[\left.\frac{\partial P}{\partial \rho}\right|_\varepsilon + \left.\frac{P}{\rho^2}\frac{\partial P}{\partial \varepsilon}\right|_\rho\right] = \frac{1}{h}\left(\chi + \frac{P}{\rho}\kappa\right). \tag{4.68}$$

Here, ρ_{ene} is the total energy density $\rho(1+\varepsilon)$ and the partial derivative of P with respect to ρ_{ene} has to be taken fixing the specific entropy s.

The eigenvalues λ^k may be derived by a more concise way [Anile (1989)] from a covariant set of hydrodynamics equations, in which we consider u^μ ($\mu = t, x, y, z$) and P as the primitive variables (remember that the eigenvalues do not depend on the variables chosen to express the basic equations). As before, we consider the perfect fluid satisfying the adiabatic condition $s = $ constant. Then, $\nabla_a T^{ab} = 0$ yields

$$\rho h u^a \nabla_a u^b + h^{ab} \nabla_a P = 0, \tag{4.69}$$

where $h^{ab} := g^{ab} + u^a u^b$. Noting that the first law of thermodynamics is written as $u^a \nabla_a P = h c_s^2 u^a \nabla_a \rho$ for $s = $ constant, we also have

$$\rho h \nabla_a u^a + \frac{1}{c_s^2} u^a \nabla_a P = 0, \tag{4.70}$$

where we used the continuity equation for obtaining this form. The set of equations (4.69) and (4.70) is written in a matrix form

$$\sum_{I=1}^{5} \left(A_I{}^J\right)^\mu \nabla_\mu V^I = 0 \quad (J = 1, 2, \cdots, 5), \tag{4.71}$$

where $V^I = (u^\mu, P)$ is a five-components vector, and $\left(A_I{}^J\right)^\mu$ is a 5×5 matrix written as

$$\left(A_I{}^J\right)^\mu = \begin{pmatrix} \rho h u^\mu \delta_I^J & h^{J\mu} \\ \rho h \delta_I^\mu & u^\mu/c_s^2 \end{pmatrix}. \tag{4.72}$$

Here, δ_I^J is the 4×4 unit matrix, and I and J denote components of columns and rows, respectively.

Now we consider the characteristics in the x-direction. Let U^μ be a vector that points to the direction of the characteristics and we write its components as $U^\mu = (1, \lambda, 0, 0)$ for (t, x, y, z). Then, a dual vector perpendicular to U^μ is written as $\phi_\mu = (-\lambda, 1, 0, 0)$, and $\left(A_I{}^J\right)^\mu$ satisfies

$$\det\left[\left(A_I{}^J\right)^\mu \phi_\mu\right] = 0. \tag{4.73}$$

The matrix $\left(A_I{}^J\right)^\mu \phi_\mu$ is written in a simple form

$$\left(A_I{}^J\right)^\mu \phi_\mu = \begin{pmatrix} \rho h a & 0 & 0 & 0 & h^{t\mu}\phi_\mu \\ 0 & \rho h a & 0 & 0 & h^{x\mu}\phi_\mu \\ 0 & 0 & \rho h a & 0 & h^{y\mu}\phi_\mu \\ 0 & 0 & 0 & \rho h a & h^{z\mu}\phi_\mu \\ \rho h \phi_t & \rho h \phi_x & \rho h \phi_y & \rho h \phi_z & a/c_s^2 \end{pmatrix}, \tag{4.74}$$

where $a := u^\mu \phi_\mu = u^t(-\lambda + v^x)$. It is easy to calculate the determinant of this matrix and the resulting eigenvalues agree with those shown in equation (4.66). This method is in particular useful for deriving the eigenvalues for magnetohydrodynamics systems (see section 4.6.3).

For the derived eigenvalues, the next task for a solution of hydrodynamics equations is to derive the right eigenvectors of the Jacobian matrix. We denote them by $\left(r_a^k\right)^{(I)}$ where $I = 1, \cdots, 5$. Then, five independent eigenvectors are present, satisfying

$$\sum_{b=1}^{5} M_{ab}^k \left(r_b^k\right)^{(I)} = \left(\lambda^k\right)^{(I)} \left(r_a^k\right)^{(I)}, \tag{4.75}$$

where we set $\left(\lambda^k\right)^{(1)} = \lambda_+^k$, $\left(\lambda^k\right)^{(2)} = \left(\lambda^k\right)^{(3)} = \left(\lambda^k\right)^{(4)} = v^k$, and $\left(\lambda^k\right)^{(5)} = \lambda_-^k$. For a 5×5 matrix defined from $\left(r_a^k\right)^{(I)}$ by

$$R_{ab}^k = \left[\left(r_a^k\right)^{(1)}, \left(r_a^k\right)^{(2)}, \left(r_a^k\right)^{(3)}, \left(r_a^k\right)^{(4)}, \left(r_a^k\right)^{(5)}\right], \tag{4.76}$$

M_{ab}^k is diagonalized as

$$M_{ab}^k = \sum_{c,d} R_{ac}^k \Lambda_{cd}^k \left(R_{bd}^k\right)^{-1}, \tag{4.77}$$

where Λ^k_{cd} is a 5×5 diagonal matrix composed of $(\lambda^k)^{(I)}$ in the following order: $[\lambda^k_+, v^k, v^k, v^k, \lambda^k_-]$. Using the decomposition (4.77), the left-hand side of equation (4.59) is written as

$$\boldsymbol{R}^k \left[\left(\boldsymbol{R}^k\right)^{-1} \partial_t \boldsymbol{Q} + \Lambda^k \left(\boldsymbol{R}^k\right)^{-1} \partial_k \boldsymbol{Q} \right], \tag{4.78}$$

where we omitted the trivial subscripts denoting the matrix components, for simplicity.

For a specific example, we here consider the advection only in the x-direction, assuming that the source term, $\boldsymbol{S}$, is vanishing. Here, the vanishing source term is equivalent to vanishing gravitational forces, and hence, we may say that we focus on local advection of a hydrodynamic flow. In this case, equation (4.78) becomes

$$(\boldsymbol{R}^x)^{-1} \partial_t \boldsymbol{Q} + \Lambda^x (\boldsymbol{R}^x)^{-1} \partial_x \boldsymbol{Q} = 0, \tag{4.79}$$

or recovering the subscripts of the matrix element, we have

$$\sum_{b=1}^{5} \left[(\boldsymbol{R}^x)^{-1}_{ab} \partial_t \boldsymbol{Q}_b + (\lambda^x)^{(a)} (\boldsymbol{R}^x)^{-1}_{ab} \partial_x \boldsymbol{Q}_b \right] = 0. \tag{4.80}$$

This shows that hydrodynamics equations (with $\boldsymbol{S}_a = 0$) locally constitute five-components simultaneous wave equations with the characteristic speeds $(\lambda^x)^{(a)}$ ($a = 1, \cdots, 5$). This implies that the general solution for each component of $\boldsymbol{Q}_a$ should be written in the superposition of functions of $x - (\lambda^x)^{(I)} t$ as

$$\boldsymbol{Q}_a = \sum_{I=1}^{5} F^I_a \left[x - (\lambda^x)^{(I)} t \right], \tag{4.81}$$

where F^I_a is a set of functions which can be in principle determined from equation (4.80). As this example shows, the general solution of the hydrodynamics equations is written in an analytic manner, if we focus only on one direction and we do not consider gravitational forces. In modern numerical hydrodynamics, this fact is used for many implementations.

Specifically, the analytic calculation of $\boldsymbol{R}^k_{ab}$ is performed in the following manner [Font et al. (1994); Banyuls et al. (1997)]. First, we define a matrix $\boldsymbol{T}^k_{ab}$ that satisfies the relation

$$\boldsymbol{R}^k_{ab} = \sum_{c=1}^{5} \boldsymbol{C}_{ac} \boldsymbol{T}^k_{cb}. \tag{4.82}$$

Because $\boldsymbol{R}^k_{ab}$ is composed of the right eigenvectors of $\boldsymbol{M}^k_{ab}$, $\boldsymbol{T}^k_{ab}$ is composed of the five vectors $\boldsymbol{t}^{(I)}_a$ ($I = 1, \cdots, 5$) that satisfy the equation

$$\sum_{b=1}^{5} \left(\boldsymbol{B}^k_{ab} - \lambda^k \boldsymbol{C}_{ab} \right) \left(\boldsymbol{t}^k_b\right)^{(I)} = 0 \quad \text{for} \quad I = 1, \cdots, 5. \tag{4.83}$$

This equation for $\left(t_a^k\right)^{(I)}$ is solved in a straightforward calculation, and then, $\boldsymbol{T}_{ab}^k = [(t_a^k)^{(1)}, (t_a^k)^{(2)}, (t_a^k)^{(3)}, (t_a^k)^{(4)}, (t_a^k)^{(5)}]$ is obtained as

$$\boldsymbol{T}_{ab}^x = \begin{bmatrix} 1 & -\kappa & 0 & 0 & 1 \\ H^{xx}(\lambda_+^x) & 0 & 0 & 0 & H^{xx}(\lambda_-^x) \\ H^{xy}(\lambda_+^x) & 0 & \rho^{-1} & 0 & H^{xy}(\lambda_-^x) \\ H^{xz}(\lambda_+^x) & 0 & 0 & \rho^{-1} & H^{xz}(\lambda_-^x) \\ \dfrac{P}{\rho^2} & \dfrac{\chi}{\rho} & 0 & 0 & \dfrac{P}{\rho^2} \end{bmatrix}, \quad (4.84)$$

$$\boldsymbol{T}_{ab}^y = \begin{bmatrix} 1 & 0 & -\kappa & 0 & 1 \\ H^{yx}(\lambda_+^y) & \rho^{-1} & 0 & 0 & H^{yx}(\lambda_-^y) \\ H^{yy}(\lambda_+^y) & 0 & 0 & 0 & H^{yy}(\lambda_-^y) \\ H^{yz}(\lambda_+^y) & 0 & 0 & \rho^{-1} & H^{yz}(\lambda_-^y) \\ \dfrac{P}{\rho^2} & 0 & \dfrac{\chi}{\rho} & 0 & \dfrac{P}{\rho^2} \end{bmatrix}, \quad (4.85)$$

$$\boldsymbol{T}_{ab}^z = \begin{bmatrix} 1 & 0 & 0 & -\kappa & 1 \\ H^{zx}(\lambda_+^z) & \rho^{-1} & 0 & 0 & H^{zx}(\lambda_-^z) \\ H^{zy}(\lambda_+^z) & 0 & \rho^{-1} & 0 & H^{zy}(\lambda_-^z) \\ H^{zz}(\lambda_+^z) & 0 & 0 & 0 & H^{zz}(\lambda_-^z) \\ \dfrac{P}{\rho^2} & 0 & 0 & \dfrac{\chi}{\rho} & \dfrac{P}{\rho^2} \end{bmatrix}, \quad (4.86)$$

where

$$H^{jk}(\lambda) = \frac{-c_s^2\left\{\alpha^2 \gamma^{jk} - V^k(\beta^j + \lambda)\right\}}{\rho w^2(v^j - \lambda)} \quad (j,k = x,y,z). \quad (4.87)$$

4.5 Hydrodynamics with microphysics

One of the important issues in numerical relativity is to clarify the mechanism of the merger of neutron-star binaries (see chapter 8) and the gravitational collapse of a massive stellar core (see chapter 9). In these phenomena, a state with high density, $\rho \gtrsim 10^{14}$ g/cm^3, and high temperature, $T \gtrsim 10$ MeV, is realized. In a high-density state with $\rho \gtrsim 10^{11}$ g/cm^3, the electron degenerate pressure could play a major role in determining the total pressure and the neutron richness. In addition, if a high-temperature state with $T \gtrsim$ a few MeV is realized, the neutrino emission and neutrino transfer could also play an important role not only in cooling and heating but also in determining the electron number density, because one of the dominant neutrino processes is electron and positron captures on nucleons and heavy nuclei. These facts imply that we need to accurately follow the evolution of the electron number density and the neutrino transfer processes for physically modeling high-density and high-temperature phenomena. Therefore, in addition to the continuity, Euler, and energy equations, we have to solve equations for them:

One is the radiation transfer equation for neutrinos, for which several formulations and methods will be described in section 4.7. The other is the evolution equation for lepton number density. In this section, we describe the latter supposing that a microphysically-motivated equation of state is employed.

Equations of state for high-density and high-temperature matter, which have been derived aiming primarily at the use for core-collapse supernova simulations, are usually constructed in a tabulated form [Lattimer and Swesty (1991); Shen et al. (1998a,b, 2011c,b,a); Hempel et al. (2012); Furusawa et al. (2013)]. In them, thermodynamical quantities such as pressure, internal energy, and entropy (or temperature) as well as fractions of protons, neutrons, light nuclei, and heavy nuclei are written as functions of the rest-mass density, ρ, the temperature, T (or the specific entropy, s), and the electron fraction per nucleon, Y_e; e.g., the pressure has the form $P = P(\rho, T, Y_e)$ [or $P = P(\rho, s, Y_e)$]: see section 1.4.7. In general relativistic hydrodynamics, ρ and T are determined from the weighted rest-mass density, ρ_*, the momentum density, S_i, and the energy density, S_0. However, an additional equation or assumption for determining Y_e is necessary.

The simplest method to determine Y_e is to impose some conditions. For example, assuming the β-equilibrium such as $p + e^- \leftrightarrow n + \nu_e$ and $n + e^+ \leftrightarrow p + \bar{\nu}_e$ yields a way for determining Y_e. However, such assumptions are valid only for the restricted cases (β-equilibrium is not always satisfied in nature). For physical simulations, in general, we have to follow the evolution of Y_e. The evolution equation for Y_e is written in the fluid rest frame as

$$u^a \nabla_a Y_e = -\gamma_e, \tag{4.88}$$

where γ_e is defined by the local capture rate of electrons by nuclei minus the local capture rate of positrons by nuclei. Multiplying ρ in equation (4.88) yields

$$\nabla_a(\rho Y_e u^a) = -\rho \gamma_e. \tag{4.89}$$

Here, we used the continuity equation, $\nabla_a(\rho u^a) = 0$. In a coordinate basis, it is written as

$$\partial_t(\rho_* Y_e) + \partial_k(\rho_* Y_e v^k) = -\rho\sqrt{-g}\gamma_e. \tag{4.90}$$

Thus, a continuity-type equation with a source term, $-\rho\sqrt{-g}\gamma_e$, becomes the basic equation.

If the fractions of neutrinos and total leptons, Y_ν and Y_l, are required to be followed in addition, we also have to solve their continuity-type equations as

$$\nabla_a(\rho Y_\nu u^a) = -\rho \gamma_\nu, \tag{4.91}$$

$$\nabla_a(\rho Y_l u^a) = -\rho \gamma_l, \tag{4.92}$$

where γ_ν and γ_l are defined by the capture rates of neutrinos and leptons, respectively. For further details, we recommend the readers to refer, e.g., to Sekiguchi (2010b).

The method for a numerical solution of u^t (or $w = \alpha u^t$) becomes complicated when we employ tabulated finite-temperature equations of state. In general relativistic hydrodynamics, the quantities numerically evolved are ρ_*, $\rho_* Y_e$ (or $\rho_* Y_l$), $S_i (= \rho_* h u_i)$, $S_0 (= \rho_* e_0)$, and geometric variables, while the argument variables of tabulated equations of state are (ρ, T, Y_e) or (ρ, T, Y_l). Here, Y_e or Y_l is easily obtained from ρ_* and $\rho_* Y_e$ or $\rho_* Y_l$, whereas ρ and T are not determined directly by ρ_*, $\hat{u}_i := h u_i$, and e_0. This fact requires us to explore an efficient method for determining ρ and T.

Since ρ is determined by $\rho = \rho_*/(w\sqrt{\gamma})$, finding a solution of ρ is equivalent to finding a solution of w. Here, from the normalization relation of u^a, w is written as

$$w^2 = 1 + h^{-2} \gamma^{ij} \hat{u}_i \hat{u}_j, \tag{4.93}$$

and h in tabulated equations of state is a function of (ρ, T, Y_e) or (ρ, T, Y_l) with $\rho = \rho_*/(w\sqrt{\gamma})$. We also have an algebraic equation

$$e_0 = hw - \frac{P}{\rho w} = hw - \frac{P\sqrt{\gamma}}{\rho_*}. \tag{4.94}$$

Here, P is also a function of (ρ, T, Y_e) or (ρ, T, Y_l), and thus, the right-hand side of this equation is also a function of these argument quantities. Since ρ_* and Y_e or Y_l are determined by the numerical evolution, the task is to find solutions of equations (4.93) and (4.94) for (w, T) in the table. However, the two-dimensional table search is highly elaborated and we have to develop a smart method in this search.

Sekiguchi (2010b) developed the following prescription. First, one should give a trial function of w. This could be determined by a guess from the values at previous time steps. Then, we will have a trial value of ρ calculated by $\rho_*/(w\sqrt{\gamma})$. For the hypothetical value of ρ, the table search reduces to a one-dimensional search for T. This T is subsequently determined by finding a solution of equation (4.94). Because this is the one-dimensional table search, the computational cost is not very expensive. For the obtained values of ρ and T, we can obtain h and P using the table, and then, can update the value of w using equation (4.93). This procedure is repeated until a convergent solution is obtained. Sekiguchi (2010b) indeed found that this method is efficient for a solution of w and h.

4.6 Electromagnetohydrodynamics

Magnetohydrodynamics and electrohydrodynamics simulations in general relativity have been an active field since 2000s. For these simulations, we have to solve Maxwell's equations as well as equations of motion for the matter fields. Maxwell's evolution equations are composed basically of evolution equations for the electric and magnetic fields. In addition, we have to solve the conservation equation of the electric charge with appropriately determining the electric current in general.

In astrophysics, however, the equations to be solved depend on the situation. In the following, we will first summarize the general electromagnetic equations to be solved. Then, we will describe a set of equations that is often employed in numerical astrophysics.

4.6.1 Definitions

For electromagnetism, the fundamental variables typically chosen are an electric field, E^a, a magnetic field, B^a, and a spacetime electric current vector, j^a. Here, E^a and B^a are defined in a laboratory frame [e.g., Thorne et al. (1986)], and in the context of the 3+1 formulation,

$$E^a n_a = 0, \quad B^a n_a = 0, \tag{4.95}$$

where n^a denotes the unit vector field normal to spatial hypersurfaces Σ_t as before. Equation (4.95) implies that E^a and B^a are purely spatial, i.e., $E^t = B^t = 0$, and

$$E_a = \gamma_{ab} E^b, \quad B_a = \gamma_{ab} B^b. \tag{4.96}$$

In other words, E^a and B^a are vector fields on spatial hypersurfaces Σ_t.

In terms of E^a and B^a, the antisymmetric electromagnetic tensor field is written as

$$F^{ab} = n^a E^b - n^b E^a + \epsilon^{abc} B_c, \tag{4.97}$$

where ϵ_{abc} is the three-dimensional Levi–Civita tensor defined by

$$\epsilon_{abc} := n^d \epsilon_{dabc}. \tag{4.98}$$

Equation (4.97) implies that E^a and B^a are defined from F^{ab} as

$$E^a = F^{ab} n_b, \quad B^a = \frac{1}{2} \epsilon^{abc} F_{bc}. \tag{4.99}$$

Here, ϵ_{ijk} satisfies the condition

$$\epsilon_{ijk} = \sqrt{\gamma} e_{ijk}, \tag{4.100}$$

where e_{ijk} is the Levi–Civita tensor in flat space, and satisfies $e_{xyz} = e^{xyz} = 1$ in Cartesian coordinates. Note that ϵ^{abc} is nonzero only for spatial indices, while ϵ_{abc} may not be zero even in the case that one index is t. Specifically, $n^a \epsilon_{abc} = 0$ gives

$$\epsilon_{tij} = \beta^k \epsilon_{kij}, \tag{4.101}$$

and thus, in the presence of $\beta^k \neq 0$, $\epsilon_{tij} \neq 0$.

j^a is also decomposed as

$$j^a = \rho_e n^a + \bar{j}^a, \tag{4.102}$$

where $\rho_e := -j^a n_a$ is the electric charge density defined on spatial hypersurfaces, and $\bar{j}^a = \gamma^a{}_b j^b$ is the electric current vector on spatial hypersurfaces Σ_t (i.e., $\bar{j}^a n_a = 0$).

The stress-energy tensor is composed of the purely hydrodynamic part, $T_{ab}^{\rm HD}$, and the electromagnetic part, $T_{ab}^{\rm EM}$, as

$$T_{ab} = T_{ab}^{\rm HD} + T_{ab}^{\rm EM}, \tag{4.103}$$

where

$$T_{ab}^{\rm HD} = (\rho + \rho\varepsilon + P)u_a u_b + P g_{ab}, \tag{4.104}$$

$$\begin{aligned} T_{ab}^{\rm EM} &= \frac{1}{4\pi}\left(F_{ac}F_b{}^c - \frac{1}{4}g_{ab}F_{cd}F^{cd}\right) \\ &= \frac{1}{4\pi}\Bigg[\frac{E^2+B^2}{2}(\gamma_{ab}+n_a n_b) - E_a E_b - B_a B_b \\ &\quad + n_a \epsilon_{bcd}E^c B^d + n_b \epsilon_{acd}E^c B^d\Bigg], \end{aligned} \tag{4.105}$$

with $E^2 = E_a E^a$ and $B^2 = B_b B^b$. For the following calculations, we define

$$^*F_{ab} := \frac{1}{2}\epsilon_{abcd}F^{cd} = \epsilon_{abc}E^c - n_a B_b + n_b B_a, \tag{4.106}$$

$$\tag{4.107}$$

and note the relations

$$F_{ab}F^{ab} = 2(B^2 - E^2), \tag{4.108}$$

$$^*F_{ab}F^{ab} = 4E^a B_a. \tag{4.109}$$

The 3+1 decomposition of T_{ab} yields

$$\rho_{\rm h} := T_{ab}n^a n^b = \rho h w^2 - P + \frac{1}{8\pi}(E^2 + B^2), \tag{4.110}$$

$$J_a := -T_{bc}n^b \gamma^c{}_a = \rho h w u_a + \frac{1}{4\pi}\epsilon_{abc}E^b B^c, \tag{4.111}$$

$$\begin{aligned} S_{ab} &:= T_{cd}\gamma^c{}_a \gamma^d{}_b \\ &= \rho h u_a u_b + P\gamma_{ab} + \frac{1}{4\pi}\left[-E_a E_b - B_a B_b + \frac{1}{2}\gamma_{ab}(E^2+B^2)\right]. \end{aligned} \tag{4.112}$$

We also define $S_0 := \sqrt{\gamma}\rho_{\rm h}$ and $S_i := \sqrt{\gamma}J_i$ as before. Using $\rho_{\rm h}$, J_a, and S_{ab}, the stress-energy tensor is written to the 3+1 form (4.25). Substituting these variables into equations (4.31) and (4.33), evolution equations for S_i and S_0 are obtained, respectively (see section 4.6.1.2). The definition of these variables indicates that $(E^2 + B^2)/8\pi$ in $\rho_{\rm h}$ denotes electromagnetic energy density and $\epsilon_{abc}E^b B^c/4\pi$ in J_a is a Poynting flux defined on spatial hypersurfaces Σ_t.

For the following manipulations, we also define new variables for the electromagnetic part as

$$\mathcal{E}^a := \sqrt{\gamma}E^a, \quad \mathcal{B}^a := \sqrt{\gamma}B^a, \quad Q := \sqrt{\gamma}\rho_{\rm e}, \quad \mathcal{J}^a := \alpha\sqrt{\gamma}\bar{j}^a.$$

Maxwell's equations and charge continuity equations are written into a simple form using these variables.

4.6.1.1 Maxwell's equations

Maxwell's equations are written as

$$\nabla_a F^{ab} = -4\pi j^b, \tag{4.113}$$

$$\nabla_{[a} F_{bc]} = 0 \quad \text{or} \quad \nabla_a {}^*F^{ab} = 0. \tag{4.114}$$

Using these Maxwell's equations, the spacetime covariant derivative of T_{ab}^{EM} is written as

$$\nabla^b T_{ab}^{\text{EM}} = -F_{ab} j^b, \tag{4.115}$$

and thus,

$$\nabla^b T_{ab}^{\text{HD}} = F_{ab} j^b. \tag{4.116}$$

The right-hand side of this equation denotes the electromagnetic forces.

Equation (4.113) together with the antisymmetric property of F^{ab} yields the continuity equation for the electric charge,

$$\nabla_a j^a = 0, \tag{4.117}$$

which determines the evolution of ρ_e. In addition to this equation, we need an equation for determining the three current $\bar{j}^a$ that appears in equations (4.113) and (4.117). The Ohm's law is the corresponding equation unless the system considered is a perfect conductor, and is written in the form

$$j^a + (j^b u_b) u^a = \sigma_c F^{ab} u_b, \tag{4.118}$$

where σ_c denotes the conductivity, for which we suppose in this section that $\sigma_c < \infty$. We also call σ_c^{-1} the resistivity. We note that for ideal magnetohydrodynamics case, $\sigma_c = \infty$ is assumed, and thus, $F^{ab} u_b = 0$ is assumed to realize a finite value of j^a. Equations in ideal magnetohydrodynamics will be described in section 4.6.2.

Now, we are going to write down explicit forms of equations for E^a and B^a. Operating n_a to equation (4.113) and ϵ^{abc} to equation (4.114), respectively, yields the constraint equations (Gauss' law and no-monopole constraint) as

$$D_a E^a = 4\pi \rho_e, \tag{4.119}$$

$$D_a B^a = 0. \tag{4.120}$$

In a coordinate basis, these are written as

$$\partial_k \mathcal{E}^k = 4\pi Q, \tag{4.121}$$

$$\partial_k \mathcal{B}^k = 0. \tag{4.122}$$

Note that equation (4.121) may be used for determining Q during the evolution of a system.

Other components of equations (4.113) and (4.114) are the evolution equations for E^a and B^a. Using the following relations,

$$\nabla_a F^{ab} = \frac{1}{\alpha\sqrt{\gamma}} \partial_a \left(\alpha\sqrt{\gamma} F^{ab} \right), \tag{4.123}$$

$$\nabla_a {}^*F^{ab} = \frac{1}{\alpha\sqrt{\gamma}} \partial_a \left(\alpha\sqrt{\gamma} {}^*F^{ab} \right), \tag{4.124}$$

the evolution equations for E^i and B^i (Ampére–Maxwell's law and Faraday's law) are written as

$$\partial_t E^i - \mathscr{L}_\beta E^i = \alpha K E^i - D_k\left(\alpha \epsilon^{kij} B_j\right) - 4\pi\alpha \bar{j}^i, \qquad (4.125)$$

$$\partial_t B^i - \mathscr{L}_\beta B^i = \alpha K B^i + D_k\left(\alpha \epsilon^{kij} E_j\right), \qquad (4.126)$$

where $\mathscr{L}_\beta$ is the Lie derivative with respect to β^i. These are written in conservative forms as

$$\partial_t \mathcal{E}^i = -\partial_k\left(\beta^i \mathcal{E}^k - \beta^k \mathcal{E}^i + \alpha \epsilon^{kij} \mathcal{B}_j\right) - 4\pi\left(\mathcal{J}^i - Q\beta^i\right), \qquad (4.127)$$

$$\partial_t \mathcal{B}^i = -\partial_k\left(\beta^i \mathcal{B}^k - \beta^k \mathcal{B}^i - \alpha \epsilon^{kij} \mathcal{E}_j\right). \qquad (4.128)$$

In numerical simulations, the conservative form is favorable because it enables us to guarantee conservation laws such as the magnetic-flux conservation and the no-monopole constraint.

The continuity equation for the electric charge density, (4.117), is written in the form

$$\partial_t Q + \partial_k\left(\mathcal{J}^k - Q\beta^k\right) = 0. \qquad (4.129)$$

Here, $\mathcal{J}^k$ should be determined by an equation derived from Ohm's law (4.118) in the following manner: Operating n_a and $\gamma_a{}^i$ to equation (4.118) yields

$$-\rho_e - (j^a u_a)w = -\sigma_c E^a u_a, \qquad (4.130)$$

$$\bar{j}^i + (j^a u_a)\gamma^{ib} u_b = \sigma_c\left(wE^i + \epsilon^{ijk} u_j B_k\right), \qquad (4.131)$$

where $w = \alpha u^t$ as before. Eliminating $j^a u_a$ from equations (4.130) and (4.131), $\bar{j}^i$ is written as

$$\bar{j}^i = w^{-1}\gamma^{ij} u_j\left(\rho_e - \sigma_c E^k u_k\right) + \sigma_c\left(wE^i + \epsilon^{ijk} u_j B_k\right), \qquad (4.132)$$

and therefore

$$\mathcal{J}^i = (v^i + \beta^i)Q + \sigma_c\left[-\left(v^i + \beta^i\right)\mathcal{E}^j u_j + \alpha\left(w\mathcal{E}^i + \epsilon^{ijk} u_j \mathcal{B}_k\right)\right], \qquad (4.133)$$

where we used $\gamma^{ij} u_j = u^t(v^i + \beta^i)$ [see equation (4.38)]. This is the Ohm's law in 3+1 form. Thus, the continuity equation for the electric charge density is written in the form

$$\partial_t Q + \partial_k\left(Qv^k\right) = \sigma_c \partial_k S_{\text{el}}^k, \qquad (4.134)$$

where

$$S_{\text{el}}^k = \left(v^k + \beta^k\right)\mathcal{E}^j u_j - \alpha\left(w\mathcal{E}^k + \epsilon^{kij} u_i \mathcal{B}_j\right), \qquad (4.135)$$

and we assumed that σ_c is constant. Equation (4.134) shows that the evolution of the electric charge density is determined by the bulk motion and a drift associated with the resistivity, σ_c^{-1}. For the infinite resistivity with $\sigma_c = 0$, the bulk motion simply determines the electric charge distribution.

To summarize, $\mathcal{E}^i$ and $\mathcal{B}^i$ are determined by solving equations (4.127) and (4.128), respectively, and Q is determined by solving either equation (4.121) or (4.134).

4.6.1.2 Electromagnetohydrodynamics equations

Electromagnetohydrodynamics equations are composed of equations of motion and energy equation which are derived by substituting equation (4.103) into (4.27) and (4.28), respectively. There are two methods for writing down these equations.

In the first method, the electromagnetic force is treated as an external force, and hence, only the fluid part is written in a conservative form. Using equation (4.115), equations (4.27) and (4.28) are written as

$$\gamma_i{}^b \nabla^a T_{ab}^{\mathrm{HD}} = F_{ia} j^a, \tag{4.136}$$

$$n^b \nabla^a T_{ab}^{\mathrm{HD}} = F_{ab} n^a j^b. \tag{4.137}$$

The left-hand sides of these equations are rewritten in the same manner as in the pure hydrodynamics case (see section 4.4), and thus,

$$\partial_t S_i^{\mathrm{HD}} + \partial_k \left(S_i^{\mathrm{HD}} v^k + P\alpha\sqrt{\gamma}\delta^k{}_i \right) = -S_0^{\mathrm{HD}} \partial_i \alpha + S_k^{\mathrm{HD}} \partial_i \beta^k - \frac{1}{2}\alpha\sqrt{\gamma} S_{jk}^{\mathrm{HD}} \partial_i \gamma^{jk}$$
$$+ \frac{1}{\sqrt{\gamma}} \left(\alpha Q \mathcal{E}_i + \varepsilon_{ijk} \mathcal{J}^j \mathcal{B}^k \right), \tag{4.138}$$

$$\partial_t S_0^{\mathrm{HD}} + \partial_i \left[S_0^{\mathrm{HD}} v^i + P(v^i + \beta^i)\sqrt{\gamma} \right] = \alpha\sqrt{\gamma} S_{ij}^{\mathrm{HD}} K^{ij} - S_i^{\mathrm{HD}} D^i \alpha + \frac{1}{\sqrt{\gamma}} \mathcal{J}_i \mathcal{E}^i, \tag{4.139}$$

where as in equation (4.36),

$$S_i^{\mathrm{HD}} := -\sqrt{\gamma} T_{ab}^{\mathrm{HD}} n^a \gamma^b{}_i = \sqrt{\gamma}\rho w h u_i = \rho_* h u_i, \tag{4.140}$$

$$S_0^{\mathrm{HD}} := \sqrt{\gamma} T_{ab}^{\mathrm{HD}} n^a n^b = \sqrt{\gamma}\left(\rho h w^2 - P\right). \tag{4.141}$$

In this case, the normalization relation for u^a is written in the same form as equation (4.46):

$$w^2 = 1 + \gamma^{ij} u_i u_j = 1 + \gamma^{ij} \frac{S_i^{\mathrm{HD}} S_j^{\mathrm{HD}}}{\rho_*^2 h^2}, \tag{4.142}$$

and thus, the equation for w has the same form as before.

In this formulation, numerical integration of the basic equations can be done using the same method as that used for the pure hydrodynamics equations. However, the conservation of the total momentum and energy (sum of the hydrodynamic and electromagnetic parts) is not guaranteed. Moreover, shocks may not be accurately captured because the information of the characteristic speeds associated with the electromagnetic field is not taken into account in the advection terms.

Next, we describe a fully conservative form, which is obtained by substituting the stress-energy tensor into equations (4.31) and (4.33):

$$\partial_t S_j + \partial_i \left[S_j v^i + \left(P + \frac{E^2 + B^2}{8\pi} \right) \alpha \sqrt{\gamma} \delta^i{}_j - \frac{E^i E_j + B^i B_j}{4\pi} \alpha \sqrt{\gamma} \right.$$
$$\left. - \frac{1}{4\pi}(v^i + \beta^i)\epsilon_{jkl} E^k B^l \sqrt{\gamma} \right]$$
$$= -S_0 \partial_j \alpha + S_i \partial_j \beta^i - \frac{\alpha}{2} \sqrt{\gamma} S_{kl} \partial_j \gamma^{kl}, \qquad (4.143)$$

$$\partial_t S_0 + \partial_i \left[S_0 v^i + \left(P - \frac{E^2 + B^2}{8\pi} \right)(v^i + \beta^i)\sqrt{\gamma} + \frac{\alpha \sqrt{\gamma}}{4\pi} \epsilon^i{}_{jk} E^j B^k \right]$$
$$= \alpha \sqrt{\gamma} S_{ij} K^{ij} - S_i D^i \alpha. \qquad (4.144)$$

Here $S_0 := \rho_h \sqrt{\gamma}$, $S_i := J_i \sqrt{\gamma}$, and S_{ij} are defined by equations (4.110)–(4.112).

The algebraic equation of the normalization relation for u^a becomes slightly complicated. First, we have to subtract the electromagnetic part from S_i and S_0 to obtain S_i^{HD} and S_0^{HD} as

$$S_i^{\mathrm{HD}} = S_i - \frac{\sqrt{\gamma}}{4\pi} \epsilon_{ijk} E^j B^k, \qquad (4.145)$$

$$S_0^{\mathrm{HD}} = S_0 - \frac{\sqrt{\gamma}}{8\pi}\left(E^2 + B^2 \right). \qquad (4.146)$$

Then, we obtain

$$w^2 = 1 + \gamma^{ij} \frac{S_i^{\mathrm{HD}} S_j^{\mathrm{HD}}}{\rho_*^2 h^2}. \qquad (4.147)$$

Here, $S_0^{\mathrm{HD}}/\rho_* = hw - P\sqrt{\gamma}/\rho_*$ [see equation (4.47)], and this gives the relation between h and w for a given equation of state $P = P(\rho, h)$.

The subtlety in numerical electromagnetohydrodynamics is that the positivity of the resulting value of S_0^{HD} is not always guaranteed within the numerical accuracy. For the case that it is negative, a physical solution for w and h is not obtained. Such pathology could be encountered when the magnitude of the electric and/or magnetic fields become accidentally large (accompanying usually with an accidentally large numerical error). For such a case, S_0 would be also very large, and thus, S_0^{HD} has to be obtained by subtracting a large number from another large number. However, the accurate operation for this type of the subtraction is usually impossible in numerical simulations. Hence, to avoid the accidental enhancement of the field magnitude, an accurate integration scheme or an artificial restriction for the resulting numerical solution is always required when the equations in the conservative form are solved.

4.6.2 Ideal magnetohydrodynamics

Basic equations for ideal magnetohydrodynamics in general relativity in a conservative form are written in a number of articles, e.g., Koide et al. (1999); Komissarov (1999); McKinney and Gammie (2004); Duez et al. (2005); Shibata and Sekiguchi (2005a); Antón et al. (2006, 2010). We here follow the derivation described in Shibata and Sekiguchi (2005a).

In the ideal magnetohydrodynamics case, $\sigma_c = \infty$, and thus, for a finite value of j^a, we have to require

$$F^{ab}u_b = 0. \tag{4.148}$$

This equation implies that the electric field in the frame comoving with fluids vanishes. The spatial component of equation (4.148) together with equation (4.97) yields

$$E^i = -\frac{1}{w}\epsilon^{ijk}u_j B_k. \tag{4.149}$$

This implies that E^i and B^i are perpendicular each other ($E_a B^a = 0$), and in addition, if B^k as well as u_j are determined by their evolution equations, E^i is automatically obtained. Thus, we do not have to solve the equations for E^i. In this case, equations (4.121) and (4.127) are used for determining the charge density and the electric current density from E^i, respectively.

Equation (4.128) determines the evolution of the magnetic field. Substituting equation (4.149) into (4.128), we have the induction equation

$$\partial_t \mathcal{B}^i = \partial_k(v^i \mathcal{B}^k - v^k \mathcal{B}^i). \tag{4.150}$$

This is only one equation to be solved for the electromagnetic part in ideal magnetohydrodynamics.

Because it satisfies equation (4.148), F^{ab} in ideal magnetohydrodynamics may be written in the form

$$F^{ab} = u_d \epsilon^{dabc} b_c, \tag{4.151}$$

and thus,

$${}^*F^{ab} = b^a u^b - b^b u^a, \tag{4.152}$$

where b^a denotes the magnetic-field vector in the frame comoving with fluids, satisfying $b^a u_a = 0$. Using b^a, the stress-energy tensor of the electromagnetic field is simply written as

$$T^{EM}_{ab} = \frac{1}{4\pi}\left(b^2 u_a u_b + \frac{b^2}{2}g_{ab} - b_a b_b\right), \tag{4.153}$$

where $b^2 = b^a b_a$. Thus, it is found that $b^2/4\pi$ is the electromagnetic energy density and $b^2/8\pi$ is the magnetic pressure. T^{EM}_{ab}, written in this simple form, helps deriving magnetohydrodynamics equations in a straightforward manner. Indeed, ρ_h, J_a, and S_{ab} are written in the following simple forms:

$$\rho_h = \left(\rho h + \frac{b^2}{4\pi}\right)w^2 - \left(P + \frac{b^2}{8\pi}\right) - \frac{1}{4\pi}(\alpha b^t)^2, \tag{4.154}$$

$$J_a = \left(\rho h + \frac{b^2}{4\pi}\right)wu_a - \frac{1}{4\pi}\alpha b^t b_a, \tag{4.155}$$

$$S_{ab} = \left(\rho h + \frac{b^2}{4\pi}\right)u_a u_b + \left(P + \frac{b^2}{8\pi}\right)\gamma_{ab} - \frac{1}{4\pi}b_a b_b. \tag{4.156}$$

Substituting these variables (with $S_0 := \rho_{\rm h}\sqrt{\gamma}$ and $S_i := J_i\sqrt{\gamma}$) into equations (4.31) and (4.33) yields the equations for S_i and S_0 as before.

Multiplying equation (4.151) by ϵ_{eab} yields the relation between B^a and b^a as

$$B^a = wb^a + u^a b^c n_c = wb^a - \alpha u^a b^t. \tag{4.157}$$

Multiplying this equation by u_a, we also have $\alpha b^t = B^k u_k$, and

$$b_a = \frac{1}{w}\left[B_a + u_a(B^k u_k)\right], \tag{4.158}$$

$$b^2 = \frac{1}{w^2}\left[B^2 + (B^k u_k)^2\right]. \tag{4.159}$$

Thus, the terms composed of b^a are immediately written in terms of B^a. For the following calculation, we write S_0 and S_i as

$$S_0 = \rho_* hw - P\sqrt{\gamma} + \frac{\sqrt{\gamma}}{4\pi}\left(B^2 - \frac{1}{2w^2}\left[B^2 + (B^k u_k)^2\right]\right), \tag{4.160}$$

$$S_i = \rho_* h u_i + \frac{\sqrt{\gamma}}{4\pi w}\left[B^2 u_i - (B^j u_j) B_i\right]. \tag{4.161}$$

The resulting equations for S_i and S_0 are

$$\partial_t S_i + \partial_k \left[S_i v^k + \alpha\sqrt{\gamma} P_{\rm tot} \delta^k_i - \frac{\alpha\sqrt{\gamma}}{4\pi w^2} B^k \left(B_i + u_i B^j u_j\right)\right]$$
$$= -S_0 \partial_i \alpha + S_k \partial_i \beta^k - \frac{\alpha\sqrt{\gamma}}{2} S_{jk} \partial_i \gamma^{jk}, \tag{4.162}$$

$$\partial_t S_0 + \partial_k \left[S_0 v^k + \sqrt{\gamma} P_{\rm tot}(v^k + \beta^k) - \frac{\alpha\sqrt{\gamma}}{4\pi w} B^k \left(u_j B^j\right)\right]$$
$$= \alpha\sqrt{\gamma} S_{ij} K^{ij} - S_k D^k \alpha. \tag{4.163}$$

where

$$P_{\rm tot} := P + \frac{b^2}{8\pi} = P + \frac{1}{8\pi w^2}\left[B^2 + (B^k u_k)^2\right]. \tag{4.164}$$

In modern numerical astrophysics in general relativity, choosing a conservative form is standard. Thus, we here restricted our attention only to the conservative form described above.

When we employ the conservative form of the ideal magnetohydrodynamics equations, (4.162) and (4.163), the normalization relation, $u^a u_a = -1$, can be written, using equation (4.161), as

$$q^2 := \rho_*^{-2} \gamma^{ij} S_i S_j = (hw + \beta_1)^2 \left(1 - w^{-2}\right) - \beta_2 (hw)^{-2}(2hw + \beta_1), \tag{4.165}$$

where $\beta_1 := \sqrt{\gamma} B^2 (4\pi \rho_*)^{-1}$ and $\beta_2 := \sqrt{\gamma}(B^i S_i)^2 (4\pi)^{-1} \rho_*^{-3}$. For each time step, ρ_*, S_i, B^2, and γ_{ij} are updated by their evolution equations, and thus, β_1 and β_2 are also updated. Hence, equation (4.165) gives the relation between hw and w^{-2}. Specifically, it is written as

$$w^{-2} = 1 + \left[q^2 + \beta_2 (hw)^{-2}(2hw + \beta_1)\right](hw + \beta_1)^{-2}. \tag{4.166}$$

In addition to this, equation (4.160) yields the relation

$$e_0 = \rho_*^{-1} S_0 = hw - \frac{P\sqrt{\gamma}}{\rho_*} + \frac{\beta_1}{2}(2 - w^{-2}) - \frac{\beta_2}{2(hw)^2}. \qquad (4.167)$$

For a given value of e_0 and an equation of state in the form, $P = P(\rho, h)$, equation (4.167) also provides a relation between hw and w^{-2}. Eliminating w^{-2} from equations (4.165) and (4.167) yields

$$e_0 = hw - \frac{P\sqrt{\gamma}}{\rho_*} + \frac{\beta_1}{2}\left[1 + (hw + \beta_1)^{-2}\left(e_0^2 + \frac{\beta_2}{(hw)^2}(2hw + \beta_1)\right)\right] - \frac{\beta_2}{2(hw)^2}. \qquad (4.168)$$

Thus, if the term $P\sqrt{\gamma}/\rho_*$ is rewritten as a function of hw, we can obtain an algebraic equation for hw. This can be in general achieved. For example, for the Γ-law equation of state, (4.48), this term is written as

$$\frac{P\sqrt{\gamma}}{\rho_*} = \frac{\Gamma - 1}{\Gamma}\left(\frac{(hw)}{w^2} - \frac{1}{w}\right). \qquad (4.169)$$

By eliminating w^{-1} and w^{-2} using equation (4.166), $P\sqrt{\gamma}/\rho_*$ is rewritten as a function of hw.

4.6.3 *Properties of the ideal magnetohydrodynamics equations*

The set of the ideal magnetohydrodynamics equations is written in a vector form as equation (4.52) for pure hydrodynamics. In the present context, $\boldsymbol{Q}_a$, $\boldsymbol{F}_a^k$, and $\boldsymbol{S}_a$ are written in an eight-components vector form as

$$\boldsymbol{Q}_a = \begin{pmatrix} \rho_* \\ S_i \\ S_0 \\ \mathcal{B}^i \end{pmatrix}, \qquad (4.170)$$

$$\boldsymbol{F}_a^k = \begin{pmatrix} \rho_* v^k \\ S_i v^k + P_{\text{tot}} \alpha\sqrt{\gamma}\delta^k{}_i - \frac{\alpha\sqrt{\gamma}}{4\pi w^2} B^k \left\{B_i + u_i(B^j u_j)\right\} \\ S_0 v^k + \sqrt{\gamma} P_{\text{tot}}(v^k + \beta^k) - \frac{\alpha\sqrt{\gamma}}{4\pi w} B^k(B^j u_j) \\ \sqrt{\gamma}\left(v^k B^i - v^i B^k\right) \end{pmatrix}, \qquad (4.171)$$

$$\boldsymbol{S}_a = \begin{pmatrix} 0 \\ -S_0 \partial_i \alpha + S_j \partial_i \beta^j - \frac{1}{2}\alpha\sqrt{\gamma} S_{jk}\partial_i \gamma^{jk} \\ \alpha\sqrt{\gamma} S_{ij} K^{ij} - S^i D_i \alpha \\ 0 \end{pmatrix}. \qquad (4.172)$$

Then, the Jacobian matrix is defined in the same manner from $\boldsymbol{F}_a^k$ and $\boldsymbol{Q}_a$.

Here, note that the induction equation for i-th component does not have the advection term in the i-th direction [the last component of $\boldsymbol{F}_a^k$ in equation (4.171) vanishes for $i = k$]. This implies that for each advection direction, actually, only 7×7 (not 8×8) Jacobian matrix is defined. For example, for the advection in the x-direction, $\boldsymbol{Q}_a$ should be chosen as $\boldsymbol{Q}_a^x = (\rho_*, S_x, S_y, S_z, S_0, \mathcal{B}^y, \mathcal{B}^z)^T$ because $F_a^x = 0$ for $\mathcal{B}^x$ in $\boldsymbol{Q}_a$, and the corresponding set of the primitive variables should be defined by $\boldsymbol{q}_a^x = (\rho, v^x, v^y, v^z, \varepsilon, B^y, B^z)^T$. In the following, we calculate the Jacobian matrix and characteristic speeds focusing on the x-direction.

As in the hydrodynamics case, the Jacobian matrix for the x^k-direction is decomposed as

$$\boldsymbol{M}_{ab}^k = \sum_{c=1}^{7} \frac{\partial F_a^k}{\partial q_c^k} \frac{\partial q_c^k}{\partial Q_b^k} = \sum_{c=1}^{7} \boldsymbol{B}_{ac}^k (\boldsymbol{C}_{bc}^k)^{-1}, \tag{4.173}$$

where $\boldsymbol{B}_{ac}^k$ and $\boldsymbol{C}_{bc}^k$ are

$$\boldsymbol{B}_{ac}^k := \frac{1}{\sqrt{\gamma}} \frac{\partial F_a^k}{\partial q_c}, \quad \boldsymbol{C}_{bc}^k := \frac{1}{\sqrt{\gamma}} \frac{\partial Q_b^k}{\partial q_c}, \tag{4.174}$$

and F_a^k is defined by the seven-components vector, neglecting the trivial term.

The eigenvalues of the matrix $\boldsymbol{M}_{ab}^k$, referred to as λ^k, correspond to the characteristic speeds of the magnetized fluid in the x^k-direction, and are derived from the eigenvalue equation

$$\det\left(\boldsymbol{B}_{ab}^k - \lambda^k \boldsymbol{C}_{ab}^k\right) = 0. \tag{4.175}$$

It is possible to derive a seventh-order equation for λ^k from equation (4.175). However, the calculation is extremely lengthy and time-consuming.

In ideal magnetohydrodynamics, the eigenvalues are often derived by employing different fundamental variables and a different set of basic equations, which are written in a matrix form as equation (4.71) [Anile (1989); Komissarov (1999); Antón et al. (2006)]. As noted in section 4.4.2, the eigenvalues do not depend on the chosen variables.

Following Anile (1989) but slightly simplifying his analysis, we choose the fundamental variables composed of nine components as $V^I := (u^\mu, b^\mu, P)$ where $I = 1$–9. Here, we assume that $s = \text{constant}$, and we do not consider its equation $u^a \nabla_a s = 0$ for simplicity. The start point to derive an equation in the matrix form like equation (4.71) is the set of the basic equations composed of the continuity equation, the conservation equation of the stress-energy tensor, and the induction equation:

$$\nabla_a(\rho u^a) = 0, \quad \nabla_a T^{ab} = 0, \quad \nabla_a(b^a u^b - b^b u^a) = 0.$$

Contracting the stress-energy conservation law and the induction equation by u_b and b_b yields, respectively,

$$\rho h \nabla_a u^a + u^a \nabla_a \rho_{\text{ene}} = \rho h \nabla_a u^a + c_s^{-2} u^a \nabla_a P = 0, \tag{4.176}$$

$$\rho h u^a u^b \nabla_a b_b - b^a \nabla_a P = 0, \tag{4.177}$$

$$\nabla_a b^a + u^a u^b \nabla_a b_b = 0, \tag{4.178}$$

$$b^a b^b \nabla_a u_b - b^2 \nabla_a u^a - \frac{1}{2} u^a \nabla_a b^2 = 0, \tag{4.179}$$

where to derive equations (4.176) and (4.177), equations (4.178) and (4.179) were used, and the sound speed is defined by equation (4.68). Using these relations, the stress-energy conservation law and the induction equation are written in the following forms:

$$\left(\rho h + \frac{b^2}{4\pi}\right) u^a \nabla_a u^b - \frac{1}{4\pi} b^a \nabla_a b^b + \frac{1}{4\pi}(h^{bc} + u^b u^c) b_a \nabla_c b^a$$
$$+ \frac{1}{\rho h}\left[\rho h h^{ab} - \frac{1}{4\pi c_s^2} b^2 u^a u^b + \frac{b^a b^b}{4\pi}\right] \nabla_a P = 0, \qquad (4.180)$$

$$u^a \nabla_a b^b - b^a \nabla_a u^b + \frac{1}{\rho h}\left(-c_s^{-2} u^a b^b + b^a u^b\right) \nabla_a P = 0. \qquad (4.181)$$

Then, the 9×9 matrix $\left(A_I{}^J\right)^\mu$, as in equation (4.71), is written as

$$\left(A_I{}^J\right)^\mu = \begin{pmatrix} [\rho h + b^2/(4\pi)] u^\mu \delta_I^J & (-b^\mu \delta_I^J + p^{\mu J} b_I)/(4\pi) & \ell^{\mu J} \\ b^\mu \delta_I^J & -u^\mu \delta_I^J & H^{\mu J} \\ \rho h \delta_I^\mu & 0^\mu{}_I & c_s^{-2} u^\mu \end{pmatrix}, \qquad (4.182)$$

where $\delta_I{}^J$ is the unit 4×4 matrix, $0^\mu{}_I$ is a 1×4 matrix for which all the components are zero, $p^{ab} := g^{ab} + 2u^a u^b$, $\ell^{ab} := h^{ab} - (4\pi\rho h)^{-1}\left[c_s^{-2} b^2 u^a u^b - b^a b^b\right]$, and $H^{\mu J} := \left(u^\mu b^J c_s^{-2} - b^\mu u^J\right)/\rho h$. We again note that I and J denote the components of columns and rows.

The purpose here is to derive the eigenvalues of $\left(A_I{}^J\right)^\mu$. Since $\left(A_I{}^J\right)^\mu$ is a 9×9 matrix, there are nine eigenvalues. However, there should exist only seven physical components as we find from 7×7 matrix equation (4.175). We do not have to worry about this point because, as we find finally, the extra two eigenvalues agree with two of seven physical eigenvalues.

For deriving the equation for the eigenvalues, λ, for the x direction, we multiply $\left(A_I{}^J\right)^\mu$ by $\phi_\mu = (-\lambda, 1, 0, 0)$ and calculate the determinant of $\left(A_I{}^J\right)^\mu \phi_\mu$ as in section 4.4.2. Here,

$$\left(A_I{}^J\right)^\mu \phi_\mu = \begin{pmatrix} (\rho h + b^2/4\pi) a \delta_I^J & m_I^J/(4\pi) & \ell^J \\ \hat{B} \delta_I^J & -a \delta_I^J & H^J \\ \rho h \phi_I & 0_I & c_s^{-2} a \end{pmatrix}, \qquad (4.183)$$

where $a := u^\mu \phi_\mu = u^t(-\lambda + v^x)$, $\hat{B} := b^\mu \phi_\mu = -\lambda b^t + b^x$, $m_I{}^J := -\hat{B} \delta_I{}^J + \left(\phi^J + 2au^J\right) b_I$, $\ell^J := \ell^{\mu J} \phi_\mu = \phi^J + \left[1 - b^2/(4\pi \rho h c_s^2)\right] au^J + \hat{B} b^J/(4\pi \rho h)$, and $H^J := H^{\mu J} \phi_\mu = (ab^J c_s^{-2} - \hat{B} u^J)/\rho h$. Note that $\det\left(m_I{}^J\right) = 0$, and this fact helps saving the labor for the following calculation. Then, a relatively simple calculation for $\det\left[\left(A_I{}^J\right)^\mu \phi_\mu\right] = 0$ yields the equation for a in the form [Anile (1989)]

$$a \left[\left(\rho h + \frac{b^2}{4\pi}\right) a^2 - \frac{1}{4\pi} \hat{B}^2\right]^2$$
$$\times \left[\rho h(1 - c_s^2) a^4 - \left(\rho h c_s^2 + \frac{b^2}{4\pi}\right) \phi_\mu \phi^\mu a^2 + \frac{1}{4\pi} \hat{B}^2 c_s^2 \phi_\mu \phi^\mu\right] = 0, \qquad (4.184)$$

where
$$\phi_\mu \phi^\mu = \gamma^{xx} - \alpha^{-2}(\lambda + \beta^x)^2. \tag{4.185}$$

Thus, nine solutions are written in the analytic form. For y and z directions, essentially the same equations are derived.

$a = 0$ and $[\rho h + b^2/(4\pi)] a^2 - \hat{B}^2/4\pi = 0$ yield three of the seven solutions for λ^k in the x^k direction, respectively, as

$$\lambda = v^k \quad \text{and} \quad \frac{b^k \pm u^k \sqrt{4\pi\rho h + b^2}}{b^t \pm u^t \sqrt{4\pi\rho h + b^2}}. \tag{4.186}$$

We note that the first and second ones are associated with the bulk motion of the fluid and Alfvén waves (which propagate to the x^k direction), respectively.

The remaining four solutions, which correspond to the slow and fast modes of magnetohydrodynamics waves, are derived by a fourth-order equation written as

$$(u^t)^4 (\lambda^k - v^k)^4 (1 - \zeta) + \left[c_s^2 \frac{(b^k - \lambda^k b^t)^2}{4\pi\rho h + b^2} - (u^t)^2 (\lambda^k - v^k)^2 \zeta \right]$$
$$\times \left[\gamma^{kk} - \frac{(\beta^k + \lambda^k)^2}{\alpha^2} \right] = 0 \quad \text{(no summation for } k\text{)}, \tag{4.187}$$

where ζ and the Alfvén velocity v_A are defined, respectively, by

$$\zeta := \frac{4\pi\rho h c_s^2 + b^2}{4\pi\rho h + b^2} = v_A^2 + c_s^2 - v_A^2 c_s^2 \quad \text{and} \quad v_A^2 := \frac{b^2}{4\pi\rho h + b^2}. \tag{4.188}$$

In ideal magnetohydrodynamics, there are in general seven independent characteristic modes if no degeneracy is present. This is in contrast to pure hydrodynamics, in which there are only three characteristic modes.

The eigenvectors can be derived by the same procedure as in the pure hydrodynamics case after an extremely lengthy calculation [e.g., Anile (1989); Komissarov (1999)]. Showing details of this calculation is beyond the scope of this volume, and if a reader is interested in the details of the eigenvectors, we recommend to refer to Antón et al. (2010).

4.6.4 Force-free electromagnetodynamics

Force-free electromagnetodynamics is the low fluid inertia limit of electromagnetohydrodynamics: In this limit, the magnitude of the stress-energy tensor components for the fluid part, T_{ab}^{HD}, is assumed to be much smaller than that for the electromagnetic part, T_{ab}^{EM} (hence we neglect T_{ab}^{HD}). The force-free approximation in general relativity often works well in a low-density region around a magnetized neutron star and black hole [see discussion, e.g., of Komissarov (2002b, 2004, 2006); McKinney (2006)].

In the force-free limit, the conservation equation of the stress-energy tensor becomes
$$\nabla^a T_{ab}^{\rm EM} = 0. \tag{4.189}$$
This implies
$$F_{ab}j^b = n_a E_b j^b + \rho_e E_a + \epsilon_{abc} \bar{j}^b B^c = 0. \tag{4.190}$$
Thus, the conditions of the force-free limit may be described by
$$\begin{array}{ll} E_b j^b = 0 & \left(F_{ab}j^b n^a = 0\right), \\ \rho_e E_a + \epsilon_{abc}\bar{j}^b B^c = 0 & \left(F_{bc}j^b \gamma^c{}_a = 0\right). \end{array} \tag{4.191}$$
The second equation of (4.191) implies that E_a and B_a are perpendicular, i.e., $E_a B^a = 0$, for $\rho_e \neq 0$. In the following, we will assume $E_a B^a = 0$ even for $\rho_e = 0$.

In general situations, E^a, B^a, ρ_e, and $\bar{j}^a$ are determined from ten independent equations (see section 4.6.1.2): the evolution equations for E^a and B^a, the electric charge conservation law for ρ_e, and Ohm's law for $\bar{j}^a$. ρ_e may be determined simply by Gauss' law, i.e.,
$$\rho_e = \frac{1}{4\pi} D_a E^a. \tag{4.192}$$
However, we do not naturally have the Ohm's law in the force-free limit, because we do not solve equations for the hydrodynamics part and cannot define the fluid rest frame in this framework. In this case, $\bar{j}^a$ has to be determined by another method. Here, we describe two methods [see, e.g., Komissarov (2004); McKinney (2006) for a detailed discussion].

Because of the relation $j^a E_a = 0$ [see the first equation of (4.191)], j^a is in general written in the form
$$j_a = C_1 \frac{\epsilon_{abc} E^b B^c}{B^2} + \rho_e n_a + \frac{j_b B^b}{B^2} B_a, \tag{4.193}$$
where C_1 is found from the following two relations. First, from the second equation of (4.191), we obtain
$$\rho_e^2 E^2 = \bar{j}^a \bar{j}_a B^2 - (\bar{j}_a B^a)^2. \tag{4.194}$$
Equation (4.193) in addition yields
$$j^a j_a = -\rho_e^2 + \bar{j}^a \bar{j}_a = C_1^2 \frac{E^2}{B^2} - \rho_e^2 + \frac{(\bar{j}_b B^b)^2}{B^2}, \tag{4.195}$$
where we used $j_a B^a = \bar{j}_a B^a$ and $E_a B^a = 0$. Comparing equations (4.194) and (4.195), we find $C_1 = \rho_e$.

The remaining task is to write $\bar{j}_a B^a$ in terms of E^a and B^a. $\bar{j}_a B^a (= j_a B^a)$ is calculated from equations (4.113) and (4.114) in the following manner. First, from equation (4.113), we have
$$\begin{aligned} -4\pi j_b B^b &= B_b \nabla_a F^{ab} = B_b \nabla_a \left(n^a E^b - n^b E^a + \epsilon^{abc} B_c\right) \\ &= B_b \left[n^a \nabla_a E^b - E^a \nabla_a n^b + \nabla_a(\epsilon^{abc} B_c)\right] \\ &= B_b n^a \nabla_a E^b + E^a B^b K_{ab} + \alpha^{-1} B_b D_a \left(\alpha \epsilon^{abc} B_c\right), \end{aligned} \tag{4.196}$$

where we used $E_a B^a = E_a n^a = B_a n^a = 0$ and equation (2.30). Equation (4.114) also yields

$$\begin{aligned}
0 = -E_b \nabla_a {}^*F^{ab} &= E_b \nabla_a (n^a B^b - n^b B^a - \epsilon^{abc} E_c) \\
&= E_b n^a \nabla_a B^b - E_b B^a \nabla_a n^b - E_b \nabla_a (\epsilon^{abc} E_c) \\
&= -B_b n^a \nabla_a E^b + E^a B^b K_{ab} - \alpha^{-1} E_b D_a (\alpha \epsilon^{abc} E_c), \quad (4.197)
\end{aligned}$$

and hence, $B_b n^a \nabla_a E^b = E^a B^b K_{ab} - \alpha^{-1} E_b D_a \left(\alpha \epsilon^{abc} E_c \right)$. Substituting it into the first term in the last line of equation (4.196), we have

$$\begin{aligned}
-4\pi j_b B^b &= 2 E^a B^b K_{ab} - \alpha^{-1} E_b D_a \left(\alpha \epsilon^{abc} E_c \right) + \alpha^{-1} B_b D_a \left(\alpha \epsilon^{abc} B_c \right) \\
&= 2 E^a B^b K_{ab} - E_b D_a \left(\epsilon^{abc} E_c \right) + B_b D_a \left(\epsilon^{abc} B_c \right), \quad (4.198)
\end{aligned}$$

and therefore,

$$j_a = \rho_e \left(n_a + \frac{\epsilon_{abc} E^b B^c}{B^2} \right)$$
$$+ \frac{B_a}{4\pi B^2} \left[-2 E^b B^c K_{bc} + E_b D_d \left(\epsilon^{dbc} E_c \right) - B_b D_d \left(\epsilon^{dbc} B_c \right) \right]. \quad (4.199)$$

Thus, j_a (and hence $\bar{j}^a$) for a given geometry are written in terms of E_a and B_a on each time slice. This implies that E^a and B^a are in principle solved by straightforwardly integrating Maxwell's equations. However, in this formulation, we need special care for guaranteeing the condition $E_a B^a = 0$ as well as the divergence-free condition for B^a [Alic et al. (2012b)]. Moreover, it is known that if the condition $E^2 < B^2$ is violated,[2] the system can become non-hyperbolic [see Komissarov (2002b, 2004) for an analysis in flat spacetime], leading to the break-down of the force-free framework. In numerical simulations, we have to prepare some physical prescription to avoid such pathological situation [see, e.g., Komissarov (2006); Alic et al. (2012b)].

McKinney (2006) also proposed an alternative method: In addition to the induction equation for evolving B^i, he proposed to solve the equation of motion (4.31) for S_i with

$$S_i = \frac{\sqrt{\gamma}}{4\pi} \epsilon_{ijk} E^j B^k, \quad (4.200)$$

$$S_0 = \frac{\sqrt{\gamma}}{8\pi} \left(E^2 + B^2 \right), \quad (4.201)$$

$$S_{ij} = -\frac{1}{4\pi} \left(E_i E_j + B_i B_j \right) + \frac{1}{8\pi} \gamma_{ij} \left(E^2 + B^2 \right). \quad (4.202)$$

Here, we note $S_i B^i = 0$. Then, B^k and S_k are updated for each time step, and subsequently, E^k is determined by

$$E^k = \frac{4\pi}{\sqrt{\gamma} B^2} \epsilon^{kij} B_i S_j. \quad (4.203)$$

---

[2]The condition $E^2 < B^2$ is not guaranteed in the force-free framework.

Thus, the condition $E_a B^a = 0$ is automatically satisfied. From E^k thus determined, we have the following relation necessary for the induction equation (4.128):

$$\epsilon^{ijk} \mathcal{E}_k = \frac{4\pi}{B^2} \left(B^i S^j - B^j S^i \right). \tag{4.204}$$

This implies that if we define an auxiliary "three velocity" as

$$\hat{v}^i = -\beta^i + \frac{4\pi\alpha S^i}{\sqrt{\gamma} B^2} = -\beta^i + \alpha \mathcal{B}^{-2} \epsilon^{ijk} \mathcal{E}_j \mathcal{B}_k, \tag{4.205}$$

the induction equation is written in the same form as in the ideal magnetohydrodynamics case,

$$\partial_t \mathcal{B}^i = \partial_k \left(\hat{v}^i \mathcal{B}^k - \hat{v}^k \mathcal{B}^i \right). \tag{4.206}$$

Thus, E^i and B^i are evolved in the same manner as that for the ideal magnetohydrodynamics case.

Here, we note that $\hat{v}^i$ does not always correspond to the real velocity because the real velocity may have a component parallel to B^a which is absent in $\hat{v}^i$. Rather, this velocity is often considered to correspond to the velocity of the so-called **E** × **B** drift motion because the main component of $\hat{v}^i$ is $\alpha B^{-2} \epsilon^{ijk} E_j B_k$. Then, what is serious is that this three velocity could exceed the speed of light for $E^2 > B^2$. In reality, for $|\hat{v}^i| \to c$, the inertia of any particle in such electromagnetic fields will diverge because the Lorentz factor diverges. Hence the force-free condition is likely to break down in such a situation. Thus, for the force-free framework to work, $|\hat{v}^i|$ has to be smaller than the speed of light and E^2 has to be smaller than B^2.

To see this point more clearly, let $\hat{u}^a$ be an auxiliary four velocity satisfying $\hat{u}^a \hat{u}_a = -1$ with which we write $\hat{v}^i = \hat{u}^i / \hat{u}^t$ to suppose that the magnitude of the three velocity, $\hat{v}^i$, does not exceed the speed of light. Then, a straightforward calculation using $(\hat{u}^t)^2 = \left[\alpha^2 - \gamma_{ij} (\hat{v}^i + \beta^i)(\hat{v}^j + \beta^j) \right]^{-1}$ yields

$$\alpha \hat{u}^t = \sqrt{\frac{B^2}{B^2 - E^2}}, \quad \hat{u}_i = \sqrt{\frac{B^2}{B^2 - E^2}} \epsilon_{ijk} E^j B^k, \tag{4.207}$$

or

$$\hat{u}^a = \sqrt{\frac{B^2}{B^2 - E^2}} \left(n^a + \frac{\epsilon^{abc} E_b B_c}{B^2} \right). \tag{4.208}$$

Thus, if $B^2 > E^2$ is satisfied, the three velocity $\hat{v}^i$ does not exceed the speed of light. By contrast, unless the condition $B^2 > E^2$ is satisfied, the spacetime never have the force-free nature. Therefore, in this formalism, we also have to prepare some physical prescription to avoid the appearance of the region with $E^2 > B^2$ [McKinney (2006)].

4.7 Radiation transfer and radiation hydrodynamics

Radiation fields and their interaction with fluids often play a crucial role in general-relativistic astrophysical phenomena (e.g., the neutrino cooling and heating play an important role in core-collapse supernovae), because they are usually accompanied with a high-density and high-temperature state, in which neutrinos (as well as photons) strongly interact with the fluids. For a *physical* simulation in numerical relativity, we are often required to take into account the neutrino transfer effects. For this, it is necessary to solve radiation transfer equations.

For strictly handling radiation transfer effects, it is necessary to numerically solve Boltzmann's equation, taking into account the absorption, emission, and scattering terms. This equation has a 3+3+1 dimensional form (3 dimensions in real space and momentum space, respectively, and 1 dimension in time), and hence, the computational domain has to cover six-dimensional phase space for a simulation (unless we do not impose any spatial symmetry). It is an extremely challenging task to perform a well-resolved numerical simulation with a sufficient grid resolution for this equation, unless a high spatial-symmetry such as spherical symmetry is imposed [e.g., see Mezzacappa and Matzner (1989); Liebendörfer *et al.* (2001a, 2004) for formulations and Liebendörfer *et al.* (2001b); Sumiyoshi *et al.* (2005) for results in spherically symmetric and general relativistic simulations]. To date, no challenge has been made on this issue [but see Sumiyoshi and Yamada (2012) in Newtonian gravity]. Even for formulations suitable for a numerical simulation, only a few attempts [Cardall and Mezzacappa (2003); Cardall *et al.* (2013); Shibata *et al.* (2014)] have been published, and there is no established formulation to date.

Efforts for developing an *approximate* method incorporating key radiation effects have been also made in numerical astrophysics. Historically, a popular method for approximate radiation hydrodynamics is a flux-limited diffusion (FLD) method [see, e.g., a review of Mihalas and Weibel-Mihalas (1999)]. In this method, radiation flux density is in general assumed to be described only by radiation energy density, and a resulting basic equation for the radiation energy density becomes a diffusion-type equation in the optically thick and grey regions. In this case, the propagation speed of characteristics is not guaranteed to be smaller than the speed of light and the causality may be violated. Another drawback of the FLD scheme in general relativity is associated with the presence of the constraint equations (Hamiltonian and momentum constraints) for the initial-value problem of general relativity: These are satisfied only for the case that $\nabla_a T^{ab} = 0$ is fully solved. However, in the FLD scheme, this is not the case, because only one component is solved for the radiation fields.

In this section, we will describe a formulation for Boltzmann's equation and approximate formulations for the radiation transfer which are suitable in numerical relativity. For an approximate formulation, we pay particular attention to a moment formalism developed by Thorne (1981), because it is well formulated in the framework of general relativity. Furthermore, computational costs for a numerical solution in this formalism should not be as expensive as those for a solution of full Boltzmann's equation, because in moment formalisms, two momentum-space dimensions (two angle directions) are integrated out and the resulting argument dimension of the transfer equations becomes four (three is for real space and one is for frequency space). There is still one additional phase space to be taken into account, but the resulting four-dimensional equations may be accurately solved using current and near-future multi-core supercomputers.

In the following, we will first outline Boltzmann's equation and present a possible formulation for it (i.e., conservation form) that is suitable for numerical-relativity simulations, based on Shibata et al. (2014) (section 4.7.1). Then, in section 4.7.2, we will describe the Thorne's covariant moment formalism [Thorne (1981)], which is the most general moment formalism in general relativity, and a numerical-relativity formulation based on it. Finally, in section 4.7.3, we outline a leakage scheme for radiation hydrodynamics, which is useful for a semi-quantitative radiation-hydrodynamics study.

4.7.1 Boltzmann's equation

4.7.1.1 Basics

First, we review Boltzmann's equation in the context of the radiation transfer for massless particles in general relativity [e.g., Shibata et al. (2014)]. Let p^a be a null vector of massless particles, and $f(x^\alpha, p^i)$ be their distribution function. Here $x^\alpha = (t, x^i)$ and p^i denotes the spatial component of particle's momentum. Then, Boltzmann's equation is written in the form [e.g., Lindquist (1966)]

$$\frac{dx^\alpha}{d\tau}\frac{\partial f}{\partial x^\alpha}\bigg|_{p^i} + \frac{dp^i}{d\tau}\frac{\partial f}{\partial p^i}\bigg|_{x^\alpha} = (-p^\alpha \hat{u}_\alpha)S_{\rm rad}(p^\mu, x^\mu, f), \tag{4.209}$$

or

$$p^\alpha \frac{\partial f}{\partial x^\alpha}\bigg|_{p^i} - \Gamma^i{}_{\alpha\beta}p^\alpha p^\beta \frac{\partial f}{\partial p^i}\bigg|_{x^\alpha} = (-p^\alpha \hat{u}_\alpha)S_{\rm rad}(p^\mu, x^\mu, f), \tag{4.210}$$

where $S_{\rm rad}$ is a source term which is determined by interaction processes between the radiation and fluids (i.e., scattering, absorption, and emission), τ is the affine parameter for a trajectory of radiation particles (i.e., $p^a = dx^a/d\tau$), and $\hat{u}^a$ may be in general any unit timelike vector field (e.g., the four velocity of a fluid or timelike unit normal to spatial hypersurfaces). Here, $(\partial f/\partial x^\alpha)|_{p^i}$ and $(\partial f/\partial p^i)|_{x^\alpha}$ are partial derivatives of f fixing p^i and x^α, respectively. Instead of p^i, we may use other sets of variables for describing the momentum phase space: Let $q_{(i)}$ ($i=1, 2, 3$) be a set of momentum-space variables. Then, Boltzmann's equation is rewritten as

$$p^\alpha \frac{\partial f}{\partial x^\alpha}\bigg|_{q_{(j)}} + \sum_{i=1}^{3}\frac{dq_{(i)}}{d\tau}\frac{\partial f}{\partial q_{(i)}}\bigg|_{x^\alpha} = (-p^\alpha \hat{u}_\alpha)S_{\rm rad}. \tag{4.211}$$

As described by Lindquist (1966) and Ehlers (1971), the number of world lines crossing an invariant three volume dV_x with four momenta in the range of an invariant momentum-phase-space volume dV_p is

$$dN = f(x^\alpha, p^i)(-p^a \hat{u}_a)dV_x dV_p, \tag{4.212}$$

where
$$dV_x = \hat{u}^a \epsilon_{abcd} dx_1^b dx_2^c dx_3^d, \qquad (4.213)$$
$$dV_p = \frac{1}{(-p^e \hat{v}_e)} \hat{v}^a \epsilon_{abcd} dp_1^b dp_2^c dp_3^d, \qquad (4.214)$$

and $x^\mu = (t, x_1, x_2, x_3)$ and $p^\mu = (p_0, p_1, p_2, p_3)$ denote real and momentum space coordinates, respectively. Again, $\hat{u}^a$ and $\hat{v}^a$ are arbitrary timelike unit vector fields ($\hat{v}^a$ may be equal to $\hat{u}^a$). dV_p may be defined by an integral with the on-shell condition as [Landau and Lifshitz (1962)]

$$2 \int \epsilon_{abcd} dp_0^a dp_1^b dp_2^c dp_3^d \delta(p^e p_e) = \frac{\sqrt{-g}}{-p_0} dp_1 dp_2 dp_3, \qquad (4.215)$$

where $p^e p_e = 0$ for massless particles and $\delta(x)$ is the delta function.

In any local orthonormal frame, dV_p is written as [Thorne (1981)]

$$dV_p = \frac{d\hat{p}^1 d\hat{p}^2 d\hat{p}^3}{\hat{p}^0}, \qquad (4.216)$$

where $\hat{p}^\alpha$ denotes four momenta of a radiation particle in a local orthonormal frame. This may be written as

$$dV_p = \nu d\nu d\bar{\Omega}, \qquad (4.217)$$

where ν denotes the frequency (i.e., energy in units of $h=1$)[3] of radiation particles measured in the local orthonormal frame, $\nu = -p_a e^a_{(0)}$ with $e^a_{(0)}$ being the timelike unit vector field in this frame: In numerical relativity, one of the simplest choices could be $e^a_{(0)} = n^a$. If $e^a_{(0)}$ denotes the four velocity of a fluid, ν is the frequency measured by the fluid rest frame. $\bar{\Omega}$ denotes the surface element over the solid angle of an unit sphere: The specific definition of a local orthonormal frame will be given below.

In numerical astrophysics, Boltzmann's equation is often rewritten for the distribution function as a function of (t, x^i) and momentum-space argument variables defined in a local orthonormal frame. This method is in particular robust in spherically symmetric spacetime [Lindquist (1966); Castor (1972)]. Basic equations along this line is derived by setting

$$p^a = \nu \left(e^a_{(0)} + \sum_{i=1}^{3} \ell_{(i)} e^a_{(i)} \right), \qquad (4.218)$$

where $e^a_{(\mu)}$ ($\mu = 0, 1, 2, 3$) denotes a set of a tetrad basis for a local orthonormal frame satisfying $g_{ab} e^a_{(\alpha)} e^b_{(\beta)} = \eta_{\alpha\beta}$ and $\eta^{\alpha\beta} e^a_{(\alpha)} e^b_{(\beta)} = g^{ab}$ with $\eta_{\alpha\beta}$ being the Minkowski metric. Following Lindquist (1966), we write $\ell_{(i)}$ as

$$\ell_{(1)} = \cos\bar{\theta}, \quad \ell_{(2)} = \sin\bar{\theta}\cos\bar{\varphi}, \quad \ell_{(3)} = \sin\bar{\theta}\sin\bar{\varphi}, \qquad (4.219)$$

where $\bar{\theta}$ and $\bar{\varphi}$ denote angles of radiation rays in the momentum space at each spatial position [see figure 4.2] and in this context, $e^a_{(1)}$, $e^a_{(2)}$, and $e^a_{(3)}$ should be unit

---
[3] We note that the Planck constant, h, is set to be unity throughout this section.

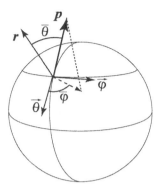

Fig. 4.2 The locally-defined angles $(\bar{\theta}, \bar{\varphi})$ of null rays, $\boldsymbol{p}(= p^i)$. $\boldsymbol{r}$, $\vec{\theta}$, and $\vec{\varphi}$ denote unit vector fields pointing to radial, θ, and φ directions, corresponding to $e^a_{(1)}$, $e^a_{(2)}$, and $e^a_{(3)}$, respectively.

vector fields pointing approximately to r, θ, and φ directions, respectively. Using $\bar{\theta}$ and $\bar{\varphi}$, the area element is written as $d\bar{\Omega} = \sin\bar{\theta} d\bar{\theta} d\bar{\varphi}$.

Choosing the argument variables as $q_{(1)} = \nu$, $q_{(2)} = \bar{\theta}$, and $q_{(3)} = \bar{\varphi}$, the second term of equation (4.211) is written as

$$\sum_{i=1}^{3} \frac{dq_{(i)}}{d\tau} \frac{\partial f}{\partial q_{(i)}} = \frac{d\nu}{d\tau} \frac{\partial f}{\partial \nu} + \frac{d\bar{\theta}}{d\tau} \frac{\partial f}{\partial \bar{\theta}} + \frac{d\bar{\varphi}}{d\tau} \frac{\partial f}{\partial \bar{\varphi}}. \tag{4.220}$$

To proceed further, we should remember the following relations,

$$\nu = -p_a e^a_{(0)} =: -p_{(0)}, \quad \tan\bar{\theta} = \frac{\sqrt{p_{(2)}^2 + p_{(3)}^2}}{p_{(1)}}, \quad \tan\bar{\varphi} = \frac{p_{(3)}}{p_{(2)}}, \tag{4.221}$$

where $p_{(j)} = p_a e^a_{(j)}$ and thus, $\sum_{i=1}^{3} p_{(i)}^2 = \nu^2 = p_{(0)}^2$. Using the geodesic equation for p^a (i.e., $p^b \nabla_b p^a = 0$) and equation (4.221), we obtain

$$\frac{d\nu}{d\tau} = -p^a p_b \nabla_a e^b_{(0)}, \tag{4.222}$$

$$\frac{d\bar{\theta}}{d\tau} = \frac{1}{\nu} \sum_{j=1}^{3} \frac{\partial \ell_{(j)}}{\partial \bar{\theta}} \frac{dp_{(j)}}{d\tau} = \frac{1}{\nu} \sum_{j=1}^{3} \frac{\partial \ell_{(j)}}{\partial \bar{\theta}} p^a p_b \nabla_a e^b_{(j)}, \tag{4.223}$$

$$\frac{d\bar{\varphi}}{d\tau} = \frac{1}{\nu \sin^2 \bar{\theta}} \sum_{j=2}^{3} \frac{\partial \ell_{(j)}}{\partial \bar{\varphi}} \frac{dp_{(j)}}{d\tau} = \frac{1}{\nu \sin^2 \bar{\theta}} \sum_{j=2}^{3} \frac{\partial \ell_{(j)}}{\partial \bar{\varphi}} p^a p_b \nabla_a e^b_{(j)}. \tag{4.224}$$

For spherically symmetric spacetime with the comoving gauge (see appendix B.4), it is easy to show that equation (4.211) reduces to the equation derived by Lindquist (1966) as

$$\nu D_t f + \bar{\mu}\nu D_r f - \nu^2 \Big[\bar{\mu} D_r \Phi + \mu^2 D_t \Lambda + (1-\bar{\mu}^2) D_t \ln R\Big] \frac{\partial f}{\partial \nu}$$

$$+ \nu(1-\bar{\mu}^2) \Big[-D_r \Phi + D_r \ln R + \bar{\mu}(D_t \ln R - D_t \Lambda) \Big] \frac{\partial f}{\partial \bar{\mu}} = (-p_\alpha \hat{u}^\alpha) S_{\text{rad}}, \tag{4.225}$$

where we assumed the line element in the form
$$ds^2 = -e^{2\Phi(t,r)}dt^2 + e^{2\Lambda(t,r)}dr^2 + R(t,r)^2 d\Omega. \tag{4.226}$$
Here, $D_t := e^{-\Phi}\partial_t$, $D_r := e^{-\Lambda}\partial_r$, $\bar{\mu} := \cos\bar{\theta}$, $u^a = e^a_{(0)}$,
$$e^a_{(0)} = e^{-\Phi}\left(\frac{\partial}{\partial t}\right)^a, \quad e^a_{(1)} = e^{-\Lambda}\left(\frac{\partial}{\partial r}\right)^a, \quad e^a_{(2)} = \frac{1}{R}\left(\frac{\partial}{\partial \theta}\right)^a, \quad e^a_{(3)} = \frac{1}{R\sin\theta}\left(\frac{\partial}{\partial \varphi}\right)^a,$$
and we note
$$p^a p_b \nabla_a e^b_{(\mu)} = p^\alpha p_\beta \partial_\alpha e^\beta_{(\mu)} + \frac{1}{2}p^\alpha p^\beta e^\sigma_{(\mu)} \partial_\sigma g_{\alpha\beta}. \tag{4.227}$$

Also, it is easy to derive Boltzmann's equation in the laboratory frame for flat spacetime as
$$\partial_t f + \bar{\mu}\partial_r f + \frac{1}{r}\sin\bar{\theta}\cos\bar{\varphi}\partial_\theta f + \frac{1}{r\sin\theta}\sin\bar{\theta}\sin\bar{\varphi}\partial_\varphi f$$
$$- \frac{1}{r}\left(\sin\bar{\theta}\partial_{\bar{\theta}}f + \sin\bar{\theta}\sin\bar{\varphi}\cot\theta\partial_{\bar{\varphi}}f\right) = S_{\rm rad}, \tag{4.228}$$
where
$$e^a_{(0)} = \left(\frac{\partial}{\partial t}\right)^a, \quad e^a_{(1)} = \left(\frac{\partial}{\partial r}\right)^a, \quad e^a_{(2)} = \frac{1}{r}\left(\frac{\partial}{\partial \theta}\right)^a, \quad e^a_{(3)} = \frac{1}{r\sin\theta}\left(\frac{\partial}{\partial \varphi}\right)^a.$$

In general curved spacetime, we have to constitute $e^a_{(\mu)}$ for a general geometry. However, in the local orthonormal frame with the choice $e^a_{(0)} = n^a$, the procedure is straightforward because $e^a_{(i)}$ has only spatial components [remember $n_\mu = (-\alpha, 0, 0, 0)$]. For example, for spherical polar coordinates (r, θ, φ), it is easy to find the following set as the tetrad bases:
$$e^\alpha_{(1)} = \left(0, \gamma_{rr}^{-1/2}, 0, 0\right),$$
$$e^\alpha_{(2)} = \left(0, -\frac{\gamma_{r\theta}}{\sqrt{\gamma_{rr}(\gamma_{rr}\gamma_{\theta\theta} - \gamma_{r\theta}^2)}}, \sqrt{\frac{\gamma_{rr}}{\gamma_{rr}\gamma_{\theta\theta} - \gamma_{r\theta}^2}}, 0\right),$$
$$e^\alpha_{(3)} = \left(0, \frac{\gamma^{r\varphi}}{\sqrt{\gamma^{\varphi\varphi}}}, \frac{\gamma^{\theta\varphi}}{\sqrt{\gamma^{\varphi\varphi}}}, \sqrt{\gamma^{\varphi\varphi}}\right). \tag{4.229}$$

In numerical astrophysics, $e^a_{(0)}$ is often chosen to be the four velocity of a fluid; $e^a_{(0)} = u^a$. In this case, all the momentum-space argument variables are defined in the local rest frame moving with the fluid. For evaluating the source term of Boltzmann's equation, it is necessary to perform integrations for the momentum-space variables in the fluid rest frame. With general argument variables for the momentum space, we have to perform a transformation of the frame to adjust the variables. What is nice in the momentum-space variables defined in the fluid rest frame is that the transformation is not necessary. A possible drawback in this scheme, however, is that we have to take a derivative of the four-velocity in general relativity, as found from equations (4.222)–(4.224), for which accurate numerical computation is not an easy task in numerical hydrodynamics. Therefore, in the following, we will not take the fluid rest frame for $e^a_{(0)}$.

4.7.1.2 Conservative form

In numerical astrophysics, it is often desirable to write Boltzmann's equation in a conservative form, in which accurate conservations of the particle number can be numerically guaranteed in the equation level [Liebendörfer et al. (2001a); Cardall and Mezzacappa (2003)]. Boltzmann's equation is written in a conservative form by a straightforward procedure, remembering the fact that dV_x and dV_p are invariant real and momentum space volumes. Starting from equation (4.210), a conservation form is derived as [e.g., Shibata et al. (2014)]

$$\frac{p_t}{(-g)}\left[\frac{\partial}{\partial x^\alpha}\left(\frac{f(-g)p^\alpha}{p_t}\right) - \frac{\partial}{\partial p^i}\left(\frac{f(-g)}{p_t}\Gamma^i{}_{\alpha\beta}p^\alpha p^\beta\right)\right] = (-p_a\hat{u}^a)S_{\rm rad}. \quad (4.230)$$

This equation was derived straightforwardly using the following relations

$$p^\alpha \frac{\partial f}{\partial x^\alpha}\bigg|_{p^j} = \frac{1}{(-g)}\frac{\partial[f(-g)p^\alpha]}{\partial x^\alpha}\bigg|_{p^j} - 2p^\alpha \Gamma^\beta{}_{\alpha\beta}f - f\frac{\partial p^t}{\partial t}\bigg|_{p^j},$$

$$\frac{\partial p^t}{\partial t}\bigg|_{p^j} = -\frac{p^\alpha p^\beta}{2p_t}\partial_t g_{\alpha\beta} = -\frac{1}{p_t}\Gamma^\beta{}_{t\alpha}p^\alpha p_\beta = -\frac{1}{p_t}\frac{dp_t}{d\tau},$$

$$-\Gamma^i{}_{\alpha\beta}p^\alpha p^\beta \frac{\partial f}{\partial p^i}\bigg|_{x^\mu} = -\frac{\partial(f\Gamma^i{}_{\alpha\beta}p^\alpha p^\beta)}{\partial p^i}\bigg|_{x^\mu} + 2fp^\alpha\left(\Gamma^i{}_{i\alpha} + \Gamma^i{}_{t\alpha}\frac{\partial p^t}{\partial p^i}\bigg|_{x^\mu}\right),$$

$$\Gamma^i{}_{t\alpha}\frac{\partial p^t}{\partial p^i}\bigg|_{x^\mu} = -\Gamma^i{}_{t\alpha}\frac{p_i}{p_t} = -\Gamma^\beta{}_{t\alpha}\frac{p_\beta}{p_t} + \Gamma^t{}_{t\alpha},$$

and remembering the definition

$$\frac{dp_t}{d\tau} = \frac{dx^\alpha}{d\tau}\frac{\partial p_t}{\partial x^\alpha}\bigg|_{p^i} + \frac{dp^i}{d\tau}\frac{\partial p_t}{\partial p^i}\bigg|_{x^\alpha} = p^\alpha \frac{\partial p_t}{\partial x^\alpha}\bigg|_{p^i} - \Gamma^i{}_{\alpha\beta}p^\alpha p^\beta \frac{\partial p_t}{\partial p^i}\bigg|_{x^\alpha}. \quad (4.231)$$

The conservative form is also derived for a local orthonormal frame. Starting from equation (4.211) with the choice of $\hat{u}^a = e^a_{(0)}$ and equations (4.220), (4.222)–(4.224), we obtain

$$\frac{1}{\sqrt{-g}}\frac{\partial(\sqrt{-g}\nu^{-1}p^\alpha f)}{\partial x^\alpha}\bigg|_{q_{(i)}} + \frac{1}{\nu^2}\frac{\partial}{\partial\nu}\left(-\nu f p^\alpha p_\beta \nabla_\alpha e^\beta_{(0)}\right)$$

$$+ \frac{1}{\sin\bar\theta}\frac{\partial}{\partial\bar\theta}\left(\nu^{-2}\sin\bar\theta f \sum_{j=1}^{3} p^\alpha p_\beta \nabla_\alpha e^\beta_{(j)}\frac{\partial \ell_{(j)}}{\partial\bar\theta}\right)$$

$$+ \frac{1}{\sin^2\bar\theta}\frac{\partial}{\partial\bar\varphi}\left(\nu^{-2}f \sum_{j=1}^{3} p^\alpha p_\beta \nabla_\alpha e^\beta_{(j)}\frac{\partial \ell_{(j)}}{\partial\bar\varphi}\right) = S_{\rm rad}, \quad (4.232)$$

and a practical form is written as

$$\frac{1}{\sqrt{-g}}\frac{\partial}{\partial x^\alpha}\bigg|_{q_{(i)}}\left[\left(e^\alpha_{(0)} + \sum_{i=1}^{3}\ell_{(i)}e^\alpha_{(i)}\right)\sqrt{-g}f\right] - \frac{1}{\nu^2}\frac{\partial}{\partial\nu}\left(\nu^3 f\omega_{(0)}\right)$$

$$+ \frac{1}{\sin\bar\theta}\frac{\partial}{\partial\bar\theta}\left(\sin\bar\theta f\omega_{(\bar\theta)}\right) + \frac{1}{\sin^2\bar\theta}\frac{\partial}{\partial\bar\varphi}\left(f\omega_{(\bar\varphi)}\right) = S_{\rm rad}, \quad (4.233)$$

where

$$\omega_{(0)} := \nu^{-2} p^\alpha p_\beta \nabla_\alpha e^\beta_{(0)} = \sum_{i=1}^{3} \ell_{(i)} \left(\gamma_{i00} + \sum_{j=1}^{3} \gamma_{i0j} \ell_{(j)} \right), \quad (4.234)$$

$$\omega_{(\bar\theta)} := \sum_{j=1}^{3} \omega_{(j)} \frac{\partial \ell_{(j)}}{\partial \bar\theta}, \quad (4.235)$$

$$\omega_{(\bar\varphi)} := \sum_{j=2}^{3} \omega_{(j)} \frac{\partial \ell_{(j)}}{\partial \bar\varphi}, \quad (4.236)$$

and

$$\omega_{(j)} := \nu^{-2} p^\alpha p_\beta \nabla_\alpha e^\beta_{(j)} = \gamma_{0j0} + \sum_{i=1}^{3} \ell_{(i)} \left\{ (\gamma_{0ji} + \gamma_{ij0}) + \sum_{k=1}^{3} \gamma_{ijk} \ell_{(k)} \right\}. \quad (4.237)$$

$\gamma_{\alpha\beta\gamma} = -\gamma_{\beta\alpha\gamma}$ is the Ricci rotation coefficients defined by $\gamma_{\alpha\beta\gamma} := e^a_{(\alpha)} e^b_{(\gamma)} \nabla_b (e_{(\beta)})_a$. We also used the following relations:

$$\nabla_a \left(e^a_{(0)} + \sum_{i=1}^{3} \ell_{(i)} e^a_{(i)} \right) = \sum_{i=1}^{3} \left(\gamma_{i0i} - \gamma_{0i0} \ell_{(i)} + \sum_{k=1}^{3} \gamma_{iki} \ell_{(k)} \right), \quad (4.238)$$

$$-\cot\bar\theta \frac{\partial \ell_{(j)}}{\partial \bar\theta} - \frac{1}{\sin^2\bar\theta} \frac{\partial^2 \ell_{(j)}}{\partial \bar\varphi^2} = \ell_{(j)}, \quad (4.239)$$

$$\frac{\partial \ell_{(i)}}{\partial \bar\theta} \frac{\partial \ell_{(j)}}{\partial \bar\theta} + \frac{1}{\sin^2\bar\theta} \frac{\partial \ell_{(i)}}{\partial \bar\varphi} \frac{\partial \ell_{(j)}}{\partial \bar\varphi} = \delta_{ij} - \ell_{(i)} \ell_{(j)}. \quad (4.240)$$

Note that the partial derivative with respect to x^α that appears in the first term of equations (4.232) and (4.233) is taken fixing ν, $\bar\theta$, and $\bar\varphi$ (*not* fixing p^i).

It is soon found that $\omega_{(0)}$ is related to $\omega_{(i)}$ by

$$\omega_{(0)} = -\sum_{i=1}^{3} \omega_{(i)} \ell_{(i)}. \quad (4.241)$$

Since $\ell_{(i)}$, $\partial \ell_{(i)}/\partial \bar\theta$, and $(\partial \ell_{(i)}/\partial \bar\varphi)/\sin\bar\theta$ constitute an orthonormal set of the unit vector in the local three-momentum space of subscript (i), we find that $\omega_{(0)}$, $\omega_{(\bar\theta)}$, and $\omega_{(\bar\varphi)}$ are the independent components of $\omega_{(i)}$: $[\omega_{(0)}, \omega_{(\bar\theta)}, \omega_{(\bar\varphi)}]$ are independent projection components of the $\omega_{(i)}$ vector, satisfying

$$\omega^2_{(0)} + \omega^2_{(\bar\theta)} + \frac{1}{\sin^2\bar\theta} \omega^2_{(\bar\varphi)} = \sum_{i=1}^{3} \omega^2_{(i)}. \quad (4.242)$$

We note that $\omega_{(0)}$ and $\omega_{(j)}$ are composed of nine bases functions of $Y_{lm}(\bar\theta, \bar\varphi)$ with $0 \leq l \leq 2$ and $0 \leq |m| \leq 2$ where Y_{lm} is the spherical harmonics function. Also, $\omega_{(\bar\theta)} \sin\bar\theta$ and $\omega_{(\bar\varphi)}$ are composed of fourteen bases functions of $Y_{lm}(\bar\theta, \bar\varphi)$ with $0 \leq l \leq 3$ and $0 \leq |m| \leq 2$. Thus, in general, $[\omega_{(0)}, \omega_{(\bar\theta)}, \omega_{(\bar\varphi)}]$ are written as functions of these bases, although with a good choice of the tetrad bases, they can be written in a simple form in particular in spacetime of a special symmetry (see below).

As an illustration, we show a conservative form of Boltzmann's equation in the Schwarzschild background [see equation (1.97) for its metric]. In this case, a natural choice of the tetrad components would be

$$e^a_{(0)} = \left(1 - \frac{2M}{r}\right)^{-1/2} \left(\frac{\partial}{\partial t}\right)^a, \quad e^a_{(1)} = \left(1 - \frac{2M}{r}\right)^{1/2} \left(\frac{\partial}{\partial r}\right)^a,$$

$$e^a_{(2)} = \frac{1}{r}\left(\frac{\partial}{\partial \theta}\right)^a, \quad e^a_{(3)} = \frac{1}{r\sin\theta}\left(\frac{\partial}{\partial \varphi}\right)^a. \tag{4.243}$$

Note that this choice is valid only for $r > 2M$ because for $r \leq 2M$, $e^a_{(0)}$ is not timelike (and $e^a_{(1)}$ is not spacelike). Then, the nonzero components of $\gamma_{\alpha\beta\gamma}$ are

$$\gamma_{122} = -\gamma_{212} = \gamma_{133} = -\gamma_{313} = -\frac{1}{r}\left(1 - \frac{2M}{r}\right)^{1/2}, \quad \gamma_{233} = -\gamma_{323} = -\frac{\cot\theta}{r},$$

$$\gamma_{100} = -\gamma_{010} = \frac{M}{r^2}\left(1 - \frac{2M}{r}\right)^{-1/2}. \tag{4.244}$$

Hence,

$$\omega_{(0)} = \frac{M}{r^2}\left(1 - \frac{2M}{r}\right)^{-1/2} \cos\bar{\theta}, \tag{4.245}$$

$$\omega_{(1)} = -\frac{M}{r^2}\left(1 - \frac{2M}{r}\right)^{-1/2} + \frac{1}{r}\left(1 - \frac{2M}{r}\right)^{1/2} \sin^2\bar{\theta}, \tag{4.246}$$

$$\omega_{(2)} = -\frac{1}{r}\left(1 - \frac{2M}{r}\right)^{1/2} \sin\bar{\theta}\cos\bar{\theta}\cos\bar{\varphi} + \frac{\cot\theta}{r}\sin^2\bar{\theta}\sin^2\bar{\varphi}, \tag{4.247}$$

$$\omega_{(3)} = -\frac{1}{r}\left(1 - \frac{2M}{r}\right)^{1/2} \sin\bar{\theta}\cos\bar{\theta}\sin\bar{\varphi} - \frac{\cot\theta}{r}\sin^2\bar{\theta}\sin\bar{\varphi}\cos\bar{\varphi}, \tag{4.248}$$

$$\omega_{(\bar{\theta})} = \frac{3M - r}{r^2}\left(1 - \frac{2M}{r}\right)^{-1/2} \sin\bar{\theta}, \tag{4.249}$$

$$\omega_{(\bar{\varphi})} = -\frac{\cot\theta}{r}\sin^3\bar{\theta}\sin\bar{\varphi}. \tag{4.250}$$

We note that $\omega_{(0)}$, $\omega_{(\bar{\theta})}$, and $\omega_{(\bar{\varphi})}$ are composed only of one basis function of $(\bar{\theta}, \bar{\varphi})$, respectively, although they may have nine functions in general. Due to this reason, Boltzmann's equation in the Schwarzschild background is written in a simple form:

$$\left(1 - \frac{2M}{r}\right)^{-1/2} \frac{\partial f}{\partial t} + \frac{1}{r^2}\frac{\partial}{\partial r}\left[f\cos\bar{\theta}r^2\left(1 - \frac{2M}{r}\right)^{1/2}\right]$$

$$+ \frac{1}{r\sin\theta}\frac{\partial}{\partial \theta}\left(f\sin\theta\sin\bar{\theta}\cos\bar{\varphi}\right) + \frac{1}{r\sin\theta}\frac{\partial}{\partial \varphi}\left(f\sin\bar{\theta}\sin\bar{\varphi}\right)$$

$$- \frac{1}{\nu^2}\frac{\partial}{\partial \nu}\left[f\nu^3 \cos\bar{\theta}\frac{M}{r^2}\left(1 - \frac{2M}{r}\right)^{-1/2}\right]$$

$$- \frac{1}{\sin\bar{\theta}}\frac{\partial}{\partial \bar{\theta}}\left[f\sin^2\bar{\theta}\frac{r - 3M}{r^2}\left(1 - \frac{2M}{r}\right)^{-1/2}\right]$$

$$- \frac{\partial}{\partial \bar{\varphi}}\left(f\frac{\cot\theta}{r}\sin\bar{\theta}\sin\bar{\varphi}\right) = S_{\text{rad}}. \tag{4.251}$$

It is found that the transport term associated with ν in equation (4.251) is present only for curved spacetime, and hence, this term is related to the gravitational redshift (for $\cos\bar{\theta} > 0$) and blueshift (for $\cos\bar{\theta} < 0$). It is also interesting to point out that the transport term associated with $\bar{\theta}$ changes the sign at the so-called photon sphere $r = 3M$: For $r > 3M$, the direction of outgoing rays tends to converge toward $\bar{\theta} \to 0$ as usual in flat spacetime, while for $r < 3M$, rays are dragged by the gravity of the black hole.

By setting $M = 0$, we can also obtain Boltzmann's equation in flat spacetime:

$$\frac{\partial f}{\partial t} + \frac{1}{r^2}\frac{\partial}{\partial r}\left(f\cos\bar{\theta}r^2\right)$$
$$+ \frac{1}{r\sin\theta}\frac{\partial}{\partial\theta}\left(f\sin\theta\sin\bar{\theta}\cos\bar{\varphi}\right) + \frac{1}{r\sin\theta}\frac{\partial}{\partial\varphi}\left(f\sin\bar{\theta}\sin\bar{\varphi}\right)$$
$$- \frac{1}{r}\frac{1}{\sin\bar{\theta}}\frac{\partial}{\partial\bar{\theta}}\left(f\sin^2\bar{\theta}\right) - \frac{\partial}{\partial\bar{\varphi}}\left(f\frac{\cot\theta}{r}\sin\bar{\theta}\sin\bar{\varphi}\right) = S_{\rm rad}. \quad (4.252)$$

It is easily checked that this equation is consistent with equation (4.228). This equation together with equation (4.251) shows that for $\omega_{(0)}$, $\omega_{(\bar{\theta})}$, and $\omega_{(\bar{\varphi})}$, $\cos\bar{\theta}$, $\sin\bar{\theta}$, and $\sin^3\bar{\theta}\sin\bar{\varphi}$ are the primary basis functions, respectively.

Non-spinning black holes may be written in the Eddington–Finkelstein form [e.g., Hawking and Ellis (1973) and equation (1.118)]. The lapse function for this is

$$\alpha = \left(\frac{r}{r+2M}\right)^{1/2}, \quad (4.253)$$

and thus it is always positive for $r > 0$. Hence, $n_a = -\alpha\nabla_a\bar{t}$ is always timelike for the entire region with $r > 0$. If we choose,

$$e^a_{(0)} = n^a, \qquad e^a_{(1)} = \left(1 + \frac{2M}{r}\right)^{-1/2}\left(\frac{\partial}{\partial r}\right)^a,$$
$$e^a_{(2)} = \frac{1}{r}\left(\frac{\partial}{\partial\theta}\right)^a, \qquad e^a_{(3)} = \frac{1}{r\sin\theta}\left(\frac{\partial}{\partial\varphi}\right)^a, \quad (4.254)$$

$e^a_{(0)}$ and $e^a_{(1)}$ are timelike and spacelike for $r > 0$, respectively. Then, we have

$$\omega_{(0)} = \frac{M}{r^3}\left(1 + \frac{2M}{r}\right)^{-3/2}\left(-M + r\cos\bar{\theta} + (2r+3M)\cos(2\bar{\theta})\right), \quad (4.255)$$

$$\omega_{(\bar{\theta})} = \frac{1}{r^3}\left(1 + \frac{2M}{r}\right)^{-3/2}\sin\bar{\theta}\left(-r(r+M) + 2M(2r+3M)\cos\bar{\theta}\right), \quad (4.256)$$

$$\omega_{(\bar{\varphi})} = -\frac{\cot\theta}{r}\sin^3\bar{\theta}\sin\bar{\varphi}. \quad (4.257)$$

In this case, $\omega_{(0)}$ and $\omega_{(\bar{\theta})}$ are different from equations (4.245) and (4.249), respectively: They are composed of three and two functions of $(\bar{\theta}, \bar{\varphi})$, respectively, and hence, it seems that they might be redundant components. However, if we choose

$$e^a_{(0)} = \left(1 - \frac{2M}{r}\right)^{-1/2} \left(\frac{\partial}{\partial t}\right)^a,$$

$$e^a_{(1)} = \left(1 - \frac{2M}{r}\right)^{-1/2} \left[\left(1 - \frac{2M}{r}\right)\left(\frac{\partial}{\partial t}\right)^a + \frac{2M}{r}\left(\frac{\partial}{\partial r}\right)^a\right],$$

$$e^a_{(2)} = \frac{1}{r}\left(\frac{\partial}{\partial \theta}\right)^a, \quad e^a_{(3)} = \frac{1}{r\sin\theta}\left(\frac{\partial}{\partial \varphi}\right)^a, \quad (4.258)$$

the resulting form of $\omega_{(0)}$, $\omega_{(\bar{\theta})}$, and $\omega_{(\bar{\varphi})}$ agree with equations (4.245), (4.249), and (4.250), respectively (although $e^a_{(0)}$ is timelike only for $r > 2M$). This example shows that the functions included in the transport term of the momentum-space variables depend in general on the choice of the tetrad basis, and implies that choice of $e^a_{(\mu)}$ corresponds to the "gauge choice" of the momentum-space coordinates composed of ν, $\bar{\theta}$, and $\bar{\varphi}$. Therefore, for the physical interpretation of the momentum-space variables, appropriate choices of the tetrad basis as well as the spacetime gauge would be necessary.

4.7.2 Moment formalism

In this section, we describe a moment formalism in numerical relativity. First of all, we review Thorne's moment formalism [Thorne (1981)], and then in section 4.7.2.2, we describe a truncated moment formalism of Shibata et al. (2011) which is formulated based on the Thorne's formalism.

4.7.2.1 Moment formalism of Thorne

In the Thorne's moment formalism, hierarchical equations for the following unprojected moments of massless particles associated with a moving medium are derived:

$$M^{\alpha_1\alpha_2\cdots\alpha_k}_{(\nu)}(x^\beta) = \int \frac{f(p'^\alpha, x^\beta)\delta(\nu - \nu')}{\nu'^{k-2}} p'^{\alpha_1} p'^{\alpha_2} \cdots p'^{\alpha_k} dV'_p. \quad (4.259)$$

Here f is the distribution function of the relevant radiation, p^α is the momentum vector of radiation particles, $\nu' = -\hat{u}^a p'_a$ is the frequency of the radiation in the rest frame of the medium (i.e., in the rest frame of the fiducial observer) with $\hat{u}^a$ being the four velocity of the medium, and dV_p is the invariant integration element on the light cone [see equation (4.216)]. k, here, is a positive integer, 1, 2, $\cdots$. Note that "ν" throughout this subsection denotes the frequency (not the subscript for spacetime components).

As pointed out by Thorne (1981), it is possible to choose any fiducial frame in the moment formalism. However, we have to keep in mind that for deriving a truncated moment formalism in a closed form, it is necessary to assume a closure relation which is determined by a physically reasonable assumption. This requires us to choose a good fiducial observer for deriving a useful truncated formalism from the moment formalism.

In the dense medium, any radiation is strongly coupled to a fluid. This implies that at the zeroth order, the radiation is in equilibrium with the medium, and radiation flow (measured by an observer comoving with the fluid) is a small correction. To reproduce this property in the closure relation, the best way is to choose the fluid rest frame as the fiducial frame. In this choice, the frequency, ν, in $M_{(\nu)}^{\alpha_1\alpha_2\cdots\alpha_k}$, always denotes the frequency *measured in the fluid rest frame*, and $\hat{u}^a = u^a$ where u^a is the four velocity of the fluid.

As in section 4.7.1, we write p^α in the form

$$\frac{dx^\alpha}{d\tau} = p^\alpha = \nu(u^\alpha + \ell^\alpha), \tag{4.260}$$

where ℓ^α is a unit normal spacelike four-vector orthogonal to u^α. Using this decomposition of p^α, equation (4.259) is rewritten in the form

$$M_{(\nu)}^{\alpha_1\alpha_2\cdots\alpha_k} = \nu^3 \int f(\nu,\bar{\Omega},x^\mu)(u^{\alpha_1}+\ell^{\alpha_1})(u^{\alpha_2}+\ell^{\alpha_2})\cdots(u^{\alpha_k}+\ell^{\alpha_k})d\bar{\Omega}, \tag{4.261}$$

where we used equation (4.217). The angular dependence in the integration is included in ℓ^α, for which the following relations hold,

$$\int d\bar{\Omega}\,\ell^\alpha = 0 = \int d\bar{\Omega}\,\ell^\alpha\ell^\beta\ell^\gamma,$$

$$\frac{1}{4\pi}\int d\bar{\Omega}\,\ell^\alpha\ell^\beta = \frac{1}{3}h^{\alpha\beta},$$

$$\frac{1}{4\pi}\int d\bar{\Omega}\,\ell^\alpha\ell^\beta\ell^\gamma\ell^\delta = \frac{1}{15}\left(h^{\alpha\beta}h^{\gamma\delta} + h^{\alpha\gamma}h^{\beta\delta} + h^{\alpha\delta}h^{\beta\gamma}\right), \tag{4.262}$$

where $h_{\alpha\beta}$ is the projection operator defined by

$$h_{\alpha\beta} := g_{\alpha\beta} + u_\alpha u_\beta. \tag{4.263}$$

Following Thorne (1981), we denote $M_{(\nu)}^{\alpha_1\alpha_2\cdots\alpha_k}$ by $M_{(\nu)}^{A_k}$. Taking the covariant derivatives of $M_{(\nu)}^{A_k\beta}$, we obtain a covariant equation with respect to real-space coordinates as [Thorne (1981)]

$$\nabla_\beta M_{(\nu)}^{A_k\beta} - \frac{\partial}{\partial\nu}\left(\nu M_{(\nu)}^{A_k\beta\gamma}\nabla_\gamma u_\beta\right) - (k-1)M_{(\nu)}^{A_k\beta\gamma}\nabla_\gamma u_\beta = S_{(\nu)}^{A_k}, \tag{4.264}$$

and

$$S_{(\nu)}^{A_k} = \nu^3\int S_{\rm rad}(\nu,\bar{\Omega},x^\mu,f)(u^{\alpha_1}+\ell^{\alpha_1})(u^{\alpha_2}+\ell^{\alpha_2})\cdots(u^{\alpha_k}+\ell^{\alpha_k})d\bar{\Omega}. \tag{4.265}$$

Here, the spacetime derivative is taken holding ν and the frequency derivative is taken holding the spacetime location. It should be noted that equation (4.264) has a coordinate-independent form as stressed by Thorne (1981). Also, the following relation is worthy to note:

$$M_{(\nu)}^{A_k\beta}u_\beta = -M_{(\nu)}^{A_k}. \tag{4.266}$$

Thus, the rank-$(k+1)$ equations include the lower-rank equations.

Since the frequency, ν, in equation (4.264) denotes the frequency *observed in a fluid rest frame*, not in the laboratory frame, $M_{(\nu)}^{A_k}$ is not directly related to the spectrum observed in the laboratory frame. However, if the fluid is assumed to be at rest in a distant zone far away from a radiation source where we observe the spectrum, the radiation moments in the rest frame agree with those in the laboratory frame. We suppose that the moment formalism will be used for the system that this assumption holds; e.g., the supernova stellar core collapse and the merger of binary compact objects. Thus, it is possible to directly compute the radiation spectrum from $M_{(\nu)}^{A_k}$, if we estimate it for $r \to \infty$.

Integrating equation (4.264) by ν, we obtain (for each species of the radiation components)

$$\nabla_\beta M^{A_k \beta} - (k-1) M^{A_k \beta \gamma} \nabla_\gamma u_\beta = S_{\rm rad}^{A_k}, \qquad (4.267)$$

where

$$M^{A_k} = \int_0^\infty d\nu\, M_{(\nu)}^{A_k} \quad \text{and} \quad S_{\rm rad}^{A_k} = \int_0^\infty d\nu\, S_{(\nu)}^{A_k}. \qquad (4.268)$$

Equation (4.267) is essentially the same as the moment formalism derived by Anderson and Spiegel (1972). We note that the second-rank tensor $M^{\alpha\beta}$ is equal to the stress-energy tensor for one of the radiation components, and thus, the total stress-energy tensor for the radiation is written as

$$(T^{\rm rad})^{\alpha\beta} = \sum \int_0^\infty d\nu\, M_{(\nu)}^{\alpha\beta}, \qquad (4.269)$$

where the summation denotes to take a sum for all the species of the radiation fields. Equation (4.267) implies that $T_{ab}^{\rm rad}$ obeys

$$\nabla_a (T^{\rm rad})^{ab} = S_{\rm rad}^b. \qquad (4.270)$$

In the following, we will analyze only the second-rank component of equation (4.264), truncating the higher-rank components. In the next section, we develop such formalism.

4.7.2.2 *Truncated moment formalism*

First of all, we define moments necessary for the following analysis as

$$J_{(\nu)} := \nu^3 \int f(\nu, \bar{\Omega}, x^\mu)\, d\bar{\Omega}, \qquad (4.271)$$

$$H_{(\nu)}^\alpha := \nu^3 \int \ell^\alpha f(\nu, \bar{\Omega}, x^\mu)\, d\bar{\Omega}, \qquad (4.272)$$

$$L_{(\nu)}^{\alpha\beta} := \nu^3 \int \ell^\alpha \ell^\beta f(\nu, \bar{\Omega}, x^\mu)\, d\bar{\Omega}, \qquad (4.273)$$

$$N_{(\nu)}^{\alpha\beta\gamma} := \nu^3 \int \ell^\alpha \ell^\beta \ell^\gamma f(\nu, \bar{\Omega}, x^\mu)\, d\bar{\Omega}, \qquad (4.274)$$

where all these integrals are assumed to be performed in the local rest frame moving with a fluid component. In terms of these new moments, the second- and third-rank unprojected moments are written as

$$M_{(\nu)}^{\alpha\beta} = J_{(\nu)}u^\alpha u^\beta + H_{(\nu)}^\alpha u^\beta + H_{(\nu)}^\beta u^\alpha + L_{(\nu)}^{\alpha\beta}, \quad (4.275)$$

$$M_{(\nu)}^{\alpha\beta\gamma} = J_{(\nu)}u^\alpha u^\beta u^\gamma + H_{(\nu)}^\alpha u^\beta u^\gamma + H_{(\nu)}^\beta u^\alpha u^\gamma + H_{(\nu)}^\gamma u^\alpha u^\beta$$
$$+ L_{(\nu)}^{\alpha\beta} u^\gamma + L_{(\nu)}^{\alpha\gamma} u^\beta + L_{(\nu)}^{\beta\gamma} u^\alpha + N_{(\nu)}^{\alpha\beta\gamma}. \quad (4.276)$$

In the truncated formalism of Shibata et al. (2011), evolution equations for the zeroth- and first-rank moments are solved, and the second- and third-rank moments are determined by closure relations. For the optically thick region, this is approximately equivalent to assuming that the degree of anisotropy of the distribution function in the local rest frame is weak and that the distribution function is approximated by lower-order harmonics as

$$f(\nu, \bar{\Omega}, x^\mu) = f_0(\nu, x^\mu) + f_1^\alpha(\nu, x^\mu)\ell_\alpha + f_2^{\alpha\beta}(\nu, x^\mu)\ell_\alpha \ell_\beta. \quad (4.277)$$

Here, f_0, f_1^α, and $f_2^{\alpha\beta}$ do not depend on the propagation angle of the radiation in the local rest frame and $f_2^{\alpha\beta}$ is a tracefree tensor with respect to $h_{\alpha\beta}$ (i.e., $f_2^{\alpha\beta} h_{\alpha\beta} = 0$). $|f_0|$ is assumed to be much larger than the absolute magnitude of f_1^α and $f_2^{\alpha\beta}$. For the expansion of equation (4.277), we obtain

$$J_{(\nu)} = 4\pi \nu^3 f_0, \quad (4.278)$$

$$H_{(\nu)}^\alpha = \frac{4\pi}{3}\nu^3 f_1^\alpha, \quad (4.279)$$

$$L_{(\nu)}^{\alpha\beta} = \frac{4\pi}{3}\nu^3 \left(f_0 h^{\alpha\beta} + \frac{2}{5}f_2^{\alpha\beta}\right) = \frac{1}{3}J_{(\nu)} h^{\alpha\beta} + \frac{8\pi}{15}\nu^3 f_2^{\alpha\beta}, \quad (4.280)$$

$$N_{(\nu)}^{\alpha\beta\gamma} = \frac{1}{5}\left(H_{(\nu)}^\alpha h^{\beta\gamma} + H_{(\nu)}^\beta h^{\alpha\gamma} + H_{(\nu)}^\gamma h^{\alpha\beta}\right), \quad (4.281)$$

where we used the relations (4.262). Thus, f_0 and f_1^α are related directly to $J_{(\nu)}$ and $H_{(\nu)}^\alpha$, and $f_2^{\alpha\beta}$ to the tracefree part of $L_{(\nu)}^{\alpha\beta}$, respectively. Because of the truncated expansion for $f(\nu, \bar{\Omega}, x^\mu)$, we naturally obtain a closure relation for $N_{(\nu)}^{\alpha\beta\gamma}$. For $L_{(\nu)}^{\alpha\beta}$, it is often assumed that $f_2^{\alpha\beta}$ is negligible for simplicity. In this case, the closure relation becomes

$$L_{(\nu)}^{\alpha\beta} = \frac{1}{3}h^{\alpha\beta} J_{(\nu)}, \quad (4.282)$$

and hence, the radiation stress tensor, $L_{(\nu)}^{\alpha\beta}$, becomes isotropic.

For the optically thin limit, by contrast, we should first give a physical assumption *in the laboratory frame* because the radiation does not interact with fluids. An acceptable assumption is that the radiation should propagate with the speed of light, and the radiation flow at each spacetime point should be pointed primarily to a particular null direction. In reality there are many radiation rays which point to different directions. However, it would be acceptable to assume that a primary direction should exist. In particular, this assumption is natural for the region far from the source of the radiation. The former assumption implies that the radiation flow points to a null direction in any frame (although the spacetime coordinate basis changes). Thus, together with the later assumption, the distribution function can be written in the form

$$f(\nu, \bar{\Omega}, x^\mu) = 4\pi f_f(\nu, x^\mu) \delta(\bar{\Omega} - \bar{\Omega}_f), \tag{4.283}$$

where $\bar{\Omega}_f$ denotes the flow direction in the fluid rest frame, and $f_f(\nu, x^\mu)$ is the partial distribution function of $\bar{\Omega} = \bar{\Omega}_f$. Then, the radiation moments are calculated to give

$$J_{(\nu)} = 4\pi \nu^3 f_f, \tag{4.284}$$

$$H_{(\nu)}^\alpha = 4\pi \nu^3 f_f \ell_f^\alpha, \tag{4.285}$$

$$L_{(\nu)}^{\alpha\beta} = 4\pi \nu^3 f_f \ell_f^\alpha \ell_f^\beta, \tag{4.286}$$

$$N_{(\nu)}^{\alpha\beta\gamma} = 4\pi \nu^3 f_f \ell_f^\alpha \ell_f^\beta \ell_f^\gamma, \tag{4.287}$$

where ℓ_f^α denotes the unit vector of the flow direction (observed in the fluid rest frame). Then, we find a natural choice of closure relations as

$$L_{(\nu)}^{\alpha\beta} = J_{(\nu)} \frac{H_{(\nu)}^\alpha H_{(\nu)}^\beta}{h_{\sigma\lambda} H_{(\nu)}^\sigma H_{(\nu)}^\lambda}, \tag{4.288}$$

$$N_{(\nu)}^{\alpha\beta\gamma} = J_{(\nu)} \frac{H_{(\nu)}^\alpha H_{(\nu)}^\beta H_{(\nu)}^\gamma}{(h_{\sigma\lambda} H_{(\nu)}^\sigma H_{(\nu)}^\lambda)^{3/2}}. \tag{4.289}$$

For optically grey regions, these two closure relations for the two limiting cases are matched by a phenomenological manner using the so-called variable Eddington factor [Livermore (1984)]. The variable Eddington factor, χ, is often defined as a function of the quantity $\bar{F}$ as $\chi = \chi(\bar{F})$ where

$$\bar{F} := \frac{h_{\sigma\lambda} H_{(\nu)}^\sigma H_{(\nu)}^\lambda}{J_{(\nu)}^2}, \tag{4.290}$$

which is zero and unity for the optically thick and thin limits, respectively. An example of the functional form of χ is [Livermore (1984)]

$$\chi(\bar{F}) = \frac{3 + 4\bar{F}^2}{5 + 2\sqrt{4 - 3\bar{F}^2}}, \tag{4.291}$$

which is 1/3 for $\bar{F} = 0$, and 1 for $\bar{F} = 1$. Using χ, $L_{(\nu)}^{\alpha\beta}$ in optically grey regions is written as

$$L_{(\nu)}^{\alpha\beta} = \frac{3\chi - 1}{2} \left(L_{(\nu)}^{\alpha\beta}\right)_{\text{thin}} + \frac{3(1-\chi)}{2} \left(L_{(\nu)}^{\alpha\beta}\right)_{\text{thick}}, \tag{4.292}$$

where $\left(L_{(\nu)}^{\alpha\beta}\right)_{\text{thin}}$ and $\left(L_{(\nu)}^{\alpha\beta}\right)_{\text{thick}}$ are defined by equations (4.288) and (4.282), respectively.

The equations for $J_{(\nu)}$ and $H^\alpha_{(\nu)}$ are derived from the second-rank component of equation (4.264) as

$$\nabla_\beta M^{\alpha\beta}_{(\nu)} - \frac{\partial}{\partial \nu}\left(\nu M^{\alpha\beta\gamma}_{(\nu)} \nabla_\gamma u_\beta\right) = S^\alpha_{(\nu)}, \qquad (4.293)$$

where

$$\begin{aligned} M^{\alpha\beta\gamma}_{(\nu)} \nabla_\beta u_\gamma &= \left(H^\gamma_{(\nu)} u^\alpha u^\beta + L^{\alpha\gamma}_{(\nu)} u^\beta + L^{\beta\gamma}_{(\nu)} u^\alpha + N^{\alpha\beta\gamma}_{(\nu)}\right) \nabla_\beta u_\gamma \\ &= \left(H^\gamma_{(\nu)} u^\alpha + L^{\alpha\gamma}_{(\nu)}\right) a_\gamma + \left(L^{\beta\gamma}_{(\nu)} u^\alpha + N^{\alpha\beta\gamma}_{(\nu)}\right) \Sigma_{\beta\gamma}. \end{aligned} \qquad (4.294)$$

The acceleration a^α and the shear $\Sigma_{\alpha\beta}$ are defined by

$$a^\alpha := u^\beta \nabla_\beta u^\alpha, \quad \Sigma_{\alpha\beta} := \frac{1}{2} h_\alpha{}^\gamma h_\beta{}^\delta \left[\nabla_\gamma u_\delta + \nabla_\delta u_\gamma\right]. \qquad (4.295)$$

Substituting equation (4.275) into (4.293), the evolution equations for $J_{(\nu)}$ and $H^\alpha_{(\nu)}$ are obtained, respectively, from

$$\nabla_\alpha Q^\alpha_{(\nu)} + Q^{\alpha\beta}_{(\nu)} \nabla_\beta u_\alpha - \frac{\partial}{\partial \nu}\left[\nu \left(Q^{\alpha\beta}_{(\nu)} \nabla_\beta u_\alpha\right)\right] = -S^\alpha_{(\nu)} u_\alpha, \qquad (4.296)$$

$$h_{k\alpha}\left[\nabla_\beta Q^{\alpha\beta}_{(\nu)} + Q^\beta_{(\nu)} \nabla_\beta u^\alpha - \frac{\partial}{\partial \nu}\left\{\nu \left(L^{\alpha\gamma}_{(\nu)} u^\beta + N^{\alpha\beta\gamma}_{(\nu)}\right) \nabla_\beta u_\gamma\right\}\right] = h_{k\alpha} S^\alpha_{(\nu)}, \qquad (4.297)$$

where

$$Q^\alpha_{(\nu)} := -M^{\alpha\beta}_{(\nu)} u_\beta = J_{(\nu)} u^\alpha + H^\alpha_{(\nu)}, \qquad (4.298)$$

$$Q^{\alpha\beta}_{(\nu)} := h^\alpha{}_\gamma M^{\gamma\beta}_{(\nu)} = H^\alpha_{(\nu)} u^\beta + L^{\alpha\beta}_{(\nu)}. \qquad (4.299)$$

The frequency-integrated equations are

$$\nabla_\alpha Q^\alpha + Q^{\alpha\beta} \nabla_\beta u_\alpha = -S^\alpha u_\alpha, \qquad (4.300)$$

$$h_{k\alpha}\left(\nabla_\beta Q^{\alpha\beta} + Q^\beta \nabla_\beta u^\alpha\right) = h_{k\alpha} S^\alpha, \qquad (4.301)$$

where

$$Q^\alpha := \int_0^\infty d\nu\, Q^\alpha_{(\nu)}, \quad Q^{\alpha\beta} := \int_0^\infty d\nu\, Q^{\alpha\beta}_{(\nu)}. \qquad (4.302)$$

It is found that equations (4.300) and (4.301) are not in a conservation form because of the presence of the second term in their left-hand sides. The reason for this result is simply that Q^t and Q^{kt} are not conservative quantities even in the absence of the source terms.

Instead of using equation (4.275), $M^{\alpha\beta}_{(\nu)}$ may be written as

$$M^{\alpha\beta}_{(\nu)} = E_{(\nu)} n^\alpha n^\beta + F^\alpha_{(\nu)} n^\beta + F^\beta_{(\nu)} n^\alpha + P^{\alpha\beta}_{(\nu)}, \qquad (4.303)$$

where n^α is the unit vector field normal to spacelike hypersurfaces Σ_t as before. $E_{(\nu)}$, $F^\alpha_{(\nu)}$, and $P^{\alpha\beta}_{(\nu)}$ may be regarded as radiation fields measured in the laboratory frame. Here, it should be noted again that ν is the frequency observed in the *fluid rest frame*. For obtaining the quantities fully defined in the laboratory frame, we need a further transformation of ν to the frequency measured in the laboratory frame. However, in the moment formalism, we do not consider such transformation, as already mentioned.

Then, $E_{(\nu)}$, $F_{(\nu)}^{\alpha}$, and $P_{(\nu)}^{\alpha\beta}$ are defined by

$$E_{(\nu)} = M_{(\nu)}^{\alpha\beta} n_\alpha n_\beta, \quad F_{(\nu)}^{i} = -M_{(\nu)}^{\alpha\beta} n_\alpha \gamma_\beta^{i}, \quad P_{(\nu)}^{ij} = M_{(\nu)}^{\alpha\beta} \gamma_\alpha^{i} \gamma_\beta^{j}, \quad (4.304)$$

where $\gamma_{\alpha\beta} = g_{\alpha\beta} + n_\alpha n_\beta$ as usual in this volume. Because $F_{(\nu)}^{\alpha} n_\alpha = P_{(\nu)}^{\alpha\beta} n_\alpha = 0$, we have the relations $F_{(\nu)}^{t} = P_{(\nu)}^{t\alpha} = 0$.

Here, we consider a formalism in which $E_{(\nu)}$ and $F_{(\nu)}^{k}$ are evolved, and $P_{(\nu)}^{ij}$ is determined by a closure relation. Then, $J_{(\nu)}$ and $H_{(\nu)}^{\alpha}$ are determined by

$$J_{(\nu)} = E_{(\nu)} w^2 - 2F_{(\nu)}^{k} w u_k + P_{(\nu)}^{ij} u_i u_j, \quad (4.305)$$

$$H_{(\nu)}^{\alpha} = \left(E_{(\nu)} w - F_{(\nu)}^{k} u_k \right) h_{\beta}^{\alpha} n^{\beta} + w h_{\beta}^{\alpha} F_{(\nu)}^{\beta} - h_{i}^{\alpha} u_j P_{(\nu)}^{ij}, \quad (4.306)$$

where $w = \alpha u^t$ as before. We note $h_{\beta}^{\alpha} n^{\beta} = n^{\alpha} - w u^{\alpha}$ and $n_\alpha h^{\alpha\beta} \gamma_{\beta k} = -w u_k$. In terms of $P_{(\nu)}^{ij}$, the closure relations for the optically thin and thick limits, described in equations (4.288) and (4.282), are rewritten as

$$\left(P_{(\nu)}^{ij} \right)_{\text{thin}} = E_{(\nu)} \frac{F_{(\nu)}^{i} F_{(\nu)}^{j}}{\gamma_{kl} F_{(\nu)}^{k} F_{(\nu)}^{l}}, \quad (4.307)$$

$$\left(P_{(\nu)}^{ij} \right)_{\text{thick}} = \frac{E_{(\nu)}}{2w^2 + 1} \left[(2w^2 - 1)\gamma^{ij} - 4V^i V^j \right] + \frac{1}{w} \left(F_{(\nu)}^{i} V^j + F_{(\nu)}^{j} V^i \right)$$

$$+ \frac{2 F_{(\nu)}^{k} u_k}{(2w^2 + 1) w} \left(-w^2 \gamma^{ij} + V^i V^j \right), \quad (4.308)$$

where we used

$$Q_{(\nu)}^{\alpha} n_\alpha = -J_{(\nu)} w + H_{(\nu)}^{\alpha} n_\alpha = -E_{(\nu)} w + F_{(\nu)}^{k} u_k, \quad (4.309)$$

$$Q_{(\nu)}^{\alpha} \gamma_{\alpha i} = J_{(\nu)} u_i + H_{(\nu)i} = w F_{(\nu)i} - P_{(\nu)i}^{k} u_k. \quad (4.310)$$

Then, the evolution equations for $E_{(\nu)}$ and $F_{(\nu)i}$ are written in the conservation forms as

$$\partial_t (\sqrt{\gamma} E_{(\nu)}) + \partial_j \left[\sqrt{\gamma} (\alpha F_{(\nu)}^{j} - \beta^j E_{(\nu)}) \right] + \frac{\partial}{\partial \nu} \left(\nu \alpha \sqrt{\gamma} n_\alpha M_{(\nu)}^{\alpha\beta\gamma} \nabla_\gamma u_\beta \right)$$

$$= \alpha \sqrt{\gamma} \left[P_{(\nu)}^{ij} K_{ij} - F_{(\nu)}^{j} \partial_j \ln \alpha - S_{(\nu)}^{\alpha} n_\alpha \right], \quad (4.311)$$

$$\partial_t (\sqrt{\gamma} F_{(\nu)i}) + \partial_j \left[\sqrt{\gamma} (\alpha P_{(\nu)i}^{j} - \beta^j F_{(\nu)i}) \right] - \frac{\partial}{\partial \nu} \left(\nu \alpha \sqrt{\gamma} \gamma_{i\alpha} M_{(\nu)}^{\alpha\beta\gamma} \nabla_\gamma u_\beta \right)$$

$$= \sqrt{\gamma} \left[-E_{(\nu)} \partial_i \alpha + F_{(\nu)k} \partial_i \beta^k + \frac{\alpha}{2} P_{(\nu)}^{jk} \partial_i \gamma_{jk} + \alpha S_{(\nu)}^{\alpha} \gamma_{i\alpha} \right]. \quad (4.312)$$

The frequency-integrated equations are

$$\partial_t (\sqrt{\gamma} E) + \partial_j \left[\sqrt{\gamma} \left(\alpha F^j - \beta^j E \right) \right]$$
$$= \alpha \sqrt{\gamma} \left[P^{ij} K_{ij} - F^j \partial_j \ln \alpha - S^\alpha n_\alpha \right], \quad (4.313)$$

$$\partial_t (\sqrt{\gamma} F_i) + \partial_j \left[\sqrt{\gamma} (\alpha P_i^{j} - \beta^j F_i) \right]$$
$$= \sqrt{\gamma} \left[-E \partial_i \alpha + F_k \partial_i \beta^k + \frac{\alpha}{2} P^{jk} \partial_i \gamma_{jk} + \alpha S^\alpha \gamma_{i\alpha} \right], \quad (4.314)$$

where

$$E := \int_0^\infty d\nu E_{(\nu)}, \quad F_j := \int_0^\infty d\nu F_{(\nu)j}, \quad P^{ij} := \int_0^\infty d\nu P_{(\nu)}^{ij}. \quad (4.315)$$

Equations (4.313) and (4.314) have full-conservation forms, because E and F_i are the conservative quantities in the absence of source terms and gravitational fields. This suggests that equations (4.311) and (4.312) will be suitable for numerical simulations.

We remark that for problems we are interested in (merger of compact objects and stellar core collapse), it is natural to assume that $u_i = 0$ ($w = 1/\alpha \approx 1$) for a distant zone far from the radiation source. Then, $E_{(\nu)}$ and $J_{(\nu)}$ agree with each other asymptotically, because the frequency ν agrees with that in the laboratory frame. Thus, if we analyze $E_{(\nu)}$ in the distant zone, it is feasible to extract a radiation energy spectrum in an observer frame.

Finally, we touch on $S_{(\nu)}^\alpha$, for which the expression depends on the interaction processes between radiation particles and medium (absorption, emission, scattering, and pair creation). If we focus only on the absorption and the emission, it is written in a familiar form [e.g., Shibata et al. (2011)]

$$S_{(\nu)}^\alpha = \kappa_{(\nu)} \left[\left(J_{(\nu)}^{\text{eq}} - J_{(\nu)} \right) u^\alpha - H_{(\nu)}^\alpha \right], \quad (4.316)$$

where $\kappa_{(\nu)}$ denotes an opacity which depends on the local condition of the medium and the cross section of radiation particles to the medium. $J_{(\nu)}^{\text{eq}}$ is the energy density for the radiation in equilibrium with the medium, which is written as a function of temperature and chemical potential of the radiation. In the presence of a scattering process, $S_{(\nu)}^\alpha$ is written in a more complicated form [see, e.g., Shibata et al. (2011)].

4.7.2.3 Radiation hydrodynamics equation

Basic equations for radiation hydrodynamics are derived from the conservation law of the stress-energy tensor, $\nabla_b T^b{}_a = 0$, and in the 3+1 formulation, they are decomposed into equations (4.27) and (4.28). Here, the stress-energy tensor is written as

$$T_{ab} = T_{ab}^{\text{HD}} + T_{ab}^{\text{rad}}, \quad (4.317)$$

where T_{ab}^{HD} is written in equation (4.104), and T_{ab}^{rad} is defined by equation (4.269).

Using equation (4.270), the conservation law of T_{ab}^{HD} is written as

$$\nabla_a \left(T^{\text{HD}} \right)^{ab} = -S_{\text{rad}}^b. \quad (4.318)$$

Then, the Euler and energy equations [corresponding to equations (4.27) and (4.28)] can be written as

$$\partial_t S_i + \partial_j \left[S_i v^j + \alpha \sqrt{\gamma} P \delta_i{}^j \right]$$
$$= -S_0 \partial_i \alpha + S_k \partial_i \beta^k + \frac{\alpha \sqrt{\gamma}}{2} S^{jk} \partial_i \gamma_{jk} - \alpha \sqrt{\gamma} S_{\text{rad}}^\alpha \gamma_{\alpha i}, \quad (4.319)$$

$$\partial_t S_0 + \partial_j \left[S_0 v^j + \sqrt{\gamma} P (v^j + \beta^j) \right]$$
$$= \alpha \left[\sqrt{\gamma} S^{ij} K_{ij} - \gamma^{ik} S_i \partial_k \ln \alpha + \sqrt{\gamma} S_{\text{rad}}^\alpha n_\alpha \right]. \quad (4.320)$$

We note that extension to the radiation magnetohydrodynamics equations is straightforward [e.g., Shibata and Sekiguchi (2012)].

In addition to the Euler and energy equations, we have to solve the continuity equation (4.44). For the neutrino-radiation hydrodynamics, it is further necessary to solve the continuity equations for the number density of leptons and electrons like equation (4.90).

To guarantee the conservation of total momentum and energy, it may be useful to solve

$$\partial_t \left[S_i + \sqrt{\gamma} F_i \right] + \partial_j \left[S_i v^j + \alpha \sqrt{\gamma} \left(P \delta_i{}^j + P_i{}^j \right) - \beta^j \sqrt{\gamma} F_i \right]$$
$$= -(S_0 + \sqrt{\gamma} E) \partial_i \alpha + (S_k + \sqrt{\gamma} F_k) \partial_i \beta^k + \frac{\alpha \sqrt{\gamma}}{2} \left(S^{jk} + P^{jk} \right) \partial_i \gamma_{jk}, \quad (4.321)$$
$$\partial_t \left[S_0 + \sqrt{\gamma} E \right] + \partial_j \left[S_0 v^j + \sqrt{\gamma} P(v^j + \beta^j) + \alpha \sqrt{\gamma} F^j - \beta^j \sqrt{\gamma} E \right]$$
$$= \alpha \left[\sqrt{\gamma} \left(S^{ij} + P^{ij} \right) K_{ij} - \gamma^{ik} (S_i + \sqrt{\gamma} F_i) \partial_k \ln \alpha \right]. \quad (4.322)$$

We note that E, F_k, and P^{ij} here denote the sum of the contribution from all the neutrino species.

4.7.3 Leakage scheme

Leakage schemes for radiation hydrodynamics are phenomenological schemes for handling radiation-transfer effects. In this scheme, equations for the radiation transfer are not essentially solved, but cooling (and heating in some case) effects are phenomenologically taken into account. Computational costs for a solution of radiation hydrodynamics by this scheme is much less expensive than those by solving Boltzmann's equation or moment equations. Nevertheless, it can still yield a semi-quantitative numerical result for radiation hydrodynamics that captures the essence of the radiation effects [see Epstein and Pethick (1981); van Riper and Lattimer (1981, 1982); Baron et al. (1985); Cooperstein (1988) for pioneer works on stellar core-collapse simulations and Ruffert et al. (1996); Rosswog and Liebendörfer (2003) for pioneer works on binary neutron star mergers by leakage schemes]. A leakage scheme in multidimensional numerical relativity was first developed by Sekiguchi (2010a,b). Here, we outline this Sekiguchi's scheme, because it is well formulated in the framework of general relativity and can yield a reliable solution for radiation hydrodynamics and neutrino luminosity. In the following, we do not consider the frequency-dependence of the radiation energy density and stress following Sekiguchi (2010a,b) although incorporating this dependence is in principle possible in his framework.

As in ordinary radiation transfer problems, in the leakage scheme of Sekiguchi (2010b), the stress-energy tensor is written in the form (4.317) and basic equations are (4.318) for radiation hydrodynamics and (4.270) for a *streaming* part of the radiation (i.e., for the radiation in the optically non-thick region). The radiation in the optically thick region escapes outwards in a diffusion time scale, which we denote as τ_{diff}. If the dynamical time scale of the system, τ_{dyn}, in the optically thick region is shorter than τ_{diff}, the corresponding region is called a *trapped* region. The radiation in the trapped region interacts frequently with fluids, and thus, it is handled as a radiation fluid in thermal equilibrium with its temperature equal to fluid temperature.

Taking into account this fact, T_{ab} is decomposed as

$$T_{ab} = \left(T_{ab}^{\text{HD}} + T_{ab}^{\text{rad,T}}\right) + T_{ab}^{\text{rad,S}}, \quad (4.323)$$

where $T_{ab}^{\text{rad,T}}$ denotes the stress-energy tensor of the radiation trapped by the fluid and $T_{ab}^{\text{rad,S}}$ denotes that of the streaming radiation which is assumed to propagate freely. The trapped part, $T_{ab}^{\text{rad,T}}$, can be written as

$$T_{ab}^{\text{rad,T}} = (e_{\text{rad}} + P_{\text{rad}}) u_a u_b + P_{\text{rad}} g_{ab}, \quad (4.324)$$

where $e_{\text{rad}}(= 3P_{\text{rad}})$ is the radiation energy density and P_{rad} is the radiation pressure, which are determined by its chemical potential and fluid temperature in the trapped region. We have to implement that $T_{ab}^{\text{rad,T}}$ vanishes in the streaming region.

Separating the stress-energy tensor of the radiation into the trapped and streaming parts, equation (4.318) is rewritten as

$$\nabla_a \left[\left(T^{\text{HD}}\right)^{ab} + \left(T^{\text{rad,T}}\right)^{ab}\right] = -S_{\text{leak}}^b, \quad (4.325)$$

$$\nabla_a \left(T^{\text{rad,S}}\right)^{ab} = S_{\text{leak}}^b, \quad (4.326)$$

where S_{leak}^b determines a dissipation effect by the radiation, which is written in a phenomenological manner by contrast to ordinary radiation transfer problems. Assuming a local cooling function, $\dot{Q}_{\text{cool}}$, defined in the fluid rest frame, it is defined by

$$S_{\text{leak}}^a = \dot{Q}_{\text{cool}} u^a. \quad (4.327)$$

This simplification significantly reduces computational costs.

The diffusion time scale, which is necessary for classifying the trapped and streaming regions and for calculating the cooling rate through the diffusion, should be formally determined by an optical depth τ_{opt}. This is locally defined by some approximate calculation in leakage schemes. In spherically symmetric simulations, τ_{opt} can be defined by [e.g., Cooperstein (1988)]

$$\tau_{\text{opt}}(r) = \int_r^\infty dr' \bar{\kappa}(r'), \quad (4.328)$$

where $\bar{\kappa}$ is an averaged opacity: see equation (4.316) for the frequency-dependent form; $\bar{\kappa}$ is obtained by an approximate average of $\kappa_{(\nu)}$. In multidimensional simulations, τ_{opt} is not defined simply by the radial integral. Ruffert et al. (1996) and Sekiguchi (2010b) proposed to perform a line integral for several ad hoc directions and then to take an average [e.g., $(\tau_1 \tau_2 \tau_3)^{1/3}$ for three values]. Once τ_{opt} is determined, the diffusion time scale is calculated by

$$\tau_{\text{diff}} \sim \tau_{\text{opt}}^2 \bar{\kappa}^{-1}, \quad (4.329)$$

and the cooling rate per volume by the diffusion is estimated by

$$\dot{Q}_{\text{cool,diff}} \sim \frac{e_{\text{rad}}}{\tau_{\text{diff}}}. \tag{4.330}$$

On the other hand, in the streaming region, the cooling rate is determined by the local emission rate of the radiation, which is determined by the local condition of the matter responsible for the emission. We write it as $\dot{Q}_{\text{cool,local}}$. This should vanish in the trapped region, because the radiation cannot escape from such region. Thus, $\dot{Q}_{\text{cool}}$ should be determined by $\dot{Q}_{\text{cool,diff}}$ in the trapped region while it is determined by $\dot{Q}_{\text{cool,local}}$ in the streaming region. In often-employed phenomenological methods, one first evaluates $\dot{Q}_{\text{cool,diff}}$ and $\dot{Q}_{\text{cool,local}}$ for the whole spatial region and then evaluates the total, $\dot{Q}_{\text{cool}}$, by

$$\dot{Q}_{\text{cool}} = \dot{Q}_{\text{cool,diff}} \left(1 - e^{-b\tau_{\text{opt}}}\right) + \dot{Q}_{\text{cool,local}} e^{-b\tau_{\text{opt}}}, \tag{4.331}$$

where b is a constant of order unity.

4.8 Numerical methods for hydrodynamics and magnetohydrodynamics: Handling transport term

In numerical hydrodynamics and magnetohydrodynamics, the time integration of equation (4.52) from n-th to $(n+1)$-th time step is typically performed using a finite-differencing algorithm written in a conservative form as

$$\boldsymbol{Q}^{n+1}_{j,k,l} = \boldsymbol{Q}^n_{j,k,l} - \frac{\Delta t}{\Delta x} \left(\boldsymbol{F}^j_{j+1/2,k,l} - \boldsymbol{F}^j_{j-1/2,k,l}\right)^n - \frac{\Delta t}{\Delta y} \left(\boldsymbol{F}^k_{j,k+1/2,l} - \boldsymbol{F}^k_{j,k-1/2,l}\right)^n$$
$$- \frac{\Delta t}{\Delta z} \left(\boldsymbol{F}^l_{j,k,l+1/2} - \boldsymbol{F}^l_{j,k,l-1/2}\right)^n + \boldsymbol{S}^n_{j,k,l} \Delta t, \tag{4.332}$$

where indices (j, k, l) label a numerical cell for three spatial directions (x, y, z), Δt is a time-step interval, and Δx, Δy, and Δz denote grid spacing of (x, y, z). $\boldsymbol{F}^j_{j\pm 1/2,k,l}$ are the numerical fluxes at cell interfaces $j \pm 1/2$ (see figure 4.3 for a one-dimensional example), and $\boldsymbol{S}^n_{j,k,l}$ denotes the sum of other source terms. We note that in actual hydrodynamics/magnetohydrodynamics, the time evolution from n-th to $(n+1)$-th time step is usually performed by employing a multi step scheme, e.g., a Runge–Kutta scheme (see section 3.1), in order to guarantee a high-order accuracy in Δt. We also note that the use of an *explicit* scheme is supposed in equation (4.332). However, for special problems (in which the equations are "stiff"), we have to employ at least partially *implicit* scheme; see section 4.9.3 for an option.

One of the most important tasks in the field of numerical hydrodynamics/ magnetohydrodynamics is to find an appropriate scheme with which an *accurate* and *stable* simulation is feasible. The evaluation of the term $\boldsymbol{S}^n_{j,k,l} \Delta t$ can be done in a straightforward manner unless *stiff* source terms are present; the spatial partial derivatives included in $\boldsymbol{S}^n_{j,k,l}$ are evaluated by a high-order accurate centered finite

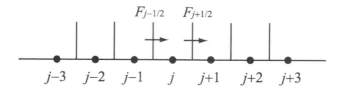

Fig. 4.3 Schematic figure for one-dimensional grid points (filled circles labeled by j) and cells (composed of small regions between $j - 1/2$ and $j + 1/2$) for numerical hydrodynamics/magnetohydrodynamics. $F_{j+1/2}$ and $F_{j-1/2}$ denote fluxes at cell interfaces.

differencing that is described in section 3.1. On the other hand, for the evaluation of the transport terms associated with $\boldsymbol{F}^{j}_{j\pm1/2,k,l}$, $\boldsymbol{F}^{k}_{j,k\pm1/2,l}$, and $\boldsymbol{F}^{l}_{j,k,l\pm1/2}$, a sophisticated numerical scheme is necessary. The straightforward finite differencing based on a Taylor expansion, described in section 3.2, cannot be employed due to the following reasons.

The unique and bothersome point in numerical compressible hydrodynamics is the possible presence of locally steep gradients or discontinuities (e.g., shocks) in hydrodynamics quantities. This is in particular the case in the context of astrophysics, because astrophysical fluids are usually highly compressible, and furthermore, supersonic motion is often realized. This implies that the Taylor expansion could often break down due to the loss of the regularity. If we use finite-differencing schemes based on the Taylor expansion in the vicinity of a discontinuity, unphysical oscillations (often called the Gibbs phenomenon) are developed in numerical solutions. By contrast, discontinuities are not present in geometric quantities (at least up to the first derivative) as far as appropriate gauge conditions are chosen, and this situation enables us to employ finite-differencing schemes based on the Taylor expansion for Einstein's equation (see section 3.1). Hence, the numerical scheme for solving hydrodynamics/magnetohydrodynamics equations should be different from that for gravitational fields: We have to employ a scheme which is not fully based on the Taylor expansion but still enables us to accurately follow the evolution of both continuous and (nearly) discontinuous fluids.

Just by this reason, one of the most important issues in numerical compressible hydrodynamics/magnetohydrodynamics becomes to discover a scheme by which discontinuities are accurately computed, avoiding the occurrence of unphysical oscillations. The numerical schemes that have such a property are often called high-resolution shock-capturing schemes [Toro (1999); Martí and Müller (2003); Font (2006)]. The purpose of this section is to outline the properties of these schemes and to introduce several well-known schemes in this category. In the following, we will first describe the fundamental ingredients for constructing high-resolution shock-capturing schemes. Then, we will introduce several popular numerical schemes in general relativistic hydrodynamics/magnetohydrodynamics that enable us to perform accurate and stable numerical simulations, capturing discontinuities.

4.8.1 Monotonicity preserving

As we mentioned above, handling transport terms is the most subtle issue in numerical compressible hydrodynamics for which partial differential equations are usually solved by a finite-differencing scheme. As described in section 3.1, a standard way for deriving finite-differencing equations is to use the Taylor expansion. In the Taylor expansion, the order of the expansion determines the order of the accuracy of a numerical solution (as far as the scheme stably yields a numerical solution and the numerical solution is regular). For obtaining a numerical solution of a high accuracy, we in principle should employ a high-order finite-differencing scheme, taking into account higher-order terms of the Taylor expansion.

In numerical simulations, we have not only to make an effort for improving the accuracy of numerical solutions but also to guarantee their *stability*. In numerical hydrodynamics in which discontinuities could be present, a high-frequency unphysical oscillation is often excited when an inappropriate numerical scheme is employed. We cannot employ any scheme in which unstable oscillations are excited, even if it is a high-order scheme that can yield an accurate numerical solution for a smooth quantity like the metric g_{ab}. The subtlety that we often encounter in numerical hydrodynamics is that a high-order (more than second-order) scheme is often more subject to numerical instability.

To clarify the origin of this problem, we often analyze the simplest linear-wave equation (4.56) that has a form, qualitatively the same as transport terms in hydrodynamics. For this equation, the general solution is written by equation (4.57), and hence, it is easy to examine how accurate a numerical solution derived is by comparing it with the exact solution. In particular, it is often examined how accurately a simple solution with a steep gradient or a discontinuity, such as a moving step function illustrated in figure 4.4, is numerically evolved. In this solution, the profile is unchanged and remains a monotonically decreasing function of x ($\partial_x \phi \leq 0$) at any given time. Thus, in any numerical evolution, this monotonicity has to be preserved. Otherwise, unphysical oscillations are excited and an inaccurate numerical solution is obtained.

In general, the *monotonicity preserving* implies that if a function ϕ is monotonic in x for a given time step t^n, it is also monotonic in x for the next time step t^{n+1}. The main subject in computational hydrodynamics has been to find a scheme that has this monotonicity-preserving property, because such schemes can avoid a spurious excitation of an unphysical oscillation in numerical solutions.

4.8.2 Godunov's theorem

The monotonicity-preserving property is desirable one for a stable numerical simulation of equation (4.56). However, Godunov's theorem strongly constrains possible finite-differencing schemes that satisfied this property. For describing Godunov's

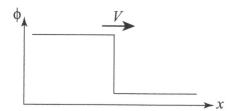

Fig. 4.4 A solution of equation (4.56) with a discontinuity which moves with velocity V.

theorem, it is convenient to write a finite-differencing form of equation (4.56) as

$$\phi_j^{n+1} = \sum_{k=-k_L}^{k_R} b_k \phi_{j+k}^n, \qquad (4.333)$$

where coefficients b_k are assumed to be *constants*, and k_L and k_R are non-negative integers. Note that all the numerical schemes introduced in section 3.2 for the same equation [see also equation (4.338)] are written in the form of (4.333).

From equation (4.333), we have

$$\Delta\phi_j^{n+1} = \sum_{k=-k_L}^{k_R} b_k \Delta\phi_{j+k}^n, \qquad (4.334)$$

where $\Delta\phi_j^n := \phi_{j+1}^n - \phi_j^n$. In monotonicity-preserving schemes, if $\Delta\phi_j^n \geq 0$, $\Delta\phi_j^{n+1} \geq 0$ has to be satisfied for any value of j. Now, we consider the case in which $b_k < 0$ for a particular value of k. Then, if the configuration at a time step n satisfies $\Delta\phi_j^n = 0$ except for $j = k$, $\Delta\phi_k^{n+1}$ has a sign opposite to $\Delta\phi_k^n$, violating the monotonicity. This implies that monotonicity-preserving schemes have to satisfy $b_k \geq 0$ for all the values of k.

Then, Godunov's theorem proves that there are no *monotone schemes* (i.e., monotonicity-preserving *linear* schemes) of the form (4.333) that are second or higher-order accurate [see, e.g., chapter 13 of Toro (1999) for more details]. The proof is based on the fact that a scheme of the form (4.333) is p-th order accurate in space and time if and only if the following conditions are satisfied

$$s_q := \sum_{k=-k_L}^{k_R} k^q b_k = (-\nu)^q, \qquad (4.335)$$

where $0 \leq q \leq p$ and $\nu := V\Delta t/\Delta x$ (note that ν is not frequency in this section). Thus, for second-order accurate schemes, one requires

$$s_0 = 1, \quad s_1 = -\nu, \quad s_2 = \nu^2. \qquad (4.336)$$

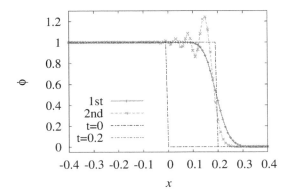

Fig. 4.5 Numerical solutions of equation (4.56) with $V = 1$ for a step function of figure 4.4 by a first-order upwind scheme (solid curve) and a Lax–Wendroff (second-order) scheme (dashed curve) at $t = 0.2$. The dash-dot step functions show the initial condition (with a discontinuity at $x = 0$) and the exact solution at $t = 0.2$ (with a discontinuity at $x = 0.2$).

However, equation (4.335) for $q = 2$ is written as

$$s_2 = \sum_{k=-k_L}^{k_R} k^2 b_k = \sum_{k=-k_L}^{k_R} (k+\nu)^2 b_k - 2\nu \sum_{k=-k_L}^{k_R} k b_k - \nu^2 \sum_{k=-k_L}^{k_R} b_k$$

$$= \sum_{k=-k_L}^{k_R} (k+\nu)^2 b_k - 2\nu s_1 - \nu^2 s_0$$

$$= \sum_{k=-k_L}^{k_R} (k+\nu)^2 b_k + \nu^2 \geq \nu^2, \qquad (4.337)$$

where we used $b_k \geq 0$, $s_0 = 1$, and $s_1 = -\nu$. The equality of equation (4.337) holds only for useless cases ($b_k = 0$ for all the values of k, or ν is an integer as $-k_0$ with $b_{k_0} \neq 0$ only for $k = k_0$ and $b_k = 0$ for other values of k). Thus, $s_2 \neq \nu^2$ for the case we concern. Therefore, Godunov's theorem shows that any monotone scheme is at most first-order accurate.

To illustrate the fact that Godunov's theorem states, we show numerical results obtained by a first-order and a second-order scheme in figure 4.5. As the first-order and second-order schemes, the upwind scheme shown in equation (3.70) and a Lax–Wendroff scheme [Press et al. (1989); Toro (1999)] are employed, respectively. Here, the finite-differencing equation for the Lax–Wendroff scheme is written as

$$\phi_j^{n+1} = \phi_j^n - \frac{\nu}{2}(\phi_{j+1}^n - \phi_{j-1}^n) + \frac{\nu^2}{2}(\phi_{j+1}^n - 2\phi_j^n + \phi_{j-1}^n), \qquad (4.338)$$

which is second-order accurate in time and space, and can yield a stable numerical solution for smooth functions as long as $\nu \leq 1$ (this fact can be easily found by the von Neumann stability analysis that is described in section 3.1.5).

Figure 4.5 shows that the second-order scheme yields an unphysically oscillatory solution, as predicted by Godunov's theorem. On the other hand, the first-order upwind scheme yields a non-oscillatory solution, although the profile of the step function cannot be captured accurately and the numerical solution is too diffusive to be of practical use near the discontinuity. These results clearly illustrate that we cannot use finite-differencing schemes of the *linear* form (4.333) in contrast to the case of computing advection terms of Einstein's evolution equations (see section 3.2). Namely, coefficients b_k cannot be constants.

4.8.3 *Circumventing Godunov's theorem*

Godunov's theorem thus shows that an appropriate choice for non-constant coefficients b_k in equation (4.333) can be the key to obtain a monotonicity-preserving and high-order scheme. In modern numerical hydrodynamics, b_k at each grid point is varied taking into account the local values and derivatives of ϕ in order to satisfy both the monotonicity-preserving property and a high-order accuracy. Specifically, for smooth regions of ϕ, coefficients b_k are basically set to be constants which are determined from the requirement of the order of accuracy, but a *flux-limiter function* has to be introduced in order to modify b_k depending on local conditions for less smooth regions.

To explain the concept of the flux-limiter function, we write the first-order upwind scheme and the Lax–Wendroff scheme in the form

$$\phi_j^{n+1} = \phi_j^n - \nu \left(\tilde{F}_{j+1/2}^n - \tilde{F}_{j-1/2}^n \right), \tag{4.339}$$

where $\tilde{F}_{j+1/2}^n$ determines the flux which is

$$\tilde{F}_{j+1/2}^n = \begin{cases} \phi_j^n & \text{first-order upwind} \\ \phi_j^n + \dfrac{1-\nu}{2}(\phi_{j+1}^n - \phi_j^n) & \text{Lax–Wendroff} \end{cases} \tag{4.340}$$

for $\nu \geq 0$, and

$$\tilde{F}_{j+1/2}^n = \begin{cases} \phi_{j+1}^n & \text{first-order upwind} \\ \phi_{j+1}^n - \dfrac{1+\nu}{2}(\phi_{j+1}^n - \phi_j^n) & \text{Lax–Wendroff} \end{cases} \tag{4.341}$$

for $\nu \leq 0$. This shows that the flux in the Lax–Wendroff scheme is composed of the sum of the first-order upwind scheme and an additional term, $(1-\nu)(\phi_{j+1}^n - \phi_j^n)/2$ or $-(1+\nu)(\phi_{j+1}^n - \phi_j^n)/2$, which is determined by the spatial derivative at $j+1/2$, $\partial_x \phi \approx (\phi_{j+1}^n - \phi_j^n)/\Delta x$. This additional term associated with the spatial derivative improves the order of accuracy but it could excite an unphysical oscillation in the presence of discontinuities.

Here, we consider to modify the Lax–Wendroff scheme by setting

$$\tilde{F}_{j+1/2}^n = \begin{cases} \phi_j^n + \dfrac{1-\nu}{2} B_{j+1/2} \left(\phi_{j+1}^n - \phi_j^n \right) & \nu \geq 0, \\ \phi_{j+1}^n - \dfrac{1+\nu}{2} B_{j+1/2} \left(\phi_{j+1}^n - \phi_j^n \right) & \nu \leq 0, \end{cases} \tag{4.342}$$

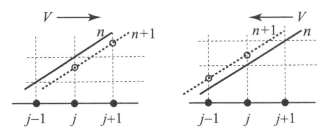

Fig. 4.6 Local behavior of exact solutions of equation (4.56). The solid and dashed curves denote the solutions at n-th and $(n+1)$-th time step levels, respectively. It is found that the values at j and at $(n+1)$-th time level are always between the values at j and $j - \text{sign}(V)$ at n-th time level.

where $B_{j+1/2}$ is a flux-limiter function, which is a function of ϕ_{j+k}^n with $k = 0, \pm 1, \cdots$. Thus, the scheme becomes *nonlinear* in ϕ_j^n.

The flux of the form (4.342) yields

$$\phi_j^{n+1} - \phi_j^n = -\nu \left[\phi_j^n - \phi_{j-1}^n + \frac{1-\nu}{2} B_{j+1/2} \left(\phi_{j+1}^n - \phi_j^n \right) \right.$$
$$\left. - \frac{1-\nu}{2} B_{j-1/2} \left(\phi_j^n - \phi_{j-1}^n \right) \right], \qquad (4.343)$$

for $\nu \geq 0$, and

$$\phi_j^{n+1} - \phi_j^n = -\nu \left[\phi_{j+1}^n - \phi_j^n - \frac{1+\nu}{2} B_{j+1/2} \left(\phi_{j+1}^n - \phi_j^n \right) \right.$$
$$\left. + \frac{1+\nu}{2} B_{j-1/2} \left(\phi_j^n - \phi_{j-1}^n \right) \right], \qquad (4.344)$$

for $\nu \leq 0$. These equations are written to the forms, respectively,

$$\frac{\phi_j^{n+1} - \phi_j^n}{\phi_{j-1}^n - \phi_j^n} = \nu \left(1 + \frac{1-\nu}{2 r_j} B_{j+1/2} - \frac{1-\nu}{2} B_{j-1/2} \right), \qquad \nu \geq 0$$

$$\frac{\phi_j^{n+1} - \phi_j^n}{\phi_{j+1}^n - \phi_j^n} = -\nu \left(1 - \frac{1+\nu}{2} B_{j+1/2} + \frac{1+\nu}{2} r_j B_{j-1/2} \right), \qquad \nu \leq 0$$

(4.345)

where

$$r_j = \frac{\Delta_{j-1}}{\Delta_j}, \qquad \Delta_j := \phi_{j+1}^n - \phi_j^n. \qquad (4.346)$$

Remembering the simply propagating nature of exact solutions for the linear-wave equation (4.56), we find that a sufficient condition for the absence of unphysical oscillations is that the left-hand sides of equation (4.345) are between 0 and 1; see figure 4.6. Assuming $0 < |\nu| < 1$, these conditions, respectively, yield the following conditions:

$$\frac{2}{1-\nu} \geq B_{j-1/2} - B_{j+1/2} \frac{\Delta_j}{\Delta_{j-1}} \geq -\frac{2}{\nu}, \qquad \nu \geq 0,$$

$$\frac{2}{1+\nu} \geq B_{j+1/2} - B_{j-1/2} \frac{\Delta_{j-1}}{\Delta_j} \geq \frac{2}{\nu}, \qquad \nu \leq 0.$$

(4.347)

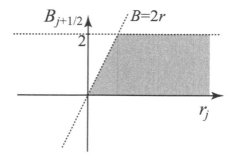

Fig. 4.7 The values that $B_{j+1/2}$ can take (shaded region) as a function of r_j. The dashed horizontal line is $B_{j+1/2} = 2$.

For further constraining the possible range of $B_{j+1/2}$, the simplest method is to find a sufficient condition. Since $|\nu|$ is assumed to be in the range between 0 and 1, the conditions in equation (4.347) are satisfied if the following conditions are satisfied:

$$2 \geq B_{j-1/2} - B_{j+1/2} \frac{\Delta_j}{\Delta_{j-1}} \geq -2, \quad \nu \geq 0,$$

$$2 \geq B_{j+1/2} - B_{j-1/2} \frac{\Delta_{j-1}}{\Delta_j} \geq -2, \quad \nu \leq 0. \quad (4.348)$$

The conditions in equation (4.348) are automatically satisfied, if the following two conditions are satisfied for any value of j:

$$0 \leq B_{j+1/2} \leq 2, \quad 0 \leq B_{j+1/2} \frac{\Delta_j}{\Delta_{j-\mathrm{sign}(\nu)}} \leq 2. \quad (4.349)$$

Therefore, the conditions shown in equation (4.349) can be employed as sufficient conditions for constructing a monotonicity-preserving scheme. Figure 4.7 plots the possible range of $B_{j+1/2}$ as a function of $r_j = \Delta_{j-\mathrm{sign}(\nu)}/\Delta_j$.

Equation (4.349) states that the value of $B_{j+1/2}$ has to be positive or zero, and hence, in the presence of a local extremum, $\Delta_j/\Delta_{j-\mathrm{sign}(\nu)} < 0$, it has to be zero. For such places, thus, the first-order scheme has to be locally employed. This is an essential point in the concept of flux limiter.

Equation (4.349) also shows that the possible range of $B_{j+1/2}$ depends on $\Delta_j/\Delta_{j-\mathrm{sign}(\nu)}$, i.e., the local values of the first and second derivatives of ϕ. Also, it depends on the sign of ν; the derivative in the upwind direction determines the condition for $B_{j+1/2}$. These are general properties of flux-limiter functions.

There are several flux-limiter functions proposed so far [see, e.g., a list in Toro (1999)]. The minmod (minimum-modulus) limiter function is one of the popular ones, and it is defined by

$$B_{j+1/2}(r) = \mathrm{minmod}(1, br), \quad (4.350)$$

where the minmod function is defined by

$$\mathrm{minmod}(x,y) = \begin{cases} x & \text{when } |x| < |y|, \text{ and } xy \geq 0, \\ y & \text{when } |x| > |y|, \text{ and } xy \geq 0, \\ 0 & \text{when } xy < 0. \end{cases} \quad (4.351)$$

b is a constant, and the possible range for it will be discussed in section 4.8.5. In sections 4.8.5, we will reconstruct numerical fluxes using this minmod limiter function.

4.8.4 *Total variation diminishing*

The flux-limiter functions have to be determined by a certain guideline for higher-order numerical schemes. The concept of the total variation diminishing (TVD) provides a robust guideline for constructing a flux-limiter function for the practical use [Harten (1983, 1984)] [see also Toro (1999) for a historical review].

The total variation is usually defined for a function that obeys a 1+1 hyperbolic equation as

$$\mathrm{TV}\left(\phi^n\right) := \sum_{j=-\infty}^{\infty} \left|\phi_{j+1}^n - \phi_j^n\right|, \quad (4.352)$$

where n and j denote a time step level and a spatial grid point as before. Here, we implicitly assume that $\phi_j = 0$ for $j \to \pm\infty$. The TVD condition is defined by

$$\mathrm{TV}\left(\phi^n\right) \geq \mathrm{TV}\left(\phi^{n+1}\right), \quad (4.353)$$

and a numerical scheme that satisfies this condition is called a TVD scheme.

Harten (1983) proved that the set of monotone schemes (monotonicity-preserving *linear* schemes) is contained in the set of TVD schemes and that the set of the TVD schemes is contained in monotonicity-preserving schemes. Therefore, if a TVD scheme is employed, the monotonicity-preserving property is guaranteed in numerical solutions.

Harten (1983) also proved a very important theorem for a class of non-linear schemes in which the finite-differencing equation is written as

$$\phi_j^{n+1} = \phi_j^n - C_{j-1/2}^n \Delta_{j-1}^n + D_{j+1/2}^n \Delta_j^n, \quad (4.354)$$

where $\Delta_j^n = \phi_{j+1}^n - \phi_j^n$, and $C_{j-1/2}^n$ and $D_{j+1/2}^n$ are functions of ϕ_j^n. He showed that for any finite-differencing equation of the form (4.354) to solve equation (4.56), a sufficient condition for schemes to be TVD is as follows:

$$C_{j+1/2}^n \geq 0, \quad D_{j+1/2}^n \geq 0, \quad 0 \leq C_{j+1/2}^n + D_{j+1/2}^n \leq 1. \quad (4.355)$$

This theorem is proven by a straightforward algebra as follows. From equation (4.354), we have

$$\phi_{j+1}^{n+1} - \phi_j^{n+1} = \left(\phi_{j+1}^n - \phi_j^n\right)\left(1 - C_{j+1/2}^n - D_{j+1/2}^n\right) \\ + C_{j-1/2}^n \left(\phi_j^n - \phi_{j-1}^n\right) + D_{j+3/2}^n \left(\phi_{j+2}^n - \phi_{j+1}^n\right), \quad (4.356)$$

and thus,

$$|\phi_{j+1}^{n+1} - \phi_j^{n+1}| \leq |\phi_{j+1}^n - \phi_j^n|\left(1 - C_{j+1/2}^n - D_{j+1/2}^n\right)$$
$$+ C_{j-1/2}^n |\phi_j^n - \phi_{j-1}^n| + D_{j+3/2}^n |\phi_{j+2}^n - \phi_{j+1}^n|, \quad (4.357)$$

where the conditions (4.355) are used. Then,

$$\mathrm{TV}\left(\phi^{n+1}\right) = \sum_j |\phi_{j+1}^{n+1} - \phi_j^{n+1}|$$
$$\leq \sum_j |\phi_{j+1}^n - \phi_j^n|\left(1 - C_{j+1/2}^n - D_{j+1/2}^n\right)$$
$$+ \sum_j C_{j+1/2}^n |\phi_{j+1}^n - \phi_j^n| + \sum_j D_{j+1/2}^n |\phi_{j+1}^n - \phi_j^n|$$
$$= \sum_j |\phi_{j+1}^n - \phi_j^n| = \mathrm{TV}\left(\phi^n\right). \quad (4.358)$$

Therefore, if one can write a finite-differencing equation in the form of (4.354), the constraints for the flux-limiter function are derived from the condition (4.355) for $C_{j+1/2}^n$ and $D_{j+1/2}^n$.

4.8.5 *Reconstruction of numerical flux at cell interfaces*

There are several high-order reconstruction schemes of numerical fluxes at cell interfaces (i.e., methods of evaluating $F_{j+1/2}$ at interfaces $j + 1/2$ of figure 4.3). Here, we focus only on the so-called MUSCL (Monotone Upstream-centered Scheme for Conservation Laws) originally proposed by van Leer (1977, 1979). In this scheme, the numerical flux is written in an upwind form as a first step, and then, high-order upwind fluxes are constructed using a high-order interpolation. In the modern context, a set of variables (not the fluxes themselves) such as primitive variables at cell interfaces are reconstructed by a high-order interpolation with an appropriate choice of a limiter function, and then, a high-order numerical flux is determined incorporating an upwind scheme.

First of all, we describe a typical method of reconstructing a variable [hereafter denoted by $q(x)$] at cell interfaces. Hydrodynamics and magnetohydrodynamics equations are usually written in a conservative form (besides external forces). This implies that the volume integral of conservative variables, such as ρ_* which derives the total rest mass, is conserved. Taking into account this fact, we usually consider that values of the hydrodynamics variables at a grid point do not represent those at the particular point but represent an average value in a cell that contains the grid point. Thus, we consider that q_j is defined by

$$q_j = \frac{1}{\Delta x} \int_{x_{j-1/2}}^{x_{j+1/2}} dx\, q(x). \quad (4.359)$$

Here, we assume for simplicity that the grid points labeled by x_j are uniformly distributed as $x_j = j\Delta x$, and $x_{j+1/2} = (x_j + x_{j+1})/2$. We note that it is straightforward to extend the following scheme for non-uniform grid profiles. The value of q at a cell interface, e.g., at $x_{j+1/2}$, has to be determined by an interpolation procedure using the discretized values q_j. A popular scheme is a piecewise parabolic reconstruction [Colella and Woodward (1984)], in which $q(x)$ is written as

$$q(x) = q_j + \frac{x - x_j}{2\Delta x}(q_{j+1} - q_{j-1})$$
$$+ \frac{3\kappa_{\text{pwp}}}{2}\left[\frac{(x-x_j)^2}{(\Delta x)^2} - \frac{1}{12}\right](q_{j+1} - 2q_j + q_{j-1}), \qquad (4.360)$$

where κ_{pwp} is a constant, and for $\kappa_{\text{pwp}} = 1/3$, the interpolation becomes third-order accurate in Δx, while for other values of κ_{pwp}, it is second-order accurate. We note that substituting equation (4.360) into (4.359), of course, yields q_j for any value of κ_{pwp}.

Using equation (4.360), the value of q at $x_{j+1/2}$ is determined, but this value, determined in the cell labeled by j, is in general different from that determined in the cell labeled by $j+1$ (see schematic figure 4.8). Thus, we denote the former and later by $(q_L)_{j+1/2}$ and $(q_R)_{j+1/2}$, respectively, which are (in the uniform grid) given by

$$(q_L)_{j+1/2} = q_j + \frac{1-\kappa_{\text{pwp}}}{4}(q_j - q_{j-1}) + \frac{1+\kappa_{\text{pwp}}}{4}(q_{j+1} - q_j), \qquad (4.361)$$

$$(q_R)_{j+1/2} = q_{j+1} - \frac{1+\kappa_{\text{pwp}}}{4}(q_{j+1} - q_j) - \frac{1-\kappa_{\text{pwp}}}{4}(q_{j+2} - q_{j+1}). \qquad (4.362)$$

Here, subscripts L and R denote the left and right sides of $x_{j+1/2}$, respectively.

Now, we revisit equation (4.56). In the MUSCL, the finite-differencing equation for this is written as

$$\phi_j^{n+1} = \phi_j^n - \frac{\Delta t}{\Delta x}\left(F_{j+1/2}^n - F_{j-1/2}^n\right), \qquad (4.363)$$

where the numerical flux, $F_{j+1/2}^n$, is determined by an upwind scheme as

$$F_{j+1/2}^n = \frac{V}{2}\left[(\phi_R)_{j+1/2}^n + (\phi_L)_{j+1/2}^n - \frac{|V|}{V}\left\{(\phi_R)_{j+1/2}^n - (\phi_L)_{j+1/2}^n\right\}\right]. \qquad (4.364)$$

Note that for $V > 0$, $F_{j+1/2}^n = V(\phi_L)_{j+1/2}^n$ and for $V < 0$, $F_{j+1/2}^n = V(\phi_R)_{j+1/2}^n$. In the first-order upwind scheme, $F_{j+1/2}^n = V\phi_j^n$ for $V > 0$ and $V\phi_{j+1}^n$ for $V < 0$. This implies that the first terms in the right-hand side of equations (4.361) and (4.362) correspond to the terms for the first-order upwind scheme, and the second and third terms are higher-order corrections.

As described and illustrated in section 4.8.2, high-order correction terms often cause unphysical oscillations (numerical instability), and to avoid their occurrence, we have to introduce a flux-limiter function. Incorporating a minmod limiter function, equations (4.361) and (4.362) are often rewritten as

$$(q_L)_{j+1/2} = q_j + \frac{1-\kappa_{\text{pwp}}}{4}\Phi\left(r_{j-1}^+\right)\Delta_{j-1} + \frac{1+\kappa_{\text{pwp}}}{4}\Phi\left(r_j^-\right)\Delta_j, \qquad (4.365)$$

$$(q_R)_{j+1/2} = q_{j+1} - \frac{1+\kappa_{\text{pwp}}}{4}\Phi\left(r_j^+\right)\Delta_j - \frac{1-\kappa_{\text{pwp}}}{4}\Phi\left(r_{j+1}^-\right)\Delta_{j+1}, \qquad (4.366)$$

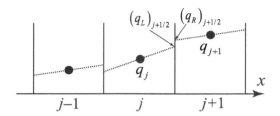

Fig. 4.8 Piecewise linear or parabolic interpolation schemes. The filled circles denote averaged values in each cell and the dotted curves denote interpolated values in each cell.

where $\Delta_j = q_{j+1} - q_j$, $\Phi(r_j) = \mathrm{minmod}(1, br_j)$ [see equation (4.351)], and

$$r_j^\pm = \frac{\Delta_{j\pm 1}}{\Delta_j}. \tag{4.367}$$

Note that when $\Phi(r_j) = 1$, the original accuracy for the reconstruction at cell interfaces is recovered. In the minmod limiter, $\Phi(r)$ is unity for $br_j \geq 1$. Thus, a large value of b is favorable for guaranteeing the desired accuracy to satisfy $\Phi(r_j) = 1$ for a wide range of r. It should be also required that $b \geq 1$; otherwise, the order of accuracy becomes poorer than second order even for a purely linear configuration of q in which $r_j = 1$.

The possible range of b is determined by the TVD condition. To see this, we again analyze equation (4.56). Using equations (4.365) and (4.366), equation (4.363) is written as

$$\phi_j^{n+1} = \phi_j^n - \frac{\Delta t}{\Delta x} V^+ \left[\Delta_{j-1} + \frac{1}{4} \left\{ (1 - \kappa_{\mathrm{pwp}}) \bar{\Delta}_{j-1}^+ + (1 + \kappa_{\mathrm{pwp}}) \bar{\Delta}_j^- \right\} \right.$$
$$\left. - \frac{1}{4} \left\{ (1 - \kappa_{\mathrm{pwp}}) \bar{\Delta}_{j-2}^+ + (1 + \kappa_{\mathrm{pwp}}) \bar{\Delta}_{j-1}^- \right\} \right]$$
$$- \frac{\Delta t}{\Delta x} V^- \left[\Delta_j - \frac{1}{4} \left\{ (1 + \kappa_{\mathrm{pwp}}) \bar{\Delta}_j^+ + (1 - \kappa_{\mathrm{pwp}}) \bar{\Delta}_{j+1}^- \right\} \right.$$
$$\left. + \frac{1}{4} \left\{ (1 + \kappa_{\mathrm{pwp}}) \bar{\Delta}_{j-1}^+ + (1 - \kappa_{\mathrm{pwp}}) \bar{\Delta}_j^- \right\} \right], \tag{4.368}$$

where $\Delta_j = \phi_{j+1}^n - \phi_j^n$, $\bar{\Delta}_j^\pm = \Phi(r_j^\pm)\Delta_j$, and

$$V^\pm = \frac{1 \pm \mathrm{sign}(V)}{2} V, \qquad (V^+ \geq 0, \quad V^- \leq 0). \tag{4.369}$$

Now, using

$$\bar{\Delta}_j^\pm = \mathrm{minmod}\left(1, b \frac{\Delta_{j\pm 1}}{\Delta_j}\right) \Delta_j = \mathrm{minmod}\left(\frac{\Delta_j}{\Delta_{j\pm 1}}, b\right) \Delta_{j\pm 1}, \tag{4.370}$$

equation (4.368) is rewritten as

$$\phi_j^{n+1} = \phi_j^n - \frac{\Delta t}{\Delta x} \left(\bar{V}^+ \Delta_{j-1} + \bar{V}^- \Delta_j \right), \tag{4.371}$$

where
$$\bar{V}^+ := V^+\left[1 + \frac{1}{4}\left\{(1-\kappa_{\text{pwp}})\Phi(r^+_{j-1}) + (1+\kappa_{\text{pwp}})\text{minmod}\left(\frac{\Delta_j}{\Delta_{j-1}},b\right)\right\}\right.$$
$$\left. - \frac{1}{4}\left\{(1-\kappa_{\text{pwp}})\text{minmod}\left(\frac{\Delta_{j-2}}{\Delta_{j-1}},b\right) + (1+\kappa_{\text{pwp}})\Phi(r^-_{j-1})\right\}\right],$$
$$\bar{V}^- := V^-\left[1 - \frac{1}{4}\left\{(1+\kappa_{\text{pwp}})\Phi(r^+_j) + (1-\kappa_{\text{pwp}})\text{minmod}\left(\frac{\Delta_{j+1}}{\Delta_j},b\right)\right\}\right.$$
$$\left. + \frac{1}{4}\left\{(1+\kappa_{\text{pwp}})\text{minmod}\left(\frac{\Delta_{j-1}}{\Delta_j},b\right) + (1-\kappa_{\text{pwp}})\Phi(r^-_j)\right\}\right].$$
(4.372)

Now, equation (4.371) has the form of equation (4.354). Thus, the TVD condition is guaranteed, if the conditions in equation (4.355) are satisfied. The required conditions are

$$\bar{V}^+ \geq 0, \quad \bar{V}^- \leq 0, \quad 0 \leq \bar{V}^+ - \bar{V}^- \leq \frac{\Delta x}{\Delta t}. \tag{4.373}$$

The first condition of equation (4.373) gives the strongest condition, when the following situation is realized:

$$\Delta_j \Delta_{j-1} < 0 \quad \text{and} \quad \frac{\Delta_{j-2}}{\Delta_{j-1}} > b. \tag{4.374}$$

Together with the requirement $b \geq 1$, these conditions yield

$$1 \leq b \leq \frac{3-\kappa_{\text{pwp}}}{1-\kappa_{\text{pwp}}}. \tag{4.375}$$

The second condition of equation (4.373) becomes strongest, when the following situation is realized:

$$\Delta_j \Delta_{j-1} < 0 \quad \text{and} \quad \frac{\Delta_{j+1}}{\Delta_j} > b. \tag{4.376}$$

These also yield the condition (4.375). Thus for the third-order scheme with $\kappa_{\text{pwp}} = 1/3$, the value of b should be in the range $1 \leq b \leq 4$.[4]

Here, the condition for the value of b is derived for equation (4.56). For more complicated, nonlinear equations such as hydrodynamics and magnetohydrodynamics equations in which fluxes at cell interfaces are nonlinear functions of primitive variables, the condition can be modified. However, it is *empirically* known that if we reconstruct the quantities at cell interfaces by equations (4.365) and (4.366) with $1 \leq b \lesssim 4$, and then, we calculate the fluxes at the cell interfaces, unphysical oscillations are efficiently suppressed. Therefore, in a popular scheme in numerical relativity, (i) the reconstruction at cell interfaces is performed for a set of primitive or similar variables multiplying a flux-limiter function and then (ii) the fluxes at cell interfaces are evaluated using the reconstructed variables. In the next section, we will describe the concept to evaluate the fluxes at cell interfaces in hydrodynamics, and then, introduce several popular schemes for evaluating the fluxes in numerical relativity.

---

[4]In this analysis, we suppose that the grid is uniformly distributed. In non-uniform grids, we need a separate analysis.

4.8.6 Approximate Riemann solvers

In numerical hydrodynamics/magnetohydrodynamics, basic equations are solved basically using schemes similar to those developed for solving the scalar-wave equation (4.56). In particular, the method of reconstructing quantities at cell interfaces, described in section 4.8.5, can be fully employed. However, a more careful and sophisticated treatment is required in numerical hydrodynamics/magnetohydrodynamics. The first reason for this is that in contrast to the scalar-wave equation, hydrodynamics/magnetohydrodynamics equations have multi components and there are multi-component characteristic speeds and characteristic curves (see sections 4.4.2 and 4.6.3). This implies that numerical fluxes for the multi-component quantities have to be determined taking into account several characteristic speeds and characteristic curves. The second reason is that hydrodynamics/magnetohydrodynamics equations are nonlinear partial differential equations. This implies that the characteristic speeds and curves depend on the local conditions, and hence, profiles of the quantities such as density, velocity, and pressure change with time, leading to formation of caustics; discontinuities are frequently formed. Thus, numerical schemes have to enable us to accurately capture such discontinuities that are formed frequently and in an irregular manner.

More specifically, the major purpose of modern numerical hydrodynamics is to develop numerical schemes that can reproduce solutions of the *Riemann problem* as accurately as possible. Thus, before going ahead, we briefly describe the Riemann problem. The Riemann problem is an initial-value problem for hydrodynamics with initial conditions of the form

$$\boldsymbol{Q}_a = \begin{cases} \boldsymbol{Q}_a^L & \text{for } x < x_0, \\ \boldsymbol{Q}_a^R & \text{for } x > x_0, \end{cases} \qquad (4.377)$$

where x_0 is a constant, and $\left(\boldsymbol{Q}_a^L, \boldsymbol{Q}_a^R\right)$ denote two uniform states for which $\boldsymbol{Q}_a^L \neq \boldsymbol{Q}_a^R$; two states are separated by a discontinuity present at $x = x_0$. Here, we assume that the initial state as well as subsequent states during the evolution do not depend on y and z. For the case that fluids are at rest initially, the problem is called the Riemann shock-tube problem. Solutions of the special relativistic Riemann shock-tube problems were analytically derived by Centrella and Wilson (1984); Hawley *et al.* (1984a) for limited cases and by Thompson (1986) for the general case. The general Riemann problem was analytically solved by Martí and Müller (1994); Pons *et al.* (2000) (see also a review in appendix E.1). As illustrated in appendix E.1, the solution of the Riemann problem is derived fully using the information of the characteristic speeds and curves. This indicates that a numerical scheme (specifically, a method for determining numerical fluxes at cell interfaces) should be implemented carefully taking into account their information.

A numerical scheme that can reproduce the solutions of the Riemann problem is often called a *Riemann solver*, and the most well-known one is the Godunov's scheme, which was originally proposed by Godunov (1959) [see Toro (1999) for a detailed description]. A scheme that can reproduce an approximate solution of the Riemann problem is called an *approximate Riemann solver*.

Probably, the most straightforward and strict method for computing fluxes at cell interfaces is to employ the Godunov's scheme. In this scheme, we first reconstruct hydrodynamics/magnetohydrodynamics variables at cell interfaces, $j+1/2$, from its left and right sides as described in section 4.8.5: We here denote the set of the variables as $\boldsymbol{q}^L_{j+1/2}$ and $\boldsymbol{q}^R_{j+1/2}$, respectively, and suppose an initial state that for $x < x_{j+1/2}$, $\boldsymbol{q}$ is uniformly equal to $\boldsymbol{q}^L_{j+1/2}$ while for $x > x_{j+1/2}$, it is uniformly equal to $\boldsymbol{q}^R_{j+1/2}$. This approximation is valid as far as we pay attention only to a local region around $x = x_{j+1/2}$. This setting of the initial state agrees with that of the Riemann problem. Thus, it is possible to obtain a solution for the subsequent local evolution by a semi-analytic calculation (see appendix E.1), and hence, we can obtain the flux at the cell interface, $x = x_{j+1/2}$.

However, employing the Godunov's method requires a large computational cost. The reason for this is that, as shown in appendix E.1, the basic equation is implicit [Martí and Müller (1994)], and an iterative calculation is necessary for obtaining a solution of the Riemann problem even for the special relativistic case. In general relativity, in addition, the effect of the curved metric has to be handled.

Thus, in numerical relativity, it is practical to employ a scheme based on an approximate Riemann solver (see section 4.4.2 for its foundation), in which the computational costs are saved significantly and a reasonably accurate approximate solution of the Riemann problem is obtained. In the following subsections, we will introduce two popular approximate Riemann solvers often employed in numerical relativity. For a wider review of the Riemann solvers in relativity, we recommend the readers to refer to Martí and Müller (2003).

4.8.6.1 *Roe-type scheme*

As described in sections 4.4.2 and 4.6.3, hydrodynamics/magnetohydrodynamics equations are written in the form of equation (4.52). Then, using the Jacobian matrix, equation (4.52) is written in the form of equation (4.59). In the Roe-type scheme, which was originally proposed by Roe (1981), the Jacobian matrix, $\boldsymbol{M}^k_{ab}$, at cell interfaces (labeled by $j+1/2$) is approximated as a constant matrix determined by the left and right states of the primitive variables, $\boldsymbol{q}^L_{j+1/2}$ and $\boldsymbol{q}^R_{j+1/2}$, which are in advance determined by an appropriate reconstruction (see section 4.8.5); we denote it as $\tilde{\boldsymbol{M}}^k_{ab} = \tilde{\boldsymbol{M}}^k_{ab}\left(\boldsymbol{q}^L_{j+1/2}, \boldsymbol{q}^R_{j+1/2}\right)$. Then, the left-hand side of equation (4.59) becomes linear in $\boldsymbol{Q}_a$, and in the absence of the right-hand side, it is solved exactly [see equation (4.81)].

In the Roe-type scheme, the numerical fluxes at cell interfaces for the x^k direction are computed by an upwind scheme as

$$(\tilde{\boldsymbol{F}}_a)^k_{j+1/2} = \frac{1}{2}\bigg[[\boldsymbol{F}_a(\boldsymbol{q}^R)]^k_{j+1/2} + [\boldsymbol{F}_a(\boldsymbol{q}^L)]^k_{j+1/2}$$
$$-\sum_{b=1}^{5}\Big(\boldsymbol{R}^k|\Lambda^k|(\boldsymbol{R}^k)^{-1}\Big)_{ab}\Big[(\boldsymbol{Q}_b)^R_{j+1/2} - (\boldsymbol{Q}_b)^L_{j+1/2}\Big]\bigg],$$
$$(a = 1, 2, \cdots, 5), \qquad (4.378)$$

where the definitions of $\boldsymbol{R}^k_{ab}$ and Λ^k_{ab} are described in section 4.4.2, and they are derived from the Jacobian matrix determined by $\boldsymbol{q}^L_{j+1/2}$ and $\boldsymbol{q}^R_{j+1/2}$ in the Roe-type method. $[\boldsymbol{F}_a(\boldsymbol{q}^L)]^k_{j+1/2}$ and $[\boldsymbol{F}_a(\boldsymbol{q}^R)]^k_{j+1/2}$ denote the fluxes at the left and right sides of cell interfaces determined from $\boldsymbol{q}$, which are evaluated using a spatial reconstruction as described in section 4.8.5. $(\boldsymbol{Q}_b)^R_{j+1/2}$ and $(\boldsymbol{Q}_b)^L_{j+1/2}$ denote the states of (ρ_*, S_i, S_0) at the right and left sides of cell interfaces. The flux given by equation (4.378) is the strict one, if the Jacobian matrix is a constant matrix.

As mentioned above, the components of matrices $\boldsymbol{R}^k_{ab}$ and Λ^k_{ab} at cell interfaces are calculated from the Jacobian matrix $\tilde{\boldsymbol{M}}^k_{ab}\left(\boldsymbol{q}^L_{j+1/2}, \boldsymbol{q}^R_{j+1/2}\right)$. Here, we have to be careful for determining the values for each component of $\tilde{\boldsymbol{M}}^k_{ab}$ by an appropriate average of $\boldsymbol{q}^L_{j+1/2}$ and $\boldsymbol{q}^R_{j+1/2}$. For this averaging, Roe (1981) required that the following three conditions should be satisfied (in the non-relativistic case): (i) The Jacobian matrix has real eigenvalues and a complete set of linearly independent right eigenvectors; (ii) The constant Jacobian matrix satisfies a consistency condition as $\tilde{\boldsymbol{M}}^k_{ab}(\boldsymbol{q}, \boldsymbol{q}) = \boldsymbol{M}^k_{ab}(\boldsymbol{q})$; (iii) Conservation across discontinuities holds as $\boldsymbol{F}^k_a\left(\boldsymbol{Q}^R_b\right) - \boldsymbol{F}^k_a\left(\boldsymbol{Q}^L_b\right) = \tilde{\boldsymbol{M}}^k_{ab}\left[\boldsymbol{Q}^R_b - \boldsymbol{Q}^L_b\right]$. Then, the average should have the form

$$\boldsymbol{q}_{j+1/2} = \frac{\sqrt{\rho_R}\,\boldsymbol{q}_R + \sqrt{\rho_L}\,\boldsymbol{q}_L}{\sqrt{\rho_R} + \sqrt{\rho_L}}, \qquad (4.379)$$

where the primitive variables, $\boldsymbol{q}$, should be composed of three velocity and specific enthalpy in the non-relativistic case.

The extension for the relativistic equations was done by Eulderink and Mellema (1995). In this case, the average should be done for a set of primitive variables composed of four velocity u^a and $P/\rho h$, and the weight, $\sqrt{\rho}$, should be replaced by a weighted density as $\sqrt{\rho_* h/u^t}$.

4.8.6.2 HLL scheme

A scheme proposed by Harten et al. (1983) (hereafter referred to as HLL scheme) is even more simplified scheme than the Roe's scheme. This scheme avoids the explicit calculation of the eigenvectors of the Jacobian matrix $\boldsymbol{M}^k_{ab}$, and is based on an approximate solution of the Riemann problem. Although there are some variants of the original HLL scheme such as HLLE scheme [Einfeldt (1988)], we will not distinguish them in the following. (Note however that the HLLC scheme is different from the HLL scheme in its policy; see below.)

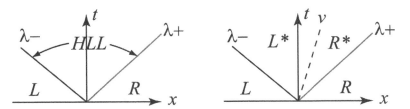

Fig. 4.9 1+1 spacetime diagram for an approximate solution of the Riemann problem supposed in the HLL (left) and HLLC (right) schemes. The thick solid lines (labeled by $\lambda_\pm$) and dashed lines (labeled by v) denote trajectories of shocks and a contact discontinuity, respectively. $\lambda_\pm$ and v are three characteristic speeds [see equation (4.66)]. L and R denote the initial uniform states, HLL denotes an approximate state assumed in the HLL scheme, and L^* and R^* in the HLLC scheme (separated by the dashed line) are uniform states between the shocks and contact discontinuity. Compare these figures with figure E.1.

For simplicity, we restrict our attention only to the Riemann problem for pure hydrodynamics. As described in appendix E.1, there are at least four different states in solutions of the Riemann problem, which are separated by two shock or rare-faction waves with the propagation speed $\lambda_\pm$ [see equation (4.67)] and by one contact discontinuity with the propagation speed $v^x(=u^x/u^t)$. In the HLL scheme (and also in the HLLE scheme), the solution of the Riemann problem is approximated assuming that there are only three different *uniform* states separated only by two waves for which the propagation speed is assumed to be $\lambda_\pm$ (see the left panel of figure 4.9). Namely, the solution is approximated by the following form

$$\boldsymbol{Q}(x,t) = \begin{cases} \boldsymbol{Q}^L & \dfrac{x-x_0}{t-t_0} < \lambda_-, \\ \boldsymbol{Q}^{HLL} & \lambda_- < \dfrac{x-x_0}{t-t_0} < \lambda_+, \\ \boldsymbol{Q}^R & \dfrac{x-x_0}{t-t_0} > \lambda_+, \end{cases} \qquad (4.380)$$

where $\lambda_\pm$ are supposed to be determined from equation (4.67). t_0 and x_0 are the initial time and the location of the discontinuity at $t=t_0$, respectively.

$\boldsymbol{Q}^{HLL}$ is a peculiar state determined by the following analysis of the equation

$$\partial_t \boldsymbol{Q} + \partial_x \boldsymbol{F} = 0. \qquad (4.381)$$

We integrate this equation for a small spacetime domain composed of $[0:T]$ for the time and $[x_L:x_R]$ for the x direction (see figure 4.10 in which we assume $\lambda_+ > 0$ and $\lambda_- < 0$). Then, we have

$$\int_{x_L}^{x_R} \boldsymbol{Q}(T,x)dx = \int_{x_L}^{x_R} \boldsymbol{Q}(0,x)dx + \int_0^T \boldsymbol{F}(t,x_L)dt - \int_0^T \boldsymbol{F}(t,x_R)dt$$
$$= \boldsymbol{Q}^R x_R - \boldsymbol{Q}^L x_L + T\left(\boldsymbol{F}^L - \boldsymbol{F}^R\right), \qquad (4.382)$$

where $\boldsymbol{F}^L$ and $\boldsymbol{F}^R$ denote $\boldsymbol{F}$ for the left and right states, respectively. Assuming that a solution of the Riemann problem is written in the form of equation (4.380), the left-hand side of equation (4.382) is written as

$$\int_{x_L}^{x_R} \boldsymbol{Q}(T,x)dx = \boldsymbol{Q}^R(x_R - \lambda_+ T) + \boldsymbol{Q}^{HLL}(\lambda_+ - \lambda_-)T + \boldsymbol{Q}^L(\lambda_- T - x_L). \quad (4.383)$$

Substituting this into equation (4.382) yields

$$\boldsymbol{Q}^{HLL} = \frac{\boldsymbol{Q}^R \lambda_+ - \boldsymbol{Q}^L \lambda_- + \boldsymbol{F}^L - \boldsymbol{F}^R}{\lambda_+ - \lambda_-}. \quad (4.384)$$

With these preparations, it is possible to determine the flux at $x = x_0$ (hereafter denoted by $\boldsymbol{F}^{HLL}$), which is employed as the numerical flux at cell interfaces in the HLL scheme. For $\lambda_- > 0$ and $\lambda_+ < 0$, it is trivially written as $\boldsymbol{F}^L$ and $\boldsymbol{F}^R$, and thus, we focus only on the case $\lambda_- < 0 < \lambda_+$. For evaluating $\boldsymbol{F}^{HLL}$, we consider the integral of equation (4.381) for a spacetime domain composed of $[0:T]$ and $[x_L : 0]$, which yields

$$\int_{x_L}^{0} \boldsymbol{Q}(T,x)dx = \int_{x_L}^{0} \boldsymbol{Q}(0,x)dx + \int_{0}^{T} \boldsymbol{F}(t,x_L)dt - \int_{0}^{T} \boldsymbol{F}(t,0)dt$$

$$= -\boldsymbol{Q}^L x_L + T\left(\boldsymbol{F}^L - \boldsymbol{F}^{HLL}\right). \quad (4.385)$$

The left-hand side of this equation is evaluated as

$$\int_{x_L}^{0} \boldsymbol{Q}(T,x)dx = \boldsymbol{Q}^L(\lambda_- T - x_L) - \boldsymbol{Q}^{HLL}\lambda_- T. \quad (4.386)$$

Substituting this into equation (4.385) yields

$$\boldsymbol{F}^{HLL} = \frac{\lambda_+ \boldsymbol{F}^L - \lambda_- \boldsymbol{F}^R + \lambda_+ \lambda_-\left(\boldsymbol{Q}^R - \boldsymbol{Q}^L\right)}{\lambda_+ - \lambda_-}. \quad (4.387)$$

We note that the integration for the spatial domain $[0 : x_R]$ yields the same result for $\boldsymbol{F}^{HLL}$. In summary, the numerical flux in the HLL scheme is written as

$$\boldsymbol{F} = \begin{cases} \boldsymbol{F}^L & 0 < \lambda_-, \\ \boldsymbol{F}^{HLL} & \lambda_- < 0 < \lambda_+, \\ \boldsymbol{F}^R & \lambda_+ < 0. \end{cases} \quad (4.388)$$

In the HLL scheme, the presence of the contact discontinuity in the Riemann problem is ignored. The HLLC scheme is a modified version of the HLL scheme, in which the effect of the contact discontinuity is taken into account [Toro (1999)]. In this scheme, thus, the solution of the Riemann problem is approximated so that the solution is composed of four uniform states (L, L^*, R^*, R), which are separated by three characteristic curves. The method for determining the numerical flux is similar to that for the HLL scheme and the details are reviewed in Toro (1999). Thus, we do not describe it any longer.

Kurganov and Tadmor (2000) scheme (hereafter KT scheme) is a even more simplified variant of the HLL scheme. This scheme is obtained by setting $\lambda_\pm = \pm \lambda$

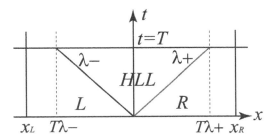

Fig. 4.10 The spacetime domain $[x_L : x_R][0 : T]$ for equation (4.382).

where $\lambda = \max(|\lambda_+|, |\lambda_-|)$ for the HLL flux. Then, the result is

$$\boldsymbol{F}^{KT} = \frac{1}{2}\left[\boldsymbol{F}^L + \boldsymbol{F}^R - \lambda\left(\boldsymbol{Q}^R - \boldsymbol{Q}^L\right)\right]. \tag{4.389}$$

Lucas-Serrano et al. (2004) performed a detained comparison among the HLL, KT, Roe, and other schemes. They found that as far as a high-order reconstruction is performed for the primitive variables at cell interfaces, the accuracy of the numerical solutions for several Riemann problems depends very weakly on the chosen schemes. Thus, although the HLL and KT schemes are simplified versions of approximate Riemann solvers, they may be employed for the practical use. Because a complicated coding is not required for them, they are recommendable schemes, in particular, for the beginner of numerical hydrodynamics/magnetohydrodynamics.

4.9 Other ingredients in numerical hydrodynamics and magnetohydrodynamics

In this section, we summarize some technical details about solving general relativistic hydrodynamics and magnetohydrodynamics equations.

4.9.1 *Solving normalization relation for four velocity*

As shown in equations (4.51) and (4.168), the normalization relation of the four velocity, $u^a u_a = -1$, is written as a single algebraic equation for w or h or hw, as far as equations of state employed are written in a simple analytic form. The single algebraic equation can be accurately solved by the Newton–Raphson method [Press et al. (1989)].

Here, we consider an equation for x in the form $f(x) = 0$ where the functional form of f is supposed to be known. Let $x = x_0$ be an approximate solution and δx be the difference between the exact solution and x_0. Then,

$$f(x_0 + \delta x) = 0. \tag{4.390}$$

Assuming that δx is small, this equation may be expanded by the first-order Taylor series as

$$0 = f(x_0 + \delta x) \approx f(x_0) + f'(x_0)\delta x, \qquad (4.391)$$

where $f' = df/dx$. Then, δx is determined approximately by

$$\delta x \approx \delta x_1 = -\frac{f(x_0)}{f'(x_0)}, \qquad (4.392)$$

and a new approximate solution is obtained as $x_0 + \delta x_1$. If this procedure is repeated until a sufficient convergence is achieved, we will finally obtain an accurate numerical solution of $f(x) = 0$. This method works well as far as equations (4.51) and (4.168) are written in a polynomial form and the initial guess for the solution would not be set in an extremely bad manner.

If a tabulated equation of state is adopted in a numerical simulation as described in section 4.5, the normalization relation does not result in a simple algebraic equation. Also, the variable to be determined from the equation is not in general h, w, and hw but one of the argument variables in the table such as temperature. In this case, the solution has to be determined referring to the table at each iteration step. If a finite differencing can be taken for the variables in the table and the resulting value is accurate enough, it is possible to use the Newton–Raphson method. However, if the finite differencing does not give an accurate result, the solution has to be found by referring to a wide range of the values in the table at each time step [see, e.g., Sekiguchi (2010b)].

4.9.2 Constrained transport

When numerically solving the induction equation for magnetic fields, we have to guarantee during the numerical simulation that the no-monopole constraint, (4.120) or (4.122), is satisfied and magnetic fluxes are conserved. These are the constraints to be satisfied among three magnetic-field components, which are not automatically satisfied even if the induction equation is written in a conservative form. Evans and Hawley (1988) proposed the robust scheme of numerical evolution, so-called *constrained transport scheme*, in which the no-monopole constraint is guaranteed to be satisfied in the machine precision level.

The constrained transport scheme of Evans and Hawley (1988) is composed of two procedures as follows. Let (j, k, l) denote a grid point (a cell center) of (x, y, z) at which density, internal energy, pressure, velocity, and geometric quantities have their representative values. Then, the first task is to put the magnetic-field components on cell interfaces as shown in figure 4.11. Specifically, B^x, B^y, and B^z are assigned at $(j \pm 1/2, k, l)$, $(j, k \pm 1/2, l)$, and $(j, k, l \pm 1/2)$, respectively.

Now, we consider the no-monopole constraint in the volume integral form. Integrating equation (4.122) by $dx\,dy\,dz$ yields

$$\oint B^k dS_k^{\rm f} = 0, \qquad (4.393)$$

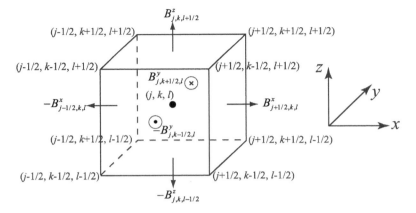

Fig. 4.11 Spatial assignment of magnetic-field components on the discretized grids in the constrained transport scheme. The square box denotes one of cells in the computational domain and the magnetic-field components are assigned on cell interfaces, while other hydrodynamics quantities are assigned at the center of the cell denoted by the filled circle.

where $dS_k^{\rm f}$ is the surface integral element in flat space. The constrained transport scheme requires us to enforce the condition (4.393) for each cubic cell in the finite-differencing level (see the cubic box of figure 4.11 for such a cell). Let (dx, dy, dz) be the side lengths of the cubic cell. Then, the discretized form of equation (4.393) is written as

$$dy\,dz\left(\mathcal{B}^x_{j+1/2,k,l} - \mathcal{B}^x_{j-1/2,k,l}\right) + dx\,dz\left(\mathcal{B}^y_{j,k+1/2,l} - \mathcal{B}^y_{j,k-1/2,l}\right)$$
$$+\,dx\,dy\left(\mathcal{B}^z_{j,k,l+1/2} - \mathcal{B}^z_{j,k,l-1/2}\right) = 0. \tag{4.394}$$

Then, the second task is to find a finite-differencing scheme for the induction equation in which equation (4.394) is automatically satisfied. This is achieved in the following manner.

The induction equation is in general written in the form

$$\partial_t \mathcal{B}^i = -e^{ijk}\partial_j \hat{E}_k, \tag{4.395}$$

where for the general case,

$$\hat{E}_j = \alpha E_j + \epsilon_{jkl}\beta^k B^l, \tag{4.396}$$

and for the ideal magnetohydrodynamics case,

$$\hat{E}_j = -\epsilon_{jkl}v^k B^l. \tag{4.397}$$

In the constrained transport scheme, equation (4.395) is written in a finite-differencing form by integrating it for surfaces of each cubic cell and by using the Stokes' theorem. To do this in a self-consistent manner, we assign $\hat{E}_k$ at the sides

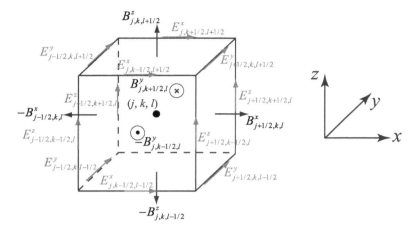

Fig. 4.12 Spatial assignment of "electric field" components, $\hat{E}_k$, in the constrained transport scheme.

of each cubic cell as shown in figure 4.12. Then, the resulting finite-differencing equations for three components are, respectively,

$$\partial_t \mathcal{B}^x_{j\pm1/2,k,l} dy\, dz = -\left(\hat{E}^z_{j\pm1/2,k+1/2,l} - \hat{E}^z_{j\pm1/2,k-1/2,l}\right) dz$$
$$+ \left(\hat{E}^y_{j\pm1/2,k,l+1/2} - \hat{E}^y_{j\pm1/2,k,l-1/2}\right) dy,$$
$$\partial_t \mathcal{B}^y_{j,k\pm1/2,l} dx\, dz = -\left(\hat{E}^x_{j,k\pm1/2,l+1/2} - \hat{E}^x_{j,k\pm1/2,l-1/2}\right) dx$$
$$+ \left(\hat{E}^z_{j+1/2,k\pm1/2,l} - \hat{E}^z_{j-1/2,k\pm1/2,l}\right) dz,$$
$$\partial_t \mathcal{B}^z_{j,k,l\pm1/2} dx\, dy = -\left(\hat{E}^y_{j+1/2,k,l\pm1/2} - \hat{E}^y_{j-1/2,k,l\pm1/2}\right) dy$$
$$+ \left(\hat{E}^x_{j,k+1/2,l\pm1/2} - \hat{E}^x_{j,k-1/2,l\pm1/2}\right) dx. \qquad (4.398)$$

If all these equations are simultaneously solved, the following relation is satisfied:

$$\partial_t \left[dy\, dz \left(\mathcal{B}^x_{j+1/2,k,l} - \mathcal{B}^x_{j-1/2,k,l}\right) + dx\, dz \left(\mathcal{B}^y_{j,k+1/2,l} - \mathcal{B}^y_{j,k-1/2,l}\right) \right.$$
$$\left. dx\, dy \left(\mathcal{B}^z_{j,k,l+1/2} - \mathcal{B}^z_{j,k,l-1/2}\right) \right] = 0. \qquad (4.399)$$

Thus, if equation (4.394) is initially satisfied and equation (4.398) is subsequently solved, the constraint (4.394) is automatically satisfied during the evolution in the machine precision level.

The remaining task is to give $\hat{E}^i$ at each side of the cell. Because quantities other than the electromagnetic fields are assigned at the cell center and the magnetic-field components are assigned at the cell interfaces, some interpolation procedures are necessary to determine $\hat{E}^i$. For achieving a high-order accuracy, a high-order reconstruction as described in section 4.8.5 is necessary. As warned in section 3.2, we here have to be careful for employing a high-order (more than second-order) interpolation procedure, because in the transport terms of equations (4.128) and (4.150), the magnetic-field components are contained: A straightforward interpolation procedures could cause numerical instability, and thus, an upwind-type interpolation is necessary. For this purpose, it is empirically known that MUSCL-type schemes work well [see, e.g., Zanna et al. (2003)].

Before closing this section, we comment on an important point for implementing adaptive mesh-refinement (AMR) algorithms for ideal magnetohydrodynamics equations. As described in section 3.3, AMR is one of the key ingredients in numerical relativity for evolving Einstein's equation. This implies that in the presence of fluids, an AMR algorithm for evolving hydrodynamics/magnetohydrodynamics equations has to be also implemented. When employing an AMR algorithm, we prepare several computational domains for which the grid resolutions and the sizes of the domains are different from each other. Then, the central tasks in AMR algorithms are to assign data for a "child" level in a refinement boundary using data of its "parent" level, and in addition, to restrict data for a parent level using its children levels. Here, the child has a finer grid resolution and smaller domain size than its parent has. As described in section 3.3, for evolving Einstein's equation, data for a child level in a refinement boundary are simply determined by interpolating the data of its parent level. Also, the restriction is done by simply copying the data in the child level to those in the parent level for overlapping points. However, this simple prescription should not be used for integrating ideal magnetohydrodynamics equations, because the simple ones in general do not guarantee the conservation of the divergence-free constraint and the magnetic flux conservation. The violation of these conservation laws in numerical computations results in an unreliable numerical result. To guarantee these conservation laws, we have to be very careful in the interpolation and restriction procedures. Describing the detailed schemes is beyond the scope of this volume. We here only refer the readers to references by Barsara (2001, 2009), which describe excellent prescriptions for precisely guaranteeing these constraints [see also Kiuchi et al. (2012) for an implementation in general relativistic magnetohydrodynamics].

4.9.3 *Implicit-explicit scheme*

As we mentioned in the first paragraph of section 4.8, time evolution of numerical hydrodynamics/magnetohydrodynamics equations is usually performed in an explicit manner, e.g., using an explicit Runge–Kutta scheme (see section 3.1.2). Although it is always possible to employ an explicit scheme in principle, expensive computational costs for the so-called *stiff* equation often prohibit employing any explicit time integration practically. A typical example of the stiff equation is a radiation transfer equation for which the interaction time scale between the medium and the radiation is much shorter than the dynamical and sound-crossing time scales of the system.

To illustrate the difficulty of explicitly time-integrating a stiff equation, we here consider the radiation transfer equations in the moment formalism described in section 4.7.2.2. For simplicity, we consider the frequency-integrated equations, (4.313) and (4.314), with the source term [see equation (4.316)]

$$S^\alpha = \bar{\kappa}\left[(J^{\text{eq}} - J)u^\alpha - H^\alpha\right], \quad (4.400)$$

where $\bar{\kappa}$ is an opacity. Here, J and H^α are written in linear functions of E and F_i, and $J \sim E$ as far as the motion of the medium is not extremely relativistic. This implies that the interaction time scale between the medium and the radiation is basically determined by the following part of the equation for E,

$$\partial_t E \sim -\bar{\kappa} E, \quad (4.401)$$

and thus, the interaction time scale is approximately $\bar{\kappa}^{-1}$. For explicitly integrating this equation, the time-step interval, Δt, has to be much smaller than $\bar{\kappa}^{-1}$ (say $\bar{\kappa}^{-1}/10$). If $\bar{\kappa}$ is extremely large, Δt has to be extremely small (i.e., much shorter than the dynamical time scale of the system), implying that the explicit time integration is practically impossible. For such cases, an implicit-explicit (or partially implicit) time integration scheme is quite useful.

To describe an implicit-explicit scheme, we schematically write equations (4.313) and (4.314) in a matrix form

$$\partial_t \boldsymbol{F} = \boldsymbol{T} + \bar{\kappa}(\boldsymbol{S}_0 - \boldsymbol{A}\boldsymbol{F}), \quad (4.402)$$

where $\boldsymbol{F}$ denotes a vector composed of

$$\boldsymbol{F} = \begin{pmatrix} \mathcal{E} \\ \mathcal{F}_x \\ \mathcal{F}_y \\ \mathcal{F}_z \end{pmatrix}. \quad (4.403)$$

$\boldsymbol{T}$ denotes the sum of transport terms $-\partial_i(\cdots)^i$ and a source term associated with gravitational fields. $\boldsymbol{S}_0$ denotes the source term composed of the thermal quantity J^{eq}, and $\boldsymbol{A}$ is a 4×4 matrix composed of hydrodynamic and geometric quantities. In each Runge–Kutta time-integration step, the term $\partial_t \boldsymbol{F}$ is discretized as $(\boldsymbol{F}^{n+1} - \boldsymbol{F}^n)/\Delta t$ where n and $n+1$ denote neighboring two time-step levels for the fourth-order Runge–Kutta time integration ($n = 0$–3). In the implicit-explicit scheme employed here, n-th quantities are assigned for $\boldsymbol{T}$, $\boldsymbol{S}_0$, $\boldsymbol{A}$, and $\bar{\kappa}$, while we assign $(n+1)$-th quantities for $\boldsymbol{F}$ in the right-hand side. Namely, we write the equation in the following form:

$$(1 + \bar{\kappa}^n \Delta t \boldsymbol{A}^n)\boldsymbol{F}^{n+1} = \boldsymbol{F}^n + \Delta t\left(\boldsymbol{T}^n + \bar{\kappa}^n \boldsymbol{S}_0^n\right), \quad (4.404)$$

where all the upperscripts n and $n+1$ denote the time-step levels. This is a simple 4×4 matrix equation and can be solved in a straightforward manner. In this scheme, only the stiff part is handled in an implicit manner.

The result of choosing this implicit-explicit scheme is found by taking the limit $\bar{\kappa}\Delta t \to \infty$, which yields

$$\boldsymbol{A}^n \boldsymbol{F}^{n+1} = \boldsymbol{S}_0^n. \tag{4.405}$$

This implies that $\boldsymbol{F}^{n+1}$ is determined by the thermal equilibrium condition where the thermal quantities at the n-th time-step level is responsible for $\boldsymbol{F}^{n+1}$. In reality, $\boldsymbol{F}^{n+1}$ should be determined by $\boldsymbol{A}^{n+1}$ and $\boldsymbol{S}_0^{n+1}$. For obtaining such a solution at least approximately, the time integration of the radiation and hydrodynamics equations is performed for several times iteratively until $\boldsymbol{F}^{n+1}$ settles to a converged solution at each Runge–Kutta time integration level.

This method is quite robust and efficient for stable time integration with a reasonably short (not extremely short) time-step interval. Indeed, numerical simulations for radiation hydrodynamics show that this is an adaptable scheme in numerical relativity [e.g., Shibata and Sekiguchi (2012)].

4.10 Testing hydrodynamics and magnetohydrodynamics codes

In this section, we introduce several standard test-bed problems for which the solutions can be obtained analytically or at most by solving algebraic equations numerically. For any new numerical code of hydrodynamics/magnetohydrodynamics, we have to examine whether it can reproduce these solutions accurately, in order to confirm the reliability of the new code.

4.10.1 *Riemann shock-tube problems in hydrodynamics*

For confirming the reliability of a hydrodynamics implementation newly developed, in particular, for confirming that it can yield an accurate numerical solution for shocks, test simulations for Riemann shock-tube and wall shock problems in 1+1 special relativistic spacetime are often performed (see appendix E.1 for a description of these problems). For both problems, left ($x < 0$) and right ($x \geq 0$) sides are initially composed of uniform but different fluid states; in the following we will denote them by (ρ_L, v_L, P_L) and (ρ_R, v_R, P_R), respectively. For some of such initial conditions, a shock and other discontinuity (such as a contact discontinuity) are formed, soon after the onset of the dynamical evolution. The exact analytic solutions for these problems are derived using the Rankine–Hugoniot conditions across discontinuities [Landau and Lifshitz (1959)] and the information of characteristic curves (see appendix E.1). Thus, it is possible to test whether a numerical code can yield an accurate solution for shocks by comparing the numerical solutions with the exact solutions.

Figures 4.13 and 4.14 show an example of the comparison between numerical and analytic solutions for a Riemann shock-tube and a wall shock problem for the choice $b = 1$ (left panels) and 2 (right panels). In these test problems, the Γ-law

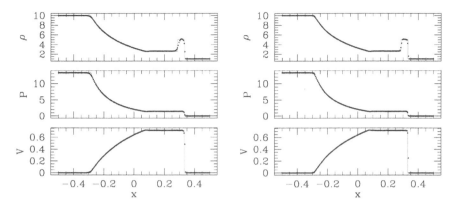

Fig. 4.13 Comparison of numerical solutions (filled circles) of a one-dimensional Riemann shock-tube problem with the analytical solution (solid curves) at $t = 0.4$ for $b = 1$ (left) and 2 (right). Here, b is the parameter in the minmod-limiter function: see equation (4.350). The grid number is 801 and the grid spacing is 1/400. Only 200 data points are plotted. The figure is taken from Shibata (2003a).

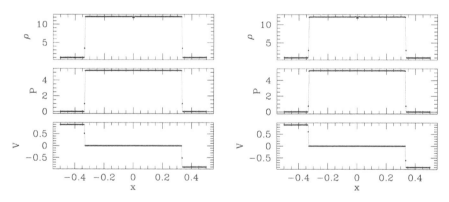

Fig. 4.14 The same as figure 4.13, but for a one-dimensional wall shock problem with $\Gamma = 4/3$ at $t = 1.6$. The figure is taken from Shibata (2003a).

equation of state, $P = (\Gamma - 1)\rho\varepsilon$, is usually employed. An often-employed initial condition for the Riemann shock-tube problem is $\rho = 10$ and $P = 13.3$ for $x < 0$ and $\rho = 1$ and $P = 10^{-6}$ for $x > 0$ with $\Gamma = 5/3$ [Hawley et al. (1984a)] [see also Font (2006)]. We note that the adiabatic constant $\kappa_\mathrm{p} := P/\rho^\Gamma$ is different between the left and right sides. For the wall shock problem of figure 4.14, the initial condition is $v^x = -0.9x/|x|$, $\rho = 1$, and $P = 10^{-6}$ with $\Gamma = 4/3$. In both tests, one-dimensional uniform grid with $N = 801$ points covering a domain $[-1 : 1]$, i.e., $\Delta x = 1/400$, is prepared.

The test simulations presented here were performed by Shibata (2003a) employing a Roe-type scheme outlined in section 4.8.6.1 with the third-order piecewise cell reconstruction described in section 4.8.5 and with the minmod limiter [equation (4.351)]. A free parameter in the minmod limiter was set to be $b = 1$ or 2. Remember that b has to be in the range $1 \leq b \leq 4$ when we employ the third-order scheme (see section 4.8.5).

As indicated in figures 4.13 and 4.14, numerical results agree well with the analytic solutions. Unphysical oscillations associated with numerical instability are absent for both problems, indicating that such instability is suppressed by the flux-limiter function. The flux-limiter function plays a stronger role and the numerical dissipation is larger when the value of b is close to unity. Indeed, the contact discontinuity at $x \sim 0.3$ in ρ is more sharply computed for $b = 2$ than for $b = 1$. This indicates that in numerical computations, the value of b should be as large as possible: In the practical use, we often employ $b = 2-3$.

4.10.2 Riemann shock-tube problems in magnetohydrodynamics

For any numerical code for magnetohydrodynamics, we also have to examine whether it can reproduce basic waves such as shock and rare-faction waves accurately. Komissarov (1999) (and its correction [Komissarov (2002a)]) proposed a suite of 1+1-dimensional test problems for ideal magnetohydrodynamics in special relativity; propagation of fast and slow shocks, fast and slow rare-faction waves, Alfvén waves, compound waves, shock-tube tests, and collision of two flows (see appendix E.4 for the description of some of them). These tests have been performed for general relativistic magnetohydrodynamics codes [e.g., Gammie et al. (2003); Villiers and Hawley (2003); Duez et al. (2005); Shibata and Sekiguchi (2005a) for early works].

Actually, nonsingular solutions for the Riemann problem in 1+1 ideal magnetohydrodynamics do not always exist. In general, a singularity (divergence) would be formed in a discontinuous region [Inoue and Inutsuka (2007)]. Nonsingular solutions can be present only for a restricted combination of left and right initial states. The solutions of Komissarov (2002a) were ingeniously derived choosing such states (see appendix E.4).

Figures 4.15–4.17 illustrate some of typical numerical results for Komissarov's test-bed problems reported by Shibata and Sekiguchi (2005a). Note that many other codes have shown similar results universally [Komissarov (1999); Gammie et al. (2003); Villiers and Hawley (2003); Duez et al. (2005)] with approximately the same grid size N and the same spacing Δx (these are described in the figure captions). In the code of Shibata and Sekiguchi (2005a), the advection terms of magnetohydrodynamics equations were handled by a KT scheme [Kurganov and Tadmor (2000)] (see also section 4.8.6.2) using the minmod limiter with $b = 2$.

The test simulations were performed for 1+1 spacetime composed of t and x. Thus, all the quantities are functions of t and x. However, this does not imply that nonzero spatial components of u^i and B^i are only u^x and B^x; other components may exist.

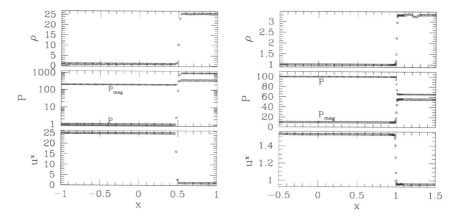

Fig. 4.15 Left: Propagation of a fast shock. The snapshot at $t = 2.5$ is shown. The numerical simulation was performed with $N = 101$ and $\Delta x = 0.02$ for the spatial domain $[-1:1]$. Right: Propagation of a slow shock. The snapshot at $t = 2.0$ is shown. The numerical simulation was performed with $N = 401$ and $\Delta x = 0.005$ for the spatial domain $[-0.5:1.5]$. Only one fourth of data points are plotted except for $[0.92, 1.08]$ in which all the data points are plotted. For both cases, the initial discontinuities were located at $x = 0$, and the shock fronts move with a constant velocity V_0 where $V_0 = 0.2$ and 0.5 for the fast and slow shocks, respectively. The figure is taken from Shibata and Sekiguchi (2005a).

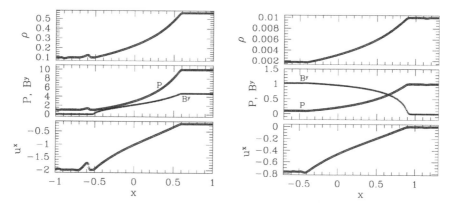

Fig. 4.16 Left: Propagation of a fast switch-off rare-faction wave. The snapshot at $t = 1.0$ is shown. Right: Propagation of a slow switch-on rare-faction wave. The snapshot at $t = 2$ is shown. For both cases, the initial discontinuities were located at $x = 0$, and the numerical simulations were performed with $N = 401$ and $\Delta x = 0.005$ for the spatial domain $[-1:1]$. Only half among all the data are plotted for both figures. The figure is taken from Shibata and Sekiguchi (2005a).

Figure 4.15 plots results for a fast shock (left panel) and a slow shock (right panel) for which there are two uniform states separated by a shock. The fast and slow shocks, respectively, satisfy a condition $B_L^y < B_R^y$ and $B_L^y > B_R^y$ for the condition $P_L < P_R$ ($B^z = 0$ in this case). In these problems, the system is stationary with respect to the frame moving with the shock front. The velocity of

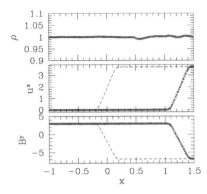
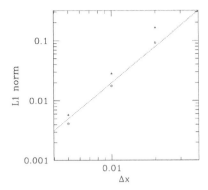

Fig. 4.17 Left: Propagation of a strong continuous Alfvén wave. The snapshot at $t = 2.0$ is shown. The initial configurations and the analytic solution at $t = 2.0$ are shown by the dashed curves. The numerical simulation was performed with $N = 501$ and $\Delta x = 0.005$ for the spatial domain $[-1 : 1.5]$. Only 1/4 of the data points are plotted. Right: L1 norm of the error for ρ (circles) and P (triangles) as functions of Δx. These variables should be constants in this problem, and the deviation from the stationary values is due to a numerical error. The dotted line denotes the slope expected for the second-order convergence. The figure is taken from Shibata and Sekiguchi (2005a).

the shocks is 0.2 and 0.5 for the fast and slow shocks, respectively. For both cases, the numerical results agree with the exact solutions besides a small numerical error. For quantifying the error, it is useful to compute an L1 norm which is defined by the spatial sum for absolute values of the difference between the numerical and exact solutions as

$$\sum |Q_{\text{numerical}} - Q_{\text{exact}}|. \tag{4.406}$$

It decreases as the grid spacing decreases, and in this case, the convergence is achieved at first order (the L1 norm decreases in proportional to Δx). This is the universal property in the presence of discontinuities, around which the advection terms of magnetohydrodynamics equations are computed with the first-order accuracy.

Figure 4.16 shows results for switch-off (left panel) and switch-on (right panel) rare-faction waves. For these tests, there are initially two uniform states separated by a discontinuous surface, and during the time evolution, the two uniform states are connected by a solution of rare-faction waves. For this case, the numerical results also agree with the exact solutions besides a small error.

Figure 4.17(a) shows results for the propagation of an Alfvén wave. In this problem, the initial condition is smooth for all the quantities, and during the evolution, the density and pressure should be unchanged (see appendix E.4). Again, the numerical results agree with the exact solutions within a small error. Since no discontinuities are present in this problem, the convergence of the numerical solution to the exact one should be achieved approximately at second order (because a second-order scheme was employed). This is confirmed by computing the L1 norm. Figure 4.17(b) indeed shows that the L1 norms for the errors of ρ and P decrease in proportional to Δx^2.

4.10.3 Bondi flow

To the previous subsections, only the test-bed simulations in special relativity have been described. The reliability of general relativistic hydrodynamics/magnetohydrodynamics codes should be also tested in curved spacetime. For this purpose, exact solutions again play an important role. A solution of spherically symmetric flow (the so-called Bondi flow) in Schwarzschild black-hole spacetime is a well-suited one.

This is a stationary and spherically symmetric solution not only in hydrodynamics but also in ideal magnetohydrodynamics in the presence only of a purely radial magnetic field. In the Bondi flow, we assume that the flow is adiabatic. The basic equations are composed of the continuity, Euler, and induction equations ($\nabla_a {}^*F^{ab} = 0$), which are, respectively, written in stationary and spherically symmetric spacetime as

$$\dot{M} = 4\pi\rho\sqrt{-g_2}u^r = \text{const}, \tag{4.407}$$

$$\partial_r \left[\sqrt{-g_2} \left\{ (\rho h + b^2) u^r u_\alpha - b^r b_\alpha + \delta^r{}_\alpha \left(P + \frac{b^2}{2} \right) \right\} \right]$$
$$- \frac{1}{2}\sqrt{-g_2} \left[(\rho h + b^2)u^\mu u^\nu - b^\mu b^\nu + \left(P + \frac{b^2}{2} \right) g^{\mu\nu} \right] \partial_\alpha g_{\mu\nu} = 0, \tag{4.408}$$

$$\partial_r \left[\sqrt{-g_2} \left(-u^r b^\mu + u^\mu b^r \right) \right] = 0, \tag{4.409}$$

where we substituted the stress-energy tensor for ideal magnetohydrodynamics and $g_2 := g/\sin^2\theta$ is a function of r ($g_2 = r^4$ for the Schwarzschild metric). For the radial flow $v^\theta = v^\varphi = 0$ and for radial magnetic fields $b^\theta = b^\varphi = 0$, the last equation is trivially satisfied for the spatial components, and for $\mu = t$, this yields

$$\partial_r \left[\sqrt{-g_2}(b^r/u_t) \right] = 0 \quad \Rightarrow \quad b^r \propto u_t(-g_2)^{-1/2}, \tag{4.410}$$

where we used $u^a b_a = 0$.

The Euler equation (4.408) is non-trivial only for the r component. However, this equation is a bit complicated for the analysis. Instead of analyzing it, we focus on the energy equation which is equivalent to the Euler equation in this context. The energy conservation equation may be derived by setting $\alpha = t$ in equation (4.408) because for stationary spacetime that satisfies $\partial_t g_{\mu\nu} = 0$, $T^r{}_t$ obeys

$$\partial_r(\sqrt{-g_2}T^r{}_t) = 0. \tag{4.411}$$

Here, $T^r{}_t$ does not depend on the magnetic field because of the relation

$$b^2 u^r u_t - b^r b_t = 0, \tag{4.412}$$

which is derived from $0 = b^a u_a = u^r b_r + u^t b_t = u_r b^r + u_t b^t$. Therefore, the Euler equation is written in the form independent of the magnetic fields, and the resulting equation is

$$\partial_r(\sqrt{-g_2}\rho h u^r u_t) = 0, \quad \text{and thus,} \quad \sqrt{-g_2}\rho h u^r u_t = \text{const}. \tag{4.413}$$

To summarize, even in the presence of a purely radial magnetic field, the basic equations are written in the same forms as those for the usual Bondi-flow problem, (4.407) and (4.413), together with an equation of state and the normalization relation for u^a, $u^a u_a = -1$.

Solutions for the Bondi flow in a non-spinning black-hole background are described in many standard textbooks, e.g. Shapiro and Teukolsky (1983), and hence, we only outline its derivation. We here use the Schwarzschild metric, (1.97), or the Kerr–Schild metric, (1.111), with $a = 0$, in both of which $\sqrt{-g_2} = r^2$ and $g_{tt} = -(1 - 2M/r)$, and thus the solution shown below is identical in these metric forms. In the following, we will assume that the equation of state has the polytropic form as $P = \kappa_p \rho^\Gamma$ (where κ_p is the adiabatic constant), and thus, $P = (\Gamma - 1)\rho\varepsilon$ for the adiabatic flow.

Using equation (4.407), equation (4.413) is written as

$$\tilde{u}_t := h u_t = \text{const.} \tag{4.414}$$

In the assumed metric forms, we have

$$u_t = -\sqrt{u^2 - g_{tt}}, \tag{4.415}$$

where we set $u = u^r$. From equations (4.407) and (4.414), we have

$$\frac{\rho'}{\rho} + \frac{2}{r} + \frac{u'}{u} = 0, \tag{4.416}$$

$$\frac{h'}{h} + \frac{u'_t}{u_t} = 0, \tag{4.417}$$

where the dash ' denotes the ordinary derivative, d/dr. The first law of thermodynamics for the adiabatic flow gives

$$\frac{h'}{h} = \frac{P'}{\rho h} = c_s^2 \frac{\rho'}{\rho}, \tag{4.418}$$

where c_s is the local sound speed. In the adiabatic equation of state, h and c_s are related to each other by

$$h = \frac{\Gamma - 1}{\Gamma - 1 - c_s^2}. \tag{4.419}$$

Using equations (4.415) and (4.418), equation (4.417) is written in the form

$$\frac{\rho'}{\rho} c_s^2 (u^2 - g_{tt}) + uu' - \frac{1}{2} g'_{tt} = 0. \tag{4.420}$$

From equations (4.416) and (4.420), we finally obtain the well-known forms

$$\frac{u'}{u} = \frac{4c_s^2(u^2 - g_{tt}) + rg'_{tt}}{2r[u^2 - c_s^2(u^2 - g_{tt})]}, \tag{4.421}$$

$$\frac{\rho'}{\rho} = -\frac{4u^2 + rg'_{tt}}{2r[u^2 - c_s^2(u^2 - g_{tt})]}. \tag{4.422}$$

For a flow solution which is regular anywhere, there is one sound point, $r = r_s$, at which the denominator and the numerator vanish simultaneously. Hence, at $r = r_s$,

$$u^2 = u_s^2 = -\frac{r_s}{4}g'_{tt}(r=r_s) = \frac{M}{2r_s}, \qquad (4.423)$$

$$c_s^2 = c_{s,s}^2 = \frac{u_s^2}{u_s^2 - g_{tt}(r=r_s)} = \frac{M}{2r_s - 3M}, \qquad (4.424)$$

and thus,

$$\dot{M} = 4\pi(\rho r^2 u)|_{r=r_s} = 4\pi\sqrt{\frac{M}{2}}\rho_s r_s^{3/2}, \qquad (4.425)$$

$$\tilde{u}_t = (hu_t)|_{r=r_s} = -\frac{(\Gamma-1)(2r_s - 3M)}{(\Gamma-1)(2r_s - 3M) - M}\sqrt{1 - \frac{3M}{2r_s}}, \qquad (4.426)$$

where ρ_s denotes the density at $r = r_s$, which should be determined by the definition of the sound velocity at $r = r_s$

$$c_{s,s}^2 = \frac{\kappa_p \Gamma}{h_s}\rho_s^{\Gamma-1} = \frac{\Gamma(\Gamma-1)\varepsilon_s}{1 + \Gamma\varepsilon_s} < \Gamma - 1. \qquad (4.427)$$

Therefore, for given values of r_s and ρ_s (or κ_p), two constants, $\dot{M}$ and $\tilde{u}_t$, are determined. Note that for $\Gamma > 2$, the constraint of equation (4.427) is replaced by $c_{s,s} \leq 1$.

The remaining task is to determine ρ and u as functions of r. This is achieved by solving simultaneous equations composed of (4.407) and (4.414). One point to be noted is that solutions do not always exist for all the values of Γ. This is found from a relation, which is derived from equations (4.414), (4.415), and (4.419) as

$$(1 + 3c_{s,s}^2)\left(1 - \frac{c_{s,s}^2}{\Gamma - 1}\right)^2 = \left(1 - \frac{c_{s,\infty}^2}{\Gamma - 1}\right)^2, \qquad (4.428)$$

where $c_{s,\infty}$ is the sound velocity at $r = \infty$. The right-hand side of equation (4.428) is always smaller than unity, and hence, the left-hand side has to be too. This provides a condition for Γ as

$$\Gamma \leq \frac{5 - 7X + 3X^2}{3(1-X)^2}, \qquad (4.429)$$

where $X := c_{s,s}^2/(\Gamma - 1)$ with $0 < X < 1$. Therefore, a large value of $\Gamma > 5/3$ is prohibited for $X \ll 1$, as is well known. For $X \lesssim 1$, Γ may be larger than 5/3. The constraint $c_{s,s} \leq 1$ yields the constraint $\Gamma \leq 3$.

Figure 4.18 plots a Bondi-flow solution for $r_s = 8M$ and $\rho_s = 1$ with $\Gamma = 4/3$ shown in Kerr–Schild coordinates of a non-spinning black hole. This solution is valid for any value of a purely radial magnetic field satisfying equation (4.410). Any numerical code has to reproduce this solution in an appropriate boundary condition.

Employing a magnetized Bondi-flow solution of figure 4.18, a test simulation has been performed. We here show an example of the numerical results obtained by an axisymmetric code with cylindrical coordinates (ϖ, z). The computational

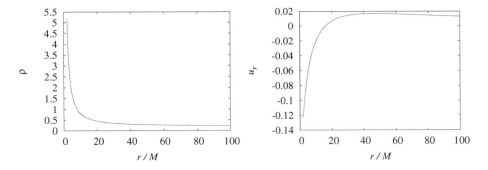

Fig. 4.18 ρ and u_r of a Bondi-flow solution with $r_\mathrm{s} = 8M$, $\rho = 1$ at $r = r_\mathrm{s}$, and $\Gamma = 4/3$.

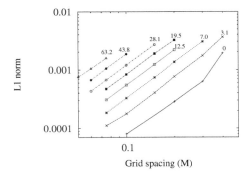

Fig. 4.19 L1 norm for the error of the weighted density ρ_* as a function of grid spacing in units of M for a magnetized Bondi flow. The numerical numbers attached for each curve denote the values of b^2/ρ at $r = 2M$. The figure is taken from Shibata and Sekiguchi (2005a).

domain was chosen to be $[0, 18M]$ for ϖ and z and was covered by a uniform grid with a variety of grid spacing $\Delta\varpi = \Delta z$. The inner boundary was set inside the event horizon at $r = 1.9M$ at which the system was imposed to be stationary. The solution for the stationary Bondi flow was prepared as the initial condition, and the simulation was performed for the time duration $100M$.

Irrespective of the grid resolution (irrespective of the values of $\Delta\varpi = \Delta z$), the system relaxes to a stationary state long before $100M$ besides small deviations from the exact stationary configuration caused by numerical errors associated with finite differencing. Because the solution is smooth and a second-order accurate scheme was employed in this example, these deviations should converge to zero at second order with improving the grid resolution. Figure 4.19 plots the L1 norm for $\rho_* - \rho_*^\mathrm{exact}$ where ρ_*^exact denotes the exact stationary solution of ρ_*. Specifically, the L1 norm is here defined by

$$\int_{r \geq 2M} \left| \rho_* - \rho_*^\mathrm{exact} \right| d^3x \bigg/ \int_{r \geq 2M} \rho_*^\mathrm{exact} \, d^3x \,. \tag{4.430}$$

For the convergence test, the grid spacing was changed from $0.06M$ to $0.4M$. Also, the strength of the radial magnetic field was varied. Figure 4.19 shows that irrespective of the various radial magnetic-field strengths, the error converges at second order. This illustrates the reliability of the numerical code for a strongly gravitational field.

4.10.4 *Stationary accretion torus around a black hole*

For stationary axisymmetric background spacetime, the Euler equation for perfect fluids is integrated to yield a first integral if $u^r = u^\theta = 0$ are assumed [see equations (5.211) and (5.215) in section 5.7.1.2]. In particular, for a torus rotating around a central body with a constant specific angular momentum j, the profile of h is determined from equation (5.217) for given constants j and C'. Then, for a given angular momentum distribution and an equation of state as well as several parameters such as locations of inner and outer edges, it is quite easy to obtain a stationary accretion torus around non-spinning and spinning black holes simply by solving an algebraic equation [Fishbone and Moncrief (1976); Abramowicz et al. (1978)]. Solutions are also easily obtained in the presence of a purely toroidal magnetic field [e.g., Shibata (2007)]. These solutions have been used for testing hydrodynamics/magnetohydrodynamics codes in a strongly general relativistic field [e.g., Hawley et al. (1984a); Font and Daigne (2002)].

4.11 Testing a numerical-relativity code with matter

Hydrodynamics codes in full general relativity have to be examined for a self-gravitating system [see, e.g., Shibata (1999b); Font et al. (2000, 2002); Shibata (2003a) for early works]. We here introduce test-bed problems for which the initial conditions are easily prepared and the test run can be executed inexpensively: Specifically, we describe tests that can be performed by simply preparing a spherical neutron star, which can be obtained by integrating the TOV equation (see section 1.4.3; note that in numerical relativity, it is more convenient to employ the equations in isotropic coordinates in which the spatial line element is conformally flat). In these test-bed problems, the analysis to examine the reliability of a numerical code can be performed using well-known physical properties of compact objects. We also describe how the degree of the reliability can be examined in these tests.

4.11.1 *Stability of neutron stars*

As outlined in section 1.4 for the properties of spherical neutron stars, there are in general two branches for sequences of neutron stars; one is stable and the other is unstable against radial perturbations. For a given equation of state, a spherical neutron star is stable if the central density, ρ_c, is smaller than a critical value (see figure 1.16). By contrast, if the central density is larger than the critical value,

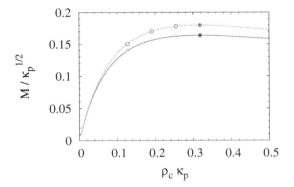

Fig. 4.20 Relations between the mass and the central density for a sequence of spherical stars in a polytropic equation of state with $\Gamma = 2$. The solid and dashed curves denote the gravitational mass and the total rest mass, respectively. The filled circles denote the star with the maximum mass, and the open circles denote the stable stars employed in the test simulations described in section 4.11.1.

the neutron star is unstable [see, e.g., Zel'dovich and Novikov (1971); Shapiro and Teukolsky (1983) for a proof]. Any numerical-relativity code has to be able to answer the stability of neutron stars. This is quite important because in numerical relativity, we often encounter the problem in which the criterion for the formation of a black hole has to be determined. The numerical-relativity code also has to be able to evolve a stable neutron star for a time scale, much longer than the dynamical time scale $\sim \rho_c^{-1/2}$. This simple test is also quite important because neutron stars are often formed in the stellar core collapse and the merger of binary neutron stars, which are long-term emitters of strong gravitational waves and neutrinos. To precisely compute them, the code has to be able to evolve any stable neutron star accurately for a long time scale.

In this section, we introduce one of the simplest test simulations along this line. We consider spherical neutron stars employing the polytropic equation of state, $P = \kappa_p \rho^\Gamma$, as in section 4.10.3. For $\Gamma \gtrsim 2$, the relation between the mass [gravitational mass M (ADM mass; see section 5.3.1.1) or rest mass $M_0 (> M)$] and the central density along a sequence of spherical stars is qualitatively similar to that for neutron stars with more realistic equations of state (compare figures 1.16 and 4.20): There are two branches along the sequence; one with $dM/d\rho_c > 0$ (stable one) and the other with $dM/d\rho_c < 0$ (unstable one).

One important property in this equation of state is that $\kappa_p^{n/2}$ and κ_p^{-n} have the dimension of mass (and radius) and density, respectively, in the geometrical units $c = 1 = G$. Here, $n := 1/(\Gamma - 1)$ is the so-called polytropic index. This implies that if we focus only on normalized dimensionless quantities such as $M/\kappa_p^{n/2}$ and $\rho \kappa_p^n$, any relation among the normalized quantities are independent of the value of κ_p; κ_p can be factored out of the basic equations. Thus, in the present and next sections, we only show such nondimensional quantities.

The purpose of this section is to show results that illustrate that a numerical-relativity code can determine the stability of spherical stars. We here choose four stars (see figure 4.20) with $(M/\kappa_{\rm p}^{1/2}, \rho_c\kappa_{\rm p}, M/R) = (0.1398, 0.1273, 0.1456)$, $(0.1561, 0.1910, 0.1781)$, $(0.1623, 0.2546, 0.1996)$, and $(0.1637, 0.3183, 0.2144)$, where R denotes the circumferential radius of the spherical stars. For all four models, the values of the compactness parameter M/R are as large as those of the typical neutron stars in the range ~ 0.13–0.25 (see section 1.4.3). The first three models are stable while the fourth model is marginally stable, i.e., the mass is maximum along the equilibrium sequence.

The following test simulations were performed in the assumption of axial symmetry, employing the Shibata–Nakamura-type BSSN formulation for the evolution of Einstein's equation (see section 2.3.3) and the scheme of Kurganov and Tadmor (2000) for integrating hydrodynamics equations. The axisymmetric conditions for gravitational-field quantities were imposed using the Cartoon method (see section 2.5). As the gauge condition, a dynamical gauge condition in the following form was employed (see section 2.2.6 for the dynamical gauge),

$$\partial_t \alpha = -2\alpha K, \qquad \partial_t \beta^i = \tilde{\gamma}^{ij}(F_j + \partial_t F_j \Delta t), \qquad (4.431)$$

where Δt is the time-step interval.

For integrating Einstein's equation, a fourth-order accurate finite differencing scheme was used, and for hydrodynamics, a third-order cell-reconstruction scheme (see section 4.8.5) was used. The time integration was performed using a fourth-order explicit Runge–Kutta scheme. These schemes are standard in numerical relativity for non-vacuum spacetime. The initial conditions were prepared by solving the equations shown in section 1.4.3; specifically, the equations in isotropic coordinates are solved. The initial data for the following simulations have to be appropriately reprocessed, because an interpolation procedure for the data obtained in the spherical coordinates has to be done for the use in non-spherical coordinates. The interpolated data do not precisely satisfy the Hamiltonian constraint, and hence, in this test, we reinforced this constraint. The numerical scheme for solving the constraint equation in this test was second-order accurate, and hence, the accuracy of the following numerical results was affected by the initial condition: The convergence order becomes smaller than four (but larger than two).

Figure 4.21 plots the evolution of the central density, ρ_c, and the central value of the lapse function, α_c, for the four models. For these simulations, the pressure was initially reduced by 0.1% uniformly to enforce slight contraction. In spite of this pressure reduction, the central values for three stable models are kept to be approximately constant (besides a quasi-periodic oscillation of a small amplitude; see below) for a time scale much longer than the dynamical time scale, $(\rho_c)^{-1/2}$. Thus, these stars are shown to be indeed stable against the radial perturbation. By contrast, the star with the maximum mass collapses to a black hole. This implies that this star is unstable against the radial perturbation by the small pressure reduction. All these results agree with the properties of spherical neutron stars,

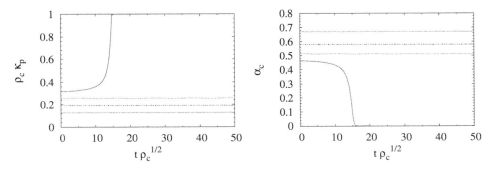

Fig. 4.21 Evolution of the central density (left) and the central value of the lapse function (right) for four chosen models of different compactness (see the circles in figure 4.20). The units of the time is $\rho_c^{-1/2}$.

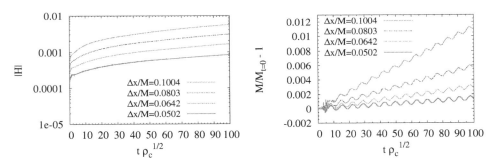

Fig. 4.22 Violations of the Hamiltonian constraint (left) and the gravitational-mass conservation (right) as functions of time for the model with $(M/\kappa_p^{1/2}, \rho_c \kappa_p, M/R) = (0.1561, 0.1910, 0.1781)$. The results for four different grid resolutions are plotted, and the magnitude of the violation converges approximately at second order.

predicted by the well-known theorem for the stability. Therefore, we conclude that this numerical-relativity code can yield a reliable result for the stability of neutron stars.

Next, we describe several methods to assess how accurate the numerical solutions are. One of the most popular methods is to analyze the magnitude for the violation of the constraint equations. Here, we focus on the violation of the Hamiltonian constraint defined by

$$|H| = \frac{1}{M_0} \int d^3x\, \rho_* \left(R - K_{ij}K^{ij} + K^2 - 16\pi\rho_h \right), \quad (4.432)$$

where M_0 denotes the total rest mass. In this case, the integrated constraint violation is defined choosing the rest-mass density as the weight.

Another popular approach is to analyze the magnitude for the violation of the conservation of the global quantities. Examining the conservation of the gravitational mass, M (ADM mass), is among the most popular tests, because it corresponds to examining the energy conservation in general relativity. In the spherical symmetric and adiabatic systems, this mass should be constant. Specifically, we here measure the violation for the conservation of the ADM mass by

$$\frac{M(t)}{M(t=0)} - 1. \quad (4.433)$$

Figure 4.22 plots the evolution for the violation of the Hamiltonian constraint and the conservation of the ADM mass for one of the stable models. This shows that the violation is maintained within 0.1% error level for ~ 100 dynamical time scales in the best-resolved run. Also shown is that the violation converges approximately at second and third orders for the violation of the Hamiltonian constraint and the ADM-mass conservation, respectively. The important fact to be pointed out is that the error increases with time but at most linearly (not exponentially) with time. This guarantees that with a sufficient grid resolution, a long-term and precise numerical simulation is in principle feasible. Even if the grid resolution is not extremely high, we can determine the (approximately) exact result by extrapolating the numerical results obtained for several grid resolutions.

The test simulations presented here were performed for spherical neutron stars. Essentially the same test simulations can be performed also for rigidly rotating neutron stars, for which the stability can be determined by the turning point theorem [Friedman et al. (1988); Cook et al. (1992, 1994b)] (see also section 9.1.1). For the rotating case, the neutron star with the maximum mass along a sequence is not in general marginally stable, and thus, a more careful analysis is necessary for determining the stable and unstable branches. However, the turning point theorem enables us to classify a neutron star into either of these two classes.

The first numerical-relativity analysis of the stability for rigidly and rapidly rotating neutron stars was performed by Shibata et al. (2000b). In this analysis, the neutron stars were modeled by the polytropic equation of state. They showed that the neutron stars in the stable branch simply oscillate for a long time scale while those in the unstable branch collapse to a black hole, as in the case of spherical neutron stars (see section 9.1.2 for this topic).

4.11.2 Oscillation of neutron stars

Any star has its characteristic oscillation modes. In particular, the fundamental pressure modes (often called the f-mode), for which the density perturbation profile does not have a node, are primarily excited and their frequency is well-known to be of order of $\rho_c^{-1/2}$. It is a good test to examine whether the f-mode oscillation can be accurately reproduced in numerical simulations [Shibata (1999b)]. This test is also quite important because accurate computation of gravitational waves emitted by a nonspherical oscillation of neutron stars is one of the primary issues in numerical relativity.

Before going ahead, we present a formula for approximately computing the

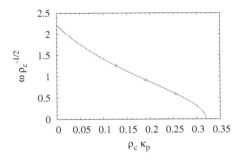

Fig. 4.23 Angular frequency of the radial f-mode of spherical neutron stars as a function of $\rho_c \kappa_p$ for a polytropic equation of state with $\Gamma = 2$. The filled circles denote the results obtained by numerical simulations.

oscillation frequency associated with the purely radial f-mode of spherical stars. For the line element in the form [see equation (1.169)],

$$ds^2 = -e^\Phi dt^2 + e^\lambda dr^2 + r^2 d\Omega, \tag{4.434}$$

the perturbation analysis shows that the angular frequency ω of the radial oscillation for a star of radius R with polytropic equations of state is written as [Chandrasekhar (1964)]

$$\omega^2 = \left[4 \int_0^R e^{(\lambda+\Phi)/2} r \frac{dP}{dr} \xi^2 dr + \int_0^R e^{(\lambda+3\Phi)/2} \frac{\Gamma P}{r^2} \left\{ \frac{d}{dr}(r^2 e^{-\Phi/2} \xi) \right\}^2 dr \right.$$
$$\left. - \int_0^R e^{(\lambda+\Phi)/2} \left(\frac{dP}{dr} \right)^2 \frac{r^2 \xi^2}{\rho h} dr + 8\pi \int_0^R e^{(3\lambda+\Phi)/2} \rho P h r^2 \xi^2 dr \right]$$
$$\times \left[\int_0^R e^{(3\lambda-\Phi)/2} \rho h r^2 \xi^2 dr \right]^{-1}, \tag{4.435}$$

where ξ denotes the r-component of the Lagrangian displacement. If we substitute an eigen function ξ for the eigenvalue equation, we can obtain a correct value of ω. However, it is empirically known that we can approximately evaluate ω for the f-mode by substituting the following approximate (nodeless) eigen function for ξ [Chandrasekhar (1964)],

$$\xi = r e^{b_f \Phi}, \tag{4.436}$$

where b_f is a constant. Here, we choose $b_f = 0$ for simplicity. It is known that the eigen angular frequency depends only weakly on the value of b_f as long as $|b_f| \leq 1/2$ for a moderately compact star with $M/R \lesssim 0.20$ and $\Gamma = 2$ [Shibata (1999b)]. Only for the stars with the ADM mass near the maximum mass with $M/R \gtrsim 0.21$, the eigen angular frequency depends rather strongly on the value of b_f, suggesting that this approximate method does not work well.

Figure 4.23 plots the approximate values of $\omega \rho_c^{-1/2}$ as a function of $\rho_c \kappa_p$ for $\Gamma = 2$ with $b_f = 0$. Beyond a critical density $\rho_c \kappa_p \approx 0.32$, ω^2 becomes negative,

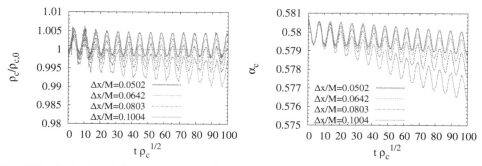

Fig. 4.24 Evolution of the central density (left) and the central value of the lapse function (right) for the model with $(M/\kappa_p^{1/2}, \rho_c \kappa_p, M/R) = (0.1561, 0.1910, 0.1781)$. The results for four different grid resolutions are plotted.

implying that the star becomes dynamically unstable for this oscillation mode. We note that the real critical density is $\rho_c \kappa_p \approx 0.318$, and hence, in this approximation, this value is approximately determined within 1% error.

Figure 4.24 plots the evolution of the central density and the central value of the lapse function for the model with $(M/\kappa_p^{1/2}, \rho_c \kappa_p, M/R) = (0.1561, 0.1910, 0.1781)$. These simulations were performed for four grid resolutions with $\Delta x/M = 0.0502$, 0.0642, 0.0803, and 0.1004. Irrespective of the grid resolutions, these local quantities oscillate in a quasi-periodic manner with an approximately constant amplitude for a time scale much longer than the dynamical time scale. As the time proceeds, the average central density and the average central value of the lapse function slightly decrease. This is due to a numerical spurious effect. This effect is suppressed with improving the grid resolution (approximately at second order), and the magnitude of such unphysical error is at most 1% of the average values even in the lowest resolution runs at $t = 100\rho_c^{-1/2}$. The behavior of ρ_c and α_c illustrated here is universally seen in the simulations of any stable star.

The oscillation frequency of these stars is determined by performing a Fourier analysis of ρ_c or α_c. Because there is a secular offset of these values which is roughly proportional to the time, we correct this offset in the Fourier transformation procedure.

Figure 4.25 plots the resulting Fourier spectrum for three stable models. The peak angular frequency for the models $\rho_c \kappa_p = 0.1273, 0.1910$, and 0.2546 is, respectively, $\omega \rho_c^{-1/2} \approx 1.26, 0.918$, and 0.586. For the first two, the numerical results agree with the perturbation results within 1% error. For the most compact models, the numerical result is by about 2% smaller than the perturbation results (see the solid circles of figure 4.23). This suggests that the perturbation results obtained for a trial function of the Lagrangian perturbation are very accurate for not-extremely general relativistic stars with $M/R < 0.18$. For highly general relativistic stars $M/R \approx 0.20$, the accuracy is slightly poor, but the agreement is still good and the error is as large as that expected above.

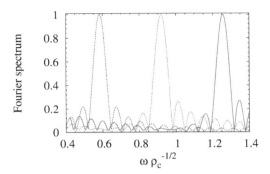

Fig. 4.25 The Fourier spectrum of the curve of $\rho_c(t)$ for three stable models. For $\rho_c \kappa_p = 0.1273$, 0.1910, and 0.2546, the peak is located at $\omega \rho_c^{-1/2} = 1.26$, 0.918, and 0.586, respectively. The amplitude is normalized so that the maximum becomes unity.

The test presented here can be easily extended to non-radial oscillations. If we add a non-spherical perturbation initially, the star oscillates in a non-radial manner which is different from that in the radial oscillation. Such test simulations have been also performed since the first work by Shibata (1999b).

4.11.3 Collapse of unstable neutron stars

Spherical neutron stars along the sequence of $dM/d\rho_c < 0$ are unstable against radial oscillations and collapse into a black hole. A spherical neutron star along the sequence at $dM/d\rho_c = 0$ is marginally stable, and thus, a small perturbation could induce the gravitational collapse. Any numerical-relativity code has to be able to find this instability and, in addition, has to be able to follow the collapse process to a black hole until a final state is reached. Here, the final state in the adiabatic collapse should be a black hole of mass equal to the initial ADM mass of the system, if all the material collapse into the black hole. Thus, examining whether the final state is the expected one is also a good test.

Here, we present a numerical result for the marginally stable neutron star with $M/\kappa_p^{1/2} = 0.1637$, $\rho_c \kappa_p = 0.3183$, and $M/R = 0.2144$ (see the solid circles in figure 4.20). For inducing the collapse, the simulation was started from for the initial condition for which the pressure is uniformly reduced by 0.1%. The formation of a black hole is confirmed by locating an apparent horizon at each time step (see section 7.1 for apparent horizon).

Figure 4.26 plots the evolution of the central value of the lapse function, α_c, for four grid resolutions. As already shown in figure 4.21, the neutron star collapses into a black hole, and hence, the value of α_c approaches zero in slicing with a singularity-avoidance property. Here, the apparent horizon is first formed when α_c reaches ~ 0.033. The numerical results for different grid resolutions show an approximate convergence. The time to the black-hole formation is shorter for the lower grid resolutions. This is due to the fact that a spurious dissipative effect associated with

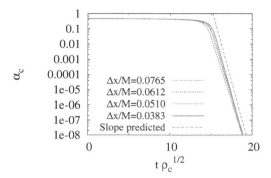

Fig. 4.26 The central value of the lapse function, α_c, as a function of time. M denotes the ADM mass of the system. The dot-dot line denotes the exponential decay $\propto e^{-t/\tau}$ where $\tau = (3/2)^{3/2} M(t=0)$ (see figure 2.8). The apparent horizon is first formed when α_c reaches ≈ 0.03 irrespective of the grid resolutions.

the numerical truncation error is larger for the poorer grid resolution. After the formation of the black hole (apparent horizon), the value of α_c steeply decreases in an exponential manner, $\propto e^{-t/\tau}$, with the time scale approximately equal to $\tau = (3/2)^{3/2} M$, as in the maximal slicing (see section 2.2.2).

The evolution of the formed black hole is analyzed calculating the area of the apparent horizons, $A_{\rm AH}$ (see section 7.1 for methods of finding apparent horizons). Figure 4.27 plots the evolution of $(A_{\rm AH}/16\pi M^2)^{1/2}$ which is unity for static black holes. This figure shows that it approaches unity after the formation of a black hole and then relaxes to a constant ~ 1. The final relaxed values are slightly larger than unity and depends on the grid resolutions. The deviation from the unity is larger for the lower grid-resolution runs. However, the deviation approaches zero at approximately second order as the grid resolution is improved, and shows the convergent behavior.

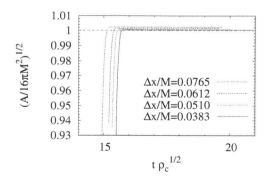

Fig. 4.27 Square root of the area of apparent horizons as a function of time in units of $\sqrt{16\pi M^2}$. Here, M denotes the ADM mass of the system.

Chapter 5

Formulations for initial data, equilibrium, and quasi-equilibrium

Initial data are the starting points for any numerical simulation. In numerical relativity, Einstein's equation constrains our choices of initial data because of the presence of the constraint equations as we described in section 2.1.5. The purpose of this chapter is to introduce several formulations used for determining initial data in the 3+1 and N+1 formulations of Einstein's equation.

5.1 Properties of initial-data equations

In the 3+1 and N+1 formulations of Einstein's equation, spatial metric γ_{ab} and extrinsic curvature K_{ab} are evolved forward in time starting from an initial condition. The initial data for them cannot be freely specified because of the presence of the constraint equations (2.48) and (2.50). Valid initial conditions have to satisfy this set of the constraint equations. However, the number of components for the constraint equations is $N+1$, and hence, it is smaller than the total number of the components of γ_{ab} and K_{ab}, $N(N+1)$, for $N \geq 2$. This implies that we have a flexibility for setting initial conditions, as found in the following.

The Hamiltonian constraint is composed of the scalar curvature of the spatial metric (i.e., second-order spatial derivatives of the spatial metric). Thus, this equation most naturally constrains one scalar component of the spatial metric among $N(N+1)/2$ components. The momentum constraint is composed of first-order spatial derivatives of the extrinsic curvature. Any spatial tensor is in general decomposed into scalar, pure vector, and pure tensor parts as (see section 2.3.1)

$$K_{ab} = \frac{K}{N}\gamma_{ab} + D_a D_b C + D_a V_b + D_b V_a + K^{tt}_{ab}, \tag{5.1}$$

where $K(= K_a{}^a)$ and C are the scalars, V_b is the pure vector that satisfies $D_a V^a = 0$, and K^{tt}_{ab} is the pure tensor (transverse-tracefree tensor) that satisfies $D^a K^{tt}_{ab} = 0 = \gamma^{ab} K^{tt}_{ab}$. Here, V^a and K^{tt}_{ab} have $N-1$ and $(N+1)(N-2)/2$ independent components, respectively. Since the momentum constraint is linear in K_{ab}, the decomposition like equation (5.1) is quite useful for finding its solution. Substituting equation (5.1) into the momentum constraint, N-components second-order partial

Table 5.1 A counting for the number of degrees of freedom in γ_{ab} and K_{ab}.

	Components	Constraints	Gauge	Dynamical degrees
In total	$N(N+1)$	$N+1$	$N+1$	$(N-2)(N+1)$
γ_{ab}	$N(N+1)/2$	1	N	$(N-2)(N+1)/2$
K_{ab}	$N(N+1)/2$	N	1	$(N-2)(N+1)/2$
γ_{ab} ($N=3$)	6	1	3	2
K_{ab} ($N=3$)	6	3	1	2

differential equations for $D_a C$ and V^a are derived. Thus, it is most natural to consider that the momentum constrain determines these N components included in the extrinsic curvature (see section 5.2 for more details).

For a more careful analysis, let us count the number of degrees of freedom that the geometric quantities have. We know that gravitational waves carry the dynamical degrees of freedom in the geometric variables, which cannot be constrained by the constraint equations. Gravitational waves are the tensor parts and this implies that $(N-2)(N+1)/2$ among $N(N+1)/2$ components in the spatial metric are not constrained (see tables 2.1 and 5.1 for the number of the degrees of freedom). Since one degree of freedom should be constrained by the Hamiltonian constraint, the number of the remaining degrees of freedom in the spatial metric is $N(N+1)/2 - (N-2)(N+1)/2 - 1 = N$. This is related to N components of the spatial gauge freedom for a given spatial hypersurface. For the extrinsic curvature that is constrained by the momentum constraint, the remaining degrees of freedom is $N(N+1)/2 - (N-2)(N+1)/2 - N = 1$. This is also related to 1 among $(N+1)$ degrees of the gauge freedom, which is usually related to K (the trace part of K_{ab}); the freedom for the choice of temporal slicing.

A note on the gauge choice is appropriate here. In the 3+1 and N+1 formulations, it is often mentioned that the lapse function and shift vector are the variables that correspond to the degrees of gauge freedom (coordinate transformation). However, this statement should be refined more precisely as follows: The lapse function and shift vector determine how the gauge should change with time (see section 2.2). This implies that the spatial metric and extrinsic curvature on a given hypersurface retain full coordinate freedom in initial-data problems. This is reflected in the remaining $N+1$ degrees of freedom for gauge as described above.

To summarize, we have two issues when specifying initial data: First, the dynamical degree of freedom in the spatial metric and extrinsic curvature cannot be determined by the constraint equations. We have to additionally impose a physical condition for controlling this degree of freedom. Second, the gauge freedom is present, and hence, we have to appropriately determine it. To resolve these things, anyway, we have to extract the dynamical degrees of freedom and gauge freedom from the geometric variables as a first step. The problem here is that we do not know which part in the geometric variables reflects the dynamical and gauge degrees of freedom exactly. However, it is possible to at least approximately decompose the geometric variables into scalar, vector, and tensor parts both for the spatial metric and extrinsic curvature. This method was developed originally by Lichnerowicz and extensively by York using a conformal decomposition method. In the next section, we will describe their formulation.

5.2 York–Lichnerowicz formulation

The York–Lichnerowicz formulation (hereafter referred to as conformal transverse-tracefree decomposition formulation) is based on the conformal decomposition described in section 2.1.7, in which the Hamiltonian and momentum constraint equations are written to equations (2.95) and (2.96), respectively. In this formulation, the Hamiltonian constraint is used for determining the conformal factor (i.e., one scalar component), ψ, which obeys an elliptic-type equation. Other components are denoted by $\tilde{\gamma}_{ij}(=\psi^{-4/(N-2)}\gamma_{ij})$, which is called conformal spatial metric. This part is considered to include dynamical degrees of freedom and has to be determined by imposing additional conditions that is different from the constraint equations.

Decomposition methods for the extrinsic curvature were extensively developed by York (1973). His method is based on the fact that any symmetric tracefree tensor is decomposed in the covariant way as

$$\bar{T}_{ij} = (LX)_{ij} + \bar{T}_{ij}^{\mathrm{tt}}, \tag{5.2}$$

where $\bar{T}_{ij}^{\mathrm{tt}}$ is a symmetric transverse-tracefree tensor, i.e., $\hat{D}^i \bar{T}_{ij}^{\mathrm{tt}} = 0 = \mathrm{tr}\left(\bar{T}_{ij}^{\mathrm{tt}}\right)$. Here, we consider a spatial metric $\hat{\gamma}_{ij}$ (such as γ_{ij} and $\tilde{\gamma}_{ij}$) with $\hat{D}_i$ being the covariant derivative associated with $\hat{\gamma}_{ij}$. $\bar{T}_{ij}^{\mathrm{tt}}$ denotes the pure tensor component of $\bar{T}_{ij}$. $(LX)_{ij}$ is referred to as a tracefree longitudinal part defined from a vector field, X^i, by

$$(LX)_{ij} = \hat{D}_i X_j + \hat{D}_j X_i - \frac{2}{N}\hat{\gamma}_{ij}\hat{D}_k X^k. \tag{5.3}$$

Here, $X_i = \hat{\gamma}_{ij} X^j$. York (1973) described this transverse-tracefree decomposition of the extrinsic curvature with respect to both the physical spatial metric γ_{ab} and the conformal spatial metric $\tilde{\gamma}_{ab}$. In numerical relativity, the latter (conformal transverse-tracefree decomposition) is quite useful for solving initial-data problems, and hence, in the following, we will describe only the latter setting $\hat{\gamma}_{ij} = \tilde{\gamma}_{ij}$ and $\hat{D}_i = \tilde{D}_i$.

The momentum constraint equation for the tracefree extrinsic curvature $\tilde{A}_{ij}$, (2.96), is rewritten as

$$\tilde{D}_i \bar{A}^i{}_j - \frac{N-1}{N}\psi^{2N/(N-2)}\tilde{D}_j K = 8\pi J_j \psi^{2N/(N-2)}, \tag{5.4}$$

where $\bar{A}^i{}_j := \psi^{2N/(N-2)}\tilde{A}^i{}_j$. We here adopt the rule that the subscripts of $\bar{A}_{ij}$ and $\bar{A}^{ij}$ are raised and lowered by $\tilde{\gamma}^{ij}$ and $\tilde{\gamma}_{ij}$ as in the case of $\tilde{A}_{ij}$. Applying the decomposition (5.2) to $\bar{A}_{ij}$ with the conformal spatial metric, we write

$$\bar{A}_{ij} = \tilde{D}_i W_j + \tilde{D}_j W_i - \frac{2}{N}\tilde{\gamma}_{ij}\tilde{D}_k W^k + \bar{A}_{ij}^{\mathrm{tt}}, \tag{5.5}$$

where $\bar{A}_{ij}^{\rm tt}$ is the transverse-tracefree part of $\bar{A}_{ij}$ satisfying $\tilde{D}^i \bar{A}_{ij}^{\rm tt} = 0 = \tilde{\gamma}^{ij}\bar{A}_{ij}^{\rm tt}$, and W^i is a vector field in the conformal space: $W_i = \tilde{\gamma}_{ij} W^j$. Substituting this decomposition into equation (5.4), the equation for W^i is derived as

$$\tilde{D}_j \tilde{D}^j W_i + \frac{N-2}{N}\tilde{D}_i(\tilde{D}_j W^j) + \tilde{R}_{ij}W^j$$
$$- \frac{N-1}{N}\psi^{2N/(N-2)}\tilde{D}_i K = 8\pi J_i \psi^{2N/(N-2)}. \tag{5.6}$$

Thus, as expected in section 5.1, the momentum constraint equation reduces to the N-component vector elliptic equations for W^i for a given set of $\tilde{\gamma}_{ij}$, K, ψ, and J_i.[1] Here, $\bar{A}_{ij}^{\rm tt}$ is free from the constraint, and thus, it does not appear in the momentum constraint.

The Hamiltonian constraint is written, in terms of $\bar{A}_{ij}$ (not K_{ij}), as

$$\tilde{\Delta}\psi = \frac{(N-2)\psi}{4(N-1)}\tilde{R} - \frac{4(N-2)\pi}{N-1}\rho_{\rm h}\psi^{(N+2)/(N-2)}$$
$$- \frac{(N-2)\psi^{(-3N+2)/(N-2)}}{4(N-1)}\bar{A}_{ij}\bar{A}^{ij} + \frac{(N-2)\psi^{(N+2)/(N-2)}}{4N}K^2. \tag{5.7}$$

This is the equation for ψ for a given set of $\tilde{\gamma}_{ij}$, K, $\bar{A}_{ij}$, and $\rho_{\rm h}$. In summary, the initial-data equations reduce to $(N+1)$-component equations composed of equations (5.6) and (5.7).

One of the important properties in the conformal transverse-tracefree decomposition formulation described above is that if we choose $K = {\rm const}$ for fixing a gauge freedom in K_{ab} and if we give $S_i = J_i \psi^{2N/(N-2)}$ (not J_i) or $J_i = 0$, equation (5.6) becomes independent of ψ and totally decouples from the Hamiltonian constraint: The equation for W_i becomes an elliptic equation for a given set of $\tilde{\gamma}_{ij}$ and S_i. Subsequently, equation for ψ, (5.7), is to be solved for a resulting solution of $\bar{A}_{ij}$. This prescription is often used for constructing initial data of black-hole spacetime, as described in section 5.5.

5.3 Mass, linear momentum, and angular momentum

For setting initial data, we often need to specify mass (energy), linear momentum, and angular momentum of the systems. In particular, for defining quasi-equilibrium states of compact binaries and for exploring their properties, it is necessary to calculate the mass and the angular momentum of the systems. In general relativity, they have to be defined from geometric quantities in a geometric manner. In this section, we first summarize the definition of these global quantities. We then introduce an important relation, a so-called virial relation that is satisfied among two masses for equilibrium and quasi-equilibrium states. The relations among the global quantities of a black hole, which are often used in initial-data problems, are also summarized in appendix G.

---

[1] Note that W^i is composed of both pure vector components ($N-1$ components) and one pure scalar component.

5.3.1 ADM mass, linear momentum, and angular momentum in the Hamiltonian formulation

5.3.1.1 ADM mass

One of the most subtle problems in general relativity is to define the mass (energy) of a system, because the mass of gravitational fields cannot be locally defined in general relativity. However, it is still possible to define global mass in asymptotically flat spacetime. The most popular mass often used in numerical relativity is the Arnowitt–Deser–Misner (ADM) mass, which is measured at spatial infinity and defines the total mass in asymptotically flat spacetime [Arnowitt et al. (1960)].

The ADM mass was defined originally by Arnowitt et al. (1960) as

$$M_{\text{ADM}} := \frac{1}{16\pi} \oint_{r\to\infty} f^{jk} f^{il} (\partial_k \gamma_{ij} - \partial_i \gamma_{jk}) d\overset{(0)}{S}_l, \tag{5.8}$$

where f_{ij} is a flat spatial metric and $d\overset{(0)}{S}_i$ denotes an element of the surface integral in the flat spatial hypersurface, $f_{ij}\bar{n}^j r^{N-1} d\Omega_{N-1}$, with $\bar{n}^i$ being a unit vector orthogonal to a closed surface at spatial infinity. Note that the ADM mass is usually defined in Cartesian coordinates.

The ADM mass corresponds to the value of the Hamiltonian in the Hamiltonian formulation of general relativity: It results from the surface terms in the Hamiltonian formulation [Arnowitt et al. (1960); DeWitt (1967); Regge and Teitelboim (1974)] [see also appendix E of Wald (1984); a brief summary on the Lagrangian and Hamiltonian formulations of general relativity is given in Appendix D]. Therefore, it can be considered as the total mass of an isolated system.

The ADM mass is not defined in a covariant way, and hence, a particular gauge choice is implicitly assumed. For its definition to be valid, in addition to asymptotic flatness, the metric components have to approach those of the flat Minkowski metric, $\eta_{\mu\nu}$, sufficiently quickly with increasing the distance toward spatial infinity as [York (1979); Gourgoulhon (2012)]

$$g_{\mu\nu} - \eta_{\mu\nu} = O\left(r^{-N+2}\right), \tag{5.9}$$

$$\partial_k \gamma_{ij} = O\left(r^{-N+1}\right), \tag{5.10}$$

$$K_{ij} = O\left(r^{-N+1}\right), \tag{5.11}$$

$$\partial_k K_{ij} = O\left(r^{-N}\right). \tag{5.12}$$

In the violation of these conditions, the ADM mass is not able to give the correct mass. A well-known example in which the ADM mass cannot be calculated is the metric of a four-dimensional non-spinning black hole in Painlevé–Gullstrand coordinates,

$$ds^2 = -dt^2 + \left(dr + \sqrt{\frac{2M}{r}} dt\right)^2 + r^2 d\Omega, \tag{5.13}$$

for which $\gamma_{ij} = f_{ij}$ and the ADM mass is zero because of the violation of the condition (5.9) [Gourgoulhon (2012)].

Together with the conditions (5.9)–(5.12), we need to assume that the matter distribution falls off sufficiently rapidly in order to guarantee the asymptotic flatness. Specifically, we assume that the stress-energy tensor satisfies the following condition written in (t, x^k) with x^k Cartesian coordinates:

$$T_{\mu\nu} = O\left(r^{-N-1}\right). \tag{5.14}$$

There are several reasons why we believe that the ADM mass denotes the mass of isolated systems. First, the ADM mass agrees with the mass of Schwarzschild and Kerr black holes as long as it is evaluated at spatial infinity. Second, the ADM mass agrees with conserved energy defined from the Landau–Lifshitz pseudo tensor at spatial infinity [Arnowitt et al. (1962); Misner et al. (1973)] (see also appendix F). Finally, the ADM mass is shown to be non-negative and hence it is reasonable to consider it to be mass [see chapter 11 of Wald (1984) for references and the historical details for the proof of the positive energy theorem].

As described in Regge and Teitelboim (1974), the ADM mass is associated with the time translation invariance of the Hamiltonian of general relativity at spatial infinity. This motivates us to define linear and angular momenta from spatial-translation and rotation invariances of the Hamiltonian of general relativity at spatial infinity (see below).

Motivated by the fact that it is the surface term of the Hamiltonian of general relativity (see appendix D), DeWitt (1967) and ÓMurchadha and York (1974) wrote the ADM mass in the form

$$M_{\mathrm{ADM}} = \frac{1}{16\pi} \oint_{r \to \infty} \gamma^{ij} \gamma^{kl} (\partial_j \gamma_{ik} - \partial_k \gamma_{ij}) dS_l. \tag{5.15}$$

Here, we basically suppose that Cartesian coordinates are used, and dS_l denotes a covariant element of the surface integral defined by

$$dS_l = \sqrt{\gamma_{N-1}} s_l d\theta^1 d\theta^2 \cdots d\theta^{N-1} =: \sqrt{\gamma_{N-1}} s_l \, d^{N-1}\theta, \tag{5.16}$$

where θ^i denote angular coordinates, s^l is a unit out-pointing radial vector field ($s_l s^l = 1$) orthogonal to the closed surface for the integral, and γ_{N-1} is the determinant of $(N-1)$-dimensional induced metric on the closed surface. We also note $dS_l = \sqrt{\gamma/\det(f_{ij})} \overset{(0)}{dS_l}$. For further manipulation, we rewrite equation (5.15) as

$$M_{\mathrm{ADM}} = \frac{1}{16\pi} \oint_{r \to \infty} \gamma^{ij} (\partial_j \gamma_{ik} - \partial_k \gamma_{ij}) \tilde{\gamma}^{kl} d\tilde{S}_l. \tag{5.17}$$

Here, we assumed that at spatial infinity, the conformal spatial metric $\tilde{\gamma}_{ij}$ is equal to γ_{ij} and f_{ij}, and $d\tilde{S}_l$ denotes an element of the surface integral defined in the conformal space. A straightforward calculation with $\gamma_{ij} = \psi^{4/(N-2)} \tilde{\gamma}_{ij}$ yields

$$M_{\mathrm{ADM}} = \frac{1}{16\pi} \oint_{r \to \infty} \left(-\frac{4(N-1)}{N-2} \tilde{\gamma}^{kl} \frac{\partial_k \psi}{\psi} - \partial_j \tilde{\gamma}^{jl} \right) d\tilde{S}_l$$

$$= \frac{1}{16\pi} \oint_{r \to \infty} \left(-\frac{4(N-1)}{N-2} \tilde{D}^l \psi - \partial_j \tilde{\gamma}^{jl} \right) d\tilde{S}_l, \tag{5.18}$$

where we used $\det(\tilde{\gamma}_{ij}) = 1$ and $\psi = 1$ at spatial infinity.

Further manipulation for M_{ADM} is simplified by introducing asymptotically quasi-isotropic gauge conditions proposed by York (1979) as [see also Gourgoulhon (2012)]

$$\partial_j \tilde{\gamma}^{jk} = O\left(r^{-N}\right), \tag{5.19}$$

where Cartesian coordinates are supposed to be used again. This gauge condition holds in a popular formulation in numerical relativity such as BSSN formulation together with moving puncture gauge conditions (see sections 2.3.3 and 2.2.6). Also, it holds for black-hole solutions in quasi-isotropic coordinates. By contrast, the Schwarzschild solution does not satisfy this condition.

Assuming that we use a coordinate system that satisfies equation (5.19) and using Gauss' law, equation (5.18) is written as

$$M_{\mathrm{ADM}} = -\frac{N-1}{4\pi(N-2)}\left[\oint_{\mathcal{H}} \tilde{D}^l \psi \, d\tilde{S}_l + \int_{\Sigma'_t} \tilde{\Delta}\psi \sqrt{f} \, d^N x\right], \tag{5.20}$$

where $\mathcal{H}$ denotes the surface of black-hole horizons, the associated surface element, $d\tilde{S}_l$, is outward pointing, and Σ'_t denotes a portion of spatial hypersurfaces Σ_t except for the inside of $\mathcal{H}$. We assumed that $\det(\tilde{\gamma}_{ij}) = \det(f_{ij}) = f$. Substituting equation (5.7) into equation (5.20) yields a volume-integral form of the ADM mass

$$M_{\mathrm{ADM}} = -\frac{N-1}{4\pi(N-2)} \oint_{\mathcal{H}} \tilde{D}^l \psi \, d\tilde{S}_l$$
$$+ \int_{\Sigma'_t} \left[-\frac{\psi}{16\pi}\tilde{R} + \rho_{\mathrm{h}} \psi^{(N+2)/(N-2)} + \frac{\psi^{(-3N+2)/(N-2)}}{16\pi} \bar{A}_{ij}\bar{A}^{ij} \right.$$
$$\left. - \frac{(N-1)\psi^{(N+2)/(N-2)}}{16\pi N}K^2 \right] \sqrt{f} \, d^N x. \tag{5.21}$$

To find that the ADM mass (which should be the total energy of asymptotically flat systems) is composed of rest-mass energy, internal energy, gravitational potential energy, and kinetic energy, we consider the Newtonian limit in four-dimensional spacetime, where $\tilde{R} \to 0$, $\bar{A}_{ij} \to 0$, $K \to 0$, $\tilde{\gamma} \to 1$, and $\psi \to 1 - \Phi_{\mathrm{N}}/2$ with Φ_{N} denoting the Newtonian potential that satisfies $\Delta_{\mathrm{f}} \Phi_{\mathrm{N}} = 4\pi\rho$. For the Newtonian limit of ρ_{h}, we assume $\varepsilon \ll 1$, $P/\rho \ll 1$, and $\alpha u^t \to 1 + v^2/2$, where v is the magnitude of the velocity. Then,

$$\rho_{\mathrm{h}} \to \rho\left(1 + \varepsilon + v^2\right). \tag{5.22}$$

We also neglect the surface integral associated with the presence of black holes. Then, we have the following equation in the Newtonian limit,

$$M_{\mathrm{ADM}} = \int d^3 x \, \rho \left(1 + \varepsilon + v^2 - \frac{5}{2}\Phi_{\mathrm{N}}\right). \tag{5.23}$$

The rest mass in the Newtonian limit is taken as [cf. equation (4.45)]

$$M_0 = \int d^3 x \, \rho \alpha u^t \psi^6 \to \int d^3 x \, \rho \left(1 + \frac{v^2}{2} - 3\Phi_{\mathrm{N}}\right). \tag{5.24}$$

Then, we obtain the expected result as

$$\begin{aligned} M_{\text{ADM}} &= M_0 + \int d^3x\, \rho \left(\varepsilon + \frac{v^2}{2} + \frac{1}{2}\Phi_{\text{N}} \right) \\ &= M_0 + U + T_{\text{kin}} + W, \end{aligned} \qquad (5.25)$$

where U, T_{kin}, and W denote the internal energy, kinetic energy, and gravitational potential energy in the Newtonian limit, defined, respectively, by

$$U = \int d^3x\, \rho\varepsilon, \quad T_{\text{kin}} = \frac{1}{2} \int d^3x\, \rho v^2, \quad W = \frac{1}{2} \int d^3x\, \rho \Phi_{\text{N}}. \qquad (5.26)$$

5.3.1.2 ADM linear momentum

The ADM linear momentum is defined also by Arnowitt et al. (1960); Regge and Teitelboim (1974) as

$$P_i^{\text{ADM}} := \frac{1}{8\pi} \oint_{r\to\infty} \left(K_i{}^j - K\gamma_i{}^j \right) dS_j. \qquad (5.27)$$

For this, we assume that $K_{ij} = O\left(r^{-N+1}\right)$ in the asymptotic region and that Cartesian coordinates are employed. As described by Regge and Teitelboim (1974), this linear momentum is associated with the spatial translation invariance of the Hamiltonian of general relativity at spatial infinity, and equation (5.27) is derived from this fact. The ADM linear momentum agrees with the linear momentum defined from the Landau–Lifshitz pseudo-tensor at spatial infinity [Arnowitt et al. (1962); Misner et al. (1973)] (see also appendix F).

In the conformally flat spatial geometry with maximal slicing, $K = 0$, equation (5.27) can be written as

$$P_i^{\text{ADM}} = \frac{1}{8\pi} \oint_{\mathcal{H}} K_i{}^j dS_j + \int_{\Sigma_t'} J_i\, dV, \qquad (5.28)$$

where dV is the volume element $\sqrt{\gamma}d^N x$, and we used the momentum constraint. Thus in the absence of black holes, the total linear momentum of the system is determined simply by integrating the linear momentum density of the matter for this case. This property will be used in section 5.6.4.

From the ADM mass and momentum, we can define a four vector as

$$\mathcal{P}_\mu = M_{\text{ADM}} n_\mu + P_\mu^{\text{ADM}}, \qquad (5.29)$$

which behaves as a spacetime vector with respect to the Poincaré transformation at spatial infinity [Arnowitt et al. (1962)]. This also indicates that the ADM mass and linear momentum are global quantities characterizing the isolated system, which is invariant under the Poincaré transformation.

5.3.1.3 Angular momentum in the Hamiltonian formulation

Arnowitt et al. (1962) did not define angular momentum and hence there is no "ADM angular momentum" [Gourgoulhon (2012)]. However, it is still possible to define it in the manner similar to that for defining ADM linear momentum, from a rotational invariance of the Hamiltonian of general relativity at spatial infinity [Regge and Teitelboim (1974); Gourgoulhon (2012)]. Let $\varphi_{(ij)}^a$ be a vector associated with the rotational invariance in the x^i-x^j plane at spatial infinity, for which we write

$$\varphi_{(ij)}^a = x^i \left(\frac{\partial}{\partial x^j}\right)^a - x^j \left(\frac{\partial}{\partial x^i}\right)^a, \tag{5.30}$$

where x^i and x^j denote two of N Cartesian coordinates. We note that (ij) here does not mean the symmetrization. Then, the angular momentum associated with the x^i-x^j plane is defined by

$$L_{(ij)} = \frac{1}{8\pi} \oint_{r \to \infty} \varphi_{(ij)}^k \left(K_k{}^l - K\gamma_k{}^l\right) dS_l. \tag{5.31}$$

In the conformally flat spatial geometry with $K = 0$ or in the case that $\varphi_{(ij)}^k$ is a rotational Killing vector field, equation (5.31) can be rewritten as

$$L_{(ij)} = \frac{1}{8\pi} \oint_{\mathcal{H}} \varphi_{(ij)}^k \left(K_k{}^l - K\gamma_k{}^l\right) dS_l + \int_{\Sigma_t'} \varphi_{(ij)}^l J_l \, dV. \tag{5.32}$$

Thus in the absence of black holes, the angular momentum for this case is determined by integrating the angular momentum density of the matter, as in equation (5.28).

As pointed out by York (1979) [see also Gourgoulhon (2012)], angular momentum defined here is not invariant under a class of coordinate transformation (the so-called supertranslations) that preserves conditions (5.9)–(5.12). For defining the angular momentum in the absence of an exact rotational spacetime Killing vector field, we have to impose further asymptotic gauge conditions. One is the asymptotically quasi-isotropic gauge (5.19), and the other is the asymptotically maximal slicing condition

$$K = O\left(r^{-N}\right). \tag{5.33}$$

These are not satisfied in general coordinate systems, but in the coordinate systems often used in numerical relativity (e.g., moving puncture gauge conditions with initially quasi-isotropic gauge and maximal slicing; see section 2.2), they are satisfied.

5.3.2 Komar mass and angular momentum

5.3.2.1 General expressions

In addition to the ADM mass described above, the mass defined by Komar (1959) (the so-called Komar mass) is often used in numerical relativity, in particular for the studies of equilibria and quasi-equilibria numerically computed (see appendix G.4). This mass is defined in a covariant way, and thus, has a gauge invariant meaning. However, it can be defined only for stationary spacetime, i.e., in the presence of a timelike Killing vector field, ξ^a, and thus, this quantity does not strictly give a mass for dynamical spacetime.

The Komar mass is defined as

$$M_{\rm K} := -\frac{1}{4\pi} \oint_{\mathcal{S}} dS_a n_b \nabla^a \xi^b, \qquad (5.34)$$

where the closed surface $\mathcal{S}$ is usually taken in an asymptotically flat region of spatial hypersurfaces Σ_t. For rewriting this equation, we often use Stokes' theorem, which holds for an antisymmetric tensor $T_{\rm ant}^{ab} = -T_{\rm ant}^{ba}$ as [see a standard textbook like Wald (1984)]

$$\oint_{\mathcal{S}} dS_a n_b T_{\rm ant}^{ab} = \oint_{\mathcal{H}} dS_a n_b T_{\rm ant}^{ab} + \int_{\Sigma_t'} dV n_b \nabla_a T_{\rm ant}^{ab}. \qquad (5.35)$$

Then, equation (5.34) is written as

$$\begin{aligned} M_{\rm K} &= -\frac{1}{4\pi} \oint_{\mathcal{H}} dS_a n_b \nabla^a \xi^b - \frac{1}{4\pi} \int_{\Sigma_t'} dV n_b \nabla_a \nabla^a \xi^b \\ &= -\frac{1}{4\pi} \oint_{\mathcal{H}} dS_a n_b \nabla^a \xi^b + \frac{1}{4\pi} \int_{\Sigma_t'} dV n_b R^b{}_a \xi^a \\ &= -\frac{1}{4\pi} \oint_{\mathcal{H}} dS_a n_b \nabla^a \xi^b + 2 \int_{\Sigma_t'} dV \left(T_{ab} - \frac{1}{2} g_{ab} T \right) \xi^a n^b, \end{aligned} \qquad (5.36)$$

where for deriving the second line, equation (A.8) was used. Thus, in the absence of black holes, the Komar mass is derived only from the stress-energy tensor T^{ab}. In the presence of a black hole, the first term in the last line of equation (5.36) contributes to the Komar mass. However, it should be noted that this term does not denote the mass of the black hole in general (e.g., in the presence of a torus surrounding the black hole). Only in the absence of any matter it agrees with the mass of the black hole [Bardeen et al. (1973); Bardeen (1973)].

The Komar angular momentum, which is associated with a rotational Killing vector field, φ^a, whose integral curves are closed, is also defined by

$$J_{\rm K} := \frac{1}{8\pi} \oint_{\mathcal{S}} dS_a n_b \nabla^a \varphi^b. \qquad (5.37)$$

Here, φ^a is defined in the same manner as equation (5.30). The same manipulation that is shown in equation (5.36) yields

$$\begin{aligned} J_{\rm K} &= \frac{1}{8\pi} \oint_{\mathcal{H}} dS_a n_b \nabla^a \varphi^b - \int_{\Sigma_t'} dV T_{ab} \varphi^a n^b \\ &= \frac{1}{8\pi} \oint_{\mathcal{H}} dS_a n_b \nabla^a \varphi^b + \int_{\Sigma_t'} dV J_a \varphi^a, \end{aligned} \qquad (5.38)$$

where we used $n_a \varphi^a = 0$ and $T_{ab} \varphi^a n^b = -J_a \varphi^a$. In contrast to the Komar mass, the first term in the right-hand side of equation (5.38) denotes the angular momentum of a black hole at least for four-dimensional spacetime even in the presence of a matter field outside the black hole [Bardeen et al. (1973); Bardeen (1973)].

The Komar angular momentum is a conserved quantity in four-dimensional axisymmetric spacetime even if the spacetime is dynamical. Thus, examining the conservation of this quantity in an axisymmetric numerical-relativity simulation is an important test to validate the reliability of the simulation.

5.3.2.2 Expressions in $N+1$ variables

In numerical relativity, it is useful to rewrite the Komar mass and angular momentum in terms of standard $N+1$ variables. First, we focus on the Komar mass, assuming that $\xi^t = 1$ and $\xi^k = 0$ (the subscript k denotes a spatial component), and thus, $\xi^a = \alpha n^a + \beta^a$. We need to rewrite $dS_a n_b \nabla^a \xi^b = \sqrt{\gamma_{N-1}}(d^{N-1}\theta) s_a n_b \nabla^a \xi^b$ in equation (5.34) where θ denotes angular coordinates. Using the following relation,

$$\begin{aligned} s_a n_b \nabla^a \xi^b &= s^a \nabla_a \left(n_b \xi^b \right) - \xi^b s^a \nabla_a n_b \\ &= -s^a D_a \alpha + \xi^b s^a (K_{ab} + n_a D_b \ln \alpha) \\ &= -s^a D_a \alpha + \beta^a s^b K_{ab}, \end{aligned} \quad (5.39)$$

we have

$$M_K = \frac{1}{4\pi} \oint_S \left(D^a \alpha - K^a_{\ b} \beta^b \right) dS_a. \quad (5.40)$$

In numerical relativity, this expression is often referred to as the Komar mass (or the Komar charge if it is not exactly the conserved quantity). In section 5.7.2, we describe laws that hold in a quasi-equilibrium binary using equation (5.40) as the definition of the Komar mass.

The Komar mass was defined in a covariant way and is independent of the coordinate choice. However, it is worthy to note that in the conditions (5.9)–(5.12), the asymptotic behavior of α is written by the Komar mass as

$$\alpha = 1 - \frac{M_K}{r} + O\left(r^{-2}\right). \quad (5.41)$$

For this case, the second term of equation (5.40) does not contribute to M_K.

Using equation (5.36), the Komar mass is also written in a volume integral form

$$M_K = \frac{1}{4\pi} \oint_{\mathcal{H}} \left(D^a \alpha - K^a_{\ b} \beta^b \right) dS_a + \int_{\Sigma'_t} dV \alpha \left(\rho_h - 2\alpha^{-1} \beta^a J_a + S^a_{\ a} \right). \quad (5.42)$$

As we mentioned above, each integral term does not have a particular physical meaning on its own in the presence of both terms.

As we did for M_{ADM}, it is instructive to take the Newtonian limit for the Komar mass for four-dimensional spacetime. In addition to the assumptions made to derive equation (5.23), we assume the Newtonian limit for α as $\alpha \to 1 + \Phi_{\text{N}}$. Then, the Newtonian limit of M_K is derived as

$$M_K \to M_0 + U + 3T_{\text{kin}} + 3\Pi + 2W, \quad (5.43)$$

where

$$\Pi := \int d^3x\, P\,. \qquad (5.44)$$

For stationary spacetime for which the Komar mass is defined, the following virial relation has to be satisfied [e.g., chapter 7 of Shapiro and Teukolsky (1983)]:

$$2T_{\text{kin}} + 3\Pi + W = 0\,. \qquad (5.45)$$

This implies that M_K in the Newtonian limit may be written as

$$M_K \to M_0 + U + T_{\text{kin}} + W, \qquad (5.46)$$

which agrees with the ADM mass in the Newtonian limit. The calculation in the derivation of equation (5.46) indicates that the condition, $M_{\text{ADM}} = M_K$, should be satisfied for stationary spacetime as a general relativistic virial relation (see section 5.3.3).

For rewriting the Komar angular momentum, we use the following relation

$$s_a n_b \nabla^a \varphi^b = -s_a \varphi_b \nabla^a n^b = K_{ab} s^a \varphi^b, \qquad (5.47)$$

where we used $n_a \varphi^a = 0$. Then, equations (5.37) and (5.38) are written as

$$J_K = \frac{1}{8\pi} \oint_S K^a{}_b \varphi^b dS_a, \qquad (5.48)$$

and

$$J_K = \frac{1}{8\pi} \oint_{\mathcal{H}} K^a{}_b \varphi^b dS_a + \int_{\Sigma'_t} dV J_a \varphi^a. \qquad (5.49)$$

In the presence of the rotational Killing vector field φ^a, it is easy to find that equation (5.31) for $N = 3$ is rewritten to equation (5.49). On the other hand, the relation between M_{ADM} and M_K [written in the form of equation (5.42)] is not trivial even in the presence of ξ^a (but see section 5.3.3).

The Komar mass and angular momentum are computed for any spacetime from equations (5.42) and (5.49) in numerical relativity. However, they are not conserved quantities in the absence of corresponding symmetry, as already mentioned. A consequence of this point will be shown in section 5.3.3.

5.3.3 Virial relation

As we described in section 5.3.2.2, the ADM mass and the Komar mass agree with each other for stationary spacetime in the Newtonian limit. Beig (1978) showed that this also holds in general relativity: The relation $M_{\text{ADM}} = M_K$ is a necessary condition for stationary spacetime, and this is a general relativistic version of the virial relation. In this section, we outline the derivation of this virial relation using 3+1 variables following Beig (1978). [The virial relations in the variables of the 3+1 formulation were also explored in Gourgoulhon and Bonazzola (1994).] We here focus only on the $N = 3$ case because the virial relation is practically used only for this case.

From the evolution equation of K_{ab}, (2.61), the following equation is derived,

$$\mathcal{L}_t K_{ij} - \frac{1}{2}\gamma_{ij}\mathcal{L}_t K = \alpha \bar{G}_{ij} - \left(D_i D_j \alpha - \frac{1}{2}\gamma_{ij} D_k D^k \alpha\right)$$
$$+ \alpha \left(K K_{ij} - 2 K_{ik} K^k{}_j - \frac{1}{2}\gamma_{ij} K^2\right)$$
$$+ \beta^k D_k \left(K_{ij} - \frac{1}{2}\gamma_{ij} K\right) + K_{ik} D_j \beta^k + K_{jk} D_i \beta^k$$
$$- 8\pi\alpha \left(S_{ij} - \frac{1}{4}\gamma_{ij}(S_k{}^k + \rho_{\mathrm{h}})\right), \qquad (5.50)$$

where $\bar{G}_{ij}$ denotes the Einstein tensor on spatial hypersurfaces. Using the prescription of Landau and Lifshitz (1962), $\bar{G}^{ij}$ is written as (see appendix F)

$$2\bar{G}^{ij} = -16\pi \bar{t}_{\mathrm{LL}}^{ij} + \gamma^{-1} \partial_k \partial_l \bar{H}^{ikjl}, \qquad (5.51)$$

where $\bar{H}^{ikjl}$ is a superpotential on spatial hypersurfaces defined here by

$$\bar{H}^{ikjl} := \gamma \left(\gamma^{ij}\gamma^{kl} - \gamma^{il}\gamma^{jk}\right). \qquad (5.52)$$

In addition, we define $\bar{h}^{ikj} := \partial_l \bar{H}^{ikjl}$. In the following manipulation, we will suppose to use Cartesian coordinates.

As in the case that we define the ADM mass, we assume that the conditions (5.9)–(5.12) hold at spatial infinity for geometric quantities with $N = 3$. In terms of the variables in the 3+1 formulation, those conditions are written (in the Cartesian coordinate basis) as

$$\alpha = 1 - \frac{M_{\mathrm{K}}}{r} + O\left(r^{-2}\right), \qquad (5.53)$$

$$\beta^i = O\left(r^{-1}\right), \qquad (5.54)$$

$$\gamma_{ij} = f_{ij} + O\left(r^{-1}\right), \qquad (5.55)$$

$$\partial_k \gamma_{ij} = O\left(r^{-2}\right), \qquad (5.56)$$

$$K_{ij} = O\left(r^{-2}\right), \qquad (5.57)$$

$$\partial_k K_{ij} = O\left(r^{-3}\right), \qquad (5.58)$$

$$T_{ab} = O\left(r^{-4}\right). \qquad (5.59)$$

Here, the leading part of K_{ij} is assumed to be composed only of the ADM linear momentum. These assumptions yield the following results at spatial infinity: (i) the evolution equation for K_{ij} implies $\partial_t K_{ij} = O\left(r^{-3}\right)$; (ii) equation (5.53) implies $D_c D^c \alpha = O\left(r^{-4}\right)$; (iii) $\bar{t}_{\mathrm{LL}}^{ij} = O\left(r^{-4}\right)$.

Following Beig (1978), we evaluate a surface integral at spatial infinity as

$$\begin{aligned}
I_1 &= \frac{1}{8\pi} \oint_{r\to\infty} f_{jk} x^k \bar{G}^{ij} \overset{(0)}{dS_i} \\
&= \frac{1}{16\pi} \oint_{r\to\infty} f_{jl} x^l \partial_k \bar{h}^{ikj} \overset{(0)}{dS_i} \\
&= \frac{1}{16\pi} \oint_{\mathcal{H}} f_{jl} x^l \partial_k \bar{h}^{ikj} \overset{(0)}{dS_i} + \frac{1}{16\pi} \int_{\Sigma'_t} \partial_i \left(f_{jl} x^l \partial_k \bar{h}^{ikj} \right) d^3 x,
\end{aligned} \quad (5.60)$$

where we used equation (5.51) with $\bar{t}^{ij}_{\mathrm{LL}} = O(r^{-4})$ for obtaining the second line. From the identity, $\partial_i \partial_k \bar{h}^{ikj} = 0$, which holds because of the antisymmetry of $\bar{h}^{ikj} = -\bar{h}^{kij}$, we have $\partial_i \left(f_{jl} x^l \partial_k \bar{h}^{ikj} \right) = f_{ij} \partial_k \bar{h}^{ikj}$. Then,

$$\begin{aligned}
I_1 &= \frac{1}{16\pi} \oint_{\mathcal{H}} f_{jl} x^l \partial_k \bar{h}^{ikj} \overset{(0)}{dS_i} + \frac{1}{16\pi} \int_{\Sigma'_t} f_{jl} \partial_i \bar{h}^{jil} d^3 x \\
&= \frac{1}{16\pi} \oint_{\mathcal{H}} f_{jl} \left(x^l \partial_k \bar{h}^{ikj} - \bar{h}^{jil} \right) \overset{(0)}{dS_i} + \frac{1}{16\pi} \oint_{r\to\infty} f_{jl} \bar{h}^{jil} \overset{(0)}{dS_i} \\
&= \frac{1}{16\pi} \oint_{\mathcal{H}} f_{jl} \partial_k (x^l \bar{h}^{ikj}) \overset{(0)}{dS_i} + \frac{1}{16\pi} \oint_{r\to\infty} f_{jl} \bar{h}^{jil} \overset{(0)}{dS_i} \\
&= \frac{1}{16\pi} \oint_{r\to\infty} f^{il} f^{jk} (\partial_i \gamma_{jk} - \partial_k \gamma_{ij}) \overset{(0)}{dS_l} =: -M_{\mathrm{ADM}},
\end{aligned} \quad (5.61)$$

where in the last line, we used the definition of the ADM mass. We also used the relation for an antisymmetric tensor T^{ij}_{ant} in flat space,

$$\oint \partial_i T^{ij}_{\mathrm{ant}} \overset{(0)}{dS_j} = 0, \quad (5.62)$$

and the asymptotic behavior of $f_{jk} \bar{h}^{jik}$

$$f_{jk} \bar{h}^{jik} \to f^{il} f^{jk} (\partial_l \gamma_{jk} - \partial_k \gamma_{jl}). \quad (5.63)$$

Using equation (5.50) together with the condition (5.53), I_1 is written as

$$\begin{aligned}
I_1 &= \frac{1}{8\pi} \oint_{r\to\infty} f_{ik} x^k \left(\mathscr{L}_t K^{ij} - \frac{1}{2} \gamma^{ij} \mathscr{L}_t K \right) \overset{(0)}{dS_j} \\
&+ \frac{1}{8\pi} \oint_{r\to\infty} x^i D_i D^j \alpha \overset{(0)}{dS_j},
\end{aligned} \quad (5.64)$$

where we discarded the terms which vanish. Using equation (5.53), the second term of equation (5.64) is calculated to yield

$$\frac{1}{8\pi} \oint_{r\to\infty} f^{jk} x^i \partial_i \partial_j \alpha \overset{(0)}{dS_k} = -M_{\mathrm{K}}. \quad (5.65)$$

Finally we obtain

$$\frac{1}{8\pi} \oint_{r\to\infty} f_{ik} x^k \left(\mathscr{L}_t K^{ij} - \frac{1}{2} \gamma^{ij} \mathscr{L}_t K \right) \overset{(0)}{dS_j} = M_{\mathrm{K}} - M_{\mathrm{ADM}}. \quad (5.66)$$

Therefore, for stationary spacetime for which $\mathscr{L}_t K^{ij} = 0 = \mathscr{L}_t K$, $M_\text{K} = M_\text{ADM}$; this is often called the virial relation in general relativity.

For non-stationary spacetime, the left-hand side of equation (5.66) is not zero in general, because $O\left(r^{-3}\right)$ parts of K_{ij} may vary. This implies that M_K is not constant in dynamical spacetime (see section 5.5.2.1 for an example), because M_ADM is independent of the time variability of K_{ab}.

5.3.4 Irreducible mass

As briefly introduced in section 1.3.2.3, the irreducible mass defined from the area of black holes is useful for guessing possible evolution processes of a system composed of black holes. Focusing only on four-dimensional spacetime, the irreducible mass is defined by

$$M_\text{irr} := \sqrt{\frac{A_\text{H}}{16\pi}}. \qquad (5.67)$$

For Schwarzschild black holes, their mass M_BH agrees with M_irr and for Kerr black holes with nonzero spin, $M_\text{irr} < M_\text{BH}$. Because A_H never decreases due to the area theorem which holds in globally hyperbolic spacetime (see appendix G), M_irr gives the lower bound of the black-hole mass; any black-hole mass cannot be reduced below M_irr by any classical processes.

It is impossible to determine event horizons and their area only from initial data. Thus, we cannot obtain the irreducible mass as well. However, it is still possible to determine apparent horizons and their area for initial data (see section 7.1 for apparent horizon). The existence of an apparent horizon means the existence of a black hole, whose event horizon lies outside the apparent horizon. The area of the apparent horizon, A_AH, is always smaller than (or equal to) the area of the event horizon (see appendix G), and hence, we can set the lower bound of the irreducible mass by

$$M_\text{irr} \geq \sqrt{\frac{A_\text{AH}}{16\pi}}. \qquad (5.68)$$

Because this also gives the lower bound of black-hole mass, we may refer to $\sqrt{A_\text{AH}/16\pi}$ as an irreducible mass.

To summarize, we have the relation

$$\sqrt{\frac{A_\text{AH}}{16\pi}} \leq \sqrt{\frac{A_\text{H}}{16\pi}} = M_\text{irr} \leq M_\text{BH} \leq M_\text{ADM}, \qquad (5.69)$$

where we can determine only $\sqrt{A_\text{AH}/16\pi}$ and M_ADM in initial-data problems.

The relation between the irreducible mass and the black-hole mass can be used for guessing a possible process in two black-hole problems [Hawking (1971)]. Let $A_\text{AH,1}$ and $A_\text{AH,2}$ be area of apparent horizons for two black holes measured on an initial spatial hypersurface. We suppose that two black holes eventually merge to be one black hole for which we denote its area and mass by $A_{\rm H,f}$ and $M_{\rm BH,f}$, respectively. Then, the following relation has to be satisfied:

$$A_{\rm AH,1} + A_{\rm AH,2} \leq A_{\rm H,f} \leq 16\pi M_{\rm BH,f}^2. \tag{5.70}$$

Now, we estimate a possible amount of gravitational waves emitted in the merger process of two black holes. The total amount should be calculated by

$$\Delta M = M_{\rm ADM} - M_{\rm BH,f}, \tag{5.71}$$

where $M_{\rm ADM}$ is the ADM mass of the system composed of two black holes and can be determined from the initial data. If the initial separation of two black holes is sufficiently large, $M_{\rm ADM} = M_{\rm BH,1} + M_{\rm BH,2}$, but we do not have to restrict our attention to such a case. $M_{\rm BH,f}$ is constrained by the following inequalities

$$M_{\rm BH,f} \geq M_{\rm irr,f} = \sqrt{\frac{A_{\rm H,f}}{16\pi}} \geq \sqrt{\frac{A_{\rm AH,1} + A_{\rm AH,2}}{16\pi}}. \tag{5.72}$$

Then, we have

$$\Delta M \leq M_{\rm ADM} - \sqrt{\frac{A_{\rm AH,1} + A_{\rm AH,2}}{16\pi}}. \tag{5.73}$$

Here, all the quantities can be measured only from the initial data, and hence, the upper bound of ΔM is obtained.

5.4 Initial data for pure gravitational waves

In general relativity, vacuum spacetime can have positive ADM mass even in the absence of black holes, if gravitational waves are present. Vacuum initial data composed of pure gravitational waves in three-dimensional asymptotically flat spacetime have been studied by a number of people [e.g., Brill (1959); Eppley (1977); Miyama (1981); Nakamura *et al.* (1987); Abrahams and Evans (1993, 1994); Shibata and Nakamura (1995); Alcubierre *et al.* (2000) for early works]. We here touch on methods for constructing initial data for such spacetime.

There are two popular methods for constructing gravitational-wave spacetime. The first method was proposed originally by Brill (1959) in which the axial symmetry is assumed and the spatial metric is chosen in the form

$$dl^2 = \psi^4 \left[e^{2q} \left(dr^2 + r^2 d\theta^2 \right) + r^2 \sin^2\theta d\varphi^2 \right], \tag{5.74}$$

where ψ and q are functions of r and θ. Setting $K_{ij} = 0$, momentarily static initial data are considered. Then, a solution of the initial data is determined by solving the Hamiltonian constraint, which is written as

$$\left[\frac{\partial^2}{\partial r^2} + \frac{2}{r}\frac{\partial}{\partial r} + \frac{1}{r^2}\left(\frac{\partial^2}{\partial \theta^2} + \cot\theta \frac{\partial}{\partial \theta} \right) \right] \psi = -\frac{\psi}{4} \left(\frac{\partial^2}{\partial r^2} + \frac{1}{r}\frac{\partial}{\partial r} + \frac{1}{r^2}\frac{\partial^2}{\partial \theta^2} \right) q. \tag{5.75}$$

Thus, if a wave packet for q is given and the elliptic equation for ψ is solved, the initial data are constructed. This initial condition is called Brill wave.

In the other typical method, the conformal flatness is assumed for the spatial metric, while a transverse-tracefree form of K_{ij} is given as a gravitational-wave component. Specifically, a nonzero function of $\bar{A}_{ij}^{\rm tt}$ [see equation (5.5)] is given. This method may be used for non-axisymmetric initial data.

Since $\bar{A}_{ij}^{\rm tt}$ is a transverse-tracefree tensor satisfying $\tilde{D}^i \bar{A}_{ij}^{\rm tt} = 0 = \tilde{\gamma}^{ij} \bar{A}_{ij}^{\rm tt}$, the general solution for it is obtained by a spherical-harmonic decomposition method, which is described in appendix C.2. In this method, a regular solution is obtained for modes with $l \geq 2$ in a straightforward calculation; for example, the solution described in equation (C.59) with $l = 2 = m$ was used in Shibata and Nakamura (1995). For that solution, $h_{ij} = \tilde{\gamma}_{ij} - f_{ij} = 0$ at $t = 0$, while $\partial_t h_{ij} \neq 0$ at $t = 0$. Thus, the corresponding extrinsic curvature, which is transverse-tracefree, is not zero initially. Employing such an extrinsic curvature component as $\bar{A}_{ij}^{\rm tt}$ that satisfies the momentum constraint equation irrespective of its magnitude, the Hamiltonian constraint in the following form should be solved:

$$\Delta_{\rm f} \psi = -\frac{1}{8\psi^7} \bar{A}_{ij}^{\rm tt} \bar{A}_{kl}^{\rm tt} f^{ik} f^{jl}. \tag{5.76}$$

Then, the initial data are obtained [see, e.g., Shibata and Nakamura (1995)].

5.5 Initial data for black holes

There are several analytic solutions for the constraint equations in vacuum where $\rho_{\rm h} = 0$ and $J_i = 0$. Initial data for black holes are the most interesting solutions, which have been extensively used in numerical relativity. In this section, we summarize these solutions.

5.5.1 Time symmetric case

5.5.1.1 One black hole

First, we focus on the time symmetric (momentarily static) case in which K_{ab} vanishes. Then, the momentum constraint equation is trivially satisfied and the Hamiltonian constraint equation reduces to (see section 2.1.7)

$$\tilde{\Delta} \psi = \frac{(N-2)\psi}{4(N-1)} \tilde{R}. \tag{5.77}$$

If we further assume the conformal flatness for the spatial hypersurface, we have $\tilde{R} = 0$ and $\tilde{\Delta} = \Delta_{\rm f}$, and hence, equation (5.77) reduces to a simple Laplace equation for ψ:

$$\Delta_{\rm f} \psi = 0. \tag{5.78}$$

This basic equation has been often employed for deriving initial data composed of non-spinning black holes. The essential reason that this simplification is justified

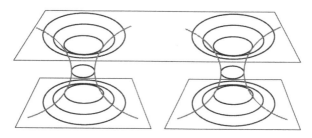

Fig. 5.1 Schematic embedding diagram for the geometry of two black holes constructed by the method of Brill and Lindquist (1963). The upper sheet denotes our world and the lower sheets are other worlds connected by wormhole throats.

is that solutions for spherically symmetric black holes in isotropic coordinates are written in this conformally flat form as [cf. equation (1.156)]

$$ds^2 = -\left[\frac{1-(r_g/r)^{N-2}}{1+(r_g/r)^{N-2}}\right]^2 dt^2 + \left[1+\left(\frac{r_g}{r}\right)^{N-2}\right]^{4/(N-2)} f_{ij}dx^i dx^j, \quad (5.79)$$

and a solution of equation (5.78),

$$\psi = 1 + \left(\frac{r_g}{r}\right)^{N-2}, \quad (5.80)$$

gives a physical initial condition for a static black hole.[2] Here r_g is the coordinate radius of the event horizon. The topology of this initial data is described in figure 1.8: There are two world sheets that are causally disconnected, and they are isometric with respect to the coordinate transformation (irrespective of the dimensionality N)

$$y^i = x^i \left(\frac{r_g}{r}\right)^2, \quad (5.81)$$

which yields

$$ds^2 = -\left(\frac{1-(r_g/y)^{N-2}}{1+(r_g/y)^{N-2}}\right)^2 dt^2 + \left[1+\left(\frac{r_g}{y}\right)^{N-2}\right]^{4/(N-2)} f_{ij}dy^i dy^j, \quad (5.82)$$

where $y = |y^i|$. This fact shows that the region for $r = 0$ denotes spatial infinity in the other world sheet defined for $r \leq r_g$.

5.5.1.2 Multiple black holes: Brill–Lindquist initial data

Because equation (5.78) is linear in ψ, it is possible to derive a new solution of ψ superimposing solutions of the form $\propto r^{-(N-2)}$ as

$$\psi = 1 + \sum_{I=1}^{n} \left(\frac{a_I}{r_I}\right)^{N-2}, \quad (5.83)$$

---

[2]Of course, this is not the case in the presence of more than two black holes.

where a_I is a constant and $r_I := |x^i - x_I^i|$ with $x_I^i (|x_I^i| < \infty)$ denoting the location of a pole that can be chosen arbitrarily. The topology of spatial hypersurfaces for this solution with $n = 2$ is schematically written in figure 5.1: For this case, there are 3 asymptotically flat world sheets, and an observer in our world sheet (the upper sheet) finds two black holes (unless a horizon encompassing two horizons of each hole is present). For general values of n, there are $n + 1$ asymptotically flat world sheets, and an observer in our world sheet could find n black holes. The region for $|x^i| \to \infty$ denotes spatial infinity in our world sheet, while the regions for $r_I \to 0$ (coordinate singularities) are spatial infinity in other world sheets. The initial data described by equation (5.83) is called Brill–Lindquist initial data for multiple black holes [Lindquist (1963); Brill and Lindquist (1963)].

Brill–Lindquist initial data are similar to one-black-hole initial data except for one major difference that the solution does not represent initial data composed of two identical world sheets joined at minimal surfaces. This fact is found by a coordinate transformation

$$y^i = (x^i - x_I^i) \left(\frac{a_I}{r_I}\right)^2. \tag{5.84}$$

By this, it is shown that a pole at $x^i = x_I^i$ corresponds to one of spatial infinity on other world sheets, as in the one-black-hole case. However, we find at the same time that an isometric relation is not satisfied among different sheets. Indeed, the transformation by equation (5.84) yields the spatial line element as

$$dl^2 = [\psi(y^i)]^{4/(N-2)} \left(dy^2 + y^2 d\Omega_{N-1}\right), \tag{5.85}$$

where $y = |y^i|$ and

$$\psi(y^i) = 1 + \left(\frac{a_I}{y}\right)^{N-2} + \sum_{J \neq I} \left(\frac{a_I a_J}{|(x_I^i - x_J^i)y + a_I^2 y^i/y|}\right)^{N-2}. \tag{5.86}$$

This functional form is different from that in equation (5.83). For constructing initial data composed of multiple black holes with an isometry, a more complicated solution of ψ is necessary (see section 5.5.1.3).

The ADM mass measured in our world sheet is derived using equation (5.18) as

$$M_{\mathrm{ADM}} = \sum_{I=1}^{N} \frac{N-1}{4\pi} a_I^{N-2} \Omega_{N-1}, \tag{5.87}$$

where $\Omega_{N-1} = 2\pi^{N/2}/\Gamma[N/2]$ with $\Gamma[x]$ being the Gamma function. The ADM mass may be measured in other world sheets, and for the I-th world sheet, it is

$$M_{\mathrm{ADM},I} = \frac{N-1}{4\pi} a_I^{N-2} \Omega_{N-1} \left[1 + \sum_{J \neq I} \left(\frac{a_J}{|x_I^i - x_J^i|}\right)^{N-2}\right]. \tag{5.88}$$

This is called a bare mass [Lindquist (1963); Brill and Lindquist (1963)]. Summation of all the bare mass is larger than the ADM mass measured in our world sheet as long as $|x_I^i - x_J^i| < \infty$. This suggests that the second term of equation (5.88) denotes the absolute value of the binding energy between the I-th black hole and others.

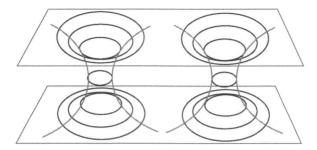

Fig. 5.2 Schematic embedding diagram for the geometry of two black holes in the Misner data. This spatial hypersurface has the inversion symmetry with respect to minimal surfaces along two wormhole throats.

5.5.1.3 *Multiple black holes: Misner initial data*

Misner constructed initial data composed of multiple black holes that satisfy an isometric condition as in the one-black-hole case [Misner (1960, 1963)]: see figure 5.2 for a two-black-hole case. For this type of initial data, two causally disconnected spatial would sheets are isometric with respect to a coordinate transformation like equation (5.84). This implies that we only need to consider one of the two world sheets for studying this system.

In the past, the Brill–Lindquist initial data were believed to be not suitable for numerical-relativity computation because they possess coordinate singularities at $x^i = x^i_I$ (although this drawback disappears after the development of puncture and moving puncture approaches described in section 2.3.3.1). By contrast, the Misner initial data are free of the coordinate singularities, if we focus only on one world sheet imposing the isometric boundary conditions at the minimal surfaces of the wormhole throats.

The method of constructing the Misner initial data is essentially the same as the method of images often used in electrostatics [Jackson (1975)]: It is a method of solving the Laplace equation with a special boundary condition. One of typical problems in electrostatics for undergraduate students is to determine the electric field for a system composed of a point charge and a perfectly conducting sphere. In this problem, the inner boundary condition is imposed on the surface of the perfectly conducting sphere where the electric potential should be constant, with the boundary condition at spatial infinity where the electric potential should be zero. For solving this problem, it is quite convenient to tentatively put two point charges inside the perfectly conducting sphere; one is located at the center of the perfectly conducting sphere and the other is at a point on the line between the original point charge and point charge at the center. The absolute values of two additional charges are the same but the sign is different. Then, we simply calculate the electric potential determined by three electric charges. For an appropriate choice of the position and values of the additional electric charge, we can obtain an electric field for which the electric potential is constant on the surface of the perfectly conducting sphere. This technique gives a required solution of this problem quite easily. If there are more than two perfectly conducting spheres, we have to add infinite number of electric charges to achieve a constant electric potential on all the spheres.

For constructing the Misner initial data, the inner boundary condition is imposed on minimal surfaces along wormhole throats. This condition is derived by the requirement that for the coordinate transformation (5.84), two world sheets have an isometric condition. Specifically, this leads to the condition

$$\psi(r_I)^{1/(N-2)} = \psi(|y^i|)^{1/(N-2)} \left(\frac{a_I}{r_I}\right), \qquad (5.89)$$

where $r_I = |x^i - x_I^i|$. We note that the minimal surfaces are denoted by $r_I = a_I = |y^i|$. Suppose that x_I^i is defined for $r_I > a_I$ and thus $|y^i| < a_I$. If r_I is chosen to be only slightly larger than a_I, equation (5.89) gives a Robin-type boundary condition, which is derived by differentiating equation (5.89) with respect to r_I as

$$\frac{1}{N-2}\frac{\partial \psi(r_I)}{\partial r_I} = -\frac{a_I^N}{(N-2)|y^i|^N}\frac{\partial \psi(|y^i|)}{\partial |y^i|} - \frac{a_I^{N-2}}{r_I^{N-1}}\psi(r_I). \qquad (5.90)$$

Then, for $r_I = a_I = |y^i|$, we obtain the required boundary condition [Bowen and York (1980); Cook and York (1990)]

$$\left.\frac{2}{N-2}\frac{\partial \psi}{\partial r_I}\right|_{r_I=a_I} + \frac{1}{a_I}\psi(a_I) = 0. \qquad (5.91)$$

This boundary condition has to be satisfied on all the minimal surfaces.

The first step toward the construction of the Misner initial data is to choose the Brill–Lindquist initial data (5.83) as a seed solution of ψ. Then, to obtain a solution that satisfies the boundary conditions (5.91) for all the minimal surfaces, we have to add infinite numbers of homogeneous solutions of the form $b_J^{N-2}/|x^i - x_J^i|^{N-2}$ with an appropriate choice of constants, b_J and x_J^i. Misner explicitly derived such a solution for two black holes of equal mass in a three dimensional hypersurface. In cylindrical coordinates (ϖ, φ, z), his solution for ψ is written as

$$\psi = 1 + \sum_{n=1}^{\infty} \frac{1}{\sinh(n\mu_0)}\left[\frac{a}{r_{+,n}} + \frac{a}{r_{-,n}}\right], \qquad (5.92)$$

where the black holes are assumed to be located along the z-axis, $r_{\pm,n} = \sqrt{\varpi^2 + [z \pm a\coth(n\mu_0)]^2}$, and (a, μ_0) are arbitrary constants, which determine the location of black holes and the total mass of the system. More specifically, the minimal surfaces are located at spheres of their coordinate centers, $\pm a\coth\mu_0$, and of radius $a/\sinh\mu_0$. Here, the total ADM mass of this system is

$$M = \sum_{n=1}^{\infty} \frac{4a}{\sinh(n\mu_0)}. \qquad (5.93)$$

Some properties of the Misner initial data were explored by Lindquist (1963) and by Smarr et al. (1976) to which the readers may refer for more details.

5.5.1.4 Remark

Before closing this section, we note the following fact. Equation for the conformal factor, (5.78), is a simple Laplace equation. This implies that a general solution is written by superimposing infinite number of terms composed of multipole moments such as $(3\cos^2\theta - 1)/r^3$ for $N = 3$. However, these terms are not positive definite and have to be absent because with its presence, zero points appear somewhere for ψ. A zero point is a physical singularity because spacelike geodesics become incomplete there. For this reason, only the monopole terms of positive coefficients are considered.

5.5.2 Time asymmetric case

For the case that black holes are moving or spinning, we have to determine the extrinsic curvature by solving the momentum constraint equation. As described in section 5.2, one of the most convenient ways for a solution of the momentum constraint equation is to employ the conformal transverse-tracefree decomposition. York and his collaborators [Bowen and York (1980); Kulkarni et al. (1983); Kulkarni (1984); Bowen et al. (1984); Cook and York (1990); Cook (1991, 1994)] [see also Cook (2000) for a review] applied this procedure for asymptotically flat spatial hypersurfaces with a set of simplifying assumptions as follows: Maximal slicing, $K = 0$, conformal flatness, $\tilde{\gamma}_{ij} = f_{ij}$, and transverse-tracefree part of $\bar{A}_{ij}$, $\bar{A}_{ij}^{tt}$, is absent. Then, equation (5.6) reduces to

$$\Delta_{\rm f} W_i + \frac{N-2}{N}\overset{(0)}{D}_i\overset{(0)}{D}_j W^j = 0. \tag{5.94}$$

In the following, we will describe procedures for a solution of this equation.

5.5.2.1 General solution for extrinsic curvature

There are several ways for deriving an analytic solution of equation (5.94) in three spatial dimensions [Bowen and York (1980); Bowen (1982); Oohara et al. (1997); Shibata (1997)]. Here, we introduce a method of Shibata (1997, 1999c) assuming the use of Cartesian coordinates. Since it is easy to derive a general solution of the extrinsic curvature for any value of N [Yoshino et al. (2006)], we do not restrict our attention to the three-dimensional case in this subsection.

For more general uses, we here consider solutions of the equation for the non-vacuum case,

$$\Delta_{\rm f} W_i + \frac{N-2}{N}\overset{(0)}{D}_i\overset{(0)}{D}_j W^j = 8\pi S_i, \quad \left(S_i = J_i \psi^{2N/(N-2)}\right). \tag{5.95}$$

Then, we introduce the following decomposition of W_i using auxiliary variables B_i and χ as

$$W_i = \frac{3N-2}{4(N-1)}B_i - \frac{N-2}{4(N-1)}\left(x^k\partial_i B_k + \partial_i \chi\right). \tag{5.96}$$

Since we assume to use Cartesian coordinates, $\overset{(0)}{D}_j = \partial_j$ will be used in the following. Subscripts of W_i and B_i are raised by f^{ij}. Imposing the auxiliary condition $\Delta_f \chi = -x^k \Delta_f B_k$, equation (5.95) is decomposed into two simple elliptic equations

$$\Delta_f B_i = 8\pi S_i, \qquad \Delta_f \chi = -8\pi S_i x^i. \tag{5.97}$$

For $S_i = 0$, these equations reduce to Laplace equations and hence it is possible to derive general solutions as

$$B_i = -\frac{p_i}{r^{N-2}} - \frac{p_{ij}\bar{n}^j}{r^{N-1}} - \frac{p_{ijk}\left(N\bar{n}^j\bar{n}^k - f^{jk}\right)}{r^N} + \cdots, \tag{5.98}$$

$$\chi = \frac{C^\chi}{r^{N-2}} + \frac{C_i^\chi \bar{n}^i}{r^{N-1}} + \frac{C_{ij}^\chi \left(N\bar{n}^i\bar{n}^j - f^{ij}\right)}{r^N} + \cdots, \tag{5.99}$$

where $p_{i\ldots}$, C^χ, and $C_{i\ldots}^\chi$ are constants, and $\bar{n}^i = x^i/r$.

Here, physical quantities of black holes that are reflected in the extrinsic curvature are their linear momentum and spin. Thus, we will later relate some of constants in equations (5.98) and (5.99) to them using equations (5.27) and (5.31).

Substituting equation (5.96) into (5.5) yields

$$\bar{A}_{ij} = \frac{N}{2(N-1)}\left(B_{i,j} + B_{j,i} - \frac{2}{N}B^l{}_{,l}f_{ij}\right) - \frac{N-2}{2(N-1)}\left(x^k B_{k,ij} + \chi_{,ij}\right). \tag{5.100}$$

Then, substituting equations (5.98) and (5.99) into equation (5.100) yields

$$\bar{A}_{ij} = \frac{N(N-2)}{2(N-1)r^{N-1}}\left[p_i\bar{n}_j + p_j\bar{n}_i + p_l\bar{n}^l\left[(N-2)\bar{n}_i\bar{n}_j - f_{ij}\right]\right]$$
$$+ \frac{N}{2(N-1)r^N}\left[p_{ik}F^k{}_j + p_{jk}F^k{}_i - \frac{2}{N}f_{ij}p_{kl}F^{kl}\right.$$
$$+ (N-2)\bar{n}^m p_{mk}f^{kl}\left\{(N+2)\bar{n}_i\bar{n}_j\bar{n}_l - (\bar{n}_i f_{jl} + \bar{n}_j f_{il} + \bar{n}_l f_{ij})\right\}$$
$$\left. - C^\chi \frac{(N-2)^2}{N}F_{ij}\right] + O\left(r^{-N-1}\right), \tag{5.101}$$

where $F_{ij} := N\bar{n}_i\bar{n}_j - f_{ij}$ and its subscript is raised by f^{ij}.

Using equation (5.101), the ADM linear momentum and angular momentum in the Hamiltonian formulation are calculated by equations (5.27) and (5.31) as

$$P_i^{\text{ADM}} = \frac{(N-2)\Omega_{N-1}}{8\pi}p_i, \tag{5.102}$$

$$L_{(ij)} = -\frac{N}{16\pi}p_{kl}^{\text{a}}\oint\left(\bar{n}^k\varphi_{(ij)}^l - \bar{n}^l\varphi_{(ij)}^k\right)r^{-1}d\Omega_{N-1} = \frac{\Omega_{N-1}}{4\pi}p_{ij}^{\text{a}}, \tag{5.103}$$

where Ω_{N-1} is the surface volume of a $(N-1)$-dimensional unit sphere [see equation (1.155)], and p_{kl}^{a} denotes the antisymmetric part of p_{kl}, i.e., $p_{kl}^{\text{a}} = -p_{lk}^{\text{a}} = (p_{kl} - p_{lk})/2$. Thus, linear and angular momenta of a black hole are related to p_i and p_{ij}^{a}, respectively.

Other constants in equation (5.101) do not have a particular physical meaning. However, C^χ and the symmetric part of p_{ij} (denoted by p_{ij}^{s}) are closely concerned with the virial relation in general relativity, which was described in section 5.3.3. We remark this fact as follows.

Using equation (5.101), the left-hand side of equation (5.66) is written as

$$\frac{1}{8\pi}\frac{d}{dt}\oint_{r\to\infty} x^i K_i{}^j dS_j = \left[N\frac{dp^s_{kk}}{dt} - (N-2)^2\frac{dC^\chi}{dt}\right]\frac{\Omega_{N-1}}{16\pi}. \quad (5.104)$$

Here, C^χ and p^s_{kk} are not conserved quantities in contrast to linear and angular momenta, and hence, they may be time-dependent. Thus, the virial relation does not hold in their presence; see equation (5.66). In other words, if they vanish, the virial relation is satisfied. However, this does not always imply that the spacetime is stationary, because higher-order coefficients, e.g., $C^\chi_{i\ldots}$ may be time-dependent. This shows that the virial relation, $M_{\text{ADM}} = M_{\text{K}}$, is a necessary condition for spacetime to be stationary, but it is not a sufficient condition.

For the further clarification, we consider the case that a matter field is present while black holes are absent. Then, the surface integral of equation (5.104) is rewritten using Gauss' law as

$$\frac{1}{8\pi}\frac{d}{dt}\oint_{r\to\infty} x^i K_i{}^j dS_j = \frac{1}{8\pi}\frac{d}{dt}\int D_j\left(x^i K_i{}^j\right) dV = \frac{d}{dt}\int J_i x^i dV, \quad (5.105)$$

where we used $K = 0$. In the Newtonian limit, $J_i \to \rho v^i$, and hence, the last term of equation (5.105) reduces to $\ddot{I}/2$, where I is the moment of inertia. This result is expected from the virial relation in Newtonian theory [see, e.g., chapter 7 of Shapiro and Teukolsky (1983)]

$$\frac{1}{2}\ddot{I} = 2T_{\text{kin}} + 3\Pi + W. \quad (5.106)$$

[Compare this with equation (5.45).]

5.5.2.2 Puncture approach

The puncture formulation was proposed by Brandt and Brügmann (1997) as a novel approach that enables us to construct initial data of multiple moving and/or spinning black holes with the same topology as that of the Brill–Lindquist initial data (see figure 5.1). Numerical computation in this approach is quite simple, and now, the puncture formulation becomes one of the most popular approaches for preparing binary-black-hole initial data.

In the puncture formulation, we adopt the terms composed of p_i and $p^a_{ij} =: s_{ij}$ in the solution of $\bar{A}_{ij}$ [see equation (5.101)] as

$$\bar{A}_{ij} = \sum_{I=1}^{n}\left[\frac{N(N-2)}{2(N-1)r_I^{N-1}}\left(p_i^I \bar{n}_j^I + p_j^I \bar{n}_i^I + f^{kl} p_k^I \bar{n}_l^I\left[(N-2)\bar{n}_i^I \bar{n}_j^I - f_{ij}\right]\right)\right.$$
$$\left. + \frac{N}{r_I^N} f^{kl}\left(s_{ik}^I \bar{n}_l^I \bar{n}_j^I + s_{jk}^I \bar{n}_l^I \bar{n}_i^I\right)\right], \quad (5.107)$$

where the subscript I denotes a label of black holes, $r_I = |x^i - x_I^i|$, and $\bar{n}^i = (x^i - x_I^i)/r_I$ as before. p_k^I and s_{jk}^I determine the values of linear and angular momenta for each black hole.

For this given form of $\bar{A}_{ij}$, equation for ψ, (5.7), is solved with $K = 0 = \tilde{R}$ and $\tilde{\Delta} = \Delta_f$ by extending equation (5.83) as

$$\psi = 1 + \sum_{I=1}^{n} \left(\frac{a_I}{r_I}\right)^{N-2} + u, \tag{5.108}$$

where a_I is a constant of mass dimension again, and u is a new function. Since the second term of equation (5.108) is a homogeneous solution of $\Delta_f \psi = 0$, the equation for u becomes

$$\Delta_f u = -\frac{N-2}{4(N-1)} \psi^{(-3N+2)/(N-2)} \bar{A}^{ij} \bar{A}_{ij}. \tag{5.109}$$

Here, for $r_I \to 0$, $\bar{A}^{ij}\bar{A}_{ij}$ diverges as $r_I^{-2(N-1)}$ for $p_i^I \neq 0$ and $s_{ij}^I = 0$, and as r_I^{-2N} for $s_{ij}^I \neq 0$. On the other hand, $\psi^{(-3N+2)/(N-2)}$ approaches zero as r_I^{3N-2}. This implies that as long as $N \geq 3$, the right-hand side of equation (5.109) is guaranteed to be finite for the whole space covered by the chosen coordinate system. Therefore, equation (5.109) can be numerically solved simply imposing asymptotically flat boundary conditions for $r \to \infty$ ($u \to 0$). We also note that this approach can be used even in the presence of non-vanishing regular functions of $\bar{A}_{ij}^{tt}$.

As already mentioned in section 5.5.1.2, the spatial hypersurface is composed of $n + 1$ distinct world sheets when ψ is written in the form of equation (5.108), and $r_I = 0$ denotes spatial infinity in the I-th world sheet. Thus, in the chosen coordinate system, such spatial infinity is compactified, and hence, any numerical solution for ψ cannot have the resolution for the region of $r_I \ll a_I$. However, this shortcoming does not matter as far as we are interested in our world, i.e., outside black-hole horizons, for which a well-resolved numerical solution can be obtained.

5.5.2.3 Bowen – York initial data

A generalization of the Misner approach for time-asymmetric initial data was achieved by York and his collaborators. They considered spatial hypersurfaces composed of two isometric world sheets like figure 5.2, and derived an analytic solution for the extrinsic curvature that satisfies the isometric condition. Their approach is called a conformal-imaging method.

The first work in this approach was achieved by Bowen and York (1980) for a single black hole. Although they focused only on the three-dimensional case, $N = 3$, it is easy to generalize their analysis for any value of N, and hence, we here describe their approach for an arbitrary value of N.

For a single black hole, the isometric condition is defined by the fact that the extrinsic curvature is invariant for the transformation (5.81) except for the parity. Accordingly, they required the condition

$$K^{kl}(y^m) = \pm J^k{}_i J^l{}_j K^{ij}(x^m), \tag{5.110}$$

where

$$J^i{}_k := \frac{\partial y^i}{\partial x^k} = \left(\frac{|y^j|}{|x^j|}\right)(\delta^i{}_k - 2\bar{n}^i \bar{n}_k). \tag{5.111}$$

For ψ, conditions (5.89) and (5.91) should be required. Note that these conditions may be written as

$$\psi\left(x^i\right)^{2/(N-2)}|x^i| = \psi\left(y^i\right)^{2/(N-2)}|y^i|. \tag{5.112}$$

Thus, for $\bar{A}^{ij} := K^{ij}\psi^{2(N+1)/(N-2)}$ [which results from $\bar{A}^i{}_j = K^i{}_j\psi^{2N/(N-2)}$], the isometric condition is written as

$$\bar{A}^{ij}(y^m) = \pm\frac{\partial y^i}{\partial x^k}\frac{\partial y^j}{\partial x^l}\bar{A}^{kl}(x^m)\left(\frac{|x^m|}{|y^m|}\right)^{N+2}. \tag{5.113}$$

At their start point, a "seed" extrinsic curvature is written by equation (5.107), which contains information of linear and angular momenta of a black hole. It is found that the term composed of s_{ij} satisfies the condition (5.113) with minus sign irrespective of the dimensionality, whereas the term composed of p_i does not. To satisfy the isometric condition, Bowen and York (1980) added a term that satisfies equation (5.94). Specifically it is composed of a nonzero value of C_i^χ in χ (or $p_{ijk}f^{jk}$ in B_i) as

$$\chi = \pm\frac{p_i\bar{n}^i}{r^{N-1}}. \tag{5.114}$$

Then, we have

$$\bar{A}_{ij} = \frac{N(N-2)}{2(N-1)r^{N-1}}\Bigg[\Big(p_i\bar{n}_j + p_j\bar{n}_i + p_l\bar{n}_l\left[(N-2)\bar{n}_i\bar{n}_j - f_{ij}\right]\Big)$$
$$\mp\frac{a^2}{r^2}\Big(p_i\bar{n}_j + p_j\bar{n}_i - p_l\bar{n}_l\left[(N+2)\bar{n}_i\bar{n}_j - f_{ij}\right]\Big)\Bigg]$$
$$+\frac{N}{r^N}f^{kl}\left(s_{ik}\bar{n}_l\bar{n}_j + s_{jk}\bar{n}_l\bar{n}_i\right), \tag{5.115}$$

which satisfies the condition (5.113). Here, a denotes the coordinate radius of the minimal surface.

An important property for the inversion asymmetric solution with plus sign in equation (5.115) is that $\bar{A}_{ij}\bar{n}^i\bar{n}^j$ vanishes at the minimal surface $r = a$. This implies that the minimal surface becomes an apparent horizon (see section 7.1), and hence, we can determine the apparent horizon without solving its equation.

For a given isometric conformal extrinsic curvature $\bar{A}^{ij}$, the equation for ψ should be numerically solved imposing the isometric boundary condition (5.91) at $r = a$ and asymptotically flat boundary condition at $r \to \infty$ ($\psi \to 1$). Here the equation for ψ is

$$\Delta_f\psi = -\frac{N-2}{4(N-1)}\psi^{(-3N+2)/(N-2)}\bar{A}^{ij}\bar{A}_{ij}. \tag{5.116}$$

An extension of this Bowen–York approach for systems of multiple black holes was carried out by Kulkarni et al. (1983); Kulkarni (1984); Bowen et al. (1984); Cook (1991, 1994). Like a solution for ψ in the Misner initial data, a solution for $\bar{A}_{ij}$ is written as a superposition of infinite series of homogeneous solutions of the functional form (5.115). The first numerical solution for ψ was derived by Cook

(1991, 1994) by solving equation (5.116) employing such an infinite-series solution. He derived numerical solutions of initial data for two spinning black holes with $L_{(xy)}$ along z axis and for binary black holes in a quasi-circular orbit for the first time (see section 8.6).

The Bowen–York-type initial data are (coordinate) singularity-free while the puncture initial data are accompanied with coordinate singularities. On the other hand, numerically constructing Bowen–York-type initial data for multiple black holes is a little bit complicated because we have to take a sum of the infinite series and have to impose isometric boundary conditions at minimal surfaces of black holes. By contrast, constructing puncture initial data in numerical computation is much simpler. Before the puncture framework was discovered by Brandt and Brügmann (1997), we did not know how to handle the coordinate singularities in numerical relativity, and hence, employing the Bowen–York initial data were well motivated. However, after the development of the puncture and moving puncture approaches in numerical relativity (see section 2.3.3.1), it is found that we no longer have to worry about the coordinate singularities. As a result of this, the puncture approach is now more popular than the Bowen–York approach.

5.5.2.4 Remark for conformally flat initial data

It should be noted that conformally flat initial data do not provide Kerr and Myers–Perry black holes. Indeed, the spatial line element of these black holes is not written in the conformally flat form even in quasi-isotropic coordinates (see section 1.3.3). This fact is rigorously shown in the three spatial dimensions case.

The spatial line element of four-dimensional Kerr black holes in quasi-isotropic coordinates, $(\bar{r}, \theta, \varphi)$, is written as [Krivan and Price (1998)] (see section 1.3.1)

$$dl^2 = \frac{\Xi}{\bar{r}^2 \Sigma} \left[\frac{\Sigma^2}{\Xi} \left(d\bar{r}^2 + \bar{r}^2 d\theta^2 \right) + \bar{r}^2 \sin^2 \theta d\varphi^2 \right]. \tag{5.117}$$

This indicates that Kerr black holes may not be written in the conformally flat form in this slicing.

For proving that a three-dimensional hypersurface of spinning black holes is not conformally flat, Bach (or Cotton–York) tensor [York (1971); Gourgoulhon (2012)] is useful. This tensor is defined on three-dimensional spatial hypersurfaces, e.g., by

$$B^{ab} := 2\epsilon^{acd} D_d \left[R^b_{\ c} - \frac{1}{4} \delta^b_{\ c} R \right], \tag{5.118}$$

and vanishing of the Bach tensor is a necessary and sufficient condition for a three-geometry to be conformally flat. Garat and Price (2000) explicitly calculated the Bach tensor for the Kerr metric and showed that this is indeed nonzero. Therefore, spatial slices of Kerr spacetime are not conformally flat, and hence, the puncture and Bowen–York approaches cannot be used for preparing stationary spinning black holes exactly; in particular for a rapidly spinning black hole, the difference between conformally flat initial data and the exact stationary solution is significant (note that the spatial metric for non-spinning black holes is written in a conformally flat form, but with increasing the black-hole spin, the degree of the deviation from the conformal flatness monotonically increases). For realistic initial data of spinning black holes, we have to prepare conformally non-flat data.

5.5.2.5 Black hole plus gravitational waves

The Brill's formulation for axisymmetric initial data composed of pure gravitational waves described in section 5.4 can be generalized for black-hole spacetime [Abrahams et al. (1992); Brandt and Seidel (1995a,b, 1996)]. Here, we focus on the $N = 3$ case although it is possible to extend this method to the higher-dimensional case.

As shown in equation (5.117), the line element for initial data of spinning black holes is written in the form of equation (5.74) as

$$dl^2 = \psi_K^4 \left[e^{2q_K} (dr_K^2 + r_K^2 d\theta^2) + r_K^2 \sin^2 \theta d\varphi^2 \right], \tag{5.119}$$

where ψ_K and q_K are determined from equation (5.117). The metric of the form (5.117) is isometric with respect to the minimal surface (event horizon) located at $\bar{r} = \sqrt{M^2 - a^2}/2$ where M and a are mass and spin of the black hole (see section 1.3.1). Thus, we may assume to impose an inner boundary condition at this surface. In this case, no coordinate singularity is present in the spatial metric. Alternatively, we may use a puncture-like approach, because the coordinate singularity at $r_K = 0$ denotes spatial infinity on the other world sheet, as in section 5.5.2.2.

First, we consider a method for a solution of the momentum constraint. For Kerr black holes, the nonzero components of $K_i{}^j$ are $K_r{}^\varphi = e^{2q_K} K^r{}_\varphi/(r_K^2 \sin^2 \theta)$ and $K_\theta{}^\varphi = e^{2q_K} K^\theta{}_\varphi/\sin^2 \theta$, which are written as (see section 1.3.1)

$$(K_r{}^\varphi)_K = \frac{1}{2\alpha_K} \partial_r \beta_K, \qquad (K_\theta{}^\varphi)_K = \frac{1}{2\alpha_K} \partial_\theta \beta_K, \tag{5.120}$$

where α_K and β_K are the lapse function and φ component of the shift vector of the Kerr black hole;

$$\alpha_K = \sqrt{\frac{\Sigma_K \Delta_K}{(r_K^2 + a^2)\Sigma_K + 2Ma^2 r_K \sin^2 \theta}}, \tag{5.121}$$

$$\beta_K = -\frac{2Mar_K}{(r_K^2 + a^2)\Sigma_K + 2Ma^2 r_K \sin^2 \theta}. \tag{5.122}$$

For this case, the φ component of the momentum constraint is only one nontrivial component, which is written as

$$\frac{1}{r_K^2} \partial_{r_K}(r_K^4 K_1) + \frac{1}{\sin^3 \theta} \partial_\theta (\sin^3 \theta K_2) = 0, \tag{5.123}$$

where $K_1 := \psi_K^6 (K_r{}^\varphi)_K$ and $K_2 := \psi_K^6 (K_\theta{}^\varphi)_K$.

Now we consider systems composed of a black hole and gravitational waves. Then, if $\psi^6 K_r{}^\varphi$ and $\psi^6 K_\theta{}^\varphi$ are given by K_1 and K_2 defined above (with replacement of r_K to r) and other components of the extrinsic curvature are supposed to be zero, the momentum constraint is automatically satisfied. Thus, by employing K_1 and K_2 as $\psi^6 K_r{}^\varphi$ and $\psi^6 K_\theta{}^\varphi$, we do not have to solve the momentum constraint.

Then, the Hamiltonian constraint equation for the line element given by equation (5.74) is written as

$$\left[\frac{\partial^2}{\partial r^2} + \frac{2}{r}\frac{\partial}{\partial r} + \frac{1}{r^2}\left(\frac{\partial^2}{\partial \theta^2} + \cot\theta \frac{\partial}{\partial \theta}\right)\right]\psi = -\frac{\psi}{4}f_q(q) - \frac{(r^2 K_1^2 + K_2^2)\sin^2\theta}{4\psi^7}, \tag{5.124}$$

where

$$f_q(q) := \left(\frac{\partial^2}{\partial r^2} + \frac{1}{r}\frac{\partial}{\partial r} + \frac{1}{r^2}\frac{\partial^2}{\partial \theta^2}\right)q. \tag{5.125}$$

The simplest choice of q is

$$q = q_K + q_{gw}, \tag{5.126}$$

where q_{gw} denotes the contribution of gravitational waves and for $q_{gw} = 0$, a Kerr black hole is recovered. Then, decomposing ψ as $\psi = \psi_K + \psi_{gw}$, we have

$$\left[\frac{\partial^2}{\partial r^2} + \frac{2}{r}\frac{\partial}{\partial r} + \frac{1}{r^2}\left(\frac{\partial^2}{\partial \theta^2} + \cot\theta \frac{\partial}{\partial \theta}\right)\right]\psi_{gw}$$
$$= -\frac{\psi_{gw}}{4}f_q(q_K + q_{gw}) - \frac{\psi_K}{4}f_q(q_{gw}) - \frac{1}{4}\left(\psi^{-7} - \psi_K^{-7}\right)(r^2 K_1^2 + K_2^2)\sin^2\theta. \tag{5.127}$$

Thus, initial data are obtained by solving equation (5.127) for a given function of q_{gw}. (If a spatial hypersurface should be isometric with respect to the minimal surface, we have to impose an additional condition $\partial_r q_{gw} = 0$ there).

5.6 Isenberg–Wilson–Mathews/Conformal thin-sandwich formulation

In the conformal transverse-tracefree formulation described in section 5.2, initial data are determined paying attention only to an initial spatial hypersurface, i.e., taking only the constraint equations into consideration. A shortcoming of this approach is that we can determine only N components among $N(N+1)$ components for (γ_{ab}, K_{ab}), although we would like to determine all these components for initial data, as well as α and β^a, in a physically motivated way. Isenberg–Wilson–Mathews (IWM) formulation [Isenberg (2008); Wilson and Mathews (1989, 1995); Wilson et al. (1996)] and/or conformal thin-sandwich (CTS) formulation [York (1999); Pfeiffer and York (2003)] are formulations that take into account a part of the evolution equations and thus can determine additional components, by specifying conditions for several components of $\partial_t \gamma_{ij}$ and $\partial_t K_{ij}$.

As in the conformal transverse-tracefree decomposition formulation, the IWM and CTS formulations employ the conformal decomposition of the spatial metric as $\gamma_{ij} = \psi^{4/(N-2)}\tilde{\gamma}_{ij}$. Then, ψ is supposed to be determined by the Hamiltonian constraint (2.95), while $\partial_t \tilde{\gamma}_{ij}$ as well as $\tilde{\gamma}_{ij}$ are supposed to be given by imposing additional conditions. Also, the extrinsic curvature is decomposed into the tracefree and trace parts as $K_{ij} = A_{ij} + \gamma_{ij} K/N$. Then, the momentum constraint is used for determining some of components in A_{ij}, and the evolution equation of K are used for determining the lapse function by imposing a slicing condition.

Although these formulations are obviously better than the conformal transverse-tracefree decomposition formulation, the full set of Einstein's equation is not still solved; the basic equations are derived basically using the constraint equations. Hence, we here classify them as formulations for initial data (not for equilibria and quasi-equilibria). Nevertheless, the IWM and CTS formulations have been extensively used for computing (approximately) quasi-equilibrium initial data for a wide variety of binary systems composed of neutron stars and black holes [see, e.g., Shibata and Taniguchi (2011); Faber and Rasio (2012) for reviews of black hole-neutron star binaries and binary neutron stars, respectively, and Gourgoulhon et al. (2002); Cook and Pfeiffer (2004); Caudill et al. (2006); Lovelace et al. (2008) for binary black holes].

As the readers find below, the IWM and CTS formulations are similar to each other. However, there are slight differences between them. Thus, we describe each formulation separately as follows.

5.6.1 Isenberg–Wilson–Mathews formulation

The IWM formulation was originally proposed as an approximate formulation of Einstein's equation by Isenberg in 1978 [the paper on this was published thirty-years later as Isenberg (2008)], and independently by Wilson and Mathews (1989, 1995); Wilson et al. (1996) aiming at computing binary neutron stars in quasi-equilibrium.

The IWM formulation is characterized by the conditions imposed as follows:

$$K = 0 = \partial_t K \quad \text{(maximal slicing condition)}, \tag{5.128}$$

$$\tilde{\gamma}_{ij} = f_{ij}, \quad \partial_t \tilde{\gamma}_{ij} = 0 \quad \text{(spatially conformal flatness)}. \tag{5.129}$$

The maximal slicing condition yields an elliptic equation for α as shown in section 2.2.2. Specifically, combining the Hamiltonian constraint equation, Wilson and Mathews (1989) employed the equation for $\alpha\psi$ [see equation (2.114)] as

$$\Delta_f(\alpha\psi) = \frac{4\pi}{N-1}\left[(N-2)\rho_h + 2 S_k{}^k\right]\alpha\psi^{(N+2)/(N-2)} + \frac{3N-2}{4(N-1)}\tilde{A}_{ij}\tilde{A}^{ij}\alpha\psi^{(N+2)/(N-2)}, \tag{5.130}$$

where $\tilde{A}_{ij}$ is given by equation (5.132), which is shown below.

The equation for ψ is derived from the Hamiltonian constraint, and in this formulation, it is written as [cf. equations (2.95) and (5.7)]

$$\Delta_\mathrm{f}\psi = -\frac{4(N-2)\pi}{N-1}\rho_\mathrm{h}\psi^{(N+2)/(N-2)}$$
$$-\frac{(N-2)}{4(N-1)}\tilde{A}_{ij}\tilde{A}^{ij}\psi^{(N+2)/(N-2)}. \quad (5.131)$$

Thus, from equations (5.130) and (5.131), α as well as ψ are determined.

The condition (5.129) determines the relation between the shift vector and $\tilde{A}_{ij}$ through the evolution equation for $\tilde{\gamma}_{ij}$, i.e., equation (2.83), as

$$\tilde{A}_{ij} = \frac{1}{2\alpha}\left(\overset{(0)}{D}_i\tilde{\beta}_j + \overset{(0)}{D}_j\tilde{\beta}_i - \frac{2}{N}f_{ij}\overset{(0)}{D}_k\tilde{\beta}^k\right), \quad (5.132)$$

where $\tilde{\beta}_i = f_{ij}\beta^j$ ($\tilde{\beta}^i = \beta^i$). Substituting equation (5.132) into the momentum constraint (5.4) with the condition (5.129) yields an elliptic equation for β^i as

$$\Delta_\mathrm{f}\beta^i + \frac{N-2}{N}\overset{(0)}{D}{}^i\overset{(0)}{D}_j\beta^j + 2\alpha\tilde{A}^{ij}\left(\frac{2N}{N-2}\overset{(0)}{D}_j\ln\psi - \overset{(0)}{D}_j\ln\alpha\right) = 16\pi\alpha J_j f^{ij}. \quad (5.133)$$

Thus, a longitudinal part of the extrinsic curvature is determined in a manner different from the conformal transverse-tracefree decomposition formulation, in which it is determined by introducing an auxiliary function W_i. A merit in the IWM approach is that β^i is automatically obtained.

To summarize, in this formulation, α, β^i, and ψ are determined by solving elliptic equations for them, while the conformal metric $\tilde{\gamma}_{ij}$ is assumed to be f_{ij}. The tracefree part of the extrinsic curvature is determined from the evolution equation for $\tilde{\gamma}_{ij}$ with the assumption $\partial_t\tilde{\gamma}_{ij} = 0$ substituting the solutions of α and β^i. The trace part of the extrinsic curvature is set to be zero (maximal slicing is employed).

5.6.2 Conformal thin-sandwich formulation

In the original version of the CTS formulation, a class of initial data, which is wider than that in the conformal transverse-tracefree decomposition formulation, is given by setting

$$\partial_t\tilde{\gamma}_{ij} = u_{ij}, \quad (5.134)$$

where u_{ij} as well as $\tilde{\gamma}_{ij}$ are supposed to be arbitrary functions. In this formulation, the conformal flatness is not always assumed. Rather, non-conformally flat spatial metric such as Kerr–Schild-like metric (see section 1.3.1) is often employed as a background metric for $\tilde{\gamma}_{ij}$ [e.g., Lovelace et al. (2008)].

For a given function of u_{ij}, equation (2.83) yields the relation between the shift vector and $\tilde{A}_{ij}$ as

$$\tilde{A}_{ij} = \frac{1}{2\alpha}\left(-u_{ij} + \tilde{D}_i\tilde{\beta}_j + \tilde{D}_j\tilde{\beta}_i - \frac{2}{N}\tilde{\gamma}_{ij}\tilde{D}_k\tilde{\beta}^k\right). \quad (5.135)$$

As in the IWM formulation, substituting equation (5.135) into the momentum constraint (5.4) yields the equation for β^i:

$$\tilde{\Delta}\beta^i + \frac{N-2}{N}\tilde{D}^i\tilde{D}_j\beta^j + \tilde{R}^i{}_k\beta^k + 2\alpha\tilde{A}^{ij}\left(\frac{2N}{N-2}\tilde{D}_j\ln\psi - \tilde{D}_j\ln\alpha\right)$$
$$- \tilde{D}_j(\tilde{\gamma}^{jk}\tilde{\gamma}^{il}u_{kl}) - \frac{2(N-1)}{N}\alpha\tilde{D}^i K = 16\pi\alpha J_j\tilde{\gamma}^{ij}. \qquad (5.136)$$

In an extended CTS formulation proposed by Pfeiffer and York (2003), the conditions for K and $\partial_t K$ are further imposed; they are supposed to be given functions and not always to vanish. Then, the evolution equation for K reduces to an elliptic equation for α. Specifically, elliptic equations for ψ and $\alpha\psi$ in this formulation are written by

$$\tilde{\Delta}\psi = \frac{(N-2)\psi}{4(N-1)}\tilde{R} - \frac{4(N-2)\pi}{N-1}\rho_{\rm h}\psi^{(N+2)/(N-2)}$$
$$- \frac{(N-2)\psi^{(N+2)/(N-2)}}{4(N-1)}\left(\tilde{A}_{ij}\tilde{A}^{ij} - \frac{N-1}{N}K^2\right), \qquad (5.137)$$

$$\tilde{\Delta}(\alpha\psi) = \frac{(N-2)\alpha\psi}{4(N-1)}\tilde{R} + \frac{4\pi}{N-1}\Big[(N-2)\rho_{\rm h} + 2S_k{}^k\Big]\alpha\psi^{(N+2)/(N-2)}$$
$$+ \left(\frac{3N-2}{4(N-1)}\tilde{A}_{ij}\tilde{A}^{ij} + \frac{N+2}{4N}K^2\right)\alpha\psi^{(N+2)/(N-2)}$$
$$- \psi^{(N+2)/(N-2)}\left(\partial_t - \beta^l\partial_l\right)K. \qquad (5.138)$$

For the case that $K = 0 = \partial_t K$ and $\tilde{\gamma}_{ij} = f_{ij}$, the basic equations in this formulation are the same as those in the IWM formulation.

5.6.3 *Handling black-hole horizon*

Many attempts have been done for constructing initial data of black-hole binary systems in quasi-equilibrium states in four-dimensional spacetime employing the IWM and/or CTS formulations, since the first works by Gourgoulhon *et al.* (2002); Grandclément *et al.* (2002); Cook (2002). For constructing realistic initial data, it is crucial to impose physical and geometrical boundary conditions on a horizon surface (see appendix G for several important properties of black holes). We here summarize the boundary conditions proposed by Cook (2002); Cook and Pfeiffer (2004), the so-called *excision boundary condition*, which are derived in the framework of an isolated horizon formulation developed and reviewed in Ashtekar *et al.* (1999, 2000a,b); Ashtekar and Krishnan (2004); Gourgoulhon and Jaramillo (2006) (see also appendix G.6), restricting our attention only to the case $D = 4$ ($N = 3$). Their method has been widely used for computing binary black holes [Caudill *et al.* (2006); Lovelace *et al.* (2008)] and black hole-neutron stars binaries in quasi-equilibrium [Grandclément (2006, 2007); Taniguchi *et al.* (2007, 2008)] (see sections 8.4 and 8.6).

Cook (2002) and Cook and Pfeiffer (2004) proposed to solve the field equations of initial-data problem in the framework of the extended CTS (or IWM) formulation described in section 5.6.2: They supposed to solve elliptic-type equations for ψ, $\alpha\psi$, and β^i, imposing inner boundary conditions on black-hole horizons. Here, a slicing condition, $\partial_t K = 0$, is imposed for fixing the equation for $\alpha\psi$. In their method, the black holes are assumed to be characterized by a non-expanding horizon (see appendix G.6 for the definition of the non-expanding horizon).[3] Then, they proposed to impose the boundary conditions that the expansion and shear of an outgoing null vector field characterizing the non-expanding horizons vanish. In addition, they required a local presence of a helical Killing vector field, k_h^a [see equation (5.228)] assuming that the spacetime in the vicinity of the black holes is momentarily in an equilibrium state in the frame comoving with the orbital angular velocity $\Omega_{\rm orb}$.[4] The local presence of k_h^a is used for imposing a boundary condition on the closed two surface of an apparent horizon $\mathcal{S}$ (which is a spacelike cross section of non-expanding horizons).

Let s^a be an outward-pointing unit spatial vector field normal to the two-surface $\mathcal{S}$. Then, using the timelike unit vector field normal to Σ_t, n^a, as well as s^a, outgoing and ingoing null vector fields are defined, respectively, by

$$k^a := \frac{1}{\sqrt{2}}(n^a + s^a), \quad l^a := \frac{1}{\sqrt{2}}(n^a - s^a). \quad (5.139)$$

Here, k^a is supposed to be normal to a non-expanding horizon, and the two null vector fields satisfy $k^a l_a = -1$. Denoting the induced metric on $\mathcal{S}$ by $H_{ab}(:= \gamma_{ab} - s_a s_b)$, the expansion and shear of k^a are defined by (see appendix G)

$$\Theta := H^{ab}\nabla_a k_b, \quad \sigma_{ab} := H_a{}^c H_b{}^d \nabla_c k_d - \frac{1}{2}\Theta H_{ab}. \quad (5.140)$$

Here, $\sigma_{ab} = \sigma_{ba}$ for the definition of equation (5.140) because k^a is normal to the non-expanding horizon (i.e., due to the absence of the rotation tensor part of $\nabla_a k_b$ according to Frobenius' theorem; see appendix A.2). The conditions, $\Theta = 0$ and $\sigma_{ab} = 0$, are used as boundary conditions for ψ and β^a, as shown in the following. Note that a closed two-surface with $\Theta = 0$ on each spatial hypersurface is an apparent horizon (see section 7.1).

The equation $\Theta = 0$ yields

$$D_a s^a - K_{ab} H^{ab} = 0. \quad (5.141)$$

Defining $\tilde{s}^a := \psi^2 s^a$ for which $\tilde{\gamma}_{ab}\tilde{s}^a\tilde{s}^b = 1$, equation (5.141) is written as

$$\tilde{s}^a \tilde{D}_a \ln \psi \Big|_{\mathcal{S}} = \frac{1}{4}\left(\psi^2 K_{ab}H^{ab} - \tilde{D}_a\tilde{s}^a\right)\Big|_{\mathcal{S}}. \quad (5.142)$$

---

[3] The non-expanding horizons do not have to agree with event horizons, but it is natural to consider that the corresponding event horizons are located outside the non-expanding horizons (see appendix G).

[4] In the global presence of the helical Killing vector field, the event horizon should be a Killing horizon satisfying $g_{ab}k_h^a k_h^b = 0$. In the formulation of Cook (2002) and Cook and Pfeiffer (2004), this condition is not required for the non-expanding horizons because they do not have to agree with event horizons. Since a non-expanding horizon agree with an the event horizon or is located inside an event horizon, $g_{ab}k_h^a k_h^b$ should be zero or positive.

This may be considered as the boundary condition of ψ on $\mathcal{S}$.

Next, we assume that the non-expanding horizons are unchanged along the integration curve of the helical Killing vector field locally defined. Then, we have $k_h^a k_a = 0$ on the non-expanding horizons. Here, the helical Killing vector field is written as

$$k_h^a = \alpha n^a + \beta^a + \Omega_{\rm orb}\varphi^a, \tag{5.143}$$

where $\varphi^a = (\partial/\partial\varphi)^a$. Thus, the condition, $k_h^a k_a = 0$, is written as

$$\alpha\big|_{\mathcal{S}} = s_a(\beta^a + \Omega_{\rm orb}\varphi^a)\big|_{\mathcal{S}}. \tag{5.144}$$

This imposes the condition between the lapse function α and a component of β^a, which is perpendicular to the two-surface $\mathcal{S}$. Following Cook and Pfeiffer (2004), we denote $s_a(\beta^a + \Omega_{\rm orb}\varphi^a)$ by $\beta_\perp$. Equation (5.144) will be used as the condition for determining $\beta_\perp$ when the condition for α on $\mathcal{S}$ is imposed (see below).[5]

It is also necessary to impose a boundary condition for the other component of the shift vector[6]

$$\beta_\parallel^a := \beta^a + \Omega_{\rm orb}\varphi^a - s^a \beta_\perp = H^a_{\ b}\left(\beta^b + \Omega_{\rm orb}\varphi^b\right). \tag{5.145}$$

The boundary condition for this is derived from the condition that the shear tensor on $\mathcal{S}$ vanishes. For deriving this, we write the extrinsic curvature in the following form

$$K_{ab} = \frac{1}{2\alpha}\left(D_a\beta_b + D_b\beta_a - \partial_t\gamma_{ab}\right)$$
$$= \frac{1}{2\alpha}\left[D_a(\beta_b + \Omega_{\rm orb}\varphi_b) + D_b(\beta_a + \Omega_{\rm orb}\varphi_a) - (\partial_t + \Omega_{\rm orb}\partial_\varphi)\gamma_{ab}\right]. \tag{5.146}$$

Since k_h^a is a helical Killing vector field in the vicinity of the non-expanding horizons, the last term of equation (5.146) vanishes near $\mathcal{S}$ as

$$(\partial_t + \Omega_{\rm orb}\partial_\varphi)\gamma_{ab} = \mathcal{L}_{k_h}\gamma_{ab} = 0. \tag{5.147}$$

An explicit form of the shear tensor is derived from the following manipulation:

$$\sqrt{2}H_a^{\ c}H_b^{\ d}\nabla_c k_d = H_a^{\ c}H_b^{\ d}\nabla_c(n_d + s_d)$$
$$= -H_a^{\ c}H_b^{\ d}K_{cd} + H_a^{\ c}D_c s_b$$
$$= -\frac{1}{2\alpha}H_a^{\ c}H_b^{\ d}\left[D_c(\beta_d + \Omega_{\rm orb}\varphi_d) + D_d(\beta_c + \Omega_{\rm orb}\varphi_c)\right] + L_{ab}$$
$$= -\frac{1}{2\alpha}H_{ac}H_{bd}\left(\overset{H}{D}{}^c\beta_\parallel^d + \overset{H}{D}{}^d\beta_\parallel^c\right) + L_{ab}\left(1 - \alpha^{-1}\beta_\perp\right), \tag{5.148}$$

where $L_{ab} := H_a^{\ c}D_c s_b$ and we used

$$H^{ac}H^{bd}D_c\beta_d = \beta_\perp L^{ab} + \overset{H}{D}{}^a\beta_\parallel^b. \tag{5.149}$$

---
[5]Cook and Pfeiffer (2004) derived their formulation in the corotating frame in which $\beta^a + \Omega_{\rm orb}\varphi^a$ in this volume is written as β^a. Here, we employ the inertial frame.

[6]We note that for $\beta_\parallel^a = 0$, k_h^a becomes null on the non-expanding horizon. Here, we do not suppose this condition.

Here, $\overset{H}{D}_a$ denotes the covariant derivative associated with H_{ab} on $\mathcal{S}$. Thus, the shear tensor on $\mathcal{S}$ is obtained as

$$\sigma_{ab} = -\frac{1}{2\sqrt{2}\alpha} H_{ac} H_{bd} \left(\overset{H}{D}{}^c \beta_\parallel^d + \overset{H}{D}{}^d \beta_\parallel^c - H^{cd} \overset{H}{D}_e \beta_\parallel^e \right)$$
$$+ \frac{1}{\sqrt{2}} \left(L_{ab} - \frac{1}{2} H_{ab} H^{cd} L_{cd} \right) (1 - \alpha^{-1}\beta_\perp), \quad (5.150)$$

where the second term vanishes due to the boundary condition $\beta_\perp = \alpha$ on $\mathcal{S}$ that results from $k_h^a k_a = 0$. Therefore, the first term yields the boundary condition for $\beta_\parallel^a$ as

$$\left. \left(\overset{H}{D}{}^c \beta_\parallel^d + \overset{H}{D}{}^d \beta_\parallel^c - H^{cd} \overset{H}{D}_e \beta_\parallel^e \right) \right|_\mathcal{S} = 0. \quad (5.151)$$

This equation implies that $\beta_\parallel^a$ has to be a conformal Killing vector field on $\mathcal{S}$, and thus, the same equation is satisfied for the conformal two metric as (see section A.1),

$$\left. \left(\overset{\tilde{H}}{D}{}^c \beta_\parallel^d + \overset{\tilde{H}}{D}{}^d \beta_\parallel^c - \tilde{H}^{cd} \overset{\tilde{H}}{D}_e \beta_\parallel^e \right) \right|_\mathcal{S} = 0, \quad (5.152)$$

where $H_{ab} = \psi^4 \tilde{H}_{ab}$. Following Cook and Pfeiffer (2004), it is convenient to set

$$\beta_\parallel^a = \Omega_\mathrm{rot} \tilde{\xi}^a, \quad (5.153)$$

where $\tilde{\xi}^a$ is a conformal Killing vector field on $\mathcal{S}$, and Ω_rot denotes an arbitrary constant used for specifying black-hole spins: For the corotating case (the angular velocity of black holes is equal to that of the orbital motion, Ω_orb), $\Omega_\mathrm{rot} = 0$, while for other cases, it does not vanish. For non-spinning black holes, $\Omega_\mathrm{rot} \approx -\Omega_\mathrm{orb}$ with $\tilde{\xi}^i$ being associated with the axis of the orbital angular momentum although the equality between Ω_rot and $-\Omega_\mathrm{orb}$ is not satisfied exactly [Caudill et al. (2006)]. Specifically, the spin of black holes is determined using a formulation in the isolated horizon framework. For the definition of the spin angular momentum, see appendix G.6.[7]

Finding $\tilde{\xi}^i$ is in general non-trivial. However, for a conformally flat spatial metric and for an excision sphere being a coordinate sphere, it is simply written by $\tilde{\xi}^i = \tilde{\epsilon}^i{}_{jk} e^j x^k$ where e^j is a unit vector and $\tilde{\epsilon}_{ijk}$ is a completely antisymmetric tensor on the flat conformal space.

The final task in this excision-boundary formulation is to impose a boundary condition for $\alpha\psi$ for which the basic equation is derived by imposing maximal slicing. As Cook and Pfeiffer (2004) perceived, the boundary condition on the horizon is not uniquely determined in maximal slicing: As described in section 2.2.2, there are

---
[7] We note that a formulation proposed by Gourgoulhon et al. (2002) is similar to that by Cook (2002); Cook and Pfeiffer (2004), but it was developed for a special class of spacetime (corotating binaries) in which $\beta_\perp = 0 = \alpha$ and $\beta_\parallel^a = 0$ on $\mathcal{S}$. In this formulation, $\mathcal{S}$ is supposed to be a Killing horizon.

infinite number of maximal slicing of a Schwarzschild black hole depending on the time parameter $T(t)$, and if $T(t)$ is different, the value of α on the event horizon is different. Thus we may choose a variety of boundary conditions for $\alpha\psi$ on $\mathcal{S}$.[8] Cook and Pfeiffer (2004) chose three conditions,

$$\tilde{s}^a \tilde{D}_a(\alpha\psi) = 0, \quad \tilde{s}^a \tilde{D}_a \left[r(\alpha\psi)^2\right] = 0, \quad \alpha\psi = 1/2, \qquad (5.154)$$

and the popular choice is the first one; see, e.g., Taniguchi et al. (2007, 2008); Lovelace et al. (2008). When α on $\mathcal{S}$ is determined by one of these conditions, $\beta_\perp$ is determined using equation (5.144).

5.6.4 IWM plus conformal transverse-tracefree formulation

Another way for handling black holes is to employ the puncture approach described in section 5.5.2.2. In this approach, we do not have to take care of boundary conditions on black-hole horizons. Shibata and Uryū (2006, 2007) developed a formulation in three spatial dimensions, in which the puncture approach is incorporated into the IWM formulation.

In this formulation, the basic equations are composed of the equations in the IWM formulation, (5.133) for β^i, (5.131) for ψ, (5.130) for $\alpha\psi$, and (5.4) for $\bar{A}_{ab}$. However, equation (5.132) is not used for determining $\tilde{A}_{ij}$. For solving equations (5.131), (5.130), and (5.4), an extended puncture approach is used. Thus, we set

$$\psi = 1 + \frac{M_P}{2r_{\text{BH}}} + u, \qquad (5.155)$$

$$\alpha\psi = 1 - \frac{M_\alpha}{r_{\text{BH}}} + v, \qquad (5.156)$$

$$\bar{A}_{ij} \left(= \bar{A}_i{}^k f_{jk}\right) = \overset{(0)}{D}_i W_j + \overset{(0)}{D}_j W_i - \frac{2}{3} f_{ij} f^{kl} \overset{(0)}{D}_k W_l + K^{\text{P}}_{ij}, \qquad (5.157)$$

where M_P and M_α are constants of mass dimension, which are related to the ADM mass and the Komar mass, respectively; $r_{\text{BH}} = |x^i - x^i_P|$ with x^i_P being the location of the puncture; u and v are scalar functions to be numerically determined; $W_i (= f_{ij} W^j)$ is an auxiliary three-dimensional function; and K^{P}_{ij} denotes a weighted extrinsic curvature component associated with the linear momentum and spin of a black hole and it is written by

$$K^{\text{P}}_{ij} = \frac{3}{2r_{\text{BH}}^2} \left(\bar{n}_i P^{\text{BH}}_j + \bar{n}_j P^{\text{BH}}_i + (\bar{n}_i \bar{n}_j - f_{ij}) P^{\text{BH}}_k \bar{n}^k \right)$$

$$+ \frac{3}{r_{\text{BH}}^3} f^{kl} \left(S^{\text{BH}}_{ik} \bar{n}_l \bar{n}_j + S^{\text{BH}}_{jk} \bar{n}_l \bar{n}_i \right). \qquad (5.158)$$

Here, $\bar{n}^i = (x^i - x^i_P)/r_{\text{BH}}$, and P^{BH}_i and S^{BH}_{ij} denote the linear and spin angular momenta of the black hole.

---
[8] However, we cannot choose, e.g., $\alpha\psi = 0 = \beta_\perp$ on $\mathcal{S} = 0$, which were employed in Gourgoulhon et al. (2002); Grandclément et al. (2002), because this condition violates the regularity on $\mathcal{S}$ [Cook (2002)].

Then, equations (5.131), (5.130), and (5.4) are rewritten to elliptic equations for u, v, and $W_i (= f_{ij} W^j)$ as

$$\Delta_f u = -2\pi \rho_h \psi^5 - \frac{1}{8} \bar{A}_i{}^j \bar{A}_j{}^i \psi^{-7}, \tag{5.159}$$

$$\Delta_f v = 2\pi \alpha \psi \left[\psi^4 \left(\rho_h + 2 S_k{}^k \right) + \frac{7}{16\pi} \psi^{-8} \bar{A}_i{}^j \bar{A}_j{}^i \right], \tag{5.160}$$

$$\Delta_f W_i + \frac{1}{3} \overset{(0)}{D}_i \overset{(0)}{D}_k W^k = 8\pi J_i \psi^6. \tag{5.161}$$

In addition, the equation for β^i is written in the following form

$$\Delta_f \beta^i + \frac{1}{3} \overset{(0)}{D}^i \overset{(0)}{D}_j \beta^j + 2\alpha \psi^{-6} \bar{A}^{ij} \left(6 \overset{(0)}{D}_j \ln \psi - \overset{(0)}{D}_j \ln \alpha \right) = 16\pi \alpha J_j f^{ij}. \tag{5.162}$$

This formulation has been applied to computing initial data of black hole-neutron star binaries in quasi-equilibrium states [Shibata and Uryū (2006, 2007); Kyutoku et al. (2009); Shibata and Taniguchi (2011)]. In this case, the density and velocity field configurations of the neutron star are determined by solving a hydrostatic equation (see section 5.7.2.2), and then, ρ_h, J_i, and $S_k{}^k$ are supplied.

For a solution of gravitational-field equations for u, v, W_i, and β^i, we have to specify M_P, M_α, P_i^{BH}, and S_{ij}^{BH}. M_P is determined when we specify the black-hole mass. M_α is determined by the virial relation: We suppose that the ADM mass and the Komar mass agree with each other:

$$\oint_{r \to \infty} \partial_i(\alpha \psi) f^{ij} \overset{(0)}{dS}_j = - \oint_{r \to \infty} \partial_i \psi f^{ij} \overset{(0)}{dS}_j = 2\pi M_{\text{ADM}}. \tag{5.163}$$

We note that the virial relation does not have to be exactly satisfied for the initial data in this formulation, but it is chosen as a condition. P_i^{BH} is determined by the requirement that the total ADM linear momentum is zero in this system, which results in

$$-P_i^{\text{BH}} = \int J_i \psi^6 d^3 x, \tag{5.164}$$

where the right-hand side denotes the linear momentum of the neutron star, and we used equation (5.28) for this. Finally, the black-hole spin, S_{ij}^{BH}, is arbitrary. More detailed numerical methods for a solution of quasi-equilibria will be described in section 5.7.2.

5.6.5 Using excision initial data of black holes in the time evolution

As we described in the previous sections, there are two approaches for preparing initial data composed of black holes. In one approach, the initial data are computed imposing boundary conditions on apparent horizons (non-expanding horizons). In this case, the domain inside the apparent horizons is excised and no data are determined there. This approach is often employed for computing initial data of binary black holes and black hole-neutron star binaries in quasi-equilibrium. The other approach is the puncture approach. In this case, the initial data are determined for the entire spatial hypersurface.

In numerical evolution, there are also two approaches for evolving black-hole spacetime: excision and moving puncture approaches (see section 2.2.4). In the GH (generalized harmonic) formulation (see section 2.3.4), the excision approach is often employed, while the moving puncture approach is usually employed in the BSSN formulation.

The initial data prepared by the excision approach can be evolved in the excision formulation in a straightforward manner. This is also the case when the initial data prepared by the puncture approach are evolved in the moving puncture formulation. Then, a question is how we evolve the initial data, prepared in the excision approach, in the moving puncture approach, because no data are present inside the apparent horizons. Etienne et al. (2007) and Brown et al. (2009) gave a simple answer for this question based on the following fact: As long as we are interested in a spacetime region outside the horizon, we can artificially fill initial data inside the apparent horizons. The reason for this is that the information inside the apparent horizons does not escape to their outside. Thus, only the necessary condition is that the artificial data are regular and regularly matched to the data outside the apparent horizon: They do not have to satisfy the constraint equations (nor any equation). Of course, we have to guarantee the causality near the horizons in the finite-differencing level, and a careful implementation of filling the horizon is necessary [Etienne et al. (2007); Brown et al. (2009)]. However, that is not a particularly difficult task and we only need to employ a sufficiently accurate interpolation procedure that is often used.

5.7 Formulation of equilibrium and quasi-equilibrium

In sections 5.2, 5.5, and 5.6, we introduced formulations for a solution of initial data in general relativity, paying particular attention to a solution of the constraint equations in general relativity. The introduced formulations, however, do not provide a full set of equations for stationary or quasi-stationary solutions, because the full set of Einstein's equation is not used. For a scientific work, we have to prepare realistic initial data that are likely to be realized in nature. For example, for exploring the properties of neutron stars such as their stability and their oscillation properties, we have to prepare an equilibrium state (only with a tiny perturbation). For exploring the *late* inspiral and merger phases of coalescing compact binaries, such as binary neutron stars and binary black holes, we have to prepare a realistic quasi-equilibrium state as the initial condition.[9] These examples indicate that we

---

[9]To avoid possible misunderstanding, we note as follows: For *distant* orbits of coalescing compact binaries, the IWM and CTS formulations described in section 5.6 could be used for generating realistic quasi-equilibrium states because the neglected terms in these formulations are likely to be

have to prepare an equilibrium or a realistic quasi-equilibrium state as the initial condition for scientific numerical-relativity works. In particular, it is crucial to prepare an initial condition as realistic as possible for deriving accurate gravitational waveforms that could be used as theoretical templates for the gravitational-wave data analysis (see section 1.6).

Motivated by the facts mentioned above, in this section, we will describe formulations for computing equilibrium and realistic quasi-equilibrium states in numerical relativity. For spherical stars, we have only to solve the TOV equations as described in section 1.4.3. In this section, we pay attention to formulations for axisymmetric equilibria (e.g., rotating neutron stars) and for non-axisymmetric quasi-equilibria (e.g., binary compact objects in close orbits), focusing only on the case $D = 4$.

5.7.1 Axisymmetric equilibria

5.7.1.1 Formulations for gravitational fields

The formulation for stationary axisymmetric stars composed of a perfect fluid with no meridian circulation flow was established in the 1970s [Carter (1970); Bardeen (1970); Bardeen and Wagoner (1971); Butterworth and Ipser (1976)]. Here, we call spacetime stationary and axisymmetric if they possess both a timelike Killing vector field $\xi^a = (\partial/\partial t)^a$ and a spacelike Killing vector field $\varphi^a = (\partial/\partial \varphi)^a$ whose integral curves are closed, and in addition, if these two Killing vector fields commute [Carter (1970)]

$$\xi^a \nabla_a \varphi^b = \varphi^a \nabla_a \xi^b. \tag{5.165}$$

Here, equation (5.165) means that t and φ can be employed as coordinates, and the spacetime metric is written as a function of other coordinates (x^2, x^3) as $g_{\mu\nu}(x^2, x^3)$.

In general, Einstein's equation is a partial differential equation for ten components of $g_{\mu\nu}$, and even if we could employ a special coordinate condition, at least six components have to be determined by the field equations for a solution of general axisymmetric spacetime [see an attempt by Gourgoulhon and Bonazzola (1993); Birkl et al. (2011)]. However, the situation can be simplified drastically if we further assume the *circularity condition* to the stress-energy tensor, which is written as [Carter (1969, 1973a)] [see also chapter 7 of Wald (1984)]

$$\xi^a T_a{}^b \xi^c \varphi^d \epsilon_{bcde} = 0, \tag{5.166}$$

$$\varphi^a T_a{}^b \xi^c \varphi^d \epsilon_{bcde} = 0. \tag{5.167}$$

This condition is satisfied if the four-velocity is restricted in the plane spanned by ξ^a and φ^a, or the stress-energy tensor is composed of a stationary and axisymmetric electromagnetic field [Carter (1969)].

---

sufficiently small (see the last paragraph of section 5.7.2.1). However, such neglected terms play a role for close orbits.

Equation (5.167) is equivalent to the conditions

$$\xi^a \overset{(4)}{R}_a{}^b \xi^c \varphi^d \epsilon_{bcde} = 0, \tag{5.168}$$

$$\varphi^a \overset{(4)}{R}_a{}^b \xi^c \varphi^d \epsilon_{bcde} = 0. \tag{5.169}$$

If these conditions are satisfied, we may choose coordinates (x^2, x^3) for a series of two surfaces which are orthogonal to ξ^a and φ^a (and thus each of the two surfaces is labeled by different values of t and φ) so that the coordinates (t, φ, x^2, x^3) cover the entire spacetime [e.g., section 7.1 of Wald (1984) for a proof].

In the presence of the circularity condition, two two-surfaces spanned by (t, φ) and (x^2, x^3) are orthogonal. Thus, the metric is written in the form

$$g_{\mu\nu} = \begin{pmatrix} g_{tt} & g_{t\varphi} & 0 & 0 \\ * & g_{\varphi\varphi} & 0 & 0 \\ * & * & g_{22} & g_{23} \\ * & * & * & g_{33} \end{pmatrix}, \tag{5.170}$$

where $*$ denotes a symmetric relation, $g_{\mu\nu} = g_{\nu\mu}$. Note that if the circularity condition is not satisfied, we have to take into account the full components [Gourgoulhon and Bonazzola (1993); Birkl et al. (2011)]. Since $g_{\mu\nu}$ is a function of (x^2, x^3), two-dimensional metric for (x^2, x^3) is written in a conformally flat form as

$$\sum_{i,j=2}^{3} g_{ij} dx^i dx^j = \Psi(x^2, x^3) \sum_{i,j=2}^{3} f_{ij} dx^i dx^j, \tag{5.171}$$

where f_{ij} denotes a flat spatial metric again. Thus, the number of the field variables reduces to four.

Bardeen and Wagoner (1971) then wrote the line element in the following form,

$$ds^2 = -e^{2\nu} dt^2 + r^2 \sin^2\theta B^2 e^{-2\nu} (d\varphi - \omega dt)^2 + e^{2\mu} (dr^2 + r^2 d\theta^2), \tag{5.172}$$

where $(x^2, x^3) = (r, \theta)$ in this line element, and (ν, B, ω, μ) are the four field variables. In the terminology of 3+1 formulations, the lapse function and the shift vector are $\alpha = e^\nu$ and $\beta^\varphi = -\omega$, respectively; $n_A = (-e^\nu, 0)$ and $n^A = (e^{-\nu}, e^{-\nu}\omega)$ with $A = t$ or φ, and $n_r = n_\theta = n^r = n^\theta = 0$.

There are two ways for writing the field equations for these four field variables. Bardeen and Wagoner (1971); Butterworth and Ipser (1976); Friedman et al. (1986); Komatsu et al. (1989); Cook et al. (1992); Shibata and Sasaki (1998) employed a formulation in which (ν, B, ω) are determined by solving elliptic-type equations, while for $\zeta := \mu + \nu$ an ordinary differential equation is solved. On the other hand, Bonazzola et al. (1993) employed a formulation in which all four components are determined by solving elliptic-type equations. In the former formulation, the basic field equations are, e.g., following Butterworth and Ipser (1976),

$$\Delta_2 \nu + B^{-1}(\boldsymbol{D}B)\boldsymbol{D}\nu = \frac{1}{2} r^2 \sin^2\theta B^2 e^{-4\nu} (\boldsymbol{D}\omega) \boldsymbol{D}\omega$$

$$+ 4\pi e^{2\mu} \left(\rho_h + S_k{}^k \right), \tag{5.173}$$

$$\boldsymbol{D}(B^3 e^{-4\nu} r^2 \sin^2\theta \boldsymbol{D}\omega) = 16\pi T_{\alpha\beta} n^\alpha \varphi^\beta B e^{2\zeta-3\nu}, \quad (5.174)$$

$$\boldsymbol{D}(r\sin\theta \boldsymbol{D}B) = 8\pi(T_{rr} + r^{-2}T_{\theta\theta})Br\sin\theta, \quad (5.175)$$

$$\partial_u \zeta = \left[\{\boldsymbol{D}(Br\sin\theta)\}\boldsymbol{D}(Br\sin\theta)\right]^{-1}$$
$$\times \left[\{\partial_r(rB)\}\{-e^{-4\nu}(\partial_r\omega)(\partial_u\omega)r^3 B^3(1-u^2)^2/2 + r(\partial_r\partial_u B)(1-u^2)\right.$$
$$\left.-ru\partial_r B + 2r(\partial_r \nu)(\partial_u \nu)B(1-u^2)\}\right.$$
$$-\frac{1}{2}\{\partial_\theta(\sin\theta B)\}\left[e^{-4\nu}B^3 r^2(1-u^2)\{(r\partial_r\omega)^2 - (1-u^2)(\partial_u\omega)^2\}/2\right.$$
$$-r^2\partial_r^2 B - r\partial_r B - 2B(r\partial_r\nu)^2$$
$$+(1-u^2)\partial_u^2 B - 3u\partial_u B + 2(\partial_u\nu)^2(1-u^2)B\right]$$
$$+4\pi\left(r^2 T_{rr} - T_{\theta\theta}\right)B\left[Bu - (1-u^2)\partial_u B\right]$$
$$\left.-8\pi T_{r\theta}Br\sin\theta(B+r\partial_r B)\right], \quad (5.176)$$

where $u := \cos\theta$, and $\boldsymbol{D}$ denotes a derivative vector for the flat spatial metric f_{ij} such that

$$(\boldsymbol{D}\omega)\boldsymbol{D}\omega = (\partial_r\omega)^2 + r^{-2}(\partial_\theta\omega)^2, \quad (5.177)$$

$$\Delta_2 := \boldsymbol{D}\boldsymbol{D} = \partial_r^2 + 2r^{-1}\partial_r + r^{-2}(\partial_\theta^2 + \cot\theta\partial_\theta). \quad (5.178)$$

Here, $\Delta_2 = \Delta_f$ in axial symmetry. Equations (5.173), (5.174), (5.175), and (5.176) are, respectively, derived from

$$R^{\mu\nu}n_\mu n_\nu = 8\pi\left(T^{\mu\nu} - \frac{1}{2}T_\alpha{}^\alpha g^{\mu\nu}\right)n_\mu n_\nu = 4\pi\left(\rho_h + S_k^k\right), \quad (5.179)$$

$$R^\mu{}_\nu n_\mu \varphi^\nu = 8\pi T^\mu{}_\nu n_\mu \varphi^\nu = -8\pi J_i \varphi^i, \quad (5.180)$$

$$G_r{}^r + G_\theta{}^\theta = 8\pi\left(T_r{}^r + T_\theta{}^\theta\right) = 8\pi\left(S_k^k - S_\varphi^\varphi\right), \quad (5.181)$$

$$R_{r\theta} = 8\pi T_{r\theta} \quad \text{and} \quad R_r{}^r - R_\theta{}^\theta = 8\pi\left(T_r{}^r - T_\theta{}^\theta\right), \quad (5.182)$$

where we used the definitions (2.43)–(2.45). The elliptic-type equations are solved imposing appropriate boundary conditions at spatial infinity and an inner boundary on a black-hole horizon or at $r = 0$ with a regular boundary condition at $\theta = 0$ and π (or at $\theta = \pi/2$ in the presence of the equatorial-plane symmetry). The ordinary differential equation for ζ is solved imposing the regularity condition (see section 2.4.2), $e^\zeta = B$, at the rotational axis ($\theta = 0$ and π).

The second version of the basic equations is derived in several ways. Because the readers now would be familiar with 3+1 formulations, we introduce a method based on it [Bonazzola et al. (1993); Shibata (2007)], writing the line element in the form

$$ds^2 = -\alpha^2 dt^2 + \psi^4\left[e^{2q}\left(dr^2 + r^2 d\theta^2\right) + r^2\sin^2\theta\left(\beta dt + d\varphi\right)^2\right], \quad (5.183)$$

where α, β, ψ, and q are functions of r and θ. These functions obey the equations in the 3+1 formulation with $\partial_t \gamma_{ij} = 0$ and $\partial_t K_{ij} = 0$. Then, the maximal slicing condition, $K = 0 = \partial_t K$, is automatically satisfied, yielding an elliptic equation for α; see equation (2.113). ψ is determined by the Hamiltonian constraint; see equation (5.7). $\partial_t \gamma_{ij} = 0$ gives the relation between K_{ij} and β^i as $2\alpha K_{ij} = D_i \beta_j + D_j \beta_i$. For the metric of equation (5.183) with $\partial_\varphi \beta = 0$, the nonzero components of K_{ij} are written as

$$K_r{}^\varphi = \frac{1}{2\alpha}\partial_r \beta, \qquad K_\theta{}^\varphi = \frac{1}{2\alpha}\partial_\theta \beta. \tag{5.184}$$

Substituting these relations into the momentum constraint equation yields an elliptic equation for β. Finally, from the combination of the evolution equation of $\partial_t K_{ij}$,

$$\gamma^{rr}\partial_t K_{rr} + \gamma^{\theta\theta}\partial_t K_{\theta\theta} - 3\gamma^{\varphi\varphi}\partial_t K_{\varphi\varphi} - \partial_t K = 0, \tag{5.185}$$

we obtain an elliptic equation for q.

A possible set of the basic field equations is summarized as follows:

$$\Delta_2(\alpha\psi) = \frac{\alpha\psi e^{2q}\tilde{R}}{8} + 2\pi(\rho_h + 2S_k{}^k)e^{2q}\alpha\psi^5 + \frac{7A^2}{4}\alpha\psi^5, \tag{5.186}$$

$$\Delta_2 \psi = \frac{\psi e^{2q}\tilde{R}}{8} - 2\pi\rho_h e^{2q}\psi^5 - \frac{A^2}{4}\psi^5, \tag{5.187}$$

$$\Delta_2 \beta + \left(\frac{2}{r} + \frac{6}{\psi}\frac{\partial\psi}{\partial r} - \frac{1}{\alpha}\frac{\partial\alpha}{\partial r}\right)\frac{\partial\beta}{\partial r} + \frac{1}{r^2}\left(2\cot\theta + \frac{6}{\psi}\frac{\partial\psi}{\partial\theta} - \frac{1}{\alpha}\frac{\partial\alpha}{\partial\theta}\right)\frac{\partial\beta}{\partial\theta}$$
$$= \frac{16\pi\alpha e^{2q}J_\varphi}{r^2 \sin^2\theta}, \tag{5.188}$$

$$\left(\frac{\partial^2}{\partial r^2} + \frac{1}{r}\frac{\partial}{\partial r} + \frac{1}{r^2}\frac{\partial^2}{\partial\theta^2}\right)q$$
$$= -8\pi\psi^4 e^{2q}[S_k{}^k - 2S_\varphi{}^\varphi] + 3A^2\psi^4 + 2\left(\frac{1}{r}\frac{\partial}{\partial r} + \frac{\cot\theta}{r^2}\frac{\partial}{\partial\theta}\right)\ln(\alpha\psi^2)$$
$$+ \frac{4}{\alpha\psi^2}\left(\frac{\partial(\alpha\psi)}{\partial r}\frac{\partial\psi}{\partial r} + \frac{1}{r^2}\frac{\partial(\alpha\psi)}{\partial\theta}\frac{\partial\psi}{\partial\theta}\right) =: S_q. \tag{5.189}$$

Here,

$$A^2 := \frac{1}{2}e^{2q}K_{ij}K^{ij} = \sin^2\theta\left[(rK_r{}^\varphi)^2 + (K_\theta{}^\varphi)^2\right], \tag{5.190}$$

$$\tilde{R} = -2e^{-2q}\left(\frac{\partial^2}{\partial r^2} + \frac{1}{r}\frac{\partial}{\partial r} + \frac{1}{r^2}\frac{\partial^2}{\partial\theta^2}\right)q. \tag{5.191}$$

Combining equations (5.186) and (5.187), and then using (5.190), the following simplified equation, instead of equation (5.187), is obtained

$$\left(\Delta_2 + \frac{1}{r}\frac{\partial}{\partial r} + \frac{\cot\theta}{r^2}\frac{\partial}{\partial\theta}\right)(\alpha\psi^2) = 8\pi e^{2q}\alpha\psi^6\left(S_k{}^k - S_\varphi{}^\varphi\right). \tag{5.192}$$

This equation combined with equation (5.189) also yields

$$\left(\frac{\partial^2}{\partial r^2} + \frac{1}{r}\frac{\partial}{\partial r} + \frac{1}{r^2}\frac{\partial^2}{\partial\theta^2}\right)\zeta$$
$$= 8\pi\psi^4 e^{2q}S_\varphi{}^\varphi + 3A^2\psi^4 - \left(\frac{\partial\ln\alpha}{\partial r}\right)^2 - \frac{1}{r^2}\left(\frac{\partial\ln\alpha}{\partial\theta}\right)^2 =: S_\zeta, \tag{5.193}$$

where $\zeta = q + \ln(\alpha\psi^2)$. Note that this equation cannot be used in the presence of a black hole with the typical boundary condition at its horizon, $\alpha = 0$ (see below). For a perfect fluid with $u^r = 0 = u^\theta$ which we consider here,

$$\rho_{\rm h} = \rho h (\alpha u^t)^2 - P = \rho h \frac{\alpha^2}{\alpha^2 - \gamma_{\varphi\varphi}(\Omega + \beta)^2} - P, \tag{5.194}$$

$$J_\varphi = \rho \alpha u^t h u_\varphi = \rho h \alpha u^t \gamma_{\varphi\varphi} \frac{\Omega + \beta}{\sqrt{\alpha^2 - \gamma_{\varphi\varphi}(\Omega + \beta)^2}}, \tag{5.195}$$

$$S_\varphi{}^\varphi = \rho h (u_\varphi)^2 \gamma^{\varphi\varphi} + P = \rho h \gamma_{\varphi\varphi} \frac{(\Omega + \beta)^2}{\alpha^2 - \gamma_{\varphi\varphi}(\Omega + \beta)^2} + P, \tag{5.196}$$

$$S_k{}^k = S_\varphi{}^\varphi + 2P, \tag{5.197}$$

where we used $u_\varphi = \gamma_{\varphi\varphi} u^t (\Omega + \beta)$ with $\Omega = u^\varphi/u^t$ being the angular velocity, and

$$u^t = \frac{1}{\sqrt{\alpha^2 - \gamma_{\varphi\varphi}(\Omega + \beta)^2}}. \tag{5.198}$$

Elliptic equations (5.186)–(5.189) are solved imposing outer boundary conditions at spatial infinity, and on a black-hole horizon in the presence of a black hole. The outer boundary conditions are

$$\alpha\psi^2 \to 1 + \frac{C_{\alpha\psi^2}}{2r^2}, \tag{5.199}$$

$$\psi \to 1 + \frac{M}{2r}, \tag{5.200}$$

$$\beta \to -\frac{2J}{r^3}, \tag{5.201}$$

$$q \to \frac{q_1 \sin^2\theta}{r^2}, \tag{5.202}$$

where M is the gravitational mass, J is the angular momentum of the system, and q_1 and $C_{\alpha\psi}$ are constants. M should agree with both the ADM mass and the Komar mass because the virial relation has to be satisfied (see section 5.3.3). Note that $q = 0$ at the rotation axis, $\theta = 0$ and π, because the regularity condition has to be satisfied there (see section 2.4.2).

The inner boundary conditions in the absence of a black hole are supplied from the regularity conditions at $r = 0$ as $\partial_r \alpha = \partial_r \psi = \partial_r q = \partial_r \beta = 0$. The inner boundary conditions in the presence of a black hole are in a sense up to our choice: We only need to guarantee that the closed boundary surface becomes the Killing horizon (see appendix G), and then, we have a degree of freedom. However, in the numerical computation, it is convenient to impose the conditions that the horizon is denoted by $r = $ const and $\alpha = 0$ on the horizon like in the Kerr metric of quasi-isotropic coordinates [Nishida and Eriguchi (1994)]. Then, in order for the extrinsic curvature written by equation (5.184) to be finite, the following boundary conditions of β on the horizon are required:

$$\beta = \text{const} \quad \text{and} \quad \partial_r \beta = 0. \tag{5.203}$$

Here, the constant value of $-\beta$ on the horizon is equal to the angular velocity of the black hole Ω_H [see equation (1.131) for its definition].

Next, we derive the boundary conditions of ψ and q on the horizon. For stationary spacetime, the event horizon has to agree with apparent horizons on which the expansion of an outgoing null vector field vanishes (see sections 5.6.3, 7.1, and appendix G). For the extrinsic curvature written by equation (5.184) and for the chosen coordinate system of the horizon, the condition of the zero expansion is written as [e.g., see equation (5.141) or (7.6)]

$$0 = D_a s^a + K_{ab} s^a s^b = \frac{1}{\sqrt{\gamma}} \partial_i \left(\sqrt{\gamma} s^i \right) + K_{ij} s^i s^j, \tag{5.204}$$

where s^i is a unit radial vector which is orthonormal to the horizon surface and has only the s^r component in the present setting. Thus, $s^r = (\psi^2 e^q)^{-1}$. Since other components of s^i vanish, $K_{ij} s^i s^j = 0$ on the horizon [see equation (5.184)]. Then, equation (5.204) becomes

$$\frac{4}{\psi} \partial_r \psi + \frac{2}{r} + \partial_r q = 0, \tag{5.205}$$

yielding the boundary condition for ψ or q on the horizon.

The remaining boundary condition comes from equation (5.189): To guarantee that the right-hand side of equation (5.189) is finite on the horizon where $\alpha = 0$, the condition $2\partial_r \psi + \psi/r = 0$ is required, and hence, we find from equation (5.205) that $\partial_r q = 0$ should be the boundary condition.

5.7.1.2 *Hydrostatics*

For stationary, axisymmetric spacetime that satisfy the circularity condition, the conservation equation for the stress-energy tensor of the perfect fluid is integrated to yield a first integral, if the fluid is isentropic. This is shown by setting the four velocity as

$$u^\mu = u^t \xi^\mu + u^\varphi \varphi^\mu. \tag{5.206}$$

Then, the continuity equation, $\nabla_\mu (\rho u^\mu) = 0$, is automatically satisfied.

The conservation equation of the stress-energy tensor is written as

$$0 = \nabla_a T^a{}_b = \nabla_a (\rho h u^a u_b) + \nabla_b P$$
$$= \rho u^a \nabla_a (h u_b) + \nabla_b P, \tag{5.207}$$

where the continuity equation was used. Using equation (5.206), the first term of equation (5.207) is calculated as follows:

$$\rho u^\mu \nabla_\mu (h u_\nu) = \rho \left(u^t \xi^\mu + u^\varphi \varphi^\mu \right) \nabla_\mu (h u_\nu)$$
$$= \rho u^t \left[\mathcal{L}_\xi (h u_\nu) - h u_\mu \nabla_\nu \xi^\mu \right] + \rho u^\varphi \left[\mathcal{L}_\varphi (h u_\nu) - h u_\mu \nabla_\nu \varphi^\mu \right]$$
$$= \rho h u_\mu \left[-\nabla_\nu (u^t \xi^\mu + u^\varphi \varphi^\mu) + \xi^\mu \partial_\nu u^t + \varphi^\mu \partial_\nu u^\varphi \right]$$
$$= \rho h \left[-u_\mu \nabla_\nu u^\mu + (u_t + \Omega u_\varphi) \partial_\nu u^t + u^t u_\varphi \partial_\nu \Omega \right]$$
$$= \rho h \left[-\frac{1}{u^t} \partial_\nu u^t + u^t u_\varphi \partial_\nu \Omega \right], \tag{5.208}$$

where we used $\mathcal{L}_\xi(hu_\nu) = 0 = \mathcal{L}_\varphi(hu_\nu)$, $\Omega := u^\varphi/u^t$, and $u_t + \Omega u_\varphi = -1/u^t$. Hence,

$$0 = \nabla_\mu T^\mu{}_\nu = \rho h \left[-\partial_\nu \ln(u^t) + u^t u_\varphi \partial_\nu \Omega + \frac{1}{\rho h} \partial_\nu P \right]. \tag{5.209}$$

The first law of thermodynamics is written in the form

$$\rho dh = dP + \rho T ds, \tag{5.210}$$

where T and s are temperature and specific entropy, respectively. Thus, assuming that $u^t u_\varphi$ is a function of Ω as $F(\Omega)$ and $s = $ const, equation (5.209) is integrated, yielding [e.g., Butterworth and Ipser (1976)]

$$-\ln u^t + \int F(\Omega) d\Omega + \ln h = C, \tag{5.211}$$

where C is a constant. An often employed relation for $F(\Omega)$ is

$$F(\Omega) = \varpi_0^2 (\Omega_0 - \Omega), \tag{5.212}$$

where ϖ_0 is a constant of length dimension and Ω_0 is the angular velocity along the rotation axis. In the Newtonian limit with $u^t \to 1$ and $u_\varphi \to \varpi^2 \Omega$ where ϖ denotes a cylindrical radius, relation (5.212) reduces to

$$\Omega = \frac{\Omega_0 \varpi_0^2}{\varpi^2 + \varpi_0^2}. \tag{5.213}$$

Thus, for the limit $\varpi_0 \to \infty$, the angular velocity reduces to a constant $\Omega = \Omega_0$, while for small values of ϖ_0 (more specifically, when ϖ_0 is smaller than the equatorial stellar radius), the degree of differential rotation becomes high.

Alternatively, equation (5.209) may be written for $ds = 0$ as

$$\rho u^t \left[h \partial_\nu \left(\frac{1}{u^t} \right) + h u_\varphi \partial_\nu \Omega + \frac{1}{u^t} \partial_\nu h \right] = 0. \tag{5.214}$$

Thus, assuming that the specific angular momentum, $j := h u_\varphi$, is a function of Ω as $j(\Omega)$, equation (5.214) is integrated, yielding

$$\frac{h}{u^t} + \int j(\Omega) d\Omega = C'(= \text{const}). \tag{5.215}$$

For the rigidly rotating case $\Omega = $ const, both equations (5.211) and (5.215) yield the same relation

$$\frac{h}{u^t} = C'. \tag{5.216}$$

Here, u^t is related to Ω by equation (5.198), and hence, for given values of C' and Ω, h is determined. This relation is often used for constructing rigidly rotating neutron stars and supermassive stars [Stergioulas (2003); Friedman and Stergioulas (2013)].

For the case $j = $ constant, equation (5.215) is written as

$$-j\beta + \alpha \sqrt{h^2 + \gamma^{\varphi\varphi} j^2} = C', \tag{5.217}$$

where we used $(\alpha u^t)^2 = 1 + \gamma^{\varphi\varphi} j^2 h^{-2}$ and $\Omega = -\beta + \gamma^{\varphi\varphi} j/(hu^t)$. Thus, for given values of C' and j, h is determined.

Before closing this section, we note the following fact: In the absence of the circularity condition, the procedure for a solution of the hydrodynamics equations is much more complicated in general. For such cases, the continuity equation is not trivial and the Euler equation has to be solved by the direct integration, unless a special velocity configuration is assumed for the four-velocity u^a [Birkl et al. (2011)].

5.7.1.3 Approximate solution by IWM formulation

It is worthy to note that axisymmetric equilibria can be approximately computed using the IWM formulation described in section 5.6. In this case, the basic gravitational-field equations are (5.186)–(5.188) with $q = 0$, and the hydrostatic equation is the same as in section 5.7.1.2. Cook et al. (1996) computed rapidly and rigidly rotating neutron stars using both the exact and approximate formulations, and showed that the approximate solutions agree well with the corresponding exact solutions within 1% error for various quantities. The reason for this is that the IWM formulation can derive exact solutions in the spherically symmetric case, and as far as the deviation from the spherical symmetry is small, it can derive a good approximate solution. Rigidly rotating neutron stars are not highly non-spherical even if they are rapidly rotating, and hence, the IWM formulation can derive an accurate approximate solution. An approximate equilibrium solution, derived in the IWM formulation, is also used as an initial condition of numerical-relativity simulations [e.g., Shibata et al. (2000b,a); Shibata (2000)].[10]

It was not clear whether the IWM formulation could derive a good *approximate* solution for quasi-equilibrium states of compact binary systems until the work by Cook et al. (1996). Thus, their analysis of rapidly rotating neutron stars provided an encouraging evidence, and now gives us a foundation for computing non-axisymmetric quasi-equilibrium states using the IWM formulation.

5.7.1.4 Virial and consistency relations

As shown in section 5.3.3, the ADM mass and the Komar mass agree with each other in stationary spacetime, and this relation is called the virial relation. The virial relation is quite useful for examining the accuracy of an equilibrium solution derived in a numerical computation as pointed out by Bonazzola (1973) and Bonazzola and Gourgoulhon (1994).

For stationary axisymmetric spacetime that satisfies the circularity condition, the virial relation, $M_K = M_{ADM}$, is trivial when these quantities are determined from the asymptotic behavior (r^{-1} parts) of $\alpha(\to 1 - M_K/r)$ and $\psi(\to 1 + M_{ADM}/2r)$

---

[10]It should be noted that for *differentially* rotating neutron stars, their central region could be highly rapidly rotating. For this case, the conformal flatness assumption may not be very good and the error in the results by the IWM formulation could not be negligible [Iosif and Stergioulas (2014)].

because $\alpha\psi^2$ obeys equation (5.192), which implies that the asymptotic behavior of $\alpha\psi^2$ has to be

$$\alpha\psi^2 \to 1 + O\left(r^{-2}\right). \tag{5.218}$$

However, M_K and M_{ADM} may be also defined by the volume integral using the elliptic equations for $\alpha\psi$ and ψ [equations (5.186) and (5.187)]. Examining the equality of these two quantities thus determined is a valuable method for finding the accuracy of the numerical results [Bonazzola and Gourgoulhon (1994)].

For this stationary axisymmetric spacetime, there is an additional consistency relation that has to be satisfied. This fact is found by observing equation (5.189) or (5.193) [Bonazzola (1973); Bonazzola and Gourgoulhon (1994)]: These equations admit a homogeneous solution that is proportional to $\ln r$ and independent of θ. The presence of such a homogeneous solution must be prohibited because g_{rr} and $g_{\theta\theta}$ diverge at $r = 0$ and $r \to \infty$ in the presence of this homogeneous solution. Its presence is realized when the following two-dimensional volume integral does not vanish,

$$\int r dr d\theta S_q, \quad \int r dr d\theta S_\zeta. \tag{5.219}$$

This fact, therefore, yields the following relations that must be satisfied:

$$\int r dr d\theta S_q = 0, \quad \int r dr d\theta S_\zeta = 0. \tag{5.220}$$

These are new consistency relations, different from the virial relation. The latter equation in (5.220) is the same as the relation, named GRV2, of Bonazzola (1973); Bonazzola and Gourgoulhon (1994), which is written as

$$\int r dr d\theta \left[8\pi\psi^4 e^{2q} S_\varphi^{\ \varphi} + 3A^2\psi^4 - \left(\frac{\partial \ln \alpha}{\partial r}\right)^2 - \frac{1}{r^2}\left(\frac{\partial \ln \alpha}{\partial \theta}\right)^2 \right] = 0. \tag{5.221}$$

For deriving GRV2 from the former equation of (5.220), we only need to combine it with equation (5.192) and use the identity [see equation (5.218)]

$$\int r dr d\theta \left(\frac{\partial^2}{\partial r^2} + \frac{1}{r}\frac{\partial}{\partial r} + \frac{1}{r^2}\frac{\partial^2}{\partial \theta^2} \right) \ln\left(\alpha\psi^2\right) = 0. \tag{5.222}$$

Then, the former equation of (5.220) reduces to equation (5.221).

We have to be careful when using equation (5.221) in the presence of a black hole with $\alpha = 0$ at the horizon surface, because the integral is not well defined. Thus, for the general use, it is better to employ the equation derived from the former equation of (5.220) as

$$\int r dr d\theta \left[-8\pi\psi^4 e^{2q}(S_k^{\ k} - 2S_\varphi^{\ \varphi}) + 3A^2\psi^4 + \frac{2}{\psi}\frac{\partial \ln(\alpha\psi)}{\partial r}\left(2\frac{\partial \psi}{\partial r} + \frac{\psi}{r}\right) + \frac{2}{r\psi}\frac{\partial \psi}{\partial r} \right.$$
$$\left. + \frac{2}{r^2}\left(\cot\theta \frac{\partial \ln(\alpha\psi^2)}{\partial \theta} + 2\frac{\partial \ln(\alpha\psi)}{\partial \theta}\frac{\partial \ln \psi}{\partial \theta}\right) \right] = 0. \tag{5.223}$$

Note here $2\partial_r \psi + \psi/r = \alpha = \partial_\theta \alpha = 0$ on the horizon surface according to the boundary conditions introduced in section 5.7.1.1.

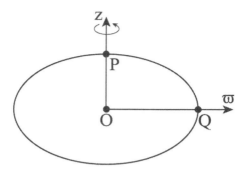

Fig. 5.3 Schematic figure for a rotating star in the ϖ-z cross section (ϖ is the cylindrical radius). P and Q are points at equatorial and polar surfaces, and O is the origin.

5.7.1.5 *Procedure for a numerical solution*

For a given set of gravitational-field variables, Ω, and $F(\Omega)$ or $j(\Omega)$, u^t is written by equation (5.198), and hence, h is determined by equation (5.211) or (5.215). If an equation of state, $P = P(\rho)$, is given (with $s = \text{const}$), $h = h(\rho)$ is also obtained from the first law of thermodynamics (5.210), and thus, ρ is determined. The source terms of the equations for gravitational-field variables are composed of the matter variables, ρ, P, h, and u^μ. Thus, a system is determined self-consistently from the four gravitational-field equations and the first integral of the matter equation for given relations $F(\Omega)$ or $j(\Omega)$ and $P(\rho)$.

The formulations presented in the previous section can be applied to the computation of equilibrium states for axisymmetric rotating stars and/or rotating black holes surrounded by a torus. For obtaining such a state, we have to satisfy all the equations self-consistently. This is carried out by iteratively solving the gravitational-field equations and the first integral of the matter equation until a sufficient convergence is achieved. For obtaining a numerical solution by an iterative method, however, a careful prescription is necessary, in particular, for obtaining a rapidly rotating object: With an inappropriate iterative method, the convergence is never achieved. A robust iterative method was developed and established by Komatsu et al. (1989) for the first time, using the so-called Hachisu's self-consistent field method. We here briefly describe this method focusing primarily on rigidly rotating stars for simplicity. We note that even for the differentially rotating case with $\Omega \neq \text{const}$, the method is qualitatively the same.

In this problem, there are two unknown constants that should be appropriately determined during the iteration procedure; C (or C') and Ω. Thus, it might be quite natural to consider the following iteration procedure fixing these two constants to physically plausible values: First, the density of the star at the origin, O, is determined as ρ_c, as well as Ω; then, using equation (5.216) and a given equation of state which is used for driving h_c (specific enthalpy at O), C' is fixed. Next, we give a trial density profile for which the density at O is given by ρ_c, and solve the gravitational-field equations; then, using equation (5.216) again, h is determined, and using the equation of state, ρ, ε, and P are determined again. Note that for this step, the central density is no longer equal to ρ_c. We then solve the gravitational-field equations again, and repeat these procedures until a sufficient convergence is achieved. Unfortunately, it is well-known that by this iteration procedure, a convergent solution cannot be obtained for rotating neutron stars. The reason for this is that inappropriate quantities were fixed during the iteration.

As an alternative (but slightly better) method, we may fix ρ_c or h_c instead of C (or C'). Then, during each step of the iteration, C is updated. Again, Ω is given and fixed during the iteration procedure. In this case, also, a convergent solution cannot be obtained for *rapidly* rotating neutron stars.

These two *inappropriate* examples show that we have to carefully consider what are fixed during the iteration. The solution to this problem was first discovered by Hachisu (1986) in his computation of rapidly rotating stars in Newtonian gravity. Subsequently, Komatsu et al. (1989) applied the Hachisu's method for a solution of rapidly rotating stars in general relativity.

The first step in the Hachisu's method is to pick up the coordinate radius at the equator, r_{eq}, as a constant that is determined during the iteration, in addition to C (or C') and Ω. Then for determining these three constants, three conditions are imposed during the iteration in the manner described below.

The next step in his method is to rescale coordinates, e.g., by $r = r_{eq}\hat{r}$, and rewrite the equations using the dimensionless coordinates like $\hat{r}$ (e.g., $\Delta_2 \psi = r_{eq}^{-2} \hat{\Delta}_2 \psi$). Then, the location on the equatorial surface is automatically fixed to be $\hat{r} = 1$, to which we refer as the point Q (see figure 5.3). The unique point in his method is to further require us to fix the location of the stellar surface along the rotation axis, which is referred to as the point P and is written as $\hat{r} = \hat{r}_p$. Note that $\hat{r}_p$ denotes the coordinate axial ratio because $\hat{r} = 1$ at Q. Finally, we have to fix the value of h at the point O as h_c. As a result, we obtain three conditions as

$$h = 1 (\rho = 0) \text{ at P and Q}, \quad h = h_c \text{ at O}. \tag{5.224}$$

These three conditions are used for determining C, Ω, and r_{eq} during the iteration procedure. We emphasize that the crucial point in his method is to fix the axial ratio of the star by determining P.

Using equation (5.211), the conditions, $h = 1$ at P and Q, are written as (for $\Omega = $ const)

$$-(\ln u^t)_P = -(\ln u^t)_Q = C, \tag{5.225}$$

where u^t in this scheme is given by

$$u^t = \frac{1}{\sqrt{\alpha^2 - r_{eq}^2 \hat{\gamma}_{\varphi\varphi}(\Omega + \beta)^2}}. \tag{5.226}$$

Here, r_{eq} appears because the dimensionless coordinates are employed for solving the gravitational-field equations. In addition, the condition at the point O results in

$$-(\ln u^t)_{\text{O}} + \ln h_c = C. \tag{5.227}$$

Thus, three conditions that determine C, Ω, and r_{eq} for given gravitational-field variables are explicitly derived.

An illustrative robust iteration procedure is described as follows. In this example, we suppose that a specific equation of state is provided.

(1) First, a trial density profile is given. Then, the pressure and specific enthalpy are also determined through the given equation of state.
(2) For the trial matter profiles, the gravitational-field equations are solved.
(3) Using the resulting solutions of the gravitational-field variables, three free parameters C, Ω, and r_{eq} are calculated from equations (5.225) and (5.227). In this procedure, the values of h at P, Q, and O are unchanged.
(4) For the new set of the free parameters, new profiles of the specific enthalpy, rest-mass density, and velocity are determined.
(5) We then go to procedure (2) and repeat these procedures until a sufficient convergence is achieved.

This iteration procedure has been proven to be quite robust for obtaining rapidly rotating stars. We note that for an efficient convergence, a good choice of the initial trial density profile is necessary.

For an arbitrary choice of h_c and axial ratio, $\hat{r}_{\text{p}}$, we in general cannot obtain a desired rotating star of specific values of mass and angular velocity. For obtaining the desired solution, we have to compute rotating stars for a number of times changing h_c and axial ratio. However, we know that the mass increases with the increase of h_c for a given axial ratio and that the angular velocity typically increases with the decrease of the axial ratio. Thus, the desired solution can be usually obtained in several straightforward iteration steps.

The Hachisu's method has been proven to be quite robust not only for the rigidly rotating case but also for the differentially rotating case. A messy point in the differentially rotating case is that Ω becomes a function of spatial coordinates. This implies that we have to determine its profile solving equations (5.211) and (5.226) self-consistently (note that in this case, the free parameter is not Ω but Ω at some point; typically Ω along the rotation axis is fixed). However, a numerical solution for the profile of Ω can be obtained in a straightforward manner using the algorithm of solving algebraic equations such as the Newton–Raphson method [Press et al. (1989)].

For the case that one wants to determine an equilibrium of a torus, we have to change the location of the points to be fixed; e.g., see, figure 5.4. In this case, it is important to fix the inner and/or outer edges of the torus instead of the points at the polar and equatorial surfaces.

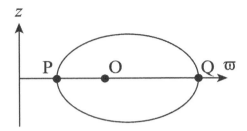

Fig. 5.4 Schematic figure for a torus in the ϖ-z cross section (ϖ is the cylindrical radius). P and Q are points at the inner and outer edges of the torus, and O is an approximate density maximum.

5.7.2 Quasi-equilibria for non-axisymmetric spacetime

One of the most important roles of numerical relativity is to clarify the evolution process of coalescing compact binary systems, such as binary neutron stars and binary black holes in close relativistic orbits. As described in section 1.4.8, these binaries evolve due to the gravitational-radiation reaction. The time scale of this reaction is longer than the orbital period for most of the orbits, except for a sufficiently close orbit (several orbits just before the merger). This implies that even for moderately close orbits, the binaries are approximately seen as stationary systems in the reference frame comoving with the orbital motion. Thus, we can approximately define a timelike Killing vector field near the binary objects for such a system. This Killing vector field is called helical Killing vector field, and in the laboratory frame, it is defined by

$$k_{\rm h}^a := \left(\frac{\partial}{\partial t}\right)^a + \Omega_{\rm orb}\left(\frac{\partial}{\partial \varphi}\right)^a, \tag{5.228}$$

where $\Omega_{\rm orb}$ denotes the orbital angular velocity measured in an observer located at spatial infinity. We note that this is (usually) timelike inside the light cylinder $r < \Omega_{\rm orb}^{-1}$ while for $r > \Omega_{\rm orb}^{-1}$, it is spacelike. (On the black-hole horizon, it should be null if the spacetime has globally helical symmetry; see Appendix G.4.2.)

The strictly helical Killing vector field cannot exist in asymptotically flat systems, because in binary systems, the energy and angular momentum are dissipated by the emission of gravitational waves, and the exact symmetry is never realized.[11] In a virtual system in which the amount of outgoing gravitational waves is equal to that of ingoing gravitational waves, standing gravitational waves of a constant wavelength could be present [Blackburn and Detweiler (1992); Detweiler (1994)]. For this case, a helical Killing vector field can be defined. However, such spacetime cannot be asymptotically flat due to the following reason: The amplitude of standing waves decreases as $A = O\left(r^{-1}\right)$ with the increase of the distance from the source, r, and hence, (average) energy density of gravitational waves, which is

---

[11]Note that if asymptotically anti-de Sitter spacetime is considered, helically symmetric spacetime composed of a binary system may exist because gravitational waves emitted are reflected back.

proportional to A^2, decreases as $O\left(r^{-2}\right)$.[12] This implies that the total energy of the system obtained by integrating over the entire space would diverge as $O(r)$, and hence, asymptotic flatness would be violated. For a real binary system, gravitational waves propagating in distant zones were emitted earlier, and their wavelength is longer. Hence, the energy density decreases as $O(r^{-n})$ with $n > 2$. Furthermore, the lifetime of real binaries is finite and the total amount of gravitational-wave energy has to be finite. Therefore, a helical Killing vector field could be defined only in the vicinity of binary systems.

All these considerations suggest that it is reasonable to suppose that in a near zone with $r \lesssim \Omega_{\rm orb}^{-1}$, a helical Killing vector field of the form (5.228) exists at least approximately, although it is not a good idea to suppose that the helical Killing vector field exists in the distant zone (or wave zone) with $r \gg \Omega_{\rm orb}^{-1}$. In the very distant zone, it is rather reasonable to suppose the approximate presence of a timelike Killing vector field neglecting the presence of gravitational waves. Thus, it is a good idea to consider a formulation assuming the presence of a Killing vector field of the form

$$k_{\rm h'}^a := \left(\frac{\partial}{\partial t}\right)^a + \Omega_{\rm orb} F(r) \left(\frac{\partial}{\partial \varphi}\right)^a, \qquad (5.229)$$

where $F(r) = 1$ in the near zone $r \lesssim \Omega_{\rm orb}^{-1}$ and $F(r) \to 0$ for $r \to \infty$. In section 5.7.2.1, we will describe a possible formulation for gravitational-field equations in the presence of a Killing vector field of the form (5.229). Subsequently, we will describe hydrostatic equations, virial relation, first law, and numerical procedure for a solution of neutron-star binary systems.

5.7.2.1 Formulation for gravitational fields

Shibata et al. (2004) proposed a formulation of Einstein's equation in the presence of a Killing vector field of the form (5.229). The basic equations were derived in the similar manner to that in the CTS formulation [York (1999)].

In the formulation of Shibata et al. (2004), a full set of Einstein's equation in the conformal 3+1 formulation (see section 2.1.7) is taken into account. The basic equations are derived for variables, ψ, $\tilde{\gamma}_{ij}$, α, and β^k. The maximal slicing condition $K = 0$ is imposed for simplicity, but we may impose only $\partial_t K = 0$ and K may be different from zero. (In the following analysis, we will set $K = 0$.)

The conformal factor and the lapse function are determined by the Hamiltonian constraint and by the equation $\partial_t K = 0$, respectively, as

$$\tilde{\Delta}\psi = \frac{\psi}{8}\tilde{R} - 2\pi\rho_{\rm h}\psi^5 - \frac{\psi^5}{8}\tilde{A}_{ij}\tilde{A}^{ij}, \qquad (5.230)$$

$$\tilde{\Delta}(\alpha\psi) = 2\pi\alpha\psi^5\left(\rho_{\rm h} + 2S_k{}^k\right) + \frac{7}{8}\alpha\psi^5\tilde{A}_{ij}\tilde{A}^{ij} + \frac{\alpha\psi}{8}\tilde{R}. \qquad (5.231)$$

Following the concept of York (1999), we set

$$u^{ij} := \partial_t\tilde{\gamma}^{ij}, \qquad v_{ij} := \partial_t\tilde{A}_{ij}, \qquad (5.232)$$

---

[12]In this section, we suppose that the spacetime dimensionality is four.

where $u_{ij} = -\tilde{\gamma}_{ik}\tilde{\gamma}_{jl}u^{kl}$. The time derivatives in u^{ij} and v_{ij} are rewritten using the conditions $\mathscr{L}_{k_\mathrm{h}}\tilde{\gamma}^{ij} = 0$ and $\mathscr{L}_{k_\mathrm{h}}\tilde{A}_{ij} = 0$. Then, the evolution equation for $\tilde{\gamma}_{ij}$ is seen as the definition of $\tilde{A}_{ij}$ as

$$2\alpha\tilde{A}_{ij} = \tilde{D}_i\tilde{\beta}_j + \tilde{D}_j\tilde{\beta}_i - \frac{2}{3}\tilde{\gamma}_{ij}\tilde{D}_k\tilde{\beta}^k - u_{ij}, \tag{5.233}$$

where $\tilde{\beta}_i = \tilde{\gamma}_{ij}\beta^j$ and $\tilde{\beta}^k = \beta^k$. Substituting this equation into the momentum constraint equation (2.96) yields

$$\tilde{\Delta}\tilde{\beta}_i + \frac{1}{3}\tilde{D}_i\tilde{D}_j\tilde{\beta}^j + \tilde{R}_{ij}\tilde{\beta}^j + \tilde{D}^j \ln\left(\frac{\psi^6}{\alpha}\right)\left[\tilde{D}_j\tilde{\beta}_i + \tilde{D}_i\tilde{\beta}_j - \frac{2}{3}\tilde{\gamma}_{ij}\tilde{D}_k\tilde{\beta}^k\right]$$
$$- \frac{\alpha}{\psi^6}\tilde{D}_k\left(\alpha^{-1}\psi^6\tilde{\gamma}^{jk}u_{ij}\right) = 16\pi\alpha J_i. \tag{5.234}$$

This elliptic equation is used for determining $\tilde{\beta}_i$, and subsequently, $\tilde{A}_{ij}$ is determined by equation (5.233).

The equation for $\tilde{\gamma}_{ij}$ is derived from the evolution equation for $\tilde{A}_{ij}$, (2.84), as

$$\Delta_\mathrm{f}\tilde{\gamma}_{ij} = 2\left[R^\psi_{ij} - \frac{1}{2}\overset{(0)}{D}_i\left(F^k\tilde{\gamma}_{jk}\right) - \frac{1}{2}\overset{(0)}{D}_j\left(F^k\tilde{\gamma}_{ik}\right) + R^\mathrm{NL}_{ij} - \frac{\psi^4}{3}\tilde{\gamma}_{ij}R\right.$$
$$\left. - \frac{1}{\alpha}\left(D_iD_j\alpha - \frac{\psi^4}{3}\tilde{\gamma}_{ij}D_kD^k\alpha\right)\right] - 4\psi^4\tilde{A}_{ik}\tilde{A}_j{}^k$$
$$+ \frac{2\psi^4}{\alpha}\left[\left(\tilde{D}_i\beta^k\right)\tilde{A}_{jk} + \left(\tilde{D}_j\beta^k\right)\tilde{A}_{ik} - \frac{2}{3}\left(\tilde{D}_k\beta^k\right)\tilde{A}_{ij} + \beta^k\tilde{D}_k\tilde{A}_{ij}\right]$$
$$- 16\pi\left(S_{ij} - \frac{\psi^4}{3}\tilde{\gamma}_{ij}S_k{}^k\right) - \frac{2\psi^4}{\alpha}v_{ij}, \tag{5.235}$$

where we decomposed the Ricci tensor as $R_{ij} = \tilde{R}_{ij} + R^\psi_{ij}$, and

$$\tilde{R}_{ij} = \frac{1}{2}\left[-\tilde{\gamma}^{kl}\overset{(0)}{D}_k\overset{(0)}{D}_l\tilde{\gamma}_{ij} - \overset{(0)}{D}_j\left(\tilde{\gamma}_{ik}F^k\right) - \overset{(0)}{D}_i\left(\tilde{\gamma}_{jk}F^k\right)\right.$$
$$\left. - \left(\overset{(0)}{D}_k\tilde{\gamma}_{jl}\right)\overset{(0)}{D}_i\tilde{\gamma}^{kl} - \left(\overset{(0)}{D}_k\tilde{\gamma}_{il}\right)\overset{(0)}{D}_j\tilde{\gamma}^{kl} + 2F^k\overset{(0)}{\Gamma}{}_{k,ij} - 2\overset{(0)}{\Gamma}{}^l{}_{jk}\overset{(0)}{\Gamma}{}^k{}_{il}\right]$$
$$= \frac{1}{2}\left[-\Delta_\mathrm{f}\tilde{\gamma}_{ij} - \overset{(0)}{D}_j\left(\tilde{\gamma}_{ik}F^k\right) - \overset{(0)}{D}_i\left(\tilde{\gamma}_{jk}F^k\right)\right] + \tilde{R}^\mathrm{NL}_{ij}. \tag{5.236}$$

R^ψ_{ij} and $\overset{(0)}{\Gamma}{}^k{}_{ij}$ are defined in equations (2.82) and (2.93), respectively, R^NL_{ij} denotes the sum of nonlinear terms of $\tilde{\gamma}_{ij} - f_{ij}$ in $\tilde{R}_{ij}$, and $F^i := \overset{(0)}{D}_j\tilde{\gamma}^{ij}$.

The equations for α, β^k, and ψ are written in elliptic equations. On the other hand, the type of the equation for $\tilde{\gamma}_{ij}$ depends on the chosen spatial gauge condition. The simplest choice for this is a transverse gauge $F^k = 0$. With this choice, $\Delta_\mathrm{f}\tilde{\gamma}_{ij}$ becomes a unique term linear in $\overset{(0)}{D}_k\overset{(0)}{D}_l\tilde{\gamma}_{ij}$. Then, this equation can be seen as an equation of mixed type. In the near zone,

$$v_{ij} \sim -\Omega_\mathrm{orb}\mathscr{L}_\varphi\tilde{A}_{ij} \sim (\Omega_\mathrm{orb}\mathscr{L}_\varphi)^2\tilde{\gamma}_{ij}. \tag{5.237}$$

Thus the equation is a Helmholtz-type equation. On the other hand, in the distant zone, $v_{ij} \to 0$, and thus, the equation becomes elliptic type.

This formulation has not been used yet for $F(r) \neq 0$. For the assumption $F(r) = 0$, Uryū et al. (2006, 2009) computed quasi-equilibrium states of binary neutron stars employing this formulation (see section 8.2). For this choice of $F(r)$, we approximate that gravitational waves are absent in the spacetime. Thus, this approximation is called a *waveless* approximation. However, even with the waveless approximation, effects of $\tilde{\gamma}_{ij} - f_{ij}$ are taken into account in the near zone.

Alternatively, when we set $\tilde{\gamma}_{ij} = f_{ij}$ and ignore the equation for $\tilde{\gamma}_{ij}$, (5.235), the formulation reduces to the IWM formulation irrespective of the choice of $F(r)$. This implies that the IWM formulation may be regarded as a formulation for computing quasi-equilibria neglecting the effects of the tensor component $\tilde{\gamma}_{ij} - f_{ij}$; the IWM formulation imposes an even simpler waveless approximation. This assumption is justified when the magnitude of $\tilde{\gamma}_{ij} - f_{ij}$ is small. This is not guaranteed in general, in particular, in a strong gravitational field. Indeed, rapidly spinning black holes are not well approximated in this formulation. However, numerical studies by Uryū et al. (2006, 2009) showed that the magnitude of $\tilde{\gamma}_{ij} - f_{ij}$ is small as far as binary neutron stars, which are not in extremely close orbits, are concerned (see section 8.2.3.2). This fact encourages us to employ the IWM formulation for constructing quasi-equilibria of compact binary systems (in moderately close orbits).

5.7.2.2 *Formulation for hydrostatics*

As shown in section 5.7.1.2, the conservation equation for the stress-energy tensor of a perfect fluid is integrated to yield a first integral, if the fluid is isentropic and the spacetime is stationary, axisymmetric, and satisfies the circularity condition. In this section, we show that the first integral can be obtained even when there is only one timelike Killing vector field at least for a special class of velocity fields [Bonazzola et al. (1997); Asada (1998); Shibata (1998); Teukolsky (1998)]. The helical Killing vector field (5.228) or (5.229) is a timelike Killing vector field outside black-hole horizons and inside the light cylinder. Thus, the following calculation is valid for the neutron-star binaries.

For perfect fluids, the conservation equation for the stress-energy tensor is written in the form of equation (5.207). Denoting a timelike Killing vector field by ξ^a and writing the four velocity as $u^a = u^t(\xi^a + v^a)$ where $v^a n_a = 0$, equation (5.207) is written in the form

$$\rho u^a \nabla_a (hu_b) + \nabla_b P$$
$$= \rho u^t \left[\mathcal{L}_\xi (hu_b) - hu_a \nabla_b \xi^a + v^a \nabla_a (hu_b) \right] + \nabla_b P$$
$$= \rho u^t \left[\mathcal{L}_\xi (hu_b) + v^a \omega_{ab} - \nabla_b (hu_a \xi^a) \right] - \rho \nabla_b h + \nabla_b P = 0, \quad (5.238)$$

where ω_{ab} is an anti-symmetric tensor defined by

$$\omega_{ab} := \nabla_a (hu_b) - \nabla_b (hu_a). \quad (5.239)$$

In equation (5.238), $\mathscr{L}_\xi(hu_b)$ vanishes because ξ^a is a Killing vector field. In addition, equation (5.207) is written in the form

$$\rho u^a \omega_{ab} - \rho \nabla_b h + \nabla_b P = 0. \tag{5.240}$$

Thus, if $\omega_{ab} = 0$, then $-\rho \nabla_b h + \nabla_b P = 0$ have to be satisfied (i.e., the fluid is isentropic), and equation (5.238) yields a first integral as

$$hu_a \xi^a = \text{const}, \tag{5.241}$$

or

$$\frac{h}{u^t} + hu_a v^a = \text{const}. \tag{5.242}$$

Therefore, for $\omega_{ab} = 0$, the first integral of the equations of motion is derived in the presence of a timelike Killing vector field.

We may rephrase the condition of $\omega_{ab} = 0$ in the presence of the isentropic condition $\nabla_a s = 0$, and thus, $u^a \nabla_a(hu_b) + \nabla_b h = 0$, because ω_{ab} in this case agrees with the definition of the vorticity tensor as

$$\hat{\omega}_{ab} = q_a{}^c q_b{}^d [\nabla_c(hu_d) - \nabla_d(hu_c)], \tag{5.243}$$

where $q_{ab} := g_{ab} + u_a u_b$. Thus, we can obtain the first integral of the Euler equation, (5.242), if the velocity field is irrotational, $\hat{\omega}_{ab} = 0$, and the isentropic condition is satisfied.

For $\omega_{ab} = 0$, hu_a is written using the velocity potential Ψ as

$$hu_a = \nabla_a \Psi. \tag{5.244}$$

Then, equation for Ψ is derived from the continuity equation which is rewritten in the form

$$\nabla_a(\rho u^a) = \nabla_a \left[\rho u^t (\xi^a + v^a)\right]$$
$$= \alpha^{-1} D_a(\rho \alpha u^t v^a) = 0, \tag{5.245}$$

where we used $\xi^a \nabla_a(\rho u^t) = 0 = \nabla_a \xi^a$. From u_a, v^a (spatial velocity vector) is calculated as

$$v^a = \gamma^{ab}\frac{u_b}{u^t} - (\bar{\xi}^a + \beta^a) = \frac{1}{hu^t} D^b \Psi - (\bar{\xi}^a + \beta^a), \tag{5.246}$$

where $\bar{\xi}^a = \gamma^a{}_b \xi^b$ and for the case that ξ^a is a helical Killing vector field, it is written by the orbital angular velocity as $\Omega_{\text{orb}}(\partial/\partial\varphi)^a$.

Substituting equation (5.246) into (5.245), an elliptic-type equation for Ψ is derived as

$$D_a[\rho\alpha\{h^{-1}D^a\Psi - u^t(\bar{\xi}^a + \beta^a)\}] = 0. \tag{5.247}$$

This equation should be solved inside stars with the boundary condition on the stellar surface that the velocity field is tangential to the stellar surface [i.e., the surface of $\rho = 0$ (or $h = 1$)].

Equation (5.238) also shows that for $v^a = 0$ (i.e., corotating velocity fields) and $\nabla_a s = 0$ (isentropy), the first integral is obtained as

$$hu_a\xi^a = \frac{h}{u^t} = \text{const.} \qquad (5.248)$$

This agrees with equation (5.216), although equation (5.248) is derived without assuming axial symmetry. Note that the continuity equation (5.245) in this case is automatically satisfied.

For neutron star-neutron star and black hole-neutron star binaries, evolution time scale due to the gravitational-radiation reaction is believed to be shorter than the time scale of the viscous dissipation [Kochanek (1992); Bildsten and Cutler (1992)] (see also section 8.2). For this case, we can consider that the vorticity of the system is conserved. Thus, if the vorticity is zero initially, it is preserved to be zero for their entire evolution. The vorticity of the neutron stars is in reality not zero because they have spins. However, the spin period is typically $0.1-1$ s for the new-born pulsars, and the period should be even longer for the older pulsars (unless they experience the accretion spin-up). One of neutron stars in binaries (the first-born neutron stars) tends to be rapidly spinning according to the observation of the binary neutron stars [e.g., Lorimer (2008)], but the spin period of the neutron stars in the binary neutron stars observed so far is longer than 20 ms. Such periods are much longer than the orbital period of close binaries [approximately estimated as $\sim 5(M/5M_\odot)$ ms where M denotes the total mass of the system] in a stage just before the merger. Thus, for the close binary systems, we may consider that the neutron-star spin does not have a strong impact, and the velocity field may be approximated by an irrotational one. For determining the velocity-field profile of such neutron stars, we should employ equation (5.242).

A corotating velocity field will be achieved if the viscous time scale is shorter than the time scale of the orbital evolution by the gravitational-radiation reaction. We believe that this will not be realized in close neutron-star binaries (see also section 8.2). However, if one is interested in such a binary and would like to determine the velocity field profile, equation (5.248) should be employed as the basic equation for hydrostatics.

5.7.2.3 Virial relation and first law

For the formulation described in section 5.7.2.1, the presence of a timelike Killing vector field is assumed for the entire spacetime (except for the inside of black-hole horizons). The spacetime obtained in that formulation is asymptotically flat (this is not the case for spacetime computed assuming the presence of a helical Killing vector field for the entire spacetime). This implies that the ADM mass and the Komar mass can be defined for the spacetime in that formulation. Then, the virial relation, $M_K = M_{ADM}$, should be satisfied, as we described in section 5.3.3. The virial relation is useful for examining the accuracy of a numerical solution in the formulation of section 5.7.2.1, as well as in the IWM formulation (see section 8.2).[13] The virial relation is, alternatively, used for defining quasi-equilibrium states of black hole-neutron stars binaries and binary black holes (see section 8.6).

The first law of binaries is a relation that holds between two slightly different solutions of quasi-equilibria. If we consider a sequence of a binary that evolves due to the gravitational-radiation reaction, a version of the first law has to be satisfied. We here touch on this fact in the following.

The general form of the first law of binaries in the presence of a helical Killing vector field of the form (5.228) is derived by Friedman *et al.* (2002) (see also appendix G.4.2) as

$$\delta M_{\rm ADM} = \Omega_{\rm orb} \delta J + \sum \frac{\kappa}{8\pi} \delta A_{\rm H} \\ + \int_{\Sigma'_t} \left[\frac{T}{u^t}(\Delta s) dM_B + \left(\frac{h}{u^t} + h u_b v^b\right)\Delta(dM_B) + v^b \Delta(h u_b) dM_B \right]. \tag{5.249}$$

Here, $A_{\rm H}$ is the area of a black hole in its presence, $dM_B = \rho u^t \sqrt{-g}\, d^3x$, and v^a is defined from $u^a = u^t(k_{\rm h}^a + v^a)$. We set $J_{\rm K} = J$ which is equal to $J_{\rm ADM}$ for $K = 0$ that is the slicing condition employed.

As we describe in section 8.2, for neutron stars in compact binaries in nature, rest mass, entropy, and vorticity are likely to be conserved during their evolution. In addition, as far as the binary separation is not extremely small (i.e., the binary is in a quasi-equilibrium state), the absorption of gravitational waves by the black hole is negligible [e.g., Mino *et al.* (1997) for a review], and $\delta A_{\rm H}$ is approximately zero. Thus, the first law should be

$$\delta M_{\rm ADM} = \Omega_{\rm orb} \delta J. \tag{5.250}$$

For a quasi-equilibrium binary that is approximately in a circular orbit with the angular velocity, $\Omega_{\rm orb}$, the dissipation rates of energy and angular momentum by the gravitational-wave emission obey the relation (see section 1.2.4)

$$\frac{dE}{dt} = \Omega_{\rm orb} \frac{dJ}{dt}. \tag{5.251}$$

This implies that the relation (5.250) is consistent with the evolution by the gravitational-radiation reaction.

The first law of the form (5.249) is derived assuming the presence of a helical Killing vector field and using Einstein's equation. For deriving the first law in the IWM formulation and in the formulation of section 5.7.2.1, we have to elaborate on another calculations. Such a calculation was performed by Friedman *et al.* (2002) and Shibata *et al.* (2004), who clarified that for the IWM formulation, the first law is written in the same form as equation (5.249). By contrast, the first law is slightly different in the presence of the Killing vector field of the form (5.229). This is also

---

[13]The virial relation is shown to be satisfied even in the IWM formulation in the presence of a helical Killing vector field [Friedman *et al.* (2002)].

the case for the waveless approximation. Specifically, the first law of binaries in the conservation of rest mass, entropy, vorticity, and horizon area are is written as

$$\delta M_{\rm ADM} = \Omega_{\rm orb}\delta J + \frac{1}{16\pi}\int_{\Sigma_t'} \left(\delta \bar{A}_{ab}\mathscr{L}_{k_{\rm h'}}\tilde{\gamma}^{ab} - \delta\tilde{\gamma}^{ab}\mathscr{L}_{k_{\rm h'}}\bar{A}_{ab}\right) d^3x, \quad (5.252)$$

where $\bar{A}_{ab} = \tilde{A}_{ab}\psi^6$. Equation (5.252) implies that it does not in general reduce to equation (5.250) [and hence, this formulation derives a binary sequence that might contradict with the evolution by the gravitational-radiation reaction]. However, the computation of quasi-equilibrium sequences in the waveless formulation by Uryū et al. (2006, 2009) found that the deviation from the first-law relation is quite small for a wide variety of binary neutron stars. This is due to the fact that the magnitude of the second term (the integral term) in the right-hand side of equation (5.252) is small. Thus, the waveless formulation is still useful for computing sequences of binary neutron stars.

5.7.2.4 *Procedure for a numerical solution of neutron-star binary systems*

As in the case for a numerical solution of rapidly rotating stars described in section 5.7.1.5, a careful iteration procedure is necessary for computing quasi-equilibria of neutron star binaries. Again, the basic equations for this problem are gravitational-field equations and hydrostatic equations, and these equations have to be self-consistently solved using an iteration method. The free parameters in this equation system are C (an integration constant of the equations of motion) and $\Omega_{\rm orb}$. We here focus only on the case that the velocity field of neutron stars is irrotational, because this is believed to be the velocity field that realistic neutron stars in close binaries would approximately have. (We note that for corotational velocity fields, the same procedure would work.) In this case, the neutron-star spin is automatically fixed to be zero and we do not have to introduce additional parameters that would be necessary for computing rapidly rotating stars (see section 5.7.1.5). However, if the system is composed of two different neutron stars, we have two different constants for C. In addition, we have to determine the center of mass of the system in an appropriate manner, if the masses of two stars are different. Thus in total, we have to fix these 2–4 parameters imposing 2–4 conditions.

In this section, we summarize the methods for fixing these parameters employed in the latest works [Bonazzola et al. (1999); Uryū and Eriguchi (2000); Gourgoulhon et al. (2001); Taniguchi and Gourgoulhon (2002, 2003); Taniguchi and Shibata (2010)]. Although a few different methods for fixing free parameters have been proposed, we will only describe the typical one here. In the following, we will always suppose that the center of mass of each binary component is located on the x-axis.

First, we outline the procedure for a solution of binaries composed of two identical neutron stars. In this case, the values of C for two neutron stars are identical, and the intersection between the rotation axis and the orbital plane trivially agrees with the center of mass. Thus, we can focus on the procedure for determining two parameters C and $\Omega_{\rm orb}$. These parameters are determined by giving the location of the center of each neutron star $x = \pm X_{\rm NS}$. If the "center" is supposed to be the maxima of the specific enthalpy, the following relations hold there

$$h|_{x=\pm X_{\rm NS}} = h_c, \qquad \left.\frac{\partial \ln h}{\partial x}\right|_{x=\pm X_{\rm NS}} = 0, \qquad (5.253)$$

where h_c is a given value, which specifies the mass of each neutron star for a given equation of state. Using equation (5.242), these two conditions are used for deriving a simultaneous equation for C and $\Omega_{\rm orb}$, and thus, these parameters are updated at each iteration step.

For unequal-mass binary neutron stars, we have four free parameters; C of stars 1 and 2, $\Omega_{\rm orb}$, and the location of the center of mass for the system along the x-axis. In this case, condition (5.253) is present for two stars, and thus, there are in total four conditions, which can determine the four parameters. Logically, this procedure is valid, but in actual numerical computation, in particular for a system of a large mass asymmetry, this method is not robust because the computation often does not lead to a convergent solution [Taniguchi and Shibata (2010)]. A better procedure is to fix the center of mass of the system by imposing the condition that the ADM linear momentum of the system vanishes, $P_i^{\rm ADM} = 0$ [see equation (5.27) for the definition of $P_i^{\rm ADM}$]. Taniguchi and Shibata (2010) showed that this method is robust irrespective of the mass ratio. A drawback in this method is that one of four equations composed of (5.253) becomes a redundant equation. Specifically, the values of $\Omega_{\rm orb}$ determined from the conditions for two different stars do not agree with each other. However, the difference is empirically quite small. Thus, Taniguchi and Shibata (2010) proposed to take a pure average of two slightly different values for determining $\Omega_{\rm orb}$.

For black hole-neutron star binaries, free parameters are C, $\Omega_{\rm orb}$, and the location of the center of mass of the system. The procedure for fixing C and $\Omega_{\rm orb}$ is the same as that for equal-mass binary neutron stars. On the other hand, the procedure for determining the location of the center of mass of the system depends on the method of handling the black-hole horizon. For the case that the excision method is employed, a black hole has to be characterized by imposing several boundary conditions on a hypothetical surface of a black-hole horizon (see section 5.6.3). In this case, the center of mass of the system is determined by the condition $P_i^{\rm ADM} = 0$. On the other hand, in the puncture approach, the condition $P_i^{\rm ADM} = 0$ was already used for determining the linear momentum parameter of the black hole (see section 5.6.4). In this case, we do not know the best way for determining the center of mass. In the works performed so far, a rather phenomenological method was used [Kyutoku et al. (2009)]; e.g., they proposed to impose that the dipole part of the conformal factor is asymptotically zero,

$$\oint_{r\to\infty} x^k f^{ij} \partial_i \overset{(0)}{\psi} dS_j = 0, \qquad (5.254)$$

or to impose that the total angular momentum of the system should agree with the results predicted by a post-Newtonian approximation for a resultant value of $\Omega_{\rm orb}$. Kyutoku et al. (2009) showed that the latter prescription can provide a fairly good quasi-equilibrium state.

5.8 Numerical methods for a solution of elliptic equations

For a numerical solution of initial data, equilibria, and quasi-equilibria, we have to solve elliptic equations such as equations (5.6) and (5.7). In this section, we will outline two typical numerical methods for deriving the solution of elliptic equations.

In the following, we will focus basically on a method of a numerical solution for the simplest elliptic equation (i.e., Poisson's equation) of the form

$$\Delta_{\rm f}\phi = S, \tag{5.255}$$

where ϕ is a smooth function and S is a function of spatial coordinates. In practice, S may be a nonlinear function of ϕ like equation (5.7). In this case, we have to solve equation (5.255) iteratively starting from a trial function of ϕ until a sufficient convergence is achieved. However, even in such a case, the first step is to solve an equation in the form of (5.255) for a given function of S. Thus, a numerical method for a solution of equation (5.255) is fundamental for a solution of any elliptic equation.

5.8.1 *Finite-differencing schemes*

In finite-differencing schemes, we first discretize equation (5.255) using the same method as that described in section 3.1.1 (i.e., based on the Taylor expansion). For example, in the second-order accurate finite differencing (assuming the uniform grid), we have

$$\partial_x^2 \phi = \frac{\phi(x + \Delta x) - 2\phi(x) + \phi(x - \Delta x)}{\Delta x^2}, \tag{5.256}$$

and the same finite differencing is used for y and z directions. Note that throughout this section, we always use these Cartesian coordinates.

Together with the finite differencing, boundary conditions have to be imposed at the outer boundaries of the computational domain. For Poisson's equation, the outer boundary condition is written as

$$\phi = -\frac{C^0}{r} - \frac{C_i^1 \bar{n}^i}{r^2} - \frac{C_{ij}^2 (3\bar{n}^i \bar{n}^j - \delta^{ij})}{2r^3} - \cdots, \tag{5.257}$$

where $\bar{n}^i = x^i/r$ and $C_{ij\ldots}^I$ (for $I = 0, 1, 2, \cdots$) are constants determined by the volume integral of S, e.g.,

$$C^0 = \frac{1}{4\pi} \int S \, d^3x,$$
$$C_i^1 = \frac{1}{4\pi} \int S x^i d^3x,$$
$$C_{ij}^2 = \frac{1}{4\pi} \int S \left(x^i x^j - \frac{1}{3} r^2 \delta^{ij} \right) d^3x. \tag{5.258}$$

Here, we assumed that S quickly approaches zero for $r \to \infty$ [in the presence of a slow fall-off term, the higher-order coefficients could not be simply defined and the boundary condition could be different from equation (5.257)].

If we take into account only the lowest-order contribution of $O\left(r^{-1}\right)$ (assuming that the outer boundaries are located at a sufficiently distant zone), the boundary conditions can be written as

$$\partial_r (r\phi) = r\partial_r \phi + \phi = 0, \qquad \partial_\theta \phi = 0, \qquad \partial_\varphi \phi = 0, \tag{5.259}$$

and thus,

$$(r^2/x)\partial_x \phi + \phi = 0, \qquad (r^2/y)\partial_y \phi + \phi = 0, \qquad (r^2/z)\partial_z \phi + \phi = 0. \tag{5.260}$$

These can be used for imposing the outer boundary conditions along x, y, and z directions, respectively. For example, the second-order finite-differencing form of the first equation of (5.260) is written as

$$\phi(x + \Delta x) = \phi(x - \Delta x) - \frac{2x\Delta x}{r^2} \phi(x). \tag{5.261}$$

By this procedure, equation (5.255) is written to a matrix equation of the form

$$M_{ab}\phi_b = S_a, \tag{5.262}$$

where M_{ab} could be written in a symmetric matrix, $M_{ab} = M_{ba}$, in the centered finite differencing together with the outer boundary conditions described above.[14] We defined $\phi_b := \phi(x_i, y_j, z_l)$ and $S_b := S(x_i, y_j, z_l)$ with $b = i + (j-1)N_x + (k-1)N_x N_y$ where N_k denotes the number of the grid points in the x^k directions. Thus, $1 \leq b \leq N_x N_y N_z =: N_{xyz}$, and M_{ab} is a $N_{xyz} \times N_{xyz}$ matrix. Finally, we note that the summation is taken for the subscript b in equation (5.262).

In the second-order finite differencing, the schematic form of M_{ab}, e.g., for $N_x = N_y = N_z = 3$ is written as

---

[14]In this section, subscripts a, b, $\cdots$ do not denote abstract indices.

$$M_{ab} = \begin{pmatrix} * * & * & & & * & & & & & \\ * * * & * & & & & * & & & & \\ * * & * & & & & * & & & & \\ * & * * & * & & & & * & & & \\ * & * * * & * & & & & * & & & \\ & * & * * & * & & & & * & & \\ & & * & * * * & & & & & * & \\ & & & * & * * & & & & & * \\ * & & & & * * & * & & & * & \\ & * & & & * * * & * & & & * \\ & & * & & & * * & * & & & * \\ & & & * & * & * * * & * & & & * \\ & & & & * & * * * & * & & & * \\ & & & & & * & * * & & & & * \\ & & & * & & & * * & * & & \\ & & & & * & & & * * * & * & \\ & & & & & * & & & * * & * \\ & & & & & & * & & * & * * & * \\ & & & & & & & * & & * * * & * \\ & & & & & & & & * & * * & \\ & & & & & & & & * & & * * \\ & & & & & & & & & * & * * * \\ & & & & & & & & & & * & * * \end{pmatrix}, \quad (5.263)$$

where $*$ denotes a nonzero component and in its absence, the components are zero. This matrix is composed of seven diagonal elements (note that zero on these diagonal elements is due to the fact that it is located at an outer boundary). If we employ the fourth-order finite differencing, the matrix is composed of thirteen diagonal elements. These facts hold irrespective of N_{xyz}, and thus, the matrix components are not densely distributed.

Although it is sparse, the matrix in numerical relativity is quite large because N_x, N_y, and N_z are typically of order 100. This implies that N_{xyz} is larger than 10^6, and a quite large matrix equation has to be solved. The primary reason for this is that in the finite-differencing scheme, the numerical error decreases as the power of N_x, e.g., as $O\left(N_x^{-2}\right)$ in the second-order scheme and as $O\left(N_x^{-4}\right)$ in the fourth-order scheme. Thus for an accurate numerical computation, large values of N_x, N_y, and N_z are necessary (this situation is in contrast to that in the spectral method; see section 5.8.2).

For a solution of such large but sparse matrix equations, preconditioned conjugate gradient (PCG) methods are quite robust [e.g., appendix A of Oohara *et al.* (1997)]. We will outline this method as follows.

As a first step, we describe a simple conjugate gradient (CG) method. The CG method is developed from the fact that the solution of equation (5.262) extremizes the following function

$$f(\phi_a) = \frac{1}{2} M_{ab} \phi_a \phi_b - \phi_a S_a. \quad (5.264)$$

Again, we assumed that the summation is taken for the subscripts a and b. In the following, we will assume that $M_{ab}\phi_a\phi_b \neq 0$ for any nonzero vector ϕ_a (this is usually satisfied). Taking the derivative of f with respect to ϕ_a, we find

$$\frac{\partial f}{\partial \phi_a} := M_{ab}\phi_b - S_a. \quad (5.265)$$

Thus, when f is extremized for all the components of ϕ_a, the solution of equation (5.262) is obtained.

In the CG method, we try to find the solution by iteratively adding a series of vectors p_a^n starting from a trial solution ϕ_a^0 as

$$\phi_a^{n+1} = \phi_a^n + \hat{\alpha}^n p_a^n = \phi_a^0 + \sum_{k=0}^{n} \hat{\alpha}^k p_a^k, \tag{5.266}$$

where the superscript n denotes the iteration step. An explicit method for giving p_a^n is described as follows. The coefficient $\hat{\alpha}^n$ is determined by the condition that $f(\phi_a^{n+1})$ should be extremized with respect to $\hat{\alpha}^n$ for given vectors of ϕ_a^n. This condition is easily derived from

$$f(\phi_a^{n+1}) = f(\phi_a^n) - \hat{\alpha}^n r_a^n p_a^n + \frac{1}{2}(\hat{\alpha}^n)^2 M_{ab} p_a^n p_b^n, \tag{5.267}$$

where r_a^n is called a residual vector defined by

$$r_a^n := -\frac{\partial f}{\partial \phi_a^n} = S_a - M_{ab} \phi_b^n. \tag{5.268}$$

From equation (5.267), $\hat{\alpha}^n$ is found as

$$\hat{\alpha}^n = \frac{p_a^n r_a^n}{M_{bc} p_b^n p_c^n}. \tag{5.269}$$

We note that r_a^{n+1} is also calculated from

$$r_a^{n+1} = r_a^n - \hat{\alpha}^n M_{ab} p_b^n. \tag{5.270}$$

Since the solution of ϕ_a is written as a linear combination of p_a^k ($k = 1, 2, 3, \cdots$), each component of p_a^k should be independent. For obtaining such an independent set, the following condition is required in the CG method:

$$M_{ab} p_a^k p_b^l = 0, \quad \text{for } k \neq l. \tag{5.271}$$

It can be proven that the condition (5.271) is satisfied if

$$p_a^0 = r_a^0 \quad \text{and} \quad p_a^{n+1} = r_a^{n+1} + \hat{\beta}^n p_a^n, \tag{5.272}$$

where

$$\hat{\beta}^n = -\frac{M_{ab} r_a^{n+1} p_b^n}{M_{cd} p_c^n p_d^n}. \tag{5.273}$$

To summarize, the algorithm of the simple CG method is:

(1) A trial solution, ϕ_a^0, is given. Then, $r_a^0 = S_a - M_{ab}\phi_b^0$ and set $p_a^0 = r_a^0$.
(2) If $|r^k| < \epsilon|S|$ ($k = 0, 1, 2, \cdots$), stop the iteration. Otherwise, repeat the following procedures until the convergence ($|r^k| < \epsilon|S|$) is achieved:

$$\hat{\alpha}^k = \frac{p_a^k r_a^k}{M_{bc} p_b^k p_c^k},$$

$$\phi_a^{k+1} = \phi_a^k + \hat{\alpha}^k p_a^k, \quad r_a^{k+1} = r_a^k - \hat{\alpha}^k M_{ab} p_b^k,$$

$$\hat{\beta}^k = -\frac{M_{ab} r_a^{k+1} p_b^k}{M_{cd} p_c^k p_d^k},$$

$$p_a^{k+1} = r_a^{k+1} + \hat{\beta}^k p_a^k. \tag{5.274}$$

Here, $|r^k| = \sum_a |r_a^k|$, $|S| = \sum_a |S_a|$, and ϵ is the tolerance, e.g., 10^{-8}.

In the simple CG method, a solution of ϕ_a should be obtained in principle by the N_{xyz} iteration steps because of the linear independence of p_a^k for different values of k. In numerical computation, however, taking the sum of all the terms is not practical at all, because N_{xyz} is extremely large and thus the computational cost is quite expensive.

It is well known that the convergence property of the CG method is significantly improved if the eigenvalues of the matrix M_{ab} are sufficiently degenerate to special values. This implies that if we can modify equation (5.262) to a form with a desired kernel matrix for which the eigenvalues are sufficiently degenerate, the CG method could be an efficient method in numerical computation. Such modified CG methods, by which an efficient computation is feasible, are called preconditioned CG (PCG) methods.

There are several PCG methods proposed so far. One of the most popular ones would be incomplete Choleskii conjugate gradient (ICCG) method [Meijerink and van der Vorst (1977)]. Here, we introduce one of the most simplest ones, called Neumann-type PCG method, to illustrate that the PCG scheme can be easily developed. The algorithm of this method is quite simple, and thus, it is recommendable to the beginners of numerical computation.

In this method, we first modify equation (5.262) as

$$(M_{ac}D_{cb}^{-1})(D_{bd}\phi_d) = M'_{ab}\phi'_b = S_a, \tag{5.275}$$

where

$$M'_{ab} = M_{ac}D_{cb}^{-1} \quad \text{and} \quad \phi'_a = D_{ab}\phi_b. \tag{5.276}$$

For D_{ab} we choose a diagonal matrix satisfying $D_{ab} = M_{ab}$ for $a = b$ and $D_{ab} = 0$ for $a \neq b$. Then, M'_{ab} has the form

$$M'_{ab} = I_{ab} - Z_{ab}, \tag{5.277}$$

where I_{ab} is the unit matrix and Z_{ab} is composed only of off-diagonal components. For the finite-differencing form of equation (5.255), the absolute magnitude of the components of Z_{ab} is smaller than unity. This implies that if we multiply the following matrices to M'_{ab},

$$I_{ab} + Z_{ab} \quad \text{or} \quad I_{ab} + Z_{ab} + Z_{ac}Z_{cb}, \tag{5.278}$$

the resulting matrix approaches I_{ab}, and many of its eigenvalues could approach unity. By this motivation, equation (5.275) is further modified, e.g., to the form,

$$(I_{ab} - Z_{ac}Z_{cb})\phi'_b = (I_{ab} + Z_{ab})S_b, \tag{5.279}$$

where, in this case, $I_{ab} + Z_{ab}$ is multiplied. Then, applying the simplest CG method could be an efficient scheme. This fact indeed has been shown to be the case in numerical computation.

5.8.2 Spectral and pseudo-spectral methods

Spectral and pseudo-spectral methods are powerful tools for a solution of partial differential equations, in particular, for the case that all the involved functions are analytic. Multi-domain spectral methods can furthermore handle discontinuous functions as far as the discontinuity is enclosed within domain boundaries. We here outline the spectral method, focusing on a solution of elliptic equations for initial data of numerical relativity, because it is now the standard method for computing initial data and quasi-equilibria. In general, the spectral method is also applicable to solving hyperbolic equations [for more detail, see, e.g., Pfeiffer et al. (2003); Grandclément and Novak (2009) and references therein]. In particular, Caltech-Cornell-CITA group has been extensively developing this approach for their simulations of black-hole binaries [e.g., Boyle et al. (2007); Duez et al. (2008); Scheel et al. (2009) for their earliest works].[15]

5.8.2.1 The idea of the spectral method

The central idea of the spectral method is to represent a function by an expansion into a complete set of basis functions, and then, to solve equations for the spectral coefficients. In numerical computation, continuous functions have to be represented by a finite degree of freedom. In the spectral method, a function $f(x)$ is written by the expansion into $L+1$ basis functions of the form

$$f(x) \simeq \sum_{n=0}^{L} \tilde{f}_n \phi_n(x), \qquad (5.280)$$

where $\{\phi_n\}$ constitutes a complete basis, such as the trigonometric functions. Coefficients $\{\tilde{f}_n\}$ are the discretized representation of the function $f(x)$ in the spectral method, and equations to be solved are rewritten as simultaneous equations for these spectral coefficients. This should be compared with the finite-differencing methods, described in sections 3.1 and 5.8.1.

A striking difference between the spectral method and the finite-differencing method is in the way to evaluate derivatives $f'(x)$. The spectral method evaluates derivatives directly using the basis functions as

$$f'(x) \simeq \sum_{n=0}^{L} \tilde{f}_n \phi'_n(x). \qquad (5.281)$$

Expanding $\phi'_n(x)$ into $\phi_n(x)$, the coefficients $\{\tilde{f}_n\}$ give the expression for $f'(x)$. Here, the error associated with the discretization results from the finite number, $L+1$, of the basis functions. It is known that this error decreases faster than any power of L^{-1} for smooth functions, and the error (the L^∞-norm of the difference between $f'(x)$ and its discretized expression) typically decreases in an exponential manner

---
[15]A draft of this section was written by K. Kyutoku.

as e^{-L}. By contrast, in finite-differencing methods, derivatives are evaluated using a local expansion into a Taylor series, such as

$$f'_n \simeq \frac{f_{n+1} - f_{n-1}}{2\Delta x}, \tag{5.282}$$

where Δx is the grid separation. Here, we employ a second-order finite differencing. In general, the discretization error with an l-th-order formula behaves as $(\Delta x)^l \propto L^{-l}$. This suggests that the spectral method is quite powerful to accurately solve the differential equations with limited computational resources. In the following, we will describe the spectral method in more detail.

5.8.2.2 *Representing functions*

The first step in the spectral method is to establish a way to project a function $f(x)$ onto spectral space composed of a set of basis functions. For simplicity, we focus on a one-dimensional problem in $x \in [-1:1]$. One of the simplest spectral methods is to use the discrete Fourier transform as

$$f(x) \simeq \tilde{a}_0 + \sum_{m=1}^{L} \tilde{a}_m \cos[m\pi(x+1)] + \sum_{n=1}^{L} \tilde{b}_n \sin[n\pi(x+1)]. \tag{5.283}$$

This choice is suited for the case that a periodic boundary condition is satisfied. However, the discrete Fourier transformation is not suited for problems with general boundary conditions. The reason for this is that the use of trigonometric functions usually results in enforcing undesired boundary conditions on the higher-order derivatives. For example, Dirichlet boundary conditions $f(\pm 1) = 0$ requires the sine-series expansion, and this enforces any even-order derivative to be zero at the boundary.

For problems with general boundary conditions, Chebyshev polynomials $T_n(x)$ are often used for the expansion basis. Here, n is zero or positive integer. The Chebyshev polynomials are defined as solutions of an ordinary differential equation,

$$\left[\sqrt{1-x^2}T'_n(x)\right]' + \frac{n^2}{\sqrt{1-x^2}}T_n(x) = 0, \tag{5.284}$$

and satisfy orthogonality relations in $[-1:1]$ written by

$$\int_{-1}^{1} \frac{T_n(x)T_m(x)}{\sqrt{1-x^2}}dx = \frac{\pi}{2}(1+\delta_{0n})\delta_{nm}. \tag{5.285}$$

The Chebyshev polynomials also satisfy the recurrence relation,

$$T_{n+1}(x) = 2xT_n(x) - T_{n-1}(x), \tag{5.286}$$

with $T_0(x) = 1$ and $T_1(x) = x$. Hence, $T_n(x)$ is found to be an n-th-order polynomial in x. Their values at $x = \pm 1$ are given by $T_n(\pm 1) = (\pm 1)^n$, and $T_n(x)$ has n zeros in $[-1:1]$. An important fact is that the Chebyshev polynomial is related to the trigonometric functions by

$$T_n(x) = \cos(n\theta), \quad x = \cos\theta. \tag{5.287}$$

This implies that the expansion by the Chebyshev polynomial is essentially the same as that by the trigonometric functions. The merit in the presence of this relation is that the fast Fourier transform (FFT) is applicable in the Chebyshev spectral method. The FFT provides a very efficient way to exchange the physical and spectral space, and is also robust for evaluating nonlinear terms as we will show later.

In the second step, we have to determine how to project a function $f(x)$ onto spectral space composed of a finite polynomial series. Specifically, we have to determine the way to obtain the coefficients $\{\tilde{f}_n\}$ for a finite-series expansion,

$$f(x) \simeq \sum_{n=0}^{L} \tilde{f}_n T_n(x). \tag{5.288}$$

Ideally (e.g., in an analytic calculation), a function $f(x)$ should admit the unique expansion onto the space of a complete basis, and the spectral coefficients may be determined by the coefficients of this expansion, discarding higher-order information for $n > L$. However, in numerical computation [i.e., for a discretized data set of $f(x)$], it is not always guaranteed that the exact coefficients $\{\tilde{f}_n\}$ are obtained. Hence, we have to define an appropriate rule for the projection of $f(x)$ onto the spectral coefficients.

This is achieved by using the orthogonality of the basis functions. Here, the orthogonality is defined by the inner product of functions $f(x)$ and $g(x)$ with a weight function $w(x)$ in the form

$$(f, g) := \int_{-1}^{1} f(x) g(x) w(x) dx. \tag{5.289}$$

In particular, for the Chebyshev spectral method, $w(x)$ should be $1/\sqrt{1-x^2}$. Using this inner product, the spectral coefficients $\tilde{f}_n$ of the basis functions $\phi_n(x)$ are determined by

$$\tilde{f}_n = \frac{(f, \phi_n)}{(\phi_n, \phi_n)}. \tag{5.290}$$

The key task in this procedure is to compute the inner product (f, ϕ_n) for a discretized data set of $f(x)$. For this purpose, the Gaussian quadrature gives a reasonable solution. By taking appropriate collocation points $x_n \in [-1:1]$ with $x_0 = -1$ and $x_L = 1$ as well as weights w_n, the integral of a function $f(x)$ with its polynomial order less than $2L$ is evaluated exactly by

$$\int_{-1}^{1} f(x) w(x) dx = \sum_{n=0}^{L} f(x_n) w_n, \tag{5.291}$$

and the right-hand side of this equation can be evaluated on computers up to the truncation error.[16] For the Chebyshev spectral method, it is known that the

---

[16]The choices $x_0 = -1$ and $x_L = 1$ are not always necessary for the Gaussian quadrature, and this choice is called the Gauss–Lobatto quadrature. Although enforcing these two conditions degrades a degree of freedom for $f(x)$, to which the Gaussian quadrature applies, it significantly simplifies the procedure for imposing the boundary condition.

appropriate choice of the collocation points and weights is given, respectively, by

$$x_n = \cos\frac{n\pi}{L}, \quad (5.292)$$

$$w_n = \frac{\pi}{2L}(2 - \delta_{0n} - \delta_{Ln}). \quad (5.293)$$

Similarly, we can also evaluate the spectral coefficients for the discrete Fourier transformation, (5.283), using the Gaussian quadrature (see also section 5.8.2.4).

5.8.2.3 Solving differential equations

The main task of the initial-data problem is to solve differential equations, schematically written by

$$\hat{L}[f](x) = S(x), \quad (5.294)$$

where $\hat{L}$ is a linear differential operator and (f, S) are fields. In the following, we will assume that the fields can be expanded by

$$f(x) \simeq \sum_{n=0}^{L} \tilde{f}_n \phi_n(x), \quad S(x) \simeq \sum_{n=0}^{L} \tilde{S}_n \phi_n(x). \quad (5.295)$$

In the initial-data problem of numerical relativity, $\hat{L}$ is a second-order, elliptic-type operator, and hence, boundary conditions are necessary for the problem to be well-posed. Here, we write the boundary conditions as

$$\hat{B}[f](x) = 0. \quad (5.296)$$

In the spectral method, discretized equations for a solution of equation (5.294) are usually derived in the idea of the weighted-residual method. For this, we first define the residual of an approximate solution of equation (5.294) by

$$R(x) := \hat{L}[f](x) - S(x), \quad (5.297)$$

and next, we require the residual to satisfy $L + 1$ simultaneous equations,

$$(R, \xi_n) = 0, \quad (5.298)$$

where $\xi_n(x)$ ($0 \leq n \leq L$) is a set of test functions. Different choices of the test functions lead to different methods. In the following, we will show three methods; tau, Galarkin, and collocation methods.

In the tau method, the test functions are the same as the basis functions. As a standard technique in numerical computation, the differential operator and fields are expressed as a matrix and vectors, respectively. In the tau method, equation (5.294) yields $L + 1$ simultaneous equations,

$$\sum_{m=0}^{L} L_{nm} \tilde{f}_m = \tilde{S}_n, \quad (5.299)$$

and the solution of this equation is obtained by a matrix inversion method. However, for this system to be invertible, appropriate boundary conditions have to be imposed to fix freely chosen homogeneous solutions. For imposing the boundary conditions, equations for $m = L - 1$ and L are often replaced by the boundary conditions.

The Galarkin method handles the boundary condition more elegantly. In this method, the function $f(x)$ is expanded by $L - 1$ Galarkin basis functions G_n, all of which are constructed from the orthogonal basis to satisfy the boundary conditions (5.296) by themselves. For example, to satisfy the boundary condition $f(\pm 1) = 0$ with the Chebyshev polynomials, the Galarkin basis is constructed by

$$G_{2j}(x) = T_{2j+2}(x) - T_0(x), \quad G_{2j+1} = T_{2j+3}(x) - T_1(x). \tag{5.300}$$

Here, we should note that they no longer satisfy the orthogonality relation. On the other hand, $S(x)$ are expanded by original orthogonal functions, such as $T_n(x)$, because they do not obey the boundary condition. As the test functions to compute equation (5.298), this Galarkin basis is also chosen, and then, we obtain the discretized equation for the spectral coefficients in the Galarkin basis. A drawback in this method may be that the discretized equations become complicated, especially when the boundary condition is complicated.

The collocation method requires equation (5.294) to be locally satisfied on the collocation points. In this method, the test functions are formally chosen such that they satisfy $\xi_m(x_n) = \delta_{mn}$ on the collocation points $\{x_n\}$. The equation to be solved in the Chebyshev spectral method is given by

$$\sum_{l=0}^{L} \sum_{m=0}^{L} L_{lm} \tilde{f}_l T_m(x_n) = S(x_n), \tag{5.301}$$

and the equations at x_0 and x_L are replaced by the boundary conditions.

5.8.2.4 Non-linearity and the pseudo-Spectral Method

In the initial-data problem of general relativity, we have to handle several nonlinear terms. Evaluation of the nonlinear terms in the spectral method would become computationally expensive, if we tried to do it straightforwardly. As a simple example, we consider the following ordinary differential equation defined in $x \in [0 : 2\pi]$:

$$f''(x) + f(x)^2 = 0. \tag{5.302}$$

Here, we consider the expansion in terms of an exponential function

$$f(x) = \sum_{n=-L}^{L} \tilde{f}_n e^{inx}, \tag{5.303}$$

where the spectral coefficients are complex. As we consider $f(x)$ to be real, the number of independent components is $2L + 1$. The spectral expression of this equation is given by

$$-\sum_{n=-L}^{L} n^2 \tilde{f}_n e^{inx} + \sum_{l=-L}^{L} \sum_{m=-L}^{L} \tilde{f}_l \tilde{f}_m e^{i(l+m)x} = 0, \tag{5.304}$$

and then, taking the exponential function as the test function, we finally obtain the discretized equation in the form

$$-n^2 \tilde{f}_n + \sum_{m=\max(-L,n-L)}^{\min(L,n+L)} \tilde{f}_m \tilde{f}_{n-m} = 0. \tag{5.305}$$

It is evident that the evaluation of the second term requires $O(L^2)$ operations, while the finite-differencing method requires only $O(L)$ operations. The computational cost could become much larger in the presence of higher-order nonlinear terms; specifically $O(L^l)$ for l-th-order terms. For such a case, the spectral method would be too computationally expensive to be employed.

The evaluation of nonlinear terms becomes computationally inexpensive, if we rely on numerical integration. In this model problem, the nonlinear term may be numerically evaluated by

$$\frac{1}{2\pi} \int_0^{2\pi} f(x)^2 e^{-inx} dx. \tag{5.306}$$

Specifically, this is approximated by a finite sum on equidistant J collocation points $x_j = 2\pi j/J$ $(0 \le j \le J-1)$ as

$$\frac{1}{J} \sum_{j=0}^{J-1} f(x_j)^2 e^{-inx_j}. \tag{5.307}$$

It is important to recall that the integral of e^{-inx} with $-L \le n \le L$ can be evaluated exactly by a finite sum as long as the number of collocation point J is larger than $L+1$. This implies that the nonlinear term above can be evaluated exactly if we take $J \ge 3L+1$ collocation points, because the highest-order term in the integrand of equation (5.306) is $e^{\pm i(3L)x}$. This property is sometimes called "the 3/2-rule," because $\approx 3L$ collocation points are required to compute second-order terms composed of terms with $\sim 2L$ independent components. As is naturally expected, for the integration of l-th-order nonlinear terms, $(l+1)L$ collocation points are necessary, and the rule becomes "$(l+1)/2$." For the practical computation, we first obtain values of $f(x_j)$ on each collocation point by the inverse transformation, and next compute the finite sum. Although one might think that these procedures still require $O(L^2)$ operations, the computational cost can be relaxed to $O(L \log L)$ by using the FFT. Therefore, the spectral method could also become an efficient tool to solve the nonlinear differential equations with the help of the FFT.

The pseudo-spectral method also evaluates the nonlinear terms in the physical space via the finite sum, but in it, collocation points taken are not as many as those in the spectral method. As is described above, for evaluating the integral with higher nonlinear terms, a large number of collocation points is necessary. To make the matter worse, we may sometimes encounter terms for which the nonlinear order is infinite, such as $\exp[f(x)]$. In the pseudo-spectral method, the computational cost is reduced by evaluating the nonlinear terms only using the finite sum with the FFT. A drawback of the pseudo-spectral method is that the spectral coefficient of the nonlinear terms is contaminated by higher-order spectral coefficients. For example, if we evaluate the second-order term using only $J \leq 3L$ collocation points, the "real" spectral coefficient $\tilde{f^2}_n$ cannot be obtained but $\tilde{f^2}_n + \tilde{f^2}_{n\pm J}$ is the result, because $-2L < n+J < 2L$ or $-2L < n-J < 2L$ is satisfied. This error is called the alias error, which always occurs due to the higher-order coefficients. If L is large and the higher-order coefficients are sufficiently small, the alias error could be small. In such cases, the pseudo-spectral method can yield a reasonably accurate numerical solution for highly nonlinear differential equations.

The collocation method can be regarded as one of the pseudo-spectral methods in which the number of collocation points is taken to be the same as the number of independent spectral components. The spectral space is equivalent to the physical space in the collocation method due to the same number of degrees of freedom, and the spectral space can be regarded as the tool to compute the derivative of functions. In spite of the unavoidable alias error, the derivatives are evaluated more accurately than those in the finite-differencing method. Therefore, the pseudo-spectral method together with the collocation method is often employed in numerical relativity.

5.8.2.5 *Multi-domain and multidimensional cases*

Up to the previous sections, we have considered only the one-dimensional problem in an interval $x \in [-1:1]$ (or $[0:2\pi]$). However, in general, the problem is multidimensional: Many problems which numerical relativity has to solve is usually multidimensional, e.g., compact binary mergers and supernova explosions. Therefore, developing a multidimensional technique is essential.

The multi-domain method paves the way to treat domains that we cannot map onto $[-1:1]$, such as $r \in [0:\infty)$, and also allows us to handle non-smooth functions, such as rest-mass density accompanying a stellar surface, by adapting the domain boundary to the location at which the value of the function or its derivative jumps.

The basic idea of the multi-domain technique is understood by division of the interval. For example, if $f(x)$ is defined in $x \in [-1:1]$ with some irregular behavior at $x = 0$, it would be natural to divide the computational domain into $x \in [-1:0]$ and $x \in [0:1]$. For this purpose, new coordinates $x_1 := 2x+1$ $(-1 \leq x \leq 0)$ and $x_2 := 2x-1$ $(0 \leq x \leq 1)$ are useful to describe these divided domains. In these two domains, the function $f(x)$ shows completely regular behavior. The boundary conditions at $x = -1$ and $x = 1$ are now regarded as new boundary conditions at $x_1 = -1$ and $x_2 = 1$. The conditions at $x_1 = 1$ and $x_2 = -1$, both of which give $x = 0$, should be replaced by a matching condition for the value and the first derivative of f. If the values of the function and its first derivative do not jump at $x = 0$, the matching condition is simply given by requiring the same values on $x_1 = 1$ and $x_2 = -1$. Appropriate junction conditions are required if the function or first derivative has a jump at $x = 0$.

The idea of the division of the domain is quite useful for computing stars that have a surface. For example, consider a spherical star with its radius R. For such a problem, it is convenient to divide the entire space, $r \in [0:\infty)$, into two regions $r \in [0:R]$ and $r \in [R:\infty)$. The inner region $r \in [0:R]$ can be mapped onto $x \in [-1:1]$ by the transformation $x = 2r/R - 1$, and the outer unbounded interval $r \in [R:\infty)$ can be mapped onto $x \in [-1:1]$ by the transformation $x = -2R/r + 1$.

Multi-dimensionality is handled in a manner similar to the separation of variables. In particular, spherical coordinates, (r, θ, φ), are the most popular in astrophysical problems. A useful method to solve the initial-data problem is to represent a function $f(r, \theta, \varphi)$ using the spherical harmonics $Y_{lm}(\theta, \varphi)$ as

$$f(r, \theta, \varphi) = \sum \tilde{f}_{klm} r^k Y_{lm}(\theta, \varphi). \tag{5.308}$$

By the expansion in terms of the spherical harmonics, the irregular behavior associated with the coordinate singularity at $r = 0$ and $\theta = 0, \pi$ is avoided easily. More importantly, in the initial-data problem of numerical relativity, we often need to solve an elliptic-type equation with the Laplacian being the major kernel, and the spherical harmonics is an eigen function of the angular part of the flat Laplacian. Thus, with the help of the spherical-harmonics decomposition, the operation of the angular Laplacian becomes only the multiplication by $-l(l+1)$.

In practical computation, it is often advantageous to expand radial functions by the Chebyshev polynomials with the multi-domain technique, and to expand polar and azimuthal functions by the trigonometric functions, because the FFT can be used for these expansions. The expansion by the spherical harmonics in the spectral space should be performed when it is required, e.g., computing the Laplacian of a function $f(r, \theta, \varphi)$.

Chapter 6

Extracting gravitational waves

One of the most important roles of numerical relativity is to accurately derive waveforms of gravitational waves emitted from strong gravitational-wave sources such as coalescing binary neutron stars and binary black holes. Here, gravitational waves are defined as time-dependent tensor parts (transverse and tracefree parts: see appendix C.2) in the spatial metric observed in a *wave zone*. The wave zone is defined as a domain satisfying $r \gtrsim \lambda$ where r and λ are the distance from the gravitational-wave source and the wavelength of gravitational waves, respectively. In asymptotically flat spacetime, each component of the metric perturbation defined by $g_{ab} - \eta_{ab}$ is much smaller than unity in the wave zone. Thus, for deriving gravitational waveforms in numerical-relativity simulations, we have to extract the tensor components of the metric perturbation in the wave zone.

There are two major methods for extracting gravitational waves from geometric quantities. One is the so-called gauge-invariant wave-extraction method and the other is the method through the calculation of an outgoing-wave component of the complex Weyl scalar. For a simulation of axisymmetric spacetime with special gauge conditions such as quasi-isotropic and quasi-radial gauges (see sections 2.4.3 and 2.4.4), it is possible to extract gravitational waves by analyzing particular metric quantities (η and ξ) [Stark and Piran (1985); Piran and Stark (1986); Abrahams and Evans (1988, 1990)]. We do not touch on these special cases in this section.

In the following two sections, we will describe the two major methods focusing only on the case that the spacetime dimension is four ($D=4$ and $N=3$). We note that a gauge-invariant wave extraction is feasible using the formulation of Kodama and Ishibashi (2003) in higher-dimensional spacetime. On the other hand, there is no established method based on the extraction of complex Weyl scalars in higher dimensions.

In the last section of this chapter, we also touch on a quadrupole formula which is often used in Newtonian and post-Newtonian simulations. By this formula, it is not possible to derive accurate gravitational waveforms. However, we will point out that it is still a useful formula to study gravitational-wave phases with a good accuracy and gravitational waveforms within the error of 20–30% in amplitude. In particular, this method is useful for the case that the wave amplitude is so tiny that it is not easy to extract gravitational waves from geometric quantities.

6.1 Gauge-invariant wave extraction

As shown in section 2.3.1, the spatial metric in the linear approximation of Einstein's equation is in general composed of two scalar, two vector, and two tensor components for $N = 3$. Here, the vector V^a has to be divergence-free, $\overset{(0)}{D}_a V^a = 0$ and the tensor $h_{ab}^{\rm tt}$ has to be transverse-tracefree, $\overset{(0)}{D}{}^a h_{ab}^{\rm tt} = 0 = f^{ab} h_{ab}^{\rm tt}$. In general relativity, metric components are not invariant and changed by a gauge transformation, $x^a \to x^a - \xi^a$ where ξ^a denotes a small variation of the coordinates composed of two scalar and two vector components, and obviously, it does not contain the tensor components. This implies that in the linear approximation, the tensor components are invariant with respect to the coordinate transformation, and thus, for extracting gravitational waves from geometric quantities, we should extract the gauge-invariant components in the metric.

The idea of gauge-invariant wave extractions is rather old. Moncrief (1974) developed a formulation of a gauge-invariant extraction of gravitational waves that propagate on the Schwarzschild background for the first time. In his method, the metric perturbation is decomposed in terms of tensor spherical harmonics derived by Regge and Wheeler (1957) (see also appendix C for a brief summary). Subsequently, Cunningham *et al.* (1978, 1979); Seidel and Moore (1987); Seidel (1990) extended the Moncrief's formulation to that in the presence of matter in their perturbation studies on a stellar core collapse to a black hole and a neutron star. In numerical relativity, the original method by Moncrief for vacuum spacetime has been used, because the wave extraction is achieved usually in an asymptotically flat region where the matter is absent. Abrahams *et al.* (1992) applied the Moncrief's method for the first time (for axisymmetric black-hole spacetime). Subsequently, their method was used for many other applications. In this section, we outline this method based on a review presented in Nagar and Rezzolla (2005).

Before describing the specific method of a gauge-invariant wave extraction, we summarize how to derive gauge-invariant metric perturbations. For simplicity, we first restrict our attention to the metric perturbation on the flat spacetime background. Let h_{ab} be the metric perturbation, $g_{ab} - \eta_{ab}$. For an infinitesimal coordinate transformation $x^a \to x^a - \xi^a$, the metric perturbation is transformed as

$$h_{ab} \to h_{ab} + \overset{(0)}{\nabla}_a \xi_b + \overset{(0)}{\nabla}_b \xi_a, \qquad (6.1)$$

where $\overset{(0)}{\nabla}_a$ is the covariant derivative associated with the Minkowski metric η_{ab}. We then decompose the spatial components of ξ_a by vector spherical harmonics (see appendix C.1.2) as

$$\xi_r = \sum_{l,m} a_{lm} Y_{lm}, \qquad \xi_\theta = \sum_{l,m} \left(b_{lm} \partial_\theta Y_{lm} - c_{lm} \frac{\partial_\varphi Y_{lm}}{\sin\theta} \right),$$
$$\xi_\varphi = \sum_{l,m} \left(b_{lm} \partial_\varphi Y_{lm} + c_{lm} \sin\theta \partial_\theta Y_{lm} \right), \qquad (6.2)$$

where a_{lm}, b_{lm}, and c_{lm} are functions of t and r, and Y_{lm} is the spherical harmonics function. In terms of the spherical harmonics decomposition, the spatial components of $\overset{(0)}{\nabla}_a \xi_b + \overset{(0)}{\nabla}_b \xi_a$ in spherical-polar coordinates are written as

$$(rr) = 2 \sum_{l,m} \partial_r a_{lm} Y_{lm},$$
$$(r\theta) = \sum_{l,m} \left[\{a_{lm} + r^2 \partial_r (r^{-2} b_{lm})\} \partial_\theta Y_{lm} - r^2 \partial_r (r^{-2} c_{lm}) \frac{\partial_\varphi Y_{lm}}{\sin\theta} \right],$$
$$(r\varphi) = \sum_{l,m} \left[\{a_{lm} + r^2 \partial_r (r^{-2} b_{lm})\} \partial_\varphi Y_{lm} + r^2 \partial_r (r^{-2} c_{lm}) \sin\theta \partial_\theta Y_{lm} \right],$$
$$(\theta\theta) = \sum_{l,m} \left[b_{lm} W_{lm} + \{2 r a_{lm} - l(l+1) b_{lm}\} Y_{lm} - c_{lm} \frac{X_{lm}}{\sin\theta} \right],$$
$$(\theta\varphi) = \sum_{l,m} \left[b_{lm} X_{lm} + c_{lm} W_{lm} \sin\theta \right],$$
$$(\varphi\varphi) = \sum_{l,m} \left[-b_{lm} W_{lm} + \{2 r a_{lm} - l(l+1) b_{lm}\} Y_{lm} + c_{lm} \frac{X_{lm}}{\sin\theta} \right] \sin^2\theta, \qquad (6.3)$$

where W_{lm} and X_{lm} are functions written by Y_{lm}; see appendix C.1.3.

We also decompose the spatial components of h_{ab} as [Regge and Wheeler (1957); Zerilli (1970)] (see also appendix C.1.4)

$$h_{ij} = \sum_{l,m} \left(A_{lm} \begin{bmatrix} Y_{lm} & 0 & 0 \\ * & 0 & 0 \\ * & * & 0 \end{bmatrix} + B_{lm} \begin{bmatrix} 0 & \partial_\theta Y_{lm} & \partial_\varphi Y_{lm} \\ * & 0 & 0 \\ * & * & 0 \end{bmatrix} \right.$$
$$+ r^2 K_{lm} \begin{bmatrix} 0 & 0 & 0 \\ * & Y_{lm} & 0 \\ * & * & \sin^2\theta Y_{lm} \end{bmatrix} + r^2 F_{lm} \begin{bmatrix} 0 & 0 & 0 \\ * & W_{lm} & X_{lm} \\ * & * & -\sin^2\theta W_{lm} \end{bmatrix}$$
$$+ C_{lm} \begin{bmatrix} 0 & -\partial_\varphi Y_{lm}/\sin\theta & \sin\theta \partial_\theta Y_{lm} \\ * & 0 & 0 \\ * & * & 0 \end{bmatrix}$$
$$\left. + r^2 D_{lm} \begin{bmatrix} 0 & 0 & 0 \\ * & -X_{lm}/\sin\theta & \sin\theta W_{lm} \\ * & * & \sin\theta X_{lm} \end{bmatrix} \right), \qquad (6.4)$$

where A_{lm}, B_{lm}, C_{lm}, K_{lm}, D_{lm}, and F_{lm} are functions of t and r. A_{lm}, B_{lm}, K_{lm}, and F_{lm} are associated with the even-parity mode, while C_{lm} and D_{lm} are associated with the odd-parity mode. These variables are calculated from the perturbed spatial metric, $\gamma_{ij} - f_{ij}$, by performing integrations over a coordinate two-dimensional sphere in a far zone (denoted by $\mathcal{S}$) as

$$A_{lm} = \oint_{\mathcal{S}} (\gamma_{rr} - f_{rr}) Y_{lm}^* d\Omega,$$

$$B_{lm} = \frac{1}{l(l+1)} \oint_{\mathcal{S}} \left(\gamma_{r\theta} \partial_\theta Y_{lm}^* + \gamma_{r\varphi} \frac{\partial_\varphi Y_{lm}^*}{\sin^2 \theta} \right) d\Omega,$$

$$K_{lm} = \frac{1}{2r^2} \oint_{\mathcal{S}} \left((\gamma_{\theta\theta} - f_{\theta\theta}) Y_{lm}^* + (\gamma_{\varphi\varphi} - f_{\varphi\varphi}) \frac{Y_{lm}^*}{\sin^2 \theta} \right) d\Omega,$$

$$F_{lm} = \frac{(l-2)!}{2(l+2)! r^2} \oint_{\mathcal{S}} \left(\gamma_{\theta\theta} W_{lm}^* - \gamma_{\varphi\varphi} \frac{W_{lm}^*}{\sin^2 \theta} + 2\gamma_{\theta\varphi} \frac{X_{lm}^*}{\sin^2 \theta} \right) d\Omega,$$

$$C_{lm} = \frac{1}{l(l+1)} \oint_{\mathcal{S}} \left(-\gamma_{r\theta} \frac{\partial_\varphi Y_{lm}^*}{\sin \theta} + \gamma_{r\varphi} \frac{\partial_\theta Y_{lm}^*}{\sin \theta} \right) d\Omega,$$

$$D_{lm} = \frac{(l-2)!}{2(l+2)! r^2} \oint_{\mathcal{S}} \left(-\gamma_{\theta\theta} \frac{X_{lm}^*}{\sin \theta} + \gamma_{\varphi\varphi} \frac{X_{lm}^*}{\sin^3 \theta} + 2\gamma_{\theta\varphi} \frac{W_{lm}^*}{\sin^2 \theta} \right) d\Omega, \quad (6.5)$$

where Y_{lm}^*, W_{lm}^*, and X_{lm}^* are the complex conjugates of Y_{lm}, W_{lm}, and X_{lm}, respectively.

Equation (6.3) shows that by the coordinate transformation, $A_{lm}, \cdots, F_{lm}$ are transformed as

$$A_{lm} \to A_{lm} + 2\partial_r a_{lm},$$
$$B_{lm} \to B_{lm} + a_{lm} + r^2 \partial_r (r^{-2} b_{lm}),$$
$$C_{lm} \to C_{lm} + r^2 \partial_r (r^{-2} c_{lm}),$$
$$K_{lm} \to K_{lm} + 2r^{-1} a_{lm} - l(l+1) r^{-2} b_{lm},$$
$$D_{lm} \to D_{lm} + r^{-2} c_{lm},$$
$$F_{lm} \to F_{lm} + r^{-2} b_{lm}. \quad (6.6)$$

A gauge-invariant quantity is a variable composed of A_{lm}–F_{lm} that is independent of a_{lm}, b_{lm}, and c_{lm}. For the even-parity mode, there are several ways for constructing such a gauge-invariant variable. Here, we focus on the quantity that is not composed of a derivative of F_{lm} because F_{lm} contains the leading order part of gravitational waves in the form $F(r-t)/r$ with F being a regular function (see appendix C.2). Such a quantity is derived by a straightforward calculation as

$$\psi_{lm}^{\text{even}} := \sqrt{\frac{2(l-2)!}{(l+2)!}} r \left[(l-1)(l+2) \left[l(l+1) F_{lm} + K_{lm} \right] \right.$$
$$\left. - 2r \partial_r K_{lm} + 2A_{lm} - 2l(l+1) \frac{B_{lm}}{r} \right], \quad (6.7)$$

where the coefficient is chosen so as to write the luminosity of gravitational waves in a wave zone in a specific form [see Abrahams et al. (1992) and/or equation (6.9)].

For the odd-parity mode, the gauge-invariant variable is uniquely defined (when we use only the spatial tensor components) as

$$\partial_t \psi_{lm}^{\text{odd}} := \sqrt{\frac{2(l+2)!}{(l-2)!}} r^{-1} \left[-C_{lm} + r^2 \partial_r D_{lm} \right]. \tag{6.8}$$

Again, the coefficient is chosen so as to write the gravitational-wave luminosity in a specific form [see equation (6.9)]. We note that in this case, it is not possible to define a variable composed of no-derivative of D_{lm} that contains the leading order part of gravitational waves in the form $F(r-t)/r$.

The gravitational-wave luminosity is calculated using equation (1.71) [or equation (F.6)] in the asymptotically flat region. Among the coefficients of the spherical harmonics expansion, F_{lm} and D_{lm} carry the primary information of gravitational waves (in the gauge often used in numerical relativity), because they behave as $F(r-t)/r$ in the wave zone (see appendix C.2). This implies that the luminosity can be written in the form

$$\begin{aligned}\left.\frac{dE}{dt}\right|_{\text{GW}} &= \frac{1}{16\pi} \sum_{l,m} \oint_{r\to\infty} \left(|\partial_t F_{lm}|^2 + |\partial_t D_{lm}|^2\right) \left(|W_{lm}|^2 + \frac{|X_{lm}|^2}{\sin^2\theta}\right) r^2 d\Omega \\ &= \frac{1}{16\pi} \lim_{r\to\infty} \sum_{l,m} \frac{(l+2)!}{(l-2)!} \left(|\partial_t F_{lm}|^2 + |\partial_t D_{lm}|^2\right) r^2 \\ &= \frac{1}{32\pi} \sum_{l,m} \left(|\partial_t \psi_{lm}^{\text{even}}|^2 + |\partial_t \psi_{lm}^{\text{odd}}|^2\right). \end{aligned} \tag{6.9}$$

Here, for calculating the luminosity of the odd-parity mode, we used the relation $\partial_r D_{lm} = -\partial_t D_{lm}$ which holds in the asymptotic region, $r \to \infty$ [Nagar and Rezzolla (2005)].

Next, we describe a gauge-invariant wave extraction method on a curved-spacetime background. For extending the method based on the spherical-harmonics decomposition described above, we have to choose a spherical background. If we further assume that the background spacetime is static, a straightforward extension of the flat-background formulation can be done. Although for general time-dependent spherical spacetime, it is still possible to construct the gauge-invariant variables [Gerlach and Sengupta (1979)], the formulation becomes slightly complicated.

Choosing a static spherical curved background, instead of flat background, is motivated by the fact that we can renormalize a spherically symmetric mode ($l = m = 0$ mass monopole part), which is the leading-order perturbation term in the far zone, in the background metric. In numerical relativity, however, the background spacetime is not always time-independent (even if it is physically static), and hence, the resulting formulation is applicable only for a restricted slicing condition in which the background metric can become at least approximately time-independent. A good news is that in the gauge condition often used in numerical relativity such as maximal slicing and dynamical slicing (see section 2.2.6.1), a static background is approximately achieved; e.g., in section 2.2.2, we show that a static limit hypersurface is soon reached in the spherically symmetric black-hole spacetime. This suggests that the formulation based on the static spherical background would be applicable for a variety of the problems.

Let h_{ab} be the metric perturbation calculated by $g_{ab} - \overset{(b)}{g}_{ab}$. Here, $\overset{(b)}{g}_{ab}$ is written in the form

$$\overset{(b)}{g}_{\mu\nu} dx^\mu dx^\nu = -\alpha_0^2 dt^2 + A_0^2 dr^2 + r^2 d\Omega, \tag{6.10}$$

where α_0 and A_0 are functions of the areal radius r. In general, the background metric is not written in the form of (6.10), but for the static case, a coordinate transformation can yield the form of (6.10).

For an infinitesimal coordinate transformation $x^a \to x^a - \xi^a$, the metric perturbation is again transformed as

$$h_{ab} \to h_{ab} + \overset{(b)}{\nabla}_a \xi_b + \overset{(b)}{\nabla}_b \xi_a, \tag{6.11}$$

where $\overset{(b)}{\nabla}_a$ is the covariant derivative with respect to $\overset{(b)}{g}_{ab}$. We then decompose the spatial components of ξ_a by vector spherical harmonics as in equation (6.2), and the spatial components of $\overset{(b)}{\nabla}_a \xi_b + \overset{(b)}{\nabla}_b \xi_a$ in spherical-polar coordinates are written as

$$(rr) = 2 \sum_{l,m} [\partial_r a_{lm} - a_{lm}(\partial_r A_0)/A_0] Y_{lm},$$

$$(r\theta) = \sum_{l,m} \left[\{a_{lm} + r^2 \partial_r(r^{-2} b_{lm})\} \partial_\theta Y_{lm} - r^2 \partial_r(r^{-2} c_{lm}) \frac{\partial_\varphi Y_{lm}}{\sin\theta} \right],$$

$$(r\varphi) = \sum_{l,m} \left[\{a_{lm} + r^2 \partial_r(r^{-2} b_{lm})\} \partial_\varphi Y_{lm} + r^2 \partial_r(r^{-2} c_{lm}) \sin\theta \partial_\theta Y_{lm} \right],$$

$$(\theta\theta) = \sum_{l,m} \left[b_{lm} W_{lm} + \{2r a_{lm}/A_0^2 - l(l+1) b_{lm}\} Y_{lm} - c_{lm} \frac{X_{lm}}{\sin\theta} \right],$$

$$(\theta\varphi) = \sum_{l,m} \left[b_{lm} X_{lm} + c_{lm} W_{lm} \sin\theta \right],$$

$$(\varphi\varphi) = \sum_{l,m} \left[-b_{lm} W_{lm} + \{2r a_{lm}/A_0^2 - l(l+1) b_{lm}\} Y_{lm} + c_{lm} \frac{X_{lm}}{\sin\theta} \right] \sin^2\theta. \tag{6.12}$$

Here, we note that ξ_t does not appear in equation (6.12) even if it is not zero. This is a result of choosing the static background: In the case that a time-dependent background is chosen, ξ_t is coupled.

We then decompose the spatial component of h_{ab} as

$$h_{ij} = \sum_{l,m} \left(A_0^2 A_{lm} \begin{bmatrix} Y_{lm} & 0 & 0 \\ * & 0 & 0 \\ * & * & 0 \end{bmatrix} + B_{lm} \begin{bmatrix} 0 & \partial_\theta Y_{lm} & \partial_\varphi Y_{lm} \\ * & 0 & 0 \\ * & * & 0 \end{bmatrix} \right.$$

$$+ r^2 K_{lm} \begin{bmatrix} 0 & 0 & 0 \\ * & Y_{lm} & 0 \\ * & * & \sin^2\theta Y_{lm} \end{bmatrix} + r^2 F_{lm} \begin{bmatrix} 0 & 0 & 0 \\ * & W_{lm} & X_{lm} \\ * & * & -\sin^2\theta W_{lm} \end{bmatrix}$$

$$+ C_{lm} \begin{bmatrix} 0 & -\partial_\varphi Y_{lm}/\sin\theta & \sin\theta \partial_\theta Y_{lm} \\ * & 0 & 0 \\ * & * & 0 \end{bmatrix}$$

$$\left. + r^2 D_{lm} \begin{bmatrix} 0 & 0 & 0 \\ * & -X_{lm}/\sin\theta & \sin\theta W_{lm} \\ * & * & \sin\theta X_{lm} \end{bmatrix} \right). \tag{6.13}$$

The gauge-invariant variables are found in the same procedure as before, which yields [Moncrief (1974)]

$$\psi_{lm}^{\text{even}} := \sqrt{\frac{2(l-1)(l+2)}{l(l+1)}} r \left[l(l+1) - 2S_0 + r\frac{dS_0}{dr} \right]^{-1}$$

$$\times \left[\left(l(l+1) - 2S_0 + r\frac{dS_0}{dr} \right) [l(l+1)F_{lm} + K_{lm}] \right.$$

$$\left. + 2S_0 \left(-r\partial_r K_{lm} + A_{lm} - l(l+1)\frac{B_{lm}}{r} \right) \right], \tag{6.14}$$

$$\partial_t \psi_{lm}^{\text{odd}} := \sqrt{\frac{2(l+2)!}{(l-2)!}} \frac{S_0}{r} \left[-C_{lm} + r^2 \partial_r D_{lm} \right], \tag{6.15}$$

where $S_0 = A_0^{-2}$. Note that the gauge invariance is not changed even if the factor S_0^n (n is arbitrary) is multiplied to ψ_{lm}^{even} and ψ_{lm}^{odd}. The choice in the above expressions is motivated by the fact that for $S_0 = 1 - 2M/r$ where M is the black-hole mass, ψ_{lm}^{even} and ψ_{lm}^{odd} satisfy the Zerilli equation [Zerilli (1970)] and Regge–Wheeler equation [Regge and Wheeler (1957)], respectively. These gauge-invariant variables are often called Moncrief variables. These variables reduce to those with $S_0 = 1$ in the far zone [compare with equations (6.7) and (6.8)], and hence, the gravitational-wave luminosity is calculated by equation (6.9) from them.

6.2 Extraction using a complex Weyl scalar

The Newman–Penrose formalism [Newman and Penrose (1962)] is a tetrad formalism in general relativity, in which four null vector fields (composed of two real and two complex vectors) are employed as the tetrad bases and then complex Weyl scalars are analyzed. This formalism is in particular robust for analyzing linear perturbation waves such as gravitational and electromagnetic waves propagating on curved background spacetime that has a special symmetry [Type D symmetry according to the Petrov's classification; see, e.g., Teukolsky (1973); Chandrasekhar (1983)].

In the following, we first describe the reason why the analysis of gravitational waves in terms of the Newman–Penrose formalism is robust. Then, we describe how a component of the complex Weyl scalar is used for extracting gravitational waves from numerically computed metric.

6.2.1 Overview of the Newman–Penrose formalism

In this section, we first outline the Newman–Penrose formalism, and then describe the reasons that it is robust for extracting gravitational waves in a local wave zone.

The Newman–Penrose formalism gives relations between Weyl tensor and Ricci rotation coefficients. The basic equations are composed of a relation between the Riemann tensor and Ricci rotation coefficients and of the Bianchi identity written in the null tetrad bases. Let $(e_\alpha)^a$ be the tetrad vector fields where the subscript α specifies a series of the tetrad bases with $\alpha = 1$–4. Here, $(e_\alpha)^a (e_\alpha)_a = 0$ and $(e_\alpha)^a (e_\beta)_a = \bar{\eta}_{\alpha\beta}$ with $\bar{\eta}_{\alpha\beta}$ being a matrix composed of constant components [see equation (6.22)] and

$$g_{ab} = \bar{\eta}^{\alpha\beta}(e_\alpha)_a (e_\beta)_b. \tag{6.16}$$

Then, the Ricci rotation coefficient, $\gamma_{\alpha\beta\sigma}$, is defined by

$$\gamma_{\alpha\beta\sigma} := (e_\alpha)^a (e_\sigma)^b \nabla_b (e_\beta)_a. \tag{6.17}$$

Here, $\gamma_{\alpha\beta\sigma} = -\gamma_{\beta\alpha\sigma}$, and thus, it has in general twenty-four independent components.

The tetrad components of the Riemann tensor can be written in terms of the Ricci rotation coefficients by

$$\begin{aligned}R_{\alpha\beta\sigma\lambda} &= R_{abcd}(e_\alpha)^a (e_\beta)^b (e_\sigma)^c (e_\lambda)^d \\ &= -\partial_\lambda \gamma_{\alpha\beta\sigma} + \partial_\sigma \gamma_{\alpha\beta\lambda} \\ &\quad -\bar{\eta}^{\mu\nu}[\gamma_{\alpha\beta\mu}(\gamma_{\sigma\nu\lambda} - \gamma_{\lambda\nu\sigma}) + \gamma_{\beta\mu\sigma}\gamma_{\nu\alpha\lambda} - \gamma_{\beta\mu\lambda}\gamma_{\nu\alpha\sigma}].\end{aligned} \tag{6.18}$$

The Bianchi identity in the tetrad bases is written by

$$\begin{aligned}0 &= (\nabla_{[e} R_{cd]ab})(e_\alpha)^a (e_\beta)^b (e_\sigma)^c (e_\lambda)^d (e_\rho)^e \\ &= \partial_{[\rho} R_{\sigma\lambda]\alpha\beta} - \bar{\eta}^{\mu\nu}\left(\gamma_{\mu\alpha[\rho} R_{\sigma\lambda]\nu\beta} + \gamma_{\mu\beta[\rho} R_{\sigma\lambda]\alpha\nu} \right. \\ &\quad \left. - \gamma_{\mu[\sigma\rho} R_{\lambda]\nu\alpha\beta} + \gamma_{\mu[\lambda\rho} R_{\sigma]\nu\alpha\beta}\right).\end{aligned} \tag{6.19}$$

Newman and Penrose (1962) chose the following set of the tetrad bases,

$$l^a = (e_1)^a, \quad k^a = (e_2)^a, \quad m^a = (e_3)^a, \quad \bar{m}^a = (e_4)^a, \tag{6.20}$$

which satisfies

$$\begin{aligned}-l_a k^a &= m_a \bar{m}^a = 1, \\ l_a m^a &= l_a \bar{m}^a = k_a m^a = k_a \bar{m}^a = 0.\end{aligned} \tag{6.21}$$

Here, $\bar{m}^a$ denotes the complex conjugate of m^a. In this case, $\bar{\eta}_{\alpha\beta}$ is written as

$$\bar{\eta}_{\alpha\beta} = \begin{pmatrix} 0 & -1 & 0 & 0 \\ -1 & 0 & 0 & 0 \\ 0 & 0 & 0 & 1 \\ 0 & 0 & 1 & 0 \end{pmatrix}. \tag{6.22}$$

Associated with the four tetrad bases, four directional derivatives are defined by [Newman and Penrose (1962)]

$$D := l^\alpha \partial_\alpha, \quad \Delta := k^\alpha \partial_\alpha, \quad \delta := m^\alpha \partial_\alpha, \quad \bar{\delta} := \bar{m}^\alpha \partial_\alpha, \tag{6.23}$$

and twelve complex functions, called spin coefficients that are associated with twenty-four real Ricci rotation coefficients, are defined as[1]

$$\kappa = \gamma_{311}, \quad \rho = \gamma_{314}, \quad \sigma = \gamma_{313}, \quad \tau = \gamma_{312},$$
$$\nu = \gamma_{242}, \quad \mu = \gamma_{243}, \quad \lambda = \gamma_{244}, \quad \pi = \gamma_{241},$$
$$\alpha = \frac{1}{2}(\gamma_{214} + \gamma_{344}), \quad \beta = \frac{1}{2}(\gamma_{213} + \gamma_{343}),$$
$$\gamma = \frac{1}{2}(\gamma_{212} + \gamma_{342}), \quad \epsilon = \frac{1}{2}(\gamma_{211} + \gamma_{341}). \tag{6.24}$$

The Newman–Penrose formalism was derived originally for non-vacuum spacetime. The Teukolsky equation, which governs the perturbations on a rotating black hole, was indeed derived from the Newman–Penrose formalism for non-vacuum spacetime [Teukolsky (1973)], combining the Bianchi identity and the relation between the Riemann tensor and the spin coefficients appropriately. For the purpose of a gravitational-wave extraction in a local wave zone, however, we may focus only on vacuum spacetime, and hence, we do not have to consider the relation between the Riemann tensor and the spin coefficients. In the following, we focus only on the vacuum spacetime analyzing the Bianchi identity.

In the vacuum case, the Newman–Penrose formalism reduces to the equations for the complex Weyl scalars and spin coefficients. Using the null tetrad-basis components, Newman and Penrose (1962) defined the following five complex Weyl scalars:

$$\Psi_0 := -C_{1313} = -C_{abcd} l^a m^b l^c m^d, \quad \Psi_1 := -C_{1213} = -C_{abcd} l^a k^b l^c m^d,$$
$$\Psi_2 := -\frac{1}{2}(C_{1212} - C_{1234}) = -\frac{1}{2} C_{abcd} \left(l^a k^b l^c k^d - l^a k^b m^c \bar{m}^d \right),$$
$$\Psi_3 := C_{1224} = C_{abcd} l^a k^b k^c \bar{m}^d, \quad \Psi_4 := -C_{2424} = -C_{abcd} k^a \bar{m}^b k^c \bar{m}^d. \tag{6.25}$$

We note that $C_{1314} = C_{2324} = C_{1323} = C_{1424} = 0$ and $C_{1212} = C_{3434}$ hold, because of the tracefree property of the Weyl tensor, i.e., $C_{abcd} g^{ac} = 0$. It is also shown that $C_{1334} = \Psi_1$, $C_{1324} = \Psi_2$, and $C_{2434} = -\Psi_3$. Then, the sixteen independent

---

[1] We note that these variables denoted in Greek are used only in this section.

components of the Bianchi identity for vacuum spacetime is written as

$$-D\Psi_1 + \bar{\delta}\Psi_0 + 3\kappa\Psi_2 - (2\epsilon + 4\rho)\Psi_1 - (\pi - 4\alpha)\Psi_0 = 0, \qquad (6.26)$$
$$-D\Psi_2 + \bar{\delta}\Psi_1 + 2\kappa\Psi_3 - 3\rho\Psi_2 - 2(\pi - \alpha)\Psi_1 + \lambda\Psi_0 = 0, \qquad (6.27)$$
$$-D\Psi_3 + \bar{\delta}\Psi_2 + \kappa\Psi_4 + 2(\epsilon - \rho)\Psi_3 - 3\pi\Psi_2 + 2\lambda\Psi_1 = 0, \qquad (6.28)$$
$$-D\Psi_4 + \bar{\delta}\Psi_3 + (4\epsilon - \rho)\Psi_4 - (4\pi + 2\alpha)\Psi_3 + 3\lambda\Psi_2 = 0, \qquad (6.29)$$
$$-\Delta\Psi_0 + \delta\Psi_1 - (4\gamma - \mu)\Psi_0 + (4\tau + 2\beta)\Psi_1 - 3\sigma\Psi_2 = 0, \qquad (6.30)$$
$$-\Delta\Psi_1 + \delta\Psi_2 - \nu\Psi_0 - 2(\gamma - \mu)\Psi_1 + 3\tau\Psi_2 - 2\sigma\Psi_3 = 0, \qquad (6.31)$$
$$-\Delta\Psi_2 + \delta\Psi_3 - 2\nu\Psi_1 + 3\mu\Psi_2 + 2(\tau - \beta)\Psi_3 - \sigma\Psi_4 = 0, \qquad (6.32)$$
$$-\Delta\Psi_3 + \delta\Psi_4 - 3\nu\Psi_2 + 2(\gamma + 2\mu)\Psi_3 + (\tau - 4\beta)\Psi_4 = 0, \qquad (6.33)$$

where these eight equations are derived from the following components of equation (6.19): $R_{13[13,4]}$, $R_{13[12,4]}$, $R_{24[13,4]}$, $R_{24[12,4]}$, $R_{13[12,3]}$, $R_{13[23,4]}$, $R_{24[12,3]}$, and $R_{24[23,4]}$.

6.2.2 The gauge freedom of the Newman–Penrose quantities and classification of spacetime

For a set of null tetrad bases, there is a six-parameters group of homogeneous Lorentz transformations, which can be performed preserving the tetrad orthogonality relations. This group can be decomposed into three Abelian subgroups [Janis and Newman (1965)]:

(1) The vector l^c is unchanged:

$$l^c \to l^c, \quad m^c \to m^c + al^c, \quad k^c \to k^c + a\bar{m}^c + \bar{a}m^c + a\bar{a}l^c. \qquad (6.34)$$

Here, a is a complex scalar.

(2) The vector k^c is unchanged:

$$k^c \to k^c, \quad m^c \to m^c + bl^c, \quad l^c \to l^c + b\bar{m}^c + \bar{b}m^c + b\bar{b}k^c. \qquad (6.35)$$

Here, b is another complex scalar.

(3) The directions of l^c and k^c are unchanged:

$$l^c \to \Lambda l^c, \quad k^c \to \Lambda^{-1} k^c, \quad m^c \to e^{i\theta} m^c. \qquad (6.36)$$

Here, Λ and θ are real scalars.

Under these transformations, the complex Weyl scalars are transformed as follows: For the transformation (1),

$$\Psi_0 \to \Psi_0, \quad \Psi_1 \to \Psi_1 + \bar{a}\Psi_0, \quad \Psi_2 \to \Psi_2 + 2\bar{a}\Psi_1 + \bar{a}^2\Psi_0,$$
$$\Psi_3 \to \Psi_3 + 3\bar{a}\Psi_2 + 3\bar{a}^2\Psi_1 + \bar{a}^3\Psi_0,$$
$$\Psi_4 \to \Psi_4 + 4\bar{a}\Psi_3 + 6\bar{a}^2\Psi_2 + 4\bar{a}^3\Psi_1 + \bar{a}^4\Psi_0. \qquad (6.37)$$

For the transformation (2),

$$\Psi_4 \to \Psi_4, \quad \Psi_3 \to \Psi_3 + b\Psi_4, \quad \Psi_2 \to \Psi_2 + 2b\Psi_3 + b^2\Psi_4,$$
$$\Psi_1 \to \Psi_1 + 3b\Psi_2 + 3b^2\Psi_3 + b^3\Psi_4,$$
$$\Psi_0 \to \Psi_0 + 4b\Psi_1 + 6b^2\Psi_2 + 4b^3\Psi_3 + b^4\Psi_4. \tag{6.38}$$

For the transformation (3),

$$\Psi_0 \to \Lambda^2 e^{2i\theta}\Psi_0, \quad \Psi_1 \to \Lambda e^{i\theta}\Psi_1, \quad \Psi_2 \to \Psi_2,$$
$$\Psi_3 \to \Lambda^{-1}e^{-i\theta}\Psi_3, \quad \Psi_4 \to \Lambda^{-2}e^{-2i\theta}\Psi_4. \tag{6.39}$$

Now, paying attention to the transformation (2), we consider to obtain $\Psi_0 = 0$ by solving the following equation for b:

$$\Psi_0 + 4b\Psi_1 + 6b^2\Psi_2 + 4b^3\Psi_3 + b^4\Psi_4 = 0. \tag{6.40}$$

Then, the resulting new direction of l^a, derived from equation (6.35), is called the principal null direction of the Weyl tensor. Assuming that $\Psi_4 \neq 0$, there are four roots of equation (6.40). The so-called Petrov's classification for a global structure of spacetime is based on the number of the roots for this equation as follows [e.g., Chandrasekhar (1983)]:

- Type I : Four distinct roots of equation (6.40) are present.
- Type II : Two coincident roots of equation (6.40) are present.
- Type D : Two distinct double roots of equation (6.40) are present.
- Type III: Three coincident roots of equation (6.40) are present.
- Type N : Four coincident roots of equation (6.40) are present.
- Type O : The Weyl scalar vanishes (the spacetime is conformally flat).

Note that the conditions classified above have to be satisfied for the entire domain of the spacetime.

Here, the number of the roots determines the number of non-trivial components of the complex Weyl scalars. In the following, we show this property, focusing in particular on the type D case because all the black-hole solutions in four dimensions are of this type [e.g., Chandrasekhar (1983)]. Let b_1 and $b_2(\neq b_1)$ be two double roots of equation (6.40), and thus, $b_1 + b_2 = -2\Psi_3/\Psi_4$. Here, the double roots satisfy equation (6.40) and

$$0 = \frac{1}{4}\frac{\partial}{\partial b}\left(\Psi_0 + 4b\Psi_1 + 6b^2\Psi_2 + 4b^3\Psi_3 + b^4\Psi_4\right)$$
$$= \Psi_1 + 3b\Psi_2 + 3b^2\Psi_3 + b^3\Psi_4. \tag{6.41}$$

This implies that $\Psi_1 = 0$ is also achieved by the same transformation for this case [see equation (6.38)].

Next, we consider the transformation (1), by which Ψ_0 and Ψ_1 remain to be zero. As in the transformation (2), $\Psi_4 = 0$ and $\Psi_3 = 0$ can be achieved for an appropriate choice of a, i.e., $\bar{a} = 1/(b_2 - b_1)$, for the type D spacetime. Therefore, for the type D spacetime, we can set $\Psi_0 = \Psi_1 = \Psi_3 = \Psi_4 = 0$ for an appropriate choice of the null tetrad. In the same procedure, it can be shown that for types I, II, III, and N, it is possible to have $\Psi_0 = \Psi_4 = 0$, $\Psi_0 = \Psi_1 = \Psi_4 = 0$, $\Psi_0 = \Psi_1 = \Psi_2 = \Psi_4 = 0$, and $\Psi_0 = \Psi_1 = \Psi_2 = \Psi_3 = 0$, respectively.

As clarified above, Ψ_2 in the principal null tetrad is only nonzero component for the type D spacetime. It denotes the so-called Coulomb part of the gravitational field and represents the curvature of the black-hole spacetime. In the Kinnersley tetrad [Kinnersley (1969)] (see section 1.3.2.5), Ψ_2 for Kerr black holes is

$$\Psi_2 = \frac{M}{(r - ia\cos\theta)^3}, \qquad (6.42)$$

where M and a are black-hole mass and spin, respectively (see section 1.3.1).

Broadly speaking, Ψ_0 and Ψ_4 are composed of ingoing and outgoing gravitational waves, respectively. For a linear perturbation on flat Minkowski spacetime, this fact is easily found (see section 6.2.5).

An important theorem that holds for the complex Weyl scalars in asymptotically flat spacetime is the so-called peeling theorem, which states the asymptotic behavior of the complex Weyl scalars for $r \to \infty$ under the quite general conditions. Newman and Penrose (1962) proved this theorem supposing the condition $\Psi_0 = O(r^{-5})$; this supposes the absence of ingoing waves at null infinity. Then, the asymptotic behavior of the other components is written as

$$\Psi_n = O(r^{-5+n}). \qquad (6.43)$$

As examples, we already saw that Ψ_2 in Kerr spacetime behaves as $O(r^{-3})$, and we will explicitly see $\Psi_n = O(r^{-5+n})$ for linear quadrupole gravitational waves propagating in the far zone in section 6.2.5.

6.2.3 Weyl tensor in the 3+1 formulation

Before going ahead, we rewrite the four-dimensional Weyl tensor defined in equation (1.10) in terms of the quantities in the 3+1 formulation. The resulting expression significantly helps calculating the complex Weyl scalar in a particular choice of the null tetrad.

Following Hawking (1966); Ellis (1971, 2009), we first define the electric and magnetic parts of the Weyl tensor as

$$E_{ac} := C_{abcd} n^b n^d, \qquad B_{ac} := \frac{1}{2} C_{abef} \epsilon^{ef}{}_{cd} n^b n^d, \qquad (6.44)$$

where C_{abcd} denotes the four-dimensional Weyl tensor in this section, and n^a is the unit timelike vector field orthonormal to spatial hypersurfaces Σ_t as before. E_{ac} and B_{ac} are symmetric, tracefree spatial tensors, i.e.,

$$g^{ac} E_{ac} = g^{ac} B_{ac} = E_{ac} n^c = B_{ac} n^c = E_{[ac]} = B_{[ac]} = 0. \qquad (6.45)$$

This implies that two tensors have five independent components, respectively, and thus, the total number of the independent components are equal to that of the four-dimensional Weyl tensor. This indicates that the four-dimensional Weyl tensor is written by E_{ac} and B_{ac}, and indeed, it is written as

$$C_{abcd} = p_{ac}E_{bd} - p_{ad}E_{bc} - p_{bc}E_{ad} + p_{bd}E_{ac}$$
$$- n_a B_{be} \epsilon^e{}_{cd} + n_b B_{ae} \epsilon^e{}_{cd} - n_c B_{de} \epsilon^e{}_{ab} + n_d B_{ce} \epsilon^e{}_{ab}, \quad (6.46)$$

where $p_{ab} := g_{ab} + 2n_a n_b$ and $\epsilon_{bcd} = n^a \epsilon_{abcd}$.

E_{ac} and B_{ac} are obtained from the same manipulations as those performed in section 2.1. Using equations (1.10), (2.53)–(2.57), and (2.41), it is straightforward to derive

$$E_{ac} = R_{ac} + KK_{ac} - K_a{}^b K_{bc} - \frac{16\pi}{3}\rho_h \gamma_{ac} - 4\pi \left(S_{ac} - \frac{1}{3}\gamma_{ac} S_b{}^b \right), \quad (6.47)$$

$$B_{ac} = \epsilon_a{}^{bd} [D_b K_{dc} - 4\pi \gamma_{cb} J_d]. \quad (6.48)$$

Now, we write the outgoing and ingoing null vectors as

$$l^a = \frac{1}{\sqrt{2}}(n^a + s^a), \quad k^a = \frac{1}{\sqrt{2}}(n^a - s^a), \quad (6.49)$$

where s^a is a unit spatial vector pointing to a radial direction and satisfying $s_a m^a = 0 = s_a n^a$. Then, the complex Weyl scalars are written in terms of E_{ac} and B_{ac} from equation (6.46) as [e.g., Gunnarsen et al. (1995)]

$$\Psi_0 = -(E_{ac} - iB_{ac})m^a m^c, \quad \Psi_4 = -(E_{ac} - iB_{ac})\bar{m}^a \bar{m}^c,$$
$$\Psi_1 = \frac{1}{\sqrt{2}}(E_{ac} - iB_{ac})s^a m^c, \quad \Psi_3 = -\frac{1}{\sqrt{2}}(E_{ac} - iB_{ac})s^a \bar{m}^c,$$
$$\Psi_2 = -\frac{1}{2}(E_{ac} - iB_{ac})s^a s^c, \quad (6.50)$$

where we used $n^a s^b m^c \bar{m}^d \epsilon_{abcd} = s^b m^c \bar{m}^d \epsilon_{bcd} = -i$, and thus $s^b \bar{m}^d \epsilon_{bcd} = -i\bar{m}_c$, $s^b m^c \epsilon_{bcd} = -im_d$, and $m^c \bar{m}^d \epsilon_{bcd} = -is_b$.

6.2.4 Perturbation on type D spacetime

Suppose that we can choose the tetrad bases with which only Ψ_2 is nonzero for type D spacetime. Then, we consider another spacetime which is slightly perturbed from the type D spacetime. For it, Ψ_0, Ψ_1, Ψ_3, and Ψ_4 are perturbed quantities. These quantities are transformed for the transformation (1)–(3) in section 6.2.2 where a, b, Λ, and θ are small constants associated with the coordinate transformation. It is found that variations of Ψ_0 and Ψ_4 is second-order in b and $\bar{a}$, respectively. This implies that in the linear order, these are gauge-invariant perturbation quantities on the type D background spacetime, and that Ψ_4 on the type D background spacetime carries the information of outgoing gravitational waves [Teukolsky (1973)]. This is one of the motivations that encourages us to employ Ψ_4 for extracting gravitational waves.

However, words of caution are necessary for employing Ψ_4 in numerical relativity. First, the background spacetime is not in general type D spacetime, although we may naively expect that the background spacetime that we generate in numerical relativity is, for most cases, similar to Kerr spacetime in a zone far from the central region where we extract gravitational waves. Second, it is not an easy task to choose the basis vectors composed of the principal null tetrad for numerically generated spacetime, even if the spacetime is of type D. For a choice of a non-principal null tetrad, Ψ_4 is not exactly gauge-invariant. However, we have the reason to believe that Ψ_4 is an approximately gauge-invariant quantity in the wave zone if we employ a reasonable null tetrad, as illustrated in the next subsection.

6.2.5 Ψ_4 and outgoing gravitational waves

To see the fact that Ψ_4 is associated with outgoing gravitational waves, we consider gravitational waves of small amplitude propagating on flat background spacetime, using a linear perturbation analysis. For the analysis of the complex Weyl scalars, equations (6.47) and (6.48) are quite useful.

In the linear perturbation analysis in vacuum, i.e., in the assumption that $h_{ab} = \gamma_{ab} - f_{ab}$ and K_{ab} are small quantities, E_{ab} and B_{ab} are written as

$$E_{ab} = R_{ab} = \partial_t \delta K_{ab} + \overset{(0)}{D}_a \overset{(0)}{D}_b \delta\alpha, \tag{6.51}$$

$$B_{ab} = \epsilon_a{}^{cd} \overset{(0)}{D}_c \delta K_{db}, \tag{6.52}$$

where δK_{ab} denotes a perturbed extrinsic curvature, and $\delta\alpha$ denotes $\alpha - 1$. In the linear perturbation analysis, it is useful to decompose δK_{ab} in terms of tensor spherical harmonics as

$$\begin{aligned}\delta K_{ij} = \sum_{l,m} &\left(\begin{bmatrix} A_{lm} & 0 & 0 \\ * & r^2 G_{lm} & 0 \\ * & * & r^2 \sin^2\theta G_{lm} \end{bmatrix} Y_{lm} + B_{lm} \begin{bmatrix} 0 & \partial_\theta Y_{lm} & \partial_\varphi Y_{lm} \\ * & 0 & 0 \\ * & * & 0 \end{bmatrix} \right.\\ &+ r^2 F_{lm} \begin{bmatrix} 0 & 0 & 0 \\ * & W_{lm} & X_{lm} \\ * & * & -\sin^2\theta W_{lm} \end{bmatrix} \\ &+ C_{lm} \begin{bmatrix} 0 & -\partial_\varphi Y_{lm}/\sin\theta & \sin\theta \partial_\theta Y_{lm} \\ * & 0 & 0 \\ * & * & 0 \end{bmatrix} \\ &\left. + r^2 D_{lm} \begin{bmatrix} 0 & 0 & 0 \\ * & -X_{lm}/\sin\theta & \sin\theta W_{lm} \\ * & * & \sin\theta X_{lm} \end{bmatrix} \right), \end{aligned}\tag{6.53}$$

where $A_{lm}, B_{lm}, C_{lm}, D_{lm}, F_{lm}$, and G_{lm} are functions of r and t. We also write

$$\delta\alpha = \sum_{l,m} a_{lm} Y_{lm}. \tag{6.54}$$

The reasonable choice of the tetrad basis vectors in the wave zone is

$$l^\alpha = \frac{1}{\sqrt{2}}(1,1,0,0), \quad k^\alpha = \frac{1}{\sqrt{2}}(1,-1,0,0), \quad m^\alpha = \frac{1}{\sqrt{2}r}\left(0,0,1,\frac{i}{\sin\theta}\right). \tag{6.55}$$

Then, each mode of the complex Weyl scalars is written in the following forms

$$\Psi_0 = -\left[\partial_t F_{lm} + \frac{1}{r}\partial_r(rF_{lm}) + \frac{a_{lm} - B_{lm}}{2r^2}\right.$$
$$\left. + i\left(\partial_t D_{lm} + \frac{1}{r}\partial_r(rD_{lm}) - \frac{C_{lm}}{2r^2}\right)\right]\left(W_{lm} + \frac{i}{\sin\theta}X_{lm}\right), \tag{6.56}$$

$$\Psi_4 = -\left[\partial_t F_{lm} - \frac{1}{r}\partial_r(rF_{lm}) + \frac{a_{lm} + B_{lm}}{2r^2}\right.$$
$$\left. - i\left(\partial_t D_{lm} - \frac{1}{r}\partial_r(rD_{lm}) + \frac{C_{lm}}{2r^2}\right)\right]\left(W_{lm} - \frac{i}{\sin\theta}X_{lm}\right), \tag{6.57}$$

$$\Psi_1 = -\frac{1}{2r}\left[-\partial_t B_{lm} + \frac{B_{lm}}{r} - \bar{\lambda}_l F_{lm} - G_{lm} + \partial_r a_{lm} - \frac{a_{lm}}{r}\right.$$
$$\left. + i\left(-\partial_t C_{lm} + \frac{C_{lm}}{r} - \bar{\lambda}_l D_{lm}\right)\right]\left(\partial_\theta Y_{lm} + \frac{i}{\sin\theta}\partial_\varphi Y_{lm}\right), \tag{6.58}$$

$$\Psi_3 = -\frac{1}{2r}\left[\partial_t B_{lm} + \frac{B_{lm}}{r} - \bar{\lambda}_l F_{lm} - G_{lm} + \partial_r a_{lm} - \frac{a_{lm}}{r}\right.$$
$$\left. - i\left(\partial_t C_{lm} + \frac{C_{lm}}{r} - \bar{\lambda}_l D_{lm}\right)\right]\left(\partial_\theta Y_{lm} - \frac{i}{\sin\theta}\partial_\varphi Y_{lm}\right), \tag{6.59}$$

$$\Psi_2 = -\frac{1}{2}\left(\partial_t A_{lm} + \partial_r^2 a_{lm} + i(l+1)l\frac{C_{lm}}{r^2}\right)Y_{lm}, \tag{6.60}$$

where $\bar{\lambda}_l := (l+2)(l-1)$. Equations (6.56)–(6.60) imply that Ψ_0 and Ψ_4 are nonzero only for $l \geq 2$, and Ψ_1 and Ψ_3 are nonzero only for $l \geq 1$.

For providing a specific form of A_{lm}–G_{lm}, we assume to choose a slicing condition in which $\delta\alpha$ is a second-order quantity. This is indeed the case, for example, for the case that we choose maximal slicing and dynamical slicing described in sections 2.2.2 and 2.2.6. In the choice of such slicing conditions, K is also second-order, because in the linear perturbation approximation, its basic equation reduces to

$$\partial_t K = -\overset{(0)}{D_a}\overset{(0)}{D^a}\delta\alpha. \tag{6.61}$$

Then, δK_{ab} in the linear order obeys a simple wave equation $\Box \delta K_{ab} = 0$. In addition, δK_{ab} has to satisfy the momentum constraint, $\overset{(0)}{D^a}\delta K_{ab} = 0$ for $K = 0$, and thus, it is a transverse-tracefree tensor.

As shown in appendix C.2, the general regular solution of the wave equation for the transverse-tracefree tensor is written analytically: The solutions of outgoing waves for $l = 2$ modes are described by equations (C.56) and (C.57). Substituting these solutions into the expressions of $\Psi_0 - \Psi_4$, we have

$$\Psi_0 = \left[\frac{1}{8}\frac{F'_e}{r^5} - \frac{3i}{4}\frac{F_o}{r^5}\right]\left(W_{2m} + \frac{i}{\sin\theta}X_{2m}\right), \tag{6.62}$$

$$\Psi_4 = \left[\frac{1}{12r}\left(F'''''_e - \frac{2F''''_e}{r} + \frac{3F'''_e}{r^2} - \frac{3F''_e}{r^3} + \frac{3F'_e}{2r^4}\right)\right.$$
$$\left. - \frac{i}{2r}\left(F''''_o - \frac{2F'''_o}{r} + \frac{3F''_o}{r^2} - \frac{3F'_o}{r^3} + \frac{3F_o}{2r^4}\right)\right]\left(W_{2m} - \frac{i}{\sin\theta}X_{2m}\right), \tag{6.63}$$

$$\Psi_1 = \left[-\frac{1}{4r^5}\left(F'_e - \frac{2F_e}{r}\right) + \frac{3i}{2r^4}\left(F'_o - \frac{2F_o}{r}\right)\right]\left(\partial_\theta Y_{2m} + \frac{i}{\sin\theta}\partial_\varphi Y_{2m}\right), \tag{6.64}$$

$$\Psi_3 = \left[\frac{1}{6r^2}\left(F''''_e - \frac{3F'''_e}{r} + \frac{9F''_e}{2r^2} - \frac{3F'_e}{2r^3}\right)\right.$$
$$\left. - \frac{i}{r^2}\left(F'''_o - \frac{3F''_o}{r} + \frac{9F'_o}{2r^2} - \frac{3F_o}{r^3}\right)\right]\left(\partial_\theta Y_{2m} - \frac{i}{\sin\theta}\partial_\varphi Y_{2m}\right), \tag{6.65}$$

$$\Psi_2 = \left[\frac{1}{2r^3}\left(F'''_e - \frac{3F''_e}{r} + \frac{3F'_e}{r^2}\right) - \frac{3i}{r^3}\left(F''_o - \frac{3F'_o}{r} + \frac{3F_o}{2r^2}\right)\right]Y_{2m}, \tag{6.66}$$

where F_e and F_o are arbitrary regular functions, and $F'_e = \partial_r F_e$. These expressions illustrate that $\Psi_n = O(r^{-5+n})$ and Ψ_4 carries the information of outgoing gravitational waves. We note that these expressions hold irrespective of the choice of the shift vector, and the gauge dependence appears only through the slicing condition. As mentioned above, however, as far as the popular slicing conditions in numerical relativity are employed, its effect may be ignored. Therefore, we can conclude that Ψ_4 is a robust tool for extracting gravitational waves in numerical relativity.

6.2.6 Waveforms and luminosity

Ψ_4 in equation (6.57) is written in terms of the extrinsic curvature. If we assume to employ a spatial gauge condition with which $2K_{ab} \approx -\partial_t \gamma_{ab}$ for $r \to \infty$, it is rewritten by the spatial metric. As usually done, we write the asymptotic metric form associated with gravitational waves as $\gamma_{ab} = f_{ab} + h_{ab}$ where

$$\frac{1}{r^2}\begin{pmatrix} h_{\theta\theta} & h_{\theta\varphi} \\ * & h_{\varphi\varphi} \end{pmatrix} = \sum_{l,m}\left[h_e^{lm}\begin{pmatrix} W_{lm} & X_{lm} \\ * & -\sin^2\theta W_{lm} \end{pmatrix}\right.$$
$$\left. + h_o^{lm}\begin{pmatrix} -X_{lm}/\sin\theta & \sin\theta W_{lm} \\ * & \sin\theta X_{lm} \end{pmatrix}\right]. \tag{6.67}$$

Then, the asymptotic form of Ψ_4 is written as

$$\Psi_4 = \sum_{l,m}\left(\ddot{h}_e^{lm} - i\ddot{h}_o^{lm}\right)\left(W_{lm} - \frac{i}{\sin\theta}X_{lm}\right), \tag{6.68}$$

where we supposed that h_e^{lm} and h_o^{lm} are functions of the form $F(r-t)/r$, and $\ddot{h}$ denotes $\partial_t^2 h$. Equation (6.68) indicates that Ψ_4 is written as

$$\Psi_4 = \ddot{h}_+ - i\ddot{h}_\times, \tag{6.69}$$

where h_+ and $h_\times$ denote the plus and cross modes of gravitational waves defined by

$$h_+ := \sum_{l,m} \left(h_e^{lm} W_{lm} - h_o^{lm} \frac{X_{lm}}{\sin\theta} \right), \tag{6.70}$$

$$h_\times := \sum_{l,m} \left(h_o^{lm} W_{lm} + h_e^{lm} \frac{X_{lm}}{\sin\theta} \right). \tag{6.71}$$

Equation (6.68) is often written as

$$\Psi_4 := \sum_{l,m} \Psi_4^{l,m} {}_{-2}Y_{lm} \tag{6.72}$$

where ${}_{-2}Y_{lm}$ is a spin-weight spherical harmonics (see appendix C.1.5). Thus,

$$\Psi_4^{l,m} = \oint d\Omega \Psi_4 {}_{-2}Y_{lm}^* = \sqrt{\frac{(l+2)!}{(l-2)!}} \left(\ddot{h}_e^{lm} - i\ddot{h}_o^{lm} \right), \tag{6.73}$$

and subsequently, h_e^{lm} and h_o^{lm} are determined by the double time integration as

$$h_e^{lm} - ih_o^{lm} = \int^t dt' \int^{t'} dt'' \Psi_4^{l,m}(r, t''). \tag{6.74}$$

For obtaining an accurate gravitational waveform, the double time integration has to be carefully performed. For such an integration method, we recommend the reader to refer, e.g., to Reisswig and Pollney (2011), in which Fourier transformation and subsequent appropriate frequency cut-off prescription are proposed. It should be also noted that Ψ_4 is usually extracted at finite radii, and thus, for obtaining the asymptotic waveform at $r \to \infty$, certain extrapolation procedure is necessary. We here introduce Nakano's method, which is based on black-hole perturbation theory [Lousto et al. (2010)] and can be easily implemented, as follows: First, we need to extract $\Psi_4^{l,m}(r, t)$ at a finite radius $r = r_0$ and then the asymptotic waveform is obtained approximately by

$$\Psi_4^{l,m}(\infty, t) = C(r_0) \left[\Psi_4^{l,m}(r_0, t) - \frac{(l-1)(l+2)}{2r_A} \int^t \Psi_4^{l,m}(r_0, t') dt' \right]. \tag{6.75}$$

Here, r_A is an appropriately determined areal radius at $r = r_0$. For example, for isotropic coordinates, we should choose $r_A = r_0[1 + M/(2r_0)]^2$ where M is the ADM mass of the system. $C(r_0)$ depends on the choice of the tetrad components; for the tetrad components of equation (6.55), we should choose $C(r_0) = 1 - 2M/r_A$.

From the asymptotic metric derived, it is straightforward to calculate the luminosity, and the angular- and linear-momentum emission rates in terms of Ψ_4 as

$$\frac{dE}{dt} = \lim_{r\to\infty} \frac{r^2}{16\pi} \oint_r \left| \int^t \Psi_4 dt' \right|^2 d\Omega, \tag{6.76}$$

$$\frac{dJ_i}{dt} = -\lim_{r\to\infty} \frac{r^2}{16\pi} \oint_r \mathrm{Re} \left[\left(\int^t \Psi_4^* dt' \right) \left(\int^t dt' \int^{t'} \epsilon_{ij}{}^k x^j \partial_k \Psi_4 dt'' \right) \right] d\Omega, \tag{6.77}$$

$$\frac{dP_i}{dt} = \lim_{r\to\infty} \frac{r^2}{16\pi} \oint_r \left|\int^t \Psi_4 dt'\right|^2 \frac{x^i}{r} d\Omega. \tag{6.78}$$

Using the multipole components, these are written as [Ruiz *et al.* (2008)]

$$\frac{dE}{dt} = \lim_{r\to\infty} \frac{r^2}{16\pi} \sum_{l,m} \left|\int^t \Psi_4^{l,m} dt'\right|^2, \tag{6.79}$$

$$\frac{dJ_x}{dt} = \lim_{r\to\infty} \frac{r^2}{32\pi} \sum_{l,m} \mathrm{Im}\left[\int^t \left(q_{l,m}\Psi_4^{l,m+1} + q_{l,-m}\Psi_4^{l,m-1}\right)^* dt'\right.$$

$$\left.\times \int^t dt' \int^{t'} \Psi_4^{l,m} dt''\right], \tag{6.80}$$

$$\frac{dJ_y}{dt} = -\lim_{r\to\infty} \frac{r^2}{32\pi} \sum_{l,m} \mathrm{Re}\left[\int^t \left(q_{l,m}\Psi_4^{l,m+1} - q_{l,-m}\Psi_4^{l,m-1}\right)^* dt'\right.$$

$$\left.\times \int^t dt' \int^{t'} \Psi_4^{l,m} dt''\right], \tag{6.81}$$

$$\frac{dJ_z}{dt} = \lim_{r\to\infty} \frac{r^2}{16\pi} \sum_{l,m} m\, \mathrm{Im}\left[\left(\int^t \Psi_4^{l,m} dt'\right)^* \int^t dt' \int^{t'} \Psi_4^{l,m} dt''\right], \tag{6.82}$$

and

$$\frac{dP_x}{dt} = \lim_{r\to\infty} \frac{r^2}{8\pi} \sum_{l,m} \mathrm{Re}\left[\int^t \Psi_4^{l,m} dt'\right.$$

$$\left.\times \int^t \left(p^1_{l,m}\Psi_4^{l,m+1} + p^2_{l,-m}\Psi_4^{l-1,m+1} - p^2_{l+1,m+1}\Psi_4^{l+1,m+1}\right)^* dt'\right], \tag{6.83}$$

$$\frac{dP_y}{dt} = \lim_{r\to\infty} \frac{r^2}{8\pi} \sum_{l,m} \mathrm{Im}\left[\int^t \Psi_4^{l,m} dt'\right.$$

$$\left.\times \int^t \left(p^1_{l,m}\Psi_4^{l,m+1} + p^2_{l,-m}\Psi_4^{l-1,m+1} - p^2_{l+1,m+1}\Psi_4^{l+1,m+1}\right)^* dt'\right], \tag{6.84}$$

$$\frac{dP_z}{dt} = \lim_{r\to\infty} \frac{r^2}{16\pi} \sum_{l,m} \mathrm{Re}\left[\int^t \Psi_4^{l,m} dt'\right.$$

$$\left.\times \int^t \left(p^3_{l,m}\Psi_4^{l,m} + p^4_{l,m}\Psi_4^{l-1,m} + p^4_{l+1,m}\Psi_4^{l+1,m}\right)^* dt'\right], \tag{6.85}$$

where

$$q_{l,m} = \sqrt{(l-m)(l+m+1)},$$

$$p^1_{l,m} = \sqrt{\frac{(l-m)(l+m+1)}{l^2(l+1)^2}}, \quad p^2_{l,m} = \frac{1}{2l}\sqrt{\frac{(l-2)(l+2)(l+m)(l+m-1)}{(2l+1)(2l-1)}},$$

$$p^3_{l,m} = \frac{2m}{l(l+1)}, \quad p^4_{l,m} = \frac{1}{l}\sqrt{\frac{(l-2)(l+2)(l-m)(l+m)}{(2l-1)(2l+1)}}.$$

6.3 Quadrupole formula

Quadrupole formulas are often used for approximately studying gravitational waves in Newtonian and post-Newtonian simulations [e.g., see Blanchet et al. (1990) for a formulation]. In these simulations, gravitational waves are computed simply from the time variation of a mass quadrupole of the system [see Oohara and Nakamura (1989); Finn and Evans (1990) for specific numerical methods for an accurate computation of gravitational waveforms and their luminosity], because the tensor part of the metric is not solved.

In this section, we point out that a quadrupole formula is useful for approximately studying gravitational waves in numerical relativity as well. In particular, when we focus mainly on the characteristic frequency of gravitational waves emitted, this formula is helpful even for a quantitative study.

In quadrupole formulas, gravitational waveforms are derived from the double time derivatives of a tracefree quadrupole moment, I_{ij}, as shown in equation (1.44). In the Newtonian order, the quadrupole moment is written as

$$I_{ij} = \int d^3x \rho x^i x^j, \qquad (6.86)$$

where ρ is the rest-mass density. In general relativity, we do not know the unique way for defining I_{ij}. Following a post-Newtonian work by Blanchet et al. (1990), Shibata (2003a) proposed to use the following definition

$$I_{ij} = \int d^3x \rho_* x^i x^j, \qquad (6.87)$$

where $\rho_* := \rho\sqrt{-g}u^t$ is the conserved rest-mass density (see section 4.4). The merit of this form is that the first time derivative is written in an analytic expression

$$\frac{dI_{ij}}{dt} = \int d^3x \rho_* \left(v^i x^j + x^i v^j\right), \qquad (6.88)$$

where $v^i = dx^i/dt$ is the three velocity written by u^μ and β^i [see equation (4.38)]. Then, we only need to numerically take one more additional time derivative.

This expression is valid up to the Newtonian order from the view point of the post-Newtonian approximation. If we focused on problems associated with a weakly gravitating system, it would be a good strategy to add post-Newtonian corrections to I_{ij}. However, the purpose of numerical relativity is to solve dynamical phenomena in strongly gravitating systems. For this case, convergence of the post-Newtonian expansion is poor in general, and hence, adding the post-Newtonian corrections does not guarantee to improve the accuracy. The quadrupole formula is useful only for a qualitative study after all, and thus, the formula as simple as possible would be better. This motivates us to employ a simple formula like equation (6.87).

Figure 6.1 displays gravitational waveforms from a merger of binary neutron stars computed by the quadrupole formula (solid curve) and a complex Weyl scalar (dashed curve). Here, the waveform observed along the axis perpendicular to the

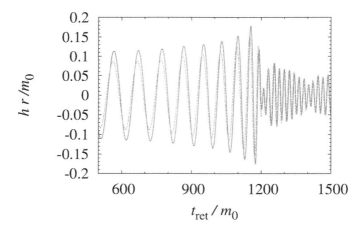

Fig. 6.1 Comparison of gravitational waveforms from a merger of binary neutron stars computed by a quadrupole formula (solid curve) and a complex Weyl scalar (dashed curve). m_0 is the total mass (sum of each mass in binary components) of the system. Note that for the quadrupole-formula case, the horizontal axis is t/m_0 (time retardation is not necessary).

orbital plane is plotted. This shows that wave phases in the waveform derived by the quadrupole formula agree fairly with those by the complex Weyl scalar. Although the wave amplitude is overestimated for the inspiral phase ($t_{\rm ret} \lesssim 1150 m_0$ with m_0 being the total mass of the system) by a factor of 1.2–1.3, the error is still in an acceptable level for a semi-quantitative study. For the merger phase ($t_{\rm ret} \gtrsim 1150 m_0$) in which gravitational waves are emitted from a hypermassive (high-mass) neutron star, the agreement of the wave amplitude is also within $\sim 20\%$ error. This example illustrates that the quadrupole formula is a good tool for extracting gravitational waves in a semi-quantitative manner. Similar comparison studies were performed for oscillating neutron stars [Shibata and Sekiguchi (2003)] and for stellar core collapse [Reisswig et al. (2011)], and the good quality of the quadrupole formula was also confirmed.

Quadrupole formulas are in particular useful when the amplitude of gravitational waves is too small to be accurately extracted from geometric quantities: For small-amplitude gravitational waves, numerical noise often prevents the accurate extraction of gravitational waves. This is in particular the case for simulations of stellar core collapses in which computational costs are so high that it is not easy to guarantee a high accuracy [Shibata and Sekiguchi (2004, 2005a); Dimmelmeier et al. (2007, 2008); Ott et al. (2011)] [but also see Reisswig et al. (2011) for the case that gravitational-wave signals are extracted geometrically].

Chapter 7

Finding black holes

One of the most important purposes in numerical relativity is to clarify the formation and evolution processes of a variety of black holes. For this purpose, we have to determine the location of the black holes and then analyze their properties. This procedure has to be performed in general for non-stationary spacetime. In this chapter, we describe methods for finding and analyzing the black holes in numerical relativity.

7.1 Apparent horizon

The apparent horizon (i.e., the marginally trapped outer surface; see, e.g., appendix G.5 for the definition) is an invaluable tool for finding black holes in numerical relativity: In numerical relativity, the existence of a black hole is usually confirmed by finding the presence of an apparent horizon. By contrast to the event horizon that is related to a global structure of spacetime, the apparent horizon can be defined on each spatial hypersurface Σ_t. Because of this property, the apparent horizon has the great advantage for locating a black hole in numerical relativity.

For four-dimensional asymptotically flat spacetime, the apparent horizon is defined by a closed two-dimensional surface on a spatial hypersurface Σ_t for which the expansion of an outgoing congruence of future-directed null geodesics orthogonal to the apparent-horizon surface vanishes (see appendix G.5 for more details). In higher dimensions with $N \geq 4$, the apparent horizon does not always have the S^{N-1} topology; for example, a ring-like apparent horizon (and event horizon) is present for black ring solutions [Emparan and Reall (2002)]. In this section, however, we describe a method for finding apparent horizons, focusing only on the case that the topology of the apparent horizon is S^{N-1}. For the general topology, we need to modify the method to numerical solutions, but it should be qualitatively identical.

We denote an future-directed outgoing null vector field as k^a and suppose that it is tangent of null geodesics. Then, we have the relations

$$k^a k_a = 0, \quad \text{and} \quad k^b \nabla_b k^a = 0. \tag{7.1}$$

Defining another null vector field, l^a, that satisfies $k^a l_a = -1$ (see figure 7.1), the

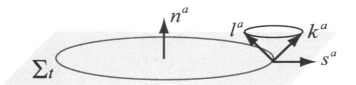

Fig. 7.1 A diagram for an apparent horizon (denoted by a circle on a spacelike hypersurface Σ_t) and the relation among vectors k^a, l^a, n^a, and s^a. k^a denotes an outgoing null vector defining the apparent horizon on Σ_t, and l^a is another null vector. s_a is a unit radial vector on Σ_t, and n^a is the unit timelike vector field normal to Σ_t. The cone, for which k^a and l^a are tangent, denotes a light cone.

spacetime metric is written as

$$g_{ab} = -k_a l_b - l_a k_b + H_{ab}, \tag{7.2}$$

where H_{ab} is a two-dimensional metric that satisfies $H_{ab}k^a = H_{ab}l^a = 0$. Then, the expansion of a congruence of the null geodesics for which k^a is tangent is defined by

$$\Theta := H^{ab}\nabla_a k_b. \tag{7.3}$$

Using the conditions (7.1), Θ may be written as $\nabla_a k^a = 0$.

Now, we write the condition $\Theta = 0$ using the variables in the $N+1$ formulation. Denoting a unit radial vector normal to the $(N-1)$-dimensional surface of an apparent horizon by s^a, k^a is written as

$$k^a = F(x^i)(n^a + s^a), \tag{7.4}$$

where $F(x^i)$ is a function which is introduced to satisfy the geodesic equation in equation (7.1) and n^a is the timelike unit vector field normal to Σ_t as before (i.e., $n_a s^a = 0$). Using s^a, H_{ab} can be written as $\gamma_{ab} - s_a s_b$, and thus, the equation, $\Theta = 0$, becomes

$$(\gamma^{ab} - s^a s^b)\nabla_a k_b = 0. \tag{7.5}$$

Substituting equation (7.4) into equation (7.5) yields

$$D_a s^a + K_{ab}s^a s^b - K = 0, \tag{7.6}$$

where we used $H^{ab}k_b = 0$ and the definition of the spatial covariant derivative. Here, the equation for determining the apparent horizon does not depend on F because of $H^{ab}k_b = 0$; we only needed to require that k^a is a null vector.

The next task is to rewrite equation (7.6) to a form by which the surface of an apparent horizon can be located. For this purpose, we denote the surface of the apparent horizon by $r = f(\theta_k)$ where f is a function to be determined and θ_k $(k = 1, 2, \cdots, N-1)$ denotes a set of angular coordinates of the apparent horizon (remember we assume that the apparent horizon has a spherical topology). We assume that the origin of the chosen coordinates is located inside the closed surface of the apparent horizon. Since s^a is the spacelike unit vector normal to the apparent horizon surface, it is denoted by

$$s_i = C\nabla_i \left[r - f(\theta_k)\right] = C\left(1, -\partial_i f\right), \quad i \neq r, \tag{7.7}$$

where C is a constant for the normalization: From $s^a s_a = 1$, we find

$$C = \left[\gamma^{rr} - 2\gamma^{rj}\partial_j f + \gamma^{jk}(\partial_j f)(\partial_k f)\right]^{-1/2}, \tag{7.8}$$

where j and k in this equation denote angular coordinates (not r): The components of s_a are written assuming the use of the coordinate system $(r, \theta_1, \theta_2, \cdots, \theta_{N-1})$. Substituting equation (7.7) into (7.6), an $(N-1)$-dimensional elliptic-type equation for f is derived.

In the following, we will describe equations for determining apparent horizons restricting our attention only to the case of $N = 3$, because the extension to the higher-dimensional case is straightforward. We will assume that spherical polar coordinates (r, θ, φ) are used in the following. We do not suppose that an actual numerical simulation is performed in this coordinate system, but it is easy to obtain the corresponding spherical-polar components of the geometrical quantities by a simple coordinate transformation even if simulations are performed in other coordinate systems.

7.1.1 Spherically symmetric case

For the spherically symmetric case, f is constant. Thus,

$$D_a s^a = \frac{1}{\sqrt{\gamma}}\partial_r\left(\sqrt{\gamma} s^r\right). \tag{7.9}$$

Because the spatial line element is written in a diagonal form, as

$$\gamma_{ij} = \mathrm{diag}(\gamma_{rr}, \gamma_{\theta\theta}, \gamma_{\theta\theta}\sin^2\theta), \tag{7.10}$$

the following relations hold; $\gamma^{rr} = 1/\gamma_{rr}$, $\sqrt{\gamma}s^r = \gamma_{\theta\theta}\sin\theta$, and $K_{ab}s^a s^b - K = -2K_\theta{}^\theta$. Thus, equation (7.6) is written as

$$\partial_r(\ln\gamma_{\theta\theta}) - 2\sqrt{\gamma_{rr}}K_\theta{}^\theta = 0. \tag{7.11}$$

As a reasonable consequence, the equation is written primarily by the components on the two-dimensional sphere. Indeed, $A(r) = 4\pi\gamma_{\theta\theta}(r)$ denotes the surface area of a radius, r, and using the equation of $\gamma_{\theta\theta}$ in the form, $-2\alpha K_\theta{}^\theta = (\partial_t - \beta^r\partial_r)\ln\gamma_{\theta\theta}$, equation (7.11) is rewritten as

$$\left[\partial_t + \left(\alpha\gamma_{rr}^{-1/2} - \beta^r\right)\partial_r\right]A(r) = 0 \quad \text{or} \quad k^a\nabla_a A(r) = 0. \tag{7.12}$$

Thus, the apparent horizon in spherical symmetry may be defined as the surface where the local variation rate of its area along outgoing light rays is zero (note this does not mean that trapping horizon has a null surface; see appendix G.5 for the definition of trapping horizon).

To check that equation (7.11) yields the correct result, we may consider a non-rotating black hole in isotropic coordinates (see, e.g., section 1.3.1). In this case, $\gamma_{rr} = \psi^4$, $\gamma_{\theta\theta} = \psi^4 r^2$, and $K_{ab} = 0$. Thus equation (7.11) is written as [see equations (1.107) and (5.79)]

$$2\frac{\partial_r \psi}{\psi} + \frac{1}{2r} = 0 \quad \rightarrow \quad -\frac{M}{r^2}\left(1 + \frac{M}{2r}\right)^{-1} + \frac{1}{2r} = 0. \tag{7.13}$$

Then, we find that the location of the apparent horizon agrees with that of the event horizon $r = M/2$.

In numerical-relativity simulations, the presence and the location of an apparent horizon are determined from equation (7.11) by finding the outermost radius that satisfies this equation, and in addition, by checking that the left-hand side of this equation (i.e., Θ) is positive and negative outside and inside the zero point, respectively. This property is confirmed from equation (7.13).

7.1.2 Axially symmetric case

For the axisymmetric case, f is a function of θ, and obeys an ordinary differential elliptic-type equation. This equation is derived in a straightforward calculation using variables in the conformal decomposition formulation, i.e., $\tilde{\gamma}_{ab} = \psi^{-4}\gamma_{ab}$, and defining

$$\tilde{C} = C\psi^{-2} = \left[\tilde{\gamma}^{rr} - 2\tilde{\gamma}^{r\theta}\partial_\theta f + \tilde{\gamma}^{\theta\theta}(\partial_\theta f)^2\right]^{-1/2}. \tag{7.14}$$

Then, we have

$$D_a s^a = \tilde{C}\psi^{-2}\left[\left(4\frac{\partial_r \psi}{\psi} + \frac{\partial_r \tilde{C}}{\tilde{C}} + \frac{2}{r}\right)\left(\tilde{\gamma}^{rr} - \tilde{\gamma}^{r\theta}\partial_\theta f\right) + \partial_r \tilde{\gamma}^{rr} - \left(\partial_r \tilde{\gamma}^{r\theta}\right)\partial_\theta f \right.$$
$$\left. + \left(4\frac{\partial_\theta \psi}{\psi} + \frac{\partial_\theta \tilde{C}}{\tilde{C}} + \cot\theta\right)\left(\tilde{\gamma}^{r\theta} - \tilde{\gamma}^{\theta\theta}\partial_\theta f\right) \right.$$
$$\left. + \partial_\theta \tilde{\gamma}^{r\theta} - \left(\partial_\theta \tilde{\gamma}^{\theta\theta}\right)\partial_\theta f - \tilde{\gamma}^{\theta\theta}\partial_{\theta\theta}f\right]. \tag{7.15}$$

It is straightforward to evaluate the terms composed of the extrinsic curvature $K_{ab}s^a s^b - K$. One point worthy to note is that this often vanishes for stationary spacetime. For example, this vanishes for Kerr spacetime in Boyer–Lindquist and quasi-isotropic coordinates because nonzero components of the extrinsic curvature are only $K_{r\varphi}$ and $K_{\theta\varphi}$ (thus $K = 0$), and in addition, the surface of the apparent horizon (i.e., event horizon) is spherical in these coordinates (see section 1.3.1), i.e., $s^\varphi = 0$. This is also the case for black hole-torus systems in equilibrium (see section 5.7.1). By contrast, $K_{ab}s^a s^b - K \neq 0$ in general dynamical spacetime.

For flat space where $\psi = 1$, $\tilde{\gamma}^{rr} = 1$, $\tilde{\gamma}^{r\theta} = 0$, and $\tilde{\gamma}^{\theta\theta} = r^{-2}$, equation (7.15) reduces to

$$D_a s^a = \tilde{C}\left[\frac{\partial_r \tilde{C}}{\tilde{C}} + \frac{2}{f} - \frac{1}{f^2}\left(\frac{\partial_\theta \tilde{C}}{\tilde{C}} + \cot\theta\right)\partial_\theta f - \frac{1}{f^2}\partial_{\theta\theta}f\right], \tag{7.16}$$

where $\tilde{C} = [1 + (\partial_\theta f)^2/r^2]^{-1/2}$, and after we evaluate the partial derivative, r should be replaced by f because the apparent horizon is defined by $r = f$. Unless the horizon is highly deformed, we should have $|\partial_\theta f| \ll f$, and thus, $|\partial_r \tilde{C}/\tilde{C}| \ll 1/r$ and $|\partial_\theta \tilde{C}/\tilde{C}| \ll 1$. This implies that the major part of equation (7.16) is

$$-\frac{1}{f^2}\left(\partial_{\theta\theta} + \cot\theta\partial_\theta - 2\right)f. \tag{7.17}$$

Thus, the equation of the apparent horizon is written as

$$\left(\partial_{\theta\theta} + \cot\theta\partial_\theta - 2 + \eta_f\right)f = \tilde{S}\left[f, \partial_\theta f, \partial_{\theta\theta} f, \gamma_{ij}, K_{ij}\right] + \eta_f f =: S, \tag{7.18}$$

where η_f is often introduced as a free parameter to accelerate the convergence to a solution in numerical computation (see below).

The elliptic equation (7.18) is solved imposing boundary conditions appropriately. The boundary conditions should be in general imposed along the symmetric axis at $\theta = 0$ and π. Because the surface of any apparent horizon has to be smooth, the boundary condition should be $\partial_\theta f = 0$ at $\theta = 0$ and π. If the equatorial plane symmetry may be imposed, the reflection symmetric boundary condition should be imposed at $\theta = \pi/2$ as $\partial_\theta f = 0$ instead of the condition at $\theta = \pi$.

Iteration is necessary for solving the equation in the form (7.18) because f is contained in its right-hand side. For example, the equation can be solved in the following manner: First, a trial function of $f(\theta)$ is prepared. Then, substituting this trial function in the right-hand side, the following linear elliptic equation for f is solved:

$$\left(\partial_{\theta\theta} + \cot\theta\partial_\theta - 2 + \eta_f\right)f = \text{given source term}. \tag{7.19}$$

We then determine a new trial function for f from the solution obtained. This procedure is repeated until a sufficient convergence for the solution of f is achieved.

There are several methods for a numerical solution of equation (7.19). A typical method is to discretize f as f_k where f_k is a value of f at $\theta_k = (k - 1/2)\Delta\theta$ with $k = 1, \cdots, N_\theta$ and with $\Delta\theta$ being the grid spacing, π/N_θ. Here, N_θ denotes the total grid number. Then, the derivatives of f are evaluated employing finite differencing; e.g., in the second-order finite differencing, we should employ

$$(\partial_\theta f)_k = \frac{f_{k+1} - f_{k-1}}{2\Delta\theta}, \quad (\partial_{\theta\theta} f)_k = \frac{f_{k+1} - 2f_k + f_{k-1}}{\Delta\theta^2}. \tag{7.20}$$

At boundaries, $k = 1$ and N_θ, we can use $f_0 = f_1$ and $f_{N_\theta+1} = f_{N_\theta}$, respectively, which are derived from the boundary conditions, $\partial_\theta f = 0$ at $\theta = 0$ and π. These conditions are used for rearranging the finite differencing at the boundaries.

As a result of the finite differencing, equation (7.19) reduces to a $N_\theta \times N_\theta$ matrix equation in the form, $M_{ij}f_j = S_i$, where the schematic form of M_{ij} is

$$M_{ij} = \begin{pmatrix} a_1 & b_1 & 0 & 0 & \cdots & 0 & 0 & 0 \\ c_2 & a_2 & b_2 & 0 & \cdots & 0 & 0 & 0 \\ 0 & c_3 & a_3 & b_3 & 0 & \cdots & 0 & 0 \\ 0 & 0 & c_4 & a_4 & b_4 & 0 & \cdots & 0 \\ \vdots & \vdots & \ddots & \ddots & \ddots & \ddots & \vdots & \vdots \\ 0 & \cdots & 0 & 0 & c_{N_\theta-2} & a_{N_\theta-2} & b_{N_\theta-2} & 0 \\ 0 & 0 & \cdots & 0 & 0 & c_{N_\theta-1} & a_{N_\theta-1} & b_{N_\theta-1} \\ 0 & 0 & 0 & \cdots & 0 & 0 & c_{N_\theta} & a_{N_\theta} \end{pmatrix}. \quad (7.21)$$

Because M_{ij} is a tridiagonal matrix, the solution for the matrix equation can be derived by the so-called LU decomposition method. In it, we define double-diagonal matrices L_{ij} and U_{ij} which satisfy $L_{ij}U_{jk} = M_{ik}$ as

$$L_{ij} = \begin{pmatrix} d_1 & 0 & 0 & 0 & \cdots & 0 & 0 & 0 \\ c_2 & d_2 & 0 & 0 & \cdots & 0 & 0 & 0 \\ 0 & c_3 & d_3 & 0 & 0 & \cdots & 0 & 0 \\ 0 & 0 & c_4 & d_4 & 0 & 0 & \cdots & 0 \\ \vdots & \vdots & \ddots & \ddots & \ddots & \ddots & \vdots & \vdots \\ 0 & \cdots & 0 & 0 & c_{N_\theta-2} & d_{N_\theta-2} & 0 & 0 \\ 0 & 0 & \cdots & 0 & 0 & c_{N_\theta-1} & d_{N_\theta-1} & 0 \\ 0 & 0 & 0 & \cdots & 0 & 0 & c_{N_\theta} & d_{N_\theta} \end{pmatrix}, \quad (7.22)$$

$$U_{ij} = \begin{pmatrix} 1 & u_1 & 0 & 0 & \cdots & 0 & 0 & 0 \\ 0 & 1 & u_2 & 0 & \cdots & 0 & 0 & 0 \\ 0 & 0 & 1 & u_3 & 0 & \cdots & 0 & 0 \\ 0 & 0 & 0 & 1 & u_4 & 0 & \cdots & 0 \\ \vdots & \vdots & \ddots & \ddots & \ddots & \ddots & \vdots & \vdots \\ 0 & \cdots & 0 & 0 & 0 & 1 & u_{N_\theta-2} & 0 \\ 0 & 0 & \cdots & 0 & 0 & 0 & 1 & u_{N_\theta-1} \\ 0 & 0 & 0 & \cdots & 0 & 0 & 0 & 1 \end{pmatrix}. \quad (7.23)$$

A straightforward calculation yields $d_k = a_k - c_k b_{k-1}/d_{k-1}$ with $d_1 = a_1$ and $u_k = b_k/d_k$. Thus, the components of L_{ij} and U_{ij} are determined simply by a recursive algebra.

Then, we write the matrix equation $M_{ij}f_j = S_i$ in the form

$$L_{ij}y_j = S_i, \quad U_{ij}f_j = y_i. \quad (7.24)$$

The former equation yields a recurrence relation for y_k as

$$c_k y_{k-1} + d_k y_k = S_k, \quad (7.25)$$

with $y_1 = S_1/d_1$. Thus, y_k is recursively determined from $k = 2$ to $k = N_\theta$. The latter equation of (7.24) yields

$$f_k + u_k f_{k+1} = y_k, \quad (7.26)$$

with $f_{N_\theta} = y_{N_\theta}$. Thus, f_k is also recursively determined from $k = N_\theta - 1$ to 1. In this method, the solution of f_k is obtained semi-analytically with a simple algebra. This new solution of f_k is then used as a new trial function.

As already mentioned, a solution of equation (7.18) is obtained by iteratively solving the equation of the form (7.19) until a sufficient convergence is achieved. The degree of convergence is checked, e.g., by calculating the following quantity,

$$\text{CONV} := \frac{\sum_{i=1}^{N_\theta} |f_i - f_i^b|}{\sum_{i=1}^{N_\theta} |f_i|}, \tag{7.27}$$

where f_i^b denotes a solution in the previous iteration step. If CONV is much smaller than unity (say 10^{-6}), we may consider that the convergence is well achieved.

To accelerate the iteration procedure, the choice of the free parameter η_f plays an important role [Nakamura et al. (1987); Shibata (1997)]. According to several numerical experiments performed by the author, the best choice of η_f is $\sim 0.5-1$.

7.1.3 Non-axially symmetric case

For the non-axisymmetric case, the procedure to a solution of f is qualitatively the same as in the axisymmetric case. However, the numerical costs increase because f is a function of θ and φ and the equation becomes a two-dimensional elliptic equation. In addition, a modified numerical method is necessary.

In this case, s_a is written by

$$s_k = \psi^2 \tilde{C}(1, -\partial_\theta f, -\partial_\varphi f), \tag{7.28}$$

where the variables in the conformal-decomposition formulation were used, and

$$\tilde{C} = [\tilde{\gamma}^{rr} - 2\tilde{\gamma}^{r\theta}\partial_\theta f - 2\tilde{\gamma}^{r\varphi}\partial_\varphi f + \tilde{\gamma}^{\theta\theta}(\partial_\theta f)^2 + 2\tilde{\gamma}^{\theta\varphi}(\partial_\theta f)(\partial_\varphi f) + \tilde{\gamma}^{\varphi\varphi}(\partial_\varphi f)^2]^{-1/2}. \tag{7.29}$$

Then,

$$\begin{aligned}D_a s^a = \tilde{C}\psi^{-2}\bigg[&\left(4\frac{\partial_r \psi}{\psi} + \frac{\partial_r \tilde{C}}{\tilde{C}} + \frac{2}{r}\right)\left(\tilde{\gamma}^{rr} - \tilde{\gamma}^{r\theta}\partial_\theta f - \tilde{\gamma}^{r\varphi}\partial_\varphi f\right) \\&+ \left(4\frac{\partial_\theta \psi}{\psi} + \frac{\partial_\theta \tilde{C}}{\tilde{C}} + \cot\theta\right)\left(\tilde{\gamma}^{r\theta} - \tilde{\gamma}^{\theta\theta}\partial_\theta f - \tilde{\gamma}^{\theta\varphi}\partial_\varphi f\right) \\&+ \left(4\frac{\partial_\varphi \psi}{\psi} + \frac{\partial_\varphi \tilde{C}}{\tilde{C}}\right)\left(\tilde{\gamma}^{r\varphi} - \tilde{\gamma}^{\theta\varphi}\partial_\theta f - \tilde{\gamma}^{\varphi\varphi}\partial_\varphi f\right) \\&+ \partial_i \tilde{\gamma}^{ir} - \left(\partial_i \tilde{\gamma}^{i\theta}\right)\partial_\theta f - \left(\partial_i \tilde{\gamma}^{i\varphi}\right)\partial_\varphi f \\&- \tilde{\gamma}^{\theta\theta}\partial_{\theta\theta}f - 2\tilde{\gamma}^{\theta\varphi}\partial_{\theta\varphi}f - \tilde{\gamma}^{\varphi\varphi}\partial_{\varphi\varphi}f\bigg]. \end{aligned} \tag{7.30}$$

As in the axisymmetric case, we first consider the flat-space case. Then, the main kernel of equation (7.30) is found to be

$$-\frac{1}{f^2}\left(\partial_{\theta\theta} + \cot\theta\partial_\theta + \sin^{-2}\theta\partial_{\varphi\varphi} - 2\right)f, \tag{7.31}$$

and thus, the equation of the apparent horizon is written in the form

$$\left(\partial_{\theta\theta} + \cot\theta\partial_\theta + \sin^{-2}\theta\partial_{\varphi\varphi} - 2 + \eta_f\right)f$$
$$= \tilde{S}\left[f, \partial_\theta f, \partial_{\theta\theta}f, \partial_\varphi f, \partial_{\theta\varphi}f, \partial_{\varphi\varphi}f, \gamma_{ij}, K_{ij}\right] + \eta_f f =: S. \quad (7.32)$$

As in the axisymmetric case, the elliptic equation (7.32) may be iteratively solved imposing boundary conditions at $\theta = 0$, π, $\varphi = 0$, and 2π. Again, the boundary condition is determined by the requirement that the surface of the apparent horizon has to be smooth anywhere. However, in the non-axisymmetric case, this does not yield a specific boundary condition such as $\partial_\theta f = 0$ at $\theta = 0$ and π. This situation stimulated people in the field of numerical relativity to rely on the pseudo-spectral expansion in 1980s and 1990s (e.g., see Nakamura et al. (1984); Baumgarte et al. (1996a); Anninos et al. (1998); Gundlach (1998) for representative works). However, the boundary condition is actually imposed in an appropriate manner, and thus, we here focus on the approach in which equation (7.32) is solved as a boundary-value problem as in the axisymmetric case [Shibata (1997); Shibata and Uryū (2000)] (see also Thornburg (1996)).

In this approach, we take the grid points for θ and φ as

$$\varphi_i = \left(i - \frac{1}{2}\right)\frac{2\pi}{N_\varphi}, \quad i = 1, 2, \cdots, N_\varphi$$

$$\theta_j = \left(j - \frac{1}{2}\right)\frac{\pi}{N_\theta}, \quad j = 1, 2, \cdots, N_\theta,$$

where N_φ and N_θ are numbers of grid points for $0 < \varphi < 2\pi$ and $0 < \theta < \pi$, respectively. Here, N_φ has to be an even number, i.e., $\varphi_i = \varphi_{i \pm N_\varphi/2} \mp \pi$ for $i \leq N_\varphi/2$ and $i > N_\varphi/2$, respectively. The subscripts i and j will always denote the grid points for φ and θ in the following.

We suppose that the partial derivatives of f are evaluated by second-order finite differencing as in section 7.1.2. Then, we have

$$(\partial_\varphi f)_{i,j} = \frac{f_{i+1,j} - f_{i-1,j}}{2\Delta\varphi}, \quad (\partial_{\varphi\varphi}f)_{i,j} = \frac{f_{i+1,j} - 2f_{i,j} + f_{i-1,j}}{\Delta\varphi^2},$$

$$(\partial_\theta f)_{i,j} = \frac{f_{i,j+1} - f_{i,j-1}}{2\Delta\theta}, \quad (\partial_{\theta\theta}f)_{i,j} = \frac{f_{i,j+1} - 2f_{i,j} + f_{i,j-1}}{\Delta\theta^2}.$$

For evaluating the derivatives at $i = 1$, $i = N_\varphi$, $j = 1$, and $j = N_\theta$, the boundary conditions are necessary. At $i = 1$ and N_φ, we should simply impose the periodic boundary condition, i.e., the dummy points $i = 0$ and $N_\varphi + 1$ are identical to $i = N_\varphi$ and 1, respectively, irrespective of the θ coordinate. At $j = 1$ and N_θ, the situation is not as straightforward as at $i = 1$ and N_φ. However, in the present setting of the grid points, we find that a dummy point $(i, 0)$ is identical to $(i + N_\varphi/2, 1)$ for $i \leq N_\varphi/2$ and to $(i - N_\varphi/2, 1)$ for $i > N_\varphi/2$. By the same procedure, a dummy point $(i, N_\theta + 1)$ can be replaced by a regular grid point of $j = N_\theta$. Therefore, all the derivatives are evaluated only in terms of the regular grid points.

Therefore, as in the axisymmetric case, equation (7.32) reduces to a matrix equation in the form, $M_{kl}f_l = S_k$, where the schematic form of M_{kl} is

$$M_{kl} = \begin{pmatrix} * * * & & * * & & * * & & \\ * * * * & & * & & * & & \\ * * * * * & & * & & * & & \\ * * * * * & & * & & * & & \\ * * * * & & * & & * & & \\ * * * & & * & & * & & \\ * & & * * * & & * & & \\ * & & * * * * & & * & & \\ * & & * * * * * & & * & & \\ * & & * * * * * & & * & & \\ * & & * * * * & & * & & \\ * & & * * * & & * & & \\ * & & * & & * * * & & * \\ * & & * & & * * * * & & * \\ * & & * & & * * * * * & & * \\ * & & * & & * * * * * & & * \\ * & & * & & * * * * & & * \\ * & & * & & * * * & & * \\ & & * & & * & & * * * \\ & & * & & * & & * * * * \\ & & * & & * & & * * * * * \\ & & * & & * & & * * * * * \\ & & * & & * & & * * * * \\ & & * & & * & & * * * \end{pmatrix}. \quad (7.33)$$

Here we supposed $N_\varphi = 8$ and $N_\theta = 4$ in this example. $*$ denotes a nonzero component and in the absence of $*$, the components are zero. The matrix is composed basically of five-diagonal elements and due to the boundary conditions imposed, additional components appear. The matrix equation is more complicated than that in the axisymmetric case, and it is necessary to employ a more sophisticated method for its numerical solution. However, the matrix is still not dense and several well established matrix-equation solvers are available for a solution of this equation [see, e.g., Shibata (1997); Shibata and Uryū (2000); Huq et al. (2002); Schnetter (2003)].

7.1.4 *Finding apparent horizons for simple initial data*

By exploring apparent horizons for initial data, we can study the properties of a black hole and a strongly gravitating object. We here describe benefits of such studies giving two examples.

7.1.4.1 *Two black holes momentarily at rest*

First, we introduce the result for an analysis of apparent horizons for two black holes that are momentarily at rest (i.e., $K_{ab} = 0$). Cadez (1974) analyzed the apparent horizon for this system for the first time. For the initial data, he employed Brill–Lindquist initial data, for which the spatial metric is written in the conformally flat form and the conformal factor is given by equation (5.83) with $n = 2$ and $N = 3$ (see section 5.5.1.2). Thus, the system is axisymmetric. Cadez (1974) focused on the equal-mass case and wrote $a_1 = a_2 = m/2 =: a$ and $x_1^i = -x_2^i = (0, 0, a)$:

$$dl^2 = \psi^4 f_{ij} dx^i dx^j, \quad \psi = 1 + \frac{m}{2r_+} + \frac{m}{2r_-}. \quad (7.34)$$

Here, $r_\pm = \sqrt{x^2 + y^2 + (z \mp a)^2}$ and m is the so-called bare mass.

The equation for apparent horizons has a quite simple form for the conformally flat axisymmetric metric with $K_{ab} = 0$ as

$$(\partial_{\theta\theta} + \cot\theta \partial_\theta - 2) f$$
$$= \frac{4\left[f^2 + (\partial_\theta f)^2\right]}{f^2 \psi} \left(f^2 \partial_r \psi - (\partial_\theta f)\partial_\theta \psi\right) + \frac{3(\partial_\theta f)^2}{f} - \frac{(\partial_\theta f)^3 \cot\theta}{f^2}. \quad (7.35)$$

Here $f(\theta)$ denotes the location of an apparent horizon for an appropriate choice of the origin. Equation (7.35) is solved for the boundary conditions $\partial_\theta f = 0$ at $\theta = 0$ and π. This is a simple ordinary differential equation, and is quite easily solved numerically, e.g., by the method described in section 7.1.2.

For a sufficiently large value of a, there are only two apparent horizons, each of which encompasses two separate black holes. For a small value of a, in addition, a new apparent horizon that encompasses two black holes appears. Cadez (1974) determined the criterion for the appearance of this common apparent horizon as $a \leq a_{\text{crit}} \approx 1.53(m/2)$, and for $a = a_{\text{crit}}$, its shape is highly deformed from the spherical shape. He also found that the non-spherical degree of this common apparent horizon decreases steeply with the decrease of a. This suggests that for values of $a < a_{\text{crit}}$, the initial data is well described by a Schwarzschild background plus a non-spherical perturbation.

This analysis performed for the simple initial data gave us a rich insight for the head-on collision of two black holes. This analysis enables us to expect that event horizons of two black holes merge only at a close separation of $a \sim 0.76m$. At the merger, a horizon of the new black hole is likely to be highly non-spherical but it will then settle to a spherical black hole when the separation of two black holes decreases subsequently. Because the horizon is deformed at the merger, gravitational waves are likely to be emitted, and after a sufficient emission of gravitational waves, the merged black hole will relax to a spherical one. Indeed, all these expectations were confirmed by numerical-relativity simulations for the head-on collision subsequently performed [Smarr (1979); Anninos et al. (1993)].

As mentioned already, the analysis of Cadez (1974) indicates that a black hole formed after the two black-hole collision could be well described by a perturbed Schwarzschild black hole. This suggests that gravitational waves emitted may be analyzed in a framework of a black-hole perturbation theory. Motivated by this idea, Price and Pullin (1994) performed a linear perturbation analysis of gravitational waves employing Misner initial data as initial conditions (see section 5.5.1.3). They solved a linear perturbation equation on a Schwarzschild background initially dividing the Misner initial data into a background part and a linear perturbation part. They showed that waveforms and total radiated energy of gravitational waves, computed in numerical-relativity simulations, are approximately reproduced by the linear perturbation analysis (see Anninos et al. (1995d) in which linear and fully nonlinear results were compared). This indicates the robustness of a linear perturbation analysis combined with initial-data problem for a particular issue. It is worthy to point out that computational costs for the linear perturbation analysis are not as expensive as those for fully nonlinear simulations.

7.1.4.2 Deformed collapsed objects

Next, we pay attention to a study concerning the presence or absence of apparent horizons for highly deformed collapsed objects. Thorne (1972) proposed a hoop conjecture which states that event horizons may not be formed for a highly deformed collapsed object even if the compactness of such object is sufficiently high. A quantitative statement of the hoop conjecture is as follows: A black hole (i.e., an event horizon) can be formed only when a mass M gets compacted into a region for which proper circumferential lengths in all the directions is $\lesssim 4\pi M$. If the hoop conjecture is correct, aspherical collapse, for which the circumferential lengths for one or two directions are sufficiently longer than the others, might lead to a naked singularity. Here, by the analysis of apparent horizons to initial data, it is possible to explore this issue.

Nakamura et al. (1988) considered initial data composed of spheroidal compact matter sources and searched for apparent horizons encompassing these objects. Specifically, they analyzed momentarily static initial data ($K_{ab} = 0$), which were analytically derived by solving the Hamiltonian constraint for the conformally flat spatial metric with a special form of the matter source as $\rho_\mathrm{h} \psi^5 = 2\rho_\mathrm{N}$. Then, the basic equation for the conformal factor ψ is

$$\Delta_\mathrm{f} \psi = -4\pi \rho_\mathrm{N}, \tag{7.36}$$

where ρ_N was set as

$$\rho_\mathrm{N} = \begin{cases} \dfrac{M_\mathrm{N}}{4\pi a^2 c/3} & \dfrac{\varpi^2}{a^2} + \dfrac{z^2}{c^2} \leq 1, \\ 0 & \text{elsewhere.} \end{cases} \tag{7.37}$$

Here, ϖ is the cylindrical coordinate, a and c are coordinate axial lengths of a spheroid, and M_N is constant. Although the choice of the density profile is rather ad hoc, we can obtain the metric analytically as follows. For the oblate case with $a \geq c$, the solution is given by

$$\psi = 1 + \frac{3M_\mathrm{N}}{2ae} \zeta + \frac{1}{2} (\varpi K_\varpi + z K_z), \tag{7.38}$$

where

$$K_\varpi = -\frac{3M_\mathrm{N}}{2(ae)^3} \varpi \left(\zeta - \sin \zeta \cos \zeta \right), \tag{7.39}$$

$$K_z = -\frac{3M_\mathrm{N}}{(ae)^3} z \left(\tan \zeta - \zeta \right), \tag{7.40}$$

and $e = \sqrt{1 - c^2/a^2}$. Here, ζ is obtained from

$$\begin{cases} \sin \zeta = e & \text{inside the spheroid,} \\ \varpi^2 \sin^2 \zeta + z^2 \tan^2 \zeta = a^2 e^2 & \text{outside the spheroid.} \end{cases} \tag{7.41}$$

For the prolate case with $a \leq c$,

$$\psi = 1 + \frac{3M_\mathrm{N}}{2ce} \zeta + \frac{1}{2} (\varpi K_\varpi + z K_z), \tag{7.42}$$

where
$$K_\varpi = \frac{3M_\mathrm{N}}{2(ce)^3}\varpi\left(\zeta - \sinh\zeta\cosh\zeta\right), \tag{7.43}$$

$$K_z = \frac{3M_\mathrm{N}}{(ce)^3}z\left(\tanh\zeta - \zeta\right), \tag{7.44}$$

and $e = \sqrt{1 - a^2/c^2}$. Here, ζ is obtained from
$$\begin{cases} \sinh\zeta = ce/a & \text{inside the spheroid,} \\ \varpi^2\sinh^2\zeta + z^2\tanh^2\zeta = c^2e^2 & \text{outside the spheroid.} \end{cases} \tag{7.45}$$

An important fact for these analytic solutions is that for the oblate spheroid, ψ at the origin is finite even for $e \to 1$ (or $c \to 0$ for a fixed value of a). This is also recognized by the fact that $\zeta \to \pi/2$ in this limit. By contrast, ψ at the origin diverges for the prolate case for $e \to 1$ (or $a \to 0$ for a fixed value of c). This is recognized by the fact that $\zeta \to \infty$ in this limit. For both spheroids, the density diverges at the limit $e \to 1$, and thus, a divergence appears. However, the divergence for the oblate case is weak (i.e., it does not cause singular behavior in the geometry).

The property of the physical singularity in general relativity should be more rigorously defined by the geodesic incompleteness [e.g., Wald (1984)]. The relevant quantity for the geodesic equation is the derivative of the metric. In this case, thus, we should pay attention to $\partial_i\psi$. For the oblate case, $\partial_i\psi$ is everywhere finite even for $e \to 1$. This shows that the oblate spheroid does not form any physical singularity. By contrast, for the prolate case, $\partial_i\psi$ diverges at the top of the prolate shape ($z = c$ and $\varpi = 0$). This indicates that a physical singularity is formed.

If the physical singularity is hidden inside a horizon, as in the case for a black hole, it is innocuous. Nakamura et al. (1988), however, showed that even for highly compact matter profiles with $c \gtrsim 1.4M$, where M denotes the ADM mass, apparent horizons are not found for $e \to 1$. This suggested that a naked singularity may be formed.

The presence of a naked singularity for rather ad hoc choice of the matter configuration in the initial data is not surprising because of the fact that general relativity admits static and stationary solutions such as Weyl and Tomimatsu–Sato solutions [Tomimatsu and Sato (1972, 1973)] that possess naked singularities. However, Nakamura et al. (1988) indicated that there are a wide variety of prolate-type matter configurations that are likely to yield a naked singularity. This suggested that a naked singularity may be formed in reality without assuming a special matter configuration. Shapiro and Teukolsky (1991) subsequently performed numerical-relativity simulations employing collisionless particles as the matter field. They indeed showed that a naked singularity with a prolate configuration may be formed even if a fairly natural initial condition is chosen. Thus, the prediction made by the analysis of initial data was confirmed. This example clearly shows that the analysis of initial data is useful for obtaining an insight about the strongly general relativistic objects.

7.2 Event horizon

Black holes are characterized by event horizons (see appendix G for more details). This implies that the best way for exploring black holes is to locate and analyze their event horizons. The event horizon of a black hole is defined as a causal boundary of the causal past of null infinity. This implies that strictly speaking, it is determined only when we know the global structure of the corresponding spacetime that is found only after the entire evolution of the spacetime is solved. However, event horizons can be located approximately after finite evolution time, provided that the spacetime has settled to a nearly stationary state. Several numerical methods for this have been developed [Hughes *et al.* (1994); Libson *et al.* (1996); Anninos *et al.* (1995a)].

First, event horizons could be directly located by evolving a large number of null geodesics [Hughes *et al.* (1994)], for which world-lines $x^a(\lambda)$ are numerically determined by the geodesic equation (1.5) written in the form

$$k^\nu \partial_\nu k_\mu + \frac{1}{2} k^\alpha k^\beta \partial_\mu g_{\alpha\beta} = 0, \qquad (7.46)$$

where k^a is the tangent vector of the null geodesics. Using the variables in the $N+1$ formulation, the spatial component of this equation is written as

$$\frac{dk_i}{dt} = -\alpha k^t \partial_i \alpha + k_j \partial_i \beta^j - \frac{k_j k_l}{2k^t} \partial_i \gamma^{jl}, \qquad (7.47)$$

where k^t is calculated from $k^a k_a = 0$ as $k^t = \sqrt{\gamma^{ij} k_i k_j}/\alpha$, and thus, only the equation for k_i (not k^i) should be solved. Then, equation (7.47) together with

$$\frac{dx^i}{dt} = \frac{k^i}{k^t} = -\beta^i + \frac{\gamma^{ij} k_j}{k^t} \qquad (7.48)$$

determines the world-lines. We note that equation (7.47) is linear in $|k_i|$ and thus we need to pay attention only to the direction of k_i (not to its absolute magnitude).

For spacetime numerically generated, α, β^i, and γ^{ij} have values on the computational grid points of each spatial hypersurface. We then can track the trajectories of light rays on such numerical spacetime by solving simultaneous ordinary differential equations (7.47) and (7.48). This task is not technically difficult, and in principle, it is possible to determine the location of an event horizon by launching a number of light rays in different directions k^i from every points in spacetime. However, this method is out of touch with reality because a huge number of the light rays have to be launched even in four spacetime dimensions; for D-dimensional spacetime, we have to take into account D-dimensional spacetime points plus $N(=D-1)$-dimensional directions for the launching direction of k^i. Thus, in total, $2N+1$ dimensions have to be taken into account. The number of points for each dimension would be at least 100, and hence, the total number of the light rays is 10^{4N+2}: This is too huge even for $N=3$.

The efficiency for the search of an event horizon may be significantly improved if we know the location of corresponding apparent horizons (or a trapping horizon;

see appendix G.5 for the definition of trapping horizon). In particular, this is the case if the apparent horizons are located for a nearly stationary final state, because they should agree with the event horizon for the stationary spacetime. In this case, we can restrict the spacetime launching points of light rays to the vicinity of the apparent horizons. Furthermore, we can use the following property of apparent horizons: If at any time a light ray passes into a trapping horizon we know with certainty that it is within an event horizon. Thus, if all light rays launched from a spacetime point, regardless initial choice of k^i, are captured by the trapping horizon, then that spacetime point is inside the black hole. On the other hand, if at least one light ray is not captured but escapes, the point is not inside the black hole. The surface separating these capture or escape points is the event horizon of the black hole.

However, the computational cost is still large using the geodesic-integration method, because we have to still consider a huge number of the direction in k^i. To overcome this inefficiency, Libson et al. (1996); Anninos et al. (1995a) proposed the *null-surface integration* method. The proposal of this approach is based on a property that the event horizon can be generally considered as a $[(N-1)+1]$-dimensional null surface. Thus, the event horizon can be represented as a level surface in terms of a function

$$f(t, x^i) = 0, \qquad (7.49)$$

for which the normal vector to the event horizon is defined by $\nabla^a f$. Since the normal vector is null, the function f satisfies

$$g^{ab}(\partial_a f)(\partial_b f) = 0, \qquad (7.50)$$

and in the $N+1$ formulation, this is written to the evolution equation for f as

$$\left(\partial_t - \beta^k \partial_k\right) f = \pm \sqrt{\alpha^2 \gamma^{ij}(\partial_i f)\partial_j f}. \qquad (7.51)$$

Here, we should choose the minus sign so that the surface is generated by future directed null geodesics ($\partial_t f < 0$ and $\nabla^t f > 0$). [1]

To further improve the efficiency in numerical computation, equation (7.51) should be integrated backwards in time [Hughes et al. (1994); Libson et al. (1996); Anninos et al. (1995a)]: Obviously, it is not easy to accurately locate the event horizon by the integration forward in time, because small numerical errors cause the ray to drift and diverge rapidly; this fact is easily understood by the fact that purely outgoing light rays launched just outside the event horizon can escape from the event horizon toward null infinity whereas outgoing light rays with a tangential momentum just outside the event horizon can be easily captured by the black hole. It should be noted, however, that when employing the backward-integration method, we have to store all the data of numerical spacetime at least for the vicinity of the event

---

[1]For flat spacetime, a null surface generated by future-directed outgoing light rays should be denoted by $f = r - t + $ const. This equation satisfies equation (7.51) with the minus sign.

horizon, and locating the event horizon has to be performed as a post process. Therefore, huge data storing is required.

In this approach, we only need to solve the first-order partial differential equation for f, (7.51), with an appropriate initial condition. This equation is essentially the same as a first-order scalar-wave equation. This fact is in particular clear for the spherically symmetric case in which the equation is written in the form

$$\partial_t f = - \left(\alpha\sqrt{\gamma^{rr}} - \beta^r\right) \partial_r f. \tag{7.52}$$

This is a wave equation with the characteristic speed $\alpha\sqrt{\gamma^{rr}} - \beta^r$: Note that for the static case, equation (7.52) can be integrated analytically as

$$f = f\left[\int_r \frac{dr'}{\alpha\sqrt{\gamma^{rr}} - \beta^r} - t\right]. \tag{7.53}$$

Thus, the equation can be solved by a numerical method similar to that used for solving Einstein's equation (see section 3.2). Another advantage in this method is that the equation is composed only of the metric itself (not its derivative). Thus, the cost for the numerical solution of this equation is less than that for the numerical solution of the geodesic equations.

As the initial condition of the backward integration of equation (7.51), we should use a property of the apparent horizons that they coincide with the corresponding event horizon in stationary spacetime. Thus, the first step in this approach is to determine an apparent horizon after a sufficiently long-term numerical computation that enables the spacetime to relax approximately to a stationary state. Then, we choose a function which is constant on the apparent horizon as f; e.g., for three-dimensional spatial hypersurfaces, we should set $f = r - r_{\rm AH}(\theta,\varphi)$ where $r_{\rm AH}$ denotes the surface of an apparent horizon. Here, the surface of the event horizon is supposed to be $r = r_{\rm AH}(\theta,\varphi)$.

7.3 Analyzing horizons

The properties of a black hole are to be extracted from its horizon. The most important quantities of black holes are mass and spin angular momentum, and hence, we here describe how to obtain (or approximately estimate) them from the quantities of the horizons. In the following discussion, we will pay attention only to four-dimensional black holes.

First, we consider the case that an event horizon is determined in vacuum spacetime (or in spacetime of a tiny amount of matter). Then, the procedure for accurately determining the mass, M, and the spin, a, of the black hole is straightforward because we know that the final state should be a Kerr black hole due to the presence of uniqueness theorem (e.g., Hawking and Ellis (1973); Wald (1984) for a review). In this case, we should measure the area, the circumferential lengths, and the Ricci scalar of the horizon, which are, respectively, written in terms of M and $\hat{a} = a/M$

by (see also section 1.3.2.3)

$$A_H = 8\pi M^2 \hat{r}_+, \quad C_{eq} = 4\pi M, \quad C_p = 4M \int_0^{\pi/2} d\theta \sqrt{\hat{r}_+^2 + \hat{a}^2 - \hat{a}^2 \sin^2\theta},$$

$$\overset{(2)}{R} = 2M^{-2}\frac{(\hat{r}_+^2 + \hat{a}^2)(\hat{r}_+^2 - 3\hat{a}^2 \cos^2\hat{\theta})}{(\hat{r}_+^2 + \hat{a}^2 \cos^2\hat{\theta})^3}, \qquad (7.54)$$

where C_{eq} and C_p are the circumferential lengths around the equatorial surface and a meridian curve, $\hat{r}_+ = 1 + \sqrt{1-\hat{a}^2}$, and $\overset{(2)}{R}$ is the Ricci scalar on the horizon surface for a particular choice of the angular coordinate $\hat{\theta}$. Note that $\hat{\theta} = 0$ and π for $\overset{(2)}{R}$ denote the spin axis direction. Then, there are several ways for determining the mass and the spin; for example, from C_{eq}, M is determined and then, $\hat{a}$ is determined either from A_H or from C_p; or from C_p/C_{eq}, $\hat{a}$ is determined and then, M is determined from A_H. Furthermore, the results obtained by several different methods are used for measuring the error in the numerical results.

For the general case, the direction of the black-hole spin axis is often nontrivial (note that $\hat{\theta}$ is in general not equal to θ of spherical polar coordinates). The Ricci scalar on horizons is a useful quantity for determining the spin axis direction as well as the magnitude of the spin [see equation (1.137)]: The spin axis can be determined by finding the locations of the minimum values of the Ricci scalar.

Even if we are able to determine only an apparent horizon, the above procedure is useful for approximately obtaining the mass and the spin of the black hole in vacuum, if the spacetime relaxes to a stationary state for which the apparent horizon should agree with the event horizon. This is likely to be also the case when we consider a black hole in a binary of quasi-equilibrium orbits (i.e., the orbital separation is much larger than the black hole radius). For these cases, we may derive the mass and the spin by calculating the area, the circumferential lengths, and the Ricci scalar of the apparent horizon and then by using the formulas like equation (7.54).

In the presence of an appreciable amount of matter and/or in dynamical spacetime, the notions of the isolated horizon and the dynamical horizon are valuable (see appendix G.5 and G.6), and the mass and the angular momentum may be extracted from the information of these horizons. For these cases, however, the physically valuable quantities to be calculated from the horizons are only the area, $\mathcal{A}$, and angular momentum, J_{H_d} (assuming that there exists a local rotational Killing vector, φ^a, that defines the angular momentum of the horizon). Here the angular momentum is derived by the surface integral on the apparent horizon

$$J_{H_d} = \frac{1}{8\pi}\oint_H K_{ab}\varphi^a s^b dA. \qquad (7.55)$$

Then, the mass is defined by (see appendix G.5)

$$M_{H_d} = \frac{1}{2R_{H_d}}\sqrt{R_{H_d}^4 + 4J_{H_d}^2}, \qquad (7.56)$$

where $R_{H_d} := \sqrt{\mathcal{A}/4\pi}$. This mass agrees with the mass of a black hole if we consider stationary black holes.

PART 2
Applications

Chapter 8

Coalescence of binary compact objects

The merger of coalescing compact binary objects, binary neutron stars or black hole-neutron star binaries or binary black holes, has not been directly observed yet (at the end of 2014). This indicates that observing the merger events is a great challenge in astronomy and astrophysics. For the direct observation of the merger events, we have to predict possible signals generated by these phenomena as accurately as possible. Therefore, not only the observational effort but also the theoretical study for the merger events has been the important subject in astronomy and astrophysics.

As we summarized in section 1.5, the inspiral and merger of binary compact objects are among the most promising sources of gravitational waves for ground-based laser-interferometric gravitational-wave detectors. As introduced in section 1.6, for an efficient detection of gravitational waves and for extracting the nature of the observed gravitational-wave sources, it is necessary to prepare accurate theoretical templates of gravitational waveforms. In the final phase of binary compact objects, general relativistic gravity plays a substantial role, and hence, accurate theoretical templates of gravitational waves can be derived only by fully solving Einstein's equation together with matter-field equations without imposing any approximation. For this purpose, numerical relativity is probably the unique approach.

The mergers of binary neutron stars and black hole-neutron star binaries are promising sources for strong transient electromagnetic signals. As briefly mentioned in section 1.5.2, predicting the properties of the electromagnetic signals such as luminosity curve and typical wavelength is as important as predicting the gravitational waveforms. This is because the observation of an electromagnetic counterpart is crucial for the confirmation of a gravitational-wave detection and also for exploring the properties of the gravitational-wave source. For such astronomical observations, the theoretical prediction is also necessary, and for this issue, numerical relativity plays an important role.

The mergers of binary neutron stars and black hole-neutron star binaries are also promising progenitors for the central engine of gamma-ray bursts with short time duration (the so-called short gamma-ray bursts). From the viewpoint of high-energy astrophysics, theoretically pursuing the merger hypothesis of short gamma-ray bursts is one of the central subjects, because their origin has been a long-standing unsolved issue. Specifically, we have to explore a variety of possible formation processes of the central engines for the huge gamma-ray emission in the first-principle theoretical study. For this issue, numerical relativity incorporating relevant microphysical processes is the unique approach.

Since 1999, a wide variety of numerical-relativity studies have been performed for understanding the nature of the final evolution phase of binary compact objects. The first successful simulation for binary neutron stars was performed by Shibata (1999b); Shibata and Uryū (2000) in 1999. The first successful simulation for binary black holes was achieved by Pretorius (2005a) in 2005. Soon after the success by Pretorius (2005a), Campanelli *et al.* (2006a) and Baker *et al.* (2006a) also succeeded in the simulation by another robust numerical-relativity formulation (BSSN-puncture formulation; see section 2.3.3.1) together with a novel numerical scheme, which is now widely used in the community of numerical relativity. Soon after the success of the binary-black-hole simulations by Campanelli *et al.* (2006a); Baker *et al.* (2006a), simulations of black hole-neutron star binaries became also feasible [see Shibata and Uryū (2006, 2007); Shibata and Taniguchi (2008); Etienne *et al.* (2008); Duez *et al.* (2008) for the early works]. Since these first successes, more accurate and more physically motivated simulations have been performed for these three phenomena, and now, we have fruitful knowledge for the merger processes of binary compact objects. In the following, we will summarize what have been achieved in the community of numerical relativity and our current understanding of the compact binary mergers based on the results derived by numerical-relativity simulations.

8.1 Binary neutron stars: Brief introduction

As described in sections 1.4.8 and 1.5.1 (see, e.g., figure 1.23), binary neutron stars are likely to be in quasi-circular orbits (with negligible eccentricity, $e \approx 0$) in a late inspiral phase. They adiabatically evolve due to the gravitational-radiation reaction [see equation (1.93)] as far as the orbital radius, a, is much larger than the radius of neutron stars, R. For the phase where $a \gg R$, each neutron star in binaries can be approximated by a point particle, and in addition, the orbital velocity is at most 20–30% of the speed of light. The orbital evolution and emitted gravitational waves in such inspiral phases are accurately and analytically determined in post-Newtonian frameworks [Blanchet (2014)] combining the point-particle and adiabatic approximations (see appendix H). However, for $R/a \gtrsim 0.2$ (i.e., $a \lesssim 50$–75 km for typical neutron-star radii, $R = 10$–15 km), neutron stars are deformed by the tidal field of their companions, and hence, the point-particle approximation becomes a poor approximation. Due to the tidal deformation of neutron stars, the gravitational field and the orbital velocity are modified from the results by the point-particle approximation. Therefore, finite-size effects have to be taken into account. In addition, for $a/6m \lesssim 2.4$ [see equation (1.93)], the ratio of the gravitational-radiation reaction time scale to the orbital period is smaller than 10. Here, for $m = 2.8 M_\odot$, $6m \approx 25$ km. Thus, for $a \lesssim 60$ km, the adiabatic approximation becomes a poor approximation (although it is *qualitatively* acceptable and does not completely break down for $a \gtrsim 6m$). For $a \lesssim 30$ km, the gravitational-radiation reaction time scale is as short as the orbital period, and hence, the system starts dynamical evolution.

The brief summary described above shows that theoretical studies of binary neutron stars should be done keeping the following facts in mind: (i) for $a \lesssim 5R (= 50-75$ km), finite-size effects have to be incorporated; (ii) for $30 \text{ km} \lesssim a \lesssim 60 \text{ km}$, nonadiabatic effects for the inspiral orbits should be incorporated, (iii) for $a \lesssim 30$ km, the merger occurs and hence a fully dynamical treatment is the unique approach for this stage. The finite-size effects can be incorporated only by fully solving gravitational-field equations together with hydrodynamics equations. Hence, for $a \lesssim 5R$, a numerical computation is necessary. However, this does not always imply that numerical-relativity simulations are necessary for this phase. For a relatively large orbital separation, e.g., $a \sim 50$ km, the adiabatic approximation may still work, because the gravitational-radiation reaction time scale is still ~ 5 times as long as the orbital period. For this phase, we may approximate the binary system in a stationary state when observing it in a frame comoving with the orbital velocity. Thus, not only numerical-relativity simulations but also constructing quasi-equilibrium sequences is one of the possible approaches for studying this evolution phase. Computing quasi-equilibrium states is also the necessary element for the studies of binary mergers, because they are used as initial data of numerical-relativity simulations. Quasi-equilibrium states of binary neutron stars have been constructed and analyzed in detail in the past two decades [Baumgarte *et al.* (1997, 1998b,a); Bonazzola *et al.* (1999); Uryū and Eriguchi (2000); Gourgoulhon *et al.* (2001); Taniguchi and Gourgoulhon (2002, 2003); Uryū *et al.* (2006, 2009); Taniguchi and Shibata (2010); Tichy (2012)]. These works have been providing us rich information for the late inspiral phase of binary neutron stars. Thus, we summarize the status and the results for this in section 8.2.

As already mentioned, for studying the merger phase in which the system evolves in a dynamical manner, numerical-relativity simulation is the unique approach. This is also the robust approach for studying a late quasi-stationary phase with $a \lesssim 5R$, if a long-term numerical simulation is performed. In section 8.3, the results and the knowledge obtained by various numerical simulations to date will be summarized.

8.2 Binary neutron stars: Quasi-equilibrium states and sequences

The gravitational-radiation reaction plays a minor role for binaries with $a \gg R$ and such binaries are considered to be approximately in a stationary state if we observe them in a frame comoving with the orbital rotation velocity. As described in section 5.7.2, such systems are considered to have a helical Killing vector field, (5.228), at least in the region inside the light cylinder, $r \lesssim 2\pi c/\Omega_{\rm orb}$ where $\Omega_{\rm orb}$ denotes the orbital angular velocity. For this case, hydrodynamics equations are integrated to yield a first integral for irrotational or corotational velocity fields, for which we do not have to fully solve hydrodynamics equations. The question is what the velocity field of the neutron stars in real binaries is. In the next section, we will show several reasons by which the velocity field can be approximated by the irrotational one.

8.2.1 *Velocity field*

In general, the velocity field inside stars is determined by their formation process (initial state at their formation) and by micro and macro-physical processes which are relevant in their subsequent evolution. For general velocity fields, we do not know how to integrate hydrodynamics equations for yielding a first integral even if the star is in a stationary state. Fortunately, for binary neutron stars (and also for black hole-neutron star binaries), it is reasonable to assume that the velocity field of each neutron star is approximated by the irrotational field [Kochanek (1992); Bildsten and Cutler (1992)].

The first reason for this is that the typical spin period of ordinary neutron stars is $0.1-1$ s [see, e.g., figure 8.8 of Lyne and Graham-Smith (2005) or figure 3 of Lorimer (2008)]. This implies that the dimensionless spin parameter, cS/GM^2, where S and M are the spin angular momentum and mass of neutron stars, is $< 10^{-2}$ for the typical radius and mass of neutron stars. Post-Newtonian spin effects on the orbital motion play a negligible role for such a tiny spin. Furthermore, the spin of neutron stars at the onset of the merger will be smaller than this ordinary value because the spin angular momentum is dissipated by the long-term emission of electromagnetic waves as a pulsar unless their magnetic-field strength is extremely small. Some of neutron stars in observed binary systems are spinning more rapidly with the period $20-100$ ms [e.g., table 1 of Lorimer (2008)]. However, the spin parameter is still much smaller than 0.1, and hence, the effect is minor.

Second, the orbital frequency of binary neutron stars in close orbits is quite high as

$$f_{\rm orb} \approx 210\,{\rm Hz}\left(\frac{a}{60\,{\rm km}}\right)^{-3/2}\left(\frac{m_0}{2.8M_\odot}\right)^{1/2}. \tag{8.1}$$

Thus, for a binary with typical total mass $m_0 = 2.8 M_\odot$, the orbital period is shorter than 5 ms for $a \leq 60$ km; at the onset of the merger, it is ~ 2 ms. This is much shorter than the plausible spin period of neutron stars in binaries. This also indicates that at the merger stage, the spin of neutron stars would only play a minor role.

Then, we have to ask the possibility whether neutron stars could spin up in close orbits due to tidal spin-up effect: If the combination of tidal torque and viscous angular-momentum transport can contribute to the spin-up, the velocity field would approach a corotating one, as in the moon around the earth (note that the moon corotates with the orbital motion). As shown by Kochanek (1992) and Bildsten and Cutler (1992), however, the corotating velocity field is unlikely to be achieved: Only a small amount of the spin-up could occur. In the following, we will describe the reason for this.

The efficiency of the tidal spin-up is determined by the torque exerted by each neutron star on its companion. The magnitude of the torque exerted on star 1 by the tidal field of star 2 is written approximately by [Bildsten and Cutler (1992)]

$$N_{\text{torq}} \approx \frac{m_2^2 R^5}{2a^6} \sin(2\alpha_{\text{torq}}), \qquad (8.2)$$

where m_2 is the mass of star 2, R is the radius of star 1, a is the orbital separation, and α_{torq} is the misalignment angle between the line connecting centers of the two stars and major axis of star 1. α_{torq} is induced by a slight violation of quasi-stationarity of the orbital motion due to the emission of gravitational waves, and is a function of a. Here, α_{torq} could increase steeply with the decrease of a for close orbits, although α_{torq} is much smaller than unity and would be at most $O(0.1)$ even at the onset of the merger, i.e., for $a \sim 3-4R$. Thus, we may approximate N_{torq} by

$$N_{\text{torq}} = \frac{m_2^2 R^5}{a^6} \alpha_{\text{torq}}. \qquad (8.3)$$

The spin-up of each neutron star is achieved through the angular-momentum transport from the orbital angular momentum to the spin angular momentum. This transport process should be proceeded by a viscous process because the generation of the spin implies the generation of the vorticity and hence a finite amount of the viscosity is necessary. In the following paragraph, we will assume that the viscous effect is strong enough that the continuous spin-up process can turn on, to show how small the tidal spin-up effect is even in such optimistic assumption. We then will show that such a large viscous effect is not realistic, and hence, the spin-up process would be a minor one.

Because binary neutron stars evolve due to the gravitational-radiation reaction, the possible maximum amount of the increased spin angular momentum, S, at a given orbital radius is estimated by $N_{\text{torq}} \times t_{\text{gw}}$ [see equation (1.92) for t_{gw}: the approximate time scale of the gravitational-radiation reaction], which gives

$$S = \frac{5R^5 m_2}{256 a^2 m_0 m_1} \alpha_{\text{torq}}, \qquad (8.4)$$

and thus, the hypothetical dimensionless spin parameter of star 1 is estimated as

$$\frac{S}{m_1^2} = \frac{5R^5 m_2}{256 a^2 m_0 m_1^3} \alpha_{\text{torq}} \approx 0.023 \left(\frac{3R}{a}\right)^2 \left(\frac{R}{6m_1}\right)^3 \left(\frac{2m_2}{m_0}\right)\left(\frac{\alpha_{\text{torq}}}{0.1}\right), \qquad (8.5)$$

where $m_0 = m_1 + m_2$ and we assumed that α_{torq} is constant at a constant orbital separation a for simplicity. As expected, the induced spin is larger for the smaller orbital separation. Equation (8.5) shows that the dimensionless spin parameter associated with the tidal spin-up would be at most of $O(0.01)$ even just before the onset of the merger and even with the optimistic assumption of a large viscous effect. This fact is also found by taking the ratio of the spin angular momentum to that of the corotating binary as

$$\frac{S}{I(m_0/a^3)^{1/2}} \approx 0.12 \left(\frac{3R}{a}\right)^{1/2} \left(\frac{R}{6m_1}\right)^{5/2} \left(\frac{8m_1 m_2^2}{m_0^3}\right)^{1/2} \left(\frac{\alpha_{\text{torq}}}{0.1}\right) \left(\frac{\alpha_{\text{inn}}}{0.3}\right)^{-1}, \quad (8.6)$$

where I is the moment of the inertia of star 1 for which we write $I = \alpha_{\text{inn}} m_1 R^2$ with $\alpha_{\text{inn}} \sim 0.3$ for realistic neutron-star models [e.g., Friedman et al. (1986)]. Thus, the corotation can be never realized by the tidal torque.

In the above, nevertheless, we found that a certain amount of spin-up may be achieved by the tidal torque. In that discussion, we assumed that the viscous effect would be large enough to allow the significant generation of the vorticity. However, we have to ask whether the viscosity of neutron stars could really play a role for such spin-up. In the following, we will find that for the spin-up, extremely high viscosity is necessary, and in reality, any efficient spin-up is unlikely to be achieved by the tidal spin-up mechanism.

The viscous dissipation could occur when the matter has shear motion. In the corotating velocity field, the shear motion is totally absent. In the irrotational velocity field, the shear motion is induced by the tidal interaction, and the magnitude of its velocity is approximately $(R/a)^3 (m_2/m_1)(a\Omega_{\text{orb}})$ where $a\Omega_{\text{orb}}$ is the approximate magnitude of the orbital velocity. Then, the order of the magnitude for the dissipation rate of the kinetic energy is written by

$$\dot{E}_{\text{vis}} \sim \frac{1}{2} \nu_{\text{vis}} m_1 \Omega_{\text{orb}}^2 \left(\frac{R}{a}\right)^6 \left(\frac{m_2}{m_1}\right)^2, \quad (8.7)$$

where ν_{vis} denotes the viscous coefficient. Here, we suppose the situation that neutron stars have an irrotational velocity field initially and due to the viscous dissipation, the irrotational velocity field subsequently approaches the corotational one by the tidal spin-up of neutron stars (i.e., the rotational kinetic energy of each neutron star increases). By the viscous dissipation, thermal energy is in general generated. Assuming that the generated thermal energy is comparable to the generated rotational kinetic energy of the spin, $I\Omega_S^2/2$ with $\Omega_S := S/I$, the viscous dissipation time scale is defined by

$$t_{\text{vis}} := \frac{I\Omega_S^2}{\dot{E}_{\text{vis}}}. \quad (8.8)$$

From this, the ratio of t_{vis} to t_{gw} is written as

$$\frac{t_{\text{vis}}}{t_{\text{gw}}} = \frac{5\alpha_{\text{torq}}^2}{128 \alpha_{\text{vis}} \alpha_{\text{inn}}} \left(\frac{R}{m_0}\right)^2 \left(\frac{a}{R}\right) \left(\frac{m_2}{m_1}\right), \quad (8.9)$$

where we rewrote ν_{vis} as $\alpha_{\text{vis}} Rc$ with α_{vis} being a coefficient smaller than unity: For $\alpha_{\text{vis}} = 1$, the momentum exchange in the microscopic process is assumed to proceed by particles with the speed of light and with the mean-free path, R. Thus, α_{vis} should be much smaller than unity in reality; for the molecular viscosity, α_{vis} is indeed by many orders of magnitude smaller than unity. Equation (8.9) shows that the viscous time scale is longer than the gravitational-wave emission time scale for distant orbits, and this process is important only for close orbits.

For the spin-up to be achieved, the viscous dissipation at a given orbital separation has to occur before the orbital radius decreases due to the gravitational-radiation reaction. Thus, for the viscous dissipation to work, we have to require the condition, $t_{\rm vis} < t_{\rm gw}$, which is written as

$$\alpha_{\rm vis} \gtrsim 0.035 \left(\frac{a}{3R}\right) \left(\frac{R}{3m_0}\right)^2 \left(\frac{m_2}{m_1}\right) \left(\frac{\alpha_{\rm inn}}{0.3}\right)^{-1} \left(\frac{\alpha_{\rm torq}}{0.1}\right)^2, \tag{8.10}$$

where we only considered close orbits because the significant spin-up can occur only for such orbits as shown in equation (8.5). Equation (8.10) implies that for the spin-up, $\alpha_{\rm vis}$ has to be of order 10^{-2} or more. As already mentioned above, such a high value is not possible by the molecular viscosity. Some process such as turbulent viscosity may achieve this value, but it seems to be quite difficult, because for such high viscous coefficient, the typical scale of the angular-momentum transport should be larger than $0.1R$ for the hypothetical maximum transport speed close to the sound velocity $\sim 0.1c$: An unnatural high-speed large-scale vortex motion is necessary.

For the corotation to be achieved, we have to require

$$\frac{I\Omega_{\rm orb}^2}{\dot{E}_{\rm vis}} < t_{\rm gw}. \tag{8.11}$$

This is rewritten as

$$\alpha_{\rm vis} \gtrsim 2.6 \left(\frac{a}{3R}\right)^2 \left(\frac{6m_1}{R}\right)^3 \left(\frac{m_0}{2m_2}\right) \left(\frac{\alpha_{\rm inn}}{0.3}\right). \tag{8.12}$$

Thus, the corotational velocity field is not realized by the tidal spin-up.

To summarize, both the tidal interaction and the viscous dissipation are unlikely to play an important role for spinning up each neutron star in binary neutron stars. Therefore, if the spin of neutron stars is not very large at a distant orbit, the slow spin is preserved throughout the entire evolution up to the onset of the merger. As mentioned in the early part of this section, the spin period of many of neutron stars is likely to be much longer than the orbital period just before the onset of the merger, and the spin can be considered to be negligible for binary neutron stars in close orbits. Therefore, it is reasonable to assume the irrotational velocity field as a good approximate one for studying binary neutron stars in close quasi-circular orbits. [However, see also Tichy (2012) for an effort to constructing quasi-equilibrium binary neutron stars with arbitrary spins.]

8.2.2 Gravitational-field equations

As described in section 5.7.2, the formulation of gravitational-field equations for non-axisymmetric quasi-stationary spacetime is non-trivial, if we try to take into account the full set of Einstein's equation. Although several candidate formulations such as those introduced in section 5.7.2 have been proposed, the basic equations for such formulations are composed of complicated nonlinear elliptic-type equations (and some additional equations), and thus, the solution to a quasi-equilibrium state is demanding because of a high cost for implementing a new code and a high cost for numerical computation. Because of this reason, this problem has been seriously tackled only by Uryū et al. (2006, 2009) to date.

Instead of using a formulation that takes into account the full set of Einstein's equation, most of quasi-equilibrium states have been derived by the IWM formulation (or a CTS formulation with $\tilde{\gamma}_{ij} = f_{ij}$; see section 5.6), although a part of the gravitational-field equations are simplified in this formulation: The spatial metric is assumed to be conformally flat (with no foundation), simply because the basic equations become significantly simplified. From the view point of post-Newtonian approximations, a part of the second post-Newtonian terms are absent in this formulation [Schäfer (1985); Damour and Schäfer (1985)], and the error of order $(m_0 \Omega_{\rm orb})^{4/3}$ relative to the leading-order term should be always present.[1] However, fortunately, the latest numerical work by Uryū et al. (2006, 2009) showed that the results in the IWM formulation and in more sophisticated formulation with $\tilde{\gamma}_{ij} \neq f_{ij}$ agree well (within $\sim 1\%$ error) with each other except for those in close orbits just before the merger, i.e., for $m\Omega_{\rm orb} \gtrsim 0.03$ (see section 8.2.3.2). This indicates that at least for binary neutron stars with a fairly distant orbit, employing the IWM formulation is acceptable. In the next section, we will first review the results obtained in the IWM formulation, and summarize the scientific outputs discovered based on them. Then, we will touch on the work by Uryū et al. (2006, 2009).

8.2.3 Numerical results

8.2.3.1 *IWM formulation case*

A numerical computation for quasi-equilibrium states of binary neutron stars was first performed by Baumgarte et al. (1997, 1998b,a) in the IWM formulation. They focused only on *corotating* binary systems. The reason for this is that numerical computation is much less demanding for the corotational velocity field than for the irrotational one: For the corotational field, we do not have to solve the equation for the velocity potential, but for the irrotational field, a special numerical technique is required for an accurate solution of the velocity potential [e.g., see Bonazzola et al. (1998) for a numerical scheme to a precise numerical solution]. Although the properties of quasi-equilibrium states of corotating binaries are qualitatively similar to those of irrotational binaries, assuming the corotational velocity field is not acceptable for the quantitative study of an accurate binary state, as already

---

[1] On the other hand, for the infinite orbital separation, the IWM formulation yields an exact solution for non-spinning binaries because this formulation can yield exact solutions for spherical neutron stars.

mentioned in the previous section.[2]

The first numerical computation for irrotational binary neutron stars in quasi-equilibrium states was performed by Bonazzola et al. (1999). Subsequently, Uryū and Eriguchi (2000); Uryū et al. (2000); Gourgoulhon et al. (2001) performed the detailed analyses for the properties of irrotational binary neutron stars [see also Taniguchi and Gourgoulhon (2002, 2003); Uryū et al. (2006, 2009); Taniguchi and Shibata (2010)]. Here, we summarize the properties of irrotational binary neutron stars in quasi-equilibrium states based on one of the pioneer works by Uryū et al. (2000).

In the assumption of the irrotational velocity field, the degree of freedom for a solution is sufficiently restricted, and thus, for a given equation of state and rest mass of each neutron star, the solution is characterized uniquely by the orbital angular velocity $\Omega_{\rm orb}$ (or an appropriately determined orbital separation). Thus, a sequence is obtained by constructing quasi-equilibrium states for a series of $\Omega_{\rm orb}$ while fixing the zero-temperature equation of state and the rest mass of the system. Here, the temperature of neutron stars in close orbits is likely to be negligible (i.e., thermal energy of each particle is much smaller than the Fermi energy of neutrons), because of the absence of appreciable heating processes, and thus, the zero entropy and zero temperature can be reasonably assumed. The rest-mass conservation comes from the fact that it is reasonable to suppose that no mass loss and no mass exchange occur in the inspiral phase of binary neutron stars.

Each quasi-equilibrium solution at each value of $\Omega_{\rm orb}$ is characterized by the total ADM mass and angular momentum of the system, $M_{\rm ADM}$ and J (see sections 5.3.1.1 and 5.3.1.3 for their definition). Thus, a sequence is characterized by $M_{\rm ADM}(\Omega_{\rm orb})$ and/or $J(\Omega_{\rm orb})$. We here note that in the IWM formulation, the ADM mass and angular momentum agree with the Komar mass and Komar angular momentum (see section 5.3.2), respectively, if the numerical solution is precisely obtained. Thus, we may use them instead of the ADM quantities. We also remind the readers that in the IWM formulation, any sequence of quasi-equilibrium states satisfies the first law of binary mechanics [see Friedman et al. (2002) and section 5.7.2.3] which is written, in the assumption of the rest-mass conservation, zero entropy, and irrotational velocity field [see, e.g., equation (G.79)], as

$$\delta M_{\rm ADM} = \Omega_{\rm orb} \delta J. \qquad (8.13)$$

This means that $M_{\rm ADM}(\Omega_{\rm orb})$ and $J(\Omega_{\rm orb})$ are not independent of each other.

Typical curves of $J(\Omega_{\rm orb})$ are depicted in figure 8.1. In the horizontal axis, an orbital separation (denoted by $d_{\rm G}$) instead of $\Omega_{\rm orb}$ is employed in this figure. In this example, sequences of irrotational binary neutron stars of equal mass are

---

[2] Remember that the orbital period of binary neutron stars is 2–3 ms just before the onset of the merger. Thus, if we employ the corotating velocity field in which the spin period is equal to the orbital one, we would assume that each neutron star in close orbits was unphysically rapidly spinning.

constructed using (cold) polytropic equations of state of the form

$$P = \kappa_\mathrm{p} \rho^{1+1/n}, \quad \left(\text{and } \varepsilon = \kappa_\mathrm{p} n \rho^{1/n}\right), \tag{8.14}$$

where κ_p and n are the polytropic constant and polytropic index, respectively. In the polytropic equation of state (with the units of $c = G = 1$), $\kappa_\mathrm{p}^{n/2}$ has the dimension of length, time, and mass, and using this fact, the figure is generated for dimensionless quantities. Thus, the curve of $J\kappa_\mathrm{p}^{-n}$ for a given value of n is invariant for the change of κ_p (this is also the case for $M_\mathrm{ADM}\kappa_\mathrm{p}^{-n/2}$).

Before going ahead, we briefly remind the readers of a property of the polytropic equations of state, which depends essentially only on the value of n. For small values of n, the pressure steeply increases with the increase of the density and the stellar density profile approaches more uniform one: In the limit of $n = 0$, the density should be constant inside the neutron star. By contrast, for larger values of n, the polytropic star has a density profile in which the degree of central condensation is high. The dependence of the density profile on the value of n is reflected in the configuration of each neutron star in binaries of close orbits as described below.

Figure 8.1 plots three curves of $J\kappa_\mathrm{p}^{-n}$ for $n = 1/2$, $2/3$, and $4/5$ for the equal-mass binaries. Here, a compact neutron star with the compactness $\mathcal{C} = 0.19$ (see section 1.4.5 for its definition) is employed. Note that for a hypothetical mass of each neutron star, $1.4 M_\odot$, the radius is about $10.9\,\mathrm{km}$ for $\mathcal{C} = 0.19$. Irrespective of the value of n, the curve terminates at a minimum orbital separation where neutron stars have a cusp (a Lagrange point) at the inner edge of their stellar surfaces. Here, the appearance of the cusps implies that at this orbital separation, the strength of the tidal force by the companion star is equal to that of its own self-gravity at the inner edge of their stellar surface. Thus, for orbits inside this minimum separation, mass shedding will turn on and no quasi-equilibrium state composed of two neutron stars exists.

For a large orbital separation, the angular momentum monotonically decreases with the decrease of d_G universally. This is also the case for the ADM mass of the system. This behavior is reasonable because binary neutron stars evolve due to the emission of gravitational waves dissipating the energy and angular momentum of the system. However, the feature of the curves depends on n for close orbits. For $n = 1/2$, the curve has a minimum value of J (and also M_ADM) along the sequence (at $d_\mathrm{G}\kappa_\mathrm{p}^{-n/2} \approx 1.0$). For $n = 2/3$, the location of the minimum of J coincides approximately with the minimum separation (at $d_\mathrm{G}\kappa_\mathrm{p}^{-n/2} \approx 1.2$). For $n = 4/5$, there is no minimum along the sequence. Uryū et al. (2000) also found that this property holds irrespective of the compactness parameter as far as $\mathcal{C}$ is in the range between ~ 0.1 and 0.2: For low values of $n \lesssim 2/3$ (for relatively uniform density profiles of the neutron star), the minimum of J (and M_ADM) always exists while it does not for high values of n. We note that the location of the minimum of J agrees with that of M_ADM according to the first-law of binary mechanics [see equation (8.13)].

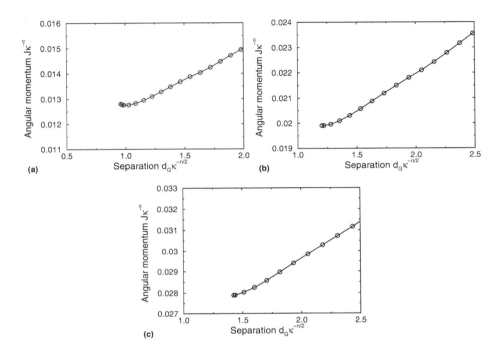

Fig. 8.1 Angular momentum in the dimensionless unit, $J\kappa_p^{-n}$, as a function of a half of orbital separation in the dimensionless unit $d_G\kappa_p^{-n/2}$ for (a) $n = 1/2$, (b) $n = 2/3$ and (c) $n = 4/5$ and for $\mathcal{C} = M_{\rm NS}/R_{\rm NS} = 0.19$. The point of the minimum value of $d_G\kappa_p^{-n/2}$ along the curves correspond to the state where each neutron star in the binary has a cusp (a Lagrange point) at the inner edge of the stellar surface; mass shedding will turn on inside the minimum value of $d_G\kappa_p^{-n/2}$. κ in the figure means κ_p in the text.

The presence of the minimum of J implies that there is actually no quasi-equilibrium state with the orbital separation smaller than this minimum point. The reason for this is that binary neutron stars in nature evolve due to the emission of gravitational waves, dissipating the energy and angular momentum, and hence, at the minimum of J (and $M_{\rm AD}$), no next quasi-equilibrium state is present. Therefore, it is natural to consider that the dynamical merger process sets in at the orbit of this minimum: The system becomes dynamically unstable. Thus, the orbit of this minimum should be referred to as the innermost stable circular orbit (ISCO).

The presence of this ISCO is determined primarily by a hydrodynamics effect (not by a general relativistic effect) as touched on in section 1.5.1. In binaries composed of fluid stars, tidal effects play an important role for close orbits because each star is tidally deformed and the deformed configuration modifies the gravitational field. At the leading order in Newtonian gravity, the gravitational potential is modified as equation (1.192) shows; the second term is the correction due to the tidal deformation. This term yields an additional attractive force and depends on

r^{-6} where r is the orbital separation. For very close orbits, the magnitude of this term steeply increases and before two stars come into contact, it could overcome the potential barrier associated with the centrifugal force that is proportional to r^{-2}. If this occurs, circular orbits could be destabilized before two stars come in contact.

We note that circular orbits for a test particle orbiting a black hole also have ISCO. For this case, a general relativistic effect yields an additional attractive potential that is proportional to r^{-3} and overcomes the centrifugal potential for close orbits. In the presence of an ISCO in binary neutron stars, the origin of the additional attractive force is different from this general relativistic one.

Figure 8.1 implies that for low values of n, an ISCO is present, and the dynamical merger process of two neutron stars sets in before they come into contact. The reason for this dependence on n is that for neutron stars of more uniform density profile, the tidal deformation is more uniformly enhanced in close orbits. However, even in the presence of the ISCO, its location is close to the mass shedding limit, and hence, it is expected that soon after the onset of the dynamical instability, a mass flow will occur prior to the collision of two neutron stars and subsequent merger process. On the other hand, for high values of n, the dynamical merger process is likely to set in after mass shedding occurs. For this case, the density profile of the neutron stars is not as uniform as that for low values of n, and hence, the tidal effect is more significantly enhanced near the inner edge of the stellar surface. These facts suggest that the merger process depends on the equation of state (in particular on the density profile of each neutron star).

8.2.3.2 Beyond IWM formulation

In their latest work, Uryū et al. (2009) constructed quasi-equilibrium sequences using a formulation beyond the conformally flat (IWM) approximation and employing equations of state more realistic than the polytropic one (a piecewise polytrope was employed: see section 1.4.4). To date this is the best work for the study of quasi-equilibrium states of binary neutron stars. They developed the so-called waveless formulation in which gravitational-field equations for all the components of the metric are determined by solving new elliptic equations for the conformal three metric, $\tilde{\gamma}_{ab}$, with $v_{ab} = 0$ [see equation (5.235)]. They clarified that the effect of the conformal three metric is not negligible for compact and close binary systems, although this effect is minor for distant orbits.

Figure 8.2 plots the binding energy and total angular momentum as functions of $\Omega_{\rm orb}$ along sequences of equal-mass binary neutron stars for given equations of state and a fixed value of the rest mass. Here, the binding energy is defined by

$$E_{\rm b} := M_{\rm ADM} - m_1 - m_2, \qquad (8.15)$$

where $m_1 = m_2$ is the ADM mass of neutron stars in isolation (for the infinite separation). The total mass is defined by $m_0 := m_1 + m_2$. The mass of each neutron star for this sequence is $m_1 = m_2 = 1.35 M_\odot$. For each sequence of a given equation of state, two curves are plotted: one is derived in the IWM formulation and the other in the waveless (WL) formulation. Equations of state, 2H, HB, and 2B, are stiff, moderately stiff, and soft ones, respectively (see table 8.1 for details). The solid curves denote the results derived in a third post-Newtonian approximation together with the point-particle approximation (in which no spin and no multipole moments are taken into account: see appendix H). This figure shows that for a distant orbit with small values of $\Omega_{\rm orb}$, all the curves agree well. This is reasonable because each neutron star in binaries may be considered to be a point particle, and thus, the effects of finite size and equations of state do not play a role, and also because higher post-Newtonian general relativistic corrections associated with the orbital motion are negligible. However, for close orbits, the curves in the WL and IWM formulations as well as the curve in the third post-Newtonian approximation do not agree with each other.

The primary reason for the fact that the curves of the WL and IWM formulations do not agree with that of the third post-Newtonian approximation is that a finite-size effect, i.e., the tidal-deformation effect, comes into play for close orbits. This effect becomes non-negligible from smaller values of $\Omega_{\rm orb} m_0$ for the case that the radius of the neutron stars is larger: Note that the radii of the spherical neutron star for 2H, HB, and 2B equations of state are $\approx$ 15.2, 11.6, and 9.7 km, respectively (see table 8.1). For example, the tidal effect becomes appreciable for $\Omega_{\rm orb} m_0 \gtrsim 0.02$ with 2H, while for $\Omega_{\rm orb} m_0 \gtrsim 0.03$ with 2B.

The curves of the WL and IWM formulations do not agree with each other for close orbits. This shows that the effects by the conformal three metric play an appreciable role for such orbits. The closer orbits are possible for systems composed of smaller-radius neutron stars, and thus, the difference of the two curves is in particular remarkable for the close orbits when the 2B equation of state is employed. The work of Uryū et al. (2009) clearly showed that the IWM formulation cannot yield accurate binary neutron stars in quasi-equilibria of close orbits. Figure 8.2 also shows that no minimum appears along the sequences both in the WL and IWM formulations irrespective of the equations of state employed. This indicates that for realistic equations of state, there is no ISCO before reaching the mass-shedding limit.

The work of Uryū et al. (2009) warns that the IWM formulation, which is the most popular formulation for constructing quasi-equilibrium states and for providing initial data of numerical-relativity simulations for the merger of binary compact objects, should not be used for close orbits. If one wants to perform a numerical-relativity simulation with the initial data prepared in the IWM formulation, the orbital separation should be taken to be sufficiently large: The simulations have to be performed with many inspiral orbits before the onset of the merger. Indeed, in the latest numerical-relativity simulations, the initial data with a sufficiently large orbital separation are standard (see section 8.3).

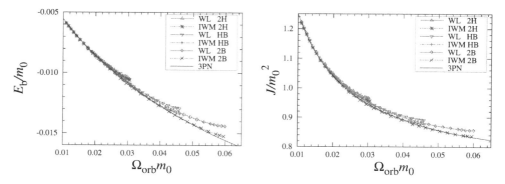

Fig. 8.2 Plots of equal-mass binary-neutron-star sequences for several equations of state. Left panel: Binding energy $E_b = M_{\rm ADM} - m_0$ normalized by $m_0 = 2m_1$ (where $m_1 = 1.35 M_\odot$ is the ADM mass of each star in isolation) as a function of $\Omega_{\rm orb} m_0$ in the waveless and IWM formulations. Right panel: Total angular momentum J normalized by m_0^2. The solid curves denote the results derived in the third post-Newtonian approximation together with the point-particle approximation. The figure is taken from Uryū et al. (2009).

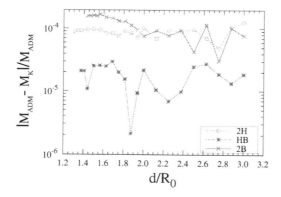

Fig. 8.3 Violation of the virial relation as a function of an orbital separation, d, for numerical solutions obtained in the waveless formulation. R_0 denotes the radius of spherical neutron star for a given equation of state and mass. The figure is taken from Uryū et al. (2009).

Finally, we comment on how to examine the reliability of the numerical solutions for quasi-equilibria and their sequences. There are at least two methods for this. The first one is to examine how accurately the virial relation, $M_{\rm ADM} = M_{\rm K}$, is satisfied (see sections 5.3.3 and 5.7.2.3): For the numerical solutions of quasi-equilibrium states, we often compute $|1 - M_{\rm K}/M_{\rm ADM}|$ as a measure of the numerical error. For example, in the work of Uryū et al. (2009), the error is at most $\sim 10^{-4}$ (see figure 8.3). This value indicates the typical error size of the numerical solutions. The second method is to examine how accurately the first-law relation is satisfied [see section 5.7.2.3 and equation (5.250)]: We should examine how accurately the relation, $dM_{\rm ADM} = \Omega_{\rm orb} dJ$, is satisfied along the quasi-equilibrium sequences. To do this, we have to take a finite differencing for the mass and angular momentum along the sequences, and hence, the accuracy is in general not as good as that for the virial relation. However, for the published numerical results, the first-law relation is satisfied in a reasonable accuracy.

8.3 Binary neutron stars: Numerical simulations

Since the first work by Shibata (1999b) and Shibata and Uryū (2000), many numerical-relativity simulations have been performed for the merger of binary neutron stars. In particular, in 2010s, a variety of physically well-modeled and accurate simulations are ongoing. Now, we can describe the merger and subsequent evolution process of the merger remnant in detail, based on the results of numerical-relativity simulations. In the following, we will summarize our understanding for the entire evolution process of this system.

8.3.1 *Equations of state for numerical simulations*

Before going ahead, we summarize hypothetical neutron-star equations of state that should be employed for a realistic simulation of binary neutron stars and black hole-neutron star binaries.

Physical modeling for the equations of state is a key ingredient for accurately clarifying the merger process, the remnant formed after the onset of the merger, and gravitational waves emitted during the late inspiral and merger stages of neutron star binaries. However, as already described in section 1.4.4, the equation of state for high density beyond the normal nuclear-matter density $\approx 2.8 \times 10^{14}\,{\rm g/cm^3}$ is still uncertain due to the lack of the strong constraints from experiments and astronomical observations. The latest discoveries of high-mass neutron stars with $M_{\rm NS} \sim 2 M_\odot$, PSR J1614-2230 and PSR J0348+432, by Demorest *et al.* (2010) and Antoniadis *et al.* (2013) certainly constrain the allowed equations of state. However, the constraint is not still sufficiently strong, primarily because the radius of neutron stars has not been strongly constrained yet.

As shown in the following subsections as well as in section 8.5, gravitational waves emitted during the merger phase of neutron star binaries and properties of the merger remnants depend strongly on the properties of the hypothetical equation of state employed: Even if the total mass and mass ratio hypothetically chosen for the binary system are identical, the outputs in the simulations can be significantly varied if different hypothetical equations of state are employed. For example, the maximum-allowed mass for the formation of a hypermassive neutron star after the merger of binary neutron stars (see section 1.4.6 for the definition of hypermassive neutron stars), which determines the remnant (black hole or massive neutron star) formed soon after the onset of the merger, depends strongly on the equation of state employed.

One of the most important purposes of numerical-relativity simulations for binary neutron stars and black hole-neutron star binaries is to explore the possibility for determining the equation of state of neutron stars from astrophysical observations. As we noted in the above paragraph and describe in the following, the equation of state will be well reflected in gravitational waves emitted during the merger stage of neutron star binaries. One interesting possibility that we then think of is to constrain the equation of state by the near-future gravitational-wave observation. For this purpose, we need to systematically perform simulations preparing a wide variety of hypothetical equations of state for deriving a wide variety of possible gravitational waveforms that could be possible sets of theoretical templates of gravitational waves.

For the simulations, we have to carefully choose the equations of state employed taking into account the possible evolution process of the binary systems. Before the onset of the merger, neutron stars are cold; their thermal energy is much lower than the Fermi energy of nucleons and electrons [see equation (1.167)]. After the onset of the merger, a strong shock is generated on an inner surface of two neutron stars. In addition, outer parts of two neutron stars form spiral arms, because the merged object usually has a non-axisymmetric shape that exerts torque to the matter surrounding it and thus plays a major role for transporting its angular momentum outward. The spiral arms thus formed subsequently collide with each other or with outer parts of the merger remnant located at the center. This results in additional shock heating of the matter located in the envelope of the merger remnant. Therefore, an inner part of the merger remnant as well as its envelope can be very hot after the onset of the merger. These facts imply that (i) for simulating binary neutron stars before the onset of the merger, cold equations of state should be employed while (ii) after the onset of the merger, the finite-temperature effects of the equation of state have to be taken into account. For a detailed study of the thermal and composition evolution, neutrino transport and evolution of the electron fraction per baryon should be also taken into account.

In the early stage of this research field, many works were done with a simple Γ-law equation of state, $P = (\Gamma - 1)\rho\varepsilon$ (in particular with $\Gamma = 2$), preparing a quasi-equilibrium state as the initial condition with the polytropic equation of state, $P = \kappa_\mathrm{p}\rho^\Gamma$. In this equation of state, the cold property of the neutron stars is preserved (i.e., κ_p remains constant) in the absence of shocks, and in the presence of shocks, the shock heating is taken into account. However, the polytropic equation of state is too simple to model the density profile of neutron stars in a realistic manner [see, e.g., figure 1 of Kyutoku et al. (2010b)]. Therefore, this equation of state should be used only for a very qualitative study of the merger of binary neutron stars and black hole-neutron star binaries.

A piecewise polytropic equation of state (or similar fitting formula), which can reproduce nuclear-theory-based cold equations of state at high density only with a small number of parameters, is a better model for describing the cold equations of state (see section 1.4.4). In the presence of shocks during the merger phase, however, the assumption of zero temperature breaks down. An often-used minimum prescription to compensate this drawback is that, in addition to the piecewise polytropic part, a correction term associated with the thermal pressure is added [Dimmelmeier et al. (2002); Shibata et al. (2005); Shibata and Taniguchi (2006); Kiuchi et al. (2009); Read et al. (2009b)]. In this method, the equation of state is written in the form

$$P = P_{\rm cold}(\rho) + P_{\rm th}(\rho, \varepsilon), \qquad (8.16)$$

where $P_{\rm cold}(\rho)$ denotes the piecewise polytropic part, and

$$P_{\rm th} = (\Gamma_{\rm th} - 1)\rho\varepsilon_{\rm th}, \qquad (8.17)$$

with $\Gamma_{\rm th}$ an adiabatic index for the thermal part, and $\varepsilon_{\rm th} = \varepsilon - \varepsilon_{\rm cold}(\rho)$ with $\varepsilon_{\rm cold}(\rho)$ determined by the first law of thermodynamics from the piecewise polytropic equation of state. Here, the effective value of $\Gamma_{\rm th}$ for neutron stars was estimated to be ~ 1.8 [Bauswein et al. (2010)], although it would be safe to take into account the uncertainty $\sim \pm 0.3$. This type of the equations of state [including a variant such as proposed in Haensel and Potekhin (2004)] has been employed by Shibata et al. (2005); Shibata and Taniguchi (2006); Kiuchi et al. (2009); Read et al. (2009b); Hotokezaka et al. (2011, 2013b,a); Takami et al. (2014) for the merger of binary neutron stars, and a variety of simulations have been performed systematically changing the parameters of the piecewise polytropic equations of state. With this equation of state, the late inspiral and early merger stages can be explored in a physical manner. However, for a more careful study of the thermal evolution of the merger remnant and its envelope, an improved treatment for the thermal effects is necessary. Moreover, with this equation of state, microphysical properties of the merger remnant such as emissivity of neutrinos, electron fraction per baryon, and baryon composition cannot be studied.

To self-consistently take into account the thermal and microphysical effects, the best way is to employ a finite-temperature equation of state which is derived based on a nuclear physics theory (see section 1.4.7). This type of equations of state is usually described in a table form as $P = P(\rho, T, Y_e)$ and $\varepsilon = \varepsilon(\rho, T, Y_e)$, where T and Y_e are the temperature and the electron fraction per baryon. Approximately speaking, ρ is determined by the continuity equation (4.42), and T by the internal energy which is determined essentially by the energy equation (4.40) in general relativity. Y_e is determined by a continuity equation for the electron fraction per baryon (4.90) (see section 4.5). For solving this equation correctly, we further have to take into account the neutrino emission/absorption, by which the electron fraction per baryon, Y_e, is varied; see, e.g., Sekiguchi (2010a,b).

Numerical-relativity simulations have been performed employing all the types of equations of state referred to in this section. In the following, we will pay attention to the numerical results obtained in these physical equations of state, piecewise polytropic ones written in the form (8.16) and finite-temperature ones, and based on these physical results, we will summarize our understanding for the merger of binary neutron stars.

8.3.2   Eccentricity reduction for initial condition

Quasi-equilibrium states described in section 8.2 are usually employed as the initial condition. These are computed assuming the presence of a helical Killing vector field in the form of equation (5.228). Actually, the spacetime concerned does not have the helical Killing vector field because of the presence of the gravitational-radiation reaction and the resulting approaching velocity between two neutron stars. The assumption for the presence of the helical Killing vector field is valid only for the case that the orbital separation of the system is so large that the gravitational-radiation reaction is negligible. Thus, for the case that a quasi-equilibrium initial condition is used, a binary of a sufficiently large orbital separation has to be employed for an astrophysically accurate simulation. If the orbital separation of the initial condition is not very large, the binary orbit becomes slightly eccentric. If we do not require highly precise numerical results, such residual eccentricity does not matter. However, for a precise simulation, in particular for a precise computation of gravitational waves, a large residual eccentricity $\gtrsim 10^{-2}$ is not acceptable.

To reduce the eccentricity in the initial data, we need to take into account the approaching velocity associated with the gravitational-radiation reaction. Such initial data can be approximately computed appropriately adding an approaching velocity for the formulation of the quasi-equilibrium and self-consistently solving the gravitational field equations and modified hydrostatic equations. Kyutoku et al. (2014b) showed that with a modified formulation, the eccentricity can be reduced to the level with $\lesssim 10^{-3}$. (See also section 8.4.1 for the related topic but for different systems.)

8.3.3   Merger process and remnants

8.3.3.1   Overview

Broadly speaking, there are two possible final fates of binary neutron stars: A massive (typically hypermassive or supramassive) neutron star is formed if the total mass of the binaries, m_0, is smaller than a critical value for the collapse to a black hole (hereafter denoted by M_{thr}), while a black hole is formed in the dynamical time scale $\lesssim 1$ ms for $m_0 > M_{\text{thr}}$. The critical mass, M_{thr}, depends on the equation of state for neutron stars: If the equation of state of the neutron-star matter is so stiff that it can yield a high value of the maximum mass of spherical neutron stars, $M_{\text{max},0}$, the value of M_{thr} is in general large. Numerical simulations by Hotokezaka et al. (2011) showed that the ratio, $M_{\text{thr}}/M_{\text{max},0}$, is much larger than unity as 1.3–1.7, although it depends strongly on the hypothetical equations of state employed. We note that if the equation of state is extremely stiff and $M_{\text{max},0} \sim 2.5 M_\odot$, a supramassive or normal neutron star (not a hypermassive neutron star) may be formed for the canonical binary mass of $m_0 = 2.6 - 2.8 M_\odot$. However, for $M_{\max,0} = 2.0 - 2.2 M_\odot$ which are the values for many of hypothetical equations of state, a hypermassive neutron star is the typical remnant for the canonical mass of binary neutron stars.

Figures 8.4–8.6 display snapshots of rest-mass density profiles for three merger simulations performed by Shibata and Taniguchi (2006) in the early stage of this field. For all the models, the hypothetical equation of state employed is APR4 (see section 1.4.4), but masses of two neutron stars are different for each model. Here, APR4 is a representative stiff equation of state with $M_{\max,0} \approx 2.2 M_\odot$, and the maximum mass for rigidly rotating neutron stars with zero temperature in this equation of state would be $M_{\max,\text{spin}} \approx 2.5 - 2.6 M_\odot$, as Cook et al. (1994a) showed that for most of nuclear-theory-based hypothetical stiff equations of state, $M_{\max,\text{spin}}$ is by 15–20% larger than $M_{\max,0}$.

Figure 8.4 shows the merger process for an equal-mass model with the total mass $m_0 = 2.6 M_\odot$. For this case, a long-lived massive (hypermassive or supramassive) neutron star is formed. The remnant massive neutron star is rapidly and differential rotating. Numerical simulations for several stiff equations of state have also found the same nature of the remnant massive neutron stars irrespective of the employed equations of state [Shibata et al. (2005); Hotokezaka et al. (2011); Sekiguchi et al. (2011b,a); Hotokezaka et al. (2013b,a)]. The rapid and differential rotation is the key property to enhance the maximum-allowed mass of massive neutron stars because the total mass of the remnant massive neutron star in this case is $\sim 2.53 M_\odot$ ($= M_{\text{ADM},0} - \Delta E_{\text{GW}}$ where $M_{\text{ADM},0}$ is the initial ADM mass and ΔE_{GW} is the energy dissipated by gravitational radiation) and hence, it is much larger than $M_{\max,0}$ and as larger as $M_{\max,\text{spin}}$. It is worthy to note that the differential rotation plays an important role for forming the hypermassive neutron star, because in its absence, a remnant of mass $\gtrsim M_{\max,\text{spin}}$ is likely to collapse to a black hole (note that thermal pressure associated with thermal energy generated by shocks during the merger could slightly increase this critical mass).

As described in section 1.4.6, hypermassive neutron stars formed cannot live forever. They will eventually collapse to a black hole if a substantial fraction of angular momentum is dissipated (e.g., by the gravitational-wave emission) or transported from its inner to outer region (e.g., by gravitational torque or viscous or magnetohydrodynamics effects; see a discussion in section 8.3.3.5). Here, the lifetime depends strongly on the total mass. If the total mass of the binary, m_0, is close to M_{thr}, the lifetime is short. By contrast, if m_0 is much smaller than M_{thr} and rather close to $M_{\max,\text{spin}}$, the lifetime can be much longer than the dynamical time scale and the rotation period of the system ~ 1 ms. For supramassive neutron stars with mass smaller than $M_{\max,\text{spin}}$, the lifetime will be even longer (see section 8.3.3.2).

Figure 8.5 displays the merger process for a high-mass equal-mass model with the total mass $m_0 = 3.0 M_\odot$. For this case, a black hole is formed promptly after the

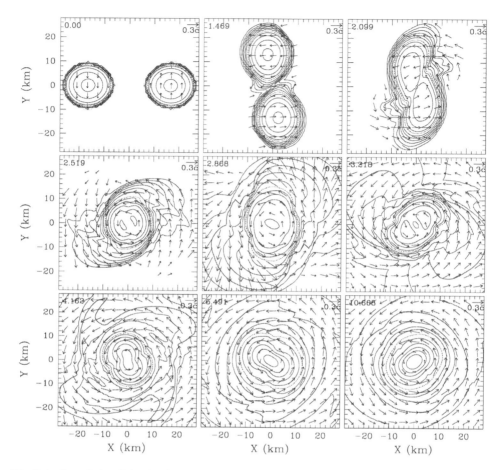

Fig. 8.4 Snapshots of the density contour curves for ρ on the equatorial plane for an equal-mass model with $m_0 = 2.6 M_\odot$ and APR4 equation of state (see section 1.4.4 for APR4). The solid contour curves are drawn for $\rho = 2 \times 10^{14} \times i$ g/cm^3 ($i = 1, 2, 3, \cdots$) and for $1 \times 10^{14-0.5i}$ g/cm^3 ($i = 1 \sim 6$). The (blue) thick dotted and solid curves denote 1×10^{14} g/cm^3 and 1×10^{12} g/cm^3, respectively. The number in the upper left-hand side denotes the elapsed time from the beginning of the simulation in units of millisecond. The vectors indicate the local velocity field (v^x, v^y), and the scale is shown in the upper right-hand corner. The figure is taken from Shibata and Taniguchi (2006).

onset of the merger, and more than 99.9% of the neutron-star matter is swallowed into the remnant black hole. Figure 8.6 also displays the merger process for $m_0 = 3.0 M_\odot$ but for an unequal-mass model with $m_1 = 1.6 M_\odot$ and $m_2 = 1.4 M_\odot$. For this case, a black hole is also formed promptly after the onset of the merger, but a fraction of the neutron-star matter stays outside the black hole and eventually forms a disk surrounding the black hole. The reason for this disk formation is that the less-massive neutron star is tidally deformed during the merger process and a

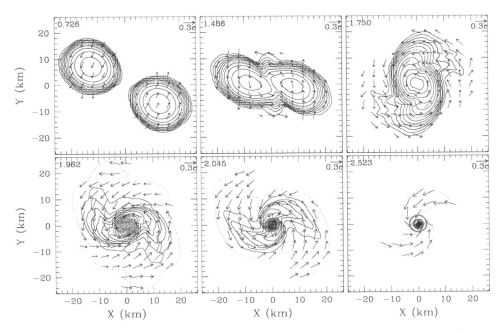

Fig. 8.5 The same as figure 8.4 but for a high-mass equal-mass model with $m_0 = 3.0 M_\odot$ and APR4 equation of state. The outermost dotted curves denote $\rho = 1 \times 10^{10}$ g/cm^3 and the thick circles around the origin in the last two panels denote apparent horizons. The figure is taken from Shibata and Taniguchi (2006).

fraction of its component located in an outer region obtains angular momentum via hydrodynamical angular-momentum transport process (i.e., via gravitational torque exerted by the merged object which has a non-axisymmetric shape). Then, a fraction of the neutron-star matter can have specific angular momentum which is larger than that of the ISCO around the remnant black hole. For given total mass, m_0, the mass of the remnant disk surrounding the remnant black hole tends to be larger for the larger mass ratio $Q := m_1/m_2 (\geq 1)$ [Shibata and Taniguchi (2006); Hotokezaka et al. (2013b,a)]. However, for making a massive disk with mass $\gtrsim 0.01 M_\odot$, the mass ratio has to be fairly large as $Q \gtrsim 1.2$ for this prompt-collapse case [Shibata and Taniguchi (2006); Hotokezaka et al. (2013b,a)].

As illustrated in the above paragraphs, the final outcome of the merger (a black hole or a massive neutron star) depends primarily on the ratio of the total mass, m_0, to $M_{\rm thr}$. Here, a key point is that the value of $M_{\rm thr}$ depends strongly on the equation of state which is still poorly known, while the observation of binary neutron stars (see table 1.2) suggests that the typical total mass of binary neutron stars is in a narrow range as $m_0 = 2.6$–$2.8 M_\odot$. Therefore, the possible outcome of the merger in nature is uncertain, because our knowledge of the equation of state for the neutron-star matter is still poor.

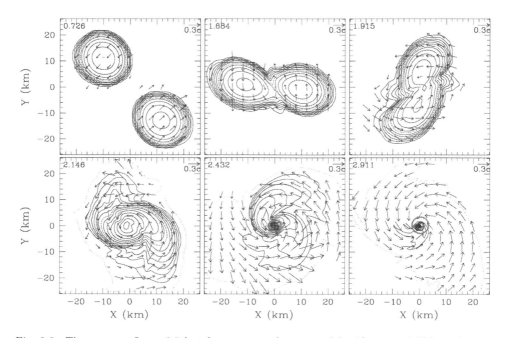

Fig. 8.6 The same as figure 8.5 but for an unequal-mass model with $m_1 = 1.6 M_\odot$ and $m_2 = 1.4 M_\odot$. The figure is taken from Shibata and Taniguchi (2006).

Nevertheless, the discoveries of two-solar mass neutron stars, PSR J1614-2230 and PSR J0348+432, by Demorest *et al.* (2010) and Antoniadis *et al.* (2013) certainly have improved this situation. These discoveries constrain that the maximum mass of spherical neutron stars, $M_{\rm max,0}$, has to be larger than $\sim 2 M_\odot$. This implies that the equation of state has to be stiff enough. Hotokezaka *et al.* (2011); Sekiguchi *et al.* (2011b); Hotokezaka *et al.* (2013b,a) performed a large number of simulations using hypothetical stiff equations of state (such as ALF2, APR4, H4, MS1, and SLy listed in table 1.1 and Shen's finite-temperature equations of state for all of which $M_{\rm max,0} \gtrsim 2 M_\odot$) and found that for very typical total mass of binary neutron stars, $m_0 = 2.7 M_\odot$, a massive neutron star is always formed after the onset of the merger at least temporarily. Therefore, we can now mention that a massive neutron star is the canonical remnant formed after the merger of binary neutron stars. In the following subsections, we will summarize a variety of possible evolution processes of the remnant massive neutron stars in more detail.

8.3.3.2 *Variety for evolution processes of remnant massive neutron stars*

Although massive neutron stars are the canonical remnants of the binary neutron star merger, their properties and evolution processes depend strongly on the total mass of the system and equations of state hypothetically employed. We describe

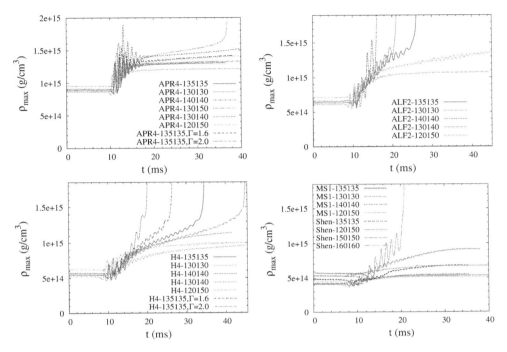

Fig. 8.7 The maximum rest-mass density as a function of time for APR4 (top left), ALF2 (top right), H4 (bottom left), and MS1 (bottom right) equations of state with several values of binary mass. Note that Γ for the panel of APR4 and H4 denotes $\Gamma_{\rm th}$ in equation (8.17) and the absence of Γ value means that $\Gamma_{\rm th}=1.8$. For the panel of MS1, some of the results for Shen equation of state are also plotted. Each model is referred to as the abbreviation "EOS"-"m_2" "m_1"; e.g., the model employing APR4, $m_1=1.50 M_\odot$, and $m_2=1.20 M_\odot$ is referred to as model APR4-120150. For the models in which the maximum rest-mass density steeply increases beyond $\sim 2\times 10^{15}\,{\rm g/cm^3}$, a black hole is formed. The figure is taken from Hotokezaka et al. (2013a).

this fact referring to numerical-relativity results by Hotokezaka et al. (2013b,a), in which they in detail explored the evolution process of the remnant massive neutron stars choosing six stiff equations of state, APR4, ALF2, H4, MS1, Shen, and SLy.

Figure 8.7 shows the evolution of the maximum rest-mass density for models with five hypothetical equations of state and with several values of binary mass in the range $m_0=2.6$–$2.8 M_\odot$. (For Shen equation of state the results with $m_0=2.7$ and $3.0 M_\odot$ are plotted.) In the following, we will first summarize the dependence of the evolution process of remnant massive neutron stars on the six equations of state, APR4, SLy, ALF2, H4, MS1, and Shen. For the properties of each equation of state employed, see section 1.4.4.

APR4: For this relatively soft equation of state ($R_{1.35}\approx 11$ km where $R_{1.35}$ is the radius of the $1.35 M_\odot$ neutron star), the pressure at $\rho=\rho_2:=5\times 10^{14}\,{\rm g/cm^3}$ (i.e., the value of p_1 in table 1.1) is rather low. However, the pressure for $\rho\geq\rho_3:=1\times 10^{15}\,{\rm g/cm^3}$ is rather high because the adiabatic index for this density range is

Fig. 8.8 Snapshots of rest-mass density profiles in the equatorial plane at selected time slices for equal-mass models ($m_1 = m_2 = 1.35 M_\odot$) with APR4 (upper three panels) and H4 (lower three panels) and with $\Gamma_{\rm th} = 1.8$. For these examples, the time at the onset of the merger is $t \sim 10\,{\rm ms}$. The figure is taken from Hotokezaka et al. (2013a).

high as $\Gamma_3 \sim 3.35$ (see figure 1.15). Reflecting the small pressure for $\rho < \rho_3$, the maximum rest-mass density increases steeply during the early stage of the merger in which the strength of the self-gravity steeply increases (see top-left panel of figure 8.7). However, also, reflecting the high pressure for $\rho > \rho_3$ as a result of the high value of Γ_3, the steep increase of the density is eventually hung up and subsequently the maximum rest-mass density oscillates with high amplitude for several oscillation periods. This is a unique feature for this type of equations of state (similar results were also found for SLy equation of state). After a subsequent relaxation process in which the matter in the envelope interacts with the remnant massive neutron star, the maximum rest-mass density eventually relaxes approximately to a constant as long as m_0 is not very high as $\lesssim 2.8 M_\odot$ (in this equation of state, $M_{\rm thr} \approx 2.9 M_\odot$).

In this relaxation process, angular momentum is transported substantially from the inner to the outer region via a hydrodynamical angular-momentum transport process (i.e., by gravitational torque), because the remnant massive neutron star has a highly non-axisymmetric structure (see figures 8.8 and 8.9) and can gravitationally exert the torque to the surrounding material during its early evolution stage. It is worthy to note that the high-amplitude oscillation of the remnant massive neutron

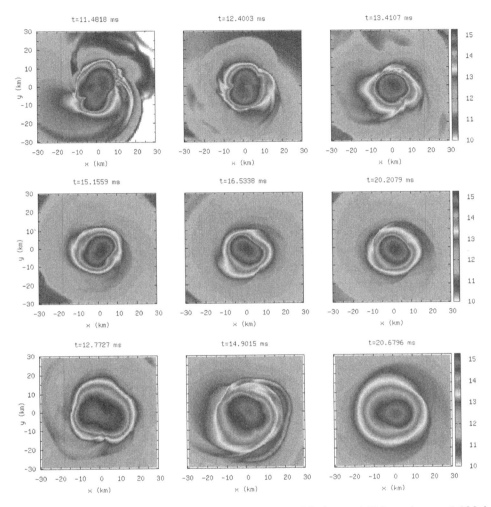

Fig. 8.9 The same as figure 8.8 but for unequal-mass models ($m_1 = 1.5 M_\odot$ and $m_2 = 1.2 M_\odot$) with APR4 (upper and middle panels) and H4 (bottom three panels). The figure is taken from Hotokezaka et al. (2013a).

stars also plays an important role for enhancing the angular-momentum transport because the non-axisymmetric central remnant interacts directly with the envelope during this oscillation.

After the relaxation, the remnant massive neutron stars result in a quasi-stationary state. For relatively small total mass with $m_0 \lesssim 2.7 M_\odot$, the maximum density in the quasi-stationary stage remains approximately constant for sufficiently long time $\gg 10$ ms. This is due to the facts that the angular momentum of the massive neutron stars is not significantly lost by the gravitational-wave emission and the hydrodynamical angular-momentum transport, and also that their mass is close to

$M_{\max,\mathrm{spin}}$; for models with $m_0 = 2.6 M_\odot$ and $2.7 M_\odot$, the final mass is $\approx 2.53 M_\odot$ and $2.60 M_\odot$, respectively. The value of $M_{\max,0}$ for this equation of state is $\approx 2.20 M_\odot$, and thus, that of $M_{\max,\mathrm{spin}}$ would be $\sim 2.6 M_\odot$ according to the numerical results by Cook et al. (1994a). Therefore, for $m_0 = 2.6 M_\odot$, the remnants are supramassive neutron stars, and even for $m_0 = 2.7 M_\odot$, the remnants may be hypermassive neutron stars with the mass close to $M_{\max,\mathrm{spin}}$.

For $m_0 \gtrsim 2.8 M_\odot$, on the other hand, the remnants are hypermassive neutron stars for which the mass is much larger than $M_{\max,\mathrm{spin}}$. For this case, their maximum density monotonically increases due to the gravitational-radiation reaction and hydrodynamical angular-momentum transport. The simulations by Hotokezaka et al. (2013b,a) showed that for $m_0 = 2.8 M_\odot$, the hypermassive neutron stars eventually collapse to a black hole surrounded by a massive torus with mass $\sim 0.1 M_\odot$ in several 10 ms after the onset of the merger. The massive torus is formed as a result of the fact that during the long-term evolution of the hypermassive neutron stars, angular momentum is continuously transported outward in their outer region: In general, a longer lifetime of the hypermassive neutron stars results in larger mass of the remnant torus surrounding the remnant black hole.

SLy: This is also a relatively soft equation of state in which $R_{1.35} \approx 11.5$ km. For this equation of state, the evolution process of remnant massive neutron stars is quite similar to that for APR4. However, the value of $M_{\max}$ for this equation of state is slightly (by $0.14 M_\odot$) smaller than that for APR4. Hence, the value of $M_{\max,\mathrm{spin}}$ should be also smaller by $\sim 0.15 M_\odot$, and thus, the plausible value of $M_{\max,\mathrm{spin}}$ would be $\sim 2.45 M_\odot$. Reflecting this fact, the threshold mass for the prompt formation of a black hole becomes $M_{\mathrm{thr}} \approx 2.8 M_\odot$ for this equation of state. For $m_0 = 2.7 M_\odot$ with which the mass of the remnant massive neutron star is $\sim 2.6 M_\odot > M_{\max,\mathrm{spin}}$ and thus it is hypermassive, a black hole is formed in a few 10 ms after the onset of the merger irrespective of the mass ratio; the lifetime of the hypermassive neutron star is not very long. For $m_0 = 2.6 M_\odot$, the mass of the remnant massive neutron star is $\gtrsim 2.5 M_\odot$, and thus, the massive neutron star is hypermassive as well. However, for this mass, the lifetime is $\gg 10$ ms.

As we mention in the following, binary neutron stars for ALF2 and H4 equations of state with $m_0 = 2.8 M_\odot$ do not result in the black-hole formation promptly after the onset of the merger by contrast to the case of SLy, although for all these equations of state, $M_{\max}$ is $\sim 2 M_\odot$. This suggests that for a given value of $M_{\max}$, the black-hole formation is more subject to the equations of state with smaller values of p_1, or in other words, with smaller neutron-star radii. As shown in section 8.3.4.3, a characteristic peak in the Fourier spectrum of gravitational waves for a high-frequency band ~ 2–4 kHz is present for the case that a remnant massive neutron star is formed after the merger. This suggests that if high-frequency gravitational waves from the mergers of binary neutron stars with particularly high total mass, say $2.8 M_\odot$, are observed, we will be able to constrain the equation of state of neutron stars only by determining whether the peak is present or not [Shibata (2005)].

ALF2: For this moderately soft equation of state ($R_{1.35} \approx 12.4$ km), not only canonical neutron stars of mass 1.2–$1.5 M_\odot$ but also remnant massive neutron stars formed just after the merger with $m_0 = 2.6$–$2.8 M_\odot$ have the maximum rest-mass density between ρ_2 and ρ_3 (see figure 1.15). Namely, the prompt black-hole formation does not occur for $m_0 \leq 2.8 M_\odot$, and hence, $M_{\text{thr}} > 2.8 M_\odot$ in this equation of state. Because the steep increase in the density is absent at the onset of the merger, the amplitude in the oscillation of the maximum rest-mass density for the remnant massive neutron stars is not as high as for APR4 and SLy, and as a result, the angular-momentum transport process in the remnant is not also as efficient as for APR4 and SLy. For this equation of state, however, the adiabatic index for the density range between ρ_2 and ρ_3 is small ($\Gamma_2 \sim 2.4$; see table 1.1), although the pressure in this range is higher than that for APR4. Due to this fact, the maximum density of the remnant massive neutron stars increases by the gravitational-radiation reaction and hydrodynamical angular-momentum transport moderately steeply. For the models with $m_0 \gtrsim 2.7 M_\odot$, the maximum density becomes eventually larger than ρ_3. For $\rho > \rho_3$, the adiabatic index is quite small due to the hypothetical phase transition to quark matter ($\Gamma_3 \sim 1.9$). Then, the increase of the maximum density by the loss of the angular momentum is enhanced, leading to the eventual gravitational collapse to a black hole. For this evolution process, the formation time scale of the black hole is determined by the time scale of the gravitational-wave emission or hydrodynamical angular-momentum transport.

For this equation of state, $M_{\text{max},0} \approx 2.0 M_\odot$ and $M_{\text{max,spin}}$ would be $\lesssim 2.4 M_\odot$. For $m_0 = 2.6 M_\odot$, the mass of the remnant massive neutron star is $\approx 2.54 M_\odot$, and thus, it is hypermassive for $m_0 \geq 2.6 M_\odot$. Hence, it is reasonable to have the results that with $m_0 \gtrsim 2.7 M_\odot$, the remnant collapses to a black hole in a short time scale ~ 10 ms. For relatively small mass $m_0 = 2.6 M_\odot$, by contrast, the emission of gravitational waves and hydrodynamical angular-momentum transport become inactive before the maximum density significantly exceeds ρ_3. In this case, the increase of the maximum density is hung up and the hypermassive neutron star relaxes to a quasi-stationary state. Subsequent evolution of such hypermassive neutron stars will be determined by other (not-purely hydrodynamical) angular-momentum transport processes or cooling (see section 8.3.3.5).

One point worthy to be noted is that the evolution process for the $m_0 = 2.7 M_\odot$ case depends on the mass ratio (compare the plots for ALF2-135135 and ALF2-120150 in figure 8.7). For a sufficiently large asymmetry (i.e., $m_1/m_2 = 1.25$), the lifetime of the hypermassive neutron star becomes much longer than that of the equal-mass model. The reason for this is that for the asymmetric case, the merger occurs at a larger orbital separation than for the equal-mass case; i.e., before a sufficient amount of angular momentum is dissipated by the gravitational-wave emission, the merger sets in. In addition, a large fraction of the matter (in particular the matter of the less-massive neutron star) obtains sufficient specific angular momentum to result in a disk/torus or ejecta, because tidal torque is exerted by the asymmetric merger product during the merger process. This reduces the mass (and hence the self-gravity) of the remnant hypermassive neutron stars, and as a result, the collapse to a black hole is delayed.

H4: The evolution process of remnant massive neutron stars in this relatively stiff equation of state ($R_{1.35} \approx 13.6$ km) is similar to that in ALF2. Because the values of $M_{\max,0}$ are approximately identical for H4 and ALF2, the criterion for the remnant massive neutron star formation is also very similar. For this equation of state, however, the adiabatic index does not decrease with the increase of the density for $\rho > \rho_2$ as drastically as for ALF2. Thus, the increase rate of the maximum density with time is relatively slow, and reflecting this fact, the configuration of the remnant massive neutron stars relaxes to a quasi-stationary one in a relatively short time scale after its formation. The resulting quasi-stationary massive neutron stars subsequently evolve through the hydrodynamical angular-momentum transport and gravitational-wave emission. However the evolution time scale is much longer than 10 ms.

A new point clearly seen in the bottom left panel of figure 8.7 is that the efficiency of the shock heating (determined by the value of $\Gamma_{\rm th}$) plays a role for changing the lifetime of the hypermassive neutron stars: By the increasing efficiency of the shock heating (for the larger values of $\Gamma_{\rm th}$), the lifetime of the hypermassive neutron stars becomes longer. It is also found that in the presence of mass asymmetry, the lifetime becomes longer due to the same reason as in the ALF2 case.

MS1: For this very stiff equation of state ($R_{1.35} \approx 14.5$ km), the maximum mass of spherical neutron stars is too high ($M_{\max,0} \approx 2.77 M_\odot$) to form supramassive or hypermassive neutron stars for $m_0 \leq 2.8 M_\odot$ because the remnant mass for such initial mass range is smaller than $2.75 M_\odot$. For this case, the remnant is a normal neutron star, which relaxes to a quasi-stationary state in a short time scale ~ 10 ms. Although a dissipation or a transport process of the angular momentum plays a role for the subsequent evolution, a black hole will not be formed for $m_0 \leq 2.8 M_\odot$ for which the remnant mass is smaller than $M_{\max,0} = 2.77 M_\odot$.

Shen: This is a tabulated equation of state and a finite-temperature effect is taken into account in a microphysical manner. The evolution process of remnant massive neutron stars in this stiff equation of state ($R_{1.35} \approx 14.5$ km) is similar to that in H4, although the critical mass for the prompt formation of a black hole is much higher than that for H4 ($M_{\rm thr} > 3.0 M_\odot$ in this case, while $M_{\rm thr} \gtrsim 2.8 M_\odot$ for H4). For this equation of state, a long-lived hypermassive neutron star is formed even for $m_0 = 3 M_\odot$ which is by 36% larger than the value of $M_{\max} \approx 2.2 M_\odot$. By contrast, for H4, a black hole is formed in several 10 ms if $m_0 \gtrsim 2.7 M_\odot \approx 1.33 M_{\max}$. This suggests that in the tabulated equation of state in which the thermal and microphysics effects are taken into account in a more strict manner, the pressure for sustaining the self-gravity of the hypermassive neutron stars could be slightly enhanced. (See section 8.3.3.4 for more details.)

We finally touch on the following aspect about the dependence of the evolution process for remnant massive neutron stars of canonical mass $m_0 \approx 2.7 M_\odot$ on hypothetical equations of state:

- For the relatively soft equations of state such as APR4 and SLy for which p_1 and $R_{1.35}$ have relatively small values, the evolution process of the remnant massive neutron stars depends primarily on the adiabatic index for $\rho > \rho_3 = 10^{15}\,\mathrm{g/cm^3}$ (i.e., Γ_3) because their central density exceeds ρ_3.
- For the moderately stiff equations of state such as ALF2 and H4, the evolution process of the remnant massive neutron stars depends on the adiabatic index for $\rho \gtrsim \rho_2 = 5 \times 10^{14}\,\mathrm{g/cm^3}$ (i.e., both on Γ_2 and Γ_3) because their central density comes in the range between ρ_2 and ρ_3 at their formation.
- For the stiff equations of state such as MS1 and Shen, the evolution process of the remnant massive neutron stars depends only on the adiabatic index for $\rho \lesssim \rho_3$ (i.e., Γ_2) because their central density does not exceed ρ_3.

These facts suggest that observations for the remnant massive neutron stars by gravitational-wave detectors and other telescopes will be used for exploring the properties of the equation of state in a specific density range.

8.3.3.3 Effects of mass asymmetry

Since remnant massive neutron stars formed after mergers are rapidly rotating and non-axisymmetric (e.g., figures 8.8 and 8.9), they emit gravitational waves of high amplitude (see section 8.3.4.3). The detailed property of the gravitational waveform depends on the density and velocity profiles of the remnant massive neutron star. The equation of state determines their characteristic radius, and hence, the frequency of gravitational waves depends strongly on the equation of state (see section 8.3.4). The merger process depends not only on the equation of state but also on the mass ratio and total mass. The mass ratio (i.e., mass asymmetry) in particular becomes a key ingredient for determining the evolution process of the density profile and the final configuration of the remnant massive neutron star. In this subsection, thus, we pay attention to the dependence for the evolution of the remnant configuration on the binary mass ratio.

First, we summarize the evolution process of the configuration and density profile of the remnants for the equal-mass case (see figure 8.8). For this case, the remnants form a dumbbell-like shape composed of two density peaks soon after the onset of the merger irrespective of the equations of state hypothetically employed. Then, due to the loss of their angular momentum by the hydrodynamical angular-momentum transport and gravitational-wave emission, the shape changes gradually to an ellipsoidal one, and subsequently, the ellipticity decreases with time. Here, the time scale of the angular momentum loss depends on the equation of state employed. For APR4 and SLy for which a violent quasi-radial oscillation occurs in an early evolution stage of the remnant massive neutron stars (see figure 8.7), the time scale of the angular momentum loss is relatively short ($\sim 10\,\mathrm{ms}$), while for stiff equations of state such as H4, MS1, and Shen, this time scale is rather long

> 10 ms. For APR4 and H4, these facts can be observed from figure 8.8.

For the unequal-mass case, the evolution process of the configuration and density profile of the remnants is different from that for the equal-mass case. To make the difference clear, we focus here on an appreciably asymmetric model with $m_1 = 1.5 M_\odot$ and $m_2 = 1.2 M_\odot$ (see figure 8.9). For this case, the configuration and density profile of the remnant change with the dynamical time scale in an early evolution stage. The reason for this is as follows: During the early merger stage, the less-massive neutron star is tidally deformed and its outer part is stripped. Then, the stripped material forms an envelope of the remnant massive neutron star, while the core of the less-massive neutron star rotates around the core of the massive companion. Hence, a massive neutron star composed of two asymmetric cores is formed (see the first two panels of figure 8.9). Because the self-gravity of the less-massive core is much weaker than that of the massive one, it behaves as a satellite that is significantly and dynamically deformed by the main core, varying its shape, like an amoeba, in a time scale of a few ms. During its evolution, the total shape of the remnant massive neutron star often becomes approximately spheroidal at a moment (see the third panel of figure 8.9). For such a moment, the emission of gravitational waves is suppressed transiently, and hence, the entire gravitational-wave amplitude has non-trivial modulation (see section 8.3.4).

However, after the substantial hydrodynamical angular-momentum transport process, which occurs via the interaction with the envelope, the remnant massive neutron star relaxes to a quasi-stationary state irrespective of the equations of state employed. Then, quasi-periodic gravitational waves are emitted. The quasi-stationary massive neutron star appears to be composed of major and minor cores which are rotating in a quasi-stationary manner (see the late-time snapshots of figure 8.9). This system looks like a hammer thrower rotating with a hammer (here the thrower is the major core and the hammer is the minor core). It subsequently loses the angular momentum primarily through the hydrodynamical angular-momentum transport process, and thus, the degree of the asymmetry decreases gradually, although the time scale of this change is much longer than the dynamical time scale.

8.3.3.4 *Thermal properties and neutrino luminosity*

To explore the evolution of hot remnant massive neutron stars in a physical manner, we have to take into account microphysical processes such as neutrino emission and evolution of matter composition. For this purpose, we have to employ a hypothetical equation of state constructed in the framework of a high-density and high-temperature nuclear-matter theory. Incorporating microphysical processes in a numerical-relativity simulation has in particular an important aspect for exploring the merger hypothesis of short gamma-ray bursts because it may be driven through the pair annihilation of neutrino-antineutrino pairs [see, e.g., Zhang and Mészáros (2004); Piran (2004) for a review, and section 1.5.2 for a brief introduction]. Taking

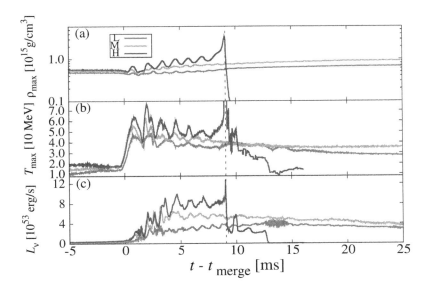

Fig. 8.10 Maximum rest-mass density, maximum matter temperature, and total neutrino luminosity as functions of time for three equal-mass models of binary neutron star mergers with different total mass. The dashed vertical line shows the time at which a black hole is formed for the highest-mass model (labeled by H). The figure is taken from Sekiguchi et al. (2011b).

the microphysical processes into account is also important for exploring the nucleosynthesis processes in the matter ejected from the system that will emit strong electromagnetic signals and thus will be observed by large-scale telescopes (see sections 1.5.2 and 8.3.5.3 for a brief introduction).

The first numerical-relativity simulation of binary neutron star mergers in the framework of full general relativity incorporating microphysical processes was performed by Sekiguchi et al. (2011b,a). They reported the thermal processes in the merger and subsequent evolution of remnant massive neutron stars, and derived neutrino-luminosity curves for the first time. In these works, they solved the evolution equations for the lepton fractions per baryons, Y_ν and Y_e (or Y_l), as well as the ordinary hydrodynamics equations, taking into account weak interaction processes [see Sekiguchi (2010b) for details] and using Shen equations of state [Shen et al. (1998a, 2011c)]. In addition, neutrino cooling was incorporated employing the so-called Sekiguchi's general relativistic leakage scheme developed in his original manner: In his leakage scheme, the emission processes of electron neutrinos (ν_e), electron antineutrinos ($\bar{\nu}_e$), and other types (μ and τ neutrinos) denoted by ν_x are taken into account [see Sekiguchi (2010b) for his scheme and the results for the calibration].

In the following, we will introduce the pioneer work of Sekiguchi et al. (2011b) in which they performed simulations for the merger of equal-mass binary neutron stars with each mass $m_1 = m_2 = 1.35 M_\odot$, $1.5 M_\odot$, and $1.6 M_\odot$ (denoted by models L,

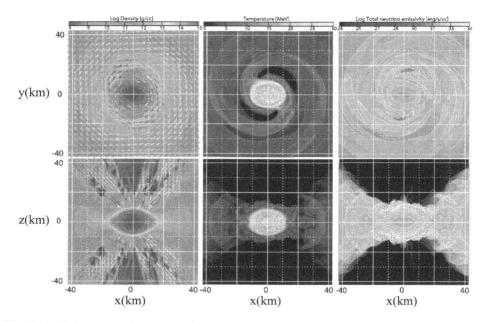

Fig. 8.11 Color maps of rest-mass density, matter temperature, and total neutrino luminosity of a hypermassive neutron star at $t \approx 15\,\mathrm{ms}$ after the onset of the merger for the model with $m_1 = m_2 = 1.5 M_\odot$. The upper and lower panels show the configuration in the x-y and x-z planes, respectively. The figure is taken from Sekiguchi et al. (2011b).

M, and H, respectively). Figure 8.10 plots the evolution of the maximum rest-mass density, $\rho_{\max}$, maximum matter temperature, $T_{\max}$, and total neutrino luminosity as functions of $t - t_{\mathrm{merge}}$ where t_{merge} is the onset time of the merger. For all the models, after the merger sets in, a massive neutron star is formed and it subsequently is evolved by the emission of gravitational waves and the hydrodynamical angular-momentum transport process, which carry energy and angular momentum from the remnant massive neutron stars. As a result of this dissipation, $\rho_{\max}$ increases with the dissipation time scale. For the models with $m_1 = m_2 = 1.35 M_\odot$ and $1.5 M_\odot$, the degree of the non-axisymmetry of the formed massive neutron stars damps gradually with time, and the dissipation time scale of the angular momentum becomes much longer than $\sim 10\,\mathrm{ms}$ for $t - t_{\mathrm{merge}} \gtrsim 10\,\mathrm{ms}$. Thus, a quasi-stationary massive neutron star is formed as described in section 8.3.3.2. These are hypermassive because the total mass is larger than $M_{\max,\mathrm{spin}}$ that would be $\sim 2.6 M_\odot$. For the highest-mass model with $m_1 = m_2 = 1.6 M_\odot$, a hypermassive neutron star is formed only temporarily and it eventually collapses to a black hole after the loss of the angular momentum by the gravitational-wave emission and hydrodynamical angular-momentum transport.

The evolution of $T_{\max}$ plotted in figure 8.10(b) shows that the remnant massive neutron stars just after their formation are quite hot with $T_{\max} \sim 40\text{--}70\,\mathrm{MeV}$. Such

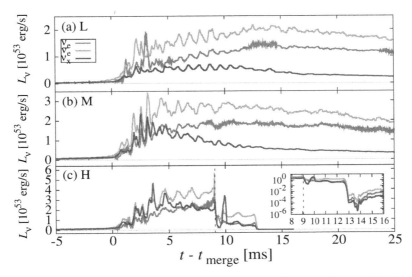

Fig. 8.12 The same as figure 8.10 but for neutrino luminosity for three flavors. The inset of the bottom panel focuses on the luminosity in the black-hole formation case. The meaning of the dashed line is the same as in figure 8.10. The figure is taken from Sekiguchi et al. (2011b).

high temperature is achieved due to the liberation of kinetic energy associated with the orbital motion at the collision of two neutron stars. Subsequently, the maximum temperature goes down due to the neutrino cooling. Here, the maximum luminosity of neutrinos is quite high as $L_\nu = 3\text{--}10 \times 10^{53}$ ergs/s [see figure 8.10(c)]. However, the cooling time scale estimated from $E_{\rm th}/L_\nu$, with $E_{\rm th}$ the total thermal energy of the remnant massive neutron stars, is quite long (of order seconds which is much longer than the dynamical time scale), and hence, the maximum temperature relaxes to a high value with 25–50 MeV for $t-t_{\rm merge} \gtrsim 10$ ms. Around the remnant massive neutron stars, spiral arms are formed and shock heating continuously occurs when the spiral arms hit the outer region of the massive neutron stars (see figure 8.11). Due to this process and because of the long neutrino cooling time scale ($\sim 1\text{--}10$ s), the temperature (and thermal energy) will be quasi-steady in seconds. For the highest-mass model for which a black hole is formed in ~ 10 ms after the onset of the merger, the temperature increases just before its formation because of violent adiabatic contraction. Associated with this, the total neutrino luminosity exceeds 10^{54} ergs/s. However, the region with the highest temperature is swallowed by the newly-formed black hole and the luminosity steeply decreases after its formation (see below). The formed black hole is surrounded by a disk/torus with mass $\sim 0.1 M_\odot$. The maximum temperature of the disk/torus is ~ 10 MeV.

Figure 8.11 plots the color maps of the rest-mass density, matter temperature, and total neutrino luminosity for the $m_1 = m_2 = 1.5 M_\odot$ model at $t - t_{\rm merger} = 15$ ms, at which the hypermassive neutron star relaxes to a quasi-steady state. This figure shows that the hypermassive neutron star is weakly spheroidal and the temperature is high in its outer region. The neutrino luminosity is also high in its outer region, in particular, near the polar surface. This is the universal feature for the thermal profile of the remnant massive neutron stars. For these massive neutron stars, the typical spin period is $\sim 1\,\mathrm{ms}$. This value is not as short as the dynamical time scale and hence the centrifugal force associated with the spin would play a mild role for sustaining the self-gravity. Exploring in detail Shen equation of state for the high-density range tells us that the effect of the thermal energy with the temperature, $20\text{--}30\,\mathrm{MeV}$, is significant and can increase $M_{\mathrm{max},0}$ by $\sim 10\text{--}20\%$ (see figure 1.19). This also suggests that the hypermassive neutron stars will be alive for a long cooling time scale, $E_{\mathrm{th}}/L_\nu \gtrsim 1\,\mathrm{s}$.

Figure 8.12 plots the neutrino luminosity as a function of time for three flavors; electron neutrinos, electron antineutrinos, and sum of other (μ- and τ-) neutrinos. It is found that antineutrinos are dominantly emitted for any model. The reason for this is as follows: The remnant massive neutron stars have high temperature $\gg 1\,\mathrm{MeV}$, and hence, electron-positron pairs are efficiently created from thermal photons, in particular in its envelope. Neutrons efficiently capture the created positrons and emit antineutrinos, whereas electrons are not captured by protons as frequently as positrons, because the fraction of protons is much smaller than that of neutrons in the remnant massive neutron stars.

Soon after the black-hole formation for the highest-mass model, the neutrino luminosity steeply decreases from $\gtrsim 10^{54}\,\mathrm{ergs/s}$ to $\sim 10^{53}\,\mathrm{ergs/s}$. In the subsequent accretion phase onto the black hole, μ- and τ-neutrino luminosity increases steeply, because the high matter temperature is achieved due to the compression during the infall and hence pair processes for leptons are enhanced. After the system relaxes to a quasi-stationary state, neutrinos are emitted only from the accretion disk/torus of mass $\sim 0.1 M_\odot$ and total luminosity falls to $\lesssim 10^{51}\,\mathrm{ergs/s}$.

The *antineutrino* luminosity for the long-lived hypermassive neutron stars is $L_{\bar\nu} \sim 1.5\text{--}3 \times 10^{53}\,\mathrm{ergs/s}$ with small time variability. It is by a factor of $\sim 1\text{--}5$ larger than that emitted from supernovae and from proto-neutron stars formed after the supernovae [e.g., Sumiyoshi *et al.* (2005)]. The average neutrino energy is quite high as $\epsilon_{\bar\nu} \sim 30\,\mathrm{MeV}$. However, it will be quite difficult to detect such neutrinos for an event at a distance larger than $10\,\mathrm{Mpc}$, even by large water-Cherenkov neutrino detectors such as Super–Kamiokande and future Hyper–Kamiokande, and thus, the neutrino detection will not be a robust tool for exploring the merger of binary neutron stars.

One remark to be done is that Shen equation of state is a quite stiff one; with it, the radius of zero-temperature spherical neutron stars is typically $14\text{--}15\,\mathrm{km}$. If the equation of state is in reality softer with the typical neutron-star radius $11\text{--}12\,\mathrm{km}$, a higher-density and higher-temperature state could be realized by a stronger compression. Remembering the fact that the neutrino luminosity depends very strongly on the temperature, it may be much larger than $\sim 10^{53}\,\mathrm{ergs/s}$ in reality. We have to keep in mind that the uncertainty of the neutrino luminosity is quite large. A large number of systematic simulations employing a variety of hypothetical equations of state in the future are awaited in this field.

8.3.3.5 Lifetime of remnant massive neutron stars

As summarized in sections 8.3.3.2–8.3.3.4, hypermassive neutron stars collapse to a black hole in a typical time scale of $O(10\,\mathrm{ms}) - O(1\,\mathrm{s})$. We described that the collapse would be triggered by the angular-momentum loss due to the gravitational-wave emission or by the hydrodynamical angular-momentum transport process from the hypermassive neutron star to its outer envelope or neutrino cooling. Hotokezaka et al. (2013b,a) found that the mass of disks/tori surrounding the remnant black holes eventually formed is in general larger for the longer lifetime of the hypermassive neutron stars. In addition, the emissivity of gravitational waves is quite low for not-young hypermassive neutron stars (with age $\gtrsim 10\,\mathrm{ms}$) as we describe in section 8.3.4. These facts show that the hydrodynamical angular-momentum transport process plays a more important role than the gravitational-radiation reaction for the black-hole formation. For the hypermassive neutron stars with relatively short lifetime $\lesssim 100\,\mathrm{ms}$, it is reasonable to conclude that the black-hole formation is determined primarily by the hydrodynamical angular-momentum transport process, and hence, we write the time scale as τ_hyd in the following.

However, for less-massive hypermassive neutron stars, neither the emission of gravitational waves nor the hydrodynamical effect are likely to trigger the collapse. For such systems, other dissipation processes, which are not purely hydrodynamical (i.e., not by gravitational torque), will play an important role, and the evolution will proceed in a different dissipation time scale. If the system is hypermassive and its degree of differential rotation is sufficiently high, there are two possible effects for the angular-momentum transport, both of which are activated in the presence of strong magnetic fields (see section 10.3 for more details). One is the magnetic winding effect [e.g., Baumgarte et al. (2000)] for which the order of the angular-momentum transport time scale is estimated by

$$\tau_\mathrm{wind} \sim \frac{R}{v_A} \sim 10^2\,\mathrm{ms}\left(\frac{\rho}{10^{15}\,\mathrm{g/cm^3}}\right)^{1/2}\left(\frac{B}{10^{15}\,\mathrm{G}}\right)^{-1}\left(\frac{R}{10^6\,\mathrm{cm}}\right), \tag{8.18}$$

where R and ρ are the typical radius and density of a hypermassive neutron star, B is the typical magnitude of the radial component of magnetic fields, and v_A is the Alfvén velocity approximately calculated by [see equation (4.188)]

$$v_A \approx \frac{B}{\sqrt{4\pi\rho}}. \tag{8.19}$$

Thus, for a sufficiently high magnetic-field strength that could be generated by the winding, compression, and magnetorotational instability (MRI; see below), the angular-momentum transport is significantly enhanced.

The other is the MRI [see Balbus and Hawley (1998) for a detailed review of this] by which an effective viscosity is likely to be generated with the effective viscous parameter

$$\nu_{\rm vis} \sim \alpha_{\rm vis} \frac{c_s^2}{\Omega}, \qquad (8.20)$$

where $\alpha_{\rm vis}$ is the so-called α parameter that would be 0.01–0.1 [Balbus and Hawley (1998)], c_s is the typical sound velocity that would be of order $\sim 0.1c$, and Ω is the typical spin angular velocity of the remnant massive neutron star $\sim 10^4$ rad/s. Thus, the MRI-driven viscous angular-momentum transport time scale in the presence of magnetic fields would be

$$\tau_{\rm mri} \sim \frac{R^2}{\nu_{\rm vis}} \sim 10^2\,{\rm ms} \left(\frac{R}{10^6\,{\rm cm}}\right)^2 \left(\frac{\Omega}{10^4\,{\rm rad/s}}\right) \left(\frac{\alpha_{\rm vis}}{10^{-2}}\right)^{-1} \left(\frac{c_s}{0.1c}\right)^{-2}, \qquad (8.21)$$

and hence, the time scale is as short as $\tau_{\rm wind}$ for a hypothetical value of $B \sim 10^{15}$ G.

These two effects work as long as differential rotation is present even in the case that the remnant massive neutron star is axisymmetric. Therefore, unless any other process stabilizes them, hypermassive neutron stars (with sufficiently high mass) are likely to collapse to a black hole in the time scale, $\tau_{\rm wind}$ or $\tau_{\rm mri}$, which would be $\sim 10^2$ ms. Indeed, numerical simulations by Duez *et al.* (2006a); Shibata *et al.* (2006); Siegel *et al.* (2013) showed that such mechanisms are likely to work for hypermassive neutron stars with sufficiently high mass (see also section 10.3).

Here, some words are appropriate: To explore the effects of magnetic fields in the merger of binary neutron stars more strictly, it would be necessary to perform a magnetohydrodynamics simulation for the entire merger process. Therefore, a magnetohydrodynamics simulation of binary neutron stars is one of the important subjects in numerical relativity. However, it is quite challenging to resolve the MRI in current computational resources, because of its short wavelength for the typical magnetic-field strength $\sim 10^{12}$ G as

$$\sim v_A P_{\rm rot} \approx 10\,{\rm cm} \left(\frac{B}{10^{12}\,{\rm G}}\right) \left(\frac{P_{\rm rot}}{1\,{\rm ms}}\right) \left(\frac{\rho}{10^{15}\,{\rm g/cm^3}}\right)^{-1/2}, \qquad (8.22)$$

where $P_{\rm rot}$ is the typical spin period of the remnant massive neutron stars. Hence, a simulation in a realistic setting has not yet been done. However, for an artificially strong magnetic field with a large-scale computation, it is feasible to resolve the MRI and to explore its effect. We will describe the details of magnetohydrodynamics effects in the binary neutron star merger in section 10.3.2.3 based on the latest work by Kiuchi *et al.* (2014) [see also Rezzolla *et al.* (2011) for an early effort].

To this paragraph, we have not considered finite-temperature (thermal) effects. This effect could be important for the evolution of low-mass hypermassive neutron stars, as already mentioned in section 8.3.3.4. The reason for this is that during the merger process, strong shocks are often generated and the maximum temperature of the remnant massive neutrons stars is increased up to 30–50 MeV [Sekiguchi *et al.* (2011b,a)]. The thermal pressure associated with this high temperature could be

$\sim 10\%$ of the cold-part pressure caused by the repulsive nuclear force, and hence, it is not negligible. Although it is not easy to quantify its effect, it is reasonable to consider that the finite-temperature effect could increase the values of $M_{\mathrm{max},0}$ and $M_{\mathrm{max,spin}}$ by $\sim 0.1 M_\odot$ (see figure 1.19).

The hot neutron stars will eventually dissipate the thermal energy in the cooling time scale $\tau_{\mathrm{cool}} \sim 1\text{--}10\,\mathrm{s}$ by the neutrino emission. Sekiguchi et al. (2011b,a) indeed estimated that the time scale of the neutrino emission is of order seconds (see section 8.3.3.4). Here, the point is that the cooling time scale is much longer than τ_{hyd}, τ_{wind}, and τ_{mri}. Thus, when we consider the possible evolution processes of remnant massive neutron stars, we have to keep in mind that in this time scale, the values of $M_{\mathrm{max},0}$ and $M_{\mathrm{max,spin}}$ could be larger by $\sim 0.1 M_\odot$. For example, consider a differentially rotating and hot remnant massive neutron star for which the mass is slightly larger than $M_{\mathrm{max,spin}}$. When an angular-momentum transport process works and the degree of the differential rotation is significantly reduced, such a remnant will be unstable against gravitational collapse, if the thermal effect is negligible. However, if the thermal effect is important even after the angular-momentum transport process works, it will be stable in the cooling time scale. It will eventually collapse to a black hole after the neutrino cooling. However, its lifetime is much longer than the angular-momentum transport time scale.

As summarized above, hypermassive neutron stars with a high degree of differential rotation could collapse to a black hole via magnetohydrodynamics effects even if they were able to survive after the dissipation by the gravitational-wave emission and hydrodynamical angular-momentum transport. The lifetime of such hypermassive neutron stars is likely to be relatively short as τ_{wind} or τ_{mri}, which is $\sim 100\,\mathrm{ms}$. However, if the degree of the differential rotation is not high and the thermal effect plays an important role for sustaining the self-gravity of the hypermassive neutron stars, the neutrino cooling will determine their lifetime: If the remnant mass is only slightly larger than $M_{\mathrm{max,spin}}$, the magnetic winding and MRI could not trigger the collapse to a black hole. For such a system, the neutrino cooling will trigger the collapse eventually.

For an even smaller-mass system with $m_0 \lesssim M_{\mathrm{max},0}$, the remnant neutron star is not hypermassive, and it evolves simply to a cold massive neutron star in τ_{cool}. This is the case for MS1 with $m_0 \lesssim 2.8 M_\odot$ and for remnant massive neutron stars with small mass irrespective of equations of state, which may be formed after the merger of low-mass binary neutron stars composed of, e.g., $1.2 M_\odot$ neutron stars.

Taking into account all the facts discussed above, we may classify the fate of remnant massive neutron stars by its evolution time scale. Figure 8.13 shows such a classification. In this figure, the horizontal axis denotes hypothetical equations of state and the vertical axis shows the total mass of the binaries. τ_{dyn} implies that a black hole is formed soon after the onset of the merger (i.e., in the dynamical time scale of the system $\lesssim 1\,\mathrm{ms}$); τ_{hyd} shows that a hypermassive neutron star is formed and its lifetime is determined by the time scale of the hydrodynamical angular-

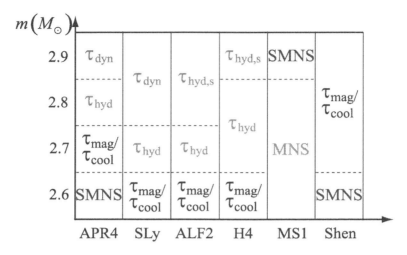

Fig. 8.13 The evolution time scale of the system in the plane composed of hypothetical equations of state and total mass. $\tau_{\rm dyn}$: A black hole is formed soon after the onset of the merger (in the dynamical time scale of $\lesssim 1\,{\rm ms}$). $\tau_{\rm hyd}$: A hypermassive neutron star is formed and its lifetime is determined by the time scale of the hydrodynamical angular-momentum transport (and partially the gravitational-wave emission) ~ 10–$100\,{\rm ms}$. $\tau_{\rm hyd,s}$: The same as for $\tau_{\rm hyd}$ but the lifetime is shorter than $10\,{\rm ms}$. $\tau_{\rm mag}/\tau_{\rm cool}$: A hypermassive neutron star is formed and its lifetime would be determined by the time scale of an angular-momentum transport process associated with some magnetohydrodynamics effects or by the neutrino cooling time scale. The evolution time scale for a given value of total mass depends weakly on the mass ratio. For MS1, only normal massive neutron stars (MNS) or supramassive neutron stars (SMNS) are formed for $m_0 \lesssim 2.9 M_\odot$. For APR4 and Shen, the remnant for the $m_0 \lesssim 2.6 M_\odot$ case is likely to be a supramassive neutron star (not a hypermassive neutron star). The figure is taken from Hotokezaka et al. (2013a).

momentum transport (and partially the gravitational-wave emission) ~ 10–$100\,{\rm ms}$. $\tau_{\rm hyd,s}$ implies that the evolution process is the same as for $\tau_{\rm hyd}$ but the lifetime is shorter than $10\,{\rm ms}$; $\tau_{\rm mag}/\tau_{\rm cool}$ shows that a hypermassive neutron star is formed and its lifetime would be determined by the time scale of an angular-momentum transport process associated with some magnetohydrodynamics effects or by the neutrino cooling time scale; "SMNS" shows that a supramassive neutron star is formed and its lifetime would be much longer than $\tau_{\rm mag}$ and $\tau_{\rm cool}(\sim 1\,{\rm s})$.

Figure 8.13 clearly shows that the evolution process and the lifetime of a remnant massive neutron star depend strongly on its equation of state and binary initial total mass m_0. Furthermore, the dependence of the lifetime of a remnant massive neutron star on the initial mass depends strongly on the hypothetical equation of state. This property is well reflected in the gravitational waveforms, as we show in section 8.3.4.

8.3.3.6 Summary for the process and fate of the binary neutron star merger

We here summarize the possible evolution paths for which binary neutron stars follow after their mergers in figure 8.14. As described in the previous subsections,

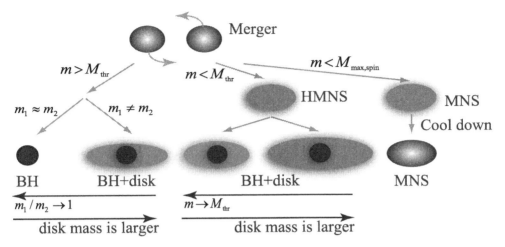

Fig. 8.14 A summary for the evolution process of the merger remnant of binary neutron stars, which depends on the total mass and mass ratio of the system for a given hypothetical equation of state. $M_{\rm thr}$ and $M_{\rm max,spin}$ denote the threshold mass for the prompt formation of a black hole and the maximum mass of supramassive neutron stars, respectively. The values of these mass depend on the equation of state, although for stiff equations of state that give $M_{\rm max,0} \gtrsim 2.0 M_\odot$, they are high values as $M_{\rm thr} \gtrsim 2.7 M_\odot$ and $M_{\rm max,spin} \gtrsim 2.4 M_\odot$. For $m_0 > M_{\rm thr}$, a black hole is formed in the dynamical time scale after the onset of the merger, and for the nearly equal-mass case, $m_1 \approx m_2$, the mass of the disk surrounding the black hole is tiny $\ll 10^{-2} M_\odot$, while it could be $\gtrsim 10^{-2} M_\odot$ for a highly asymmetric system with $m_1/m_2 \gtrsim 1.1$. For $M_{\rm max,spin} < m_0 < M_{\rm thr}$, a hypermassive neutron star is formed, and it subsequently evolves due to the loss of the angular momentum through some angular-momentum transport processes and the gravitational-wave emission, leading to eventual collapse to a black hole surrounded by a disk or torus. When m_0 is close to $M_{\rm thr}$, the lifetime of the hypermassive neutron star is relatively short, while for smaller values of m_0 toward $M_{\rm max,spin}$, the lifetime is longer. For the longer lifetime, the angular-momentum transport process works for a longer time scale, and the mass of the disk/torus could be $\gtrsim 0.1 M_\odot$, whereas for a short lifetime, it could be $\sim 10^{-2} M_\odot$ or less. For $m_0 < M_{\rm max,spin}$, a supramassive neutron star is formed and it will be alive for a dissipation time scale of the angular momentum which will be much longer than the cooling time scale $\sim 1\,{\rm s}$ (see section 1.4.6). HMNS and BH denote a hypermassive neutron star and a black hole, respectively. MNS denotes either a supramassive neutron star or a normal neutron star.

for a given hypothetical equation of state, the evolution path depends on the total mass and mass ratio of the binary system.

In this figure, $M_{\rm thr}$ denotes the threshold mass for the prompt formation of a black hole. This mass depends strongly on the equation of state, but for stiff equations of state that give $M_{\rm max,0} \gtrsim 2.0 M_\odot$, numerical-relativity results have shown that $M_{\rm thr} \gtrsim 2.7 M_\odot$ and $M_{\rm max,spin} \gtrsim 2.4 M_\odot$. For $m_0 > M_{\rm thr}$, a black hole is formed in the dynamical time scale ($\lesssim 1\,{\rm ms}$) after the onset of the merger, and for the nearly equal-mass case, $m_1 \approx m_2$, the mass of the disk/torus surrounding the black hole is tiny $\ll 10^{-2} M_\odot$, while it could be $\gtrsim 10^{-2} M_\odot$ for a highly asymmetric system with $m_1/m_2 \gtrsim 1.1$. For $M_{\rm max,spin} < m_0 < M_{\rm thr}$, a hypermassive neutron star is formed, subsequently evolves due to the loss of the angular momentum through some angular-momentum transport processes and the gravitational-wave emission, and eventually collapses to a black hole surrounded by a disk/torus. When m_0 is close to $M_{\rm thr}$, the lifetime of the hypermassive neutron star is relatively short, while for smaller values of m_0, the lifetime is longer. In particular for m_0 close to $M_{\rm max,spin}$, the collapse to a black hole may occur in a neutrino cooling time scale of order seconds. For the longer lifetime of the hypermassive neutron stars, the angular-momentum transport process works for a longer time scale, and the mass of the disk/torus could be $\gtrsim 0.1 M_\odot$, whereas for a short lifetime, it could be $\sim 10^{-2} M_\odot$ or less. For $m_0 < M_{\rm max,spin}$, a supramassive neutron star is formed and it will be alive for a dissipation time scale of angular momentum, which will be much longer than the cooling time scale $\sim 1\,{\rm s}$ (see section 1.4.6), although it should eventually collapse to a slowly spinning black hole.

8.3.4 *Gravitational waves*

The feature of gravitational waves emitted by coalescing binary neutron stars is summarized as follows.

In their inspiral stage in which the orbital separation gradually decreases with time, the so-called chirp signal is emitted. For the early inspiral stage in which the orbital separation is much larger than the neutron-star radii, i.e., each neutron star is well approximated by a point particle, and also the general relativistic two-body interaction is not very strong, the chirp waveform is accurately calculated by post-Newtonian formalisms (see appendix H). For the late inspiral stage in which the orbital separation is smaller than ~ 5 times of the neutron-star radii, tidal-deformation effects of each neutron star modify the orbital motion as already described in section 8.2.3 (see also appendix H). For such a stage, the gravitational waveform is still of a chirp-signal type but the tidal-deformation effects have to be taken into account for accurately computing the gravitational waveforms. For this purpose, a numerical-relativity simulation is necessary: We will describe the feature of gravitational waves in this stage in section 8.3.4.1.

In the merger stage, we have two possibilities; a black hole is promptly formed soon after the onset of the merger or a massive neutron star is formed. For the former case, the merger waveform is characterized universally by a ringdown waveform as illustrated in figure 1.25 and section 8.3.4.2, irrespective of the total mass and mass ratio of the binary and equations of state of the neutron stars. For the latter case, quasi-periodic gravitational waves are emitted from the remnant massive neutron stars that are rapidly rotating and have a non-axisymmetric structure, as illustrated in figure 1.26. Gravitational waves from the remnant massive neutron stars appreciably reflect the equation of state of the neutron stars that is still poorly understood. This implies that the detection of these gravitational waves will provide us an invaluable opportunity for constraining the equation of state, and therefore, clarifying the dependence of the gravitational waveform on the equations of state is one of the most important subjects in numerical relativity. In section 8.3.4.3, we will summarize the results of the numerical-relativity simulations on this topic in detail.

8.3.4.1 Late inspiral stage

The tidal effects on the late inspiral waveforms have been explored by Baiotti et al. (2010, 2011); Bernuzzi et al. (2012b,a); Hotokezaka et al. (2013c) in numerical-relativity simulations. In the following, we will summarize their finding.

As far as it is not extremely large, the degree of the tidal deformation of inspiraling neutron stars is characterized simply by a quadrupole moment induced by the tidal field of their companion stars. Here, the magnitude of the quadrupole moment is proportional to the strength of the (lowest-order) tidal field, and the coefficient of the proportionality is called the tidal deformability (see also appendix H). For a weakly tidal-deformation stage, the corrections to the orbital motion and gravitational waveforms by the tidal effect can be described only by this tidal deformability.

The tidal deformability is proportional to R_{NS}^5 where R_{NS} denotes the neutron-star radius. Thus, it depends strongly on the equation of state of the neutron star. This implies that the orbital motion in the late inspiral stage reflects the equation of state of the neutron star through the degree of the tidal deformation: For neutron stars with a larger radius (i.e., for stiffer equations of state), the effect of the tidal deformation is more appreciable.

The adiabatic evolution of the inspiraling orbit due to the gravitational-radiation reaction is accurately determined in a post-Newtonian approximation (see appendix H) for the case that the orbital separation is sufficiently large. For example, the Taylor T4 approximant, described in equation (H.20), is a well-known robust formalism for accurately determining the orbital motion of binaries composed of nearly equal-mass non-spinning objects [Boyle et al. (2007)]. In the Taylor T4 approximant (and other similar approximants), the evolution of $x(t) := [m_0 \Omega_{\rm orb}(t)]^{2/3}$ is determined by integrating an ordinary differential equation for $dx(t)/dt = F[x]$ where F is a function of x. The corrections (up to the first post-Newtonian order in the tidal effect) associated with the tidal deformation were derived by Vines et al. (2011) [see equation (H.21)] and it is found that the corrections in the source term for the evolution equation of $x(t)$ are proportional to the tidal deformability and thus positive definite; it accelerates the evolution of $\Omega_{\rm orb}$. The correction terms start from the fifth post-Newtonian order in the viewpoint of the post-Newtonian approximation, and thus, one might think that the effect should be quite small. However, because the coefficient could be quite large [which is proportional to the tidal deformability, and hence, of order $(R_{NS}/M_{NS})^5$, where M_{NS} is the neutron-star mass], such a high-order correction could play an important role in close orbits. Indeed, the evolution of $\Omega_{\rm orb}$ could be significantly accelerated in the presence of the tidal deformation in particular for the last few

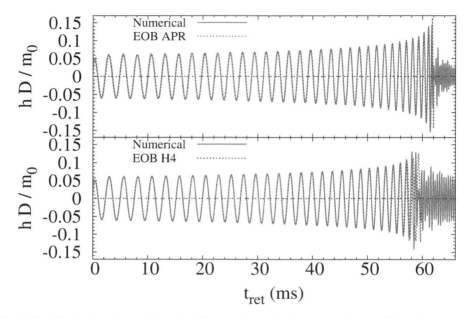

Fig. 8.15 Gravitational waveforms (solid curves) in the late inspiral stage (through the early merger stage) of binary neutron stars with $m_1 = m_2 = 1.35 M_\odot (= m_0/2)$ for APR4 (upper panel) and H4 (lower panel) equations of state. The vertical axis shows hD/m_0 where h and D are the gravitational-wave amplitude observed along a direction perpendicular to the orbital plane and the hypothetical distance from the gravitational-wave source. The horizontal axis shows an appropriately-defined retarded time. The dotted curves show the results by an effective-one-body formalism (see appendix H). For the comparison of the numerical waveforms with the waveforms by the effective one-body formalism, the gravitational waveforms for $5\,\mathrm{ms} \leq t_\mathrm{ret} \leq 20\,\mathrm{ms}$ are matched. Note that for $t_\mathrm{ret} = 10\,\mathrm{ms}$, the gravitational-wave frequency is $\approx 395\,\mathrm{Hz}$ for both cases.

orbits, and the acceleration rate is higher for the larger neutron-star radius (for stiffer equations of state).

The property indicated by the post-Newtonian analysis is based on the assumption that the degree of the tidal deformation is not very large. In reality, the neutron stars in close orbits could be highly deformed. Thus, the prediction by the post-Newtonian calculations has to be calibrated by comparing it with the result derived by an accurate numerical-relativity simulation, in which fully non-linear effects on the tidal deformation are taken into account. Such comparison was performed by Baiotti *et al.* (2010, 2011); Bernuzzi *et al.* (2012b,a); Hotokezaka *et al.* (2013c). In particular, Bernuzzi *et al.* (2012b,a) and Hotokezaka *et al.* (2013c) performed the first detailed analysis of their numerical waveforms. Specifically, they performed careful extrapolation of the numerical waveforms to the continuous limit ($\Delta x \to 0$ where Δx is the grid spacing), and compared the extrapolated results with the predictions by the post-Newtonian and effective one-body calculations [see, e.g., Buonanno and Damour (1999); Damour and Nagar (2009, 2010); Bini

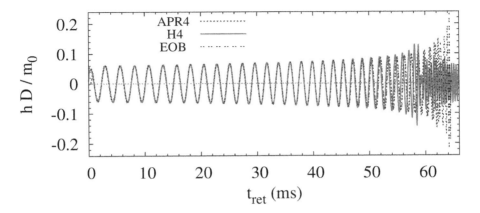

Fig. 8.16 The same as figure 8.15 but for gravitational waves from binary neutron stars with $m_1 = m_2 = 1.35 M_\odot$ for APR4 (dotted curve) and H4 (solid curve) equations of state together with a gravitational waveform calculated by an effective one-body formalism for binary black holes (dot-dot curve).

et al. (2012) for the effective one-body formulations]. Then, they found that these semi-analytic predictions are accurate up to a few orbits before the onset of the merger. Therefore, for most of the inspiral orbits, the semi-analytic formalisms can yield an accurate gravitational waveform.

Figure 8.15 displays an unpublished result of gravitational waveforms derived by a long-term numerical-relativity simulation by Hotokezaka, Kyutoku, and Shibata. In this simulation, they employed equal-mass binary neutron stars with $m_1 = m_2 = 1.35 M_\odot$ for APR4 and H4 equations of state (see section 1.4.4 for these equations of state). The gravitational waveforms were determined by an extrapolation procedure using numerical results with several different grid resolutions, and the estimated error of the resulting extrapolated waveforms is $\sim 1\%$ in amplitude and within $\sim 1\,\mathrm{radian}$ in phase during $\approx 15\text{--}16$ orbits ($\approx 30\text{--}32$ gravitational-wave cycles). The numerical waveforms are compared with the results by an effective one-body formalism [Bini et al. (2012)] (see appendix H). For the comparison of the numerical waveforms with the waveforms by the effective one-body formalism, a cross correlation between two gravitational waveforms for $5\,\mathrm{ms} \leq t_{\mathrm{ret}} \leq 20\,\mathrm{ms}$ is calculated, and the best-matched time range is determined to compare the two waveforms. Figure 8.15 indeed illustrates that the results by the effective one-body approach agree well with the numerical waveforms up to a few orbits before the onset of the merger. In particular, for the more compact case (for the APR4 equation of state), the agreement is quite good. (We remind the reader that the stellar radii of APR4 and H4 for $m_1 = m_2 = 1.35 M_\odot$ are 11.1 and 13.6 km, respectively.)[3]

---

[3] We note that for the less compact case (for H4 equation of state), the agreement between numerical and effective one-body waveforms in the last cycle becomes poor: For the numerical result, the phase evolution is faster. This would be due to the fact that in the employed version of

Figure 8.16 plots gravitational waves from binary neutron stars with $m_1 = m_2 = 1.35 M_\odot$ for APR4 and H4 equations of state, together with a gravitational waveform calculated by an effective one-body formalism for binary black holes (i.e., a gravitational waveform in the absence of tidal deformation). This figure shows that for binaries composed of more compact neutron stars (i.e., for APR4), gravitational waves in the inspiral phase continue for a longer duration, while for binaries composed of less-compact binaries, the inspiral stage ends at earlier time. Comparing these waveforms with the result by the binary black hole case, the duration in the inspiral stage for H4 equation of state is by ~ 6 ms shorter than that for the binary black hole, while for APR4, the overlap with the binary-black-hole result is fairly good even for the late inspiral stage up to a few orbits before the onset of the merger. It is also found that, as predicted by the post-Newtonian analysis, the frequency of gravitational waves increases more rapidly for binaries composed of less-compact neutron stars; the deviation from the waveform in the binary-black-hole case occurs for an earlier inspiral stage.

Figure 8.16 also indicates the fact that for binaries of less-compact neutron stars, the maximum gravitational-wave amplitude (i.e., the amplitude at the onset of the merger) is smaller [Read *et al.* (2013)]. Thus, if the maximum amplitude is measured in a gravitational-wave observation, we will be also able to use this information for constraining the equation of state.

8.3.4.2 *Gravitational waves in the prompt black hole formation*

Figure 8.17 plots a typical gravitational waveform for the case that a black hole is promptly formed after the onset of the merger. In addition to the waveform, the gravitational-wave frequency as a function of the retarded time is shown in the bottom panel. Here, the frequency is determined by extracting the phase of the outgoing part of the complex Weyl scalar, $\arg(\Psi_4)$, and by taking the time derivative as $2\pi f := d[\arg(\Psi_4)]/dt$.

It is found that the gravitational waveform is composed of the inspiral, short-term merger, and ringdown waveforms for this case. Before the ringdown waveform is excited, a merger waveform is emitted, but the duration for this wave emission is quite short $\lesssim 1$ ms. These are the universal features for the gravitational waveform in the prompt formation of a black hole.

The gravitational-wave frequency monotonically increases with time. During the inspiral stage, the frequency is smaller than ~ 1 kHz and it increases adiabatically with the decrease of the orbital separation. In the merger stage, the frequency increases steeply from ~ 1 kHz to ~ 6 kHz. This reflects the fact that the compactness of the system steeply increases during the collapse of the merged object to a black hole. Finally, ringdown gravitational waves are emitted, and the gravitational-

---

the effective one-body formalism, any effect except for the lowest-order tidal deformation effect is not taken into account, and hence, for the significantly deformed case, the approximation would be poor. Overcoming this weak point is an issue in the future.

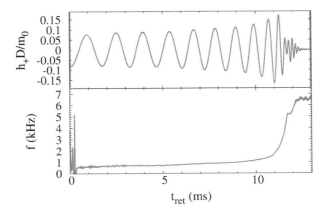

Fig. 8.17 Gravitational waves (of plus mode) and their frequency for coalescing binary neutron stars leading to prompt formation of a black hole for $(m_1, m_2) = (1.6M_\odot, 1.3M_\odot)$ with APR4 equation of state. Here, gravitational waves shown are those observed along the rotational axis which is perpendicular to the binary orbital plane, and the vertical axis is $h_+ D/m_0$ where D is the distance from the source and $m_0 = 2.9M_\odot$. The bumps found in the early stage of the frequency evolution curve are artifacts caused by the initial relaxation of the numerical system. The figure is taken from Hotokezaka et al. (2013b).

wave frequency eventually reaches a high value which characterizes the ringdown waveform. This ringdown gravitational waveform is determined primarily by the fundamental quasi-normal mode of the formed black hole [see equation (1.203)]. For the model shown in figure 8.17, the real-part frequency of the quasi-normal mode is $\approx 6.54\,\mathrm{kHz}$. Since the mass and spin of the formed black holes for this case are $\approx 2.81 M_\odot$ and ≈ 0.77, respectively (which are derived by analyzing the circumferential lengths of the apparent horizon; see section 7.3), the frequency of these ringdown gravitational waves agrees approximately with the frequency of the quasi-normal mode, analytically derived using equation (1.203). This is also the universal feature found in the simulations in which a black hole is formed as the final remnant.

8.3.4.3 Gravitational waves from remnant massive neutron stars

As already described in section 8.3.3.2, massive neutron stars are likely to be often formed after the onset of the merger of binary neutron stars in nature. The remnant massive neutron stars are in general rapidly rotating and have a non-axisymmetric configuration, and hence, they are the strong emitters of quasi-periodic gravitational waves. Such gravitational waves have been studied using a variety of nuclear-theory-based equations of state in general relativity by Shibata et al. (2005); Shibata and Taniguchi (2006); Kiuchi et al. (2009, 2010); Sekiguchi et al. (2011b,a); Hotokezaka et al. (2011, 2013a); Takami et al. (2014) [see also the works by Bauswein and Janka (2012); Bauswein et al. (2012) which were performed in an approximate framework of general relativity]. These works have clarified that irrespective of hypothetically employed equations of state, the remnant massive neutron stars emit quasi-periodic gravitational waves for a long duration of $O(10\,\text{ms})$. They have also found that the characteristic frequency of gravitational waves and the gravitational waveform depend on the equations of state employed. This fact suggests that if these quasi-periodic gravitational waves are detected, they will be used for constraining the equation of state for the neutron-star matter that is still poorly understood. Thus, in the following, we will summarize the dependence of gravitational waveforms on the hypothetical equations of state, based on the results in the numerical-relativity simulations.

Figures 8.18 and 8.19 display the plus-mode gravitational waves, h_+, and the corresponding frequency of gravitational waves emitted by remnant massive neutron stars for four representative hypothetical equations of state (APR4, ALF2, H4, and MS1) and several binary masses. Note that for the cross mode, essentially the same waveforms are obtained. In the left and right columns of figure 8.18, waveforms for equal-mass and unequal-mass binaries with $m_0 = 2.7 M_\odot$ are shown, respectively. In the left and right columns of figure 8.19, waveforms for equal-mass binaries with $m_0 = 2.6 M_\odot$ and $2.8 M_\odot$ are shown, respectively. Below, we will describe the properties for the evolution curves of the amplitude and frequency of these gravitational waveforms separately.

Amplitude: Broadly speaking, the amplitude of quasi-periodic gravitational waves emitted by remnant massive neutron stars decreases with time because the angular momentum for them is lost by the hydrodynamical angular-momentum transport process and gravitational-wave emission, and as a result, the degree of their non-axisymmetric deformation is lowered. However, the way of its time variation depends on the hypothetically employed equation of state and mass ratio of the binary.

There are two patterns for the damping process of the gravitational-wave amplitude. (i) For equal-mass or approximately equal-mass (weakly asymmetric) binaries, the amplitude decreases approximately monotonically with time (besides small modulation) irrespective of the equation of state employed, although the damping time scale depends on the equation of state: For relatively soft equations of state (that yield small-radius neutron stars, e.g., for APR4 in figures 8.18 and 8.19), the damping time scale is relatively short $\sim 10\,\text{ms}$, while for stiff equations of state (e.g., for H4 and MS1), the damping time scale is longer. This reflects the fact that for the softer equations of state, remnant massive neutron stars more quickly lose their angular momentum by the hydrodynamical angular-momentum transport and gravitational-wave emission.

(ii) For unequal-mass binaries with a sufficiently high asymmetry, e.g., $m_1/m_2 \gtrsim 1.2$, the gravitational-wave amplitude damps with a unique modulation as found from the right column of figure 8.18. The origin of this modulation is closely related to the evolution process of the remnant massive neutron stars, in particular, the

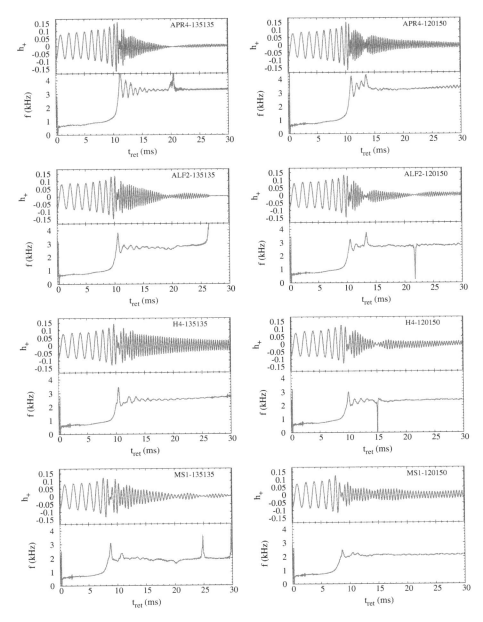

Fig. 8.18 Gravitational waves (plus mode; $h_+ D/m_0$) and the frequency of gravitational waves f (kHz) as functions of a retarded time for models $m_1 = m_2 = 1.35 M_\odot$ (left column) and $(m_1, m_2) = (1.50 M_\odot, 1.20 M_\odot)$ (right column) with APR4 (top row), ALF2 (second row), H4 (third row), and MS1 (bottom row). Here, $m_0 = 2.7 M_\odot$. For the equal-mass model with ALF2, a black hole is formed at $t_{\rm ret} \approx 27$ ms. For all the models, the piecewise polytropic equation of state with $\Gamma_{\rm th} = 1.8$ is employed [see equation (8.16)]. The figure is taken from Hotokezaka et al. (2013a).

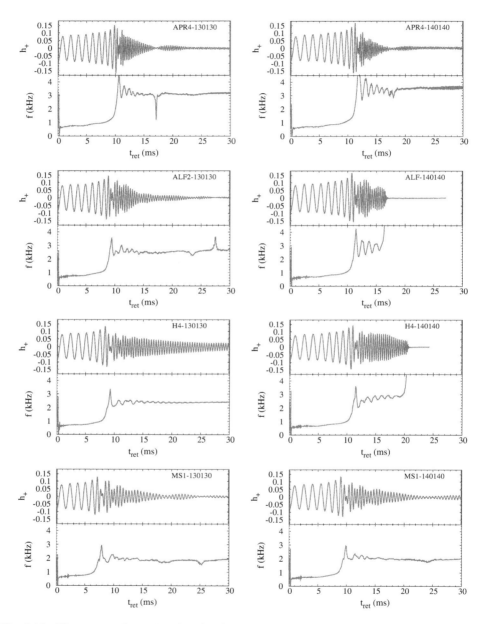

Fig. 8.19 The same as figure 8.18 but for the equal-mass models with $m_1 = m_2 = 1.3M_\odot$ (left column) and $m_1 = m_2 = 1.4M_\odot$ (right column) and with APR4 (top row), ALF2 (second row), H4 (third row), and MS1 (bottom row). For ALF2 and H4 with $m_1 = m_2 = 1.4M_\odot$, a black hole is formed at $t_{\rm ret} \approx 17$ and 21 ms, respectively. For all the models, the piecewise polytropic equation of state with $\Gamma_{\rm th} = 1.8$ is employed [see equation (8.16)]. The figure is taken from Hotokezaka et al. (2013a).

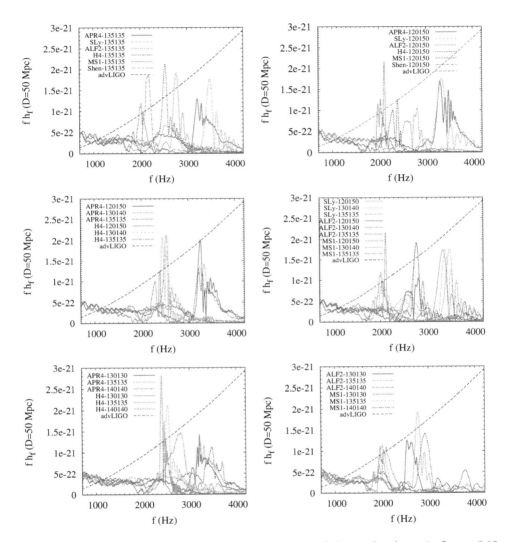

Fig. 8.20 Fourier spectra of gravitational waves for some of the results shown in figures 8.18 and 8.19. h_f in this figure denotes $|\tilde{h}(f)|$. Top left: For $m_1 = m_2 = 1.35M_\odot$ with $\Gamma_{\rm th} = 1.8$ and with five piecewise polytropic (APR4, ALF2, H4, MS1, SLy) and Shen equations of state. Top right: The same as top left but for $m_1 = 1.5M_\odot$ and $m_2 = 1.2M_\odot$. Middle left: For three mass ratios with $m_0 = 2.7M_\odot$, $\Gamma_{\rm th} = 1.8$, and with APR4 and H4. Middle right: The same as middle left but for ALF2, MS1, and SLy. Bottom left: For equal-mass models with $m_0 = 2.6, 2.7,$ and $2.8M_\odot$, $\Gamma_{\rm th} = 1.8$, and with APR4 and H4. Bottom right: The same as bottom left but for ALF2 and MS1. The amplitude is shown for a hypothetical event at a distance of $D = 50$ Mpc along the direction perpendicular to the orbital plane (the most optimistic direction). The black dot-dot curve is a planned noise spectrum of the advanced LIGO with a configuration for the detection of gravitational waves from inspiraling binary neutron stars (see https://dcc.ligo.org/cgi-bin/DocDB/ShowDocument?docid=2974). The figure is taken from Hotokezaka *et al.* (2013a).

dynamical evolution of a less-massive core resulting from the less-massive binary companion: During the early stage of their evolution, the central region of the remnant massive neutron stars appears to be composed of a massive core and a deformed satellite for which the shape dynamically varies (see section 8.3.3.2). Here, the massive core and satellite stem from the massive and less-massive neutron stars of the binary, respectively. Two asymmetric cores rotate around the center of mass, and high-amplitude quasi-periodic gravitational waves are emitted for $\sim 3-5\,\mathrm{ms}$ after the onset of the merger. Then, the amplitude once damps to be very small at a moment at which the satellite core has a highly elongated shape surrounding the massive core; not double cores but a nearly spheroidal core is transiently formed (see the third panel of figure 8.9). Subsequently, long-term quasi-periodic gravitational waves are again emitted when a quasi-steady binary-like structure composed of two asymmetric cores is recovered. This feature of gravitational waves is clearly seen for APR4, ALF2, and H4 in figure 8.18.

Frequency: As already mentioned, the frequency of gravitational waves emitted by remnant massive neutron stars is approximately constant (see figures 8.18 and 8.19). The exception could occur for the very early stage just after the formation of the remnant massive neutron stars in which the frequency oscillates with a dynamical time scale (this can be observed for all the models to a greater or lesser degree) and/or for the stage just before the formation of a black hole in which the frequency increases steeply with time (see the results with labels ALF2-135135, ALF2-140140, H4-135135, and H4-140140). These qualitative features hold irrespective of the equation of state employed.

Figure 8.20 shows the Fourier spectra for some of gravitational waveforms displayed in figures 8.18 and 8.19 (as well as for SLy and Shen equations of state). Here, plotted is the effective amplitude defined by $|f\tilde{h}(f)|$ as a function of f, where $\tilde{h}(f)$ is the Fourier spectrum of $h_+ - ih_\times$ with $h_\times$ the cross mode [see equation (1.197)], for a hypothetical distance of $D = 50\,\mathrm{Mpc}$. This shows that there are indeed characteristic peaks at $2\,\mathrm{kHz} \lesssim f \lesssim 4\,\mathrm{kHz}$ irrespective of the models with different equations of state. For relatively soft equations of state for which $R_{1.35} \lesssim 12\,\mathrm{km}$ (i.e., for APR4 and SLy in these examples), the characteristic peak frequency is relatively high with $f \gtrsim 3\,\mathrm{kHz}$, while for stiff equations of state (H4, MS1, and Shen for which $R_{1.35} \gtrsim 13.5\,\mathrm{km}$), it is lower as $f \sim 2-2.5\,\mathrm{kHz}$.

Because the spin angular velocity of the remnant massive neutron stars is close to the Keplerian velocity, the characteristic peak frequency of gravitational waves is qualitatively proportional to $(M_{\mathrm{MNS}}/R_{\mathrm{MNS}}^3)^{1/2}$ where M_{MNS} and R_{MNS} denote the typical mass and radius of the remnant massive neutron stars. This is the reason that for softer equations of state (i.e., for smaller values of R_{MNS}), the characteristic frequency is higher. Due to the same reason, the characteristic frequency is higher for the lower values of Γ_{th} for the case that the piecewise polytropic equations of state are employed, because, for this case, the effect of shock heating is weaker and the remnant massive neutron stars become more compact.

Hotokezaka et al. (2013a) reported that a certain correlation exists between the characteristic frequency and a stellar radius [see also Bauswein and Janka (2012); Bauswein et al. (2012) for the original proposal]. Figure 8.21 shows the frequency of the Fourier spectrum peaks as a function of the neutron-star radius of mass $1.8M_\odot$ for six employed equations of state (denoted by $R_{1.8}$ in units of km) for $m_0 = 2.7M_\odot$ (left panel) and $2.6M_\odot$ (right panel). Here, for some Fourier spectrum, there are several peaks among which the difference in the spectrum amplitude is less than 20%. For such cases, the several values for the frequency are plotted together to indicate a possible dispersion of the characteristic frequency. The dotted curves are

$$f = (4.0 \pm 0.3)\,\text{kHz}\left(\frac{(R_{1.8}/\text{km}) - 2}{8}\right)^{-3/2}, \qquad (8.23)$$

for the left panel and

$$f = (3.8 \pm 0.2)\,\text{kHz}\left(\frac{(R_{1.8}/\text{km}) - 2}{8}\right)^{-3/2}, \qquad (8.24)$$

for the right panel. These curves approximately show the upper and lower limits of the characteristic frequency for a given value of $R_{1.8}$. The subtraction factor of -2 for $R_{1.8}$ in these curves is empirically needed to capture the upper and lower limits for $R_{1.8} \lesssim 11\,\text{km}$. This would be due to the fact that general relativistic corrections come into play for the small values of $R_{1.8}$. The reason why $R_{1.8}$ is chosen is simply that Hotokezaka et al. (2013a) found it a relatively good choice to get a clear correlation between the peak frequency and a neutron-star radius [see also Bauswein and Janka (2012); Bauswein et al. (2012) for other choice].

Figure 8.21 shows that a clear correlation indeed exists. In particular for the lower-mass ($m_0 = 2.6M_\odot$) models, the dispersion is small. This figure suggests that if the characteristic peak frequency can be determined accurately in the gravitational-wave detection, we will be able to constrain the radius of a high-mass neutron star with the uncertainty of $\sim 1\,\text{km}$. However, it is also found that it is not easy to reduce the estimation error to $\ll 1\,\text{km}$, because of the presence of the dispersion of the peak frequency.

The origin of the dispersion in the peak frequency is explained as follows: As already mentioned, the (non-axisymmetric) oscillation frequency of remnant massive neutron stars vary during the early stage of their evolution due to a quasi-radial oscillation and angular-momentum dissipation processes such as gravitational-radiation reaction and hydrodynamical angular-momentum transport. This results in the time variation of the major frequency and hence in the broadening of the peak or appearance of the multi peaks (see the spectra, e.g., for APR4-135135, H4-120150, ALF2-130130, and ALF2-120150). This broadening is not very large for particular models such as equal-mass models of some equations of state (see a discussion in the last paragraph of this section), and thus, for these particular cases, the characteristic frequency may be determined with a small dispersion. However, in general, the broadening, Δf, is $\sim 10\%$ of the peak frequencies, i.e., is

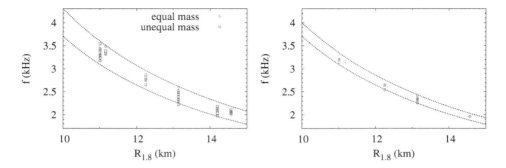

Fig. 8.21 The frequency of the Fourier spectrum peak as a function of a neutron-star radius of mass $1.8 M_\odot$ with several hypothetical equations of state for $m_0 = 2.7 M_\odot$ (left) and $2.6 M_\odot$ (right). The figure is taken from Hotokezaka et al. (2013a).

$\Delta f \sim \pm 0.2$–0.3 kHz [Hotokezaka et al. (2013b,a)]. This magnitude of the dispersion agrees approximately with the dispersion found in figure 8.21.

It should be also mentioned that there are several interesting properties found in the Fourier spectrum. The first one is that the peak frequency varies with the mass ratio even for the same total mass m_0 and that the way of this variation depends on the equation of state employed. For APR4, SLy, and ALF2, the frequency depends only weakly on the mass ratio. For H4 and Shen, the frequency is lower for the higher values of $Q = m_1/m_2 \geq 1$, i.e., for more asymmetric systems. By contrast, for MS1, the frequency tends to be higher for more asymmetric systems. This property causes the dispersion in the relation between the peak frequency and $R_{1.8}$, as displayed in figure 8.21. The second one is that the peak amplitude of the Fourier spectrum is higher for the symmetric binaries for ALF2, H4, and Shen, while it is higher for the asymmetric binaries for APR4 and MS1 (see figure 8.20). In other words, for ALF2, H4, and Shen, the spectrum peak is sharper for the equal-mass binaries, while for APR4 and MS1, it is sharper for the asymmetric binaries. These properties will be used for constraining the equation of state if the frequency and shape of the spectrum peaks are determined for a variety of merger events.

8.3.4.4 Overall spectrum shapes

As already mentioned in section 8.3.4.3, the Fourier spectrum reflects the properties of merger remnants. Here, we summarize the general features of the shapes of the gravitational-wave spectrum.

For the case that a black hole is formed promptly after the merger, the spectrum is quite simple, like figure 8.22 and the upper panel of figure 8.23 [Kiuchi et al. (2009, 2010); Hotokezaka et al. (2011)]. During the inspiral stage, the amplitude decreases with the increase of the frequency in proportional to $f^{-1/6}$ for a low-frequency band [i.e., for distant binary inspiral orbits; see equation (1.199)], and in proportional to $f^{-\alpha_f}$ with $\alpha_f \gtrsim 1/6$ for a high-frequency band (i.e., for close binary inspiral orbits).

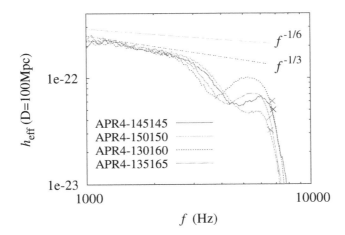

Fig. 8.22 Fourier spectra (effective amplitude; $|f\tilde{h}(f)|$) of gravitational waves emitted by binary neutron star mergers that form a black hole promptly after the onset of the merger. The equation of state employed in this example is APR4 and each mass in binary is $(m_1, m_2) = (1.45 M_\odot, 1.45 M_\odot)$, $(1.50 M_\odot, 1.50 M_\odot)$, $(1.60 M_\odot, 1.30 M_\odot)$, and $(1.65 M_\odot, 1.35 M_\odot)$. The cross symbols denote the quasi-normal mode frequency ($\sim 6.5\,\mathrm{kHz}$) for the formed black holes. The figure is taken from Kiuchi et al. (2009).

Even for the merger stage in which $f \gtrsim 1\,\mathrm{kHz}$, the amplitude approximately obeys the power law with $1/6 \lesssim \alpha_f \lesssim 1/3$ for $f \lesssim 3\,\mathrm{kHz}$. This is due to the fact that after the onset of the merger, the merged object forms a double-core structure that emits gravitational waves similar to inspiral waves. However, for $f \gtrsim 3\,\mathrm{kHz}$, the amplitude steeply damps and a dip is seen at $f \sim 3$–$5\,\mathrm{kHz}$. This is due to the fact that the double-core structure disappears in the merger remnant, and changes to a spheroidal object for which the gravitational-wave emissivity decreases. For $\sim 5\,\mathrm{kHz} \lesssim f \lesssim 7\,\mathrm{kHz}$, a bump associated with ringdown gravitational waves then appears, and for $f > f_{\mathrm{QNM}}$ where f_{QNM} is the frequency of a quasi-normal mode of the formed black hole ≈ 6–$7\,\mathrm{kHz}$, the amplitude damps exponentially. Kiuchi et al. (2010) explored this type of gravitational waves changing the total mass, binary mass ratio, and hypothetically employed equations of state, and found that the spectrum shape is universal, although the frequency at the dip and bump, and the width and amplitude of the bump depend on the mass and equations of state employed.

For the case that a remnant massive neutron star is formed after the merger of binary neutron stars, the spectrum is not as simple as that for the prompt black-hole formation case, as already described in section 8.3.4.3. For this case, several peaks including the major peak appear in particular in the frequency range, 2–4 kHz, associated with quasi-periodic gravitational waves emitted by remnant massive neutron stars, and the spectrum shape schematically becomes like the bottom panel of figure 8.23. As shown in section 8.3.4.3, the height, width, and frequency of the peaks

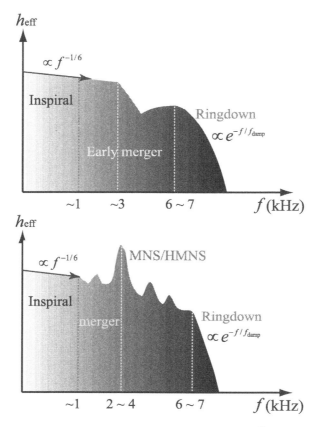

Fig. 8.23 Summary of the Fourier spectrum (effective amplitude, $|f\tilde{h}(f)|$) of gravitational waves emitted by binary neutron stars. The upper and lower panels show the schematic spectra for the cases that a black hole is formed promptly after the onset of the merger and that a massive neutron star is formed. MNS and HMNS denote the primary peaks generated by gravitational waves from a massive neutron star and/or a hypermassive neutron star, respectively.

depend strongly on the total mass, binary mass ratio, and hypothetically employed equation of state. However, the shape is qualitatively universal.

8.3.4.5 Modeling gravitational waveforms from massive neutron stars

Gravitational waves emitted by remnant massive neutron stars formed after the merger have a universal feature irrespective of the mass of binary components and hypothetical equations of state, as depicted in figure 8.23. This implies that it may be feasible to model these gravitational waveforms by a universal fitting formula in terms of a particular functional form. As shown in figures 8.18 and 8.19, the gravitational waveforms are characterized universally by a quasi-periodic oscillation with a gradual damping of the amplitude and with certain variation of the characteristic frequency. The damping of the amplitude may be described approximately by a steeply decreasing function plus a slowly decreasing term (or a constant). The time variation of the frequency is described by a slowly varying function plus a damping oscillation around an average one in an early stage of the remnant massive neutron stars.

Based on these considerations, Hotokezaka et al. (2013a) proposed the following fitting formula for $h := h_+ - ih_\times$,

$$h_{\text{fit}}(t) := A(t)\exp[2\pi i f(t)], \tag{8.25}$$

where

$$A(t) = \left[a_1 e^{-(t-t_i)/a_d} + a_0\right] \frac{1-\exp\{-(t-t_i)/a_{\text{ci}}\}}{1+\exp\{(t-a_{\text{co}})/t_{\text{cut}}\}}, \tag{8.26}$$

$$f(t) = p_0 + p_1(t-t_i) + p_2(t-t_i)^2 + p_3(t-t_i)^3$$
$$+ p_4 e^{-(t-t_i)/p_d}\left[p_c\cos\{p_f(t-t_i)\} + p_s\sin\{p_f(t-t_i)\}\right]. \tag{8.27}$$

Here, t_i is an initial time, i.e., only for $t \geq t_i$, the fitting is considered. Hotokezaka et al. (2013a) chose it the time at which the peak gravitational-wave amplitude is reached. Because high-amplitude gravitational waves are emitted from remnant massive neutron stars for a time scale of ~ 10 ms irrespective of the equations of state employed, Hotokezaka et al. (2013a) considered the fitting only for $t_i \leq t \leq t_f$ where $t_f = t_i + 10$ ms.

As equation (8.26) shows, the fitting function for the amplitude is modeled by three parts: (i) $a_1 e^{-(t-t_i)/a_d} + a_0$ denotes the evolution of the amplitude, which is composed of an exponentially damping term and a constant term. a_0, a_1, and a_d are free parameters that should be determined by a fitting procedure: (ii) $[1+\exp\{(t-a_{\text{co}})/t_{\text{cut}}\}]^{-1}$ denotes a cutoff term, which specifies a time interval of high-amplitude stage with $a_{\text{co}} < 10$ ms being the center of the cutoff time and t_{cut} being the time scale for the shutdown. a_{co} should be determined in the fitting process while t_{cut} is a free parameter that should be arbitrarily chosen as far as it is short enough; Hotokezaka et al. (2013a) chose it to be 0.1 ms: (iii) $1 - \exp[-(t-t_i)/a_{\text{ci}}]$ is a steeply increasing function for $t \gtrsim t_i$ with a_{ci} being the growth time scale. The reason for introducing this term is that at a moment of $t \gtrsim t_i$, the amplitude of gravitational waves is universally low, but after this moment, the amplitude steeply increases (see figures 8.18 and 8.19).

Equation (8.27) defines that the fitting function for the frequency is modeled by a secularly evolving term $p_0+p_1(t-t_i)+p_2(t-t_i)^2+p_3(t-t_i)^3$ and a damping oscillation term $p_4 e^{-(t-t_i)/p_d}\left[p_c\cos\{p_f(t-t_i)\} + p_s\sin\{p_f(t-t_i)\}\right]$. Here, eight constants p_0, p_1, p_2, p_3, p_4, p_d, p_f, and p_s are free parameters that should be determined by a fitting procedure. Thus, in total, there are 13 parameters to be determined in their model.

The parameters should be chosen so as to reproduce each numerical waveform, h_{NR}, as accurately as possible. To determine the parameter set, Hotokezaka et al.

(2013a) considered to maximize the following function,

$$C(h_{\rm fit}) := -\mathscr{M}(h_{\rm fit}) + \frac{[(h_{\rm NR}, h_{\rm NR}) - (h_{\rm fit}, h_{\rm fit})]^2}{(h_{\rm NR}, h_{\rm NR})^2}, \quad (8.28)$$

where

$$\mathscr{M}(h_{\rm fit}) := \frac{(h_{\rm NR}, h_{\rm fit})}{\sqrt{(h_{\rm NR}, h_{\rm NR})(h_{\rm fit}, h_{\rm fit})}}, \quad (8.29)$$

and $(\cdot, \cdot)$ is an inner product defined in the time domain by

$$(a, b) := \int_{t_i}^{t_f} a(t) b^*(t) \, dt. \quad (8.30)$$

The first term of equation (8.28), $\mathscr{M}(h_{\rm fit})$, is an inner product between a numerical-relativity waveform and the fitting formula. The second term is a normalization factor that keeps the power of $h_{\rm NR}$ and $h_{\rm fit}$ the same: In the absence of the second term, the absolute amplitude cannot be determined in the fitting procedure.

Hotokezaka et al. (2013a) found that $C(h_{\rm fit})$ is always larger than 0.9 for more than 50 numerical-relativity waveforms for a variety of binary models of different total mass, mass ratio, and hypothetical equations of state. For the best fit model, $C(h_{\rm fit})$ can be $\gtrsim 0.98$. This suggests that the fitting waveform composed of equations (8.26) and (8.27) could capture the features of gravitational waves emitted by remnant massive neutron stars. They also found that even when the parameters such as p_s and $a_{\rm co}$ are fixed to be particular values, the value of $C(h_{\rm fit})$ is maintained to be $\gtrsim 0.9$. This implies that if we require the minimum-allowed fitting factor to be ~ 0.9, we may choose the 11-parameter fitting. However, for making a template of gravitational waves of sufficiently small systematic errors, the fitting value of 0.9 does not seem to be sufficiently high. Deriving a better fitting formula perhaps with more parameters is one of the issues for the future.

8.3.5 *Dynamical mass ejection and electromagnetic counterparts*

During the merger of binary neutron stars, a fraction of neutron-rich matter obtains a large amount of kinetic energy and escapes from the system. Because such ejecta are dense and neutron-rich, a substantial fraction of r-process heavy elements are likely to be produced [Lattimer and Schramm (1974); Symbalisty and Schramm (1982); Eichler et al. (1989)]. As briefly introduced in section 1.5.2, the ejecta will subsequently emit electromagnetic signals of a high luminosity via a radioactive decay of the produced unstable r-process heavy nuclei. This is the so-called kilonova or macronova scenario.

The ejecta will also emit electromagnetic signals during the interaction with ambient interstellar medium and subsequent shock heating [Nakar and Piran (2011)]. The luminosity of the electromagnetic signals depends primarily on the total mass of the ejecta for the radioactive decay scenario and on the total kinetic energy for the interaction with ambient interstellar medium. To accurately derive these key quantities of the ejecta, a numerical-relativity simulation is needed.

For exploring the properties of the ejecta, the numerical-relativity simulation has to be performed preparing a wide computational domain, to follow the motion of the ejecta for a long time scale until they relax to a free-expansion stage and to confirm that the ejecta component is indeed unbound by the merger remnant. This can be determined by confirming that the ejected matter satisfies the condition $u_t < -1$ in a zone far from the central object where u_t is the time component of u_μ.

A detailed study for the dynamical mass ejection[4] in numerical relativity was first performed by Hotokezaka et al. (2013b) [see also the works in other frameworks by Goriely et al. (2011); Bauswein et al. (2013); Piran et al. (2013); Rosswog et al. (2013); Grossman et al. (2014)]. Hotokezaka et al. (2013b) showed that the dynamical mass ejection indeed occurs in a mildly relativistic manner with the average and maximum velocities $\sim 0.15c$–$0.3c$ and $\sim 0.5c$–$0.8c$, respectively. They also showed that the total mass and kinetic energy of the ejecta depend strongly on the binary total mass, binary mass ratio, and hypothetically employed equations of state: The total ejected mass is in a wide range between $\sim 10^{-4}$ and $\sim 10^{-2} M_\odot$, and the total kinetic energy is also in a wide range between $\sim 10^{49}$ and $\sim 10^{51}$ ergs (see figure 8.28), depending on the parameters. These values are in general larger for the case that a massive neutron star is formed as a remnant after the merger. In the following, we will summarize the mass and kinetic energy of the dynamical ejecta in more detail.

8.3.5.1 *Massive neutron star formation case*

First, we focus on the case that a massive neutron star is formed as a remnant after the merger. Figures 8.24 and 8.25 display snapshots of rest-mass density profiles for the merger of equal-mass and unequal-mass binary neutron stars, respectively, with the total mass $m_0 = 2.7 M_\odot$ and with APR4 equation of state. Figures 8.26 and 8.27 also display snapshots of the thermal part of the specific internal energy, $\varepsilon_{\rm th}$, in the vicinity of remnant massive neutron stars during their early evolution stage, for the equal-mass and unequal-mass models, respectively. All these figures indicate that there are two important processes for the dynamical mass ejection. The first one is the heating by shocks formed both at the onset of the merger for the contact surface of two neutron stars and during the subsequent quasi-periodic oscillation of the remnant massive neutron star (see figure 8.7). Figures 8.26 and 8.27 clearly show that hot matter with $\varepsilon_{\rm th} \lesssim 0.1$ (i.e., $\lesssim 100\,{\rm MeV}$) is ejected from the remnant

---

[4]Here, "dynamical mass ejection" means the mass ejection during the merger stage of binary systems. Mass ejection could also occur via electromagnetic effects or neutrino wind in a quasi-stationary manner after the system relaxes to a quasi-stationary stage. Here, we do not consider such possibility, although it could play an important role in a mass ejection process [Fernández and Metzger (2013)].

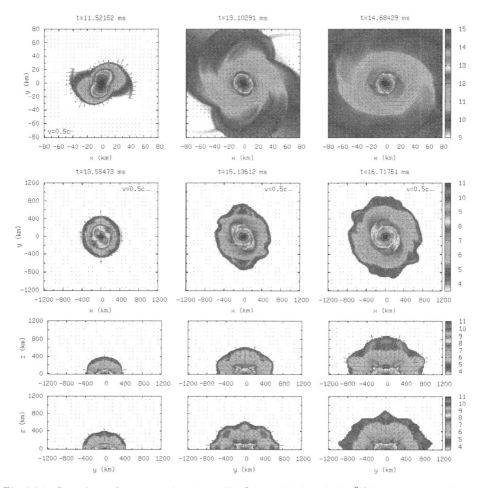

Fig. 8.24 Snapshots of rest-mass density profiles [plotted for $\log \rho(\mathrm{g/cm^3})$] for the merger of binary neutron stars for an equal-mass model ($m_1 = m_2 = 1.35 M_\odot$) with APR4 equation of state. The first row shows the profiles in the equatorial plane and in the central region, and second–fourth ones show rest-mass density profiles for a wide region in the x-y, x-z, and y-z planes, respectively. The merger time is $t_{\mathrm{merge}} \approx 11.3$ ms for this model. The figure is taken from Hotokezaka *et al.* (2013b).

massive neutron stars, in particular, toward the polar and equatorial regions. The mass ejection toward the polar region occurs in an appreciable manner in particular for the equal-mass (and only slightly asymmetric) binaries. The heated-up matter is pushed outwards by the thermal pressure generated by the shock approximately toward the plane parallel to the (rotating) shock surface. Subsequently, the hot matter *quasi-spherically* expands outwards with rotation, and the equatorial component forms hot spiral arms around the remnant massive neutron star.

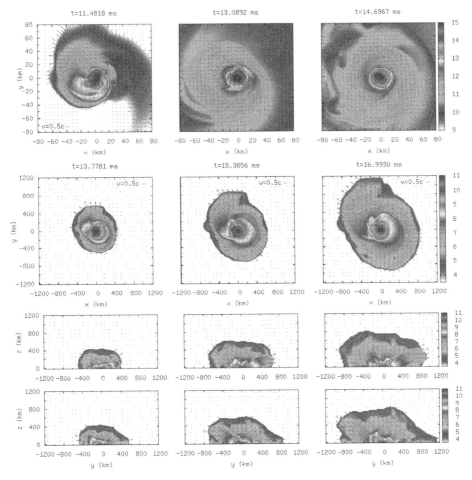

Fig. 8.25 The same as figure 8.24 but for an unequal-mass model ($m_1 = 1.5 M_\odot$ and $m_2 = 1.2 M_\odot$) with APR4 equation of state. The figure is taken from Hotokezaka *et al.* (2013b).

This component subsequently gains angular momentum (and hence kinetic energy) due to the torque gravitationally exerted by the remnant massive neutron star of a non-axisymmetric configuration. By these two mechanisms, a fraction of the matter eventually gains the kinetic energy that is large enough for them to escape from the system. Therefore, both the shock heating and the tidal torque play a primary role for the dynamical mass ejection. In the following, we will describe these two mechanisms in detail.

One evidence for the fact that the shock is one of the two key mechanisms for the dynamical mass ejection is found as follows: A shock formed at the onset of the merger is stronger for softer equations of state that yield compact neutron stars (e.g., APR4 and SLy in the context of canonical-mass neutron stars), because the

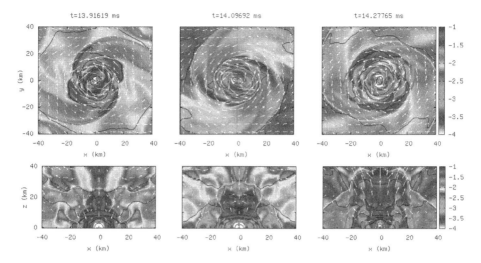

Fig. 8.26 Snapshots for profiles of the thermal part of the specific internal energy [plotted for $\log(\varepsilon_{\rm th})$] in the vicinity of a remnant massive neutron star on the equatorial (top) and x-z (bottom) planes for an equal-mass model ($m_1 = m_2 = 1.35 M_\odot$) with APR4 equation of state. The rest-mass density contours are over-plotted for every decade from 10^{15} g/cm^3. The figure is taken from Hotokezaka et al. (2013b).

binary system for such soft equations of state can achieve a significantly compact state, and hence, the collision velocity of two neutron stars is quite high. For these equations of state, the amount of the heated-up matter is high and the dynamical mass ejection is enhanced, leading to the enhancement of the ejecta mass (see figure 8.28).

Besides those at the onset of the merger, shocks are also formed continuously in the outer part of the remnant massive neutron stars during their evolution through the interaction with their envelope. Here it should be noted that the remnant massive neutron stars, which have a non-axisymmetric configuration, are oscillating and rapidly rotating. Hence, they interact violently with the envelope and, as a result, kinetic energy is transported outwards. This mechanism could play a substantial role for the quasi-spherical mass ejection in particular in their early stage in which the amplitude of quasi-radial oscillation is high. This also plays an important role in a long-term gradual mass ejection with the duration of $O(10\,{\rm ms})$.

The second important effect for the dynamical mass ejection is the tidal torque gravitationally exerted by the remnant massive neutron stars of a non-axisymmetric configuration: The matter surrounding the remnant massive neutron stars obtains angular momentum by the tidal torque. Because they are rapidly rotating, the remnant massive neutron stars work as an efficient engine for supplying the torque. Hotokezaka et al. (2013a) showed that this process also plays a key role for the dynamical mass ejection in the early stage of the merger in the following manner.

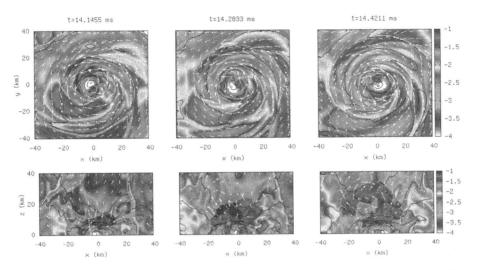

Fig. 8.27 The same as figure 8.26 but for an unequal-mass model ($m_1 = 1.5 M_\odot$ and $m_2 = 1.2 M_\odot$) with APR4 equation of state. The figure is taken from Hotokezaka et al. (2013b).

For the nearly equal-mass binaries, a fraction of the matter, which first spreads outwards by shocks formed at the merger, subsequently gains angular momentum from the remnant massive neutron star by the tidal torque and eventually obtains kinetic energy large enough for it to escape from the system. For sufficiently asymmetric binaries, a less-massive neutron star is tidally elongated during the early stage of the merger, and a fraction of its matter forms spiral arms. A part of the spiral-arm component subsequently gains large angular momentum from the remnant massive neutron star by the torque enough to escape from the system. In the dynamical mass ejection caused by the torque exerted by the remnant massive neutron stars, the matter is primarily ejected in the anisotropic manner near the equatorial plane.

The typical velocity of the ejecta is quite high $\sim 0.5c$–$0.8c$ in the very early stage (following the locations of the head of the ejecta in figures 8.24 and 8.25, this fact is found). The maximum velocity is larger for the cases with soft equations of state; e.g., for APR4 and SLy, it is $\sim 0.8c$ while for stiffer one such as MS1, it is $\sim 0.5c$. This value also depends on the mass ratio for models in particular with less-compact neutron stars (e.g., for models of H4 and MS1).

In the subsequent stage followed by the very early stage, the dynamical mass ejection also appears to occur by the combination of the shock heating and the torque exerted by the remnant massive neutron stars. For this stage, the continuous (weak) shock heating occurs in the envelope of the remnant massive neutron stars. Due to this, a fraction of the matter gains certain kinetic energy. In addition, the matter in the outer region gains angular momentum by the torque exerted by the remnant massive neutron stars. These two effects supply escape velocity to a

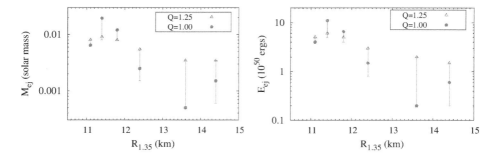

Fig. 8.28 Total rest mass (left panel) and kinetic energy (right panel) of dynamical ejecta as functions of $R_{1.35}$ for $m_0 = 2.7 M_\odot$ with several mass ratios $1 \leq Q \leq 1.25$ and with six equations of state (APR4, SLy, Steiner, ALF2, H4, and MS1 from small to large values of $R_{1.35}$). The solid circles and open triangles denote the data for the equal-mass binary and for $Q = 1.25$, respectively. The error bars show the range of the results for several mass ratios of $1 \leq Q \leq 1.25$. The data are taken from Hotokezaka et al. (2013b) and our private computations.

fraction of the matter. By this process, the matter is gradually ejected from the system in a quasi-spherical manner; the anisotropy of the ejecta configuration is not as large as that of the matter ejected in the early stage. This indicates that the shock heating plays a relatively important role. The average velocity of the ejecta in this process is sub-relativistic $\sim 0.15c - 0.25c$ [Hotokezaka et al. (2013b)].

Total mass and kinetic energy of the ejecta depend strongly on hypothetically employed equations of state. Figure 8.28 plots the value of the ejecta mass, $M_{\rm ej}$, and kinetic energy, $E_{\rm ej}$, as functions of $R_{1.35}$ for $m_0 = 2.7 M_\odot$ with several mass ratios $1 \leq Q \leq 1.25$ and with six equations of state. Here, $R_{1.35}$ is the radius of spherical neutron stars of mass $1.35 M_\odot$. The data are taken from Hotokezaka et al. (2013b). This figure shows that, broadly speaking, the ejecta mass and kinetic energy *increase* with the decrease of $R_{1.35}$. In particular, for soft equations of state with $R_{1.35} \lesssim 12$ km, these quantities are much larger than those for other stiffer equations of state. For the soft equations of state, it is also found that these quantities depend weakly on the mass ratio. This is due to the fact that the shock heating during the early stage of the merger primarily determines the mass of the dynamical mass ejection. By contrast, for stiff equations of state, the mass asymmetry of the binary is necessary to enhance the ejecta mass and kinetic energy. This is due to the fact that the tidal effect plays a relatively important role in the dynamical mass ejection for such equations of state.

8.3.5.2 Prompt black-hole formation case

We then summarize the properties of dynamical ejecta for the case that a black hole is formed promptly after the merger. This occurs for the case that total mass of the binaries is sufficiently high. Figure 8.29 displays snapshots of rest-mass density profiles for the merger of unequal-mass binary neutron stars with $m_1 = 1.6 M_\odot$ and

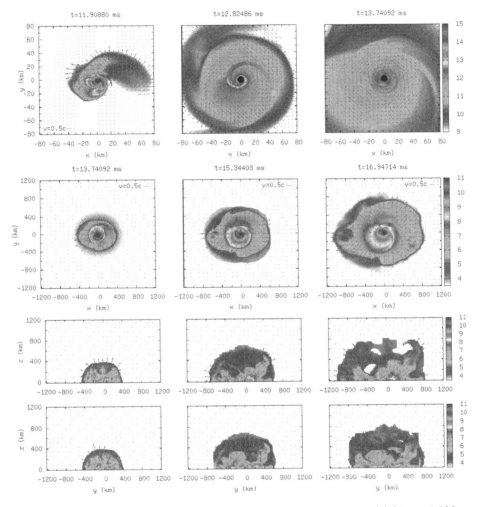

Fig. 8.29 The same as figure 8.24 but for a high-mass unequal-mass model ($m_1 = 1.6 M_\odot$ and $m_2 = 1.3 M_\odot$) with APR4 equation of state that leads to a black hole promptly after the onset of the merger. The figure is taken from Hotokezaka et al. (2013b).

$m_2 = 1.3 M_\odot$, and with APR4 equation of state. For this model, the dynamical mass ejection proceeds primarily at the instance of the merger, i.e., during a short duration before the formation of a black hole. Because a black hole is promptly formed, a region shock-heated at the collision of two neutron stars is soon swallowed by the black hole, and thus, the shock heating does not play a primary role in the mass ejection. A significant mass ejection occurs only for the case that a mass asymmetry is present, and the mass ejection is induced primarily by a tidal torque: In the presence of the mass asymmetry, the less-massive neutron star is tidally elongated during the merger, and a fraction of the tidally elongated neutron-star matter receives a sufficient torque from the merged object just before the formation of a black hole and gets an escape velocity. In the example of figure 8.29, the gain of the angular momentum is large enough that the ejecta rest mass becomes $\sim 2 \times 10^{-3} M_\odot$ with the kinetic energy $\sim 10^{50}$ ergs. The average velocity of the ejecta for this case is $\sim 0.3c$ and larger than that in the case of the massive neutron star formation. The reason for this is that the mass ejection is caused primarily by the tidal interaction at the onset of the merger, and for this case, the induced velocity is larger than that by long-term shock heating. Because the tidal interaction induces the mass ejection, the matter is ejected primarily in the direction of the equatorial plane. The motion to the z-direction is also induced by shock heating that occurs when spiral arms surrounding the formed black hole collide each other. However, this is a secondary effect.

For the equal-mass or nearly equal-mass binaries, the total rest mass of ejecta is quite small, because of the absence of the asymmetry and of the lack of the time during which the matter located in the outer region receives sufficient torque from the merged object (note that most of the fluid elements of binary neutron stars just before the onset of the merger do not have the specific angular momentum large enough to escape even from the infall into the formed black hole [Shibata and Uryū (2000)]). Therefore, for the prompt black-hole formation case, a substantial asymmetry is needed for producing large-mass ejecta. Note that a large asymmetry is also necessary for the formation of appreciable disks surrounding the remnant black hole (see section 8.3.3.1).

8.3.5.3 *Possible electromagnetic counterparts*

In this section, we discuss possible electromagnetic signals emitted by dynamical ejecta of the binary neutron star merger, referring to the results in numerical relativity. As already mentioned, the discovery of high-mass neutron stars, PSR J1614-2230 and PSR J0348+432, suggests that the maximum mass of spherical neutron stars should be larger than $\sim 2 M_\odot$ [Demorest et al. (2010); Antoniadis et al. (2013)]. This indicates that long-lived massive neutron stars with their lifetime $\gtrsim 10$ ms would be the canonical remnants of the binary neutron star merger, if the binaries were composed of neutron stars of canonical mass of $1.3-1.4 M_\odot$ (i.e., the total mass is $\sim 2.6-2.8 M_\odot$). A number of numerical-relativity results indicate that from the long-lived massive neutron stars, a fraction of matter could be ejected with large kinetic energy, in particular, for the soft equations of state that give a small neutron-star radius $\lesssim 12$ km.

There are several pioneer works [Li and Paczyński (1998); Kurkarni (2005); Metzger et al. (2010)] that discussed an electromagnetic (ultraviolet-optical-infrared) signal by the radioactive decay of unstable r-process nuclei that could be produced in the neutron-rich ejecta. The signal may be observable by current and planned optical telescopes suitable for searching transient phenomena such as PTF,

Pan-STARRs, SUBARU and LSST. In this scenario, the ejecta are initially so dense that photons are optically thick in their early expansion stage: In this stage, the emission occurs in a long-term diffusion process, and the emissivity is not very high. Thus, the cooling of the ejecta occurs primarily through the adiabatic expansion. During the subsequent expansion stage, however, the optical depth decreases and eventually the ejecta become optically thin against the electromagnetic emission. The peak luminosity is reached approximately at the moment that the diffusion time scale of photons in the ejecta is equal to the expansion time scale, which is approximately equal to the elapsed time after the merger (hereafter referred to as $t_{\rm elap}$). In the following, we will estimate the time at which the peak luminosity is reached and the resulting peak luminosity.

Assuming random walks of photons in spherically symmetric and uniform ejecta for which the expansion velocity is $v_{\rm ave}$, the total mass is $M_{\rm ej}$, and the average value of the opacity is $\bar{\kappa}$, the diffusion time scale is approximately estimated as [Li and Paczyński (1998)]

$$t_{\rm diff} \sim \frac{\bar{\kappa}\rho_{\rm ej}R_{\rm ej}^2}{c} = \frac{3\bar{\kappa}M_{\rm ej}}{4\pi c R_{\rm ej}}, \tag{8.31}$$

where $\rho_{\rm ej}$ is the average density and $R_{\rm ej}$ is the radius of the spherical ejecta which is approximately equal to $v_{\rm ave}t_{\rm elap}$. The peak time is determined by the relation, $t_{\rm diff} = t_{\rm elap}$, which yields

$$t_{\rm peak} \approx \left(\frac{3\bar{\kappa}M_{\rm ej}}{4\pi c v_{\rm ave}}\right)^{1/2} = 6\,{\rm days}\left(\frac{\bar{\kappa}}{10\,{\rm cm}^2/{\rm g}}\right)^{1/2}\left(\frac{M_{\rm ej}}{0.01 M_\odot}\right)^{1/2}\left(\frac{\beta_0}{0.2}\right)^{-1/2}, \tag{8.32}$$

and thus, $R_{\rm ej} \approx 10^{-3}$ pc at the peak. Here, $\beta_0 := v_{\rm ave}/c$. If a fraction $f_{\rm eff}$ of $M_{\rm ej}c^2$ is radiated, the peak luminosity is expected to be

$$L_{\rm peak} \approx \frac{f_{\rm eff} M_{\rm ej}c^2}{t_{\rm peak}}$$

$$\approx 1 \times 10^{41}\,{\rm ergs/s}\left(\frac{f_{\rm eff}}{3 \times 10^{-6}}\right)\left(\frac{\bar{\kappa}}{10\,{\rm cm}^2/{\rm g}}\right)^{-1/2}\left(\frac{M_{\rm ej}}{10^{-2} M_\odot}\right)^{1/2}\left(\frac{\beta_0}{0.2}\right)^{1/2}, \tag{8.33}$$

and the temperature at the peak is

$$T_{\rm peak} \approx \left(\frac{L_{\rm peak}}{4\pi v_{\rm ave}^2 t_{\rm peak}^2 \sigma_{\rm SB}}\right)^{1/4}$$

$$\approx 2 \times 10^3\,{\rm K}\left(\frac{f_{\rm eff}}{3 \times 10^{-6}}\right)^{1/4}\left(\frac{\bar{\kappa}}{10\,{\rm cm}^2/{\rm g}}\right)^{-3/8}\left(\frac{M_{\rm ej}}{10^{-2} M_\odot}\right)^{-1/8}\left(\frac{\beta_0}{0.2}\right)^{-1/8}, \tag{8.34}$$

where $\sigma_{\rm SB}$ is the Stefan–Boltzmann constant. $f_{\rm eff}$ is of order 10^{-6} according to the results of Metzger et al. (2010). $\bar{\kappa}$ is predicted to be rather high $\sim 10\,{\rm cm}^2/{\rm g}$ for the r-process heavy elements [Kasen et al. (2013); Tanaka and Hotokezaka (2013)].

Note that the luminosity of $\sim 10^{41}$ ergs/s is more than hundred times larger than the Eddington luminosity for the remnant of mass $\sim 2.7 M_\odot$.

Numerical-relativity results indicate, for canonical-mass binaries with $m_0 \sim 2.7 M_\odot$, that the typical average velocity of ejecta is $\beta_0 = 0.15\text{--}0.25$, and the total dynamical ejecta mass is $M_{\rm ej} \sim 5 \times 10^{-3}\text{--}2 \times 10^{-2} M_\odot$ for binaries composed of neutron stars with a small radius $\sim 11\text{--}12.5$ km, and $\sim 0.3 \times 10^{-3}\text{--}5 \times 10^{-3} M_\odot$ for binaries composed of neutron stars with a larger radius $\sim 13.5\text{--}14.5$ km (see figure 8.28). Thus, if the neutron-star equation of state is soft, an observable signal due to the radioactive decay can be expected with a duration of several days. If the equation of state is stiff, the strength of the signal will be weaker and the duration shorter, although it would be still possible to detect the signal in particular for the merger of unequal-mass (sufficiently asymmetric) neutron stars.

More careful radiative-transfer studies of Barnes and Kasen (2013) and Tanaka and Hotokezaka (2013), which were based on a systematic study for the opacity by r-process elements, suggested that if the total ejected mass is $\gtrsim 10^{-3} M_\odot$ with a hypothetical value of $\bar{\kappa} \sim 10\,\text{cm}^2/\text{g}$, the signal will be detected by a large telescope such as LSST for a typical distance to the sources ~ 200 Mpc. They also showed that for the ejecta mass $\sim 10^{-2} M_\odot$, the ejecta would be most luminous for a near-infrared band at $\sim 5\text{--}10$ days after the onset of the merger.

There is also another possible channel for the electromagnetic emission: Synchrotron radio flare. According to Nakar and Piran (2011); Metzger and Berger (2012); Piran et al. (2013), the ejecta in the free expansion will sweep up the interstellar matter and form blast waves. During this process turning on, the shocked matter could generate magnetic fields and accelerate electrons that emit synchrotron radiation, for a hypothetical amplification of the electromagnetic field and a hypothetical electron injection. The emission will reach a peak luminosity when the total swept-up mass approaches the ejecta mass, because the blast waves are decelerated and transit to the phase in which the matter motion is described by the (non-relativistic) Sedov-Taylor's self-similar solution.

Assuming that the interstellar matter is composed of hydrogens, the deceleration radius, $R_{\rm dec}$, for spherical ejecta is defined by $M_{\rm ej} = 4\pi n_0 m_{\rm H} R_{\rm dec}^3/3$, and thus,

$$R_{\rm dec} := \left(\frac{3 M_{\rm ej}}{4\pi m_{\rm H} n_0}\right)^{1/3} \approx 0.46 \text{ pc} \left(\frac{M_{\rm ej}}{10^{-2} M_\odot}\right)^{1/3} \left(\frac{n_0}{1\,\text{cm}^{-3}}\right)^{-1/3}. \quad (8.35)$$

Here, n_0 is the number density of the interstellar matter and $m_{\rm H}$ is the hydrogen mass. Then, the deceleration time is given by Nakar and Piran (2011) as

$$t_{\rm dec} = \frac{R_{\rm dec}}{v_{\rm ave}} \approx 7.5 \text{ yrs} \left(\frac{M_{\rm ej}}{10^{-2} M_\odot}\right)^{1/3} \left(\frac{n_0}{1\,\text{cm}^{-3}}\right)^{-1/3} \left(\frac{\beta_0}{0.2}\right)^{-1}, \quad (8.36)$$

or

$$t_{\rm dec} \approx 7.7 \text{ yrs} \left(\frac{E_0}{10^{51}\,\text{ergs}}\right)^{1/3} \left(\frac{n_0}{1\,\text{cm}^{-3}}\right)^{-1/3} \left(\frac{\beta_0}{0.2}\right)^{-5/3}. \quad (8.37)$$

By the synchrotron radiation, a radio signal could be emitted as in the late phase of supernovae and the afterglow of gamma-ray bursts [Nakar and Piran (2011)]. Numerical-relativity results indicate that the typical velocity of the ejecta is $\beta_0 = 0.15-0.25$ irrespective of the equations of state employed and the mass of neutron stars in binaries. However, E_0 is in a wide range between $\sim 10^{49}$ ergs and 10^{51} ergs, depending strongly on the equation of state employed, mass ratio, and total mass of the binaries, and its value is highly uncertain. Thus the predicted value of $t_{\rm dec}$ is in a wide range $\sim 1-10$ yrs, even for an optimistic value of $n_0 = 1\,{\rm cm}^{-3}$. For smaller values of n_0 which are likely when the merger occurs outside the galactic plane, the value of $t_{\rm dec}$ will be longer.

For the typical value of the ejecta velocity $\beta_0 \sim 0.2$, the peak flux for the observed frequency is obtained at the deceleration time described in equation (8.37). Specifically, the peak flux may be obtained at the self-absorption frequency, $\sim 1-2$ hundreds MHz, and the typical synchrotron frequency is sub-MHz. The peak flux for a given observed radio-band frequency $\nu_{\rm obs}$ is estimated as [Nakar and Piran (2011)]

$$F_\nu \approx 90\,\mu{\rm Jy} \left(\frac{E_0}{10^{50}\,{\rm ergs}}\right) \left(\frac{n_0}{1\,{\rm cm}^{-3}}\right)^{0.9} \left(\frac{\beta_0}{0.2}\right)^{2.8} \left(\frac{D}{200\,{\rm Mpc}}\right)^{-2} \left(\frac{\nu_{\rm obs}}{1.4\,{\rm GHz}}\right)^{-0.75}, \tag{8.38}$$

where we assumed the power-law distribution of the electron's Lorentz factor with the power 2.5. Equation (8.38) is applicable as long as the observed frequency is higher than the typical synchrotron and self-absorption frequency at the deceleration time, $t_{\rm dec}$. Equation (8.38) indicates that for a hypothetical event at a distance of 200 Mpc, $E_0 \sim 10^{50}$ ergs with $n_0 = 1\,{\rm cm}^{-3}$ is strong enough to be observed by future-planned radio instruments such as EVLA, ASKAP, and MeerKAT for which the root-mean square value of the background noise for one hour observation is smaller than 50 μJy as shown in Nakar and Piran (2011). Therefore, if the equation of state is rather soft (i.e., the neutron-star radius is fairly small) or the binary is significantly asymmetric, the synchrotron radio flare will be observable.

In this scenario, the duration to the peak luminosity and the strength of the radio signal depend strongly on the value of n_0. In nature, the value of n_0 will depend strongly on the site where the merger of binary neutron stars happens. If it is in a galactic disk, n_0 would be typically $\sim 1\,{\rm cm}^{-3}$, whereas if it is outside a galaxy, the value may be much smaller as $\sim 10^{-3}\,{\rm cm}^{-3}$. Equation (8.38) shows that for a small value of $n_0 \sim 10^{-3}\,{\rm cm}^{-3}$, $F_\nu \sim 1\mu$Jy even for $E_0 = 10^{51}$ ergs. Numerical-relativity simulations have shown that the maximum value of E_0 would be $\sim 10^{51}$ ergs. Therefore, for a low value of $n_0 \sim 10^{-3}\,{\rm cm}^{-3}$, this type of electromagnetic signals may not be observable as a counterpart of the gravitational-wave signal [Metzger and Berger (2012)].

8.4 Black hole-neutron star binaries: Quasi-equilibrium states and sequences

8.4.1 *Formulations*

8.4.1.1 *Brief summary*

As in the case of binary neutron stars, for constructing quasi-equilibrium states of black hole-neutron star binaries, we have to consider the velocity field of neutron stars as a first step. Due to the same reason described in section 8.2.1, it is reasonable to suppose that an irrotational velocity field is approximately realized. Quasi-equilibrium states for such black hole-neutron star binaries have been constructed by Grandclément (2006); Taniguchi et al. (2006, 2007, 2008); Foucart et al. (2008); Kyutoku et al. (2009); Foucart et al. (2011).

Broadly classifying, two different formulations have been proposed for constructing black hole-neutron star binaries in quasi-equilibrium. For these two formulations, the basic equations for the gravitational field and the method for handling the black hole are different. Grandclément (2006); Taniguchi et al. (2006, 2007, 2008); Foucart et al. (2008, 2011) employed the CTS formulation (see section 5.6.2) together with the excision boundary condition (see section 5.6.3) for handling the black hole. For non-spinning black holes, they set the conformal three metric $\tilde{\gamma}_{ij}$ to be flat, $\tilde{\gamma}_{ij} = f_{ij}$, and the trace-part of the extrinsic curvature to be zero, respectively, while for spinning black holes, the Kerr-Schild geometry for $\tilde{\gamma}_{ij}$ and K with an appropriate cutoff for a region far from the black hole was also employed [Foucart et al. (2011)]. On the other hand, Kyutoku et al. (2009) employed the IWM plus transverse-tracefree formulation (see section 5.6.4) together with the moving puncture approach (see section 5.5.2.2) for handling the black hole.

In both approaches, all the components of Einstein's equation are not solved. Hence, they would not yield a good quasi-equilibrium state for close orbits (see section 8.2.3).

8.4.1.2 *Eccentricity reduction*

If we employ quasi-equilibrium data with not-distant orbits as the initial condition of a numerical-relativity simulation, the trajectories obtained in the numerical simulation usually result in a slightly eccentric orbit because of the lack of the approaching radial velocity caused by the gravitational radiation reaction. If one desires to obtain a high-precision gravitational waveform, such an eccentricity should be suppressed to be as small as possible. One prescription to suppress it might be to choose an initial condition in which the binary separation is large enough that the effect of the radial velocity is not serious. However, for such a simulation, we have to follow a large number of inspiral orbits and it is not very practical.

To overcome this problem, Foucart et al. (2008, 2011) implemented an eccentricity reduction technique for constructing a realistic quasi-circular initial condition.

This was originally developed for simulating binary black holes by Pfeiffer et al. (2007); Boyle et al. (2007). Their method is summarized as follows:

(1) First of all, one has to prepare a quasi-equilibrium state that has zero approaching velocity with an orbital angular velocity Ω_0.
(2) Then, one needs to evolve the quasi-equilibrium initial data for at least one and a half orbits. During the time evolution, one needs to record the time derivative of the measured coordinate separation between the centers of the compact objects, $\dot{d}(t)$, and then, one fits $\dot{d}(t)$ in the form

$$\dot{d} = A_0 + A_1 t + A_2 \sin(\omega_0 t + \phi_0), \tag{8.39}$$

where the parameters A_0, A_1, A_2, ω_0, and ϕ_0 are all determined by the fitting. The $A_0 + A_1 t$ part denotes the smooth inspiral, while the $A_2 \sin(\omega t + \phi)$ part denotes the undesirable oscillations due to an eccentricity of the orbit. For a nearly circular orbit in Newtonian gravity, the eccentricity e of the orbit can be written as

$$e = \frac{A_2}{\omega_0 d_0}, \tag{8.40}$$

where $d_0 = d(t=0)$. This implies that reducing the orbital eccentricity is equivalent to reducing A_2.
(3) Finally, one needs to add the correction of the approaching velocity $\dot{a}_0$ and the correction to the orbital angular velocity $\Omega_{\rm orb}$ using the parameters which appear in equation (8.39). Such corrections should be chosen so that the eccentricity-induced initial radial velocity and radial acceleration can be removed. Hence, the initial radial velocity is changed as

$$\delta \dot{d} = -A_2 \sin \phi_0, \tag{8.41}$$

and the initial orbital angular velocity as

$$\delta \Omega_0 = -\frac{A_2 \omega_0 \cos \phi_0}{2 d_0 \Omega_0}. \tag{8.42}$$

This was derived from the assumption that the modulation part of the acceleration, $\ddot{d}$, would stem from the error in the centrifugal force $\approx \Omega_0^2 d_0$. Using the corrected approaching radial velocity and the orbital angular velocity for the initial step (1), one should perform the same procedure until a convergence is achieved. Pfeiffer et al. (2007); Foucart et al. (2008) showed that by twice integrations, the orbital eccentricity is reduced by about an order of magnitude.

Although there are many works for constructing the initial data of numerical simulations, only a few works have been done for the analysis of quasi-equilibrium sequences. Only Taniguchi et al. (2007, 2008) performed an analysis for such sequences. Although their analysis was performed based on the conformally flat approximation (IWM formulation) and only for a simple polytropic equation of state with $n = 1$ and non-spinning black holes, they reported several important properties that characterize the sequences of black hole-neutron star binaries. Thus,

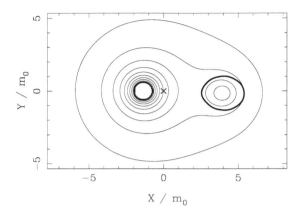

Fig. 8.30 Contours of the conformal factor in the equatorial plane for a binary of a close orbit with $Q = 3$ and $\mathcal{C} = 0.145$ shown in Taniguchi *et al.* (2008). The thick solid circle in the left-hand side denotes the position of the excised surface of the black hole (the apparent horizon), while that in the right-hand side denotes the location of the deformed neutron-star surface. The cross indicates the position of the rotation axis of the system.

in the following, we will summarize the properties of quasi-equilibrium sequences based on their results.

8.4.2 *Two types of quasi-equilibrium sequences*

As in the case of binary neutron stars, a sequence of black hole-neutron star binaries for a given equation of state is uniquely determined if we specify mass and spin (or area and spin) of the black hole, and rest mass (or compactness) of the neutron star, for the case that the irrotational velocity field of the neutron star is assumed. In the following, we will implicitly assume that the black-hole mass is larger than the neutron-star mass, because this is likely to be the case in nature. A quasi-equilibrium state along a sequence is specified by the orbital angular velocity, $\Omega_{\rm orb}$ (or an appropriately determined orbital separation), and the corresponding total ADM mass and angular momentum are determined for each value of $\Omega_{\rm orb}$. For distant orbits, the total ADM mass and angular momentum decrease monotonically with the increase of $\Omega_{\rm orb}$. This is natural because the binary evolves due to the gravitational-radiation reaction dissipating the energy and angular momentum.

For close orbits, this monotonic relation for the total ADM mass and angular momentum could be modified because of two important ingredients. One is general relativistic gravity of the black hole, around which orbits of a test particle have the innermost stable circular orbit (ISCO). This suggests that at a close orbit of a neutron star around the black hole, the ADM mass and angular momentum along quasi-equilibrium sequences could have a minimum point. The other possibility is as follows: For close orbits, the neutron star is tidally deformed by the black-hole tidal field, and for the sufficiently strong tidal field, a cusp (a Lagrange point) could

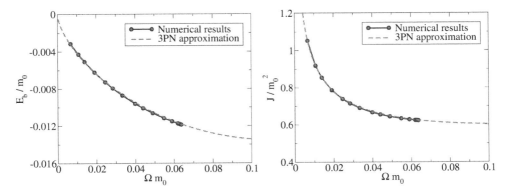

Fig. 8.31 Binding energy E_b/m_0 (left panel) and total angular momentum J/m_0^2 (right panel) in dimensionless units as functions of Ωm_0 for binaries of mass ratio $Q = 3$ and neutron-star mass $\bar{M}_B = M_B \kappa_p^{-1/n} = 0.15$ ($\mathcal{C} = 0.145$). Here, Ω denotes $\Omega_{\rm orb}$, and M_B and κ_p denote the rest mass and polytropic constant with $n = 1$. The solid curve with the filled circles show numerical results, and the dashed curve denotes the results in a third post-Newtonian approximation with the point-particle approximation [Blanchet (2014)]. The figure is taken from Taniguchi et al. (2008).

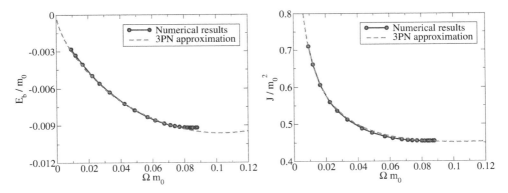

Fig. 8.32 The same as figure 8.31 but for the sequence of mass ratio $Q = 5$. The minima of these curves (i.e., ISCO) appear at $\Omega m_0 \approx 0.085$. The figure is taken from Taniguchi et al. (2008).

appear at an inner edge of the neutron star. When the cusp appears, mass shedding sets in and the quasi-equilibrium sequence terminates at this orbit (i.e., the mass shedding occurs before the ISCO is reached). Taniguchi et al. (2008) showed that either of these possibilities is realized for any binaries they studied and the type of the end point depends primarily on the mass ratio, $Q := M_{\rm BH}/M_{\rm NS}$, and the compactness of the neutron star, $\mathcal{C} := M_{\rm NS}/R_{\rm NS}$, where $M_{\rm BH}$ and $M_{\rm NS}$ denote the ADM mass of the black hole and neutron star in isolation, and $R_{\rm NS}$ is the circumferential radius of the neutron star in isolation.

Figure 8.30 displays contour curves of the conformal factor $[\det(\gamma_{ij})]^{1/12}$ for a binary of a close orbit with $Q = 3$ and $\mathcal{C} = 0.145$. A saddle point of the conformal factor is found in the vicinity of the neutron-star surface along the X-axis, suggesting that the Lagrange point is present in the vicinity of an inner edge of the neutron star and that the neutron star is close to the mass-shedding limit. If the mass-shedding limit is encountered before the binary reaches an ISCO, the quasi-equilibrium sequence terminates at this orbit. By contrast, if this is not the case, the sequence will terminate at an ISCO.

Figure 8.31 shows the binding energy (E_b/m_0) and the total angular momentum (J/m_0^2) in dimensionless units as functions of the orbital angular velocity in the dimensionless units ($\Omega_{\text{orb}} m_0$) for $Q = 3$ and $\mathcal{C} = 0.145$. Here, m_0 is $M_{\text{BH}} + M_{\text{NS}}$ and the binding energy is defined by $E_b := M_{\text{ADM}} - m_0$. The solid curves with the filled circles denote the numerical results, and the dashed curves the results in a third post-Newtonian approximation together with the point-particle approximation [Blanchet (2014)] (see appendix H). In this case, the sequence terminates at an orbit of the mass-shedding limit before the ISCO is reached (i.e., before the minima of the binding energy and angular momentum appear).

As described in section 1.5.4, the mass shedding is less subject for the larger black-hole mass (for the larger value of Q) for a given compactness of the neutron star. Figure 8.32 shows the binding energy and the total angular momentum as functions of the orbital angular velocity for $\mathcal{C} = 0.145$ and for $Q = 5$; the mass ratio is larger than that shown in Figure 8.31. In this model, the sequence of the binary reaches an ISCO before the onset of mass shedding; i.e., minima in the binding energy and angular momentum appear just before the end of the sequence at which mass shedding occurs.

8.4.3 Endpoint of sequences

As shown in section 8.4.2, there are two possible final fates of black hole-neutron star binaries. Then, one of the most important issues for this system is to determine the condition for the onset of mass shedding before reaching an ISCO. For the case that the mass-shedding occurs, the neutron star could be subsequently tidally disrupted by the companion black hole. The disrupted neutron-star matter could form an accretion disk/torus surrounding the black hole and also could be ejecta. A system composed of a black hole surrounded by a disk is a candidate for the central engine of gamma-ray bursts. Also, the ejecta could emit luminous electromagnetic signals (see section 1.5.2 for a review). As reviewed in section 1.5.4, if the tidal disruption occurs, resulting gravitational waves will have a characteristic property, which can be used for exploring the properties of the neutron star such as its radius and nuclear equation of state. Thus, exploring the condition for the onset of mass shedding is an important subject in high-energy astrophysics and gravitational-wave physics.

For exploring this issue quantitatively, a numerical-relativity simulation is probably the unique approach. However, the study of quasi-equilibrium sequences can also provide a guide for this issue approximately. Taniguchi *et al.* (2008) performed

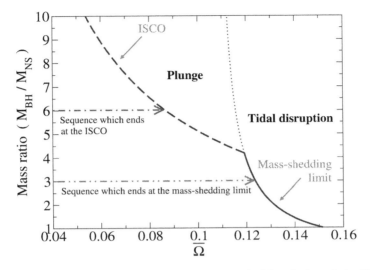

Fig. 8.33 An example of the boundary between the mass-shedding limit and the ISCO derived by Taniguchi et al. (2008). The vertical and horizontal axes denote the mass ratio, Q, and normalized angular velocity $\Omega_{\rm orb} m_0 (= \bar{\Omega})$. For this selected model, $\mathcal{C} = 0.145$. The solid and long-dashed curves, respectively, denote the mass-shedding limit and the ISCO for each mass ratio as functions of $\bar{\Omega}$. The dotted curve denotes the mass-shedding limit for unstable quasi-equilibrium sequences.

a smart analysis of quasi-equilibrium sequences to classify the final fate. In the following, we will outline their analysis.

Taniguchi et al. (2008) first derived a fitting formulation for the condition of mass shedding from their numerical results as [compare it with equation (1.209)][5]

$$\Omega_{\rm ms} m_0 = 0.270\, \mathcal{C}^{3/2} (1+Q) \left(1 + Q^{-1}\right)^{1/2}, \qquad (8.43)$$

where $\Omega_{\rm ms}$ denotes the angular velocity at the onset of mass shedding. Taniguchi et al. (2008) then derived a simple empirical fitting formula that predicts the angular velocity at the ISCO, $\Omega_{\rm ISCO}$, for an arbitrary companion orbiting a black hole. They search for an expression, composed of three free parameters, that expresses $\Omega_{\rm ISCO}$ as a function of the mass ratio Q and the compactness $\mathcal{C}$ of the neutron star. Then, the three parameters are determined using the following three known values of $\Omega_{\rm ISCO}$; (i) $\Omega_{\rm ISCO}$ of a test particle orbiting a Schwarzschild black hole, i.e., $\Omega_{\rm ISCO} m_0 = 6^{-3/2}$ for $Q = \infty$; (ii) $\Omega_{\rm ISCO}$ of an equal-mass binary black hole computed by Caudill et al. (2006), i.e., $\Omega_{\rm ISCO} m_0 = 0.1227$ for $Q = 1$ and $\mathcal{C} = 0.5$; (3) $\Omega_{\rm ISCO}$ of a black hole-neutron star binary computed by Taniguchi et al. (2008), i.e., for $Q = 5$ and $\mathcal{C} = 0.1452$, $\Omega_{\rm ISCO} m_0 = 0.0854$. An additional requirement is that for a test particle (with $Q = \infty$), the expression should be independent of the

---

[5] Equation (8.43) is valid only for $n = 1$ polytropic equation of state

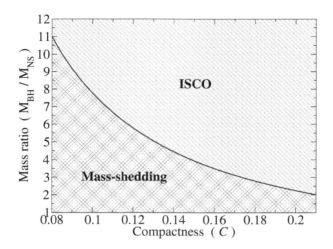

Fig. 8.34 The critical mass ratio that separates the final fates of black hole-neutron star binaries as a function of the neutron-star compactness for $n = 1$ polytropic equation of state. Black hole-neutron star binaries either encounter an ISCO before reaching mass shedding or undergo the mass shedding limit. The figure is taken from Taniguchi *et al.* (2008).

companion's compactness. The resulting equation by Taniguchi *et al.* (2008) is

$$\Omega_{\text{ISCO}} m_0 = 0.0680 \left[1 - \frac{0.444}{Q^{0.25}} \left(1 - 3.54 \, \mathcal{C}^{1/3} \right) \right]. \tag{8.44}$$

Combining equations (8.43) and (8.44), one can determine the critical binary parameters that separate the final fates. Figure 8.33 illustrates the final fate for $\mathcal{C} = 0.145$. The solid and long-dashed curves denote the orbital angular velocity at the mass-shedding limit and at the ISCO, respectively, for a given mass ratio, Q. As found from equations (8.43) and (8.44), these curves depend on the value of Q in different ways, and they intersect at a point, $\bar{\Omega} := \Omega m_0 \approx 0.12$. During the evolution of an inspiraling binary, $\bar{\Omega}$ increases along a horizontal line until reaching either the ISCO or the mass-shedding limit. After the binary reaches the ISCO (i.e., for a sufficiently large mass ratio), we cannot predict the fate of the neutron star by this quasi-equilibrium study, because it is subsequently in a dynamical plunge phase. As shown in figure 8.33, the model with mass ratio $Q \gtrsim 4$ (e.g., dotted-dashed line for $Q = 6$) encounters the ISCO, while that of $Q \lesssim 4$ (e.g., dot-dot-dashed line for $Q = 3$) ends up at a mass-shedding limit. The intersection between the mass-shedding and ISCO curves marks a critical point that separates the two distinct fates of the binary inspiral.

By setting $\Omega_{\text{ms}} = \Omega_{\text{ISCO}}$ for equations (8.43) and (8.44), we can draw a curve of the critical mass ratio that separates the fate of black hole-neutron star binaries. The resulting critical curve is determined by

$$0.270 \, \mathcal{C}^{3/2} (1+Q) \left(1 + Q^{-1} \right)^{1/2} = 0.0680 \left[1 - \frac{0.444}{Q^{0.25}} \left(1 - 3.54 \, \mathcal{C}^{1/3} \right) \right], \tag{8.45}$$

and the curve that separates those two regions is depicted in figure 8.34. The solid curve denotes the critical mass ratio for each compactness. If the mass ratio of a binary is larger than the critical one, the quasi-equilibrium sequence terminates at an ISCO, while if smaller, it ends up at a mass-shedding limit.

Before closing this section, we note that the result of equation (8.44) holds only for binaries composed of a non-spinning black hole. In the presence of a black-hole spin, $\Omega_{\rm ISCO}$ changes significantly; for example, for $Q \to \infty$, the value of $\Omega_{\rm ISCO} m_0 = 0.0680$ for the non-spinning case can be changed into a wide range between ≈ 0.0385 and 0.5 [see equations (1.129) and (1.130)]. In particular, if the black hole is corotating with the orbital motion, this value can be significantly large for a rapidly spinning black hole. For such a case, the solid curve of figure 8.34 would be significantly shifted to the high mass-ratio side (see figure 8.43 for a related topic).

8.5 Black hole-neutron star binaries: Numerical simulations

Since the first works by Shibata and Uryū (2006, 2007); Shibata and Taniguchi (2008); Etienne *et al.* (2008); Duez *et al.* (2008), many numerical simulations have been performed for the merger of black hole-neutron star binaries. The primary interest in this problem is summarized as follows:

(1) What is the condition for tidal disruption of neutron stars ?
(2) What is the property of the outcome if tidal disruption occurs ? Specifically, how large is the mass of the remnant disk/torus and what is the property of the disk/torus ?
(3) How large fraction of the neutron-star matter is ejected from the system for the case that tidal disruption occurs ?
(4) What is the feature of gravitational waves emitted in the late inspiral, merger, and ringdown phases ?

In the following subsections, we will summarize our understanding for these questions, which have been obtained by numerical-relativity simulations.

8.5.1 *Methodology of simulations and equations of state*

First, we summarize the methodology of numerical-relativity simulations employed for this problem. Two formulations have been employed for a solution of Einstein's equation: BSSN-puncture formulation together with the moving puncture gauge (see section 2.3.3) and GH formulation together with the excision boundary condition for handling black holes (see section 2.3.4). Both formulations work well, and for some particular models, it has been shown that the numerical results by these two formulations agree well with each other. In both approaches, an adaptive mesh refinement algorithm or a multi-domain algorithm, in which the simulations are performed on multi computational domains, is employed (except in the early works performed before 2007). For hydrodynamics or magnetohydrodynamics, employing high-resolution shock capturing schemes (see section 4.8) is standard.

As already mentioned in section 8.3.1, physical modeling of equations of state for the neutron-star matter is a key ingredient for accurately exploring the remnant formed after the onset of the merger and gravitational waves emitted during the late inspiral and merger stages. The primary reason for this on this problem is that the merger remnant and gravitational waveforms in the merger stage depend strongly on the relation between the mass and the radius (i.e., equation of state) of the neutron star, because the sensitivity of the neutron star to the tidal force exerted by its companion black hole depends strongly on its compactness as outlined in section 1.5.4. Specifically, a neutron star of smaller compactness (with a stiffer equation of state) will be disrupted at a larger orbital separation (or a lower orbital frequency). When the tidal disruption of a neutron star occurs at a large distance, a large amount of the neutron-star matter could be widely spread around its companion black hole, and consequently, a high-mass remnant disk/torus will be formed. Such a system could be a central engine of gamma-ray bursts, and hence, its formation process deserves a detailed study. In addition, the gravitational-wave frequency at tidal disruption is the key frequency and deserves a detailed analysis when it is detected. This frequency depends on the compactness of the neutron star: It would be in general lower for a neutron star of smaller compactness. All these features have to be quantified employing physical equations of state.

Again, hypothetical equations of state have to be carefully employed for simulations, taking into account the evolution process. Before the onset of the merger, neutron stars are cold; their thermal energy is much lower than the Fermi energy of nucleons. After the onset of the merger, a neutron star may be tidally disrupted by its companion black hole. For this case, the disrupted neutron-star matter can form spiral arms, and subsequently, the spiral arms collide with each other. This results in shock heating of the matter in the spiral arms, and hence, the disk or torus formed subsequently could be hot. These facts imply that (i) for simulating the system before the onset of the merger, cold equations of state should be employed; (ii) for the case that tidal disruption does not occur (a neutron star is simply swallowed by its companion black hole), employing cold equations of state is acceptable; (iii) for following long-term evolution of the tidally disrupted matter which will form spiral arms and a disk/torus around the remnant black hole, the finite-temperature effects of equations of state have to be taken into account. The finite-temperature equation of state is also a key ingredient for exploring the physical condition of dynamical ejecta. For a detailed thermodynamics study, microphysics effects such as neutrino cooling and heating, nucleon composition, and electron fraction per baryon have to be also taken into account. We note that the density of the disk/torus (typically $\sim 10^{11}$–$10^{12}\,\mathrm{g/cm^3}$) is much lower than that of neutron stars, and thus, the Fermi energy of electron and nucleon is significantly lowered during the merger process. Hence, the role of the thermal gas pressure becomes increasingly important.

Table 8.1 The parameters and key ingredients for some of a simplified version of piecewise polytropic equations of state employed in Kyutoku et al. (2010b, 2011); Lackey et al. (2012, 2014). $\Gamma_1 (= \Gamma_2 = \Gamma_3)$ is the adiabatic index for the neutron-star core region and p_1 is the pressure at the fiducial density $\rho_{\text{fidu}} = 10^{14.7}$ g/cm^3 (see also table 1.1). M_{max} is the maximum mass of spherical neutron stars for a given equation of state. $R_{1.35}$ and $\mathcal{C}_{1.35}$ are the circumferential radius and the compactness of spherical neutron stars with $M_{\text{NS}} = 1.35\,M_\odot$.

Model	Γ_1	$\log_{10} p_1/c^2$ (g/cm^3)	$M_{\text{max}}(M_\odot)$	$R_{1.35}$ (km)	$\mathcal{C}_{1.35}$
2H	3.0	13.95	2.835	15.23	0.1309
1.5H	3.0	13.75	2.525	13.69	0.1456
H	3.0	13.55	2.249	12.27	0.1624
HB	3.0	13.45	2.122	11.61	0.1718
B	3.0	13.35	2.003	10.96	0.1819

In the early history of this field, many works were carried out with a simple Γ-law equation of state, $P = (\Gamma - 1)\rho\varepsilon$, with $\Gamma = 2$ preparing quasi-equilibrium states with the polytropic equation of state, $P = \kappa_{\text{p}}\rho^\Gamma$, as the initial condition. As we already noted for several times, this equation of state is not very physical, although for a qualitative study, it may be used.

With the progress of this field, more physical equations of state such as piecewise polytropic equations of state in the form of equation (8.16) [Kyutoku et al. (2010b, 2011); Lackey et al. (2012, 2014); Kyutoku et al. (2013)] have been employed. With this equation of state, tidal disruption and subsequent short-term evolution can be explored in a realistic manner. In the following, we will often refer to the works in which a simplified version of the piecewise polytropic equation of state, for which $\Gamma_1 = \Gamma_2 = \Gamma_3$ is assumed, was used: see table 8.1 (compare with table 1.1). The motivation for setting $\Gamma_1 = \Gamma_2 = \Gamma_3$ stems from the fact that for black hole-neutron star binaries, the maximum density monotonically decreases during the merger process, and hence, simplifying the equation of state for the high-density range is an acceptable approximation.

To self-consistently take into account thermal and microphysics effects, which play an important role for the evolution of a disk/torus formed after tidal disruption and for dynamical ejecta, the best way is to employ finite-temperature equations of state derived based on a nuclear physics theory together with taking into account neutrino transfer effects (see section 1.4.7). With the progress of this field, simulations with such equations of state are getting popular [Deaton et al. (2013); Foucart et al. (2014)]. We will introduce some of the pioneer works in section 8.5.3.3.

8.5.2 Merger process

8.5.2.1 Zero black-hole spin case: $\hat{a} = 0$

Broadly speaking, the final fates of black hole-neutron star binaries are classified into two classes: the neutron star is tidally disrupted before it is swallowed by its companion black hole or the neutron star is simply swallowed by the black hole.

Figures 8.35 and 8.36 display the snapshots of rest-mass density profiles and the location of apparent horizons of the black hole on the orbital plane at selected time slices for two typical cases. For these results, the neutron stars are modeled by one of the piecewise polytropic equations of state listed in Table 8.1. Figure 8.35 illustrates the evolution process in which a neutron star is tidally disrupted before the binary orbit reaches the ISCO and then a disk is formed surrounding a remnant black hole. For this model, $M_{\rm BH} = 2.7\,M_\odot$, $\hat{a} = 0$, $M_{\rm NS} = 1.35\,M_\odot$, and $R_{\rm NS} = 15.2$ km (2H equation of state in table 8.1 was used). In this case, the mass ratio is small ($Q = 2$) and the neutron-star radius is large; it is larger than the extent of the black-hole horizon. For this setting, the neutrons star is significantly tidally deformed in close orbits, and eventually, mass shedding from an inner cusp (Lagrange point) of the neutron star sets in far outside the ISCO. During the mass-shedding phase, a substantial amount of fluid elements are removed from the inner cusp, the neutron star is highly elongated, and eventually tidally disrupted. It should be emphasized that tidal disruption does not occur immediately after the onset of mass shedding in this case. The tidal disruption occurs *for an orbital separation which is smaller than that at the onset of mass shedding*. The reason for this is that at the mass-shedding limit, the neutron star already had appreciable approaching velocity that was induced by the gravitational-radiation reaction and it continued to approach the black hole after the onset of the mass shedding.

Even for the case that tidal disruption occurs, most of the neutron-star matter (for the zero spin case, more than 80% of the total rest mass) falls into the black hole. Because the radius of the neutron star is as large as or larger than the extent of the black-hole horizon, the infall of the neutron-star matter proceeds for a wide region of the black hole (see left panel of figure 1.30). On the other hand, a fraction of the neutron-star matter located in a far region from the black hole forms a spiral arm after the onset of the tidal disruption. Due to the angular-momentum transport in this spiral arm, a large amount of material spreads outward, and after the spiral arm is wound due to the differential rotation, a disk of approximately axisymmetric configuration is formed around the black hole. Because of the presence of a non-axisymmetric structure at its formation, however, the disk does not completely relax to an axisymmetric state in the rotation period ~ 10 ms. Rather, non-axisymmetric perturbations of small amplitude are present in the disk for a long time scale, and help gradually transporting angular momentum outward that results in a gradual mass accretion into the black hole [Hawley (1991); Kiuchi *et al.* (2011a)]. However, the mass accretion time scale is much longer than the orbital period around the black hole, and hence, the disk remains to be quasi-stationary for $\gg 10$ ms. In reality, the subsequent evolution of the disk will be determined by the time scale of viscous accretion, which is probably induced by the viscous stress associated with the MRI [Balbus and Hawley (1998)] (see also section 10.3).

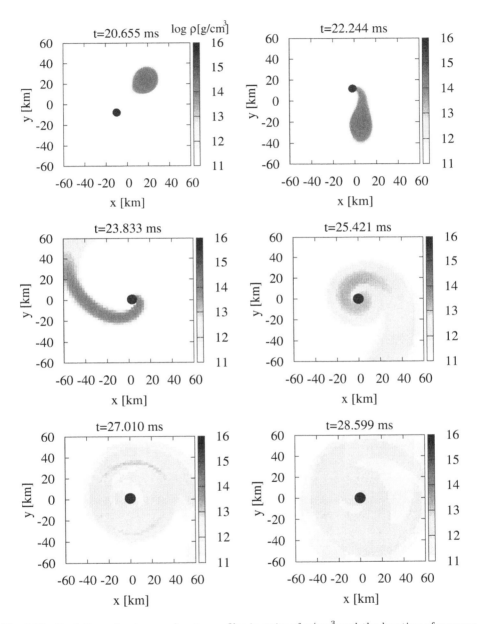

Fig. 8.35 Evolution of rest-mass density profiles in units of g/cm^3 and the location of apparent horizons on the equatorial (orbital) plane for a model with $M_{\rm BH} = 2.7\,M_\odot$, $\hat{a} = 0$, $M_{\rm NS} = 1.35\,M_\odot$, and $R_{\rm NS} = 15.2$ km (2H equation of state was used). The filled circle denotes the region inside the apparent horizons of the black hole. The color panel in the right-hand side of each figure shows $\log_{10}(\rho)$. The figure is taken from Kyutoku et al. (2010b).

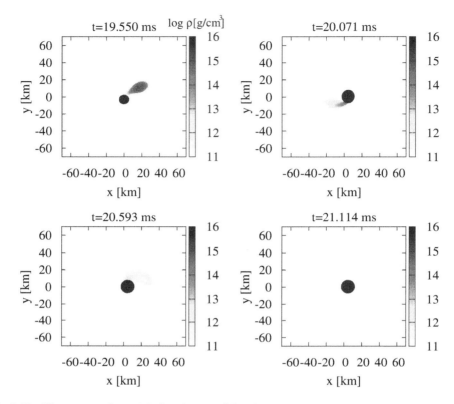

Fig. 8.36 The same as figure 8.35 but for a model with $M_{\rm BH} = 4.05\,M_\odot$, $\hat{a} = 0$, $M_{\rm NS} = 1.35\,M_\odot$, and $R_{\rm NS} = 11.0$ km (B equation of state was used).

Figure 8.36 illustrates the case in which a neutron star is not tidally disrupted before it is swallowed by a black hole. For this model, $M_{\rm BH} = 4.05\,M_\odot$, $\hat{a} = 0$, $M_{\rm NS} = 1.35\,M_\odot$, and $R_{\rm NS} = 11.0$ km (B equation of state in table 8.1 was used). In this case, the neutron star is tidally deformed only for a very close orbit. Although mass shedding sets in eventually, it occurs at an orbit which is too close to induce subsequent tidal disruption outside the ISCO. Hence, most of the neutron-star matter falls into the black hole in a short time scale (within an orbital period). Consequently, mass of the disk formed is negligible (much smaller than $0.01\,M_\odot$).

In addition, the infall occurs from a narrow region of the black-hole horizon (see middle panel of figure 1.30). These processes help coherently exciting a fundamental non-axisymmetric quasi-normal mode of the remnant black hole.

Generally speaking, the final outcome depends on the orbit at which mass shedding of the neutron star sets in. If the orbit is in the vicinity of or inside the ISCO, most of the neutron-star matter falls into the companion black hole, and hence, the remnant black hole is surrounded by negligible matter. With the increase of the orbital separation at the onset of mass shedding, the mass of the matter surrounding the remnant black hole increases. However, we note again that mass shedding has to set in sufficiently outside the ISCO for inducing tidal disruption, because the tidal disruption occurs only after a substantial amount of the matter is striped from the the inner part of the neutron star.

It is interesting to note that for the zero black-hole spin case, roughly speaking, tidal disruption occurs if the size of the neutron star is larger than the size of the black-hole horizon (compare figures 8.35 and 8.36). Thus, when tidal disruption occurs, the subsequent infall of the neutron-star matter into the black hole proceeds from a wide region of the black-hole surface (see the left panel of figure 1.30). On the other hand, when tidal disruption does not occur, the infall of the neutron star proceeds from a narrow region of the black-hole surface (see the middle panel of figure 1.30) as already mentioned.

8.5.2.2 Nonzero black-hole spin case: $\hat{a} \neq 0$

The effect of the black-hole spin significantly modifies the orbital evolution process in the late inspiral phase and the merger dynamics. Figure 8.37 shows the trajectories of a black hole and a neutron star for models with $Q = 3$, $C = 0.145$, and $\hat{a} = 0$ (left) and $\hat{a} = 0.75$ (right) computed by Etienne et al. (2009). In this example, the neutron star is modeled by the Γ-law equation of state with $\Gamma = 2$, and the black-hole spin vector aligns with the orbital angular-momentum vector. The initial orbital angular velocity is the same for these two models. For the binary composed of a non-spinning ($\hat{a} = 0$) black hole, the merger occurs after about 4 orbits while for the case with a spinning black hole ($\hat{a} = 0.75$), it occurs after about 6 orbits. Qualitatively, and in the terminology of the post-Newtonian approximation, these differences may be explained primarily by the presence of a spin-orbit coupling effect, which exerts the additional repulsive (attractive) force for $\hat{a} > 0$ ($\hat{a} < 0$) [e.g., Kidder et al. (1993); Kidder (1995) and section 1.3.2.1 for the post-Newtonian analysis]. In the presence of this additional force, the centrifugal force may be weaker (for $\hat{a} > 0$) or stronger (for $\hat{a} < 0$) for a given orbital separation. This slows down the orbital velocity for $\hat{a} > 0$, and therefore, the luminosity of gravitational waves is decreased and orbital evolution due to the gravitational-radiation reaction is delayed (the lifetime of the binary becomes longer). An even more important fact is that the orbital radius at the ISCO around the black hole is decreased (increased) for $\hat{a} > 0$ ($\hat{a} < 0$) due to the spin-orbit coupling. This implies that the absolute value of the binding energy at the ISCO around the black hole is increased (decreased) due to the spin-orbit coupling effect for $\hat{a} > 0$ ($\hat{a} < 0$) [e.g., Bardeen et al. (1972) and section 1.3.2.1]. This further helps increasing (decreasing) the lifetime of the binary that evolves due to the gravitational-radiation reaction because for $\hat{a} > 0$ ($\hat{a} < 0$) the binary system has to emit more (less) gravitational waves until reaching the ISCO.

Since closer orbits are allowed for a binary with a spinning (corotating) black hole, the possibility of tidal disruption is enhanced. Also, due to the spin-orbit

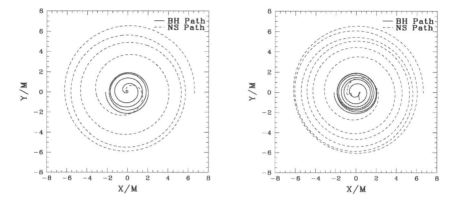

Fig. 8.37 Trajectories of the black-hole and neutron-star coordinate centroids for models with $Q = 3$, $\mathcal{C} = 0.145$, and $\hat{a} = 0$ (left) and $\hat{a} = 0.75$ (right). The neutron star is modeled by the Γ-law equation of state with $\Gamma = 2$. The coordinate centroids of the black hole and neutron star correspond to the centroids of the apparent horizon and the neutron-star density maximum, respectively. The figure is taken from Etienne et al. (2009).

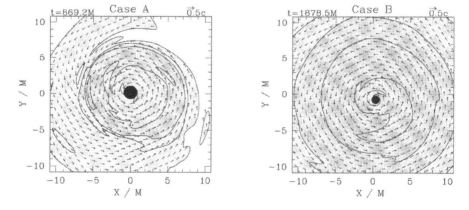

Fig. 8.38 Remnant black hole-disk system for models shown in figure 8.37 (left panel is for $\hat{a} = 0$ and right panel is for $\hat{a} = 0.75$). The contour curves, velocity fields (arrows), and black hole (solid circles) are plotted. The figure is taken from Etienne et al. (2009).

coupling effect, the final outcome after the onset of tidal disruption is modified. Figure 8.38 displays density contour curves for the remnant disk and the location of the remnant black hole for the same models as those shown in figure 8.37. For both models, the neutron stars are tidally disrupted outside the ISCO and a disk is formed. For the spinning black-hole case ($\hat{a} = 0.75$), more extended, more massive, and denser disk is the outcome. For the non-spinning case ($\hat{a} = 0$), the disk mass is only $\approx 4\%$ of the total rest mass while for $\hat{a} = 0.75$, it is $\approx 13\%$. This is due to the effects that the physical radius of the ISCO (or specific angular momentum of a particle orbiting the ISCO) around the spinning (corotating) black hole is smaller than that for the non-spinning black hole and also that the radial approaching velocity at tidal disruption is smaller for the spinning black hole because of the repulsive nature of the spin-orbit coupling effect.

Subsequently, more systematic works by Duez et al. (2010); Foucart et al. (2011); Kyutoku et al. (2011); Foucart et al. (2012, 2013); Lovelace et al. (2013); Kyutoku et al. (2013) also have shown that the corotating spin of black holes systematically enhances the possibility of tidal disruption and the resulting increase of the remnant disk mass with the increase of the magnitude of the spin. Foucart et al. (2011, 2012, 2013); Lovelace et al. (2013) performed simulations for $\hat{a} = 0.9$–0.97 with several values of $\mathcal{C}$ and Q in the Γ-law equation of state (and a slightly modified one) and showed that tidal disruption occurs far outside the ISCO for the rapidly spinning (and spin-aligned) case, and the resulting disk mass is quite high, $\gtrsim 30\%$ of the total rest mass. In particular, Lovelace et al. (2013) performed a simulation with $\hat{a} = 0.97$ and $\mathcal{C} = 0.144$, and illustrated the strong impact of the spin effect on the final remnant: They showed that $\sim 50\%$ of the total rest mass can escape from swallowing by the black hole and indicated that a disk of large mass and a large amount of mass ejection may be expected.

Kyutoku et al. (2011, 2014a) also found in their study employing neutron stars of mass $1.35 M_\odot$ with a variety of equations of state that for binaries composed of a high-spin black hole ($\hat{a} = 0.75$), tidal disruption occurs for a variety of the neutron-star compactness with $\mathcal{C} \lesssim 0.20$ for $Q = 4$, for $\mathcal{C} \lesssim 0.18$ for $Q = 5$, and for $\mathcal{C} \lesssim 0.16$ for $Q = 7$. Thus, in the presence of a black-hole spin, tidal disruption of a neutron star is possible for large black-hole mass (i.e., for a wide horizon area). In such high-mass case, the tidally disrupted neutron-star matter may be swallowed from a relatively narrow region of the black-hole surface (note that this is never the case for the non-spinning black-hole case). This mechanism helps exciting a non-axisymmetric fundamental quasi-normal mode of the remnant black hole (see section 8.5.4).

This mechanism is indeed observed in figure 8.39. For this case, $M_{\rm NS} = 1.35 M_\odot$, $M_{\rm BH} = 4.05 M_\odot$ ($Q = 3$), $\hat{a} = 0.5$, and HB equation of state in table 8.1 was used. The evolution process shown in figures 8.39 is similar to that in figures 8.35: mass shedding of the neutron star occurs at an orbit sufficiently far from the black hole, and subsequently, the neutron star is extensively elongated and tidally disrupted. Then, most of the neutron-star matter in the vicinity of the black hole falls into the black hole, while a spiral arm is formed in the region far from the black hole, and then the spiral arm composed of dense matter is wound around the black hole, leading to the formation of a disk surrounding the black hole. The difference is found in the infall process onto the black hole: The infall of the dense matter proceeds from a wide region of the black-hole surface for the non-spinning case (see, e.g., the fourth panel of figure 8.35). By contrast, for $\hat{a} = 0.5$ (see the second–fourth panels of figure 8.39), the infall proceeds from a relatively narrow region of the black-hole surface. This is due to the fact that the tidal disruption occurs relatively in the vicinity of the black hole, although the site is outside the ISCO. The essence here is that the relative distance between black-hole horizon and ISCO radius is closer than that for the non-spinning black-hole case (see the right panel of figure 1.30 in chapter 1 that schematically shows the situation). For such cases, the matter infall could occur from a narrow region of the black hole.

The significant impact of the black-hole spin on the merger process is also found from figure 8.40. For this case, the mass and equation of state are the same as those for figure 8.39 but for $\hat{a} = -0.5$ (counter-rotating spin): The spin-orbit coupling effect exerts the attractive force. In this case, the ISCO is located at a more distant orbit than that for the non-spinning case, and tidal disruption does not occur; more than 99.99% of the neutron-star matter falls into the black hole from a narrow region of the black-hole horizon in a short time scale.

To summarize, numerical-relativity simulations have clarified that there are three types of the merger processes for the canonical-mass ($\sim 1.35 M_\odot$) neutron star depending on the black-hole spin and binary mass ratio:

I   For the case that the black-hole mass is low or the black-hole spin is sufficiently high, the neutron star is tidally disrupted for an orbit far from the ISCO. The final remnant in this case is composed of a black hole surrounded by a massive disk (or torus).

II For the case that the black-hole mass is not low and the black-hole spin is small or counter-rotating ($\hat{a} < 0$), the neutron star is not tidally disrupted.

III For the case that the black-hole mass is not low and the black-hole spin is corotating ($\hat{a} > 0$) and high, the neutron star is tidally disrupted for an orbit close to the ISCO.

These three types of the merger processes are schematically described in figure 1.30 of chapter 1. These differences in the the late inspiral and merger processes are well reflected in gravitational waveforms and their spectra (see section 8.5.4).

The orientation of the black-hole spin with respect to the orbital plane is also an important parameter for determining the merger remnant as first demonstrated by Foucart et al. (2011, 2013). They performed various simulations for $\hat{a} = 0.5$ and 0.9 and for $Q = 3$ and 7 with a Γ-law equation of state ($\Gamma = 2$), and found that the remnant disk mass decreases sensitively with the increase of the inclination angle (misalignment angle between the spin vector $\mathbf{S}$ and orbital angular momentum vector $\mathbf{L}$) for given values of $\hat{a}$ and Q. This is due to the fact that the spin-orbit coupling force is proportional to $\mathbf{S} \cdot \mathbf{L}$, and also, the radius of the ISCO around the black hole approaches the radius for $\hat{a} = 0$ with the increase of the inclination angle. Hence, the effects of the black-hole spin become less important with the increase of the inclination angle: Even when the black-hole spin is large, its effect could be minor if the inclination angle is close to $90°$ ($\mathbf{S} \cdot \mathbf{L} \sim 0$).

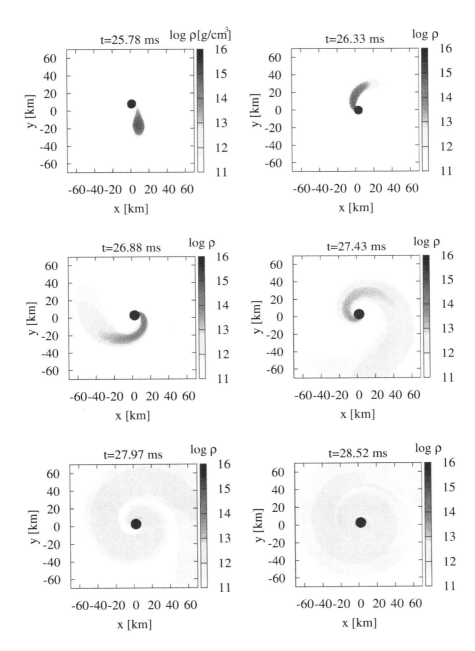

Fig. 8.39 The same as figure 8.35 but for a model with $M_{\rm BH} = 4.05\,M_\odot$, $\hat{a} = 0.5$, $M_{\rm NS} = 1.35\,M_\odot$, and $R_{\rm NS} = 11.6$ km (HB equation of state was used). The figure was taken from Kyutoku et al. (2011).

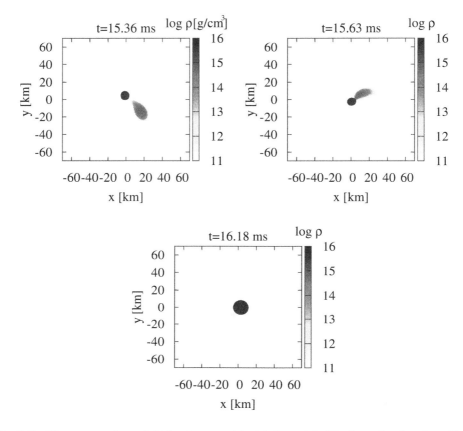

Fig. 8.40 The same as figure 8.39 but for a model with $\hat{a} = -0.5$. The figure is taken from Kyutoku et al. (2011).

Foucart et al. (2013) also explored the configuration of the remnant disk/torus formed after tidal disruption. Figure 8.41 shows some results by them. For this case, $Q = 7$, $\hat{a} = 0.9$, $\mathcal{C} = 0.144$, and the inclination angle is 20° (left) and 40° (right), respectively. In the presence of the inclination angle, the binary has a precessing motion during the inspiral phase. In the same manner, each fluid component of the spiral arm formed after tidal disruption has a precessing motion around the remnant black hole. Reflecting these facts, the remnant material surrounding the black hole does not constitute a disk but a geometrically-thick torus of a scattered density distribution around the black hole. Such torus would subsequently evolve due to an angular-momentum transport process perhaps associated with the magneto-viscous effect in a unique way spending a long time scale [Fragile et al. (2007)]. Eventually, the torus may relax to a state in which the angular-momentum axis of the torus approximately aligns with the axis of the remnant black-hole spin [Bardeen and Petterson (1975)], if the lifetime of the torus is sufficiently long.

Fig. 8.41 The tidal-disruption process of precessing binaries with $Q = 7$, $\hat{a} = 0.9$, and $\mathcal{C} = 0.144$. For left and right panels, the initial inclination angle between the orbital angular momentum and black-hole spin is 20° and 40°, respectively. The figure is taken from Foucart et al. (2013).

8.5.2.3 Criteria of tidal disruption

Tidal disruption occurs *after* the onset of mass shedding and subsequent substantial mass strip of the neutron-star matter. Thus, it is worthy to emphasize again that *the condition of tidal disruption is in general different from that of mass shedding for black hole-neutron star binaries* and that tidal disruption could occur in a more restricted condition than that for mass shedding. For determining the condition of tidal disruption, numerical-relativity simulation is necessary: Studies of quasi-equilibrium sequences are not sufficient for this issue.

However, it is not easy to strictly determine the condition for tidal disruption, because its concept is not as clear as that for mass shedding. The simplest way is to use the property of the merger remnant; we may recognize that tidal disruption occurs if the mass of a remnant disk surrounding a remnant black hole is substantially large, say larger than 1% of the total rest mass at $t \sim 10$ ms after the onset of mass shedding. We here employ this criterion.

Broadly speaking, the condition for tidal disruption depends on the mass ratio, the compactness (i.e., equation of state) of the neutron star, and the spin of the black hole. First, we summarize the criterion for the case that the black-hole spin is zero ($\hat{a} = 0$). The most detailed study for this issue was done employing the Γ-law equation of state with $\Gamma = 2$. Thus, first of all, we summarize the results for this case.

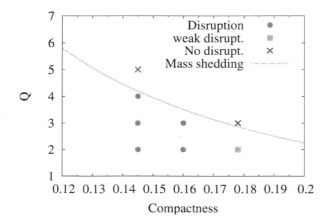

Fig. 8.42 A summary for the conditions of tidal disruption and mass shedding for $\hat{a} = 0$ and for Γ-law equation of state with $\Gamma = 2$ in the plane of $(\mathcal{C}, Q)$. If the value of Q or $\mathcal{C}$ is smaller than that of the threshold curve shown here, mass shedding occurs [see figure 8.34 and Taniguchi et al. (2008)]. The circles and crosses denote that tidal disruption occurs and does not occur, respectively, for the corresponding values of $\mathcal{C}$ and Q. The square denotes that tidal disruption only weakly occurs; tidal disruption occurs but the resulting disk mass is smaller than 1% of the total rest mass. On the other hand, the disk mass is between 1% and 10% of the total rest mass for the models marked by the circles.

Figure 8.42 summarizes of the numerical results by Shibata and Taniguchi (2008) [corrected in Shibata et al. (2012)], Duez et al. (2010), and Etienne et al. (2009) on tidal disruption together with the mass-shedding limit [Taniguchi et al. (2008)] in the plane of $(\mathcal{C}, Q)$. We note that the $Q = 3$ case was studied independently by these three groups and their conclusions agree with each other. The circles and crosses denote that tidal disruption occurs and does not occur, respectively, for the corresponding values of $(\mathcal{C}, Q)$. This indicates that for a small compactness of the neutron stars, $\mathcal{C} \lesssim 0.15$, the condition for tidal disruption agrees approximately with that for mass shedding. By contrast, for a large value of the compactness with $\mathcal{C} \gtrsim 0.17$, two conditions do not agree well: Tidal disruption occurs for the more restricted case, i.e., for the large compactness with $\hat{a} = 0$, tidal disruption occurs only for a small value of Q. For such a system, the time scale for the gravitational-radiation reaction is as short as the orbital period in close orbits [see equation (1.93)]. This implies that at the onset of mass shedding, the radial approaching velocity induced by the gravitational-radiation reaction is high enough to significantly decrease (increase) the orbital radius (angular frequency) during the subsequent mass-shedding phase. If the fraction of this decrease in the orbital radius is large enough to enforce the orbit to go inside the ISCO, tidal disruption is prohibited.

For the case that the value of $\mathcal{C}$ is small, tidal disruption could occur for a large value of Q. Here, with the increase of Q (i.e., with the decrease of the ratio of the reduced mass to the total mass), the ratio of the gravitational-radiation reaction time scale to the orbital period at the ISCO increases [see equation (1.93)]. For such a high-Q case, the effect of the orbital shrinking by the gravitational-radiation reaction after the onset of mass shedding becomes relatively minor, and therefore, the critical curves of tidal disruption and mass shedding approach each other.

The criterion of tidal disruption for the non-spinning black-hole case was also studied by Kyutoku et al. (2010b) for a variety of realistic equations of state (based on the piecewise polytrope) with $Q = 2$ and 3. Their results agree approximately with the result shown in figure 8.42 (see figure 8.43). One point to be added is as follows. The density profile of neutron stars depends on the equation of state hypothetically employed, even if the mass and compactness for two models are identical. Thus, the condition for tidal disruption could depend not only on the compactness but also on the density profile. Kyutoku et al. (2010b) showed that this is indeed the case. Specifically, for neutron stars of a centrally condensed density profile, tidal disruption becomes less subject.

In the presence of a spinning black hole, in particular for the case that its spin axis aligns with the orbital angular-momentum axis, the condition for tidal disruption is highly relaxed. Here, we focus on this aligned case in which the spin effect is most significantly reflected: If these two axes misalign, the spin effect is less appreciable, as already mentioned in the last two paragraphs of section 8.5.2.2.

As in the non-spinning case, the condition for tidal disruption can be approximately determined by the rest mass of the remnant disk. Kyutoku et al. (2011, 2014a) explored it by the simulations changing the mass ratio, black-hole spin, and equations of state of neutron stars systematically (but with the fixed value of the neutron-star mass $M_{NS} = 1.35 M_\odot$). In their work, piecewise polytropic equations of state shown in tables 1.1 and 8.1 were employed. Figure 8.43 displays a summary of their results. This shows that for $\hat{a} = 0.75$ and $\hat{a} = 0.5$, tidal disruption always occurs for realistic neutron stars with $\mathcal{C} \leq 0.18$ as long as $Q \leq 5$ and $Q \leq 3$, respectively. We note that Duez et al. (2010) also performed a number of simulations for $\hat{a} = 0.5$ and $Q = 3$, and also found that tidal disruption occurs irrespective of equations of state for which $\mathcal{C} < 0.18$. These results are in clear contrast with the results for $\hat{a} = 0$ and demonstrate that with the increase of the spin magnitude, the allowed parameter range of tidal disruption steeply increases. By contrast, for a counter-rotating spin ($\hat{a} = -0.5$), the allowed range of tidal disruption is significantly restricted; even for $Q = 2$, $\mathcal{C} \lesssim 0.16$ is necessary.

In the presence of a high black-hole spin, say, $\hat{a} \gtrsim 0.9$, tidal disruption universally occurs even for a high mass ratio, $Q \sim 7$ ($M_{BH} \sim 10 M_\odot$ for $M_{NS} = 1.4 M_\odot$) for a variety of the neutron-star compactness (see also section 1.5.4). Foucart et al. (2013) showed that tidal disruption indeed occurs for such cases and the mass of the resulting disk surrounding the remnant black hole is larger than 10% of the total rest mass for $\mathcal{C} \lesssim 0.17$. A binary composed of a high-mass high-spin black hole is the promising system for observing the tidal-disruption event.

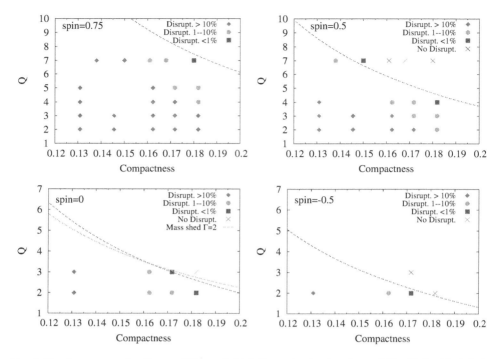

Fig. 8.43 A summary for the conditions of tidal disruption in the plane of ($\mathcal{C}$, Q) for black-hole spins $\hat{a} = 0.75$ (top left), $\hat{a} = 0.5$ (top right), $\hat{a} = 0$ (bottom left), and $\hat{a} = -0.5$ (bottom right) for the case that several piecewise polytropic equations of state listed in tables 1.1 and 8.1 are employed. The diamonds, circles, and squares denote the cases in which tidal disruption occurs and the remnant disk mass are $> 10\%$, $1-10\%$, and $< 1\%$ of the total rest mass, respectively. On the other hand, the crosses denote that tidal disruption does not occur or the remnant disk mass is smaller than 0.1% of the total rest mass. The dashed curve for $\hat{a} = 0$ denotes the mass shedding limit for the Γ-law equation of state with $\Gamma = 2$ (the same as in figure 8.42). The dot-dot curves drawn for all the panels denote the formula in equation (8.47). The numerical results are taken from Kyutoku et al. (2011, 2014a).

Figure 8.43 shows that the condition for tidal disruption has a systematic feature irrespective of $\hat{a}$. This suggests that the condition may be written by an analytic fitting formula. Foucart (2012) suggested a fitting formula for the condition of tidal disruption as

$$\mathcal{C} \lesssim \left(A_0 + A_1 Q^{2/3} \hat{r}_{\text{ISCO}} \right)^{-1}, \qquad (8.46)$$

where $\hat{r}_{\text{ISCO}}$ is defined in equation (1.130). A_0 and A_1 are constants for which Foucart (2012) suggested to employ 2 and ≈ 2.14, respectively. Numerical-relativity results shown in figure 8.43 indicate that A_1 should be slightly smaller as ≈ 1.9 with which the dotted curve of figure 8.42 is approximately reproduced, although the fitting formula of the form (8.46) captures a qualitative feature for reproducing the numerical results.

There would be a lot of possible alternative formulas if we restrict our attention to a realistic range of $\mathcal{C}$ as $0.1 \lesssim \mathcal{C} \lesssim 0.2$. For example, based on equation (8.43), we may employ the following formula

$$\mathcal{C} = \left[(2.1 - 0.45\hat{a})(Q+1)Q^{-1/3}(\hat{r}_{\text{ISCO}}/6)\right]^{-1}. \tag{8.47}$$

This can approximately reproduce the threshold. The dot-dot curves in figure 8.43 denote the result of this formula.

Finally, it should be noted that for the high-mass high-spin black-hole case, the conditions for tidal disruption and mass shedding would agree approximately with each other. The reason for this is that these occur at an orbit at which the gravitational-radiation reaction time scale is not as short as the orbital period: The approaching velocity of the binary component is much smaller than the orbital velocity. This point has not been explored quantitatively, and should be clarified by the future study.

8.5.3 *Properties of the remnants*

8.5.3.1 *Properties of remnant black hole*

During the merger, the neutron-star matter falls into its companion black hole, and then, the mass and spin of the black hole vary. The final black-hole mass may be approximately estimated by

$$M_{\text{BH,f}} = M_{\text{BH}} + M_{\text{NS}} - M_{r>\text{AH}} - E_{\text{GW}}, \tag{8.48}$$

where $M_{r>\text{AH}}$ denotes the mass of fluid elements which do not fall into the black hole: It is the total mass of the disk and ejecta (the disk mass is dominant: see section 8.5.6). E_{GW} denotes the total energy dissipated by the gravitational-wave emission. Typically, $M_{r>\text{AH}}$ and E_{GW} are less than 10% and $\sim 1\%$ of the initial total mass m_0, respectively. Hence, the final black-hole mass will be between $0.9m_0$ and m_0. This final mass depends strongly on how the neutron star is tidally disrupted: When the value of $M_{r>\text{AH}}$ is large as a result of tidal disruption, the final black-hole mass is suppressed.

The final black-hole spin depends sensitively on the mass ratio and initial black-hole spin. This can be understood by the following simple analysis: In Newtonian gravity, the total orbital angular momentum for two point particles in a circular orbit with an angular velocity Ω_{orb} is

$$J_{\text{orb}} = \frac{M_{\text{BH}} M_{\text{NS}}}{(\Omega_{\text{orb}} m_0)^{1/3}}. \tag{8.49}$$

Thus, the dimensionless spin parameter of the system is written approximately as

$$\hat{a}_f = \frac{J_{\text{orb}} + M_{\text{BH}}^2 \hat{a}}{m_0^2} = \frac{(\Omega_{\text{orb}} m_0)^{-1/3} Q + \hat{a} Q^2}{(1+Q)^2}, \tag{8.50}$$

where we assumed that the black-hole spin aligns with the orbital angular-momentum axis. If the mass and angular momentum of the remnant disk and

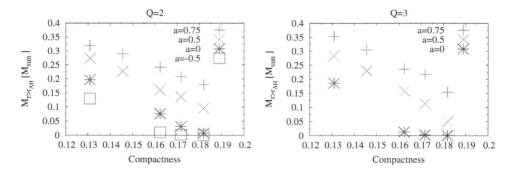

Fig. 8.44 Left: Disk mass at 10 ms after the onset of the merger as a function of the neutron-star compactness $\mathcal{C}$ with various piecewise polytropic equations of state and with various values of $\hat{a}$ for $Q = 2$. Right: The same as the left panel but for $Q = 3$. The figure is taken from Kyutoku et al. (2011).

ejecta as well as loss by gravitational waves could be neglected, $\hat{a}_f$ would be equal to the spin of the remnant black hole. At the onset of the merger or at tidal disruption, the angular velocity becomes $\Omega_{\rm orb} m_0 \sim 0.05$–$0.1$, and thus, $(\Omega_{\rm orb} m_0)^{-1/3}$ is in a narrow range ~ 2.1–2.7. This implies that $\hat{a}_f$ is primarily determined by Q and $\hat{a}$.

Equation (8.50) is a rather qualitative estimate for the spin of the remnant black hole. Nevertheless, it can still yield a good approximate value of the final spin with the choice of $(\Omega_{\rm orb} m_0)^{-1/3} = 3$ as far as the mass of the remnant disk and ejecta is much smaller than the total mass. With this choice, $\hat{a}_f = 0.67, 0.56, 0.48$, and 0.42 for $Q = 2, 3, 4$, and 5, respectively, for $\hat{a} = 0$. These values agree with several results derived by Etienne et al. (2009); Shibata et al. (2009); Kyutoku et al. (2010b,a); Foucart et al. (2011) within the error of $\Delta \hat{a} = 0.01$–0.02.

For a large black-hole spin with $\hat{a} \gtrsim 0.5$, mass of the remnant disk and ejecta is often large ($\gtrsim 0.1\,M_\odot$: see section 8.5.3.2) even for a large value of $Q \sim 5$–7, in particular for the case that $\mathcal{C}$ is small. In such cases, equation (8.50) overestimates the final black-hole spin. However, this equation still captures the qualitative property for the final spin; e.g., for a small initial black-hole spin, the final black-hole spin is determined by the value of Q and the larger values of Q result in the smaller final black-hole spin; for the larger values of Q with a large initial black-hole spin $\hat{a} \gtrsim 0.5$, the final black-hole spin is determined primarily by the initial black-hole spin.

8.5.3.2 Mass of remnant disk

The mass and characteristic density of a remnant disk surrounding a remnant black hole depend sensitively on the mass ratio, the black-hole spin, and the equation of state (or the compactness) of the neutron star. Figures 8.44–8.46 illustrate this fact

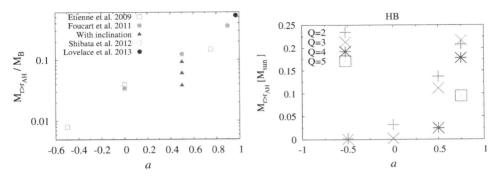

Fig. 8.45 Left: Summary of the remnant disk mass as a function of the initial black-hole spin for a fixed equation of state (Γ-law one with $\Gamma = 2$), neutron-star compactness ($\mathcal{C} \approx 0.145$), and mass ratio ($Q = 3$), computed by three different groups. The vertical axis shows the fraction of the disk mass $M_{r>r_{\rm AH}}/M_{\rm B}$ where $M_{\rm B}$ is the baryon rest mass of the neutron star. "With inclination" shows the results by Foucart et al. (2011) in the presence of an inclination angle of the black-hole spin with respect to the orbital angular-momentum vector, 40, 60, and 80 degrees (from upper to lower points). Right: The same as the left panel but for the disk mass in the solar mass unit for more compact neutron star ($\mathcal{C} \approx 0.172$) with a piecewise polytropic equation of state (HB listed in table 8.1). The simulation was performed by Kyutoku et al. (2011). For both panels, the disk mass is measured at $t \approx 5$–10 ms after the onset of the merger (for the Γ-law equation of state, $M_{\rm NS}$ is assumed to be $1.4\,M_\odot$ to fix the units).

clearly.[6] Figure 8.44 displays the remnant disk mass as a function of the neutron-star compactness for $Q = 2$ (left) and $Q = 3$ (right) for various piecewise polytropic equations of state and for a variety of black-hole spins [Kyutoku et al. (2011)]. This shows that the remnant disk mass decreases steeply and systematically with the increase of the compactness irrespective of the black-hole spin and mass ratio.

Comparison between the left and right panels of figure 8.44 as well as figures 8.42 and 8.43 also shows basically that the remnant disk mass monotonically decreases with the increase of Q for many cases. However, it is also shown in figure 8.44 that for the case that a large-mass disk is formed, this rule is not clear. For example, for a small mass ratio $Q \lesssim 3$, for relatively small compactness ($\mathcal{C} \lesssim 0.16$), and for high black-hole spins ($\hat{a} = 0.5$ and 0.75), the remnant disk mass depends only weakly on the value of Q for given values of $\hat{a}$ and $\mathcal{C}$ (see also figure 8.46).

The left panel of figure 8.45 plots together the results obtained by different groups [Etienne et al. (2009); Foucart et al. (2011); Shibata et al. (2012); Lovelace et al. (2013)] for the $\Gamma = 2$ equation of state with $\mathcal{C} \approx 0.145$ and $Q = 3$. This shows that the remnant disk mass increases steeply with the black-hole spin, $\hat{a}$, for given values of $\mathcal{C}$ and Q. The results by the three groups [Etienne et al. (2009); Foucart et al. (2011); Shibata et al. (2012)] agree approximately with each other for $\hat{a} = 0$. Foucart et al. (2011) in addition showed for the first time that the remnant disk mass decreases with the increase of the inclination angle of the black-hole spin

---

[6]In the "remnant disk mass" presented in this section, the mass of the ejecta is partly included. However, the fraction of the ejecta mass is always smaller than the disk fraction.

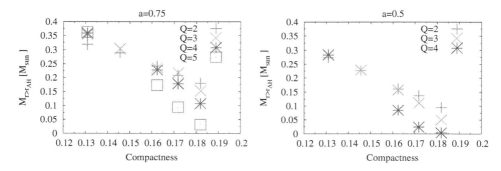

Fig. 8.46 Left: Remnant disk mass as a function of the neutron-star compactness for various values of Q with $\hat{a} = 0.75$. Right: The same as the left panel but for $\hat{a} = 0.5$. The figure is taken from Kyutoku et al. (2011).

with respect to the orbital angular-momentum axis, and toward the limit to 90° ($\mathbf{S} \cdot \mathbf{L} = 0$), the remnant disk mass approaches the value for $\hat{a} = 0$. It is also worthy to note that Lovelace et al. (2013) showed for the first time that the remnant disk mass steeply increases for $\hat{a} \geq 0.9$.

The right panel of figure 8.45 also plots the remnant disk mass as a function of the black-hole spin for different compactness ($\mathcal{C} \sim 0.172$ using HB equation of state) with $M_{\mathrm{NS}} = 1.35\,M_\odot$ [Kyutoku et al. (2011)]. This shows again that the remnant disk mass increases with the increase of the black-hole spin, and also that for a high black-hole spin (e.g., $\hat{a} = 0.75$), the remnant disk mass is as large as $0.1\,M_\odot$ even for $Q = 5$. Foucart et al. (2012, 2013) also showed that the remnant disk mass could be larger than $0.1\,M_\odot$ even for $\mathcal{C} \sim 0.17$ and $Q = 7$, illustrating the strong spin effect.

Figure 8.46 displays the remnant disk mass as a function of the neutron-star compactness for $\hat{a} = 0.75$ and 0.5 [Kyutoku et al. (2011)]. A steep decrease of the remnant disk mass with the increase of the compactness is found irrespective of the values of $\hat{a}$ and Q. The dependence is in particular strong for a high black-hole spin ($\hat{a} = 0.75$) and for a high mass ratio ($Q = 5$).

An interesting property found in figure 8.46 is that the remnant disk mass does not decrease steeply with the increase of the mass ratio for the high black-hole spin cases with small values of $\mathcal{C} \lesssim 0.15$. This indicates that the remnant disk mass is likely to be large even for a higher value of $Q (\geq 5)$ with a high black-hole spin ($\hat{a} \geq 0.5$).

We also note that the remnant disk mass depends not only on the compactness but weakly on the density profile as found by Kyutoku et al. (2010b,a). For neutron stars with more centrally concentrated density profiles, the remnant disk mass is smaller. The reason for this is that if the degree of the central mass concentration is larger, tidal deformation becomes less subject, and hence, the onset of tidal disruption after the onset of mass shedding is delayed.

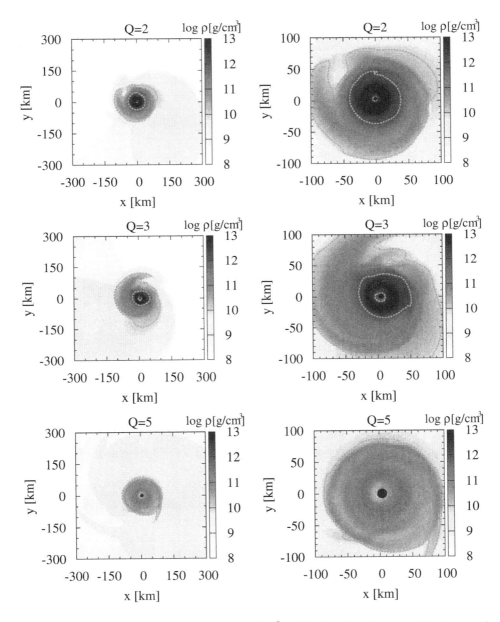

Fig. 8.47 Rest-mass density profiles in units of g/cm^3 and the location of apparent horizons on the equatorial plane at 10 ms after the onset of the merger for models with $\hat{a} = 0.75$, $M_{\rm NS} = 1.35\,M_\odot$, HB equation of state, and $Q = 2$ (top panels), 3 (middle panels), and 5 (bottom panels). The left and right columns show the profiles for square regions of $(600\,{\rm km})^2$ and $(200\,{\rm km})^2$, respectively. The filled circles around the origin denote the region inside apparent horizons, and the dotted contour curves are for $\rho = 10^{10}$ and $10^{12}\,{\rm g/cm}^3$. The color panel in the right-hand side of each figure shows $\log_{10}(\rho)$. The figure is taken from Kyutoku *et al.* (2011).

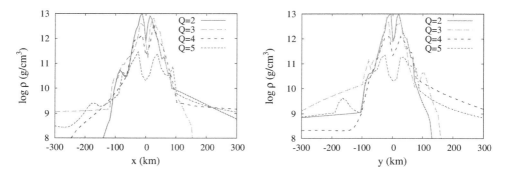

Fig. 8.48 Radial profiles of the rest-mass density at 10 ms after the onset of the merger for different values of Q. The left and right panels show the profiles along x and y axes, respectively. For these models, $(M_{NS}, \hat{a}) = (1.35 M_\odot, 0.75)$ and HB equation of state are employed. The figure is taken from Kyutoku et al. (2011).

8.5.3.3 Density, extent, and temperature of remnant disk/torus

The structure of the remnant disk and its time evolution process depend on the mass ratio (or black-hole mass or total mass) of the binary. Figure 8.47 displays the rest-mass density profile at 10 ms after the onset of the merger for binaries with $\hat{a} = 0.75$, $M_{NS} = 1.35 M_\odot$, HB equation of state, and different values of $Q = 2, 3$, and 5. Figure 8.48 also displays the corresponding radial profiles of ρ along x and y axes for the models with $Q = 2$–5. Irrespective of the mass ratio, a massive disk is formed for these cases because of the high black-hole spin. The left and right columns of figure 8.47 show the profiles for an extended region (left) and a close-up region (right). Although the remnant disk is approximately axisymmetric for a region $r \lesssim 100$ km, non-axisymmetric structures still remain reflecting the merger and tidal-disruption processes. Also for a far region, a spiral arm is still expanding, and thus, the degree of non-axisymmetry is still high.

We find that for the smaller values of Q, the remnant disk is denser; for $Q = 2$, its maximum density is $\sim 10^{13}$ g/cm^3 while for $Q = 5$, it is $\sim 10^{11}$ g/cm^3. This is clearly seen in figure 8.48. This is mainly due to the fact that for the smaller values of Q (with the fixed neutron-star mass), the mass of the remnant black hole is smaller, and hence, the ISCO radius of the black hole is smaller. As a consequence of this, the material of the remnant disk can concentrate at a small radius for the smaller values of Q, resulting in the enhancement of the density.

On the other hand, the region with $\rho \sim 10^{10}$ g/cm^3 universally extends to ~ 100 km. Also, the profiles in the left column of figure 8.47 suggest that the region of $\rho > 10^8$ g/cm^3 extends to a large distance for the large values of Q. Taking these facts into account, we conclude that a typical profile of $\rho(r)$ is steeper for smaller values of Q around the remnant black hole. All these features are qualitatively the same for binaries with other equations of state.

For the remnant disk of mass larger than $0.1 M_\odot$, the maximum density is larger than $\gtrsim 10^{12}\,\mathrm{g/cm^3}$ for $Q \lesssim 4$ and $\sim 10^{11}\,\mathrm{g/cm^3}$ for $Q = 5$. Hence, a high-mass remnant disk formed after the merger with a relatively small value of $Q \lesssim 5$ is likely to be universally opaque against the thermal-neutrino emission for a typical geometrical thickness of the disk because of the following reasons: The cross section of neutrinos with nucleons is

$$\sigma \sim \sigma_0 \left(\frac{E_\nu}{m_e c^2}\right)^2, \tag{8.51}$$

where $\sigma_0 = 1.76 \times 10^{-44}\,\mathrm{cm^2}$ denotes the characteristic cross section of weak interactions [Shapiro and Teukolsky (1983)], $m_e c^2$ is the electron rest-mass energy, 511 keV, and E_ν is the typical energy of neutrinos which is likely to be $\sim 10\,\mathrm{MeV}$ (see section 8.5.3.4). Assuming that the remnant disk is composed primarily of neutrons, the optical depth is estimated by

$$\sigma \frac{\rho}{m_n} R \approx 3.6 \left(\frac{\rho}{10^{11}\,\mathrm{g/cm^3}}\right)\left(\frac{E_\nu}{10\,\mathrm{MeV}}\right)^2 \left(\frac{H}{10 M_{\mathrm{BH}}}\right)\left(\frac{M_{\mathrm{BH}}}{6 M_\odot}\right), \tag{8.52}$$

where we supposed that the geometrical thickness of the remnant disk denoted by H is approximately $10 M_{\mathrm{BH}}$ where M_{BH} is the mass of the remnant black hole. Thus, for $Q \lesssim 5$, neutrinos cannot escape freely and they escape by diffusion. This implies that the cooling time scale is rather long and high disk temperature can be preserved. Therefore, a neutrino-dominated *hot* accretion disk surrounding a spinning black hole is the outcome. Such a remnant emits high-energy neutrinos for a time scale longer than the dynamical time scale of the system and hence it could be a promising candidate for the central engine of short-hard gamma-ray bursts (see the final paragraph of this section).

It should be noted here that not only the total mass but also the typical density is an important factor for determining the property of the remnant disk: As equation (8.52) shows, if the typical density is smaller than $\sim 10^{10}\,\mathrm{g/cm^3}$, neutrinos generated inside the remnant disk would soon escape from the system, and its temperature, T, would be significantly lowered.

According to the standard accretion disk model [Shakura and Sunyaev (1973); Novikov and Thorne (1973); Shapiro and Teukolsky (1983)], the remnant disk will evolve due to viscous dissipation. Using the so-called α-viscous law in which a viscous stress component is proportional to the pressure with a dimensionless coefficient α_v, the mass accretion rate into the black hole, $\dot{M}$, is roughly written as

$$\dot{M} \sim 4\pi \rho c_s^3 \Omega^{-2} \alpha_v, \tag{8.53}$$

where ρ, c_s, and Ω denote the typical density, sound speed, and angular velocity in the inner region of the remnant disk, and $\alpha_v \sim 0.01$–0.1. Here, $\rho \propto M_{\mathrm{disk}}/M_{\mathrm{BH}}^3$ and $\Omega \propto M_{\mathrm{BH}}^{-1}$. Thus, $\dot{M}$ is proportional to M_{BH}^{-1}. Approximately c_s^2 is proportional to the temperature T. For a low-density disk from which neutrinos freely escape, T goes down to a value much lower than that for the neutrino-dominated hot accretion disk, as mentioned above. Thus, $\dot{M}$ for the low-density disk would be much smaller than that for the high-density disk. Broadly speaking, the neutrino luminosity is proportional to

$$\frac{M_{\rm BH}\dot{M}}{R}, \qquad (8.54)$$

where R is the inner edge of the disk. Because $M_{\rm BH}/R(\sim 0.1)$ does not (or only weakly) depend on the disk structure, the luminosity is essentially proportional to $\dot{M}$. Therefore, the decrease in the density of the disk below a threshold value will result in a significant decrease of the neutrino luminosity.[7]

One of the promising models for generating a fireball of a gamma-ray burst is the pair annihilation of neutrinos and antineutrinos [Zhang and Mészáros (2004); Piran (2004)]. The efficiency of this model depends totally on the neutrino luminosity; the required neutrino luminosity is inferred to be 10^{53} ergs/s [e.g., Setiawan et al. (2006); Zalamea and Beloborodov (2011)]. Such a huge neutrino luminosity is likely to be achieved only for the neutrino-dominated hot accretion disk [e.g., Di Matteo et al. (2002); Kohri and Mineshige (2002); Lee et al. (2005); Shibata et al. (2007)]. This suggests that a binary composed of a high-mass black hole with $Q > 5$ might not be a good candidate for the progenitor that subsequently forms a central engine of gamma-ray bursts.

8.5.3.4 *Simulations with finite-temperature equations of state and neutrino transfer*

To explore the thermal properties of the remnant disk and neutrino luminosity, numerical-relativity simulations have to be performed incorporating finite-temperature equations of state together with a neutrino-transfer process. Only a few works have been done in this topic.

Duez et al. (2010) performed the first simulation incorporating a tabulated finite-temperature equation of state [of Shen et al. (1998a)] in this field. To omit both taking into account the effects of neutrino emission and solving the evolution equation of the electron fraction per baryon, Y_e, they assumed that the system is in β-equilibrium or that Y_e is unchanged in the fluid-moving frame (it is simply advected). In the former and latter, the entire system is assumed to be in either of the following two limiting cases; the weak interaction time scale is much shorter and longer than the dynamical time scale, respectively (neither limits are not realized in reality for the entire system, although they are partly realized). They performed simulations focusing on the case $\hat{a} = 0.5$ and $Q = 3$, and irrespective of choice of the two rules for Y_e, they found that the remnant disk is neutron-rich with $Y_e \sim 0.1$ (the simple advection case) – 0.2 (the β-equilibrium case) and temperature is only moderately high (maximum is ~ 10 MeV with the average ~ 3 MeV) with the maximum density $\sim 10^{12}$ g/cm^3 and disk mass $\sim 0.1\,M_\odot$. The typical density and mass

---

[7]The emissivity of neutrinos is determined primarily by the electron and positron capture by nucleons and approximately proportional to T^6 [e.g., appendix B of Shibata et al. (2007)]. Thus, even for a slight lowering of T, the luminosity is significantly reduced.

of the remnant disk agree approximately with those derived in piecewise polytropic equations of state with approximately the same neutron-star compactness for the zero-temperature state ($\mathcal{C} = 0.147$).

The typical value of Y_e for the neutron stars before the onset of the merger is ~ 0.1, and thus, for the simple advection case, the value of Y_e is essentially unchanged. For the β-equilibrium case, with the decrease of the density and the increase of the temperature, the production of electrons and protons from neutrons is enhanced, and thus, the typical value of Y_e is increased.

Subsequently, Deaton *et al.* (2013); Foucart *et al.* (2014) performed simulations using an equation of state of Lattimer and Swesty (1991) and employing a relatively simple version of the Leakage scheme for the neutrino cooling. In particular, Foucart *et al.* (2014) performed a detailed study. For their work, they employed $M_{\rm NS} = 1.2 M_\odot$ and $1.4 M_\odot$, $M_{\rm BH} = 7 M_\odot$ and $10 M_\odot$, and $\hat{a} = 0.7$, 0.8, and 0.9. Because of the high black-hole spin, the remnant disk mass becomes reasonably high as $\sim 0.1 M_\odot$. They found that the temperature for the relaxed state of the disk is in a relatively narrow range as $\sim 3-7\,{\rm MeV}$ while the value of Y_e is in a wide range $\sim 0.1-0.4$ depending on the physical condition of the disk: For a high-density region in which the degree of electron degeneracy is high, Y_e has small values. The high value of Y_e stems from the high temperature for which the degeneracy of electrons becomes weaker. Their result reflects the fact that the value of Y_e is sensitive to the physical conditions which depend on the location of the disk. They found that total neutrino luminosity is $\sim 10^{53}\,{\rm ergs/s}$ irrespective of the models. This order of magnitude agrees with that found in the study of accretion disks around black holes [e.g., Di Matteo *et al.* (2002); Kohri and Mineshige (2002); Lee *et al.* (2005); Shibata *et al.* (2007)]. In the high-density region of the disk, neutrinos do not freely escape as already mentioned in section 8.5.3.3.

8.5.4 *Gravitational waves*

Gravitational waves emitted by black hole-neutron star binaries have unique features, in particular in their merger stage. In the following, we will summarize them referring mainly to the numerical results by Etienne *et al.* (2009) and Kyutoku *et al.* (2010b, 2011).

8.5.4.1 *Zero black-hole spin case*

Figure 8.49 displays the typical gravitational waveforms for $\hat{a} = 0$, which clearly reflect the features of the late-stage orbital evolution and subsequent merger processes (tidal disruption or not). In the inspiral stage, the so-called chirp gravitational waves are emitted. Figure 8.49 shows that the gravitational waveforms in the late inspiral stage before the onset of the merger (or tidal disruption) are reproduced approximately by a modified post-Newtonian formula (Taylor-T4 formula: see appendix H). Note that in the first few wave cycles, the agreement is not very

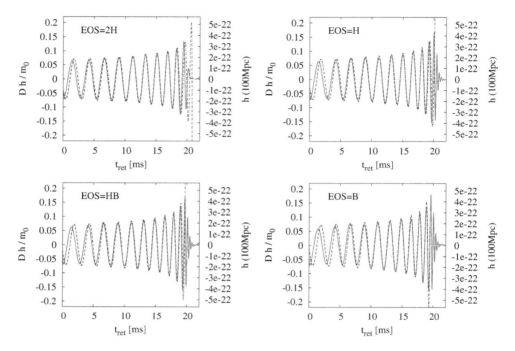

Fig. 8.49 Gravitational waveforms observed along the axis perpendicular to the orbital plane for $Q = 3$ and $\hat{a} = 0$ with 2H, H, HB, and B piecewise-polytropic equations of state (see table 8.1). The solid and dashed curves denote the numerical results and results derived by the Taylor-T4 formula (see appendix H). h is the amplitude, D is the distance from the source and $m_0 = M_{\rm BH} + M_{\rm NS}$. The left and right axes show the normalized amplitude (Dh/m_0) and physical amplitude for $D = 100$ Mpc, respectively. The figure is taken from Kyutoku et al. (2010b).

good. This is because the initial condition given for these simulations included a residual eccentricity.

For close orbits, the waveforms may deviate from the prediction by the Taylor-T4 formula, in particular for a small value of Q or for a large neutron-star radius. This is because the neutron star for such binaries is significantly deformed by the tidal force exerted by the companion black hole. For the case that the tidal effect becomes significant, the neutron star is tidally disrupted. At the tidal disruption, the amplitude damps suddenly in the middle of the inspiral waveform (see the top left panel of figure 8.49). The features of such gravitational waveforms are described later in more detail.

By contrast, for a sufficiently large value of Q or for a sufficiently compact neutron star, the tidal-deformation effect is negligible up to the merger, and hence, the waveforms in the merger stage are characterized by the ringdown oscillation associated with a fundamental quasi-normal mode of the remnant black hole, as in the waveform of binary black holes (see section 8.7.2). For this case, the highest frequency of gravitational waves is determined by the quasi-normal mode. Even for a small degree of tidal deformation and mass shedding, most of the neutron-star matter falls into the black hole simultaneously (this case corresponds to the type-II according to the definition of section 8.5.2.2: see figure 1.30). In such a case, a fundamental quasi-normal mode of the remnant black hole is also excited in the final merger stage (see the bottom two panels in figure 8.49).

The degree of the quasi-normal mode excitation depends strongly on the degree of tidal disruption of the companion neutron star. The primary reason for this is that a phase cancellation suppresses the excitation of the quasi-normal mode. Here, the phase cancellation is the effect that gravitational waves emitted in a non-coherent manner (with different phases) interfere with each other, and as a result, the amplitude of gravitational waves is suppressed [Nakamura and Sasaki (1981); Shapiro and Wassermasn (1982); Nakamura and Oohara (1983)]: With the increasing degree of tidal disruption, the phase cancellation effect plays an increasingly important role and the amplitude of ringdown gravitational waves decreases. For the case that a neutron star is tidally disrupted far outside the ISCO, this effect is significant, because the neutron-star matter does not simultaneously fall into the black hole. Rather, a widely spread material surrounding the black hole, which has the density much smaller than the typical density of neutron stars, falls from a wide region of the black-hole horizon spending a relatively long time duration (this case corresponds to the type-I according to the definition of figure 1.30). Here, the reason that the infall into a wide region of the black-hole horizon occurs is as follows: The black-hole mass has to be small enough for mass shedding to occur for $\hat{a} = 0$, and thus, the areal radius of the black hole is smaller than or as small as the neutron-star radius. Both the low density and the wide extension of the infalling material are discouraging for efficiently exciting a quasi-normal mode. Therefore, the amplitude of ringdown gravitational waves is strongly suppressed for the case that tidal disruption occurs (see top left panel of figure 8.49).

For the case that tidal disruption occurs, the highest frequency of gravitational waves is determined approximately by the orbital frequency at the tidal disruption, not by the frequency of a quasi-normal mode. One important remark appropriate here is that this highest characteristic frequency is not in general determined by the frequency at the onset of mass shedding. Even after the onset of mass shedding, the neutron star could remain to be a self-gravitating star for a while and gravitational waves associated with an approximately inspiral motion could be emitted. After the orbital separation appreciably decreases due to the gravitational-wave emission, tidal disruption would occur. At such a moment, the amplitude of gravitational waves damps steeply, and hence, the highest frequency of gravitational waves should be determined by the tidal-disruption event.

The qualitative features summarized above could be modified by the black-hole spin; for binaries composed of a high-spin black hole, tidal disruption may occur for a high mass ratio, and hence, the infall process of the tidally disrupted neutron-star matter into the black hole may be qualitatively modified. This is well reflected in gravitational waveforms, as described in section 8.5.4.2.

8.5.4.2 Nonzero black-hole spin case

Gravitational waveforms in the merger stage are quantitatively and qualitatively modified in the presence of a black-hole spin. In this subsection, we will show this fact focusing on the case that the axis of the black-hole spin aligns with the orbital angular momentum. Figure 8.50 plots gravitational waveforms for $Q = 3$ with a stiff equation of state (HB in table 8.1) and with an initial angular velocity ($\Omega_{\rm orb} m_0 = 0.030$) but with different values of the black-hole spin. This obviously shows that with the increase of the black-hole spin, the lifetime of the binary system increases and hence the number of the wave cycle increases. This is explained primarily by the spin-orbit coupling effect (see section 8.5.2.2), which brings the repulsive force into the binary for the corotating black-hole spin (see also section 8.7.2.3 for the same effect in binary black holes). Due to the presence of this repulsive force, first, the orbital separation at the ISCO can be smaller than that for the non-spinning black hole (we may say that the absolute value of the binding energy at the ISCO becomes larger in the presence of the corotating spin). This effect increases the lifetime of the binary, and furthermore, enhances the possibility for tidal disruption of a neutron star because a circular orbit with a closer orbital separation is allowed.

The repulsive force also reduces the orbital velocity for a given orbital separation, because the centrifugal force for maintaining a quasi-circular orbit may be reduced due to this. The slow-down of the orbital velocity results in the decrease of the gravitational-wave luminosity, and this decelerates the orbital evolution due to the gravitational-radiation reaction, making the lifetime of the binary longer and increasing the number of cycle of gravitational waves. We note that all these effects are also clearly reflected in the gravitational-wave spectrum (see section 8.5.5).

Figure 8.50 shows that for $\hat{a} \leq 0$, a ringdown waveform associated with a quasi-normal mode of the remnant black hole is excited, while for $\hat{a} = 0.75$, such a feature is absent. This reflects the fact that tidal disruption of the neutron star occurs for $\hat{a} = 0.75$ far outside the ISCO, while it does not for $\hat{a} \leq 0$. Interestingly, for $\hat{a} = 0.5$, tidal disruption occurs but a ringdown waveform associated with a quasi-normal mode is also excited. This is a new type of the gravitational waveform. In this case, tidal disruption occurs near the ISCO and a large fraction of the neutron-star matter falls into the black hole. The infall occurs approximately simultaneously and proceeds from a narrow region of the black-hole horizon. This new type appears for the case that the black-hole mass (or mass ratio Q) is large enough that the surface of the event horizon is wider than the extent of the infalling material, as explained in section 8.5.2.2 (see also the right row of Figure 1.30).

For binaries composed of a spinning black hole, the inspiral waveform does not match well to that by the Taylor-T4 formula for a few orbits just before the onset of the merger. The top panels of figure 8.50 shows that the wave cycles are underestimated in the Taylor-T4 model and illustrates the limitation of this formula; it cannot be used for binary systems composed of high-spin black holes. Hence, a better approximate formulation is necessary for modeling the waveforms for binaries

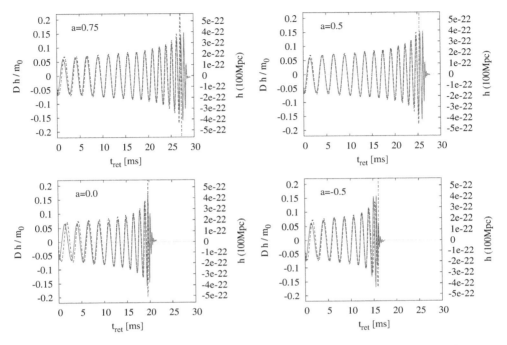

Fig. 8.50 The same as figure 8.49 but for $Q=3$ and $\hat{a}=0.75$ (top left), 0.5 (top right), 0 (bottom left), and -0.5 (bottom right) with HB equation of state. The figure is taken from Kyutoku et al. (2011).

composed of spinning black holes. This issue is described in section 8.7.3 for the context of binary black holes in more detail.

8.5.5 Fourier spectrum of gravitational waves

8.5.5.1 General feature

The final fate of neutron stars in black hole-neutron star binaries is reflected clearly in the spectrum of gravitational waves. General qualitative features of the gravitational-wave spectrum are summarized as follows. For the inspiral stage, during which the gravitational-wave frequency is $\lesssim 1\,\text{kHz}\,(R_{\text{NS}}/12\,\text{km})^{-3/2}$ and the post-Newtonian and point-particle approximations work well, the gravitational-wave spectrum is reproduced approximately by the post-Newtonian-based resummation formula such as Taylor-series formulas (e.g., appendix H). For this stage, the spectrum amplitude of $h_{\text{eff}} := f|\tilde{h}(f)|$ decreases as f^{-n_i} where $n_i \approx 1/6$ for $f \ll 1\,\text{kHz}$ and the value of n_i increases with f for $f \lesssim 1\,\text{kHz}\,(R_{\text{NS}}/12\,\text{km})^{-3/2}$. With the decrease of the orbital separation, both the nonlinear effect of general relativity and the finite-size effect of the neutron star come into play, and thus, both the post-Newtonian and point-particle approximations break down. When tidal disruption (not mass shedding) occurs for a relatively large separation (e.g., for a neutron star composed of a stiff equation of state or for a small value of Q), the amplitude of the gravitational-wave spectra damps above a *cutoff* frequency $f_{\rm cut}$. The cutoff frequency exists in a late inspiral stage and satisfies $f_{\rm cut} \sim 1$–3 kHz (it is lower than the frequency at the ISCO). The cutoff frequency depends on the binary parameters as well as the equation of state of the neutron star. A more strict definition for it will be described below.

On the other hand, if tidal disruption does not occur or occurs at a close orbit near the ISCO, the spectrum amplitude for a high-frequency band ($f \gtrsim 1$ kHz) is larger than that predicted by the post-Newtonian-based formulas (i.e., the value of n_i decreases and even can become negative). In this case, an inspiral-like motion may continue even inside the ISCO for a dynamical time scale and gravitational waves with a high amplitude are emitted. This property holds even in the presence of mass shedding. The evolution process described above is reflected in the fact that $h_{\rm eff}$ becomes a slowly varying function of f for $1 \text{ kHz} \lesssim f \lesssim f_{\rm cut}$, where $f_{\rm cut} \sim 2$–3 kHz is determined by the infall process of the neutron star into the black hole in this case. We note that the similar property is originally discovered by Buonanno *et al.* (2007a) in the context of binary black-hole mergers (see section 8.7.3).

An exponential damping of the spectra for $f \gtrsim f_{\rm cut}$ is universally observed, and for softer equations of state hypothetically employed (for smaller neutron-star radii), the frequency of $f_{\rm cut}$ becomes higher for given mass of the black hole and the neutron star and for a given black-hole spin. This cutoff frequency is determined by the frequency of gravitational waves emitted at a moment that the neutron star is tidally disrupted (for the case of large neutron-star radii) or by the frequency of a fundamental quasi-normal mode of the remnant black hole (for the case of small neutron-star radii). Therefore, the cutoff frequency will provide potential information for the equations of state through the tidal-disruption event of neutron stars.

8.5.5.2 *Zero black-hole spin case*

Figure 8.51, plotted by Etienne *et al.* (2009) for their simulation with the Γ-law equation of state ($\Gamma = 2$) and $\mathcal{C} \approx 0.145$, clearly illustrates the facts described in section 8.5.5.1. The top panel (case E) shows the spectrum for $Q = 1$, in which the neutron star is tidally disrupted far outside the ISCO. In this case, the spectrum damps for $f \sim 1$ kHz at which tidal disruption occurs. The bottom panel (case D) shows the spectrum for $Q = 5$, in which the neutron star is not tidally disrupted. In this case, the steep damping of the spectrum for $f \sim 2$ kHz is determined by the event that the neutron star falls into the companion black hole, and thus, the cutoff frequency is characterized by ringdown gravitational waves associated with a fundamental quasi-normal mode of the remnant black hole. Because the finite-size effect of the neutron star is not very important in this case, the gravitational-wave spectrum is similar to that of the merger of binary black holes with the same mass

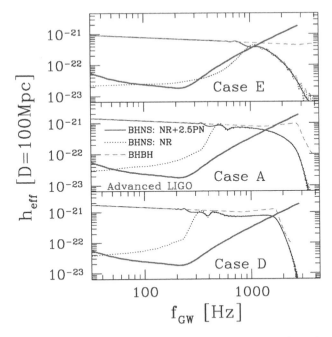

Fig. 8.51 Gravitational-wave spectrum for $Q = 1$ (Case E), 3 (Case A), and, 5 (Case D) with $\hat{a} = 0$ and with the Γ-law equation of state ($\Gamma = 2$). The solid curve shows the spectrum of a hybrid waveform composed of post-Newtonian and numerical waveforms, while the dotted curve shows the contribution from the numerical waveform only. The dashed curve is an analytic fit derived in Ajith et al. (2008) from analysis of gravitational waves from binary black holes composed of a nonspinning black hole with the same values of Q as the black hole-neutron star. The heavy solid curve is the effective strain of the advanced LIGO. To set physical units, a rest mass is assumed to be $M_B = 1.4\,M_\odot$ ($R_{NS} \approx 13$ km) for the neutron star with a source distance of $D = 100$ Mpc. The figure is taken from Etienne et al. (2009).

ratio ($Q = 5$; see the dashed curve). In the middle panel (case A), the cutoff frequency, at which the steep damping of h_{eff} sets in, is different from that for the binary black holes with the same mass ratio. This shows that tidal deformation and disruption play an important role in the merger process and in determining the gravitational-wave spectrum.

As described above, the cutoff frequency at which the steep damping of the spectrum amplitude is found will bring us the information for the degree of tidal deformation and for the frequency at the onset of tidal disruption. As emphasized for several times to this time, these tidal effects depend on the equation of state of the neutron star. This suggests that the cutoff frequency should have the information of the equation of state. Motivated by this idea, Kyutoku et al. (2011) performed a wide variety of simulations, changing the mass ratio, hypothetical equation of state, and black-hole spin for the first time, and then, systematically analyzed the resulting gravitational waveforms [see also subsequent studies by Lackey et al. (2012);

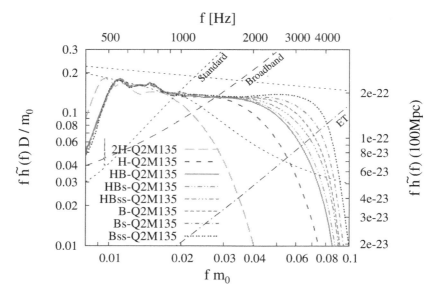

Fig. 8.52 Spectra of gravitational waves from black hole-neutron star binaries for $Q=2$, $\hat{a}=0$, and $M_{\rm NS}=1.35\,M_\odot$ (i.e., $m_0 = 4.05\,M_\odot$) with various equations of state. The bottom axis denotes the normalized dimensionless frequency fm_0 and the left axis the normalized amplitude $f|\tilde{h}(f)|D/m_0$ with D the hypothetical distance from the source. Note that $\tilde{h}(f)$ in this figure denotes $|\tilde{h}(f)|$. The top axis denotes the physical frequency f in units of Hz and the right axis denotes the effective amplitude $f|\tilde{h}(f)|$ observed at a distance of $D=100$ Mpc from the binaries. The short-dashed slope line plotted in the upper left region denotes a planned noise curve of the advanced-LIGO [Abbott et al. (2009)] optimized for $1.4\,M_\odot$ neutron star-neutron star inspiral detection ("Standard"), the long-dashed slope line denotes a noise curve optimized for the burst detection ("Broadband") and the dot-dashed slope line plotted in the lower right region denotes a planned noise curve of the Einstein Telescope ("ET") [Hild et al. (2008, 2010)]. The upper transverse dashed line is the spectrum derived by the quadrupole formula and lower one is the spectrum derived by the Taylor-T4 formula, respectively. The figure is taken from Kyutoku et al. (2010b).

Pannarale et al. (2013); Lackey et al. (2014)]. Figure 8.52 shows the spectrum as a function of the frequency for $Q=2$, $M_{\rm NS}=1.35\,M_\odot$, and with a variety of hypothetical equations of state for $\hat{a}=0$. Irrespective of the equations of state, the spectrum has the qualitatively universal shape with different values of the cutoff frequency, as already mentioned.

The features of gravitational-wave spectra for $\hat{a}=0$, are schematically summarized in figure 8.53. Here, four curves are plotted assuming that the mass of black hole and neutron star are given, $\hat{a}=0$, and a relatively small value of Q is employed, but the neutron-star equation of state is different. The three curves labeled by (i) schematically denote the gravitational-wave spectra for stiff equations of state (for the large neutron-star radius case). The curve labeled by (ii) denotes the spectrum for a soft equation of state (for a small neutron-star radius). For the curves of (i), the damping of the spectrum is determined by tidal disruption. In this case, the

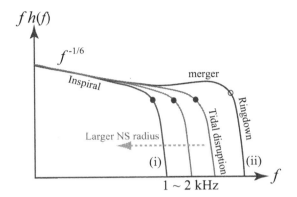

Fig. 8.53 Schematic figure of the gravitational-wave spectrum for $\hat{a} = 0$ and for given mass of black hole and neutron star but with different neutron-star equations of state. For the three curves labeled by (i), tidal disruption occurs far outside the ISCO. With the increase of the neutron-star radius, the cutoff frequency shifts to the lower frequency side. For the curve labeled by (ii), tidal disruption does not occur and the neutron star is swallowed simply by the black hole. We refer to the spectra (i) as type-I and the spectrum (ii) as type-II. For $\hat{a} = 0$ and $\mathcal{C} \gtrsim 0.13$, type-I spectrum is seen only for small values of the mass ratio with $Q \lesssim 3$. The filled and open circles denote the cutoff frequency associated with tidal disruption and the ringdown oscillation of a fundamental quasi-normal mode, respectively.

spectrum is characterized simply by the exponential damping for $f \gtrsim f_{\text{cut}}$, and the value of f_{cut} is lower for the larger neutron-star radius. We refer to the spectrum of this type as type-I. For the case (ii), on the other hand, tidal disruption does not occur, and the cutoff frequency is determined by a quasi-normal mode of the remnant black hole. In this spectrum, for $f \lesssim f_{\text{cut}}$, the amplitude of the spectrum slightly increases with the frequency; this is a characteristic feature also seen for the spectrum of gravitational waves from binary black holes [see figure 8.51; see also Buonanno et al. (2007a) and section 8.7.3]. We refer to the spectrum of this type as type-II.

To quantify the cutoff frequency and to strictly study its dependence on the hypothetical equations of state, a systematic analysis for the numerical data is necessary. Shibata et al. (2009); Kyutoku et al. (2010b, 2011) proposed to fit all the spectra by a function with seven free parameters [see also Lackey et al. (2012); Pannarale et al. (2013) for other modeling]

$$\tilde{h}_{\text{fit}}(f) = \tilde{h}_{\text{3PN}}(f) e^{-(f/f_{\text{ins}})^{\sigma_{\text{ins}}}} + \frac{Am_0}{Df} e^{-(f/f_{\text{cut}})^{\sigma_{\text{cut}}}} \left[1 - e^{-(f/f_{\text{ins2}})^{\sigma_{\text{ins2}}}}\right], \quad (8.55)$$

where $\tilde{h}_{\text{3PN}}(f)$ is the amplitude of the Fourier spectrum calculated by the Taylor-T4 formula and f_{ins}, f_{ins2}, f_{cut}, σ_{ins}, σ_{ins2}, σ_{cut}, and A are free parameters. The first and second terms in the right-hand side of equation (8.55) denote the spectrum models for the inspiral and merger stages, respectively. These free parameters are determined by searching the minimum for a weighted norm defined by

$$\sum_i \left\{ [f_i h(f_i) - f_i h_{\text{fit}}(f_i)] f_i^{1/3} \right\}^2, \qquad (8.56)$$

where i denotes the data point for the spectrum.

Among these seven free parameters, f_{cut} depends most strongly on the compactness $\mathcal{C}$ and the neutron-star equation of state. Figure 8.54 plots $f_{\text{cut}} m_0$, obtained in this fitting procedure, as a function of $\mathcal{C}$ for $\hat{a} = 0$. Also the values of the typical quasi-normal mode frequency, f_{QNM}, of the remnant black hole for $Q = 2$ and 3 are plotted by the two horizontal lines, which show that the values of $f_{\text{cut}} m_0$ for compact models ($\mathcal{C} \gtrsim 0.16$) with $Q = 3$ agree approximately with f_{QNM} and indicate that f_{cut} for these models are irrelevant to tidal disruption: For $Q = 3$, $f_{\text{cut}} m_0$ depends on the equation of state only for $\mathcal{C} \lesssim 0.16$. By contrast, $f_{\text{cut}} m_0$ for $Q = 2$ depends strongly on the neutron-star compactness, $\mathcal{C}$, irrespective of M_{NS} for a wide range of $\mathcal{C} \lesssim 0.19$.

An interesting finding in Kyutoku et al. (2010b) is that the following relation approximately holds for $Q = 2$ and $\Gamma_1 = 3$:

$$\ln(f_{\text{cut}} m_0) = (3.87 \pm 0.12) \ln \mathcal{C} + (4.03 \pm 0.22). \qquad (8.57)$$

Thus, $f_{\text{cut}} m_0$ is approximately proportional to $\mathcal{C}^4$. This is a note-worthy point because the power of $\mathcal{C}$ is much larger than a well-known factor 1.5 that is expected from the relation for the mass-shedding limit presented in section 1.5.4. (For the binaries composed of a spinning black hole, this power is smaller than 3.9, but it is still much larger than 1.5 [Kyutoku et al. (2011)]). This difference indicates that the cutoff frequency is not determined simply by mass shedding: Remember the fact that at the onset of mass shedding, neutron stars have an appreciable approaching velocity, in particular for the small value of $Q \lesssim 3$, and tidal disruption occurs at an orbital separation smaller than that at mass shedding. Qualitatively, this high power is natural because the lifetime for the survival against tidal disruption after the onset of mass shedding is in general longer for more compact neutron stars, and hence, the cutoff frequency for the larger values of $\mathcal{C}$ tends to be higher.

Figure 8.52 illustrates that f_{cut} is rather high, 1–3 kHz, for a variety of hypothetical equations of state. The reason for this is that for $\hat{a} = 0$, tidal disruption can occur only for a small mass ratio (and thus for small total mass) with a typical neutron-star mass 1.3–$1.4 \, M_\odot$; see equation (1.211). The effective amplitude of gravitational waves at $f = 2$ kHz is $\sim 2 \times 10^{-22}$ for a hypothetical distance to the source of $D = 100$ Mpc. The amplitude is smaller than the planned noise level of the advanced LIGO, VIRGO, and KAGRA, and unless the event occurs near our Galaxy with $D \ll 100$ Mpc, the observation of the cutoff frequency will be difficult by the near-future detectors, as illustrated by Lackey et al. (2012) in their Fisher-matrix analysis. For the detection of such a signal, we have to wait for next-generation detectors such as Einstein Telescope [Hild et al. (2008, 2010)]. However, the situation is improved in the presence of black-hole spins, as we describe in section 8.5.5.3 [see also Lackey et al. (2014)].

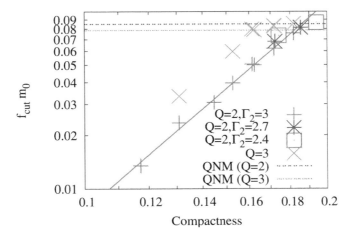

Fig. 8.54 $f_{\mathrm{cut}} m_0$ as a function of $\mathcal{C}$ in logarithmic scales for $\hat{a} = 0$ with several hypothetical equations of state. The solid line is obtained by a linear fitting of the data for $Q = 2$ and $\Gamma_1 = 3$ [see equation (8.57)]. The short-dashed and long-dashed horizontal lines show approximate frequency of the fundamental quasi-normal mode of the remnant black hole for $Q = 2$ and $Q = 3$, respectively. The figure is taken from Kyutoku et al. (2010b).

8.5.5.3 Nonzero black-hole spin case

The gravitational-wave spectrum is quantitatively and qualitatively modified by the presence of a black-hole spin. In the following, we will focus only on the case that the axes of the black-hole spin and orbital angular momentum align, because gravitational waves for the misaligned case have not been studied in detail yet [but see Foucart et al. (2013)]. Figure 8.55 is essentially the same as figure 8.52 but for $Q = 3$, $M_{\mathrm{NS}} = 1.35\,M_\odot$, and $\hat{a} = 0.75$, 0.5, 0, and -0.5 with HB equation of state (in table 8.1). This figure shows that the spectrum shapes for $\hat{a} \leq 0$, $\hat{a} = 0.5$, and $\hat{a} = 0.75$ are qualitatively different. For $\hat{a} \leq 0$, tidal disruption does not occur and the exponential damping above cutoff frequency is determined by a fundamental quasi-normal mode of the remnant black hole. This is the type-II spectrum according to the classification in figure 8.53. On the other hand, for $\hat{a} = 0.75$, the cutoff frequency ($f_{\mathrm{cut}} \sim 1.5$ kHz) is determined by the frequency at which tidal disruption occurs. This is the type-I spectrum according to the classification in figure 8.53. The spectrum for $\hat{a} = 0.5$ is neither type-I nor type-II: In this case, there are two characteristic frequencies. One is at $f \sim 2$ kHz, above which the spectrum amplitude slowly declines, and the other is at $f \sim 3$ kHz, above which the spectrum amplitude steeply damps. The first frequency is determined primarily by the frequency at which tidal disruption occurs, and the second one corresponds to the fundamental quasi-normal mode frequency of the remnant black hole. We call this new spectrum type-III (according to the definition of section 8.5.2.2; see also figure 1.30). In the left panel of figure 8.56, we summarize three types of the gravitational-wave spectrum. For the type-III spectrum, we will refer to the first

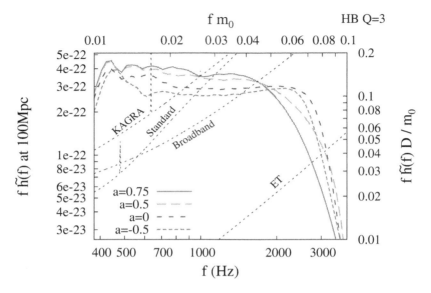

Fig. 8.55 The same as figure 8.52 but for $Q = 3$, $M_{\rm NS} = 1.35\,M_\odot$, and $\hat{a} = 0.75$, 0.5, 0, and -0.5 with a moderately stiff equation of state (HB in table 8.1). The thin dashed curves show planned noise curves for the KAGRA, advanced LIGO, broadband-designed advanced LIGO, and Einstein telescope (from upper to lower). The figure is taken from Kyutoku et al. (2011).

(lower) characteristic frequency as the cutoff frequency in the following.

Figure 8.55 shows that with the increase of black-hole spins (for $\hat{a} \geq 0$), the cutoff frequency, $f_{\rm cut}$, goes down, and in addition, the amplitude of the gravitational-wave spectrum for $f \lesssim f_{\rm cut}$ increases. These two effects are favorable for the gravitational-wave detection by near-future advanced detectors as found from figure 8.55. These quantitative changes stem again from the spin-orbit coupling effect (see also sections 8.5.2.2 and 8.5.4.2), as explained in the following manner.

Due to the spin-orbit coupling effect that brings a repulsive force into black hole-neutron star binaries for the corotating spin, $\hat{a} > 0$, the orbital velocity for a given separation is reduced, because the centrifugal force may be weaker than for $\hat{a} = 0$ for maintaining a quasi-circular orbit. Due to the slow-down of the orbital velocity, (i) the orbital angular velocity at a given separation is lowered and, as a result, (ii) the luminosity of gravitational waves is decreased. The effect (i) results in lowering the cutoff frequency at which tidal disruption occurs. The effect (ii) decelerates the orbital evolution by the gravitational-radiation reaction, resulting in a longer lifetime of the binary system and in the increase of the number of the gravitational-wave cycle. This effect increases the amplitude of the gravitational-wave spectrum of $f < f_{\rm cut}$ for $\hat{a} > 0$. These two effects are schematically described in the right panel of figure 8.56.

More quantitative description is as follows: From the luminosity formula of

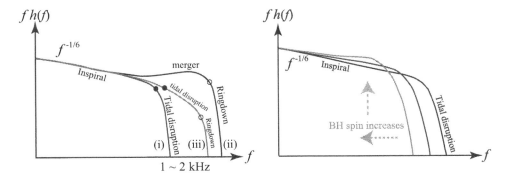

Fig. 8.56 Left: A schematic figure for three types of the gravitational-wave spectrum. Type-I (i) and type-II (ii) are the same as those shown in figure 8.53. Type-III (iii) is often seen for the case that the black-hole spin is high and the mass ratio (and thus the area of the black-hole horizon) is large. The filled and open circles denote the cutoff frequency associated with tidal disruption and with a quasi-normal mode, respectively. Right: A schematic figure for explaining the property of figure 8.55. With the increase of the black-hole spin but for given mass of black hole and neutron star and for given neutron-star equation of state, the cutoff frequency goes down and the amplitude below the cutoff frequency increases. The figures are taken from Kyutoku et al. (2011).

gravitational waves, the power spectrum is written as

$$\frac{dE}{df} \propto \left[f\tilde{h}(f)\right]^2. \quad (8.58)$$

In the one-and-half post-Newtonian approximation in which the lowest-order spin-orbit coupling effect is taken into account (see appendix H), dE/df in the inspiral stage is written as

$$\frac{dE}{df} = \frac{1}{G(\pi f)^2} \frac{Q}{(1+Q)^2} (\pi m_0 f)^{5/3} \left[1 + (\pi m_0 f)\hat{\mathbf{S}} \cdot \hat{\mathbf{L}} \frac{20Q^2 + 15Q}{3(1+Q)^2}\right], \quad (8.59)$$

where we supposed vanishing neutron-star spin, and also, we omitted other post-Newtonian terms. $\hat{\mathbf{S}}$ and $\hat{\mathbf{L}}$ are unit vectors of the black-hole spin and orbital angular momentum, respectively. Equation (8.59) shows that for a given value of frequency f, $|dE/df|$ is larger in the presence of the corotating spin ($\hat{\mathbf{S}} \cdot \hat{\mathbf{L}} > 0$) than for zero black-hole spin. Therefore, the effective amplitude, $f|\tilde{h}(f)|$, for a given value of frequency is larger in the presence of the corotating spin than for zero spin.

For the binaries composed of a high-spin black hole, tidal disruption can occur outside the ISCO even for a high-mass black hole for a variety of equations of state (see section 8.5.2.3). Thus, the dependence of f_{cut} on the equation of state is clearly seen even for a high value of Q. Figure 8.57 shows the spectrum for $Q = 4$, $M_{\text{NS}} = 1.35\,M_\odot$, and $\hat{a} = 0.75$ with four different equations of state. For all the models, tidal disruption occurs, and the values of f_{cut} depend strongly on the hypothetical equations of state employed. For 2H and H (stiff equations of state), the spectra are of type-I, but for HB and B (soft equations of state), they are of type-III. Thus, for a high black-hole spin, the type-I or type-III spectrum is often seen even for a high value of Q.

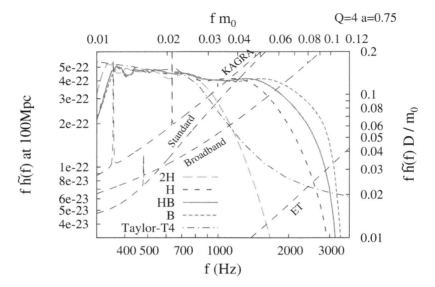

Fig. 8.57 The same as figure 8.52 but for $Q = 4$, $\hat{a} = 0.75$, and $M_{\rm NS} = 1.35\,M_\odot$ with four hypothetical equations of state. The figure is taken from Kyutoku et al. (2011).

With the increase of Q (for the canonical mass of a neutron star ~ 1.3–$1.4\,M_\odot$), the value of $f_{\rm cut}$ goes down and the effective amplitude at $f = f_{\rm cut}$ increases. These are also favorable properties for the gravitational-wave detection. Figure 8.57 indeed illustrates that $f_{\rm cut}$ is lower than $2\,{\rm kHz}$ irrespective of the hypothetical equations of state, and also, the effective amplitude at $f \lesssim 2$ kHz is as large as planned noise curves of the advanced detectors for a hypothetical distance of $D = 100\,{\rm Mpc}$ [see also Lackey et al. (2014) for a more detailed analysis]. For an even higher spin, say $\hat{a} \gtrsim 0.9$, tidal disruption is likely to occur for a higher black-hole mass with $Q \gtrsim 7$ and $M_{\rm BH} \gtrsim 10 M_\odot$ [e.g., Lovelace et al. (2013)]. For such cases, the amplitude of gravitational waves at tidal disruption is likely to be high enough to be observable irrespective of the equation of state for $D = 100\,{\rm Mpc}$. This indicates that a black hole-neutron star binary with a high-mass high-spin black-hole will be a promising experimental site for constraining the equation of state of the high-density nuclear matter, when advanced gravitational-wave detectors are in operation.

8.5.6 Dynamical mass ejection and electromagnetic counterparts

8.5.6.1 Dynamical mass ejection

For the case that a neutron star is tidally disrupted by its companion black hole, a part of the neutron-star matter gains specific angular momentum due to tidal torque exerted by the black hole during the tidal disruption process. Then, the fluid elements that obtain a sufficient amount of the specific angular momentum (and hence kinetic energy) escape from the system. As described in section 8.3.5, such ejecta can shine as a strong electromagnetic counterpart of gravitational waves.

A detailed study for this dynamical mass ejection by the merger of black hole-neutron star binaries in numerical relativity was first performed by Kyutoku et al. (2013). We here summarize the properties of the ejecta, based on their results.

For the case that a neutron star is disrupted by the black-hole tidal field, a one-armed spiral structure (the tidal tail) is formed around the black hole (see figures 8.35 and 8.39). Although a large part of the tidal tail eventually falls back onto the remnant disk and black hole, its outermost part gains a sufficient amount of angular momentum and kinetic energy and becomes unbound via hydrodynamical angular-momentum transport processes. This dynamical mass ejection is driven dominantly by the tidal effect. Some material in the vicinity of the black hole becomes unbound when the tidal tail hits itself and is shock-heated as it is wound around the black hole. However, this shock-driven component is always sub-dominant.

For most cases, ejecta exhibit a crescent-like or boomerang shape as depicted in figure 8.58, and the typical opening angle of the ejecta in the equatorial plane is $\varphi_{\rm ej} \approx \pi$. Such a non-axisymmetric shape arises from the fact that the sound-crossing time scale inside the neutron star is shorter than the orbital period at the onset of tidal disruption. Furthermore, ejecta spread dominantly in the equatorial plane, and expand only slowly in the direction perpendicular to the equatorial plane (hereafter referred to as z-direction). The reason for this is that this mass ejection is driven primarily by the tidal effect, which is most efficient in the equatorial plane.[8] Thus, only a portion of circumferential material will be subsequently swept by the ejecta. A typical half opening angle of the ejecta in the meridian direction around the equatorial plane is $\theta_{\rm ej} \approx 1/5$ radian, and hence, the ejecta velocity in the equatorial plane $v_\parallel$ is larger by a factor of $1/\theta_{\rm ej} \approx 5$ than that in the z-direction $v_\perp$. Here, $v_\parallel$ may be identified with the radial velocity, and the azimuthal velocity is negligible soon after the ejection due to the angular momentum conservation. Indeed, the azimuthal velocity is very small as found in figure 8.58. Aside from the ejecta themselves, the region above the remnant black hole is much less dense than that for the merger of binary neutron stars (compare figure 8.58 with figures 8.24 and 8.25).

The ejecta mass, $M_{\rm ej}$, depends on binary parameters and hypothetical equations of state employed for neutron stars, and is typically in the range of ~ 0.01–$0.1 M_\odot$ for the case that tidal disruption occurs and the disk mass $M_{\rm disk}$ exceeds $\sim 0.1 M_\odot$. The value of $M_{\rm ej}$ is generally larger for stiffer hypothetical equations of state, for given values of mass ratio, Q, and black-hole spin, $\hat{a}$, because the mass ejection is driven primarily by the tidal effect. This dependence on the equation of state

---

[8] As shown in section 8.5.3.2, after tidal disruption, a disk/torus is formed around the remnant black hole. During the subsequent evolution of the disk/torus due to viscous dissipation or magnetohydrodynamics effect, mass ejection may be driven by disk wind [Fernández and Metzger (2013)]. Such mass ejection may proceed in a quasi-spherical manner, and the degree of the anisotropy of the ejecta may be reduced in the presence of this component.

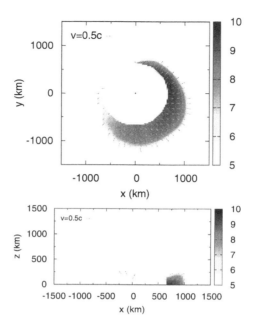

Fig. 8.58 Rest-mass density profiles and velocity vectors of ejecta at ≈ 10 ms after the onset of the merger for a model with $Q = 5$, $\hat{a} = 0.75$, and H4 equation of state (see table 1.1). Only unbound matter is shown to elucidate geometry of the ejecta. The blank region between the ejecta and black hole is filled with unshown bound material. The top and bottom panels are for the xy- and xz-planes, respectively. The color panel on the right of each plot shows $\log_{10}[\rho\,(\mathrm{g\,cm^{-3}})]$. The region above ~ 500 km in the bottom panel is much less dense than that in a typical binary neutron star merger (compare it with figures 8.24 and 8.25).

(and thus on the neutron-star radius) is opposite to the case of binary neutron star mergers, for which shock heating as well as rapid rotation and oscillation of remnant massive neutron stars play an important role for driving mass ejection (see section 8.3.5). For the case that the neutron star is not disrupted prior to the merger, the ejecta mass is negligible for the astrophysical interest. Kyutoku et al. (2013) found that the ejecta mass is always smaller than the disk mass, and a relation $M_{\rm ej} \approx 0.05 - 0.25 M_{\rm disk}$ approximately holds as far as a range in their study is concerned.

The ejecta from the mergers of black hole-neutron star binaries have a bulk linear momentum, $P_{\rm ej}$, and resulting large bulk velocity $v_{\rm ej} := P_{\rm ej}/M_{\rm ej} \sim 0.2c$, in a particular direction. By contrast, these bulk linear momentum and velocity would essentially vanish for nearly spherical ejecta such as one from binary neutron star mergers. The value of $v_{\rm ej} \sim 0.2c$ depends only weakly on the binary parameters and equation of state as far as the mass ejection is appreciable.

Typical values of kinetic energy $T_{\rm ej}$ are in the range of $\sim 5 \times 10^{50} - 5 \times 10^{51}$ ergs. This value is larger than that for binary neutron star mergers for which it is $\lesssim 5 \times 10^{50}$ ergs. The average velocity of the ejecta, $v_{\rm ave} \approx (2T_{\rm ej}/M_{\rm ej})^{1/2}$, is typically

$0.25-0.3c$ and is naturally larger than $v_{\rm ej}$. The contribution of $v_\perp$ to $T_{\rm ej}$ is smaller by a factor of $\theta_{\rm ej}^2$ than that of $v_\parallel$, and thus a relation $v_{\rm ave} \approx v_\parallel$ holds. For a partially axisymmetric outflow truncated at an opening angle $\varphi_{\rm ej}$, a relation $v_{\rm ej}/v_\parallel \approx \sin(\varphi_{\rm ej}/2)/(\varphi_{\rm ej}/2)$ should hold, and thus the ejecta opening angle in the equatorial plane is estimated to be $\varphi_{\rm ej} \approx (0.7-1.3)\pi$ radian. This is consistent with figure 8.58.

Kyutoku et al. (2013) suggested several possible unique phenomena associated with the dynamical mass ejection from black hole-neutron star mergers. One of them is that the remnant black hole should receive a back reaction from the ejecta with total mass $M_{\rm ej}$ and total linear momentum $P_{\rm ej}$, resulting in substantial *ejecta kick* velocity [see also Rosswog et al. (2000)]. The total mass of the remnant black hole and surrounding disk is determined approximately by the mass of the binary at infinite separation $m_0 = (1+Q)M_{\rm NS}$, neglecting $M_{\rm ej}$ and energy carried by gravitational waves $\lesssim 0.05 m_0 c^2$. Then, the ejecta kick velocity of the remnant system would be

$$v_{\rm kick} \approx \frac{M_{\rm ej}}{m_0} v_{\rm ej} \approx 220 \text{ km s}^{-1} \left(\frac{M_{\rm ej}}{0.03 M_\odot}\right) \left(\frac{\beta_{\rm ej}}{0.2}\right) \left(\frac{M_{\rm NS}}{1.35 M_\odot}\right)^{-1} \left(\frac{1+Q}{1+5}\right)^{-1}, \quad (8.60)$$

where $\beta_{\rm ej} := v_{\rm ej}/c$.

The ejecta kick velocity, $v_{\rm kick}$, can be larger than the kick velocity due to the gravitational-radiation reaction because the gravitational-wave kick velocity is at most ~ 150 km s^{-1} for the case that the neutron star is disrupted prior to the merger [Shibata et al. (2009); Kyutoku et al. (2010b, 2011); Foucart et al. (2012, 2013); Lovelace et al. (2013)]. The reason for this is that the linear momentum is radiated most efficiently when the binary is about to merge, and therefore, earlier disruption, which is necessary for the efficient mass ejection, significantly suppresses the gravitational-wave kick velocity. Thus, the ejecta kick velocity can dominate the velocity of the remnant for the case that tidal disruption occurs.

Because the ejecta and remnant black hole-disk travel in the opposite direction after the merger, non-oscillatory gravitational-wave emission, i.e., gravitational-wave memory, is expected. Assuming that the remnant mass $\approx m_0$ is much larger than $M_{\rm ej}$, magnitude of this (linear) ejecta memory is given approximately by [Braginski and Thorne (1987)]

$$\delta h \approx \frac{2 M_{\rm ej} v_{\rm ej}^2}{D} \approx 1.2 \times 10^{-24} \left(\frac{M_{\rm ej}}{0.03 M_\odot}\right) \left(\frac{\beta_{\rm ej}}{0.2}\right)^2 \left(\frac{D}{100 \text{ Mpc}}\right)^{-1}, \quad (8.61)$$

where D is a distance from the binary to the observer. Taking into account the fact that the expected rise time of ejecta memory is much shorter than the inverse of frequency at which ground-based detectors are most sensitive, $(\sim 100 \text{ Hz})^{-1}$, it may be possible to detect ejecta memory by the Einstein Telescope [Hild et al. (2010)] if the ejecta mass is larger than $\sim 0.1 M_\odot$.

8.5.6.2 Possible electromagnetic counterparts

As in the case of binary neutron star mergers, the ejecta can be a source of the strong transient electromagnetic emission by essentially the same mechanisms (see section 8.3.5.3). One mechanism for the emission is based on the production of unstable r-process heavy nuclei and subsequent radioactive decay [Li and Paczyński (1998); Kurkarni (2005); Metzger et al. (2010)].

This emission could be modified by ejecta geometry for black hole-neutron star mergers because the ejecta is highly asymmetric.[9] We here approximate the ejecta geometry by an axisymmetric cylinder with velocity $v_\parallel$ and $v_\perp$ in the radial and z-directions, respectively, and then truncate at an opening angle $\varphi_{\rm ej}$ in the φ-direction. Assuming that photons escape primarily to the z-direction due to a short distance, the peak time is estimated as

$$t_{\rm peak} \approx \left(\frac{\bar{\kappa} M_{\rm ej} v_\perp}{c \varphi_{\rm ej} v_\parallel^2} \right)^{1/2}$$

$$\approx 4\,{\rm days} \left(\frac{\bar{\kappa}}{10\,{\rm cm}^2/{\rm g}} \right)^{1/2} \left(\frac{M_{\rm ej}}{0.03 M_\odot} \right)^{1/2} \left(\frac{\beta_0}{0.3} \right)^{-1/2} \left(\frac{\theta_{\rm ej}}{0.2} \right)^{1/2} \left(\frac{\varphi_{\rm ej}}{\pi} \right)^{-1/2},$$
(8.62)

where $\beta_0 := v_{\rm ave}/c$. Accordingly, the peak luminosity is

$$L_{\rm peak} \approx \frac{f_{\rm eff} M_{\rm ej} c^2}{t_{\rm peak}}$$

$$\approx 4 \times 10^{41}\,{\rm ergs} \left(\frac{f_{\rm eff}}{3 \times 10^{-6}} \right) \left(\frac{\bar{\kappa}}{10\,{\rm cm}^2/{\rm g}} \right)^{-1/2} \left(\frac{M_{\rm ej}}{0.03 M_\odot} \right)^{1/2}$$

$$\times \left(\frac{\beta_0}{0.3} \right)^{1/2} \left(\frac{\theta_{\rm ej}}{0.2} \right)^{-1/2} \left(\frac{\varphi_{\rm ej}}{\pi} \right)^{1/2},$$
(8.63)

and finally the temperature at the peak is

$$T_{\rm peak} \approx \left(\frac{L_{\rm peak}}{\varphi_{\rm ej} v_\parallel^2 t_{\rm peak}^2 \sigma_{\rm SB}} \right)^{1/4}$$

$$\approx 4 \times 10^3\,{\rm K} \left(\frac{f_{\rm eff}}{3 \times 10^{-6}} \right)^{1/4} \left(\frac{\bar{\kappa}}{10\,{\rm cm}^2/{\rm g}} \right)^{-3/8} \left(\frac{M_{\rm ej}}{0.03 M_\odot} \right)^{-1/8}$$

$$\times \left(\frac{\beta_0}{0.3} \right)^{-1/8} \left(\frac{\theta_{\rm ej}}{0.2} \right)^{-3/8} \left(\frac{\varphi_{\rm ej}}{\pi} \right)^{3/8}.$$
(8.64)

These approximate estimates suggest that characteristics of the radioactively-powered emission associated with anisotropic ejecta from the black hole-neutron star merger will be modified by a factor of a few compared to those associated with the isotropic one when the values of the ejecta mass and velocity are the same [compare the above equations with equations (8.32)–(8.34)].

---
[9] We note again that in the presence of disk/torus wind, the ejecta may become less asymmetric.

As pointed out by Tanaka *et al.* (2014), the morphology plays an important role for determining the light curves for different colors. Indeed, as found from equations (8.34) and (8.64), the bluer spectrum is expected at the peak time for the merger of black hole-neutron star binaries.

The synchrotron radio flare, proposed by Nakar and Piran (2011) and introduced in section 8.3.5.3, is also one of the most promising electromagnetic counterparts if a number density of the ambient interstellar matter is high as $n_0 \gtrsim 0.1\,\mathrm{cm}^{-3}$ [see equations (8.38)]. For the mergers of black hole-neutron star binaries, the ejecta are not likely to be spherical, and hence, equation (8.37) has to be modified. Specifically, the anisotropic ejecta will accumulate only a small portion of the ambient interstellar matter because of their anisotrophy. Thus, the anisotropic ejecta have to travel a longer distance $R_{\mathrm{dec,as}} > R_{\mathrm{dec}}$ to be decelerated than spherical ejecta. Here, we assume that values of $v_\parallel$ and $v_\perp$ do not change significantly before the deceleration. This would lead to constant values of θ_{ej} and φ_{ej} before the deceleration, which is expected to occur at

$$R_{\mathrm{dec,as}} \approx R_{\mathrm{dec}} \left(\frac{\pi/2}{\theta_{\mathrm{ej}}}\right)^{1/3} \left(\frac{2\pi}{\varphi_{\mathrm{ej}}}\right)^{1/3}$$

$$\approx 1.7\,\mathrm{pc}\,\left(\frac{M_{\mathrm{ej}}}{0.03 M_\odot}\right)^{1/3} \left(\frac{n_0}{1\,\mathrm{cm}^{-3}}\right)^{-1/3} \left(\frac{\theta_{\mathrm{ej}}}{0.2}\right)^{-1/3} \left(\frac{\varphi_{\mathrm{ej}}}{\pi}\right)^{-1/3}, \quad (8.65)$$

and, therefore, the peak time is

$$t_{\mathrm{dec,as}} = \frac{R_{\mathrm{dec}}}{v_\parallel}$$

$$\approx 18\,\mathrm{yrs}\,\left(\frac{M_{\mathrm{ej}}}{0.03 M_\odot}\right)^{1/3} \left(\frac{n_0}{1\,\mathrm{cm}^{-3}}\right)^{-1/3} \left(\frac{\beta_0}{0.3}\right)^{-1} \left(\frac{\theta_{\mathrm{ej}}}{0.2}\right)^{-1/3} \left(\frac{\varphi_{\mathrm{ej}}}{\pi}\right)^{-1/3}. \quad (8.66)$$

The peak time should be closer to (but still longer than due to the energy conservation) t_{dec} if the ejecta approaches a spherical state. The reality would be in between these two limits.

8.6 Binary black holes: Quasi-equilibrium states

The first successful numerical-relativity simulations for binary black holes were performed in 2005 [Pretorius (2005a); Campanelli *et al.* (2006a); Baker *et al.* (2006a)] (see also Tichy and Brügmann (2004) for the prelude to these works). Before this year, it was not feasible to explore the dynamics of binary black holes. In such a situation, studying binary black holes in quasi-equilibrium and its sequence was only one possible approach for understanding the properties of binary black holes in close orbits. Also, computing a quasi-equilibrium state of the binary black hole is an important task because it is necessary as the initial condition of numerical-relativity simulations. In this section, we summarize the studies for the quasi-equilibrium states of binary black holes performed so far.

8.6.1   *Effective-potential method*

The first attempt for constructing binary black holes in quasi-equilibrium states (quasi-circular orbits) was done by Cook (1994) [see also Pfeiffer *et al.* (2000) and Baumgarte (2000) for approaches similar to that by Cook (1994)]. Although we now know that realistic quasi-equilibrium states in close circular orbits cannot be obtained by his method, his novel idea based on an *effective-potential method* for constructing a quasi-equilibrium sequence is memorable.

Cook (1994) employed the conformal transverse-tracefree decomposition formulation (see section 5.2) together with the Bowen–York type approach (see section 5.5.2.3). Specifically, the method of isometry on the conformally flat spatial hypersurface (see sections 5.5.1.3 and 5.5.2.3) is employed. He focused only on the equal-mass and non-spinning black holes. His first step in this framework was to construct a large number of initial data changing the orbital separation between two black holes (denoted by l) and the linear momentum of each black hole. Then, he plotted many curves of given values of constant total angular momentum, J, as a function of l for a given mass (area) of black holes. He discovered that as far as the chosen angular momentum is not extremely small, each curve has a minimum of binding energy of the system, E_b, along the $J = $ const sequence. Here, the binding energy is defined by equation (8.15). In practice, it is not possible to strictly define the black-hole mass in the initial-data problem, and thus, irreducible mass, $\sqrt{A_{\rm AH}/16\pi}$, was used as the mass where $A_{\rm AH}$ is the area of apparent horizons (see section 5.3.4). Cook (1994) then defined the initial data at the minimum of the binding energy as a quasi-circular orbit, and the sequence of the minimum points for a series of J was identified as a sequence of the quasi-circular orbits. In addition, for the quasi-circular sequence of E_b and J, he defined the orbital angular velocity, according to the first-law relation, by (see section 5.7.2.3)

$$\Omega_{\rm orb} = \frac{dE_b}{dJ}. \qquad (8.67)$$

Cook (1994) called this method effective-potential method.

This approach was motivated by the following fact: Consider two equal-mass point particles of each mass m_1 at a separation l in Newtonian gravity. For this system, the effective potential energy is written as $E_b = -m_1^2/l + J^2/l^2 m_1$ where J is the total angular momentum. Minimizing E_b for constant values of m_1 and J with respect to l,

$$\left.\frac{\partial E_b}{\partial l}\right|_{m_1,J} = 0, \qquad (8.68)$$

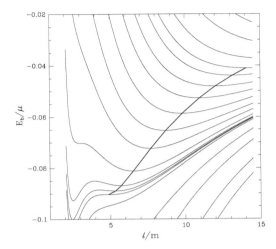

Fig. 8.59 The binding energy denoted by E_b/μ as a function of separation denoted by l/m for several values of angular momentum (thin curves). Here $m(=m_0)$ and μ are total and reduced masses of two black holes, respectively. Plotted as a bold curve is the sequence of quasi-circular orbits which cross the effective-potential curves at local minima. The figure is taken from Cook (1994).

yields $J^2 = m_1^3 l/2$ for circular orbits which is the correct value in Newtonian gravity. The angular velocity is derived from equation (8.67), yielding $\sqrt{2m_1/l^3}$ which also agrees with the Keplerian angular velocity.

The thin solid curves of figure 8.59 denote E_b as a function of l for several values of J, derived by Cook (1994). This shows that there is (at least) one minimum of E_b along each curve of a constant value of J. The thick solid curve denotes the sequence of the minimum points, and in Cook (1994), this curve was identified as the sequence of the quasi-circular orbits.

His result is qualitatively reasonable. However, we now know that it was not quantitatively accurate. Figure 8.60 shows the binding energy and the angular momentum as functions of the orbital angular velocity. Together with the numerical results, the curves derived in Newtonian, first post-Newtonian, and second post-Newtonian approximations are shown together. It is found that the numerical sequence of the angular momentum agrees well with the second post-Newtonian curve, although the numerical sequence of the binding energy does not (except for distant orbits). This is not consistent. Here, we do not mean that the numerical results have to agree with the second post-Newtonian results because the system is highly general relativistic. However, the numerical results of both the binding energy and the angular momentum should self-consistently agree or disagree with the second post-Newtonian ones. Thus, the results of figure 8.60 indicated that the framework by Cook (1994) would have a lack for an accurate study of binary black holes in close quasi-circular orbits.

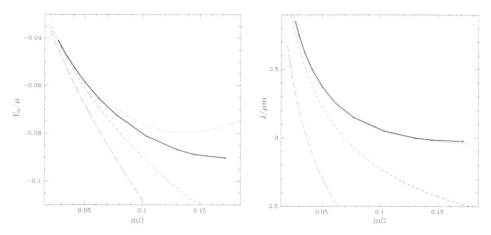

Fig. 8.60 The binding energy denoted by E_b/μ (left) and the angular momentum denoted by $J/m\mu$ (right) as functions of the orbital angular velocity denoted by $m\Omega(=m_0\Omega_\mathrm{orb})$ (solid curves). The dotted, dashed, and long-dashed curves denote the results in Newtonian, first post-Newtonian, and second post-Newtonian approximations. The figure is taken from Cook (1994).

The other question in this work was that the minima of the binding energy and angular momentum along the sequence (i.e., ISCO) was located at extremely large angular velocity, $m_0\Omega_\mathrm{orb} \approx 0.17$. This value is much larger than the angular velocity of the ISCO for test particles around non-spinning black holes, $1/6^{3/2} \approx 0.068$ (see section 1.3.2.1). It should be also mentioned that the compactness of the system defined by $(m_0\Omega_\mathrm{orb})^{2/3}$ is extremely large as ≈ 0.31 for $m_0\Omega_\mathrm{orb} \approx 0.17$. Here, assuming that Ω_orb is written in the Keplerian form, $(m_0\Omega_\mathrm{orb})^{2/3}$ should denote m_0/ℓ where ℓ is a proper length of the orbital separation. Thus, this result suggested that the compactness of the ISCO might be about twice as large as that for the test-particle case in which it is $1/6$. However, figure 8.59 shows $m_0/\ell \sim 1/5$ at the ISCO, and suggests a huge deviation from the usual Keplerian law. We did not know the effect that could explain this huge derivation.

Now, there are several evidences that the framework of Cook (1994) was short of a theoretical piece for constructing binary black holes in quasi-equilibrium, although the idea of the effective-potential method would be viable [see Caudill et al. (2006) for a supporting evidence of the effective-potential method; see also figure 8.61]. For example, any *quasi-equilibrium condition for black-hole horizons* was not taken into account in this framework [see also a discussion and an analysis of Pfeiffer et al. (2000)]: Stationary black-hole horizons have several unique properties; e.g., a Killing horizon is present, its shear and expansion vanish, and the angular velocity on the horizon is constant (see, e.g., appendix G). However, none of these conditions were imposed in the formulation of Cook (1994). Obviously, a more sophisticated formulation had to be applied for computing a quasi-equilibrium state in close quasi-circular orbits of binary black holes. Cook of course well understood this point and in his subsequent works, he resolved this issue (see the next subsection).

8.6.2 Quasi-equilibrium with physical horizon boundary conditions

For deriving more physical quasi-equilibrium sequences, subsequently, Gourgoulhon *et al.* (2002); Cook (2002); Cook and Pfeiffer (2004) proposed more physical formulations in the IWM/CTS frameworks together with a geometrical boundary condition on black-hole horizons (see section 5.6.3). Based on their new formulations together with introducing spectral methods in numerical computations (see section 5.8.2), accurate and physical numerical computations have been performed for binary black holes in quasi-equilibrium since the first work by Grandclément *et al.* (2002). In the following, we will introduce pioneer numerical works performed so far in this field.

8.6.2.1 *The work of Grandclément et al. (2002)*

The work by Grandclément *et al.* (2002) developed and proposed many of basic ideas and procedures for a numerical solution of binary black holes in quasi-equilibrium that have become standard. In their work, they employed the IWM formulation (see section 5.6), and assumed the presence of a helical Killing vector field, denoted by equation (5.228). A unique point of their work is that they considered a Misner-bridge type topology for two black holes (see section 5.5.1.3), and in addition, they focused only on corotating binary black holes.

With these setting, they solved the basic equations imposing asymptotically flat outer boundary conditions and an excision boundary condition at minimal surfaces of wormhole throats; specifically, the conditions employed there are (in the terminology of section 5.6.3)

$$\alpha = 0, \qquad \beta_\perp = \beta_\parallel^a = 0, \qquad \frac{\partial \psi}{\partial r_I} + \frac{1}{2r_I}\psi = 0, \tag{8.69}$$

where $r_I = |x^i - x_I^i|$ with $I = 1$ and 2 and with x_I^i being coordinate centers of the minimal surfaces. This boundary condition is derived based on the the consideration that the properties of Killing horizons should be satisfied (see, e.g., appendix G). With the condition, $\beta_\perp = \beta_\parallel^a = 0$, the angular velocity of the black holes is guaranteed to be a constant which is equal to the orbital one. Later, Cook (2002) pointed out that the boundary condition (8.69) yields a slight irregularity on the minimal surfaces (for a solution of $\tilde{A}_{ij}$), and thus, the formulation of Grandclément *et al.* (2002) needed to be modified. However, the influence of this irregularity was subsequently found to be small [Cook and Pfeiffer (2004)], and the basic idea of Grandclément *et al.* (2002) is acceptable.[10]

For a solution of the basic equations, $\Omega_{\rm orb}$ is a free parameter, which has to be determined by a physical condition for constructing quasi-equilibria. Grandclément *et al.* (2002) proposed to employ the general relativistic virial relation, $M_{\rm ADM} = M_{\rm K}$

---

[10]Specifically, $\alpha = 0$ at the horizon cannot be used for the regularity to be satisfied. Cook (2002); Cook and Pfeiffer (2004) developed a Killing horizon boundary condition in which $\alpha \neq 0$ on the horizon (see section 5.6.3).

(see section 5.3.3), for identifying quasi-equilibria, assuming that the solution is in equilibrium in the frame corotating with the orbital motion. Friedman et al. (2002) later showed that the virial relation is indeed satisfied for quasi-equilibria in the IWM formulation in the presence of a helical Killing vector field, and hence, imposing the condition, $M_{\rm ADM} = M_{\rm K}$, is now strictly justified.

When a set of quasi-equilibrium states is obtained, a sequence has to be constructed from them. Binary black holes evolve simply due to the emission of gravitational waves, (approximately) conserving the area of each black hole. Here, the conservation of the area is not strictly satisfied in the presence of the absorption of gravitational waves, but black-hole perturbation studies suggest that this absorption effect would be quite minor [e.g., Poisson and Sasaki (1995); Mino et al. (1997)]. As described in section 5.7.2.3, for such a system, the first law is satisfied. In the present context, it is written as

$$\delta M_{\rm ADM} = \Omega_{\rm orb} \delta J + \frac{1}{8\pi} \sum_{i=1}^{2} \kappa_i \delta A_{{\rm H}_i} \,. \tag{8.70}$$

Here, it is reasonable to assume that the black-hole area is constant, and hence, $\delta A_{{\rm H}_i} = 0$ for each black hole, resulting in

$$\delta M_{\rm ADM} = \Omega_{\rm orb} \delta J \,. \tag{8.71}$$

Thus, there are two ways for constructing a sequence; one is to find a sequence for a fixed value of $A_{{\rm H}_i}$, and the other is to find a sequence that satisfies the first law in the form (8.71). Unfortunately, we cannot determine the event horizon from the quasi-equilibrium data, and can only determine an apparent horizon and its area $A_{\rm AH}$. Thus, the most strict way in the study of quasi-equilibrium sequences is to construct a sequence that satisfies equation (8.71).

Alternatively, we may infer that $A_{\rm AH}$ is approximately equal to $A_{\rm H}$ for a solution of quasi-equilibria. Thus, as an approximate condition, we may choose that $A_{\rm AH}$ of each black hole is constant along a sequence. Grandclément et al. (2002) employed the condition (8.71) to construct the sequence. However, they also confirmed that along the obtained sequence, $A_{\rm AH}$ is approximately constant; this was reconfirmed by subsequent works by Cook and Pfeiffer (2004).

One of the most remarkable discoveries in the numerical results of Grandclément et al. (2002) was that an ISCO (the minima of the energy and the angular momentum) is present along the sequence of equal-mass (and corotating) black holes, and the resulting angular velocity of the ISCO is

$$m_0 \Omega_{\rm orb} \approx 0.10, \tag{8.72}$$

where each mass ($m_0/2$) is defined when they are in isolation (i.e., $\Omega_{\rm orb} \to 0$). This value is slightly larger than the angular velocity of the ISCO for test particles around non-spinning black holes, $1/6^{3/2} \approx 0.068$ (see section 1.3.2.1), but it is not as larger as that by Cook (1994). In this case, the compactness of the ISCO is $(m_0 \Omega_{\rm orb})^{2/3} \approx 0.22$, which agrees approximately with $\ell/m_0 \sim 5$ [see figure 8.61

which is essentially the same as that by Grandclément *et al.* (2002)], and thus, this result indicates that the Keplerian law is not significantly and only reasonably modified even in the strongly general relativistic gravity.

In Damour *et al.* (2002), the numerical result of Grandclément *et al.* (2002) such as $M_{\rm ADM}$ as a function of $\Omega_{\rm orb}$ was compared with analytic results of post-Newtonian and effective one-body approaches (see appendix H.4). Damour *et al.* (2002) showed that the numerical result of Grandclément *et al.* (2002) agrees with those derived by these analytic approaches in a manner much better than the numerical result of Cook (1994). In particular, for non-close orbits with $m_0 \Omega_{\rm orb} \lesssim 0.02$, their numerical result agrees well with the analytic results (see also figure 8.62).

8.6.2.2 *The works of Cook and Pfeiffer (2004) and Caudill* et al. *(2006)*

Subsequently, Cook and Pfeiffer (2004) and Caudill *et al.* (2006) performed numerical computations for a solution of quasi-equilibria in the framework of the excision boundary condition (see section 5.6.3), which is more general than the condition employed by Grandclément *et al.* (2002) and enables to compute binary black holes for a wide variety of black-hole spins. In particular, Caudill *et al.* (2006) reported sophisticated numerical results both for corotating and irrotational (non-spinning) binary black holes. This work represents the most advanced achievement in this field. In the rest of this section, we summarize the finding by Caudill *et al.* (2006).

Figure 8.61 shows the binding energy as a function of an appropriately determined orbital separation for equal-mass binary black holes of corotating spin (left) and irrotational spin (no spin, right). The thick solid and dashed curves are sequences of binary black holes with quasi-circular orbits. The thin solid curves labeled by $J/m_0\mu = 4.5, \cdots$ denote the binding energy as a function of the orbital separation for (generally) non-circular orbits with given values of J. The quasi-equilibrium circular states denoted by the solid curves are determined by the virial relation $M_{\rm ADM} = M_{\rm K}$ while those denoted by the dashed curves are determined by the effective-potential approach. These two curves agree approximately with each other, and this indicates that the effective-potential method is a good approach for approximately determining quasi-equilibrium circular states. This figure shows that the ISCO is reached at $\ell/m_0 \sim 5.5$ for the corotating binary and $\ell/m_0 \sim 5$ for the irrotational binary.

Figure 8.62 plots the binding energy and the angular momentum as functions of the orbital angular velocity for the irrotational black-hole binary obtained by Caudill *et al.* (2006), together with various other results including the results by Cook (1994) (labeled by IVP) and the results by an effective one-body approach [Damour *et al.* (2002)] (labeled by EOB). All the results agree approximately with each other for non-close orbits with $m_0 \Omega_{\rm orb} \lesssim 0.03$. However, the degree of disagreement is quantitatively significant for close orbits. This disagreement is caused by the degree of the sophistication for the chosen formulation.

In particular, the predicted location of the ISCO is significantly different among

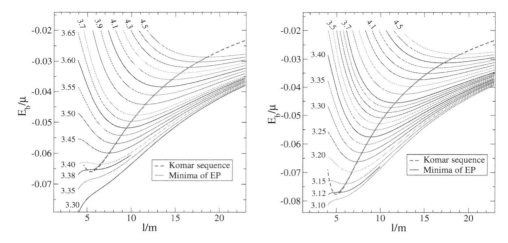

Fig. 8.61 The binding energy (E_b/μ) as a function of an appropriately determined orbital separation in units of total mass m_0 (denoted by m) for corotating binary black holes (left) and for irrotational binary black holes (right). The thin curves with labels $J/m_0\mu = 4.5, 4.3, 4.1, \cdots$ denote the sequences of constant angular momentum. The thick solid curves denote sequences of quasi-equilibria (quasi-circular orbits) determined by the virial relation $M_{\rm ADM} = M_{\rm K}$, and the dashed curves denote those determined by the effective potential method, which agree approximately with the solid curves. The figure is taken from Caudill et al. (2006).

these results, as well as the post-Newtonian results [Blanchet (2002)] in which $m_0\Omega_{\rm orb}$ at the ISCO for the irrotational binary is 0.137 and 0.129 in the second and third post-Newtonian approximations. This fact suggests that the numerical solutions obtained by Caudill et al. (2006) (as well as other solutions obtained by other methods) might not be accurate for binary black holes in close quasi-circular orbits. On the work of Caudill et al. (2006), a disadvantage was that the conformal flatness for the spatial metric was assumed. As we pointed out in section 8.2.3, this approximation should not be imposed for the study of compact binaries in close orbits, because the magnitude of the conformal three metric increases in proportional to $(m_0\Omega_{\rm orb})^{4/3}$.

Another serious problem for the study of quasi-equilibrium states in close orbits is that the quasi-equilibrium is unlikely to be realized for such orbits. As described in section 1.2.5, the ratio of the gravitational-radiation reaction time scale to the orbital period is written approximately by [see equation (1.93)]

$$\frac{t_{\rm gw}}{P_0} \approx 1.1 \left(\frac{m_0\Omega_{\rm orb}}{6^{-3/2}}\right)^{-5/3}. \tag{8.73}$$

This is derived in Newtonian gravity together with the quadrupole formula for the gravitational-wave luminosity, and hence, is not quantitatively very accurate. However, for an orbit with $m_0\Omega_{\rm orb} \sim 6^{-3/2}$, any general relativistic correction is not likely to modify this estimation by a factor of more than 2. Thus, it is natural to consider that for $m_0\Omega_{\rm orb} \gtrsim 6^{-3/2} \approx 0.07$, the orbit does not evolve in an adiabatic

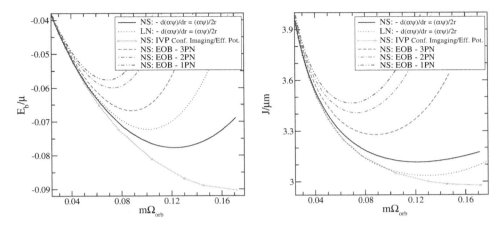

Fig. 8.62 The binding energy (E_b/μ) and the angular momentum ($J/m_0\mu$) as functions of the orbital angular velocity ($m_0\Omega_{\rm orb}$) for irrotational binary black holes. "NS" and "LN" denote that each black hole is non-spinning and slightly spinning (choosing $\Omega_{\rm rot} = -\Omega_{\rm orb}$; see section 5.6.3). Note that for $\Omega_{\rm rot} = -\Omega_{\rm orb}$ the black-hole spin is not fixed in a physical way and this unphysical degree is reflected in the fact that the location of minima of the binding energy and angular momentum disagree. "NS:IVP" denotes the work of Cook (1994). "EOB" denotes the results by several versions of effective one-body approach [Damour et al. (2002)] (see, e.g., appendix H.4). The figure is taken from Caudill et al. (2006).

manner but in a dynamical manner. Therefore, quasi-equilibrium solutions for such close orbits would not be very physical.

Recently, quasi-equilibrium states in close orbits have not been studied in the initial-data problem. Rather, binary black holes in close orbits are studied in numerical-relativity simulations with an initial condition in which the orbital separation is large enough to guarantee $t_{\rm gw}/P_0 \gg 1$, because it is now feasible to perform long-term simulations of binary black holes evolving through the inspiral orbits to the merger (see section 8.7).

8.7 Binary black holes: Numerical simulations

Soon after the first successful simulations for the merger of binary black holes performed in 2005 [Pretorius (2005a); Campanelli et al. (2006a); Baker et al. (2006a)], a variety of numerical simulations for the inspiral, merger, and ringdown of binary black holes have been performed. We now know how binary black holes evolve and merge in their final stage. Furthermore, highly accurate numerical results (as accurate as or more accurate than the results by post-Newtonian calculations for close circular orbits) have been derived since the work by Boyle et al. (2007, 2008). The resulting gravitational waveforms by these works can be used for constructing theoretical templates of gravitational waves [e.g., Ajith et al. (2008)] and for calibrating several approximations such as post-Newtonian approximations [e.g., Boyle et al.

(2007); Buonanno *et al.* (2007a)] and effective one-body approaches [e.g., Buonanno *et al.* (2007b); Pan *et al.* (2011)]. In this section, we would like to summarize the evolution process of coalescing binary black holes and resulting gravitational waves emitted, based on the results of the numerical-relativity simulations. Because a large number of works have been done and it is not possible for us to describe all of them, we will only pick up representative works (in particular early works) in this field in the following.

8.7.1 Two approaches for evolving binary black holes

There are two representative approaches for numerical-relativity simulations of binary black holes. For both approaches, using an adaptive mesh refinement (AMR) algorithm or a multi-domain algorithm, in which the simulations are performed on multi computational domains, is standard. One approach is based on the GH formulation (see section 2.3.4) together with the black-hole excision scheme for handling black holes. This approach was adopted first by Pretorius (2005a) and subsequently have been developed by Caltech–Cornell–CITA group extensively [Boyle *et al.* (2007, 2008)]. Caltech–Cornell–CITA group employs a multi-domain pseudo-spectral method (see section 5.8.2) for their numerical simulations, which enables them to yield a high-precision numerical result (much more precise than the results obtained by a standard finite-differencing scheme for a given computational resource). However, implementing such a sophisticated code requires a large research effort and development time, and thus, for beginners of numerical relativity, it would not be easy to develop a similar code in a short time scale.

The other approach is based on the BSSN formulation together with the moving puncture scheme for handling black holes (see section 2.3.3.1) and with the moving puncture gauge condition (see section 2.2.6). Since the first discovery of this method by Campanelli *et al.* (2006a) and Baker *et al.* (2006a), a large number of people in the community of numerical relativity has been employing this approach [e.g., Brügmann *et al.* (2008) and references therein]. This formulation is not suitable for the use of pseudo-spectral methods, and to date, numerical simulations have been performed only with the finite-differencing method. For a given computational resource, thus, it is not possible for this approach to yield a numerical result of coalescing binary black holes which is as precise as that by the GH-excision-spectral approach of Caltech–Cornell–CITA group [Boyle *et al.* (2007, 2008)]. However, implementing a new numerical code in this approach is quite straightforward,[11] and hence, for beginners of numerical relativity and for a subject in which an extremely high precision is not required, this is a recommendable approach.

---

[11]For example, the code described in Yamamoto *et al.* (2008) was developed by a beginner (one of my former students, T. Yamamoto) of numerical relativity.

8.7.2 Inspiral and merger processes

A variety of simulations performed so far in this field have made our knowledge about the inspiral and merger processes of binary black holes quite rich. We now confirm that the coalescence process is composed of three stages, inspiral, merger, and ringdown stages, irrespective of the mass ratio and spins of black holes. In the merger stage, a deformed black hole encompassing two black holes is formed when the separation of two black holes is small enough, and then, after the emission of ringdown gravitational waves associated primarily with the fundamental quasi-normal mode of the remnant black hole, a new stationary black hole is formed (see section 1.5.3). These processes proceed with no additional complicated process even in the presence of black-hole spins. However, the details of each stage and the state of the remnant black hole depend quantitatively on the mass ratio, magnitude and orientation of black-hole spins. In the following, we will describe generic features for the following three cases; equal-mass non-spinning black-holes case, unequal-mass non-spinning black-holes case, and spinning black-holes case.

8.7.2.1 Equal-mass and no spin case

This case was studied most deeply in particular in the first three years after the success of Pretorius [Pretorius (2005a,b); Campanelli *et al.* (2006a); Baker *et al.* (2006a); Pretorius (2006); Baker *et al.* (2006b, 2007c,b); Boyle *et al.* (2007); Scheel *et al.* (2009); Brügmann *et al.* (2008)]). In the absence of spins or in the presence of only spins which align with the orbital angular momentum vector, the orbital plane is unchanged throughout the entire evolution of the system. For the equal-mass and no spin case, furthermore, after the onset of the merger, a remnant black hole is formed at the center of mass of the system with no kick velocity because of no linear momentum loss by gravitational waves. This is in contrast to other cases in which the orbital plane precesses or nonzero kick velocity is excited. Although the final state of the remnant black hole depends on the mass of each black hole, the dimensionless parameter, such as dimensionless spin parameter and the frequency of gravitational waves in units of $M_{\rm BH}^{-1}$ where $M_{\rm BH}$ is the mass of the remnant black hole, is universal because of the scale-free nature of this vacuum system.

Among many simulations, Scheel *et al.* (2009) performed a high-precision and long-term simulation for the first time using the multi-domain pseudo-spectral method, and derived the final black-hole mass and the dimensionless spin parameter most accurately as

$$\frac{M_{\rm BH}}{m_0} = 0.95162 \pm 0.00002, \quad \frac{J_{\rm BH}}{M_{\rm BH}^2} = 0.68646 \pm 0.00004, \qquad (8.74)$$

where m_0 is the initial total mass and $J_{\rm BH}$ is the angular momentum of the remnant black hole. In their simulation as well as in the simulation by Boyle *et al.* (2007), 16 inspiral orbits were accurately evolved before the onset of the merger [see figure 8.63 and Mroué *et al.* (2013) for longer-term simulations]. Although the accuracy is not

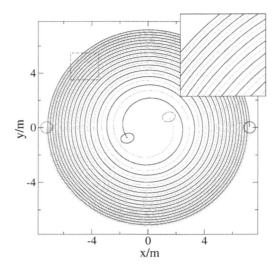

Fig. 8.63 Coordinate trajectories of the centers of inspiraling two black holes in a high-precision simulation by Boyle et al. (2007). The small circles/ellipsoids show the apparent horizons at the initial time and at the time when the simulation ends. The inset shows an enlargement of the dashed box.

as high as that by Scheel et al. (2009), other simulations [Baker et al. (2006b); Buonanno et al. (2007a)], before the work by Scheel et al. (2009), also derived approximately the same results for the remnant black hole in their early work.

Equation (8.74) implies that about 4.84% of the initial total mass energy is dissipated by the gravitational-wave emission. This value is much smaller than the maximum possible value of 29% allowed by the area theorem [see equation (1.144)]. Nevertheless, $\sim 5\%$ of the total mass energy is still huge, and the energy generation rate is much higher than those in other processes such as nuclear burning in which the conversion efficiency is at most $\sim 1\%$.

Equation (8.74) also shows that a moderately rapidly spinning black hole is formed in the end. This implies that a large fraction of the orbital angular momentum is carried into the remnant black hole.

For the coalescence of equal-mass non-spinning binary black holes, the gravitational waveform has a quite simple structure, which is much less complicated than that emitted by binary neutron stars. Figure 8.64 shows a gravitational waveform (a component of the complex Weyl scalar, Ψ_4) derived in a high-resolution run of Scheel et al. (2009). The left panel displays the inspiral waveform, in which the amplitude and the frequency of gravitational waves increase monotonically with time. The right panel displays the merger and inspiral waveforms. The peak of the amplitude is reached around the formation of a common apparent horizon, and then, the amplitude exponentially decreases to zero in the ringdown stage during which the new black hole formed relaxes to a final stationary state.

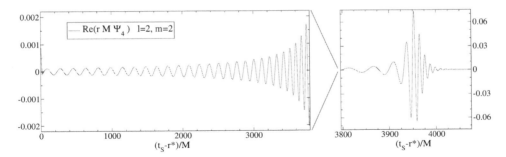

Fig. 8.64 Gravitational waveform ($l = m = 2$ mode of Ψ_4 multiplied by the initial binary mass and distance) for an equal-mass non-spinning binary black hole derived in a high-resolution run of Scheel et al. (2009). The left panel zooms in on the inspiral waveform, and the right panel zooms in on the merger and ringdown waveforms.

8.7.2.2 Unequal-mass and no spin case

This asymmetric case has been studied focusing in particular on kick (recoil) motion of the remnant black hole induced by the anisotropic gravitational-wave emission during the merger and ringdown stages [e.g., Baker et al. (2006c); González et al. (2007b); Schnittman et al. (2008); Hannam et al. (2010); Buchman et al. (2012)]. Because of the absence of spins, the orbital plane is unchanged throughout the entire evolution of the system, as in the case of equal-mass non-spinning binaries. On the other hand, kick motion is always excited as a result of the back reaction of the anisotropic gravitational-wave emission.

As described in section 1.2.4, gravitational waves carry away the linear momentum from binaries in circular orbits in the presence of the mass asymmetry, and as the back reaction of this emission, the binary system always has a kick velocity in the inspiral stage. However, because of the quasi-periodic nature of the binary orbits, the kick velocity changes in a quasi-periodic manner, and is not accumulated in a particular direction during the inspiral stage. By contrast, through the last inspiral to the merger and ringdown stages, the linear momentum emission may be dynamically accumulated in a particular direction in a short time scale and, as a result, a large kick velocity for the remnant black hole could be excited.

After several preliminary results were derived in numerical relativity[12], González et al. (2007b) derived the first accurate results for the kick velocity; see figure 8.65 for their results. Here the kick velocity is estimated from the total linear momentum carried away by gravitational waves divided by the mass of the remnant black hole. González et al. (2007b) performed the simulations in the BSSN-puncture

---

[12] In those days, the competition among several groups working in the simulation of binary black holes was quite severe and many short-term simulations with initial data of quite close separation were performed to accelerate the publication. However, in many of such simulations, the initial condition was not very realistic, and hence, the result for the kick velocity was not obtained as accurately as that in the subsequent work by González et al. (2007b).

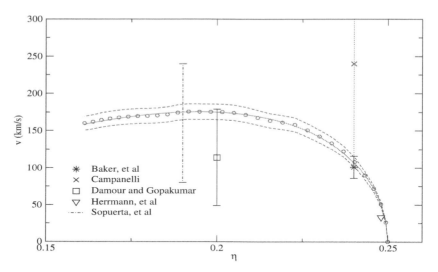

Fig. 8.65 The kick velocity of remnant black holes as a function of the symmetric mass ratio $m_1 m_2/(m_1+m_2)^2 (=: \eta)$. The solid circles and solid curve denote the numerical results and the fitting for them derived by González et al. (2007b). The dashed curves indicate the possible error range. Other plots (including the error bars) denote the earlier results [see González et al. (2007b) for details].

formulation for mass ratio $q := m_2/m_1 (= 1/Q)$ in the range between 0.25 and 1 (and thus, $0.16 \leq \eta \leq 0.25$ where $\eta := m_1 m_2/m_0^2$). The results were derived within the error of $\sim 10\,\mathrm{km/s}$. They found that the maximum kick velocity of $\approx 175\,\mathrm{km/s}$ is achieved for $q \approx 0.36$. It is interesting to note that the extreme for $f_{\mathrm{fitchett}}(q)$ in equation (1.96) is located at $q \approx 0.38$ and coincides approximately with the numerical-relativity result for the merger and ringdown stages. This maximum velocity was subsequently confirmed by a high-precession numerical-relativity study of Buchman et al. (2012) which was performed using a multi-domain pseudo-spectral method.

The mechanism for the emission of the linear momentum by gravitational waves was in detail analyzed, e.g., in Schnittman et al. (2008). They performed a multipolar expansion analysis of the gravitational-wave emission, in which gravitational waves numerically obtained are decomposed into spherical harmonics components [denoted by (l, m) in the following] and the contribution of each component is determined (see chapter 6). They clarified that the linear momentum is emitted primarily by three coupling modes (see figure 8.66); between the mass quadrupole ($l = |m| = 2$) and the mass octapole mode ($l = |m| = 3$), between the mass quadrupole ($l = |m| = 2$) and the current quadrupole mode ($l = 2, |m| = 1$), and between the mass octapole ($l = |m| = 3$) and the mass hexadeca-pole mode ($l = |m| = 4$). They also showed that other coupling modes give only minor contribution for the linear momentum emission. It should be stressed that the linear

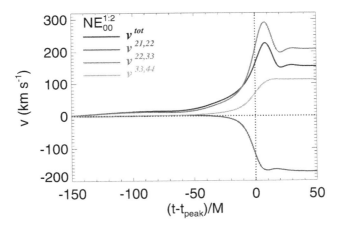

Fig. 8.66 The kick velocity as a function of time for a binary black hole with $q(= Q^{-1}) = m_2/m_1 = 1/2$ and no spins, derived by Schnittman et al. (2008). The total kick velocity as well as the contributions from the coupling between $l = |m| = 2$ and $(l, |m|) = (2, 1)$ modes, between $l = |m| = 2$ and $l = |m| = 3$ modes, and between $l = |m| = 3$ and $l = |m| = 4$ modes are plotted. t_{peak} denotes the time at which the gravitational-wave amplitude of $l = |m| = 2$ mode reaches its maximum.

momentum is emitted only by the coupling between two different modes in contrast to the energy and angular momentum for which the dominant mode is universally the quadrupole mode ($l = |m| = 2$ mode): Thus, there is no quadrupole formula for the emission of the linear momentum [see equation (1.78)]. Schnittman et al. (2008) indeed clarified that there is no absolutely primary mode for the linear momentum emission, and all the three coupling modes are leading terms for the emission as long as the binaries of no spins are concerned.

For the kick velocity and the parameters of the remnant black hole, approximate fitting formulas have been derived. Based on the Fitchett's formula, (1.95), González et al. (2007b) gave a fitting formula for the kick velocity as (see the solid curve of figure 8.65)

$$V_{\text{kick}} = A f_{\text{fitchett}}(q)(1 + B\eta), \tag{8.75}$$

where $A = 1.2 \times 10^4$ km/s and $B = -0.93$.

Fitting formulas for the dimensionless spin parameter of the remnant black-hole were first studied by González et al. (2007b) and by a follow-up analysis of Berti et al. (2007). Subsequently, Buonanno et al. (2008); Tichy and Marronetti (2009); Barausse and Rezzolla (2009) also derived fitting formulas. González et al. (2007b) gave for $1 \leq Q \leq 4$

$$\frac{J_{\text{BH}}}{M_{\text{BH}}^2} \approx 0.089(\pm 0.003) + 2.4(\pm 0.025)\eta. \tag{8.76}$$

For the equal-mass case with $\eta = 1/4$, the result of this fitting formula agrees well with the value of equation (8.74). However, for $\eta \to 0$, this is not a good formula, because for this limit, the spin of the remnant black hole should vanish. Berti et al. (2007) corrected this drawback and derived a better fitting formula

$$\frac{J_{\rm BH}}{M_{\rm BH}^2} \approx 3.272\eta - 2.075\eta^2. \tag{8.77}$$

The work of Barausse and Rezzolla (2009) further improved this result relying thoroughly on many numerical-relativity results as

$$\frac{J_{\rm BH}}{M_{\rm BH}^2} \approx 2\sqrt{3}\eta - 3.5171(\pm 0.1208)\eta^2 + 2.5763(\pm 0.4833)\eta^3, \tag{8.78}$$

which reproduces the result for $\eta = 1/4$ and 0 well. This fitting formula also fits well with the latest high-precision results by Buchman et al. (2012).

The fitting formulas for the mass of the remnant black hole have a diversity. This is due to the fact that the results of numerical-relativity simulations for the total energy emitted by gravitational waves depended rather strongly on which group performed the simulations. If we believe the latest work by Buchman et al. (2012), the following fitting formula derived by Buonanno et al. (2007b) would be a good formula:

$$\frac{M_{\rm BH}}{m_0} = 1 - 0.024(\pm 0.057)\eta - 0.641(\pm 0.249)\eta^2. \tag{8.79}$$

This shows that (for the best fit formula) the total energy emitted by gravitational waves monotonically decreases with the increase of the mass asymmetry, and the equal-mass system is the most efficient energy generator.

Gravitational waveforms for the asymmetric systems are qualitatively similar to those for the equal-mass binary black holes. Figure 8.67 displays the results for $q = 1\text{-}1/6$ ($Q = 1\text{-}6$) derived by Buchman et al. (2012), and shows that the waveforms are composed of the inspiral, merger, and ringdown parts irrespective of the mass ratio. However, quantitatively, significant dependence on the mass ratio is found. For example, the maximum amplitude of the quadrupole mode (in units of the total mass) decreases with the decrease of the mass ratio q. Also, the relative amplitude among different multipole modes depends strongly on the mass ratio: e.g., the ratio of the amplitudes of the octapole mode to the quadrupole mode increases with the decrease of q.

The remaining challenge for the simulation of non-spinning black holes is to explore the case where q is much smaller than unity. To date, for the mass ratio less than 0.1 ($q < 0.1$), no long-term simulation has been done yet [but see González et al. (2009) for the run with several orbits and Lousto and Zlochower (2011); Nakano et al. (2012) for the first attempt to $q = 1/100$ case]. The reason for this status is that evolving small-mass black holes in highly asymmetric binaries requires a high grid resolution (a small grid spacing) and hence the time step interval is severely limited by the CFL condition (see sections 3.1.4 and 3.4.3; in the later section, we showed that for evolving a black hole, the grid spacing should be smaller than $\sim M_{\rm BH}/20$ where $M_{\rm BH}$ is the mass of the black hole.) Thus, a

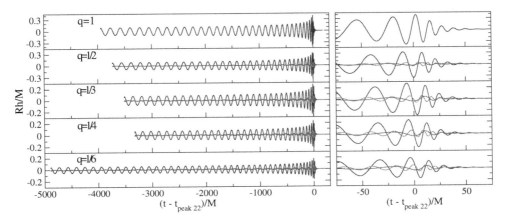

Fig. 8.67 Gravitational waveforms emitted by unequal-mass non-spinning binary black holes. The simulations were performed by Buchman et al. (2012). $t_{\text{peak 22}}$ denotes the time at which the gravitational-wave amplitude of the quadrupole mode reaches its maximum. R and M are the radius of extraction and initial total mass with h being the gravitational-wave amplitude for the $l = |m| = 2$ mode (highest amplitude), $l = |m| = 3$ mode (middle-amplitude), and $(l, |m|) = (2, 1)$ mode (lowest-amplitude).

large computational resource is necessary. For $q \ll 1$ (and $\eta = \mu/m \ll 1$), the ratio of the gravitational-radiation reaction time scale to the orbital period is much larger than unity even near the ISCO [see equation (1.93)]. This implies that many highly relativistic orbits will be maintained before the orbital motion changes to the plunging motion toward the merger. This is the special property of the extremely small mass-ratio case. Extremely small mass-ratio binaries are likely to exist in nature; e.g., binary supermassive black holes located in the center of a galaxy and a binary composed of a massive black hole with mass $\gtrsim 100 M_\odot$ and a stellar-mass black hole $\lesssim 10 M_\odot$. For clarifying the merger process of such binaries, a long-term high-resolution numerical simulation is awaited.

8.7.2.3 Spinning binary black hole case

The effects of spins significantly modify the orbital motion of inspiraling binary black holes. In the presence of spins, the orbital plane in general precesses due to the spin-orbit and spin-spin coupling effects [e.g., Kidder et al. (1993); Kidder (1995)]. In addition, the remnant black hole obtains in general a kick velocity even for the equal-mass case [Kidder (1995)] due to the anisotropic gravitational-wave emission. Because there are seven degrees of freedom for the free parameters of the system in the presence of generic spins (for six components of two spin vectors and mass ratio), a huge number of simulations is required for the complete survey of the parameter space that has not been done yet. In particular, it is a challenging task to perform simulations for the case that black holes have spins close to the extreme

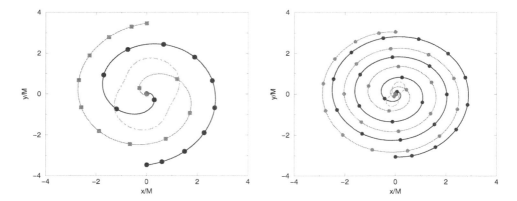

Fig. 8.68 The trajectories of punctures (centers of two black holes) on the orbital plane for inspiraling two black holes for the case that spins of two black holes anti-align (left) and align (right) with the axis of the orbital angular momentum vector. The dimensionless spin parameter of each black hole is 0.757. The dashed closed curves denote the apparent horizons of the remnant black holes formed at $t \approx 106 M_{\mathrm{ADM},0}$ (left) and $233 M_{\mathrm{ADM},0}$ (right) where $M_{\mathrm{ADM},0}$ is the initial ADM mass. The figure is taken from Campanelli et al. (2006c).

limit [Lovelace et al. (2011, 2012); Hemberger et al. (2013)] and that the spin axes of two black holes significantly precess [Mroué et al. (2013)]. Nevertheless, a variety of knowledge has been obtained by numerical-relativity simulations.

First, we summarize the inspiral and merger processes for the case that the spin axes of two black holes are parallel to the axis of the orbital angular momentum. In this special case, the orbital plane and directions of spins are unchanged, and thus, there are only three parameters; magnitudes of two spins and mass ratio. The spin effects for such a case were first studied by Campanelli et al. (2006c) and subsequently by Campanelli et al. (2006b); Baker et al. (2007a); Herrmann et al. (2007); Koppitz et al. (2007); Pollney et al. (2007); Marronetti et al. (2008); Lovelace et al. (2011, 2012); Hemberger et al. (2013). Among them, Marronetti et al. (2008); Lovelace et al. (2011, 2012); Hemberger et al. (2013) focused in particular on the simulation for rapidly spinning black holes, and demonstrated it feasible to perform simulations for highly spinning black holes in the range of $\hat{a}$ between 0.90 and 0.97.

Campanelli et al. (2006c) performed the first simulation of spinning binary black holes for two cases: The spins of two black holes were anti-aligned or aligned with the axis of the orbital angular momentum. Here, the dimensionless spin parameter of the two black holes and initial orbital angular velocity were fixed to be 0.757 and $m_0 \Omega_{\mathrm{orb}} = 0.05$. They employed the BSSN formulation together with the moving puncture scheme. Figure 8.68 displays the trajectory of punctures (centers of two black holes) for the anti-aligned (left) and aligned cases (right). It is found that for the anti-aligned case, the two black holes merge in one orbit. On the other hand, for the aligned case, the two black holes experience three orbits: The lifetime for the aligned case is much longer than that for the anti-aligned case. This is primarily due to the spin-orbit coupling effect; in the terminology of the post-Newtonian approximation [Kidder (1995)], a repulsive or attractive force in addition to Newtonian gravity is exerted to black holes in the presence of spins and its magnitude is approximately proportional to $\mathbf{S} \cdot \mathbf{L}$ where $\mathbf{S}$ and $\mathbf{L}$ denote the total black-hole spin and orbital angular momentum vector, respectively (see, e.g., appendix H). As already described in section 8.5.2.2 (in the context of black hole-neutron star binaries), the aligned (anti-aligned) spin effect slows down (speeds up) the orbital velocity, and hence, the luminosity of gravitational waves is decreased (increased) and orbital evolution due to the gravitational-radiation reaction is decelerated (accelerated). Even more importantly, the orbital radius at the ISCO is decreased (increased) for the aligned (anti-aligned) case. This enhances increasing (decreasing) the lifetime of the aligned (anti-aligned) binary. In addition to the spin-orbit coupling, the spin-spin coupling could play an important role because if two spins align and anti-align, repulsive and attractive forces are exerted, respectively. Since the magnitude of the spins employed was high, the spin-spin coupling would also play a certain role in the result of Campanelli et al. (2006c).

All these facts can be more clearly found in the simulation with high black-hole spins. Lovelace et al. (2011, 2012) performed two long-term simulations for equal-mass binary black holes. One is for the aligned case in which the dimensionless spins of both black holes are 0.97 and the other is for the anti-aligned case in which the dimensionless spins are 0.95. For both cases, the initial orbital separation is sufficiently large ($m_0 \Omega_{\rm orb} \approx 0.0138$ and 0.0145 for the aligned and anti-aligned cases, respectively). For the aligned case, ≈ 25.5 orbits are experienced (see figure 8.69), and more than 10% of the initial total mass energy of the system is dissipated by the gravitational-wave emission during the entire coalescence process. On the other hand, for the anti-aligned case, only ≈ 12.5 orbits are experienced and only $\sim 3\%$ of the initial total mass energy is dissipated. The primary reason for the difference in the energy dissipated is that for the aligned case, the absolute value of the binding energy near the ISCO is quite large (i.e., the orbital separation at the ISCO is small) and a large amount of the energy has to be dissipated for reaching the ISCO. On the other hand, for the anti-aligned case, the absolute value of the binding energy near the ISCO is rather small because the orbital separation at the ISCO would be quite large.

Hemberger et al. (2013) subsequently performed a number of additional simulations for the equal-mass and equal-spin (aligned or anti-aligned) case, varying the spin of the black hole. They analyzed the final spin of the remnant black hole and the energy radiated by gravitational waves during the entire coalescence, and derived interesting fitting formulas: For the dimensionless spin of the remnant black hole, they derived a fitting formula of a fourth-order polynomial form as

$$\hat{a}_f = a_0 + a_1 \hat{a}_i + a_2 \hat{a}_i^2 + a_3 \hat{a}_i^3 + a_4 \hat{a}_i^4, \tag{8.80}$$

where $a_0 = 0.686402$, $a_1 = 0.30660$, $a_2 = -0.02684$, $a_3 = -0.00980$, and $a_4 = -0.00499$ with $\hat{a}_i$ being the initial spin magnitude of the black holes. This shows

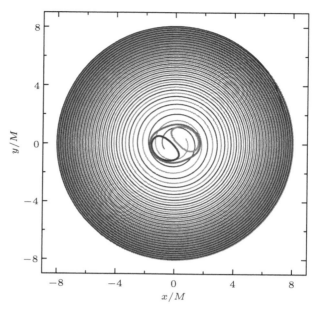

Fig. 8.69 Trajectories of the centers of the individual apparent horizons for two equal-mass equal-spin black holes with the individual spin $\hat{a} = 0.97$. The individual and common apparent horizons at the final time are also shown by the thick closed curves. The figure is taken from Lovelace et al. (2012).

that even if the extreme black holes ($\hat{a}_i = 1$) were initial components, the resulting final spin would be $\hat{a}_f \approx 0.95$. Therefore, the merger of the extreme black holes will not result in the extreme black hole.

Hemberger et al. (2013) also derived a fitting formula for the total radiated energy in a simple form

$$\frac{\Delta E_{\text{GW}}}{m_0} = b_0 + \frac{b_1}{b_2 + \hat{a}_i}, \tag{8.81}$$

where $b_0 = 0.00258$, $b_1 = -0.07730$, and $b_2 = -1.6939$ with m_0 being the initial total mass. This shows that for $\hat{a}_i = 1, 0$, and -1, $\Delta E_{\text{GW}}/m_0 \approx 11.4\%$, 4.8%, and 3.1%, and illustrates the importance of the initial black-hole spin for determining the remnant black-hole mass.

In the presence of a misalignment between axes of spins and angular momentum vector, the orbital plane and spin axes precess as shown in post-Newtonian studies [e.g., Kidder (1995)]. Campanelli et al. (2007a) explicitly showed the significance of this effect for the first time in numerical relativity. Figure 8.70 displays one of the results reported by Campanelli et al. (2007a). This clearly shows that the orbital plane and spin axes precess.

By this precession effect, gravitational waveforms are significantly modulated if we observe gravitational waves from a particular direction. Although such wave-

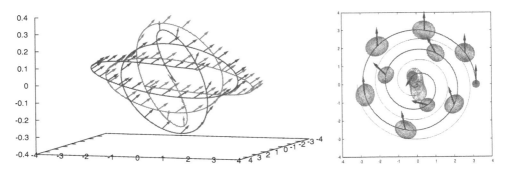

Fig. 8.70 Left: The trajectories of punctures and directions of spins in the inspiraling and precessing two spinning black holes. Note that the scale of z-axis in the left panel is by 10 times enlarged. Right: The x-y plane projection of the orbital track, apparent horizons, and spin directions for the result of the left panel. The dimensionless spin parameter of each black hole is 0.5. The figure is taken from Campanelli *et al.* (2007a).

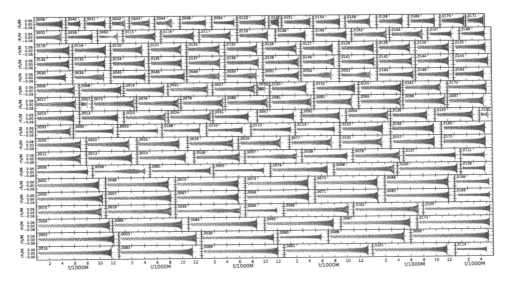

Fig. 8.71 The 174 gravitational waveforms computed by Mroué *et al.* (2013).

forms were found in Campanelli *et al.* (2009), a systematic survey for this issue has not been completed yet. Caltech–Cornell–CITA group has been performing a large number of long-term high-precision simulations for the spinning binary black holes, improving the computational efficiency [Mroué and Pfeiffer (2012); Mroué *et al.* (2013)]. In particular, Mroué *et al.* (2013) published more than 100 waveforms (see figure 8.71). This type of systematic works will play an important role in preparing gravitational-wave templates for the future gravitational-wave detection.

The effect of the black-hole spins significantly can enhance the emissivity of the linear momentum of gravitational waves and hence the kick velocity. This effect is partly understood from the analysis of Favata et al. (2004) for the inspiral orbits [which is based on the result of Kidder (1995)]. Here, we focus on the case that two black-hole spins align or anti-align with the axis of the orbital angular momentum vector (no precession case). According to a post-Newtonian formula of Kidder (1995) [see also Favata et al. (2004)], the magnitude of the kick velocity for the inspiraling binary black holes in circular orbits is written as

$$|V_{\text{kick}}| \approx \frac{q^2}{(1+q)^5}\left[c_1(1-q) - c_2(q\hat{a}_2 - \hat{a}_1)\left(\frac{2m_0}{r_{\text{term}}}\right)^{1/2}\right]\left(\frac{2m_0}{r_{\text{term}}}\right)^4, \quad (8.82)$$

where $\hat{a}_1$ and $\hat{a}_2$ are the magnitudes of dimensionless spins of two black holes and r_{term} is the orbital radius for which we calculate the kick velocity. Note that $q = m_2/m_1 < 1$. c_1 and c_2 are numerical values as $1480\,\text{km/s}$ and $883\,\text{km/s}$, respectively. This suggests that a kick velocity may be excited even for equal-mass binaries in the presence of spins with $\hat{a}_1 \neq \hat{a}_2$. Note that in the presence of the spin effects, the orbital radius of the ISCO can be smaller than that in their absence. Hence, the kick velocity could be enhanced by the presence of the spins.

The kick velocity for non-precession binaries was first studied by Herrmann et al. (2007); Baker et al. (2007a); Koppitz et al. (2007); Pollney et al. (2007). All these groups employed the BSSN formulation with the moving puncture approach. They indeed found that a moderately large kick velocity with up to $\sim 500\,\text{km/s}$ can be excited even for equal-mass binaries in the presence of large anti-parallel spins. However, it was later found that this is in a sense a minor effect by spins, because in the presence of a precession orbit, the spin effect becomes much more significant.

The possibility for an extremely large kick velocity with $\gtrsim 2000\,\text{km/s}$ was first discovered by González et al. (2007a) and Campanelli et al. (2007c). González et al. (2007a) studied the merger of an equal-mass binary black hole for which the spin axes are initially parallel to the orbital plane and the two spins are anti-aligned. The magnitude of each black-hole spin was chosen to be ~ 0.8. They found that for this case, the kick velocity becomes $\sim 2500\,\text{km/s}$. Figure 8.72 displays the trajectories of two black-hole punctures found in González et al. (2007a). This shows that the black holes move out of the original orbital plane, and after the merger, the remnant black hole receives a kick in the negative z direction. This is qualitatively understood by the fact that in the presence of spins which misalign with the orbital angular momentum vector, the post-Newtonian formula for the kick velocity (8.82) is modified as [see also Campanelli et al. (2007c)]

$$V_{\text{kick}}^i \approx \frac{q^2}{(1+q)^5}\left[c_1(1-q)\mathbf{e}_1^i - \left\{c_2\left(q\hat{a}_2^{\parallel} - \hat{a}_1^{\parallel}\right)\mathbf{e}_1^i \sin\xi\right.\right.$$
$$\left.\left. + 2c_2\left(q\hat{a}_2^{\perp} - \hat{a}_1^{\perp}\right)\mathbf{e}_3^i\right\}\left(\frac{2m}{r_{\text{term}}}\right)^{1/2}\right]\left(\frac{2m}{r_{\text{term}}}\right)^4, \quad (8.83)$$

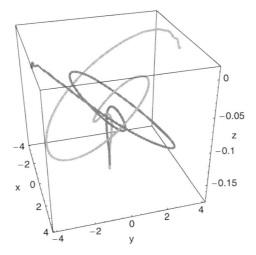

Fig. 8.72 The trajectories of two spinning black-hole punctures. The black holes move out of the original orbital plane, and after the merger, the remnant black hole receives a kick velocity in the negative z direction. Note that the scale of the z-direction is about 40 time larger. The figure is taken from González et al. (2007a).

where $\perp$ and $\parallel$ refer to perpendicular and parallel to the orbital angular momentum vector, $\mathbf{e}_1^i$ is a unit orthogonal vector in the orbital plane and $\mathbf{e}_3^i$ is a unit vector normal to $\mathbf{e}_1^i$. Thus, for a large value of spin components in the orbital plane, the *super* kick could be explained. A possible origin of such a super kick was also analyzed in Brügmann et al. (2007). They pointed out that the large kick was excited due to the asymmetry in the gravitational-wave emission by the two modes, $l = m = 2$ and $l = -m = 2$.

8.7.3 *Modeling gravitational waveforms and their spectrum*

As we mentioned in section 1.6, theoretical templates are necessary for an efficient detection of gravitational waves. Modeling gravitational waves computed in numerical relativity is probably the most promising approach for constructing the theoretical templates in the late inspiral, merger, and ringdown stages of binary black holes. Such works have been actively done for many years [see, e.g. Baker et al. (2007d); Buonanno et al. (2007a,b); Boyle et al. (2007); Ajith et al. (2008); Pan et al. (2008); Hannam et al. (2008b,a); Gopakumar et al. (2008); Santamaría et al. (2010); Hannam et al. (2010) for early works].

Here, the ringdown gravitational waveform is easily modeled because it is determined by the oscillation of a remnant black hole associated with its quasi-normal mode. Thus the main task is to model the gravitational waveform emitted during the inspiral and merger stages.

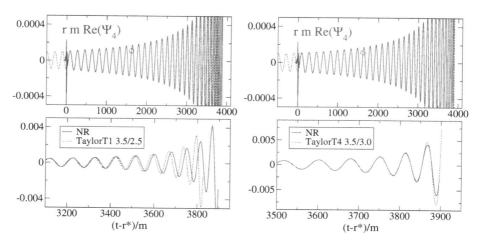

Fig. 8.73 Comparison between a numerical-relativity waveform and waveforms derived in the Taylor T1 approach (left) and in the Taylor T4 approach (right) for the equal-mass non-spinning binary black hole. The figure is taken from Boyle et al. (2007).

Broadly speaking, there are two analytic (or semi-analytic) approaches for modeling gravitational waveforms. One is the Taylor expansion approach (Taylor T1–T4 approaches) and the other is the effective one-body approach developed originally by Buonanno and Damour (1999) [e.g., Buonanno et al. (2009) for the overview of two approaches; see also appendix H]. Both approaches rely on the results derived in the currently-best post-Newtonian calculations together with a resummation prescription to phenomenologically incorporate higher-order post-Newtonian terms that have not been derived yet. In the Taylor expansion approach, the adiabatic approximation is usually adopted for determining the orbital evolution of binaries, while in the effective one-body approach, an equation of motion derived in this framework is directly integrated for determining the orbital motion.

The Taylor T1–T4 formulas were examined by Buonanno et al. (2007a) and Boyle et al. (2007) for gravitational waves from the equal-mass non-spinning black hole binary. Figure 8.73 displays the results of comparison performed by Boyle et al. (2007), who compared the numerical waveform derived in their high-precision numerical-relativity simulation with those obtained by Taylor T1–T4 formulas (see appendix H for their definition). They discovered that the Taylor T4 can reproduce the numerical waveform quite accurately for this particular model. This opened a window to construct a precise template of gravitational waves in analytic forms at least for equal-mass compact binaries.

Subsequently, Hannam et al. (2010) examined the robustness of the Taylor T1 and T4 formulas for unequal-mass non-spinning binaries and equal-mass spinning binaries. They found that for the unequal-mass non-spinning cases, the Taylor T4 formula is still robust (although it is not as robust as for the equal-mass case) while

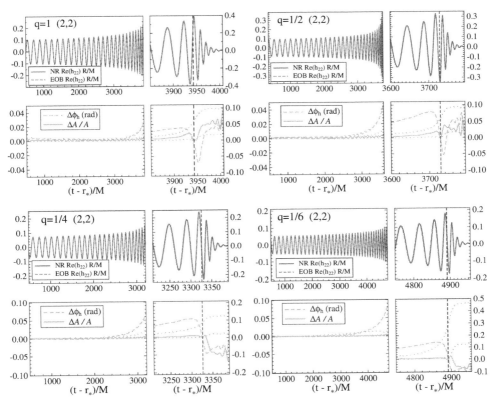

Fig. 8.74 Comparison between numerical-relativity waveforms and model waveforms derived in an effective one-body approach. The top panels of each figure show the gravitational waveforms by numerical-relativity simulations and by an effective one-body approach, and the bottom panels show the amplitude and phase differences between them. The results for equal-mass ($q = 1 = Q$) and unequal-mass ($q = 1/2, 1/4, 1/6$ or $Q = 2, 4, 6$) non-spinning binary black holes are displayed. The figure is taken from Pan *et al.* (2011).

for the equal-mass spinning case, not the Taylor T4 but the Taylor T1 formula works relatively well.

The modeling of numerical-relativity waveforms of coalescing binary black holes in terms of effective one-body approaches has been also extensively performed by Damour, Buonanno, and their collaborators. Figure 8.74 is one of the results for such comparison. In this figure the numerical waveforms derived by Buchman *et al.* (2012) are compared with the waveforms by an effective one-body formalism. In the effective one-body approach, equations of motion are numerically integrated (without the adiabatic approximation), and hence, not only the inspiral motion but also the plunge motion can be derived (see appendix H.4). Then, the waveform is derived using a resummed post-Newtonian formula until the merger sets in. For the merger and ringdown stages, the waveform is appropriately modeled

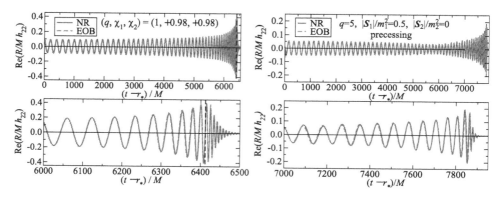

Fig. 8.75 Left: Numerical-relativity (NR) and effective one-body (EOB) waveforms for a binary black hole with $Q = 1$ and $\hat{a}_1 = \hat{a}_2 = 0.98$. The spin axes of the black holes are parallel to the orbital angular-momentum axis. Right: The same as the left panel but with $Q = 5$, $\hat{a}_1 = 0.5$, and $\hat{a}_2 = 0$. The spin and orbital angular-momentum axes are misaligned in this case. For both plots, the lower panels show the zoom-up of the final merger state. The figures are taken from Taracchini et al. (2014).

using the information of expected quasi-normal modes of the remnant black hole. Figure 8.74 clearly shows that with such a relatively simple procedure, numerical-relativity waveforms can be well modeled irrespective of the mass ratio within the phase error ~ 0.1 radian and the amplitude error of $\lesssim 10\%$.

In their subsequent works, Taracchini et al. (2014) performed a modeling of gravitational waves for generic mass ratios and spins of binary black holes. They showed that for the aligned spin cases, the numerical-relativity gravitational waveforms are well reproduced by an effective one-body approach even for $\hat{a} = 0.98$ (see the left panel of figure 8.75). They furthermore showed that even for the precessing binary cases, the numerical-relativity gravitational waveforms are well reproduced if the spin is not extremely high (see the right panel of figure 8.75). All these works demonstrate that the effective one-body approach is powerful for constructing a template family.

The spectrum of gravitational waves from coalescing binary black hole has a unique feature. As shown in section 1.5.1, for a distant orbit in which the quadrupole formula together with Newtonian gravity describes accurately the binary motion and emitted gravitational waves, the amplitude of the Fourier spectrum is proportional to $f^{-7/6}$ [and the effective amplitude $f|\tilde{h}(f)|$ is proportional to $f^{-1/6}$; see equation (1.199)]. For close orbits, this relation is modified by general relativistic effects. Buonanno et al. (2007a) and Baker et al. (2007c) independently discovered that for equal-mass non-spinning binary black holes, the power changes from $-7/6$ approximately to $-n$ with $n = 0.6-0.8$ (see the upper panel of figure 8.76; note that the bumps are due to the presence of a small residual eccentricity of the orbital motion). The transition occurs near the ISCO at $m_0 \Omega_{\rm orb} \approx 0.10-0.15$. The amplitude of the Fourier spectrum subsequently damps for a high-frequency band; the

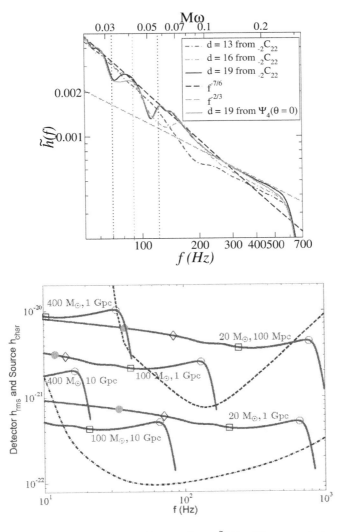

Fig. 8.76 Top: The amplitude of the Fourier transform, $\tilde{h}(f)$, of the numerical waveforms by Buonanno et al. (2007a) for an equal-mass binary with hypothetical total mass $30M_\odot$ at a hypothetical distance to the source, $D = 100\,\mathrm{Mpc}$. Three numerical results with different values of d are shown where d denotes a coordinate separation of two black holes for the initial conditions of the numerical simulation. The three vertical lines mark the frequency at which each run starts. Bottom: The same as the top panel but for the numerical waveforms by Baker et al. (2007c) and for a variety of black-hole mass and distance. In this plot, $h_{\mathrm{char}} := 2f|\tilde{h}(f)|$ are shown. The mass labeled is the total mass of binary black holes. The two convex curves denote planned noise levels of the initial (upper curve) and advanced LIGO (lower curve).

cutoff frequency agrees approximately with the frequency of the fundamental quasi-normal mode. Using equation (1.204) with $M_{\rm BH} = 30 \times 0.95 M_\odot$ and $\hat{a} = 0.69$ [see equation (8.74)], the frequency of the quasi-normal mode is approximately 600 Hz, and is written as

$$f_{\rm QNM} \approx 600\,{\rm Hz} \left(\frac{M_{\rm BH}}{28.5 M_\odot}\right)^{-1}. \tag{8.84}$$

The cutoff frequency found in the upper panel of figure 8.76 agrees approximately with this value.

Using the bottom panel of figure 8.76, Baker *et al.* (2007c) compared the amplitude of gravitational waves from equal-mass non-spinning binary black holes with the noise levels of the initial and advanced LIGO. This clearly demonstrates that by the advanced LIGO, it will be feasible to detect late inspiral, merger, and ringdown gravitational waves with total mass $m_0 \gtrsim 20 M_\odot$, even if the merger occurs at a cosmological distance of 1 Gpc.

For unequal-mass non-spinning binary black holes, the Fourier spectrum was analyzed in detail in Pan *et al.* (2011). The Fourier spectra for a variety of waveforms from both non-spinning and spinning binary black holes are also found in Santamaría *et al.* (2010). These show that the shape of the Fourier spectrum is similar to those in figure 8.76 irrespective of the mass ratio and the magnitude of spins. This simpleness helps modeling the Fourier spectrum of gravitational waves from coalescing binary black holes with a small number of parameters [e.g., Santamaría *et al.* (2010); Hannam *et al.* (2014)].[13]

---

[13] Note however that in the presence of orbital precession due to the presence of misaligned spins and orbital angular momentum vector, the spectrum amplitude depends on the direction of the observation and usually has a modulation.

Chapter 9

Gravitational collapse to a black hole

One of the most important roles of numerical relativity is to clarify the formation process of a variety of black holes. Stellar-mass black holes of mass in the range from $\sim 3M_\odot$ to $\lesssim 1000M_\odot$ will be formed by the core collapse of massive stars and massive neutron stars. Supermassive black holes of mass larger than $\sim 10^5 M_\odot$ may be formed as a result of the collapse of a supermassive star. In addition, in the very early epoch of the universe, primordial black holes may be formed from a highly nonlinear density fluctuation. For clarifying the formation processes of these black holes, numerical relativity is probably the unique approach. In the following, we will introduce some of the numerical-relativity studies for the formation process of these black holes performed so far.

9.1 Collapse of a supramassive neutron star to a black hole

This section focuses on the collapse of rapidly and rigidly rotating supramassive neutron stars to a black hole (see section 1.4.6 for the definition of the supramassive neutron star). A supramassive neutron star can be formed when a substantial amount of matter falls onto a neutron star from a normal binary companion, as in the formation of recycled millisecond pulsars for which the spin velocity is believed to be increased by the mass accretion [see, e.g., Cook *et al.* (1994c) for the possible evolution scenario]. The supramassive neutron stars and recycled pulsars are believed to have a rigid-rotation velocity profile because the accretion time scale is likely to be much longer than the viscous time scale inside the neutron stars,[1] and hence, the viscous angular-momentum transport process drives these neutron stars to a rigid-rotation state. In addition, the supramassive neutron stars are likely to be rapidly rotating because they are formed as a result of the accretion of a large amount of the matter from the companion, and hence, a significant amount of angular momentum is also increased. Probably, the supramassive neutron stars with mass close to the maximum mass $M_{\text{max,spin}}$ will be close to mass-shedding

---

[1] Assuming that the mass accretion rate is equal to the Eddington luminosity divided by c^2, it is $\approx 2.4 \times 10^{-9} M_\odot/\text{yrs}$, and hence, to increase the mass of $\sim 0.2 M_\odot$, $\sim 10^8$ yrs are necessary.

limit at which their equators rotate with the Keplerian velocity, and hence, any further spin-up is prohibited [Cook et al. (1994c)]. The dynamical stability (against the quasi-radial collapse) of such high-mass supramassive stars for which the mass is close to $M_{\mathrm{max,spin}}$ was studied by Shibata et al. (2000b); Shibata (2003b); Baiotti et al. (2005). They also followed the collapse process of the dynamically unstable supramassive neutron stars leading to a spinning black hole. Although they employed simple polytropic equations of state for modeling the neutron stars, their study clarified the essential mechanism for the onset of the dynamical instability and the formation process of the black hole. Thus, in the following, we will describe the process for the black hole formation from rapidly rotating supramassive neutron stars based on their results.

9.1.1 The criterion for the quasi-radial collapse

Before introducing numerical-relativity works, we briefly describe an important theorem proven by Friedman et al. (1988), who showed that for rigidly rotating axisymmetric stars, the condition of *secular* instability can be determined quite easily by applying a *turning-point method* for equilibrium sequences of the rotating stars [see Friedman and Stergioulas (2013) for a detailed description]. This method has been applied for finding the branches of secularly unstable stars with respect to gravitational collapse for numerical models of rigidly rotating neutron stars [e.g., Cook et al. (1992, 1994b,a) for a pioneer work]. In this subsection, we outline the turning-point method and summarize the results for the stability property of supramassive neutron stars.

We consider a family of rigidly rotating axisymmetric stellar models based on a one-parameter equation of state in the form $P = P(\rho)$ and $\varepsilon = \varepsilon(\rho)$. We suppose that the gravitational mass (ADM mass and Komar mass) of the stars, M, is determined as a function of angular momentum, J, and the central density ρ_c; $M = M(J, \rho_c)$. The angular velocity is denoted by Ω. The primarily important fact to be noted is that a turning point of M along equilibrium sequences for a fixed value of J marks the location of a zero-frequency (static) mode (see figure 9.1), as it is well known for the sequence of spherical stars with $J = 0$. This is due to the fact that we can find two equilibria in the close neighborhood of the turning points, which have the same rest mass, gravitational mass, and angular momentum, and hence, by a static perturbation, the transition from one to the other is possible. The presence of the static mode usually implies that one of the two branches around the turning point is a stable sequence and the other branch is an unstable one [Zel'dovich and Novikov (1971)]. Thus, the turning point marks the onset of some instability along the equilibrium sequence. For the neutron stars, the sequence in the lower-density side is known to be a stable branch. Thus, we can determine that the sequence in the higher-density side is an unstable branch.

As mentioned in the above paragraph, for rigidly rotating axisymmetric neutron

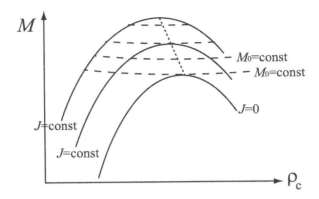

Fig. 9.1 Schematic figure for sequences of rotating neutron stars in the plane of the gravitational mass M and the central density ρ_c. The solid curves denote the sequences of constant angular momenta. The long dashed curves are the sequences of constant rest mass. The dotted curve shows the sequence of the marginally stable stars.

stars that are specified by J and ρ_c for a given one-parameter equation of state, the turning point is determined by [Friedman et al. (1988)]

$$\left.\frac{\partial M}{\partial \rho_c}\right|_J = 0 . \qquad (9.1)$$

Cook et al. (1992) showed an alternative formula as

$$\left.\frac{\partial J}{\partial \rho_c}\right|_{M_0} = 0 , \qquad (9.2)$$

where M_0 is the rest mass of the star. This is derived from the first law (see appendix G.4)

$$dM = \Omega dJ , \qquad (9.3)$$

which is satisfied for the sequence of M_0 =const, together with the relation

$$\Omega \left.\frac{\partial J}{\partial \rho_c}\right|_{M_0} = \left.\frac{\partial M}{\partial \rho_c}\right|_{M_0} = \left.\frac{\partial M}{\partial \rho_c}\right|_J + \left.\frac{\partial M}{\partial J}\right|_{\rho_c} \left.\frac{\partial J}{\partial \rho_c}\right|_{M_0} . \qquad (9.4)$$

Cook et al. (1992, 1994b,a) numerically derived supramassive neutron stars for a wide variety of hypothetical equations of state. They then applied the turning-point method to the numerical solutions for identifying the sequence of marginally stable stars that separates the stable and unstable regions. As a result, they found the universal feature for neutron-star sequences as follows.

For all the equations of state employed by them, the diagram becomes like figure 9.1. First, we remind the readers of the relation between M and ρ_c for a sequence of spherical neutron stars (see section 1.4.4): For a low-density and low-mass region, M monotonically increases with ρ_c. Then, M eventually reaches the maximum (turning point) at a critical value of ρ_c, and it decreases with the further increase of ρ_c. For sequences of constant values of angular momentum (the solid curves in figure 9.1), the shape of the curve of $M(J = \text{const}, \rho_c)$ is essentially the same as that for the spherical neutron stars. Hence, turning points can be determined for each curve of constant values of J quite easily. Then, the curve connecting these turning points separates the stable and unstable branches (the dotted curve in figure 9.1). As shown by Cook *et al.* (1992), this (dotted) curve is also drawn by connecting another set of turning points which are located by finding the minimum of angular momentum along sequences of constant values of the rest mass (see the dashed curves in figure 9.1). The turning point along these (dashed) curves is always located on the dotted curve.

Unfortunately, the turning-point method can only identify the marginal points of the secular stability, and thus, does not tell us any information on the dynamical stability. This is because in this method we only compare neighboring, rigidly rotating configurations with the same angular momentum. Maintaining rigid rotation during perturbations tacitly assumes high viscosity. In reality, the star will preserve circulation as well as angular momentum in a dynamical time scale, and thus, rigid rotation is not preserved. It is thus possible that a secularly unstable star may be dynamically stable: For sufficiently small viscosity, the star may change to a differentially rotating, stable configuration instead of collapsing to a black hole. Ultimately, the presence of viscosity will bring the star back into rigid rotation, driving the star to an unstable state. However, the secular instability evolves on a dissipative, viscous time scale, which is usually much longer than the dynamical time scale. As Friedman *et al.* (1988) showed, along a sequence of rigidly rotating axisymmetric stars, the secular instability always occurs *before* the dynamical instability (implying that all secularly stable stars are also dynamically stable or marginally stable).

For spherical stars, the onset of the secular and dynamical instability coincides because for a non-rotating star a radial perturbation conserves both circulation and uniform rotation. This suggests that for rigidly rotating stars for which the rotational kinetic energy T is typically a small fraction of the gravitational potential energy W, the marginally stable point of the dynamical instability is close to that of the secular instability. One goal in numerical relativity is to examine this hypothesis and to identify the onset of the dynamical instability in rigidly rotating neutron stars.

The other goal is to follow the nonlinear growth of the quasi-radial instability and to determine the final fate of unstable configurations. Unstable non-rotating neutron stars simply collapse to a spherical black hole, but unstable rotating neutron stars could form a spinning black hole surrounded by disks.

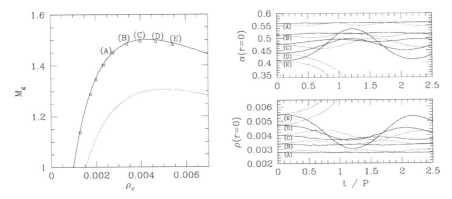

Fig. 9.2 Left: The gravitational mass (denoted by M_g in this figure instead of M in the text) as a function of the central density ρ_c for rotating neutron stars with $\Gamma = 2$ and $\kappa_p = 200/\pi$ in units of $c = G = M_\odot = 1$. The solid and dashed curves denote equilibrium sequences of rotating neutron stars at the mass-shedding limit and of spherical neutron stars, respectively. The open circles denote a sequence of initial data that were employed. The data used in the dynamical evolution simulations are marked with (A)–(E). Right: α and ρ at $r = 0$ as functions of t/P in the evolution of supramassive neutron stars (A)–(E). Here, P denotes the rotation period. The solid and dotted curves denote the results for the case that the pressure is uniformly reduced initially by 0% (solid curves) and 1% (dashed curves), respectively. The figures are taken from Shibata *et al.* (2000b).

9.1.2 *The dynamical stability and final fate of gravitational collapse*

The first numerical-relativity simulation for the collapse of rapidly rotating supramassive neutron stars was performed by Shibata *et al.* (2000b). Although they employed a simple polytropic equation of state with the polytropic index $n = 1$ (i.e., the adiabatic index $\Gamma = 2$, and the equation of state is written by $P = \kappa_p \rho^2$ with κ_p the polytropic constant) for modeling neutron stars and thus their study was rather qualitative, they for the first time illustrated the stability property and collapse process of rapidly rotating supramassive neutron stars that have approximately the maximum-allowed mass. Thus, we first introduce their work.

They studied the dynamical stability of supramassive neutron stars approximately at the mass-shedding limit near the turning point (see the plots marked with (A)–(E) in the left panel of figure 9.2). According to the turning-point theorem, the supramassive neutron stars (A)–(C) are secularly stable, while (D) and (E) are secularly unstable. Shibata *et al.* (2000b) evolved these five supramassive neutron stars in numerical-relativity simulations. To investigate the stability of these stars, the simulations were performed by slightly reducing the pressure uniformly at the initial stage. The right panel of figure 9.2 shows the evolution of the lapse function and rest-mass density at the center of the stars. This shows that for (A)–(C), the neutron stars are in a quasi-radial oscillation state, implying that they are dynamically stable. By contrast, for (D) and (E), the oscillation amplitude is quite high even in the case that the pressure was not reduced initially, and further-

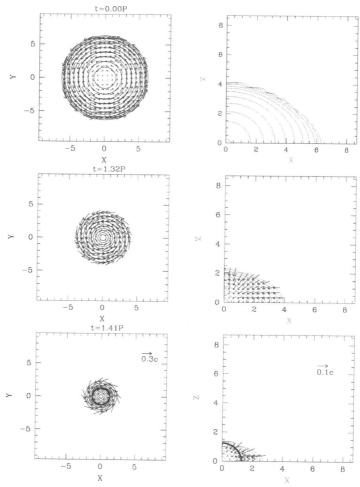

Fig. 9.3 Snapshots of density contours for the weighted rest-mass density $\rho_*(=\rho\alpha u^t\psi^6)$ and the velocity vector for (v^x, v^y) in the equatorial plane (left) and in the $y = 0$ plane (right) for the collapse of star (D) with $\Delta\kappa_{\rm p}/\kappa_{\rm p} = 1\%$. The contour curves are drawn for $\rho_*/\rho_{*c} = 10^{-0.3j}$ for $j = 0, 1, 2, \cdots, 10$ where ρ_{*c} is the central value of ρ_* which is 0.034, 0.64 and 2.04 (in units of $c = G = M_\odot = 1$) for the three different time slices. The time is shown in units of the rotation period of the given neutron star, denoted by P. The lengths of the velocity vectors are normalized to $0.3c$ (left) and $0.1c$ (right). The thick solid curves in the bottom panels denote the location of the apparent horizon. The figure is taken from Shibata et al. (2000b).

more, the stars collapse to a black hole in the case that the pressure was initially reduced by 1%. These results suggest that the supramassive neutron stars (D) and (E) are close to the marginally stable point for dynamical instability. Remembering the fact that stars (A)–(C) are secularly stable and (D) and (E) are secularly unstable, it is reasonable to conclude the criterion of the dynamical instability along the mass-shedding sequence coincides closely with that of the secular instability.

The next issue in this subject is to determine the final state of the collapsing supramassive neutron stars. Figure 9.3 displays the snapshots of density contours for the collapse of star (D) with the 1% initial pressure reduction. This figure clearly shows that within a few rotation periods, the supramassive neutron star collapses to a black hole. At the final stage of the collapse, more than 99.9% of the matter is swallowed by the black hole, and as this figure shows, there is no evidence for the formation of a disk with a substantial amount of mass. The result shown here indicates that the universal outcome formed after the collapse of rapidly and rigidly rotating supramassive neutron stars approximately at the marginally stable state is a black hole surrounded by a disk of negligible mass. Note that the formation of a disk is less subject for more slowly rotating supramassive neutron stars.

The final fate of unstable supramassive neutron stars was subsequently explored by Shibata (2003b) in more detail. He performed higher-resolution simulations assuming the axial symmetry (that was performed in a cartoon method described in section 2.5) and employing a wide variety of polytropic equations of state with the polytropic indices $n = 2/3, 4/5, 1, 3/2$, and 2. Here, the equation of state for neutron stars (above nuclear density) may be approximated by a polytropic equation of state with $1/2 \lesssim n \lesssim 1$; only with such stiff equations of state, neutron stars at the marginally stable state against gravitational collapse have the compactness $M/R \gtrsim 0.2$ [Cook et al. (1994b)] where R denotes the circumferential radius of spherical neutron stars. Hence, by employing such a wide range of the polytropic indices, the conclusion for the gravitational collapse of rotating neutron stars could be universally derived. For the initial conditions, marginally stable supramassive neutron stars with the approximately maximum mass along the sequence at the mass-shedding limit were employed. Shibata (2003b) then found that irrespective of the values of n, the final state after the collapse is a rotating black hole surrounded by a disk of negligible mass.

The reason for this result can be explained in the following analysis [Shibata (2003b)]. First, for the corresponding supramassive neutron star, we should calculate a rest-mass distribution as a function of specific angular momentum, j, of fluid elements, which is defined by

$$M_0(j) = \int_{j>j'} \rho_*(x') d^3x', \qquad (9.5)$$

where the integral should be performed for the fluid elements with $j > j'$ for a given value of $j := hu_\varphi$. For axisymmetric systems, the specific angular momentum distribution is preserved with time in the absence of angular momentum transport processes, and during the gravitational collapse that proceeds in the dynamical time scale, we may consider that it is approximately preserved in realistic situations. Figure 9.4 displays the specific angular momentum distribution for marginally stable supramassive neutron stars at the mass-shedding limit for several values of the polytropic index n. This shows that irrespective of the values of n, only a tiny fraction of fluid elements have the specific angular momentum larger than $2M$.

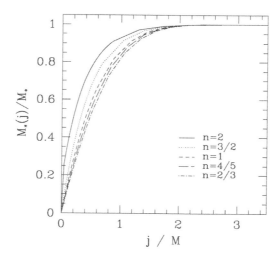

Fig. 9.4 Rest-mass distribution as a function of specific angular momentum, j, for several values of n for marginally stable supramassive neutron stars at the mass-shedding limit. M_* in this plot denotes M_0 (total rest mass). The horizontal axis is normalized by the gravitational mass M. The figure is taken from Shibata (2003b).

This is the reasonable consequence found in the following simple analysis: At the mass-shedding limit, the angular velocity is $\sim \sqrt{M/R_{\rm eq}^3}$, and hence, the specific angular momentum of a fluid element at the equator is $\sim \sqrt{MR_{\rm eq}}$. Here, $R_{\rm eq}$ denotes the circumferential radius of the neutron stars at the equator. Therefore, for supramassive neutron stars with $M/R_{\rm eq} > 0.1$, the specific angular momentum even at the equator is smaller than $\sim 3M$.

Now, we suppose that after the collapse of a supramassive neutron star, most of the fluid elements is swallowed by the formed black hole for which the final mass and angular momentum are approximately the initial mass M and initial angular momentum J, respectively. The numerical results of Shibata (2003b) indeed showed that this is the case. In this assumption, we can approximately predict the final dimensionless spin parameter of the black hole. Shibata (2003b) estimated it as ≈ 0.67, 0.63, 0.56, 0.45, and 0.39 for $n = 2/3$, $4/5$, 1, $3/2$, and 2, respectively. For these spin parameters, we can determine the specific angular momentum of a test particle orbiting the Kerr black hole at ISCO as $j/M = 2.64$, 2.71, 2.82, 2.97, and 3.05. If the specific angular momentum of a fluid element is smaller than these values, it has to be swallowed by the black hole. As figure 9.4 shows, the specific angular momentum of most of the fluid elements for the marginally stable supramassive stars at the mass-shedding limit is smaller than these values. This implies that as long as a highly efficient angular momentum transport process is absent, the final state of the collapse should be a black hole surrounded by a disk of negligible mass. Because any efficient angular momentum transport process was absent in his simulation, the numerical results by Shibata (2003b) are consistent with this prediction.

In real systems, it is also reasonable to expect that there is no efficient angular momentum transport process that works in the dynamical time scale of about a few milliseconds. Therefore, for the collapse of a marginally stable supramassive neutron star, the final state should be a Kerr black hole surrounded by a disk of negligible mass.

Finally, we note the following fact: The universal final fate (a black hole surrounded by a tiny disk) obtained for $n \leq 2$ is the result only for stiff equations of state. As we describe in section 9.2.1, the equation of state for massive stars could be approximated by $n = 3$ polytropic equation of state. In the collapse of such massive stars, the final fate could be significantly different from those found for $n \leq 2$ and for neutron stars. The subject in the next section is to illustrate this fact.

9.2 Collapse of a rotating supermassive star to a supermassive black hole

A supermassive star with mass $\sim 10^5 - 10^6 M_\odot$ is a possible end product of very massive gas cloud [Hoyle and Fowler (1963)], although it has not been discovered yet. A supermassive star, if it is formed, could eventually collapse to a supermassive black hole following quasi-stationary contraction via radiative cooling and the subsequent onset of general-relativistic quasi-radial instability [Iben, Jr. (1963); Chandrasekhar (1964); Fowler (1966); Zel'dovich and Novikov (1971)] that occurs before the nuclear burning sets in. This suggests that the supermassive stars are possible progenitors or seeds of supermassive black holes with mass $\sim 10^6 M_\odot - 10^{10} M_\odot$ for which the formation process is still unknown. Thus, clarifying the formation process of a supermassive black hole through the collapse of a supermassive star is an important subject. For the rotating supermassive stars, a multi-dimensional numerical-relativity study is necessary. In this section, we describe the stability of rapidly rotating supermassive stars and the final fate after the collapse of supermassive stars, which have been studied in numerical relativity [Baumgarte and Shapiro (1999a); Shibata and Shapiro (2002); Shibata (2004); Liu et al. (2007); Montero et al. (2012)].

9.2.1 Equation of state for massive stars

Before going ahead, we outline the equation of state for massive stars with their initial mass larger than $\sim 100 M_\odot$. As briefly mentioned in section 1.5.5, there are in general three major pressure sources for ordinary stars; gas pressure, radiation pressure, and pressure by degenerate electrons. For most stages of massive stars, only the first two sources could play a major role because the density cannot be so high that the degree of degeneracy of electrons is low. In particular, for very massive stars with mass $\gg 10 M_\odot$, the radiation pressure becomes dominant. This is found from the following order estimate.

For simplicity, we first assume that the gas pressure is the dominant pressure for supporting the self-gravity of a massive star. This assumption is justified when the mass of the star is not very high ($\lesssim 10 M_\odot$). Then, the order estimate for the force balance equation in Newtonian gravity (i.e., the pressure gradient is equal to the gravitational acceleration) yields

$$\frac{\bar{\rho} k_B \bar{T}}{\bar{m}} \sim \frac{GM}{R} \bar{\rho}, \tag{9.6}$$

where $\bar{T}$, $\bar{m}$, $\bar{\rho}$, and R are the average temperature, average mass of the gas particles, average density, and stellar radius, respectively. Here, we recover G to clarify the units, and M is the total mass in Newtonian gravity written by

$$M = \frac{4\pi}{3} \bar{\rho} R^3. \tag{9.7}$$

Equations (9.6) and (9.7) yield (erasing R)

$$\bar{T} \sim \frac{\bar{m}}{k_B} GM^{2/3} \left(\frac{4\pi}{3} \bar{\rho}\right)^{1/3}. \tag{9.8}$$

Now, we consider the ratio of the radiation pressure to the gas pressure as

$$\beta'_p := \frac{\bar{P}_{\text{rad}}}{\bar{P}_{\text{gas}}} = \frac{a_{\text{rad}} \bar{T}^4/3}{\bar{\rho} k_B \bar{T}/\bar{m}} \sim \frac{4\pi}{9} k_B^{-4} \bar{m}^4 a_{\text{rad}} G^3 M^2 \approx 1.6 \left(\frac{M}{10 M_\odot}\right)^2, \tag{9.9}$$

where we set $\bar{m}$ to be half of the hydrogen atomic mass assuming that the gas is composed of free electrons and free protons. If the massive star is composed of heavy nuclei, the numerical value of 1.6 is much larger. (We also note that in a more strict analysis, the numerical factor is slightly different from 1.6 [Fowler and Hoyle (1964); Shapiro and Teukolsky (1983)].) These facts imply that for the massive stars with $M \gtrsim 10 M_\odot$, the radiation pressure is the dominant pressure.

Because equation (9.9) can be used only for $\beta'_p \ll 1$, we next derive a relation for the case that the radiation pressure supports the self-gravity of a massive star. In this case, equation (9.6) is replaced by

$$\frac{a_{\text{rad}} \bar{T}^4}{3} \sim \frac{GM}{R} \bar{\rho}, \tag{9.10}$$

and hence,

$$\bar{T} \sim \left(\frac{3 GM \bar{\rho}}{a_{\text{rad}} R}\right)^{1/4}. \tag{9.11}$$

Then,

$$\beta_p := \frac{\bar{P}_{\text{rad}}}{\bar{P}_{\text{gas}}} \sim \left(\frac{4\pi a_{\text{rad}}}{9}\right)^{1/4} \bar{m} k_B^{-1} G^{3/4} M^{1/2} \approx 1.1 \left(\frac{M}{10 M_\odot}\right)^{1/2}. \tag{9.12}$$

We note again that in a more strict analysis, the numerical factor is slightly different from 1.1 but 2.7 if we assume that the gas is composed of protons and electrons [Fowler and Hoyle (1964); Shapiro and Teukolsky (1983)]. We again find that the radiation pressure is dominant for the massive stars with $M \gtrsim 10 M_\odot$. Equation (9.12) also shows that with the increase of the stellar mass, the role of the radiation pressure is enhanced (in proportional to $M^{1/2}$), and hence, we may consider that supermassive stars with $M \gtrsim 10^5 M_\odot$ are primarily supported by the radiation pressure; the gas pressure would be $\lesssim 1\%$ of the radiation pressure.

The next hypothesis often supposed is that massive stars are entirely convectively unstable, because the radiation transport is not efficient enough to transport the thermal energy from their inner to outer region, in spite of the situation that the massive stars emit a large amount of radiation from their surface approximately with the Eddington luminosity. In the presence of active convection, the massive stars should be isentropic; the specific entropy inside the massive stars should be constant. The photon entropy per baryon is written as [see the standard textbook like Shapiro and Teukolsky (1983)]

$$s_r = \frac{8}{3}\frac{\bar{m}a_{\rm rad}T^3}{\rho}, \tag{9.13}$$

and s_r is approximately constant. This implies that ρ should be approximately proportional to T^3. Since the massive stars are supported primarily by the radiation pressure, the equation of state for them can be written approximately by the polytropic equation of state with $n=3$ as

$$P = \frac{a_{\rm rad}T^4}{3} = \kappa_{\rm p}\rho^{4/3}, \tag{9.14}$$

where the polytropic constant $\kappa_{\rm p}$ is determined by s_r ($\kappa_{\rm p} \propto s_r^{4/3}$). Because s_r is written by $8k\beta_p$ in this case, $\kappa_{\rm p}$ is a monotonically increasing function of the stellar mass M ($\kappa_{\rm p} \propto M^{2/3}$). Although the gas pressure slightly modifies these relations, the equation of state is close to the $\Gamma = 4/3$ polytropic equation of state, i.e., the equation of state is quite soft in contrast to that for neutron stars.

The spherical stars with $\Gamma = 4/3$ are marginally stable against gravitational collapse in Newtonian gravity and unstable in general relativity [Chandrasekhar (1964)]. In reality, the correction by the gas pressure plays an important role for stabilizing the massive stars. In the presence of the gas pressure, the total pressure and the internal energy density are written as

$$P = \frac{1}{3}aT^4 + \frac{\rho k_{\rm B}T}{\bar{m}}, \tag{9.15}$$

$$\rho\varepsilon = aT^4 + \frac{3\rho k_{\rm B}T}{2\bar{m}}. \tag{9.16}$$

Then, the effective adiabatic index for massive stars for $\beta_p^{-1} \ll 1$ is estimated as [Bond et al. (1984)]

$$\Gamma_{\rm eff} \approx 1 + \frac{P}{\rho\varepsilon} \approx \frac{4}{3} + \frac{\beta_p^{-1}}{6}. \tag{9.17}$$

Thus, the effective adiabatic index is slightly larger than 4/3, and hence, in Newtonian gravity, stable spherical stars in equilibrium can be obtained. In general relativity, on the other hand, the critical value of the adiabatic index for the stability against gravitational collapse is modified as [Chandrasekhar (1964)]

$$\Gamma_{\text{crit}} = \frac{4}{3} + 2.249 \frac{M}{R} + O\left[(M/R)^2\right]. \tag{9.18}$$

Thus, only when the following condition is satisfied, the spherical stars are stable:

$$\frac{\beta_p^{-1}}{6} > 2.249 \frac{M}{R}. \tag{9.19}$$

Supermassive stars with $M \gtrsim 10^5 M_\odot$ formed from a gas cloud could contract during a radiative cooling. For the stage where M/R is sufficiently small, they are stable. However, they will eventually reach the marginally stable state of $\beta_p^{-1}/6 = 2.249 M/R$. Then, if the effect of rotation is negligible, they start collapsing at this critical point. During the collapse, the nuclear burning could set in, and if the energy generation efficiency is high enough, the collapse could be hung up [Fricke (1973); Fuller et al. (1986); Montero et al. (2012)]. However, for a sufficiently high-mass and metal-poor star, the self-gravity is strong enough to overcome the increase of the thermal energy by the nuclear burning and the collapse will continue, leading to black hole formation. The black-hole formation in spherical symmetry was first explored in numerical relativity by Shapiro and Teukolsky (1979).

However, supermassive stars are likely to be rapidly rotating because it is reasonable to consider that angular momentum is increased during the formation process of a massive gas cloud; the merger of small gas clouds and the accretion of gas will spin up the massive cloud. In the presence of the rapid rotation, the density profile and the resulting stability property of the massive star with the soft equation of state (with $\Gamma \sim 4/3$) can be significantly modified as shown by Fowler (1966); Baumgarte and Shapiro (1999a); Shibata (2004). In the next subsection, we describe the equilibrium and the stability of rapidly rotating supermassive stars, focusing on the star with the $\Gamma = 4/3$ equation of state.

9.2.2 Equilibrium and stability of rotating supermassive stars

A detailed numerical analysis for the structure and the stability of a rapidly rotating supermassive star in equilibrium was performed by Baumgarte and Shapiro (1999a). Supposing that the viscous time scale for angular momentum transport is shorter than the evolution time scale (Kelvin–Helmholtz time scale) which would be determined by the radiative cooling, the supermassive star will settle into a rigidly rotating state and evolve to the mass-shedding limit following the cooling and resulting contraction. Baumgarte and Shapiro (1999a) surveyed equilibrium states of rigidly rotating supermassive stars employing the $\Gamma = 4/3$ ($n = 3$ polytropic) equation of state. If we focus on the stability of rapidly rotating supermassive stars of sufficiently high mass against quasi-radial collapse, the stabilized effect by the rotation

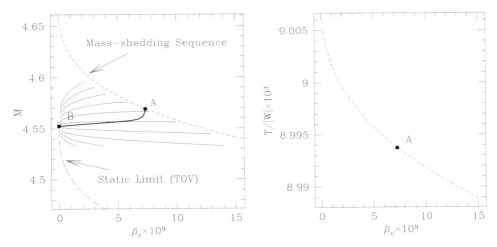

Fig. 9.5 Left: Mass versus central density plot for rigidly rotating supermassive stars with $\Gamma = 4/3$ in general relativity. The long dashed curve is the solution for non-rotating, static configurations (TOV configuration; see section 1.4.3), and the short dashed curve marks the mass-shedding limit. The thin solid curves denote sequences of constant angular momentum, ranging from $\bar{J} = 15$ (lowest curve) to $\bar{J} = 24$ (highest curve) in increments of $\Delta \bar{J} = 1$. Turning points of these curves mark the onset of instability. The thick solid curve connects these turning points and hence separates a region of stable configurations (above this curve) from a region of unstable configurations (below this curve). In particular, all non-rotating stars are unstable to radial perturbations. A configuration evolving along the mass-shedding sequence with increasing central density becomes unstable at the critical point A. All sequences of constant angular momentum connect the mass-shedding limit with point B for $\bar{\rho}_c \to 0$. Right: $T_{\rm rot}/|W|$ as a function of central density along the mass-shedding curve. The filled circle marks the critical configuration A (see the left panel). All the quantities (with bar) are shown in the polytropic units of $c = G = \kappa_{\rm p} = 1$. The figure is taken from Baumgarte and Shapiro (1999a).

is significant, and hence, it is acceptable to neglect the effect of the gas pressure which only slightly increases the critical value of the adiabatic index.

Figure 9.5 summarizes the key results of Baumgarte and Shapiro (1999a). The left panel shows the gravitational mass as a function of the central density. The long dashed curve is the solution for non-rotating, static configurations (TOV configuration; see section 1.4.3), and the short dashed curve marks the mass-shedding limit. The thin solid curves denote sequences of constant angular momentum, ranging from $\bar{J} = 15$ (lowest curve) to $\bar{J} = 24$ (highest curve) in increments of $\Delta \bar{J} = 1$. Here, $\bar{J}$ is the angular momentum in the polytropic unit (in units of $c = G = \kappa_{\rm p} = 1$).[2] Turning points of these curves mark the onset of the secular instability for the quasi-

---

[2]In the polytropic equations of state, physical units enter the problem only through the polytropic constant $\kappa_{\rm p}$, which can be chosen arbitrarily or else completely scaled out of the problem. Dimensional quantities scale with the change of $\kappa_{\rm p}$; e.g., mass, radius, and time scale with $\kappa_{\rm p}^{n/2}$ in units of $c = G = 1$ while ratios of the mass to radius and the mass to time are invariant. Here, n is the polytropic index. Thus, the results obtained in the so-called polytropic units, $c = G = \kappa_{\rm p} = 1$, can be universally used for translating results for any value of mass.

radial collapse [see Friedman et al. (1988) and section 9.1.1]. The thick solid curve connects these turning points and hence separates a region of stable configurations (above this curve) from a region of unstable configurations (below this curve). We note that all the non-rotating stars with $\Gamma = 4/3$ are unstable for radial perturbations in general relativity, and thus, the stabilizing effect by rapid rotation is significant in this case. A configuration evolving along the mass-shedding sequence with increasing the central density becomes unstable at the critical point A.

The right panel of figure 9.5 plots $T_{\rm rot}/|W|$ as a function of the central density along the mass-shedding curve. Here, $T_{\rm rot}$ and W are rotational kinetic energy and gravitational potential energy in general relativity, defined, respectively, by

$$T_{\rm rot} = \frac{1}{2} \int \Omega J_\varphi \sqrt{\gamma}\, d^3x, \tag{9.20}$$

$$W = M - (T_{\rm rot} + E_{\rm int} + M_0), \tag{9.21}$$

where $J_\varphi = -T_{ab} n^a \varphi^b = -T_{\mu\varphi} n^\mu$,

$$M_0 = \int \rho_* d^3x, \tag{9.22}$$

$$E_{\rm int} = \int \rho_* \varepsilon\, d^3x, \tag{9.23}$$

and $\rho_* := \rho\sqrt{-g}u^t$ (see section 4.4.1). The filled circle marks the critical configuration A where $T_{\rm rot}/|W| \approx 0.009$. This implies that the rotational kinetic energy at the onset of the instability is not very large for supermassive stars with $n = 3$ ($\Gamma = 4/3$); e.g., for neutron-star models with $n = 2/3, 1$, and $3/2$, $T_{\rm rot}/|W|$ at the onset of the instability along the mass-shedding curves are $\approx 0.12, 0.10$, and 0.08, respectively. However, the dimensionless spin parameter, J/M^2, at point A is quite large as ≈ 0.97. This is in contrast to the values for $n = 2/3$–$3/2$ for which the dimensionless spin parameter is in the range 0.67–0.45. This difference is reflected in the significant difference in the final remnant formed after the collapse of rapidly rotating unstable supermassive star (see section 9.2.3).

For strictly determining the final outcome after the onset of the collapse of supermassive stars, we have to perform a numerical simulation. However, as described by Shibata and Shapiro (2002); Shapiro and Shibata (2002); Shibata (2004), it can be predicted from the specific angular momentum distribution of the corresponding supermassive star, defined in equation (9.5), if we suppose that angular momentum transport processes during the collapse would not play an important role and hence the relation $M_0(j)$ is preserved during the collapse.

Now, we suppose that a seed black hole is formed during the collapse and the subsequent infall into the black hole proceeds sequentially from the ambient matter with lower values of j. Then, we consider the ISCO around the hypothetically growing black hole at the center. If j of an ambient element is smaller than the value at the ISCO of the black hole (hereafter denoted by $j_{\rm ISCO}$), the element will fall into the black hole eventually. Now the possibility exists that some fluid

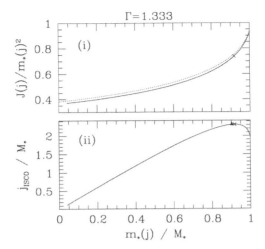

Fig. 9.6 Upper panel: Dimensionless spin parameter defined by $J(j)/M_0(j)^2$ as a function of $M_0(j)/M_0$. Lower panel: The specific angular momentum at ISCO, $j_{\rm ISCO}$, around a hypothetical black hole as a function of $M_0(j)/M_0$. M_* and $m_*(j)$ in this plot should be replaced by M_0 and $M_0(j)$ in the text. The horizontal axis is normalized by the total rest mass. The crosses denote the point at which $j_{\rm ISCO}$ becomes maximum, and the triangle denotes the point of $j = j_{\rm ISCO}$. The solid curves are derived from the numerical solution of a marginally stable equilibrium star at the mass-shedding limit and the dotted curve is from an analytic model of Shapiro and Shibata (2002) and Shibata (2004). The figures are taken from Shibata (2004).

elements can be captured even for $j > j_{\rm ISCO}$, if it is in a non-circular orbit. Ignoring these trajectories yields the minimum amount of mass that will fall into the black hole at each moment. The value of $j_{\rm ISCO}$ changes as the black hole grows. If $j_{\rm ISCO}$ increases, additional mass will fall into the black hole. However, if $j_{\rm ISCO}$ decreases, ambient fluid elements for all of which j should be larger than $j_{\rm ISCO}$ will no longer be captured. This expectation implies that when $j_{\rm ISCO}$ reaches a maximum value (strictly speaking, if $j_{\rm ISCO}$ is equal to the minimum value of j for the ambient medium), the dynamical growth of the black hole will terminate.

For analyzing the hypothetical growth of the black hole spin during collapse, we should calculate

$$J(j) = \int_{j>j'} \rho_*(x')j'd^3x', \qquad (9.24)$$

in addition to $M_0(j)$, and approximately define a dimensionless spin parameter of the hypothetical black hole by

$$a(j) := J(j)/M_0(j)^2. \qquad (9.25)$$

Because we assume that the infall of the ambient matter into the black hole proceeds sequentially from the matter with smaller values of j, we can calculate the (approximate) hypothetical mass and dimensionless spin of the black hole as $M_0(j)$ and $a(j)$ at a given moment that the matter with the angular momentum smaller than j constitutes the black hole. For these hypothetical values, we can also calculate the specific angular momentum at ISCO around the hypothetical black hole, $j_{\rm ISCO}(j)$, at the given moment, if we assume that the spacetime may be approximated instantaneously by a Kerr metric [see equation (1.130)].

The solid curves of figure 9.6 show the spin, $J(j)/M_0(j)^2$, and the specific angular momentum at ISCO, $j_{\rm ISCO}/M_0$, as functions of $M_0(j)/M_0$ for a marginally stable equilibrium star at the mass-shedding limit with $\Gamma = 1.333$. This result is approximately identical with that for $\Gamma = 4/3$. The solid curves in figure 9.6 may be interpreted as an approximate evolutionary track of mass and angular momentum for the growing black hole. As shown by Shibata and Shapiro (2002), the numerical results indicated that this is close to the real track, and thus, we can suppose that our assumptions made in this analysis are reasonable.

The solid curve in the upper panel of figure 9.6 shows that with the increase of the black hole mass, $M_0(j)$, the dimensionless spin parameter $[a(j) := J(j)/M_0(j)^2]$ also increases. The specific angular momentum at ISCO also increases with $M_0(j)$ in the early evolution stage of the hypothetical black hole, i.e., for $M_0(j) \lesssim 0.9 M_0$. This suggests that the mass of the black hole will increase to $\sim 0.9 M_0$ by the matter infall. However, at $M_0(j)/M_0 \approx 0.9$, $j_{\rm ISCO}$ reaches the maximum (which agrees approximately with the point where $j = j_{\rm ISCO}$), and $j_{\rm ISCO}/M_0(j)$ steeply decreases above this mass fraction. Thus, once the black hole reaches this point, it will stop growing dynamically. The upper panel of figure 9.6 shows that at this stage, the value of the dimensionless spin parameter is ~ 0.7. Therefore, at the end of this collapse, (1) $\sim 90\%$ of the total rest mass of the supermassive star will form a black hole and (2) the dimensionless spin parameter of the remnant black hole will be ~ 0.7. Thus, the prediction for the supermassive-star collapse is that the remnant is a supermassive black hole surrounded by a torus with $\sim 0.1 M_0$. In the next section, we will introduce numerical-relativity simulations that confirmed this prediction.

9.2.3 The final state of rapidly rotating supermassive stars

Shibata and Shapiro (2002); Liu et al. (2007) performed numerical-relativity simulations for the collapse of a marginally stable rotating supermassive star at the mass-shedding limit, employing a Γ-law equation of state with $\Gamma = 4/3$, and clarified that the final remnant is indeed a supermassive black hole surrounded by a massive torus, as predicted from the pre-collapse configuration (see the previous section). In the following, we will describe their results of axisymmetric simulations in which the collapse was followed for a supermassive star with $R_{\rm eq}/M = 622$, $T/|W| = 0.0088$, $J/M^2 = 0.96$, $M = 4.566$, and $\rho_{\rm c} = 7.8 \times 10^{-9}$ in the polytropic unit $c = G = \kappa_{\rm p} = 1$.[3]

---

[3] See footnote 1 of this chapter. The results in the polytropic units can be translated to the results of any value of the stellar mass.

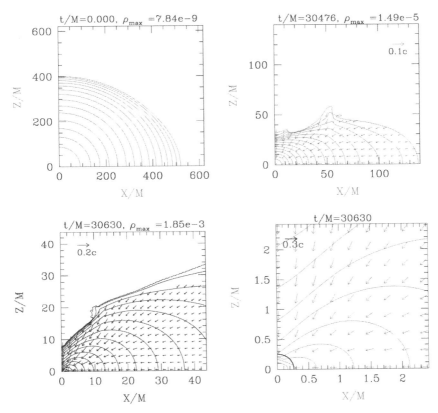

Fig. 9.7 Snapshots of density contours and velocity vectors in the x-z plane at selected time slices for the collapse of a marginally stable supermassive star at the mass-shedding limit. The contours are drawn for $\rho/\rho_{\max} = 10^{-0.4j}$ ($j = 0 \sim 15$), where $\rho_{\max}$ denotes the maximum rest-mass density at each time. The fourth figure is a blow-up of the third one in the central region: The thick solid curve at $r \approx 0.3M$ shows the location of the apparent horizon. The figure is taken from Shibata and Shapiro (2002).

A technical issue in this simulation is that the dynamical range is quite large: Initially the stellar radius is $\sim 600M$, and thus, the computational domain has to cover at least the entire structure of this star. On the other hand, a supermassive black hole with its horizon radius $\lesssim M$ is eventually formed, and hence, the grid spacing in the final stage has to be smaller than $\sim 0.1M$ to resolve the black hole. If we performed a simulation employing a uniform grid for the entire stage of the collapse, the total number of the grid point would have to be larger than $\sim 10^4$ for one direction. Even for an axisymmetric simulation, this number is too large to perform a simulation within reasonable computational time. Thus, for this problem, a mesh refinement algorithm such as AMR algorithm (see section 3.3) or a regridding technique is required. Shibata and Shapiro (2002) and Liu et al. (2007) employed a regridding technique in their axisymmetric simulations.

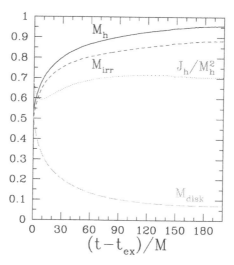

Fig. 9.8 Post-excision evolution of the mass in units of the initial mass M (denoted by M_h), dimensionless spin parameter (denoted by J_h/M_h^2), and the irreducible mass $M_{\rm irr}$ (in units of M) of the supermassive black hole, and the rest mass of the torus in units of M (denoted by $M_{\rm disk}$). Time is measured from the beginning of excision ($t_{\rm ex} = 28284M$). The figure is taken from Liu et al. (2007).

Figure 9.7 displays snapshots of density contours and velocity vectors in the x-z plane at selected time slices for the collapse of the supermassive star. It is found that the collapse proceeds nearly homologously in the central region during the early stage in which the effects of the rotation and general relativity are not still significant. However, this homologous property is modified by these effects in the late stage: By the general relativistic effect, the collapse near the central region is accelerated significantly and by the centrifugal force caused by the rotation, the collapsing supermassive star is flattened and a torus is formed in the equator. In the central region, the collapse proceeds in a runaway manner, leading to a black-hole formation; the increase speed of the density near the central region is much faster than other regions. This is due to the general relativistic gravity and the soft equation of state by which the supermassive star should have centrally condensed density profiles.

For this simulation, an apparent horizon is first found at $t \approx 30630M$. At the first formation of the apparent horizon, $\sim 60\%$ of the total rest mass has already swallowed into the black hole. However, this is not the final state and the mass of the black hole subsequently increases gradually by the matter infall.

The simulation by Shibata and Shapiro (2002) was not able to follow the subsequent growth of the black hole until the final state is reached, because of the lack of the technique for computing the growth of the black hole at that time. In a subsequent work by Liu et al. (2007), the evolution of the black hole due to the accretion of the infalling matter was followed using the so-called excision algorithm for the black-hole horizon and clarified until the black hole reaches approximately the final stage. Figure 9.8 displays the evolution of mass, irreducible mass, and dimensionless spin parameter of the supermassive black hole as well as the torus mass. This shows that the final mass and dimensionless spin parameter of the black hole are $\sim 0.95M$ and ~ 0.7, respectively. These values agree approximately with the prediction that was made by analyzing the initial state of the supermassive star.

Before closing this section, we note the collapse process clarified by Shibata and Shapiro (2002) and Liu et al. (2007) holds only for the case that the mass of a supermassive star is sufficiently high (say, larger than $10^6 M_\odot$ for the solar-metallicity) or metallicity is sufficiently low, $\lesssim 5 \times 10^{-3}$ of the solar metallicity (for $M \sim 10^5 M_\odot$) [Fuller et al. (1986)]. If the mass is not high enough $\lesssim 10^6 M_\odot$ and the metallicity is not low, the effect of the nuclear burning (primarily hydrogen burning) could play an important role during the collapse: The nuclear burning supplies a huge amount of thermal energy during the collapse, and as a result, the collapse is hung up, possibly leading to explosion instead of black-hole formation, although the critical (upper-threshold) mass and the critical (lower-threshold) metallicity for the explosion depend strongly on the angular momentum of the supermassive star [see the simulations of Montero et al. (2012)].

9.3 Stellar core collapse of massive stars to a black hole

This section focuses on the formation of black holes through the collapse of ordinary massive stars with their initial mass smaller than $\sim 10^3 M_\odot$. The ordinary massive stars satisfying certain conditions, which depend on the metallicity and initial mass, [see, e.g., figure 9.9 drawn by Heger et al. (2003) or Eldridge and Tout (2004)] will form a black hole after the stellar core collapse. For clarifying the formation process of such black holes and the properties of the resulting final remnant, a numerical-relativity simulation taking into account relevant microphysics and realistic equation of state is probably the unique approach. In this section, we will introduce some of numerical-relativity studies for the black-hole formation.

Broadly speaking, there are two possible paths for the formation of a stellar-mass black hole. If the core of the massive star is large enough (for this, metallicity of the primordial gas is low enough to form a very massive star), a black hole is promptly (or approximately promptly) formed after the collapse. This occurs for the case that a high-density object formed during the core collapse has mass larger than the maximum-allowed value for the formation of a stable proto-neutron star [Sekiguchi and Shibata (2005)]. On the other hand, if the core mass is not sufficiently large, a black hole is not promptly formed after the core collapse [Sumiyoshi et al. (2006); Nakazato et al. (2007)]. In this case, a shock wave will be generated at the core bounce in the formation of a proto-neutron star and subsequently the shock stalls

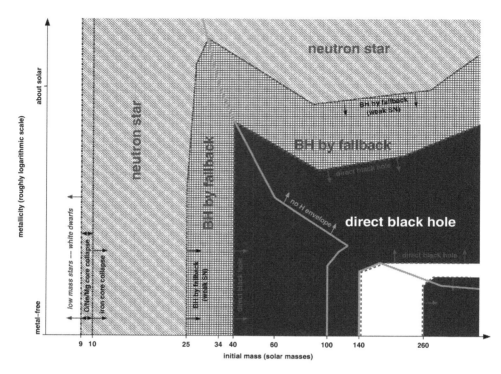

Fig. 9.9 A diagram which (qualitatively) predicts remnants of massive single stars as a function of initial metallicity (vertical axis) and initial mass (horizontal axis). Figure is taken from Heger et al. (2003). See also Eldridge and Tout (2004) for a related issue.

at the middle of the propagation. Then, the material in the outer region falls onto the transient massive proto-neutron star spending a long time scale. The black-hole formation will be achieved after a substantial amount of the material falls onto the massive proto-neutron star and its mass exceeds the maximum-allowed mass.

The former path occurs for the case that the initial mass of the massive star is sufficiently high, say, more than several hundred solar mass, which could be realized only for a metal-poor star such as population III stars. On the other hand, for other less massive stars, a black hole will be formed in the latter path. The collapse to a black hole from population III stars will be triggered by the instability associated with electron-positron pair production [Zel'dovich and Novikov (1971); Bond et al. (1984)]. On the other hand, for other cases (for the collapse of population I and II stars), the collapse will be triggered by the partial photo-dissociation of the iron core (see, e.g., section 1.5.5). Basically, this difference comes from the initial mass and metallicity. In the next section, we will outline the reason for this.

9.3.1 Path to the gravitational collapse of massive stars

Massive stars quasi-stationarily evolve due to sequential nuclear burning processes, and eventually come to the end of their lives at which some instability that induces gravitational collapse sets in. The final fate of the massive stars composed of heavy nuclei depends on the mechanism for this instability, which is determined by the initial stellar mass and metallicity. As described below, there are three different instability mechanisms, by all of which the effective adiabatic index for the core of the massive stars becomes smaller than 4/3 and the stars are destabilized.

Figure 9.10 illustrates how these three mechanisms trigger gravitational collapse for massive stars in a high-density or high-temperature state for which the adiabatic index is $\Gamma \approx 4/3$ [see, e.g., Zel'dovich and Novikov (1971) for details].

- Cores of massive stars in a high-density ($\rho_c \gtrsim 10^{10} \,\mathrm{g/cm^3}$) and relatively low-temperature ($T_c \lesssim 5 \times 10^9 \,\mathrm{K}$) state become unstable by the *electron-capture* process of heavy nuclei. Since the primary pressure source of the massive-star cores in such a state is the degenerate pressure of electrons, the depletion of this pressure destabilizes the massive stars.
- Iron cores of massive stars in a high-temperature state ($T_c \gtrsim 5 \times 10^9 \,\mathrm{K}$) with its central density $10^6 \,\mathrm{g/cm^3} \lesssim \rho_c \lesssim 10^{10} \,\mathrm{g/cm^3}$ become unstable through the *partial photo-dissociation* process of iron: This consumes a large amount of thermal energy, leading to sufficient reduction of the adiabatic index.
- Cores of very massive stars in a high-temperature ($T_c \gtrsim 10^9 \,\mathrm{K}$) and low-density ($\rho_c \lesssim 10^6 \,\mathrm{g/cm^3}$) state become unstable through the creation of *electron-positron pairs* from thermal photons. This also consumes a large amount of thermal energy, leading to sufficient reduction of the adiabatic index.

Massive stars are eventually destabilized by some of these three mechanisms, and the final fate is determined basically by the mass of the massive-star core that is composed of heavy nuclei. Here, the mass is determined approximately by the specific entropy of the core as $M_\mathrm{core} \propto s_r^{3/2}$ as briefly described in section 9.2.1. Here, from equation (9.13), s_r is approximately proportional to T_c^3/ρ_c. This implies that for a given value of the central density, the temperature is higher for the core of higher mass. Therefore, the onset of gravitational collapse is determined (i) by the electron-capture process for low-mass massive stars, (ii) by the partial photo-dissociation of iron for middle-mass massive stars, and (iii) by the creation of electron-positron pairs from the thermal photons for high-mass massive stars. Paths of the stellar evolution for representative massive stars derived by stellar evolution calculations are shown in figure 9.10.

Detailed studies for the stellar evolution have been clarifying that there are many possible evolution tracks for massive stars with initial mass $9M_\odot \lesssim M_\mathrm{init} \lesssim 10^3 M_\odot$ [see, e.g., Heger et al. (2003); Eldridge and Tout (2004)]. The evolution process also depends strongly on the metallicity of the massive star. Here, we focus on the cases

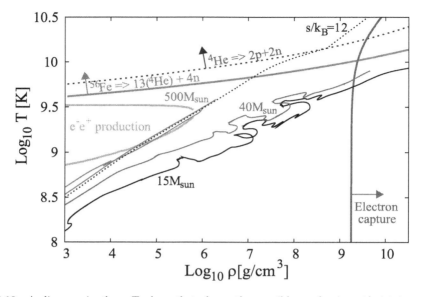

Fig. 9.10 A diagram in the ρ-T plane that shows the possible mechanisms that trigger gravitational collapse for massive-star cores in a high-density or high-temperature state. "Electron capture" denotes that for such high-density region, the star is unstable due to the electron capture by heavy nuclei. "e^-e^+ production" denotes that for the enclosed region, the star is so hot (but less dense) that electron-positron pairs are produced in a catastrophic manner and the star is unstable due to the significant loss of thermal energy. "^{56}Fe $\Rightarrow$ 13^{4}He + 4n" and "^{4}He $\Rightarrow$ 2p + 2n" show that above these curves (for high-density regions), 50% of iron and helium are photo-dissociated, respectively. Three curves denoted by $15M_\mathrm{sun}$, $40M_\mathrm{sun}$, and $500M_\mathrm{sun}$ denote the evolution tracks for massive stars of initial mass $15M_\odot$, $40M_\odot$, and $500M_\odot$, respectively. The curve labeled by $s/k_\mathrm{B} = 12$ denotes the approximate evolution track for a very massive star with the entropy per baryon of $12k_\mathrm{B}$. This figure was generated by Y. Sekiguchi with a help of T. Yoshida.

that the metallicity is quite low (close to zero) or that the metallicity is as high as the solar one.

For the case of the negligible metallicity, there are five possibilities as follows:

i) For $\sim 9M_\odot \lesssim M_\mathrm{init} \lesssim 10M_\odot$, a massive-star core composed of oxygen, neon, and magnesium (not iron) is eventually destabilized due to the electron capture processes, and then, collapses. In this case, a neutron star is likely to be formed after a supernova explosion that occurs soon after the core bounce without shock stall [Kitaura et al. (2006)].

ii) For $10M_\odot \lesssim M_\mathrm{init} \lesssim M_\mathrm{crit}$ where M_crit denotes a threshold value of the initial mass between $\sim 10M_\odot$ and $140M_\odot$, a massive-star core composed primarily of iron is eventually destabilized due to the partial photo-dissociation, and then, collapses. For this relatively smaller-mass case, a neutron star is believed to be formed after a supernova explosion, although the explosion mechanism is still poorly understood (see, e.g., section 1.5.5).

iii) For $M_{\rm crit} \lesssim M_{\rm init} \lesssim 140 M_\odot$, also, a massive-star core composed primarily of iron is eventually destabilized due to the partial photo-dissociation. For this case, a massive proto-neutron star will be transiently formed after a core bounce. However, the shock launched at that time could not explode the entire star of sufficiently large mass, and hence, a substantial amount of the material located outside the core eventually would fall onto the massive proto-neutron star, leading to the formation of a black hole. For this case, a black hole is not likely to be formed directly during the gravitational collapse, as explored in spherical symmetric studies by Sumiyoshi et al. (2006) and Nakazato et al. (2007). The reason for this is that the collapsing core has high entropy, which can increase the maximum-allowed mass of hot proto-neutron stars significantly.

iv) For $140 M_\odot \lesssim M_{\rm init} \lesssim 260 M_\odot$, the temperature of the core becomes quite high whereas the density is low $< 10^6 \, {\rm g/cm^3}$ in the oxygen burning stage at which the core is composed of oxygen. Then, the electron-positron pair creation destabilizes the oxygen core, leading to gravitational collapse. However, the collapse is likely to be halted during the early contraction because of the explosive nuclear burning of oxygen, and then, the star will be exploded as indicated, e.g., by Fryer et al. (2001). Such explosion is called pair-instability supernova explosion. For this case, no remnant object is left.

v) For $260 M_\odot \lesssim M_{\rm init}$, the oxygen core destabilized by the electron-positron pair creation collapses to a black hole directly because the explosive oxygen burning does not generate thermal energy enough to explode the heavy core.

A black hole is formed for the cases iii) and v). For iii), i.e., a relatively less-massive case, a black hole is likely to be formed in the fall-back of the material onto a massive proto-neutron star and for the massive case, v), a black hole is likely to be promptly formed during the gravitational collapse.

For a high metallicity (e.g., solar metallicity), it is expected that a massive star with mass larger than $\sim 150 M_\odot$ would not be formed. Thus, for this case, the black-hole formation by the direct gravitational collapse of very massive core is unlikely to occur, and a black hole will be formed by a fall-back process onto a massive proto-neutron star.

The first detailed studies for the black-hole formation in general relativity were performed by Sumiyoshi et al. (2006) and Nakazato et al. (2007) in their spherically symmetric simulations, in which Einstein's equation and hydrodynamics equations were solved together with neutrino-radiation-transfer equations incorporating nuclear-theory-based equations of state. They explored the criterion for the prompt formation of a black hole and neutrino luminosity in detail. However, they were not able to follow the evolution process of the black holes because they performed the simulations using Misner–Sharp coordinates (see appendix B.4), which do not allow us to explore the evolution of the black holes due to the inappropriate choice of the time coordinate.

Moreover, the effect of angular momentum in massive stars cannot be explored by spherically symmetric simulations. High-energy phenomena such as gamma-ray bursts are believed to be associated with the formation of a black hole surrounded by a massive and hot torus (or disk), which can be formed only for the case that the massive star has sufficiently large angular momentum. To clarify the formation process of the black hole-torus systems, we have to perform a long-term multi-dimensional simulation and follow not only the black-hole formation but also the subsequent evolution of the black hole and surrounding torus, taking into account relevant microphysical processes.

The history of this field in numerical relativity is still limited, although a couple of pioneering works were already performed. In the next section, we will introduce the pioneering works of long-term and physical numerical simulations for the formation of black hole-torus systems by Sekiguchi and his collaborators.

9.3.2 *Numerical-relativity simulation of rotating massive stars to a black hole*

9.3.2.1 *Fall-back collapse to a black hole: $M_{\text{init}} \sim 100 M_\odot$*

Iron cores formed from ordinary massive stars with the initial mass between M_{crit} and $\sim 100 M_\odot$ are believed to collapse to a black hole by the fall-back process onto a massive and hot proto-neutron star that is formed at the core bounce of the gravitational collapse. The formation process of the black hole in the presence of stellar rotation and neutrino cooling was first computed by Sekiguchi *et al.* (2012) in numerical relativity.[4] They explored the process for the formation of a spinning black hole surrounded by a hot torus in long-term simulations. In addition, emissivity of neutrinos from such a system was quantitatively explored for the first time. This numerical-relativity field is still in an early stage of the research and only a few results have been reported. Thus, we here refer only to their results. The deeper knowledge of this issue will be obtained by a wide variety of systematic simulations changing the initial stellar models and the equations of state in the future.

As the initial condition, they employed a core model of the initial stellar mass $100 M_\odot$ derived by a stellar evolution calculation of Umeda and Nomoto (2008). For this model, the mass and radius of the iron core are $\approx 3.2 M_\odot$ and $\approx 2500\,\text{km}$, respectively, with the central density $\rho_c \approx 10^{9.5}\,\text{g/cm}^3$ and the central temperature $T_c \approx 10^{10}\,\text{K}$. The central value of entropy per baryon is $\approx 4 k_B$ which is much larger than that of ordinary supernova cores $\lesssim k_B$, and this is the key property of the very massive iron core.

Because Umeda and Nomoto (2008) computed a non-rotating stellar core model, Sekiguchi *et al.* (2012) superimposed angular momentum with a rotation profile

---
[4]See also Cerda-Duran *et al.* (2013).

according to the rule

$$\Omega(\varpi) = \Omega_0 \frac{R_0^2}{R_0^2 + \varpi^2} \mathcal{F}_{\text{cut}}, \qquad (9.26)$$

where $\varpi := \sqrt{x^2 + y^2}$, Ω_0 is angular velocity on the rotation axis, and R_0 is a radius that determines the degree of differential rotation: Ω_0 and R_0 are parameters that can be chosen arbitrarily. For the outer region, the angular velocity is artificially reduced by a cut-off factor $\mathcal{F}_{\text{cut}}$ to avoid mass-shedding. Sekiguchi et al. (2012) considered moderately rapidly rotating models by choosing $\Omega_0 = 1.2\,\text{rad/s}$ which is much smaller than the Keplerian value at the core surface $R = R_{\text{core}}$, $(M_{\text{core}}/R_{\text{core}}^3)^{1/2} \approx 5.2\,\text{rad/s}$. For R_0, a rigid-rotation model ($R_0 = \infty$) and a differential-rotation model ($R_0 = R_{\text{core}}$) were employed. Simulations were performed assuming the axial symmetry together with the equatorial-plane symmetry by using a cartoon method described in section 2.5. A finite-temperature (Shen) equation of state together with a neutrino leakage scheme [see Sekiguchi (2010b) and section 4.7.3] for the neutrino cooling was employed for exploring the microphysical process of stellar core collapse and subsequent formation of a hot massive proto-neutron star and a massive torus surrounding a black hole.

Dynamical features

Figure 9.11 (a) shows the evolution path (the middle solid curve) of the central values of the rest-mass density and temperature in the ρ-T plane for the rigid-rotation model (for the differential-rotation model, the results are essentially the same before the core bounce; see figure 9.12). As in the core collapse of ordinary supernovae for which the central value of entropy per baryon is $s \lesssim k_B$, gravitational collapse is induced by both the electron capture and the partial photo-dissociation of heavy nuclei. Because of the higher value of the entropy per baryon for this case ($s \sim 4k_B$), the partial photo-dissociation plays a primary role in the destabilization. This is found from the fact that a substantial amount of heavy nuclei are resolved into helium by the photo-dissociation (see figure 9.11 (a)). The collapse proceeds in a homologous and adiabatic manner after the neutrino trapping occurs at $\rho_c \sim 10^{11}\,\text{g/cm}^3$ (see, e.g., section 1.5.5). Thereafter, as the collapse proceeds, the temperature steeply increases, and helium are eventually resolved into free nucleons (p, n).

Figure 9.11 (b) shows the evolution path for the central values of the rest-mass density and Y_e in the ρ-Y_e plane for the rigid-rotation model. Because the temperature for the present model is higher than that for ordinary supernovae for a given value of the density, the partial photo-dissociation of helium and heavy nuclei into free nucleons sets in for lower density. Here, free protons are more subject to the electron capture than helium and heavy nuclei, and consequently, by an efficient electron capture on the free protons, the electron fraction becomes smaller than that for ordinary supernovae in the collapse phase [compare two curves in figure 9.11 (b)].

The time evolution of the central values of the rest-mass density, electron fraction, temperature, and lapse function for the two different initial rotation profiles

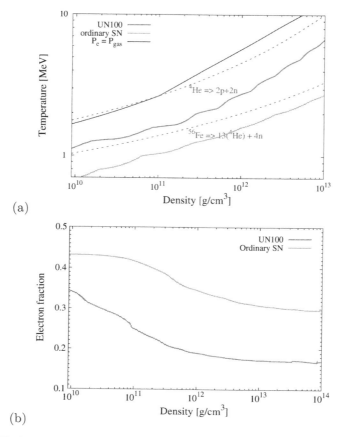

Fig. 9.11 (a) Evolution paths for the central values of the rest-mass density and temperature in the ρ-T plane. The middle solid curve shows the evolution path for the rigid-rotation model employed in section 9.3.2.1. The top solid curve shows the boundary at which the condition $P_e = P_{\text{gas}}$ is satisfied ($P_e > P_{\text{gas}}$ for the higher density side). The two dashed curves denote the values of (ρ, T) with which ^{56}Fe or ^{4}He will be half by mass due to the partial photo-dissociation. An evolution path for an ordinary supernova core [Sekiguchi (2010b)] is shown together for comparison (bottom solid curve). (b) Evolution paths for the central values of the rest-mass density and the electron fraction in the ρ-Y_e plane (bottom curve). For both panels, "UN100" denotes the model concerned. The figure is taken from Sekiguchi et al. (2012).

is shown in figure 9.12. As in the collapse of ordinary supernova cores, the collapsing core experiences a bounce when the central density exceeds the nuclear density $\rho_{\text{nuc}} \sim 10^{14}\,\text{g/cm}^3$ above which the pressure steeply increases due to the repulsive nuclear force, and then, shock waves are formed and propagate outwards. Because the electron fraction at the core bounce is small ($Y_e \approx 0.17$) and hence the core is neutron-rich, the nuclear force starts playing a role at relatively low density. After the core bounce, a massive proto-neutron star, which is supported partly by centrifugal force associated with rotation and by thermal pressure, is formed. At the

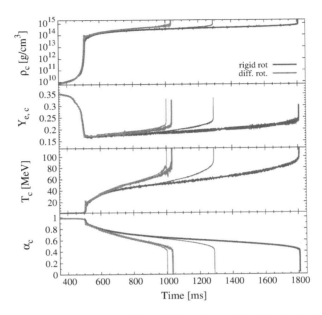

Fig. 9.12 Time evolution of the central values of the rest-mass density, electron fraction, temperature, and lapse function for the differential-rotation (red curves) and rigid-rotation (blue curves) models. The results for the low grid resolution (thin curves) are shown together with those for the high grid resolution (thick curves). The collapsing core experiences a core bounce at $t \approx 510$ ms and collapse to a black hole at $t \approx 1000$ and 1810 ms, respectively. The figure is taken from Sekiguchi et al. (2012).

moment of the core bounce, the central temperature was already high as ~ 20 MeV, and subsequently increases with the matter accretion that induces shock heating and compression; the central temperature increases to ~ 40 MeV in ~ 100 ms. This suggests that the thermal pressure plays an important role for supporting the strong self-gravity of the massive proto-neutron star by increasing the maximum-allowed mass.

The shock wave formed at the core bounce propagates outwards but eventually stalls at $r \approx 100$ km due to the neutrino cooling and photo-dissociation of heavy nuclei contained in the infalling matter (see the top panels of figures 9.13 and 9.14). Then, a standing accretion shock is formed (see, e.g., section 1.5.5). As in the case of the ordinary supernova, convection is activated between the massive proto-neutron star and the standing accretion shock. However, it is not strong enough to push the stalled shock outward. As the matter accretion proceeds, the central density and temperature of the hot massive proto-neutron star increase gradually, and eventually, it collapses to a black hole when its mass exceeds the maximum-allowed value. In Shen equation of sate, it is $\sim 3 M_\odot$.

The formation time of the black hole depends on the grid resolution (compare the thin and thick curves in figure 9.12). This is because the massive proto-neutron

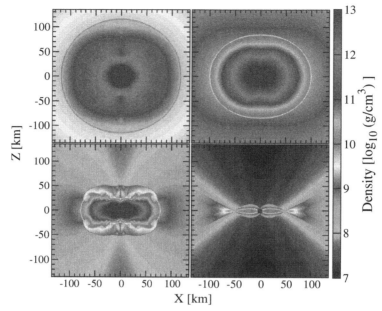

Fig. 9.13 Rest-mass density profiles in the x-z plane at $t = 570$ ms (top left), 645 ms (top right), 890 ms (bottom left), and 1150 ms (bottom right) for the differential-rotation model (see figure 9.12 for the state at each time slice). The figure is taken from Sekiguchi et al. (2012).

star just before the onset of the collapse is close to a marginally stable state, and hence, a small change in thermodynamical quantities and rotation profile results in a significant change in the central density. In general, for a finer grid resolution, the lifetime of the massive proto-neutron star is longer. The short lifetime in poorer grid resolutions is often seen and unavoidable (spurious) result in numerical computation.

One of the interesting discoveries of Sekiguchi et al. (2012) was that the rotation profiles in the central region of collapsing cores depend only weakly on the initial rotation profile, and their evolution process agrees well with each other up to the stage soon after the core bounce. However, the subsequent evolution processes of the massive proto-neutron stars are significantly different from each other: In the differential-rotation model, the matter in the outer region falls primarily onto the hot massive proto-neutron star while the infalling matter could be accumulated to be a torus or a disk in the rigid-rotation model.

Figure 9.13 displays snapshots of rest-mass density profiles in the x-z plane for the differential-rotation model, at selected time slices until the hot massive proto-neutron star collapses to a black hole. For this model, a large fraction of collapsing matter falls only onto the central region because the matter in the outer region has the angular velocity slower than that in the central region. As the accretion of the matter proceeds, the shock front of the standing accretion shock gradually recedes, and the shape of the shock wave is deformed by the rotation to be *spheroidal* (see

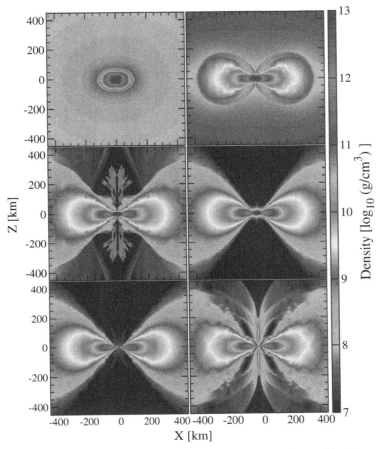

Fig. 9.14 Rest-mass density profiles in the x-z plane at $t = 617$ ms (top left), 863 ms (top right), 1256 ms (middle left), 1437 ms (middle right), 1822 ms (bottom left), and 2225 ms (bottom right) for the rigid-rotation model (see figure 9.12 for the state at each time slice). In the middle left panel, outflow velocity vectors larger than $0.15c$ are plotted together (red arrows). Note that the scale of the figure is different from that of figure 9.13. In the bottom two panels, a black hole is formed at the center and surrounded by a hot torus. The figure is taken from Sekiguchi et al. (2012).

second and third panels of figure 9.13). Convection is also activated near a neutrino sphere due to the presence of negative gradients of the entropy per baryon, s, and of the total-lepton (electron) fraction (see the bottom left panel of figure 9.13). Here, the negative gradients are generated by the fact that neutrinos carry away both energy and lepton number, as in the case of an ordinary pre-supernova. However, the convection does not play an important role for the explosion in this case.

The hot massive proto-neutron star eventually collapses to a black hole due to the mass accretion. Accompanied with the black hole formation, a geometrically

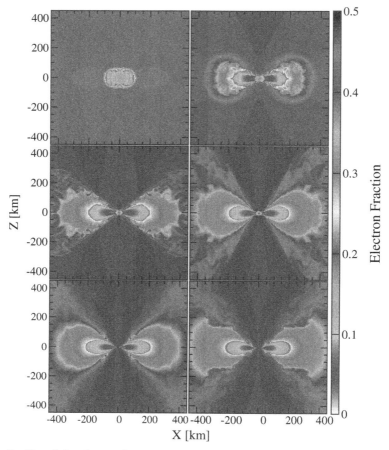

Fig. 9.15 Profiles of the electron fraction in the x-z plane at the same time slices as figure 9.14. The figure is taken from Sekiguchi et al. (2012).

thin disk is formed around the black hole (see the last panel of figure 9.13). For this stage, the mass of the disk is quite small $< 0.01 M_\odot$ because most part of the massive proto-neutron star collapses into the black hole.

Figures 9.14, 9.15, 9.16, and 9.17, respectively, display profiles of the rest-mass density, electron fraction, entropy per baryon, and temperature in the x-z plane at selected time slices for the rigid-rotation model. For this case, the matter located in the outer region has more specific angular momentum than in the central region. As in the differential-rotation model, the shock wave formed after the core bounce stalls at $r \sim 100$ km (the top left panel of figures 9.14–9.17). Due to the faster rotation of the outer region, the shock wave is deformed to be a torus-like configuration (the top right panel of figures 9.14–9.17). The formation of this torus-shaped shock is the key ingredient that characterizes the dynamics of this model (compare

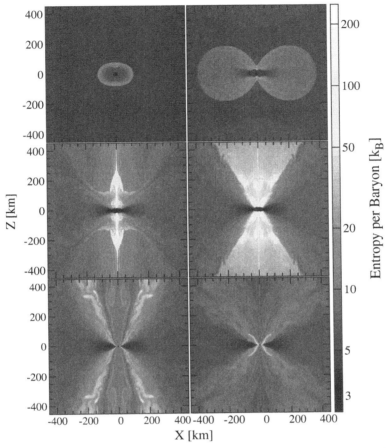

Fig. 9.16 Profiles of the entropy per baryon in the x-z plane at the same time slices as figure 9.14. The figure is taken from Sekiguchi et al. (2012).

the second panel of figure 9.14 with 9.13). At the shock, kinetic energy associated with the motion perpendicular to the shock surface is dissipated but the momentum associated with the parallel component is preserved; the amount of the kinetic energy dissipated at the shock is not as large as that for the differential-rotation model. Due to the torus-like configuration, the infalling matter is eventually accumulated in the central region and their kinetic energy is dissipated at the surface of the hot massive proto-neutron star. Figure 9.18, which displays the velocity field in the x-z plane at a time slice, clearly shows this mechanism. During this process, oscillations of the hot massive proto-neutron star are excited as the infalling matter hits it. Also, shock waves generated at the stellar surface obtain thermal energy and propagate outwards intermittently.

Due to the accumulation of the matter onto the massive proto-neutron star

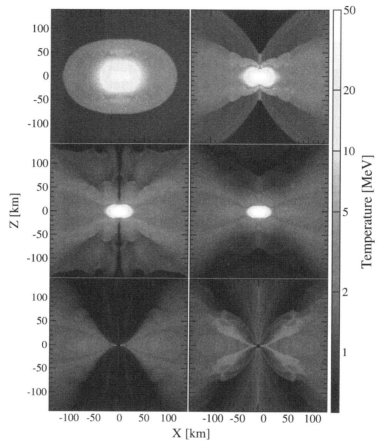

Fig. 9.17 Profiles of the temperature in the x-z plane at the same time slices as figure 9.14 but with a different (zooming) scale. The figure is taken from Sekiguchi et al. (2012).

and the resulting shock heating, thermal energy is stored in the pole region of the massive proto-neutron star, increasing the gas pressure, $P_{\rm gas}$, there. On the other hand, the ram pressure, $P_{\rm ram}$, of the infalling matter decreases with the elapse of time because its density decreases. When the condition, $P_{\rm ram} < P_{\rm gas}$, is realized, outflows are launched from the polar stellar surface, forming strong shocks (see the middle left panel of figures 9.14–9.17). It can be seen that the entropy around the rotation axis is significantly enhanced due to the shock heating associated with the outflows (see the middle right panel of figure 9.16).

The outflows eventually lose the driving power by the neutrino cooling and shock dissipation, and a fraction of the matter again falls back onto the polar region of the massive proto-neutron star. Due to the continuous mass accretion, the massive proto-neutron star eventually collapses to a black hole surrounded by a geometrically

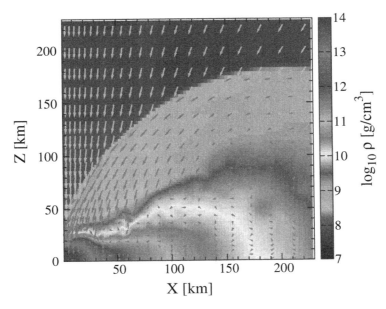

Fig. 9.18 Velocity vector fields outside the shock surface (blue arrows) and inside the shock (red arrows) together with a rest-mass density profile in the x-z plane at $t = 645$ ms. It is found that only the velocity component perpendicular to the shock surface is dissipated at the shock. As a result, the infalling matter is accumulated toward the massive proto-neutron star located at the center. The figure is taken from Sekiguchi et al. (2012).

thick, dense, and massive *torus* (see the bottom left panel of figures 9.14–9.17). This result is significantly different from that for the differential-rotation model.

The torus surrounding the central black hole is massive and hot (~ 5 MeV), and hence, has the following unique properties. First, its central region is dense enough that copious neutrinos are emitted, playing a dominant role for cooling [Popham et al. (1999); Narayan et al. (2001); Di Matteo et al. (2002); Kohri and Mineshige (2002)]. Namely, a neutrino-dominated accretion torus is the outcome (see, e.g., section 1.5.2 for a description). Second, because the density is high with their maximum values $\sim 10^{12}$ g/cm^3 while the temperature is only moderately high with several MeV, the degree of electron degeneracy is quite high in the central region of the torus, resulting in $Y_e < 0.1$ there (see figure 9.15). Consequently, the torus is composed mostly of neutrons. In addition, the specific entropy in its high-density region is not very high with $s \lesssim 5k_B$ while for a low-density region surrounding the torus, it is quite high $> 10k_B$. This state agrees semi-quantitatively with that of an analytically simple model, e.g., by Di Matteo et al. (2002); Kohri and Mineshige (2002); Chen and Beloborodov (2007).

One of the interesting discoveries of Sekiguchi et al. (2012) [see also Sekiguchi and Shibata (2011)], which cannot be found in the analytic models, is that the black hole-torus system shows violent time variability (see the bottom right panel

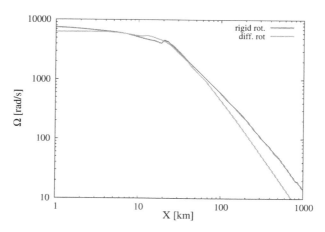

Fig. 9.19 Profiles of the rotation angular velocity along the equator just before the black-hole formation for the differential-rotation (green steeply decreasing curve) and rigid-rotation (red curve) models. The figure is taken from Sekiguchi et al. (2012).

of figures 9.14–9.17). The reason for this violent activity is that the torus is heated up continuously by the shock caused by the accretion of the matter infalling from the outer region, and in addition, the neutrino cooling enhances the inhomogeneity on the profiles of specific entropy and lepton number per baryon leading to convection. This time variability is also reflected in the strong variability of neutrino luminosity (see below).

The final point to be stressed is that a slight difference in the initial velocity profile results in a large difference in the formation and evolution processes of the torus or disk. Figure 9.19 compares profiles of the rotation angular velocity on the equator just before the black-hole formation for two models. The rotation profiles of the massive proto-neutron star ($r \lesssim 40$ km) are similar in the central region, and even in the outer region, the difference is at most by a factor of 2–3. This result indicates that the final outcome depends strongly on the rotation profile of progenitor stars. This obviously implies that a number of systematic simulations changing the initial rotation profile, that have not been done yet, are required for totally clarifying the formation process of a black hole surrounded by a torus (or disk).

Neutrino Luminosity

Figures 9.20 (a) and 9.20 (b) plot time evolution of the neutrino luminosity for the differential-rotation model. In the pre-bounce phase, electron neutrinos are dominantly emitted and the emissivity of electron anti-neutrinos is much smaller. This is because electrons are (mildly) degenerate (i.e., the chemical potential of electrons μ_e is higher than $k_B T$) blocking the β-decay of neutrons and also the positron fraction, which is responsible for the anti-neutrino emission, is small. Soon

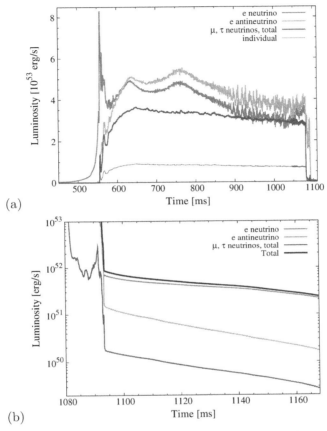

Fig. 9.20 Luminosity of neutrinos for the differential-rotation model (a) before the black-hole formation and (b) after the black-hole formation. The red, green, and blue thick curves correspond to the luminosity of ν_e, $\bar{\nu}_e$, total of μ and τ pair neutrinos, respectively. The thin blue curve in the upper panel shows the luminosity of *individual* μ/τ neutrino (namely the quarter of thick blue curve). The thick black curve in the lower panel shows the total neutrino luminosity. The figure is taken from Sekiguchi *et al.* (2012).

after the core bounce, the so-called neutrino burst occurs at the time that the shock wave passes through the neutrino-sphere (see the sharp peaks at $t \sim 560$ ms), as in the ordinary supernova case.

After the neutrino burst, the emission of electron anti-neutrinos is enhanced and their luminosity becomes larger than that of electron neutrinos. This property is different from that in the ordinary supernova [Bethe (1990)], and the reason for this is described as follows. During the post neutrino burst phase, a large number of positrons are produced via electron-positron pair creation process because the degeneracy parameter becomes low as $\eta_e := \mu_e/k_B T \sim 1$ due to the high temperature with $T \gtrsim 20$ MeV, which is higher than the temperature in the ordinary

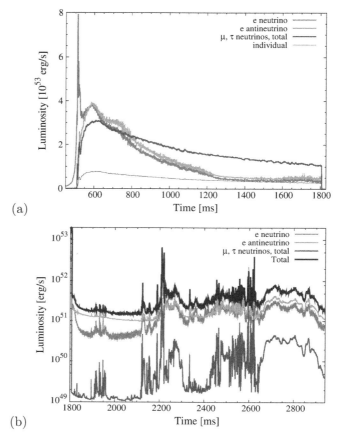

Fig. 9.21 Luminosity of neutrinos for the rigid-rotation model (a) before the black-hole formation and (b) after the black-hole formation. Meanings of all curves are the same as those of figure 9.20. The figure is taken from Sekiguchi et al. (2012).

supernova for which $T \sim 5\,\mathrm{MeV}$ [Bethe (1990)]. Then, because the neutron fraction is much larger than the proton fraction, the positron capture on neutrons occurs more efficiently, and therefore, the electron anti-neutrino luminosity becomes larger than the electron neutrino luminosity. This dominant emission of electron anti-neutrinos is also found for hot massive neutron stars formed after the merger of binary neutron stars (see section 8.3) and in a black hole-torus system formed in the collapse of a more massive stellar core [Sekiguchi and Shibata (2011)] (see section 9.3.2.2). Both systems have a higher-temperature and lower-electron-fraction state than the system formed in the collapse of the ordinary supernova core.

The luminosity of μ and τ neutrinos is smaller than that of electron neutrinos and anti-neutrinos. This is simply due to the absence of the neutrino production channel mediated by the charged weak current for them. At later phases of the

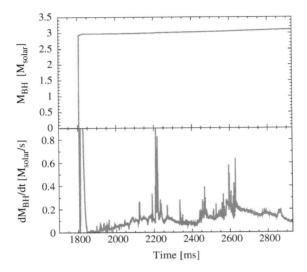

Fig. 9.22 Time evolution of the irreducible mass of the black hole (the upper panel) and the mass accretion rate (the lower panel) for the rigid-rotation model.

hot massive proto-neutron star evolution (800 ms $\lesssim t \lesssim$ 1100 ms), the electron neutrino and anti-neutrino luminosity shows weak time variability. This is due to the convective activities that occur near the neutrino sphere. The luminosity of μ and τ neutrinos does not show the variability because they are emitted primarily from the hot central regions that do not suffer from the convection.

Soon after the black-hole formation that occurs at $t \approx 1090$ ms, the neutrino luminosity decreases drastically because the main neutrino-emission region is swallowed into the black hole. After the black-hole formation, a geometrically thin accretion torus emits neutrinos with luminosity $\sim 10^{51}$–10^{52} ergs/s in its early evolution phase with the duration ~ 100 ms, and then, the luminosity decreases monotonically with time. Here, the luminosity from the accretion torus is, at zeroth order, determined by shock heating due to the infall of the material onto the accretion torus. The generation rate of thermal energy by the liberation of kinetic energy of the infalling material is estimated as

$$\frac{M_{\rm BH}}{r}\dot{M} \sim 0.1\dot{M} \approx 2 \times 10^{52}\left(\frac{\dot{M}}{M_\odot\,{\rm s}^{-1}}\right){\rm ergs/s}, \tag{9.27}$$

where $M_{\rm BH}$ is the black-hole mass, r is the typical location of the accretion torus, and $\dot{M}$ is the mass infall rate. We set $r \sim 10 M_{\rm BH}$ for the typical accretion torus scale. The neutrino luminosity is roughly equal to several ten percent of this value. In this phase, electron neutrinos are dominantly emitted because the torus is in a low-temperature state with $T \lesssim$ a few MeV, and hence, positron production by the pair process is not efficient.

Figures 9.21 (a) and 9.21 (b) show time evolution of the neutrino luminosity for the rigid-rotation model. The features of the evolution is similar to those for the differential-rotation model before the neutrino burst ($t \lesssim 700$ ms). After that time, the luminosity of electron neutrinos and anti-neutrinos gradually decreases. This is because the optical depth (diffusion time) increases as the torus grows. The luminosity of electron neutrinos and anti-neutrinos shows only weak time variability, reflecting the weaker convective activity for this stage.

At $t \approx 800$ ms, the *total* luminosity of μ and τ neutrinos ($L_{\nu_\mu} + L_{\bar\nu_\mu} + L_{\nu_\tau} + L_{\bar\nu_\tau}$) becomes larger than the luminosity of electron neutrinos and anti-neutrinos. This is basically due to the fact that the neutrino sphere has so high temperature that pair neutrino production processes are enhanced. Another partial reason is that the optical depth along the rotation axis becomes small due to the flattened shape of the massive proto-neutron star: Thermal neutrinos from the hot massive proto-neutron star will be almost directly *seen* due to the low density along the rotation axis. Indeed, after $t \gtrsim 1300$ ms, all flavors of neutrinos and anti-neutrinos are almost equally emitted, indicating the dominant emission of thermal neutrinos from the very hot massive proto-neutron star. This feature is not seen in the remnants of the binary neutron star mergers (cf. section 8.3) and the collapse of the more massive stellar core [Sekiguchi and Shibata (2011)]. This is because the continuous mass accretion is absent in the merger of binary neutron stars and the massive proto-neutron star very quickly collapses to a black hole in the collapse of the more massive stellar core. By contrast, at the final phase in the fall-back collapse of an ordinary core (a failed supernova) [Sumiyoshi *et al.* (2006)], an enhancement of the emission of $\nu_{\mu/\tau}$ and $\bar\nu_{\mu/\tau}$ is also seen.

Note that the above result implies that observational signals of neutrinos could depend on the viewing angle: If we see the system from the direction along the rotation axis, we will observe a brighter emission of neutrinos with a higher value of average energy from the hot massive proto-neutron star. By contrast we will observe smaller neutrino luminosity if we see the system from the direction along the equator.

The neutrino luminosity shows a sudden decrease when the black hole is formed at $t \approx 1805$ ms. The total luminosity of neutrinos emitted from the torus around the black hole amounts to $L_{\nu,\text{tot}} \sim 10^{51}$ ergs/s, which is much smaller than the luminosity emitted by the hot massive proto-neutron star. This value is also smaller than that for the differential-rotation model. This is reasonable because the black hole in the rigid-rotation model is formed after a long-term mass infall from the outer region, and as a result, the mass accretion rate onto the torus for this model is relatively small $\sim 0.1 M_\odot/$s [see equation (9.27)]. Nevertheless, this luminosity is maintained for $\gtrsim 1$ s because the accretion time scale is much longer than 1 s. In addition, the neutrino luminosity shows violent time variability. Such long-term high luminosity and time variability may be associated with the time variability that some gamma-ray bursts show.

The numerical results for the neutrino luminosity may be used for an order estimate of the energy deposition rate ($\dot{E}_{\nu\bar{\nu}}$) by the neutrino pair-annihilation, which is one of the possible processes to drive relativistic jets required to produce gamma-ray bursts. Note that the energy from the neutrino pair annihilation should be deposited in a baryon-poor region for generating highly relativistic outflows. The funnel region near the rotation axis above the central black hole is a promising site for this.

According to the estimate by Beloborodov (2008) [see also Birkl et al. (2007) for more careful computation], the deposition rate in the black hole-torus system would be proportional to $\dot{M}^{9/4} M_{\rm BH}^{-3/2}$. In this estimation, the neutrino luminosity is assumed to arise from viscous heating. For the simulations of Sekiguchi et al. (2012), the neutrino luminosity is determined by the dynamical infalling rate of the matter that experiences the shock heating at the surface of the torus and increases thermal energy of the torus. However, the dependence of the pair-annihilation rate on the mass infall rate $\dot{M}$ should be essentially the same for the thick torus phase. Figure 9.22 shows the evolution of irreducible mass of the black hole (the upper panel) and the mass accretion rate (the lower panel). Due to the strong dependence of the pair deposition rate on $\dot{M}$, the energy deposition by the neutrino pair-annihilation would be important only for a phase in which $L_{\nu,\rm tot} \gtrsim 10^{51}$ ergs/s [see figures 9.20 (b) and 9.21 (b)]. Figure 9.21 (b) shows that the duration of the neutrino emission (in the black-hole phase) with $L_{\nu,\rm tot} \gtrsim 10^{51}$ ergs/s is longer than 1 s, and thus, a part of the energy deposition for a gamma-ray burst may be explained. Taking into account the dependence of the neutrino pair annihilation rate on the geometry of the torus [Beloborodov (2008); Zalamea and Beloborodov (2011)], $\dot{E}_{\nu\bar{\nu}}$ would be given by [Beloborodov (2008)],

$$\dot{E}_{\nu\bar{\nu}} \sim 10^{48} \text{ ergs/s} \left(\frac{100 \text{ km}}{R_{\rm fun}}\right) \left(\frac{0.1}{\theta_{\rm fun}}\right)^2 \left(\frac{E_\nu + E_{\bar{\nu}}}{10 \text{ MeV}}\right)$$
$$\times \left(\frac{L_\nu}{10^{51} \text{ ergs/s}}\right) \left(\frac{L_{\bar{\nu}}}{10^{51} \text{ ergs/s}}\right) \cos^2 \Theta, \qquad (9.28)$$

where $R_{\rm fun}$ and $\theta_{\rm fun}$ are the characteristic radius and the opening angle of the funnel region. Θ denotes the collision angle of the neutrino pair. Thus, only low-luminosity gamma-ray burst could be explained (however, note that the value could be enhanced by a factor of ~ 10 if the black hole is rapidly spinning [Zalamea and Beloborodov (2011)]). The primary reason for the low values of $\dot{E}_{\nu\bar{\nu}}$ and the neutrino luminosity in this simulation is that the mass infall rate onto the torus surrounding the central black hole is rather small, of order $0.1 M_\odot/$s [see equation (9.27)].

In the massive-proto-neutron-star phase, by contrast, the neutrino luminosity is huge as $L_\nu \gtrsim 10^{53}$ ergs/s [see figures 9.20 (a) and 9.21 (a)], and hence, the deposition rate would be very large as

$$\dot{E}_{\nu\bar{\nu}} \sim 3 \times 10^{52} \text{ ergs/s} \left(\frac{100 \text{ km}}{R_{\rm fun}}\right) \left(\frac{0.1}{\theta_{\rm fun}}\right)^2 \left(\frac{E_\nu + E_{\bar{\nu}}}{30 \text{ MeV}}\right)$$
$$\times \left(\frac{L_\nu}{10^{53} \text{ ergs/s}}\right) \left(\frac{L_{\bar{\nu}}}{10^{53} \text{ ergs/s}}\right) \cos^2 \Theta. \qquad (9.29)$$

If the outflows launched due to the mass accumulation mechanism can penetrate the stellar envelope and construct a low-density funnel region, a system composed of a long-lived hot massive proto-neutron star and a geometrically thick torus may be a candidate of the central engine of gamma-ray bursts for a relatively short duration.

9.3.2.2 Collapse to a black hole soon after the bounce: $M_{\text{init}} \sim 150 M_\odot$

For the case that iron core mass is $\sim 10 M_\odot$ (i.e., entropy per baryon of the core is high as $s/k_B \sim 6$–10), the core should collapse promptly to a black hole. Although it is not clear whether such an iron core is formed in nature, it is interesting to explore the formation process of a black hole from the massive iron cores of this mass range.

A simple order estimate can clarify one important property for the collapse of such massive iron cores. Let M_c be the mass of a part of the core that will form a black hole in the collapse. For the formation of a black hole, the mass has to be enclosed inside an event horizon for which the circumferential radius is $\sim 2M_c (= 2GM_c/c^2)$. Thus, the required average density is estimated by

$$\frac{M_c}{4\pi(2M_c)^3/3} \approx 1.5 \times 10^{16} \left(\frac{M_c}{M_\odot}\right)^{-2} \text{g/cm}^3. \tag{9.30}$$

This suggests that for $M_c \gtrsim 10 M_\odot$, a black hole is likely to be formed before the average density exceeds the nuclear-matter density $\sim 2.8 \times 10^{14}$ g/cm^3. Thus, for the collapse of such a massive core, a neutron star is never formed.

Nevertheless, a black hole is not formed directly after the core collapse. Nakazato et al. (2007) discovered in their spherically symmetric simulations in general relativity that the iron core collapses to a black hole after a weak bounce; a *proto-black hole* is formed [Fryer et al. (2001)]. This weak bounce occurs at a relatively low central density $\sim 10^{12}$–10^{13} g/cm^3, and hence, it is not associated with the stiffening of the equation of state caused when the matter density exceeds the nuclear-matter density. Rather, this is caused by the fact that the gas pressure (pressure of protons and neutrons) dominates the degenerate pressure of electrons during the core collapse: The bounce occurs when the average adiabatic index of the core changes from $\sim 4/3$ to $\sim 5/3$. This situation is realized due to the facts that a high-temperature state with $T \gtrsim 10$ MeV is reached and primary baryon components become free nucleons [i.e., gas pressure is enhanced; see equations (1.216) and (1.218)]. This weak bounce is the unique feature for the collapse of a high-mass and high-entropy iron core.

Sekiguchi and Shibata (2011) performed simulations for the collapse of rotating iron cores with $s = 8 k_B$ using the same formulation and numerical method as those described in section 9.3.2.1. In their simulation, iron-core mass and total mass are $\approx 13 M_\odot$ and $\approx 23 M_\odot$, respectively. They confirmed that after a weak bounce, a black hole is formed as far as the initial degree of rotation is not extremely high. By

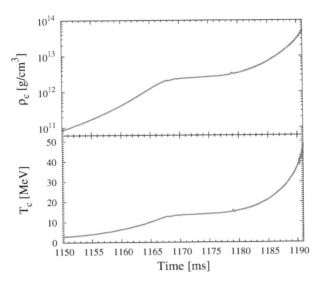

Fig. 9.23 Time evolution for the central values of the rest-mass density and temperature for the core collapse with a model of $s/k_B = 8$ in spherical symmetry. The collapsing core experiences a weak bounce at $t \approx 1168$ ms and apparent horizon is formed at $t \approx 1193$ ms. The figure is taken from Sekiguchi and Shibata (2011).

contrast to the simulation by Nakazato et al. (2007), they performed simulations for a long time scale after the formation of the black hole and followed the formation and evolution of a torus surrounding the black hole. They discovered that after the formation of the black hole, a hot torus is formed for a certain range of initial angular momentum and that such a torus shows long-term violent time variability with quite high neutrino luminosity.

Figure 9.23 displays time evolution of the central values of the rest-mass density and temperature for the collapse of the massive iron core with no angular momentum. For this case, at $\rho_c \sim 2 \times 10^{12}$ g/cm^3, a weak bounce occurs and a proto-black hole is formed. As already mentioned, for this stage, the temperature is high $\gtrsim 10$ MeV and the baryon is composed primarily of free nucleons. Hence, the gas pressure is significantly enhanced and dominates over the degenerate pressure by electrons, leading to the sufficient increase of the adiabatic index and to the resulting weak bounce during the collapse. However, by subsequent mass accretion onto it, the proto-black hole can be destabilized by the general relativistic effect, and eventually, it collapses to a black hole. Because of its high mass, the black hole is formed before the density exceeds 10^{14} g/cm^3 as shown in equation (9.30). For this model, the initial mass of the black hole is $\sim 6M_\odot$. This indicates that the maximum-allowed mass for the high-entropy and high-mass proto-black hole is $\sim 6M_\odot$, which is much larger than that for proto-neutron stars.

Figure 9.24 displays profiles of the rest-mass density, electron fraction, entropy

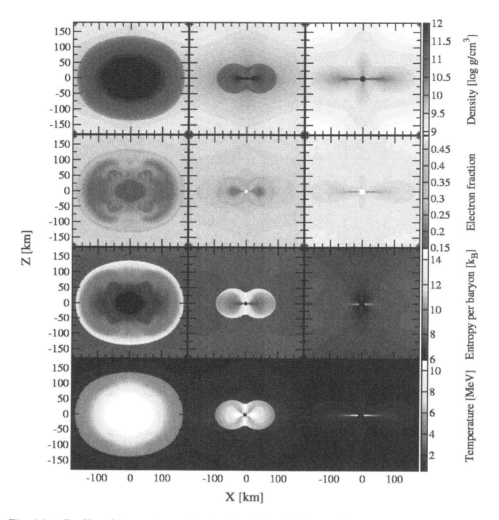

Fig. 9.24 Profiles of the rest-mass density (panels in the first row), electron fraction (panels in the second row), entropy per baryon (panels in the third row), and temperature (panels in the fourth row) at a weak bounce (left column), at the black-hole formation (middle column), and at 70 ms after the formation of the black hole (right column) for the moderately rotating progenitor model. The black regions in the contours of the rest-mass density and entropy per baryon, and the white regions in the contours of the electron fraction for middle and right panels are inside the apparent horizon. The figure is taken from Sekiguchi and Shibata (2011).

per baryon, and temperature at the weak bounce (left column), at the black-hole formation (middle column), and at 70 ms after the formation of the black hole (right column) for the moderately rotating progenitor model. Here, "moderately rotating" implies that the dimensionless spin parameter $a(j)$ [see equation (9.25) for its definition] of the progenitor in the central region is ≈ 0.5, and that for the

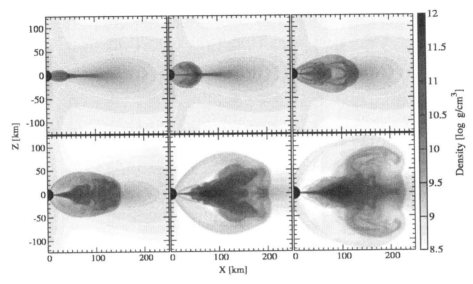

Fig. 9.25 Profiles of the rest-mass density after the formation of the system composed of a black hole and torus for the moderately rotating progenitor model. The elapsed time after the formation of the black hole is ≈ 205 ms (upper left), 211 ms (upper middle), 218 ms (upper right), 271 ms (lower left), 333 ms (lower middle), and 427 ms (lower right). The figure is taken from Sekiguchi and Shibata (2011).

inner region that encloses 70% of the mass element is smaller than unity; thus, it is a model that forms a moderately spinning black hole at its formation stage. The black regions in the contours of the rest-mass density and entropy per baryon, and the white regions in the profiles of the electron fraction for middle and right columns denote the inside of the apparent horizon.

This figure captures the characteristic properties for the formation mechanism of a spinning black hole during the collapse of a rotating high-entropy and high-mass iron core. At the weak bounce, a slightly deformed spheroid is formed. At this moment, the central temperature was already quite high as ~ 10 MeV and the electron fraction per baryon is low as ~ 0.15. By contrast to the temperature, the central entropy per baryon is relatively low as $s \sim 6.5 k_{\rm B}$, while that in the shocked surface becomes quite high as $s \sim 13 k_{\rm B}$, indicating a strong effect of the bounce.

After the bounce, a black hole is formed, and most of the material inside the shock surface is swallowed by it. Although a fraction of the material forms a disk surrounding the black hole at its formation, the disk mass $\sim 0.1 M_\odot$ is much smaller than the black-hole mass $\sim 6 M_\odot$. The dimensionless spin parameter of the black hole is initially moderately large ~ 0.6 and subsequently increases with the infall of the rotating material from the outer region.

The disk mass subsequently and gradually increases with the infall of the material from the outer part. In particular, in a late stage of the infall for which the

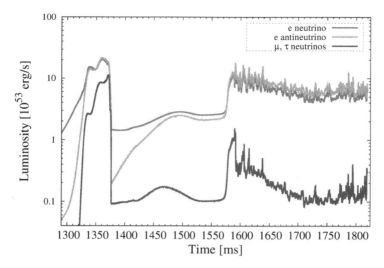

Fig. 9.26 Time evolution of neutrino luminosity for the moderately rotating progenitor model. A black hole is formed at $t \approx 1373\,\mathrm{ms}$. The figure is taken from Sekiguchi and Shibata (2011) with a minor modification.

material of relatively large specific angular momentum falls onto the disk, the disk mass steeply increases and the disk changes to a torus during the evolution. Figure 9.25 displays profiles of the rest-mass density for the evolution process of the torus surrounding the central black hole; the upper left panel of this figure displays the density profile just before the disk mass steeply increases. For the next $\sim 200\,\mathrm{ms}$, the disk/torus mass increases from $\lesssim 0.1 M_\odot$ to $\sim 0.8 M_\odot$, and during this rapid accretion phase, a torus is formed (found from second to sixth panels). The maximum rest-mass density of the torus is $\sim 10^{12}\,\mathrm{g/cm^3}$ with the maximum temperature $\gtrsim 10\,\mathrm{MeV}$, and hence, neutrinos are copiously emitted for cooling while some fraction is trapped; neutrino-dominated accretion torus is formed (see section 9.3.2.1 for the similar outcome). What is remarkable for this model is that the neutrino luminosity is quite high.

Figure 9.26 displays time evolution of the neutrino luminosity for each flavor (electron neutrinos, anti-electron neutrinos, and others). For this model, a black hole is formed at $t \approx 1373\,\mathrm{ms}$ and a torus is formed at $\sim 200\,\mathrm{ms}$ after the black-hole formation. This shows that the total neutrino luminosity is of order of $10^{53}\,\mathrm{ergs/s}$ before the formation of the torus, but it becomes $\gtrsim 10^{54}\,\mathrm{ergs/s}$ after the formation of the massive torus. Also remarkable is that the high luminosity is maintained for more than 300 ms.

This long-term high neutrino luminosity comes from the fact that the mass infall onto the torus proceeds with a high rate for a long time scale: see equation (9.27). One reason that the high-mass infall accretion rate is maintained for a long time scale is that a black hole is formed soon after the collapse due to the high mass of the progenitor, and furthermore, at the formation of the black hole, its mass is much smaller than the total mass of the collapsing progenitor. Therefore, there are a lot of accreting material for subsequently producing the torus in the outer region surrounding the black hole. This situation is by significant contrast to the collapse of ordinary massive stars for which the iron core mass is at most $\sim 3M_\odot$, and hence, most of iron in the core fall into a black hole at its formation. The other reason to be mentioned is that for this model, angular momentum distribution is suitable for forming a massive torus. If the specific angular momentum of the material in the outer region is not large enough, a massive torus cannot be formed.

The high-luminosity torus surrounding a spinning black hole is a promising candidate for the central engine of gamma-ray bursts through the pair-annihilation process of neutrinos and anti-neutrinos. Substituting the typical luminosity into equation (9.28) yields the possible pair-annihilation luminosity as [Beloborodov (2008)],

$$\dot{E}_{\nu\bar{\nu}} \sim 10^{54} \, \text{ergs/s} \left(\frac{100 \, \text{km}}{R_{\text{fun}}}\right) \left(\frac{0.1}{\theta_{\text{fun}}}\right)^2 \left(\frac{E_\nu + E_{\bar{\nu}}}{10 \, \text{MeV}}\right)$$
$$\times \left(\frac{L_\nu}{10^{54} \, \text{ergs/s}}\right) \left(\frac{L_{\bar{\nu}}}{10^{54} \, \text{ergs/s}}\right) \cos^2 \Theta. \quad (9.31)$$

This suggests that the collapse of a high-entropy and massive iron core could be a promising progenitor for the central engine of gamma-ray bursts.

9.4 Formation of a primordial black hole

As we already introduced in the beginning of this chapter, primordial black holes may be formed from a highly nonlinear density fluctuation in the very early epoch of the universe. For exploring their formation process, we have to provide a plausible initial condition that is indicated by the standard cosmological evolution scenario. Shibata and Sasaki (1999) proposed a formulation in numerical relativity that is well suited for exploring the formation of primordial black holes in such realistic setting even for the non-spherical case. In this section, we summarize their method and discovery.[5]

First, they developed a numerical-relativity formulation that is well suited for the use in asymptotically Friedmann spacetime. The essence of their method is to extract a uniformly expanding part from the full metric and matter field, and only to solve the other non-uniform parts (see section 2.6.2 for this formulation). As a result of this formulation, the set of the basic equations settles to a form similar to that in asymptotically flat spacetime.

---

[5]There are several works for the study of primordial black-hole formation assuming the spherical symmetry. Among them, we would like to refer to the latest work by Nakama et al. (2014) in which the condition for the formation of primordial black holes is systematically surveyed.

The next task is to prepare a realistic initial condition in the cosmological context [see textbooks, e.g., Mukhanov (2005) for standard cosmology and inflation scenarios]. The standard formation scenario of primordial black holes is as follows: In a very early epoch of the universe just after an inflation epoch, a scalar-mode perturbation (the so-called super-horizon-scale perturbation) generated from quantum fluctuations of hypothetical inflaton fields has the length scale much longer than the Hubble horizon scale. Some of these metric perturbations may have large amplitude. This is allowed as long as such a high-amplitude perturbation does not contradict with the current cosmological observational results. The super-horizon-scale metric perturbation does not change the amplitude in the radiation-dominated post-inflation epoch (see below). However, the length scale of the perturbation increases with the scale factor $a \propto t^{2/3\Gamma}$ with $1/3 \leq 2/3\Gamma \leq 2/3$ (see section 2.6.2), while the Hubble horizon scale increases in proportional to t. Therefore, the Hubble horizon scale eventually exceeds the length scale of the metric perturbation.[6] Once this occurs, the amplitude of the metric perturbation starts changing. If this amplitude is high enough, in other words, if the total mass M enclosed in a small region is comparable to or larger than the radius of the region, the high-density region would collapse to a primordial black hole.

As summarized above, primordial black holes are formed from a perturbation for which the length scale, L, is much longer than the Hubble horizon scale in an early epoch. For analyzing such a state, the so-called long-wavelength approximation is robust. Here the long wavelength implies $L \gg a/\dot{a} = H^{-1}$ where H is the Hubble parameter. In the long-wavelength approximation, one introduces a small parameter ϵ, and assumes that $\delta := E/E_0 - 1$ is of order ϵ^2 and L is of order ϵ^{-1}. Here, E is energy density and E_0 is the energy density in a hypothetically homogeneous universe (see section 2.6.2). The assumption that $L = O(\epsilon^{-1})$ is equivalent to assuming that the magnitude of spatial gradients of any quantities (denoted by Q) is given by

$$\partial_i Q \sim \frac{Q}{L} = Q \times O(\epsilon). \tag{9.32}$$

In this approximation, the order of both geometric and hydrodynamical quantities in the context of the formulation described in section 2.6.2 should be as follows:

$$\text{Quantities of } O\left(\epsilon^0\right) : \Psi - 1,$$
$$O\left(\epsilon^1\right) : u^i \text{ and } v^i,$$
$$O\left(\epsilon^2\right) : \tilde{A}_{ij}, h_{ij}(:= \tilde{\gamma}_{ij} - \eta_{ij}), \chi(:= \alpha - 1), \text{ and } \delta,$$
$$O\left(\epsilon^3\right) : u_i \text{ and } v^i + \beta^i, \tag{9.33}$$

where $v^i := u^i/u^t$. Here, we chose $\tilde{K} = O\left(\epsilon^4\right)$ as the slicing condition for simplicity, and assumed that the amplitude of primordial gravitational-wave (tensor) perturbation is smaller than the scalar perturbation according to standard inflationary scenarios so that h_{ij} is of order ϵ^2. In addition, we assumed that the vector

---

[6] In this section, a denotes the scale factor, not orbital separation nor black-hole spin.

part (i.e., vorticity part) is absent in any quantity, because it is composed only of decaying modes.

Substituting the variables shown in equation (9.33) into the basic equations described in section 2.6.2, the equations in the lowest order in ϵ are derived as

$$\frac{1}{\Gamma}\dot{\delta} + \frac{6\dot{\Psi}}{\Psi} + \nabla_k v^k = O(\epsilon^4), \tag{9.34}$$

$$\partial_t(E_0 a^3 u_i) + \frac{6\dot{\Psi}}{\Psi}E_0 a^3 u_i = -E_0 a^3\left(\partial_i\chi + \frac{\Gamma-1}{\Gamma}\partial_i\delta\right) + O(\epsilon^6), \tag{9.35}$$

$$\frac{6\dot{\Psi}}{\Psi} - 3\frac{\dot{a}}{a}\chi = \nabla_k\beta^k + O(\epsilon^4), \tag{9.36}$$

$$\Delta_f\Psi = -2\pi a^2\Psi^5 E_0\delta + O(\epsilon^4), \tag{9.37}$$

$$\nabla^2\chi = 4\pi E_0 a^2\{(3\Gamma-2)\delta + 3\Gamma\chi\} + O(\epsilon^4), \tag{9.38}$$

$$\partial_t h_{ij} = -2\tilde{A}_{ij} + \delta_{ik}\beta^k{}_{,j} + \delta_{jk}\beta^k{}_{,i} - \frac{2}{3}\delta_{ij}\beta^k{}_{,k} + O(\epsilon^4), \tag{9.39}$$

$$\partial_t\tilde{A}_{ij} + 3\frac{\dot{a}}{a}\tilde{A}_{ij} = \frac{1}{\Psi^4 a^2}\left[-\frac{2}{\Psi}\left(\overset{(0)}{D}_i\overset{(0)}{D}_j\Psi - \frac{\eta_{ij}}{3}\Delta_f\Psi\right)\right.$$
$$\left.+ \frac{6}{\Psi^2}\left((\overset{(0)}{D}_i\Psi)\overset{(0)}{D}_j\Psi - \frac{\eta_{ij}}{3}(\overset{(0)}{D}_k\Psi)\overset{(0)}{D}_k\Psi\right)\right] + O(\epsilon^4), \tag{9.40}$$

where

$$\nabla_k := \frac{1}{\Psi^6\eta^{1/2}}\partial_k\Psi^6\eta^{1/2},$$

$$\nabla^2 := \frac{1}{\Psi^6\eta^{1/2}}\partial_k\left(\Psi^2\eta^{1/2}\eta^{kl}\partial_l\right), \tag{9.41}$$

with $\eta = \gamma/\Psi^{12}$. The first two equations, (9.34) and (9.35), are derived from hydrodynamics equations and other five, (9.36)–(9.40), are derived from equations for geometric variables (see section 2.6.2).

From equations (9.34) and (9.36), we find that the following relations have to be satisfied;

$$\dot{\Psi} = O(\epsilon^2), \tag{9.42}$$

$$\frac{1}{\Gamma}\dot{\delta} + 3\frac{\dot{a}}{a}\chi = -\nabla_k(v^k + \beta^k) = O(\epsilon^4). \tag{9.43}$$

Also, we find that the right-hand side of equation (9.38) has to be of $O(\epsilon^4)$, i.e.,

$$(3\Gamma-2)\delta + 3\Gamma\chi = O(\epsilon^4). \tag{9.44}$$

From these relations with a reasonable assumption that $\delta \to 0$ for $t \to 0$, we finally find the time dependence for each variable at the leading order in ϵ as

$$\delta = -\frac{3\Gamma}{3\Gamma-2}\chi \propto t^{2-4/3\Gamma}, \tag{9.45}$$

$$\nabla_k(v^k + \beta^k) \propto t^{3-8/3\Gamma}, \tag{9.46}$$

$$u_k \propto t^{3-4/3\Gamma}, \tag{9.47}$$

$$\Psi \propto t^0, \tag{9.48}$$

$$\tilde{A}_{ij} \propto t^{1-4/3\Gamma}. \tag{9.49}$$

Time dependence of v^i, β^i, $O(\epsilon^2)$ part of Ψ, and h_{ij} is found when we determine a spatial gauge condition for β^i. For example, in the minimal distortion gauge (see section 2.2.5), it is found that

$$v^k \text{ and } \beta^k \propto t^{1-4/3\Gamma}, \tag{9.50}$$

$$h_{ij} \propto t^{2-4/3\Gamma}, \tag{9.51}$$

$$O(\epsilon^2) \text{ part of } \Psi \propto t^{2-4/3\Gamma}. \tag{9.52}$$

The initial condition for a numerical-relativity simulation has to be provided taking into account all these relations and time-dependence.

An important point to be emphasized is that we do not restrict our attention to the case of $\Psi - 1 \ll 1$ in the long-wavelength approximation. Therefore, even when the scalar part of the metric is non-linear, we can still find the analytic solution as long as $L \gg H^{-1}$.

Shibata and Sasaki (1999) performed simulations for the formation of primordial black holes employing this solution as the realistic initial condition with the most plausible value of the adiabatic index $\Gamma = 4/3$. Although their simulation was performed assuming spherical symmetry, the derived result captures the general property for the formation of primordial black holes. Hence, we will introduce their finding in the following.

In their simulation, Shibata and Sasaki (1999) provided initial conditions in the following manner (note that in their spherical symmetric study $h_{ij} = 0$ was always chosen by the spatial gauge condition that determines β^i). First, δ is written as

$$\delta = f(\boldsymbol{r})t^{2-4/3\Gamma}, \tag{9.53}$$

where $f(\boldsymbol{r})$ is a function of spatial coordinates. From equations (9.45), χ is soon found as

$$\chi = -\frac{3\Gamma - 2}{3\Gamma} f(\boldsymbol{r}) t^{2-4/3\Gamma}. \tag{9.54}$$

Subsequently, u_i is derived from equation (2.408) as

$$u_i = \frac{1}{3\Gamma + 2} \partial_i f(\boldsymbol{r}) t^{3-4/3\Gamma}. \tag{9.55}$$

Thus, if we specify the function $f(\boldsymbol{r})$ on the initial time slice $t = t_0$, where t_0 is much smaller than the length scale of δ, and subsequently solve constraint equations for Ψ and $\tilde{A}_{ij}$, an initial condition is derived.

In the non-spherical case, we have to be more careful for determining h_{ij} which still depends on the spatial gauge condition. If we can ignore the tensor mode of the metric perturbation, the simplest way is to choose $h_{ij} = 0 = \partial_t h_{ij}$. Then, equation (9.39) yields in the long-wavelength approximation,

$$\tilde{A}_{ij} = \frac{1}{2}\left(\delta_{ik}\beta^k{}_{,j} + \delta_{jk}\beta^k{}_{,i} - \frac{2}{3}\delta_{ij}\beta^k{}_{,k}\right). \tag{9.56}$$

Substituting this equation into the momentum constraint, which has to be solved, we can determine $\tilde{A}_{ij}$ and β^i.

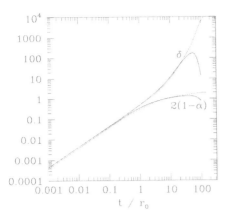
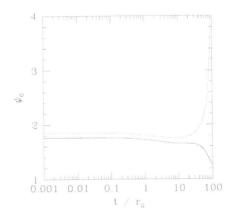

Fig. 9.27 Left: Time evolution of δ and $2(1-\alpha)$ at the center for the black-hole formation case (dotted curves) and no formation case (solid curves), respectively. Right: The same as the left panel but for Ψ at the center. The figure is taken from Shibata and Sasaki (1999).

In the spherically symmetric work of Shibata and Sasaki (1999), δ is given in a Gaussian-like form as

$$\delta \cdot \psi^6 = C_\delta \left[\exp\left(-\frac{r^2}{r_0^2}\right) - \sigma^{-3} \exp\left(-\frac{r^2}{\sigma^2 r_0^2}\right) \right] \left(\frac{t}{r_0}\right)^{2-4/3\Gamma}, \qquad (9.57)$$

where C_δ, $r_0 (\gg t_0)$, and σ are constants. r_0 determines the epoch for which the density fluctuation scale becomes comparable to the Hubble horizon scale. C_δ and σ specify the amplitude and shape of the density fluctuation, respectively.

As shown in equation (9.45), the amplitude of δ for the early epoch ($t \ll r_0$) has to be sufficiently small. However, Ψ can still be sufficiently large. This can be found from equation (2.405) with the fact that $E_0 a^2$ is a large quantity for $t \ll L$ ($E_0 a^2 \to \infty$ for $t \to 0$). Such a large scalar perturbation for the metric is the key for the formation of primordial black holes.

As the gauge condition during the numerical simulation, the constant mean curvature slicing ($K = -3H$) together with the minimal distortion gauge condition were employed. Since the conformally flat form for the spatial metric was chosen, it is easy to impose the minimal distortion gauge in spherical symmetry (see appendix B.1).

Figure 9.27 plots the general features for the evolution of δ, -2χ, and Ψ at $r = 0$ taking the case $\sigma = \infty$. Essentially the same features are found also for $\sigma < \infty$. The left panel of figure 9.27 plots the evolution of δ and -2χ at $r = 0$ for a black-hole formation case ($C_\delta = 15$, dotted curves) and no formation case ($C_\delta = 13$, solid curves), respectively. Note that $-2\chi = 2(1-\alpha)$ agrees with δ in the long-wavelength approximation (i.e., for $t \ll r_0$), and in the numerical results, this is satisfied. In addition, δ and $\alpha - 1$ are proportional to t, as the solution in the long-wavelength approximation shows. The difference between δ and $2(1-\alpha)$

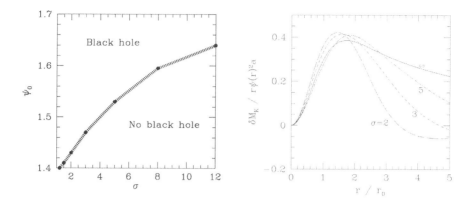

Fig. 9.28 Left: The criterion for the formation of primordial black holes in the plane of σ and initial value of Ψ at $r = 0$ (ψ_0) in the study of Shibata and Sasaki (1999). The thick curve shows the threshold. Right: A compactness function (mass enclosed at each radius divided by the circumferential radius there) as a function of r for several models with different values of σ along the threshold curve of the left panel.

gradually becomes appreciable for $t/r_0 \gtrsim 0.1$ and as large as unity at $t \sim r_0$. This is reasonable because the amplitude of the density fluctuation is non-linear at the time that its length scale becomes comparable to the Hubble horizon scale.

The right panel of figure 9.27 plots the evolution of Ψ at $r = 0$ (denoted by ψ_0). As the solution in the long-wavelength approximation shows, it is approximately constant for $t \ll r_0$. For $t \gtrsim 0.1 r_0$, however, it gradually changes the value, and for $t \gtrsim 10 r_0$, we have two possibilities: For the case that ψ_0 is smaller than a threshold value at $t = 0$, it eventually settles toward unity. By contrast, if the initial value is larger than the threshold value, ψ_0 monotonically increases in the late time: For this case, a primordial black hole is formed at $t \sim 100 r_0$.

Primordial black holes are formed only when the initial amplitude of Ψ is sufficiently high. The left panel of figure 9.28 shows a summary on the criterion of the primordial black-hole formation in the plane composed of σ and initial value of ψ_0 in the study of Shibata and Sasaki (1999). Although the threshold value of ψ_0 at $t = 0$ depends on the value of σ (i.e., the initial configuration of δ and Ψ), it is found that only for the case that $\Psi(r = 0) - 1$ is non-linear, a primordial black hole is formed. In spherical symmetry, it is possible to locally define the mass [Kodama (1980)]. Using this fact, Shibata and Sasaki (1999) also defined a compactness function (mass enclosed at each radius divided by the circumferential radius there) and found that if the maximum value of the compactness function is larger than ~ 0.4, a primordial black hole could be formed irrespective of the value of σ (see the right panel of figure 9.28).[7]

---

[7] See Nakama et al. (2014) for more systematic survey for the condition of the formation of primordial black holes.

Chapter 10

Non-radial instability and magnetohydrodynamics instability

Rapidly rotating objects in nature are often subject to non-axisymmetric deformation by some mechanism of instability [see, e.g., Chandrasekhar (1969); Shapiro and Teukolsky (1983)]. Rapidly rotating and non-axially deformed neutron stars emit gravitational waves that could be important targets for the detection by grand-based laser-interferometric detectors such as advanced LIGO, advanced VIRGO, and KAGRA (see section 1.2.6). This fact has stimulated a detailed study for the instability of rapidly rotating neutron stars.

Accretion tori surrounding a black hole can be also subject to non-axisymmetric deformation due to the so-called Papaloizou–Pringle instability [Papaloizou and Pringle (1984)]. If the mass of the tori is sufficiently high, gravitational waves with high amplitude could be emitted. These gravitational waves could also be targets for the detection of gravitational-wave detectors.

Differentially rotating neutron stars and accretion tori are also subject to a variety of magnetohydrodynamics instability in the presence of magnetic fields [see, e.g., Spruit (1999) for a review]. The instability could be the primary engine for the angular-momentum transport and also could enhance the dynamical activity of these compact objects, which may be the origin of high-energy phenomena such as gamma-ray burst and high-energy outflow.

For exploring the nature of these dynamical instability for compact objects, numerical relativity has been playing an important role. The purpose of this chapter is to introduce some of such numerical-relativity simulations and their outputs.

10.1 Non-axisymmetric instability of rapidly rotating neutron stars

As introduced in section 1.5.5.4, it is well-known that rapidly rotating neutron stars with $\beta := T_{\rm rot}/|W| \gtrsim 0.27$ are dynamically unstable against non-axisymmetric bar-mode deformation in Newtonian gravity. Here, $T_{\rm rot}$ is the rotational kinetic energy and W is the gravitational potential energy. In general relativity, these are defined by equations (9.20) and (9.21), respectively.

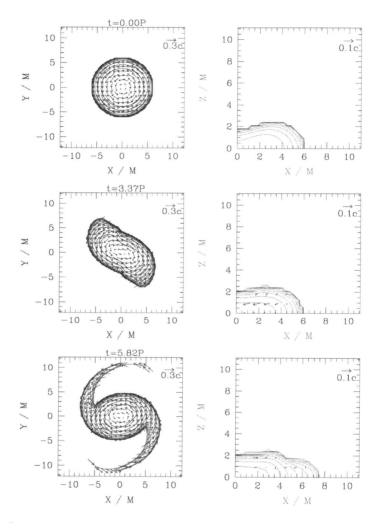

Fig. 10.1 Snapshots of density contours for the weighted rest-mass density $\rho_*(=\rho\alpha u^t\psi^6)$ and the velocity flow for $v^i(=u^i/u^t)$ in the equatorial plane (left) and in the $y=0$ plane (right) for an unstable differentially rotating star. The contour lines are drawn for $\rho_*/\rho_{*\mathrm{max}}=10^{-0.3j}$ for $j=0,1,2,\cdots,10$ where $\kappa_\mathrm{p}\rho_{*\mathrm{max}}$ is 0.126, 0.172 and 0.264 for the three different time slices (the corresponding values of $\bar{\rho}_{\mathrm{max}}$ are 0.045, 0.059, and 0.065, respectively) with κ_p the polytropic constant. The lengths of arrows are normalized to $0.3c$ (left) and $0.1c$ (right). The time is shown in units of the rotation period along the z-axis (denoted by P in the figure). The figure is taken from Shibata et al. (2000a).

The threshold value for $\beta \sim 0.27$ has been found in Newtonian gravity for a variety of equations of state and angular-velocity profiles (for which the degree of differential rotation is not very high). However, it was not known whether this is also the case in general relativity until the work by Shibata et al. (2000a), who performed a numerical-relativity simulation for rapidly and differentially rotating neutron stars

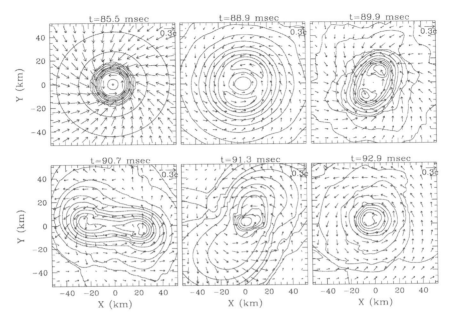

Fig. 10.2 Snapshots of the density contour curves for ρ in the equatorial plane for a dynamically unstable proto-neutron star model. The solid contour curves are drawn for $\rho/\rho_{\max} = 1$, 0.8, 0.6, 0.4, 0.2, and $10^{-j/2}$ for $j = 2, \cdots, 8$ with $\rho_{\max}$ the maximum value of ρ. Vectors indicate the local velocity field (v^x, v^y), and the scale is shown in the upper right-hand corner. The figure is taken from Shibata and Sekiguchi (2005b).

for a wide range of β for the first time. In their simulation, neutron stars were modeled by a simple $\Gamma = 2$ polytropic equation of state. However, the compactness of the neutron stars was set to be in a realistic range $\sim 0.13\text{--}0.20$, and thus, general relativistic effects were fully explored. As the initial condition, rapidly rotating axisymmetric equilibrium states with moderate degree of differential rotation were prepared, superposing a bar-mode perturbation for the density profile.

Shibata et al. (2000a) discovered in their simulations that for general relativistic stars with compactness > 0.1, the dynamical bar-mode instability can set in even for $\beta \sim 0.25$ [see also a work by Baiotti et al. (2007)]. The critical value for the onset of the dynamical instability is by $\sim 10\%$ smaller than 0.27. This is due to a general relativistic effect as clarified by Saijo et al. (2002) in their post-Newtonian computation performed as a follow-up study of Shibata et al. (2000a).

Figure 10.1 displays the evolution of density contours and velocity vectors for an unstable neutron star. In the early stage of the evolution, the amplitude of the bad-mode perturbation evolves exponentially in time. This implies that the neutron star is indeed dynamically unstable. The amplitude of the perturbation subsequently becomes so high that a nonlinear effect of the bar-mode deformation plays an important role. In a sufficiently deformed stage, the non-axisymmetric neutron star exerts a torque to the material in its outer part by a gravitational effect, and hence, an angular-momentum transport process becomes active. As a result of this process turning on, the specific angular momentum in the central region of the neutron star decreases and the growth of the bad-mode perturbation is eventually saturated. At the same time, the outer part of the neutron star obtains the specific angular momentum through the angular-momentum transport process, resulting in expansion and eventually in forming spiral arms. After the spiral-arm formation, a fraction of the material spreads outwards carrying the mass and angular momentum. On the other hand, the central region of the neutron star reduces rotational kinetic energy and β, and hence, it relaxes to a new state which is dynamically stable.

We note that after the formation of a stable neutron star, the bar-mode perturbation of small amplitude is still present. This often becomes the primary mechanism for continuously transporting angular momentum outwards, as in massive neutron stars formed after the merger of binary neutron stars (see section 8.3.3).

The next question is whether a neutron star with $\beta \gtrsim 0.25$ could be really formed in nature. A possible formation process for rapidly rotating neutron stars is the stellar core collapse for which their progenitor may have large angular momentum before the onset of the collapse. In the context of numerical relativity, Shibata and Sekiguchi (2005b) first explored this possibility. They performed a simulation employing a piecewise polytropic equation of state without taking into account realistic equations of state and detailed microphysical processes. They found that a rapidly rotating proto-neutron star with $\beta \gtrsim 0.25$ may be formed if the progenitor core is rapidly and *highly differentially* rotating before the collapse, and in addition, the equation of state for the proto-neutron star satisfies certain conditions: Specifically, the cooling effect in the collapsing core has to be efficient enough to soften the equation of state. Their simulations showed that not only the bar-mode ($m = 2$) but also $m = 1$ mode could set in soon after the core bounce, if the high value of β is achieved (see figure 10.2).

However, it is not very clear whether such a high degree of differential rotation is realized in the real progenitor core. Moreover, the simulation by Shibata and Sekiguchi (2005b) were performed employing a phenomenological equation of state without taking into account realistic equations of state and microphysics. Subsequent numerical simulations with a more realistic equation of state and microphysics by Dimmelmeier et al. (2007); Ott et al. (2007a,b); Dimmelmeier *et al.* (2008) showed that for a plausible rotation profile of the progenitor core, the formation of high-β proto-neutron stars is not very likely for the ordinary supernova collapse (see section 1.5.5.4). The primary reason for this is that in the realistic equation of state, the cooling is not enhanced enough to form a proto-neutron star of a high degree of differential rotation.

Alternatively, Ott *et al.* (2007b) showed in their numerical-relativity simulations that the dynamical instability could set in even for a low value of $\beta \sim 0.1$, because the outer region of formed proto-neutron stars could have a high degree of differential rotation and thus may be unstable as shown in Newtonian simulations by Tohilne and Hachisu (1990); Centrella et al. (2001); Shibata et al. (2003). For this type of the dynamical instability, the deformation occurs not only for the bar-mode ($m = 2$ mode) but also for the $m = 1$ mode, and the growth of non-axisymmetric deformation modes is slow; the growth time scale is by a factor of $\gtrsim 10$ longer than the dynamical time scale that is sub-milliseconds. Furthermore, the mode growth is saturated at a weakly nonlinear stage. However, after the saturation, non-axisymmetric deformation modes are preserved for a long time scale, and hence, quasi-periodic gravitational waves are emitted for a long term. Figure 10.3 shows the evolution of non-axisymmetric deformation modes (modes with $m = 1$–4) and gravitational waves emitted by the bar-mode ($l = m = 2$ mode), obtained by a simulation of Ott et al. (2007b). The upper panel displays the evolution of mode amplitudes for the non-axisymmetric density perturbation, and shows that modes with $m = 1$–3 are unstable because the density perturbation grows exponentially with time. The exponential growth is saturated at a weakly nonlinear deformation stage, and subsequently, the degree of the deformation is preserved for a long time scale. The lower panel displays gravitational waves observed along the most optimistic direction (i.e., along the rotation axis) at a hypothetical distance of 10 kpc. This shows that quasi-periodic gravitational waves with a typical frequency (~ 930 Hz for this case) are emitted for a time scale that is much longer than the rotation period. Ott et al. (2007b) concluded that gravitational waves from such a proto-neutron star at a hypothetical distance of 10 kpc could be detected by advanced LIGO with the signal-to-noise ratio $\sim 10^2$.

As a completely alternative model, Zink et al. (2007) considered the case that a progenitor is extremely rapidly rotating, and hence, a toroidal object (not proto-neutron star) of entirely *subnuclear-matter* density is formed after the collapse. Namely, they focused on the case that before the nuclear-matter density is reached, the collapse is halted by the large centrifugal force. Because such a remnant should have a soft equation of state, they performed a simulation simply employing the $\Gamma = 4/3$ equation of state. In their study, they also discovered that not only the bar-mode instability but also $m = 1$ mode instability set in. These modes grow to a nonlinear stage, and subsequently, govern the process of angular momentum transport from the inner to the outer region. As a result of the development of the $m = 1$ or $m = 2$ mode density peaks together with the angular momentum transport, the toroidal configuration changes to a configuration composed of a single star surrounded by a torus or a system composed of two stars.

Zink et al. (2006); Reisswig et al. (2013) also proposed a scenario for formation of a supermassive black hole through the fragmentation of toroidal supermassive stars. The mechanism is essentially the same as that of Zink et al. (2007). For this scenario to be realized, highly differentially rotating supermassive stars have to be formed, and thus, the scenario is based on the assumption which is different from

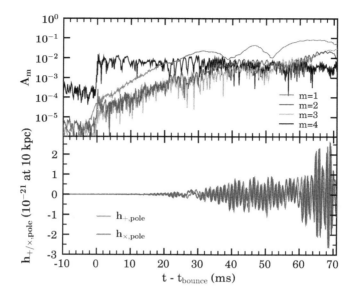

Fig. 10.3 Upper panel: Normalized mode amplitudes of non-axisymmetric density perturbation at a post-bounce epoch. The high amplitude for $m = 4$ mode is spuriously excited because the simulation was performed in the Cartesian coordinates. Lower panel: Gravitational waves from a non-axisymmetric proto-neutron star. The figure is taken from Ott et al. (2007b).

that described in section 9.1 [see also Saijo and Hawke (2009)].

As described in section 9.3.2, the central engine of gamma-ray bursts is believed to be composed typically of a stellar-mass black hole surrounded by a hot and massive torus, and it is reasonable to consider that such a system is formed by a fall-back of the material onto a rapidly rotating proto-neutron star through a long-term accretion in the stellar core collapse. During the evolution of the proto-neutron star, the material of high specific angular momentum could fall onto it, leading to spin-up. The resulting system composed of a rapidly rotating proto-neutron star surrounded by a massive torus could be unstable against the dynamical non-axisymmetric instability. The simulation for this issue has not been done yet. This is an issue for the future.

10.2 Non-axisymmetric instability of torus surrounding black hole

A torus surrounding a black hole is universally found in the universe. For example, the central region of active galactic nuclei is believed to consist of a supermassive black hole surrounded by a torus. The merger of binary neutron stars and black hole-neutron star binaries could often result in a black hole surrounded by a hot and massive torus as described in chapter 8. A black hole-torus systems could be also formed after the core collapse of massive stars, as described in section 9.3.

It has been long suggested that the central engine of short- and long-duration gamma-ray bursts is formed by the merger of binary neutron stars (or black hole-neutron star binaries) and core collapse of massive stars, respectively [Narayan et al. (1992); Woosley (1993)]. In these hypotheses, a system composed of a spinning black hole surrounded by a massive torus is supposed to be formed. However, their formation and evolution processes have not yet been observed directly, because such sites are opaque to electromagnetic signals due to their intrinsic high density. Gravitational waves are much more transparent than electromagnetic waves against the absorption and scattering with matter. If black hole-torus systems emit strong gravitational waves, it will be possible to explore their formation and evolution processes by detecting these gravitational waves, along with the popular hypotheses that associate them to gamma-ray burst engines.

Black hole-torus systems are known to be subject to the non-axisymmetric Papaloizou–Pringle instability [Papaloizou and Pringle (1984)] [see also, e.g., Kojima (1986); Narayan et al. (1987)]. Exploring the nonlinear growth, saturation, and subsequent evolution of this instability for black hole-torus systems requires long-term three-dimensional general relativistic simulations. Such a simulation was first performed by Hawley (1991) in a fixed background geometry of a black hole. If the mass of the torus is a substantial fraction of the black-hole mass and the self-gravity of the torus plays an important role, we need numerical-relativity simulations. Such simulations have been performed in the recent years.

In the following, we will first summarize our knowledge on the stability and nonlinear growth of the non-axisymmetric instability for systems composed of a black hole and a self-gravitating massive torus. Then, some of numerical-relativity simulations will be introduced.

Our knowledge on the instability of a non-self-gravitating torus (or disk) around a central object against non-axisymmetric deformation is summarized as follows:

- Tori of constant specific angular momentum are always dynamically unstable for a variety of non-axisymmetric deformation modes with $m = 1, 2, 3, \cdots$ [Kojima (1986)]. Here, the fastest-growing mode is determined by the thickness (the slenderness in the equatorial plane) of the torus. Higher-m modes are increasingly important for more slender tori while for radially-wide tori that are likely to be astrophysically realistic, $m = 1$ mode is often the fastest-growing mode.
- By contrast, tori for which the velocity profile is close to the Keplerian one (i.e., far from the profile of constant specific angular momentum) could be stable against non-axisymmetric deformation modes [Blaes and Hawley (1988)].
- After the growth of non-axisymmetric deformation modes for the unstable tori, angular momentum is transported from the inner region to the outer one because the non-axisymmetric modes exert the torque. As a result of the angular momentum redistribution, the velocity profile approaches a stable configuration (which is away from the profile of constant specific angular momentum) [Blaes and Hawley (1988); Hawley (1991)].

The purpose of numerical-relativity simulations for black hole-torus systems is to clarify the effects of general relativistic self-gravity. Such numerical-relativity studies were performed by Oleg et al. (2011) and Kiuchi et al. (2011a) for the first time. A detailed study for the non-axisymmetric stability of the torus of constant specific angular momentum was performed by Oleg et al. (2011) employing a $\Gamma = 4/3$ polytropic equation of state. They confirmed that the conditions of the non-axisymmetric dynamical instability satisfied for non-self-gravitating tori are basically satisfied for self-gravitating general relativistic tori.

Subsequently, Kiuchi et al. (2011a) performed long-term numerical-relativity simulations employing two rotation profiles of tori and two ratios of torus mass to mass of the central black hole with a $\Gamma = 4/3$ polytropic equation of state. They also confirmed that the conditions of the non-axisymmetric dynamical instability satisfied for non-self-gravitating tori are also satisfied for self-gravitating general relativistic tori. In addition, they found that the self-gravity of the torus (i.e., increasing the ratio of the torus mass to the black-hole mass) enhances the growth rate of the dynamical instability. In particular, they clarified that even for the case that the velocity profile of the torus is significantly different from that of constant specific angular momentum profile, the massive torus can be dynamically unstable.

Oleg et al. (2011) and Kiuchi et al. (2011a) also found that in the presence of self-gravity of the torus, the central black hole moves after the growth of the $m = 1$ mode (see figure 10.4). This is reasonable because a density peak is developed in the torus and the black hole receives asymmetric force from the torus. Since the black hole has to move to the opposite direction of the density peak for the preservation of the mass center, the growth of the density peak is enhanced. Therefore, in the presence of the self-gravity, the growth rate of the $m = 1$ mode instability is enhanced. Kiuchi et al. (2011a) also showed that after the perturbation of this mode becomes nonlinear, mass accretion from the torus to the black hole sets in. If the inner edge of the torus is located in the vicinity of the innermost stable circular orbit of the black hole, a significant fraction of the mass in the torus could fall into the black hole.

Kiuchi et al. (2011a) performed long-term simulations that were continued after the saturation of the growth of the dynamically unstable modes in the torus. They found that after the saturation, the torus relaxes to a new quasi-steady state. However, the non-axisymmetric modes are still preserved for a time scale much longer than the orbital period of the torus. This suggests that quasi-periodic gravitational waves of small amplitude are continuously emitted. Since their emissivity is low and the radiation reaction time scale is quite long, gravitational waves will be emitted for a time scale much longer than the rotation period, unless other dissipation processes govern the dissipative evolution. Kiuchi et al. (2011a) then estimated the effective amplitude of gravitational waves for a hypothetical accumulation of the gravitational-wave cycles, and found that gravitational waves from black hole-torus systems may be observed by planned detectors such as advanced LIGO for

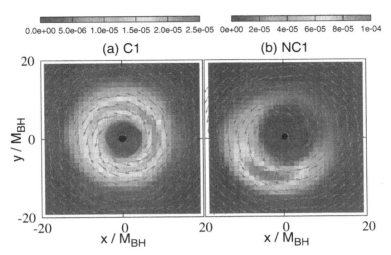

Fig. 10.4 Snapshots of the rest-mass density distribution on the equatorial plane for two models after the saturation of the growth of dynamically unstable modes. The left and right panels show the results for the models in which the specific angular momentum of the tori is constant and non-constant (closer to the Keplerian motion), respectively. For both models, it is seen that the $m=1$ mode is developed. Vectors show the velocity fields, and black filled circles around the centers show the region inside the apparent horizons. The figure is taken from Kiuchi et al. (2011a).

the black-hole mass $\sim 10 M_\odot$ with the torus mass $\sim 1 M_\odot$ for an event at a distance of several 10 Mpc.

10.3 Magnetohydrodynamical instability of neutron stars

Magnetohydrodynamics instability often govern the evolution of general relativistic compact objects; neutron stars and torus (or disk) surrounding a central compact object. In the following, we will first summarize a variety of instability and outline the possible effects by them. We then introduce several numerical-relativity simulations that explored the possible effects of these instability.

10.3.1 *Magnetohydrodynamics instability*

There are many known magnetohydrodynamics instability that play a crucial role in relativistic astrophysics. In this subsection, we outline some of them briefly.

10.3.1.1 *Magnetic winding*

In the presence of a radial component of magnetic fields in a differentially rotating object, the toroidal component of magnetic fields grow linearly with time. This process is often called magnetic winding. The mechanism for this can be seen from an analysis of the induction equation (4.150).

If the magnetic field is weak and has a negligible back-reaction on the fluid, the velocity components of the fluid will remain constant with time. Here, we consider an object in which the rotational velocity is much larger than other velocity components. Then, in cylindrical coordinates (ϖ, φ, z), the induction equation, assuming axial symmetry, is approximately written as

$$\partial_t \mathcal{B}^I \approx 0, \tag{10.1}$$

$$\partial_t \mathcal{B}^\varphi \approx \frac{1}{\varpi} \partial_I \left(\varpi \Omega \mathcal{B}^I \right), \tag{10.2}$$

where the z-axis is the rotation axis, $\Omega[= \Omega(\varpi)]$ denotes the angular velocity, and the subscript I denotes ϖ or z. We assumed that $|v^\varpi| \ll \varpi\Omega$ and $|v^z| \ll \varpi\Omega$. Using the no-monopole constraint, $\partial_I \left(\varpi \mathcal{B}^I \right) = 0$, we have

$$\partial_t \mathcal{B}^\varphi \approx \mathcal{B}^I \partial_I \Omega = \mathcal{B}^\varpi \partial_\varpi \Omega, \tag{10.3}$$

Equation (10.3) indicates that the toroidal component of the magnetic field, $\mathcal{B}^T (= \varpi \mathcal{B}^\varphi)$, grows linearly with time according to

$$\left| \mathcal{B}^T(t; \varpi, z) \right| \approx \left| \mathcal{B}^T(0; \varpi, z) \right| + t\varpi \left| \mathcal{B}^\varpi(0; \varpi, z) \partial_\varpi \Omega(0; \varpi) \right|$$

$$\approx \left| \mathcal{B}^T(0; \varpi, z) \right| + \frac{3\pi t}{P_{\rm rot}} \left| \mathcal{B}^\varpi(0; \varpi, z) \right|$$

$$\approx \left| \mathcal{B}^T(0; \varpi, z) \right| + 10^{15} \left(\frac{t}{100 \, {\rm ms}} \right) \left(\frac{P_{\rm rot}}{1 \, {\rm ms}} \right)^{-1} \left(\frac{|\mathcal{B}^\varpi(0; \varpi, z)|}{10^{12} \, {\rm G}} \right) {\rm G}, \tag{10.4}$$

where $t = 0$ is an initial time, $P_{\rm rot}$ denotes the local rotation period, and we have assumed a Keplerian angular velocity profile $\Omega \propto \varpi^{-3/2}$. For the last line, we substituted the value of $P_{\rm rot}$ supposing rapidly and differentially rotating neutron stars.

After the strength of the toroidal magnetic field is sufficiently amplified by winding, the toroidal magnetic-field energy will become comparable to the rotational kinetic energy: The relation is approximately written as

$$\rho R^2 \Omega^2 \sim \frac{1}{4\pi} \left(\mathcal{B}^T \right)^2 \sim \frac{1}{4\pi} \left(\Omega t \mathcal{B}^\varpi \right)^2, \tag{10.5}$$

where R is the characteristic radius of the system such as neutron-star radius. Thus, at $t \sim (4\pi\rho)^{1/2} R/\mathcal{B}^\varpi \sim R/v^A$, the magnetic tension will be large enough to change the angular velocity profile of the fluid and specific angular momentum will be redistributed by a magnetic force [e.g., Spruit (1999)]. This mechanism is called magnetic braking. Here, v^A is the local Alfvén velocity, and R/v^A is called the Alfvén time scale. For neutron stars with strong magnetic fields, it is

$$t_A \sim \frac{R}{v_A} \approx 10 \, {\rm ms} \left(\frac{|\mathcal{B}^\varpi|}{10^{15} \, {\rm G}} \right)^{-1} \left(\frac{R}{10 \, {\rm km}} \right) \left(\frac{\rho}{10^{14} \, {\rm g/cm}^3} \right)^{1/2}. \tag{10.6}$$

Thus, equation (10.4) can be used only for $t \lesssim t_A$. For $t \gtrsim t_A$, specific angular momenta of the fluid elements are redistributed, and the degree of differential rotation of the system will be reduced.

The state realized after the onset of the magnetic braking is determined by nonlinear magnetohydrodynamics processes. For clarifying such a state, numerical simulations are required.

10.3.1.2 *Magnetorotational instability*

The magnetorotational instability (MRI) often plays an important role in a weakly magnetized, rotating fluid wherever $\partial_\varpi \Omega < 0$ [Balbus and Hawley (1991a,b, 1998)]. By this instability, the magnetic-field strength exponentially grows, and eventually, the strength of magnetic forces becomes as large as gas pressure. When the instability reaches such nonlinear regime, the distortions in the magnetic-field lines and resulting highly non-uniform velocity field induced by the high magnetic forces lead to turbulence. The turbulence subsequently could play a crucial role for transporting angular momentum.

For estimating the growth time scale, t_MRI, and the wavelength of the fastest-growing mode, λ_max, of the MRI, we here use a Newtonian local linear analysis given in Balbus and Hawley (1998). Linearizing magnetohydrodynamics equations for a local patch of a rotating fluid and imposing a plane-wave dependence, $e^{i(\boldsymbol{k}\cdot\boldsymbol{x}-\omega t)}$, on the perturbations lead to a dispersion relation given in equation (125) of Balbus and Hawley (1998). Specializing this equation for a constant-entropy medium leads to the dispersion relation in the form

$$\omega^4 - \left[2(\boldsymbol{k}\cdot\boldsymbol{v}_A)^2 + \kappa_e^2\right]\omega^2 + (\boldsymbol{k}\cdot\boldsymbol{v}_A)^2\left[(\boldsymbol{k}\cdot\boldsymbol{v}_A)^2 + \kappa_e^2 - 4\Omega^2\right] = 0, \quad (10.7)$$

where $\boldsymbol{v}_A = \boldsymbol{B}/\sqrt{4\pi\rho}$ is the (Newtonian) Alfvén velocity vector and κ_e is the epicyclic frequency of Newtonian theory defined by

$$\kappa_e^2 := \frac{1}{\varpi^3}\frac{\partial(\varpi^4\Omega^2)}{\partial\varpi}. \quad (10.8)$$

Equation (10.7) is modified for a medium of inhomogeneous entropy [Balbus and Hawley (1998)], and the stability associated with convective motion comes into play in this case. We here neglected any entropy gradient and focused on the effects of magnetorotational shear. We further simplify the analysis by considering only the vertical modes ($\boldsymbol{k} = k\boldsymbol{e}_z$) since these are likely to be the dominant modes (i.e., modes with other directions in $\boldsymbol{k}$ have typically smaller growth rates). The value of ω^2 can be found by solving equation (10.7) and then, it is extremized with respect to k to obtain the frequency of the fastest-growing mode, ω_max as

$$\omega_\text{max}^2 = -\frac{1}{4}\left(\frac{\partial\Omega}{\partial\ln\varpi}\right)^2. \quad (10.9)$$

This maximum growth rate corresponds to

$$(kv_A^z)_\text{max}^2 = \Omega^2 - \frac{\kappa_e^4}{16\Omega^2}. \quad (10.10)$$

The growth time (*e*-folding time) and wavelength of the fastest-growing mode are then given, respectively, by

$$t_\text{MRI} = |\omega_\text{max}|^{-1} = 2\left|\frac{\partial\Omega}{\partial\ln\varpi}\right|^{-1}, \quad (10.11)$$

$$\lambda_\text{max} = \frac{2\pi}{k_\text{max}} = \frac{2\pi v_A^z}{\Omega}\left[1-\left(\frac{\kappa_e}{2\Omega}\right)^4\right]^{-1/2}. \quad (10.12)$$

For a Keplerian angular velocity distribution $\Omega \propto \varpi^{-3/2}$, we have

$$t_{\text{MRI}} = \frac{4}{3\Omega} \approx 1.3\,\text{ms} \left(\frac{\Omega}{1000\,\text{rad s}^{-1}}\right)^{-1}, \tag{10.13}$$

$$\lambda_{\max} = \frac{8\pi v_A^z}{\sqrt{15}\Omega}$$

$$\approx 0.21\,\text{km} \left(\frac{\Omega}{1000\,\text{rad s}^{-1}}\right)^{-1} \left(\frac{B^z}{10^{14}\,\text{G}}\right) \left(\frac{\rho}{10^{15}\,\text{g/cm}^3}\right)^{-1/2}. \tag{10.14}$$

If B^z is comparable to the field strength of a canonical pulsar ($B^z \sim 10^{12}$ G), $\lambda_{\max}$ is much smaller than the typical radius R of neutron stars. This illustrates the fact that resolving the fastest-growing mode of the MRI is one of the challenges in numerical relativity. Since $\lambda_{\max} \propto v_A$, larger magnetic fields will result in longer MRI wavelengths, and hence, less computational resources would be necessary for resolving the fastest-growth mode for the highly magnetized case.

Unlike $\lambda_{\max}$, t_{MRI} does not depend on the magnetic-field strength but on the angular velocity profile. The Newtonian local analysis indicates that the MRI always grows on the time scale of a rotation period for a configuration with $-\partial \ln \Omega / \partial \ln \varpi = O(1)$. Hence, the MRI is expected to play an important role for rapidly and differentially rotating object such as massive neutron stars formed after mergers of binary neutron stars. The resulting strong magnetic fields and turbulence would transport angular momentum from their rapidly rotating inner region to the more slowly rotating outer layers. This causes the inner part to contract and the outer layers to expand. One of the important purposes of magnetohydrodynamics simulations in numerical relativity is to clarify this process.

10.3.1.3 *Tayler, interchange, and Parker instability*

In differentially rotating astrophysical objects such as remnants of binary neutron star mergers and proto-neutron stars formed after the stellar core collapse of massive stars, toroidal magnetic fields are often significantly amplified by the magnetic winding process described in section 10.3.1.1. Stars with toroidal-dominant magnetic fields are often unstable against the so-called Tayler instability and interchange instability [Tayler (1973); Acheson and Gibbons (1978); Spruit (1999)]. Here, the Tayler instability and interchange instability are associated with the exchange of neighboring magnetic-field lines by the magnetic hoop stress and the magnetic buoyancy, respectively.

The nature of these instability for axisymmetric perturbations are different from that for non-axisymmetric perturbations. This point is clarified in particular by Acheson and Gibbons (1978) in their local linear perturbation analysis. Thus in the following, we will review the stability property of stars with purely toroidal magnetic fields following Acheson and Gibbons (1978). We will neglect the effect of general relativity, magnetic diffusion, thermal diffusion, and viscous effects for simplicity.

Axisymmetric case: First, we focus on axisymmetric perturbations. Acheson and Gibbons (1978) derived a dispersion relation for this in their local linear perturbation analysis, and then, showed three independent conditions for the onset of the instability; if any of these three conditions are satisfied, the star is dynamically unstable. Here, two of them are often referred to and they are written in cylindrical coordinates (ϖ, φ, z) as

$$\left(1 + \frac{v_A^2}{\Gamma c_s^2}\right) \Omega^2 \frac{\partial \ln(\Omega^2 \varpi^4)}{\partial \ln \varpi} + \left(g_\varpi + \frac{2v_A^2}{\varpi}\right) \frac{1}{\Gamma} \frac{\partial E}{\partial \varpi}$$
$$+ \left(g_\varpi - \frac{2c_s^2 \Gamma}{\varpi}\right) \frac{v_A^2}{c_s^2 \Gamma} \frac{\partial F}{\partial \varpi} < 0, \quad (10.15)$$

$$g_z \left(\frac{\partial E}{\partial z} + \frac{v_A^2}{c_s^2} \frac{\partial F}{\partial z}\right) < 0, \quad (10.16)$$

where $E := \ln(P/\rho^\Gamma)$ and $F := \ln(B/\rho\varpi)$ with B being the toroidal magnetic-field strength. E may be considered as the profile of specific entropy. Γ is the adiabatic index, and v_A and c_s are Alfvén and sound velocities as defined in Chapter 4. Ω is the local rotational angular velocity and is assumed to be a function of ϖ. g_ϖ and g_z denote magnitudes of gravitational acceleration in the ϖ and z directions, and they are defined as positive constants.

Equation (10.15) shows that the stability is determined by the rotational effects (associated with centrifugal force and Coriolis force), the thermal effects (associated with entropy gradient), and the magnetic effects. In the absence of the gradients of E and F, the condition of the instability becomes $\partial_\varpi (\Omega \varpi^2) < 0$, which is the so-called Rayleigh criterion [Chandrasekhar (1961)]. In the absence of rotation and magnetic fields, the condition for the instability is $g_\varpi \partial_\varpi E < 0$ or $g_z \partial_z E < 0$, which is the condition for the onset of the convective instability [e.g., Kippenhahn and Weigert (1994)]. The term associated with the gradient of F determines the instability induced by the magnetic fields. In the absence of the rotation and the gradients of E (entropy gradients), the condition for the instability becomes

$$\left(g_\varpi - \frac{2c_s^2 \Gamma}{\varpi}\right) \frac{\partial F}{\partial \varpi} < 0, \quad (10.17)$$

$$g_z \frac{\partial}{\partial z}\left(\frac{B}{\rho}\right) < 0. \quad (10.18)$$

Equation (10.18) implies that the so-called interchange instability sets in for $\partial_z(B/\rho) < 0$ in the isentropic fluid irrespective of ϖ. Note that for the convectively stable fluid with $g_z \partial_z E > 0$, this interchange instability could be stabilized as found from equation (10.16).

Equation (10.17) shows that the condition for the instability depends on the cylindrical radius. For $\varpi > \varpi_{\text{crit}} := 2c_s^2\Gamma/g_\varpi$, the condition for the instability is similar to the condition for the interchange instability as $g_\varpi \partial_\varpi F < 0$. On the other hand, for $\varpi < \varpi_{\text{crit}}$, i.e., near the z-axis, the condition becomes $g_\varpi \partial_\varpi F > 0$:

This is the so-called condition for the axisymmetric Tayler instability. For the axisymmetric system, near the z-axis, B/ρ is written as

$$\frac{B}{\rho\varpi} = a_0 + \frac{a_2}{2}\varpi^2 + O\left(\varpi^4\right), \tag{10.19}$$

where $a_0(>0)$ and a_2 are constant. Thus, $\partial_\varpi F = a_2\varpi/a_0 + O\left(\varpi^3\right)$, and hence, in the presence of a positive value of a_2, the system becomes unstable, while the linear term of B/ρ in ϖ does not contribute to the instability. We note that the stability property does not depend on the magnitude of the magnetic-field strength. Rather, it depends on the magnetic-field profile. The magnetic-field strength primarily determines the growth time scale of the instability: For both the interchange and Tayler instability, the growth time scale is of order Alfvén time scale [see equation (10.6)]: Denoting the stellar radius by R, it is $\sim R/v_A$. For a typical field strength $B \sim 10^{13}$ G with the typical density and radius of neutron stars, t_A is of order 1 s. For the strongly magnetized case, $B \sim 10^{15}$ G, t_A becomes ~ 10 ms.

The meaning of the critical radius, ϖ_{crit}, may be described as follows. The hoop stress of toroidal magnetic fields and magnetic buoyancy force are approximately written, respectively, as $B^2/(4\pi\varpi)$ and $g_\varpi B^2/(8\pi c_s^2)$. Thus at the critical radius, the magnitude of these two forces is identical. Inside the critical radius, the hoop stress is stronger. For such cases, the Tayler instability is the primary player. On the other hand, for the case that the magnetic buoyancy force is dominant, the interchange instability plays a role.

Note that the interchange and Tayler instability described above can be stabilized by the entropy gradient and rotational effects. For $g_\varpi \partial_\varpi E > 0$ or $\partial_\varpi(\Omega\varpi^2) > 0$, these instability can be stabilized. For rigidly rotating stars, the rotational effect always play a role for the stabilization: High angular velocity will contribute to the stabilization.

Non-axisymmetric case: Next, we consider the instability associated with non-axisymmetric perturbations. Acheson and Gibbons (1978) also derived several dispersion relations for the non-axisymmetric case. First, we focus on the non-rotating case $\Omega = 0$ in the low-frequency limit in which the mode angular frequency ω is much smaller than $mv_A/\varpi \sim m/t_A$ where m is a quantum number of perturbation in the φ direction. Thus, in this condition imposed, we cannot consider the $m = 0$ case. Then, there are also three independent conditions for the onset of the instability; if any of these three conditions are satisfied, the star is dynamically unstable. Here, two of them are often referred to and they are written as

$$\frac{g_\varpi}{\Gamma}\frac{\partial E}{\partial \varpi} + \left(g_\varpi - \frac{2c_s^2\Gamma}{\varpi}\right)\frac{v_A^2}{c_s^2\Gamma}\frac{\partial F'}{\partial \varpi} + \frac{m^2 v_A^2}{\varpi^2} < 0, \tag{10.20}$$

$$\frac{g_z}{\Gamma}\left(\frac{\partial E}{\partial z} + \frac{v_A^2}{c_s^2}\frac{\partial F'}{\partial z}\right) + \frac{m^2 v_A^2}{\varpi^2} < 0, \tag{10.21}$$

where $F' := \ln(B\varpi)$.

The conditions for the non-axisymmetric instability are similar to those for the axisymmetric ones, but there is one crucial difference for the instability associated with the presence of the magnetic fields: The non-axisymmetric instability is not determined by the gradient of B/ρ but by B itself. Thus, the interchange instability does not come into play. Rather, the Parker instability [Parker (1966)] which is associated with the bending of the magnetic-field lines could occur in the non-axisymmetric case.

However, apart from this difference, other properties are similar between axisymmetric and non-axisymmetric instability. For example, there is also a critical radius $\varpi_{\text{crit}} = 2c_s^2\Gamma/g_\varpi$, and for $\varpi > \varpi_{\text{crit}}$, the Parker instability could set in if the condition, $g_\varpi \partial_\varpi F' < 0$, is satisfied. The Parker instability could set in in the z-direction irrespective of ϖ if the condition $g_z\partial_z \ln B + \Gamma m^2 c_s^2/\varpi^2 < 0$ is satisfied. Equations (10.20) and (10.21) show that near the stellar surface where $c_s^2/(\varpi g_\varpi)$ would be much smaller than unity, the Parker instability could set in if $g_\varpi \partial_\varpi \ln B < 0$ or $g_z \partial_z \ln B < 0$ is satisfied.

For $\varpi < \varpi_{\text{crit}}$, non-axisymmetric Tayler instability could set in if $\partial_\varpi \ln(B\varpi) > 0$. We note that $B \propto \varpi$ near the z-axis for axisymmetric equilibrium states. This implies that the necessary condition for this Tayler instability is always satisfied: In particular for $m = 1$, the Tayler instability is always subject because the absolute value of $2v_A^2 \varpi \partial_\varpi \ln(B\varpi)$ is larger than v_A^2, and hence, larger than the last term of equation (10.20). We note that $g_\varpi \partial_\varpi E \to 0$ for $\varpi \to 0$ and hence the entropy gradient does not play a role near the z-axis. Thus, stars with purely toroidal magnetic fields are always dynamically unstable against the non-axisymmetric Tayler instability, and will change the magnetic-field profile after the onset of this instability. Finding a state subsequently realized is one of the important tasks in numerical relativity.

Acheson and Gibbons (1978) also derived the conditions for non-axisymmetric instability in the isentropic and rapidly rotating case with $\Omega \gg v_A/\varpi$. According to their analysis, if either of the following two criteria is satisfied, the system is unstable:

$$\frac{\partial \Omega^2}{\partial \ln \varpi} + \left(g_\varpi - \frac{2c_s^2\Gamma}{\varpi}\right)\frac{v_A^2}{c_s^2\Gamma}\frac{\partial F}{\partial \varpi} + \frac{m^2 v_A^2}{\varpi^2} < 0, \qquad (10.22)$$

$$\frac{g_z v_A^2}{\Gamma c_s^2}\frac{\partial F}{\partial z} + \frac{m^2 v_A^2}{\varpi^2} < 0. \qquad (10.23)$$

For this case, F (not F') determines the stability associated with the magnetic field. Thus, the non-axisymmetric Tayler instability and interchange instability could come into play for $\varpi\Omega \gg v_A$.

10.3.2 Numerical simulations

10.3.2.1 Evolution of differentially rotating magnetized neutron stars

As described in section 8.3, massive neutron stars are likely to be often formed after mergers of binary neutron stars. These massive neutron stars subsequently evolve due to several processes as discussed in section 8.3.3.5. Among them, magnetohydrodynamical processes could play a central role for their evolution. In particular, angular-momentum transport processes by magnetic winding and MRI could be key players, because specific angular momenta of the fluid elements inside the massive neutron stars are redistributed by such transport processes.

As described in section 1.4.6, the *hypermassive* neutron stars should have differential and rapid rotation profile, which is the key for sustaining their high mass and strong self-gravity. Here, the differential rotation implies that the angular velocity near the rotational axis is larger than that of the outer region. By magnetohydrodynamical angular-momentum transport processes, the degree of the differential rotation and the angular velocity near the rotational axis are reduced. Then, the centrifugal force in the central region, which plays a crucial role for sustaining the self-gravity of the hypermassive neutron stars, is reduced, and this reduction could trigger the collapse of a hypermassive neutron star to a black hole. Because a long-term efficient outward-transport process of the angular momentum works before the onset of the collapse, a disk or torus surrounding the formed black hole may have large mass. Furthermore, magnetohydrodynamical processes often induce turbulent motion, which could generate shocks and result in a significant increase of thermal energy. If these processes occur, the remnant black hole surrounded by the hot disk/torus may be a central engine of a gamma-ray burst.

Motivated by this consideration, magnetohydrodynamics simulations for differentially rotating hypermassive neutron stars were performed by Duez et al. (2006a); Shibata et al. (2006); Duez et al. (2006b); Siegel et al. (2013). In their series of papers, Duez et al. performed axisymmetric simulations both in a simple Γ-law equation of state with $\Gamma = 2$ and in a piecewise polytropic equation of state. Irrespective of the equations of state employed, they found that the hypermassive neutron stars eventually collapse to a black hole surrounded by a massive torus after a substantial amount of angular momentum is transported from the central region to the outer layer.

Figure 10.5 plots snapshots of the meridional density contours, velocity vectors, and magnetic-field lines at selected time slices derived in the simulation of Shibata et al. (2006). For this simulation, a piecewise polytropic equation of state composed of two pieces is employed for the cold part: $P = P_{\text{cold}} = K_1 \rho^{\Gamma_1}$ for $\rho \leq \rho_{\text{nuc}}$ and $P_{\text{cold}} = K_2 \rho^{\Gamma_2}$ for $\rho \geq \rho_{\text{nuc}}$. Γ_1 and Γ_2 are set to be 1.3 and 2.75, respectively. $K_1 = 5.16 \times 10^{14}$ cgs, $K_2 = K_1 \rho_{\text{nuc}}^{\Gamma_1 - \Gamma_2}$, and $\rho_{\text{nuc}} = 1.8 \times 10^{14}\,\text{g/cm}^3$. With this equation of state, the maximum gravitational mass (baryon rest mass) are $2.01 M_\odot$ ($2.32 M_\odot$) for spherical neutron stars and $2.27 M_\odot$ ($2.60 M_\odot$) for rigidly rotating neutron stars,

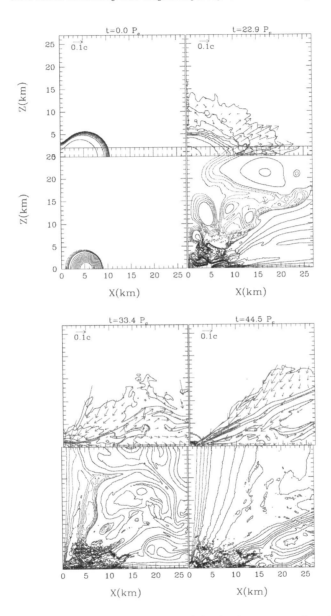

Fig. 10.5 First- and third-row panels: Snapshots of the density contours for ρ (solid curves) and velocity vectors. The contours are drawn for $\rho = 10^{15}$ g/cm$^3 \times 10^{-0.4i}$ g/cm^3 ($i = 0$–9). In the last panel, a curve with $\rho = 10^{11}$ g/cm^3 is also drawn. The (red) circle near the center in last two panels denotes an apparent horizon. The scale of the velocity is indicated in the upper left corner. Second- and fourth-row panels denote the magnetic field (contours of the toroidal component of the vector potential A_φ) at the same time slices as the corresponding upper panels. The solid contour curves are drawn for $A_\varphi = 0.8(1 - 0.1i)A_{\varphi,\mathrm{max},0}$ ($i = 0$–9) and the dotted curves are for $A_\varphi = 0.08(1 - 0.2i)A_{\varphi,\mathrm{max},0}$ ($i = 1$–4). Here, $A_\varphi = A_{\varphi,\mathrm{max},0}$ is the maximum value of A_φ at $t = 0$. P_c denotes the rotation period along the rotation axis and in this case it is 0.202 ms. The figure is taken from Shibata et al. (2006).

respectively. For such setting, a hypermassive neutron star with the following characteristics was employed for the simulation: gravitational mass $M = 2.65 M_\odot$, rest mass $M_0 = 2.96 M_\odot$, maximum density $\rho_{\max} = 9.0 \times 10^{14}\,\mathrm{g/cm^3}$, angular momentum $J = 0.82 G M^2/c$, rotation period along the rotation axis $P_c = 15.5 M = 0.202\,\mathrm{ms}$, ratio of polar to equatorial radius $1/3$, and rotation period at the equatorial surface $5.4 P_c$. This hypermassive neutron star is qualitatively similar to that found in the simulations of binary neutron star mergers.

For the simulation, a hybrid equation of state $P = P_{\mathrm{cold}}(\rho) + (\Gamma_{\mathrm{th}} - 1)\rho(\varepsilon - \varepsilon_{\mathrm{cold}})$ is used. The conversion efficiency of kinetic energy to thermal energy in shocks is determined by Γ_{th}, which is set to be 1.3 to conservatively account for shock heating. A seed poloidal magnetic field is added to the hypermassive neutron star by specifying the φ-component of the vector potential as $A_\varphi = A_b \varpi^2 \max[(P - P_{\mathrm{cut}}), 0]$ where P_{cut} is 0.04 times the maximum pressure and A_b denotes a constant that determines the initial magnetic-field strength. The value of A_b is chosen so that the maximum value of $C := B^2/8\pi P$ at $t = 0$ is 3.42×10^{-3}. Here, $B^2/8\pi$ is the magnetic pressure. This implies that the typical magnetic-field strength is $\sim 10^{17}$ G. Such a large value had to be chosen to see the onset of the MRI in a restricted computational resource at that time [see equation (10.12)]: By this setting, the wavelength of the fastest-growing mode of the MRI becomes a few km, and hence, with the grid spacing of $\sim 100\,\mathrm{m}$, this mode can be resolved.

Figure 10.5 shows that following an initial stage of linear growth of the toroidal magnetic field by winding ($t \lesssim t_A \approx 13 P_c \approx 2.6\,\mathrm{ms}$ for this case), the magnetic braking begins to transport angular momentum from the inner to the outer regions of the hypermassive neutron star, inducing its quasi-stationary contraction. At $t \sim t_A$, the growth of the toroidal magnetic field is saturated. The subsequent evolution is dominated by MRI, which distorts the poloidal magnetic-field lines and leads to the formation of turbulent eddies on a scale much smaller than the stellar radius (see the second panel of the magnetic-field configuration of figure 10.5). Because the turbulence is excited, the matter located near the stellar surface is blown outward (see the second panel of the density configuration of figure 10.5).

The hypermassive neutron star collapses at $t \approx 33 P_c$, forming a black hole composed of $\sim 85\%$ of the total rest mass (see the third panel of the density profile of figure 10.5). Material with high enough specific angular momentum remains outside the newly formed black hole and forms an accretion torus. However, the torus is secularly unstable and continuous matter infall is seen, since magnetically-induced turbulence transports angular momentum outward. The accretion rate from the torus to the black hole, $\dot{M}$, gradually decreases and eventually settles down to $\dot{M} \sim 10 M_\odot/\mathrm{s}$. At $t \sim 40 P_c$, the total rest mass of the torus is $\sim 0.1 M_\odot$, and the accretion time scale is thus $\approx 0.1 M_\odot/\dot{M} \sim 10\,\mathrm{ms}$. It was also found that a collimated magnetic field has formed along the rotation axis (see the last panel of figure 10.5).

Shibata *et al.* (2006) also explored the thermal property of the torus, and found that the maximum density and typical temperature are $\sim 10^{12}\,\mathrm{g/cm^3}$ and $\gtrsim 10\,\mathrm{MeV}$. Namely, the torus is dense and hot enough to emit a substantial amount of neutrinos. This suggests that such a remnant formed after the collapse of a hypermassive neutron star could be a central engine of gamma-ray bursts.

In the works of Duez et al. (2006a); Shibata et al. (2006); Duez et al. (2006b), they simply considered the evolution of isolated hypermassive neutron stars. However, after the merger of binary neutron stars, in reality, a dense atmosphere is formed surrounding the hypermassive neutron stars due to the mass ejection that occurs during the early stage of the merger (see, e.g., section 8.3.5). For the formation of a magnetosphere composed of coherently collimated magnetic fields in the system of a black hole and accretion torus finally formed, the atmosphere has to be blown away (the mass-energy density of the atmosphere matter should be smaller than the magnetic energy). However, the process for this is non-trivial. For clarifying the formation process of the magnetosphere, we need a self-consistent study from the merger of binary neutron stars through the formation of a hypermassive neutron star and subsequent collapse to a black hole surrounded by an accretion torus and dense atmosphere. This is an issue to be solved in the future (see section 10.3.2.3 for a preliminary work on this).

10.3.2.2 *Evolution of neutron stars with purely toroidal magnetic fields*

As described in section 10.3.1.3, linear perturbation analyses have shown that there are several magnetohydrodynamical instability for neutron stars with purely toroidal (and probably toroidal-dominant) magnetic fields. However, the perturbation analyses cannot tell us the evolution process of the unstable stars in the nonlinear growth stage of the instability that follows from the linear growth stage. To clarify this, numerical simulations are obviously necessary.

The nonlinear growth of the axisymmetric Tayler instability in neutron stars with purely toroidal magnetic fields was first explored by Kiuchi et al. (2008) in the framework of numerical relativity. As the initial condition, they prepared both non-rotating and rotating equilibrium states by solving magnetohydrostatic equations employing the polytropic equations of state with $\Gamma = 2$. They focused on the isentropic and rigidly rotating (or non-rotating) neutron stars. Magnetic-field configurations were given by $b_\varphi = B_0 u^t \left(\rho h a^2 \gamma_{\varphi\varphi}\right)^k \gamma_{\varphi\varphi}^{-1/2}$ where B_0 is a constant that determines the magnetic-field strength and k is a positive integer, $1, 2, 3, \cdots$. The linear perturbation analysis described in section 10.3.1.3 indicates that for $k = 1$, the neutron stars are stable against the axisymmetric Tayler instability irrespective of the degree of the rotation. On the other hand, for $k \geq 2$, the non-rotating neutron stars will be always unstable against the Tayler instability, while for the rotating case, the stability property will depend on the angular velocity and the profile of the magnetic field.

In their numerical-relativity simulations, Kiuchi et al. (2008) confirmed all the stability properties described in section 10.3.1.3. They then showed that for the

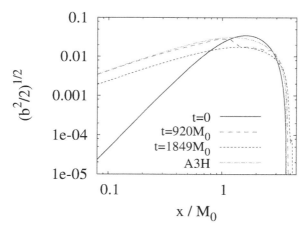

Fig. 10.6 Profiles of square root of magnetic-field energy along ϖ-axis at selected time slices, $t/M_0 = 0$, 920, and 1849 with M_0 the initial gravitational mass for an unstable non-rotating neutron star. For this model, the averaged Alfvén time scale is $\approx 73 M_0$. The profile labeled by A3H is a configuration of a stable neutron star. The figure is taken from Kiuchi et al. (2008).

unstable non-rotating neutron stars, the magnetic field is redistributed approximately in the time scale of order $10 t_A$ (see figure 10.6). The resulting profile of the magnetic fields is similar to that for (axisymmetrically) stable neutron stars with $k = 1$ (see the curve labeled by A3H in figure 10.6), indicating that neutron stars with the magnetic-field profile of $k = 1$ are attractors if only the axisymmetric perturbation is taken into account. In addition, they showed that during the growth of the dynamically unstable modes in the non-rotating neutron stars, electromagnetic energy is converted to kinetic energy via the Tayler instability, and as a result, convective motion is excited (see figure 10.7). The magnitude of the convective kinetic energy becomes approximately as large as the electromagnetic energy at the time when the growth of the instability is saturated.

Kiuchi et al. (2008) also confirmed that in the presence of a sufficiently rapid rotation, the axisymmetric Tayler instability is suppressed as predicted by equation (10.15): Even with a magnetic-field profile that leads to the Tayler instability for the non-rotating case (i.e., $k \geq 2$), the neutron star may be stable for the rapidly rotating case. Specifically, they indicated that if the rotational kinetic energy is several times larger than the electromagnetic energy, the stabilization occurs. They also found that even for the unstable rotating models, convective motion is not induced as remarkably as for the non-rotating case. For the rapidly rotating case in which the rotational kinetic energy is larger than the electromagnetic energy by a factor of $\gtrsim 2$, the maximum convective energy is smaller than the electromagnetic energy by two or three orders of magnitude. This indicates that rotation in general plays a role in stabilizing the axisymmetric Tayler instability.

Subsequently, Kiuchi et al. (2011b) explored the non-axisymmetric Tayler instability of neutron stars with purely toroidal magnetic fields in numerical relativity.

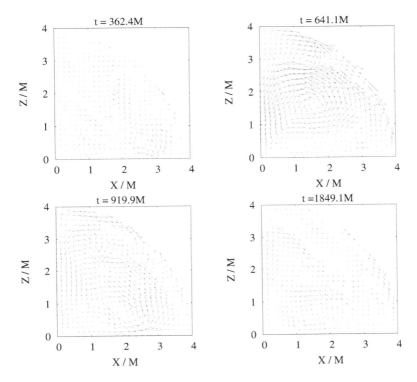

Fig. 10.7 Snapshots of velocity fields for an unstable neutron star (which is the same as that in figure 10.6) at selected time slices, $t/M_0 = 362.4$, 641.1, 919.9, and 1849.1. For each time, the maximum velocity [the maximum value of $\sqrt{(v^\varpi)^2 + (v^z)^2}$] is 0.039, 0.047, 0.041, and 0.037, respectively. The figure is taken from Kiuchi et al. (2008).

Again, they prepared both non-rotating and rigidly rotating equilibrium states employing the polytropic equations of state with $\Gamma = 2$ as the initial condition, and focused on the isentropic neutron star. Magnetic-field configurations were given by $b_\varphi = B_0 u^t \rho h \alpha^2 \gamma_{\varphi\varphi}^{1/2}$, for which the neutron stars are stable against the axisymmetric Tayler instability as already mentioned. However, they are unstable against the non-axisymmetric Tayler instability, as predicted in the linear perturbation analysis described in 10.3.1.3.

Figures 10.8 and 10.9 plot the evolution of the rest-mass density and electromagnetic energy density on the equatorial plane and in one of meridian (x-z) planes, respectively, for a non-rotating neutron star. The panels (a)–(c) in these figures show that the magnetic field near the stellar surface is disturbed by the Parker instability, as predicted by the linear perturbation analysis, and then, the magnetic flux leaks out of the stellar surface. The growth time scale is again several Alfvén time scale (for this model, the averaged Alfvén time scale is $\approx 37M_0$ where M_0 is the initial gravitational mass). Because the plasma beta defined by $P/(B^2/8\pi)$ is small and thus the matter inertia is small near the stellar surface, the matter is

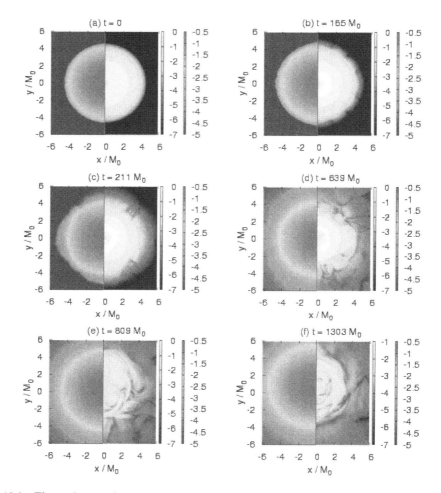

Fig. 10.8 The evolution of the rest-mass density (left in each panel) and electromagnetic energy density (right in each panel) on the equatorial plane for a non-rotating neutron star with purely toroidal magnetic fields at $t = 0$. Both of them are plotted in the logarithmic scale (zero denotes the maximum). The coordinate time at each slice is shown in units of M_0, which is the initial gravitational mass. The figure is taken from Kiuchi et al. (2011b).

dragged by the magnetic force and consequently the stellar surface is significantly distorted. The minimum value of the plasma beta is initially ≈ 2 at the stellar surface, and after the onset of the Parker instability, the leak-out magnetic-field loop produces even lower beta plasma. As a result, a weak wind expanding outward is driven. On the other hand, the ingoing magnetic-field loop enhances turbulent motion inside the neutron star as in the axisymmetric case. During the transition between the states shown in panels (c) to (d) in figures 10.8 and 10.9, the initial magnetic-field profile is completely destroyed and a random magnetic-field profile is

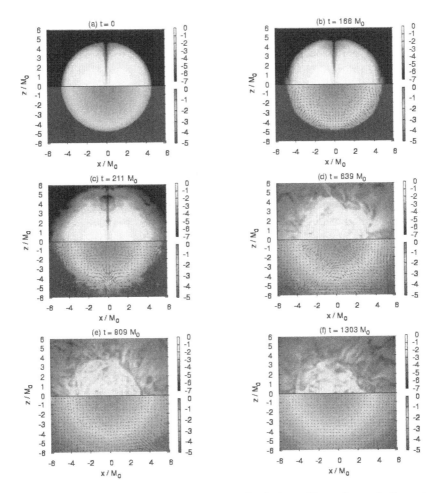

Fig. 10.9 The same as figure 10.8 but on a meridian (x-z) plane. For each panel, the rest-mass density (lower) and electromagnetic energy (upper) are shown. The arrow indicates the velocity field. The figure is taken from Kiuchi et al. (2011b).

produced. During the development of the turbulence, the non-axisymmetric Tayler instability does not appear to play a primary role. However, the region near the axis of $\varpi = 0$ is not stable against this instability and thus no mechanism seems to help stabilizing there.

The toroidal magnetic fields initially prepared behave like a rubber belt, which fastens the "waist" of neutron stars by a magnetic tension. The disappearance of the coherent toroidal magnetic fields by the growth of the Parker instability, therefore, results in the expansion of the star as shown in the panel (d) of figures 10.8 and 10.9. After the magnetic field becomes turbulent, the star stably oscillates around the hypothetical quasi-stationary state. Although the density profile relaxes to a quasi-stationary state, the turbulent motion is maintained. This indicates that a stationary state is not reached at least in several 10 Alfvén time scales.

The linear perturbation analysis in section 10.3.1.3 indicates that the $m = 1$ mode would be most relevant for the onset of the Parker instability. However, the region near the stellar surface may be also unstable against higher m modes because c_s approaches zero there. Indeed, figures 10.8 and 10.9 illustrate that this is the case (see the second panel of these figures).

During the growth of the instability and consequent development of the turbulent motion, convective motion is enhanced and shock heating continuously occurs. As a result, not only kinetic energy but also thermal energy increase while electromagnetic energy decreases. The results of Kiuchi et al. (2011b) indicate that the thermal energy becomes as high as the electromagnetic energy as the development of the turbulent motion. This suggests that this type of non-axisymmetric instability could be an efficient mechanism for converting the electromagnetic energy into the thermal energy, as the MRI.

For their study, Kiuchi et al. (2011b) considered isentropic and chemically homogeneous neutron stars. In reality, composite gradients such as gradients of the number density of neutrons, protons, and electrons could exist in the real neutron stars. Thus, its presence as well as the entropy gradients could stabilize the Parker instability near the surface in the real neutron stars. It should be also mentioned that in the absence of these stabilizing ingredients, the unstable magnetized neutron stars may never reach a stable state. Indeed, in the simulations of Kiuchi et al. (2011b), such a new state was not determined. To determine the final state, it might be necessary to take into account more realistic physics. Thus, it is appropriate to caution here that for exploring the stability of realistic magnetized neutron stars, simulations should be performed in more physical setting taking into account composite gradients and entropy gradients. These have not been yet performed in this field, and hence, remain an issue for the future.

10.3.2.3 *Merger of binary magnetized neutron stars*

The merger of binary neutron stars and the remnant subsequently formed are the sites in which a wide variety of magnetohydrodynamical instability plays an essential role. Clarifying the role of these instability has been one of the most challenging subjects in numerical relativity, because a high-resolution study is necessary in this subject.

There are three types of the instability that could come into play in the merger of binary neutron stars. One is the Kelvin–Helmholtz instability [Rasio and Shapiro (1999); Price and Rosswog (2006); Kiuchi et al. (2014)] [see Chandrasekhar (1961) for the elementary analysis of the Kelvin–Helmholtz instability]. This instability can set in whenever a shear layer in the velocity field is present, i.e., in the presence of a velocity discontinuity. During the early stage of the merger, two neutron stars with opposite velocity vectors collide near the mass center, and at the contact surface, the velocity discontinuity can be formed. Then, the Kelvin–Helmholtz instability could be significantly activated.

In the absence of gravitational effects, the growth rate of this instability is proportional to the wave number of the relevant mode, k, and the relative difference of the velocity between two discontinuous regions, Δv, as [Chandrasekhar (1961)]

$$\omega_i = k \Delta v, \qquad (10.24)$$

where ω_i is the imaginary part of the angular velocity in a linear-perturbation mode (i.e., the growth rate). We assumed that the density is continuous for simplicity (in the presence of the density discontinuity, the equation is quantitatively modified). In the presence of gravity, the equation for the growth rate is qualitatively modified. However, for a sufficiently large value of k (i.e., for a sufficiently short wavelength), equation (10.24) is approximately valid. Unique properties of the Kelvin–Helmholtz instability are that it is activated for any short-wavelength mode and the growth rate is larger for shorter wavelengths. This implies that for the numerical study of this instability, sufficiently high-resolution simulations together with varying the grid resolution for the convergence check is essential. Without such a series of simulations, we cannot draw any conclusion on this instability and on its effects for the merger process.

After the onset of the Kelvin–Helmholtz instability, vortex motion is generated in the shear layer. Associated with this vortex motion, magnetic fields are wound and the magnetic-field strength is amplified in an exponential manner with the time scale ω_i^{-1}. This is quite short (much shorter than the dynamical time scale, $\rho^{-1/2} \sim 0.1\,\mathrm{ms}$) because Δv is $\sim 0.5c$ and the wavelength, $\lambda = 2\pi/k$, is smaller than 1 km:

$$\omega_i^{-1} \sim 10^{-3}\,\mathrm{ms} \left(\frac{\lambda}{1\,\mathrm{km}}\right) \left(\frac{\Delta v}{0.5c}\right)^{-1}. \qquad (10.25)$$

As already mentioned, the amplification is in general enhanced more significantly for smaller length scales.

However, during the merger of binary neutron stars, the Kelvin–Helmholtz instability is likely to come into play only for a short time scale, as short as the dynamical one. The reason for this is that after the onset of the merger, shocks are continuously formed at the shear layer due to the oscillation of the merger remnant. The shocked layer oscillates violently due to the combination of the shock-induced expansion and the contraction by strong self-gravity. As a result, the vortexes could be dissipated by the shocks. Nevertheless, the growth time scale for the Kelvin–Helmholtz instability is significantly short for the short-wavelength modes. Thus, in quite short time scale, the magnetic-field energy could be comparable to thermal and rotational energies in the shear layer [Price and Rosswog (2006); Kiuchi et al. (2014)].

After the onset of the merger, a massive neutron star is likely to be formed as far as the total mass of the system is smaller than $\sim 2.8 M_\odot$ (see section 8.3.3).

As found by many numerical simulations, the formed massive neutron stars will be differentially rotating. Thus, the magnetic winding (see section 10.3.1.1) and MRI (see section 10.3.1.2) can play an important role for amplifying the magnetic-field strength and for the subsequent angular-momentum transport. If the magnetic-field strength is not extremely high (smaller than 10^{15} G) at the formation of the massive neutron star, the MRI should play a dominant role because its growth time scale is shorter than the amplification time scale by the magnetic winding [see equation (10.13)]. The MRI also could trigger turbulent motion in the massive neutron star [Kiuchi et al. (2014)].

After a substantial amount of angular momentum is transported from the central to the outer region, the massive neutron star would be destabilized against gravitational collapse, if it is hypermassive, and then, a black hole surrounded by a torus will be formed. The formed torus is in general differentially rotating. Thus, in this torus, the magnetic winding and MRI again can play an important role for its evolution governing the angular-momentum transport process.

The primary purpose of the simulation for magnetized binary neutron stars is to clarify the role of these magnetic processes in the merger and subsequent evolution of the merger remnants. As already mentioned above, the primarily unstable modes of the Kelvin–Helmholtz instability and MRI have short wavelengths. This implies that a high-resolution simulation is inevitable for this issue. In particular, for MRI, the wavelength of the fastest-growing mode has a particular value for given magnetic-field strength and density. For, e.g., $B \sim 10^{14}$ G, the wavelength is ~ 0.1 km for the typical density of neutron stars. We have to resolve at least this wavelength in the simulation to draw a physical conclusion on the effect of the MRI.

Since the required grid resolution is quite high, the simulation for magnetized binary neutron stars has been a quite challenging issue. To date, there is no completely physical simulation (i.e., a simulation in which the MRI is completely resolved) in this issue. However, with the development of supercomputers, our understanding for the magnetic processes becomes richer. In particular, Kiuchi et al. (2014) performed the highest-resolution simulation among those performed so far, by which we can confirm the important roles of the magnetohydrodynamics instability summarized above. We here outline the magnetohydrodynamics processes in the merger of binary neutron stars and subsequent evolution of the merger remnants following the work by Kiuchi et al. (2014).

In their simulations, an adaptive mesh-refinement algorithm was employed covering a wide region up to ≈ 2300 km by 7 refinement levels. In the finest domain, the grid spacing is $\Delta x_7 = 70$ m, which is $\sim 1/150$ of the neutron-star radius. For each refinement domain, $1025 \times 1025 \times 513$ grid points that covers the region $[-512\Delta x_l : 512\Delta x_l]$ for x and y, and $[0 : 512\Delta x_l]$ for z are assigned for x-, y-, and z-directions (the equatorial plane symmetry was imposed). Here, Δx_l is the grid spacing for l-th refinement level with $l = 1, 2, \cdots, 7$. The finest domain covers the region of $[-36 \text{ km} : 36 \text{ km}]$ for x- and y-directions and of $[0 : 36 \text{ km}]$ for z-

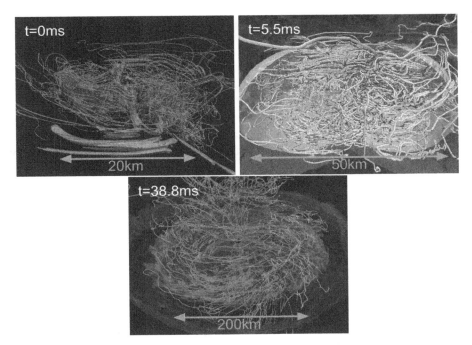

Fig. 10.10 Snapshot of the rest-mass density and magnetic-field configurations for a magnetized neutron star merger at the onset of the merger (top-left panel), at $t - t_{\rm mrg} = 5.5$ ms (top-right panel), and at $t - t_{\rm mrg} = 38.8$ ms (bottom panel). Here, $t_{\rm mrg}$ denotes the merger time. The top right and bottom panels show the configuration for a hypermassive neutron star phase and for a black hole surrounded by an accretion torus, respectively. For each panel, the white curves denote the magnetic-field lines. In the top-left panel, the cyan denotes the region for which the magnetic-field strength is greater than $10^{15.6}$ G. In the top-right panel, the yellow, green, and blue show the density iso-surface of 10^{14}, 10^{12}, and 10^{10} g/cm^3, respectively. In the bottom panel, the light blue and blue show the density iso-surfaces of $10^{10.5}$ and 10^{10} g/cm^3, respectively. The figure is taken from Kiuchi et al. (2014).

direction. Therefore, each neutron star in the binary motion, hypermassive neutron stars formed after the merger, and inner region of an accretion torus surrounding a black hole formed after the collapse of the hypermassive neutron stars are all covered by the finest refinement level. The H4 equation of state (see table 1.1) was employed in this example. Each mass of the binary is $1.4 M_\odot$.

As the initial condition for the magnetic field, a poloidal field confined inside the neutron stars are superimposed with the maximum magnetic-field strength 10^{15} G. Although this field strength is rather high, the resulting magnetic pressure does not affect the inspiral orbit because the magnetic stress-energy is much weaker than the matter pressure. It should be noted that such an artificially high magnetic field is necessary to explore the growth of the MRI in the merger remnant because the wavelength of its fastest-growing mode is proportional to the magnetic-field strength [see equation (10.14)]. For the practical purpose, the maximum magnetic-field strength in the numerical simulations is often determined by the finest grid resolution that can be employed.

Figure 10.10 shows the profiles of the rest-mass density and magnetic-field lines at selected time slices during the merger. This illustrates that after the onset of the merger, the instability associated with the presence of magnetic fields plays a substantial role. The top-left panel of figure 10.10 shows a snapshot just after two neutron stars come into contact. In this phase, the vortexes developed by the Kelvin–Helmholtz instability curl the magnetic-field lines very quickly, generating the strong toroidal fields. This significantly enhances the magnetic stress-energy in the shear layer (see also figure 10.11). The unstable shear layer then disappears in a dynamical time scale of ~ 1 ms, because the compression associated with a violent oscillation of the formed hypermassive neutron star suppresses the continuous generation of the vortices. Nevertheless, the magnetic-field energy is increased by two orders of magnitude in a short time scale for this example.

It is important to note that the growth rate of the Kelvin–Helmholtz instability is proportional to the relevant wavelength. This implies that for higher-resolution runs, we could follow the faster-growing mode, and hence, the amplification factor of the magnetic-field energy is increased with the improvement of the grid resolution. Because of the limited grid resolution, the magnetic-field energy is increased *only* by two orders of magnitude in this example. However, in reality, the growth rate could be much larger. By contrast, if the resolution was poor, we could not see the growth of the Kelvin–Helmholtz instability. Kiuchi *et al.* (2014) also performed less-resolved simulations with $\Delta x_7 = 110$ and 150 m. Indeed, for their lowest-resolution run, the amplification factor of the magnetic-field energy is less than 10, and thus, they could not clearly identify the onset of the Kelvin–Helmholtz instability (see figure 10.11). This illustrates the fact that the convergence study employing a wide range of the grid resolution is crucial for this issue.

The top-right panel of figure 10.10 plots a snapshot in the hypermassive-neutron-star phase. This shows that a large-scale toroidal magnetic field, enhanced primarily by magnetic winding, is generated. A detailed analysis also elucidates that the magnetic field is also amplified by the MRI (see below) because the resolution of their simulation is high enough to resolve the MRI.

The hypermassive neutron star eventually collapses to a black hole in this model. Then, a part of the matter in the hypermassive neutron star forms an accretion torus surrounding the black hole. For this model, the dimensionless black-hole spin is $\hat{a} \approx 0.69$ and the torus mass is $\approx 0.06 M_\odot$ at 10 ms after the black-hole formation.

The magnetohydrodynamical instability such as MRI generate the turbulent flow and vortices inside the accretion torus, and they enhance the mass accretion to the black hole by the outward angular-momentum transport. Due to this effect, the density of the accretion torus gradually decreases from $\sim 10^{12}$ g/cm^3 to lower values in their evolution process. The magnetic field remains to be toroidal-field dominant, and any coherent poloidal field is not formed in a short time scale (see the right panel of figure 10.10). The reason for this is as follows: A large amount of the matter is ejected and blown outwards in the early merger phase, and subsequently, forms a dense and hot atmosphere. The resulting mass-energy density is typically $\sim 10^{29} \left(\rho/10^8\,{\rm g\,cm}^{-3}\right)$ erg/cm^3. Since the magnetic-field lines are frozen in the fluid elements, an outflow component, which should blow away the atmosphere, is necessary to generate a coherent poloidal magnetic field. To establish such coherent magnetic field overcoming the high matter inertia, the magnetic-field strength should be higher than 10^{15} G, but it is not achieved in the end of this simulation. This suggests that for forming a coherent magnetic field, we have to wait for $\gg 10$ ms until the density of the atmosphere is sufficiently reduced perhaps due to cooling and subsequent matter infall onto the central black hole or torus.

Figure 10.11 plots the magnetic-field energy as a function of time for runs with three grid resolutions, $\Delta x_7 = 70, 110$, and 150 m. For $\Delta x_7 = 70$ m, the results with different magnetic-field strength initially given are also shown. The time coordinate is shifted to align the time at the onset of the merger to be zero. Soon after the onset of the merger, both the toroidal and poloidal components are steeply amplified because the Kelvin–Helmholtz vortexes developed in the shear layer amplify the magnetic fields. The growth rate of the magnetic-field energy is higher for the higher-resolution runs as already mentioned. In fact, they showed that the amplification factor of the field strength in the merger depends strongly on the grid resolution. This is consistent with the amplification mechanism due to the Kelvin–Helmholtz instability. The magnetic-field energy at $t - t_{\rm mrg} \approx 5$ ms in the high-resolution run is 40–50 times as large as that of the the low-resolution run. Here $t_{\rm mrg}$ denotes the merger time.

Figure 10.11 also shows that the amplification factor of the magnetic-field energy at the onset of the merger does not depend on the magnetic-field strength initially given (compare three results for $\Delta x_7 = 70$ m). This result is also consistent with the fact that the amplification is driven by the Kelvin–Helmholtz instability.

In the hypermassive-neutron-star phase, the magnetic-field strength increases significantly in the high- and middle-resolution runs but not in the low-resolution run. This growth is primarily due to the non-axisymmetric mode of the MRI associated with the strong and coherent toroidal magnetic field inside the hypermassive neutron star. Kiuchi et al. (2014) checked that their high- and middle-resolution runs satisfy the criterion $\lambda_{\rm MRI}^{\varphi}/\Delta x_7 \geq 10$ where $\lambda_{\rm MRI}^{\varphi}$ is the MRI wavelength of the fastest-growing mode for the *toroidal* magnetic field. On the other hand, the low-resolution run does not satisfy this criterion, and hence, we do not clearly see the exponential growth for this run.

The Kelvin–Helmholtz instability, which occurs soon after the onset of the merger, and the MRI, which occurs in the hypermassive-neutron-star phase, significantly amplify the magnetic-field strength. At the black-hole formation, the magnetic-field strength was already saturated in the high- and middle-resolution runs as found in figure 10.11, and thus, it does not much increase in the accretion

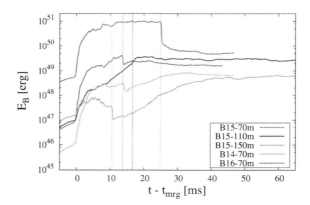

Fig. 10.11 The magnetic-field energy as a function of time with three grid resolutions $\Delta x_7 = 70$, 110, and 150 m. For $\Delta x_7 = 70$ m, the results with different magnetic-field strength initially given are also plotted. The thin vertical lines show the time at which a black hole is formed. Note that the magnetic-field energy is calculated only outside the apparent horizon. This figure is taken from Kiuchi et al. (2014).

torus formed after the hypermassive neutron star collapses to a black hole. Namely, the accretion torus was already in an equipartition situation at its formation, and hence, it was in a highly turbulent state. On the other hand, the magnetic field was still amplified in the accretion torus in the low-resolution run because the amplification in the previous phases was not sufficient. This is attributed to the insufficient grid resolution to capture the magnetohydrodynamical instability during the merger and subsequent hypermassive neutron star phases. Their work tells us the caution that in the magnetohydrodynamics simulation with an *insufficient* grid resolution, we would draw an incorrect conclusion: An in-depth resolution study is crucial for deriving essential physics from the magnetohydrodynamics simulation.

Chapter 11

Higher-dimensional simulations

This section is devoted to introducing several representative simulations in higher-dimensional numerical relativity performed in the early stage of this field. Higher-dimensional numerical relativity has been an active field in numerical relativity since a possibility of production of mini black holes in large particle accelerators such as CERN Large Hardon Collider (LHC) was pointed out (see Preface). We here pick up three representative research topics; studies of the Gregory–Laflamme instability for a black string [Lehner and Pretorius (2010)], two black-hole collisions [Zilhao et al. (2010); Witek et al. (2010, 2011); Okawa et al. (2011)], and the instability for rapidly spinning Myers–Perry (MP) black holes with one spin parameter [Shibata and Yoshino (2010b,a)].[1]

11.1 Gregory–Laflamme instability of black string

The five-dimensional black string is one of the simplest solutions of the black objects in higher-dimensional general relativity. The metric of this solution is written as

$$ds^2 = -f(r)dt^2 + \frac{dr^2}{f(r)} + r^2 d\Omega_2 + dz^2, \qquad f(r) = 1 - \frac{2G_4 M}{r}, \qquad (11.1)$$

where G_4 is the four-dimensional gravitational constant that is related to the five-dimensional one as $G_5 = G_4 L$ if the z direction is compactified with the scale L. Here, we note that this spacetime is uniform in the z-direction and has a cylindrical horizon located for $r = 2G_4 M$ encompassing the z-axis. Thus, if we focus on the z =constant cross section, this spacetime looks the same as a Schwarzschild black hole. The black string solution can exist only for spacetime with the dimensionality $D \geq 5$.

Any black string is shown to be unstable against a long-wavelength deformation along the z direction [Gregory and Laflamme (1993)]. This instability is called the Gregory–Laflamme instability. There is a critical wavelength L_c for this instability; if the wavelength of a perturbation is longer than L_c, the instability sets in. The values of L_c are evaluated numerically by Gregory and Laflamme (1993), and L_c

---
[1] This chapter was written with H. Yoshino.

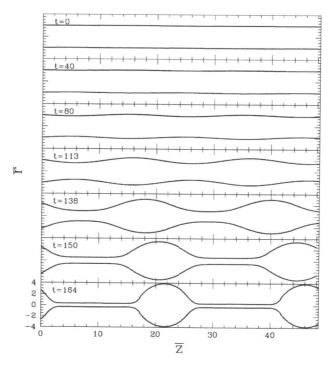

Fig. 11.1 Embedding diagrams of apparent horizons, with the two angular dimensions θ and ϕ suppressed, in the evolution of a perturbed black string with $L = 1.4L_c$. These plots describe the intrinsic geometry of the apparent horizons at selected time slices, in a coordinate system with metric $ds^2 = d\bar{r}^2 + d\bar{z}^2$ where $\bar{z}$ is a periodic coordinate, and $\bar{r}$ is an areal radius of $\bar{z}$ =constant sections of the horizon. The figure is taken from Choptuik et al. (2003c).

was found to be of order $4\pi G_4 M$. This implies that the black string is unstable if the compactification scale is longer than L_c.

The end state of this instability has been inferred by many speculative works, but it is impossible to obtain the definitive answer without numerical-relativity simulations. Choptuik et al. (2003c) first performed a numerical-relativity simulation for the evolution of a black string. This was the first multidimensional and higher-dimensional simulation in numerical relativity. Their simulation was performed assuming a spherical symmetry for the three-dimensional subspace, composed of (r, θ, φ), and adding a perturbation along the z direction. An excision algorithm was employed for removing a deep interior region inside the black-string horizon. They found that the black string that satisfies the unstable condition is indeed dynamically unstable (they chose $L = 1.4L_c$). They showed an embedding diagram of apparent horizons (see figure 11.1), and the resulting configuration looks like a system composed of large spherical black holes connected by thin black strings. The radius of the thin strings becomes narrower and narrower with the evolution.

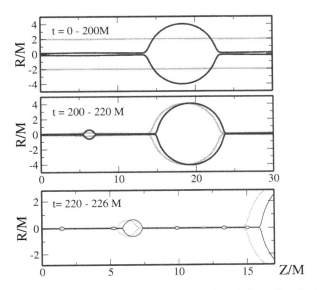

Fig. 11.2 Embedding diagrams of apparent horizons at selected time slices in the evolution of a perturbed black string. R is the areal radius, and the embedding coordinate Z is defined so that the proper length of the horizon in the z direction (for fixed coordinates t, θ, ϕ) is exactly equal to the Euclidean length of $R(Z)$. The light and dark curves, respectively, denote the first- and last-time locations of the apparent horizons for the time segment depicted in the corresponding panel. The figure is taken from Lehner and Pretorius (2010).

This indicates that a grid resolution has to be improved with time for a long-term simulation in this problem. Unfortunately, they were not able to evolve the system for a sufficiently long time scale until the end state was reached, because of the lack of the computational resources and appropriate numerical methods, at that time.

Subsequently a more sophisticated simulation was performed by Lehner and Pretorius (2010). In this simulation, they employed a modern numerical scheme and formulations including the fourth-order accurate finite differencing, generalized-harmonic formulation (section 2.3.4), a modified version of the cartoon method (section 2.5), an AMR scheme (section 3.3), and a black-hole excision algorithm (section 2.2.4). The initial data were prepared in the same method as in Choptuik et al. (2003c). The improved scheme and formulations enabled them to perform the simulation for a time scale much longer than that in Choptuik et al. (2003c).

Figure 11.2 shows snapshots of an embedding diagram of apparent horizons reported in Lehner and Pretorius (2010). For $t \lesssim 200 G_4 M$, the results are essentially the same as those in Choptuik et al. (2003c). However, for $t \geq 220 G_4 M$, it is observed that a part of the thin string segments continuously generates small black-hole-like objects while other parts continue to shrink to be thinner strings. Lehner and Pretorius called the black-hole-like objects "satellites". The satellite black holes are formed in a self-similar cascade manner. Their result also suggests

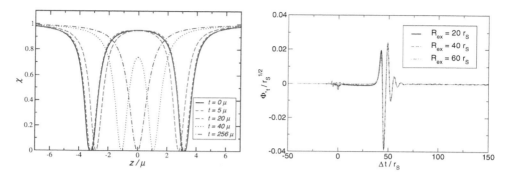

Fig. 11.3 Left panel: Snapshots of the conformal factor χ for $t/r_S(M) = 0, 5, 20, 40$ and 256 in a head-on collision of two equal-mass black holes initially at rest. Right panel: The gravitational waveform for the $l = 2$ scalar mode extracted in a gauge-invariant wave-extraction method. The results with three extract positions $R_{\rm ex}/r_S = 20, 40$, and 60 are shown. The figures are taken from Zilhao et al. (2010); Witek et al. (2010).

that the radius of the black-string horizon may become zero eventually.

An important and interesting question in this problem is whether the black-string singularity is visible, i.e., naked from observers located outside, when the horizon radius becomes zero. The result of Lehner and Pretorius (2010) suggested that the singularity is naked, because the time scale for the formation of the next-generation satellite/string-segments is proportional to the local string radius, and becomes shorter and shorter as the generation is increased; the horizon radius should reach zero in an finite time scale as a result of the self-similar cascade. Thus, the result of Lehner and Pretorius (2010) strongly supports that the cosmic censorship may be violated in five-dimensional general relativity. (This is likely to be the case also for $D > 5$.)

11.2 Black hole collisions and scattering

Now, we turn our attention to simulations of black-hole collisions in higher dimensions. In the following, we will first review simulations for head-on collisions of five-dimensional black holes and then those of off-axis collisions.

11.2.1 *Black hole head-on collision in $D = 5$*

The simulations for head-on collisions of two black holes in five dimensions were first reported in Zilhao et al. (2010); Witek et al. (2010, 2011). Their simulations were performed employing the so-called Brill–Lindquist initial data for two black holes as the initial condition (see section 5.5.1.2).

The left panel of figure 11.3 shows the snapshots of the conformal factor χ along the z axis for $t/r_S(M) = 0, 5, 20, 40$, and 256 reported in Zilhao et al. (2010). Here,

$r_S(M)$ denotes the horizon radius for a given mass parameter with M being the total mass of the system. As in the four-dimensional simulations, the positions of the punctures, at which $\chi = 0$, approach each other and eventually merge. The right panel of figure 11.3 shows the gravitational waveforms for the $l = 2$ scalar mode extracted by a gauge-invariant wave-extraction method developed by Witek et al. (2010). This shows it feasible to accurately compute the gravitational waveform from the initial burst throughout the ringdown phase. They estimated that the total radiated energy is $E_{\rm rad}/M \approx 9 \times 10^{-4}$. Simulations for the head-on collision of two unequal-mass black holes were also performed by Witek et al. (2011), and similar results to those for the equal-mass case were reported. All these results are very similar to those in the four-dimensional case. However, this is not the case for the off-axis collision, as illustrated in the next subsection.

11.2.2 Off-axis collision of two black holes

High-velocity, off-axis collisions of two black holes for $D = 5$ were first studied by Okawa et al. (2011). Their results suggested the possible formation of a naked singularity, and hence, gave a possible evidence for the violation of cosmic censorship hypothesis in higher-dimensional spacetime. This result is completely different from that in four-dimensional gravity. Their results together with the results by Lehner and Pretorius (2010) suggest that the nature of nonlinear phenomena in higher-dimensional gravity is likely to be highly different from that in four-dimensional gravity. In the following, we will outline the work by Okawa et al. (2011).

For preparing initial data composed of two boosted black holes, the puncture formulation developed by Brandt and Brügmann (1997) (see section 5.5.2.2) is popular not only for three spatial dimensions but also for more than four spatial dimensions [Yoshino et al. (2006)]. However, the initial data generated in this formulation is known to contain an unphysical radiation component, in particular for high-velocity cases. An alternative method, suitable for the high-velocity collision of black holes, was first proposed by Shibata et al. (2008) for the four-dimensional case. In this method, the initial data of one black hole in motion are prepared by boosting a Schwarzschild black hole of mass m_0, and initial data composed of two boosted black holes are prepared by superposing two boosted black-hole solutions. Because a nonlinear interaction between two black holes is present, just superposing two solutions leads to the violation of the Hamiltonian and momentum constraints. However, if the initial distance between two black holes is sufficiently large, the magnitude of the constraint violation is so small that we may safely ignore the interaction terms. Okawa et al. (2011) employed this idea in their exploration for the high-velocity collision of two black holes in five-dimensional spacetime.

Specifically, the method of Shibata et al. (2008); Okawa et al. (2011) is described as follows: First, a Schwarzschild–Tangherlini black hole solution (see section 1.3.3)

in isotropic coordinates is prepared,
$$ds^2 = -\alpha^2(\bar{r})d\bar{t}^2 + \psi^2(\bar{r})(d\bar{w}^2 + d\bar{x}^2 + d\bar{y}^2 + d\bar{z}^2), \tag{11.2}$$
where [see equation (1.156)]
$$\psi(\bar{r}) = 1 + \frac{\mu}{4\bar{r}^2} \quad \text{and} \quad \alpha(\bar{r}) = \frac{\bar{r}^2 - \mu/4}{\bar{r}^2 + \mu/4}, \tag{11.3}$$
and μ is a mass parameter. For this seed metric, the Lorentz transformation
$$\bar{t} = \gamma(\bar{t} \mp v\bar{w}), \quad w = \gamma(\mp v\bar{t} + \bar{w}), \quad x = \bar{x}, \quad y = \bar{y} \tag{11.4}$$
is first performed, and then, the spatial translations $w \to w \mp d/2$ and $x \to x \mp b/2$ are done for each black hole. Here, d is the coordinate separation between two black-hole centers along the w direction and b is the impact parameter of two black-hole collision. Then, we have two forms of the line elements as
$$ds_\pm^2 = -\gamma^2(\alpha_\pm^2 - v^2\psi_\pm^2)dt^2 \pm 2\gamma^2 v(\alpha_\pm^2 - \psi_\pm^2)dtdw + \psi_\pm^2(B_\pm^2 dw^2 + dx^2 + dy^2 + dz^2), \tag{11.5}$$
where $\alpha_\pm = \alpha(\bar{r}_\pm)$, $\psi_\pm = \psi(\bar{r}_\pm)$, and $B_\pm^2 = \gamma^2(1 - v^2\alpha_\pm^2/\psi_\pm^2)$ with
$$\bar{r}_\pm = \sqrt{\gamma^2(w \mp d/2 \pm vt)^2 + (x \mp b/2)^2 + y^2 + z^2}. \tag{11.6}$$
Equation (11.5) describes two black holes (denoted by $\pm$) located at $(w, x, y, z) = (\pm d/2, \pm b/2, 0, 0)$ with the boost velocity $v^i = (\mp v, 0, 0)$.

From these metric forms, the extrinsic curvature $K_{ij}^\pm$ can be determined for each black hole using equation (2.21). Then, the extrinsic curvature for initial data of two moving black holes is set to be
$$K_{ij} = K_{ij}^+ + K_{ij}^- + \delta K_{ij}, \tag{11.7}$$
where δK_{ij} is a correction due to the mutual nonlinear interaction between two black holes. Because it is sufficiently small for a large value of $d \gg \sqrt{\mu}/2$, it may be approximated as $\delta K_{ij} = 0$ as far as the truncation error is larger than it.

The initial spatial metric γ_{ij} is written in the form
$$\gamma_{ij}dx^i dx^j = (\Psi + \delta\Psi)^2(B^2 dw^2 + dx^2 + dy^2 + dz^2), \tag{11.8}$$
where
$$\Psi = \psi_+(\bar{r}_+) + \psi_-(\bar{r}_-) - 1 \quad \text{and} \quad B^2 = \gamma^2\left[1 - \frac{v^2}{\Psi^4}(2 - \Psi)^2\right]. \tag{11.9}$$
Here, $\delta\Psi$ is a correction due to the mutual nonlinear interaction and it may be also set to be zero for a choice $d \gg \sqrt{\mu}/2$.

Adopting the initial data described above, numerical simulations were performed using a SACRA-ND code with a dynamical gauge (moving puncture gauge) condition (suitable for high-dimensional numerical relativity) [Shibata and Yoshino (2010a)]. In the higher-dimensional case, the final fate of the off-axis collision is simply classified into two cases: two black holes merge at the instance of the first contact or after the first contact, they are scattered away to infinity. Figure 11.4 shows a summary

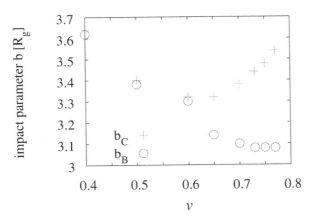

Fig. 11.4 The values of threshold impact parameters, b_B and b_C, as functions of the initial velocity v. Here, for $b \leq b_B$, two black holes are confirmed to merge to a single black, and for $b \geq b_C$, two black holes are scattered away. The figure is taken from Okawa et al. (2011).

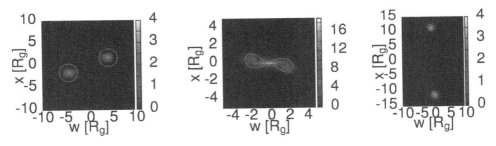

Fig. 11.5 Color maps of a curvature invariant $\mathcal{K}$ in the scattering of two black holes with $v = 0.7$ and $b = 3.38 R_g$ before scattering (left), at the instant of the collision (middle), and after scattering (right). Here, R_g is the gravitational radius of each incoming black hole $R_g = \sqrt{\mu}/2$ (therefore the gravitational radius of the system is $\sqrt{2\gamma} R_g$) where $\gamma = 1/\sqrt{1-v^2}$. The figure is taken from Okawa et al. (2011).

of the numerical results. Here, the implications for b_B and b_C are as follows: For $b \leq b_B$, the numerical simulations confirmed that two black holes merge to be a single black hole, while for $b \geq b_C$, two black holes are scattered away to infinity after the first contact. For $v \leq 0.6$, the values of b_B and b_C are identical, implying that the simulations were successfully performed for any values of the impact parameter for determining the final fate. By contrast, for $v \geq 0.65$, b_B and b_C are different, because for $b_B < b < b_C$, the simulations crashed soon after the contact: The largest value of b for the two-black-holes merger would be in a range between b_B and b_C, but the simulations were not able to determine the threshold value.

Then, Okawa et al. (2011) paid attention to the behavior of a curvature invariant $\mathcal{K} := (6\sqrt{2} E_P^2)^{-1} |R^{abcd} R_{abcd}|^{1/2}$ in the scattering process for $b(> b_C) \to b_C$. Here, E_P is the hypothetical five-dimensional Planck energy $E_P := \sqrt{3\pi/8G}$ and the normalization factor $(6\sqrt{2}E_P^2)$ of $\mathcal{K}$ is adopted as the value of $|R^{abcd}R_{abcd}|^{1/2}$ at the horizon of a black hole with mass E_P (assuming that the mini black holes have energy close to the five-dimensional Planck energy). Figure 11.5 displays the color maps of $\mathcal{K}$ for various stages (before the contact, at the instant of the collision, after the scattering, from left to right) in the scattering process of two black holes for $v = 0.7$ and $b = 3.38R_g$. At the instant of the collision, the curvature invariant $\mathcal{K}$ steeply increases around the center of mass, and it subsequently decreases with increasing the separation of two black holes. The maximum value of $\mathcal{K}$, $\mathcal{K}_{\max}$, is estimated to be

$$\mathcal{K}_{\max} \approx 50 \left(\frac{E_P}{2E}\right), \tag{11.10}$$

where E is the total energy of each incoming black hole. For the scattering of two black holes with energy $E \approx E_P$, the value of $\mathcal{K}$ becomes $O(10)$. Furthermore, the value of $\mathcal{K}_{\max}$ steeply increases with decreasing the value of b toward b_C. This suggests that in a region outside the horizons, the curvature is so large that the effect of quantum gravity may play an important role in this trans-Planckian scattering.

If this happens, the hypothetical mini-black-hole formation scenario at accelerators, frequently discussed, may have to be changed: The black hole production rate may be smaller, and instead, we may observe phenomena of quantum gravity more frequently. Unfortunately, the simulations have not provided a totally reliable evidence for/against the naked singularity formation for the scattering with $b \lesssim b_C$. Obviously, further analyses for $b_B < b < b_C$ with $v \geq 0.65$ are necessary.

11.3 Bar-mode instability of Myers–Perry black holes

Exploring the stability property of rapidly spinning Myers–Perry (MP) black holes with one spin parameter (see section 1.3.3 for the MP black holes) [Myers and Perry (1986)] in numerical relativity is an interesting subject in higher-dimensional numerical relativity. This was first performed by Shibata and Yoshino (2010b,a). They for the first time clarified that the rapidly spinning MP black holes (for $D \geq 6$) are dynamically unstable primarily against a non-axisymmetric, bar-mode deformation, and derived the condition for the onset of the bar-mode instability. They also followed the evolution of the unstable black holes until a new state is reached. In the following, we will outline this work.

11.3.1 *Studies for the instability of Myers–Perry black holes*

The stability analysis of the MP black hole has a long history. Here, we briefly summarize it (for the works done before 2013).

A standard method for the stability analysis is a linear perturbation study. If the linear perturbation equations are totally separable, the resulting equation reduces to an ordinary differential equation and its analysis may be done analytically or semi-analytically. Although the linear perturbation equations for a metric perturbation in the MP spacetime have been extensively analyzed, the separation was succeeded only for a tensor-mode perturbation [Oota and Yasui (2010); Kodama *et al.* (2010)]. For other modes, thus, the stability has not been found yet by this analysis.

The next-best method may be to numerically solve the partial differential equations for the linear perturbation. The first numerical analysis in this line was performed by Dias *et al.* (2009). In their study, an axisymmetric perturbation [i.e. the perturbation that keeps the $U(1) \times O(D-4)$ symmetry] was analyzed by solving two-dimensional simultaneous partial differential equations. Dias *et al.* (2009) discovered that the MP black hole with an ultra high spin [$a \gg \mu^{1/(D-3)}$ where $\mu^{1/(D-3)}$ has the dimension of the length; see section 1.3.3] can be unstable against axisymmetric deformation for $D \geq 6$. However, no numerical study had not been done for non-axisymmetric perturbation that breaks the $U(1)$ symmetry until the work by Shibata and Yoshino (2010b,a) [see also the latest work by Dias *et al.* (2014)].

Alternatively, Emperan and Myers (2003) analyzed the stability of MP black holes using two different methods. In one analysis, they took the so-called black-membrane limit for ultra spinning MP black holes. The ultra spinning MP black holes for $D \geq 6$ with $a \gg \mu^{1/(D-3)}$ becomes extremely oblate. For such an extremely oblate black object, they predicted that the instability analogous to the Gregory–Laflamme instability would set in. This discussion was applied to the axisymmetric instability, and subsequently, the numerical analysis of Dias *et al.* (2009) confirmed this prediction.

The other analysis was based on black-hole thermodynamics: They compared the horizon area of a spinning MP black hole with that of two identical boosted Schwarzschild–Tangherlini black holes, which recede from each other, fixing the total gravitational mass and angular momentum of the system. The horizon area of a MP black hole is [see equation (1.160)]

$$A_{\rm H} = \Omega_{D-2} \mu r_+, \tag{11.11}$$

while sum of the area of two boosted (non-spinning) black holes is

$$2A_{\rm H,S} = 2\Omega_{D-2} \mu^{(D-2)/(D-3)}. \tag{11.12}$$

As described in section 5.3.1, the total ADM mass M and the bare mass m of each black hole are related by $M = \sqrt{m^2 + p^2}$ where p is the magnitude of the linear momentum of each black hole. Thus, the total mass of the system in this analysis is $2\sqrt{m^2 + p^2}$. The total angular momentum of the system J may be given by $J = bp$, where b is an "impact parameter", i.e., the distance between two black holes in the direction orthogonal to the momenta. Here, b was chosen to be $b \sim \mu^{1/(D-3)}$ as a typical value in their analysis. Remember that r_+ approaches zero for an rapidly spinning black hole. If $A_{\rm H} < 2A_{H,S}$ is satisfied, then, the configuration of two boosted black holes may be preferred to the rapidly spinning MP black hole in a thermodynamical sense. If so, we may expect that the MP black hole would be unstable against non-axisymmetric deformation, leading to the pinch-off of its event horizon to form a system of two boosted black holes. By this discussion, the MP black holes are predicted to be unstable for

$$q := a/\mu^{1/(D-3)} \gtrsim \begin{cases} 0.85 & (D=5), \\ 0.96 & (D=6), \\ 0.99 & (D=7), \\ 1.00 & (D=8). \end{cases} \quad (11.13)$$

Here, we introduced a dimensionless spin parameter q which is used in the next subsection. In contrast to the discussion based on the black-membrane limit, this discussion can be applied to the $D = 5$ case as well as $D \geq 6$, and the predicted critical parameter for the onset of the instability is much smaller than that for the Gregory–Laflamme-like axisymmetric instability (i.e., the non-axisymmetric instability can set in for a smaller black-hole spin). Therefore, the non-axisymmetric mode could be the primary instability mode.

11.3.2 Numerical simulations

Although the prediction by Emperan and Myers (2003) appeared to be qualitatively correct, for strictly verifying that this non-axisymmetric instability indeed sets in for MP black holes and for quantitatively clarifying the criterion for the onset of this instability, we have to fully solve Einstein's equation. This can be done only by a numerical-relativity simulation.

Shibata and Yoshino (2010b,a) performed simulations for MP black holes in five, six, seven, and eight-dimensions in the following procedures. First, the MP black hole was written in quasi-isotropic coordinates [see equation (1.165)]. Then, the initial spacelike hypersurface possesses two asymptotically flat regions and one throat (see figure 1.8), and hence, the spacelike hypersurface does not cross the physical curvature singularity of the MP spacetime. The initial data are written in Cartesian coordinates (x, y, z, w_i), which are related to spherical-polar coordinates in quasi-isotropic coordinates by

$$x = r \cos\theta \cos\phi, \quad y = r \cos\theta \sin\phi, \quad \sqrt{z^2 + \sum_i w_i^2} = r \sin\theta, \quad (11.14)$$

where the (x, y)-plane corresponds to the plane of the black-hole spin.

Shibata and Yoshino (2010b,a) focused on the bar-mode instability, and a small non-axisymmetric perturbation was initially added to the conformal factor, χ, (see section 2.3.3.1) as

$$\chi = \chi_0 \left[1 + A\mu^{-1}(x^2 - y^2)\exp\left(-r^2/2r_h^2\right)\right], \quad (11.15)$$

where χ_0 is the value of unperturbed initial data, and A is a small number $\ll 1$. r_h denotes the coordinate radius of the event horizon for the unperturbed black hole.

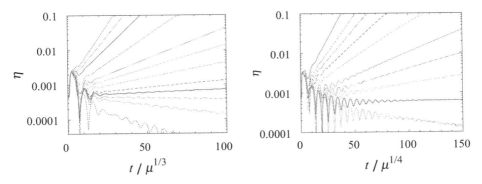

Fig. 11.6 Left panel: Time evolution of distortion parameter η for $D = 6$ and for the initial spin $q_i = a/\mu^{1/3} \approx 1.039$, 0.986, 0.933, 0.878, 0.821, 0.801, 0.781, 0.761, 0.750, 0.740, 0.718, and 0.674 (from the upper to lower curves) with $A = 0.005$. Right panel: The same as the left panel but for $D = 7$ and for $q_i = a/\mu^{1/4} = 0.960$, 0.903, 0.844, 0.813, 0.783, 0.767, 0.751, 0.735, and 0.719 (from the upper to lower curves). The figures are taken from Shibata and Yoshino (2010a).

This perturbation breaks the $U(1)$ symmetry with respect to (x, y)-plane while it keeps the $O(D-4)$ symmetry with respect to z and w_i directions.

For the initial data described above, Shibata and Yoshino (2010b,a) evolved the spacetime employing SACRA-ND code, which is the generalized version of SACRA code [Yamamoto et al. (2008)]. The modified version of the cartoon method described in section 2.5.2 was employed to impose the $O(D-4)$ symmetry. For stable simulations, the parameters of dynamical gauge conditions (see section 2.2.6) were carefully chosen [see Shibata and Yoshino (2010b,a) for details]. Their important finding on the dynamical gauge in higher-dimensional numerical relativity was that a large value of η_B is favored for performing a long-term simulation of rapidly spinning unstable black holes until the growth of the instability is saturated and subsequently the non-axisymmetric deformation damps.

Shibata and Yoshino (2010b,a) performed simulations for $D = 5-8$. The numerical results were most successfully obtained for $D = 6$ and 7, and thus, we here review their results focusing only on these cases. Note that also for other dimensions, similar numerical results were derived in their simulations.[2]

Figure 11.6 shows the time evolution of a distortion parameter η for $D = 6$ (left panel) and 7 (right panel). Here, η is defined by

$$\eta := \frac{2[(l_0 - l_{\pi/2})^2 + (l_{\pi/4} - l_{3\pi/4})^2]^{1/2}}{(l_0 + l_{\pi/2})}, \quad (11.16)$$

where l_φ denotes the proper circumferential length measured from $\theta = 0$ to $\pi/2$

---

[2] After the work of Shibata and Yoshino (2010b,a), Dias et al. (2014) performed an extensive linear-perturbation analysis for the stability of the Myers–Perry black hole. Their results on the stability agree well with those by Shibata and Yoshino (2010b,a) for $D = 6$ and 7. However, for $D = 5$, Dias et al. (2014) shows that the black hole is stable irrespective of the black-hole spin, although the damping rate of the bar-mode perturbation is extremely low for rapidly spinning black holes.

Table 11.1 The values of critical spin parameter q_{crit} for the onset of the bar-mode instability for $D = 5-8$. Note that the value of q_{crit} for $D = 5$ could be modified if a better-resolved numerical simulation is performed in the future.

D	5	6	7	8
q_{crit}	(0.91)	0.74	0.73	0.77

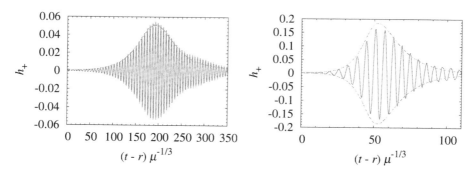

Fig. 11.7 Left panel: + modes of gravitational waveform (solid curve) emitted from an unstable black hole for $D = 6$ and for $q_i = 0.801$ as a function of a retarded time defined by $t - r$ where r is the coordinate distance from the center. $\eta/2$ is also plotted as a function of t (dashed curve). Right panel: The same as the left panel but for $q_i = 0.986$. The figures are taken from Shibata and Yoshino (2010a).

for a fixed value of φ evaluated on the surface of apparent horizons. The value of η exponentially damps in time if the black hole is not rapidly spinning (i.e., if the value of the dimensionless spin parameter q is not very large, the black holes are dynamically stable). On the other hand, the value of η grows exponentially in time for the case that the value of q is larger than a certain critical value, q_{crit}: $q_{crit} \approx 0.74$ and 0.73 for $D = 6$ and 7, respectively. These critical values of q_{crit} for the onset of the bar-mode instability is much smaller than that for the onset of the axisymmetric instability reported in Dias et al. (2009). Shibata and Yoshino (2010b,a) also determined the values of q_{crit} for $D = 5$ and 8. The results are summarized in Table 11.1.[3]

The solid curves of figure 11.7 plot gravitational waveforms in the long-term simulations for $D = 6$ for which the initial value of q is $q_i = 0.801$ (left panel) and 0.986 (right panel). The amplitude of gravitational waves initially grows in time and then is saturated when the distortion parameter becomes of order 0.1 at peak time, $t = t_{peak}$. After the saturation is reached, the amplitude exponentially damps. The reason for this behavior is explained as follows: Associated with the growth of the non-axisymmetric deformation, the emissivity of gravitational waves

---

[3]Note that Shibata and Yoshino (2010b) originally published the critical value for $D = 5$ as 0.87. Subsequent better-resolved numerical simulations by them (unpublished work) indicate that it is ≈ 0.91. However, this result does not agree with the linear-perturbation analysis by Dias et al. (2014) (thus the new value is still not reliable). The reason for this disagreement is not clear at present.

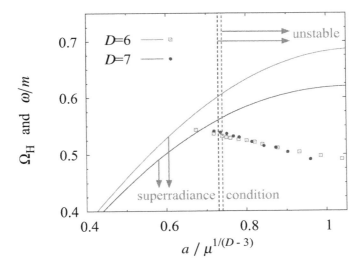

Fig. 11.8 Real part of gravitational-wave frequencies ω/m (where $m=2$) for selected values of the spin parameter for $D=6$ and 7 (points) together with Ω_H as a function of $q = a/\mu^{1/(D-3)}$ (solid curves). The units of the vertical axis are $\mu^{-1/(D-3)}$. The dotted lines denote $q = q_{\rm crit}$ for the onset of the bar-mode instability. The figure is taken from Shibata and Yoshino (2010a) with modification.

is enhanced, and energy and angular momentum are significantly extracted from the black hole, resulting in the decrease of the value of q. Eventually, q becomes smaller than $q_{\rm crit}$ (the black hole becomes stable), and the growth of the deformation amplitude is saturated at $t = t_{\rm peak}$. Because of the presence of the deformation even at $t = t_{\rm peak}$, gravitational waves continue to carry energy and angular momentum from the stable black hole, and the value of q is further decreased. Therefore, the final state is a stable state with the value of $q = q_f < q_{\rm crit}$. Here, the time, $t = t_{\rm peak}$, for $q_i = 0.986$ is smaller than that for $q_i = 0.801$. This is because the growth rate of the instability for $q_i = 0.986$ is larger than that for $q_i = 0.801$.

The dashed curves of figure 11.7 plot half of the distortion parameter η of apparent horizons. It agrees approximately with the amplitude of gravitational waves, indicating that the distortion of the apparent horizons is not due to a gauge effect: Gravitational waves are indeed generated by the distortion of the system.

Figure 11.8 shows the real part of the gravitational-wave frequency extracted from the gravitational waveform as a function of q for $D = 6$ and 7. The super-radiance condition, $\omega \leq m\Omega_H$ [Teukolsky and Press (1974)], is also shown for each value of D by the arrows, where $m = 2$ and Ω_H is the angular velocity of the horizon [see equation (1.132)]. It is well known that the super-radiance is a phenomena such that waves can extract energy and angular momentum from a spinning black hole without violating the area theorem of black holes (see section 1.3.2.4). The super-radiance condition is a necessary condition for extracting energy from the black

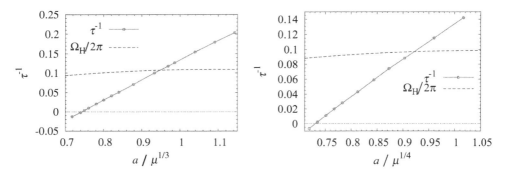

Fig. 11.9 Left panel: The growth rate $1/\tau$ of η in units of $\mu^{-1/(D-3)}$ as a function of q (solid curve) for $D=6$. The dashed curve denotes $\Omega_{\rm H}/2\pi$. Right panel: The same as the left panel but for $D=7$. The figures are taken from Shibata and Yoshino (2010a).

hole by waves. However, it is only a necessary condition and not the sufficient condition for the onset of the dynamical instability found in Shibata and Yoshino (2010b,a). In the super-radiance often discussed, one considers to inject rather an artificial ingoing wave for which the frequency satisfies this condition. For such an artificial wave, the reflected waves are amplified. For the dynamical instability to occur, however, gravitational waves have to be spontaneously excited by an unstable quasi-normal mode. Therefore, such a mode has to satisfy not only the super-radiance condition but also the condition that the imaginary part of the quasi-normal mode is negative.

Figure 11.9 shows the inverse of the growth time scale of the bar-mode instability, τ^{-1}, for $D=6$ (left panel) and 7 (right panel), which corresponds to the imaginary part of the quasi-normal mode. It indeed becomes negative for $q > q_{\rm crit}$ (i.e., τ^{-1} becomes positive), showing that the rapidly spinning black holes are dynamically unstable. Figure 11.9 also shows that for $q \gtrsim 0.9$, τ^{-1} is larger than $\Omega_{\rm H}/2\pi$. This implies that the growth time scale of the bar-mode deformation is shorter than the spin period of the black hole.

The final state eventually reached after the onset of the bar-mode instability was also clarified for $D=6$ and 7 in Shibata and Yoshino (2010a). For this purpose, the time evolution of the value of q was approximately followed by evaluating the degree of oblateness of the horizon, $C_{\rm p}/C_{\rm eq}$, where $C_{\rm p} = 2(l_0 + l_{\pi/2})$ and $C_{\rm eq}$ is the proper circumferential length measured from $\varphi=0$ to 2π along the equatorial plane $\theta=\pi/2$ on apparent horizons. For a spherically symmetric surface, the value of $C_{\rm p}/C_{\rm eq}$ is unity, and it monotonically decreases as the spin of the MP black hole increases (as the oblateness of the horizon surface increases; see section 1.3.2.3). Shibata and Yoshino (2010a) approximately determined the value of q using the relation of $C_{\rm p}/C_{\rm eq}(q)$.

Figure 11.10 shows the value of $C_{\rm p}/C_{\rm eq}$ as a function of time for $D=6$. The value of $C_{\rm p}/C_{\rm eq}$ increases with time, indicating that the black-hole spin decreases.

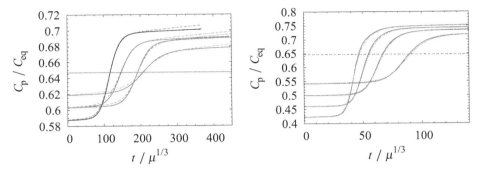

Fig. 11.10 Left panel: Time evolution of C_p/C_{eq} for $D=6$ and for dimensionless spin parameters not much greater than q_{crit}, $q_i = a/\mu^{1/3} = 0.821$, 0.801, and 0.781. The corresponding initial values of C_p/C_{eq} are ≈ 0.587, 0.602, and 0.618, respectively. The results with $A = 0.02$ and 0.005 are plotted for $q_i = 0.801$, and the results with $A = 0.02$ are plotted for $q_i = 0.821$ and 0.781. The solid and dashed curves denote the results for $N = 50$ and 40, respectively, where N is grid numbers. The thin dotted line denote $C_p/C_{eq} = 0.647$ which is the value of C_p/C_{eq} for $q = q_{crit}$. For $q_i = 0.821$, the simulation was stopped at $t/\mu^{1/3} \approx 370$ because the black hole reaches an approximately stationary state. Right panel: The same as the left panel but for the large initial spins $q_i = 0.878$, 0.933, 0.986, and 1.039 with $A = 0.005$. $C_p/C_{eq} \approx 0.542$, 0.499, 0.460, and 0.422 at $t = 0$, respectively. The figures are taken from Shibata and Yoshino (2010a).

Now, we focus on the curve starting from $C_p/C_{eq} \approx 0.62$, shown in the left panel. The initial value of q is $q_i = 0.781$, and the dotted horizontal line with $C_p/C_{eq} \approx 0.647$ corresponds to $q = q_{crit} = 0.74$. The curve crosses this dotted line at $t/\mu^{1/3} \approx 200$, which agrees approximately with the time at which the growth of the gravitational-wave amplitude is saturated. The final value of C_p/C_{eq} is ≈ 0.68, and the corresponding value of q is $q_f \approx 0.705$. Thus, a stable and moderately rapidly spinning black hole is the final outcome. Next, we focus on the curve starting from $C_p/C_{eq} \approx 0.42$ shown in the right panel. For $C_p/C_{eq} = 0.42$, the initial value of q is $q_i = 1.04$. The value of C_p/C_{eq} increases with time also in this case, and crosses the line for $q = q_{crit}$. Then, it relaxes to a stable state with the value $C_p/C_{eq} \approx 0.75$, which corresponds to $q_f = 0.61$. Again, a stable black hole is the final outcome. It is interesting to note that for higher initial spins, the final spins are smaller.

To summarize, the MP black holes are unstable against non-axisymmetric bar-mode deformation if they are spinning sufficiently rapidly. Due to the onset of this instability, their energy and angular momentum are extracted through the emission of gravitational waves that are spontaneously excited by an unstable quasi-normal mode.

It should be noted that Shibata and Yoshino (2010b,a) were not able to follow the evolution of black holes for $q_i \gg 1$ because the simulation for them is technically not easy. The reason for this is perhaps that the spatial hypersurface near the horizon for large values of q has a very long throat and it is not easy to resolve this throat numerically. Also, the growth of the instability would be very rapid. This fact also requires a higher resolution in time. For the ultra spinning case, the growth time scale of the bar-mode instability is so short that the value of η may approach unity. Namely, the horizon pinch-off may happen as discussed by Emperan and Myers (2003). It is an interesting remaining issue to clarify the evolution of the bar-mode instability for ultra-spinning black holes.

Chapter 12

Conclusion

I have described a variety of topics in numerical relativity including basic formulations, methods for a solution of the basic equations, numerical schemes, and the latest scientific results of the applications. I wish the readers to find that numerical relativity is one of the exciting fields in theoretical physics and astrophysics. I will be happy if a graduate student, motivated by this volume, participates in this field and contributes to accelerating its progress.

The readers may also find that numerical relativity has been a mature field. Before the 21st century, numerical relativity was still in the infancy: Much effort used to be paid primarily to looking for a formulation of Einstein's evolution equations that could be available in the practical numerical computation and to developing a numerical code that could be solved Einstein's equation accurately. As described in this volume, however, we now have robust tools: It is now feasible to accurately perform a wide variety of scientific numerical simulations and the numerical results obtained, such as gravitational waveforms, are accurate enough to be used for the comparison with observational and experimental data in the near future. In 1990s, to be honest, I was not able to expect that such happy future would come.

Numerical relativity has already become a robust field that can contribute significantly to gravitational-wave astronomy, high-energy astrophysics, and gravitational physics. The requirement to numerical relativity will be further enhanced in the next decade. The primary reason for this is that advanced gravitational-wave detectors will be in operation, they will detect gravitational waves, and new phenomena and hopefully new objects in our universe will be observed. For the theoretical interpretation of the forthcoming observational data, modeling with help of numerical-relativity simulations will play a special role. However, for this purpose, more detailed physical modeling and more accurate numerical simulations will be necessary. One of the near-future goals in numerical relativity is to accurately describe the observational results by a numerical-relativity simulation.

There are still challenging issues in numerical relativity. The most challenging task is to develop a numerical implementation for a solution of radiation transfer problems. Namely, a solver for Boltzmann's equation in general relativity has to be developed. This is in particular important for issues in which neutrino cooling and heating play a crucial role, such as supernova explosions and gamma-ray bursts. To date, almost no effort has been paid yet in this field. The reason for this is that fully solving Boltzmann's equation requires to solve six-dimensional problems, and hence, a computational resource, which is much more powerful than the current one, is necessary. Even if we develop a numerical code, it cannot be practically used for scientific issues right now. However, the progress of the computer power is very rapid. In the next five years, exa-flops scale, supermulti-core computers will be available for the scientists. With such a supercomputer, it may be able to solve six-dimensional problems with a good accuracy. While foreseeing the near future, we have to continue to develop a new implementation.

Numerical relativity will also play a crucial role for predicting new phenomena in the universe and in the high-energy physics as well as for discovering new phenomena in general relativity. It is in general quite difficult to observe phenomena that occur in strongly gravitating environments. The reason for this is that such phenomena are usually covered with a dense matter field, and hence, it becomes quite difficult or impossible to directly observe them through the observation of electromagnetic waves. Gravitational-wave detectors will directly observe such phenomena because gravitational waves are transparent against the dense matter. However, the phenomena, for which the emissivity of gravitational waves is weak, will not be able to be observed. There will be much more unknown phenomena in the universe. Predicting such unknown ones in computers is one of the most important roles in numerical relativity. In addition, numerical relativity will play an important role for detailed understanding of the nature of general relativity and gravity theory that are highly nonlinear theories. I hope that some of the readers who are interested in numerical relativity will work in this field and achieve a great discovery in the future.

Appendix A

Killing vector and Frobenius' theorem

A.1 Killing vector fields

In this appendix, we summarize important equations associated with a Killing vector field ξ^a, which satisfies

$$\mathscr{L}_\xi g_{ab} = 0, \tag{A.1}$$

where $\mathscr{L}_\xi$ is a Lie derivative associated with ξ^a and g_{ab} is the spacetime metric. Using the covariant derivative associated with g_{ab}, equation (A.1) is written as

$$\xi^c \nabla_c g_{ab} + g_{ac} \nabla_b \xi^c + g_{bc} \nabla_a \xi^c = 0$$
$$\Rightarrow \quad \nabla_a \xi_b + \nabla_b \xi_a = 0. \tag{A.2}$$

This last equation is called Killing's equation.

From the definition of the Riemann tensor, we have

$$(\nabla_a \nabla_b - \nabla_b \nabla_a) \xi_c = R_{abc}{}^d \xi_d, \tag{A.3}$$

where in this appendix, R_{abcd} denotes the spacetime Riemann tensor. Using Killing's equation, equation (A.3) is written as

$$\nabla_a \nabla_b \xi_c + \nabla_b \nabla_c \xi_a = R_{abc}{}^d \xi_d. \tag{A.4}$$

By cyclic permutations of the indices (a, b, c), we also have

$$\nabla_b \nabla_c \xi_a + \nabla_c \nabla_a \xi_b = R_{bca}{}^d \xi_d, \tag{A.5}$$

$$\nabla_c \nabla_a \xi_b + \nabla_a \nabla_b \xi_c = R_{cab}{}^d \xi_d. \tag{A.6}$$

Then, [(A.4)+(A.6)−(A.5)]/2 gives

$$\nabla_a \nabla_b \xi_c = \frac{1}{2} \left(R_{abc}{}^d - R_{bca}{}^d + R_{cab}{}^d \right) \xi_d = -R_{bca}{}^d \xi_d, \tag{A.7}$$

where we used a symmetry relation $R_{[abc]}{}^d = 0$. Contraction of equation (A.7) [or from $\nabla^a (\nabla_a \xi_c + \nabla_c \xi_a) = 0$] yields

$$\nabla^a \nabla_a \xi_c = -R_c{}^d \xi_d. \tag{A.8}$$

Thus, in vacuum, ξ^a obeys a vector Laplacian equation.

Because the Killing's equation means $\nabla_a \xi^a = 0$, we may rewrite equation (A.8) in the following often-seen form:

$$\nabla^a \nabla_a \xi_c + \frac{D-2}{D} \nabla_c \nabla_a \xi^a = -R_c{}^d \xi_d, \tag{A.9}$$

which is also derived from

$$\nabla^a \left(\nabla_a \xi_c + \nabla_c \xi_a - \frac{2}{D} g_{ac} \nabla_b \xi^b \right) = 0. \tag{A.10}$$

Here, D denotes the spacetime dimensionality.

In the presence of a Killing vector field, it is possible to derive a conservation law for the stress-energy tensor, T^{ab}, that satisfies $\nabla_a T^{ab} = 0$. This is because

$$\nabla_a (T^{ab} \xi_b) = \xi_b \nabla_a T^{ab} + (\nabla_a \xi_b) T^{ab}$$

$$= \xi_b \nabla_a T^{ab} + \frac{1}{2} (\nabla_a \xi_b + \nabla_b \xi_a) T^{ab} = 0, \tag{A.11}$$

where we used the conservation relation, $\nabla_a T^{ab} = 0$, and Killing's equation. Thus, vector fields such as $T^{ab} \xi_b$ and $(T^{ab} - g^{ab} T_c{}^c / 2) \xi_b$ obey a conservation equation, and for an isolated system in which the matter is confined in a finite volume, the following global quantities become conserved charges like Komar mass (see section 5.3.2):

$$-\int d^N x \sqrt{-g} T^{tb} \xi_b = \int d^N x \sqrt{\gamma} T^{ab} n_a \xi_b, \tag{A.12}$$

$$-\int d^N x \sqrt{-g} \left(T^{tb} \xi_b - \frac{1}{2} T_c{}^c g^{tb} \xi_b \right) = \int d^N x \sqrt{\gamma} \left(T^{ab} - \frac{1}{2} g^{ab} T_c{}^c \right) n_a \xi_b, \tag{A.13}$$

where n^a is the timelike unit normal vector field and N is the spatial dimensionality. It is also easy to show the fact that along geodesics for which u^a is the tangent vector, $u^a \xi_a$ is constant:

$$u^b \nabla_b (u^a \xi_a) = u^a u^b \nabla_b \xi_a = \frac{1}{2} u^a u^b (\nabla_a \xi_b + \nabla_b \xi_a) = 0. \tag{A.14}$$

A generalized notion of the Killing vector field is a conformal Killing vector field, ψ^a, which satisfies

$$\mathscr{L}_\psi g_{ab} = C_\psi g_{ab}, \tag{A.15}$$

where C_ψ is a function. This equation is written as

$$\nabla_a \psi_b + \nabla_b \psi_a = C_\psi g_{ab}. \tag{A.16}$$

Thus, by operating g^{ab}, C_ψ is determined as

$$C_\psi = \frac{2}{D} \nabla_a \psi^a. \tag{A.17}$$

A property worthy to note is that conformal Killing vector fields satisfy conformal Killing's equation for the conformal spacetime as well. A straightforward manipulation starting from equation (A.16) yields

$$\tilde{\nabla}_a \tilde{\psi}_b + \tilde{\nabla}_b \tilde{\psi}_a = \frac{2}{D} \tilde{g}_{ab} \tilde{\nabla}_c \tilde{\psi}^c, \tag{A.18}$$

where $\tilde{g}_{ab} = \Psi^{-n} g_{ab}$, $\tilde{\nabla}_a$ is the covariant derivative associated with $\tilde{g}_{ab}$, and $\tilde{\psi}_a = \tilde{g}_{ab}\psi^b$ ($\tilde{\psi}^a = \psi^a$). Ψ is a conformal factor and n is an arbitrary constant.

In the presence of a conformal Killing vector field only, the conservation law like equation (A.11) is not satisfied, except for the case that $T_a{}^a = 0$, because

$$\nabla_a(T^{ab}\psi_b) = \psi_b \nabla_a T^{ab} + \frac{1}{2}(\nabla_a\psi_b + \nabla_b\psi_a)T^{ab}$$
$$= \frac{1}{D}(\nabla_a\psi^a)T_b{}^b. \tag{A.19}$$

Also, $u^a\psi_a$ is not constant along timelike geodesics of tangent u^a, but it becomes constant if u^a is null, because

$$u^b\nabla_b(u^a\psi_a) = u^a u^b \nabla_b\psi_a = \frac{1}{2}u^a u^b(\nabla_a\psi_b + \nabla_b\psi_a) = \frac{C_\psi}{2}u^a u^b g_{ab}. \tag{A.20}$$

A.2 Frobenius' theorem

Frobenius' theorem is clearly described in standard textbooks [Wald (1984); Poisson (2004)], and hence, we only summarize the important consequence of this theorem.

Let u^a be a vector field in D-dimensional spacetime and consider the congruence of u^a. Frobenius' theorem states that a congruence of curves, for which the tangent vector is u^a, is orthogonal to a hypersurface of the dimensionality $D-1$, if and only if u^a satisfies the condition

$$u_{[a}\nabla_b u_{c]} = 0. \tag{A.21}$$

This is understand by the fact that the corresponding hypersurface is defined by $\Phi(x^\mu) = C$, where Φ is a function composed of spacetime coordinates and C is a constant specific to the hypersurface. u_μ is then written as

$$u_\mu = f(x^\alpha)\partial_\mu\Phi, \tag{A.22}$$

where $f(x^\alpha)$ is a function of spacetime coordinates. Equation (A.22) results in

$$\nabla_{[\mu}u_{\nu]} = (\partial_{[\mu}f)\partial_{\nu]}\Phi, \tag{A.23}$$

and hence, equation (A.21) is satisfied.

Next, we consider the case that u^a is a timelike unit vector filed satisfying $u^a u_a = -1$. Then, multiplying u^a to equation (A.21) yields

$$\omega_{ab} := h_a{}^c h_b{}^d (\nabla_c u_d - \nabla_d u_c) = 0, \tag{A.24}$$

where $h_a{}^c := g_a{}^c + u_a u^c$ and ω_{ab} denotes a rotation (or vorticity) tensor which satisfies $\omega_{ab}u^b = 0$ and $\omega_{ab} = -\omega_{ba}$. This rotation-free condition guarantees that the extrinsic curvature defined from timelike unit normal vector fields n^a is symmetric and spatial.

If we assume, in addition, that u^a obeys the geodesic equation, the following relation is derived from equation (A.21),

$$\nabla_c u_d - \nabla_d u_c = 0, \tag{A.25}$$

and hence, $u_a = \nabla_a \Phi$ where Φ is a scalar function.

Finally, we note that the same result holds for null vector fields satisfying $k_a k^a = 0$: If k^a satisfies $k_{[a} \nabla_b k_{c]} = 0$, the rotation tensor defined for k^a vanishes. This fact provides an important basis for deriving the laws of a black hole which is characterized by null surfaces (e.g., appendix G).

Appendix B

Numerical relativity in spherical symmetry

In this section, we describe formulations in spherically symmetric spacetime with 3+1 dimensions. In this case, the general form of the metric in spherical polar coordinates, (r, θ, φ), is written as [e.g., Nakamura *et al.* (1980); Bernstein *et al.* (1989)]

$$ds^2 = -\alpha^2 dt^2 + A^2(dr + \beta dt)^2 + B^2 r^2 d\Omega, \qquad (B.1)$$

where α is the lapse function, β is the radial component of the shift vector, and (A, B) denote the nontrivial components of the spatial metric. In addition to these four functions, we have to evolve the rr and $\theta\theta$ components of the extrinsic curvature, K_{rr} and $K_{\theta\theta}$. These six functions of (t, r) are variables to be solved in general. However, there are a variety of the choices for basic equations in spherical symmetry proposed so far.

The first reason for this variety is that the form of the metric components depends on the choice of the slicing and gauge conditions in the 3+1 formulation. The second reason stems from a unique property of the spherically symmetric spacetime: gravitational waves are absent. Because this dynamical degree of freedom is absent, we do not have to solve the dynamical equations for the gravitational field, if we solve the constraint equations at each time step. This increases the number of possible way of choosing the basic equations to be solved in numerical simulations.

In spherically symmetric hydrodynamics, there is an additional specialty that also increases the degree of freedom: In multi-dimensional numerical relativity, hydrodynamics simulations are usually performed in Eulerian coordinates. By contrast, Lagrangian coordinates may be used in spherical symmetry because the fluid moves only along the radial direction and the sequence of the positions of spherical shells never change. Note that in hydrodynamics, when two spherical shells collide, a shock is formed and one shell never passes through the other shell.

As outlined above, there could be a variety of possible choices in the coordinate condition and in the basic equations. In this appendix, we introduce several popular formulations. For the following, we define the nonzero components of the extrinsic curvature by $K_1 := K_r{}^r$ and $K_2 := K_\theta{}^\theta = K_\varphi{}^\varphi$. Thus, $K = K_1 + 2K_2$.

B.1 Spatially conformally flat gauge

In this case, the line element is written in the form

$$ds^2 = -\alpha^2 dt^2 + \psi^4 \left[(dr + \beta dt)^2 + r^2 d\Omega\right]. \tag{B.2}$$

In the minimal distortion gauge condition (see section 2.2.5), the equation for the shift vector is derived from the condition $\tilde{D}_i(\gamma^{1/2}\partial_t \tilde{\gamma}^{ij}) = 0$. In spherical symmetry, this condition is automatically satisfied in the conformally flat spatial metric, because $\partial_t \tilde{\gamma}^{ij} = 0$.

The equation for ψ is derived either from the evolution equation for it,

$$(\partial_t - \beta \partial_r)\psi = \frac{\psi}{6}\left(-\alpha K + \partial_r \beta + \frac{2\beta}{r}\right), \tag{B.3}$$

or from the Hamiltonian constraint, which is written in the form

$$(\partial_r^2 + 2r^{-1}\partial_r)\psi = -2\pi \rho_h \psi^5 - \frac{1}{8}\psi^5 \left(\frac{3}{2}K_1^2 - KK_1 - \frac{1}{2}K^2\right). \tag{B.4}$$

Combination of the evolution equations $\partial_t \gamma_{rr} = r^{-2}\partial_t \gamma_{\theta\theta}$, derived from the conformal flatness condition, gives a first-order ordinary differential equation for β as

$$\partial_r \beta - \frac{1}{r}\beta = \frac{\alpha}{2}(3K_1 - K). \tag{B.5}$$

It is also possible to derive an elliptic equation for β as in the multi-dimensional case (see section 2.2.5). However, such a complicated equation is not necessary in spherical symmetry.

Equation for K_1 is derived either from the evolution equation, or from the momentum constraint written as

$$\partial_r \left(K_1 \psi^6 r^3\right) - \psi^4 r^2 \partial_r \left(K\psi^2 r\right) = 8\pi \psi^6 r^3 J_r. \tag{B.6}$$

In numerical simulations performed so far, the momentum constraint has been always employed (with maximal slicing $K = 0$).

Equation for K depends on the slicing condition. In general case, the evolution equation for K has to be solved. However, for maximal slicing $K = 0 = \partial_t K$ (see section 2.2.2), the lapse function obeys an elliptic equation

$$\left[\partial_r^2 + 2r^{-1}\partial_r + 2\psi^{-1}(\partial_r \psi)\partial_r\right]\alpha = 4\pi\alpha\psi^4(\rho_h + S) + \frac{3}{2}\alpha\psi^4 K_1^2, \tag{B.7}$$

where $S = S_k{}^k$.

If we choose the Hamiltonian constraint for the equation of ψ and the momentum constraint for the equation of K_1, a fully constrained system with no evolution equation is constructed for $K = 0$. The fully constrained system was popular in numerical relativity, because a stable evolution with no constraint violation is possible with such a formulation. For example, Shapiro and Teukolsky performed a variety of simulations in spherical symmetry in 1980s using the fully constrained formulation [Shapiro and Teukolsky (1979, 1980, 1985a,b, 1986)] (see also Scheel et al. (1995a,b)).

B.2 Zero shift

For the zero shift vector $\beta = 0$, we have to evolve A and B or evolve one of two variables and the other is determined by the Hamiltonian constraint. The method for the evolution of the extrinsic curvature depends also on the slicing condition and on whether we employ the momentum constraint or not. In the following, we describe two interesting schemes.

B.2.1 Evolving γ_{rr} and $\gamma_{\theta\theta}$

This method was employed in Nakamura et al. (1980). Their policy in those days was *not* to employ the constraint equation for determining the spatial metric and extrinsic curvature components because they wanted to use these constraint equations to check the accuracy of numerical results obtained by solving the evolution equations. Thus, in their formulation, A, B, K_1, and K_2 have to be evolved. For this choice, the formulation becomes much more complicated than that in other choices, and moreover, we have to be careful for guaranteeing the accuracy (in particular the regularity at origin) in numerical computation. Thus, we do not recommend the readers to use this formulation. However, the readers will also learn a lesson from this formulation that it is not straightforward to perform a numerical-relativity simulation in curvilinear coordinates even for the spherically symmetric case.

All the following messy procedures in this formulation stem from the facts that we have to guarantee the regularity of geometric quantities near the coordinate singularity at $r = 0$. To explain why the regularity has to be guaranteed, we write the evolution equation for K_2 with $\beta = 0$ [cf. equation (2.62)]:

$$\partial_t K_2 = \alpha R_\theta{}^\theta - 8\pi\alpha \left(S_\theta{}^\theta - \frac{S - \rho_h}{2} \right) - D_\theta D^\theta \alpha + \alpha K K_2, \tag{B.8}$$

where

$$\begin{aligned} R_\theta{}^\theta = \big[&r^{-2}\left(B^{-2} - A^{-2}\right) \\ &+ A^{-2}\{r^{-1}\partial_r \ln A - 4r^{-1}\partial_r \ln B \\ &+ (\partial_r \ln A)(\partial_r \ln B) - (\partial_r \ln B)^2 - B^{-1}\partial_r^2 B\}\big]. \end{aligned} \tag{B.9}$$

It is found that there are several terms which look singular at $r = 0$ as

$$\frac{B^{-2} - A^{-2}}{r^2}, \quad \frac{\partial_r A}{r}, \quad \frac{\partial_r B}{r}. \tag{B.10}$$

These terms may diverge for general functions of A and B. However, in the presence of the regularity condition, they do not diverge.

As we described in section 2.4.2, the regularity conditions near the origin are

$$\partial_r A = O(r^1), \quad \partial_r B = O(r^1), \quad A - B = O(r^2), \quad K_1 - K_2 = O(r^2). \tag{B.11}$$

Thus, if these conditions are guaranteed to be satisfied in numerical simulations, the terms shown in equation (B.10) do not diverge. However, in numerical computation in which truncation errors are always present, the regularity conditions are not satisfied exactly. With an error that violates the equality of A and B at $r=0$, the term as $(B^{-2} - A^{-2})/r^2$ diverges at $r=0$. To avoid this trouble, Nakamura et al. (1980) proposed to define variables such as

$$a := \frac{A-B}{r^2}, \quad k := \frac{K_2 - K_1}{r^2}. \tag{B.12}$$

In addition, they rewrote $\partial_r B/r$ and $\partial_r a/r$ as $2\partial_{r^2} B$ and $2\partial_{r^2} a$ because B and a are functions of r^2 near the origin and such a definition of the derivative is useful to guarantee the regularity condition in the finite differencing level. Then, the evolution equation for k contains $(R_\theta{}^\theta - R_r{}^r)/r^2$ that might look singular at $r=0$. However, with the above prescription, this term is written as

$$\frac{R_\theta{}^\theta - R_r{}^r}{r^2} = \frac{1}{A^3}\left[\frac{3a^2}{B} + \frac{a^3 r^2}{B^2} - 2\partial_{r^2} a + 4(\partial_{r^2})^2 B + \frac{a}{B}\left\{4r^2(\partial_{r^2})^2 B - 2\partial_{r^2} B\right\}\right.$$
$$\left. - \frac{4r^2(\partial_{r^2} a)\partial_{r^2} B}{B} - \frac{8(\partial_{r^2} B)^2}{B} - \frac{4ar^2(\partial_{r^2} B)^2}{B^2}\right]. \tag{B.13}$$

Thus, although we need messy manipulations, all the terms are explicitly written in the regular form.

B.2.2 Maximal slicing

By contrast to the above formulation, it is possible to derive a quite simple formulation with certain choice of the gauge condition and basic equations to be solved. Shibata et al. (1994a) proposed to use the maximal slicing condition $K = 0$ (see section 2.2.2) together with the zero shift vector. For $\beta = 0$, the evolution equation of γ is written as

$$\partial_t \sqrt{\gamma} = -\alpha K \sqrt{\gamma}. \tag{B.14}$$

Thus, for $K = 0$, γ is time-independent. This implies that AB^2 is time-independent (determined at the initial stage) and we only need to solve A or B (in the following A is chosen).

In this setting, the equations to be solved are the evolution equation for A, an elliptic equation for α (derived from the maximal slicing condition), and an equation (evolution equation or momentum constraint) for K_1: In Shibata et al. (1994a), momentum constraint was employed. Then the basic equations are quite concise as

$$\partial_t A = -\alpha K_1 A, \tag{B.15}$$

$$\left[\partial_r^2 + \left(\frac{2}{r} + \frac{2\partial_r B}{B} - \frac{\partial_r A}{A}\right)\partial_r\right]\alpha = 4\pi\alpha A^2(\rho_h + S) + \frac{3}{2}\alpha A^2 K_1^2, \tag{B.16}$$

$$\partial_r(K_1 r^3 B^3) = 8\pi r^3 B^3 J_r. \tag{B.17}$$

B.3 Radial gauge

In the radial gauge, we set $B = 1$. Thus, r becomes the areal radius from which the proper area of a sphere with radius $r = r_0$ is calculated as $4\pi r_0^2$. The merit of the radial gauge is that the equations become quite simple. For example, nonzero components of R_{ij} are

$$R_{rr} = \frac{2\partial_r A}{Ar}, \quad R_{\theta\theta} = \frac{R_{\varphi\varphi}}{\sin^2\theta} = 1 - \frac{1}{A^2} + \frac{r\partial_r A}{A^3}. \tag{B.18}$$

Hence, the three-dimensional Ricci scalar becomes a first-order equation,

$$R = \frac{2}{r^2 A^2}\left(A^2 - 1 + 2r\frac{\partial_r A}{A}\right) = \frac{2}{r^2}\partial_r\left[r\left(1 - A^{-2}\right)\right]. \tag{B.19}$$

This simplifies the Hamiltonian constraint equation into the first-order ordinary differential equation.

The basic equations in the radial gauge become particularly simple when the polar slicing condition is imposed as

$$K_2 = 0. \tag{B.20}$$

In this case, we immediately have $\beta = 0$ because the evolution equation for $\gamma_{\theta\theta}$ is

$$0 = \partial_t \gamma_{\theta\theta} = -2\alpha K_{\theta\theta} + 2D_\theta \beta_\theta = -2\alpha r^2 K_2 + 2r\beta. \tag{B.21}$$

In addition, relations $K_{ij}K^{ij} = K_1^2 = K^2$ hold. This implies that the extrinsic curvature does not appear in the Hamiltonian constraint. Thus we only need to solve equations of α and A which are derived from the evolution equation of K_2 ($\partial_t K_2 = 0$) and the Hamiltonian constraint:

$$\partial_r \ln \alpha = \frac{A^2}{2r}\left[1 - A^{-2} - 16\pi r^2 \left(S_\theta{}^\theta - \frac{S}{2}\right)\right], \tag{B.22}$$

$$\partial_r\left[r\left(1 - A^{-2}\right)\right] = 8\pi \rho_h r^2. \tag{B.23}$$

Here, the first equation is derived by combining the equation of $\partial_t K_2 = 0$ and the Hamiltonian constraint. By rewriting $A^{-2} = 1 - 2m(r)/r$ and $\alpha = e^\Phi$, these equations are written in the forms

$$\partial_r \Phi = \frac{m}{r^2}\left(1 - \frac{2m}{r}\right)^{-1}\left[1 + \frac{4\pi r^3}{m}(S - 2S_\theta{}^\theta)\right], \tag{B.24}$$

$$\partial_r m = 4\pi \rho_h r^2. \tag{B.25}$$

Thus, the basic equations are essentially the same as the field equations of the Tolman–Oppenheimer–Volkoff equations (see section 1.4.3). This formulation was employed by Choptuik (1993) in his study of gravitational collapse of scalar fields which led to his discovery of critical phenomena.

A demerit in this coordinate system is that minimal surface never appears. Hence, hypersurfaces with a wormhole topology, like figure 1.8, are not constructed in this system. Since polar slicing is equal to Schwarzschild slicing in vacuum, it is also expected that apparent horizon is found only after the elapse of infinite proper time, and thus, evolution of black holes cannot be studied.

B.4 Misner–Sharp–May–White and Hernandez–Misner formulations

Misner and Sharp (1964) and May and White (1966) independently derived a formulation of general relativistic Lagrangian hydrodynamics for an ideal fluid in spherically symmetric spacetime. This formulation has been extensively employed for studying stellar core collapse and supernova explosion in spherical symmetry.

In the Misner–Sharp–May–White (MSMW) formulation, the line element is written in the form

$$ds^2 = -\alpha^2 dt^2 + e^\lambda dr^2 + R^2 d\Omega, \tag{B.26}$$

where α, λ, and R are functions of r and t. Note that R is not Ricci scalar here. Thus, they assumed the zero-shift gauge condition in the terminology of the 3+1 formulation. e^λ is often rewritten as

$$e^\lambda = (\partial_r R)^2 \Gamma^{-2}, \tag{B.27}$$

where

$$\Gamma^2 := 1 + U^2 - \frac{2m}{R}, \tag{B.28}$$

and $U := \alpha^{-1} \partial_t R$. Here, U may be related to the extrinsic curvature as $-K_{\theta\theta}/R$. m is a function of r and t, and in vacuum (outside the star), it agrees with the gravitational mass of the system. With this form, we easily find that the line element in the static case settles to the Tolman–Oppenheimer–Volkoff form [cf. equation (1.169)], because $\partial_r R dr = dR$ and $U = 0$ for the static case. On the other hand, in the absence of gravity (i.e., $m = 0$), Γ denotes the factor of Lorentz contraction by the motion of spherical shells with a velocity U.

Misner and Sharp (1964); May and White (1966) considered hydrodynamics in Lagrangian coordinates. This implies that the fluid four velocity u^μ has components

$$u^\mu = (u^t, 0, 0, 0), \quad \text{i.e.} \quad u^i = 0 \quad \text{and} \quad u^t = \alpha^{-1}. \tag{B.29}$$

Because the zero-shift gauge condition is imposed, u^a agrees with n^a. Thus, the relation $n^a = u^a$ determines the slicing condition. In the 3+1 formulation, the trace part of the extrinsic curvature is related to n^a by $K = -\nabla_a n^a$. Using $u^a = n^a$ with $u^i = 0$, we have

$$K = -\nabla_a u^a = -\alpha^{-1} \partial_t \ln \sqrt{\gamma}, \tag{B.30}$$

where we used $\alpha u^t = 1$. Remember that in the geodesic gauge with $\alpha = 1$ and $\beta^i = 0$ (see section 2.2.1.1), we have

$$K = -\nabla_a n^a = -\partial_t \ln \sqrt{\gamma}. \tag{B.31}$$

Thus, the magnitude of $|K|$ increases during the collapse of a star more rapidly than in the geodesic slicing [because $\alpha < 1$ in the comoving gauge; see equation (B.32)]. If a black hole is formed, K is likely to diverge as in the geodesic slicing case. This suggests that the comoving gauge condition does not have singularity avoidance properties, and is not suitable for studying the evolution of a black hole. On the other hand, if the collapse is halted and the collapsed core relaxes to a stable state, K will settle to zero because γ should relax to a steady value. Hence, stellar core collapse and subsequent formation of a neutron star will be followed in this gauge condition.

In the MSMW formulation, we have to solve equations for α, R, U, K_1, m, ρ (rest-mass density), and ε (specific internal energy) for a given equation of state, $P = P(\rho, \varepsilon)$. The evolution equation for R comes from the definition of U. Equation for α is derived from the r component of the Euler equation $\nabla_\mu T^\mu_r = 0$ as

$$\partial_r \ln \alpha = -\frac{1}{\rho h} \partial_r P, \tag{B.32}$$

where h is the specific enthalpy.

r component of the momentum constraint is written as

$$(\partial_r R) K_1 = \partial_r (K_2 R) = -\partial_r U. \tag{B.33}$$

Thus $K_1 = -\partial_r U / \partial_r R$ (with no time derivative of m and U). This relation will be often used in the following.

Equation for m is derived either from the evolution equation for γ_{rr} or the Hamiltonian constraint. The popular method is to employ the Hamiltonian constraint in the form

$$8\pi \rho (1+\varepsilon) R^2 = 1 + U^2 + 2RU \frac{\partial_r U}{\partial_r R} - \left[2R \partial_r^2 R + (\partial_r R)^2 \right] e^{-\lambda} - R(\partial_r R)(\partial_r e^{-\lambda})$$

$$= 2 \frac{\partial_r m}{\partial_r R}, \tag{B.34}$$

which gives at each time slice

$$m = 4\pi \int_0^R \rho (1+\varepsilon) R'^2 dR'. \tag{B.35}$$

The evolution equation for U is derived from the evolution equation for $K_{\theta\theta} (= -RU)$ in the form

$$\partial_t U = -\frac{\alpha \Gamma^2}{\rho h} \frac{\partial_r P}{\partial_r R} - \alpha \frac{m + 4\pi R^3 P}{R^2}. \tag{B.36}$$

Here, we used equations (B.33) and (B.34).

Equations for ρ and ε are derived from the continuity and energy equations as

$$\rho R^2 \Gamma^{-1} \partial_r R = \text{time independent}, \tag{B.37}$$

$$\partial_t [\rho (1+\varepsilon)] = -\rho h \alpha R^{-2} \partial_R (R^2 U). \tag{B.38}$$

In the MSMW formulation, the equation for apparent horizon, (7.6), becomes

$$\ell^\mu \partial_\mu R = 0, \tag{B.39}$$

where ℓ^μ is an outgoing null vector with $\ell^t = \alpha^{-1}$ and $\ell^r = e^{-\lambda/2}$. Then, equation (B.39) is written as [Hernandez and Misner (1966)]

$$\Gamma + U = \left(1 - \frac{2m}{R}\right)(\Gamma - U)^{-1} = 0. \tag{B.40}$$

For the collapsing star, $U < 0$ and Γ should be positive in the normal choice, and hence, $\Gamma - U > 0$. Therefore, the location of the apparent horizon is denoted by $R = 2m$.

As mentioned above, the singularity-avoidance property is absent in the MSMW formulation. Thus, although the black-hole formation can be determined by finding apparent horizons in these coordinates, it is not possible to follow the evolution of the black-hole spacetime for a long time scale. One of the most important tasks of general relativistic simulations for gravitational collapse in spherical symmetry is to explore the luminosity curves for radiation such as photons and neutrinos. To compute the luminosity curves, it is necessary to follow the propagation of radiation along null rays and to calculate the luminosity in a distant zone. To determine the luminosity of radiation emitted just before the black-hole formation, it is necessary to continue the simulation for a long time scale even after its formation until the radiation propagates to a distant zone. However, in the coordinates of the MSMW formulation, such tasks cannot be done, and thus, the MSMW formulation is not suitable for analyzing the radiation in black-hole spacetime.

To overcome this drawback, Hernandez and Misner (1966) proposed to employ an "observer time", u, as the time coordinate, defined by

$$e^{\psi} du = \alpha dt - e^{\lambda/2} dr, \tag{B.41}$$

where ψ is a function of r and u. Here, outgoing radial null rays satisfy $du = 0$, and hence, each curve of $u =$const denotes a null surface. u is the so-called retarded time, and can be scaled in such a way that it agrees with the proper time of a stationary observer at infinity. Hence, it is well suited for observing the phenomena, which occur in the strong field zone, in the distant zone. If we choose that $u = \infty$ agrees with the event horizon, we can avoid the black-hole physical singularity and also we can cover the entire region outside the event horizon. It is understood that such a choice is possible, by considering the Schwarzschild spacetime with $\alpha = e^{\psi}$: In this case, $u = t - r_*$ where r_* is the so-called tortoise coordinate defined by $r_* = r + 2M \ln(r/2M - 1)$ where r is the Schwarzschild radial coordinate, and for $r \to 2M$, u goes to infinity. Thus, for exploring stellar core collapse to a black hole, Hernandez–Misner formulation is better suited than the MSMW one [see, e.g., Baumgarte et al. (1996c,b); Harada (1998) for successful applications].

The basic equations in the Hernandez–Misner formulation are derived from those in the MSMW formulation, simply by transforming the differential operators as

$$e^{-\psi} \frac{\partial}{\partial u}\bigg|_r = \alpha^{-1} \frac{\partial}{\partial t}\bigg|_r, \quad e^{-\lambda/2} \frac{\partial}{\partial r}\bigg|_u = e^{-\lambda/2} \frac{\partial}{\partial r}\bigg|_t + \alpha^{-1} \frac{\partial}{\partial t}\bigg|_r. \tag{B.42}$$

A detailed description for the derivation of the basic equations is found in Baumgarte et al. (1995).

Appendix C

Decomposition by spherical harmonics

For solving linear partial differential equations, spherical harmonics decomposition often becomes a robust tool. In this appendix, we first summarize the scalar, vector, and tensor spherical harmonics on a unit two-dimensional sphere, and then, their applications for finding a solution of tensor and vector equations are outlined.

C.1 Spherical harmonics

Spherical harmonics are eigen functions on a unit two-dimensional sphere labeled by θ and φ. We write the line element of this sphere as

$$dl^2 = \bar{f}_{ij} dx^i dx^j = d\theta^2 + \sin^2\theta \, d\varphi^2. \tag{C.1}$$

In the following, we describe scalar, vector, and tensor harmonics on this sphere. For the manipulation, it is useful to remind us of the Riemann and Ricci tensors on this sphere:

$$\overset{(2)}{R}_{abcd} = \bar{f}_{ac}\bar{f}_{bd} - \bar{f}_{ad}\bar{f}_{bc}, \quad \overset{(2)}{R}_{ab} = \bar{f}_{ab}, \quad \overset{(2)}{R} = 2. \tag{C.2}$$

C.1.1 Scalar spherical harmonics

Scalar spherical harmonics, $Y_{lm}(\theta,\varphi)$, are the solutions of the equation

$$\Delta_2 Y_{lm} = -l(l+1) Y_{lm}, \tag{C.3}$$

where $\Delta_2 := \bar{f}^{ab}\nabla_a \nabla_b$. As a normalized solution of this equation, $Y_{lm}(\theta,\varphi)$ are written as

$$Y_{lm} = (-1)^m \sqrt{\frac{2l+1}{4\pi}} \sqrt{\frac{(l-m)!}{(l+m)!}} P_l^m(\cos\theta) e^{im\varphi}, \tag{C.4}$$

where $P_l^m(\cos\theta)$ denote Legendre polynomials, and Y_{lm} satisfy the normalization and orthogonality conditions

$$\oint Y_{lm}^* Y_{l'm'} d\Omega = \delta_{ll'} \delta_{mm'}, \tag{C.5}$$

with Y_{lm}^* the complex conjugate of Y_{lm}.

The scalar spherical harmonics are often used for finding the general solution of linear partial differential equations. Here, we consider the four-dimensional scalar-wave equation

$$\Box \phi = 0. \tag{C.6}$$

Using the decomposition

$$\phi = \sum_{l,m} \phi_l(t,r) Y_{lm}, \tag{C.7}$$

equation (C.6) reduces to 1+1 dimensional partial differential equations for each value of l as

$$\partial_t^2 \phi_l = \partial_r^2 \phi_l + \frac{2}{r} \partial_r \phi_l - \frac{l(l+1)}{r^2} \phi_l. \tag{C.8}$$

The general regular solution of this equation is written as

$$\phi_l = \int d\omega \, f_l(\omega) e^{i\omega t} j_l(\omega r), \tag{C.9}$$

where j_l is the spherical Bessel function of order l and $f_l(\omega)$ is an arbitrary function. Then, the regular solution is written in the form (see section C.2 for the derivation),

$$\phi_l = r^l \left(\frac{1}{r} \frac{\partial}{\partial r} \right)^l \frac{F(t-r) - F(t+r)}{r}, \tag{C.10}$$

where F is an arbitrary regular function.

C.1.2 *Vector spherical harmonics*

Next, we describe vector spherical harmonics. For them, there are two types, even- and odd-parity harmonics, which change the sign, respectively, as $(-1)^l$ and $(-1)^{l+1}$ for the transformation $(\theta, \varphi) \to (\pi - \theta, -\varphi)$.

The basic equation for the even-parity vector harmonics is derived simply by taking the derivative of equation (C.3), yielding

$$\Delta_2 \nabla_a Y_{lm} = -[l(l+1) - 1] \nabla_a Y_{lm}, \tag{C.11}$$

where we used equation (C.2). The normalization condition of $\nabla_a Y_{lm}$ is

$$\oint \bar{f}^{ab} (\nabla_a Y_{lm}^*) \nabla_b Y_{l'm'} d\Omega = -\oint (\Delta_2 Y_{lm}^*) Y_{l'm'} d\Omega$$

$$= l(l+1) \delta_{ll'} \delta_{mm'}. \tag{C.12}$$

Thus, one of the vector harmonics is defined by

$$V_a^{\text{even}} = \sqrt{\frac{(l-1)!}{(l+1)!}} \nabla_a Y_{lm}. \tag{C.13}$$

The odd-parity vector harmonics are defined from V_a^{even} as

$$V_a^{\text{odd}} = \sqrt{\frac{(l-1)!}{(l+1)!}} \epsilon_{ba} \nabla^b Y_{lm}, \tag{C.14}$$

where ϵ_{ab} is the completely anti-symmetric tensor on the unit sphere with $\epsilon_{\theta\varphi} = \sin\theta$. We note that ϵ_{ab} satisfies $\epsilon_{ab} \epsilon_c{}^b = \bar{f}_{ac}$.

C.1.3 Tensor spherical harmonics

Finally, we describe tensor spherical harmonics, for which there are three components. The first one is conformally flat one

$$T_{ab}^{\text{iso}} = \frac{1}{\sqrt{2}} \bar{f}_{ab} Y_{lm}. \tag{C.15}$$

This satisfies

$$\oint \bar{f}^{ac} \bar{f}^{bd} T_{ab}^{\text{iso}} T_{cd}^{\text{iso}} d\Omega = \delta_{ll'} \delta_{mm'}. \tag{C.16}$$

The other two components are the so-called even- and odd-parity components. The even-parity one is derived first by taking the covariant derivative of equation (C.11) and subtracting the trace components, yielding

$$\Delta_2 \left(\nabla_a \nabla_b - \frac{1}{2} \bar{f}_{ab} \Delta_2 \right) Y_{lm} = -[l(l+1) - 4] \left(\nabla_a \nabla_b - \frac{1}{2} \bar{f}_{ab} \Delta_2 \right) Y_{lm}, \tag{C.17}$$

where

$$(2\nabla_i \nabla_j - \bar{f}_{ij} \Delta_2) Y_{lm} = \begin{bmatrix} W_{lm} & X_{lm} \\ * & -\sin^2\theta W_{lm} \end{bmatrix}, \tag{C.18}$$

$*$ denotes a relation of symmetry, and

$$W_{lm} = \left[(\partial_\theta)^2 - \cot\theta \partial_\theta - \frac{1}{\sin^2\theta} (\partial_\varphi)^2 \right] Y_{lm}, \tag{C.19}$$

$$X_{lm} = 2\partial_\varphi [\partial_\theta - \cot\theta] Y_{lm}. \tag{C.20}$$

The normalization relation is calculated in a straightforward manner as

$$\oint \left[(2\nabla_a \nabla_b - \bar{f}_{ab}\Delta_2) Y_{lm}^* \right] (2\nabla^a \nabla^b - \bar{f}^{ab}\Delta_2) Y_{l'm'} d\Omega$$
$$= 2(l-1)l(l+1)(l+2) \delta_{ll'} \delta_{mm'}. \tag{C.21}$$

Thus, the even-parity tensor harmonics are written as

$$T_{ab}^{\text{even}} = \sqrt{\frac{(l-2)!}{2(l+2)!}} \left(2\nabla_a \nabla_b - \bar{f}_{ab}\Delta_2 \right) Y_{lm}$$
$$= \sqrt{\frac{(l-2)!}{2(l+2)!}} \begin{bmatrix} W_{lm} & X_{lm} \\ * & -\sin^2\theta W_{lm} \end{bmatrix}. \tag{C.22}$$

The odd-parity tensor harmonics are defined from T_{ab}^{even} by multiplying the antisymmetric tensor as

$$T_{ab}^{\text{odd}} = -\sqrt{\frac{(l-2)!}{2(l+2)!}} \left(\epsilon_{cb} \nabla_a \nabla^c + \epsilon_{ca} \nabla_b \nabla^c \right) Y_{lm}, \tag{C.23}$$

which satisfy

$$\oint \bar{f}^{ac} \bar{f}^{bd} T_{ab}^{\text{odd}} T_{cd}^{\text{odd}} d\Omega = \delta_{ll'} \delta_{mm'}. \tag{C.24}$$

Explicitly, the odd-parity tensor spherical harmonics are written by

$$T_{ij}^{\rm odd} = \sqrt{\frac{(l-2)!}{2(l+2)!}} \begin{bmatrix} -X_{lm}/\sin\theta & \sin\theta W_{lm} \\ * & \sin\theta X_{lm} \end{bmatrix}. \quad (C.25)$$

In numerical relativity, the explicit forms for the lower-multipole components of W_{lm} and X_{lm} are often used. For $2 \le l \le 4$, they are written as

$$W_{2\pm2} = e^{\pm 2i\varphi}\sqrt{\frac{15}{32\pi}}[3+\cos(2\theta)], \quad W_{2\pm1} = \pm e^{\pm i\varphi}\sqrt{\frac{15}{2\pi}}\cos\theta\sin\theta,$$

$$W_{20} = \sqrt{\frac{45}{4\pi}}\sin^2\theta,$$

$$W_{3\pm3} = \mp\frac{3}{8}e^{\pm 3i\varphi}\sqrt{\frac{35}{\pi}}[3+\cos(2\theta)]\sin\theta,$$

$$W_{3\pm2} = e^{\pm 2i\varphi}\sqrt{\frac{105}{32\pi}}\cos\theta[1+3\cos(2\theta)],$$

$$W_{3\pm1} = \mp\frac{5}{16}e^{\pm i\varphi}\sqrt{\frac{21}{\pi}}[\sin\theta - 3\sin(3\theta)], \quad W_{30} = \frac{15}{2}\sqrt{\frac{7}{\pi}}\cos\theta\sin^2\theta,$$

$$W_{4\pm4} = \frac{9}{8}e^{\pm 4i\varphi}\sqrt{\frac{35}{2\pi}}[3+\cos(2\theta)]\sin^2\theta, \quad W_{4\pm3} = \mp\frac{9}{2}e^{\pm 3i\varphi}\sqrt{\frac{35}{\pi}}\cos^3\theta\sin\theta,$$

$$W_{4\pm2} = \frac{9}{16}e^{\pm 2i\varphi}\sqrt{\frac{5}{2\pi}}[5+4\cos(2\theta)+7\cos(4\theta)],$$

$$W_{4\pm1} = \pm\frac{9}{8}e^{\pm i\varphi}\sqrt{\frac{5}{\pi}}[-1+7\cos(2\theta)]\sin(2\theta),$$

$$W_{40} = \frac{45}{8\sqrt{\pi}}[5+7\cos(2\theta)]\sin^2\theta \quad (C.26)$$

$$X_{2\pm2} = \pm i e^{\pm 2i\varphi}\sqrt{\frac{15}{2\pi}}\cos\theta\sin\theta, \quad X_{2\pm1} = i e^{\pm i\varphi}\sqrt{\frac{15}{2\pi}}\sin^2\theta,$$

$$X_{3\pm3} = -i\frac{3}{2}e^{\pm 3i\varphi}\sqrt{\frac{35}{\pi}}\cos\theta\sin^2\theta, \quad X_{3\pm2} = \pm i e^{\pm 2i\varphi}\sqrt{\frac{105}{2\pi}}\sin\theta\cos(2\theta),$$

$$X_{3\pm1} = i\frac{5}{2}e^{\pm i\varphi}\sqrt{\frac{21}{\pi}}\cos\theta\sin^2\theta,$$

$$X_{4\pm4} = \pm i\frac{9}{2}e^{\pm 4i\varphi}\sqrt{\frac{35}{2\pi}}\cos\theta\sin^3\theta,$$

$$X_{4\pm3} = i\frac{9}{4}e^{\pm 3i\varphi}\sqrt{\frac{35}{\pi}}\sin^2\theta\left(-2\cos^2\theta + \sin^2\theta\right),$$

$$X_{4\pm2} = \pm i\frac{3}{2}e^{\pm 2i\varphi}\sqrt{\frac{5}{2\pi}}\cos\theta\sin\theta\left(-1+7\cos^2\theta - 14\sin^2\theta\right),$$

$$X_{4\pm1} = i\frac{9}{4}e^{\pm i\varphi}\sqrt{\frac{5}{\pi}}\left(-1+7\cos^2\theta\right)\sin^2\theta, \quad (C.27)$$

where $X_{l0} = 0$ irrespective of the value of l. Note that for $l \le 1$, W_{lm} and X_{lm} vanish.

C.1.4 Decomposition of spatial tensor by spherical harmonics

An arbitrary three-dimensional spatial tensor, Q_{ab}, in the spherical-polar coordinate basis is decomposed in terms of spherical harmonics by classifying the components of Q_{ab} into three categories, Q_{rr}, Q_{rI}, and Q_{IJ}, where I and J denote θ or φ. On a unit two-dimensional sphere, Q_{rr}, Q_{rI}, and Q_{IJ} behave as scalar, vector, and tensor components, respectively. Thus, they are decomposed in the following manner using spherical harmonics:

$$Q_{rr} = \sum_{l,m} A_{lm} Y_{lm}, \tag{C.28}$$

$$Q_{rI} = \sum_{l,m} \sqrt{\frac{(l+1)!}{(l-1)!}} \left(B_{lm} V_I^{\text{even}} + C_{lm} V_I^{\text{odd}} \right), \tag{C.29}$$

$$Q_{IJ} = r^2 \sum_{l,m} \left[\sqrt{2} K_{lm}(t,r) T_{IJ}^{\text{iso}} + \sqrt{\frac{2(l+2)!}{(l-2)!}} \left(D_{lm} T_{IJ}^{\text{odd}} + F_{lm} T_{IJ}^{\text{even}} \right) \right], \tag{C.30}$$

where A_{lm}, B_{lm}, C_{lm}, K_{lm}, D_{lm}, and F_{lm} are functions of t and r. Hence, Q_{ab} is decomposed as [Regge and Wheeler (1957); Zerilli (1970)]

$$\begin{aligned} Q_{ij} = \sum_{l,m} \Bigg(& A_{lm} \begin{bmatrix} Y_{lm} & 0 & 0 \\ * & 0 & 0 \\ * & * & 0 \end{bmatrix} + B_{lm} \begin{bmatrix} 0 & \partial_\theta Y_{lm} & \partial_\varphi Y_{lm} \\ * & 0 & 0 \\ * & * & 0 \end{bmatrix} \\ & + r^2 K_{lm} \begin{bmatrix} 0 & 0 & 0 \\ * & Y_{lm} & 0 \\ * & * & \sin^2\theta Y_{lm} \end{bmatrix} + r^2 F_{lm} \begin{bmatrix} 0 & 0 & 0 \\ * & W_{lm} & X_{lm} \\ * & * & -\sin^2\theta W_{lm} \end{bmatrix} \\ & + C_{lm} \begin{bmatrix} 0 & -\partial_\varphi Y_{lm}/\sin\theta & \sin\theta \partial_\theta Y_{lm} \\ * & 0 & 0 \\ * & * & 0 \end{bmatrix} \\ & + r^2 D_{lm} \begin{bmatrix} 0 & 0 & 0 \\ * & -X_{lm}/\sin\theta & \sin\theta W_{lm} \\ * & * & \sin\theta X_{lm} \end{bmatrix} \Bigg). \end{aligned} \tag{C.31}$$

This decomposition corresponds to the extension of the decomposition of scalar functions, (C.7), to that of tensor functions.

C.1.5 Spin-weighted spherical harmonics

Vector and tensor spherical harmonics are closely related to spin-weighted spherical harmonics [Goldberg et al. (1967)], which are defined by

$$\begin{aligned} {}_s Y_{lm}(\theta,\varphi) &= \sqrt{\frac{(l-s)!}{(l+s)!}} \eth^s Y_{lm}(\theta,\varphi), & 0 \le s \le l, \\ &= \sqrt{\frac{(l+s)!}{(l-s)!}} (-1)^s \bar\eth^{-s} Y_{lm}(\theta,\varphi), & -l \le s \le 0, \end{aligned} \tag{C.32}$$

where the differential operators $\eth$ and $\bar{\eth}$, acting on a quantity η of spin-weight s, are defined by

$$\eth\eta = -(\sin\theta)^s \left[\frac{\partial}{\partial\theta} + \frac{i}{\sin\theta}\frac{\partial}{\partial\varphi}\right](\sin\theta)^{-s}\eta, \tag{C.33}$$

$$\bar{\eth}\eta = -(\sin\theta)^{-s} \left[\frac{\partial}{\partial\theta} - \frac{i}{\sin\theta}\frac{\partial}{\partial\varphi}\right](\sin\theta)^{s}\eta. \tag{C.34}$$

$\eth$ and $\bar{\eth}$ are differential operators that appear in the Newman–Penrose formalism of general relativity [Newman and Penrose (1962, 1963, 1966)]. For each value of s, the spin-weighted spherical harmonics obey the normalization relation

$$\oint {}_sY^*_{l'm'}{}_sY_{lm}d\Omega = \delta_{ll'}\delta_{mm'}. \tag{C.35}$$

Using the spin-weighted spherical harmonics, the vector and tensor spherical harmonics are written as

$$V_a^{\text{even}} + iV_a^{\text{odd}} = {}_{-1}Y_{lm}(1,\ i\sin\theta), \tag{C.36}$$

$$T_{ab}^{\text{even}} + iT_{ab}^{\text{odd}} = \frac{{}_{-2}Y_{lm}}{\sqrt{2}}\begin{bmatrix} 1 & i\sin\theta \\ * & -\sin^2\theta \end{bmatrix}, \tag{C.37}$$

where

$${}_{-1}Y_{lm}(\theta,\varphi) = \sqrt{\frac{(l-1)!}{(l+1)!}}\left(\frac{\partial}{\partial\theta} - \frac{i}{\sin\theta}\frac{\partial}{\partial\varphi}\right)Y_{lm}, \tag{C.38}$$

$${}_{-2}Y_{lm}(\theta,\varphi) = \sqrt{\frac{(l-2)!}{(l+2)!}}\left(W_{lm} - i\frac{X_{lm}}{\sin\theta}\right). \tag{C.39}$$

C.2 Tensor harmonics decomposition of spatial tensors and linear gravitational waves

Let h_{ab} be a three-dimensional tracefree tensor in a flat background and a function of t, r, θ, and φ. Using the decomposition (C.31) with the tracefree condition $A_{lm} = 2K_{lm}$, h_{ab} is decomposed as

$$h_{ij} = \sum_{l,m}\left(\frac{A_{lm}}{2}\begin{bmatrix} 2Y_{lm} & 0 & 0 \\ * & -r^2 Y_{lm} & 0 \\ * & * & -r^2\sin^2\theta Y_{lm} \end{bmatrix} + B_{lm}\begin{bmatrix} 0 & \partial_\theta Y_{lm} & \partial_\varphi Y_{lm} \\ * & 0 & 0 \\ * & * & 0 \end{bmatrix}\right.$$

$$+ r^2 F_{lm}\begin{bmatrix} 0 & 0 & 0 \\ * & W_{lm} & X_{lm} \\ * & * & -\sin^2\theta W_{lm} \end{bmatrix}$$

$$+ C_{lm}\begin{bmatrix} 0 & -\partial_\varphi Y_{lm}/\sin\theta & \sin\theta\partial_\theta Y_{lm} \\ * & 0 & 0 \\ * & * & 0 \end{bmatrix}$$

$$\left.+ r^2 D_{lm}\begin{bmatrix} 0 & 0 & 0 \\ * & -X_{lm}/\sin\theta & \sin\theta W_{lm} \\ * & * & \sin\theta X_{lm} \end{bmatrix}\right), \tag{C.40}$$

where A_{lm}, B_{lm}, C_{lm}, D_{lm}, and F_{lm} are functions of r and t. The metric components associated with A_{lm}, B_{lm}, and F_{lm} are even-parity parts, while those associated with C_{lm} and D_{lm} are odd-parity parts.

The components of the Laplacian of h_{ij} are written as

$$\Delta_\mathrm{f} h_{rr} = \sum_{l,m} \left[\frac{\partial^2 A_{lm}}{\partial r^2} + \frac{2}{r}\frac{\partial A_{lm}}{\partial r} - \frac{\lambda_l + 6}{r^2} A_{lm} + \frac{4\lambda_l}{r^3} B_{lm} \right] Y_{lm}$$

$$=: \sum_{l,m} H^a_{lm} Y_{lm}, \tag{C.41}$$

$$\Delta_\mathrm{f} h_{r\theta} = \sum_{l,m} \left[\frac{\partial^2 B_{lm}}{\partial r^2} - \frac{\lambda_l + 4}{r^2} B_{lm} + \frac{3}{r} A_{lm} + \frac{2\bar{\lambda}_l}{r} F_{lm} \right] \partial_\theta Y_{lm}$$

$$- \sum_{l,m} \left[\frac{\partial^2 C_{lm}}{\partial r^2} - \frac{\lambda_l + 4}{r^2} C_{lm} + \frac{2\bar{\lambda}_l}{r} D_{lm} \right] \frac{\partial_\varphi Y_{lm}}{\sin\theta}$$

$$=: \sum_{l,m} \left(H^b_{lm} \partial_\theta Y_{lm} + H^c_{lm} \frac{\partial_\varphi Y_{lm}}{\sin\theta} \right), \tag{C.42}$$

$$\Delta_\mathrm{f} h_{r\varphi} = \sum_{l,m} \left(H^b_{lm} \partial_\varphi Y_{lm} - H^c_{lm} \sin\theta \partial_\theta Y_{lm} \right), \tag{C.43}$$

$$\frac{\Delta_\mathrm{f} h_{\theta\varphi}}{r^2} = \sum_{l,m} \left[\frac{\partial^2 F_{lm}}{\partial r^2} + \frac{2}{r}\frac{\partial F_{lm}}{\partial r} - \frac{\bar{\lambda}_l}{r^2} F_{lm} + \frac{2}{r^3} B_{lm} \right] X_{lm}$$

$$+ \sum_{l,m} \left[\frac{\partial^2 D_{lm}}{\partial r^2} + \frac{2}{r}\frac{\partial D_{lm}}{\partial r} - \frac{\bar{\lambda}_l}{r^2} D_{lm} + \frac{2}{r^3} C_{lm} \right] \sin\theta W_{lm}$$

$$=: \sum_{l,m} \left(H^f_{lm} X_{lm} + H^d_{lm} \sin\theta W_{lm} \right), \tag{C.44}$$

$$\frac{\Delta_\mathrm{f} h_{\theta\theta}}{r^2} = \sum_{l,m} \left(-\frac{1}{2} H^a_{lm} Y_{lm} + H^f_{lm} W_{lm} - H^d_{lm} \frac{X_{lm}}{\sin\theta} \right), \tag{C.45}$$

$$\frac{\Delta_\mathrm{f} h_{\varphi\varphi}}{r^2 \sin^2\theta} = \sum_{l,m} \left(-\frac{1}{2} H^a_{lm} Y_{lm} - H^f_{lm} W_{lm} + H^d_{lm} \frac{X_{lm}}{\sin\theta} \right), \tag{C.46}$$

where $\lambda_l = l(l+1)$ and $\bar{\lambda}_l = (l+2)(l-1)$. $f^{jk} \overset{(0)}{D}_k h_{ij} = 0$ are also calculated to yield

$$\frac{\partial A_{lm}}{\partial r} + \frac{3}{r} A_{lm} - \lambda_l \frac{B_{lm}}{r^2} = 0, \tag{C.47}$$

$$\frac{\partial B_{lm}}{\partial r} + \frac{2}{r} B_{lm} - \frac{A_{lm}}{2} - \bar{\lambda}_l F_{lm} = 0, \tag{C.48}$$

$$\frac{\partial C_{lm}}{\partial r} + \frac{2}{r} C_{lm} - \bar{\lambda}_l D_{lm} = 0. \tag{C.49}$$

A tracefree tensor that satisfies these three conditions has the transverse-tracefree property for $l \geq 2$ and is often called the pure tensor. Note that for $l = 0$ and 1, the solution, which is regular everywhere, does not exist and cannot describe gravitational waves. Equations (C.47)–(C.49) show that a pure tensor with $l \geq 2$ is obtained if A_{lm} and C_{lm} are determined for the even and odd parity modes, respectively. Subsequently, B_{lm}, F_{lm}, and D_{lm} are obtained from the above three equations.

Next, we describe the general solution of gravitational waves that propagate in flat spacetime. Such solutions are derived by Teukolsky (1982) for $l = 2$ and $m = 0$ and by Nakamura (1984) for any value of l and m. The basic equations are derived from $\Box h_{ij} = 0$ and the transverse-tracefree condition for h_{ab}, equations (C.47)–(C.49). We pay attention only to the modes with $l \geq 2$. Because B_{lm} and F_{lm} are written using A_{lm} and because D_{lm} is written using C_{lm}, there are only two independent equations in this case. The equations for A_{lm} and C_{lm} are derived from rr and $r\theta$ components of $\Box h_{ij} = 0$, respectively, as

$$\frac{\partial^2 E_{lm}}{\partial t^2} = \frac{\partial^2 E_{lm}}{\partial r^2} + \frac{2}{r}\frac{\partial E_{lm}}{\partial r} - \frac{\lambda_l}{r^2} E_{lm}, \tag{C.50}$$

$$\frac{\partial^2 C_{lm}}{\partial t^2} = \frac{\partial^2 C_{lm}}{\partial r^2} + \frac{2}{r}\frac{\partial C_{lm}}{\partial r} - \frac{\lambda_l}{r^2} C_{lm}, \tag{C.51}$$

where $E_{lm} = A_{lm} r^2$. Thus E_{lm} and C_{lm} obey the same equations. The general regular solution for these equations is written in the form

$$E_{lm} = \int d\omega f_l(\omega) j_l(\omega r) e^{i\omega t}, \tag{C.52}$$

where $f_l(\omega)$ is an arbitrary function of ω and j_l is the spherical Bessel function of order l, which is written by

$$j_l(\omega r) = (-1)^l (\omega r)^l \left(\frac{1}{\omega r}\frac{d}{d(\omega r)}\right)^l \frac{\sin(\omega r)}{\omega r}. \tag{C.53}$$

Thus, the regular solutions for A_{lm} and C_{lm} are in general written in the forms

$$A_{lm} = r^{l-2} \left(\frac{1}{r}\frac{\partial}{\partial r}\right)^l \frac{F_e(r-t) - F_e(r+t)}{r}, \tag{C.54}$$

$$C_{lm} = r^l \left(\frac{1}{r}\frac{\partial}{\partial r}\right)^l \frac{F_o(r-t) - F_o(r+t)}{r}, \tag{C.55}$$

where F_e and F_o denote arbitrary regular functions.

If we focus on outgoing waves with $l = 2$, the even-parity modes are written as

$$A_{lm} = \frac{F_e''}{r^3} - \frac{3F_e'}{r^4} + \frac{3F_e}{r^5},$$

$$B_{lm} = \frac{1}{6}\left(\frac{F_e'''}{r} - \frac{3F_e''}{r^2} + \frac{6F_e'}{r^3} - \frac{6F_e}{r^4}\right),$$

$$F_{lm} = \frac{1}{24}\left(\frac{F_e''''}{r} - \frac{2F_e'''}{r^2} + \frac{3F_e''}{r^3} - \frac{3F_e'}{r^4} + \frac{3F_e}{r^5}\right), \tag{C.56}$$

and the odd-parity modes are

$$C_{lm} = \frac{F_o''}{r} - \frac{3F_o'}{r^2} + \frac{3F_o}{r^3},$$

$$D_{lm} = \frac{1}{4}\left(\frac{F_o'''}{r} - \frac{2F_o''}{r^2} + \frac{3F_o'}{r^3} - \frac{3F_o}{r^4}\right), \tag{C.57}$$

where F_e' and F_o' denote $\partial_r F_e$ and $\partial_r F_o$. Equations (C.56) and (C.57) show that terms of order $1/r$ in F_{lm} and D_{lm} are the primary parts of gravitational waves in the far zone $r \to \infty$.

In the presence of a characteristic angular velocity, ω, F' is of order ωF where F denotes F_e and F_o. An accurate extraction of gravitational waves is feasible when the first term of F_{lm} and D_{lm} is much larger than other terms. The condition may be written as $|F''''| \sim \omega |F'''| \gg 2|F'''|/r$, and hence, $r \gg 2/\omega$. This defines the wave zone which is the region that satisfies $r \gtrsim 2/\omega \sim \lambda/\pi$ with λ being the gravitational wavelength. Thus, in numerical relativity, it is required that gravitational waves are extracted for a far region that satisfies $r \gtrsim \lambda$.

If we are interested in the propagation of a wave packet, it is a good idea to set $f_l(\omega)$ in equation (C.52) as

$$f_l(\omega) = \frac{2}{i\sqrt{2\pi}} r_0 (-1)^l \omega^{l+1} \exp\left[-\frac{(r_0\omega)^2}{2}\right]. \tag{C.58}$$

Then,

$$E_{lm} = \frac{2}{i\sqrt{2\pi}} r^l \left(\frac{1}{r}\frac{d}{dr}\right)^l \frac{1}{r} \int d(r_0\omega) e^{-(r_0\omega)^2/2 + i\omega t} \sin(\omega r)$$

$$= r^l \left(\frac{1}{r}\frac{d}{dr}\right)^l \frac{1}{r}\left(e^{-(r-t)^2/2r_0^2} - e^{-(r+t)^2/2r_0^2}\right), \tag{C.59}$$

and hence, an exact solution both for the even and odd parities is derived. This type of exact solutions is often used for a test simulation; we can test whether propagation of gravitational waves is followed by a numerical-relativity code (see section 3.4.2). By this test, the validity of a new numerical-relativity code as well as the robustness of the chosen formulation has been often confirmed [Nakamura et al. (1987); Shibata and Nakamura (1995); Baumgarte and Shapiro (1999b)].

C.3 Vector harmonics decomposition of spatial vectors

Let V_a be a three-dimensional vector in a flat background and a function of r, θ, and φ. Then, it is decomposed in terms of vector spherical harmonics as

$$V_i = \sum_{l,m}\left[a_{lm}(Y_{lm},\ 0,\ 0) + b_{lm}(0,\ \partial_\theta Y_{lm},\ \partial_\varphi Y_{lm})\right.$$

$$\left. + c_{lm}\left(0,\ \frac{-\partial_\varphi Y_{lm}}{\sin\theta},\ \sin\theta\, \partial_\theta Y_{lm}\right)\right], \tag{C.60}$$

where a_{lm}, b_{lm}, and c_{lm} are functions of r. Then, the components of the Laplacian of V_a are written as

$$\Delta_{\mathrm{f}} V_r = \sum_{l,m} \left[a''_{lm} + \frac{2}{r} a'_{lm} - \frac{\lambda_l + 2}{r^2} a_{lm} + \frac{2\lambda_l}{r^3} b_{lm} \right] Y_{lm}$$

$$\Delta_{\mathrm{f}} V_\theta = \sum_{l,m} \left[\left(b''_{lm} + \frac{2}{r} a_{lm} - \frac{\lambda_l}{r^2} b_{lm} \right) \partial_\theta Y_{lm} - \left(c''_{lm} - \frac{\lambda_l}{r^2} c_{lm} \right) \frac{\partial_\varphi Y_{lm}}{\sin\theta} \right],$$

$$\Delta_{\mathrm{f}} V_\varphi = \sum_{l,m} \left[\left(b''_{lm} + \frac{2}{r} a_{lm} - \frac{\lambda_l}{r^2} b_{lm} \right) \partial_\varphi Y_{lm} + \left(c''_{lm} - \frac{\lambda_l}{r^2} c_{lm} \right) \sin\theta \partial_\theta Y_{lm} \right], \quad \text{(C.61)}$$

and the often-seen terms are

$$\Delta_{\mathrm{f}} V_r + \frac{1}{3} \overset{(0)}{D}_r (\overset{(0)}{D}_i V^i) = \sum_{l,m} \left[\frac{4}{3} a''_{lm} + \frac{8}{3r} a'_{lm} - \frac{3\lambda_l + 8}{3r^2} a_{lm} - \frac{\lambda_l}{3r^2} b'_{lm} + \frac{8\lambda_l}{3r^3} b_{lm} \right] Y_{lm},$$

$$\Delta_{\mathrm{f}} V_\theta + \frac{1}{3} \overset{(0)}{D}_\theta (\overset{(0)}{D}_i V^i) = \sum_{l,m} \left[\left(b''_{lm} + \frac{1}{3} a'_{lm} + \frac{8}{3r} a_{lm} - \frac{4\lambda_l}{3r^2} b_{lm} \right) \partial_\theta Y_{lm} \right.$$
$$\left. - \left(c''_{lm} - \frac{\lambda_l}{r^2} c_{lm} \right) \frac{\partial_\varphi Y_{lm}}{\sin\theta} \right],$$

$$\Delta_{\mathrm{f}} V_\varphi + \frac{1}{3} \overset{(0)}{D}_\varphi (\overset{(0)}{D}_i V^i) = \sum_{l,m} \left[\left(b''_{lm} + \frac{1}{3} a'_{lm} + \frac{8}{3r} a_{lm} - \frac{4\lambda_l}{3r^2} b_{lm} \right) \partial_\varphi Y_{lm} \right.$$
$$\left. + \left(c''_{lm} - \frac{\lambda_l}{r^2} c_{lm} \right) \sin\theta \partial_\theta Y_{lm} \right]. \quad \text{(C.62)}$$

Here we only consider the case $l \geq 1$.

Equation (C.62) is helpful for finding the asymptotic behavior of a vector obeying the vector Laplacian equation in the form

$$\Delta_{\mathrm{f}} V_a + \frac{1}{3} \overset{(0)}{D}_a \overset{(0)}{D}_b V^b = 8\pi J_a. \quad \text{(C.63)}$$

Here, J_a denotes a source function. Assuming $J_a = 0$ for a distance zone, the asymptotic behavior of V_a is found from

$$\frac{4}{3} a''_{lm} + \frac{8}{3r} a'_{lm} - \frac{3\lambda_l + 8}{3r^2} a_{lm} - \frac{\lambda_l}{3r^2} b'_{lm} + \frac{8\lambda_l}{3r^3} b_{lm} = 0, \quad \text{(C.64)}$$

$$b''_{lm} + \frac{1}{3} a'_{lm} + \frac{8}{3r} a_{lm} - \frac{4\lambda_l}{3r^2} b_{lm} = 0, \quad \text{(C.65)}$$

$$c''_{lm} - \frac{\lambda_l}{r^2} c_{lm} = 0. \quad \text{(C.66)}$$

Substituting the relations, $a_{lm} = a r^n$, $b_{lm} = b r^{n+1}$, and $c_{lm} = c r^p$, into these three equations, it is found that for $l \geq 2$,

$$n = l - 1, \ l + 1, \ -l, \ -l - 2,$$
$$p = l - 1, \ -l, \quad \text{(C.67)}$$

and for $l = 1$, $n = -1$ and -3, and $p = -1$.

Appendix D

Lagrangian and Hamiltonian formulations of general relativity

Lagrangian and Hamiltonian formulations are clearly described in several textbooks, e.g., Wald (1984); Poisson (2004); Gourgoulhon (2012). Here, we briefly outline these formulations.

D.1 Action principle for gravitational fields

It is well known that Einstein's equation is derived in a variation principle starting from an appropriate action. The action for the gravitational field part is composed of a Hilbert term, S_H, and boundary terms, S_B and S_0, as

$$S_G[g] = S_H[g] + S_B[g] - S_0, \tag{D.1}$$

where

$$S_H[g] = \int_\mathcal{V} \mathcal{L}_G \, d^D x = \frac{1}{16\pi} \int_\mathcal{V} \overset{(D)}{R} \sqrt{-g} \, d^D x,$$
$$S_B[g] = \frac{1}{8\pi} \oint_{\partial\mathcal{V}} \nabla_a q^a (q_c q^c) \sqrt{|h|} \, d^N x,$$
$$S_0 = \frac{1}{8\pi} \oint_\mathcal{B} \mathcal{K}_0 \sqrt{|h|} \, d^N x, \tag{D.2}$$

and h_{ab} is an induced metric on a closed boundary $\partial\mathcal{V}$ defined by

$$h_{ab} = g_{ab} - (q_c q^c) q_a q_b, \tag{D.3}$$

with q^a being the (non-null) unit vector field orthogonal to $\partial\mathcal{V}$. Here, if the boundary is spacelike, $q_c q^c = -1$, and if it is timelike, $q_c q^c = 1$. The schematic figure for $\partial\mathcal{V}$ is shown in figure D.1. $\mathcal{K}_0$ is a function independent of the spacetime metric which is introduced for the surface integral at spatial infinity $\mathcal{B}$ to be well defined [see Poisson (2004) for a detailed and comprehensive description]. Note that a nonzero value of $\mathcal{K}_0$ does not affect the field equations because h_{ab} is fixed on the boundary (see section D.5 for the procedure to determine $\mathcal{K}_0$).

The variation of the Hilbert action with respect to g_{ab} yields

$$\delta S_H = \frac{1}{16\pi} \left[\int_\mathcal{V} G_{ab} \delta g^{ab} \sqrt{-g} \, d^D x + \int_\mathcal{V} g^{ab} \delta \overset{(D)}{R}_{ab} \sqrt{-g} \, d^D x \right], \tag{D.4}$$

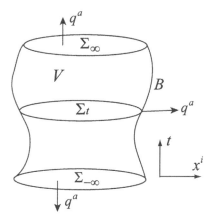

Fig. D.1 Schematic figure for the boundary $\partial \mathcal{V}$. $\Sigma_{\pm\infty}$ denote the future and past spacelike boundaries and B denotes the timelike boundary.

where

$$\begin{aligned}
\delta(\overset{(D)}{R}\sqrt{-g}) &= \delta(\overset{(D)}{R_{ab}}g^{ab})\sqrt{-g} + \overset{(D)}{R}\delta\sqrt{-g} \\
&= (\delta \overset{(D)}{R_{ab}})g^{ab}\sqrt{-g} + \overset{(D)}{R_{ab}}(\delta g^{ab})\sqrt{-g} - \frac{1}{2}\overset{(D)}{R}g_{ab}(\delta g^{ab})\sqrt{-g} \\
&= (\delta \overset{(D)}{R_{ab}})g^{ab}\sqrt{-g} + G_{ab}(\delta g^{ab})\sqrt{-g}.
\end{aligned} \quad (\text{D}.5)$$

δg^{ab} is the variation of g^{ab}, and it is fixed on the boundaries $\partial \mathcal{V}$. This implies that h_{ab} and q_a are also fixed on the boundaries. On the other hand, $\nabla_c \delta g_{ab}$ is not always zero on the boundaries, because $q^c \nabla_c \delta g_{ab}$ is not determined only by the information given on the boundaries.

The first term in the right-hand side of equation (D.4) is eventually related to Einstein's equation, whereas the second term is an extra unwanted term. This term is rewritten to a surface integral by calculating $\delta \overset{(D)}{R_{ab}}$ in the same manner as that used for deriving equation (2.17). We will find that the resulting surface term cancels out δS_B.

Let $\overset{(0)}{g_{ab}}$ and $\overset{(0)}{\nabla_a}$ be a non-perturbed original metric and its associated covariant derivative, respectively. Then, the relation between the Riemann tensors associated with g_{ab} and $\overset{(0)}{g_{ab}}$ is written as [cf. equation (2.17)]

$$\overset{(D)}{R_{abc}}{}^d = \overset{(D)}{R_{abc}}{}^d(0) + \overset{(0)}{\nabla}_b C^d{}_{ac} - \overset{(0)}{\nabla}_a C^d{}_{bc} + C^e{}_{ac} C^d{}_{be} - C^e{}_{bc} C^d{}_{ae}, \quad (\text{D}.6)$$

where $\overset{(D)}{R_{abc}}{}^d(0)$ is the Riemann tensor associated with $\overset{(0)}{g_{ab}}$. $C^c{}_{ab}$ defined here is written using equation (2.18) as

$$C^c{}_{ab} = \frac{1}{2}g^{cd}\left(\overset{(0)}{\nabla}_a g_{bd} + \overset{(0)}{\nabla}_b g_{ad} - \overset{(0)}{\nabla}_d g_{ab}\right)$$

$$= \frac{1}{2}g^{cd}\left(\overset{(0)}{\nabla}_a \delta g_{bd} + \overset{(0)}{\nabla}_b \delta g_{ad} - \overset{(0)}{\nabla}_d \delta g_{ab}\right), \quad (D.7)$$

where we used $\overset{(0)}{\nabla}_a \overset{(0)}{g}_{bc} = 0$. This shows that $C^c{}_{ab}$ is a first-order quantity of δg_{ab}, and up to the first-order variation, it is written as

$$C^c{}_{ab} = \frac{1}{2}g^{cd}\left(\nabla_a \delta g_{bd} + \nabla_b \delta g_{ad} - \nabla_d \delta g_{ab}\right). \quad (D.8)$$

Then, by taking the contraction of the subscripts b and d, equation (D.6) yields

$$\delta R^{(D)}_{ac} = R^{(D)}_{ac} - R^{(D)}_{ac}(0) = \nabla_b C^b{}_{ac} - \nabla_a C^b{}_{bc}, \quad (D.9)$$

and thus,

$$g^{ac}\delta R^{(D)}_{ac} = \nabla_b \left(g^{ac}C^b{}_{ac} - g^{bc}C^a{}_{ac}\right)$$

$$= \nabla_a \left[g^{ad}g^{bc}(\nabla_b \delta g_{cd} - \nabla_d \delta g_{bc})\right], \quad (D.10)$$

where only the first-order variation is kept, discarding more than second-order variations. Then, the second term in the right-hand side of equation (D.4) is written by the terms of total divergence, which can be rewritten to a surface integral as

$$\frac{1}{16\pi}\int_{\partial V} q^a g^{bc}(\nabla_b \delta g_{ac} - \nabla_a \delta g_{bc})(q_d q^d)\sqrt{|h|}\,d^N x. \quad (D.11)$$

Because the integrand is antisymmetric for the subscripts a and b [i.e., $q^a q^b q^c(\nabla_b \delta g_{ac} - \nabla_a \delta g_{bc}) = 0$] and because of the fact that $h^{bc}\nabla_b \delta g_{ac} = 0$ ($\delta g_{ac} = 0$ on the boundaries), the integrand of equation (D.11) reduces to

$$q^a g^{bc}(\nabla_b \delta g_{ac} - \nabla_a \delta g_{bc}) = -q^a h^{bc}\nabla_a \delta g_{bc}. \quad (D.12)$$

Thus, the surface integral resulting from the variation of S_H is written as

$$-\frac{1}{16\pi}\int_{\partial V} q^a h^{bc}(\nabla_a \delta g_{bc})(q_d q^d)\sqrt{|h|}\,d^N x. \quad (D.13)$$

On the other hand, the variation of S_B is written as

$$\delta S_B = \frac{1}{8\pi}\int_{\partial V}\delta(\nabla_a q^a)(q_c q^c)\sqrt{|h|}\,d^N x, \quad (D.14)$$

where on the boundaries

$$\delta(\nabla_a q^a) = \delta\left[h^{ab}\nabla_a q_b\right] = -h^{ab}C^c{}_{ab}q_c$$

$$= -\frac{1}{2}h^{ab}q^c\left(\nabla_a \delta g_{bc} + \nabla_b \delta g_{ac} - \nabla_c \delta g_{ab}\right)$$

$$= \frac{1}{2}h^{ab}q^c \nabla_c \delta g_{ab}. \quad (D.15)$$

Again, we used $h^{ab}\nabla_a \delta g_{bc} = 0$. Thus, $-\delta S_B$ is equal to equation (D.13), and hence, the second term in the right-hand side of equation (D.4) cancels out δS_B.

D.2 Action principle for general relativistic fluids

Lagrangian formulations and Lagrangian perturbation theories for general relativistic perfect fluids have been developed by Taub (1954, 1969); Carter (1973b); Schutz and Sorkin (1978); Friedman (1978) [see also Friedman et al. (2002) and references therein]. Here, we outline how to derive a stress-energy tensor of perfect fluids in the Lagrangian formulation, restricting our attention only to the four-dimensional spacetime (although the following calculation would be valid for any dimensionality). The basic equations for hydrodynamics and the notation for the variables are the same as those shown in section 4.4.

The action for the perfect fluid is

$$S_M = \int_\mathcal{V} \mathcal{L}_M \, d^4x = -\int_\mathcal{V} \rho_{\text{ene}} \sqrt{-g} \, d^4x, \tag{D.16}$$

where $\rho_{\text{ene}} := \rho(1+\varepsilon)$. Let ζ^a be a Lagrangian displacement vector $(\partial/\partial\lambda)^a$ where λ characterizes a perturbed fluid trajectory. Also we denote the Lagrangian and Euler displacements by Δ and δ, which have the following relation for tensor quantities, Q, as $\Delta Q = \delta Q + \mathscr{L}_\zeta Q$. Then, several relations associated with the Lagrangian perturbation, which is necessary for the following calculations, are derived as [see Carter (1973b) for the comprehensive descriptions]

$$\Delta\sqrt{-g} = \frac{1}{2}\sqrt{-g}g^{ab}\Delta g_{ab} = \frac{1}{2}\sqrt{-g}g^{ab}(\delta g_{ab} + \nabla_a\zeta_b + \nabla_b\zeta_a)$$
$$= \delta\sqrt{-g} + \sqrt{-g}\nabla_a\zeta^a, \tag{D.17}$$

$$\Delta u^a = \frac{1}{2}u^a u^b u^c \Delta g_{bc}, \tag{D.18}$$

$$\Delta(\rho u^a d\Sigma_a) = 0, \tag{D.19}$$

$$\Delta\rho = -\frac{1}{2}\rho q^{ab}\Delta g_{ab}, \tag{D.20}$$

where

$$q^{ab} := g^{ab} + u^a u^b, \tag{D.21}$$

and $d\Sigma_a$ is a volume element, i.e., $(\nabla_a t)\sqrt{-g}d^3x [= n_a (n_b n^b) \sqrt{\gamma} \, d^3x]$. Equation (D.18) is derived from the normalized condition $u^a u_a = -1$ and from the fact that in the Lagrangian frame, the direction of the fluid velocity is always that of u^a [Taub (1969); Carter (1973b)]. Equation (D.19) is an alternative expression of the conservation of the rest-mass density, equation (4.42). Equation (D.20) is derived from equations (D.18) and (D.19) by a straightforward calculation with

$$\Delta(d\Sigma_a) = \frac{1}{2}g^{bc}\Delta g_{bc} d\Sigma_a. \tag{D.22}$$

Note that the variation of the volume element for perturbed fluids comes from the variation of $\sqrt{-g}$ in this calculation.

Then, the variation of $\mathcal{L}_M$ is calculated as

$$\begin{aligned}
\delta \mathcal{L}_M &= -(\delta \rho_{\text{ene}})\sqrt{-g} - \rho_{\text{ene}} \delta \sqrt{-g} \\
&= -\left[\Delta \rho_{\text{ene}} - \zeta^a \nabla_a \rho_{\text{ene}}\right] \sqrt{-g} - \rho_{\text{ene}}[\Delta \sqrt{-g} - \sqrt{-g}\, \nabla_a \zeta^a] \\
&= -\left[\Delta(\rho_{\text{ene}} \sqrt{-g}) - \nabla_a \left(\rho_{\text{ene}} \zeta^a \sqrt{-g}\right)\right].
\end{aligned} \qquad (D.23)$$

The first law of thermodynamics is written in the form

$$\Delta \varepsilon = -P \Delta \rho^{-1} + T \Delta s = \rho^{-2} P \Delta \rho + T \Delta s, \qquad (D.24)$$

where T and s are temperature and specific entropy. Then,

$$\Delta \rho_{\text{ene}} = (\Delta \rho)(1 + \varepsilon) + \rho \Delta \varepsilon = h \Delta \rho + \rho T \Delta s, \qquad (D.25)$$

and using equations (D.17) and (D.20), we obtain

$$\begin{aligned}
\Delta \left(\rho_{\text{ene}} \sqrt{-g}\right) &= -\frac{1}{2} \sqrt{-g} \left(\rho h q^{ab} - \rho_{\text{ene}} g^{ab}\right) \Delta g_{ab} + \rho T \sqrt{-g}\, \Delta s \\
&= -\frac{1}{2} \sqrt{-g} T^{ab} \Delta g_{ab} + \rho T \sqrt{-g}\, \Delta s,
\end{aligned} \qquad (D.26)$$

where $T^{ab} = \rho h u^a u^b + P g^{ab}$ is the stress-energy tensor for the perfect fluid. In the following, we assume $\Delta s = 0$ because we consider the perfect fluid.

Substituting equation (D.26) into equation (D.23) with equation (D.17), we have

$$\begin{aligned}
\delta \mathcal{L}_M &= \frac{1}{2} \sqrt{-g} T^{ab} \delta g_{ab} + \sqrt{-g} T^{ab} \nabla_a \zeta_b + \nabla_a \left(\rho_{\text{ene}} \sqrt{-g} \zeta^a\right) \\
&= -\frac{1}{2} \sqrt{-g} T_{ab} \delta g^{ab} - (\nabla_a T^{ab}) \zeta_b \sqrt{-g} + \nabla_a \left[\rho h q^{ab} \zeta_b \sqrt{-g}\right].
\end{aligned} \qquad (D.27)$$

Thus, the variation of a total action, $S = S_H + S_B + S_M$, is written as

$$\delta S = \int_{\mathcal{V}} \left[\frac{1}{16\pi} (G_{ab} - 8\pi T_{ab})\, \delta g^{ab} + \nabla_a \left(\rho h q^{ab} \zeta_b\right) - (\nabla_a T^{ab})\, \zeta_b\right] \sqrt{-g}\, d^4x, \qquad (D.28)$$

and therefore, we obtain Einstein's equation $G_{ab} = 8\pi T_{ab}$ and the conservation equation for the matter, $\nabla_a T^{ab} = 0$. The total divergence term vanishes in the usual assumption that $\zeta^a = 0$ on the boundary $\partial \mathcal{V}$.

D.3 Gravitational-field action in $N+1$ variables

As a first step toward constructing the Hamiltonian, we have to write the action defined in the previous section in terms of the variables in the $N+1$ formulation. The start point is the relation $\overset{(D)}{R} = 2(G_{ab} - \overset{(D)}{R}_{ab}) n^a n^b$ where $2 G_{ab} n^a n^b = R - K_{ab} K^{ab} + K^2$ [see equation (2.47)] and

$$\begin{aligned}
\overset{(D)}{R}_{ab} n^a n^b &= n^a \overset{(D)}{R}_{acb}{}^c n^b = -n^a (\nabla_a \nabla_c - \nabla_c \nabla_a) n^c \\
&= -\nabla_a (n^a \nabla_c n^c) + \nabla_c (n^a \nabla_a n^c) + (\nabla_a n^a)(\nabla_c n^c) - (\nabla_c n^a)(\nabla_a n^c) \\
&= -\nabla_a (n^a \nabla_c n^c - n^c \nabla_c n^a) + K^2 - K_{ab} K^{ab}.
\end{aligned} \qquad (D.29)$$

Here, we used equation (2.32). Hence,

$$\overset{(D)}{R} = R + K_{ab}K^{ab} - K^2 + 2\nabla_a(n^a\nabla_c n^c - n^c\nabla_c n^a), \tag{D.30}$$

and

$$S_H = \frac{1}{16\pi}\int_{\mathcal{V}}\left(R + K_{ab}K^{ab} - K^2\right)\alpha\sqrt{\gamma}\,d^D x$$
$$+ \frac{1}{8\pi}\int_{\partial\mathcal{V}}(n^a\nabla_c n^c - n^c\nabla_c n^a)\,q_a\left(q_d q^d\right)\sqrt{|h|}\,d^N x. \tag{D.31}$$

Here, the boundary is composed of three elements; two spacelike hypersurfaces at past and future infinity, $\Sigma_{\mp\infty}$, and one timelike hypersurface, $\mathcal{B}$; see figure D.1. On $\Sigma_{\mp\infty}$, q^a is equal to $\mp n^a$, while on $\mathcal{B}$, q^a is a spacelike radial vector, r^a ($r^a r_a = 1$) and we suppose $r_a n^a = 0$; i.e., r_a is a vector on Σ_t. Taking into account these facts, the second integral term of equation (D.31) is written as

$$\frac{1}{8\pi}\int_{\partial\mathcal{V}}(n^a\nabla_c n^c - n^c\nabla_c n^a)\,q_a(q_d q^d)\sqrt{|h|}\,d^N x$$
$$= \frac{1}{8\pi}\left[-\int_{\Sigma_\infty}K\sqrt{\gamma}\,d^N x + \int_{\Sigma_{-\infty}}K\sqrt{\gamma}\,d^N x - \int_{\mathcal{B}}(n^c\nabla_c n^a)r_a\sqrt{-h}\,d^N x\right], \tag{D.32}$$

where h_{ab} in the second line and hereafter is $h_{ab} = g_{ab} - r_a r_b$.

In the same manner, S_B is rewritten as

$$S_B[g] = \frac{1}{8\pi}\left[\int_{\Sigma_\infty}K\sqrt{\gamma}\,d^N x - \int_{\Sigma_{-\infty}}K\sqrt{\gamma}\,d^N x + \int_{\mathcal{B}}\nabla_a r^a\sqrt{-h}\,d^N x\right]. \tag{D.33}$$

Thus, the surface terms on $\Sigma_{\pm\infty}$ resulting from S_H and S_B cancel out, and remaining terms are in total [Poisson (2004)]

$$\frac{1}{8\pi}\int_{\mathcal{B}}\mathcal{K}\sqrt{-h}\,d^N x, \tag{D.34}$$

where

$$\mathcal{K} = -(n^c\nabla_c n^a)r_a + \nabla_a r^a = n^a n^b \nabla_a r_b + \nabla_a r^a = (h^{ab} + n^a n^b)\nabla_a r_b. \tag{D.35}$$

In summary, the action is written as

$$S_G[g] = \frac{1}{16\pi}\int_{\mathcal{V}}\left(R + K_{ab}K^{ab} - K^2\right)\alpha\sqrt{\gamma}\,d^D x$$
$$+ \frac{1}{8\pi}\int_{\mathcal{B}}(\mathcal{K} - \mathcal{K}_0)\sqrt{-h}\,d^N x. \tag{D.36}$$

Now, we find that $\mathcal{K}_0$ should be equal to $\mathcal{K}$ on the flat spatial hypersurface, which ensures the value of $\delta S_G/\delta t$ to be finite.

D.4 ADM formulation from the variation principle

Here, we derive the ADM formulation from the variation principle. For simplicity, we restrict our attention to the $N = 3$ case. As a first step, $\overset{(4)}{R}$ is written in a form different from equation (D.30), explicitly rewriting the total divergence terms [Misner et al. (1973)]. As shown in section 2.1, $\nabla_a n^a = -K$ and $n^c \nabla_c n^a = D^a \ln \alpha$. Thus,

$$\nabla_a(n^a \nabla_c n^c - n^c \nabla_c n^a) = -\nabla_a(K n^a) - \nabla_a(\alpha^{-1} D^a \alpha)$$
$$= K^2 - n^a \nabla_a K - \alpha^{-1} D_a D^a \alpha, \qquad (D.37)$$

and

$$16\pi \mathcal{L}_G = \alpha \sqrt{\gamma} \left(R + K_{ab} K^{ab} + K^2 \right) - 2\sqrt{\gamma} \left(\partial_t K - \beta^a D_a K \right) - 2\sqrt{\gamma} D_a D^a \alpha. \qquad (D.38)$$

Using the definition of the extrinsic curvature in the form

$$2\alpha K_{ab} = -\partial_t \gamma_{ab} + D_a \beta_b + D_b \beta_a, \qquad (D.39)$$

we have

$$16\pi \mathcal{L}_G = \alpha \sqrt{\gamma} \left(R - K_{ab} K^{ab} + K^2 \right) - \left(K^{ab} \partial_t \gamma_{ab} + 2\partial_t K \right) \sqrt{\gamma}$$
$$- 2\sqrt{\gamma} \beta^b D_a \left(K^a{}_b - K \gamma^a{}_b \right) + 2\sqrt{\gamma} D_a \left(K^a{}_b \beta^b - D^a \alpha \right)$$
$$= 2\alpha \sqrt{\gamma} G_{ab} n^a n^b - \gamma_{ab} \partial_t \pi^{ab} + 2\sqrt{\gamma} G^a{}_b n_a \beta^b$$
$$+ D_a \left(-2\pi^a{}_b \beta^b + \pi^b{}_b \beta^a - 2\sqrt{\gamma} D^a \alpha \right), \qquad (D.40)$$

where we defined

$$\pi^{ab} := -\left(K^{ab} - K \gamma^{ab} \right) \sqrt{\gamma}, \qquad (D.41)$$

[see equation (D.49) for the original definition], and rewrote the terms of time derivative using π^{ab} as

$$\left(K^{ab} \partial_t \gamma_{ab} + 2\partial_t K \right) \sqrt{\gamma} = \gamma_{ab} \partial_t \pi^{ab}. \qquad (D.42)$$

Using total Lagrangian density $\mathcal{L} = \mathcal{L}_G - \rho_{\text{ene}} \sqrt{-g}$, the variation in the Lagrangian density with respect to α, β^a, γ_{ab}, and π^{ab} is given by [e.g., Friedman et al. (2002)]

$$\delta \mathcal{L} = -\rho T \sqrt{-g} \Delta s + \frac{h}{u^t} \Delta \left(\rho u^a n_a \sqrt{\gamma} \right)$$
$$+ \frac{1}{16\pi} \Bigg[-\delta \alpha \mathcal{H} - \delta \beta^a \mathcal{C}_a$$
$$+ \delta \pi^{ab} \left\{ \partial_t \gamma_{ab} - D_a \beta_b - D_b \beta_a - \frac{2\alpha}{\sqrt{\gamma}} \left(\pi_{ab} - \frac{1}{2} \gamma_{ab} \pi_c{}^c \right) \right\}$$
$$- \delta \gamma_{ab} \left(\bar{G}^{ab} - 8\pi S^{ab} \right) \alpha \sqrt{\gamma} \Bigg]$$
$$- \zeta_a \nabla_b T^{ab} \sqrt{-g} + D_a \tilde{\Theta}^a \sqrt{\gamma} - \frac{1}{16\pi} \partial_t \left(\delta \pi^{ab} \gamma_{ab} \right) + \partial_t \left(J_a \zeta^a \sqrt{\gamma} \right), \qquad (D.43)$$

where

$$\bar{G}_{ab} := \gamma_a{}^c \gamma_b{}^d G_{cd}, \tag{D.44}$$

$$\mathcal{H} := 2(G_{ab} - 8\pi T_{ab})n^a n^b \sqrt{\gamma} = (R + K^2 - K_{ab}K^{ab} - 16\pi\rho_{\rm h})\sqrt{\gamma}, \tag{D.45}$$

$$\mathcal{C}_c := 2(G_{ab} - 8\pi T_{ab})n^a \gamma^b{}_c \sqrt{\gamma} = 2\sqrt{\gamma}(D_a K^a{}_c - D_c K - 8\pi J_c), \tag{D.46}$$

$$\tilde{\Theta}^a := \frac{1}{16\pi}\left[\frac{1}{\sqrt{\gamma}}\left\{-2\delta\left(D^a\alpha\sqrt{\gamma}\right) + \left(\beta^a \gamma_{bc}\delta\pi^{bc} + \pi\delta\beta^a - 2\pi^a{}_b\delta\beta^b\right)\right\} \right.$$
$$\left. + \left(\gamma^{ac}\gamma^{bd} - \gamma^{ab}\gamma^{cd}\right)(\alpha D_b \delta\gamma_{cd} - D_b \alpha\, \delta\gamma_{cd})\right]$$
$$+ \alpha\rho h \gamma^a{}_c q^c{}_b \zeta^b - \beta^a J^b \zeta_b. \tag{D.47}$$

Requiring that $\delta\mathcal{L} = 0$ for an arbitrary variation of α, β^a, γ_{ab}, and π^{ab}, we obtain the Hamiltonian constraint $\mathcal{H} = 0$, momentum constrain $\mathcal{C}_a = 0$, evolution equation $\bar{G}_{ab} = 8\pi S_{ab}$, and evolution equation for γ_{ab}.

D.5 Hamiltonian formulation

The Hamiltonian density for gravitational fields, $\mathcal{H}_G$, is defined from the Lagrangian density, $\mathcal{L}_G$, by

$$\mathcal{H}_G = \frac{1}{16\pi}\pi^{ab}\dot{\gamma}_{ab} - \mathcal{L}_G, \tag{D.48}$$

where $\dot{\gamma}_{ab} = \partial_t \gamma_{ab}$. $\mathcal{L}_G$ is defined from equation (D.36) and π^{ab} is a momentum conjugate to γ_{ab} defined by

$$\pi^{ab} := 16\pi \frac{\partial \mathcal{L}_G}{\partial \dot{\gamma}_{ab}}. \tag{D.49}$$

$\dot{\gamma}_{ab}$ is contained in $\mathcal{L}_G$ through the extrinsic curvature as in equation (D.39).

Since the surface integral part of equation (D.36) is independent of $\dot{\gamma}_{ab}$, π^{ab} is defined only through the volume integral part. Defining a part of the Lagrangian density as

$$\mathcal{L}_1 := \frac{1}{16\pi}\left(R + K_{ab}K^{ab} - K^2\right)\alpha\sqrt{\gamma}, \tag{D.50}$$

π^{ab} is calculated as

$$\pi^{ab} = 16\pi \frac{\partial \mathcal{L}_1}{\partial \dot{\gamma}_{ab}} = -\frac{8\pi}{\alpha}\frac{\partial \mathcal{L}_1}{\partial K_{ab}} = -\sqrt{\gamma}\left(K^{ab} - \gamma^{ab}K\right). \tag{D.51}$$

Then, the volume integral part of the Hamiltonian density is calculated as

$$16\pi\mathcal{H}_1 := \pi^{ab}\dot{\gamma}_{ab} - 16\pi\mathcal{L}_1$$
$$= -\alpha\sqrt{\gamma}R + \frac{\alpha}{\sqrt{\gamma}}\left(\pi^{ab}\pi_{ab} - \frac{1}{N-1}(\pi^a{}_a)^2\right) + 2\pi^{ab}D_a\beta_b, \tag{D.52}$$

or in terms of (γ_{ab}, K_{ab}),

$$16\pi \mathcal{H}_1 = \sqrt{\gamma}\bigl[-\alpha\left(R - K_{ab}K^{ab} + K^2\right) + 2\beta^a D_b\left(K_a{}^b - \gamma_a{}^b K\right) \\ - 2D_a\left(K^a{}_b \beta^b - K\beta^a\right)\bigr]. \quad \text{(D.53)}$$

The full Hamiltonian is obtained by integrating $\mathcal{H}_1$ over a spatial hypersurface Σ_t, adding the contribution coming from the surface integral in equation (D.36) as

$$H_G = \frac{1}{16\pi}\int_{\Sigma_t}\left[-\alpha\left(R - K_{ab}K^{ab} + K^2\right) + 2\beta^a D_b\left(K_a{}^b - \gamma_a{}^b K\right)\right]dV \\ -\frac{1}{8\pi}\oint_{r\to\infty}(K^a{}_b - K\gamma^a{}_b)\beta^b dS_a \\ -\frac{1}{8\pi}\oint_{r\to\infty}\alpha(\mathcal{K} - \mathcal{K}_0)\sqrt{\gamma_{N-1}}\,d^{N-1}\theta, \quad \text{(D.54)}$$

where $dV = \sqrt{\gamma}d^N x$, and we used the relation $\sqrt{-h} = \alpha\sqrt{\gamma_{N-1}}$ with γ_{N-1} being the determinant of the metric on a $(N-1)$-dimensional closed surface [see equation (5.16)].

In vacuum, the value of the volume integral of equation (D.54) vanishes because of the presence of the constraint equations. Only the surface term associated with $\mathcal{K} - \mathcal{K}_0$ does not vanish, and this gives the ADM mass as found in the following calculation.

The explicit form of $\mathcal{K}$ is calculated to yield

$$\mathcal{K} = D_k r^k = \frac{1}{\sqrt{\gamma}}\partial_k\left(\sqrt{\gamma}r^k\right). \quad \text{(D.55)}$$

For further calculation, we assume $r^k = r_0^k + \delta r^k$ and $\gamma_{ij} = f_{ij} + \delta\gamma_{ij}$ in the asymptotically flat region where δr^k and $\delta\gamma_{ij}$ are perturbed quantities and we consider the perturbation up to the first order. Here $f_{ij}r_0^i r_0^j = 1$ with f_{ij} the flat spatial metric and r_0^i the radial unit vector at $r\to\infty$. Because $\gamma_{ij}r^i r^j = 1$, the relation between the perturbed quantities are

$$\delta r^j = -\frac{1}{2}\gamma^{jk}\delta\gamma_{ik}r_0^i, \quad \text{(D.56)}$$

where Cartesian coordinates are supposed to be used as in the calculation of the ADM mass in section 5.3. Then, we find

$$\mathcal{K} = \frac{1}{2}f^{ij}r_0^k\left(\partial_k \delta\gamma_{ij} - \partial_i \delta\gamma_{jk}\right) - \frac{1}{2}\delta\gamma_{ij}\left(f^{ij} - r_0^i r_0^j\right) + \frac{N-1}{r}. \quad \text{(D.57)}$$

In addition, we have to calculate $\sqrt{\gamma_{N-1}}$ in equation (D.54). This may be written as

$$\sqrt{\gamma_{N-1}} \to 1 + \frac{1}{2}\delta\gamma_{ij}f_{N-1}^{ij}. \quad \text{(D.58)}$$

Here, f_{N-1}^{ij} is the $(N-1)$-dimensional flat metric which is equal to $f^{ij} - r_0^i r_0^j$. Thus,

$$\mathcal{K} = \frac{1}{2}f^{ij}r_0^k(\partial_k\delta\gamma_{ij} - \partial_i\delta\gamma_{jk}) + \frac{N-1}{r}\sqrt{\gamma_{N-1}}. \quad \text{(D.59)}$$

This shows that $\mathcal{K}_0$ should be $(N-1)/r$, and then,

$$\mathcal{K} - \mathcal{K}_0 = \frac{1}{2} f^{ij} r_0^k (\partial_k \delta\gamma_{ij} - \partial_i \delta\gamma_{jk}). \tag{D.60}$$

Therefore, the last term of equation (D.54) with $\alpha \to 1$ agrees with the definition of the ADM mass [see equation (5.8)].

Appendix E

Solutions of Riemann problems in special relativistic hydrodynamics

As introduced in section 4.8.6, the Riemann problem is an initial-value problem for 1+1 hydrodynamics and/or magnetohydrodynamics with the initial condition composed of two uniform states separated by a discontinuity [cf. equation (4.377)]. Riemann problems in special relativistic hydrodynamics are solved in an analytic or semi-analytic manner (by solving an algebraic equation) [Martí and Müller (1994); Pons et al. (2000)]. The derived exact solutions should be reproduced by newly developed hydrodynamics codes in test simulations, as first performed by Centrella and Wilson (1984); Hawley et al. (1984a,b) for a Riemann shock-tube and a wall shock problem. For magnetohydrodynamics, it is also possible to derive solutions by solving algebraic equations for some special cases. In this appendix, we summarize the procedure for these solutions.

E.1 Riemann problems for pure hydrodynamics in special relativity

First, we outline the general properties for solutions of Riemann problems in hydrodynamics. The key property of the solutions is a self-similarity, which implies that the solutions depend only on $(x-x_0)/(t-t_0)$ where x_0 and t_0 are the initial location of a discontinuity and the initial time, respectively. The self-similarity comes from the facts that hydrodynamical waves propagating in a uniform medium have a constant velocity and that the adiabatic and uniform fluid is scale free. For $t > t_0$, the discontinuity splits into two elementary nonlinear waves, a *shock wave* and/or an expanding set of *rare-faction waves*, two of which move in opposite directions into the left and right states initially given (which is denoted by L and R in the following; see figure E.1). Between these two different waves, two new uniform states (which is denoted by L^* and R^*) are formed and they are separated by a *contact discontinuity* which moves with a local fluid velocity v. An expanding set of rare-faction waves is characterized by a nonuniform state.

Shock waves are formed when the pressure ahead an initial discontinuity, P_a, is smaller than the pressure behind the discontinuity, P_b, while a set of rare-faction

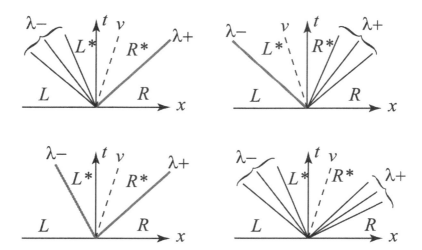

Fig. E.1 1+1 spacetime diagrams for four types of solutions of the Riemann problem in hydrodynamics. The upper two, lower left, and lower right panels show the solutions composed of one shock and rare-faction waves, two shock waves, and two sets of rare-faction waves, respectively. The thick solid line (labeled by $\lambda_\pm$), the fan-shaped three solid lines (labeled by $\lambda_\pm$), and the dashed lines (labeled by v) denote trajectories of the shocks, the rare-faction waves, and the contact discontinuity, respectively. $\lambda_\pm$ and v are three characteristic speeds [see equation (4.66)]. L and R denote the initial uniform states and L^* and R^* are states between the shocks and contact discontinuity or between the rare-faction waves and contact discontinuity, respectively.

waves are formed for $P_\mathrm{a} > P_\mathrm{b}$. At the shock, two neighboring states are discontinuous on any time slice, whereas in the rare-faction waves, all the quantities are continuous. At the contact discontinuity, the density is not continuous although the velocity and pressure are continuous (see figure E.2).

There are four types of solutions for the Riemann problem. The type is determined by the relation among the pressure for the states L, L^*, R^*, and R. For $P(L) > P(L^*) = P(R^*) > P(R)$ and $P(L) < P(L^*) = P(R^*) < P(R)$, both a shock and rare-faction waves are formed (see upper two panels of figure E.1, respectively). In this case, the pressure continuously varies in the region composed of the rare-faction waves. For $P(L) < P(L^*) = P(R^*) > P(R)$, two shock waves are formed (see the bottom left panel of figure E.1), and for $P(L) > P(L^*) = P(R^*) < P(R)$, two sets of rare-faction waves are formed (see the bottom right panel of figure E.1). In the last case, no shocks are present.

A solution of the Riemann problem is derived using a self-similar relation held for the solution [Johnson and McKee (1971); Eltgroth (1971)] together with jump conditions across discontinuities [Taub (1948); Thorne (1973)]. The procedure for obtaining general solutions is described in Martí and Müller (1994); Pons et al. (2000). In the following, we will focus only on two typical Riemann problems, shock-tube problem and wall shock problem, because they are popular solutions obtained in a simple calculation and thus have been often adopted in testbed simulations that are always performed when one wants to confirm the reliability of a new numerical code.

The basic equations in the next two sections are composed of the continuity, Euler, and energy equations in 1+1 dimension with $u^y = u^z = 0$. In special relativity, they are, respectively, written as

$$\partial_t(\rho w) + \partial_x(\rho w v) = 0, \quad \text{(E.1)}$$
$$\partial_t(\rho h w^2 v) + \partial_x(\rho h w^2 v^2 + P) = 0, \quad \text{(E.2)}$$
$$\partial_t(\rho h w^2 - P) + \partial_x(\rho h w^2 v) = 0, \quad \text{(E.3)}$$

where $v = u^x/u^t$ and $w = u^t = 1/\sqrt{1-v^2}$. In the following, we will assume that the Γ-law equation of state is used; $P = (\Gamma - 1)\rho\varepsilon$ and $h = 1 + \Gamma P/[(\Gamma - 1)\rho]$. This implies that for the left and right isentropic states, the pressure is written in the polytropic form, $P = \kappa_\mathrm{p}\rho^\Gamma$, using the first law of thermodynamics for the isentropic fluid, $d\varepsilon = -Pd(1/\rho)$. Here the values of κ_p for the left and right states are different in general.

E.2 Solutions of shock-tube problem in hydrodynamics

In the shock-tube problem often considered, fluids are supposed to be initially at rest, and the solutions illustrated in the upper panels of figure E.1 are realized. The typical configuration for the solutions are depicted in figure E.2 [Centrella and Wilson (1984); Hawley et al. (1984a); Thompson (1986)]. In this example, $P(L) > P(R)$, and thus, a shock and rare-faction waves are formed for $x > 0$ and $x < 0$, respectively. Below, we outline the method for deriving this solution following Martí and Müller (1994). In this section, we focus only on the Riemann shock-tube problem in which the condition $P(L) > P(L^*) = P(R^*) > P(R)$ holds as in figure E.2.

E.2.1 *Riemann invariants and solution for rare-faction waves*

As shown in figure E.2, the fluid state in the region of rare-faction waves is not uniform, although other regions are composed of separated uniform states. Thus, first of all, we derive the solution of rare-faction waves. As already mentioned, any fluid in the Riemann problem behaves in a self-similar manner, and hence, all the fluid quantities depend only on $\xi := x/t$ where we set $x_0 = t_0 = 0$ for simplicity. Then, equations (E.1) and (E.2) are written as

$$(v - \xi)\rho' + [\rho w^2 v(v - \xi) + \rho]v' = 0, \quad \text{(E.4)}$$
$$\rho h w^2 (v - \xi)v' + (1 - \xi v)P' = 0. \quad \text{(E.5)}$$

where the dash $'$ denotes $d/d\xi$. The definition of the sound velocity yields

$$P' = hc_s^2\rho'. \quad \text{(E.6)}$$

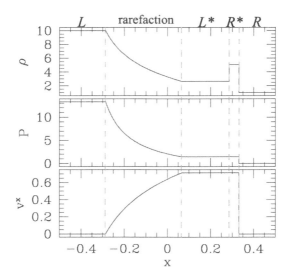

Fig. E.2 Snapshot of a solution (density, pressure, and velocity) of a Riemann shock-tube problem. For L and R states, $(\rho, P, v^x) = (10, 13.3, 0)$ and $(1, 10^{-6}, 0)$, respectively, and the solution at $t = 0.4$ with $t_0 = x_0 = 0$ is plotted. The spacetime diagram of this solution is illustrated at the upper left panel of figure E.1.

Eliminating P' and ρ' from equations (E.4) and (E.5) using equation (E.6), we obtain

$$\rho v'[(1 + c_s v)\xi - (v + c_s)][(1 - c_s v)\xi - (v - c_s)] = 0, \tag{E.7}$$

and thus

$$\xi = \frac{v \mp c_s}{1 \mp v c_s}. \tag{E.8}$$

Here, the solutions with $\mp$ denote the cases that rare-faction waves propagate in the minus and plus x directions, respectively. For the solution shown in figure E.2, the minus sign should be chosen. Substituting equation (E.8) into (E.4) yields

$$w^2 dv \mp \frac{c_s}{\rho} d\rho = \frac{1}{1 - v^2} dv \mp \frac{c_s}{\rho} d\rho = 0. \tag{E.9}$$

From this equation, the Riemann invariants are derived as [Taub (1948)]

$$J_\pm = \frac{1}{2} \ln\left(\frac{1+v}{1-v}\right) \pm \int \frac{c_s}{\rho} d\rho = \text{const}, \quad \text{along } \xi = \frac{v \mp c_s}{1 \mp v c_s}. \tag{E.10}$$

For the solution shown in figure E.2, J_+ is the relevant quantity.

The next step is to derive relations that hold among two uniform states L, L^*, and the state in rare-faction waves by using the general solution of rare-faction waves. Because the entropy is uniform for these regions, the equation of state is commonly written by $P = \kappa_p \rho^\Gamma$ where κ_p is a constant ($\kappa_p = P_L/\rho_L^\Gamma$). Then, the sound velocity in these regions is written as

$$c_s^2 = \frac{1}{h} \frac{dP}{d\rho} = \frac{\kappa_p \Gamma(\Gamma - 1)\rho^{\Gamma-1}}{\Gamma - 1 + \kappa_p \Gamma \rho^{\Gamma-1}}. \tag{E.11}$$

Thus, c_s is regarded as a function of P (or ρ). Then, the second term of equation (E.10) is integrated to yield

$$J_\pm = \frac{1}{2} \ln \left[\frac{1+v}{1-v} F(P) \right] = \text{const}, \quad (E.12)$$

where

$$F(P) := \left[\frac{\sqrt{\Gamma-1} + c_s(P)}{\sqrt{\Gamma-1} - c_s(P)} \right]^{\pm 2/\sqrt{\Gamma-1}}. \quad (E.13)$$

Note again that in the present context, the plus sign is relevant. Equation (E.12) implies that two states L and L^* are related by

$$\frac{1+v_L}{1-v_L} F(P_L) = \frac{1+v(\xi)}{1-v(\xi)} F[P(\xi)] = \frac{1+v_{L^*}}{1-v_{L^*}} F(P_{L^*}), \quad (E.14)$$

and thus, v_{L^*} (three velocity in L^*) is written as a function of P_{L^*} (pressure in L^*) for a given set of (P_L, v_L). We will use this relation between v_{L^*} and P_{L^*} in the next section for determining these values. We note that because the polytropic equation of state for the same value of κ_p is satisfied both in L and L^*, ρ_{L^*} (density in L^*) is written simply by

$$\rho_{L^*} = \rho_L \left(\frac{P_{L^*}}{P_L} \right)^{1/\Gamma}. \quad (E.15)$$

The solution in rare-faction waves is obtained in the following manner. From equation (E.8), c_s is written as a function of ξ as

$$c_s(\xi) = \pm \frac{v(\xi) - \xi}{1 - \xi v(\xi)}. \quad (E.16)$$

In the present context, the plus sign is relevant. Equation (E.14) yields,

$$v(\xi) = \frac{F[(c_s)_L](1+v_L) - F[c_s(\xi)](1-v_L)}{F[(c_s)_L](1+v_L) + F[c_s(\xi)](1-v_L)}. \quad (E.17)$$

Thus, from the simultaneous equations (E.16) and (E.17), $c_s(\xi)$ and $v(\xi)$ for rarefaction waves are determined. When $c_s(\xi)$ is obtained, $\rho(\xi)$ and $P(\xi)$ are subsequently derived from equation (E.11).

E.2.2 Jump conditions at shock waves

Next, we consider the relations that are satisfied at shocks. Specifically, we need to determine the state R^* for a given state R.

In the frame moving with a shock, the states in the vicinity of the shock are stationary, and thus, equations (E.1)–(E.3) are rewritten as

$$\partial_{x'}[\rho w(v - v_s)] = 0, \quad (E.18)$$
$$\partial_{x'}[\rho h w^2 v(v - v_s) + P] = 0, \quad (E.19)$$
$$\partial_{x'}[\rho h w^2 (v - v_s) + P v_s] = 0, \quad (E.20)$$

where v_s is the speed of the shock measured in the laboratory frame, and $x' = x - v_s t$; we used $(\partial/\partial t)_x = (\partial/\partial t)_{x'} - v_s(\partial/\partial x')_t$. As a result, we obtain the following jump conditions at the shock [Taub (1948); Thorne (1973)]

$$[\rho w(v - v_s)] = 0, \tag{E.21}$$
$$[\rho h w^2 v(v - v_s) + P] = 0, \tag{E.22}$$
$$[\rho h w^2 (v - v_s) + P v_s] = 0, \tag{E.23}$$

where $[\cdots]$ denotes the difference between the right and left states. Thus, in the present context,

$$\rho_R w_R (v_R - v_s) = \rho_{R^*} w_{R^*} (v_{R^*} - v_s) =: j, \tag{E.24}$$
$$\rho_R h_R w_R^2 v_R (v_R - v_s) + P_R = \rho_{R^*} h_{R^*} w_{R^*}^2 v_{R^*} (v_{R^*} - v_s) + P_{R^*}, \tag{E.25}$$
$$\rho_R h_R w_R^2 (v_R - v_s) + P_R v_s = \rho_{R^*} h_{R^*} w_{R^*}^2 (v_{R^*} - v_s) + P_{R^*} v_s, \tag{E.26}$$

where j denotes the mass flux. Using equation (E.24), equations (E.25) and (E.26) are rewritten, respectively, as

$$[P] = P_R - P_{R^*} = -j(h_R w_R v_R - h_{R^*} w_{R^*} v_{R^*}), \tag{E.27}$$
$$(P_R - P_{R^*}) v_s = -j(h_R w_R - h_{R^*} w_{R^*}). \tag{E.28}$$

Eliminating $h_{R^*} w_{R^*}$ from these equations, we obtain

$$v_{R^*} = \frac{h_R w_R v_R j + [P]}{h_R w_R j + [P]\{v_R - j/(\rho_R v_R)\}}, \tag{E.29}$$

where we used $j = \rho_R w_R (v_R - v_s)$ to eliminate v_s. Equation (E.29) gives a relation of v_{R^*} as a function of P_{R^*} and v_s for a given state R.

The relation of v_{R^*} as a function only of P_{R^*} is obtained if a condition between v_s and P_{R^*} is derived. Such a condition is derived from equations (E.25) and (E.26) by a straightforward calculation as

$$j^2 = -\frac{[P](1 - v_s^2)}{h_R/\rho_R - h_{R^*}/\rho_{R^*}}, \tag{E.30}$$
$$h_R^2 - h_{R^*}^2 = [P] (h_R/\rho_R + h_{R^*}/\rho_{R^*}). \tag{E.31}$$

These equations together with $j = \rho_R w_R (v_R - v_s)$ and $h = 1 + \Gamma P/[(\Gamma - 1)\rho]$ yield the relation between v_s and P_{R^*}, and thus, we have v_{R^*} as a function of P_{R^*}.

The final task is to match the solution for the state R^* with the solution for the state L^* at the contact discontinuity. As already shown in equation (E.14), v_{L^*} is written as a function of P_{L^*}. At the contact discontinuity, the pressure and velocity are continuous, i.e.,

$$P_{L^*} = P_{R^*}, \quad v_{L^*} = v_{R^*}. \tag{E.32}$$

These two relations provide simultaneous equations for the pressure and velocity at the contact discontinuity for given states L and R. Although the resulting algebraic equations are complicated and a numerical calculation is needed for a solution, the procedure for it is straightforward. Figure E.2 is a solution of the shock-tube problem that is obtained by solving the equations described in this section.

E.3 Wall shock problem in hydrodynamics

The wall shock problem is one of the Riemann problems, in which two shock waves are formed (see the bottom left panel of figure E.1). We here consider the simplest initial condition as follows [Hawley et al. (1984a)]: The density ρ_0, pressure P_0, and absolute magnitude of the velocity V_0 in the left and right states are the same, while the direction of the velocity is opposite as $v = -\text{sign}(x)V_0$. Thus, at $t = t_0$, shocks are formed at $x = 0$, and then, two shock waves propagate both to the plus and minus directions in a reflection symmetric manner. Because of the presence of the symmetry, the fluid behind the shocks is at rest. Thus, for $t > t_0$, there are essentially two regions: For $|x| > v_s(t-t_0)$, $(\rho, P, v) = (\rho_0, P_0, -\text{sign}(x)V_0)$ and for $|x| < v_s(t-t_0)$, ρ and P have new values, denoted by ρ_2 and P_2, respectively, while $v = 0$. Here v_s is the absolute magnitude of the shock propagation velocity.

ρ_2, P_2, and v_s are determined by equations (E.21)–(E.23), which are written as

$$\rho_0 w_0 (V_0 + v_s) = \rho_2 v_s, \tag{E.33}$$

$$\rho_0 h_0 w_0^2 V_0 (V_0 + v_s) + (\Gamma - 1)\rho_0 \varepsilon_0 = (\Gamma - 1)\rho_2 \varepsilon_2, \tag{E.34}$$

$$-\rho_0 h_0 w_0^2 (V_0 + v_s) + (\Gamma - 1)\rho_0 \varepsilon_0 v_s = -\rho_2 (1 + \varepsilon_2) v_s, \tag{E.35}$$

where $h_0 = 1 + \Gamma \varepsilon_0$ and the specific internal energy is determined by the equation of state, $P = (\Gamma - 1)\rho\varepsilon$. Eliminating ρ_2 and ε_2 from these equations yields a second-order algebraic equation for v_s as

$$[(\Gamma - 1)(h_0 w_0 - 1) - h_0 w_0 V_0 v_s](V_0 + v_s)w_0 = \frac{\Gamma P_0}{\rho_0} v_s, \tag{E.36}$$

and the solution is easily obtained. In particular, for the initially negligible specific internal energy $P_0/\rho_0 \ll 1$ and $h_0 \approx 1$, the solution for v_s is written in a simple form

$$v_s \approx \frac{(\Gamma - 1)(w_0 - 1)}{w_0 V_0}. \tag{E.37}$$

If v_s is obtained, ρ_2 and ε_2 are subsequently obtained from equations (E.33) and (E.35), respectively.

E.4 Solutions of Riemann problems in ideal magnetohydrodynamics

As described in section 4.6.3, there are seven characteristic speeds and characteristic curves for each spatial direction in ideal magnetohydrodynamics. Hence, the situation is much more complicated than that for pure hydrodynamics. What is serious in ideal magnetohydrodynamics is, moreover, that non-singular solutions of Riemann problems are not always guaranteed to be derived. In general, for given left and right states, a singular solution may be the result. However, it is still possible to derive solutions from algebraic equations for special cases. The solutions, which are quite helpful for testing ideal magnetohydrodynamics codes in special relativity, were derived by, e.g., Komissarov (1997, 1999): These have been indeed used as described in section 4.10.2. In this section, we outline the method of derivation for these solutions.

E.4.1 Solutions for shocks

If we assume that all the quantities depend only on t and x, ideal magnetohydrodynamics equations in special relativity are written as [see equations (4.162) and (4.163) for their original forms]

$$\partial_t S_i + \partial_x \left[S_i v^x + P_{\rm tot} \delta^x{}_i - B^x(B_i + u_i B^k u_k) w^{-2} \right] = 0, \tag{E.38}$$

$$\partial_t S_0 + \partial_x \left[S_0 v^x + P_{\rm tot} v^x - w^{-1} B^x u_k B^k \right] = 0, \tag{E.39}$$

where

$$w = u^t = \sqrt{1 + u_i u^i}, \quad (u^i = u_i), \tag{E.40}$$

$$S_i = \rho h w u_i + w^{-1} \left[B^2 u_i - B_i (B^k u_k) \right], \tag{E.41}$$

$$S_0 = \rho h w^2 - P_{\rm tot} + B^2, \tag{E.42}$$

and $P_{\rm tot} = P + \left[B^2 + (B^k u_k)^2 \right]/2w^2$ [see equation (4.164)]. In addition, the continuity and induction equations have to be solved:

$$\partial_t(\rho w) + \partial_x(\rho w v^x) = 0, \tag{E.43}$$

$$\partial_t B^i + \partial_x(B^i v^x - B^x v^i) = 0. \tag{E.44}$$

The no-monopole constraint $\partial_x B^x = 0$ together with the x-component of the induction equation implies that B^x has to be uniform and time-independent. By contrast, u^y, u^z, B^y, and B^z are not necessarily uniform and are time-varying. This situation significantly enhances the degree of the complexity for deriving solutions in ideal magnetohydrodynamics.

Nevertheless, there are several cases in which solutions can be derived from algebraic equations. The first case is that only one shock, which separates the left and right uniform states and moves with a uniform velocity V_0, is present. Denoting the difference of a quantity between the left and right states, Q_L and Q_R, as $[Q] = Q_R - Q_L$, the jump conditions in the rest frame of the shock are written as [see equations (E.38), (E.39), (E.43), and (E.44)]

$$\left[S_i(v^x - V_0) + P_{\rm tot} \delta^x{}_i - B^x(B_i + u_i B^k u_k) w^{-2} \right] = 0, \tag{E.45}$$

$$\left[S_0(v^x - V_0) + P_{\rm tot} v^x - w^{-1} B^x (u_j B^j) \right] = 0, \tag{E.46}$$

$$\left[\rho w(v^x - V_0) \right] = 0, \tag{E.47}$$

$$\left[B^i(v^x - V_0) - B^x v^i \right] = 0 \quad \text{for } i = y \text{ and } z. \tag{E.48}$$

For simplicity, we here restrict our attention to the cases $u^z = S_z = B^z = 0$, and consider to give (ρ, P, u^x, u^y, B^y) for the left state as well as a uniform value of B^x and velocity of the shock, V_0. Equations (E.45)–(E.48) are composed of five equations for S_x, S_y, S_0, ρ, and B^y for five unknowns (ρ, P, u^x, u^y, B^y) of the right state: Note that h is supposed to be determined from ρ and P using an equation of state assumed. Thus, for an appropriate choice of the left state as well as B^x and V_0, a solution for a shock is determined, although a numerical calculation is necessary for solving the algebraic equations. Solutions of shocks used in section 4.10.2 can be obtained using such procedure [see, e.g., appendix C of Villiers and Hawley (2003) for more specific description of a solution procedure].

E.4.2   Solutions for Riemann problems

For special initial conditions of the left and right states, for which some of characteristic speeds coincide with each other, the Riemann problems are simplified and their solution reduces to one of the four solutions schematically shown in figure E.1. For such cases, it is possible to derive solutions of the Riemann problem in the same procedure as described in section E.2 using the basic equations shown in section E.4.1. Komissarov (1999) derived several solutions including those of rare-faction waves (lower right panel of figure E.1) and shock-tube problems (upper panels of figure E.1).

We here outline the Komissarov's solutions, denoting the characteristic speeds of fast and slow waves by μ_f and μ_s. The sound speed, total Alfvén speed, and Alfvén speed for waves propagating to the x direction are denoted by c_s, v_A, and v_{A_x}, respectively [see equations (4.68), (4.188), and (4.186), respectively].

Komissarov (1999) derived two solutions for rare-faction waves which connect two separate uniform states (left and right states initially given). One is the so-called switch-off type solution for which $v_A = v_{A_x} = \mu_f > c_s$ and $B^y = B^z = 0$ in the down-stream side. The other is the so-called switch-on type solution for which $v_A = v_{A_x} = \mu_s < c_s$ and $B^y = B^z = 0$ in the upper-reaches side. Thus, for one of two sides, a degeneracy among some of the characteristic speeds is assumed. For appropriate choices of the left and right states initially given, rare-faction waves are produced between these left and right states. Using the self-similar nature of rare-faction waves, the solutions are obtained in the same procedure as those shown in section E.2.1.

Komissarov (1999) also derived two solutions for shock-tube problems. First, he considered the case in which $u^y = u^z = 0$, $B^y = B^z = 0$, and B^x is constant. In this case, the induction equation becomes trivial, and furthermore,

$$S_0 = \rho h w^2 - P_{\text{tot}}, \quad S_x = \rho h w u_x, \quad S_y = S_z = 0, \quad P_{\text{tot}} = P + \frac{(B^x)^2}{2}. \quad \text{(E.49)}$$

Because $\partial_t B^x = \partial_x B^x = 0$, the basic equations reduce essentially to purely hydrodynamics equations for ρw, $\rho h w u_x$, and $\rho h w^2 - P_{\text{tot}}$. Thus, solutions are obtained by the same procedure as in section E.2 (this is the case even in the presence of a nonzero uniform value of B^x): For this special case, $v_A = v_{A_x} = \mu_f$ or $v_A = v_{A_x} = \mu_s$, and remaining characteristic speeds, μ_s or μ_f as well as v^k agree with those shown in equation (4.66). Because the transverse mode associated with

v_{A_x} is assumed to vanish in this setting, there are only three relevant modes. This situation agrees totally with that in pure hydrodynamics.

In the second case, the magnetic field is assumed to be parallel to the initial discontinuity; without loss of generality, we may set $B^x = B^z = 0$ and $B^y \neq 0$. In addition, it is assumed that $u_y = u_z = 0$. In this case, $v_{A_x} = \mu_s = v^x$, and hence, there are only three characteristic speeds as in the pure hydrodynamics case. In the following, we will show this fact setting $B^y = B$ for simplicity.

Because of the presence of nonzero B, S_0 and S_x still depend on B, but in terms of the effective specific enthalpy and pressure defined by

$$\hat{h} := h + \frac{B^2}{\rho w^2}, \quad \hat{P} := P + \frac{B^2}{2w^2}, \tag{E.50}$$

they are written in the same form as those of pure hydrodynamics as,

$$S_0 = \rho \hat{h} w^2 - \hat{P}, \quad S_x = \rho \hat{h} w u_x. \tag{E.51}$$

For these new variables, the Euler and energy equations are written as

$$\partial_t \left(\rho \hat{h} w u_x \right) + \partial_x \left(\rho \hat{h} u_x^2 + \hat{P} \right) = 0, \tag{E.52}$$

$$\partial_t \left(\rho \hat{h} w^2 - \hat{P} \right) + \partial_x \left(\rho \hat{h} w u_x \right) = 0, \tag{E.53}$$

which also have the same forms as pure hydrodynamics equations (E.2) and (E.3).

In addition to these equations, the y component of the induction equation is non-trivial as

$$\partial_t B + \partial_x (B w u_x) = 0. \tag{E.54}$$

However, combining it with the continuity equation (E.1), the induction equation can be written as

$$u^\mu \partial_\mu \left(\frac{B}{\rho w} \right) = 0. \tag{E.55}$$

Using this equation, the first law of thermodynamics for the isentropic fluid, $dh = dP/\rho$, can be rewritten in the form

$$d\hat{h} = \frac{d\hat{P}}{\rho}, \tag{E.56}$$

with the effective sound velocity

$$(\hat{c}_s)^2 := \frac{\rho}{\hat{h}} \frac{\partial \hat{h}}{\partial \rho} = \frac{\rho h w^2 c_s^2 + B^2}{\rho h + B^2} = \mu_f. \tag{E.57}$$

This implies that the magnetohydrodynamics equations are rewritten in the same forms as pure hydrodynamics equations (E.1)–(E.3) for the specific enthalpy $\hat{h}$, pressure $\hat{P}$, and sound velocity $\hat{c}_s$. Therefore, if a solution for pure hydrodynamics is known, it can be also used for constructing a solution of ideal magnetohydrodynamics with $B^x = B^z = 0$ and $B^y \neq 0$ by replacing (h, P, c_s) to $(\hat{h}, \hat{P}, \hat{c}_s)$.

E.4.3 Solutions for nonlinear Alfvén waves

Komissarov (1997) derived a solution for the propagation of nonlinear Alfvén waves in isentropic fluids, which is also useful for testing ideal magnetohydrodynamics codes in special relativity. This solution is qualitatively different from solutions of the Riemann problem for which two discontinuous states are initially prepared. The Alfvén-wave solution is totally analytic, and thus, it enables us to examine whether the numerical solution shows the correct convergence behavior for a chosen numerical scheme. In the following, we will outline the procedure for this solution following Komissarov (1997).

For the analysis, the best way is probably to start from

$$(A_I{}^J)^\mu \phi_\mu l^I = 0, \tag{E.58}$$

where $(A_I{}^J)^\mu \phi_\mu$ is defined in equation (4.183), $\phi_\mu = (-\lambda, 1, 0, 0)$ with λ being an eigenvalue (characteristic speed) of the Jacobian metric $A_I{}^J$, and l^I is the corresponding right eigenvectors. Here, λ is the Alfvén speed and the solution of

$$(\rho h + b^2)a^2 - \hat{B}^2 = 0, \tag{E.59}$$

where $a = u^\mu \phi_\mu$ and $\hat{B} = b^\mu \phi_\mu$ [cf. equation (4.186)]. The solution of equation (E.58) for l^I is written as [Komissarov (1997)]

$$l^I = \left(d^\alpha, \frac{\hat{B}}{a} d^\alpha, 0\right), \tag{E.60}$$

where d^α is a four-vector, which satisfies the following constraints

$$d^\alpha \phi_\alpha = 0, \quad d^\alpha b_\alpha = 0. \tag{E.61}$$

Because the right eigenvectors determine the variation of independent variables along the corresponding characteristic waves, we have the relations $\delta u^\alpha \propto d^\alpha$ and $\delta b^\alpha \propto (\hat{B}/a) d^\alpha$. Together with $u^\alpha u_\alpha = -1$ and $u_\alpha \delta u^\alpha = 0$, we obtain

$$d^\alpha u_\alpha = 0. \tag{E.62}$$

Then, d^α is written, besides a constant factor, as

$$d^\alpha = e^{\alpha\beta\gamma\delta} \phi_\beta u_\gamma b_\delta. \tag{E.63}$$

The solution (E.60) implies that the pressure variation vanishes. The first law of thermodynamics with the constant specific entropy is written as

$$u^\alpha \nabla_\alpha P = h c_s^2 u^\alpha \nabla_\alpha \rho. \tag{E.64}$$

Thus, the density variation also vanishes. In addition, the absolute value of the magnetic field does not change because

$$\delta(b^2) = 2b_\alpha \delta b^\alpha \propto b_\alpha \frac{\hat{B}}{a} d^\alpha = 0. \tag{E.65}$$

This implies that the variation of $\rho h + b^2$ also vanishes.

Taking the variation of equation (E.59) yields

$$0 = \delta\left[(\rho h + b^2)a^2 - \hat{B}^2\right] = 2\left[(\rho h + b^2)a\delta a - \hat{B}\delta\hat{B}\right]$$
$$= 2\delta\lambda\left[(\rho h + b^2) au^\alpha t_\alpha - \hat{B}b^\alpha t_\alpha\right], \quad \text{(E.66)}$$

where $t_\alpha = (-1, 0, 0, 0)$. The terms in $[\cdots]$ do not cancel in general, and thus, the variation of λ has to vanish. This implies

$$\delta a = (\delta u^\alpha)\phi_\alpha + \delta\lambda u^a t_a = 0, \quad \text{(E.67)}$$
$$\delta\hat{B} = (\delta b^\alpha)\phi_\alpha + \delta\lambda b^a t_a = 0, \quad \text{(E.68)}$$

where we used $(\delta u^\alpha)\phi_\alpha \propto d^\alpha\phi_\alpha = 0$ and $(\delta b^\alpha)\phi_\alpha \propto (\hat{B}/a)d^\alpha\phi_\alpha = 0$.

Therefore, if two states L and R are connected via a simple Alfvén wave, then

$$[P] = [\rho] = [b^2] = [\lambda] = [a] = [\hat{B}] = 0, \quad [u^\alpha] = \frac{a}{\hat{B}}[b^\alpha], \quad \text{(E.69)}$$

where $[Q] = Q_R - Q_L$. Then, it is straightforward to derive a solution for the propagation of simple Alfvén waves.

Following Komissarov (1997), solutions of the R state can be obtained for a given L state in the frame moving with Alfvén waves. This is achieved by setting $\phi^\mu = (0, 1, 0, 0)$. Then, equation (E.69) becomes

$$[P] = [\rho] = [b^2] = [\lambda] = 0, \quad [a] = [u^x] = 0, \quad [\hat{B}] = [b^x] = 0,$$
$$[u^\alpha] = \chi[b^\alpha] \quad (\alpha \neq x),$$

where $\chi = u^x/b^x$. Setting $u^\alpha = u_L^\alpha + \chi[b^\alpha]$ for $\alpha \neq x$, b^t is written as

$$b^t = a_y b^y + a_z b^z + a_t, \quad \text{(E.70)}$$

where

$$a_y = \frac{u_L^y - \chi b_L^y}{u_L^t - \chi b_L^t}, \quad a_z = \frac{u_L^z - \chi b_L^z}{u_L^t - \chi b_L^t}, \quad a_t = \frac{\chi b^2}{u_L^t - \chi b_L^t}. \quad \text{(E.71)}$$

Because $[b^2] = [(b^x)^2] = 0$, we also have $[-(b^t)^2 + (b^y)^2 + (b^z)^2] = 0$, and thus,

$$-(b^t)^2 + (b^y)^2 + (b^z)^2 = C_b (= \text{const}). \quad \text{(E.72)}$$

Substituting equation (E.70) into (E.72) yields

$$a_{11}(b_y)^2 + a_{22}(b_z)^2 + 2a_{12}b^y b^z + 2a_{13}b^y + 2a_{23}b^z + a_{33} = 0, \quad \text{(E.73)}$$

where we used $a_{ij} = a_{ji}$ and

$$a_{11} = 1 - a_y^2, \quad a_{22} = 1 - a_z^2, \quad a_{33} = -C_b - a_t^2,$$
$$a_{12} = -a_y a_z, \quad a_{13} = -a_y a_t, \quad a_{23} = -a_z a_t.$$

Thus the transverse components, b^y and b^z, lie on a conic section; actually, they lie on the ellipse as shown in Komissarov (1997). Thus, the solution is determined for an arbitrary rotation angle, $\arctan(b^z/b^y)$. After the solution is determined, we should perform a Lorentz transformation to obtain the corresponding solution in the laboratory frame.

Appendix F

Landau–Lifshitz pseudo tensor

The Landau–Lifshitz pseudo tensor [Landau and Lifshitz (1962); Misner et al. (1973)] is useful for defining total mass-energy and linear momentum in isolated systems and energy flux of gravitational waves in a far zone. This quantity is also useful in numerical relativity for analyzing numerical results. Thus, we briefly describe the Landau–Lifshitz pseudo tensor in this appendix. For more detailed description, we recommend the readers to refer to chapter 20 of Misner et al. (1973).

For deriving their pseudo tensor, Landau and Lifshitz first defined the following quantity from the spacetime metric $g_{\mu\nu}$

$$\mathcal{G}^{\mu\nu} := \sqrt{-g}\, g^{\mu\nu}, \tag{F.1}$$

where g is the determinant of $g_{\mu\nu}$. Using $\mathcal{G}^{\mu\nu}$, the following superpotential is defined

$$H^{\mu\alpha\nu\beta} := \mathcal{G}^{\mu\nu}\mathcal{G}^{\alpha\beta} - \mathcal{G}^{\nu\alpha}\mathcal{G}^{\mu\beta}, \tag{F.2}$$

where $H^{\mu\alpha\nu\beta} = -H^{\alpha\mu\nu\beta} = -H^{\mu\alpha\beta\nu}$. Using this potential, terms composed of second derivatives of $g_{\mu\nu}$ in the Einstein tensor $G^{\mu\nu}$ is written by $(-2g)^{-1}\partial_\alpha\partial_\beta H^{\mu\alpha\nu\beta}$. Then, the Landau–Lifshitz pseudo tensor, $t_{LL}^{\mu\nu}$, is defined by

$$\begin{aligned}16\pi t_{LL}^{\mu\nu} &= (-g)^{-1}\partial_\alpha\partial_\beta H^{\mu\alpha\nu\beta} - 2G^{\mu\nu} \\ &= (-g)^{-1}\partial_\alpha\partial_\beta H^{\mu\alpha\nu\beta} - 16\pi T^{\mu\nu}.\end{aligned} \tag{F.3}$$

By definition, the Landau–Lifshitz pseudo tensor is composed only of the first derivatives of $g_{\mu\nu}$ [see equation (F.5)].

From equation (F.3), the conservation law is derived as

$$[(-g)(T^{\mu\nu} + t_{LL}^{\mu\nu})]_{,\nu} = 0. \tag{F.4}$$

This is useful for defining total energy and linear momentum of isolated systems as shown in the following.

In Landau and Lifshitz (1962), two expressions for $t_{LL}^{\mu\nu}$ are given. The first one is the expression in terms of connection coefficients, which is valid for arbitrary dimensionality D. The second one is the expression in terms of the metric functions, which depends on D as

$$16\pi(-g)t^{\mu\nu}_{LL} = \mathcal{G}^{\mu\nu}{}_{,\alpha}\mathcal{G}^{\alpha\beta}{}_{,\beta} - \mathcal{G}^{\mu\alpha}{}_{,\alpha}\mathcal{G}^{\nu\beta}{}_{,\beta} + \frac{1}{2}g^{\mu\nu}g_{\alpha\beta}\mathcal{G}^{\alpha\rho}{}_{,\sigma}\mathcal{G}^{\sigma\beta}{}_{,\rho}$$
$$- \left(g^{\mu\alpha}g_{\beta\rho}\mathcal{G}^{\nu\rho}{}_{,\sigma}\mathcal{G}^{\beta\sigma}{}_{,\alpha} + g^{\nu\alpha}g_{\beta\rho}\mathcal{G}^{\mu\rho}{}_{,\sigma}\mathcal{G}^{\beta\sigma}{}_{,\alpha}\right) + g_{\alpha\beta}g^{\rho\sigma}\mathcal{G}^{\mu\alpha}{}_{,\rho}\mathcal{G}^{\nu\beta}{}_{,\sigma}$$
$$+ \frac{1}{4}\left(2g^{\mu\alpha}g^{\nu\beta} - g^{\mu\nu}g^{\alpha\beta}\right)\left[g_{\rho\sigma}g_{\gamma\delta} - \frac{g_{\sigma\gamma}g_{\rho\delta}}{D-2}\right]\mathcal{G}^{\rho\delta}{}_{,\alpha}\mathcal{G}^{\sigma\gamma}{}_{,\beta}. \quad (F.5)$$

This shows that $t^{\mu\nu}_{LL}$ is a second-order quantity in $\partial_\mu \mathcal{G}^{\alpha\beta}$.

Because the Landau–Lifshitz pseudo tensor is not a tensor, it does not have a covariant meaning locally. However, equation (F.4) implies that a conservation law holds for global quantities. A spatial integration of equation (F.4) yields

$$\partial_t E_{LL} + \oint_S (-g)\left(T^{it} + t^{it}_{LL}\right) dS^f_i = 0, \quad (F.6)$$

$$\partial_t P^i_{LL} + \oint_S (-g)\left(T^{ij} + t^{ij}_{LL}\right) dS^f_j = 0, \quad (F.7)$$

where

$$E_{LL} := \frac{1}{16\pi}\int d^N x (-g)\left(T^{tt} + t^{tt}_{LL}\right), \quad (F.8)$$

$$P^i_{LL} := \frac{1}{16\pi}\int d^N x (-g)\left(T^{it} + t^{it}_{LL}\right). \quad (F.9)$$

dS^f_j is a surface-integral element simply defined by $(D-2)$-dimensional coordinates (see section 5.3) and S denotes a closed $(D-2)$-dimensional surface.

Here, we focus only on asymptotically flat spacetime. Then, if the second terms of equations (F.6) and (F.7) are evaluated at spatial infinity ($r \to \infty$), they vanish, and hence, E_{LL} and P^i_{LL} are conserved quantities [Landau and Lifshitz (1962)]. These are called Landau–Lifshitz energy and linear momentum. These quantities may be defined by a surface integral: Using equation (F.3), we have

$$E_{LL} = \frac{1}{16\pi}\oint_S \partial_j H^{tjtk} dS^f_k, \quad (F.10)$$

$$P^i_{LL} = \frac{1}{16\pi}\oint_S \partial_\mu H^{i\mu tk} dS^f_k, \quad (F.11)$$

where antisymmetric relations of $H^{\mu\alpha\nu\beta}$ are used for their derivations.

Using the variables of the $N+1$ formulation,

$$\partial_j H^{tjtk} = -\partial_j\left(\gamma\gamma^{jk}\right), \quad (F.12)$$
$$\partial_\mu H^{i\mu tk} = \partial_t\left(\gamma\gamma^{ik}\right) + \partial_j\left[\gamma\left(\beta^i\gamma^{jk} - \beta^j\gamma^{ik}\right)\right]$$
$$= 2\alpha\gamma\left(K^{ik} - K\gamma^{ik}\right) + \gamma\left(\gamma^{ik}\partial_j\beta^j - \gamma^{ij}\partial_j\beta^k\right) + \beta^i\partial_j\left(\gamma\gamma^{jk}\right). \quad (F.13)$$

Thus, in asymptotically flat spacetime that satisfies the conditions (5.9)–(5.12) for which the ADM mass and linear momentum are defined, $E_{LL} = M_{ADM}$ and $P^i_{LL} = P^{ADM}_i$ [Arnowitt et al. (1962); Misner et al. (1973)]. Although the Landau–Lifshitz pseudo tensor itself is not, the Landau–Lifshitz energy and linear momentum are, therefore, gauge-invariant, if it is evaluated at spatial infinity.

Landau and Lifshitz (1962) also showed that angular momentum is defined from the following conservation law:

$$[(-g)\{x^\alpha(T^{\mu\nu}+t_{LL}^{\mu\nu})-x^\mu(T^{\alpha\nu}+t_{LL}^{\alpha\nu})\}]_{,\nu}=0. \quad (F.14)$$

This yields angular momentum tensor

$$L_{LL}^{ij} = \frac{1}{16\pi}\int d^N x (-g)\left\{x^i\left(T^{jt}+t_{LL}^{jt}\right)-x^j\left(T^{it}+t_{LL}^{it}\right)\right\}. \quad (F.15)$$

If we evaluate the second terms of equations (F.6) and (F.7) at finite radius, they do not vanish; they denote the energy and linear momentum flux dissipated from an isolated system. The term composed of $T^{\mu\nu}$ is associated with the ejection of matter. On the other hand, the term composed of $t_{LL}^{\mu\nu}$ is associated with gravitational radiation. This fact is widely used for evaluating gravitational-wave flux in a wave zone in numerical relativity.

The Landau–Lifshitz energy may be evaluated for any closed surfaces. Consider to evaluate it at a finite (but sufficiently large) radius for an isolated system, and write it as $\tilde{E}_{LL}$ which is in general different from those evaluated at spatial infinity, E_{LL}, because of the possible existence of gravitational radiation. Because the asymptotic behavior of the integrands for the ADM mass and the Landau–Lifshitz energy is essentially the same, $\tilde{E}_{LL}$ is likely to agree approximately with the ADM mass, defined by equation (5.8), if they are evaluated at the same finite radius in a distant zone where the distance from the isolated system is sufficiently large and the spacetime metric is written in a linear perturbation form, $g_{\mu\nu} = \eta_{\mu\nu} + h_{\mu\nu}$ with $|h_{\mu\nu}| \ll 1$. Then, equation (F.6) suggests that for the approximate ADM mass evaluated at a finite radius (written as $\tilde{M}_{ADM}$), the following conservation relation should approximately hold:

$$\tilde{M}_{ADM}(t+\Delta t) = \tilde{M}_{ADM}(t) + \Delta E_{GW}. \quad (F.16)$$

Here, Δt denotes a time interval and ΔE_{GW} is an amount of energy emitted by gravitational waves in Δt computed by the second term of equation (F.6). In actual numerical-relativity simulations, the "approximate" ADM mass is usually evaluated at a finite distance from the isolated source, and it decreases with time. For such approximate ADM mass, the conservation law (F.16) should hold. Checking that this conservation law is satisfied is one of the important tests to confirm the reliability of numerical-relativity simulations.

Appendix G

Laws of black hole and apparent horizon

We first outline the four laws of black holes, which are defined following the remarkable similarity between the laws of thermodynamics and those of black-hole mechanics (see [Bardeen et al. (1973)] for the original summary).

The zeroth law of black-hole mechanics states that the surface gravity of stationary black holes, κ, is constant on the black-hole horizon, if a certain reasonable condition for the stress-energy tensor is satisfied (see section G.3). This corresponds to the zeroth law of thermodynamics which states that temperature is constant in thermal equilibrium. This law was shown by Bardeen et al. (1973) [see also Carter (1973a); Wald (1984); Poisson (2004) for comprehensive reviews].

The first law of black-hole mechanics determines the relation among variations in mass, spin angular momentum, and area of stationary black holes. This corresponds to the first law of thermodynamics. This law was also discovered by Bardeen et al. (1973). The extended first law, which is satisfied for a system that contains not only black holes but also matter fields, was also derived by them.

The second law of black-hole mechanics states that the black-hole area never decreases with time if the null energy condition is satisfied [see equation (G.10) for this condition]. This is often referred to as the area theorem for which the proof was shown by Hawking (1972) [see also Hawking and Ellis (1973)].

The third law of black-hole mechanics states that it is impossible to achieve $\kappa = 0$ by a physical process. This corresponds to the third law of thermodynamics that states it is impossible to achieve zero temperature by a physical process.

In numerical relativity, the first and second laws often play an important role because it gives theoretical foundations for the analysis of numerical-relativity results; e.g., the first law plays an important role when we explore how mass, spin angular momentum, and area of black holes evolve as a result of quasi-stationary evolution by mass accretion and/or gravitational radiation reaction in binary systems. The second law and related theorems such as theorems on apparent horizons provide an important baseline when we analyze dynamical black holes in numerical relativity. In this appendix, we in particular describe the first law in detail, paying attention to four-dimensional spacetime ($D = 4$), because it gives a foundation for the study of evolution sequences of binary compact objects in quasi-circular orbits.

Fig. G.1 Penrose–Carter diagram of Schwarzschild spacetime. The future null infinity, $\mathscr{I}^+$, and the boundary of its causal past, $\dot{J}^-(\mathscr{I}^+)$, are denoted by the thick lines, and black-hole physical singularities are described by the zigzags.

G.1 Raychaudhuri's equation

Loosely speaking, we say that spacetime contains a black hole if there exist outgoing null geodesics that never reach null infinity, $\mathscr{I}^+$ (see, e.g., figure G.1). This is because a null ray which originates from a black-hole interior (i.e., inside of its event horizon) cannot escape from the black hole and cannot reach $\mathscr{I}^+$. Therefore, the event horizon is generated by null geodesics with no future end point (i.e., the event horizon never disappear and never hit physical singularities).

The event horizon of a black hole is thus defined as a causal boundary of the causal past of null infinity, often written as $\dot{J}^-(\mathscr{I}^+)$, and thus, it must be a null surface (see figure G.1). Here, a null surface is characterized by a congruence of null geodesics. This implies that properties of a black hole are found by analyzing the congruence of null geodesics that characterize the event horizon (this congruence of null geodesics is often referred to as generators of the event horizon). In this section, we summarize tools for analyzing such null surfaces.

Let k^a be a null vector field which is tangent to null geodesics satisfying $k^a k_a = 0$ and $k^b \nabla_b k^a = 0$. Let l^a be another null vector field which satisfies $l^a l_a = 0$ and $k^a l_a = -1$. Then, the spacetime metric is written as

$$g_{ab} = -k_a l_b - l_a k_b + H_{ab}, \tag{G.1}$$

where H_{ab} is a two-dimensional metric which satisfies $H_{ab} k^a = H_{ab} l^a = 0$. In the following, we will project several quantities onto the two-dimensional surface of the metric H_{ab} orthogonal to k^a. The reason that a two-dimensional (not three-dimensional) surface is considered is that k^a is null and $g_{ab} \pm k_a k_b$ cannot be a projection operator.

Then, we introduce the following tensor field

$$B_{ab} := \nabla_b k_a. \tag{G.2}$$

Because k^a is null and a tangent vector field of null geodesics, $B_{ab} k^a = B_{ab} k^b = 0$. We then project B_{ab} onto the two-dimensional surface of H_{ab} and decompose it as

$$\hat{B}_{ab} := H_a{}^c H_b{}^d B_{cd} = \frac{1}{2}\Theta H_{ab} + \sigma_{ab} + \omega_{ab}, \tag{G.3}$$

where Θ, σ_{ab}, and ω_{ab} are called the expansion, shear, and rotation of the null geodesic congruence defined by

$$\Theta := H^{ab}B_{ab} = g^{ab}B_{ab}, \tag{G.4}$$

$$\sigma_{ab} := \frac{1}{2}\left(\hat{B}_{ab} + \hat{B}_{ba} - \Theta H_{ab}\right), \tag{G.5}$$

$$\omega_{ab} := \frac{1}{2}\left(\hat{B}_{ab} - \hat{B}_{ba}\right). \tag{G.6}$$

Here, to derive the last term of equation (G.4), we used the geodesic equation for k^a. The scalar expansion satisfies Raychaudhuri's equation, which is derived from an identity [e.g., see Poisson (2004)]

$$k^c \nabla_c \nabla_a k^a = \nabla_a(k^c \nabla_c k^a) - (\nabla_a k_c)\nabla^c k^a - R_{ac}k^a k^c, \tag{G.7}$$

as

$$k^a \nabla_a \Theta = -\frac{1}{2}\Theta^2 - \sigma^{ab}\sigma_{ab} + \omega_{ab}\omega^{ab} - R_{ab}k^a k^b, \tag{G.8}$$

where R_{ab} in this appendix denotes the four-dimensional Ricci tensor. To derive equation (G.8), we used the relation

$$B_{ab}B^{ba} = \hat{B}_{ab}\hat{B}^{ba} - (l^a k^c \nabla_c k_a)^2 - 2(\nabla_b k_a)l^a k^c \nabla_c k^b$$
$$= \frac{1}{2}\Theta^2 + \sigma_{ab}\sigma^{ab} - \omega_{ab}\omega^{ab} - (l^a k^c \nabla_c k_a)^2 - 2(\nabla_b k_a)l^a k^c \nabla_c k^b, \tag{G.9}$$

and the geodesic equation, $k^c \nabla_c k^a = 0$.

Einstein's equation yields $R_{ab}k^a k^b = 8\pi T_{ab}k^a k^b$. In the following, we will assume that the null energy condition is satisfied for T_{ab}:

$$T_{ab}k^a k^b \geq 0. \tag{G.10}$$

For a perfect fluid, this leads to $\rho h = \rho_{\text{ene}} + P \geq 0$ with $\rho_{\text{ene}} := \rho(1+\varepsilon)$, which is satisfied by normal fluids. We in addition will assume that the dominant energy condition is satisfied, which states that for a timelike vector V^a, $-T_{ab}V^b$ should be a future-directed timelike or null vector [see, e.g., chapter 9 of Wald (1984)]. The null dominant energy condition is an associated energy condition, which states that for a null vector k^a, $-T_{ab}k^b$ should be a future-directed timelike or null vector. This is a weaker condition than the dominant energy condition.

G.2 Expansion, shear, and rotation on event horizons

For stationary and axisymmetric spacetime that contains a black hole, we have a Killing vector field, which is null on the horizon and tangent to the generators of the horizon, as

$$\chi^a = \xi^a + \Omega_H \varphi^a, \tag{G.11}$$

where ξ^a and φ^a are timelike and spacelike Killing vector fields, respectively,

$$\xi^a := \left(\frac{\partial}{\partial t}\right)^a, \quad \varphi^a := \left(\frac{\partial}{\partial \varphi}\right)^a, \tag{G.12}$$

and Ω_H is the angular velocity of a black hole, as defined in section 1.3.1. Note that χ^a is null only on the horizon (and on the light cylinder) and it does not obey the geodesic equation.

In the above, we assumed both the stationarity and the axial symmetry for defining χ^a. However, for the calculation in this section, we only need to assume the presence of a Killing vector field χ^a that is null on the horizon and tangent to the generators of this Killing horizon. Assuming that a helical Killing vector field (see section 5.7.2) satisfies such a requirement, we will also see the identical property of black holes in binary systems in section G.4.2.

Now we derive the properties of a black hole associated with χ^a. In the following, let ℓ^a be a null vector field which satisfies $\ell^a \ell_a = 0$ and $\ell^a \chi_a = -1$ on the horizon. To avoid misunderstanding, we first define a null vector field $\hat{\chi}^a$ which agrees with χ^a only on the horizon (i.e., $\hat{\chi}^a$ is not a Killing vector field outside the horizon), and write the spacetime metric as

$$g_{ab} = -\hat{\chi}_a \ell_b - \ell_a \hat{\chi}_b + H_{ab}, \tag{G.13}$$

where H_{ab} is a two-dimensional metric, which satisfies $H_{ab}\hat{\chi}^b = H_{ab}\ell^b = 0$. Because $\hat{\chi}^a$ is a null vector, it satisfies equation (G.7) with the replacement $k^a = \hat{\chi}^a$. However, we have to keep in mind that $\hat{\chi}^a$ does not obey the geodesic equation because it does not have to be parametrized by an affine parameter. In this case, it obeys

$$\hat{\chi}^b \nabla_b \hat{\chi}^a = \hat{\kappa} \hat{\chi}^a, \tag{G.14}$$

where $\hat{\kappa}$ is a function on the null curve (not the surface gravity) for which $\hat{\chi}^a$ is tangent, and hence,

$$\hat{\kappa} = -\ell^a \hat{\chi}^b \nabla_b \hat{\chi}_a. \tag{G.15}$$

Thus, the Raychaudhuri's equation is replaced by

$$\hat{\chi}^a \nabla_a \Theta = \hat{\kappa}(\Theta + \hat{\kappa}) - (\nabla_a \hat{\chi}_c)\nabla^c \hat{\chi}^a - 8\pi T_{ac}\hat{\chi}^a \hat{\chi}^c, \tag{G.16}$$

where we used

$$\nabla_a \hat{\chi}^a = -\ell^a \hat{\chi}^b \nabla_b \hat{\chi}_a + H^{ab}\nabla_a \hat{\chi}_b = \hat{\kappa} + \Theta. \tag{G.17}$$

Here, equation (G.9) yields,

$$(\nabla_a \hat{\chi}_c)\nabla^c \hat{\chi}^a = \frac{1}{2}\Theta^2 + \sigma_{ab}\sigma^{ab} - \omega_{ab}\omega^{ab} + \hat{\kappa}^2. \tag{G.18}$$

Hence,

$$\hat{\chi}^c \nabla_c \Theta = \hat{\kappa}\Theta - \frac{1}{2}\Theta^2 - \sigma_{ab}\sigma^{ab} + \omega_{ab}\omega^{ab} - 8\pi T_{ac}\hat{\chi}^a \hat{\chi}^c. \tag{G.19}$$

We note that this equation can be used even if $\hat{\kappa}$ is not constant.

In the rest of this subsection, we only consider the quantities on the horizon setting $\hat{\chi}^a = \chi^a$. We immediately find that the expansion of χ^a, $\Theta = H^{ab}\nabla_a \chi_b$, becomes zero on the horizon because χ^a is a Killing vector field, i.e., $\nabla_{(a}\chi_{b)} = 0$. Because χ^a is orthogonal to the generators of the horizon surface, Frobenius' theorem tells us that the relation, $\chi_{[a}\nabla_b\chi_{c]} = 0$, is satisfied on the horizon (see, e.g., appendix A.2). A simple calculation from this (multiply another null vector ℓ^a and then project onto the horizon surface by multiplying $H_d{}^b H_e{}^c$) yields $\omega_{ab} = 0$.

Because $\Theta = 0 = \omega_{ab}$ on the horizon, equation (G.19) becomes

$$0 = \sigma_{ab}\sigma^{ab} + 8\pi T_{ab}\chi^a\chi^b \tag{G.20}$$

on the horizon. Here, $\sigma_{ab}\sigma^{ab} \geq 0$, because H_{ab} is a spatial tensor and $\det(H_{ab})$ is positive. Thus, the stress-energy tensor has to satisfy $T_{ab}\chi^a\chi^b \leq 0$. However, the null energy condition requires $T_{ab}\chi^a\chi^b \geq 0$. Therefore, equation (G.20) implies that $T_{ab}\chi^a\chi^b = 0 = \sigma_{ab}\sigma^{ab}$. In conclusion, the expansion, shear, and rotation of a Killing vector field χ^a are all zero on the event horizon. This is one of the important properties of stationary black holes.

G.3 Zeroth law

Here we show that the zeroth law is satisfied for stationary black holes. For the zeroth law to be satisfied, we only need to require the presence of one Killing vector field χ^a that is null on the horizon. As before, let ℓ^a be another null vector field orthogonal to the horizon surface, satisfying $\chi^a\ell_a = -1$ on the horizon. Then, the surface gravity, κ, is defined by

$$\kappa = -\ell^a\chi^b\nabla_b\chi_a, \tag{G.21}$$

which results from equation (1.139).

To show that κ is constant over the horizon, we define complex null vector fields lying on the horizon surface as m^a and $\bar{m}^a$, which satisfy the normalization relation $m^a\bar{m}_a = 1$. Then, the spacetime metric is written in the form

$$g_{ab} = -\chi_a\ell_b - \ell_a\chi_b + m_a\bar{m}_b + \bar{m}_a m_b. \tag{G.22}$$

Namely, $H_{ab} = m_a\bar{m}_b + \bar{m}_a m_b$.

Then, the purpose here becomes to show that $m^a\nabla_a\kappa = 0$ (and $\bar{m}^a\nabla_a\kappa = 0$). Using equation (G.21), we have on the horizon

$$\begin{aligned}m^a\nabla_a\kappa &= -m^c\nabla_c(\ell^a\chi^b\nabla_b\chi_a) \\ &= -m^c\ell^a\chi^b\nabla_c\nabla_b\chi_a - (\chi^b\nabla_b\chi_a)m^c\nabla_c\ell^a - \ell^a(\nabla_b\chi_a)m^c\nabla_c\chi^b \\ &= -m^c\ell^a\chi^b R_{abc}{}^d\chi_d - \kappa\chi_a m^c\nabla_c\ell^a - \ell^a(\nabla_b\chi_a)m^c\nabla_c\chi^b,\end{aligned} \tag{G.23}$$

where equations (A.7) and (G.14) were used to rewrite the first and second terms in the second line, and R_{abcd} in this appendix denotes the four-dimensional Riemann tensor. To rewrite the third term in the last line, we first use the conditions that the expansion, shear, and rotation of the generators of the horizon vanish for stationary black holes. These conditions imply that $m^b\nabla_b\chi_a$ does not have the components for the directions of m^a and $\bar{m}^a$. In addition, the ℓ_a component vanishes because of $\ell^a\ell_a = 0$. Thus, we have

$$m^b\nabla_b\chi_a = B\chi_a, \tag{G.24}$$

where $B = -\ell^a m^b \nabla_b \chi_a$. Then, the third term of equation (G.23) results in $B\kappa$, which cancels out the second term that is rewritten as $\kappa \ell^a m^b \nabla_b \chi_a = -\kappa B$. Hence,

$$m^c \nabla_c \kappa = -\ell^a \chi^b m^c \chi^d R_{abcd}. \tag{G.25}$$

The right-hand side of equation (G.25) is rewritten again using the fact that the expansion, shear, and rotation of the generators of the horizon vanish, and hence, $m^a \bar{m}^b \nabla_a \chi_b = 0$ and $m^c \nabla_c (m^a \bar{m}^b \nabla_a \chi_b) = 0$. The latter equation is written as

$$\begin{aligned} 0 &= m^c \nabla_c (m^a \bar{m}^b \nabla_b \chi_a) \\ &= m^a \bar{m}^b m^c \nabla_c \nabla_b \chi_a + (m^c \nabla_c \bar{m}^b) m^a \nabla_b \chi_a + (m^c \nabla_c m^a) \bar{m}^b \nabla_b \chi_a \\ &= R_{abcd} m^a \bar{m}^b m^c \chi^d - (m^c \nabla_c \bar{m}^b) B \chi_b + (m^c \nabla_c m^a) \bar{B} \chi_a \\ &= R_{abcd} m^a \bar{m}^b m^c \chi^d, \end{aligned} \tag{G.26}$$

where $\bar{B}$ is the complex conjugate of B and we used $\nabla_b \chi_a = -\nabla_a \chi_b$ for the second term in the second line. For deriving the last line, we used $(m^c \nabla_c \bar{m}^a) \chi_a = -m^c \bar{m}^a \nabla_c \chi_a = 0$ and $(m^c \nabla_c m^a) \chi_a = -m^c m^a \nabla_c \chi_a = 0$ with equation (G.24).

The further calculation is proceeded using equation (G.22) as

$$\begin{aligned} 0 = R_{abcd} m^a \bar{m}^b m^c \chi^d &= R_{abcd} m^a \chi^d (g^{bc} + \chi^b \ell^c + \chi^c \ell^b - m^b \bar{m}^c) \\ &= -R_{ad} m^a \chi^d + R_{abcd} m^a \chi^b \chi^d \ell^c, \end{aligned} \tag{G.27}$$

and thus, equation (G.25) is written as

$$m^c \nabla_c \kappa = -R_{cd} m^c \chi^d = -8\pi T_{cd} m^c \chi^d. \tag{G.28}$$

Since the dominant energy condition is assumed to be satisfied, $-T_{cd} V^d$ is a future-directed timelike or null vector for a timelike vector V^d. Then, $-T_{cd} \chi^d$ is also a future-directed timelike or null vector. On the other hand, Raychaudhuri's equation for the null vector shows that $T_{cd} \chi^c \chi^d = 0$ on the horizon (see section G.2). This means that $T_{cd} \chi^d$ is zero or parallel to χ_c. Therefore, if the dominant energy condition holds, $T_{cd} \chi^d$ is parallel to χ_c, and $T_{cd} m^c \chi^d = 0$, resulting in $m^c \nabla_c \kappa = 0$.

G.4 First law

The derivation of the first law is started from the Komar mass, defined in equation (5.36) [Bardeen et al. (1973)]. Using the null coordinates on the horizon, the Komar mass is written as [see, e.g., Poisson (2004) for the description of the surface element of null surfaces]

$$M_K = -\frac{1}{8\pi} \oint_{\mathcal{H}} dA (\chi_a \ell_b - \ell_a \chi_b) \nabla^a \xi^b + 2 \int_{\Sigma'_t} dV \left(T_{ab} - \frac{1}{2} g_{ab} T_c{}^c \right) \xi^a n^b, \tag{G.29}$$

where ξ^a is a timelike Killing vector field, dA is the area element of the horizon, $\sqrt{\gamma_2} d^2\theta$, with θ^i being two-dimensional angular coordinates, and Σ'_t denotes a region of the spatial hypersurface except for the inside of the event-horizon surface $\mathcal{H}$.

In the following, we will derive the first law for stationary and axisymmetric spacetime and helically symmetric spacetime, separately.

G.4.1 Stationary and axisymmetric case

For the derivation of the first law in stationary and axisymmetric spacetime, we basically follow Bardeen et al. (1973). We here focus only on spacetime in which the circularity condition (see section 5.7.1.1) is satisfied [Carter (1973a)].

Rewriting $\xi^a = \chi^a - \Omega_H \varphi^a$ and using the definition of κ, (G.21), the surface integral of equation (G.29) is written to a well-known form

$$-\frac{1}{8\pi} \oint_{\mathcal{H}} dA (\chi_a \ell_b - \ell_a \chi_b) \nabla^a \xi^b = \frac{1}{4\pi} \kappa A_H + 2\Omega_H J_H, \qquad (G.30)$$

where A_H and J_H are the area and (Komar) angular momentum of the black hole

$$A_H := \oint_{\mathcal{H}} dA, \qquad (G.31)$$

$$J_H := \frac{1}{16\pi} \oint_{\mathcal{H}} dA (\chi_a \ell_b - \ell_a \chi_b) \nabla^a \varphi^b. \qquad (G.32)$$

In the absence of T_{ab}, the black hole mass is defined by equation (G.30), and this is called the Smarr relation [Smarr (1973a)].

Denoting the four velocity of fluids as,

$$u^a = u^t (\xi^a + v^a), \qquad (G.33)$$

and substituting the perfect-fluid form of the stress-energy tensor, $T_{ab} = (\rho_{\text{ene}} + P) u_a u_b + P g_{ab}$, we have

$$\begin{aligned} T_{ab} \xi^a n^b &= T_{ab} (\xi^a + v^a) n^b - T_{ab} v^a n^b \\ &= T_{ab} \frac{u^a}{u^t} n^b - T_{ab} v^a n^b \\ &= -\rho_{\text{ene}} \xi^a n_a - (\rho_{\text{ene}} + P) u_a u_b v^a n^b \\ &= \alpha \rho_{\text{ene}} - \rho h u_a u_b v^a n^b, \end{aligned} \qquad (G.34)$$

where v^a is a spatial vector satisfying $v^a n_a = 0$ and $\xi^a n_a = -\alpha$. Then, using $T_c{}^c = -R/8\pi$, M_K is written as

$$\begin{aligned} M_K &= \frac{1}{4\pi} \kappa A_H + 2\Omega_H J_H + \int_{\Sigma'_t} \left(2\rho_{\text{ene}} \sqrt{-g} - 2\rho h u_a u_b v^a n^b \sqrt{\gamma} - \frac{1}{8\pi} R \sqrt{-g} \right) d^3 x \\ &= \frac{1}{4\pi} \kappa A_H + 2\Omega_H J_H - 2 \int_{\Sigma'_t} \left(\mathcal{L} + \rho h u_a u_b v^a n^b \sqrt{\gamma} \right) d^3 x, \end{aligned} \qquad (G.35)$$

where $\mathcal{L} := (R/16\pi - \rho_{\text{ene}}) \sqrt{-g}$ is an often-used Lagrangian density (see appendix D.1 and D.2).

Now we consider the variation of M_K. The variation here means comparing two slightly different solutions. Then, there is a certain gauge freedom; when two different solutions are compared, we have to determine the gauge (i.e., coordinates). The gauge freedoms are fixed by taking the following condition [Bardeen et al. (1973)]

$$\delta \xi^a = 0, \quad \delta \varphi^a = 0, \quad \text{for entire spacetime}, \qquad (G.36)$$

and
$$\delta\chi_a = C\chi_a, \quad \text{on the horizon}, \tag{G.37}$$

where C is some function on the horizon. The first two conditions fix the gauge conditions in t and φ (i.e., t and φ are, respectively, chosen to be identical in the two different solutions). Because of this choice, we can compare two solutions in the same spatial hypersurface Σ_t. The last condition fixes the gauge in r and θ on the horizon but does not constrain the gauge in t and φ because $\chi_\varphi = g_{t\varphi} + \Omega_H g_{\varphi\varphi}$ and $\chi_t = g_{tt} + \Omega_H g_{t\varphi}$ always vanish on the horizon due to the definition of $\Omega_H (= -g_{t\varphi}/g_{\varphi\varphi}$ on the horizon) and due to the fact that $\chi_a \chi^a = 0$ on the horizon. In this gauge condition, we have

$$\delta\chi^a = \delta\Omega_H \varphi^a, \tag{G.38}$$
$$\delta\chi_a = \delta g_{ab}\chi^b + g_{ab}\delta\Omega_H \varphi^b. \tag{G.39}$$

The variation of fluid quantities is described either by an Euler displacement or by a Lagrangian displacement in terms of a displacement vector ζ^a (see appendix D.2). The description of fluid perturbations in terms of the Lagrangian displacement also has a gauge freedom; a class of trivial displacement, $\zeta^a \propto u^a$, which yields no Euler change in the fluid variables [Friedman (1978); Friedman et al. (2002)]. We use this freedom for setting $\Delta t = 0$ which is equivalent to $\zeta^t = 0$ because δt trivially vanishes. Namely, comparison of two slightly different solutions is performed for the same time coordinate.

The variation of $-R\sqrt{-g}/8\pi$ is described in appendix D.1 and written by

$$-\frac{1}{8\pi}\delta\left(R\sqrt{-g}\right) = -\frac{1}{8\pi}\Big[G_{ab}\delta g^{ab}\sqrt{-g} + \nabla_a\left[g^{ad}g^{bc}(\nabla_b\delta g_{cd} - \nabla_d\delta g_{bc})\right]\sqrt{-g}\Big]. \tag{G.40}$$

Using the relation in the presence of the timelike Killing vector field, $\mathscr{L}_\xi V^a = 0$, for a given vector field V^a, we have

$$(\nabla_a V^a)\sqrt{-g} = D_a(\alpha \tilde{V}^a)\sqrt{\gamma}, \tag{G.41}$$

where $\tilde{V}^a = \gamma^a{}_b V^b + \alpha^{-1}\beta^a n_b V^b$, and thus,

$$\int_{\Sigma'_t}\left(\nabla_a \tilde{V}^a\right)\sqrt{-g}\,d^3x = \int_{\Sigma'_t} D_a(\alpha\tilde{V}^a)\sqrt{\gamma}\,d^3x$$
$$= \left(\oint_{r\to\infty} - \oint_{\mathcal{H}}\right)\alpha\tilde{V}^a n^b(s_a n_b - n_a s_b)(n_c n^c)dA$$
$$= \left(\oint_{r\to\infty} - \oint_{\mathcal{H}}\right)\alpha\tilde{V}^a n^b(\chi_a \ell_b - \ell_a \chi_b)(\chi_c \ell^c)dA. \tag{G.42}$$

Here, $V^a = g^{ad}g^{bc}(\nabla_b\delta g_{cd} - \nabla_d\delta g_{bc})$ satisfies the Killing relation $\mathscr{L}_\xi V^a = 0$. Then, the volume integral of the second term of equation (G.40) is written as

$$-\frac{1}{8\pi}\int_{\Sigma_t'}\nabla_a\left[g^{ad}g^{bc}\left(\nabla_b\delta g_{cd}-\nabla_d\delta g_{bc}\right)\right]\sqrt{-g}\,d^3x$$

$$=-\frac{1}{8\pi}\left[\oint_{r\to\infty}\alpha g^{ad}g^{bc}\left(\nabla_b\delta g_{cd}-\nabla_d\delta g_{bc}\right)s_a dA\right.$$

$$\left.+\oint_{\mathcal{H}}\alpha g^{ad}g^{bc}\left(\nabla_b\delta g_{cd}-\nabla_d\delta g_{bc}\right)n^e\left(\chi_a\ell_e-\ell_a\chi_e\right)dA\right]. \quad (G.43)$$

The definitions of the ADM mass given in equation (5.15) and of the Komar mass given in equation (5.40) immediately yield $-2\delta M_{\rm ADM}+\delta M_{\rm K}$ for the first integral term (in the second line), because at spatial infinity,

$$\alpha g^{ad}g^{bc}(\nabla_b\delta g_{cd}-\nabla_d\delta g_{bc})s_a \to s^k\gamma^{ij}(\partial_i\delta\gamma_{jk}-\partial_k\delta\gamma_{ij})-2s^k\partial_k\delta\alpha. \quad (G.44)$$

Here we used $g^{ab}\delta g_{ab}=\delta\ln(\alpha^2\gamma)$.

Using $\alpha n^b\ell_b=(\xi^b-\beta^b)\ell_b=\chi^b\ell_b=-1$ and $\alpha n^b\chi_b=(\xi^b-\beta^b)\chi_b=0$ on the horizon (note $\beta^b=-\Omega_{\rm H}\varphi^b$ on the horizon), the second integral term of equation (G.43) is written as

$$\frac{1}{8\pi}\oint_{\mathcal{H}}\chi^d g^{bc}\left(\nabla_b\delta g_{cd}-\nabla_d\delta g_{bc}\right)dA=\frac{1}{8\pi}\oint_{\mathcal{H}}\chi^d g^{bc}\nabla_b\delta g_{cd}dA$$

$$=-\frac{1}{8\pi}\oint_{\mathcal{H}}\chi_d\nabla_b\delta g^{bd}dA, \quad (G.45)$$

where we used $g^{ab}\chi^d\nabla_d\delta g_{ab}=g^{ab}\mathscr{L}_\chi\delta g_{ab}=0$. The last term of equation (G.45) is related to the variation of κ as follows. First, $\delta\kappa$ on the horizon is calculated as

$$\delta\kappa=-\delta(\ell^a\chi^b\nabla_b\chi_a)$$
$$=\delta(\ell^a\chi^b\nabla_a\chi_b)$$
$$=\delta\ell^a\chi^b\nabla_a\chi_b+\ell^a\delta\chi^b\nabla_a\chi_b+\ell^a\chi^b\nabla_{[a}\delta\chi_{b]}, \quad (G.46)$$

where $\delta(\nabla_a\chi_b)=\delta(\nabla_{[a}\chi_{b]})=\nabla_{[a}\delta\chi_{b]}$ from the Killing relation for χ_a. The first term in the last line is rewritten as

$$\delta\ell^a\chi^b\nabla_a\chi_b=-\delta\ell^a\kappa\chi_a=\kappa\ell^a\delta\chi_a=-\kappa C \quad \text{[we used equation (G.37)]}$$
$$=-\ell^a(C\chi_b)\nabla_a\chi^b=-\ell^a\delta\chi_b\nabla_a\chi^b$$
$$=\chi^a\ell^b\nabla_a\delta\chi_b, \quad (G.47)$$

where we used $\mathscr{L}_\chi\delta\chi_a=0$ on the horizon to derive the last term. Substituting equation (G.47) into (G.46) yields,

$$\delta\kappa=\frac{1}{2}\left(\chi^a\ell^b+\ell^a\chi^b\right)\nabla_a\delta\chi_b+\ell^a\delta\chi^b\nabla_a\chi_b$$
$$=-\frac{1}{2}\left(g^{ab}-m^a\bar{m}^b-\bar{m}^a m^b\right)\nabla_a\delta\chi_b+\ell^a\delta\Omega_{\rm H}\varphi^b\nabla_a\chi_b$$
$$=\frac{1}{2}\chi_a\nabla_b\delta g^{ab}+\ell^a\delta\Omega_{\rm H}\chi^b\nabla_a\varphi_b, \quad (G.48)$$

where we used the gauge condition on the horizon, (G.37), together with equation (G.24), and $\varphi^b \nabla_a \chi_b = -\varphi^b \nabla_b \chi_a = -\chi^b \nabla_b \varphi_a = \chi^b \nabla_a \varphi_b$ to derive the last line. Using equation (G.48), the surface integral of (G.45) is finally written as

$$-\frac{1}{8\pi} \oint_{\mathcal{H}} \chi_d \nabla_b \delta g^{bd} dA = -\frac{1}{8\pi} \oint_{\mathcal{H}} \left[2\delta\kappa + \delta\Omega_{\mathrm{H}} (\chi^a \ell^b - \ell^a \chi^b) \nabla_a \varphi_b \right] dA$$

$$= -\frac{1}{4\pi} \delta\kappa A_{\mathrm{H}} - 2\delta\Omega_{\mathrm{H}} J_{\mathrm{H}}. \qquad (G.49)$$

We next evaluate the variation of the fluid terms in equation (G.35), defining

$$I_{\mathrm{m}} := \int_{\Sigma_t'} \left(\rho_{\mathrm{ene}} \sqrt{-g} - \rho h u_a u_b v^a n^b \sqrt{\gamma} \right) d^3x. \qquad (G.50)$$

The variation of $\rho_{\mathrm{ene}} \sqrt{-g}$ is described in the same manner as in section D.2. Here we explicitly incorporate changes in baryon number and entropy. In the presence of a change in baryon mass, equation (D.20) is replaced by

$$\Delta \rho = -\frac{1}{2} \rho q^{ab} \Delta g_{ab} + \frac{\Delta(\rho u^a d\Sigma_a)}{u^b d\Sigma_b}, \qquad (G.51)$$

where $q^{ab} := g^{ab} + u^a u^b$. Note that $\rho u^a d\Sigma_a = -\rho u^a n_a \sqrt{\gamma} d^3x$. Using equation (D.25) with $\Delta s \neq 0$, we have [compare with equation (D.27)]

$$\frac{1}{\sqrt{-g}} \delta \left(\rho_{\mathrm{ene}} \sqrt{-g} \right) = \rho T \Delta s + \frac{h}{u^a d\Sigma_a} \Delta (\rho u^a d\Sigma_a)$$

$$- \frac{1}{2} T^{ab} \delta g_{ab} + \zeta_a \nabla_b T^{ab} - \nabla_a \left[\rho h q^{ab} \zeta_b \right]. \qquad (G.52)$$

The last term of this equation is rewritten as

$$\sqrt{-g} \nabla_a \left[\rho h q^{ab} \zeta_b \right] = \rho h u^t u_b \sqrt{-g} \mathcal{L}_\xi \zeta^b + \sqrt{\gamma} D_a \left[\alpha \rho h q^{ab} \zeta_b \right]$$

$$= -\rho h u^a n_a u_b \sqrt{\gamma} \mathcal{L}_\xi \zeta^b + \sqrt{\gamma} D_a \left[\alpha \rho h q^{ab} \zeta_b \right], \qquad (G.53)$$

where we used the fact that fluid quantities except for the Lagrangian displacement are time-independent in the presence of the timelike Killing vector field ξ^a.

For the second term of equation (G.50), the variation is evaluated by the Lagrangian method to yield

$$\Delta \left(\rho h u_a u_b v^a n^b \sqrt{\gamma} d^3x \right) = \Delta \left(\rho u_b n^b \sqrt{\gamma} d^3x \right) h u_a v^a + (\Delta h u_a) \rho v^a u_b n^b \sqrt{\gamma} d^3x$$

$$+ (\Delta v^a) \rho h u_a u_b n^b \sqrt{\gamma} d^3x$$

$$= -\Delta \left(\rho u^b d\Sigma_b \right) h u_a v^a - (\Delta h u_a) v^a \rho u^b d\Sigma_b$$

$$- (\Delta v^a) h u_a \rho u^b d\Sigma_b. \qquad (G.54)$$

Here, in the gauge $\zeta^t = 0$, equation (D.18) is written as

$$\Delta u^t = \frac{1}{2} u^t u^b u^c \Delta g_{bc}. \qquad (G.55)$$

This with equations (D.18) and (G.33) yields $\Delta u^a = (\Delta u^t)(\xi^a + v^a)$. On the other hand, $\Delta u^a = \Delta[u^t(\xi^a + v^a)]$. Thus, $\Delta(\xi^a + v^a) = 0$, and hence, for the last term of equation (G.54), $\Delta v^a = -\Delta \xi^a = \mathscr{L}_\xi \zeta^a$ for $a \neq t$, and $\Delta v^t = 0$ because $\zeta^t = 0$. Note that we here used the gauge condition, $\delta \xi^a = 0$.

Thus, in total,

$$\delta I_{\rm m} = \int_{\Sigma'_t} \left[\frac{T}{u^t}(\Delta s)\rho u^a d\Sigma_a + \left(\frac{h}{u^t} + h u_b v^b\right) \Delta(\rho u^a d\Sigma_a) + v^b \Delta(h u_b) \rho u^a d\Sigma_a \right.$$
$$\left. + \left(-\frac{1}{2}\sqrt{-g} T^{ab} \delta g_{ab} + \sqrt{-g} \zeta_a \nabla_b T^{ab} - \sqrt{\gamma} D_a \left[\alpha \rho h q^{ab} \zeta_b \right] \right) d^3 x \right]. \tag{G.56}$$

The last term of this equation can be rewritten to a surface integral using Stokes' theorem, and then, assuming that $\zeta^a = 0$ on boundary surfaces, this term vanishes.

We finally obtain the variation of $M_{\rm K}$ as

$$\delta M_{\rm K} = \frac{1}{4\pi}\kappa \delta A_{\rm H} + 2\Omega_{\rm H} \delta J_{\rm H} - 2\delta M_{\rm ADM} + \delta M_{\rm K}$$
$$+ 2 \int_{\Sigma'_t} \left[\frac{\rho T}{u^t}(\Delta s) u^a d\Sigma_a + \left(\frac{h}{u^t} + h u_b v^b\right) \Delta(\rho u^a d\Sigma_a) + v^b \Delta(h u_b) \rho u^a d\Sigma_a \right.$$
$$\left. + \left(\frac{1}{16\pi}\sqrt{-g}\left(G^{ab} - 8\pi T^{ab}\right) \delta g_{ab} + \sqrt{-g}\zeta_a \nabla_b T^{ab} \right) d^3 x \right]. \tag{G.57}$$

Using $G^{ab} = 8\pi T^{ab}$ and $\nabla_a T^{ab} = 0$, we arrive at

$$\delta M_{\rm ADM} = \Omega_{\rm H} \delta J_{\rm H} + \frac{\kappa}{8\pi}\delta A_{\rm H}$$
$$+ \int_{\Sigma'_t} \left[\frac{T}{u^t}(\Delta s) dM_B + \left(\frac{h}{u^t} + h u_b v^b\right) \Delta(dM_B) + v^b \Delta(h u_b) dM_B \right], \tag{G.58}$$

where $dM_B := \rho u^a d\Sigma_a$. Interestingly, $M_{\rm K}$ does not appear in the final result. We also note that equation (G.58) is satisfied irrespective of the virial relation, $M_{\rm K} = M_{\rm ADM}$.

Before closing this subsection, we pay attention to two special cases. First, using a thermodynamic relation, $h = Ts + \mu$, where μ is chemical potential, and assuming a purely rotational velocity field, $v^a = \Omega \varphi^a$, where Ω is the angular velocity of fluid elements, the second line of equation (G.58) is rewritten to yield [Bardeen et al. (1973)]

$$\delta M_{\rm ADM} = \Omega_{\rm H} \delta J_{\rm H} + \frac{\kappa}{8\pi}\delta A_{\rm H}$$
$$+ \int_{\Sigma'_t} \left[\frac{T}{u^t}\Delta(s dM_B) + \frac{\mu}{u^t}\Delta(dM_B) + \Omega \Delta(j dM_B) \right], \tag{G.59}$$

where $j = h u_a \varphi^a$ is the specific angular momentum of fluid elements and we used $\mathscr{L}_\varphi \zeta^a = 0$. This is the differential mass formula obtained by Bardeen et al. (1973). Therefore, in the conservation of entropy, rest mass, and specific angular momentum for each fluid element, we have

$$\delta M_{\rm ADM} = \Omega_{\rm H} \delta J_{\rm H} + \frac{\kappa}{8\pi}\delta A_{\rm H}. \tag{G.60}$$

Next, we consider the case that the vorticity and entropy are conserved for fluid elements. The last term of equation (G.58) is related to the conservation of the fluid vorticity which is defined by

$$\tilde{\omega}_{ab} := q_a{}^c q_b{}^d [\nabla_c(hu_d) - \nabla_d(hu_c)]$$
$$= \nabla_a(hu_b) - \nabla_b(hu_a), \tag{G.61}$$

where we used the Euler equation and the first law of thermodynamics with $ds = 0$ to derive the second line. Thus, if $\tilde{\omega}_{ab} = 0$, hu_a is written using a scalar potential, Ψ, as $hu_a = \nabla_a \Psi$.

Taking a Lagrangian displacement of $\tilde{\omega}_{ab}$ yields

$$\Delta\tilde{\omega}_{ab} = \nabla_a \Delta(hu_b) - \nabla_b \Delta(hu_a). \tag{G.62}$$

This implies that $\Delta(hu_a)$ is written in the form

$$\Delta(hu_a) = \nabla_a \Delta\Psi, \tag{G.63}$$

where $\Delta\Psi$ is a perturbed scalar potential. Substituting equation (G.63) into the last term of equation (G.58) yields,

$$\int_{\Sigma'_t} v^b \nabla_b(\Delta\Psi) \rho u^a d\Sigma_a$$
$$= \int_{\Sigma'_t} \nabla_b \left[v^b(\Delta\Psi) \rho u^t \right] \sqrt{-g}\, d^3x - \int_{\Sigma'_t} \nabla_b \left(\rho v^b u^t \right) (\Delta\Psi) \sqrt{-g}\, d^3x. \tag{G.64}$$

The first term vanishes because it can be rewritten to a surface integral (note $v^t = 0$) which vanishes because $\Delta\Psi = 0$ in the surfaces. The second term is calculated using

$$\nabla_b \left(\rho v^b u^t \right) = \nabla_b \left[\rho(u^b - u^t \xi^b) \right] = -\xi^a \nabla_a(\rho u^t) = 0, \tag{G.65}$$

where we used the continuity equation and Killing relations of ξ^a. Thus, equation (G.64) vanishes, if the vorticity is conserved. Therefore, in the conservation of entropy, rest mass, and vorticity for each fluid element, equation (G.60) is also satisfied.

G.4.2 *Helically symmetric case*

The first law in the helically symmetric case can be derived in a calculation similar to (but slightly modified from) that of the stationary and axisymmetric case described in the previous subsection. We here denote the helical Killing vector field by k^a which satisfies $k^a = t^a + \Omega \varphi^a$ where

$$t^a = \left(\frac{\partial}{\partial t}\right)^a, \quad \varphi^a = \left(\frac{\partial}{\partial \varphi}\right)^a, \tag{G.66}$$

and Ω is an angular velocity which is equal to the orbital one for binary systems in quasi-circular orbits. t^a and φ^a are not Killing vector fields in the present context, although t^a is assumed to be asymptotically a timelike Killing vector at $r \to \infty$.

k^a is assumed to be a null vector on the black hole horizon, $k^a k_a = 0$, i.e., null geodesics for which k^a is tangent are the generators of an event horizon.

Using k^a, the expression for the Komar mass, (5.34), is rewritten as

$$M_{\rm K} = -\frac{1}{4\pi} \oint_S dS_a n_b \nabla^a k^b + 2\Omega J_{\rm K}. \tag{G.67}$$

Thus, as in equation (G.29), we have

$$M_{\rm K} - 2\Omega J_{\rm K} = -\frac{1}{8\pi} \oint_{\mathcal{H}} dA (k_a \ell_b - \ell_a k_b) \nabla^a k^b$$
$$+ \int_{\Sigma'_t} dV \left(2T_{ab} - g_{ab} T_c{}^c \right) k^a n^b, \tag{G.68}$$

where ℓ^a is again an auxiliary null vector field that satisfies $\ell^a k_a = -1$ on the horizon. The surface integral term reduces to $\kappa A_{\rm H}/4\pi$ in this case [Friedman et al. (2002)] where κ is defined by $-\ell^a k^b \nabla_b k_a$ on the horizon. We note that the Komar mass is not always equal to the ADM mass in the present context and thus it is not always "the mass". However, it is a useful quantity for deriving the first law.

Using $u^a = u^t(k^a + v^a)$ where v^a is again a spatial vector that satisfies $v^a n_a = 0$, the first perfect fluid term of equation (G.68) is written as

$$T_{ab} k^a n^b = \alpha \rho_{\rm ene} - \rho h u_a u_b v^a n^b, \tag{G.69}$$

where we used $k^a n_a = -\alpha$. Here, we assume that the fluid exists only outside the event horizon and inside the light cylinder $r < \Omega^{-1}$, in which k^a is timelike as $k_a k^a < 0$. With the relation $T_c{}^c = -R/8\pi$, $M_{\rm K}$ is written in the form

$$M_{\rm K} = \frac{1}{4\pi} \kappa A_{\rm H} + 2\Omega J_{\rm K}$$
$$- 2 \int_{\Sigma'_t} \left(\frac{R\sqrt{-g}}{16\pi} - \rho_{\rm ene} \sqrt{-g} + \rho h u_a u_b v^a n^b \sqrt{\gamma} \right) d^3 x. \tag{G.70}$$

As a gauge choice, it is convenient to choose $\delta k^a = 0$ and $\delta k_a = C k_a$ on the horizon, where C is some function on the horizon [Friedman et al. (2002)]; compare these conditions with equations (G.36) and (G.37). $\delta k_a = C k_a$ on the horizon constrains the choice for δk_r and δk_θ, but it does not constrain the condition for δk_t and δk_φ because it is automatically satisfied if the condition $k^a k_a = 0$ is satisfied on the horizon. It is also worthy to note that the gauge $\delta k^a = 0$ results in $\mathscr{L}_k \delta k_a = 0$ and $\mathscr{L}_k \delta g_{ab} = 0$ on the horizon.

There is one additional degree of freedom for the gauge condition. For the naturalness, we fix this freedom by choosing $\delta \varphi^a = 0$. In this case, $\delta t^a \neq 0$, but $\delta t^a = -(\delta \Omega) \varphi^a$.

Using the gauge conditions and resulting relations, it is straightforward to show that the variation of the surface integral is calculated to yield [compare it with equation (G.48)]

$$\delta \kappa = \frac{1}{2} k_a \nabla_b \delta g^{ab}. \tag{G.71}$$

For a detailed derivation, see Friedman et al. (2002).

Now, we take the variation of M_K. As we mentioned in the previous subsection, the variation means comparing two slightly different solutions. When we compare quantities that appear in some integral on the spatial hypersurface, the time should be identical for the entire spacetime. However, in our choice of the gauge condition, $\delta t^a \neq 0$. Thus, this difference in the time slicing has to be corrected by slightly changing the time as a first step. Since M_K is originally written by

$$M_K = -\frac{1}{4\pi} \oint_S dS_a n_b \nabla^a t^b, \tag{G.72}$$

the correction for δM_K should be

$$(\delta M_K)_{\text{correction}} = -\frac{1}{4\pi} \oint_S dS_a n_b \nabla^a \delta t^b$$

$$= \frac{1}{4\pi} \delta\Omega \oint_S dS_a n_b \nabla^a \varphi^b = 2(\delta\Omega) J_K \tag{G.73}$$

In the following, the calculation will be performed supposing that two time slices are identical in two solutions. We then subtract the correction term in the final moment for deriving the correct variation of δM_K.

If two time slices are identical, for the fluid part, I_m, shown in equation (G.50), the result of the variation is the same as in equation (G.56). The variation of the Hilbert term is written in equation (G.40) and we here set

$$V^a = g^{ad}g^{bc}(\nabla_b \delta g_{cd} - \nabla_d \delta g_{bc}). \tag{G.74}$$

For this

$$\sqrt{-g}\nabla_a V^a = \sqrt{\gamma} D_a(\alpha \tilde{V}^a), \tag{G.75}$$

where

$$\tilde{V}^a = \gamma^a_{\ b} V^b - \alpha^{-1} \Omega \varphi^a n_b V^b. \tag{G.76}$$

This expression is different from equation (G.41), but the term proportional to φ^a does not contribute to the surface integrals because $\varphi^a k_a = \varphi^a \ell_a = 0$ on the horizon and $\varphi^a s_a = \varphi^a n_a = 0$ for $r \to \infty$. Thus, the volume integral yields the same answer as equation (G.43), i.e.,

$$-\frac{1}{8\pi}\int_{\Sigma'_t}(\nabla_a V^a)\sqrt{-g}\,d^3x = -2\delta M_{\text{ADM}} + \delta M_K - \frac{1}{4\pi}\delta\kappa A_H, \tag{G.77}$$

where we used equation (G.71).

Finally, we obtain the variation of M_K as

$$\delta M_K = \frac{1}{4\pi}\kappa \delta A_H - 2\delta M_{\text{ADM}} + \delta M_K + 2\delta(\Omega J_K) - (\delta M_K)_{\text{correction}}$$

$$+ 2\int_{\Sigma'_t}\left[\frac{\rho T}{u^t}(\Delta s)u^a d\Sigma_a + \left(\frac{h}{u^t} + hu_b v^b\right)\Delta(\rho u^a d\Sigma_a) + v^b \Delta(hu_b)\rho u^a d\Sigma_a\right.$$

$$\left. + \left(-\frac{1}{16\pi}\sqrt{-g}\left(G^{ab} - 8\pi T^{ab}\right)\Delta g_{ab} + \sqrt{-g}\zeta_a \nabla_b T^{ab}\right)d^3x\right], \tag{G.78}$$

and using $G^{ab} = 8\pi T^{ab}$ and $\nabla_a T^{ab} = 0$ with equation (G.73), we obtain

$$\delta M_{\text{ADM}} = \Omega \delta J_{\text{K}} + \frac{\kappa}{8\pi} \delta A_{\text{H}}$$
$$+ \int_{\Sigma'_t} \left[\frac{T}{u^t}(\Delta s) dM_B + \left(\frac{h}{u^t} + h u_b v^b \right) \Delta(dM_B) + v^b \Delta(h u_b) dM_B \right]. \text{(G.79)}$$

As before, the matter term vanishes if the rest mass, entropy, and vorticity are conserved for each fluid element. We note again that M_{K} does not appear in the final result and the first law is satisfied irrespective of the virial relation, $M_{\text{K}} = M_{\text{ADM}}$.

Before closing this section, we remark that the ADM mass and angular momentum are not always defined in the helically symmetric spacetime, although we implicitly assumed that they could be defined. As mentioned in section 5.7.2, spacetime of a binary system which is exactly helically symmetric cannot be asymptotically flat because of the presence of standing gravitational waves. In reality, helically symmetric binary systems never exist because of the presence of gravitational-wave emission and its radiation reaction. However, realistic, asymptotically-flat binary systems should be approximately helical symmetric as far as we focus only on the region inside the light cylinder. Therefore, it is reasonable to consider that for sequences of realistic binary systems, the first law written by equation (G.79) is approximately satisfied.

G.5 Apparent horizon and dynamical horizon

Apparent horizon (outer marginally trapped surface) plays a very important role in numerical relativity because the existence of a black hole is usually confirmed by the presence of an apparent horizon in numerical relativity: In contrast to event horizon which is associated with a global structure of spacetime, apparent horizons can be defined on each spatial hypersurface Σ_t.

To strictly define apparent horizon, we first have to define *trapped surface*: A trapped surface on Σ_t is a closed two-dimensional surface $\mathcal{T}$, having the property that the expansion of both ingoing and outgoing congruences of future-directed null geodesics orthogonal to $\mathcal{T}$ is everywhere negative. Then, *trapped region* is defined as a portion of Σ_t that contains trapped surfaces. Based on these preparations, the apparent horizon is defined by the boundary of the trapped region on Σ_t, and hence, the expansion for one (outgoing) congruence of null geodesics originating from the apparent horizon is zero, $\Theta = 0$. Because of its nature, apparent horizons are also called outer marginally trapped surfaces.

If a present hypersurface Σ_t contains an apparent horizon, its nearby future and past hypersurfaces also contain apparent horizons in general. Hence, the surface of the apparent horizon can be extended toward the future (and past). The union of all the apparent-horizon surfaces form a three-dimensional hypersurface. This is called the *trapping horizon* [Hayward (1994)]. The trapping horizon is either null or spacelike. By this property, the trapping horizon (and apparent horizon) can be distinguished by event horizon, as well as isolated horizon, which is introduced in the next section.

By analyzing Raychaudhuri's equation, it is straightforward to show that null geodesics orthogonal to a trapped surface never reach null infinity if the spacetime concerned is globally hyperbolic and if the null energy condition is satisfied [see chapter 12 of Wald (1984) for a proof]. This is simply due to the fact that the expansion of null geodesics on the trapped surface is initially negative $\Theta_0 < 0$ and hence within a finite affine length $\leq 2/|\Theta_0|$, a physical singularity is reached, never reaching null infinity. This implies that the trapped surface is contained in an event horizon. To show that apparent horizons with zero expansion are also contained inside event horizons, a more careful and mathematical proof is necessary, but it is still possible to prove that null geodesics originating from the apparent horizons never reach null infinity [see Wald (1984) for a concise proof]. Therefore, the existence of an apparent horizon means the existence of an event horizon encompassing the apparent horizon. In relation with this fact, the area of the apparent horizon $A_{\rm AH}$ gives the lower bound for the area of the event horizon $A_{\rm H}$, i.e., $A_{\rm AH} \leq A_{\rm H}$.

Apparent horizons coincide with corresponding surfaces of an event horizon for stationary black holes. This is quite natural because both horizons are defined as the surfaces with vanishing expansion of a null geodesics congruence. This also suggests that if the spacetime is nearly stationary and if we choose time slicing in which the spacetime is observed to be nearly stationary, an apparent horizon is likely to agree approximately with the corresponding event horizon. (See below for more sophisticated definitions of horizons.)

A shortcoming of apparent horizons is that it is not gauge-invariant. The location and property of apparent horizons depend on the chosen time slice. In the worst case, any apparent horizon may be absent on Σ_t even if an event horizon is present for the corresponding spacetime. This indicates that it may not be easy to confirm the absence of a black hole, if we search only for apparent horizons; even if any apparent horizon is absent, a black hole may be present. However, in numerical relativity performed with popular coordinate conditions so far, such a situation has been very rarely encountered; finding apparent horizons is likely to be a robust strategy for finding black holes in numerical relativity.

We then touch on *dynamical horizon*. This is a quite similar (and in the context of numerical relativity essentially identical) notion to the trapping horizon proposed and defined by Ashtekar and Krishnan (2003). As already mentioned, we can consider the union of apparent horizons which may be either null or spacelike. The dynamical horizon is defined as the *spacelike* three-dimensional portion for the union of apparent horizons. When it relaxes to a null surface after a dynamical phase is over, it is supposed to relax to an *isolated horizon* (see next section). By definition, an event horizon has to be located outside a dynamical horizon.

On the dynamical horizon, even if it is dynamical, it is possible to define angular

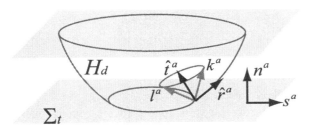

Fig. G.2 A spacetime diagram for a dynamical horizon H_d and vectors $\hat{t}^a$, $\hat{r}^a$, n^a, and s^a. k^a denotes an outgoing null vector field defining the apparent horizon on Σ_t, and l^a is another null vector field. Circles on every spatial hypersurface Σ_t denote apparent horizons, and the cone for which k^a and l^a are tangent denotes a light cone.

momentum, area, mass, angular velocity, and surface gravity as in the case of stationary black holes, if a certain symmetry is present on the horizon surface. Among these quantities, the first law is also satisfied. In the context of numerical relativity, important quantities are mass and angular momentum [see Schnetter et al. (2006) for a concise description to people in the field of numerical relativity]. These are computed in the following manner. First, a timelike unit vector field normal to (spacelike) surfaces of the dynamical horizon H_d is defined as $\hat{t}^a$. Then the induced metric on H_d is defined by $q_{ab} := g_{ab} + \hat{t}_a \hat{t}_b$. In addition, let $\hat{r}^a$ be a unit spacelike vector field orthogonal to two-dimensional surfaces on H_d (see figure G.2 for the relation among H_d, $\hat{t}^a$, $\hat{r}^a$, n^a, and s^a, where n^a is the timelike unit vector field normal to Σ_t and s^a is the unit radial vector field on Σ_t, which is orthogonal to the horizon surface; see also section 7.1).

The area, $\mathcal{A}$, is simply computed by the surface integral of a two-dimensional cross section on H_d for which the induced metric is $q_{ab} + \hat{r}_a \hat{r}_b$. From this, an areal radius is defined by $R_{H_d} := \sqrt{\mathcal{A}/4\pi}$. Angular momentum is defined in the manner similar to the Komar angular momentum (5.48) as

$$J_{H_d} = \frac{1}{8\pi} \oint_{H_d} \bar{K}_{ab} \varphi^a \hat{r}^b dA, \tag{G.80}$$

where $\bar{K}_{ab}$ is the extrinsic curvature on H_d, $\bar{K}_{ab} := -q_a{}^c q_b{}^d \nabla_c \hat{t}_d$. We note that the interpretation of J_{H_d} as angular momentum is valid only for the case $\mathcal{L}_\varphi \bar{K}_{ab} = 0 = \mathcal{L}_\varphi q_{ab}$. To express J_{H_d}, we remind the readers

$$q_{ab} = \gamma_{ab} + \hat{t}_a \hat{t}_b - n_a n_b, \tag{G.81}$$

$$k^a = \frac{1}{\sqrt{2}} \left(\hat{t}^a + \hat{r}^a \right) = \frac{f}{\sqrt{2}} \left(n^a + s^a \right), \tag{G.82}$$

$$l^a = \frac{1}{\sqrt{2}} \left(\hat{t}^a - \hat{r}^a \right) = \frac{1}{f\sqrt{2}} \left(n^a - s^a \right), \tag{G.83}$$

where f is a boost factor. Then, equation (G.80) is rewritten to give the angular momentum in the 3+1 quantities as

$$J_{H_d} = \frac{1}{8\pi} \oint_{H_d} \left(K_{ab}\varphi^a s^b - \mathscr{L}_\varphi \ln f \right) dA. \tag{G.84}$$

Again, if φ^a is a Killing vector field on H_d, the second term vanishes.

The mass (as well as angular velocity and surface gravity) is not directly defined in the framework of dynamical horizon. However, the first law is derived based on the Hamiltonian framework [Ashtekar and Krishnan (2003)]. For stationary and axisymmetric black holes (Kerr black holes), the angular velocity and surface gravity can be written as functions of area and angular momentum (see section 1.3.2.3). In this case,

$$\kappa = \frac{R_{H_d}^4 - 4J_{H_d}^2}{2R_{H_d}^3\sqrt{R_{H_d}^4 + 4J_{H_d}^2}}, \quad \Omega_H = \frac{2J_{H_d}}{R_{H_d}\sqrt{R_{H_d}^4 + 4J_{H_d}^2}}. \tag{G.85}$$

It is reasonable to assume that their functional forms would be the same as those of Kerr black holes. Then, the first law can be integrated for giving mass as

$$M_{H_d} = \frac{1}{2R_{H_d}}\sqrt{R_{H_d}^4 + 4J_{H_d}^2}. \tag{G.86}$$

Needless to say, this mass has the same dependence on the area and angular momentum as in Kerr black holes.

G.6 Isolated horizon

Isolated horizon was proposed by Ashtekar et al. (1999, 2000a) [see also Ashtekar and Krishnan (2004); Gourgoulhon and Jaramillo (2006) for a review]. The key idea of the isolated horizon is to approximate an event horizon of a stationary black hole by replacing it with that of a local three null hypersurface in quasi-equilibrium which has the topology of $\mathbb{R} \times S^2$. Isolated horizons are useful for exploring black holes in quasi-equilibrium which is just relaxing to a stationary black hole. It also plays a key role for defining black holes in quasi-equilibrium binary systems (see section 5.6.3).

As a first step toward defining an isolated horizon, a *non-expanding horizon* has to be defined. Here, an isolated horizon is a restricted class of the non-expanding horizon which is defined as a null hypersurface, $\mathcal{H}$, satisfying the following properties: (i) $\mathcal{H}$ has the topology $\mathbb{R} \times S^2$, (ii) the expansion Θ vanishes on $\mathcal{H}$, and (iii) on $\mathcal{H}$, the null dominant energy condition is satisfied for the stress-energy tensor.

Here, the expansion, Θ, shear σ_{ab}, and rotation ω_{ab} are defined as in section G.1: Let k^a be a null vector field which is normal to $\mathcal{H}$ and let l^a be another null vector field satisfying $k^a l_a = -1$. Then, $H_{ab} = g_{ab} + k_a l_b + l_a k_b$, and Θ, σ_{ab}, and ω_{ab} are defined by equations (G.4)–(G.6). By definition, Θ has to be zero on $\mathcal{H}$. Because

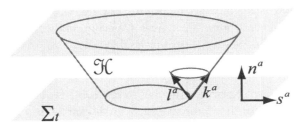

Fig. G.3 A spacetime diagram for a non-expanding horizon or an isolated horizon.

k^a is orthogonal to the null hypersurface $\mathcal{H}$, the rotation is automatically zero due to Frobenius' theorem.

Since k^a is a null vector field normal to $\mathcal{H}$, it is written as $\propto \nabla^a u$ where u is a function of spacetime coordinates and is constant on $\mathcal{H}$. Then, k^a satisfies

$$k^b \nabla_b k_a = \kappa k_a, \tag{G.87}$$

where κ is a scalar field on $\mathcal{H}$ given by

$$\kappa = -l^a k^b \nabla_b k_a. \tag{G.88}$$

Note that κ is not always constant in contrast to surface gravity.

Because equation (G.87) is satisfied, we have Raychaudhuri-like equation (G.19). Then, $\Theta = 0 = \omega_{ab}$ and the null dominant energy condition for T_{ab} yield $\sigma_{ab} = 0$ and $T_{ab} k^a k^b = 0$. Therefore, on $\mathcal{H}$,

$$\hat{B}_{ab} = H_a{}^c H_b{}^d B_{cd} = 0, \tag{G.89}$$

where $B_{cd} = \nabla_d k_c$. By a straightforward calculation, this may be written as

$$\hat{B}_{ab} = \frac{1}{2} H_a{}^c H_b{}^d \mathscr{L}_k H_{cd} = 0. \tag{G.90}$$

Thus, $\mathscr{L}_k H_{cd} = 0$ on $\mathcal{H}$. This means that the area of two-dimensional closed surfaces on $\mathcal{H}$ orthogonal to k^a, $\mathcal{A}$, is constant. All these facts show that the non-expanding horizon has the properties similar to event horizons in stationary spacetime.

However, on non-expanding horizons, κ does not have any invariant meaning like the surface gravity, by contrast to on event horizons. The reason for this is that we here do not have any Killing vector field like χ^a defined in section G.2, and we have infinite degrees of scaling freedom for choosing k^a as $C k^a$ where C is some function on $\mathcal{H}$. With this scaling l^a should be rescaled as l^a/C and, as a result, κ changes as

$$\kappa + k^b \nabla_b \ln C. \tag{G.91}$$

This shows an evidence that the definition of the non-expanding horizon is not tight enough to determine some properties of a black-hole horizon; the notion of the non-expanding horizon provides a restricted tool to extract information about black-hole spacetime.

Isolated horizon is an advanced notion. For defining this, it is necessary to impose an additional condition, which gives a specific condition for κ. Because there is no definite definition, several possibilities have been proposed. For a detailed description, we recommend the readers to refer to Gourgoulhon and Jaramillo (2006).

As in the dynamical horizon, angular momentum is defined for the isolated horizon if a Killing vector field presents on $\mathcal{H}$. Ashtekar *et al.* (2001) further shows that an isolated horizon (called weakly isolated horizon) satisfies the first law; the same relation among mass, angular momentum, area, surface gravity, and angular velocity as that for Kerr black holes holds. As in the dynamical-horizon case, if we assume that the functional forms of the surface gravity and angular velocity as functions of angular momentum and area are the same as those for Kerr black holes as in equation (G.85), the functional form of the mass is derived in the same manner as equation (G.86).

Appendix H

Post-Newtonian results for coalescing compact binaries

Post-Newtonian approximation is a robust approach for deriving an analytic solution of coalescing compact binaries in close orbits [Blanchet (2014)]. The solutions obtained are approximate but they are highly accurate as long as the orbit is not extremely relativistic. By this reason, the post-Newtonian solutions are often used to check the reliability of results of numerical-relativity simulations.

In the post-Newtonian approach, solutions of Einstein's equation and equations of motion are approximately obtained by iteratively performing a perturbation expansion of the field equation and equations of motion adopting v/c as the expansion parameter, where v is the typical velocity of the system and it is assumed to be much smaller than the speed of light c. The post-Newtonian calculations are performed using the harmonic gauge condition [Blanchet (2014)] or an ADM gauge condition [Schäfer (1985); Hartung et al. (2013)]. This approximation is in particular robust when each star of binaries can be approximated by a point particle (point-particle approximation) or by a particle characterized by a small number of parameters (mass, spin, quadrupole moment, $\cdots$). To date, the solution of the orbital motion for binary stars in circular orbits is determined up to the fourth post-Newtonian (4PN) order in the absence of spin of each star [i.e., to the order of $(v/c)^8$ where v here denotes an orbital velocity of the relative motion]. The spin effect is self-consistently taken into account up to the order of $(v/c)^5$.

Gravitational waveforms and gravitational-wave luminosity are also derived by integrating Einstein's equation iteratively using the post-Minkowski approximation. Here, the post-Minkowski approximation is an approach that Einstein's equation is iteratively integrated using gravitational constant, G, as the expansion parameter and usually using the harmonic gauge condition [Thorne (1980); Blanchet and Damour (1986)]. For the calculation of the waveform and luminosity of gravitational waves emitted by binary systems, equations of motion are solved using the post-Newtonian approximation and field equations are integrated using the post-Minkowski approach. To date, the waveform and luminosity are obtained up to third and half post-Newtonian (3.5PN) order ($O[(v/c)^7]$) in the absence of spins, and second and half post-Newtonian (2.5PN) order ($O[(v/c)^5]$) in the presence of spins. In sections H.1 and H.2, we summarize these solutions [for the summary of updated solutions, we recommend the readers to refer to Blanchet (2014); Ajith et al. (2007)].

For highly relativistic systems, the higher-order post-Newtonian results are not still accurate enough for modeling the orbital motion and emitted gravitational waves of coalescing compact binaries. To compensate the lack of higher-relativistic corrections, resummation techniques are often used. These are rather phenomenological techniques and the quality of the corrections has to be examined by comparing the results with those derived in numerical-relativity simulations. The representative resummation techniques are Taylor-T1–T4 approximants and effective one-body approach. These techniques are outlined in section H.3 and H.4.

H.1 Solutions for circular orbits

Using post-Newtonian equations of motion for binary systems of two point masses derived up to the 4PN order, angular velocity, $\Omega_{\rm orb}$, total energy, E, and total angular momentum, J, as functions of an orbital separation for circular orbits can be derived. Because the orbital separation is gauge-dependent, physical information is found from the relations of E and J as functions of $\Omega_{\rm orb}$. These relations are usually written as functions of $x := (m\Omega_{\rm orb})^{2/3}(= O[(v/c)^2])$ where $m = m_1 + m_2$ with m_1 and m_2 being individual masses as

$$
\begin{aligned}
E = -\frac{\mu x}{2}\bigg[&1 - \left(\frac{3}{4} + \frac{\eta}{12}\right)x + \left(-\frac{27}{8} + \frac{19\eta}{8} - \frac{\eta^2}{24}\right)x^2 \\
&+ \left(-\frac{675}{64} + \left[\frac{34445}{576} - \frac{205\pi^2}{96}\right]\eta - \frac{155\eta^2}{96} - \frac{35\eta^3}{5184}\right)x^3 \\
&+ \left\{-\frac{3969}{128} + \left[-\frac{123671}{5760} + \frac{9037\pi^2}{1536} + \frac{1792\ln 2}{15} + \frac{896}{15}\gamma_{\rm E}\right]\eta \right.\\
&\left. + \left(-\frac{498449}{3456} + \frac{3157\pi^2}{576}\right)\eta^2 + \frac{301\eta^3}{1728} + \frac{77\eta^4}{31104} + \frac{448}{15}\eta\ln x\right\}x^4 \\
&+ O(x^5)\bigg],
\end{aligned}
\tag{H.1}
$$

$$
\begin{aligned}
J = m\mu x^{-1/2}\bigg[&1 + \left(\frac{3}{2} + \frac{\eta}{6}\right)x - \left(-\frac{27}{8} + \frac{19\eta}{8} - \frac{\eta^2}{24}\right)x^2 \\
&- \left(-\frac{135}{16} + \left[\frac{6889}{144} - \frac{41\pi^2}{24}\right]\eta - \frac{31\eta^2}{24} - \frac{7\eta^3}{1296}\right)x^3 \\
&+ \left\{\frac{2835}{128} + \left[\frac{98869}{5760} - \frac{6455\pi^2}{1536} - \frac{256\ln 2}{3} - \frac{128}{3}\gamma_{\rm E}\right]\eta \right.\\
&\left. + \left(\frac{356035}{3456} - \frac{2255\pi^2}{576}\right)\eta^2 - \frac{215\eta^3}{1728} - \frac{55\eta^4}{31104} - \frac{64}{3}\eta\ln x\right\}x^4 \\
&+ O(x^5)\bigg],
\end{aligned}
\tag{H.2}
$$

where μ is the reduced mass defined by $m_1 m_2/m$, $\eta := \mu/m$ is the symmetric mass ratio, and γ_E is the Euler number. We note that E and J satisfy the first law

$$dE = \Omega_{\text{orb}} dJ. \tag{H.3}$$

In the presence of spins of binary components, additional terms appear in E and J. For the so-called spin-orbit coupling, the correction is currently known up to the 2.5PN order [e.g., Kidder (1995); Faye et al. (2006); Blanchet et al. (2006)], which is written as

$$E_{\text{SL}} = -\frac{\mu x}{2} \left[\frac{x^{3/2}}{m^2} \left(\frac{14}{3} \hat{S}_\ell + 2 \frac{\delta m}{m} \hat{\Sigma}_\ell \right) \right.$$
$$\left. + \frac{x^{5/2}}{m^2} \left\{ \left(11 - \frac{61}{9} \eta \right) \hat{S}_\ell + \left(3 - \frac{10}{3} \eta \right) \frac{\delta m}{m} \hat{\Sigma}_\ell \right\} + O(x^3) \right], \tag{H.4}$$

where $\delta m := m_1 - m_2$, $\hat{S}_\ell := \sum_i (S_1^i + S_2^i) \ell^i$, $\hat{\Sigma}_\ell := m \sum_i (S_2^i/m_2 - S_1^i/m_1) \ell^i$, and ℓ^i denotes the unit vector parallel to the angular momentum vector in the Newtonian order. S_1^i and S_2^i are spin vectors of two binary components for which the norm conserves with time; e.g., $\sum_i S_1^i S_1^i =$const.

For the spin-spin coupling, the correction (currently known up to the 2PN order) is written as [e.g., Kidder (1995)]

$$E_{\text{SS}} = -\frac{1}{2m^3} x^3 \left[\sum_i S_1^i S_2^i - 3 \left(\sum_i \ell^i S_1^i \right) \sum_j \ell^j S_2^j \right]. \tag{H.5}$$

In the presence of a quadrupole moment of binary components (which is induced by centrifugal force associated with the spins of the individual components), additional 2PN component is present. To take into account this effect, equation (H.5) has to be modified. For example, the quadrupole terms associated with black-hole spins are incorporated as [e.g., Buonanno et al. (2006)]

$$E_{\text{SS}} = -\frac{\eta}{4m^3} x^3 \left[\sum_i S_0^i S_0^i - 3 \left(\sum_i \ell^i S_0^i \right) \sum_j \ell^j S_0^j \right], \tag{H.6}$$

where

$$S_0^i = m \left(\frac{S_1^i}{m_1} + \frac{S_2^i}{m_2} \right). \tag{H.7}$$

If the binary component is not a black hole but a rotating star, equation (H.6) should have in general a different form. We note the corresponding correction to J in the presence of spins is derived using the first law (H.3).

One of the most remarkable properties for the binary motion in the presence of spins is that the orbital plane does not remain to be unchanged in general. Rather, it precesses with time unless S_1^i, S_2^i, and ℓ^i are all aligned (or anti-aligned). For the misaligned case, the orbital motion is not determined only by the relation among E, J, and Ω_{orb} because S_1^i, S_2^i, and ℓ^i process. For such a case, evolution equations for S_1^i, S_2^i, and ℓ^i have to be solved.

Neutron stars in close orbits are tidally deformed by the tidal field of their companion stars, and quadrupole moments are induced for them. In the presence of the quadrupole moments, the gravitational field is also modified, and hence, the correction terms to E and J appear. If the degree of the tidal deformation is small and hence each quadrupole moment is assumed to be proportional to the tidal field of each companion star, the correction to E is written as [Damour and Nagar (2010); Vines et al. (2011)]

$$E_{\text{tidal}} = \mu x \left[\frac{9}{2m^5} x^5 \left(\frac{m_1}{m_2} \lambda_2 + \frac{m_2}{m_1} \lambda_1 \right) \right.$$
$$+ \frac{11}{4m^5} x^6 \left\{ \frac{m_1}{m_2} \lambda_2 \left(3 + 2\left(\frac{m_2}{m}\right) + 3\left(\frac{m_2}{m}\right)^2 \right) \right.$$
$$\left. \left. + \frac{m_2}{m_1} \lambda_1 \left(3 + 2\left(\frac{m_1}{m}\right) + 3\left(\frac{m_1}{m}\right)^2 \right) \right\} + O(x^7) \right], \quad \text{(H.8)}$$

where λ_1 and λ_2 are the so-called tidal deformability of stars 1 and 2, and are defined by the ratio of the quadrupole moment to the tidal field by the companion star. Note that the tidal deformability has the dimension of (radius)^5, and thus, its order of the magnitude is proportional to $\mathcal{C}_k^{-5} m_k^5$ where $\mathcal{C}_k$ ($k = 1, 2$) is the compactness of each neutron star. Since $\mathcal{C}_k$ is 0.1–0.2, the coefficients of the tidal correction terms are quite large, although it is the fifth post-Newtonian term. This implies that the tidal correction term plays an important role for close orbits.

H.2 Luminosity, emission rate of linear and angular momenta, amplitude of gravitational waves

In the absence of spins, the gravitational-wave luminosity is calculated up to 3.5PN order as

$$\left(\frac{dE}{dt}\right)_{\text{GW}} = \frac{32}{5} \eta^2 x^5 \left[1 - \left(\frac{1247}{336} + \frac{35}{12} \eta \right) x + 4\pi x^{3/2} \right.$$
$$+ \left(-\frac{44711}{9072} + \frac{9271}{504} \eta + \frac{65}{18} \eta^2 \right) x^2$$
$$- \left(\frac{8191}{672} + \frac{583}{24} \eta \right) \pi x^{5/2}$$
$$+ \left\{ \frac{6643739519}{69854400} + \frac{16\pi^2}{3} - \frac{1712}{105} \gamma_E - \frac{856}{105} \ln(16x) \right.$$
$$\left. + \left(-\frac{134543}{7776} + \frac{41\pi^2}{48} \right) \eta - \frac{94403}{3024} \eta^2 - \frac{775}{324} \eta^3 \right\} x^3$$
$$\left. + \left(-\frac{16285}{504} + \frac{214745}{1728} \eta + \frac{193385}{3024} \eta^2 \right) \pi x^{7/2} + O(x^4) \right]. \quad \text{(H.9)}$$

The angular momentum emission rate for circular orbits, $(dJ/dt)_{\text{GW}}$, is derived simply from the relation

$$\left(\frac{dE}{dt}\right)_{\text{GW}} = \Omega_{\text{orb}} \left(\frac{dJ}{dt}\right)_{\text{GW}}. \tag{H.10}$$

In the presence of the spin-orbit coupling, the following correction term (currently known up to the 2.5PN order) is added to $(dE/dt)_{\text{GW}}$ [Kidder (1995); Blanchet et al. (2006, 2010)]

$$\left(\frac{dE}{dt}\right)_{\text{GW,SO}} = \frac{32}{5}\eta^2 x^5 \left[-\left(4\hat{S}_\ell + \frac{5}{4}\frac{\delta m}{m}\hat{\Sigma}_\ell\right)\frac{x^{3/2}}{m^2} \right.$$
$$\left. + \left\{\left(-\frac{9}{2} + \frac{272\eta}{9}\right)\hat{S}_\ell + \left(-\frac{13}{16} + \frac{43\eta}{3}\right)\frac{\delta m}{m}\hat{\Sigma}_\ell\right\}\frac{x^{5/2}}{m^2} \right.$$
$$\left. + O(x^3) \right]. \tag{H.11}$$

In the presence of the spin-spin coupling, the correction term (currently known up to the 2PN order) is [Kidder (1995); Buonanno et al. (2006)]

$$\left(\frac{dE}{dt}\right)_{\text{GW,SS}} = \frac{32}{5}x^7 \frac{\eta}{48m^4} \left[-103\sum_i S_1^i S_2^i + 289\left(\sum_i \ell^i S_1^i\right)\sum_j \ell^j S_2^j \right.$$
$$\left. - \frac{89}{2}\left(\frac{m_2}{m_1}\sum_i S_1^i S_1^i + \frac{m_1}{m_2}\sum_i S_2^i S_2^i\right) \right.$$
$$\left. + \frac{287}{2}\left\{\frac{m_2}{m_1}\left(\sum_i \ell^i S_1^i\right)^2 + \frac{m_1}{m_2}\left(\sum_i \ell^i S_2^i\right)^2\right\} \right]. \tag{H.12}$$

In the presence of quadrupole moments of neutron stars induced by the tidal deformation, the correction terms are known up to the 1PN order from the leading term as [Vines et al. (2011)]

$$\left(\frac{dE}{dt}\right)_{\text{GW,tidal}} = \frac{32}{5}\frac{\eta^2 x^{10}}{m^5}\left[\frac{6(3m-2m_2)}{m_2}\lambda_2 + \frac{6(3m-2m_1)}{m_1}\lambda_1 \right.$$
$$\left. + \frac{mx}{28}\left[\left\{-704 - 1803\left(\frac{m_2}{m}\right) + 4501\left(\frac{m_2}{m}\right)^2 - 2170\left(\frac{m_2}{m}\right)^3\right\}\frac{\lambda_2}{m_2} \right.\right.$$
$$\left.\left. + \left\{-704 - 1803\left(\frac{m_1}{m}\right) + 4501\left(\frac{m_1}{m}\right)^2 - 2170\left(\frac{m_1}{m}\right)^3\right\}\frac{\lambda_1}{m_1}\right]\right]. \tag{H.13}$$

The linear-momentum emission rate is known up to the 2PN order in the absence of spins as [Blanchet et al. (2005)]

$$\left(\frac{dP^i}{dt}\right)_{\rm GW} = \frac{464}{105}\eta^2 x^{11/2}\frac{\delta m}{m}\left[1 - \left(\frac{452}{87} + \frac{1139}{522}\eta\right)x + \frac{309}{58}\pi x^{3/2}\right.$$
$$+\left(-\frac{71345}{22968} + \frac{36761}{2088}\eta + \frac{147101}{68904}\eta^2\right)x^2$$
$$\left. + O(x^3)\right]\hat{v}^i, \qquad (\text{H.14})$$

where $\hat{v}^i$ is a unit tangential vector pointing to the same direction as the relative orbital velocity $v_1^i - v_2^i$.

In the presence of spins, the linear-momentum emission rate is not simply written only in terms of x because the magnitude depends on the inclination angles among spins and orbital angular momentum vector. Kidder (1995) derived a formula up to half post-Newtonian (0.5PN) order, and for circular orbits, the correction by the spins is written as

$$\left(\frac{dP^i_{\rm spin}}{dt}\right)_{\rm GW} = \frac{16}{15}\frac{\eta^2}{m^2}x^6 \sum_{j,k}\epsilon^{ijk}\bar{n}^j\left[\hat{\Sigma}^k + \hat{v}^k\sum_l \hat{v}^l\hat{\Sigma}^l\right], \qquad (\text{H.15})$$

where $\bar{n}^i = (x_1^i - x_2^i)/|x_1^i - x_2^i|$ and $\Sigma^i := m(S_2^i/m_2 - S_1^i/m_1)$. Thus, linear momentum is emitted in the presence of a spin even for equal-mass system in which $(dP^i/dt)_{\rm GW}$ in equation (H.14) vanishes.

The amplitude of the quadrupole component of gravitational waves (in the absence of spins) is often referred to in numerical relativity [compare the following equation with equation (1.193)]:

$$h^{22} = 4Q_h\frac{\eta m}{D}\exp[-i\Phi(t)]x\left[1 - \left(\frac{107}{42} - \frac{55}{42}\eta\right)x + 2\pi x^{3/2}\right.$$
$$-\left(\frac{2173}{1512} + \frac{1069}{216}\eta - \frac{2047}{1512}\eta^2\right)x^2$$
$$-\left\{\left(\frac{107}{21} - \frac{34}{21}\eta\right)\pi + 24i\eta\right\}x^{5/2}$$
$$+\left\{\frac{27027409}{646800} - \frac{856}{105}\gamma_{\rm E} + \frac{2}{3}\pi^2\right.$$
$$-\frac{428}{105}\ln(16x) - \left(\frac{278185}{33264} - \frac{41}{96}\pi^2\right)\eta$$
$$\left.\left. -\frac{20261}{2772}\eta^2 + \frac{114635}{99792}\eta^3 + \frac{428}{105}i\pi\right\}x^3\right]. \qquad (\text{H.16})$$

Here, the meaning of Φ, $Q_h(>0)$, and D is the same as that in equation (1.193). Note that Φ here is the gravitational-wave phase and it is twice as large as the orbital phase.

H.3 Adiabatic approximations for the orbital evolution, and Taylor approximants

Using the luminosity together with the total energy of binary systems such as equation (H.1), we can derive an evolution equation of $\Omega_{\rm orb}$ and the phase of the orbital motion as functions of time in the adiabatic approximation, as done in section 1.2.5. (Here the time is the proper time of an observer who is far from the source.) Integrating the equation $dx/dt = -(dE/dt)_{\rm GW}/(dE/dx)$ yields (in the absence of spins) [Blanchet (2014)]

$$
\begin{aligned}
x = \frac{1}{4}\Theta^{-1/4} \Bigg[& 1 + \left(\frac{743}{4032} + \frac{11\eta}{48}\right)\Theta^{-1/4} - \frac{\pi}{5}\Theta^{-3/8} \\
& + \left(\frac{19583}{254016} + \frac{24401}{193536}\eta + \frac{31}{288}\eta^2\right)\Theta^{-1/2} \\
& + \left(-\frac{11891}{53760} + \frac{109}{1920}\eta\right)\pi\Theta^{-5/8} \\
& + \left\{-\frac{10052469856691}{6008596070400} + \frac{\pi^2}{6} + \frac{107}{420}\gamma_{\rm E} - \frac{107}{3360}\ln\left(\frac{\Theta}{256}\right)\right. \\
& + \left(\frac{3147553127}{780337152} - \frac{451\pi^2}{3072}\right)\eta - \frac{15211}{442368}\eta^2 + \left.\frac{25565}{331776}\eta^3\right\}\Theta^{-3/4} \\
& + \left(-\frac{113868647}{433520640} - \frac{31821}{143360}\eta + \frac{294941}{3870720}\eta^2\right)\pi\Theta^{-7/8} + O(x^4) \Bigg], \quad (\text{H}.17)
\end{aligned}
$$

and integrating the equation $d\Phi/dt = 2\Omega_{\rm orb}$ yields [compare it with equation (1.196)]

$$
\begin{aligned}
\Phi = \Phi_0 - \frac{2}{\eta}\Theta^{5/8} \Bigg[& 1 + \left(\frac{3715}{8064} + \frac{55\eta}{96}\right)\Theta^{-1/4} - \frac{3\pi}{4}\Theta^{-3/8} \\
& + \left(\frac{9275495}{14450688} + \frac{284875}{258048}\eta + \frac{1855}{2048}\eta^2\right)\Theta^{-1/2} \\
& + \left(-\frac{38645}{172032} + \frac{65}{2048}\eta\right)\pi\Theta^{-5/8}\ln\left(\frac{\Theta}{\Theta_0}\right) \\
& + \left\{\frac{831032450749357}{57682522275840} - \frac{53\pi^2}{40} - \frac{107}{56}\gamma_{\rm E} + \frac{107}{448}\ln\left(\frac{\Theta}{256}\right)\right. \\
& + \left(-\frac{126510089885}{4161798144} + \frac{2255\pi^2}{2048}\right)\eta \\
& + \left.\frac{154565}{1835008}\eta^2 - \frac{1179625}{1769472}\eta^3\right\}\Theta^{-3/4} \\
& + \left(\frac{188516689}{173408256} + \frac{488825}{516096}\eta - \frac{141769}{516096}\eta^2\right)\pi\Theta^{-7/8} + O(x^4) \Bigg],
\end{aligned}
$$
(H.18)

where Φ_0 is a constant phase,

$$\Theta := \frac{\eta}{5m}(t_0 - t), \qquad (\text{H}.19)$$

and t_0 denotes the time at the onset of the merger. Θ_0 is a constant of integration that can be fixed by the initial condition. In the definition of Boyle et al. (2007), these results for $\Omega_{\rm orb}$ and Φ are called the formulas in the Taylor-T3 approximant. Note that 4PN terms are not still known, because the luminosity in the 4PN order has not been derived yet [Blanchet (2014)].

Although the analytic solutions of the phase Φ (called Taylor-T3 approximant) written above are easy to deal with, it is not guaranteed that they do always yield a good approximate solution: The error associated with the truncation of $O(x^4)$ terms could be significant for close orbits. When deriving these analytic solutions, one discards higher-order terms of x [all the terms of $O(x^4)$] in $(dE/dt)_{\rm GW}/(dE/dx)$. It may be better to use the original forms of $(dE/dt)_{\rm GW}$ and dE/dx without any discarding for obtaining $x(t)$ and $\Phi(t)$ or $t(x)$ and $\Phi(x)$ by the direct numerical integration or by the combination of an analytic calculation and numerical integration (for the former and latter methods, the resulting approximants are called Taylor-T1 and Taylor-T2, respectively [e.g., Buonanno et al. (2009)]). However, no theoretical foundation exists for this issue and for a more strict analysis, more than fourth-order post-Newtonian corrections, which have not been derived yet, are necessary.

Buonanno et al. (2007a) pointed out, in their comparison between analytical results and numerical-relativity results of gravitational waves emitted by *equal-mass non-spinning* binary black holes, that it could indeed be a poor idea to use the analytic formula obtained by the expansion (i.e., Taylor-T3 approximant). Instead, it could be better to use a formula obtained by (numerically) integrating the following equation [for the latest version, see also Blanchet et al. (2010)]:

$$\frac{dx}{dt} = \frac{64}{5}\frac{\eta}{m}x^5 \left[1 - \left(\frac{743}{336} + \frac{11}{4}\eta\right)x + 4\pi x^{3/2} \right.$$
$$- \left(\frac{47}{3}\hat{S}_\ell + \frac{25}{4}\frac{\delta m}{m}\hat{\Sigma}_\ell\right)\frac{x^{3/2}}{m^2}$$
$$+ \left(\frac{34103}{18144} + \frac{13661}{2016}\eta + \frac{59}{18}\eta^2\right)x^2$$
$$- \left(\frac{4159}{672} + \frac{189}{8}\eta\right)\pi x^{5/2}$$
$$+ \left\{\left(-\frac{5861}{144} + \frac{1001}{12}\eta\right)\hat{S}_\ell + \left(-\frac{809}{84} + \frac{281}{8}\eta\right)\frac{\delta m}{m}\hat{\Sigma}_\ell\right\}\frac{x^{5/2}}{m^2}$$
$$+ \left\{\frac{16447322263}{139708800} + \frac{16\pi^2}{3} - \frac{1712}{105}\gamma_E - \frac{856}{105}\ln(16x)\right.$$
$$+ \left.\left(-\frac{56198689}{217728} + \frac{451\pi^2}{48}\right)\eta + \frac{541}{896}\eta^2 - \frac{5605}{2592}\eta^3\right\}x^3$$
$$\left. + \left(-\frac{4415}{4032} + \frac{358675}{6048}\eta + \frac{91495}{1512}\eta^2\right)\pi x^{7/2} + O(x^4)\right]. \quad (\text{H.20})$$

Here, the gravitational-wave phase should be subsequently obtained by numerically integrating $d\Phi/dt = 2\Omega_{\rm orb}$ for the numerical solution of $\Omega_{\rm orb}$. The resulting approximate solution is often called Taylor-T4 approximant [Boyle et al. (2007)] [see, e.g., Buonanno et al. (2009) for the overview of several types of approximations based on the 3.5PN approximate formulas]. Boyle et al. (2007) showed that the Taylor-T4 approximant can (coincidently) reproduce the numerical results of the orbital evolution very accurately for equal-mass non-spinning binary black holes [see section 8.7].

In the presence of quadrupole moments induced by the tidal deformation, the following correction terms to the right-hand side of equation (H.20) should be added:

$$\left.\frac{dx}{dt}\right|_{\text{tidal}} = \frac{32x^{10}}{5m^7}\left[12\frac{m_1}{m}\lambda_2(m+11m_1) + 12\frac{m_2}{m}\lambda_1(m+11m_2)\right.$$
$$+ m_1\lambda_2 x\left\{\frac{4421}{28} - \frac{12263m_2}{28m} + \frac{1893m_2^2}{2m^2} - 661\left(\frac{m_2}{m}\right)^3\right\}$$
$$\left. + m_2\lambda_1 x\left\{\frac{4421}{28} - \frac{12263m_1}{28m} + \frac{1893m_1^2}{2m^2} - 661\left(\frac{m_1}{m}\right)^3\right\}\right]. \quad (\text{H.21})$$

This correction is quite appreciable for the evolution of x and Φ in the late inspiral phase of binary neutron stars (see section 8.3.4.1).

The stationary phase approximation can be used to derive the inspiral waveform in the frequency domain, when the gravitational radiation time scale is much longer than the orbital period [see equations (1.198) and (1.199)]. The Taylor-F2 approximant is such a solution, and the Fourier transform of the waveform for the non-spinning binary in the 3.5PN approximation is written as [e.g., Ajith et al. (2007)]

$$\tilde{h}(f) = \left(\frac{5}{24\pi^3}\right)^{1/2}\frac{Q_h}{D}\tilde{h}(x)f^{-7/6}(\pi\mathcal{M})^{5/6}e^{i\psi(f)}, \quad (\text{H.22})$$

where $\mathcal{M}$ is the chirp mass and

$$\psi(f) = 2t_0 x^{3/2}/m + \psi_0 + \frac{3}{128\eta}x^{-5/2}\left[1 + \left(\frac{3715}{756} + \frac{55}{9}\eta\right)x - 16\pi x^{3/2}\right.$$
$$+ \left(\frac{15293365}{508032} + \frac{27145}{504}\eta + \frac{3085}{72}\eta^2\right)x^2$$
$$+ \left(\frac{38645}{756} - \frac{65}{9}\eta\right)\left\{1 + \frac{3}{2}\log x\right\}\pi x^{5/2}$$
$$+ \left[\frac{11583231236531}{4694215680} - \frac{640\pi^2}{3} - \frac{6848\gamma_{\text{E}}}{21} - \frac{3424\gamma_{\text{E}}}{21}\log(16x)\right.$$
$$+ \left(-\frac{15737765635}{3048192} + \frac{2255\pi^2}{12}\right)\eta + \frac{76055}{1728}\eta^2 - \frac{127825}{1296}\eta^3\bigg]x^3$$
$$\left.+ \left(\frac{77096675}{254016} + \frac{378515}{1512}\eta - \frac{74045}{756}\eta^2\right)\pi x^{7/2} + O(x^4)\right], \quad (\text{H.23})$$

with ψ_0 being a constant phase and t_0 being the time at the onset of the merger. $\tilde{h}(x)$ denotes the correction to the amplitude appropriately determined; e.g., for the quadrupole mode, the polynomial function in the large bracket of h^{22}, $[1 - (107 - 55\eta)x/42 + \cdots]$, in equation (H.16) is one candidate. The Taylor-F2 approximant is also written in the presence of spins. For this equation, see, e.g., Ajith et al. (2007).

H.4 Effective one-body approach

Effective one-body approach for two-body problems in general relativity was originally proposed by Buonanno and Damour (1999), and this has been subsequently developed by many researchers [see Damour and Nagar (2009) for a review]. We here outline this approach following the description of Buonanno et al. (2009), focusing only on the case that spins and quadrupole deformation of binary components are absent.

In the effective one-body approach, the problem of the relative motion of two bodies is replaced into the problem of the particle motion in an effective metric of the form

$$ds^2_{\text{eff}} = -A(r)dt^2 + \frac{D(r)}{A(r)}dr^2 + r^2\left(d\theta^2 + \sin^2\theta d\varphi^2\right), \tag{H.24}$$

and then, the relative motion is explored by analyzing a test-particle motion in this effective metric written in terms of polar coordinates (r, φ) with (p_r, p_φ) being their conjugate momenta. Here, r, p_r, and p_φ are normalized to be dimensionless with respect to the total mass m, reduced mass μ, and $m\mu$, respectively. The effective-one-body Hamiltonian is defined by

$$H^{\text{real}}(r, p_r, p_\varphi) := \mu \hat{H}^{\text{real}} := m\sqrt{1 + 2\eta\left(\frac{H^{\text{eff}} - \mu}{\mu}\right)}, \tag{H.25}$$

where $\eta = \mu/m$ denotes the symmetric mass ratio as before. H^{eff} is an effective Hamiltonian

$$H^{\text{eff}}(r, p_r, p_\varphi) := \mu \hat{H}^{\text{eff}}$$
$$:= \mu\sqrt{A(r)\left[1 + \frac{A(r)}{D(r)}p_r^2 + \frac{p_\varphi^2}{r^2} + 2(4 - 3\eta)\eta\frac{p_r^4}{r^2}\right]}. \tag{H.26}$$

The Taylor approximants to the functions $A(r)$ and $D(r)$ are written as

$$A_k(r) = \sum_{i=0}^{k+1} \frac{a_i(\eta)}{r^i}, \qquad D_k(r) = \sum_{i=0}^{k} \frac{d_i(\eta)}{r^i}, \tag{H.27}$$

where k is a positive integer. In the absence of spins and multipole components, the functions $A(r)$ and $D(r)$ [and thus, $A_k(r)$, and $D_k(r)$] depend only on η through the η-dependent coefficients $a_i(\eta)$ and $d_i(\eta)$. These approximants are determined so that the resulting equations of motion are consistent with the 3PN equations of motion in the Taylor expanded form. Specifically, they are written as [e.g., Buonanno et al. (2007b)]

$$A_3(r) = 1 - \frac{2}{r} + \frac{2\eta}{r^3} + \frac{1}{r^4}\left[\left(\frac{94}{3} - \frac{41\pi^2}{32}\right)\eta - z_1\right], \tag{H.28}$$

$$D_3(r) = 1 - \frac{6\eta}{r^2} + \frac{1}{r^3}\left[7z_1 + z_2 + 2\eta(3\eta - 26)\right], \tag{H.29}$$

where z_1 and z_2 are arbitrary constants and with the choice of the effective Hamiltonian (H.26), they vanish. For the last stage of inspiral orbits and plunge stage, the equations of motion with the Taylor expansion forms of $A_3(r)$ and $D_3(r)$ cannot accurately reproduce results of numerical-relativity simulations because the lack of the higher post-Newtonian terms gives a serious damage. To compensate this lack, a resummation technique is usually employed. A popular prescription is to use the Padé approximation and the following resummation is often used,

$$A_3^{\text{resum}} = \frac{r^2[a_4(\eta, 0) + 8\eta - 16 + r(8 - 2\eta)]}{r^3(8 - 2\eta) + (r^2 + 2r)[a_4(\eta, 0) + 4\eta] + 4[\eta^2 + a_4(\eta, 0)]}, \tag{H.30}$$

$$D_3^{\text{resum}} = \frac{r^3}{r^3 + 6\eta r + 2\eta(26 - 3\eta)}, \tag{H.31}$$

where

$$a_4(\eta, z_1) = \left(\frac{94}{3} - \frac{41\pi^2}{32}\right)\eta - z_1. \tag{H.32}$$

In addition to using the resummation technique, higher-order terms, associated with a_5, a_6, d_4, and d_5, are often added to improve the accuracy near the last orbit and subsequent plunging [Buonanno et al. (2007b); Pan et al. (2011)]. These coefficients are determined by comparing the obtained results with the numerical-relativity results.

The Hamilton equations in the effective one-body approach are written in terms of the reduced quantities, $\hat{H}^{\text{real}}$, $\hat{t} := t/m$, and $\hat{\omega} := m\omega = d\varphi/d\hat{t}$ as

$$\frac{dr}{d\hat{t}} = \frac{\partial \hat{H}^{\text{real}}}{\partial p_r}, \tag{H.33}$$

$$\frac{d\varphi}{d\hat{t}} = \frac{\partial \hat{H}^{\text{real}}}{\partial p_\varphi}, \tag{H.34}$$

$$\frac{dp_r}{d\hat{t}} = -\frac{\partial \hat{H}^{\text{real}}}{\partial r} + \mathcal{F}_\varphi \frac{p_r}{p_\varphi}, \tag{H.35}$$

$$\frac{dp_\varphi}{d\hat{t}} = \mathcal{F}_\varphi, \tag{H.36}$$

where $\mathcal{F}_\varphi$ denotes the term of the gravitational radiation reaction which is written as

$$\mathcal{F}_\varphi = -\frac{1}{\eta\hat{\omega}}\left(\frac{dE}{dt}\right)_{\text{GW}}. \tag{H.37}$$

$(dE/dt)_{\rm GW}$ is basically chosen so that it can reproduce the 3.5PN luminosity formula for circular orbits in the post-Newtonian approximation. However it is well-known in a black-hole perturbation study [e.g., Mino *et al.* (1997)] that the 3.5PN formula is not accurate enough for the luminosity formula of close circular orbits. To compensate this deficit, a resummation technique based on Padé approximation [Damour *et al.* (1997)] is also employed.

Effective one-body approaches have been developed for many years and effects by spins and tidal deformation of neutron stars are partly incorporated to date [see, e.g., Damour (2001); Taracchini *et al.* (2012) in the presence of spins, and Damour and Nagar (2010); Bini *et al.* (2012) in the presence of tidal deformation]. However, the formulation is still in progress. Also, this formulation relies strongly on the post-Newtonian results and on the calibration by comparing with the results in numerical-relativity simulations. Hence, with the progress of the higher-order post-Newtonian calculation and with the improvement in the accuracy of the numerical-relativity results, more sophisticated equations of motion will be developed in the future.

Bibliography

Abbott, B. P., Abbott, R., and 499 authors (2009). LIGO: The laser interferometer gravitational-wave observatory, *Rep. Prog. Phys.* **72**, pp. 076901–1.

Abrahams, A. M., Bernstein, D., Hobill, D., Seidel, E., and Smarr, L. (1992). Numerically generated black-hole spacetimes: Interaction with gravitational waves, *Phys. Rev. D* **45**, p. 3544.

Abrahams, A. M., Cook, G. B., Shapiro, S. L., and Teukolsky, S. A. (1994a). Solving Einstein's equations for rotating spacetimes: Evolution of relativistic star clusters, *Phys. Rev. D* **49**, p. 5153.

Abrahams, A. M. and Evans, C. R. (1988). Reading off gravitational radiation waveforms in numerical relativity calculations: Matching to linearized gravity, *Phys. Rev. D* **37**, p. 318.

Abrahams, A. M. and Evans, C. R. (1990). Gauge-invariant treatment of gravitational radiation near the source: Analysis and numerical simulations, *Phys. Rev. D* **42**, p. 2585.

Abrahams, A. M. and Evans, C. R. (1992). Trapping a geon: Black hole formation by an imploding gravitational wave, *Phys. Rev. D* **46**, p. 4117.

Abrahams, A. M. and Evans, C. R. (1993). Critical behavior and scaling in vacuum axisymmetric gravitational collapse, *Phys. Rev. Lett.* **70**, p. 2980.

Abrahams, A. M. and Evans, C. R. (1994). Universality in axisymmetric vacuum collapse, *Phys. Rev. D* **49**, p. 3998.

Abrahams, A. M., Shapiro, S. L., and Teukolsky, S. A. (1994b). Disk collapse in general relativity, *Phys. Rev. D* **50**, p. 7282.

Abrahams, A. M., Shapiro, S. L., and Teukolsky, S. A. (1995). Calculation of gravitational waveforms from black hole collisions and disk collapse: Applying perturbation theory to numerical spacetimes, *Phys. Rev. D* **51**, p. 4295.

Abramowicz, M., Jaroszynski, M., and Sikora, M. (1978). Relativistic accretion disks, *Astron. Astrophys.* **63**, p. 221.

Accadia, T., Acernese, T., and 181 authors (2011). Status of the VIRGO project, *Class. Quantum Grav.* **28**, pp. 114002–1.

Acheson, D. J. and Gibbons, M. P. (1978). On the instability of toroidal magnetic fields and differential rotation in stars, *Phil. Trans. Roy. Soc. Lond. A* **289**, p. 459.

Ajith, P., Babak, S., Chen, Y., and 15 authors (2008). Template bank for gravitational waveforms from coalescing binary black holes: Nonspinning binaries, *Phys. Rev. D* **77**, pp. 104017–1.

Ajith, P., Boyle, M., Brown, D. A., Fairhurst, S., Hannam, M., Hinder, I., Husa, S., Hrishnan, B., Mercer, R. A., Ohme, F., Ott, C. D., Read, J. S., Santamaría, L., and Whelan, T. (2007). Data formats for numerical relativity, *arXiv:/0709.0093*, p. 1.

Akmal, A., Pandharipande, V. R., and Ravenhall, D. G. (1998). Equation of state of nucleon matter and neutron star structure, *Phys. Rev. C* **58**, p. 1804.

Alcubierre, M. (1995). Appearance of coordinate shocks in hyperbolic formalisms of general relativity, *Phys. Rev. D* **55**, p. 5981.

Alcubierre, M. (2003). Hyperbolic slicings of spacetime: Singularity avoidance and gauge shocks, *Class and Quantum Grav.* **20**, p. 607.

Alcubierre, M. (2008). *Introduction to 3+1 Numerical Relativity* (Oxford University Press).

Alcubierre, M., Allen, G., Brügmann, B., Lanfermann, G., Seidel, E., Suen, W.-M., and Tobias, M. (2000). Gravitational collapse of gravitational waves in 3D numerical relativity, *Phys. Rev. D* **61**, pp. 041501–1.

Alcubierre, M., Brügmann, B., Diener, P., Koppitz, M., Pollney, D., Seidel, E., and Takahashi, R. (2003). Gauge conditions for long-term numerical black hole evolutions without excision, *Phys. Rev. D* **67**, pp. 084023–1.

Alcubierre, M., Brügmann, B., Holz, D., Takahashi, R., Brandt, S., and Seidel, E. (2001a). Symmetry without symmetry, *Int. J. Mod. Phys. D* **10**, p. 273.

Alcubierre, M., Brügmann, B., Pollney, D., Seidel, E., and Takahashi, R. (2001b). Black hole excision for dynamic black holes, *Phys. Rev. D* **64**, pp. 061501–1.

Alcubierre, M. and Massó, J. (1997). Pathologies of hyperbolic gauges in general relativity and other field theories, *Phys. Rev. D* **57**, p. R4511.

Alic, D., Bona-Casas, C., Nona, C., Rezzolla, L., and Palenzuela, C. (2012a). Conformal and covariant formulation of the Z4 system with constraint-violation damping, *Phys. Rev. D* **85**, pp. 064040–1.

Alic, D., Mösta, P., Rezzolla, L., Zanotti, O., and Jaramillo, L. (2012b). Accurate simulations of binary black-hole mergers in force-free electrodynamics, *Astrophys. J. Lett.* **754**, p. 36.

Anderson, A. and York, J. W. (1999). Fixing Einstein's equations, *Phys. Rev. Lett.* **82**, p. 4384.

Anderson, J. L. and Spiegel, E. A. (1972). The moment method in relativistic radiative transfer, *Astrophys. J.* **171**, p. 127.

Anderson, M., Hirschmann, E. W., Lehner, L., Liebling, S. L., Motl, P. M., Neilsen, D., Palenzuela, C., and Tohline, J. E. (2008). Simulating binary neutron stars: Dynamics and gravitational waves, *Phys. Rev. D* **77**, pp. 024006–1.

Anile, A. M. (1989). *Relativistic Fluids and Magneto-Fluids* (Cambridge University Press).

Anninos, P., Bernstein, D., Brandt, S., Libson, J., Massó, J., Seidel, E., Smarr, L., Suen, W.-M., and Walker, P. (1995a). Dynamics of apparent and event horizons, *Phys. Rev. Lett.* **74**, p. 630.

Anninos, P., Camarda, K., Libson, J., Massó, J., Seidel, E., and Suen, W.-M. (1998). Finding apparent horizons in dynamic 3D numerical spacetimes, *Phys. Rev. D* **58**, pp. 024003–1.

Anninos, P., Hobill, D., Seidel, E., Smarr, L., and Suen, W.-M. (1993). Collision of two black holes, *Phys. Rev. Lett.* **71**, p. 2851.

Anninos, P., Hobill, D., Seidel, E., Smarr, L., and Suen, W.-M. (1995b). Head-on collision of two equal mass black holes, *Phys. Rev. D* **52**, p. 2044.

Anninos, P., Massó, J., Seidel, E., Suen, W.-M., and Towns, J. (1995c). Three-dimensional numerical relativity: The evolution of black holes, *Phys. Rev. D* **52**, p. 2059.

Anninos, P., Price, R. H., Pullin, J., Seidel, E., and Suen, W.-M. (1995d). Head-on collision of two black holes: Comparison of different approaches, *Phys. Rev. D* **52**, p. 4462.

Antón, L., Miralles, J. A., Martí, J.-M., Ibáñez, J.-M., Aloy, M. A., and Mimica, P. (2010). Numerical 3+1 general relativistic magnetohydrodynamics: A local characteristic approach, *Astrophys. J. Suppl* **188**, p. 1.

Antón, L., Zanotti, O., Miralles, J. A., Martí, J.-M., Ibáñez, J.-M., Font, J. A., and Pons, J. A. (2006). Numerical 3+1 general relativistic magnetohydrodynamics: A local characteristic approach, *Astrophys. J.* **637**, p. 296.

Antoniadis, I., Arkani-Hamed, N., Dimopoulos, S., and Dvali, G. R. (1998). New dimensions at a millimeter to a fermi and superstrings at a TeV, *Phys. Lett. B* **436**, p. 257.

Antoniadis, J., Freire, P. C. C., Wex, N., Tauris, T. M., Lynch, R. S., van Kerkwijk, M. H., Kramer, M., Bassa, C., Dhillon, V. S., Driebe, T., Hessels, J. W. T., Kaspi, V. M., Kondratiev, V. I., Langer, N., Marsh, T. R., McLaughlin, M. A., Ransom, T. T. P. S. M., Stairs, I. H., van Leeuwen, J., Verbiest, J. P. W., and Whelan, D. G. (2013). A massive pulsar in a compact relativistic binary, *Science* **340**, p. 448.

Apostolatos, T. A., Cutler, C., Sussman, G. J., and Thorne, K. S. (1994). Spin-induced orbital precession and its modulation of the gravitational waveforms from merging binaries, *Phys. Rev. D* **49**, p. 6274.

Argyres, P. C., Dimopoulos, S., and March-Russell, J. (1998). Black holes and submillimeter dimensions, *Phys. Lett. B* **441**, p. 96.

Arkani-Hamed, N., Dimopoulos, S., and Dvali, G. R. (1998). The hierarchy problem and new dimensions at a millimeter, *Phys. Lett. B* **429**, p. 263.

Arnowitt, R., Deser, S., and Misner, C. W. (1960). Energy and the criteria for radiation in general relativity, *Phys. Rev.* **118**, p. 1100.

Arnowitt, R., Deser, S., and Misner, C. W. (1962). The dynamics of general relativity, in L. Witten (ed.), *Gravitation*.

Asada, H. (1998). Formulation for the internal motion of quasiequilibrium configurations in general relativity, *Phys. Rev. D* **57**, p. 7292.

Ashtekar, A., Beetle, C., Dreyer, O., Fairhurst, S., Krishnan, B., Lewandowski, J., and Wisniewski, J. (2000a). Generic isolated horizons and their applications, *Phys. Rev. Lett.* **85**, p. 3564.

Ashtekar, A., Beetle, C., and Fairhurst, S. (1999). Isolated horizons: A generalization of black hole mechanics, *Class. Quantum Grav.* **16**, p. L1.

Ashtekar, A., Beetle, C., and Lewandowski, J. (2001). Mechanics of rotating isolated horizons, *Phys. Rev. D* **64**, pp. 044016–1.

Ashtekar, A., Fairhurst, S., and Krishnan, B. (2000b). Isolated horizons: Hamiltonian evolution and the first law, *Phys. Rev. D* **62**, pp. 104025–1.

Ashtekar, A. and Krishnan, B. (2003). Dynamical horizons and their properties, *Phys. Rev. D* **68**, pp. 104030–1.

Ashtekar, A. and Krishnan, B. (2004). Isolated and dynamical horizons and their applications, *Living Review Relativity* **7-10**.

ASKAP (). http://www.atnf.csiro.au/projects/askap/.

Baiotti, L., Damour, T., Giacomazzo, B., Nagar, A., and Rezzolla, L. (2010). Analytic modelling of tidal effects in the relativistic inspiral of binary neutron stars, *Phys. Rev. Lett.* **105**, pp. 261101–1.

Baiotti, L., Damour, T., Giacomazzo, B., Nagar, A., and Rezzolla, L. (2011). Accurate numerical simulations of inspiralling binary neutron stars and their comparison with effective-one-body analytical models, *Phys. Rev. D* **84**, pp. 024017–1.

Baiotti, L., de Pieri, R., Manca, G. M., and Rezzolla, L. (2007). Accurate simulations of the dynamical bar-mode instability in full general relativity, *Phys. Rev. D* **75**, pp. 044023–1.

Baiotti, L., Hawke, W., Rezzolla, L., and Schnetter, E. (2005). Gravitational-wave emission from rotating gravitational collapse in three dimensions, *Phys. Rev. Lett.* **94**, pp. 131101–1.

Baker, J. G., Boggs, W. D., Centrella, J., Kelly, B. J., McWilliams, S. T., Miller, M. C., and van Meter, J. R. (2007a). Modeling kicks from the merger of nonprecessing black hole binaries, *Astrophys. J.* **668**, p. 1140.

Baker, J. G., Campanelli, M., Pretorius, F., and Zlochower, Y. (2007b). Comparisons of binary black hole merger waveforms, *Class. Quantum Grav.* **24**, p. S25.

Baker, J. G., Centrella, J., Choi, D.-I., Koppitz, M., and van Meter, J. (2006a). Accurate evolutions of orbiting black-hole binaries without excision, *Phys. Rev. Lett.* **96**, pp. 111102–1.

Baker, J. G., Centrella, J., Choi, D.-I., Koppitz, M., and van Meter, J. (2006b). Binary black hole merger dynamics and waveforms, *Phys. Rev. D* **73**, pp. 104002–1.

Baker, J. G., Centrella, J., Choi, D.-I., Koppitz, M., van Meter, J., and Miller, M. C. (2006c). Getting a kick out of numerical relativity, *Astrophys. J.* **653**, p. L93.

Baker, J. G., McWilliams, S. T., van Meter, J. R., Centrella, J., Choi, D.-I., Kelly, B. J., and Koppitz, M. (2007c). Binary black hole late inspiral: Simulations for gravitational wave observations, *Phys. Rev. D* **75**, pp. 124024–1.

Baker, J. G., van Meter, J. R., McWilliams, S. T., Centrella, J., and Kelly, B. J. (2007d). Consistency of post-Newtonian waveforms with numerical relativity, *Phys. Rev. Lett.* **99**, pp. 181101–1.

Balbus, S. A. and Hawley, J. F. (1991a). A powerful local shear instability in weakly magnetized disks. I — Linear analysis, *Astrophys. J.* **376**, p. 214.

Balbus, S. A. and Hawley, J. F. (1991b). A powerful local shear instability in weakly magnetized disks. II — Nonlinear evolution, *Astrophys. J.* **376**, p. 223.

Balbus, S. A. and Hawley, J. F. (1998). Instability, turbulence, and enhanced transport in accretion disks, *Rev. Mod. Phys.* **70**, p. 1.

Banks, T. and Fischler, W. (1999). A model for high energy scattering in quantum gravity, *hep-th:/9906038*, p. 1.

Banyuls, F., Font, J. A., Ibáñez, J.-M., Martí, J.-M., and Miralles, J. A. (1997). Numerical 3+1 general relativistic hydrodynamics: A local characteristic approach, *Astrophys. J.* **476**, p. 221.

Barausse, E. and Rezzolla, L. (2009). Predicting the direction of the final spin from the coalescence of two black holes, *Astrophys. J* **704**, p. L40.

Bardeen, J. M. (1970). A variational principle for rotating stars in general relativity, *Astrophys. J.* **162**, p. 71.

Bardeen, J. M. (1973). Rapidly rotating stars, disks, and black holes, in C. DeWitt and B. S. DeWitt (eds.), *Black Holes* (Gordon and Breach Science Publishers), p. 241.

Bardeen, J. M. (1983). Gauge and radiation conditions in numerical relativity, in N. Deruelle and T. Piran (eds.), *Gravitational radiation* (North-Holland Publishing Company), p. 433.

Bardeen, J. M., Carter, B., and Hawking, S. W. (1973). The four laws of black hole mechanics, *Commun. Math. Phys.* **31**, p. 161.

Bardeen, J. M. and Petterson, J. A. (1975). The Lense-Thirring effect and accretion disks around Kerr black holes, *Astrophys. J. Lett.* **195**, p. L68.

Bardeen, J. M. and Piran, T. (1983). General relativistic axisymmetric rotating systems: Coordinates and equations, *Phys. Rep.* **96**, p. 205.

Bardeen, J. M., Press, W. H., and Teukolsky, S. A. (1972). Rotating black holes: Locally nonrotating frames, energy extraction, and scalar synchrotron radiation, *Astrophys. J.* **178**, p. 347.

Bardeen, J. M. and Wagoner, R. (1971). Relativistic disks. I. Uniform rotation, *Astrophys. J.* **167**, p. 359.

Barnes, J. and Kasen, D. (2013). Effect on a high opacity of the light curves of radioactively powered transients from compact object mergers, *Astrophys. J.* **775**, p. 18.

Baron, E. A., Cooperstein, J., and Kahana, S. (1985). Supernovae and the nuclear equation of state at high densities, *Nucl. Phys. A* **440**, p. 744.

Barsara, D. S. (2001). Divergence-free adaptive mesh refinement for magnetohydrodynamics, *J. Comp. Phys.* **174**, p. 614.

Barsara, D. S. (2009). Divergence-free reconstruction of magnetic fields and WENO schemes for magnetohydrodynamics, *J. Comp. Phys.* **228**, p. 5040.

Baumgarte, T. W. (2000). Innermost stable circular orbit of binary black holes, *Phys. Rev. D* **62**, pp. 024018–1.

Baumgarte, T. W., Cook, G. B., Scheel, M. A., Shapiro, S. L., and Teukolsky, S. A. (1996a). Implementing an apparent-horizon finder in three dimensions, *Phys. Rev. D* **54**, p. 4849.

Baumgarte, T. W., Cook, G. B., Scheel, M. A., Shapiro, S. L., and Teukolsky, S. A. (1997). Binary neutron stars in general relativity: Quasiequilibrium models, *Phys. Rev. Lett.* **79**, p. 1182.

Baumgarte, T. W., Cook, G. B., Scheel, M. A., Shapiro, S. L., and Teukolsky, S. A. (1998a). General relativistic models of binary neutron stars in quasiequilibrium, *Phys. Rev. D* **57**, p. 7299.

Baumgarte, T. W., Cook, G. B., Scheel, M. A., Shapiro, S. L., and Teukolsky, S. A. (1998b). Stability of relativistic neutron stars in binary orbit, *Phys. Rev. D* **57**, p. 6181.

Baumgarte, T. W., Janka, H.-T., Keil, W., Shapiro, S. L., and Teukolsky, S. A. (1996b). Delayed collapse of hot neutron stars to black holes via hadronic phase transitions, *Astrophys. J.* **468**, p. 823.

Baumgarte, T. W., Montero, P., Cordero-Carrión, I., and Müller, E. (2013). Numerical relativity in spherical polar coordinates: Evolution calculations with the BSSN formulation, *Phys. Rev. D* **87**, pp. 044027–1.

Baumgarte, T. W. and Naculich, S. G. (2007). Analytical representation of a black hole puncture solution, *Phys. Rev. D* **75**, pp. 067502–1.

Baumgarte, T. W. and Shapiro, S. L. (1999a). Evolution of rotating supermassive stars to the onset of collapse, *Astrophys. J.* **526**, p. 941.

Baumgarte, T. W. and Shapiro, S. L. (1999b). Numerical integration of Einstein's field equation, *Phys. Rev. D* **59**, pp. 024007–1.

Baumgarte, T. W. and Shapiro, S. L. (2010). *Numerical relativity* (Cambridge University Press).

Baumgarte, T. W., Shapiro, S. L., and Shibata, M. (2000). On the maximum mass of differentially rotating neutron stars, *Astrophys. J. Lett.* **528**, p. L28.

Baumgarte, T. W., Shapiro, S. L., and Teukolsky, S. A. (1995). Computing supernova collapse to neutron stars and black holes, *Astrophys. J.* **443**, p. 717.

Baumgarte, T. W., Shapiro, S. L., and Teukolsky, S. A. (1996c). Computing the delayed collapse of hot neutron stars to black holes, *Astrophys. J.* **458**, p. 680.

Bauswein, A., Goriely, S., and Janka, H.-T. (2013). Systematics of dynamical mass ejection, nucleosynthesis, and radioactively powered electromagnetic signals from neutron-star mergers, *Astrophys.* **773**, p. 78.

Bauswein, A. and Janka, H.-T. (2012). Measuring neutron-star properties via gravitational waves from neutron-star mergers, *Phys. Rev. Lett.* **108**, pp. 011101–1.

Bauswein, A., Janka, H.-T., Hebeler, K., and Schwenk, A. (2012). Equation-of-state dependence of the gravitational-wave signal from the ring-down phase of neutron-star mergers, *Phys. Rev. D* **86**, pp. 063001–1.

Bauswein, A., Janka, H.-T., and Oechslin, R. (2010). Testing approximations of thermal effects in neutron star merger simulations, *Phys. Rev. D* **82**, pp. 084043–1.

Beig, R. (1978). Arnowitt-Deser-Misner energy and g_{00}, *Phys. Lett. A* **69**, p. 153.
Bekenstein, J. D. (1973). Extraction of energy and charge from a black hole, *Phys. Rev. D* **7**, p. 949.
Belczynski, K., Kalogera, V., and Bulik, T. (2002). A comprehensive study of binary compact objects as gravitational wave sources: Evolutionary channels, rates, and physical properties, *Astrophys. J.* **572**, p. 407.
Beloborodov, A. M. (2008). Hyper-accreting black holes, *AIP Conferece Proc.* **1054**, p. 51.
Berger, M. and Oliger, J. (1984). Adaptive mesh refinement for hyperbolic partial differential equations, *J. Comp. Phys.* **53**, p. 484.
Bernstein, D., Hobill, D., Seidel, E., Smarr, L., and Towns, J. (1994). Numerically generated axisymmetric black hole spacetimes: Numerical methods and code tests, *Phys. Rev. D* **50**, p. 5000.
Bernstein, D. H., Hobill, D. W., and Smarr, L. L. (1989). Black hole spacetimes: Testing numerical relativity, in C. R. Evans, L. S. Finn, and D. W. Hobill (eds.), *Frontiers in Numerical Relativity* (Cambridge University Press), p. 57.
Bernuzzi, S. and Hilditch, D. (2010). Constraint violation in free evolution schemes: Comparing the BSSNOK formulation with a conformal decomposition of the Z4 formulation, *Phys. Rev. D* **81**, pp. 084003–1.
Bernuzzi, S., Nager, A., Thierfelder, M., and Brügmann, B. (2012a). Tidal effects in binary neutron star coalescence, *Phys. Rev. D* **86**, pp. 044030–1.
Bernuzzi, S., Thierfelder, M., and Brügmann, B. (2012b). Accuracy of numerical relativity waveforms from binary neutron star mergers and their comparison with post-Newtonian waveforms, *Phys. Rev. D* **85**, pp. 104030–1.
Berti, E., Cardoso, V., Gonzalez, J. A., Sperhake, U., Hannam, M., Husa, S., and Brügmann, B. (2007). Inspiral, merger, and ringdown of unequal mass black hole binaries: A multipolar analysis, *Phys. Rev. D* **76**, pp. 064034–1.
Berti, E., Cardoso, V., and Starinets, A. O. (2009). Quasinormal modes of black holes and black branes, *Class. Quantum Grav.* **26**, pp. 163001–1.
Bethe, H. A. (1990). Supernova mechanisms, *Rev. Mod. Phys.* **62**, p. 801.
Bildsten, L. and Cutler, C. (1992). Tidal interactions of inspiraling compact binaries, *Astrophys. J.* **400**, p. 175.
Bini, D., Damour, T., and Nagar, A. (2012). Effective action approach to higher-order relativistic tidal interactions in binary systems and their effective one body description, *Phys. Rev. D* **85**, pp. 124034–1.
Binney, J. and Tremaine, S. (1987). *Galactic Dynamics* (Princeton University Press).
Bionta, R. M., Blewitt, G., Bratton, C. B., and 34 authors (1987). Observation of a neutrino burst in coincidence with supernova 1987A in the Large Magellanic Cloud, *Phys. Rev. Lett.* **58**, p. 1494.
Birkl, R., Aloy, M. A., Janka, H.-T., and Müller, E. (2007). Neutrino pair annihilation near accreting, stellar-mass black holes, *Astron. and Astrophys.* **463**, p. 51.
Birkl, R., Stergioulas, N., and Müller, E. (2011). Stationary, axisymmetric neutron stars with meridional circulation in general relativity, *Phys. Rev. D* **84**, pp. 023003–1.
Bizoń, P. and Rostworowski, A. (2011). Weakly turbulent instability of anti-de Sitter spacetime, *Phys. Rev. Lett.* **107**, pp. 031102–1.
Blackburn, J. K. and Detweiler, S. (1992). Close black-hole binary systems, *Phys. Rev. D* **46**, p. 2318.
Blaes, O. M. and Hawley, J. F. (1988). Nonaxisymmetric disk instabilities — A linear and nonlinear synthesis, *Astrophys. J.* **326**, p. 277.
Blanchet, L. (2002). Innermost circular orbit of binary black holes at the third post-Newtonian approximation, *Phys. Rev. D* **65**, pp. 124009–1.

Blanchet, L. (2014). Gravitational radiation from post-Newtonian sources and inspiraling compact binaries, *Living Review Relativity* **17-2**.

Blanchet, L., Buonanno, A., and Faye, G. (2006). Higher-order spin effects in the dynamics of compact binaries. II. Radiation field, *Phys. Rev. D* **74**, pp. 104034–1.

Blanchet, L., Buonanno, A., and Faye, G. (2010). Erratum: Higher-order spin effects in the dynamics of compact binaries. II. Radiation field, *Phys. Rev. D* **81**, pp. 089901(E)–1.

Blanchet, L. and Damour, T. (1986). Radiative gravitational fields in general relativity I: General structure of the field outside the source, *Phil. Trans. Royal Soc. Lond.* **A320**, p. 379.

Blanchet, L., Damour, T., and Schäfer, G. (1990). Post-Newtonian hydrodynamics and post-Newtonian gravitational wave generation for numerical relativity, *Mon. Not. R. Astron. Soc.* **242**, p. 289.

Blanchet, L., Qusailah, M. S. S., and Will, C. M. (2005). Gravitational recoil of inspiraling black hole binaries to second post-Newtonian order, *Astrophys. J.* **635**, p. 508.

Blondin, J. M., Mezzacappa, A., and DeMarino, C. (2003). Stability of standing accretion shocks, with an eye toward core-collapse supernovae, *Astrophys. J.* **584**, p. 971.

Bona, C., Ledvinka, T., Palenzuela, C., and Zácek, M. (2003). General-covariant evolution formalism for numerical relativity, *Phys. Rev. D* **67**, pp. 104005–1.

Bona, C., Ledvinka, T., Palenzuela, C., and Zácek, M. (2004). Symmetry-breaking mechanism for the Z4 general-covariant evolution system, *Phys. Rev. D* **69**, pp. 064036–1.

Bona, C. and Massó, J. (1988). Harmonic synchronizations of spacetime, *Phys. Rev. D* **38**, p. 2419.

Bona, C. and Massó, J. (1992). Hyperbolic evolution system for numerical relativity, *Phys. Rev. Lett.* **68**, p. 1097.

Bona, C., Massó, J., Seidel, E., and Stela, J. (1995). New formalism for numerical relativity, *Phys. Rev. Lett.* **75**, p. 600.

Bonazzola, S. (1973). A virial theorem in general relativity, *Astrophys. J* **182**, p. 335.

Bonazzola, S., Frieben, J., and Gourgoulhon, E. (1996). Spontaneous symmetry breaking of rapidly rotating stars in general relativity, *Astrophys. J.* **460**, p. 379.

Bonazzola, S. and Gourgoulhon, E. (1994). A virial identity applied to relativistic stellar models, *Class. Quantum Grav.* **11**, p. 1775.

Bonazzola, S., Gourgoulhon, E., Grandclément, P., and Novak, J. (2004). Constrained scheme for the Einstein equations based on the Dirac gauge and spherical coordinates, *Phys. Rev. D* **70**, pp. 104007–1.

Bonazzola, S., Gourgoulhon, E., and Marck, J.-A. (1997). Relativistic formalism to compute quasiequilibrium configurations of nonsynchronized neutron star binaries, *Phys. Rev. D* **56**, p. 7740.

Bonazzola, S., Gourgoulhon, E., and Marck, J.-A. (1999). Numerical models of irrotational binary neutron stars in general relativity, *Phys. Rev. Lett.* **82**, p. 892.

Bonazzola, S., Gourgoulhon, E., Salgado, M., and Marck, J.-A. (1993). Axisymmetric rotating relativistic bodies: A new numerical approach for exact solutions, *Astron. Astrophys.* **278**, p. 421.

Bonazzola, S., Gourhoulhon, E., and Marck, J.-A. (1998). Numerical approach for high precision 3D relativistic star models, *Phys. Rev. D* **58**, pp. 104020–1.

Bond, J. R., Arnett, W. D., and Carr, B. J. (1984). The evolution and fate of very massive objects, *Astrophys. J.* **280**, p. 825.

Bowen, J., Rauber, J., and York, J. W. (1984). Two black holes with axisymmetric parallel spins — Initial data, *Class. Quantum Grav.* **1**, p. 591.

Bowen, J. M. (1982). General solution for flat-space longitudinal momentum, *General Relativity and Gravitation* **14**, p. 1183.

Bowen, J. M. and York, J. W. (1980). Time-asymmetric initial data for black holes and black-hole collisions, *Phys. Rev. D* **21**, p. 2047.
Boyer, R. H. and Lindquist, R. W. (1967). Maximal analytic extension of the Kerr metric, *J. Math. Phys.* **8**, p. 265.
Boyle, M., Brown, D. A., Kidder, L. E., Mroué, A. H., Pfeiffer, H. P., Scheel, M. A., Cook, G. B., and Teukolsky, S. A. (2007). High-accuracy comparison of numerical relativity simulations with post-Newtonian expansions, *Phys. Rev. D* **76**, pp. 124038-1.
Boyle, M., Buonanno, A., Kidder, L. E., Mroué, A. H., Pan, Y., Pfeiffer, H. P., and Scheel, M. A. (2008). High-accuracy numerical simulation of black-hole binaries: Computation of gravitational-wave energy flux and comparisons with post-Newtonian approximants, *Phys. Rev. D* **78**, pp. 104020-1.
Braginski, V. B. and Thorne, K. S. (1987). Gravitational-wave bursts with memory and experimental prospects, *Nature* **327**, p. 123.
Brandt, S. and Brügmann, B. (1997). A simple construction of initial data for multiple black holes, *Phys. Rev. Lett.* **78**, p. 3606.
Brandt, S. R. and Seidel, E. (1995a). Evolution of distorted rotating black holes. I. Methods and tests, *Phys. Rev. D* **52**, p. 856.
Brandt, S. R. and Seidel, E. (1995b). Evolution of distorted rotating black holes. II. Dynamics and analysis, *Phys. Rev. D* **52**, p. 870.
Brandt, S. R. and Seidel, E. (1996). Evolution of distorted rotating black holes. III. Initial data, *Phys. Rev. D* **54**, p. 1403.
Brill, D. R. (1959). On the positive definite mass of the Bondi-Weber-Wheeler time-symmetric gravitational waves, *Ann. Phys.* **7**, p. 466.
Brill, D. R. and Lindquist, R. W. (1963). Interaction energy in geometrostatics, *Phys. Rev.* **131**, p. 471.
Brown, D., Diener, P., Sarbach, O., Schnetter, E., and Tiglio, M. (2009). Turduckening black holes: An analytical and computational study, *Phys. Rev. D* **79**, pp. 044023-1.
Bruenn, S. W. (1985). Stellar core collapse: Numerical model and infall epoch, *Astrophys. J. Suppl.* **58**, p. 771.
Bruenn, S. W., Nisco, K. R. D., and Mezzacappa, A. (2001). General relativistic effects in the core collapse supernova mechanism, *Astrophys. J.* **560**, p. 326.
Brügmann, B., González, J. A., Hannam, M., Husa, S., and Sperhake, U. (2007). Exploring black hole superkicks, *Phys. Rev. D* **77**, pp. 124047-1.
Brügmann, B., González, J. A., Hannam, M., Husa, S., Sperhake, U., and Tichy, W. (2008). Calibration of moving puncture simulations, *Phys. Rev. D* **77**, pp. 024027-1.
Buchman, L. T., Pfeiffer, H. P., Scheel, M. A., and Szilágyi, B. (2012). Simulations of non-equal mass black hole binaries with spectral methods, *Phys. Rev. D* **86**, pp. 084033-1.
Buonanno, A., Chen, Y., and Damour, T. (2006). Transition from inspiral to plunge in precessing binaries of spinning black holes, *Phys. Rev. D* **74**, pp. 104005-1.
Buonanno, A., Cook, G. B., and Pretorius, F. (2007a). Inspiral, merger, and ring-down of equal-mass black-hole binaries, *Phys. Rev. D* **75**, pp. 124018-1.
Buonanno, A. and Damour, T. (1999). Effective one-body approach to general relativistic two-body dynamics, *Phys. Rev. D* **59**, pp. 084006-1.
Buonanno, A., Iyer, B. R., Ochsner, E., Pan, Y., and Sathyaprakash, B. S. (2009). Comparison of post-Newtonian templates for compact binary inspiral signals in gravitational-wave detectors, *Phys. Rev. D* **80**, pp. 084043-1.
Buonanno, A., Kidder, L. E., and Lehner, L. (2008). Estimating the final spin of a binary black hole coalescence, *Phys. Rev. D* **77**, pp. 026004-1.

Buonanno, A., Pan, Y., Baker, J. G., Centrella, J., Kelly, B. J., McWilliams, S. T., and van Meter, J. R. (2007b). Approaching faithful templates for nonspinning binary black holes using the effective-one-body approach, *Phys. Rev. D* **76**, pp. 104049–1.

Buras, R., Janka, H.-T., Rampp, M., and Kifonidis, K. (2006). Two-dimensional hydrodynamic core-collapse supernova simulations with spectral neutrino transport. II. Models for different progenitor stars, *Astron. Astrophys.* **457**, p. 281.

Burbidge, E. M., Burdidge, G. R., Fowler, W. A., and Hoyle, F. (1957). Synthesis of the elements in stars, *Rev. Mod. Phys.* **29**, p. 547.

Burrows, A. and Fryxell, B. A. (1992). An instability in neutron stars at birth, *Science* **258**, p. 430.

Burrows, A. and Hayes, J. (1996). Pulsar recoil and gravitational radiation due to asymmetrical stellar collapse and explosion, *Phys. Rev. Lett.* **76**, p. 352.

Burrows, A., Livne, E., Dessart, L., Ott, C., and Murphy, J. (2007). Features of the acoustic mechanism of core-collapse supernova explosions, *Astrophys. J.* **655**, p. 416.

Butterworth, E. M. and Ipser, J. R. (1976). On the structure and stability of rapidly rotating fluid bodies in general relativity. I — The numerical method for computing structure and its application to uniformly rotating homogeneous bodies, *Astrophys. J.* **204**, p. 200.

Cadez, A. (1974). Apparent horizons in the two-black-hole problem, *Annals of Physics* **83**, p. 449.

Campanelli, M., Lousto, C. O., Marronetti, P., and Zlochower, Y. (2006a). Accurate evolutions of orbiting black-hole binaries without excision, *Phys. Rev. Lett.* **96**, pp. 111101–1.

Campanelli, M., Lousto, C. O., Nakano, H., and Zlochower, Y. (2009). Comparison of numerical and post-Newtonian waveforms for generic precessing black-hole binaries, *Phys. Rev. D* **79**, pp. 084010–1.

Campanelli, M., Lousto, C. O., and Zlochower, Y. (2006b). Spin-orbit interactions in black-hole binaries, *Phys. Rev. D* **74**, pp. 084023–1.

Campanelli, M., Lousto, C. O., and Zlochower, Y. (2006c). Spinning-black-hole binaries: The orbital hang-up, *Phys. Rev. D* **74**, pp. 041501(R)–1.

Campanelli, M., Lousto, C. O., Zlochower, Y., Krishnan, B., and Merritt, D. (2007a). Spin flips and precession in black-hole-binary mergers, *Phys. Rev. D* **75**, pp. 064030–1.

Campanelli, M., Lousto, C. O., Zlochower, Y., and Merritt, D. (2007b). Large merger recoils and spin flips from generic black hole binaries, *Astrophys. J. Lett.* **659**, p. L5.

Campanelli, M., Lousto, C. O., Zlochower, Y., and Merritt, D. (2007c). Maximum gravitational recoil, *Phys. Rev. Lett.* **98**, pp. 231102–1.

Cardall, C. Y., Endeve, E., and Mezzacappa, A. (2013). Conservative 3+1 general relativistic Boltzmann equation, *Phys. Rev. D* **88**, pp. 023011–1.

Cardall, C. Y. and Mezzacappa, A. (2003). Conservative formulations of general relativistic kinetic theory, *Phys. Rev. D* **68**, pp. 023006–1.

Cardoso, V., Gualtieri, L., Herdeiro, C., and 29 authors (2012). HEP/NR: Roadmap for the future, *Class. Quantum Grav.* **29**, pp. 244001–1.

Cardoso, V., Lemos, P. S., and Yoshida, S. (2003). Scalar-gravitational perturbations and quasinormal modes in the five dimensional Schwarzschild black hole, *JHEP* **12**, p. 041.

Carr, B. J. and Hawking, S. W. (1974). Black holes in the early universe, *Mon. Not. R. Astron. Soc.* **399**, p. 168.

Carter, B. (1968). Hamilton-Jacobi and Schrödinger separable solutions of Einstein's equations, *Comm. Math. Phys.* **10**, p. 280.

Carter, B. (1969). Killing horizons and orthogonally transitive groups in space-time, *J. Math. Phys.* **10**, p. 70.

Carter, B. (1970). The commutation property of a stationary, axisymmetric system, *Commun. Math. Phys.* **17**, p. 233.

Carter, B. (1973a). Black hole equilibrium states, in C. DeWitt and B. S. DeWitt (eds.), *Black Holes* (Gordon and Breach Science Publishers), p. 57.

Carter, B. (1973b). Elastic perturbation theory in general relativity and a variation principle for a rotating solid star, *Commun. Math. Phys.* **30**, p. 261.

Castor, J. I. (1972). Radiative transfer in spherically symmetric flows, *Astrophys. J.* **178**, p. 779.

Caudill, M., Cook, G. B., Grigsby, J. D., and Pfeiffer, H. P. (2006). Circular orbits and spin in black-hole initial data, *Phys. Rev. D* **74**, pp. 064011–1.

Centrella, J. and Wilson, J. R. (1984). Planar numerical cosmology. II — The difference equations and numerical tests, *Astrophys. J. Suppl.* **54**, p. 229.

Centrella, J. M., New, K. C. B., Lowe, L. L., and Brown, J. D. (2001). Dynamical rotational instability at low T/W, *Astrophys. J.* **550**, p. 193.

Cerda-Duran, P., DeBrye, N., M. A. Aloy, J. A. F., and Obergaulinger, M. (2013). Gravitational wave signatures in black hole forming core collapse, *Astrophys. J. Lett.* **779**, p. L18.

Chandrasekhar, S. (1961). *Hydrodynamic and Hydromagnetic Stability* (Oxford University Press).

Chandrasekhar, S. (1964). The dynamical instability of gaseous masses approaching the Schwarzschild limit in general relativity, *Astrophys. J.* **140**, p. 417.

Chandrasekhar, S. (1969). *Ellipsoidal Figures of Equilibrium* (Dover Publications, Inc., New York).

Chandrasekhar, S. (1970). Solutions of two problems in the theory of gravitational radiation, *Phys. Rev. Lett.* **24**, p. 611.

Chandrasekhar, S. (1983). *The Mathematical Theory of Black Holes* (Oxford University Press).

Chandrasekhar, S. and Detweiler, S. (1975). The quasi-normal modes of the Schwarzschild black hole, *Proc. Roy. Soc. London A* **344**, p. 441.

Chandrasekhar, S. and Esposito, F. P. (1970). The $2\frac{1}{2}$-post-Newtonian equations of hydrodynamics and radiation reaction in general relativity, *Astrophys. J.* **160**, p. 153.

Chen, W.-Z. and Beloborodov, A. M. (2007). Neutrino-cooled accretion disks around spinning black holes, *Astrophys. J.* **657**, p. 383.

Choptuik, M. W. (1993). Universality and scaling in gravitational collapse of a massless scalar field, *Phys. Rev. Lett.* **70**, p. 9.

Choptuik, M. W., Hirschmann, E. W., Liebling, S. L., and Pretorius, F. (2003a). An axisymmetric gravitational collapse code, *Class. Quantum Grav.* **20**, p. 1857.

Choptuik, M. W., Hirschmann, E. W., Liebling, S. L., and Pretorius, F. (2003b). Critical collapse of the massless scalar field in axisymmetry, *Phys. Rev. D* **68**, pp. 044007–1.

Choptuik, M. W., Hirschmann, E. W., Liebling, S. L., and Pretorius, F. (2004). Critical collapse of a complex scalar field with angular momentum, *Phys. Rev. Lett.* **93**, pp. 131101–1.

Choptuik, M. W., Lehner, L., Olabarrueta, I., Petryk, R., Pretorius, F., and Villegas, H. (2003c). Towards the final fate of an unstable black string, *Phys. Rev. D* **68**, pp. 044001–1.

Christodoulou, D. (1970). Reversible and irreversible transformations in black-hole physics, *Phys. Rev. Lett.* **25**, p. 1596.

Colella, P. and Woodward, P. R. (1984). The piecewise parabolic method (ppm) for gas dynamical simulation, *J. Comput. Phys.* **54**, p. 174.

Colgate, S. A. and White, R. H. (1966). The hydrodynamic behavior of supernovae explosions, *Astrophys. J* **143**, p. 626.

Cook, G. B. (1991). Initial data for axisymmetric black-hole collisions, *Phys. Rev. D* **44**, p. 2983.

Cook, G. B. (1994). Three-dimensional initial data for the collision of two black holes. II. Quasicircular orbits for equal-mass black holes, *Phys. Rev. D* **50**, p. 5025.

Cook, G. B. (2000). Initial data for numerical relativity, *Living Review Relativity* **3-5**.

Cook, G. B. (2002). Corotating and irrotational binary black holes in quasi-circular orbits, *Phys. Rev. D* **65**, pp. 084003-1.

Cook, G. B. and Pfeiffer, H. P. (2004). Excision boundary conditions for black hole initial data, *Phys. Rev. D* **70**, pp. 104016-1.

Cook, G. B. and Scheel, M. A. (1997). Well-behaved harmonic time slices of a charged, rotating, boosted black hole, *Phys. Rev. D* **56**, p. 4775.

Cook, G. B., Shapiro, S. L., and Teukolsky, S. A. (1992). Spin-up of a rapidly rotating star by angular momentum loss — effects of general relativity, *Astrophys. J.* **398**, p. 203.

Cook, G. B., Shapiro, S. L., and Teukolsky, S. A. (1994a). Rapidly rotating neutron stars in general relativity: Realistic equations of state, *Astrophys. J.* **423**, p. 823.

Cook, G. B., Shapiro, S. L., and Teukolsky, S. A. (1994b). Rapidly rotating polytropes in general relativity, *Astrophys. J.* **422**, p. 227.

Cook, G. B., Shapiro, S. L., and Teukolsky, S. A. (1994c). Recycling pulsars to millisecond periods in general relativity, *Astrophys. J. Lett.* **423**, p. 117.

Cook, G. B., Shapiro, S. L., and Teukolsky, S. A. (1996). Testing a simplified version of Einstein's equations for numerical relativity, *Phys. Rev. D* **53**, p. 5533.

Cook, G. B. and York, J. W. (1990). Apparent horizons for boosted or spinning black holes, *Phys. Rev. D* **41**, p. 1077.

Cooperstein, J. (1988). Neutrinos in supernova, *Phys. Rep.* **163**, p. 95.

Cordero-Carrión, I., Cerdá-Durán, P., Dimmelmeier, H., Jaramillo, J. L., Novak, J., and Gourgoulhon, E. (2009). Improved constrained scheme for the Einstein equations: An approach to the uniqueness issue, *Phys. Rev. D* **79**, pp. 024017-1.

Creighton, J. D. E. and Anderson, W. G. (2011). *Gravitational-Wave Physics and Astronomy: An Introduction to Theory, Experiment and Data Analysis* (Wiley-VCH).

Cunningham, C. T., Price, R. H., and Moncrief, V. (1978). Radiation from collapsing relativistic stars. I — Linearized odd-parity radiation, *Astrophys. J.* **224**, p. 643.

Cunningham, C. T., Price, R. H., and Moncrief, V. (1979). Radiation from collapsing relativistic stars. II — Linearized even-parity radiation, *Astrophys. J.* **230**, p. 870.

Damour, T. (2001). Coalescence of two spinning black holes: An effective one-body approach, *Phys. Rev. D* **64**, pp. 124013-1.

Damour, T., Grandclément, P., and Gourgoulhon, E. (2002). Circular orbits of corotating binary black holes: Comparison between analytical and numerical results, *Phys. Rev. D* **66**, pp. 024007-1.

Damour, T., Iyer, B., and Sathyaprakash, B. S. (1997). Improved filters for gravitational waves from inspiraling compact binaries, *Phys. Rev. D* **57**, p. 885.

Damour, T. and Nagar, A. (2009). The effective one body description of the two-body problem, *arXiv:/0906.1769*, p. 1.

Damour, T. and Nagar, A. (2010). Effective one body description of tidal effects in inspiralling compact binaries, *Phys. Rev. D* **81**, pp. 084016-1.

Damour, T. and Schäfer, G. (1985). Lagrangians for n point masses at the second post-Newtonian approximation of general relativity, *General Relativity and Gravitation* **17**, p. 879.

Deaton, M. B., Duez, M. D., Foucart, F., O'Connor, E., Ott, C. D., Kidder, L. E., Lawrence, E., Muhlberger, C. D., Scheel, M. A., and Szilágyi, B. (2013). Black hole-neutron star mergers with a hot nuclear equation of state: Outflow and neutrino-cooled disk for a low-mass, high-spin case, *Astrophys. J* **776**, p. 47.

Demorest, P., Pennucci, T., Ransom, S., Roberts, M., and Hessels, J. (2010). Shapiro delay measurement of a two solar mass neutron star, *Nature* **467**, p. 1081.

Detweiler, S. (1994). Periodic solutions of the Einstein equations for binary systems, *Phys. Rev. D* **50**, p. 4929.

DeWitt, B. S. (1967). Quantum theory of gravity. I. The canonical theory, *Phys. Rev.* **160**, p. 1113.

Di Matteo, T., Perna, R., and Narayan, R. (2002). Neutrino trapping and accretion models for gamma-ray bursts, *Astrophys. J.* **579**, p. 706.

Dias, O. J. C., Figueras, P., Monteiro, R., Santos, J. E., and Emparan, R. (2009). Instability and new phases of higher-dimensional rotating black holes, *Phys. Rev. D* **80**, pp. 111701–1.

Dias, O. J. C., Harnett, G. S., and Santos, J. E. (2014). Quasinormal modes of asymptotically flat rotating black holes, *JHEP* .

Dimmelmeier, H., Font, J. A., and Müller, E. (2002). Relativistic simulations of rotational core collapse II. Collapse dynamics and gravitational radiation, *Astron. Astrophys.* **393**, p. 523.

Dimmelmeier, H., Ott, C. D., Janka, H.-T., Marek, A., and Müller, E. (2007). Generic gravitational-wave signals from the collapse of rotating stellar cores, *Phys. Rev. Lett.* **98**, pp. 251101–1.

Dimmelmeier, H., Ott, C. D., Marek, A., and Janka, H.-T. (2008). The gravitational wave burst signal from core collapse of rotating stars, *Phys. Rev. D* **78**, pp. 064056–1.

Dimopoulos, S. and Landsberg, G. (2001). An alternative to compactification, *Phys. Rev. Lett.* **87**, pp. 161602–1.

Dominik, M., Belczynski, K., Fryer, C., Holz, D. E., Berti, E., Bulik, T., Mandel, I., and O'Shaughnessy, R. (2012). Double compact objects. I. The significance of the common envelope on merger rates, *Astrophys. J.* **759**, p. 52.

Douchin, F. and Haensel, P. (2001). A unified equation of state of dense matter and neutron star structure, *Astron. Astrophys.* **380**, p. 151.

Duez, M. D., Foucart, F., Kidder, L. E., Ott, C. D., and Teukolsky, S. A. (2010). Equation of state effects in black hole-neutron star mergers, *Class. Quantum Grav.* **27**, pp. 114106–1.

Duez, M. D., Foucart, F., Kidder, L. E., Pfeiffer, H. P., Scheel, M. A., and Teukolsky, S. A. (2008). Evolving black hole-neutron star binaries in general relativity using pseudospectral and finite difference methods, *Phys. Rev. D* **78**, pp. 104015–1.

Duez, M. D., Liu, Y.-T., Shapiro, S. L., Shibata, M., and Stephens, B. C. (2006a). Collapse of magnetized hypermassive neutron stars in general relativity, *Phys. Rev. Lett.* **96**, pp. 031101–1.

Duez, M. D., Liu, Y.-T., Shapiro, S. L., Shibata, M., and Stephens, B. C. (2006b). Evolution of magnetized, differentially rotating neutron stars: Simulations in full general relativity, *Phys. Rev. D* **73**, pp. 104015–1.

Duez, M. D., Liu, Y.-T., Shapiro, S. L., and Stephens, B. C. (2004). General relativistic hydrodynamics with viscosity: Contraction, catastrophic collapse, and disk formation in hypermassive neutron stars, *Phys. Rev. D* **69**, pp. 104030–1.

Duez, M. D., Liu, Y.-T., Shapiro, S. L., and Stephens, B. C. (2005). Relativistic magneto-hydrodynamics in dynamical spacetimes: Numerical methods and tests, *Phys. Rev. D* **72**, pp. 024028–1.

Eardley, D. M. and Smarr, L. (1978). Time functions in numerical relativity: Marginally bound dust collapse, *Phys. Rev. D* **19**, p. 2239.
Ehlers, J. (1971). General relativity and kinetic theory, in B. K. Sachs (ed.), *General Relativity and Cosmology* (Academic Press: New York and London), p. 1.
Eichler, D., Livio, M., Piran, T., and Schramm, D. N. (1989). Nucleosynthesis, neutrino bursts and gamma-rays from coalescing neutron stars, *Nature* **340**, p. 126.
Einfeldt, B. (1988). On Godunov-type methods for gas dynamics, *SIAM J. Numerical Analysis* **25**, p. 294.
Eldridge, J. J. and Tout, C. A. (2004). The progenitors of core-collapse supernovae, *Mon. Not. R. Astron. Soc.* **353**, p. 87.
eLISA Consorthium (2013). The gravitational universe, *arXiv:/1305.5720*, p. 1.
Ellis, G. F. R. (1971). Relativistic cosmology, in B. K. Sachs (ed.), *General Relativity and Cosmology* (Academic Press: New York and London), p. 104.
Ellis, G. F. R. (2009). Relativistic cosmology: Republication of an article in 1971, *Gen. Relativ. Gravit.* **41**, p. 581.
Eltgroth, P. G. (1971). Similarity analysis for relativistic flow in one dimension, *Phys. of Fluids* **14**, p. 2631.
Emparan, R. and Reall, H. S. (2002). A rotating black ring solution in five dimensions, *Phys. Rev. Lett.* **88**, pp. 101101-1.
Emperan, R. and Myers, R. C. (2003). Instability of ultra-spinning black holes, *JHEP* **09**, p. 025.
Emperan, R. and Reall, H. S. (2008). Black holes in higher dimensions, *Living Review Relativity* **11-6**.
Eppley, K. R. (1977). Evolution of time-symmetric gravitational waves: Initial data and apparent horizon, *Phys. Rev. D* **16**, p. 1609.
Epstein, R. (1978). The generation of gravitational radiation by escaping supernova neutrinos, *Astrophys. J.* **223**, p. 1037.
Epstein, R. I. and Pethick, C. J. (1981). Lepton loss and entropy generation in stellar collapse, *Astrophys. J.* **243**, p. 1003.
Estabrook, F., Wahlquist, H., Christensen, S., DeWitt, B., Smarr, L., and Tsiang, E. (1973). Maximally slicing a black hole, *Phys. Rev. D* **7**, p. 2814.
Etienne, Z. B., Faber, J. A., Liu, Y. T., Shapiro, S. L., and Baumgarte, T. W. (2007). Filling the holes: Evolving excised binary black hole initial data with puncture techniques, *Phys. Rev. D* **76**, pp. 101503-1.
Etienne, Z. B., Faber, J. A., Liu, Y. T., Shapiro, S. L., Taniguchi, K., and Baumgarte, T. W. (2008). Fully general relativistic simulations of black hole-neutron star mergers, *Phys. Rev. D* **77**, pp. 084002-1.
Etienne, Z. B., Liu, Y. T., Shapiro, S. L., and Baumgarte, T. W. (2009). Relativistic simulations of black hole-neutron star mergers: Effects of black hole spin, *Phys. Rev. D* **79**, pp. 044024-1.
Eulderink, F. and Mellema, G. (1995). General relativistic hydrodynamics with a Roe solver, *Astron. Astrophys. Suppl.* **110**, p. 587.
Evans, C. R. (1985). A method for numerical simulation of gravitational collapse and gravitational radiation generation, in J. Centrella, J. LeBlanc, and R. Bowers (eds.), *Numerical Astrophysics* (Jones and Bartlett Publishers), p. 216.
Evans, C. R. (1986). An approach for calculating axisymmetric gravitational collapse, in J. M. Centrella (ed.), *Dynamical spacetimes and numerical relativity* (Cambridge University Press), p. 3.
Evans, C. R. and Hawley, J. F. (1988). Simulation of magnetohydrodynamic flows — a constrained transport method, *Astrophys. J.* **332**, p. 659.

EVLA (). http://www.aoc.nrao.edu/evla/.

Faber, J. A. and Rasio, F. (2000). Post-Newtonian SPH calculations of binary neutron star coalescence: Method and first results, *Phys. Rev. D* **62**, pp. 064012–1.

Faber, J. A. and Rasio, F. (2002). Post-Newtonian SPH calculations of binary neutron star coalescence. III. Irrotational systems and gravitational wave spectra, *Phys. Rev. D* **65**, pp. 084042–1.

Faber, J. A., Rasio, F., and Manor, J. B. (2001). Post-Newtonian smoothed particle hydrodynamics calculations of binary neutron star coalescence. II. Binary mass ratio, equation of state, and spin dependence, *Phys. Rev. D* **63**, pp. 044012–1.

Faber, J. A. and Rasio, F. A. (2012). Binary neutron star mergers, *Living Review Relativity* **15-8**.

Faulkner, A. J., Kramer, M., Lyne, A. G., and 9 authors (2005). PSR J1756-2251: A new relativistic double neutron star system, *Astrophys. J. Lett.* **618**, p. L119.

Favata, M., Hughes, S. A., and Holz, D. E. (2004). How black holes get their kicks: Gravitational radiation recoil revisited, *Astrophys. J. Lett.* **607**, p. L5.

Faye, G., Blanchet, L., and Buonanno, A. (2006). Higher-order spin effects in the dynamics of compact binaries. I. Equations of motion, *Phys. Rev. D* **74**, pp. 104033–1.

Ferdman, R. D., Stairs, I. H., Kramer, M., Breton, R. P., MacLaughlin, M. A., Freire, P. C. C., Possenti, A., Stappers, B. W., Kaspi, V. M., and Manchester, R. N. (2013). The double pulsar: Evidence for neutron star formation without an iron core-collapse supernova, *Astrophys. J.* **767**, p. 85.

Ferdman, R. D., Stairs, I. H., Kramer, M., Janssen, G. H., Bassa, C. G., Stappers, B. W., Demorest, P. B., Cognard, I., Desvignes, G., Theureau, G., Burgay, M., Lyne, A. G., Manchester, R. N., and Possenti, A. (2014). PSR J1756-2251: A pulsar with a low-mass neutron star companion, *Mon. Not. R. Astron. Soc.* **443**, p. 2183.

Fernández, B. and Metzger, B. D. (2013). Delayed outflows from black hole accretion tori following neutron star binary coalescence, *Mon. Not. R. Astron. Soc.* **435**, p. 502.

Ferrarese, L. and Ford, H. (2005). Supermassive black holes in galactic nuclei: Past, present and future research, *Space Sci. Rev.* **116**, p. 523.

Finkelstein, D. (1958). Past-future asymmetry of the gravitational field of a point particle, *Phys. Rev.* **110**, p. 965.

Finn, L. S. and Evans, C. (1990). Determining gravitational radiation from Newtonian self-gravitating systems, *Astrophys. J.* **351**, p. 588.

Fishbone, L. G. and Moncrief, V. (1976). Relativistic fluid disks in orbit around Kerr black holes, *Astrophys. J.* **207**, p. 962.

Fitchett, M. J. (1983). The influence of gravitational wave momentum losses on the centre of mass motion of a Newtonian binary system, *Mon. Not. R. Astron. Soc.* **203**, p. 1049.

Flanagan, É. É. and Hinderer, T. (2008). Constraining neutron star tidal love numbers with gravitational wave detectors, *Phys. Rev. D* **77**, pp. 021502–1.

Font, J. A. (2006). Numerical hydrodynamics and magnetohydrodynamics in general relativity, *Living Review Relativity* **11-7**.

Font, J. A. and Daigne, F. (2002). The runaway instability of thick discs around black holes - I. The constant angular momentum case, *Mon. Not. R. Astron. Soc.* **334**, p. 383.

Font, J. A., Goodale, T., Iyer, S., Miller, M., Rezzolla, L., Seidel, E., Stergioulas, N., Suen, W.-M., and Tobias, M. (2002). Three-dimensional numerical general relativistic hydrodynamics II. Long-term dynamics of single relativistic stars, *Phys. Rev. D* **65**, pp. 084024–1.

Font, J. A., Ibáñez, J.-M., Marquina, A., and Martí, J. M. (1994). Multidimensional relativistic hydrodynamics: Characteristic fields and modern high-resolution shock-capturing schemes, *Astron. Astrophys.* **282**, p. 304.

Font, J. A., Miller, M., Suen, W.-M., and Tobias, M. (2000). Three-dimensional numerical general relativistic hydrodynamics: Formulations, methods, and code tests, *Phys. Rev. D* **61**, pp. 044011-1.

Foucart, F. (2012). Black-hole-neutron-star mergers: Disk mass predictions, *Phys. Rev. D* **86**, pp. 124007-1.

Foucart, F., Deaton, M. B., Duez, M. D., Kidder, L. E., Lawrence, E., MacDonald, I., Ott, C. D., Pfeiffer, H. P., Scheel, M. A., Szilágyi, B., and Teukolsky, S. A. (2013). Black-hole-neutron-star mergers at realistic mass ratios: Equation of state and spin orientation effects, *Phys. Rev. D* **87**, pp. 084006-1.

Foucart, F., Deaton, M. B., Duez, M. D., O'Connor, E., Ott, C. D., Haas, R., Kidder, L. E., Lawrence, E., Pfeiffer, H. P., Scheel, M. A., and Szilágyi, B. (2014). Neutron star-black hole mergers with a nuclear equation of state and neutrino cooling: Dependence in the binary parameters, *Phys. Rev. D* **90**, pp. 024026-1.

Foucart, F., Duez, M. D., Kidder, L. E., Scheel, M. A., Szilágyi, B., and Teukolsky, S. A. (2012). Black hole-neutron star mergers for 10 solar mass black holes, *Phys. Rev. D* **85**, pp. 044015-1.

Foucart, F., Duez, M. D., Kidder, L. E., and Teukolsky, S. A. (2011). Black hole-neutron star mergers: Effects of the orientation of the black hole spin, *Phys. Rev. D* **83**, pp. 024005-1.

Foucart, F., Kidder, L. E., Pfeiffer, H. P., and Teukolsky, S. A. (2008). Initial data for black hole-neutron star binaries: A flexible, high-accuracy spectral method, *Phys. Rev. D* **77**, pp. 124051-1.

Fowler, W. A. (1966). The stability of supermassive stars, *Astrophys. J.* **144**, p. 180.

Fowler, W. A. and Hoyle, F. (1964). Neutrino processes and pair formation in massive stars and supernova, *Astrophys. J. Suppl.* **9**, p. 201.

Fragile, P. C., Blaes, O. M., Anninos, P., and Salmonson, J. D. (2007). Global general relativistic magnetohydrodynamic simulation of a tilted black hole accretion disk, *Astrophys. J.* **668**, p. 417.

Fricke, K. J. (1973). Dynamical phases of supermassive stars, *Astrophys. J.* **183**, p. 941.

Friedman, J. L. (1978). Generic instability of rotating relativistic stars, *Commun. Math. Phys.* **62**, p. 247.

Friedman, J. L., Ipser, J. R., and Sorkin, R. D. (1988). Turning-point method for axisymmetric stability of rotating relativistic stars, *Astrophys. J.* **325**, p. 722.

Friedman, J. L., Parker, L., and Ipser, J. R. (1986). Rapidly rotating neutron star models, *Astrophys. J.* **304**, p. 115.

Friedman, J. L. and Schutz, B. F. (1978). Lagrangian perturbation theory of nonrelativistic fluids, *Astrophys. J.* **221**, p. 937.

Friedman, J. L. and Stergioulas, N. (2013). *Rotating Relativistic Stars* (Cambridge University Press).

Friedman, J. L., Uryū, K., and Shibata, M. (2002). Thermodynamics of binary black holes and neutron stars, *Phys. Rev. D* **65**, pp. 064035-1.

Fryer, C. L., Woosley, S. E., and Heger, A. (2001). Pair-instability supernovae, gravity waves, and gamma-ray transients, *Astrophys. J.* **550**, p. 372.

Fuller, G. M., Woosley, S. E., and Weaver, T. A. (1986). The evolution of radiation-dominated stars. I — Nonrotating supermassive stars, *Astrophys. J.* **307**, p. 675.

Furusawa, S., Sumiyoshi, K., Yamada, S., and Suzuki, H. (2013). New equations of state based on the liquid drop model of heavy nuclei and quantum approach to light nuclei for core-collapse supernova simulations, *Astrophys. J.* **772**, p. 95.

Gammie, C. F., McKinney, J. C., and Tóth, G. (2003). Harm: A numerical scheme for general relativistic magnetohydrodynamics, *Astrophys. J.* **589**, p. 444.

Garat, A. and Price, R. H. (2000). Nonexistence of conformally flat slices of the Kerr spacetime, *Phys. Rev. D* **61**, pp. 124011–1.

Garfinkle, D. (2002). Harmonic coordinate method for simulating generic singularities, *Phys. Rev. D* **65**, pp. 064015–1.

Garfinkle, D. and Duncan, G. C. (2001). Numerical evolution of Brill waves, *Phys. Rev. D* **63**, pp. 044011–1.

Gerlach, U. H. and Sengupta, U. K. (1979). Gauge-invariant perturbations on most general spherically symmetric space-times, *Phys. Rev. D* **19**, p. 2268.

Geroch, R. (1971). A method for generating solutions of Einstein's equations, *J. Math. Phys.* **12**, p. 918.

Glendenning, N. K. and Moszkowski, S. A. (1991). Reconciliation of neutron-star masses and binding of the lambda in hypernuclei, *Phys. Rev. Lett.* **67**, p. 2414.

Godunov, S. K. (1959). A finite difference method for the numerical computation of discontinuous solutions of the equations of fluid dynamics, *Mathematicheskii Sbornik* **47**, p. 271.

Goldberg, J. N., Macfarlane, A. J., Newman, E. T., Rohrlich, F., and Sudarshan, E. C. G. (1967). Spin-s spherical harmonics and ð, *J. Math. Phys.* **8**, p. 2155.

Goldwirth, D. S. and Piran, T. (1989). Spherical inhomogeneous cosmologies and inflation: Numerical methods, *Phys. Rev. D* **40**, p. 3263.

González, J. A., Hannam, M., Sperhake, U., Brügmann, B., and Husa, S. (2007a). Supermassive recoil velocities for binary black-hole mergers with antialigned spins, *Phys. Rev. Lett.* **98**, pp. 231101–1.

González, J. A., Sperhake, U., and Brügmann, B. (2009). Black-hole binary simulations: The mass ratio 10:1, *Phys. Rev. D* **79**, pp. 124006–1.

González, J. A., Sperhake, U., Brügmann, B., Hannam, M., and Husa, S. (2007b). Maximum kick from nonspinning black-hole binary inspiral, *Phys. Rev. Lett.* **98**, pp. 091101–1.

Gopakumar, A., Hannam, M., Husa, S., and Brügmann, B. (2008). Where post-Newtonian and numerical-relativity waveforms meet, *Phys. Rev. D* **78**, pp. 064026–1.

Goriely, S., Bauswein, A., and Janka, H.-T. (2011). r-process nucleosynthesis in dynamically ejected matter of neutron star mergers, *Astrophys. J. Lett.* **738**, p. L32.

Gourgoulhon, E. (2012). *3+1 formalism in general relativity* (Springer).

Gourgoulhon, E. and Bonazzola, S. (1993). Noncircular axisymmetric stationary spacetimes, *Phys. Rev. D* **48**, p. 2635.

Gourgoulhon, E. and Bonazzola, S. (1994). A formulation of the virial theorem in general relativity, *Class. Quantum Grav.* **11**, p. 443.

Gourgoulhon, E., Grandclément, P., and Bonazzola, S. (2002). Binary black holes in circular orbits. I. A global spacetime approach, *Phys. Rev. D* **65**, pp. 044020–1.

Gourgoulhon, E., Grandclément, P., Taniguchi, K., Marck, J.-A., and Bonazzola, S. (2001). Quasiequilibrium sequences of synchronized and irrotational binary neutron stars in general relativity. I. Methods and tests, *Phys. Rev. D* **63**, pp. 064029–1.

Gourgoulhon, E. and Jaramillo, J. L. (2006). A 3+1 perspective on null hypersurfaces and isolated horizons, *Phys. Rep.* **423**, p. 159.

Grandclément, P. (2006). Accurate and realistic initial data for black hole-neutron star binaries, *Phys. Rev. D* **74**, pp. 124002–1.

Grandclément, P. (2007). Erratum: Accurate and realistic initial data for black hole-neutron star binaries, *Phys. Rev. D* **75**, pp. 129903(E)–1.

Grandclément, P., Gourgoulhon, E., and Bonazzola, S. (2002). Binary black holes in circular orbits. II. Numerical methods and first results, *Phys. Rev. D* **65**, pp. 044021–1.

Grandclément, P. and Novak, J. (2009). Spectral methods for numerical relativity, *Living Review Relativity* **12-1**.

Gregory, L. and Laflamme, R. (1993). Black strings and p-branes are unstable, *Phys. Rev. Lett.* **70**, p. 2837.

Grossman, D., Korobkin, O., Rosswog, S., and Piran, T. (2014). The longterm evolution of neutron star merger remnants II: Radioactively powered transients, *Mon. Not. R. Astron. Soc.* **439**, p. 757.

Gundlach, C. (1998). Pseudospectral apparent horizon finders: An efficient new algorithm, *Phys. Rev. D* **57**, p. 863.

Gundlach, C., Calabrese, G., Hinder, I., and Martin-Garcia, J. M. (2005). Constraint damping in the Z4 formulation and harmonic gauge, *Class. Quantum Grav.* **22**, p. 3767.

Gunnarsen, L., Shinkai, H., and Maeda, K. (1995). 3+1 method for finding principal null directions, *Class. Quantum Grav.* **12**, p. 133.

Hachisu, I. (1986). A versatile method for obtaining structures of rapidly rotating stars, *Astrophys. J. Suppl.* **61**, p. 479.

Haensel, P. and Potekhin, A. Y. (2004). Analytical representations of unified equations of state of neutron-star matter, *Astron. Astrophys.* **428**, p. 191.

Hanke, F., Müller, B., Wongwathanarat, A., Marek, A., and Janka, H.-T. (2013). SASI activity in three-dimensional neutrino-hydrodynamics simulations of supernova cores, *Astrophys. J.* **770**, p. 66.

Hannam, M., Husa, S., Brügmann, B., and Gopakumar, A. (2008a). Comparison between numerical-relativity and post-Newtonian waveforms from spinning binaries: The orbital hang-up case, *Phys. Rev. D* **78**, pp. 104007–1.

Hannam, M., Husa, S., Ohme, F., Müller, D., and Brügmann, B. (2010). Simulations of black-hole binaries with unequal masses or nonprecessing spins: Accuracy, physical properties, and comparison with post-Newtonian results, *Phys. Rev. D* **82**, pp. 124008–1.

Hannam, M., Husa, S., Pollney, D., Brügmann, B., and ÓMurchadha, N. (2007). Geometry and regularity of moving punctures, *Phys. Rev. Lett.* **99**, pp. 241102–1.

Hannam, M., Husa, S., Sperhake, U., Brügmann, B., and González, J. A. (2008b). Where post-Newtonian and numerical-relativity waveforms meet, *Phys. Rev. D* **77**, pp. 044020–1.

Hannam, M., Schmidt, P., Bohé, A., Haegel, L., Husa, S., Ohme, F., Pratten, G., and Pürrer, M. (2014). Simple model of complete precessing black-hole-binary gravitational waveforms, *Phys. Rev. Lett.* **113**, pp. 15110–1.

Harada, T. (1998). Final fate of the spherically symmetric collapse of a perfect fluid, *Phys. Rev. D* **58**, pp. 104015–1.

Harten, A. (1983). High resolution scheme for hyperbolic conservation laws, *J. Comput. Phys.* **49**, p. 357.

Harten, A. (1984). On a class of high resolution total-variation-stable finite-differencing schemes, *SIAM Journal* **21**, p. 1.

Harten, A., Lax, P. D., and van Leer, B. (1983). On upstream differencing and Godunov-type schemes for hyperbolic conservation laws, *SIAM Review* **25**, p. 35.

Hartung, J., Steinhoff, J., and Schäfer, G. (2013). Next-to-next-to-leading order post-newtonian linear-in-spin binary hamiltonians, *Ann. Phys. (Berlin)* **525**, p. 359.

Hawking, S. W. (1966). Perturbations of an expanding universe, *Astrophys. J.* **145**, p. 544.

Hawking, S. W. (1971). Gravitational radiation from colliding black holes, *Phys. Rev. Lett.* **26**, p. 1344.

Hawking, S. W. (1972). Black holes in general relativity, *Commun. Math. Phys.* **25**, p. 152.

Hawking, S. W. and Ellis, G. F. R. (1973). *The large scale structure of space-time* (Cambridge University Press).

Hawley, J. F. (1991). Three-dimensional simulations of black hole tori, *Astrophys. J.* **381**, p. 496.

Hawley, J. F., Smarr, L. L., and Wilson, J. R. (1984a). A numerical study of nonspherical black hole accretion. I. Equations and test problems, *Astrophys. J.* **277**, p. 296.

Hawley, J. F., Smarr, L. L., and Wilson, J. R. (1984b). A numerical study of nonspherical black hole accretion. II — Finite differencing and code calibration, *Astrophys. J. Suppl.* **55**, p. 211.

Hayward, S. A. (1994). General laws of black-hole dynamics, *Phys. Rev. D* **49**, p. 6467.

Heger, A., Fryer, C. L., Woosley, S. E., Langer, N., and Hartmann, D. H. (2003). How massive single stars end their life, *Astrophys. J.* **591**, p. 288.

Hemberger, D. A., Lovelace, G., Loredo, T. J., Kidder, L. E., Scheel, M. A., Szilágyi, B., Taylor, N. W., and Teukolsky, S. A. (2013). Final spin and radiated energy in numerical simulations of binary black holes with equal masses and equal, aligned or antialigned spin, *Phys. Rev. D* **88**, pp. 064014–1.

Hempel, M., Fischer, T., Schaffner-Bielich, J., and Liebendörfer, M. (2012). New equation of state in core-collapse supernova simulations, *Astrophys. J.* **748**, p. 70.

Herant, M., Benz, W., Hix, W. R., Fryer, C. L., and Colgate, S. A. (1994). Inside the supernova: A powerful convective engine, *Astrophys. J.* **435**, p. 339.

Hernandez, W. C. and Misner, C. W. (1966). Observer time as a coordinate in relativistic spherical hydrodynamics, *Astrophys. J.* **143**, p. 452.

Herrmann, F., Hinder, I., Shoemaker, D., Laguna, P., and Matzner, R. A. (2007). Gravitational recoil from spinning binary black hole mergers, *Astrophys. J.* **661**, p. 430.

Hewish, A., Bell, S. J., Pilkington, J. D. H., Scott, P. F., and Collins, R. A. (1968). Observation of a rapidly pulsating radio source, *Nature* **217**, p. 709.

Hild, S., Abernathy, M., and 137 authors (2010). Sensitivity studies for third-generation gravitational wave observatories, *arXiv:gr-qc/1012.0908*, p. 1.

Hild, S., Chelkowski, S., and Freise, A. (2008). Pushing towards the ET sensitivity using 'conventional' technology, *arXiv:gr-qc/0810.0604*, p. 1.

Hilditch, D., Bernuzzi, S., Thierfelder, M., Cao, Z., Tichy, W., and Brügmann, B. (2013). Compact binary evolutions of the Z4c formulation, *Phys. Rev. D* **88**, pp. 084057–1.

Hirata, K., Kajita, T., Koshiba, K., Nakahata, M., and 19 authors (1987). Observation of a neutrino burst from the supernova SN1987A, *Phys. Rev. Lett.* **58**, p. 1490.

Hockney, R. W. and Eastwood, J. W. (1988). *Computer Simulation Using Particles* (Institute of Physics Publishing, Bristol and Philadelphia).

Hosokawa, T., Yorke, H. W., Inayoshi, K., Omukai, K., and Yoshida, N. (2013). Formation of primordial supermassive stars by rapid mass accretion, *Astrophys. J.* **778**, p. 178.

Hotokezaka, K., Kiuchi, K., Kyutoku, K., Muranushi, T., Sekiguchi, Y., Shibata, M., and Taniguchi, K. (2013a). Remnant massive neutron stars of binary neutron star mergers: Evolution process and gravitational waves, *Phys. Rev. D* **88**, pp. 044026–1.

Hotokezaka, K., Kiuchi, K., Kyutoku, K., Okawa, H., Sekiguchi, Y., Shibata, M., and Taniguchi, K. (2013b). The mass ejection from the merger of binary neutron stars, *Phys. Rev. D* **87**, pp. 024001–1.

Hotokezaka, K., Kyutoku, K., Okawa, H., Shibata, M., and Kiuchi, K. (2011). Binary neutron star mergers: Dependence on the nuclear equation of state, *Phys. Rev. D* **83**, pp. 124008–1.

Hotokezaka, K., Kyutoku, K., and Shibata, M. (2013c). Exploring tidal effects of coalescing binary neutron stars in numerical relativity, *Phys. Rev. D* **87**, pp. 044001–1.

Houser, J. L., Centrella, J. M., and Smith, S. C. (1994). Gravitational radiation from nonaxisymmetric instability in a rotating star, *Phys. Rev. Lett.* **72**, p. 1314.

Hoyle, F. and Fowler, W. A. (1963). On the nature of strong radio sources, *Mon. Not. R. Astron. Soc.* **125**, p. 169.

Hughes, S. A., Keeton, C. R., Walker, P., Walsh, K. T., Shapiro, S. L., and Teukolsky, S. A. (1994). Finding black holes in numerical spacetimes, *Phys. Rev. D* **49**, p. 4004.

Hulse, R. A. and Taylor, J. H. (1975). Discovery of a pulsar in a binary system, *Astrophys. J. Lett.* **195**, p. L51.

Huq, M. F., Choptuik, M. W., and Matzner, R. A. (2002). Locating boosted Kerr and Schwarzschild apparent horizons, *Phys. Rev. D* **66**, pp. 084024–1.

Ibáñez, J.-M. and Martí, J.-M. (1999). Riemann solvers in relativistic astrophysics, *J. Comp. Phys.* **109**, p. 173.

Iben, Jr., I. (1963). Massive stars in quasi-static equilibrium, *Astrophys. J.* **138**, p. 1090.

Inoue, T. and Inutsuka, S. (2007). Evolutionary conditions in dissipative MHD systems revisited, *Prog. Theor. Phys.* **118**, p. 47.

Iosif, P. and Stergioulas, N. (2014). On the accuracy of the iwm-cfc approximation in differentially rotating relativistic stars, *Gen. Relativ. Gravit.* **46**, p. 1800.

Isaacson, R. A. (1968a). Gravitational radiation in the limit of high frequency. I. The linear approximation and geometrical optics, *Phys. Rev.* **166**, p. 1263.

Isaacson, R. A. (1968b). Gravitational radiation in the limit of high frequency. II. Nonlinear terms and the effective stress tensor, *Phys. Rev.* **166**, p. 1272.

Isenberg, J. A. (2008). Waveless approximation theories of gravity, *Int. J. Mod. Phys. D* **17**, p. 265.

Jackson, J. D. (1975). *Classical Electrodynamics, 2nd edition* (Wiley, New York).

Jacoby, B. A., Cameron, P. B., Jenet, F. A., Anderson, S. B., Murty, R. N., and Kulkarni, S. R. (2006). Measurement of orbital decay in the double neutron star binary PSR B2127+11C, *Astrophys. J. Lett.* **644**, p. L113.

Janis, A. I. and Newman, E. T. (1965). Structure of gravitational sources, *J. Math. Phys.* **6**, p. 902.

Janka, H.-T. (2001). Conditions for shock revival by neutrino heating in core-collapse supernovae, *Astron. Astrophys.* **368**, p. 527.

Johnson, M. H. and McKee, C. F. (1971). Relativistic hydrodynamics in one dimension, *Phys. Rev. D* **3**, p. 858.

KAGRA (). http://gwcenter.icrr.u-tokyo.ac.jp/en/.

Kalogera, V., Belczynski, K., Kim, C., O'Shaughnessy, R., and Willems, B. (2007). Formation of double compact objects, *Phys. Rep.* **442**, p. 75.

Kalogera, V., Kim, C., and 11 authors (2004). The cosmic coalescence rates for double neutron star binaries, *Astrophys. J.* **601**, p. 179.

Karino, S. and Eriguchi, Y. (2003). Linear stability analysis of differentially rotating polytropes: New results for the $m=2$ f-mode dynamical instability, *Astrophys. J.* **592**, p. 1119.

Kasen, D., Badnell, N. R., and Barnes, J. (2013). Opacities and spectra of the r-process ejecta from neutron star mergers, *Astrophys. J.* **774**, p. 25.

Kasian, L. E. (2012). Radio observations of two binary pulsars (PhD. Thesis), .

Kerr, R. P. (1963). Gravitational field of a spinning mass as an example of algebraically special metrics, *Phys. Rev. Lett.* **11**, p. 237.

Kidder, L. E. (1995). Coalescing binary systems of compact objects to (post)5/2-Newtonian order. V. Spin effects, *Phys. Rev. D* **52**, p. 821.

Kidder, L. E., Scheel, M. A., and Teukolsky, S. A. (2001). Extending the lifetime of 3D black hole computations with a new hyperbolic system of evolution equations, *Phys. Rev. D* **64**, pp. 064017–1.

Kidder, L. E., Will, C. M., and Wiseman, A. G. (1993). Spin effects in the inspiral of coalescing compact binaries, *Phys. Rev. D* **47**, p. 4183.

Kinnersley, W. (1969). Type D vacuum metrics, *J. Math. Phys.* **10**, p. 1195.

Kippenhahn, R. and Weigert, A. (1994). *Stellar Structure and Evolution* (Springer-Verlag).

Kitaura, F. S., Janka, H.-T., and Hillebrand, W. (2006). Explosions of O-Ne-Mg cores, the crab supernova, and subluminous type II-P supernova, *Astron. Astrophys.* **450**, p. 345.

Kiuchi, K., Kyutoku, K., Sekiguchi, Y., and Shibata, M. (2014). High-resolution numerical-relativity simulations for the merger of binary magnetized neutron stars, *Phys. Rev. D* **90**, pp. 041502(R)–1.

Kiuchi, K., Kyutoku, K., and Shibata, M. (2012). Three dimensional evolution of differentially rotating magnetized neutron stars, *Phys. Rev. D* **86**, pp. 064008–1.

Kiuchi, K., Sekiguchi, Y., Shibata, M., and Taniguchi, K. (2009). Longterm general relativistic simulations of binary neutron stars collapsing to a black hole, *Phys. Rev. D* **80**, pp. 064037–1.

Kiuchi, K., Sekiguchi, Y., Shibata, M., and Taniguchi, K. (2010). Exploring binary-neutron-star-merger scenario of short-gamma-ray bursts by gravitational-wave observation, *Phys. Rev. Lett.* **104**, pp. 141101–1.

Kiuchi, K., Shibata, M., Montero, P. J., and Font, J. A. (2011a). Gravitational waves from the Papaloizou-Pringle instability in black hole-torus systems, *Phys. Rev. Lett.* **106**, pp. 251102–1.

Kiuchi, K., Shibata, M., and Yoshida, S. (2008). Evolution of neutron stars with toroidal magnetic fields: Axisymmetric simulation in full general relativity, *Phys. Rev. D* **78**, pp. 024029–1.

Kiuchi, K., Shibata, M., and Yoshida, S. (2011b). Non-axisymmetric instabilities of neutron stars with purely toroidal magnetic fields, *Astron. Astrophys.* **532**, pp. A30–1.

Kochanek, C. S. (1992). Coalescing binary neutron stars, *Astrophys. J.* **398**, p. 234.

Kochanek, C. S. and Piran, T. (1993). Gravitational waves and gamma-ray bursts, *Astrophys. J. Lett.* **417**, p. L17.

Kodama, H. (1980). Conserved energy flux for the spherically symmetric system and the backreaction problem in the black hole evaporation, *Prog. Theor. Phys.* **63**, p. 1217.

Kodama, H. and Ishibashi, A. (2003). A master equation for gravitational perturbations of maximally symmetric black holes in higher dimensions, *Prog. Theor. Phys.* **110**, p. 701.

Kodama, H., Konoplya, R. A., and Zhidenko, A. (2010). Gravitational stability of simply rotating Myers-Perry black holes: Tensorial perturbations, *Phys. Rev. D* **81**, pp. 044007–1.

Kohri, K. and Mineshige, S. (2002). Can neutrino-cooled disks be an origin of gamma-ray bursts ? *Astrophys. J.* **577**, p. 311.

Koide, S., Shibata, K., and Kudoh, T. (1999). Relativistic jet formation from black hole magnetized accretion disks: Method, tests, and applications of a general relativistic magnetohydrodynamic numerical code, *Astrophys. J.* **522**, p. 727.

Kojima, Y. (1986). The dynamical stability of a fat disk with constant specific angular momentum, *Prog. Theor. Phys.* **75**, p. 251.

Komar, A. (1959). Covariant conservation laws in general relativity, *Phys. Rev.* **113**, p. 934.

Komatsu, H., Eriguchi, Y., and Hachisu, I. (1989). Rapidly rotating general relativistic stars. I — numerical method and its application to uniformly rotating polytropes, *Mon. Not. R. Astron. Soc.* **237**, p. 355.

Komissarov, S. S. (1997). On the properties of Alfvén waves in relativistic magnetohydrodynamics, *Phys. Lett. A* **232**, p. 435.

Komissarov, S. S. (1999). A Godunov-type scheme for relativistic magnetohydrodynamics, *Mon. Not. R. Astron. Soc.* **303**, p. 343.

Komissarov, S. S. (2002a). Test problems for relativistic magnetohydrodynamics, *arXiv:astro-ph/0209213*, p. 1.

Komissarov, S. S. (2002b). Time-dependent, force-free, degenerate electrodynamics, *Mon. Not. R. Astron. Soc.* **336**, p. 759.

Komissarov, S. S. (2004). Electrodynamics of black hole magnetospheres, *Mon. Not. R. Astron. Soc.* **350**, p. 427.

Komissarov, S. S. (2006). Simulations of the axisymmetric magnetospheres of neutron stars, *Mon. Not. R. Astron. Soc.* **367**, p. 19.

Koppitz, M., Pollney, D., Reisswig, C., Rezzolla, L., Thornburg, J., Diener, P., and Schnetter, E. (2007). Recoil velocities from equal-mass binary-black-hole mergers, *Phys. Rev. Lett.* **99**, pp. 041102–1.

Krivan, W. and Price, R. H. (1998). Initial data for superposed rotating black holes, *Phys. Rev. D* **58**, pp. 104003–1.

Kruskal, M. D. (1960). Maximal extension of Schwarzschild metric, *Phys. Rev.* **119**, p. 1743.

Kulkarni, A. D. (1984). Time-asymmetric initial data for the N black hole problem in general relativity, *J. Math. Phys.* **25**, p. 1028.

Kulkarni, A. D., Shepley, L. C., and York, J. W. (1983). Initial data for N black holes, *Phys. Lett. A* **96**, p. 228.

Kurganov, A. and Tadmor, E. (2000). New high-resolution central schemes for nonlinear conservation laws and convection-diffusion equations, *J. Comp. Phys.* **160**, p. 241.

Kurkarni, S. R. (2005). Modeling supernova-like explosions associated with gamma-ray bursts with short duration, *astro-ph/051025*, p. 1.

Kuroda, K. (2010). State of LCGT, *Class. Quantum Grav.* **27**, pp. 084004–1.

Kuroda, T., Kotake, K., and Takiwaki, T. (2012). Fully general relativistic simulations of core-collapse supernovae with an approximate neutrino transport, *Astrophys. J.* **755**, p. 11.

Kyutoku, K., Ioka, K., and Shibata, M. (2013). Anisotropic mass ejection from black hole-neutron star binaries: Diversity of electromagnetic counterparts, *Phys. Rev. D* **88**, pp. 041503(R)–1.

Kyutoku, K., Ioka, K., and Shibata, M. (2014a). Ultra-relativistic counterparts to binary neutron star mergers in every direction, X-ray-to-radio bands and second-to-day timescale, *Mon. Not. R. Astron. Soc.* **437**, p. L6.

Kyutoku, K., Okawa, H., Shibata, M., and Taniguchi, K. (2011). Gravitational waves from spinning black hole-neutron star binaries: Dependence on equations of state, *Phys. Rev. D* **84**, pp. 064018–1.

Kyutoku, K., Shibata, M., and Taniguchi, K. (2009). Quasiequilibrium states of black hole-neutron star binaries in moving-puncture framework, *Phys. Rev. D* **79**, pp. 124018–1.

Kyutoku, K., Shibata, M., and Taniguchi, K. (2010a). Erratum: Gravitational waves from nonspinning black hole-neutron star binaries: Dependence on equations of state, *Phys. Rev. D* **84**, pp. 049902(E)–1.

Kyutoku, K., Shibata, M., and Taniguchi, K. (2010b). Gravitational waves from nonspinning black hole-neutron star binaries: Dependence on equations of state, *Phys. Rev. D* **82**, pp. 044049–1.

Kyutoku, K., Shibata, M., and Taniguchi, K. (2014b). Reducing orbital eccentricity in initial data of binary neutron stars, *Phys. Rev. D* **90**, pp. 064006–1.

Lackey, B. D., Kyutoku, K., Shibata, M., Brady, P. R., and Friedman, J. L. (2012). Extracting equation of state parameters from black hole-neutron star mergers. I. nonspinning black holes, *Phys. Rev. D* **85**, pp. 044061–1.

Lackey, B. D., Kyutoku, K., Shibata, M., Brady, P. R., and Friedman, J. L. (2014). Extracting equation of state parameters from black hole-neutron star mergers. Aligned-spin black holes and a preliminary waveform model, *Phys. Rev. D* **89**, pp. 043009–1.

Lackey, B. D., Nayyar, M., and Owen, B. J. (2006). Observational constraints on hyperons in neutron stars, *Phys. Rev. D* **73**, pp. 024021–1.

Lai, D., Rasio, F. A., and Shapiro, S. L. (1993a). Ellipsoidal figures of equilibrium: Compressible models, *Astrophys. J. Suppl.* **88**, p. 205.

Lai, D., Rasio, F. A., and Shapiro, S. L. (1993b). Hydrodynamic instability and coalescence of close binary systems, *Astrophys. J. Lett.* **406**, p. 63.

Lai, D., Rasio, F. A., and Shapiro, S. L. (1994a). Equilibrium, stability, and orbital evolution of close binary systems, *Astrophys. J.* **423**, p. 344.

Lai, D., Rasio, F. A., and Shapiro, S. L. (1994b). Hydrodynamic instability and coalescence of binary neutron stars, *Astrophys. J.* **420**, p. 811.

Lai, D., Rasio, F. A., and Shapiro, S. L. (1994c). Hydrodynamics of rotating stars and close binary interactions: Compressible ellipsoid models, *Astrophys. J.* **437**, p. 742.

Lai, D. and Shapiro, S. L. (1995). Gravitational radiation from rapidly rotating nascent neutron stars, *Astrophys. J.* **442**, p. 259.

Lai, D. and Wiseman, A. G. (1996). Innermost stable circular orbit of inspiraling neutron-star binaries: Tidal effects, post-Newtonian effects, and the neutron-star equation of state, *Phys. Rev. D* **54**, p. 3958.

Landau, L. D. and Lifshitz, E. M. (1959). *Fluid Mechanics* (Pergamon Press, London).

Landau, L. D. and Lifshitz, E. M. (1962). *The Classical Theory of Fields* (Pergamon, Oxford).

Lattimer, J. M. (2012). The nuclear equation of state and neutron star masses, *Ann. Rev. Nucl. Part. Sci.* **62**, p. 485.

Lattimer, J. M. and Prakash, M. (2001). Neutron star structure and the equation of state, *Astrophys. J.* **550**, p. 426.

Lattimer, J. M. and Prakash, M. (2004). The physics of neutron stars, *Science* **304**, p. 536.

Lattimer, J. M. and Prakash, M. (2007). The physics of neutron stars, *Phys. Rep.* **442**, p. 109.

Lattimer, J. M. and Schramm, D. N. (1974). Black-hole-neutron-star collisions, *Astrophys. J.* **192**, p. L145.

Lattimer, J. M. and Swesty, D. F. (1991). A generalized equation of state for hot, dense matter, *Nucl. Phys. A* **535**, p. 331.

Leaver, E. W. (1985). An analytic representation for the quasi-normal modes of Kerr black holes, *Proc. Roy. Soc. London A* **402**, p. 285.

Lee, W. H., Ramirez-Ruiz, E., and Page, D. (2005). Dynamical evolution of neutrino-cooled accretion disks: Detailed microphysics, lepton-driven convection, and global energetics, *Astrophys. J.* **632**, p. 421.

Lehner, L. and Pretorius, F. (2010). Black strings, low viscosity fluids, and violation of cosmic censorship, *Phys. Rev. Lett.* **105**, pp. 101102–1.

Li, L. and Paczyński, B. (1998). Transient events from neutron star mergers, *Astrophys. J. Lett.* **507**, p. L59.

Libson, J., Massó, J., Seidel, E., Suen, W.-M., and Walker, P. (1996). Event horizons in numerical relativity: Methods and tests, *Phys. Rev. D* **53**, p. 4335.

Lichnerowicz, A. (1944). L'integration des équations de la gravitation relativiste et le probleme des n corps, *J. Math. Pures Appl.* **23**, p. 37.
Liebendörfer, M., Messer, O. E. B., Mezzacappa, A., Bruenn, S. W., Cardall, C. Y., and Thieleman, F.-K. (2004). A finite difference representation of neutrino radiation hydrodynamics in spherically symmetric general relativistic spacetime, *Astrophys. J. Suppl.* **150**, p. 263.
Liebendörfer, M., Mezzacappa, A., and Thieleman, F.-K. (2001a). Conservative general relativistic radiation hydrodynamics in spherical symmetry and comoving coordinates, *Phys. Rev. D* **63**, pp. 103003–1.
Liebendörfer, M., Mezzacappa, A., Thieleman, F.-K., Messer, O. E. B., Hix, W. R., and Bruenn, S. W. (2001b). Probing the gravitational well: No supernova explosion in spherical symmetry with general relativistic Boltzmann neutrino transport, *Phys. Rev. D* **63**, pp. 103004–1.
LIGO (). http://www.ligo.caltech.edu/.
Lindblom, L., Scheel, M. A., Kidder, L. E., Owen, R., and Rinne, O. (2006). A new generalized harmonic evolution system, *Class. Quantum Grav.* **23**, p. S447.
Lindquist, R. W. (1963). Initial-value problem on Einstein-Rosen manifolds, *J. Math. Phys.* **4**, p. 938.
Lindquist, R. W. (1966). Relativistic transport theory, *Annals of Physics* **37**, p. 487.
Liu, Y. T., Etienne, Z. B., and Shapiro, S. L. (2010). Evolution of near-extremal-spin black holes using the moving puncture technique, *Phys. Rev. D* **80**, pp. 121503(R)–1.
Liu, Y.-T., Shapiro, S. L., and Stephens, B. C. (2007). Magnetorotational collapse of very massive stars to black holes in full general relativity, *Phys. Rev. D* **76**, pp. 084017–1.
Livermore, C. D. (1984). Relating Eddington factors to flux limiters, *J. Quant. Spectrosc. Radiat.* **31**, p. 149.
Lorimer, D. R. (2008). Binary and millisecond pulsars, *Living Review Relativity* **11-8**.
Lorimer, D. R., Stairs, I. H., Freire, P. C. C., and 33 authors (2006). The yound, highly relativistic binary pulsar J1906+0746, *Astrophys. J.* **640**, p. 428.
Lousto, C. O., Nakano, H., Zlochower, Y., and Campanelli, M. (2010). Intermediate-mass-ratio black hole binaries: Intertwining numerical and perturbative techniques, *Phys. Rev. D* **82**, pp. 104057–1.
Lousto, C. O. and Zlochower, Y. (2011). Orbital evolution of extreme-mass-ratio black-hole binaries with numerical relativity, *Phys. Rev. Lett.* **106**, pp. 041101–1.
Lovelace, G., Boyle, M., Scheel, M. A., and Szilágyi, B. (2012). High-accuracy gravitational waveforms for binary black hole mergers with nearly extremal spins, *Class. Quantum Grav.* **29**, pp. 045003–1.
Lovelace, G., Duez, M. D., Foucart, F., Kidder, L. E., Pfeiffer, H. P., Scheel, M. A., and Szilágyi, B. (2013). Massive disk formation in the tidal disruption of a neutron star by a nearly extremal black hole, *Class. Quantum Grav.* **30**, pp. 135004–1.
Lovelace, G., Owen, R., Pfeiffer, H. P., and Chu, T. (2008). Binary-black-hole initial data with nearly extremal spins, *Phys. Rev. D* **78**, pp. 084017–1.
Lovelace, G., Scheel, M. A., and Szilágyi, B. (2011). Simulating merging binary black holes with nearly extreme spins, *Phys. Rev. D* **83**, pp. 024010–1.
LSST (). http://www.lsst.org/lsst/.
Lucas-Serrano, A., Font, J. A., Ibáñez, J.-M., and Martí, J. M. (2004). Assessment of a high-resolution central scheme for the solution of the relativistic hydrodynamics equations, *Astron. Astrophys.* **428**, p. 703.
Lyne, A. and Graham-Smith, F. (2005). *Pulsar Astronomy (third edition)* (Cambridge University Press).
Lyne, A. G., Burgay, M., Kramer, M., and 9 authors (2004). A double-pulsar system: A rare laboratory for relativistic gravity and plasma physics, *Science* **303**, p. 1153.

Maeda, K., Sasaki, M., Nakamura, T., and Miyama, S. (1980). A new formalism of the Einstein equations for relativistic rotating systems, *Prog. Theor. Phys.* **63**, p. 719.

Magorrian, J., Tremaine, S., Richstone, D., Bender, R., Bower, G., Dressler, A., Faber, S. M., Gebhardt, K., Green, R., Grillmair, C., Kormendy, J., and Lauer, T. (1998). The demography of massive dark objects in galaxy centres, *Astron. J.* **115**, p. 2285.

Marek, A. and Janka, H.-T. (2009). Delayed neutrino-driven supernova explosions aided by the standing accretion-shock instability, *Astrophys. J.* **694**, p. 664.

Marronetti, P., Tichy, W., Brügmann, B., González, J. A., and Sperhake, U. (2008). High-spin binary black hole mergers, *Phys. Rev. D* **77**, pp. 064010-1.

Martí, J. M., Ibáñez, J.-M., and Miralles, J. A. (1991). Numerical relativistic hydrodynamics: Local characteristic approach, *Phys. Rev. D* **43**, p. 3794.

Martí, J.-M. and Müller, E. (1994). Analytical solution of the Riemann problem in relativistic hydrodynamics, *J. Fluid Mechanics* **258**, p. 317.

Martí, J.-M. and Müller, E. (2003). Numerical hydrodynamics in special relativity, *Living Review Relativity* **6-7**.

May, M. M. and White, R. H. (1966). Hydrodynamic calculations of general-relativistic collapse, *Phys. Rev.* **141**, p. 1232.

McClintock, J. E. and Remillard, R. A. (2006). Black hole binaries, in W. H. G. Lewin and M. van der Klis (eds.), *Compact Stellar X-ray Sources* (Cambridge University Press), p. 157.

McKinney, J. C. (2006). General relativistic force-free electrodynamics: A new code and applications to black hole magnetospheres, *Mon. Not. R. Astron. Soc.* **367**, p. 1797.

McKinney, J. C. and Gammie, C. F. (2004). A measurement of the electromagnetic luminosity of a Kerr black hole, *Astrophys. J.* **611**, p. 977.

MeerKAT (). http://www.ska.ac.za/meerkat/.

Meijerink, J. and van der Vorst, H. A. (1977). An iterative solution method for linear systems of which the coefficient matrix is a symmetric M-matrix, *Math. Comp.* **31**, p. 148.

Metzger, B. D. and Berger, E. (2012). What is the most promising electromagnetic counterpart of a neutron star binary merger ? *Astrophys. J.* **746**, p. 48.

Metzger, B. D., Martínez-Pinedo, G., Darbha, S., Quataert, E., Arcones, A., Kasen, D., Thomas, R., Nugent, P., Panov, I. V., and Zinner, N. T. (2010). Electromagnetic counterparts of compact object mergers powered by the radioactive decay of r-process nuclei, *Mon. Not. R. Astron. Soc.* **406**, p. 2650.

Mezzacappa, A., Liebendörfer, M., Messer, O. E. B., Hix, W. R., Thielemann, F.-K., and Bruenn, S. W. (2001). Simulation of the spherically symmetric stellar core collapse, bounce, and postbounce evolution of a star of 13 solar masses with Boltzmann neutrino transport, and its implications for the supernova mechanism, *Phys. Rev. Lett.* **86**, p. 1395.

Mezzacappa, A. and Matzner, R. A. (1989). Computer simulation of time-dependent, spherically symmetric spacetimes containing radiating fluids — formalism and code tests, *Astrophys. J.* **343**, p. 853.

Mihalas, D. and Weibel-Mihalas, B. (1999). *Foundations of Radiation Hydrodynamics* (Dover Publication, Inc.).

Mino, Y., Sasaki, M., Shibata, M., Tagoshi, H., and Tanaka, T. (1997). Black hole perturbation, *Prog. Theor. Phys. Suppl.* **128**, p. 1.

Misner, C. W. (1960). Whormhole initial conditions, *Phys. Rev.* **118**, p. 1110.

Misner, C. W. (1963). The method of images in geometrostatics, *Ann. Phys.* **24**, p. 102.

Misner, C. W. and Sharp, D. H. (1964). Relativistic equations for adiabatic, spherically symmetric gravitational collapse, *Phys. Rev.* **136**, p. 571.

Misner, C. W., Thorne, K. S., and Wheeler, J. A. (1973). *Gravitation* (W.H. Freeman and Company, New York).

Miyama, S. M. (1981). Time evolution of pure gravitational waves, *Prog. Theor. Phys.* **65**, p. 894.

Miyoshi, M., Moran, J., Herrnstein, J., Greenhill, L., Nakai, N., Diamond, P., and Inoue, M. (1995). Evidence for a black hole from high rotation velocities in a sub-parsec region of NGC4258, *Nature* **373**, p. 127.

Moncrief, V. (1974). Gravitational perturbations of spherically symmetric systems. I. The exterior problem, *Annals of Physics* **88**, p. 323.

Montero, P. J., Janka, H.-T., and Müller, E. (2012). Relativistic collapse and explosion of rotating supermassive stars with thermonuclear effects, *Astrophys. J.* **749**, p. 37.

Mroué, A. H. and Pfeiffer, H. P. (2012). Precessing binary black holes simulations: Quasicircular initial data, *arXiv:/1210:2958*, p. 1.

Mroué, A. H., Scheel, M. A., Szilágyi, B., Pfeiffer, H. P., Boyle, M., Hemberger, D. A., Kidder, L. E., Lovelace, G., Ossokine, S., Taylor, N. W., Zenginoglu, A., Buchman, L. T., Chu, T., Foley, E., Giesler, M., Owen, R., and Teukolsky, S. A. (2013). A catalog of 174 high-quality binary black-hole simulations for gravitational-wave astronomy, *Phys. Rev. Lett.* **111**, pp. 241104–1.

Mukhanov, V. (2005). *Cosmology* (Cambridge University Press).

Müller, B., Janka, H.-T., and Marec, A. (2013). A new multi-dimensional general relativistic neutrino hydrodynamics code of core-collapse supernovae. III. Gravitational wave signals from supernova explosion models, *Astrophys. J.* **766**, p. 43.

Müller, E. (1982). Gravitational radiation from collapsing rotating stellar cores, *Astron. Astrophys.* **114**, p. 53.

Müller, E. and Janka, H.-T. (1997). Gravitational radiation from convective instabilities in Type II supernova explosions, *Astron. Astrophys.* **317**, p. 140.

Müller, E., Janka, H.-T., and Wongwathanarat, A. (2012). Parametrized 3D models of neutrino-driven supernova explosions. Neutrino emission asymmetries and gravitational-wave signals, *Astron. Astrophys.* **537**, p. A63.

Müller, E., Rampp, M., Buras, R., Janka, H.-T., and Shoemaker, D. H. (2004). Toward gravitational wave signals from realistic core-collapse supernova models, *Astrophys. J.* **603**, p. 221.

Müller, H. and Serot, B. D. (1996). Relativistic mean-field theory and the high-density nuclear equation of state, *Nucl. Phys. A* **606**, p. 508.

Murphy, J. W., Ott, C., and Burrows, A. (2009). A model for gravitational wave emission from neutrino-driven core-collapse supernovae, *Astrophys. J.* **707**, p. 1173.

Myers, R. C. and Perry, M. J. (1986). Black holes in higher dimensional space-times, *Ann. Phys.* **172**, p. 304.

Nagar, A. and Rezzolla, L. (2005). Gauge-invariant non-spherical metric perturbations of Schwarzschild black hole spacetimes, *Class. Quantum Grav.* **22**, p. 2005.

Nakama, T., Harada, T., Polnarev, A. G., and Yokoyama, J. (2014). Identifying the most crucial parameters of the initial curvature profile for primordial black hole formation, *JCAP* **01**, p. 037.

Nakamura, T. (1981). General relativistic collapse of axially symmetric stars leading to the formation of rotating black holes, *Prog. Theor. Phys.* **65**, p. 1876.

Nakamura, T. (1983). General relativistic collapse of accreting neutron stars with rotation, *Prog. Theor. Phys.* **70**, p. 1144.

Nakamura, T. (1984). General solutions to the linearized Einstein equations and initial data for three dimensional time evolution of pure gravitational waves, *Prog. Theor. Phys.* **72**, p. 746.

Nakamura, T., Kojima, Y., and Oohara, K. (1984). A method of determining apparent horizons in three-dimensional numerical relativity, *Phys. Lett.* **106A**, p. 235.

Nakamura, T., Maeda, K., Miyama, S. M., and Sasaki, M. (1980). General relativistic collapse of an axially symmetric star. I — The formulation and the initial value equations—, *Prog. Theor. Phys.* **63**, p. 1229.

Nakamura, T. and Oohara, K. (1983). Gravitational radiation emitted by N particles in circular orbits, *Phys. Lett. A* **98**, p. 483.

Nakamura, T. and Oohara, K. (1989). Methods in 3D numerical relativity, in C. R. Evans, L. S. Finn, and D. W. Hobill (eds.), *Frontiers in Numerical Relativity* (Cambridge University Press), p. 281.

Nakamura, T., Oohara, K., and Kojima, Y. (1987). General relativistic collapse to black holes and gravitational waves from black holes, *Prog. Theor. Phys. Suppl. (Chapter 1)* **90**, p. 1.

Nakamura, T. and Sasaki, M. (1981). Is collapse of a deformed star always effectual for the gravitational radiation ? *Phys. Lett. B* **106**, p. 69.

Nakamura, T. and Sato, H. (1981). General relativistic collapse of rotating supermassive stars, *Prog. Theor. Phys.* **66**, p. 2038.

Nakamura, T. and Sato, H. (1982). General relativistic collapse of non-rotating, axisymmetric stars, *Prog. Theor. Phys.* **67**, p. 1396.

Nakamura, T., Shapiro, S. L., and Teukolsky, S. A. (1988). Naked singularities and the hoop conjecture: An analytic exploration, *Phys. Rev. D* **38**, p. 2972.

Nakano, H., Zlochower, Y., Lousto, C. O., and Campanelli, M. (2012). Intermediate-mass-ratio black hole binaries. II. Modeling trajectories and gravitational waveforms, *Phys. Rev. D* **84**, pp. 124006–1.

Nakao, K., Abe, H., Yoshino, H., and Shibata, M. (2009). Maximal slicing of D-dimensional spherically symmetric vacuum spacetimes, *Phys. Rev. D* **80**, pp. 084028–1.

Nakao, K., Maeda, K., Nakamura, T., and Oohara, K. (1991a). Constant-mean-curvature slicing of the Schwarzschild-de Sitter space-time, *Phys. Rev. D* **44**, p. 1326.

Nakao, K., Maeda, K., Nakamura, T., and Oohara, K. (1991b). Numerical study of cosmic no-hair conjecture: Formalism and linear analysis, *Phys. Rev. D* **43**, p. 1788.

Nakar, E. (2007). Short-hard gamma-ray bursts, *Phys. Rep.* **442**, p. 166.

Nakar, E. and Piran, T. (2011). Detectable radio flares following gravitational waves from mergers of binary neutron stars, *Nature* **478**, p. 82.

Nakazato, K., Sumiyoshi, K., and Yamada, S. (2006). Gravitational collapse and neutrino emission of population III massive stars, *Astrophys. J.* **645**, p. 519.

Nakazato, K., Sumiyoshi, K., and Yamada, S. (2007). Numerical study of stellar core collapse and neutrino emission: Probing the spherically symmetric black hole progenitors with 3-30 $M_\odot$ iron cores, *Astrophys. J.* **666**, p. 1140.

Namtilan, H., Pretorius, F., and Gubser, S. G. (2012). Simulation of asymptotically AdS$_5$ spacetimes with a generalized harmonic evolution scheme, *Phys. Rev. D* **85**, pp. 084038–1.

Narayan, R., Goldreich, P., and Goodman, J. (1987). Physics of modes in a differentially rotating system — Analysis of the shearing sheet, *Mon. Not. R. Astron. Soc.* **228**, p. 1.

Narayan, R., Paczyński, B., and Piran, T. (1992). Gamma-ray bursts as the death throes of massive binary stars, *Astrophys. J.* **395**, p. L83.

Narayan, R., Piran, T., and Kumar, P. (2001). Accretion models of gamma-ray bursts, *Astrophys. J.* **557**, p. 949.

Narayan, R., Piran, T., and Shemi, A. (1991). Neutron star and black hole binaries in the galaxy, *Astrophys. J.* **379**, p. 17.

New, J. C. B., Centrella, J. M., and Tohline, J. E. (2000). Gravitational waves from long-duration simulations of the dynamical bar instability, *Phys. Rev. D* **62**, pp. 064019–1.

Newman, E. and Penrose, R. (1962). An approach to gravitational radiation by a method of spin coefficients, *J. Math. Phys.* **3**, p. 566.

Newman, E. and Penrose, R. (1963). Errata: An approach to gravitational radiation by a method of spin coefficients, *J. Math. Phys.* **4**, p. 998.

Newman, E. T. and Penrose, R. (1966). Note on the Bondi-Metzner-Sachs group, *J. Math. Phys.* **7**, p. 863.

Nishida, S. and Eriguchi, Y. (1994). A general relativistic toroid around a black hole, *Astrophys. J.* **427**, p. 429.

Novikov, I. D. and Thorne, K. S. (1973). Black hole astrophysics, in C. DeWitt and B. S. DeWitt (eds.), *Black Holes* (Gordon and Breach Science Publishers), p. 343.

Okawa, H., Nakao, K., and Shibata, M. (2011). Is super-Planckian physics visible ?: Scattering of black holes in 5-dimension, *Phys. Rev. D* **83**, pp. 121501–1.

Oleg, K., Abdikamolov, E. B., Schnetter, E., Stergioulas, N., and Zink, B. (2011). Stability of general-relativistic accretion disks, *Phys. Rev. D* **83**, pp. 043007–1.

ÓMurchadha, N. and York, J. W. (1974). Gravitational energy, *Phys. Rev. D* **10**, p. 2345.

Oohara, K. and Nakamura, T. (1989). Three-dimensional initial data of colliding neutron stars, *Prog. Theor. Phys.* **81**, p. 360.

Oohara, K., Nakamura, T., and Shibata, M. (1997). A way to 3D numerical relativity, *Prog. Theor. Phys. Suppl.* **128**, p. 183.

Oota, T. and Yasui, Y. (2010). Separability of gravitational perturbation in generalized Kerr-NUT-de Sitter spacetime, *Int. J. Mod. Phys. A* **25**, p. 3055.

Oppenheimer, J. R. and Snyder, H. (1939). On continued gravitational contraction, *Phys. Rev.* **56**, p. 455.

Oppenheimer, J. R. and Volkoff, G. M. (1939). On massive neutron cores, *Phys. Rev.* **55**, p. 374.

O'Shaughnessy, R., Kalogera, V., and Belczynski, K. (2010). Binary compact object coalescence rates: The role of elliptical galaxies, *Astrophys. J.* **715**, p. 1453.

O'Shaughnessy, R., Kim, C., Kalogera, V., and Belczynski, K. (2008). Constraining population synthesis models via empirical binary compact object merger and supernova rates, *Astrophys. J.* **672**, p. 479.

Ott, C. D., Abdikamalov, E., Mösta, P., Haas, R., Drasco, S., O'Connor, E. P., Reisswig, C., Meakin, C. A., and Schnetter, E. (2013). General-relativistic simulations of three-dimensional core-collapse supernovae, *Astrophys. J.* **768**, p. 115.

Ott, C. D., Dimmelmeier, H., Marek, A., Janka, H.-T., Hawke, I., Zink, B., and Schnetter, E. (2007a). 3D collapse of rotating stellar iron cores in general relativity including deleptonization and a nuclear equation of state, *Phys. Rev. Lett.* **98**, pp. 261101–1.

Ott, C. D., Dimmelmeier, H., Marek, A., Janka, H.-T., Zink, B., Hawke, I., and Schnetter, E. (2007b). Rotating collapse of stellar iron cores in general relativity, *Class Quantum Grav.* **24**, p. S139.

Ott, C. D., Ou, S., Tohline, J. E., and Burrows, A. (2005). One-armed spiral instability in a low-$T/|W|$ postbounce supernova core, *Astrophys. J.* **625**, p. 119.

Ott, C. D., Reisswig, C., Schnetter, E., O'Connor, E., Sperhake, U., Loeffler, F., Diener, P., Abdikamolow, E., Hawke, I., and Burrows, A. (2011). Dynamics and gravitational wave signature of collapsar formation, *Phys. Rev. Lett.* **106**, pp. 161103–1.

Ou, S., Tohline, J. E., and Lindblom, L. (2004). Nonlinear development of the secular bar-mode instability in rotating neutron stars, *Astrophys. J.* **617**, p. 490.

Özel, F. and Psaltis, D. (2009). Reconstructing the neutron-star equation of state from astrophysical measurements, *Phys. Rev. D* **80**, pp. 103003–1.

Özel, F., Psaltis, D., Narayan, R., and McClintock, J. E. (2010). The black hole mass distribution in the galaxy, *Astrophys. J.* **725**, p. 1918.

Pan, Y., Buonanno, A., Baker, J. G., Centrella, J., Kelly, B. J., McWilliams, S. T., Pretorius, F., and van Meter, J. R. (2008). Data-analysis driven comparison of analytic and numerical coalescing binary waveforms: Nonspinning case, *Phys. Rev. D* **77**, pp. 024014–1.

Pan, Y., Buonanno, A., Boyle, M., Buchman, L. T., Kidder, L. E., Pfeiffer, H. P., and Scheel, M. A. (2011). Inspiral-merger-ringdown multipolar waveforms of nonspinning black-hole binaries using the effective-one-body formalism, *Phys. Rev. D* **84**, pp. 124052–1.

Pan-STARRs (). http://pan-starrs.ifa.hawaii.edu/public/.

Pandhripande, V. R. and Smith, R. A. (1975). A model neutron solid with π^0 condensate, *Nucl. Phys. A* **237**, p. 507.

Pannarale, F., Berti, E., Kyutoku, K., and Shibata, M. (2013). Nonspinning black hole-neutron star mergers: A model for the amplitude of gravitational waveforms, *Phys. Rev. D* **88**, pp. 084011–1.

Papaloizou, J. C. B. and Pringle, J. E. (1984). The dynamical stability of differentially rotating discs with constant specific angular momentum, *Mon. Not. R. Astron. Soc.* **208**, p. 721.

Parker, E. N. (1966). The dynamical state of the interstellar gas and field, *Astrophys. J.* **145**, p. 811.

Peters, P. C. (1964). Gravitational radiation and the motion of two point masses, *Phys. Rev.* **136**, p. B1224.

Petrich, L. I., Shapiro, S. L., and Teukolsky, S. A. (1985). Oppenheimer-Snyder collapse with maximal time slicing and isotropic coordinates, *Phys. Rev. D* **31**, p. 2459.

Petrich, L. I., Shapiro, S. L., and Teukolsky, S. A. (1986). Oppenheimer-Snyder collapse in polar time slicing, *Phys. Rev. D* **33**, p. 2100.

Pfeiffer, H. P., Brown, D. A., Kidder, L. E., Lindblom, L., Lovelace, G., and Scheel, M. A. (2007). Reducing orbital eccentricity in binary black hole simulations, *Class. Quantum Grav.* **24**, p. S59.

Pfeiffer, H. P., Kidder, L. E., Scheel, M. A., and Teukolski, S. A. (2003). A multidomain spectral method for solving elliptic equations, *Comput. Phys. Commun* **152**, p. 253.

Pfeiffer, H. P., Teukolsky, S. A., and Cook, G. B. (2000). Quasicircular orbits for spinning binary black holes, *Phys. Rev. D* **62**, pp. 104018–1.

Pfeiffer, H. P. and York, J. W. (2003). Extrinsic curvature and the Einstein constraints, *Phys. Rev. D* **67**, pp. 044022–1.

Phinney, E. S. (1991). The rate of neutron star binary mergers in the universe — Minimal predictions for gravity wave detectors, *Astrophys. J.* **380**, p. 17.

Piran, T. (2004). The physics of gamma-ray bursts, *Rev. Mod. Phys.* **76**, p. 1143.

Piran, T., Nakar, E., and Rosswog, S. (2013). The electromagnetic signals of compact binary mergers, *Mon. Not. R. Astron. Soc.* **430**, p. 2121.

Piran, T. and Stark, R. F. (1986). Numerical relativity, rotating gravitational collapse, and gravitational radiation, in J. M. Centrella (ed.), *Dynamical spacetimes and numerical relativity* (Cambridge University Press), p. 40.

Poisson, E. (2004). *A relativist's toolkit* (Cambridge University Press).

Poisson, E. and Sasaki, M. (1995). Gravitational radiation from a particle in circular orbit around a black hole. v. black-hole absorption and tail corrections, *Phys. Rev. D* **51**, p. 5753.

Pollney, D., Reisswig, C., Rezzolla, L., Szilágyi, B., Ansorg, M., Deris, B., Diener, P., Dorband, E. N., Koppitz, M., Nagar, A., and Schnetter, E. (2007). Recoil velocities from equal-mass binary-black hole mergers: A systematic investigation of spin-orbit aligned configuration, *Phys. Rev. D* **76**, pp. 124002–1.

Pons, J. A., Martí, J.-M., and Müller, E. (2000). The exact solution of the Riemann problem with non-zero tangential velocities in relativistic hydrodynamics, *J. Fluid Mechanics* **422**, p. 125.

Popham, R., Woosley, S. E., and Fryer, C. (1999). Hyperaccreting black holes and gamma-ray bursts, *Astrophys. J.* **518**, p. 356.

Postnov, K. A. and Yungelson, L. R. (2006). The evolution of compact binary star systems, *Living Review Relativity* **9-6**.

Press, W. H., Flannery, B. P., Teukolsky, S. A., and Vetterling, W. T. (1989). *Numerical Recipes* (Cambridge University Press).

Press, W. H. and Teukolsky, S. A. (1973). Perturbations of a rotating black hole. II. Dynamical stability of the Kerr metric, *Astrophys. J.* **185**, p. 649.

Pretorius, F. (2005a). Evolution of binary black-hole spacetimes, *Phys. Rev. Lett.* **95**, pp. 121101–1.

Pretorius, F. (2005b). Numerical relativity using a generalized harmonic decomposition, *Class. Quantum Grav.* **22**, p. 425.

Pretorius, F. (2006). Simulation of binary black hole spacetimes with a harmonic evolution scheme, *Class. Quantum Grav.* **23**, p. S529.

Price, D. J. and Rosswog, S. (2006). Producing ultrastrong magnetic fields in neutron star mergers, *Science* **312**, p. 729.

Price, R. H. and Pullin, J. (1994). Colliding black holes: The close limit, *Phys. Rev. Lett.* **72**, p. 3297.

PTF (). http://www.astro.caltech.edu/ptf/.

Rampp, M. and Janka, H.-T. (2000). Spherically symmetric simulation with boltzmann neutrino transport of core collapse and postbounce evolution of a 15 $M_\odot$ star, *Astrophys. J. Lett* **539**, p. L33.

Rampp, M., Müller, E., and Ruffert, M. (1998). Simulations of non-axisymmetric rotational core collapse, *Astron. Astrophys.* **332**, p. 969.

Randall, L. and Sundrum, R. (1999a). An alternative to compactification, *Phys. Rev. Lett.* **83**, p. 4690.

Randall, L. and Sundrum, R. (1999b). Large mass hierarchy from a small extra dimension, *Phys. Rev. Lett.* **83**, p. 3370.

Rasio, F. A. and Shapiro, S. L. (1992). Hydrodynamic evolution of coalescing binary neutron stars, *Astrophys. J.* **401**, p. 226.

Rasio, F. A. and Shapiro, S. L. (1999). TOPICAL REVIEW: Coalescing binary neutron stars, *Class. Quantum Grav.* **16**, p. R1.

Read, J. S., Baiotti, L., Creighton, J. D. E., Friedman, J. L., Giacomazzo, B., Kyutoku, K., Markakis, C., Rezzolla, L., Shibata, M., and Taniguchi, K. (2013). Matter effect on binary neutron star waveforms, *Phys. Rev. D* **88**, pp. 044042–1.

Read, J. S., Lackey, B. D., Owen, B. J., and Friedman, J. L. (2009a). Constraints on a phenomenologically parametrized neutron-star equation of state, *Phys. Rev. D* **79**, pp. 124032–1.

Read, J. S., Markakis, C., Shibata, M., Uryū, K., Creighton, J. D. E., and Friedman, J. L. (2009b). Measuring the neutron star equation of state with gravitational wave observations, *Phys. Rev. D* **79**, pp. 124033–1.

Rees, M. (1984). Black hole models for active galactic nuclei, *Annu. Rev. Astron. Astrophys.* **22**, p. 445.

Rees, M. J. (1996). Astrophysical evidence for black holes, in R. M. Wald (ed.), *Black Holes and Relativistic Stars* (The University of Chicago Press), p. 79.

Regge, T. and Teitelboim, C. (1974). Role of surface integrals in the Hamiltonian formulation of general relativity, *Ann. Phys.* **88**, p. 286.

Regge, T. and Wheeler, J. A. (1957). Stability of a Schwarzschild singularity, *Phys. Rev.* **108**, p. 1063.

Reisswig, C., Ott, C., Sperhake, U., and Schnetter, E. (2011). Gravitational wave extraction in simulations of rotating stellar core collapse, *Phys. Rev. D* **83**, pp. 064008–1.

Reisswig, C., Ott, C. D., Abdikamalov, E., Haas, R., Mösta, P., and Schnetter, E. (2013). Formation and coalescence of cosmological supermassive black hole binaries in supermassive star collapse, *Phys. Rev. Lett.* **111**, pp. 151101–1.

Reisswig, C. and Pollney, D. (2011). Notes on the integration of numerical relativity waveforms, *Class. Quantum Grav.* **28**, pp. 195015–1.

Reula, O. (1998). Hyperbolic methods for Einstein's equations, *Living Review Relativity* **1-3**.

Rezzolla, L., Giacomazzo, B., Baiotti, L., Granot, J., Kouveliotou, C., and Aloy, M. A. (2011). The missing link: Merging neutron stars naturally produce jet-like structures and can power short gamma-ray bursts, *Astrophys. J. Lett.* **732**, p. L6.

Rinne, O. (2008). Constrained evolution in axisymmetry and the gravitational collapse of prolate Brill waves, *Class. Quantum Grav.* **25**, pp. 135009–1.

Rinne, O. (2010). An axisymmetric evolution code for the Einstein equations on hyperboloidal slices, *Class. Quantum Grav.* **27**, pp. 035014–1.

Roe, P. L. (1981). Approximate Riemann solvers, parameter vectors, and difference schemes, *J. Comp. Phys.* **43**, p. 357.

Rosswog, S., Davies, M. B., Thielemann, F.-K., and Piran, T. (2000). Merging neutron stars: Asymmetric systems, *Astron. Astrophys.* **360**, p. 171.

Rosswog, S. and Liebendörfer, M. (2003). High-resolution calculations of merging neutron stars - II. Neutrino emission, *Mon. Not. R. Astron. Soc.* **342**, p. 673.

Rosswog, S., Piran, T., and Nakar, E. (2013). The multimessenger picture of compact object encounters: Binary mergers versus dynamical collisions, *Mon. Not. R. Astron. Soc.* **430**, p. 2585.

Ruffert, M., Janka, H.-T., and Schäfer, G. (1996). Coalescing neutron stars - a step towards physical models. I. Hydrodynamic evolution and gravitational-wave emission, *Astron. Astrophys.* **311**, p. 532.

Ruiz, M., Alcubierre, M., Núñez, D., and Takahashi, R. (2008). Multiple expansions for energy and momenta carried by gravitational waves, *Gen. Relativ. Gravit.* **40**, p. 1705.

Saijo, M. and Hawke, I. (2009). Collapse of differentially rotating supermassive stars: Post black hole formation, *Phys. Rev. D* **80**, pp. 064001–1.

Saijo, M., Shibata, M., Baumgarte, T. W., and Shapiro, S. L. (2002). Dynamical bar instability in rotating stars: Effect of general relativity, *Astrophys. J.* **548**, p. 919.

Santamaría, L., Ohme, F., Ajith, P., Brëmann, B., Dorband, N., Hannam, M., Husa, S., Moesta, P., Pollney, D., Reisswig, C., Robinson, E. L., Seiler, J., and Krishnan, B. (2010). Matching post-Newtonian and numerical relativity waveforms: Systematic errors and a new phenomenological model for non-precessing black hole binaries, *Phys. Rev. D* **82**, pp. 064016–1.

Sato, K. (1975a). Neutrino degeneracy in supernova cores and neutral current of weak interaction, *Prog. Theor. Phys.* **53**, p. 595.

Sato, K. (1975b). Supernova explosion and neutral currents of weak interaction, *Prog. Theor. Phys.* **54**, p. 1325.

Schäfer, G. (1985). The gravitational quadrupole radiation-reaction force and the canonical formalism of adm, *Ann. Phys.* **161**, p. 81.

Scheel, M. A., Boyle, M., Chu, T., Kidder, L. E., Matthews, K. D., and Pfeiffer, H. P. (2009). High-accuracy waveforms for binary black hole inspiral, merger, and ringdown, *Phys. Rev. D* **79**, pp. 024003–1.

Scheel, M. A., Shapiro, S. L., and Teukolsky, S. A. (1995a). Collapse to black holes in Brans-Dicke theory. I. Horizon boundary conditions for dynamical spacetimes, *Phys. Rev. D* **51**, p. 4208.

Scheel, M. A., Shapiro, S. L., and Teukolsky, S. A. (1995b). Collapse to black holes in Brans-Dicke theory. II. Comparison with general relativity, *Phys. Rev. D* **51**, p. 4236.

Scheidegger, S., Käppeli, R., Whitehouse, S., Fischer, T., and Liebendörfer, M. (2010). The influence of model parameters on the prediction of gravitational wave signals from stellar core collapse, *Astron. Astrophys.* **514**, p. A51.

Schnetter, E. (2003). Finding apparent horizons and other 2-surfaces of constant expansion, *Class. Quantum Grav.* **57**, p. 863.

Schnetter, E., Hawley, S. H., and Hawke, I. (2004). Evolutions in 3D numerical relativity using fixed mesh refinement, *Class. Quantum Grav.* **21**, p. 1465.

Schnetter, E., Krishnan, B., and Beyer, F. (2006). Introduction to dynamical horizons in numerical relativity, *Phys. Rev. D* **74**, pp. 024028–1.

Schnittman, J. D., Buonanno, A., van Meter, J. R., Baker, J. G., Boggs, W. D., Centrella, J., Kelly, B. J., and McWilliams, S. T. (2008). Anatomy of the binary black hole recoil: A multipolar analysis, *Phys. Rev. D* **77**, pp. 044031–1.

Schödel, R., Ott, T., Genzel, R., Hofmann, R., Lehnert, M., Eckart, A., and 17 authros (2002). A star in a 15.2-year orbit around the supermassive black hole at the center of the Milky Way, *Nature* **419**, p. 694.

Schutz, B. F. (1985). *A first course of general relativity* (Cambridge University Press).

Schutz, B. F. (1991). Data processing analysis and storage for interferometric antennas, in D. G. Blair (ed.), *The Detection of Gravitational Waves* (The University of Chicago Press), p. 406.

Schutz, B. F. and Sorkin, R. D. (1978). Variational aspects of relativistic field theories with application to perfect fluids, *Commun. Math. Phys.* **62**, p. 247.

Schutz, B. F. and Tinto, M. (1987). Antenna patterns of interferometric detectors of gravitational waves. I — Linearly polarized waves, *Mon. Not. R. Astron. Soc.* **224**, p. 131.

Schutz, B. F. and Will, C. M. (1985). Black hole normal modes — A semianalytic approach, *Astrophys. J. Lett.* **291**, p. L33.

Seidel, E. (1990). Gravitational radiation from even-parity perturbations of stellar collapse: Mathematical formalism and numerical methods, *Phys. Rev. D* **42**, p. 1884.

Seidel, E. and Moore, T. (1987). Gravitational radiation from realistic relativistic stars: Odd-parity fluid perturbations, *Phys. Rev. D* **35**, p. 2287.

Seidel, E. and Suen, W.-M. (1992). Towards a singularity-proof scheme in numerical relativity, *Phys. Rev. Lett.* **69**, p. 1845.

Sekiguchi, Y., Kiuchi, K., Kyutoku, K., and Shibata, M. (2011a). Effects of hyperons in binary neutron star mergers, *Phys. Rev. Lett.* **107**, pp. 211101–1.

Sekiguchi, Y., Kiuchi, K., Kyutoku, K., and Shibata, M. (2011b). Gravitational waves and neutrino emission from the merger of binary neutron stars, *Phys. Rev. Lett.* **107**, pp. 051102–1.

Sekiguchi, Y., Kiuchi, K., Kyutoku, K., and Shibata, M. (2012). Current status of numerical-relativity simulations in Kyoto, *Prog. Theor. Expe. Phys.* **01**, pp. A304–1.

Sekiguchi, Y. and Shibata, M. (2005). Axisymmetric collapse simulations of rotating massive iron cores in full general relativity: Numerical study for prompt black hole formation, *Phys. Rev. D* **71**, pp. 084013–1.

Sekiguchi, Y. and Shibata, M. (2011). Formation of black hole and accretion disk in a massive high-entropy stellar core collapse, *Astrophys. J.* **737**, p. 6.

Sekiguchi, Y.-I. (2010a). An implementation of the microphysics in full general relativity: A general relativistic neutrino leakage scheme, *Class. Quantum Grav.* **27**, pp. 114107-1.

Sekiguchi, Y.-I. (2010b). Stellar core collapse in full general relativity with microphysics: Formulation and spherical collapse test, *Prog. Theor. Phys.* **124**, p. 331.

Setiawan, S., Ruffert, M., and Janka, H.-T. (2006). Three-dimensional simulations of nonstationary accretion by remnant black-holes of compact object mergers, *Astron. Astrophys.* **458**, p. 553.

Shakura, N. I. and Sunyaev, R. A. (1973). Black holes in binary systems. Observational appearance, *Astron. Astrophys.* **24**, p. 337.

Shapiro, S. L. and Shibata, M. (2002). Collapse of a rotating supermassive star to a supermassive black hole : Analytic calculation, *Astrophys. J. Lett.* **577**, p. 904.

Shapiro, S. L. and Teukolsky, S. A. (1979). Gravitational collapse of supermassive stars to black holes — Numerical solution of the Einstein equations, *Astrophys. J. Lett.* **234**, p. 177.

Shapiro, S. L. and Teukolsky, S. A. (1980). Gravitational collapse to neutron stars and black holes — Computer generation of spherical spacetimes, *Astrophys. J.* **235**, p. 199.

Shapiro, S. L. and Teukolsky, S. A. (1983). *Black holes, white dwarfs, and neutron stars: The physics of compact objects* (Wiley-Interscience Publication).

Shapiro, S. L. and Teukolsky, S. A. (1985a). Relativistic stellar dynamics on the computer. I — Motivation and numerical method, *Astrophys. J.* **298**, p. 34.

Shapiro, S. L. and Teukolsky, S. A. (1985b). Relativistic stellar dynamics on the computer. II — Physical applications, *Astrophys. J.* **298**, p. 58.

Shapiro, S. L. and Teukolsky, S. A. (1986). Relativistic stellar dynamics on the computer. IV — Collapse of a star cluster to a black hole, *Astrophys. J.* **307**, p. 575.

Shapiro, S. L. and Teukolsky, S. A. (1991). Formation of naked singularities — The violation of cosmic censorship, *Phys. Rev. Lett.* **66**, p. 994.

Shapiro, S. L. and Teukolsky, S. A. (1992a). Collisions of relativistic clusters and the formation of black holes, *Phys. Rev. D* **45**, p. 2739.

Shapiro, S. L. and Teukolsky, S. A. (1992b). Gravitational collapse of rotating spheroids and the formation of naked singularities, *Phys. Rev. D* **45**, p. 2006.

Shapiro, S. L., Teukolsky, S. A., and Winicour, J. (1995). Toroidal black holes and topological censorship, *Phys. Rev. D* **52**, p. 6982.

Shapiro, S. L. and Wassermasn, I. (1982). Gravitational radiation from nonspherical infall into black holes, *Astrophys. J.* **260**, p. 838.

Shen, G., Horowitz, C., and O'Connor, E. (2011a). Second relativistic mean field and virial equation of state for astrophysical simulations, *Phys. Rev. C* **83**, pp. 065808-1.

Shen, G., Horowitz, C., and Teige, S. (2011b). New equation of state for astrophysical simulations, *Phys. Rev. C* **83**, pp. 035802-1.

Shen, H., Toki, H., Oyamatsu, K., and Sumiyoshi, K. (1998a). Relativistic equation of state of nuclear matter for supernova and neutron star, *Nucl. Phys. A* **637**, p. 435.

Shen, H., Toki, H., Oyamatsu, K., and Sumiyoshi, K. (1998b). Relativistic equation of state of nuclear matter for supernova explosion, *Prog. Theor. Phys.* **100**, p. 1013.

Shen, H., Toki, H., Oyamatsu, K., and Sumiyoshi, K. (2011c). Relativistic equation of state for core collapse supernova simulations, *Astrophys. J. Suppl.* **197**, p. 20.

Shibata, M. (1996). Instability of synchronized binary neutron stars in the first post-Newtonian approximation of general relativity, *Prog. Theor. Phys.* **96**, p. 317.

Shibata, M. (1997). Apparent horizon finder for a special family of spacetimes in 3D numerical relativity, *Phys. Rev. D* **55**, p. 2002.

Shibata, M. (1998). Relativistic formalism for computation of irrotational binary stars in quasiequilibrium states, *Phys. Rev. D* **58**, pp. 024012–1.

Shibata, M. (1999a). 3D numerical simulation of black hole formation using collisionless particles: Triplane symmetric case, *Prog. Theor. Phys.* **101**, p. 251.

Shibata, M. (1999b). Fully general relativistic simulation of coalescing binary neutron stars: Preparatory tests, *Phys. Rev. D* **60**, pp. 104052–1.

Shibata, M. (1999c). Fully general relativistic simulation of merging binary clusters: Spatial gauge condition, *Prog. Theor. Phys.* **101**, p. 1199.

Shibata, M. (2000). Axisymmetric simulations of rotating stellar collapse in full general relativity: Criteria for prompt collapse to black holes, *Prog. Theor. Phys.* **104**, p. 325.

Shibata, M. (2003a). Axisymmetric general relativistic hydrodynamics: Longterm evolution of neutron stars and collapse to neutron stars and black holes, *Phys. Rev. D* **67**, pp. 024033–1.

Shibata, M. (2003b). Collapse of rotating supramassive neutron stars to black holes: Fully general relativistic simulations, *Astrophys. J.* **595**, p. 992.

Shibata, M. (2004). Stability of rigidly rotating relativistic stars with soft equations of state against gravitational collapse, *Astrophys. J.* **605**, p. 350.

Shibata, M. (2005). Constraining nuclear equations of state using gravitational waves from hypermassive neutron stars, *Phys. Rev. Lett.* **94**, pp. 201101–1.

Shibata, M. (2007). Rotating black hole surrounded by self-gravitating torus in the puncture framework, *Phys. Rev. D* **76**, pp. 064035–1.

Shibata, M., Baumgarte, T. W., and Shapiro, S. L. (2000a). The bar-mode instability in differentially rotating neutron stars: Simulations in full general relativity, *Astrophys. J.* **542**, p. 453.

Shibata, M., Baumgarte, T. W., and Shapiro, S. L. (2000b). Stability and collapse of rapidly rotating, supramassive neutron stars: 3D simulations in general relativity, *Phys. Rev. D* **61**, pp. 044012–1.

Shibata, M., Duez, M. D., Liu, Y.-T., Shapiro, S. L., and Stephens, B. C. (2006). Magnetized hypermassive neutron star collapse: A central engine of short gamma-ray bursts, *Phys. Rev. Lett.* **96**, pp. 031102–1.

Shibata, M. and Karino, S. (2004). Numerical evolution of secular bar-mode instability induced by the gravitational radiation reaction, *Phys. Rev. D* **70**, pp. 084022–1.

Shibata, M., Karino, S., and Eriguchi, Y. (2003). Dynamical bar-mode instability of differentially rotating stars: Effects of equations of state and velocity profiles, *Mon. Not. R. Astron. Soc.* **343**, p. 619.

Shibata, M., Kiuchi, K., Sekiguchi, Y., and Suwa, Y. (2011). Truncated moment formalism for radiation hydrodynamics in numerical relativity, *Prog. Theor. Phys.* **125**, p. 1255.

Shibata, M., Kyutoku, K., Yamamoto, T., and Taniguchi, K. (2009). Gravitational waves from black hole-neutron star binaries: Classification of waveforms, *Phys. Rev. D* **79**, pp. 044030–1.

Shibata, M., Kyutoku, K., Yamamoto, T., and Taniguchi, K. (2012). Erratum and addendum: Gravitational waves from black hole-neutron star binaries: Classification of waveforms, *Phys. Rev. D* **85**, pp. 127502–1.

Shibata, M., Nagakura, H., Sekiguchi, Y., and Yamada, S. (2014). A conservative form of Boltzmann's equation in general relativity, *Phys. Rev. D* **89**, pp. 084073–1.

Shibata, M. and Nakamura, T. (1995). Evolution of three-dimensional gravitational waves: Harmonic slicing case, *Phys. Rev. D* **52**, p. 5428.

Shibata, M., Nakao, K., and Nakamura, T. (1994a). Scalar type gravitational waves from gravitational collapse in Brans-Dicke theory: Detectability by a laser interferometer, *Phys. Rev. D* **50**, p. 7304.

Shibata, M., Nakao, K., Nakamura, T., and Maeda, K. (1994b). Dynamical evolution of gravitational waves in asymptotic de Sitter spacetime, *Phys. Rev. D* **50**, p. 708.

Shibata, M., Okawa, H., and Yamamoto, T. (2008). High-velocity collision of two black holes, *Phys. Rev. D* **78**, pp. 101501(R)–1.

Shibata, M. and Sasaki, M. (1998). Innermost stable circular orbits around relativistic rotating stars, *Phys. Rev. D* **58**, pp. 104011–1.

Shibata, M. and Sasaki, M. (1999). Black hole formation in the Friedmann universe: Formulation and computation in numerical relativity, *Phys. Rev. D* **60**, pp. 084002–1.

Shibata, M. and Sekiguchi, Y. (2003). Gravitational waves from axisymmetrically oscillating neutron stars in general relativistic simulations, *Phys. Rev. D* **68**, pp. 104020–1.

Shibata, M. and Sekiguchi, Y. (2004). Gravitational waves from axisymmetric rotating stellar core collapse to a neutron star in full general relativity, *Phys. Rev. D* **69**, pp. 084024–1.

Shibata, M. and Sekiguchi, Y. (2005a). Magnetohydrodynamics in full general relativity: Formulation and tests, *Phys. Rev. D* **72**, pp. 044014–1.

Shibata, M. and Sekiguchi, Y. (2005b). Three-dimensional simulations of stellar core collapse in full general relativity: Nonaxisymmetric dynamical instabilities, *Phys. Rev. D* **71**, pp. 024014–1.

Shibata, M. and Sekiguchi, Y. (2012). Radiation magnetohydrodynamics for black hole-torus system in full general relativity: A step toward physical simulation, *Prog. Theor. Phys.* **127**, p. 535.

Shibata, M., Sekiguchi, Y.-I., and Takahashi, R. (2007). Magnetohydrodynamics of neutrino-cooled accretion tori around a rotating black hole in general relativity, *Prog. Theor. Phys.* **118**, p. 257.

Shibata, M. and Shapiro, S. L. (2002). Collapse of a rotating supermassive star to a supermassive black hole: Fully relativistic simulations, *Astrophys. J. Lett.* **572**, p. 39.

Shibata, M. and Taniguchi, K. (2006). Merger of binary neutron stars to a black hole: Disk mass, short gamma-ray bursts, and quasinormal mode ringing, *Phys. Rev. D* **73**, pp. 064027–1.

Shibata, M. and Taniguchi, K. (2008). Merger of black hole and neutron star in general relativity: Tidal disruption, torus mass, and gravitational waves, *Phys. Rev. D* **77**, pp. 084015–1.

Shibata, M. and Taniguchi, K. (2011). Coalescence of black hole-neutron star binaries, *Living Review Relativity* **14-6**.

Shibata, M., Taniguchi, K., and Uryū, K. (2005). Merger of binary neutron stars with realistic equations of state in full general relativity, *Phys. Rev. D* **71**, pp. 084021–1.

Shibata, M. and Uryū, K. (2000). Simulation of merging binary neutron stars in full general relativity: $\Gamma = 2$ case, *Phys. Rev. D* **61**, pp. 064001–1.

Shibata, M. and Uryū, K. (2002). Gravitational waves from the merger of binary neutron stars in a fully general relativistic simulation, *Prog. Theor. Phys.* **107**, p. 265.

Shibata, M. and Uryū, K. (2006). Merger of black hole-neutron star binaries: Nonspinning black hole case, *Phys. Rev. D* **74**, pp. 121503(R)–1.

Shibata, M. and Uryū, K. (2007). Merger of black hole-neutron star binaries in full general relativity, *Class. Quantum Grav.* **24**, p. S125.

Shibata, M., Uryū, K., and Friedman, J. L. (2004). Deriving formulations for numerical computation of binary neutron stars in quasicircular orbits, *Phys. Rev. D* **70**, pp. 044044–1.

Shibata, M. and Yoshino, H. (2010a). Bar-mode instability of rapidly spinning black hole in higher dimensions, *Phys. Rev. D* **81**, pp. 104035–1.

Shibata, M. and Yoshino, H. (2010b). Nonaxisymmetric instability of rapidly rotating black hole in five dimensions, *Phys. Rev. D* **81**, pp. 021501(R)–1.

Siegel, D. M., Ciolfi, R., Harte, A. I., and Rezzolla, L. (2013). On the magnetorotational instability in relativistic hypermassive neutron stars, *Phys. Rev. D* **87**, pp. 121302–1.

Smarr, L. (1973a). Mass formula for Kerr black holes, *Phys. Rev. Lett.* **30**, p. 71.

Smarr, L. (1973b). Surface geometry of charged rotating black holes, *Phys. Rev. D* **7**, p. 289.

Smarr, L. (1979). Gauge conditions, radiation formulae, and the two black hole collisions, in L. L. Smarr (ed.), *Sources of gravitational radiation* (Cambridge University Press), p. 245.

Smarr, L., Cadez, A., DeWitt, B., and Eppley, K. (1976). Collision of two black holes: Theoretical framework, *Phys. Rev. D* **14**, p. 2443.

Smarr, L. and York, J. W. (1978a). Kinematical conditions in the construction of spacetime, *Phys. Rev. D* **17**, p. 2529.

Smarr, L. and York, J. W. (1978b). Radiation gauge in general relativity, *Phys. Rev. D* **17**, p. 1945.

Smartt, S. J. (2009). Progenitors of core-collapse supernova, *Annu. Rev. Astron. Astrophys.* **47**, p. 63.

Sorkin, E. (2010). An axisymmetric generalized harmonic evolution code, *Phys. Rev. D* **81**, pp. 084062–1.

Spruit, H. C. (1999). Differential rotation and magnetic fields in stellar interiors, *Astron. Astrophys.* **349**, p. 189.

Stairs, I. H. (2003). Testing general relativity with pulsar timing, *Living Review Relativity* **6-5**.

Stairs, I. H. (2004). Pulsars in binary systems: probing binary stellar evolution and general relativity, *Science* **304**, p. 547.

Stark, R. F. (1989). Non-axisymmetric rotating gravitational collapse and gravitational radiation, in C. R. Evans, L. S. Finn, and D. W. Hobill (eds.), *Frontiers in Numerical Relativity* (Cambridge University Press), p. 281.

Stark, R. F. and Piran, T. (1985). Gravitational-wave emission from rotating gravitational collapse, *Phys. Rev. Lett.* **55**, p. 891.

Stark, R. F. and Piran, T. (1987). A general relativistic code for rotating axisymmetric configuration and gravitational radiation: Numerical methods and tests, *Comp. Phys. Rep.* **5**, p. 221.

Steiner, A. W., Lattimer, J. M., and Brown, E. F. (2013). The neutron star mass-radius relation and the equation of state of dense matter, *Astrophys. J. Lett.* **765**, p. L5.

Stergioulas, N. (2003). Rotating stars in relativity, *Living Review Relativity* **6-3**.

Stergioulas, N. and Friedman, J. L. (1998). Nonaxisymmetric neutral modes in rotating relativistic stars, *Astrophys. J.* **492**, p. 301.

Sumiyoshi, K. and Yamada, S. (2012). Neutrino transfer in three dimensions for corecollapse supernovae. I. static configurations, *Astrophys. J. Suppl.* **199**, p. 17.

Sumiyoshi, K., Yamada, S., Suzuki, H., and Chiba, S. (2006). Neutrino signals from the formation of a black hole: A probe of the equation of state of dense matter, *Phys. Rev. Lett.* **97**, pp. 091101–1.

Sumiyoshi, K., Yamada, S., Suzuki, H., Shen, H., Chiba, S., and Toki, H. (2005). Postbounce evolution of core-collapse supernovae: Long-term effects of the equation of state, *Astrophys. J.* **629**, p. 922.

Suwa, Y., Kotake, K., Takiwaki, T., Whitehouse, S. C., Liebendörfer, M., and Sato, K. (2010). Explosion geometry of a rotating $13 M_\odot$ star driven by the SASI-aided neutrino-heating supernova mechanism, *Pub. Astron. Soc. Japan* **62**, p. L49.

Symbalisty, E. and Schramm, D. N. (1982). Neutron star collisions and the r-process, *Astrophys. Lett.* **22**, p. 143.

Szilágyi, B., Lindblom, L., and Scheel, M. A. (2009). Simulations of binary black hole mergers using spectral methods, *Phys. Rev. D* **80**, pp. 124010–1.

Takami, K., Rezzolla, L., and Baiotti, L. (2014). Constraining the equation of state of neutron stars from binary mergers, *Phys. Rev. Lett.* **113**, pp. 0911014–1.

Takiwaki, T., Kotake, K., and Suwa, Y. (2012). Three-dimensional hydrodynamic core-collapse supernova simulations for an $11.2\, M_\odot$ star with spectral neutrino transport, *Astrophys. J.* **749**, p. 98.

Tanaka, M. and Hotokezaka, K. (2013). Radiative transfer simulations of neutron star merger ejecta, *Astrophys. J.* **775**, p. 113.

Tanaka, M., Hotokezaka, K., Kyutoku, K., Wanajo, S., Kiuchi, K., Sekiguchi, Y., and Shibata, M. (2014). Radioactively powered emission from black hole-neutron star mergers, *Astrophys. J.* **780**, p. 31.

Tangherlini, F. R. (1963). Schwarzschild field in n dimensions and the dimensionality of space problem, *Nuovo Cimento* **27**, p. 636.

Taniguchi, K., Baumgarte, T. W., Faber, J. A., and Shapiro, S. L. (2006). Quasiequilibrium sequences of black-hole-neutron-star binaries in general relativity, *Phys. Rev. D* **74**, pp. 041502(R)–1.

Taniguchi, K., Baumgarte, T. W., Faber, J. A., and Shapiro, S. L. (2007). Quasiequilibrium black hole-neutron star binaries in general relativity, *Phys. Rev. D* **75**, pp. 084005–1.

Taniguchi, K., Baumgarte, T. W., Faber, J. A., and Shapiro, S. L. (2008). Relativistic black hole-neutron star binaries in quasiequilibrium: Effects of the black hole excision boundary condition, *Phys. Rev. D* **77**, pp. 044003–1.

Taniguchi, K. and Gourgoulhon, E. (2002). Quasiequilibrium sequences of synchronized and irrotational binary neutron stars in general relativity. III. Identical and different mass stars with $\gamma = 2$, *Phys. Rev. D* **66**, pp. 104019–1.

Taniguchi, K. and Gourgoulhon, E. (2003). Various features of quasiequilibrium sequences of binary neutron stars in general relativity, *Phys. Rev. D* **68**, pp. 124025–1.

Taniguchi, K. and Shibata, M. (2010). Binary neutron stars in quasi-equilibrium, *Astrophys. J. Suppl.* **188**, p. 187.

Taracchini, A., Buonanno, A., Pan, Y., and 14 authors (2014). Effective-one-body model for black-hole-binaries with generic mass ratios and spins, *Phys. Rev. D* **89**, pp. 061502(R)–1.

Taracchini, A., Pan, Y., Buonanno, A., Barausse, E., Boyle, M., Chu, T., Lovelace, G., Pfeiffer, H. P., and Scheel, M. A. (2012). Prototype effective-one-body model for nonprecessing spinning inspiral-merger-ringdown waveforms, *Phys. Rev. D* **86**, pp. 024011–1.

Taub, A. (1948). Relativistic Rankine-Hugoniot equations, *Phys. Rev.* **74**, p. 328.

Taub, A. H. (1954). General relativistic variational principle for perfect fluids, *Phys. Rev.* **94**, p. 1468.

Taub, A. H. (1969). Stability of general relativistic gaseous masses and variational principles, *Commun. Math. Phys.* **15**, p. 235.

Tayler, R. J. (1973). The adiabatic stability of stars containing magnetic fields-I. Toroidal fields, *Mon. Not. R. Astron. Soc.* **161**, p. 365.

Teukolsky, S. A. (1973). Perturbations of a rotating black hole. I. Fundamental equations for gravitational, electromagnetic, and neutrino-field perturbations, *Astrophys. J.* **185**, p. 635.

Teukolsky, S. A. (1982). Linearized quadrupole waves in general relativity and the motion of test particles, *Phys. Rev. D* **26**, p. 745.

Teukolsky, S. A. (1998). Irrotational binary neutron stars in quasi-equilibrium in general relativity, *Astrophys. J.* **504**, p. 442.

Teukolsky, S. A. (2000). Stability of the iterated Crank-Nicholson method in numerical relativity, *Phys. Rev. D* **61**, pp. 087501-1.

Teukolsky, S. A. and Press, W. H. (1974). Perturbations of a rotating black hole. II. Interaction of the hole with gravitational and electromagnetic radiation, *Astrophys. J.* **193**, p. 443.

Thompson, K. W. (1986). The special relativistic shock tube, *J. Fluid Mechanics* **171**, p. 363.

Thornburg, J. (1987). Coordinates and boundary conditions for the general relativistic initial data problem, *Class. Quantum Grav.* **4**, p. 1191.

Thornburg, J. (1996). Finding apparent horizons in numerical relativity, *Phys. Rev. D* **54**, p. 4899.

Thorne, K. S. (1972). Nonspherical gravitational collapse – A short review, in J. A. Wheeler (ed.), *Magic Without Magic* (San Francisco: W. H. Freeman), p. 231.

Thorne, K. S. (1973). Relativistic shocks: The Taub adiabat, *Astrophys. J.* **179**, p. 897.

Thorne, K. S. (1980). Multipole expansions of gravitational radiation, *Rev. Mod, Phys.* **52**, p. 299.

Thorne, K. S. (1981). Relativistic radiative transfer: Moment formalism, *Mon. Not. R. Astron. Soc.* **194**, p. 439.

Thorne, K. S. (1987). Gravitational radiation, in S. Hawking and W. Israel (eds.), *Three hundred years of gravitation* (Cambridge University Press), p. 330.

Thorne, K. S., Price, R. H., and Macdonald, D. A. (1986). *Black holes: The Membrane Paradigm* (Yale University Press).

Tichy, W. (2012). Constructing quasi-equilibrium initial data for binary neutron stars with arbitrary spins, *Phys. Rev D.* **86**, pp. 064024-1.

Tichy, W. and Brügmann, B. (2004). Quasiequilibrium binary black hole sequences for puncture data derived from helical Killing vector conditions, *Phys. Rev. D* **69**, pp. 024006-1.

Tichy, W. and Marronetti, P. (2009). Final mass and spin of black-hole merger, *Astrophys. J* **704**, p. L40.

Tohilne, J. E. and Hachisu, I. (1990). The breakup of self-gravitating rings, tori, and thick accretion disks, *Astrophys. J.* **361**, p. 394.

Tohline, J. E., Durisen, R. H., and McCollough, M. (1985). The linear and nonlinear dynamic stability of rotating N=3/2 polytropes, *Astrophys. J.* **298**, p. 220.

Tolman, R. C. (1939). Static solutions of Einstein's field equations for spheres of fluid, *Phys. Rev.* **55**, p. 374.

Tomimatsu, A. and Sato, H. (1972). New exact solution for the gravitational field of a spinning mass, *Phys. Rev. Lett.* **29**, p. 1344.

Tomimatsu, A. and Sato, H. (1973). New series of exact solutions for gravitational fields of spinning masses, *Prog. Theor. Phys.* **50**, p. 95.

Toro, E. F. (1999). *Riemann Solvers and Numerical Methods for Fluid Dynamics (2nd Edition)* (Springer).

Ugliano, M., Janka, H.-T., Marec, A., and Arcones, A. (2012). Progenitor-explosion connection and remnant birth masses for neutrino-driven supernovae of iron-core progenitors, *Astrophys. J.* **757**, p. 69.

Umeda, H. and Nomoto, K. (2008). How much ^{56}Ni can be produced in core-collapse supernovae? evolution and explosions of 30–100$M_\odot$ stars, *Astrophys. J.* **673**, p. 1014.

Uryū, K. and Eriguchi, Y. (2000). New numerical method for constructing quasiequilibrium sequences of irrotational binary neutron stars in general relativity, *Phys. Rev. D* **61**, pp. 124023–1.
Uryū, K., Limousin, F., Friedman, J. L., Gourgoulhon, E., and Shibata, M. (2006). Binary neutron stars: Equilibrium models beyond spatial conformal flatness, *Phys. Rev. Lett.* **97**, pp. 171101–1.
Uryū, K., Limousin, F., Friedman, J. L., Gourgoulhon, E., and Shibata, M. (2009). Nonconformally flat initial data for binary compact objects, *Phys. Rev. D* **80**, pp. 124004–1.
Uryū, K., Shibata, M., and Eriguchi, Y. (2000). Properties of general relativistic, irrotational binary neutron stars in close quasiequilibrium orbits: Polytropic equations of state, *Phys. Rev. D* **62**, pp. 104015–1.
van Leer, B. (1977). Toward the ultimate conservative difference scheme IV. A new approach to numerical convection, *J. Comput. Phys.* **23**, p. 276.
van Leer, B. (1979). Toward the ultimate conservative difference scheme V. A second-order sequel to Godunov's theorem, *J. Comput. Phys.* **32**, p. 101.
van Meter, J. R., Baker, J. G., Koppitz, M., and Choi, D.-I. (2006). How to move a black hole without excision: Gauge conditions for the numerical evolution of a moving puncture, *Phys. Rev. D* **73**, pp. 124011–1.
van Riper, K. A. and Lattimer, J. M. (1981). Stellar core collapse. I. Infall epoch, *Astrophys. J.* **249**, p. 270.
van Riper, K. A. and Lattimer, J. M. (1982). Stellar core collapse. II. Inner core bounce and shock propagation, *Astrophys. J.* **257**, p. 793.
Villiers, J.-P. D. and Hawley, J. F. (2003). A numerical method for general relativistic magnetohydrodynamics, *Astrophys. J.* **589**, p. 458.
Vines, J., Flanagan, E. E., and Hinderer, T. (2011). Post-1-Newtonian tidal effects in the gravitational waveform from binary inspirals, *Phys. Rev. D* **83**, pp. 084051–1.
VIRGO (). http://www.virgo.infn.it/.
Voss, R. and Tauris, T. M. (2003). Galactic distribution of merging neutron stars and black holes — prospects for short gamma-ray burst progenitors and LIGO/VIRGO, *Mon. Not. R. Astron. Soc.* **342**, p. 1169.
Wald, R. M. (1984). *General Relativity* (The University of Chicago Press).
Walker, M. and Penrose, R. (1970). On quadratic first integrals of the geodesic equations for type {22} spacetimes, *Commun. Math. Phys.* **18**, p. 265.
Weisberg, J. M., Nice, D. J., and Taylor, J. H. (2010). Timing measurements of the relativistic binary pulsar PSR B1913+16, *Astrophys. J.* **722**, p. 1030.
Whiting, B. F. (1989). Mode stability of the Kerr black hole, *J. Math. Phys.* **30**, p. 1301.
Will, C. M. (1993). *Theory and experiment in gravitational physics: revised version* (Cambridge University Press).
Will, C. M. (2014). The confrontation between general relativity and experiment, *Living Review Relativity* **17-4**.
Wilson, J. R. (1985). Supernovae and post-collapse behavior, in J. Centrella, J. LeBlanc, and R. Bowers (eds.), *Numerical Astrophysics* (Jones and Bartlett Publishers), p. 422.
Wilson, J. R. and Mathews, G. J. (1989). Relativistic hydrodynamics, in C. Evans, L. Finn, and D. Hobill (eds.), *Frontiers in Numerical Relativity* (Cambridge University Press), p. 306.
Wilson, J. R. and Mathews, G. J. (1995). Instabilities in close neutron star binaries, *Phys. Rev. Lett.* **75**, p. 4161.

Wilson, J. R., Mathews, G. J., and Marronetti, P. (1996). Relativistic numerical model for close neutron-star binaries, *Phys. Rev. D* **54**, p. 1317.

Wiseman, A. G. (1992). Coalescing binary systems of compact objects to (post)5/2-Newtonian order. II. Higher-order wave forms and radiation recoil, *Phys. Rev. D* **46**, p. 1517.

Witek, H., Cardoso, V., Gualtieri, L., Herdeiro, C., Sperhake, U., and Zilhao, M. (2011). Head-on collisions of unequal mass black holes in D=5 dimensions, *Phys. Rev. D* **83**, pp. 044017–1.

Witek, H., Zilhao, M., Gualtieri, L., Cardoso, V., Herdeiro, C., Nerozzi, A., and Sperhake, U. (2010). Numerical relativity for D dimensional space-times: Head-on collisions of black holes and gravitational wave extraction, *Phys. Rev. D* **82**, pp. 104014–1.

Woods, P. M. and Thompson, C. (2006). Soft gamma repeaters and anomalous X-ray pulsars, in W. H. G. Lewin and M. van der Klis (eds.), *Compact Stellar X-ray Sources* (Cambridge University Press), p. 547.

Woosley, S. E. (1993). Gamma-ray bursts from stellar mass accretion disks around black holes, *Astrophys. J.* **405**, p. 273.

Woosley, S. E., Heger, A., and Spruit, H. C. (2005). Presupernova evolution of differentially rotating massive stars including magnetic fields, *Astrophys. J.* **626**, p. 350.

Woosley, S. E., Heger, A., and Weaver, T. A. (2002). The evolution and explosion of massive stars, *Rev. Mod. Phys.* **74**, p. 1015.

Yakunin, K. N., Marronetti, P., Mezzacappa, A., Bruenn, S. W., Lee, C.-T., Chertkow, M. A., Hix, W. R., Blondin, J. M., Lentz, E. J., Messer, E. B., and Yoshida, S. (2010). Gravitational waves from core collapse supernovae, *Class. Quantum Grav.* **27**, pp. 194005–1.

Yamada, Y. and Shinkai, H. (2011). Formation of naked singularities in five-dimensional space-time, *Phys. Rev. D* **83**, pp. 064006–1.

Yamamoto, T., Shibata, M., and Taniguchi, K. (2008). Simulating coalescing compact binaries by a new code (SACRA), *Phys. Rev. D* **78**, pp. 064054–1.

York, J. W. (1971). Gravitational degrees of freedom and the initial-value problem, *Phys. Rev. Lett.* **26**, p. 1656.

York, J. W. (1972). Role of conformal three-geometry in the dynamics of gravitation, *Phys. Rev. Lett.* **28**, p. 1082.

York, J. W. (1973). Conformally invariant orthogonal decomposition of symmetric tensors on Riemannian manifolds and the initial-value problem of general relativity, *J. Math. Phys.* **14**, p. 456.

York, J. W. (1979). Kinematics and dynamics of general relativity, in L. L. Smarr (ed.), *Sources of gravitational radiation* (Cambridge University Press), p. 83.

York, J. W. (1999). Conformal 'thin-sandwich' data for the initial-value problem of general relativity, *Phys. Rev. Lett.* **82**, p. 1350.

Yoshino, H. and Shibata, M. (2009). Higher-dimensional numerical relativity: Formulation and code tests, *Phys. Rev. D* **80**, pp. 084025–1.

Yoshino, H., Shiromizu, T., and Shibata, M. (2006). Close-slow analysis for head-on collision of two black holes in higher dimensions: Bowen-York initial data, *Phys. Rev. D* **74**, pp. 124022–1.

Zalamea, I. and Beloborodov, A. M. (2011). Neutrino heating near hyper-accreting black holes, *Mon. Not. R. Astron. Soc.* **410**, p. 2302.

Zanna, L. D., Bucciantini, N., and Londrillo, P. (2003). An efficient shock-capturing central-type scheme for multidimensional relativistic flows. II. Magnetohydrodynamics, *Astron. Astrophys.* **400**, p. 397.

Zel'dovich, Y. B. and Novikov, I. D. (1971). *Stars and Relativity* (The University of Chicago Press).

Zerilli, F. J. (1970). Gravitational field of a particle falling in a Schwarzschild geometry analyzed in tensor harmonics, *Phys. Rev. D* **2**, p. 2141.

Zhang, B. and Meszáros, P. (2004). Gamma-ray bursts: Progress, problems, and prospects, *Int. J. Mod. Phys. A* **19**, p. 2385.

Zilhao, M., Witek, H., Sperhake, U., Cardoso, V., Gualtieri, L., Herdeiro, C., and Nerozzi, A. (2010). Numerical relativity for d dimensional axially symmetric space-times: Formalism and code tests, *Phys. Rev. D* **81**, pp. 084052–1.

Zink, B., Stergioulas, N., Hawke, I., Ott, C. D., Schnetter, E., and Müller, E. (2006). Formation of supermassive black holes through fragmentation of toroidal supermassive stars, *Phys. Rev. Lett.* **96**, pp. 161101–1.

Zink, B., Stergioulas, N., Hawke, I., Ott, C. D., Schnetter, E., and Müller, E. (2007). Nonaxisymmetric instability and fragmentation of general relativistic quasitoroidal stars, *Phys. Rev. D* **76**, pp. 024019–1.

Index

$N+1$ formulation (in conformal formulation), 120
Γ-law equation of state, 251
g-mode, 92, 93
g-mode frequency, 93
r-process, 74, 502, 510–512, 562
2+1+1 formulation, 167, 180–186
3+1 formulation (axisymmetric case), 166, 167
3+1 formulation (cosmology), 192
3+1 formulation (in conformal formulation), 120

acceleration of unit timelike normal, 114
action for perfect fluid, 714
adaptive mesh-refinement (AMR), 219–229
advection term (for hydrodynamics), 252
Alfvén time scale, 650
Alfvén velocity, 272
Alfvén wave (ideal magnetohydrodynamics), 323, 732
Ampére–Maxwell's law, 264
angular-momentum emission rate (of gravitational waves), 14, 425
angular momentum (for time-asymmetric initial data), 359
angular momentum (of asymptotically flat system), 345
angular momentum (of dynamical horizon), 444, 753, 754
angular velocity (for circular orbits around Kerr black hole), 32
angular velocity (for circular orbits around Schwarzschild black hole), 31

angular velocity (of black hole), 34, 380, 683
angular velocity (of dynamical horizon), 754
apparent horizon (basic equation), 429–431
apparent horizon (definition), 751
apparent horizon (for excision), 139
apparent horizon (head-on collision of two black holes), 437, 438
apparent horizon (in axial symmetry), 432, 433
apparent horizon (in nonaxial symmetry), 435, 436
apparent horizon (in spherical symmetry), 431, 432
apparent horizon (of spinning black hole initial data), 362
apparent horizon (schematic figure), 430, 753
apparent horizon (spheroidal object), 439, 440
approximate Riemann solver, 309
areal radius, 25, 38, 43
area (of black hole), 25, 26, 34, 39, 443, 679
area theorem, 34, 36, 737
Arnowitt–Deser–Misner (ADM) formulation ($N+1$ formulation), 109–119
Arnowitt–Deser–Misner (ADM) formulation (derived from variational principle), 717
Arnowitt–Deser–Misner (ADM) four-momentum, 344

Arnowitt–Deser–Misner (ADM) linear momentum (definition), 344
Arnowitt–Deser–Misner (ADM) linear momentum (for time-asymmetric initial data), 359
Arnowitt–Deser–Misner (ADM) mass (as the boundary term of Hamiltonian), 719
Arnowitt–Deser–Misner (ADM) mass (definition), 341, 342
Arnowitt–Deser–Misner (ADM) mass (for TOV case), 44, 45
Arnowitt–Deser–Misner (ADM) mass (Misner initial data), 357
Arnowitt–Deser–Misner (ADM) mass (multiple black holes), 355
Arnowitt–Deser–Misner (ADM) mass (Newtonian limit), 343
Arnowitt–Deser–Misner (ADM) mass (volume-integral form), 343
asymptotically quasi-isotropic gauge, 343

Bach tensor, 363
bar-mode instability (for Myers–Perry black hole), 678–685
bar-mode instability (for rotating star), 54, 94–96, 98, 641–646
bare mass (Brill–Lindquist initial data), 355
Baumgarte–Shapiro–Shibata–Nakamura (BSSN) formulation, 147, 148, 154–157
Baumgarte–Shapiro–Shibata–Nakamura (BSSN) formulation (cosmological case), 193, 194
Baumgarte–Shapiro–Shibata–Nakamura (BSSN) formulation (Friedmann universe), 195
Baumgarte–Shapiro–Shibata–Nakamura (BSSN) formulation [SO$(n+1)$ symmetric case], 191
Bianchi identity, 3, 119, 416
Bianchi identity (in Newman–Penrose formalism), 418
binary neutron star (in quasi-equilibrium), 460
binary black hole (evolution scenario), 65
binary black hole (in quasi equilibrium), 567–571
binary black hole (merger process), 573, 574, 577, 578, 580–583, 585
binary black hole (precessing), 77

binary black hole (predicted merger rate), 60–62
binary neutron star (evolution scenario), 64, 65
binary neutron star (formation scenario), 59
binary neutron star (in quasi-equilibrium), 454–461
binary neutron star (lifetime), 58
binary neutron star (merger process), 464–468
binary neutron star (observed data), 57
binary neutron star (predicted merger rate), 60, 61
black-hole excision, 139, 145, 146, 162
black hole-neutron star binary (evolution scenario), 65
black hole-neutron star binary (in quasi-equilibrium sequence), 516–518, 520
black hole-neutron star binary (merger process), 523–533
black hole-neutron star binary (merger remnant), 77–81, 537–545
black hole-neutron star binary (predicted merger rate), 60–62
black hole (as remnant of binary neutron star merger), 467, 468, 485
black hole (introduction), 22, 24
black membrane, 39, 679
black string, 671–673
Boltzmann's equation, 277
Boltzmann's equation (conservation form), 281
Boltzmann's equation (conservation form in flat spacetime), 284
Boltzmann's equation (conservation form in Schwarzschild spacetime), 283
Boltzmann's equation (flat spacetime), 280
Boltzmann's equation (spherical symmetry), 279
Bondi flow, 324–328
Bowen–York initial data, 361–363
Boyer–Lindquist coordinates, 25–28, 37, 125
Brill–Lindquist initial data, 354, 355
Brill wave, 352
Brunt–Väisälä frequency, 92
BSSN-puncture formulation, 158, 159, 162

canonical orthonormal tetrad, 37
Carter constant, 31, 32
cartoon method, 186–188
CCZ4 formulation, 162, 165
cell-centered grid, 223
characteristic speed (for hydrodynamics), 255
characteristic speed (ideal magnetohydrodynamics), 270–272
chirp mass, 66
chirp signal, 67
Christodoulou's formula, 34
circularity condition, 375
circumferential length (black-hole horizon), 34, 39, 239, 443, 491, 681, 684
Codazzi equation, 114–116
code test (Bondi flow), 324–328
code test (collapse of neutron star), 335, 336
code test (collision of black holes), 239, 240
code test (collision of black holes in five dimension), 242
code test (evolution of black hole), 234–238
code test (evolution of Myers–Perry black hole), 241, 242
code test (linear gravitational waves), 232–234
code test (oscillating neutron star), 332–335
code test (Riemann shock-tube problem), 319–321
code test (special relativistic magnetohydrodynamics), 321–323
code test (stable neutron star), 329–332
code test (wall-shock problem), 319–321
cold neutron-star equation of state, 45, 48, 49
collision of two black holes (five dimension), 674, 675
compactness parameter, 15, 51, 78
complex null tetrad, 416
conductivity, 263
conformal-imaging method, 361
conformally flat gauge (in spherical symmetry), 694
conformal decomposition, 120
conformal decomposition formulation, 120, 121
conformal decomposition formulation (linearized), 153
conformal factor (cosmological case), 192
conformal factor (definition), 120
conformal factor (quasi-isotropic formulation), 172
conformal spatial metric (definition), 120
conformal thin-sandwich (CTS) formulation, 367
conformal thin-sandwich (CTS) formulation (extended), 368
conformal transformation, 3
conformal transverse-tracefree decomposition formulation, 339, 340, 358
conjugate gradient (CG) method, 398–400
connection coefficient (shortwave formalism), 12
connection coefficient (spacetime), 1, 2
connection coefficient (spatial), 111, 112
connection tensor (for flat metric), 122
connection tensor (shortwave formalism), 11
connection tensor (spatial), 112, 121
conservation equation (for radiation hydrodynamics), 294
conservation equation (for stress-energy tensor), 3, 243, 249
conservation law (of Landau–Lifshitz), 733
consistency relation (axisymmetric spacetime), 383
constants of motion (for circular orbits around Kerr black hole), 31, 32
constants of motion (for circular orbits around Schwarzschild black hole), 31
constant mean curvature slicing, 193, 639
constrained formulation, 162
constrained transport scheme, 314–316
constraint equation, 118, 119
constraint equation (D-dimensional), 160
constraint equation (definition), 115
constraint equation (GH formulation), 161
constraint equation (in BSSN formulation), 156
constraint propagation, 162, 166
constraint violation, 162–164
contact discontinuity (in Riemann problem), 721–723

continuity equation (for electric charge density), 263, 264
continuity equation (for lepton fraction), 259
continuity equation (for rest mass), 250, 251
continuity equation (in special relativity), 723
convergence property, 149, 230, 231
convergence property (numerical example), 212, 233, 234, 236, 238, 323, 327, 330, 331, 334, 336
coordinate distortion, 141
coordinate distortion (numerical example), 126, 127
coordinate singularity (for black hole in isotropic coordinates), 27, 158, 355, 356
coordinate singularity (for black hole in quasi-isotropic coordinates), 364
coordinate singularity (in curvilinear coordinates), 166, 168
coordinate singularity (in geodesic gauge), 124
coordinate singularity (in horizon stretching), 139
coordinate singularity (in zero-shift condition), 126
coordinate singularity (of black hole horizon), 25
corotating velocity field, 392
cosmological constant, 193
could-in-cell (CIC) scheme, 247
Courant–Friedrichs–Lewy (CFL) condition, 147, 209–211, 220
Courant–Friedrichs–Lewy (CFL) number, 209–211, 221, 233, 236, 237, 240
covariant derivative (2+1+1 formulation), 181, 184
covariant derivative (for conformal metric), 120
covariant derivative (shortwave formalism), 11
covariant derivative (spacetime), 1
covariant derivative (spatial), 111
curvature invariant (definition), 2
curvature invariant (Kerr black hole), 26
curvature invariant (Schwarzschild black hole), 25

cutoff frequency (of gravitational wave spectrum), 550, 552–558

de Sitter spacetime, 192
dominant energy condition, 739
double-null coordinates, 29
dynamical gauge, 142, 148
dynamical horizon, 751–754, 756
dynamical shift, 146–148
dynamical slicing, 143, 144, 146, 148

eccentricity reduction of quasi-equilibrium, 514, 515
Eddington–Finkelstein coordinates, 30
effective amplitude of gravitational waves, 68, 97, 99
effective Einstein's equation (shortwave formalism), 11
Effective one-body approach, 766–768
effective one-body approach (comparison with numerical waveform), 488, 489, 571, 587, 588
effective potential method (for binary black hole in quasi-equilibrium), 564–566
effective stress-energy tensor (of gravitational waves), 11, 13
Einstein's equation, 1, 3
Einstein's equation (2+1+1 formulation), 184
Einstein's equation (3+1 formulation), 116, 117
Einstein's equation (derived by action principle), 712
Einstein's equation (dimensional reduction), 183
Einstein's equation (GH formulation), 161
Einstein's equation (hyperbolic nature), 4
Einstein's equation (linearized), 5
Einstein's equation (shortwave formalism), 11
Einstein's equation (Z4 formulation), 163
Einstein-Rosen bridge, 27
Einstein telescope, 21
Einstein tensor (definition), 1
Einstein tensor (linearized), 5
Einstein tensor (spatial), 349
electric charge density, 261
electric current vector, 261
electric field vector (3+1 form), 261

electromagnetic counterpart of gravitational waves, 73, 74, 510–513, 558, 562, 563
electromagnetic energy density, 262
electromagnetohydrodynamics equations, 265, 266
electron capture (of heavy nuclei), 41, 83, 84, 611
eLISA, 22, 23
energy equation (3+1 form), 249
energy equation (hydrodynamics), 250
energy equation (in Friedmann universe), 196
energy equation (in special relativity), 723
energy equation (radiation hydrodynamics), 292
equation of motion (3+1 form), 249
equation of state (for neutron star), 47
equation of state (for neutron star in binary), 461–464
Euler displacement, 714
Euler equation (hydrodynamics), 250
Euler equation (in Friedmann universe), 196
Euler equation (in special relativity), 723
Euler equation (radiation hydrodynamics), 292
event horizon (definition), 738
event horizon (in five-dimensional Myers–Perry black hole), 39
event horizon (in Hernandez–Misner formulation), 700
event horizon (in higher-dimensional isotropic coordinates), 38
event horizon (numerical method), 441–443
event horizon (of Schwarzschild–Tangherlini solution), 38
event horizon (radius in Boyer–Lindquist coordinates), 26, 33
event horizon (radius in isotropic coordinates), 26, 354
event horizon (radius in Kerr–Schild coordinates), 28
event horizon (radius in quasi-isotropic coordinates), 27
event horizon (Schwarzschild radius), 25
evolution equation (2+1+1 formulation), 184

evolution equation (for constraints), 119, 164
evolution equation (for extrinsic curvature), 116–119, 131
evolution equation (for spatial geometries), 118, 119
excision boundary condition (for black hole), 368–371
exotic particles, 43, 46
expansion (non-expanding horizon), 369
expansion of null geodesics, 739
extrinsic curvature (2+1+1 formulation), 184
extrinsic curvature (definition), 112
extrinsic curvature (geometrical meaning), 113
extrinsic curvature (of Kerr black hole), 125, 364

Faraday's law, 264
Fermi energy, 42
finite-temperature equation of state, 55, 56, 259
first law of binary, 393, 394, 455, 748–751
first law of black hole, 36, 737, 742–748
first law of thermodynamics, 715
flat spatial metric (definition), 120
flux-limiter function, 300–302, 305
focusing singularity, 125
Force-free electromagnetodynamics, 272–275
force-free limit (condition), 273
fourth-order centered finite-differencing scheme, 199, 200
four laws of black hole, 737
four velocity (for perfect fluid), 250
Friedmann equation, 193, 195
Friedmann spacetime, 194, 195
Frobenius' theorem, 25, 112, 691

gain radius, 90
gamma-ray burst, 72, 73
gauge-invariant variable, 412, 413, 415
gauge-invariant wave extraction, 410–414
gauge condition, 123
gauge freedom, 118
Gauss' law, 263
Gauss equation, 114, 115
generalized harmonic (GH) formulation, 160–162

generalized harmonic (GH) formulation (axisymmetric), 168
generalized harmonic (GH) gauge, 161, 162
generator of event horizon, 738
geodesic deviation equation, 2
geodesic deviation equation (for gravitational waves), 7
geodesic equation, 2, 124, 128, 245
geodesic gauge, 124, 125, 128
geodesic slicing, 143
Geroch's formulation, 180
Godunov's scheme, 309
Godunov's theorem, 297–299
gravitational-field action (in $N+1$ formulation), 715
gravitational-radiation reaction time scale, 17, 98
gravitational-wave detector, 18
gravitational-wave spectrum (merger of binary black hole), 585, 588–590
gravitational-wave spectrum (merger of black hole-neutron star binary), 549–558
gravitational-wave spectrum (merger remnant of binary neutron star), 495, 498–500
gravitational potential energy (in general relativity), 604
gravitational waves (amplitude), 10
gravitational waves (by anisotropic neutrino emission), 93, 94
gravitational waves (from binary black hole merger), 74–76, 575–579, 583, 586–588
gravitational waves (from binary neutron star merger to black hole), 70, 71, 490, 491
gravitational waves (from black-hole formation), 100, 101
gravitational waves (from black hole-neutron star merger), 81, 82, 546–549
gravitational waves (from gravitational collapse and supernova), 87, 88, 91–93
gravitational waves (from inspiraling binary), 16, 66–68
gravitational waves (from inspiraling binary neutron star), 487–490

gravitational waves (from remnant massive neutron star), 71, 72, 491–494, 496, 497
gravitational waves (from rotating proto-neutron star), 96–100
gravitational waves (from supermassive binary black hole), 76
gravitational waves (from supermassive black-hole formation), 101, 102
gravitational waves (generation), 7–9
gravitational waves (propagation), 6
gravitational waves (relation with Weyl scalar, 425
gravitational waves (solution for linearized Einstein's equation), 708, 709
Gregory–Laflamme instability, 671–673, 679

Hachisu's self-consistent field method, 384–386
Hamiltonian constraint, 116
Hamiltonian constraint (2+1+1 formulation), 184
Hamiltonian constraint (Brill wave), 352
Hamiltonian constraint (cosmological case), 193
Hamiltonian constraint (de Sitter universe), 194
Hamiltonian constraint (for time-symmetric vacuum initial data), 353
Hamiltonian constraint (Friedmann universe), 196
Hamiltonian constraint (in conformally flat gauge), 694
Hamiltonian constraint (in conformal decomposition formulation), 122
Hamiltonian constraint (in conformal transverse-tracefree decomposition), 340
Hamiltonian constraint (in IWM formulation), 366
Hamiltonian constraint (in linear approximation), 150
Hamiltonian constraint (in radial gauge), 697
Hamiltonian constraint (quasi-isotropic formulation), 175
Hamiltonian constraint (quasi-radial formulation), 178

Hamiltonian constraint (Z4 formulation), 163
Hamiltonian density, 718
harmonic gauge, 4, 5, 137
harmonic gauge (in linear approximation), 152
harmonic slicing, 138, 143
Hernandez–Misner formulation, 700
high-resolution shock-capturing scheme, 296
Hilbert action, 711
HLL scheme, 310–313
hoop conjecture, 439
horizon stretching, 139, 144
Hubble horizon, 636, 640
Hubble parameter, 193, 195
hydrodynamics equation (derived by action principle), 714
hydrostatic equations (for axisymmetric equilibrium), 380–382
hydrostatic equations (for equilibrium of neutron star binary), 390–392
hyperbolic formulation, 108, 109
hypermassive neutron star, 53–55
hypermassive neutron star (as remnant of binary neutron star merger), 472–474, 478–485
hypermassive neutron star (magnetohydrodynamical collapse), 656–659, 668

ideal magnetohydrodynamics, 266–268
ideal magnetohydrodynamics equations (in special relativity), 728
ideal magnetohydrodynamics equations (property), 269
implicit-explicit scheme, 318, 319
induced metric (2+1+1 formulation), 181, 184
induced metric (3+1 formulation), 109, 110
induction equation, 267, 315
initial data (for pure gravitational wave), 352, 353
initial data (puncture formulation), 360, 361
innermost stable circular orbit (ISCO) (of binary black hole), 568
innermost stable circular orbit (ISCO) (of binary neutron star), 457, 458

innermost stable circular orbit (ISCO) (of black hole), 31–34
innermost stable circular orbit (ISCO) (of black hole-neutron star binary), 517, 518, 520
interchange instability, 652–655
irreducible mass, 34, 351
irrotational velocity field, 391
irrotational velocity field (for binary neutron star), 450–453
Isenberg–Wilson–Mathews (IWM) formulation, 366, 367
isolated horizon, 754, 756
isotropic coordinates (general form), 144
isotropic coordinates (of non-spinning black hole), 26

Jacobian matrix (for hydrodynamics), 253, 254
Jacobian matrix (for ideal magnetohydrodynamics), 269, 270
jump conditions of shock, 726, 728

KAGRA, 19, 20, 556
Kelvin–Helmholtz instability, 664, 665
Kelvin–Helmholtz instability (in binary neutron star merger), 666, 668, 669
Kerr–Schild coordinates, 28
Kerr black hole, 25, 26, 28, 31, 34, 125
kick in binary merger, 77, 561, 576, 577, 584, 585
Killing's equation, 180, 689, 690
Killing horizon, 740
Killing tensor field, 31
Killing vector field (conformal), 371, 690, 691
Killing vector field (conservation law), 690
Killing vector field (definition), 180, 689
Killing vector field (helical), 370, 387, 740, 748
Killing vector field (helical+timelike), 388
Killing vector field (rotational), 26, 31, 169, 739, 754
Killing vector field (tangent of horizon generator), 35, 739–741
Killing vector field (timelike), 25, 26, 31, 742, 744, 746, 748
kilonova, 74, 502
Kinnersley's null tetrad, 37
Komar angular momentum, 346

Komar angular momentum (in $N+1$ form), 348
Komar angular momentum (in null coordinates), 743
Komar mass (definition), 346
Komar mass (for TOV case), 45
Komar mass (in $N+1$ form), 347
Komar mass (in null coordinates), 742
Komar mass (Newtonian limit), 347
Komar mass (relation with lapse function), 347
Kreiss–Oliger dissipation, 229, 230
Kruskal–Szekeres coordinates, 28, 29
Kruskal–Szekeres diagram, 29, 30, 134
Kurganov-Tadamor scheme, 312

Lagrangian density, 717, 743
Lagrangian displacement, 714
Lagrangian interpolation, 224
Lagrangian point, 518
Landau–Lifshitz energy, 734, 735
Landau–Lifshitz linear momentum, 734, 735
Landau–Lifshitz pseudo tensor, 4, 733–735
lapse function, 37
lapse function (2+1+1 formulation), 184
lapse function (definition), 109, 110
lapse function (maximal slicing), 130
lapse function (of Kerr black hole), 26, 364
Large Hadron Collider (LHC), 24
laser interferometric gravitational-wave detector, 19
Lax–Wendroff scheme, 299, 300
leakage scheme, 293–295
leapfrog scheme, 205, 217
LIGO, 19–21, 495, 551, 556, 589
limit hypersurface, 130, 134–139, 143–146, 148, 235
linear-momentum emission rate (of gravitational waves), 14, 18, 425
LISA, 21
locally non-rotating frame, 33, 37
Lorentz gauge, 5, 6
luminosity (by pair-annihilation of neutrinos), 629, 635
luminosity (of gravitational waves), 14, 15, 17, 98, 412, 413, 415, 424, 425
luminosity (of neutrinos in binary neutron star merger), 477, 479
luminosity (of neutrinos in stellar collapse to black hole), 625–630, 634
luminosity (of radioactive ejecta), 511, 562
LU decomposition, 434

macronova, 74, 502
magnetic braking, 650
magnetic field vector (3+1 form), 261
magnetic winding, 649
magnetic winding (in binary neutron star merger), 481, 668
magnetorotational instability (in binary neutron star merger), 666, 668, 669
magnetorotational instability (MRI), 482, 651, 652
massive neutron star (as remnant of binary neutron star merger), 465, 466, 468–472, 474–478, 484, 485
massive star (evolution scenario), 611–613
massive star (property), 599–602
mass (of dynamical horizon), 444, 754
mass ejection (in binary neutron star merger), 502–510
mass ejection (in black hole-neutron star merger), 558–561
mass shedding, 78–81
mass shedding (condition for black hole-neutron star binary), 519–521
mass shedding (of binary neutron star), 456
mass shedding limit (of black hole-neutron star binary), 518
mass shedding limit (rotating star), 592
matched filtering, 103, 104
maximal slicing, 129–131, 134–139, 142–146
maximal slicing (asymptotically), 345
maximal slicing (for initial data of black holes), 358
maximal slicing (in IWM formulation), 366
maximal slicing (in spherical symmetry), 696
maximal slicing (quasi-isotropic formulation), 173
maximal slicing (quasi-radial formulation), 178
Maxwell's equations, 263
minimal distortion gauge, 137, 140–142, 146–148

minimal distortion gauge (in conformal flat gauge), 638, 639, 694
minimal surface, 131
minimal surface (absence of), 177
minimal surface (as apparent horizon), 362
minimal surface (its location for black hole), 136, 144
mini black hole, 24
Minkowski metric, 2
Minkowski spacetime, 5
minmod limiter function, 302
Misner–Sharp–May–White (MSMW) formulation, 698–700
Misner initial data, 356, 357
mixed slice, 178
modified cartoon method, 189–191
momentum constraint, 116
momentum constraint (2+1+1 formulation), 184
momentum constraint (cosmological case), 193
momentum constraint (de Sitter universe), 194
momentum constraint (Friedmann universe), 196
momentum constraint (in conformally flat gauge), 694
momentum constraint (in conformal decomposition formulation), 122
momentum constraint (in conformal transverse-tracefree decomposition), 340
momentum constraint (in IWM formulation), 367
momentum constraint (in linear approximation), 150
momentum constraint (method for a solution), 358
momentum constraint (seen as evolution equation), 157
momentum constraint (Z4 formulation), 163
Moncrief variables, 415
monotone scheme, 298, 299, 303
monotone upstream-centered scheme for conservation laws (MUSCL), 304
monotonicity preserving, 297, 298, 302
moving puncture gauge, 142, 148, 159
Myers–Perry black hole, 38, 678

Nakano's method, 425
naked singularity (5D off-axis collision of black holes), 675, 678
naked singularity (Myers–Perry metric), 39
naked singularity (of spheroidal object), 439, 440
naked singularity (unstable black string), 674
NCSA formulation, 179
neutrino-dominated accretion torus, 623, 634
neutrino-driven convection, 86, 90
neutrino sphere, 86, 89, 90
neutrino trapping, 84
neutron star (basic property), 42, 43, 50, 51
neutron star (formation scenario), 41
neutron star (introduction), 40
neutron star (structure), 46, 48, 50
Newman–Penrose formalism, 415–417
Newton–Raphson method, 313, 314
no-monopole constraint, 263
non-expanding horizon, 369, 370, 754, 755
nonlinear Alfvén wave (ideal magnetohydrodynamics), 731
null dominant energy condition, 739
null energy condition, 739

off-axis collision of two black holes (five dimension), 675–678
Ohm's law, 263, 264
outer boundary condition (for Einstein's evolution equation), 207–209

Padé approximation, 767
Painlevé–Gullstrand coordinates, 341
pair-instability explosion, 101, 613
pair instability, 101, 611
Papaloizou–Pringle instability, 641
Papaloizou–Pringle instability (of torus surrounding black hole), 647–649
Parker instability, 655
Parker instability (in neutron star with toroidal magnetic field), 661, 663, 664
particle-mesh method, 246
peeling theorem, 420
Penrose–Carter diagram, 30, 738
Petrov's classification, 416, 419
phase cancellation effect, 547

photo-dissociation, 41, 83, 84, 611, 612, 616
physical singularity (of Kerr–Schild black hole), 28
physical singularity (of Kerr black hole), 26
physical singularity (of Schwarzschild black hole), 25, 29, 30
piecewise parabolic reconstruction, 305
piecewise polytropic equation of state, 47–50, 523
piecewise polytropic equation of state (modified), 463
polar slicing, 130, 697
population synthesis, 61, 62
post-Minkowski approximation (definition), 757
post-Newtonian approximation (amplitude of gravitational waves), 762
post-Newtonian approximation (angular momentum of circular orbit), 758–760
post-Newtonian approximation (definition), 757
post-Newtonian approximation (energy of circular orbit), 758–760
post-Newtonian approximation (linear-momentum emission rate of gravitational waves), 761
post-Newtonian approximation (luminosity of gravitational waves), 760, 761
post-Newtonian approximation (orbit in adiabatic evolution), 763
Poynting flux, 262
preconditioned conjugate gradient (CG) method, 400
pressure (for perfect fluid), 250
primitive recovery (electromagnetohydrodynamics), 266
primitive recovery (for simple hydrodynamics), 252
primitive recovery (for tabulate equation of state), 259, 260
primitive recovery (ideal magnetohydrodynamics), 268, 269
primordial black hole, 24, 635–640
principal null direction of Weyl tensor, 419
prompt convection, 90, 92
proto-black hole, 630, 631
proto-neutron-star convection, 90

proto-neutron star (as source of gravitational waves), 87, 92, 96, 97, 100
proto-neutron star (formation scenario), 83, 84
PSR B1534+12, 57
PSR B1913+16, 57
PSR B2127+11C, 57
PSR J0348+432, 42
PSR J0737-3039, 57, 59, 63
PSR J1614-2230, 42
PSR J1756-2251, 57
puncture plus IWM formulation, 372, 373

quadrupole formula, 9, 14, 427, 428
quadrupole moment, 9, 427
quasi-isotropic coordinates (Kerr black hole), 26
quasi-isotropic coordinates (Myers–Perry black hole), 40, 680
quasi-isotropic formulation, 172–176
quasi-normal mode (non-spinning black hole), 240
quasi-normal mode (4D black hole), 70, 102
quasi-normal mode (5D non-spinning black hole), 242
quasi-normal mode (for merger of black hole-neutron star binary), 82, 547, 548, 550, 553–555, 557
quasi-normal mode (for merger remnant of binary black hole), 75, 588, 590
quasi-normal mode (for merger remnant of binary neutron star), 70, 491, 499
quasi-normal mode (higher-dimensional black hole), 683, 684
quasi-normal mode (non-spinning black hole), 239
quasi-radial formulation, 176–179

radial gauge, 697
radiative variable, 172, 175, 177
rare-faction wave (in Riemann problem), 721–723
Raychaudhuri's equation, 738
Rayleigh criterion, 653
regularity condition, 168–170
regularity condition (violation), 176
regularized variable, 171
resistivity, 263

rest-mass density (for particles), 245
rest-mass density (for perfect fluid), 250
Ricci rotation coefficient, 282, 416, 417
Ricci scalar (horizon surface), 35, 443, 444
Ricci scalar (spacetime), 1
Ricci tensor (2D), 185
Ricci tensor (conformal), 121, 122
Ricci tensor (for GH formulation), 161
Ricci tensor (shortwave formalism), 11, 12
Ricci tensor (spacetime), 1–3
Ricci tensor (spatial), 112, 114, 121
Riemann invariant (in Riemann problem), 724
Riemann problem (ideal magnetohydrodynamics), 729
Riemann problem (in special relativistic hydrodynamics), 308, 721, 722
Riemann shock-tube problem, 308, 723–726
Riemann shock-tube problem (ideal magnetohydrodynamics), 730
Riemann solver, 309
Riemann solver (basic concept), 252
Riemann tensor (2+1+1 formulation), 183
Riemann tensor (relation with Ricci rotation coefficient), 416
Riemann tensor (shortwave formalism), 12
Riemann tensor (spacetime), 2, 3
Riemann tensor (spatial), 111
right eigenvectors of Jacobian matrix, 256
Robin-type boundary condition, 357
Roe-type scheme, 309
rotational kinetic energy (in general relativity), 604
rotation of null geodesics, 739
Runge–Kutta scheme (fourth order), 203, 204
Runge–Kutta scheme (general), 202
Runge–Kutta scheme (second order), 202
Runge–Kutta scheme (stability), 204–206
Runge–Kutta scheme (third order), 203

scalar spherical harmonics, 701, 702
scalar wave equation ($N+1$ form), 244
scale factor, 192, 195
Schwarzschild–Tangherlini solution, 38
Schwarzschild black hole, 25, 26, 145
second-order centered finite-differencing scheme, 199, 436
second law of black hole, 34, 36, 737

shear (non-expanding horizon), 369
shear of null geodesics, 739
shift vector (2+1+1 formulation), 184
shift vector (definition), 109, 123
shift vector (dynamical gauge), 147, 148
shift vector (minimal distortion gauge), 140, 141
shift vector (NCSA formulation), 180
shift vector (of Kerr black hole), 364
shift vector (quasi-isotropic formulation), 172
shift vector (quasi-radial formulation), 177
shock problem (ideal magnetohydrodynamics), 728
shock wave (in Riemann problem), 721–723
shortwave formalism, 10
signal-to-noise ratio of gravitational waves, 104
singularity avoidance property, 129, 130, 134, 136–140, 143
singularity avoidance property (numerical example), 335
Smarr relation, 35, 743
Sommerfeld boundary condition, 208
sound velocity, 255
spatial gauge (definition), 123
spatial gauge (NSCA formulation), 180
spatial gauge (quasi-isotropic formulation), 172
spatial gauge (quasi-radial formulation), 177
spatial gauge (role), 128
spatial metric (definition), 109
specific energy, 251
specific enthalpy (for perfect fluid), 250
specific internal energy (for perfect fluid), 250
specific momentum, 251
spectral and pseudo-spectral methods, 401–408
spin-orbit coupling effect, 32
spin-orbit coupling effect (for binary black hole), 581
spin-orbit coupling effect (for black hole-neutron star binary), 527, 530, 533
spin-weighted spherical harmonics, 425, 705
spin coefficient, 417
stable mass transfer, 78

standard 3+1 formulation, 118
standard 3+1 formulation (linearized), 149
standing accretion shock, 85–87, 89, 90
standing accretion shock instability (SASI), 86, 87, 90–92
stationary-phase approximation, 67
stationary axisymmetric equilibrium (formulation), 375–380
stellar-mass black hole (formation by fall-back process), 614–627
stellar-mass black hole (formation from high-entropy core), 630–634
stellar-mass black hole (formation scenario), 41, 610, 612, 613
stellar-mass black hole (observational evidence), 22
Stokes' theorem, 346
stress-energy tensor (3+1 decomposition), 115, 248
stress-energy tensor (3+1 form for hydrodynamics), 250
stress-energy tensor (definition), 1
stress-energy tensor (de Sitter), 192
stress-energy tensor (electromagnetic field), 262
stress-energy tensor (Friedmann universe), 194
stress-energy tensor (ideal magnetohydrodynamics), 267
stress-energy tensor (particles), 245
stress-energy tensor (perfect fluid), 249
stress-energy tensor (radiation), 287
stress-energy tensor (radiation hydrodynamics), 292
stress-energy tensor (scalar field), 243
stress-energy tensor (Z4 formulation), 163
strong hyperbolicity, 150–152
super-radiance, 36, 683, 684
supermassive black hole (observational evidence), 23, 24
supermassive black hole (source of gravitational waves), 22, 23
supermassive star (hypothetical collapse scenario), 605, 606
supermassive star (numerical simulation for collapsing to black hole), 606–609
supermassive star (stability against gravitational collapse), 602–604

supernova explosion (neutrino-driven scenario), 89, 90
supernova explosion (standard scenario), 83–87
superpotential, 733
superpotential (spatial), 349
supramassive neutron star, 52–54
supramassive neutron star (as remnant of binary neutron star merger), 485
supramassive neutron star (collapse to black hole), 597, 598
supramassive neutron star (numerical analysis of dynamical stability), 595, 596
surface gravity, 35, 741
surface gravity (of dynamical horizon), 754
synchrotron radio flare (of merger ejecta), 512, 513, 563

Tayler instability, 652–655
Tayler instability (in neutron star with toroidal magnetic field), 659–661, 663, 664
Taylor-F2 approximant, 765
Taylor-T1 approximant, 764
Taylor-T2 approximant, 764
Taylor-T3 approximant, 764
Taylor-T4 approximant, 765
Taylor-T4 formula (comparison with numerical waveform), 586
tensor spherical harmonics, 703, 704
tensor spherical harmonics decomposition, 411, 415, 422, 705–707
third law of black hole, 36, 737
Thorne's moment formalism, 285–287
three velocity (definition), 250
three velocity (force-free electromagnetodynamics), 275
tidal deformability, 487, 760
tidal disruption, 78–81
tidal disruption (condition for black hole-neutron star binary), 533–537
tidal force, 2, 6, 65, 78
tidal spin-up, 450–453
tidal tail, 559
time slicing, 123
Tolman–Oppenheimer–Volkoff (TOV) equation, 43–45
total rest mass (definition), 251

total variation diminishing (TVD), 303, 304
total variation diminishing (TVD) condition, 306, 307
tracefree longitudinal part (of symmetric tensor), 339
transverse gauge, 13
trapped region, 751
trapped surface, 751
trapping horizon, 751, 752
truncated moment formalism, 287–292
turning-point method, 592–594
twist, 181

upwind scheme (first-order finite-differencing), 217, 300
upwind scheme (for MUSCL), 305
upwind scheme (fourth-order finite-differencing), 218
upwind scheme (second-order finite-differencing), 218

vector spherical harmonics, 411, 702
vector spherical harmonics decomposition, 709, 710
vertex grid, 223
VIRGO, 19
virial relation, 348–351
virial relation (a remark), 360
virial relation (axisymmetric spacetime), 382
virial relation (for spherical star), 45
virial relation (helically symmetric spacetime), 392
virial relation (in Newtonian gravity), 348
virial relation (numerical accuracy), 460
virial relation (used for quasi-equilibrium formulation), 373, 567
von Neumann analysis, 213–215, 217, 218
vorticity tensor, 391, 748

wall shock problem, 727
waveless approximation (for binary quasi-equilibrium), 390
wave zone, 7, 409
weakly hyperbolic system, 152
weighted extrinsic curvature, 117, 175
weighted rest-mass density (definition), 251
weighted rest-mass density (particles), 245
Weyl scalar, 409, 415–418
Weyl scalar (asymptotic behavior), 420
Weyl scalar (for linear gravitational waves), 423
Weyl scalar (in 3+1 formulation), 420, 421
Weyl tensor, 3, 416, 417
Weyl tensor (in 3+1 formulation), 420
wormhole throat, 27, 29, 131

Z4c formulation, 162, 163, 165
Z4 formulation, 162–164
zero-shift gauge, 125–128, 138, 695
zeroth law of black hole, 737, 741